HANDBOOK OF MILK COMPOSITION

EDITED BY

Robert G. Jensen

University of Connecticut

Storrs, Connecticut

ACADEMIC PRESS

San Diego New York Boston

London Sydney Tokyo Toronto

This book is printed on acid-free paper. ♾

Academic Press, Inc.
A Division of Harcourt Brace & Company
525 B Street, Suite 1900, San Diego, California 92101-4495

United Kingdom Edition published by
Academic Press Limited
24-28 Oval Road, London NW1 7DX

Library of Congress Cataloging-in-Publication Data

Handbook of milk composition / edited by Robert G. Jensen.
p. cm. -- (Food science and technology international series)
Includes index.
ISBN 0-12-384430-4 (case)
1. Milk-Composition--Handbooks, manuals, etc. 2. Breast milk--Composition--Handbooks, manuals, etc. I. Jensen, Robert G. II. Series.
TX556.M5H33 1995
613.2'6--dc20 95-2194
CIP

PRINTED IN THE UNITED STATES OF AMERICA
95 96 97 98 99 00 MM 9 8 7 6 5 4 3 2 1

Contents

CHAPTER 3

Determinants of Milk Volume and Composition

A. Lactogenesis in Women: A Cascade of Events Revealed by Milk Composition

MARGARET C. NEVILLE

B. Volume and Caloric Density of Human Milk

MARGARET C. NEVILLE

C. Volume and Caloric Density of Bovine Milk

ROBERT G. JENSEN

D. Regional Variations in the Composition of Human Milk

ANN PRENTICE

E. Effects of Gestational Stage at Delivery on Human Milk Components

STEPHANIE A. ATKINSON

F. Miscellaneous Factors Affecting Composition and Volume of Human and Bovine Milks

ROBERT G. JENSEN

CHAPTER 4

Carbohydrates in Milks: Analysis, Quantities, and Significance

DAVID S. NEWBURG AND SUZANNE H. NEUBAUER

CHAPTER 6

Milk Lipids

A. Human Milk Lipids

ROBERT G. JENSEN, JOEL BITMAN, SUSAN E. CARLSON, SARAH C. COUCH, MARGIT HAMOSH, AND DAVID S. NEWBURG

B. Bovine Milk Lipids

ROBERT G. JENSEN AND DAVID S. NEWBURG

CHAPTER 7

Minerals, Ions, and Trace Elements in Milk

A. Ionic Interactions in Milk

MARGARET C. NEVILLE, PEIFANG ZHANG, AND JONATHAN C. ALLEN

D. Infant Formulas

ROBERT G. JENSEN, SARAH C. COUCH, JAMES W. HANSEN, ERIC L. LIEN, KARIN M. OSTROM, UMBERTO BRACCO, AND ROGER A. CLEMENS

CHAPTER 11

Contaminants in Milk

A. Drugs and Contaminants in Human Milk

RUTH A. LAWRENCE AND LINDA R. FRIEDMAN

B. Contaminants in Bovine Milk

ROBERT G. JENSEN

Contributors

Numbers in parentheses indicate the pages on which the authors' contributions begin.

Jonathan C. Allen (577), Department of Food Science, North Carolina State University, Raleigh, North Carolina 27695

Brenda P. Alston-Mills (467, 593, 827), Department of Animal Science, North Carolina State University, Raleigh, North Carolina 27695

Stephanie A. Atkinson (221, 351, 369, 593), Department of Pediatrics, Faculty of Health Sciences, McMaster University, Hamilton, Ontario Canada L8N 3Z5

Craig R. Baumrucker (475), Department of Animal and Dairy Sciences, Pennsylvania State University, University Park, Pennsylvania 16801

Joel Bitman (495), Milk Secretion and Mastitis Laboratory, USDA, Beltsville, Maryland 20705

Bernard Blanc (49), Genie Biochimique EPFS, CH-1015 Lausanne, Switzerland

Umberto Bracco (835), Nestle Ltd., CH-1000 Lausanne, Switzerland

Wendy M. Campana (475), Department of Neurosciences, Center for Molecular Genetics, School of Medicine, University of California, San Diego, La Jolla, California 92093

Louise M. Canfield (693), Department of Biochemistry, College of Medicine, University of Arizona, Tucson, Arizona 85721

Susan E. Carlson (543), Department of Pediatrics and OB/GYN, Newborn Center, University of Tennessee, Memphis, Tennessee 38163

Clare E. Casey (621), Public Health Commission, Wellington, New Zealand

Roger A. Clemens (835), Carnation Nutritional Products Division, Glendale, California 92103

Sarah C. Couch (495, 835), Department of Nutritional Sciences, University of Connecticut, Storrs, Connecticut 06269

Linda R. Friedman (857), Department of Pediatrics, University of Rochester Medical Center, Rochester, New York 14642

Angel Gil (435), Department of Biochemistry and Molecular Biology, Universidad de Granada and Uniasa, Granada, Spain

Anna R. Giuliano (693), Department of Family and Community Medicine, University of Arizona, Tucson, Arizona 85721

Randall M. Goldblum (727), Department of Pediatrics, University of Texas Medical Branch, Galveston, Texas 77555

Armond S. Goldman (727), Division of Immunology/Allergy, Department of Pediatrics, University of Texas Medical Branch, Galveston, Texas 77550

Ellen J. Graver (693), Arizona Cancer Center, University of Arizona, Tucson, Arizona 85721

Margit Hamosh (387, 495), Division of Developmental Biology and Nutrition, Department of Pediatrics, Georgetown University Medical Center, Washington, D.C. 20007

James W. Hansen (835), Department of Nutrition and Medical Affairs, Mead Johnson Nutritional Group, Evansville, Indiana 47721

Sara J. Iverson[1] (749, 789), Canadian Institute of Fisheries Technology, Techical University of Nova Scotia, Halifax, Nova Scotia, Canada B3J 2X4

Robert G. Jensen (1, 49, 79, 81, 113, 237, 495, 543, 687, 717, 745, 835, 887, 903), Department of Nutritional Sciences, The University of Connecticut, Storrs, Connecticut 06269

Thomas W. Keenan (5), Department of Biochemistry and Anaerobic Microbiology, Virginia Polytechnic Institute and State University, Blacksburg, Virginia 24061

Otakar Koldovský (427), Department of Pediatrics, and the Steele Memorial Children's Research Center, University of Arizona College of Medicine, Tucson, Arizona 85724

Carol J. Lammi-Keefe (705), Department of Nutritional Sciences, The University of Connecticut, Storrs, Connecticut 06269

Ruth A. Lawrence (857), Department of Pediatrics and OB/GYN, University of Rochester Medical Center, Rochester, New York 14627

Eric L. Lien (835), Wyeth Laboratorys Inc., Philadelphia, Pennsylvania 19104

Bo Lönnerdal (351, 369, 593), Department of Nutrition, University of California, Davis, California 95616

Suzanne H. Neubauer (273), Framingham State College, Framingham, Massachusetts 01701

Margaret C. Neville (63, 81, 87, 99, 577, 593), Departments of Physiology and Cell and Structural Biology, University of Colorado, School of Medicine, Denver, Colorado 80262

David S. Newburg (273, 495, 543), Shriver Center for Mental Retardation, Waltham, Massachusetts 02254; and Harvard Medical School, Cambridge, Massachusetts 02138

[1] Present Address: Department of Biology, Dalhousie University, Halifax Nova Scotia, Canada B3H 4J1

Olav T. Oftedal (749, 789), Department of Zoological Research, National Zoological Park, Smithsonian Institution, Washington, D.C. 20008

Karin M. Ostrom (835), Ross Products Division, Abbott Laboratories, Columbus, Ohio 43216

Stuart Patton (5, 49), Department of Neurosciences, Center for Molecular Genetics, University of California, San Diego, La Jolla, California 92093

Mary Frances Picciano (675), Department of Nutrition, College of Health and Human Development, Pennsylvania State University, University Park, Pennsylvania 16802

Ann Prentice (115), MRC Dunn Nutrition Unit, Cambrige CB4 1XJ, United Kingdom

Anne Smith (621), Department of Human Nutrition and Food Management, Ohio State University, Columbus, Ohio 43210

Vladimir Strbák (427), Institute of Experimental Endocrinology, Slovak Academy of Sciences, 83306 Bratislava, Slovak Republic

Harold E. Swaisgood (463, 471), Department of Food Science, North Carolina State University, Raleigh, North Carolina 27695

Ricardo Uauy (435), Institute of Nutrition, INTA, University of Chile, Casilla 138-11 Santiago, Chile

Peifang Zhang (577, 621), Department of Physiology, University of Colorado, School of Medicine, Denver, Colorado 80262

Foreword

It is possible to believe that all the past is but the beginning of a beginning.
H. G. Wells

The 70 pages of the revised (1953) edition of *The Composition of Milks*, edited by Macy, Kelly, and Stone and published by the National Academy of Sciences–National Research Council, consist almost entirely of five tables of elemental and organic constituents. A modern publication in equivalent detail probably could not be hefted. In their book, Macy *et al.* extracted data from 1500 publications, citing 278 key references; the contributors to the present volume have prepared a distillate from perhaps ten times that number.

A significant part of the exponential expansion of knowledge about some of the complex substances in milk over the past fifteen years has resulted from the activities of the National Institute of Child Health and Human Development. As a result of the insight of Drs. Norman Kretchmer and Thorsten J. Fjellstedt, the Institute staff made a sustained effort to increase research on human milk composition. They established a program of human milk research and through targeted grants and contracts encouraged the development of new methods for the examination of milk components. They transmitted their excitement about this subject to some of the leading investigators in the field; this was an important influence on the establishment of the International Society for Research on Human Milk and Lactation, the organization sponsoring this volume.

Icie Macy Hoobler, in her brief preface to the 1953 publication, made this insightful observation: "[I]n infant feeding, quantitative interrelationships among various components of milk may be even more significant than levels of intake of specific factors." These interrelationships have proven difficult to study using the classical scientific method of maintaining constant all variables but one. As a result, the exponential increase in our knowledge of what is in milk has led to little understanding of how things get there and why. Because of the obvious potential benefit to animal husbandry and the metabolic support of ill or premature infants, it is time for the best investigators to address those questions. When they do, they will doubtless start from this *Handbook*.

Ephraim Y. Levin

Preface

This book is the first comprehensive compilation of data on the composition of milk published in over 40 years. It contains, in the opinions of the contributors, the most reliable data available and obtained by modern analytical methods for each topic. Anyone who needs information on milk composition, including dairy scientists and processors, those who work with human milk and lactation, and those who investigate the milks of other mammals, should find it here.

This book is the only publication of its type that is currently available. It features human and bovine milk, but also contains chapters on other milks used by humans as foods and those that are not. It has chapters on sampling, determination of volume, the major components of milks, immune factors, the milk lipid globule membrane, particulate matter, factors affecting composition, and contaminants. In addition, there is a section on milk banking with in-depth discussions by many authors. Tabular data are presented.

The major area of concern in this book is nutrition, not only of the human infant but also of anyone who uses milk and milk products as foods. The book is also concerned with substances in human milk that protect infants from disease and act as messengers that provide metabolic and developmental information to the infant.

I am grateful to our many contributors, to Dr. Ephraim Y. Levin, who obtained a contract for me from NIH to help prepare the book, and to The International Society for Research on Human Milk and Lactation for its informal support. I also thank Dr. Ann M. Ferris, Head of the Department of Nutritional Sciences, University of Connecticut, for the use of departmental facilities, and Mrs. Sandra J. Beaupre, who typed my sections. I am particularly grateful to our editor, Ms. Charlotte Brabants, and her assistant, Ms. Leslie O'Brien, of Academic Press for their patience, to Drs. Margaret (Peggy) Neville and Margit Hamosh for their help, and to my contributors, who tolerated and eventually responded to my incessant nagging. I dedicate this book to Drs. Icie G. Macy and Robert Jenness, pioneers in the analysis of milk.

Robert G. Jensen

CHAPTER 1

Introduction

ROBERT G. JENSEN

I. Purpose

Macy *et al.* (1953) in their classic publication, "The Composition of Milks," summarized the data then available on the composition and properties of bovine, human, and goat milks. The effects of time postpartum (stage of lactation) were included. Macy's (later Hoobler) results from her extensive research were published (Macy and Kelly, 1961). The composition of milks from many animals were collected by Jenness (1974). Listings of the components in infant formulas are available (Tsang and Nichols, 1988). However, since the publications of Macy and colleagues, there has been no effort to compile and summarize in one volume the composition and properties of the milks and infant formulas used for food nor of the milks from other mammals for which reliable data exist. There is no single source of this information for the workers who are interested in any aspect of milk. My primary purpose for preparing our book is to provide this source.

Advances in analytical methods provide another reason for publication of the "Handbook of Milk Composition." Milk can be analyzed with sensitivity, resolving power, and speed that were impossible in years past. One of the results of improvement of sensitivity is that hitherto unknown or unrecognized compounds are detected. Thus, the complexity of milk as a system designed to deliver nutrients and nonnutritive messages to the neonate has increased.

An example of the impact of a new analytical procedure is the determination of bovine milk fatty acids by gas–liquid chromatography (GLC). The analysis which required weeks in the past is now routinely done in about 2 hr (Jensen *et al.*, 1991). The identities and amounts of fatty acids

HANDBOOK OF MILK COMPOSITION

in many samples can be quickly obtained. The new data are much more reliable and comprehensive.

My contributing authors were asked to report the data which in their opinions were the most reliable and to discuss problems with sampling, storage, and analysis which might influence composition. They were instructed to use, when possible, hours or days instead of colostrum, transitional, and mature to describe age postpartum. We prefer this term to the clumsy phrase, stage of lactation. They were required to report their data at wt/dl with use of SI units optional. Those who were gathering information on bovine milk were reminded that much of this is consumed in the pasteurized homogenized form although few data are available on this product. The length of the contributions varies, primarily because this was left to the discretion of the authors. Some sections are short, e.g., "Bovine Milk Proteins," because comprehensive, current texts are available. Other sections, notably "Carbohydrates," are long because there is no single source of information available on the subject. We have tried to provide the best data on composition that are now in the literature.

Since imitation is the highest form of flattery we will paraphrase from the preface of Macy *et al.* (1953). She recognized the importance of milk as a food for all age groups and mentioned the difficulties involved in gathering information (her group examined 1500 references) and the general inadequacy of knowledge of milk components. It was anticipated that the survey would be useful to investigators and those working directly on infant nutrition either with human milk or formulas. All of these reasons are valid today, perhaps even more so. This is because of the increased incidence of breastfeeding in Western countries, the recognition that human milk provides protection against diseases, e.g., diarrhea, endemic in the Third World, and that it may be needed for optimal growth and development of infants and their performance as adults.

II. General Description of Milks

Milks contain, with some exceptions, the nutrients required for the growth and development of the neonate. If the development time is short then the milk is nutrient dense. All milks contain specific proteins, fats designed to be easily digested, most have lactose, minerals, vitamins, and other components which may have important roles. These are organized as follows: lipids in emulsified globules coated with a membrane, proteins in colloidal dispersion as micelles, and most minerals and all lactose in true solution (Jensen *et al.,* 1991).

For the guidance of the reader we present, in Table 1, proximate analyses of bovine, human, goat, and sheep milks.

TABLE I
Proximate Composition (WT%) of Bovine, Human, Goat, and Sheep Milks[a]

Component	Bovine	Human	Goat	Sheep
Protein	3.4	1.0	2.9	5.5
Casein	2.8	0.4	2.5	4.6
Fat	3.7	3.8	4.5	7.4
Lactose	4.6	7.0	4.1	4.8
Ash	0.7	0.2	0.8	1.0

[a]Jenness (1974).

References

Jenness, R. (1974). The composition of milk. *In* "Lactation" (B. L. Larson and V. R. Smith, eds.), Vol. III, pp. 3–107. Academic Press, New York.

Jensen, R. G., Ferris, A. M., and Lammi-Keefe, C. J. (1991). The composition of milk fat. *J. Dairy Sci.* **74,** 3228–3243.

Macy, I. G., and Kelly, H. J. (1961). Human milk and cow's milk in human nutrition. *In* "The Mammary Gland and Its Secretion" (S. K. Kon and A. T. Cowie, eds.), Vol. II, pp. 265–304. Academic Press, New York.

Macy, I. G., Kelly, H. J., and Sloan, R. E. (1953). "The Composition of Milks." National Academy of Science and National Research Council, Publication 254, Washington, DC.

Tsang, R. C., and Nichols, B. L. (1988). Nutrient content of infant formulas. *In* "Nutrition during Infancy" (R. C. Tsang and B. L. Nichols, eds.), pp. 418–424. Hanley and Belfus, Philadelphia.

CHAPTER 2

The Structure of Milk: Implications for Sampling and Storage

A. The Milk Lipid Globule Membrane

THOMAS W. KEENAN
STUART PATTON

I. Intracellular Origin and Growth of Milk Lipid Globules

Membrane and membrane-associated material which surrounds the triacylglycerol-rich milk lipid globules commonly is referred to as the milk fat or milk lipid globule membrane (MLGM hereafter). This material originates from specialized regions of apical plasma membrane of mammary epithelial cells, and from endoplasmic reticulum (ER) and perhaps other intracellular compartments. That portion of the MLGM derived from apical plasma membrane, termed the primary membrane, has a typical bilayer or unit membrane appearance, with an electron-dense material on the inner membrane face. That component derived from ER lacks bilayer membrane structure, primarily is composed of proteins and polar lipids, and covers the surface of the lipid droplets within the cell. Constituents of this coat material mediate intracellular fusions through which droplets grow in volume and also may be involved in interaction of droplets with plasma membrane.

HANDBOOK OF MILK COMPOSITION

A. Droplet Formation

Earliest intracellular precursors of milk lipid globules appear to originate from ER. Triacylglycerols appear to accumulate at focal points on or in the ER membrane (Dylewski *et al.*, 1984). Whether this accumulation of triacylglycerols is due to localized synthesis or accretion is unknown. It has been suggested that triacylglycerols accumulate between the halves of the bilayer membrane and are released from ER into the cytoplasm as droplets coated with the outer or cytoplasmic half of the ER membrane (Long and Patton, 1978; Scow *et al.*, 1980). Some morphological evidence supporting this suggestion has been obtained (Patton and Keenan, 1975; Zaczek and Keenan, 1990), but information that would prove or disprove this hypothesis is lacking.

B. Growth of Droplets

By whatever mechanism they originate, milk lipid globule precursors first appear in the cytoplasm as small (diameters < 0.5 μm) droplets that have a triacylglycerol-rich core surrounded by a granular coat material lacking unit-like (or bilayer membrane structure, but that in localized regions appears thickened, with tripartite-like structure (Dylewski *et al.*, 1984; Deeney *et al.*, 1985). Small lipid droplets, termed microlipid droplets, appear to grow in volume by fusions with each other. Fusions give rise to larger droplets, termed cytoplasmic lipid droplets, operationally defined as those droplets with diameters > 1 μm.

In addition to observations made by electron microscopic examination of fixed and sectioned material, the nature of the surface coat material on intracellular lipid droplets has been explored through isolation and compositional analysis of droplets (Dylewski *et al.*, 1984; Deeney *et al.*, 1985). Droplets can be isolated by density gradient centrifugation, taking advantage of the fact that they have lower densities than do organelles and vesicles derived from components of the endomembrane system. Droplets ranging from < 1 to 1.12 g/cc in density have been characterized; density was inversely related to volume. Droplets of different density classes from cow mammary gland had lipid to protein ratios ranging from about 1.5:1 to 40:1.

Triacylglycerols were the major lipid class in droplets of all sizes. Surface coat material of droplets contained cholesterol and the same five major phospholipid classes found in milk: sphingomyelin and the phosphoglycerides of choline, ethanolamine, inositol, and serine. Lipid droplets also had monohexosyl- and dihexosylceramides and gangliosides in their surface coat material; these glycosphingolipids are known constituents of milk lipid globules.

When separated in sodium dodecylsulfate–polyacrylamide gels (SDS–PAGE), micro- and cytoplasmic lipid droplets had complex and virtually identical polypeptide patterns. Many polypeptides with electrophoretic mobilities identical to those of intracellular lipid droplets are found in MLGM material. Several polypeptides of MLGM and intracellular lipid droplets share antigenic reactivity.

In summary, morphological observations and biochemical data are consistent with an ER origin of intracellular lipid droplet precursors of milk lipid globules. The material on the surface of lipid droplets within the cell appears to remain associated with the droplets, at least in part, when they are secreted as milk lipid globules.

II. Role of Intracellular Lipid Droplet Coat Material

Coat material on surfaces of intracellular lipid droplets undoubtedly is required to stabilize the triacylglycerol-rich core of droplets and prevent their coalescence in the cytoplasm. Beyond this stabilization role, the coat material appears to participate also in droplet fusions, and in droplet–plasma membrane interactions. If cytoskeletal elements function in guiding lipid droplets from their sites of origin to their sites of secretion from the cell, coat constituents may participate in interaction with elements of the cytoskeleton. Mechanisms responsible for unidirectional transit of lipid droplets through the cytoplasm to apical cell regions, from which they are secreted, are not known with certainty. Evidence that microtubes or microfilaments may be involved in this process has been obtained, but evidence contradicting these interpretations also has been obtained (discussed in Mather and Keenan, 1983). As yet, we have no clear, definitive information on what is responsible for this unidirectional transfer of lipid droplets. In the milk of cows, lipid globules range in size from under 0.2 to over 10 μm in diameter. Small globules (below 1 μm) are most numerous, accounting for 80% or more of the total number of globules, but these small globules account for less than 10% of the total volume of milk fat. Globules with diameters between 1 and 8 μm contain 90% or more of the total volume of milk fat. Large droplets are few in number, but account for 1 to 3% of the fat volume of milk (reviewed in Brunner, 1965; Mulder and Walstra, 1974). Data available suggest a similar size range of globules in milks of humans (Rüegg and Blanc, 1981). Globule size distribution is discussed under Section V.

Within the cell, one mechanism for growth of lipid droplets appears to be fusions of microlipid droplets with each other to form larger droplets. Microlipid droplets can also fuse with larger, cytoplasmic lipid droplets, providing triacylglycerols for continued growth in volume of larger droplets (Dylewski *et al.*, 1984); this growth is pronounced in globules in apical cell regions (Stemberger and Patton, 1981, 1984). While images interpreted as microlipid droplet–microlipid droplet and microlipid droplet–cytoplasmic lipid droplet fusions are commonly seen in electron micrographs, several investigators have failed to find morphological evidence for fusions between larger, cytoplasmic lipid droplets (Wooding, 1971a; Stemberger and Patton, 1981, 1984; Dylewski *et al.*, 1984).

From research to date the size range of lipid globules in milk can be accounted for by the fusion process. Smaller milk lipid globules arise most probably from secretion of microlipid droplets that have undergone no or only a few fusions. Larger droplets originate by continued fusions with microlipid droplets. Morphological and kinetic evidence in support of this interpretation has been obtained. However, this evidence is insufficient to allow the interpretation that fusion of droplets is the sole or major mechanism for droplet growth. Other possible mechanisms for this growth, for example, lipid transfer proteins which convey triacylglycerols from their site of synthesis to growing lipid droplets, cannot be excluded (Patton, 1973).

The process of microlipid droplet fusion has been reconstituted in a cell-free system (Valivullah *et al.*, 1988). Fusion appears to involve constituents of the surface coat of lipid droplets. As droplets grow, excess coat material is lost from the surface, as would be expected from geometric consideration of surface area (area = $4\pi r^2$) to volume (volume = $4/3\pi r^3$) ratios of what are essentially spherical particles. The fate of coat material shed from droplet surfaces during fusion within cells is unknown. In the cell-free system, fusion was promoted by calcium, by a protein fraction of cytosol, and by gangliosides of the surface coat of lipid droplets. In agreement with morphological observations of sections from cells, in the cell-free system microlipid droplet–microlipid droplet and microlipid droplet–cytoplasmic lipid droplet fusions occurred, but cytoplasmic lipid droplet–cytoplasmic lipid droplet fusions did not. The reasons why cytoplasmic lipid droplets do not fuse with each other is not apparent. Within the scope of compositional analyses performed to date, coat materials on micro- and cytoplasmic lipid droplets largely are indistinguishable, except for the increased level of gangliosides per unit protein in cytoplasmic lipid droplets.

III. Milk Lipid Globule Secretion

The process of lipid droplet secretion has been described repeatedly since Bargmann and Knoop (1959) originally observed that droplets became surrounded by apical plasma membrane as they were budded from cells (reviewed in Patton and Keenan, 1975; Mather and Keenan, 1983; Keenan *et al.*, 1988). Wooding (1971a) provided morphological evidence for an alternative mechanism, one in which fat droplets contacting the apical plasma membrane also become surrounded with secretory vesicles that fuse with each other and the plasma membrane. This resulted in formation of intracellular vacuoles containing membrane-coated lipid droplets. Release of droplets, surrounded partially in apical plasma membrane, was envisioned to occur by emptying of the vacuolar contents. Morphological evidence for which of these alternative processes occurs or predominates is equivocal. Most biochemical evidence favors the interpretation that the major mechanism for secretion of milk lipid globules involves envelopment of droplets directly in plasma membrane. A minor contribution from Golgi apparatus-derived secretory vesicle membrane cannot be excluded (reviewed in Mather and Keenan, 1983; Keenan *et al.*, 1988).

Plasma membrane regions with which lipid droplets associate are characterized by the appearance of an electron-dense material on the inner (cytoplasmic) face of the membrane (Wooding, 1971a, 1977; Freudenstein *et al.*, 1979). Droplets do not contact the plasma membrane directly, but rather this material. Which constituents of this electron-dense material recognize and interact with constituents on the droplet surface remains to be elucidated. Immunomicroscopic (Franke *et al.*, 1981; Jarash *et al.*, 1981) and biochemical studies (Freudenstein *et al.*, 1979; Mather and Keenan, 1983; Niera and Mather, 1990) have shown butyrophilin and xanthine oxidase, two prominent proteins associated with the MLGM, to be major constituents of the electron-dense material on the cytoplasmic face of apical plasma membrane. Butyrophilin, a hydrophobic transmembrane glycoprotein, is highly concentrated at the apical surface of milk-secreting cells (Franke *et al.*, 1981; Niera and Mather, 1990). Xanthine oxidase is distributed throughout the cytoplasm, but appears to be enriched at the apical cell surface (Jarasch *et al.*, 1981). Butyrophilin, which is acylated (Keenan *et al.*, 1982) and binds phospholipids tightly (Freudenstein *et al.*, 1979), has been believed to be involved in mediating interaction between lipid droplets and apical plasma membrane. Recently, the gene for butyrophilin was cloned and sequenced (Jack and Mather, 1990). From the inferred primary amino acid sequence, it was not apparent how butyrophilin could interact with lipid droplets. Since butyrophilin has an exoplasmic N-terminus and a single membrane-spanning domain, interaction with lipid droplets must occur with the 257-residue C-terminal domain, if in fact this protein does interact with lipid droplets. From the primary sequence, the C-terminal domain has no obvious hydrophobic regions, but hydrophobic domains could result from acylation of serine-threonine residues. One possibility for interaction of butyrophilin is with proteins of the lipid droplet surface rather than with lipids (Jack and Mather, 1990). This could be with proteins of the lipid droplet surface directly, or through complexes formed with cytoplasmic proteins. Since butyrophilin and xanthine oxidase show a propensity to associate with each other, a butyrophilin–xanthine oxidase complex could be involved in lipid droplet interaction. The function that xanthine oxidase may play in the recognition or envelopment process remains obscure.

IV. Nature and Frequency of Cytoplasmic Crescents

A small compartment of milk that has received limited attention is crescents of cytoplasm on milk fat globules. This truly is a unique component because technically crescents are part of the mammal that made the milk rather than a true secretory product of the gland. Thus, crescents make their own distinctive contribution to the composition and properties of milk.

A. *Mode of Crescent Formation*

In the process of their secretion, milk fat globules usually are enveloped smoothly and compactly by apical membrane of the lactating cell. In the electron microscope, the droplet of fat undergoing secretion appears to be in close association with membrane over its entire surface. However, on occasion, the closure of the membrane behind the projecting fat droplet appears to occur by a route through the apical cytoplasm rather than along the droplet surface. The result is that a fat globule is secreted with a piece of cytoplasm, a so-called crescent or signet, attached. These can vary from thin slivers of cellular material to situations in which the crescent dominates the globule. A typical human globule with crescent is shown in Figure 1.

A second mechanism by which crescents may form has been proposed by Wooding (1977). The apical region of the lactating cell is populated with secretory vesicles containing the skim phase of milk, which is secreted by exocytosis. Wooding proposed that on occasion these vesicles gather behind a fat droplet that is in the process of secretion and trap cytoplasm into secretion with the droplet. It is not known whether one or the other of these mechanisms predominate. One might expect the entrapment mechanism to occasionally produce membrane-bound vesicles of cytoplasm free of a fat droplet.

On the assumption that some significant proportion of crescents on milk fat globules form as a result of a closure of apical plasma membrane through cytoplasm behind the fat droplet, Huston and Patton (1990) suggested that an abnormality in the protein coat, reputed to bind the membrane to the fat droplet (Franke *et al.*, 1981), is responsible. The principal proteins of this coat are butyrophilin and xanthine oxidase (Freudenstein *et al.*, 1979). Inadequate production or distribution of this coat complex might interfere with adhesion of the membrane to the droplet. Butyrophilin is reported to be acylated with long-chain fatty acids (Keenan *et al.*, 1982). Variation in this acylation also may alter the surface properties of the protein and its function as a coat (adhesive) substance. Determining what causes crescent formation at the molecular level seems very challenging. For example, the phenomenon appears to be subject to

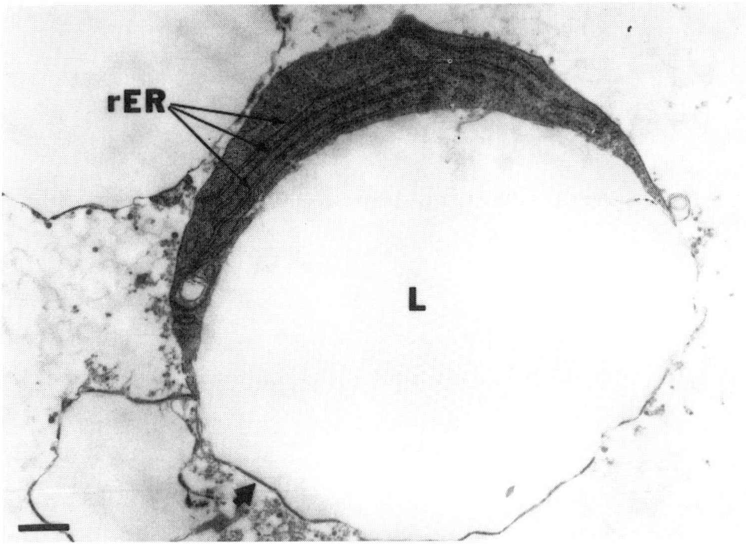

Figure 1 Electron micrograph of a typical human milk fat globule bearing a crescent of cytoplasm. Note array of well-preserved rough endoplasmic reticulum (rER) in cytoplasm and membrane (arrow, lower left) surrounding lipid droplet (L) and cytoplasm. Scale bar denotes 0.5 μm.

diurnal variation in the human, with greatest numbers of crescents being formed in the evening (Patton and Huston, 1988). The proteins and fat of milk are also subject to diurnal variations. This makes possible a situation in which synthesis and secretion of milk fat maybe favored at a time when structural elements needed for globule secretion are in short supply. It is also possible that nature favors the production of crescent material for some benefit it has in the nursling (discussed further under Section IV, D).

B. Properties of Cytoplasmic Crescents

Cytoplasmic crescents have been observed to contain all the various membranes and organelles of the lactating cell except the nucleus. As a result, they take up fluorescent dyes, such as acridine orange, in the same manner as do cells. This is a useful property in quantifying crescents (see following). It is said that the age of cells is revealed by the fluorescent hue, orange being young and yellow being older. Such variations can be seen in crescents. Using acridine staining to visualize crescents, Patton and Huston (1988) observed their gradual disappearance over 36 hr in human milk stored at 4°C. This means that some substances originally associated with milk fat globules progressively become part of the skim milk. Milk accumulates in the mammary gland following its secretion from the cell, and removal of milk by nursing or machine is never complete. Thus, prior to milk removal, deterioration of crescents can proceed in the gland. The difference in age of crescents can be detected in thin sections of plastic-embedded milk viewed in the electron microscope. "Fresh" crescents look like fresh specimens of tissue; the ER is laminar and well-defined. In older crescents, the ER is open, rounded, and swollen looking but with ribosomes still attached. Still later, ER is not recognizable as such. The cytoplasm then has a totally disorganized, granular appearance (Patton and Huston, 1988).

Globule populations with a high proportion of crescents exhibit a more complex pattern of proteins by SDS–PAGE than do low-crescent populations (Patton and Huston, 1988). Presumably, the many additional minor bands that are evident arise from the cytoplasmic components in the crescent. It should be possible to identify some of these minor proteins by immunostaining of Western blots.

When membrane is released from human milk fat globules by churning, one can isolate a crescent-rich fraction by low-speed centrifugation (1200*g* for 10 min) of the buttermilk. Viewing of this fraction in the electron microscope reveals that, despite all the manipulation of the sample, the shapes of crescents are preserved, suggesting that the structure of cell cytoplasm is not amorphous but somewhat defined (Patton and Huston, 1988), most probably by elements of the cytoskeleton (Schliwa, 1986).

C. *Detection and Quantification of Crescents*

Jannsen and Walstra (1982) originated use of the fluorescent dye, acridine orange, for staining crescents. They also devised a method of quantifying the amount of crescent material in milk based on enumeration of crescents in a defined volume of milk and calculating their mass. Patton and Huston (1988) utilized this procedure for staining and viewing globules with crescents but pursued data on the basis of what proportion of globules contain crescents. Both kinds of data are useful, e.g., the Jannsen–Walstra approach emphasizes how relatively small the amount of crescent material is in milks of the various species and how little lactating cellular material is being lost from the gland, and the Patton–Huston evaluation yields some insight into globules with crescents as a proportion of secretion events. In one sample of human milk, Patton and Huston (1988) found that there were almost as many globules with crescents (44%) as without. Both approaches suffer the limitations of light microscopy with respect to viewing very small globules. However, control experiments (Patton and Huston, 1988) to assess crescents on such globules (< 3 μm) indicated that they account for about 12% of the total crescents. Since small globules make up 80 to 90% of the population, this implies that there is a much stronger tendency for cresents to form during secretion of the larger globules. While it is also possible to enumerate crescents using electron microscopy (Patton and Huston, 1988), there are limitations with this approach. The milk must be given structure with agarose or other suitable agents; at higher magnifications, the globules need to be concentrated somehow so that there are a sufficient number of them in the field of view. In doing this there is a danger that the population selected will not be representative and that crescents may be destroyed in the process.

Crescents have been identified in association with the milk fat globules of all species examined to date. The proportion of globules with crescents varies between and within species. Wooding *et al.* (1970) estimated that 1–5% of the globules in goat and guinea pig milk had crescents, while cow's milk had relatively few (about 1%). Janssen and Walstra (1982), in a comparative study, measured crescent quantity in milks of several species, including cow, goat, rat, pig, sheep, rabbit, and human, and reported cow to have the least (7.1 mg/kg milk) and rabbit to have the most (131 mg/kg milk). In a study of 50 human milk donors (Huston and Patton, 1990), incidence of crescents on fat globules ranged from 1 to 29%. Most (80%) fell between 3 and 10% and the mean (±SD) was 7.2 ± 4.2%. Two pooled bovine milk samples, both representing over 100 animals, contained 1% or less of fat globules with crescents. The apparent evolutionary persistence of cytoplasmic crescents on human milk fat globules suggests that they may have beneficial effects in the young.

In the course of their study of 50 lactating women, Huston and Patton noted that diurnal and genetic factors seemed to be involved in crescent production. There was definite evidence that evening samples of milk had higher crescent numbers than those collected in the morning. Two sisters consistently showed much higher levels of globules with crescents (25–44%) than others in the study. Moreover, this characteristic persisted during their following lactation. One possible hypothesis is that these sisters have only one copy of the butyrophilin gene. Frequent milking and administration of oxytocin have been reported to favor secretion of globules with crescents (Wooding, 1977). Such treatments would tend to make for lactating cells with cytoplasm distended into the alveolar lumen along with projecting fat droplets, a configuration that might facilitate crescent formation.

In the Huston–Patton study, a number of factors that did not seem to be involved in the crescent phenomenon were age of the donor, stage of lactation, which breast, volume of milk expressed, a particular fraction during a complete expression of a breast, and milk lipid or protein content. They also compared crescent incidence in dairy cows and beef cows and found no difference, which suggests that selection for milk production in cattle may not be a relevant variable.

D. Significance of Crescents

It may seem unlikely that something as quantitatively limited as crescents of cytoplasm on milk fat globules can have much significance. However, it has to be remembered that the newborn is receiving milk around the clock day after day. Further, it is now well appreciated that many biologically important molecules are effective at very low levels including enzymes, hormones, receptors in and on cells, growth factors, and trace elements. In addition, crescents can be viewed as a source of foreign antigens (the mother's) that may have some conditioning effect on the developing immune system of the newborn.

Even if crescents have no nutritional value, there are a number of other significant considerations concerning them. An understanding of crescent formation at the molecular level will do much to further clarify the mechanism of milk fat globule secretion. The implication that butyrophilin is essential to the latter and may be insufficient at times, thereby resulting in crescent formation, needs research. In the preparation of milk fat globule membrane by churning or freeze-thaw, crescents tend to concentrate in the membrane faction and constitute a significant impurity in some species. Methods which involve a further purification step, such as centrifuging in a density gradient, may eliminate this problem. The discovery of heretofore unreported biological molecules in milk raises such questions as where are they in the milk?, i.e., in what compartment and how did they gain entry? Answering the former can help to define the latter. Crescents represent an important route of cellular, as opposed to secretory, substances into milk. In that connection, crescents represent a means of sampling specifically the lactating cell without having to biopsy the animal and isolate the particular cells from their tissue matrix. This is an important consideration with the human. For this purpose, human milk fat globules can be isolated (Patton and Huston, 1986), churned, the butter removed, and the buttermilk used as a suspension solution of cytoplasmic constituents from the lactating cell. Of course, this system will contain MLGM, but substances originating from the cytoplasm, such as mRNAs, can be readily distinguished from this "contaminant."

V. Size and Membrane Area Distribution of Milk Lipid Globules

A. *Globule Size*

Milk lipid globules of species examined to date fall into three overlapping size distributions: small with diameters centered below 1 μm, intermediate with diameters in the 3 to 5 μm range, and large globules with a mean diameter of about 8 to 10 μm. Some of the latter approach 20 μm and it is felt that this group, for the most part, is formed by postsecretion merging of globules. Thus, milk fat globules that are most typical as secretory products belong to the two smaller groups. However, a surprisingly large proportion of the total globule population, 70 to 90% in the bovine and human, lies in the first group below 1 μm in diameter. While these globules must account for most of the secretory events, they contain $< 5\%$ of the total milk lipid. Globules of the large-diameter group make up a very small part of the total globule population, estimated to be 0.01% in the human, but represent 1 to 4% of the lipid. Therefore, the intermediate group, comprising roughly 10 to 30% of the globule numbers, accounts for 90% or more of the total lipid. These findings are from studies by Walstra (1969) and Rüegg and Blanc (1981).

The average diameter of milk lipid globules of species examined (cow, human, buffalo, goat, ewe, sow) is in the range of 3 to 5 μm. However, there are some complications in arriving at precise figures for mean diameter. The populations of small globules, < 1 μm in diameter, are beyond enumerating limits possible with the light microscope or Coulter counter. As a consequence, their numbers and sizes often have been ignored. Two studies employing the Coulter counter, one of the cow (Walstra, 1969) and the other for the human (Rüegg and Blanc, 1981), have derived values for small globules by extrapolation with low relative uncertainty. Using this approach, Walstra determined the volume/surface average diameter (d_{vs}) for Jersey and Fresien cows to be 4.5 and 3.34 μm, respectively. Using similar methodology, Rüegg and Blanc calculated d_{vs} for mature human milk to be 4.0 μm. For details of the theory and procedures used in these studies, the original literature should be consulted.

It would be interesting to know at what, if any, lower diameter the globule population falls off sharply. This would indicate whether there is a size below which free intracellular lipid droplets do not exist or globule secretion from the cell does not occur. Resolution of this problem will require an electron microscopy approach. Presumably, the large number of very small lipid globules in milk are secreted by the same mechanism and carry the same membrane as larger globules (Deeney *et al.*, 1985). Globule size ranges in alveolar lumens were observed to be the same as those for lipid droplets inside lactating cells (Wooding, 1971a). Stemberger and Patton (1981) also observed that the lipid droplet populations in lactating tissue of cow, cat, rabbit, rat, and mouse were composed of two groups, similar to those in the milk, i.e., smaller ones < 1.5 μm and intermediate in the 1.5 to 8 μm range. No evidence of very large droplets was taken by them to indicate that the latter must result from postsecretation fusion of smaller globules.

B. *Membrane Surface Area*

Rüegg and Blanc (1981) estimate that the lipid globule surface area in 1 ml of mature human milk is 500 cm^2. In a liter of milk, this would be 500,000 cm^2 of surface or roughly the floor space in a room 23 ft^2. Whether one thinks of milk in terms of processing and storage effects or digestion and behavior in the gut, this is a large amount of surface. We can assume that the surface area of lipid globules in milk is roughly equivalent to that of globule membrane (i.e., one side of that membrane). However, there is, as mentioned in the section on membrane stability (Section VII, A), the question of how much membrane has sluffed from the globules into the skim milk. This will vary, as will the proportion of (small) globules left in the skim milk by centrifugal separation. So it is not possible to derive very precise estimates of the membrane surface area in skim milk. Using such criteria as the amount of membrane proteins and activities of membrane enzymes, it appears that about one-third to one-half of the membrane materials in milk is recovered in the skim milk. It seems reasonable to think of this as surface area like that on globules, although it may represent various configurations and even be membrane sheets with both exo- and endoplasmic faces exposed. The membrane material in skim milk often is referred to as the "fluff" fraction because it is a loose layer that rests on the casein pellet following centrifugation. For further characteristics of this fraction, especially its appearance in the electron microscope, see Stewart *et al.* (1972).

VI. Nature of the Milk Lipid Globule Membrane

A. *Isolation of Milk Lipid Globule Membrane*

The membrane surrounding lipid globules in milk closely resembles plasma membrane in ultrastructure in that it has a typical bilayer appearance, with the space between bilayers being comparable to that of plasma membrane, and has an externally disposed glycocalyx (Monis *et al.*, 1975; Freudenstein *et al.*, 1979; Sasaki and Keenan, 1979; Franke *et al.*, 1981). This membrane is characterized, as are differentiated regions of apical plasma membrane, by the appearance of the electron-dense material associated with the inner face of the membrane (Figure 2) (Wooding, 1971b; Freudenstein *et al.*, 1979; Franke *et al.*, 1981). Some of the plasma membrane initially surrounding globules may be lost following secretion, within alveolar lumina, or in expressed milk (Wooding, 1974), but estimates of the extent of this loss vary widely (reviewed in Mather and Keenan, 1983). In regions of globules where the bilayer membrane appears to have been lost, a granular material covers the surface. This material may be the coating which was on the surface of lipid droplets within milk-secreting cells. Stability of MLGM is discussed in more detail under Section VII.

Membranes can be released from milk lipid globule suspensions by several processes, including freezing and thawing, vigorous agitation (churning) (Keenan *et al.*, 1988), exposure to nonionic detergents like Triton X-100 (Patton, 1982) or to conjugated bile salts like taurodeoxycholate (Patton *et al.*, 1986), or by suspension in polar, aprotic solvents (Dapper *et al.*, 1987). After release, membranous material normally is

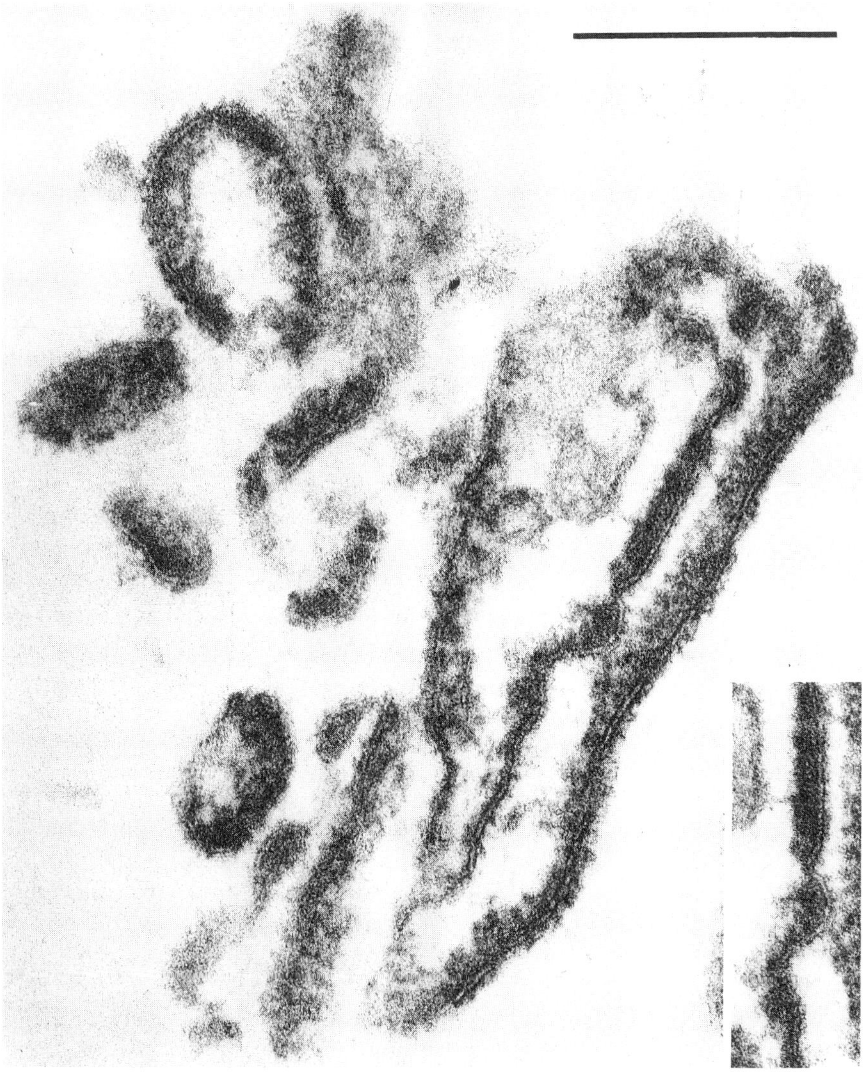

Figure 2 Electron micrograph of a cow MLGM preparation. This preparation consists largely of sheets of membrane which display typical bilayer membrane structure. The inner (originally cytoplasmic) face of the membrane is coated with a densely staining material of variable thickness. The coat material on the inner face of the membrane largely is amorphous, but regularly arranged particulate or globular structure sometimes is observed (insert). Scale bar denotes 0.2 μm. This plate was reproduced from Keenan *et al.* (1977) with permission of the publisher.

collected by centrifugation at *g* forces up to or exceeding 100,000. Alternatively, membrane material can be caused to aggregate by reducing the pH or by adding ammonium sulfate to the suspension, and can be collected by low-speed centrifugation (Brunner, 1965; McPherson and Kitchen,

1983). Milk lipid globules and membranes derived therefrom are identical or nearly so in distribution of phospholipids, but there are major quantitative differences in polypeptide composition of intact globules and isolated MLGM (Mather and Keenan, 1975; Keenan *et al.*, 1988). Depending upon the method and temperature of globule disruption, appreciable amounts of proteins and polar lipids remain associated with the congealed lipid (butter) or are dispersed in the aqueous phase entrained in the congealed lipid. When suspensions of washed lipid globules are churned at low temperatures, a considerable amount of proteinaceous material remains associated with the surfaces of congealed lipid droplets (Deeney *et al.*, 1985). The method and extent of washing of lipid globules to remove milk serum constituents, prior to release of the membrane, also can alter composition of the material recovered ultimately as MLGM. The extent of removal of peripheral (extrinsic) proteins under various washing conditions has not been quantified. The method used for collection of MLGM, and any subsequent steps used to remove entrained materials from the membrane, can alter composition as well. Relatively large proportions of xanthine oxidase and some membrane glycoproteins can be selectively removed from MLGM (Mather *et al.*, 1977; Keenan *et al.*, 1977a).

B. Gross Composition

Over 95% of the total lipids in milk from cows (Huang and Kuksis, 1967; Jenness, 1974; Patton and Jensen, 1976) and humans (Bracco *et al.*, 1972; Jensen *et al.*, 1980; Blanc, 1981; Jensen, 1989) is recovered in the globule fraction upon centrifugal fractionation of milk into globule and milk serum or skim milk phases, and 95% or more of the globule lipids are triacylglycerols. Much of the lipid in milk serum is present in a heterogeneous fraction of membrane and membrane fragments which, from the milk of cows, has been characterized thoroughly (reviewed in Patton and Keenan, 1975; Keenan *et al.*, 1988). Phospholipids (30 to 45%) and triacylglycerols (40 to 55%) comprise the bulk of the lipids in milk serum (Huang and Kuksis, 1967; Patton and Keenan, 1971; Patton *et al.*, 1973, 1980b). The percentage of the total globule mass accounted for by membrane material has not been determined with certainty. From data summarized by Brunner (1965) and by Mulder and Walstra (1974), membrane-associated materials may comprise from about 2 to more than 6% of the mass of globules. This range of values must be considered as a crude estimate at best, as differences in globule and membrane preparation methods have a major effect on the value obtained. Globules must be washed sufficiently to remove adsorbed or adherent milk serum constituents, yet extensive washing also will remove loosely associated but true constituents of the membrane. An additional complication is that membrane material will be entrained in the congealed lipids (butter) when globules are treated to release MLGM material at temperatures below the solidification point of the triacylglycerol mass.

Available information on gross composition of MLGM from cow and human milks is provided in Table I. Compositional data for human MLGM largely is from a limited number of studies. In most analyses, proteins plus lipids together have accounted for over 90% of the membrane dry weight, but relative proportions of proteins and lipids vary widely. This variation can be due to differences in methods for release and recovery of membrane and to other factors such as breed, age, and stage of lactation. Once isolated, MLGM can be subfractionated by isopycnic density gradient centrifugation into a range of density classes. Density of fractions is correlated inversely with both phospholipid and total lipid contents. All fractions have similar polypeptide patterns, as judged from patterns observed upon separation of polypeptides by electrophoresis under denaturing conditions. In one study, various density subfractions also were found to have similar or identical specific activities (units of activity/unit protein) of certain MLGM-associated enzymes (Mather *et al.*, 1977). However, Kitchen (1977), using a different procedure to prepare and subfractionate MLGM according to density, noted large differences in enzyme specific activities in different density classes.

The greatest variation in membrane composition is the content of neutral lipids, principally triacylglycerols. What part of the triacylglycerol associated with isolated MLGM represents a true membrane constituent, in contrast to an adsorbed or entrained contaminant, is not known. Since cell

TABLE I
Gross Composition of Cow and Human Milk Lipid Globule Membrane Preparations

Constituent class	Unit	Cow[a]	Human[b]
Protein	weight %	25 to 60	—
Total lipids	mg/mg protein	0.5 to 1.1	1.46
Phospholipids	mg/mg protein	0.13 to 0.34	0.35
Neutral lipids	mg/mg protein	0.25 to 0.88	1.1
Glycosphingolipids[c]	μg/mg protein	13	32
Hexoses	μg/mg protein	108	45
Hexosamines	μg/mg protein	66	44
Sialic acids	μg/mg protein	20	18
Glycosaminoglycans	μg/mg protein	0.1	—
RNA	μg/mg protein	20	15

[a]Taken from compilation in Keenan *et al.* (1988).
[b]Data from human MLGM compiled from Martel *et al.* (1973), Bouhours and Bouhours (1979), Takamizawa *et al.* (1986b), and Jensen (1989).
[c]Calculated from published data assuming average molecular weights of neutral glycosphingolipids and gangliosides of 850 and 1470, respectively.

surface membranes, isolated from homogenates of mammary gland as well as other tissues, contain only small amounts of triacylglycerols, one assumes that the triacylglycerols associated with MLGM in part originate from the core lipid. On a protein basis the phospholipid content of MLGM is more constant than is the neutral lipid content; an average value for phospholipid is about 0.25 mg per milligram protein for the cow, and a similar value was obtained for human. Martel *et al.* (1973) reported a much lower value for phospholipid (0.085 mg/mg protein) for human MLGM, but this value undoubtedly is erroneously low. MLGM, like plasma membranes, is enriched in glycosphingolipids in relationship to intracellular membrane systems. The total amount of glycosphingolipids (comprising neutral glycosphingolipids and gangliosides) in cow MLGM is about 13 μg per milligram protein. Bouhours and Bouhours (1979) reported a total neutral glycosphingolipid content of human MLGM of 11.5 μg/mg protein. The ganglioside content of human MLGM has not been measured directly, but Otnaess *et al.* (1983) reported a content of 11 mg/l milk, and Takamizawa *et al.* (1986) found 10 to 23 μmol of ganglioside sialic acid per liter of human milk. Using average values for mass of MLGM in milk, and for mass of MLGM which is protein, the approximate amount of gangliosides would be about 20 μg/mg protein. Several authors have found RNA in MLGM preparations; Swope and Brunner (1965) made careful measurements and found about 20 μg RNA/mg protein in cow MLGM, and Martel *et al.* (1973) found 15 μg RNA/mg protein in human MLGM. Extraction of membranes with high ionic strength buffers reduced RNA levels in MLGM to about 10 μg per milligram protein (Jarasch *et al.*, 1977). This RNA may be a constituent of the primary MLGM; alteratively, it may originate from ribosomes associated with the surfaces of lipid droplets within the cell (Dylewski *et al.*, 1984) or from entrainment of ER membranes or ribosomes in cytoplasmic crescents. DNA has not been detected in cow (Jarasch *et al.*, 1977) and human (Martel *et al.*, 1973) MLGM preparations. Hexose (glucose, galactose, mannose, and fucose), hexosamine (glucosamine and galactosamine), and sialic acid (*N*-acetyl- and *N*-glycoylneuraminic acids) contents of MLGM have been measured (Table I). In aggregate total, these carbohydrates amount to just under 0.2 mg per milligram protein. Most of the MLGM-associated carbohydrates are covalently bound to proteins and lipids; it is not likely that much free carbohydrate is associated with the MLGM. Glycosaminoglycans, normally associated with basement membranes, have been isolated from cow and human MLGM preparations. Lis and Monis (1978) identified hyaluronic acid, chondroitin sulfate, and heparin sulfate in the glycosaminoglycan fraction from cow MLGM. However, the value of 58 μg glycosaminoglycans per milligram protein, calculated from the data of Lis and Monis (1978), seems to be unrealistically high. Shimizu *et al.* (1981) confirmed the presence of heparin sulfate and chondroitin sulfate in glycosaminoglycans from MLGM; they found a total glycosaminoglycan content of about 0.1 μg per milligram protein in cow MLGM and a 5- to 10-fold higher amount in human MLGM.

C. *Lipid Composition*

1. *Neutral Lipids*

Values reported for the amounts of most of the lipid classes of human MLGM fall within the range of values reported for cow MLGM (Table II). In membranes from both sources, triacylglycerols are the most abundant lipid class. Since preparative method has a major influence on the amount of triacylglycerol associated with MLGM, it is likely that some of the triacylglycerols originate from the core lipids and adsorb onto or partition into the membrane material. Fatty acid composition of MLGM-associated triacylglycerols differs from that of milk fat in that MLGM triacylglycerols contain considerably higher proportions of long-chain, saturated fatty acids (principally palmitate and stearate) (Kitchen, 1977). Vasic and DeMan (1966) and Bracco *et al.* (1972) found that when fat globules were destabilized at temperatures above 37°C, isolated MLGM was not enriched in high-melting triacylglycerols. Walstra (1974) suggested that these high-melting triacylglycerols may be derived from fat crystals which "contaminate" the membrane during the cooling and churning process. Results from microelectrophoretic characterization of lipid globules led Newman and Harrison (1973) to conclude that the outer surface of the MLGM contains little neutral lipid. Trace to substantial quantities of mono- and diacylglycerols usually are found in MLGM lipids. Whether these partial glycericdes are true constituents of membranes or are products of lipolytic degradation of triacylglycerols or phosphoglycerides is not known. The amounts of unesterified fatty acids found in MLGM preparations, at least from cow, vary widely. This variation may be due to variation in lipolytic activity. Sterols and sterol esters invariably are found in MLGM lipids; however, there is extensive variation in sterol content judged from values reported for cow MLGM. Some of this variation may be the result of preparative method-induced differential partitioning of sterols between core and membrane lipids (discussed in Keenan *et al.*, 1988). Cholesterol is present in lipid droplets before secretion as milk lipid globules, but its distribution between the core lipids and material of the surface coat is unknown (Dylewski *et al.*, 1984). Cholesterol is a known and abundant lipid constituent of plasma membrane, from mammary gland (Keenan *et al.*, 1970; Kanno *et al.*, 1987), as well as other tissues (reviewed in Van Meer, 1989). An ultrastructural approach, in which cholesterol–filipin complexes were visualized in freeze-fracture replicas, provided evidence for the presence of cholesterol both in or at the surface of core lipids, and in the MLGM in intact milk lipid globules (Martin, 1989). This observation does not rule out partitioning of cholesterol between membrane and core lipids before globules are harvested and destabilized. In MLGM from both cows and humans, sterol esters account for a small proportion, 10% or less, of the total sterols. Fat globules, but not necessarily MLGM, from human milk contain a much higher amount of cholesterol than do those in cow milk

TABLE II
Lipid Composition of Cow and Human Milk Lipid Globule Membrane Preparations

Constituent class	Cow[a]	Human[b]
	% of total lipid	
Triacylglycerols	62	58
Diacylglycerols	9	8
Monoacylglycerols	Trace	0.6
Sterols	0.2 to 2	0.7
Sterol esters	0.1 to 0.3	Trace
Unesterified fatty acids	0.6 to 6	7.3
Hydrocarbons	1.2	Trace
Phospholipids	26 to 31	23
	% of total phospholipid	
Sphingomyelin	22	26
Phosphatidyl choline	36	30
Phosphatidyl ethanolamine	27	37
Phosphatidyl inositol	11	5
Phosphatidyl serine	4	1
Lysophosphatidyl choline	2	2

[a]As compiled in Keenan *et al.* (1988); Patton and Keenan (1975).
[b]As compiled in Jensen (1989).

(Braco *et al.*, 1972). Cholesterol is the major sterol in human and cow milks, accounting for over 90% of the total sterol fraction. About 17 different sterols have been isolated from cow milk; those which have been identified include 7-dehydrocholesterol, campesterol, stigmasterol, and β-sitosterol (reviewed in Blanc, 1979). In addition to cholesterol, 7-dehydrocholesterol and phytosterols have been found in human milk. Which of these sterols are in MLGM has yet to be determined with either species. Squalene, the abundant hydrocarbon of cow and human milk fat (Bracco *et al.*, 1972), has been identified as a constituent of cow MLGM (Thompson *et al.*, 1961). β-Carotene also is present in the hydrocarbon fraction of cow MLGM (Thompson *et al.*, 1961), but most of the carotenoid of the globule appears to be in the core lipid (Patton *et al.*, 1980). Carotenes also are associated with human milk lipid globules, but the distribution between core lipids and the MLGM has yet to be determined.

2. *Phospholipids*

Phospholipids of milk are mainly present in lipid globules (about 60% of the total) and in the heterogeneous membrane fraction of skim milk (about 40%) (Huang and Kuksis, 1967; Patton and Keenan, 1971). The phospholipids of lipid globules are mainly, if not exclusively, recovered with MLGM when globules are destabilized at temperatures of 40°C. The same five major phospholipids, with a similar pattern of distribution, are present in MLGM from cow and human (Table II), as well as in milk or MLGM from ass, camel, Indian buffalo, pig (Morrison 1968), goat (Patton and Keenan, 1971), mouse (Calberg-Bacq *et al.*, 1976), and rat (Keenan *et al.*, 1971). Human MLGM has relatively less phosphatidyl choline, but relatively more phosphatidyl ethanolamine and sphingomyelin, than does cow MLGM. Phosphatidyl serine and phosphatidyl inositol account for a higher proportion of phospholipids in cow MLGM than in human MLGM, but this difference was not observed upon comparison of phospholipid distribution in whole milk specimens from these species (Morrison, 1968). The phospholipid distribution pattern seen with MLGM is similar to that found with plasma membranes from mammary gland (reviewed in Keenan *et al.*, 1988; Kanno, 1990) and other organs in that sphingomyelin is high and phosphatidyl choline is low. Intracellular membranes have a much lower sphingomyelin to phosphatidyl choline ratio than that found with plasma membranes and MLGM. In addition to the five major phospholipids, lyso-derivatives of the major phospholipids also are found in MLGM, but these are relatively minor constituents in fresh samples handled so as to minimize lipolytic activity. What proportions of the various phosphoglyceride classes are alkyl or alkenyl ethers is not known as analytical methods used to establish distribution of phospholipids were unable to make such separations. Alkyl and alkenyl ethers were found in choline and ethanolamine phosphoglyceride fractions from whole milk of cows (Hay and Morrison, 1971).

Several workers have noted extensive similarity in distribution of phospholipids of skim milk, or membranes isolated therefrom, and lipid globules and in distribution of major fatty acids within each phospholipid class. Given this apparent unity of milk phospholipids, it has been assumed that those of whole milk and MLGM originate from a common cellular source and, by inference, are identical or nearly so in fatty acid composition. Actual data to validate this assumption are lacking. A very large number of minor fatty acids have been identified in milk lipids, but we have no information on the distribution of most of these fatty acids within individual lipid classes and cannot assume that there is not a preferential occurrence of any given minor fatty acid in a particular lipid class of MLGM or of skim milk.

3. Glycosphingolipids

Glycosphingolipids, relatively minor constituents of the MLGM (Table I), have been the subject of a number of investigations over the past several years. This attention has been due to the recognition that certain glycosphingolipids and products of glycosphingolipid catabolism have functional roles in a number of biological phenomena, such as cell–cell interaction, differentiation, proliferation, immune recognition, transmembrane signaling, and as receptors for certain hormones, growth factors, and toxins (reviewed in Hakomori, 1981, 1990; Fishman, 1986; Hannun and Bell, 1989; Karlsson, 1989). Two general classes of glycosphingolipids have been found in milk and MLGM: neutral glycosphingolipids (with uncharged sugars, commonly called cerebrosides) and acidic glycosphingolipids (containing sialic acid and called gangliosides). These lipids have a ceramide (a sphingosine base to which a fatty acid is attached via an amide bond) to which is linked, through the ceramide primary hydroxyl group, a mono- or oligosaccharide. Cow MLGM contains two major neutral glycosphingolipids, glucosyl- and lactosylceramides, in nearly equimolar proportions. Neutral glycosphingolipids with more complex carbohydrate structures have not yet been found in cow MLGM. Glucosyl- and lactosylceramides of MLGM are enriched in long-chain fatty acids, especially 16:0, 18:0, 22:0, 23:0, 24:0, and 24:1, but appear to lack hydroxylated fatty acids. These lipids also occur in skim milk, but those from skim milk have relatively lesser amounts of the 22-, 23-, and 24-carbon fatty acids (Kayser and Patton, 1970). In human MLGM, there is a total neutral glycosphingolipid content of about 11 μg/mg protein (Bouhours and Bouhours, 1979). In human MLGM, there is about double the amount of monohexosylceramides as dihexosylceramides. Galactosylceramide comprises about 88%, and glycosylceramide about 12%, of the total monohexosylceramides. The dihexosylceramide of human MLGM is lactosylceramide (Bouhours and Bouhours, 1979). Glucosyl- and lactosylceramides of human MLGM did not contain hydroxylated fatty acids, but galactosylceramide contained both hydroxylated and nonhydroxylated fatty acids. Among nonhydroxylated fatty acids, 16:0, 22:0, 24:0, and 24:1 were most abundant, accounting for about 70% of the total fatty acids in neutral glycosphingolipids. Over 80% of the hydroxy fatty acid was accounted for as 22:0, 23:0, and 24:0.

Nine different gangliosides have been detected and characterized structurally, at least partially, in cow MLGM or buttermilk. Those identified to date have been GM_3, GM_2, GM_1, GD_3, GD_2, GD_{1b} (Huang, 1973; Keenan, 1974; Bushway and Keenan, 1978), GT_3, and novel branched-chain mono- and trisialogangliosides (Takamizawa *et al.*, 1986a). In most investigations, GD_3 emerged as the major ganglioside of cow MLGM, and GM_3 was the next most abundant ganglioside homolog. The aggregate total of the other gangliosides amounted to perhaps no more than 20% of the total ganglioside content of the membrane. GD_3 and GM_3 are the major gangliosides of human MLGM; trace amounts of gangliosides tentatively identified as GM_2 and GM_1 also have been detected in human MLGM (Takamizawa *et al.*, 1986b; Laegreid and Otnaess, 1986). In human MLGM there is a major difference in ganglioside distribution with stage of lactation (Takamizawa *et al.*, 1986b). In milk samples collected 2–6 days postpartum, the GM_3:GD_3 ratio was 0.2 to 0.3, but in samples collected

60–390 days postpartum, this ratio was greater than 3. Milk collected at 8–40 days postpartum had GM_3:GD_3 ratios intermediate between these extremes. Gangliosides of MLGM have fatty acid compositions similar to those of neutral glycosphingolipids and sphingomyelin, perhaps indicating that all of these sphingosine-containing lipids originate from a common pool of ceramides.

4. Fat-Soluble Vitamins and the Membrane

Certain of the vitamins, A, D, E, and K are said to be fat soluble. When foods, such as milk and milk products, are extracted with so-called fat solvents, these vitamins are obtained in the solvents along with the lipids. From this, one might conclude that these vitamins in the native state are dissolved in the triacylglycerol core of the milk lipid globule. While this may be true of ester forms, such as of vitamins A and E, it would not be true of those containing polar groups, and all four vitamins exist at least in some proportion in forms containing unesterified hydroxyl groups. Under physiological conditions, these latter forms are hydrated and would resist solution in lipid droplets. More likely, they occur in membranes including the MLGM, and are oriented in the surface of lipid droplets. However, heat connected with food processing (pasteurization and sterilization) might overcome these distributions and actually extract some of the fat-soluble vitamins into lipid droplets.

There are suggestions in the literature that one or more of the fat-soluble vitamins of milk reside in part in the MLGM and are accounted for more adequately on a surface distribution basis rather than on fat content. There is also the possibility that these vitamins are associated with carrier proteins and membrane material in the skim milk. But the data are far too sketchy at this time to enable presentation of concentrations of the vitamins, either in the MLGM or in the intact globule, of either human or cow. To achieve that goal will require systematic fractionation and analysis of milk and its compartments. At the moment, the best guidance is from the data on fat-soluble vitamin concentrations in whole milk.

Related to the forgoing consideration is the question of carotenoids. Some of these, principally α- and β-carotene, are precursors of vitamin A. While there are reports of carotenoids in the MLGM (Thompson *et al.*, 1961), they apparently do not occur there with consistency (Brunner, 1965). In a study of the distribution of carotenoids in bovine mammary tissue and milk lipid globules (Patton *et al.*, 1980a), MLGM was found to be devoid of carotenoids.

D. Protein Composition

Consideration is given here mainly to proteins of the human and cow milk lipid globules because they account for most of the information and they are of greatest interest and significance. At the same time, it should be borne in mind that the major human and cow globule proteins, such as butyrophilin and xanthine oxidase, seem to show commonality across species. This suggests that while these proteins have been evolving among species to yield some variations in size, sequence, etc., they retain properties essential to their functions in the lactating cell and the nursling.

Protein represents only about 1% of the globule mass or 0.3 to 0.4 g/liter of human or bovine milk (Patton and Huston, 1986). Few of the individual proteins have been isolated and extensively characterized. However, their behavior and staining properties on SDS–gels give useful indications of their concentrations, relative molecular weights (M_r's), and whether or not they contain carbohydrate. Resolution of proteins for the two species by SDS–PAGE is shown in Figure 3. The principal bands are identified but there are many minor bands that remain unknown. The number of these will depend on the size of sample and staining technique. No doubt, some are enzymes (Section VI, E). Western blotting coupled with immunostaining affords a very sensitive technique for investigating the identities of these bands.

1. Compartmentation of Globule Proteins

The question arises as to which proteins of the globule are associated with the membrane, which with the original fat droplet surface, and which with the intervening cytoplasmic space. Because of the potential for producing artifacts in isolating these compartments, these questions cannot be answered with certainty. In fact, the process of globule secretion may effect change in the position of the proteins. If one uses the evidence of proteins released by globules subjected to churning to identify those residing in the cytoplasmic space, two principal bands are seen on gels. One at M_r 155,000 corresponding to xanthine oxidase and the other at M_r 15,000 (Patton *et al.*, 1986). Since xanthine oxidase is a component of the coat complex on the cytoplasmic (inner) face of the globule membrane, it is reasonable that some of it might be released by churning. Excepting the contribution of proteins in cytoplasmic crescents, as discussed under Section IV, the secretory mechanism effectively expels the cytoplasm from most globules and one would not expect much contribution to total globule protein from this compartment. A comparison of the butter phase from the churning process with the isolated MLGM reveals essentially identical gel protein patterns (Patton *et al.*, 1986). This suggests that the proteins of intact globules are largely components of the MLGM. However, there is evidence that lipid droplets isolated from within the lactating cell carry a spectrum of proteins originating in the endoplasmic reticulum (Deeney *et al.*, 1985). Carryover of these proteins to the secreted globules is discussed under Section II.

2. The Principal Globule Proteins

Resolution of the principal globule proteins by SDS–PAGE is shown in Fig. 3. Note the relatively strong staining of the bands for xanthine oxidase

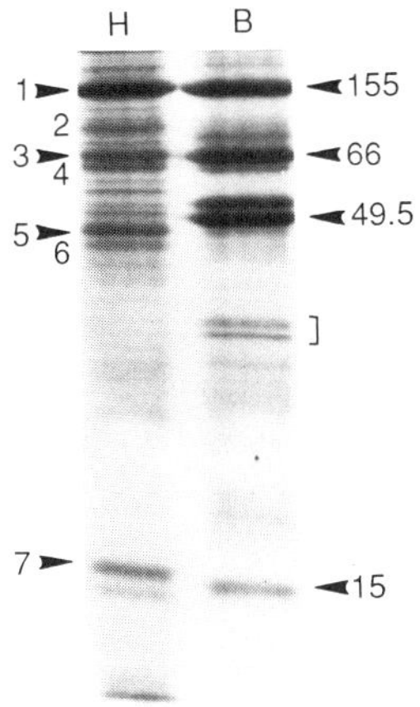

Figure 3 An SDS gel comparing proteins of human (H) and bovine (B) milk lipid globules. Major matching bands for the two species are indicated by arrowheads. Relative molecular weights (M_r) in kDa are at right. The numbered positions at left correspond to: 1, xanthine oxidase (monomer); 2, PAS-IV (M_r 80 kDa); 3, butyrophilin; 4, butyrophilin-related protein (M_r 62.5 kDa); 5, PAS-VI (glycoprotein B); 6, actin–keratin (?) band; 7, component 21. The bracket (right) locates two contaminating casein bands. The samples each contained 50 μg of protein. The gel (12.5% acrylamide) was stained with Coomassie blue.

and butyrophilin in both the human and the cow samples. This is generally the case in other species as well. These relatively high concentrations, coupled with the occurrence of the two proteins across species lines, suggest a role of fundamental importance for them in the MLGM. In comparing the patterns for the human and cow samples, it is evident that the former has many more minor bands. These are contributed at least in part by cytoplasmic crescents (see Section IV and Figure 1). The mucins are not observed in Figure 3 because of their limited penetration into a 12.5% acrylamide gel and their poor staining with Coomassie blue. However, see Figure 4 for the bovine mucin. Available information on the globule proteins is summarized below and in Table III.

a. Mucin(s). High molecular weight glycoproteins, now known as epithelial mucins, have been found in milk lipid globules of all species studied to date. In the human there are actually two such proteins detectable by SDS–PAGE; one that resolves in the stacking gel (3% acrylamide) and the other that enters a 4% acrylamide running gel. These proteins have many names and abbreviations. They are designated here as mucin A and mucin B, respectively. The former is approximately 80% and the latter about 50% carbohydrate. Both are very high molecular weight (> 400,000). Cow globules only exhibit one mucin composed of about 50% carbohydrate and ranging in M_r from 170,000 to 205,000. We refer to this mucin as periodic acid-Schiff reagent I (PAS-I) in accord with the nomenclature of Mather

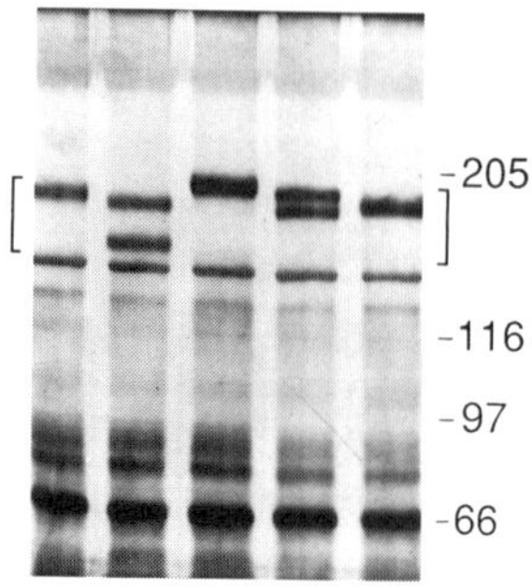

Figure 4 An SDS gel showing polymorphism of the bovine milk lipid globule mucin, PAS-I, among five animals (in brackets). Note variable number and mobility of bands. Position of molecular weight references is indicated in kDa at left. The gel contained 6% acrylamide and was stained with silver reagents. Each of the five samples were 10 μg of total globule protein.

and Keenan (1975) for the cow MLGM proteins. It appears to be of the same family of glycoproteins as human mucin B. There is considerable variation in the M_r of the milk lipid globule mucins among species. A further unique aspect of these glycoproteins is their polymorphism (Patton and Huston, 1987; Swallow *et al.*, 1987) which results from variable numbers of a tandemly repeated 20-amino acid segment (Gendler *et al.*, 1988; Spicer *et al.*, 1991). Thus, the individual alleles, one from the mother and one from the father, usually express proteins of different sizes. This is manifest on SDS–gels by two bands which vary in mobility from one milk sample to another. Both the stacking gel and running gel mucins exhibit this type of polymorphism (Patton and Huston, 1987; Patton *et al.*, 1989). However, some species, such as the cow, goat, and mouse, have only the running gel mucin, and in the case of the mouse, the polymorphism appears to have been lost although mutated tandem repeats remain (Spicer *et al.*, 1991). The polymorphism of cow PAS-I, as revealed by SDS–PAGE, is shown in Figure 4. Under the conditions in Figure 4, the human B mucin would be at the very top of the gel. Characterizational work has been done on human mucin A (Shimizu and Yamauchi, 1982) and B (Shimizu *et al.*, 1986) and on cow PAS-I (Snow *et al.*, 1977). Human mucin B (Gendler *et al.*, 1990) and the mouse mucin that resolves in the running gel on SDS–PAGE (Spicer *et al.*, 1991) have been sequenced via the cDNAs. The 20-amino acid repeating unit is rich in serines and threonines, which serve as O-glycosylation sites. On SDS–gels, the mucins stain readily with PAS, but very weakly if at all with Coomassie blue (CB).

b. Xanthine oxidase (monomer, 155,000; band 1 of Figure 3). On gels, this protein stains strongly with CB but does not stain with PAS. Thus, it contains little or no carbohydrate. However, there is a human carbohydrate-containing protein which occurs in the 155,000 M_r-region of the gel which has been isolated and characterized regarding amino acid

TABLE III
Information on the Major Milk Lipid Globule Proteins Detected by SDS–PAGE

Protein	M_r[a]	Staining[b]		Isolation	Sequence	Composition		References[g]
		CB	PAS			Amino acid	Carbohydrate	
Mucin A (human)	[c]	–	+	Yes	No	Yes	Yes	A,B
Mucin B (human)	> 400	–	+	Yes	Yes	Yes	Yes	C,D
PAS-I (cow)	165 to 205	–	+	Yes	No	Yes	Yes	E,F
Xanthine oxidase	155	+	–	Yes[d]	No[e]	Yes	No	G,H
PAS-IV	80	±	+	Yes[d]	Partial	Yes[d]	Yes[d]	I,P,Q
Butyrophilin	66	+	+	Yes	Yes	Yes	Yes	G,J,K
Butyrophilin	62.5	+	+	No	No	No	No	G,J,K
PAS-IV								
Human	46	+	+	Yes	No	No	No	L
Cow	49.5	+	+	Yes	No	Yes	Yes	M
?	ca.42	+	–	No	No	No	No	—
PAS-VII	39	±	+	Yes[f]	No	Yes[f]	Yes[f]	N,O
?	15	+	–	No	No	No	No	—

[a] Approximate molecular weights, kDa. For location in SDS gels, see Fig. 2.
[b] CB, Coomassie blue; PAS, periodic acid—Schiff's reagent.
[c] This mucin resolves in a 3% acrylamide stacking gel, but does not enter the running gel.
[d] Cow only.
[e] The primary sequence of rat liver xanthine oxidase has been reported (Amaya *et al.*, 1990).
[f] Human only.
[g] Reference key: A, Shimizu *et al.*, (1986); B, Patton *et al.* (1989); C, Shimizu and Yamauchi (1982); D, Gendler *et al.* (1990); E, Snow *et al.* (1977); F, Patton and Patton (1990); G, Freudenstein *et al.* (1979); H, Cheng *et al.* (1988); I, Mather *et al.* (1980); J, Jack and Mather (1990); K, Imam *et al.* (1981); L, Ceriani *et al.* (1983); M, Basch *et al.* (1976); N, Wiman *et al.* (1979); O, Imam *et al.* (1982); P, Greenwalt *et al.* (1990); Q, Greenwalt and Mather (1985).

and carbohydrate composition (Imam *et al.*, 1981). In the native state, xanthine oxidase exists as a homodimer, molecular weight 283,000. The M_r observed on gels for the monomer of 155,000 is thus somewhat high. Its properties as an enzyme and its unique composition are discussed under Section VI, E. The most perplexing aspect of this membrane component concerns its function in MLGM or in lipid globule secretion (discussed below).

c. Butyrophilin (band 3 of Figure 3). This protein, with M_r of 66,000, stains both with CB and PAS. It is the major protein component of the MLGM but it seems likely that xanthine oxidase may at times constitute nearly as much of the total globule protein. In the isolation of the membrane from the globule a considerable amount of xanthine oxidase is shed because it is not a true integral membrane component, but is complexed rather loosely, at least in part, with butyrophilin in the membrane coat A butyrophilin-like protein has been detected in MLGM of a variety of other species (Heid *et al.*, 1983). Sequencing and other evidence indicates that cow butyrophilin is composed of an extracellular domain which is glycosylated, a transmembrane region, and a cytoplasmic tail (Jack and Mather, 1990). A glycoprotein of M_r 70,000 isolated from human MLGM (Imam *et al.*, 1981; Ceriani *et al.*, 1983) appears to be butyrophilin. Its amino acid and carbohydrate composition have been determined (Imam *et al.*, 1981) and monoclonal antibodies to it have been prepared (Ceriani *et al.*, 1983).

d. M_r 62,500 (band 4 of Figure 3). This protein may be a close structural relative of butyrophilin.

e. M_r 46,000 to 52,000 (band 5 of Figure 3; also known as PAS-VI and glycoprotein B). This protein stains with both CB and PAS. The human protein appears to be somewhat smaller at M_r 46,000 than that of the cow (M_r 49,500) and a similar goat protein is even slightly higher (M_r 52,000) (Patton and Hubert, 1983). All three bind the lectin, concanavalin A. The amino acid and carbohydrate compositions of the cow protein have been determined (Basch *et al.*, 1976). Because some breast tumors express this protein, monoclonal antibodies to it have been prepared and used to screen for breast cancer (Ceriani *et al.*, 1983).

f. M_r 42,000 (band 6 of Figure 3). Although the band for this protein is broad on SDS–gels, it does not appear to contain carbohydrate. The molecular weight range is that of actins and some keratins, which would be plausible since proteins of the filamentous network may function in connecting the membrane to the cell's cytoskeleton.

g. M_r 39,000. There is a relatively broad, PAS-positive band in this region on SDS–gels. The human glycoprotein has been isolated and partially characterized including amino acid and carbohydrate composition (Imam *et al.*, 1982). It is not known whether this protein is the same as the HLA-DR-like antigen at M_r 35,000 detected by Wiman *et al.* (1979).

h. M_4 15,000 (band 7 of Figure 3). This band is seen on gels resolving human, cow, and goat globule proteins. It appears to be carbohydrate free. Since it is released readily during churning or freezing-thawing of globules, it is probably a peripheral rather than an integral membrane protein, and it may reside on the inner (cytoplasmic) face of the membrane.

Suggested relationships between proteins of the human and bovine globule must be considered tentative until further evidence is provided such as cross-reactivity with antibodies or homology of amino acid sequences. There is still much characterizational work to be done on the globule proteins. For example, the bovine protein in the M_r 54,000 region (Figure 3) seems to have received no attention to date, nor has the M_r 15,000 component evident in the patterns of both species (band 7, Figure 3).

3. *Lectin Binding*

Lectins have been used extensively in the characterization and purification of MLGM proteins. A study by Farrar *et al.* (1980) is particularly useful in that binding of 12 different lectins to intact human and cow globules was evaluated. Another informative study on binding of lectins to MLGM proteins is that of Murray *et al.* (1979). Peanut lectin, which binds to β-D-galactopyranosyl-(1,3)-*N*-acetyl-galactosamine, was used to purify human mucin B (Shimizu and Yamauchi, 1982). The bovine mucin, PAS-I, also binds peanut lectin strongly (Patton *et al.*, 1989). Imam *et al.* (1981) employed concanavalin A in the isolation of their M_r 70,000 human glycoprotein (butyrophilin ?); the goat glycoprotein with M_r 52,000 binds concanavalin A strongly (Patton and Hubert, 1983).

4. *Functions of the Milk Lipid Globule Membrane and Its Proteins*

The proteins of milk, including those of the lipid globules, can be viewed as having functions in the mammary gland of the mother or in her nursling. Further, it is conceivable that a particular milk protein might have functions in both locations. In this regard, it is helpful to view milk in an evolutionary sense. Biologically, milk is a very crucial element in survival of the mammalian species. Exclusive of man, with his capacity for sophisticated manipulation of food requirements, no milk means death of the newborn. Limited or poor quality milk means undernourished and sickly young who may not mature or reproduce. No doubt there has been selection of milk proteins to ensure adequate milk quality and quantity, thus perpetuating species. By the same token, in the evolution of a mammal, it is possible that a milk component that was indispensable at one time may have become nonessential at a later stage, and if harmless, carried along genetically as the species further evolved. This is not quite applicable to a protein in the sense that proteins will have some nutritional value in any event. But the likelihood is that most, if not all, milk proteins have been honed by evolution to facilitate survival of the particular species.

The nutritive value of milk for the young is fairly straightforward in that a very large number of required nutrients and their amounts can be specified; milk of a given species, in the main, meets the requirements for young of that species. But beyond this there are many other complex considerations which are now coming to the fore. Such things as cell growth factors, hormones, enzymes, antibodies, and immunogenic substances in milk may also benefit the newborn. In addition, the control of gut microflora is of critical importance to health and development of the young human. Being minor if not trace components of milk, the MLGM proteins make very little contribution to the classic nutritive value of milk, but there is likelihood that they may benefit the young in the same manner as some of these foregoing factors that aid well being.

Because of the dynamic needs and relative underdevelopment of metabolic systems in the newborn, the question of what happens to milk in the young has become a matter or rising interest. It is a difficult area of research because of its invasive requirements. The infant brain doubles in size in the first 6 months postpartum on an exclusive diet of human milk. A difficult question is what milk components, if any, are used intact in the synthesis of brain during this period? In connection with possible contribution by the MLGM, it is pertinent that the infant brain is an extensive system of evolving membranes. This research area becomes even more challenging by reports, such as Lucas *et al.* (1992) and others cited therein, that human-milk-fed infants develop greater intelligence on average than do those fed formula.

a. The MLGM. A component of milk that has been an integral structure in the lactating cell, as is the case of the MLGM, obviously will have had important functions in that cell. There is evidence of MLGM protein involvement in enzymatic, receptor, transport, immunologic, and milk-secretory functions. Much of this is by inference from the composition of the membrane, its close resemblance to plasma membranes in general, and what is known about their functions. However, it is firmly established that the MLGM plays the central role in the secretion of milk lipid globules.

With respect to possible functions of the MLGM in the nursling, we propose that one function of this material is to serve as a decoy for pathologic bacteria and their toxins in the infant gut. Like the intestinal mucosa, the lactating cells are epithelial cells and their membranes are epithelial membranes. For many enteric bacteria of the type that cause sickness in the young, binding to the intestinal mucosa is an essential first step to infection. The same is true of the diarrhea-producing toxins of the type elaborated by *Vibrio cholerae* and *Escherichia coli*. To be effective, they must first bind to ganglioside in the mucosal membrane. If a constant supply of extraneous epithelial membrane material is coursing through the lumen of the gut, the chances that these bacteria and their toxins may be tied up by such membrane are favorable, thus minimizing binding to and invasion of the intestinal mucosa. It is a well-known fact that breast-fed infants are much less afflicted with diarrheal disorders than are those given formula. We contend that in the human, MLGM and its components play an important part in this difference.

b. Mucin(s). The following observations and speculations apply particularly to the human B mucin that enters a 4% running gel in SDS–PAGE and is variously named MUC-1, PAS-0, and the B-C mucin, among others. Comparison of the tandem repeats for the human and mouse mucins reveals that positions of the serines and threonines are conserved, whereas amino acids in other positions have undergone extensive mutation (Spicer *et al.*, 1991). This strongly implies that the oligosaccharide chains are an important functional element of these mucins, since in these locations the serines and threonines are glycosylation sites. The fact that there is far more mucin and it is of higher molecular weight in human milk lipid globules than in those of the cow (Patton *et al.*, 1989) or mouse (Spicer *et al.*, 1991) suggests that the mucins have evolved to be of greater importance in the human. The oligosaccharide chains of the mucins first are exteriorized on the apical surface of the lactating cell. A plausible function for them in this location is to serve as a barrier against invading (mastitic) microorganism. It is also reported that in this location, they are connected on the cytoplasmic side of the plasma membrane to elements of the cytoskeleton (Parry *et al.*, 1990). It has been suggested that the mucins function in the immunorecognition system (Patton and Huston, 1987). The human mucin tandem repeats carry blood group antigens (Dion *et al.*, 1990) and B and T cell epitopes (Gendler *et al.*, 1990). The T cell epitope is recognized by cytotoxic T cells (Barnd *et al.*, 1989). Human mucin B inhibits the growth of BALB/c 3T3 cells (Shimizu *et al.*, 1990). It has been suggested that the human mucins facilitate digestion of milk lipid globules (Shimizu and Yamauchi, 1982; Buchheim *et al.*, 1988).

c. Xanthine oxidase. In the lactating cell this protein tends to concentrate on the cytoplasmic face of the apical plasma membrane although it is dispersed throughout the cytoplasm (Jarasch *et al.*, 1981). In the apical position it becomes complexed with butyrophilin in what is known as the protein coat of the MLGM. When a mature milk lipid droplet reaches the apical membrane, the coated membrane surface commences to envelop the droplet in the secretion process. It is reasoned that properties of the coat bind the membrane to the droplet (Keenan and Dylewski, 1991). This appears more likely in that the coat proteins are reported to be acylated with long-chain fatty acids (Keenan *et al.*, 1982) which would enhance their hydrophobic attraction to lipid droplets. Thus, assisting in the secretion of milk lipid globules appears to be a plausible function of xanthine oxidase.

Xanthine oxidase occurs in remarkably high concentrations in milk (approx 35 mg/liter). It is the richest known source of the enzyme. Doubt is cast on its functioning as an enzyme in that it is nearly, if not completely, inactive in the milk of some species, e.g., human and goat. Molybdenum deficiency may explain the lack of activity in human xanthine oxidase. At this time, there is no good evidence why the lactating cell or the newborn would require such high levels of xanthine oxidase. A nutritional role for xanthine oxidase as a purveyor to the young of trace elements (iron and molybdenum), sulfur, and intact flavin–adenine–dinucleotide complex is another possibility.

d. Butyrophilin. This protein was so named because of its apparent affinity for the milk lipid droplet at secretion (Franke *et al.*, 1981). This seems to be a well-accepted working hypothesis, discussed above, on functions of the xanthine oxidase–butyrophilin coat complex. From its sequence (Jack and Mather, 1990), butyrophilin appears to be a classical integral membrane protein with a single transmembrane region and a sizeable cytoplasmic tail. This tail apparently is a receptor for or is interactive with xanthine oxidase. Butyrophilin does not appear to be expressed in any other tissue of the body and it is only evident in and from mammary tissue during lactation (Jack and Mather, 1990). This makes it likely that the role of butyrophilin is exclusively related to lactation. Butyrophilin is glycosylated in its exoplasmic segment and possibly in its cytoplasmic tail (Jack and Mather, 1990). The function(s) of this glycosylation is not known.

In addition, the human globule membrane contains histocompatibility antigens in the form of two glycoproteins at M_r 35,000 and 28,000 (Wiman *et al.*, 1979). There are no known functions for the other major proteins of the milk lipid globule listed in the foregoing, i.e., those at M_r 80,000, 46,000, 42,000, 39,000, and 15,000.

D. Enzymes of Milk Lipid Globule Membranes

Over 25 enzymatic activities have been detected and measured in MLGM from cow milk (Table IV). In some cases, these different activities may be catalyzed by the same enzyme using different substrates or acceptors. Well over half of the enzymes detected in cow MLGM are members of the hydrolase class. Oxidoreductases are next in abundance, followed by transferases. Aldolase, the only lyase reported, is a well-known cytosolic enzyme, and its presence may be indicative of cytoplasmic crescents. Acetyl-CoA carboxylase, the only ligase reported, was present in enzymatically inactive form (Shriver *et al.*, 1989). To date, isomerases have not been reported as constituents of MLGM preparations. With human MLGM, about 11 different enzymic activities have been reported (Table IV). Several enzymes with high specific activities in MLGM, such as adenosine

TABLE IV
Enzymatic Activities Detected in Cow and Human Milk Lipid Globule Membrane Preparations

Enzyme[a]		Cow[b]	Human[b]
Alkaline phosphatase	3.1.3.1	A	
Acid phosphatase	3.1.3.2	A	
5′-Nucleotidase	3.1.3.5	A	C
Phosphodiesterase I	3.1.4.1	A	C
Inorganic pyrophosphatase	3.6.1.1	A	
Nucleotide pyrophosphatase	3.6.1.9	A	
Phosphatidic acid phosphatase	3.1.3.4	A	
Adenosine triphosphatase	3.6.1.3	A	C
Cholinesterase	3.1.1.8	A	
UDP-glycosyl hydrolases	3.2.1._	A	
Glucose-6-phosphatase	3.1.3.9	A	C
Plasmin	3.4.21.7	A	
β-Glucosidase	3.2.1.21	A	
β-Galactosidase	3.2.1.23	A	
Ribonuclease I	3.1.4.22	A	C
Thiamine pyrophosphatase	3.6.1.6		C
Lipoamide dehydrogenase	1.6.4.3	A	
Xanthine oxidase	1.2.3.2	A	D
Thiol oxidase	1.8.3.2	A	
NADH oxidase	1.6.99.3	A	E
NADPH oxidase	1.6.99.1	A	
Catalase	1.11.1.6	A	
γ-Glutamyl transpeptidase	2.3.2.1	A	
Galactosyl transferase	2.4.1._	A	F,G
Glycosyl transferases	2.4._._		H
Aldolase	4.1.2.13	A	
Acetyl-CoA carboxylase	6.4.1.2	B	

[a]Common or trivial name of enzyme is followed by the Enzyme Commission (EC) reference number.

[b]Letter indicates that enzyme has been reported in MLGM of that species, and keys the reference: A, reviewed in Keenan *et al.* (1988); B, Shriver *et al.* (1989) found acetyl-CoA carboxylase to be present in enzymatically inactive form; C, Martel-Pradal and Got (1972); D, Zikakis *et al.* (1976); E, Burder *et al.* (1978); F, Martel *et al.* (1973); G, Martel and Got (1976a); H, Parodi *et al.* (1984) found enzymes of synthesis of dolichol monophosphomannose and dolichol monophosphoglucose, as well as those involved in transfer of glycosyl residue from these dolichol derivative to dolichol diphosphooligosaccharides.

triphosphatase, phosphodiesterase I, and 5′-nucleotidase, are enriched in plasma membranes from several tissues and frequently are used as marker enzymes for plasma membranes. As discussed under a previous section, xanthine oxidase is an abundant protein associated with milk lipid globules, and this enzyme can be in the oxidase or dehydrogenase form (Nakamura and Yamazaki 1982; Cheng *et al.*, 1988). A number of enzymic activities present in MLGM preparations are associated with intracellular membranes such as, for example, Golgi apparatus (glycosyltransferases and thiamine pyrophosphatase), endoplasmic reticulum (glucose-6-phosphatase, glycosyltransferases using dolichol as acceptor), lysosomes (acid phosphatase, glycosidases), and cytoplasm (aldolase). In most cases, whether these activities are constituents of the MLGM proper or are present in cytoplasmic materials entrained between the membrane and the core lipid has not been determined. Nevertheless, these activities are associated with milk lipid globules, irrespective of their precise localization within the globule. Some activities, concentrated in one cellular membrane, are also present in other locations. For example NADH oxidases (NADH-cytochrome c and ferricyanide reductases) are enriched in endoplasmic reticulum, but also are constituents of Golgi apparatus and plasma membranes (Jarasch *et al.*, 1979). In cow and human MLGM, Bruder *et al.* (1978) found the NADH oxidase system to be linked to b_5 and P420 cytochromes. Another source which can potentially contribute enzymes to milk lipid globule membrane preparations is cells in milk, which may be entrained with globules during recovery of cream and not be removed by the washing procedures. Anderson and Cawston (1975; cf. also Anderson, 1977) have found leucocytes entrained in milk lipid globules to be the source of at least portions of certain enzymic activities measured in MLGM preparations. Martel and Got (1976) found that MLGM from human milk would incorporate free amino acids into proteins. This observation implies that several enzymes and protein factors necessary for polypeptide synthesis are present in human milk lipid globules, but as yet there have been no reports on individual components of the protein synthetic apparatus in MLGM.

VII. Reorganization of the Membrane During Storage and Processing

As might be expected of a complex biological structure, the MLGM is fragile and in a state of flux. There are many ways in which its stability may be viewed from the standpoints of physical and chemical change, such as its physical organization and continuity on globules, its composition (proteins and lipids), activity of its enzymes, binding phenomena, denaturation of proteins, oxidation of lipids, and subtle biochemical changes.

There is good evidence of a qualitative nature that change occurs, but there is a paucity of quantitative data as to how durable the membrane is in various circumstances.

A. Milk in the Gland

Milk as it is removed from the human breast or the cow's udder already has a history. That is, time has elapsed to varying degrees since secretion from the lactating cell of every lipid globule, casein micelle, and lactose or other molecule in the milking. A very small part of the milking can be very old in that a fraction, often estimated at 15–20% of total milk in the cow udder, is left in the gland at each milking. This residual milk is occluded in small ducts and alveoli. For experimental purposes, such as to obtain very freshly secreted milk, it can be largely removed by a series of short-interval milkings subsequent to a complete milking. These milking manipulations stimulate additional oxytocin release, which in turn forces the residual milk out of the fine apertures. Administration of oxytocin and additional milking is also used to achieve the same end. However, this residual milk normally remains behind and mixes with the new milk that accumulates prior to the next milking. Again at the next milking, a certain proportion of the milk in the gland is left behind. Thus, aging of the milk and associated changes begin in the gland and, depending on the conditions, continue in various ways after the milk has been expressed. These are cogent considerations when so-called freshness of milk is concerned and they are relevant to the nature of the MLGM.

Wooding (1971) has concluded from studies with the electron microscope that the MLGM is unstable, and soon after secretion of the globule from the cell it loses membrane by vesiculation into the skim milk. On the other hand, secreted globules exhibit the characteristic structure of a biomembrane on their surface, and membrane isolated from milk fat globules shows this structure as well (Figure 2). Unfortunately, there is no way of seeing this membrane other than by chemical fixation, embedding in plastic and thin sectioning, followed by very high-magnification viewing. Not only is there the danger of generating artifacts in the process of preparing the specimen for observation, but the data need a statistical-morphometric approach to have convincing significance. In other words, on how many globules was the surface structure well resolved and over what proportion of their surfaces? And of these globules, how many showed alteration in MLGM and over what proportion of their surfaces? Such studies are tedious and seldom done. In this situation, in which it is certain that some of the globules are quite old and that even fresh ones may not reflect their condition at secretion accurately, it is difficult to draw firm conclusions regarding MLGM structural stability. We suggest that a marker for this purpose might be xanthine oxidase. This enzyme is located largely, if not completely, on the inner surface of the MLGM, in the so-called coat layer (Freudenstein *et al.*, 1979). To the extent that it increases in the milk serum, disruption of the membrane is indicated. By this we do not mean enzymic activity of xanthine oxidase, which can be misleading because it is redox sensitive, but rather shifts in distribution of actual mass of this protein.

In support of Wooding's contention, it was found that short-interval milking of goats to produce fresh milk resulted in a progressive reduction in the amount of membrane material in the skim milk (Patton *et al.*, 1973). Two goats were milked at hourly intervals for 10 hr after a complete milking. Concentrations of phospholipid and cholesterol in the skim milk were used as criteria of membrane concentration and this was supported by the fact that there was close correlation in the changes in concentration of the two lipids. During the first five hourly milkings, the amount of membrane in the skim milk dropped to approximately 60% of the concentration in skim milk of the complete milking. After five hourly milkings, the membrane concentrations tended to plateau in one goat and became somewhat erratic in the other. Thus, there is the implication that in the udder of the goat, MLGM is to some extent unstable and is in part shed into the skim milk phase. Of course, milk fat globules may not be the only source of membrane in the skim milk phase, and there is some evidence that there could be species differences, as is discussed for incidence and stability of cytoplasmic crescents on milk fat globules (Section IV) and on effect of refrigerated storage on MLGM (Section VII, B).

B. Changes after Expression from the Gland

1. Cooling

It is reasonable to expect that changes in the temperature of milk will produce expansion and contraction of milk fat globules and changes in the MLGM. For reasons of keeping quality, the first procedure applied to industrial (cow) milk is cooling. This results in a phenomenon which illustrates the delicacy of MLGM changes; namely, the adsorbtion onto milk fat globules of a substance called "agglutinin" which is important to the creaming phenomenon. The deepest cream layers are formed rapidly on milk that is cooled (0–5°C). It became possible, by separating milk at different temperatures, to make cream and skim milks that were agglutinin rich or agglutinin poor. When recombined, the agglutinin poor components formed a cream layer very slowly and inadequately. The precise nature of agglutinin has not been established, but it appears to be a member of the immunoglobin M class of antibodies (Euber and Brunner, 1984). From a practical standpoint, the phenomenon of creaming is mainly of historical interest in that the depth of the cream layer on a bottle of milk (richness) lost its significance when homogenization, opaque containers, and dietary concern about fat became prevalent. It is interesting that creaming is greatly delayed in goat's milk. While a cream layer will form within 15 or 20 min on cow's milk, it requires hours of holding for goat's milk. This is not so much due to the smaller size of goat globules, which is true, as it is due to slowness of the globules in clumping together. For this reason, goat milk is said to be naturally homogenized. The earlier literature on creaming of cow milk has been reviewed extensively by Brunner (1965).

No doubt there are many temperature-sensitive adsorbtion–desorbtion phenomenon in addition to agglutinin binding transpiring in milk. Another factor related to this is the progressive crystallization of the glycerides in the globule core and of the lipids in the MLGM as milk cools. We expect this to change the position and configuration of some membrane components: in some cases, irreversibly. Moreover, we expect an accompanying loss of membrane fluidity if such fluidity exists.

As cow milk is held at 0–5°C, the amount of lipid recovered in the skim milk on centrifugal separation increases gradually over 48 hr (Patton *et al.*, 1980b). This lipid was measured by solvent extraction and weighing. It is most likely membrane derived and indicates the sluffing of MLGM. In this connection, it is of interest that skim milk of lowest cholesterol content should result by separating the freshest milk possible.

2. *Agitation, Aeration, and Off-Flavor Development*

A natural accompanying factor in removing milk from the bovine gland is a certain amount of turbulence. The splashing and foaming in the glass-walled receiving vessel of a milking line is an obvious example. In addition, when the milk is pumped from the receiver to the refrigerated holding tank, there will be further agitation by the pump and often air incorporation as well. These events set the stage for further changes in the MLGM including additional loss of membrane, some denuding of the globule core, possible adsorbtion of lipase from the milk serum, and resulting lipolysis of glycerides. The incorporation of air is conducive to oxidation of the membrane lipids. These and other factors influencing the MLGM are discussed below.

The (hydrolytically) rancid off flavor in which MLGM plays an important role results from the action of lipoprotein lipase on the triacylglycerols of the globule, with release of butyric and other short-chain acids. From the olfactory character of the off flavor it is clear that butyric acid plays the dominant role and only trace amounts need be released. For rancid flavor to develop, the triacylglycerols bearing butyrate in the lipid core of the globule must be accessible. Since the core normally is protected by a layer of MLGM, access of the lipase in the milk serum is denied. Moreover, it is not clear how much of a barrier the native surface of the fat droplet, lying under the MLGM, may present to the lipase. This appears to represent a monolayer of polar lipids and proteins (see Section III). But this surface, as well as the MLGM, can be disrupted by agitation and aeration, thus allowing the lipase to bind to its substrate. These forces are particularly pernicious because they generate air bubbles that dissipate their free surface energy by stripping membrane from lipid globules. The globule surface can be further fractured by pumps which can have an homogenizing action under the conditions of their operation. For this reason, much effort has gone into design of pipeline milking systems that will minimize agitation and aeration of milk.

Another dimension of the rancidity problem is variation in susceptibility of the milk among individual cows. Some milk tastes rancid as it comes from the udder. Whether this is simply a matter of "not enough MLGM to go around" is not known. However, it is evident that nearly all milk normally presents enough of a barrier to lipase action on its fat globules that rancidity is no problem. An exception to this is the effect of warming cold milk (0–5°C) to approximately 30°C and then recooling. On standing, many individual milk samples so treated will become rancid (pertinent literature reviewed by Brunner, 1965). This sounds very much like a desorbtion of surface substances from the globules in the warming and a selective adsorbtion of lipase onto the globules from the milk serum on recooling. Momentary heating to 60°C substantially inhibits milk lipase and pasteurization (61.8°C for 30 min or 71.8°C for 15 sec) completely inactivates it.

3. Oxidized Flavor

Oxidized flavor results from oxidative degradation of polyunsaturated fatty acids contained in lipids of the MLGM, particularly linoleate and arachidonate. A complex, free radical mechanism involving these fatty acids, oxygen, ascorbic acid, and cupric ions leads to fragmentation of the hydrocarbon chains. Ascorbic acid is present in milk from the udder at a concentration of 20–30 mg/liter. Milk is saturated with air as a result of the milking and pumping processes. Ferric ions and light also can play catalytic roles in development of this off flavor. Resulting compounds causing the flavor are alk-2-enals, particularly 2-nonenal, 2-octenal, and 2-heptenal. Flavor of any one of these is detectable at a few ppm. In view of the high flavor potency of these aldehydes, oxidized flavor can be produced with very limited destruction of the lipids involved. The use of copper-bearing metal alloys in milk plants earlier in this century led to widespread oxidized flavor in the milk supply, even to the point that in some communities, it was not only accepted, but understood to be the normal flavor of milk. The introduction of stainless steel into the dairy industry largely overcame this problem. However, exposure of milk to light for long periods in supermarket cabinets can induce this off flavor at times, and spontaneous outbreaks of the problem, presumably due to feed and metabolic effects in animals, are encountered. A more detailed discussion of oxidized flavor is provided by Badings (1984).

Because of a lack of commercial processing and consumer complaints, human milk has few flavor problems. As is well known, human milk fat contains no short-chain fatty acids, so rancid flavor would not be a problem in any event. However, it seems likely that food flavors and metabolites may gain access to human milk just as easily as in the cow because of the direct monogastric route to the circulation in the human. But whether the consumers object to the flavor at times is not known.

4. *Freezing*

Freezing disrupts the structure of milk lipid globules. Whether this is caused by expansion of the globule core due to crystallization of the triacylglycerols or to penetration of the MLGM by ice crystals, or both, is not known. It is commonly observed that when frozen milk (nonhomogenized) is thawed, the fat globules have clumped. If these clumps are further warmed, oil droplets form. Thus, freezing must create breaks in the membrane so that the core triacylglycerols can emerge and coalesce with those released from other globules. This forms the basis of the freeze-thaw method of preparing MLGM (Section VI, A).

5. *Heating*

As milk is heated from 37 to 50°C changes in the MLGM, if any, are minor. Above this temperature, inactivation of the more sensitive enzymes begins. For example, pasteurization (71.8°C for 15 sec) brings about complete inactivation of milk lipase and alkaline phosphatase. A major fraction of this latter enzyme is in the MLGM. At about 72°C, release of hydrogen sulfide from milk fat globules commences. The specific source of this volatile sulfide has not been identified but xanthine oxidase, a principal component of the MLGM and one rich in sulfur, is a logical candidate. The so-called agglutinin, which promotes creaming, is progressively denatured as heat treatment proceeds beyond pasteurization. On cooling, such milk slowly forms a shallow, indistinct cream layer. Unlike freezing, heating of itself, even to boiling, does not disrupt the physical integrity of milk lipid globules such as to bring about significant oiling off of the triacylglycerols.

6. *Effects of Surface Active Agents*

The forces which hold the MLGM together and bound to the globule core can be overcome by various surfactants with or without heat treatment. Even low-molecular-weight organic molecules showing solubility in both oil and water can release the membrane (Dapper *et al.*, 1987). Two methods for preparing MLGM based on detergent treatment, one with sodium deoxycholate (Hayashi and Smith, 1965) and the other with Triton X-100 (Patton, 1982), have been developed. The latter is capable of retaining membrane-bound concanavilin A, a lectin with strong affinity for mannosyl- and glucosyl-containing glycoproteins, during the isolation. Conjugated bile salts, such as those involved in the physiology of digestion, will also remove membrane from milk lipid globules (Patton *et al.*, 1986). Methods of preparing MLGM using surfactants are discussed under Section VI, A. Higher concentrations of surfactants, especially when coupled with heat treatment, cause both release of the membrane and dissociation of its components. The powerful detergent, SDS, is used to completely dissociate proteins from membrane to facilitate their electrophoretic analysis.

C. Changes Due to Processing

Processing of milk and milk products has effects on the MLGM due to combined effects of heating, cooling, freezing, and agitation, both with and without air incorporation. There has been little research in this area dealing specifically with the membrane because of the complex variables and changes involved. Examples of these are in the production of butter, homogenized milk, and ice cream. A brief summary of the effects such processing is understood to have on the MLGM, together with possible functions of the membrane in the products involved, follows. More extensive discussions of the physics of these processes are given by Walstra and Jenness (1984).

1. Churning

No doubt the production of butter happened by accident thousands of years ago when some beast of burden carried a container of milk over some distance. Fundamentally, agitation is all that is needed, and if it is maintained long enough, butter will be produced. The process is also aided by the incorporation of air, high fat content, such as in heavy cream, and a suitable temperature. When air bubbles are suspended in a liquid, such as milk or cream, they will tend to take up surface active materials in order to lower their free surface energy and become physically more stable. One of the substances that binds to the air cells under these conditions is MLGM. This tends to denude lipid globules and expose their underlying triacylglycerols. At the churning temperature (approximately 12°C) these patches of glycerides are semisolid (sticky) and tend to adhere to one another, especially under the rigorous physical agitation of the churn. Eventually the growing globule aggregates destabilize the air cells (foam) and, with continued churning, these aggregates become particles of butter. Much of the membrane, but not all, goes into the buttermilk. Membrane is still retained by some of the fat globules in the butter as seen in freeze-fracture replicas in the electron microscope (W. Buchheim, personal communication). It is felt that the MLGM remaining in the butter helps with the incorporation of moisture during the working process and the creation of a product that feels smooth on the tongue.

People who work with many human and cow milk samples notice differences in tendencies of samples to churn spontaneously, even when the samples are all handled in the same way. The reasons for the differences are not known. Relative softness of the fat at a given temperature is known to be one factor in churning. Thus, liquid to solid ratio of milk triacylglycerols may be a factor in the human in which diet varies greatly, but differences can be perceived in the tendency to churn among milk samples of cows receiving the same ration. Thus, more subtle individual differences must also be involved.

2. *Homogenization*

The discovery that homogenization stabilized the suspension of fat in milk, made the milk taste creamier, increased milk's resistance to oxidized flavor, and lowered the curd tension made for rapid adoption of the process. Homogenization is effected by pumping milk at high pressure through narrow orifices at temperatures approximating those for pasteurization (65–80°C). This produces drastic turbulent and cavitational forces which physically disintegrate the globules, particularly the larger ones, leading to a reduction in their mean diameter from 3 or 4 μm to less than 1 μm. The exposed new triacylglycerol surfaces adsorb milk proteins, particularly casein micelles. About 30% of the casein in homogenized milk is associated with the lipid globules. The resulting globule suspension is probably stabilized in two ways. The adsorbed protein increases the density of the globules, thus overcoming their tendency to rise, and the negative charge of adsorbed casein micelles probably makes the globules more self-repellent. It is also possible that such clumping factors as agglutinin are denatured in the homogenizing. In any event, the normal tendency of the globules to clump seems to be destroyed. However, a study (Keenan *et al.*, 1983) has made it clear that a large proportion of globules in homogenized milk still retain membrane material, both protein and lipid, and that unit (normal biological) membrane structure can be observed on them in the electron microscope.

3. *Ice Cream Manufacture*

From formulation, processing, and physicochemical standpoints, ice cream is one of the most complex foods. While detailed consideration of the product is beyond the scope of this article, we should note that MLGM plays a role in its processing and properties. After blending of ingredients, heating, homogenizing, and cooling the basic ice cream mix, it is frozen while being whipped to incorporate air. This produces a system of sugar syrup containing fat globules, proteins, ice crystals, and air cells. It is important that the air cells be small and stable. As an emulsifying agent, MLGM is considered to be an important component in achieving this condition. Sweet cream buttermilk, which is rich in MLGM and sometimes used as a source of milk solids in ice cream, is said to impart excellent whipping properties to the mix. So-called emulsifying–whipping agents are often added to the ice cream mix. They include mono- and diglycerides, Tweens and Spans, and lecithin (phospholipid)-containing products. These have properties like those of the polar lipids in MLGM. To our knowledge, no one has evaluated the specific contribution of MLGM to the body and texture of ice cream. However, in this era of decreasing the fat content, with the hope that palatability can be maintained, it seems worthy of investigation.

Acknowledgements

Research of S. P. is supported by funds from the National Dairy Promotion and Research Board. Research of T. W. K. is supported by Grant GM31244 from the National Institute of General Medical Science, NIH, and by a grant from the National Dairy Promotion and Research Board.

References

Amaya, Y., Yamazaki, K., Sato, M., Nada, K., Nishino, T., and Nishino, T. (1990). Protelytic conversion of xanthine dehydrogenase from the NAD-dependent type to the O_2-dependent type. *J. Biol. Chem.* **265,** 14170–14175.

Anderson, M. (1977). Source and significance of lysosomal enzymes in bovine milk fat globule membrane. *J. Dairy Sci.* **60,** 1217–1222.

Anderson, M., and Cawston, T. E. (1975). Reviews on the progress of dairy science. The milk-fat globule membrane. *J. Dairy Res.* **42,** 401–417.

Badings, H. T. (1984). Flavors and off-flavors. *In* "Dairy Chemistry and Physics" (P. Walstra and R. Jenness, authors and eds.), pp. 336–357. Wiley, New York.

Bargmann, W., and Knoop, A. (1959). Über die morphologie der Milchsekretion. Licht- und electronenmikroskopische Studien an der Milchdrüse der Ratte. *Z. Zellforsch.* **49,** 344–388.

Barnd, D. L., Lan, M. S., Metzgar, R. S., and Finn, O. S. (1989). Specific, major histocompatibility complex-unrestricted recognition of tumor-associated mucins by human cytotoxic T cells. *Proc. Natl. Acad. Sci. USA* **86,** 7159–7163.

Baumrucker, C. R., and Keenan, T. W. (1973). Membranes of mammary gland. VII. Stability of milk fat globule membrane in secreted milk. *J. Dairy Sci.* **56,** 1092–1094.

Blanc, B. (1981). Biochemical aspects of human milk—Comparison with bovine milk. *Wld. Rev. Nutr. Diet.* **36,** 1–89.

Bouhours, J-F., and Bouhours, D. (1979). Galactosylceramide is the major cerebroside of human milk fat globule membrane. *Biochem. Biophys. Res. Commun.* **88,** 1217–1222.

Bracco, U., Hidalgo, J., and Bohren, H. (1972). Lipid composition of the fat globule membrane of human and bovine milk. *J. Dairy Sci.* **55,** 165–172.

Bruder, G., Fink, A., and Jarasch, E-D. (1978). The b-type cytochrome in endoplasmic reticulum of mammary gland epithelium and milk fat globule membranes consists of two components, cytochrome b_5 and cytochrome P-420. *Exp. Cell Res.* **117,** 207–217.

Brunner, J. R. (1965). Physical equilibria in milk: The lipid phase. *In* "Fundamentals of Dairy Chemistry" (B. H. Webb and A. H. Johnson, eds.), pp. 403–505. Avi Publishing, Westport, CT.

Buchheim, W., Welsch, U., Huston, G. E., and Patton, S. (1988). Glycoprotein filament removal from human milk fat globules by heat treatment. *Pediatrics* **81,** 141–146.

Calberg-Bacq, C-M., Francois, C., Gosselin, L., Osterrieth, P. M., and Rentier-Delrue, F. (1976). Comparative study of the milk fat globule membrane and the mouse mammary tumor virus prepared from the milk of an infected strain of Swiss albino mice. *Biochim. Biophys. Acta* **419,** 458–478.

Ceriani, R. L., Peterson, J. A., Lee, J. Y., Moncada, R., and Blank, E. W. (1983). Characterization of cell surface antigens of human mammary epithelial cells with monoclonal antibodies prepared against human milk-fat-globule membranes. *Somatic Cell Genet.* **9,** 415–427.

Cheng, S. G., Koch, U., and Brunner, J. R. (1988). Characteristics of purified cow's milk xanthine oxidase and its submolecular characteristics. *J. Dairy Sci.* **71,** 901–916.

Dapper, C. H., Valivullah, H. M., and Keenan, T. W. (1987). Use of polar aprotic solvents to release membranes from milk fat globules. *J. Dairy Sci.* **70,** 760–765.

Deeney, J. T., Valivullah, H. M., Dapper, C. H., Dylewski, D. P., and Keenan, T. W. (1985). Microlipid droplets in milk secreting mammary epithelial cells: Evidence that they originate from endoplasmic reticulum and are precursors of milk lipid globules. *Eur. J. Cell Biol.* **38,** 16–26.

Dion, A. S., Williams, J., Herlyn, M., and Major, P. (1990). Human milk fat globule membrane glycoproteins express blood group-related determinants primarily on mucin-like epithelial membrane antigens and gp 70. *Biochem. Int.* **22,** 295–302.

Dylewski, D. P., Dapper, C. H., Valivullah, H. M., Deeney, J. T., and Keenan, T. W. (1984). Morphological and biochemical characterization of possible intracellular precursors of milk lipid globules. *Eur. J. Cell Biol.* **35,** 99–111.

Euber, J. R., and Brunner, J. R. (1984). Reexamination of fat globule clustering and creaming in cow milk. *J. Dairy Sci.* **67,** 2821–2832.

Farrar, G. H., Harrison, R., and Mohanna, N. A. (1980). Comparison of lectin receptors on the surface of human and bovine milk fat globule membranes. *Comp. Biochem. Physiol. B* **67,** 265–270.

Fishman, P. H. (1986). Recent advances in identifying the functions of gangliosides. *Chem. Phys. Lipids* **42,** 137–151.

Franke, W. W., Heid, H. W., Grund, C., Winter, S., Freudenstein, C., Schmid, E., Jarasch, E-D., and Keenan, T. W. (1981). Antibodies to the major insoluble milk fat globule membrane associated protein: Specific location in apical regions of lactating epithelial cells. *J. Cell Biol.* **89,** 485–494.

Freudenstein, C., Keenan, T. W., Eigel, W. N., Sasaki, M., Stadler, J., and Franke, W. W. (1979). Preparation and characterization of the inner coat material associated with fat globule membranes from bovine and human milk. *Exp. Cell Res.* **118,** 277–294.

Gendler, S. J., Lancaster, C. A., Taylor-Papadimitriou, J., Duhig, T., Peat, N., Burchell, J., Pemberton, L., Lalani, E.-N., and Wilson, D. (1990). Molecular cloning and expression of human tumor-associated polymorphic epithelial mucin. *J. Biol. Chem.* **265,** 15286–15293.

Gendler, S. J., Taylor-Papadimitriou, J., Duhig, T., Rothbard, J., and Burchell, J. (1988). A highly immunogenic region of a human polymorphic epithelial mucin expressed by carcinomas is made up of tandem repeats. *J. Biol. Chem.* **263,** 12820–12823.

Greenwalt, D. E., and Mather, I. H. (1985). Characterization of an apically derived epithelial membrane glycoprotein from bovine milk, which is expressed in capillary endothelia in diverse tissues. *J. Cell Biol.* **100,** 397–408.

Greenwalt, D. E., Watt, K. W. K., So, O. Y., and Jiwani, N. (1990). PAS-IV, an integral membrane protein of mammary epithelial cells, is related to platelet and endothelial cell CD36 (GP IV). *Biochemistry* **29,** 7054–7059.

Hakomori, S-I. (1981). Glycosphingolipids in cellular interaction, differentiation, and oncogenesis. *Annu. Rev.Biochem.* **50,** 733–764.

Hakomori, S-I. (1990). Bifunctional role of glycosphingolipids. *J. Biol. Chem.* **265,** 18713–18716.

Hanun, Y., and Bell, R. M. (1989). Function of sphingolipids and sphingolipid breakdown products in cellular regulation. *Science* **243,** 500–507.

Hay, J. D., and Morrison, W. R. (1971). Polar lipids in bovine milk. III. Isomeric *cis* and *trans* monoenoic and dienoic fatty acids, and alkyl and alkenyl ethers in phosphatidyl choline and phosphatidyl ethanolamine. *Biochim. Biophys. Acta* **248,** 71–79.

Hayashi, S., and Smith, L. M. (1965). Membranous material of bovine milk fat globules. I. Comparison of membranous fractions released by deoxycholate and by churning. *Biochemistry* **4,** 2550–2559.

Heid, H. W., Winter, S., Bruder, G., Keenan, T. W., and Jarasch, E.-D. (1983). Butyrophilin, an apical plasma membrane-associated glycoprotein characteristic of lactating mammary glands of diverse species. *Biochim. Biophys. Acta* **728,** 228–238.

Huang, T. C., and Kuksis, A. (1967). A comparative study of the lipids of globule membrane and fat core and of the milk serum of cows. *Lipids* **2,** 453–470.

Huston, G. E., and Patton, S. (1990). Factors related to the formation of cytoplasmic crescents on milk fat globules. *J. Dairy Sci.* **73,** 2061–2066.

Imam, A., Laurence, D. J. L., and Neville, A. M. (1981). Isolation and characterization of a major glycoprotein from milk-fat-globule membrane of human breast milk. *Biochem. J.* **193,** 47–54.

Imam, A., Laurence, D. J. R., and Neville, A. M. (1982). Isolation of two individual glycoprotein components from human milk-fat-globule membranes. *Biochem. J.* **207,** 37–41.

Jack, L. J. W., and Mather, I. H. (1990). Cloning and analysis of cDNA encoding bovine butyrophilin, an apical glycoprotein expressed in mammary tissue and secreted in association with the milk-fat globule membrane during lactation. *J. Biol. Chem.* **265,** 14482–14486.

Janssen, M. M. T., and Walstra, P. (1983). Cytoplasmic remnants in milk of certain species. *Netherlands Milk Dairy J.* **36,** 365–368.

Jarasch, E-D., Bruder, G., Keenan, T. W., and Franke, W. W. (1977). Redox constituents in milk fat globule membranes and rough endoplasmic reticulum from lactating mammary gland. *J. Cell Biol.* **73,** 223–241.

Jarasch, E-D., Kartenbeck, J., Bruder, G., Fink, A., Morre', D. J., and Franke, W. W. (1979). B-type cytochromes in plasma membranes isolated from rat liver, in comparison with those of endomembranes. *J. Cell Biol.* **80,** 37–52.

Jarasch, E-D., Bruder, G., Heid, H. W., Keenan, T. W., and Franke, W. W. (1981). Localization of xanthine oxidase in mammary gland epithelium and capillary endothelium. *Cell* **25,** 67–82.

Jenness, R. G. (1974). The composition of milk. *In* "Lactation" (B. L. Larson and V. R. Smith, eds.), Vol. **3,** pp. 3–107. Academic Press, New York.

Jensen, R. G. (1989). "The Lipids of Human Milk." CRC Press, Boca Raton, FL.

Jensen, R. G., Clark, R. M., and Ferris, A. M. (1980). Composition of lipids in human milk: A review. *Lipids* **15,** 345–355.

Kanno, C. (1990). Secretory membranes of the lactating mammary gland. *Protoplasma* **159,** 184–208.

Kanno, C., Hattori, H., and Yamauchi, K. (1987). Lipid composition of plasma membranes isolated from lactating bovine mammary gland. *Agric. Biol. Chem. (Japan)* **51,** 2995–3001.

Karlsson, K-A. (1989). Animal glycosphingolipids as membrane attachment sites for bacteria. *Annu. Rev. Biochem.* **58,** 309–350.

Kayser, S. G, and Patton, S. (1970). The function of very long chain fatty acids in membrane structure: Evidence from milk cerebrosides. *Biochem. Biophys. Res. Commun.* **41,** 1572–1578.

Keenan, T. W., Morre', D. J., Olson, D. E., Yunghans, W. N., and Patton, S. (1970). Biochemical and morphological comparison of plasma membrane and milk fat globule membrane from bovine mammary gland. *J. Cell Biol.* **44,** 80–93.

Keenan, T. W., Huang, C. M., and Morre', D. J. (1971). Lipid composition of Golgi apparatus from rat mammary gland. *J. Dairy Sci.* **55,** 51–57.

Keenan, T. W., Freudenstein, C., and Franke, W. W. (1977a). A lipoprotein complex derived from bovine milk fat globule membrane with some preparative characteristics resembling those of actin. *Cytobiologie* **14,** 259–278.

Keenan, T. W., Powell, K. M., Sasaki, M., Eigel, W. N., and Franke, W. W. (1977b). Isolation and partial characterization of a high molecular weight glycoprotein fraction from bovine milk fat globule membrane. *Cytobiologie* **15,** 96–115.

Keenan, T. W., Heid, H. W., Stadler, J., Jarasch, E.-D., and Franke, W. W. (1982). Tight attachment of fatty acids to proteins associated with milk lipid globule membrane. *Eur. J. Cell Biol.* **26,** 270–276.

Keenan, T. W., Moon, T.-W., and Dylewski, D. P. (1983). Lipid globules retain globule membrane material after homogenization. *J. Dairy Sci.* **66,** 196–203.

Keenan, T. W., Mather, I. H., and Dylewski, D. P. (1988). Physical equilibria: Lipid phase. *In* "Fundamentals of Dairy Chemistry" (N. P. Wong, R. Jenness, M. Keeney, and E. H. Marth, eds.), 3rd Ed., pp. 511–582. Van Nostrand–Reinhold, New York.

Kitchen, B. J. (1977). Fractionation and characterization of the membranes for bovine milk fat globules. *J. Dairy Res.* **44,** 469–482.

Laegreid, A., and Otnaess, A-B. K. (1986). Purification of human milk gangliosides by silica gel chromatography and analysis of trifluoroacetate derivatives by gas chromatography. *J. Chromatogr.* **377,** 59–67.

Linzell, J. L. (1970). Milk secretion. *Nature (London)* **228,** 1007.

Lis, D., and Monis, B. (1978). Glycosaminoglycans of the fat globule membrane of cow milk. *J. Supramol. Struct.* **8,** 173–176.

Long, C. A., and Patton, S. (1978). Formation of intracellular fat droplets: Interrelation of newly synthesized phosphatidyl choline and triglyceride in milk. *J. Dairy Sci.* **61,** 1392–1399.

Martel, M. B., and Got, R. (1976a). Transfer du galactose dans les membranes des globules lipidiques du lait humain. *Biochim. Biophys. Acta* **436,** 789–799.

Martel, M. B., and Got, R. (1976b). Incorporation d'acides amines radioactifs dans les membranes des globules lipidiques du lait humain. *Experientia* **32,** 330–331.

Martel, M. B., Dubois, P., and Got, R. (1973). Membranes des globules lipidiques du lait humain. Preparation, etude morphologique et composition chimique. *Biochim. Biophys. Acta* **311,** 565–575.

Martel-Pradal, M. B., and Got, R. (1972). Presence d'enzymes marqueurs des membranes plasmiques, de l'appariel de Golgi et du reticulum endoplasmique dans les membranes des globules lipidiques de lait maternel. *FEBS Lett.* **21,** 220–222.

Martin, R. W. (1989). Electron microscopic localization of cholesterol in bovine milk fat globules. *Food Microstruct.* **8,** 3–9.

Mather, I. H., and Keenan, T. W. (1975). Studies on the structure of milk fat globule membrane. *J. Membrane Biol.* **21,** 65–85.

Mather, I. H., and Keenan, T. W. (1983). Function of endomembranes and the cell surface in the secretion of organic milk constituents. *In* "Biochemistry of Lactation" (T. B. Mepham, ed.), pp. 231–283. Elsevier, Amsterdam.

Mather, I. H., Weber, K., and Keenan, T. W. (1977). Closely associated proteins and compositional heterogeneity of bovine milk fat globule membrane. *J. Dairy Sci.* **60,** 394–402.

Mather, I. H., Tamplin, C. B., and Irving, M. G. (1980). Separation of the proteins of the bovine milk-fat-globule membrane by electrofocusing with retention of enzymatic and immunologic activity. *Eur. J. Biochem.* **110,** 327–336.

McPherson, A. V., and Kitchen, B. J. (1983). Reviews of the progress of dairy science: The bovine milk fat globule membrane—Its formation, composition, structure and behaviour in milk and dairy products. *J. Dairy Res.* **50,** 107–133.

Monis, B., Rovasio, R. A., and Valentich, M. A. (1975). Ultrastructural characterization by ruthenium red of the surface of the fat globule membrane of human and rat milk with data on carbohydrates of fractions of rat milk. *Cell Tissue Res.* **157,** 17–24.

Morrison, W. R. (1968). The distribution of phospholipids in some mammalian milks. *Lipids* **3,** 101–103.

Mulder, H., and Walstra, P. (1974). "The Milk Fat Globule." Center for Agricultural Publishing and Documentation, Wageningen, The Netherlands.

Murray, L. R., Powell, K. M., Sasaki, M., Eigel, W. N., and Keenan, T. W. (1979). Comparison of lectin receptor and membrane coat-associated glycoproteins of milk lipid globule membranes. *Comp Biochem. Physiol. B* **63,** 137–145.

Nakamura, M., and Yamazaki, I. (1982). Preparation of bovine milk xanthine oxidase as a dehydrogenase form. *J. Biochem. (Japan)* **92,** 1279–1286.

Neira, L. M., and Mather, I. H. (1990). Biochemical and immunological comparison of bovine butyrophilin with a butyrophilin-like glycoprotein in guinea pig milk-fat-globule membrane. *Protoplasma* **159,** 168–178.

Newman, R. A., and Harrison, R. (1973). Characterisation of the surface of bovine milk fat globule membrane using microelectrophoresis. *Biochim. Biophys. Acta* **298,** 798–809.

Otnaess, A-B. K., Laegreid, A., and Ertresvag, K. (1983). Inhibition of enterotoxin from *Escherichia coli* and *Vibrio cholerae* by gangliosides from human milk. *Infect. Immunol.* **40,** 563–569.

Parodi, A. J., Blank, E. W., Peterson, J., and Ceriani, R. (1984). Glycosyl transferases in mouse and human milk fat globule membranes. *Mol. Cell. Biochem.* **58,** 157–163.

Patton, S. (1952). Preparation of milk fat. I. A study of some organic compounds as de-emulsifying agents. *J. Dairy Sci.* **35,** 324–328.

Patton, S. (1973). Origin of the milk fat globule. *J. Am. Oil Chem. Soc.* **50,** 178–185.

Patton, S. (1982). Release of remnant plasma membrane from milk fat globules by Triton X-100. *Biochim. Biophys. Acta* **688,** 724–734.

Patton, S., and Hubert, J. (1983). Binding of concanavalin A to milk fat globule membrane and release of the lectin-membrane complex by Triton-X 100. *J. Dairy Sci.* **66,** 2312–2319.

Patton, S., and Huston, G. E. (1986). A method for isolation of milk fat globules. *Lipids* **21,** 170–174.

Patton, S., and Huston, G. E. (1987). Differences between individuals in high-M_r glycoproteins from human mammary epithelia. *FEBS Lett.* **216,** 151–154.

Patton, S., and Huston, G. E. (1988). Incidence and characteristics of cell pieces on human milk fat globules. *Biochim Biphys Acta* **965,** 146–153.

Patton, S., and Huston, G. E. (1989). Fat globule proteins. *In* "Protein and Non-protein Nitrogen in Human Milk" (S. A. Atkinson and B. Lonnerdal, eds.), pp. 94–114. CRC Press, Boca Raton, FL.

Patton, S., and Jensen, R. G. (1976). "Biomedical Aspects of Lactation." Pergamon Press, New York.

Patton, S., and Keenan, T. W. (1971). The relationship of milk phospholipids to membranes of the secretory cell. *Lipids* **6,** 58–61.

Patton, S., and Keenan, T. W. (1975). The milk fat globule membrane. *Biochim. Biophys. Acta* **415,** 273–309.

Patton, S., Plantz, P. E., and Thoele, C. A. (1973). Factors influencing phospholipids and cholesterol in skim milk: Effect of short interval milkings. *J. Dairy Sci.* **56,** 1473–1476.

Patton, S., Kelly, J. J., and Keenan, T. W. (1980a). Carotene in bovine milk fat globules: Observations on origin and high content in tissue mitochondria. *Lipids* **15,** 33–38.

Patton, S., Long, C., and Sokka, T. (1980b). Effect of storing milk on cholesterol and phospholipid of skim milk. *J. Dairy Sci.* **63,** 697–700.

Patton, S., Borgstrom, B., Stemberger, B. H., and Welsch, U. (1986). Release of membrane from milk fat globules by conjugated bile salts. *J. Pediatr. Gastroenterol. Nutr.* **5,** 262–267.

Patton, S., Huston, G. E., Jenness, R., and Vaucher, Y. (1989). Differences between individuals in high-molecular weight glycoproteins from mammary epithelia of several species. *Biochim. Biophys. Acta* **980,** 333–338.

Rüegg, M., and Blanc, B. (1981). The fat globule size distribution in human milk. *Biochim. Biophys. Acta* **666,** 7–14.

Sasaki, M., and Keenan, T. W. (1979). Ultrastructural characterization of carbohydrate distribution on milk lipid globule membrane. *Cell Biol. Int. Rep.* **3,** 67–74.

Schliwa, M. (1986). "The Cytoskeleton: An Introductory Survey." Cell Biology Monographs, Vol. 13. Springer-Verlag, Berlin.

Scow, R. O., Blanchette-Mackie, E. J., and Smith, L. C. (1980). Transport of lipid across capillary endothelium. *Fed. Proc.* **39,** 2610–2617.

Shimizu, M., and Yamauchi, K. (1982). Isolation and characterization of mucin-like glycoprotein in human milk fat globule membrane. *J. Biochem.* **91,** 515–524.

Shimizu, M., Uryu, N., and Yamauchi, K. (1981). Presence of heparin sulfate in the fat globule membrane of bovine and human milk. *Agric. Biol. Chem. (Japan)* **45,** 741–7465.

Shimizu, M., Yamauchi, K., Miyauchi, Y., Sakurai, T., and McIlhinney, R. A. J. (1986). High-M_r profiles in human milk serum and fat-globule membrane. *Biochem. J.* **233,** 725–730.

Shimizu, M., Tanimoto, H., Azuma, N., and Yamauchi, K. (1990). Growth inhibition of BALB/c 3T3 cells by a high-molecular-weight mucin-like glycoprotein of human milk fat globule membrane. *Biochem. Int.* **20,** 147–154.

Shriver, B. J., Allred, J. B., and Roman-Lopez, C. R. (1989). Bovine milk-fat-globule membrane contains an enzymatically inactive form of acetyl-CoA carboxylase. *Biochem. J.* **257,** 925–927.

Snow, L. D., Colton, D. G., and Carraway, K. L. (1977). Purification and properties of the major sialoglycoprotein of the milk fat globule membrane. *Arch. Biochem. Biophys.* **179,** 690–697.

Spicer, A. P., Parry, G., Patton, S., and Gendler, S. J. (1991). Molecular cloning and analysis of the mouse homologue of the tumor-associated mucin, MUC-1, reveals conservation of potential O-glycosylation sites, transmembrane, and cytoplasmic domains and a loss of minisatellite-like polymorphism. *J. Biol. Chem.* **266,** 15099–15109.

Stemberger, B. H., and Patton, S. (1981). Relationship of size, intracellular location, and time required for secretion of milk fat droplets. *J. Dairy Sci.* **64,** 422–426.

Stemberger, B. H., Walsh, R. M., and Patton, S. (1984). Morphometric evaluation of lipid droplet associations with secretory vesicles, mitochondria and other components in the lactating cell. *Cell Tissue Res.* **236,** 471–474.

Stewart, P. S., Puppione, D. L., and Patton, S. (1972). The presence of microvilli and other membrane fragments in the non-fat phase of bovine milk. *Z. Zellforsch.* **123,** 161–167.

Swallow, D. M., Gendler, S., Griffiths, B., Corney, G., Taylor-Papadimitriou, J., and Bramwell, M. E. (1987). The human tumor-associated epithelial mucins are coded by an expressed hypervariable gene locus PUM. *Nature (London)* **328,** 82–84.

Swope, F. C., and Brunner, J. R. (1965). Identification of ribonucleic acid in the fat-globule membrane. *J. Dairy Sci.* **48,** 1705–1707.

Takamizawa, K., Iwamori, M., Mutai, M, and Nagai, Y. (1986a). Gangliosides of bovine buttermilk. *J. Biol. Chem.* **261,** 5625–5630.

Takamizawa, K., Iwamori, M., Mutai, M., and Nagai, Y. (1986b). Selective changes in gangliosides of human milk during lactation: A molecular indicator of the period of lactation. *Biochim. Biophys. Acta* **879,** 73–77.

Thompson, M. P., Brunner, J. R., Stine, C. M., and Lindquist, K. (1961). Lipid components of the fat globule membrane. *J. Dairy Sci.* **44,** 1589–1596.

Valivullah, H. M., and Keenan, T. W. (1989). Butyrophilin of milk lipid globule membrane contains N-linked carbohydrate and cross-links with xanthine oxidase. *Int. J. Biochem.* **21,** 103–107.

Valivullah, H. M., Bevan, D. R., Peat, A., and Keenan, T. W. (1988). Milk lipid globules: Control of their size distribution. *Proc. Natl. Acad. Sci. USA* **85,** 8775–8779.

van Meer, G. (1989). Lipid traffic in animal cells. *Annu. Rev. Cell Biol.* **5,** 247–275.

Vasic, J, and DeMan, J. M. (1966). High melting glycerides and the milk fat globule membrane. *Proc. 17th Int. Dairy Congr.* **C,** 167–172.

Walstra, P. (1969). Studies on milk fat dispersion. II. The globule size distribution of cows' milk. *Netherlands Milk Dairy J.* **23,** 99–110.

Walstra, P. (1974). High-melting triglycerides in the milk fat globule membrane; an artifact? *Netherlands Milk Dairy J.*, **28,** 3–9.

Walstra, P., and Jenness, R. (1984). "Dairy Chemistry and Physics." Wiley, New York.

Wiman, K., Curman, B., Tragardh, L., and Peterson, P. A. (1979). Demonstration of HLA-DR-like antigens on milk fat globule membranes. *Eur. J. Immunol.* **9,** 190–195.

Wooding, F. B. P. (1971a). The mechanism of secretion of the milk fat globule. *J. Cell Sci.* **9,** 805–821.

Wooding, F. B. P. (1971b). The structure of the milk fat globule membrane. *J. Ultrastruct. Res.* **37,** 388–400.

Wooding, F. B. P. (1974). Milk fat globule membrane material in skim milk. *J. Dairy Res.* **41,** 331–337.

Wooding, F. B. P. (1977). Comparative mammary fine structure. *In* "Comparative Aspects of Lactation" (M. Peaker, ed.), pp. 1–41. Academic Press, New York.

Wooding, F. B. P., Peaker, M., and Linzell, J. L. (1970). Theories of milk secretion: Evidence from the electron microscopic examination of milk. *Nature (London)* **226,** 762–764.

Zaczek, M., and Keenan, T. W. (1990). Morphological evidence for an endoplasmic reticulum origin of milk fat globules obtained using lipid-selective staining procedures. *Protoplasma* **159,** 179–182.

Zikakis, J. P., Dougherty, T. M., and Biasotto, N. O. (1976). The presence and some properties of xanthine oxidase in human milk and colostrum. *J. Food Sci.* **41,** 1408–1412.

B. Particulate Constituents in Human and Bovine Milks

ROBERT G. JENSEN
BERNARD BLANC
STUART PATTON

I. Introduction

Milks are biological fluids of exceptional complexity containing thousands of compounds. These are located in several compartments, directed there by the biological and physicochemical forces acting during milk synthesis, secretion, and thereafter. Compartmentation is one of the factors which control the flow of milk components and their products into and through the digestive and absorptive systems of the consumer. The compounds in milk provide nutriture, structural components for cellular membranes, and nonnutritive messages, e.g., immunological systems for host defense. The compartments in bovine milk, which is produced and consumed in the largest quantity, are altered by processing. Most of the milk is processed, i.e., pumped, agitated, pooled, cooled, clarified (centrifugation to remove cells, etc), the fat content standardized, usually to a lower amount by controlled separation (centrifugation), pasteurized, and homogenized to reduce the size of the fat globules. Unfortunately, compartmentation in

HANDBOOK OF MILK COMPOSITION

processed milk has received little attention. However, it is known that about 30% of the casein is associated with the fat phase in homogenized milk and that the whey proteins associate with the casein and each other as milk is heat-treated beyond pasteurization.

If human or bovine milk is centrifuged in a tube, milk components will be separated into the several compartments (see Chapter 2A). At low speeds, e.g., 300*g*, cells and any tissue debris are sedimented as a soft pellet and the fat globules rise and form a distinct cream layer. As centrifugation continues and the force is increased, smaller fat globules enter the cream layer which is becoming more compact. Casein micelles also begin to sediment. In order to attain more or less complete sedimentation of the casein into a pellet, about 100,000*g* for 1 hr is required. A so-called fluff layer will be layered on top of the casein pellet. The fluff layer contains membrane fragments, small vesicles, sloughed microvilli, etc. (Stewart *et al.*, 1972). This produces a clear infranatatant compartment which contains the soluble constituents of milk including the whey proteins. Thus, it is possible, using centrifugation, to prepare the individual compartments for further observation and analysis. Because human milk contains only about one-tenth as much casein as bovine milk, the casein pellet obtained by centrifugation is much smaller for human compared to bovine milk. It is possible to isolate the fat globule compartments which contain nearly all the fat, but not all the small globules, by a relatively simple centrifugation procedure (Patton and Huston, 1986).

In dealing with milk from an analytical standpoint, it should be borne in mind that it is progressively changing from the amount of secretion by the lactating cell. In the case of an animal that is milk fed, some of the milk is already "old" as a result of accumulation in the gland. The changes are mostly subtle, but large shifts from one compartment to another, and sometimes destruction of minor or trace constituents, can occur. Simply cooling and holding milk in the cold can cause some substances to redistribute into other compartments. In particular, the equilibria of substances absorbed on fat globules and casein micelles will change, membranes will fragment, and membrane components will tend to dissociate.

The compartments are defined by the amount, size, and solubility of milk constituents therein as shown in Table I (Jensen *et al.*, 1990). The compartments that we will discuss are cells and membrane fragments, emulsified lipid globules, and casein micelles. Information about the formation and properties of the milk lipid globule membrane and the globules is given in Chapter 2A.

TABLE I
Compartments and Their Constituents in Mature Bovine and Human Milks[a]

Compartment description	Content (%) Bovine (B)	Content (%) Human (H)	Major constituents: Name	Major constituents: Content (%)
Aqueous phase True solution (1 nm)	87.3	87.6	1. Compounds of Ca, Mg, PO_4, Na, K, Cl, CO_2, citrate, casein	0.7 as ash, B 0.2 as ash, H
Whey proteins (3 to 9 nm)			2. Whey proteins: α-lactalbumin, lactoferrin, IgA, lysozyme, and serum albumin B, 20% of total N;H, 70%	0.6 B, H
			3. Lactose and oligosaccharides: 4.8 and 0.1 B; 7.0 and 1.0% H	4.9 B 8.0 H
			4. Nonprotein nitrogen compounds: glycocyamine, urea, amino acids. B, 5% of total N; H, 25%	30 mg N/dl B 50 mg N/dl H
			5. Miscellaneous: B vitamins, ascrobic acid	
Colloidal dispersion (11 to 55 nm, 10^{16}/dl)		2.6	6. Caseins: B- and, a for B, Ca, PO_4	2.6 B 0.3 H
Emulsion Fat globules (4 μm; 1.1 × 10^{10}/dl)	3.7	4.0	7. Fat globules: triacylglycerols, vitamins	3.7 B 4.0 H
Fat globule membrane Absorbed layer Cells and fragments (8 to 40 μm; 10^4 to 10^5/dl)			8. Milk fat globule membrane: proteins phospholipids, cholesterol, enzymes, trace minerals	2% of total lipid
			9. Macrophages, neutrophils, lymphocytes, epithelial cells, leukocytes, cytoplasmic fragments	

[a]All figures are approximate. Adapted from Jensen *et al.* (1990).

II. Cells and Membrane Fragments

A. Human Milk

The cells which have been identified in human colostrum and milk and approximate numbers are listed in Table II. During early lactation, macrophages predominate with numbers of all types, except epithelial cells, decreasing markedly as lactation progresses. The numbers are representative and vary considerably among individuals (Brooker, 1980; Rüegg and Blanc, 1982; Hayward, 1983; Ogra and Ogra, 1988; Lawrence, 1989). Beneficial immunological functions for the infant and the mammary gland have been attributed to some of those cells, e.g., the macrophages and lymphocytes (Rüegg and Blanc, 1982; Hayward, 1983; Ogra and Ogra, 1988; Lawrence, 1989). The functions of leukocytes were described by Mandyla and Xanthou (1986), Buescher and Pickering (1986), and the IOM (1991).

Brooker (1980) observed membrane-bound cytoplasmic remnants in the sedimentation pellet of centrifuged human milk. There were more fragments than cells at all times postpartum studied. Most of the fragments came from secretory cells in the mammary gland. They contained vesicles of rough endoplasmic reticulum, lipid droplets, and Golgi vesicles containing casein micelles. These membranes when folded or spherical are probably the particles named milk microsomes (Rüegg and Blanc, 1982). The membranes, which in general are called milk lipoproteins, range in size from 10 to 400 nm (Rüegg and Blanc, 1982). Some of the fragments were probably associated with the cytoplasmic crescents seen on about 7% of human and 1% of bovine lipid droplets (Huston and Patton, 1990), with the array of glycoprotein filaments seen on human but not bovine globules (Buchheim *et al.*, 1986), and were displaced from their original sites by the processes of isolation (see Chapter 2A for more information). Relatively little displacement of even loosely bound material would be expected to occur during the short period of transit from the breast to the breast-fed infant's stomach. Sucking by the infant is likely to be more vigorous than hand expression of the milk, but not as much as vacuum pumping, to obtain milk samples.

Bacteria, usually innocuous skin species, are present. The numbers are low, but extremely variable, and are often attached to squamous epithelial cells (Brooker, 1980). Neubauer *et al.* (1995) found that the numbers of bacteria and leukocytes were related to the incidence and severity of mastitis. With no mastitis, leukocytes were $< 1 \times 10^6$ and bacteria $< 1 \times 10^3$/ml. For noninfectious mastitis, the figures were $\geq 10^6$ and $< 10^3$/ml and for infectious mastitis $\geq 10^6$ and $\geq 10^3$/ml. The major species present in milks from women with no mastitis (52%, $n = 89$) were skin types. An increased incidence of *Staphylococcus aureus* was associated with mastitis. In developing countries, many of the women who are breastfeeding will have

TABLE II
Cell Types and Numbers in Human Milk (per μl) before and during Lactation

Time	Total	Macrophages	Neutrophils	Lymphocytes	Epithelial[b]
Antepartum	3430	2140	360	240	—
Postpartum					
Days 0–4	2840	1490	1375	250	About 1×10^4 throughout lactation
Days 5–8	450	320	100	27	
Weeks 1 or 2	69	52	4	1	
Weeks 2–4	51	52	8	1	
Months 1 or 2	17	4	3	1	
Months 2–4	16	3	2	1	
Months 4–6	10	1	1	1	

[a]Adapted from Hayward (1983).
[b]Brooker (1980).

mastitis (Prentice *et al.*, 1985). The importance is that severe mastitis alters the composition of milk (see Chapter 3F) and destroys lactating tissue, thus reducing the volume of milk available for future lactations. While natural defenses are operative, improved hygiene would be helpful (Prentice *et al.*, 1983).

The relationship in dairy cattle with mastitis, lower milk production, and high somatic cell counts (leukocytes) in their milk has apparently not been studied in humans (See below).

B. *Bovine*

1. *Cells*

Bovine milk contains about 10^4 to 10^7 cells/ml (Lipkin *et al.*, 1993), although individual variation is large. The numbers are usually reported as somatic cell counts which are a mixture of epithelial cells (2%) and leukocytes (98%). Several enzymes are found in leukocytes, e.g., catalase, proteases, etc. Most of the nucleic acids in milk originate from these cells. These milk cells have been utilized as a source of deoxyribonucleic acid and as a substrate for the polymerase chain reaction (Lipkin *et al.*, 1993).

Somatic cell counts are routinely determined in the U.S. Dairy Herd Improvement Association Programs using electronic cell counters (Heald, 1985). High counts are associated with reduced milk yields and increased incidence of mastitis. In one study, the milk from 81% of 139,421 cows contained 18,000 ml 565,000 cells/ml. The average milk yields were 26 to

21.6 kg; milk production dropped 0.68 kg for the average of all cows each time the somatic cell count doubled. Somatic cell counts of 200,000 to 400,000/ml were associated with lower milk yields and greater mastitic infection rates (Jones *et al.,* 1984). They mentioned that somatic cell counts above 500,000/ml have been used as an indicator of significant incidence of mastitis in a herd or nonspecific mastitis if pathogenic microorganisms had not been detected. Guidry (1985) noted that leukocyte counts in excess of 200,000/ml in an individual cow sample suggested mastitis with the need for diagnosis by a chemical method such as determination of chloride. In the United States, the Federal regulatory limit has been 750,000/ml, but will be lowered to 500,000/ml in 1994 (Bennett, 1993). This will be done to align the United States with the requirements in the EEC. The regulatory limit will drop to 400,000 ml in 1998.

We mentioned earlier that somatic cells carry enzymes into the milk. Verdi and Barbano (1991) found proteolytic activity in somatic cells isolated from milk by ultracentrifugation. The proteases hydrolyzed β-casein. They could cause proteolysis in aged cheeses if not destroyed during processing. Neither cells nor enzymes, with the possible exception of bacterial proteases, are likely to survive processing. When samples of raw milk are frozen for subsequent analytical or research purposes, it should be remembered that freezing and thawing will disrupt cells. Degradative enzymes will be released. This can be prevented by a preliminary heating to 60°C to inactivate the enzymes.

Verdi *et al.* (1984) observed that milk with higher somatic cell counts had lower casein contents than milk with lower cell counts. They attributed this to proteolysis of casein. This group (Senyk *et al.,* 1985) later found that proteolysis increased when somatic cell counts increased from 50,000 to 1,000,000/ml. Some protelysis was detected in pasteurized milks with high cell counts. The proteases associated with the cells damage raw milk quality during storage and have an adverse effect on pasteurized fluid milk and milk during cheese making. Continuing their work, this group (Verdi *et al.,* 1987) found that proteolysis of caseins increased with incubation with either high or low cell counts.

Milk producers are required to and do attempt to exclude bacteria from milk; however, some microorganisms gain entry. In the United States, the bacterial count in Grade A raw milk may not exceed 300,000/ml, in the EEC, 100,000/ml, and in Switzerland, 80,000/ml. When the cow has mastitis, microorganisms associated with infections such as *S. aureus, Streptococcus uberis,* and *Streptococcus agalactiae* are found (Jones *et al.,* 1984; Guidry, 1985). Milk from cows with mastitis must be excluded from any commercial processing for human consumption, but some subclinical cases may not be detected. Pasteurization destroys most of the microorganisms in milk and all of the pathogens. In the United States the upper limit of bacteria in pasteurized milk is 20,000/ml. In Switzerland, the limit is 20,000/ml.

2. Membrane Fragments

These components are similar to those found in human milk except that there will be less derived from lipid globule cytoplasmic crescents (see Section II, A). Only 1% of the globules in bovine milk have the crescents (Huston and Patton, 1990). However, bovine milk has well-characterized fluff fractions in the skim milk phase (Stewart *et al.*, 1972). Bovine globules do not have glycoprotein filaments clustered on their surfaces as are seen on the human globules (Buchheim *et al.*, 1988). As mentioned, the membrane fragments have been termed milk lipoproteins (Rüegg and Blanc, 1982). These components may be irreversibly denatured or otherwise altered by processing although this has apparently not been reported.

III. Lipid Globule Emulsion

A. Introduction

Almost all of the lipid in human and bovine milks is found in dispersed globules. The stability of the emulsion is maintained by the amphiphilic components in the globule membrane, particularly the strong negative change carried by some of the glycolipids and proteins. Triacylglycerols make up 98% or more of the lipid with polar compounds in the membrane and the nonpolar components in the core. The core is almost totally triacylglycerol, while all of the phospholipids and most of the cholesterol are in the membrane. The purpose of the emulsion appears to be to provide a unit amount of dispersed lipid globules with large total surface area uniformly dispersed in a unit volume of milk. After ingestion, the milk forms a gel (curd) in the stomach and the globule surfaces are accessible to enzymatic action by gastric lipase and other enzymes. Rapid lipolysis occurs. The globules are resistant to digestion in the small intestine by pancreatic lipase and the bile–salt-stimulated lipase in human milk unless first conditioned by exposure to gastric lipase in the human or probably to pregastric lipase in the calf (see Chapters 6A and 6B). If the lipids were not dispersed as globules, the fat would rise and merge into a layer. It could not be secreted nor digested.

B. Size Distribution

The average diameter of globules in all species examined (cow, human, goat, ewe, sow) ranges from 3 to 5 μm. These figures are not precise since there are problems with determination of mean diameter. These problems are primarily because populations of small globules (< 1 μm) cannot be counted with the light microscope or Coulter counter [see Chapter 2A and Rüegg and Blanc (1981; 1982)]. A relatively new instrument, the Coulter

LS130 photon correlation spectrometer, which uses laser light, can determine particles with diameters ranging from 1 to 10,000 nm. Measurements are based on laser diffraction and scattering and polarization intensify differential scattering. Diameters of particles in milk ranging from 0.1 to 900 μm have been determined (Blanc, unpublished data). A broad distribution below 0.6 μm was observed which was probably due to casein micelles. Blanc noted that the casein micelle distribution was probably not correctly measured because the calculations were based on the refractive index of milk fat. Cyr *et al.* (1989) used the instrument to determine the size distribution of fat globules in intravenous fat emulsions. The mean globule diameter was 0.3 μm. There are three overlapping size distributions of human (Rüegg and Blanc, 1981, 1982) and bovine milk globules (Walstra and Jenness, 1984): small with diameters below 1 μm, intermediate with 3 to 5 μm diameters, and large with a diameter range of 8 to 10 μm. The small globules make up about 70 to 90% of the total number, but only a small portion of the total fat. The intermediate group has the largest amount of fat, but only about 10 to 30% of the globule numbers. The larger population ranges from 8 to 12 μm, but has only 0.01% of the fat.

Some important parameters of the globule dispersions in human and bovine milks are presented in Table III. In human colostrum, there is less fat, but also more globules, so the surface area of fat is 3.3 m^2/g compared to 1.4 m^2 in mature milk. This may be an adaptation in response to the neonate's relative inability to digest fats. Colostrum is excluded from commercial bovine milk. Most of the bovine milk is homogenized, at least in the United States, where the fat content is standardized to about 3.3 to 3.4%. Except for the association of casein with globules mentioned earlier, we have very little information on the redistribution of components into compartments as a result of homogenization and none on the digestion of bovine milk lipids in humans by the gastric–pancreatic lipase system (Jensen *et al.*, 1990, 1992). The globule parameters influence many factors in the processing of milk, e.g., creaming, separation formation of butter, clustering of globules, etc. For more information see Mulder and Walstra (1974) and Walstra and Jenness (1984) and Chapter 2A.

The globule size distribution in human milk is affected by gestational age of the infant. The average diameters in preterm and term milks were identical and increased 2.2 to 2.7 μm to 40 days postpartum (Simonin *et al.*, 1984). The number of globules with diameters of 1 to 1.5 and 8 to 13 μm, respectively, decreased as gestational age increased. A similar decline was observed in term milk, but the numbers of globules were lower. The numbers of larger globules leveled off to those in term milk 50 days postpartum. The fatty acids in the diet influence their profiles in the lipids of human and bovine milks. The profiles influence the liquidity of the triacylglycerols in the globules in bovine milk. Timmen and Patton (1988) observed that cows fed certain rations produced globules in skim milk with smaller diameters and altered fatty acid profiles compared to those in

TABLE III
Some Parameters of the Fat Globule Dispersion in Human Colostrum and Mature Human and Bovine Milks[a]

	Human average (+ SD)				Bovine (range)	
Parameter	Colostrum		Mature		Milk	Homogenized
Fat content (g/100 g)	2.6	1.0	3.3	0.6	3.7–4.1	3.7–4.1
Globules (approx No./ml)	6×10^{10}	(2×10^{10})	1.1×10^{10}	(3×10^{9})	$1.5–10^{10}$	$10^{12}–10^{14}$
Surface area of 1 g fat in milk (m^2)	3.3	0.5	1.4	0.1	1.4–2.9	10–30
Volume/surface average diameter ($d_{vs.}$ μm)	1.5	0.3	4.0	0.3	2.5–4.6	0.2–0.7

[a]Adapted from Rüegg and Blanc (1982).

cream. The authors suggested that the mammary gland regulates the fatty acid composition in the globules to maintain liquidity at body temperatures and that this may affect the diameters of the small globules. This influence may be irrelevant industrially, since herd milks are pooled unless most of the producers in a region are feeding the same diet to their cows. Although changes in diet markedly and rapidly influence the fatty acid profiles in human milk (Chapter 6A), there are no reports of effects on globule sizes and distributions.

IV. Casein Micelles

A. Introduction

These particles exist as complexes of protein and salts, in a colloidal, making up 20 to 40% of the protein in human milk and about 80% in bovine milk (Rüegg and Blanc, 1982). Classically and of importance in cheese making, caseins are precipitated from bovine milk by acidification to pH 4.6 at 20°C (Eigel *et al.*, 1984). The caseins in human milk are more difficult to isolate requiring acidification to pH 4.3 and addition of $CaCl_2$ (Kunz and Lonnerdal, 1989a,b). The amounts of the caseins in human and bovine milks are shown in Table IV.

The casein micellar systems in the milks differ considerably. The micelles in human milk are about 43 nm in diameter (Carroll *et al.*, 1985). The average diameters of casein micelles in bovine milk are about 83 nm (Donnelly *et al*, 1984) or 120–180 nm (Farrell, 1990). In diameter, they are about 1/50 that of a fat globule.

TABLE IV
Amounts of Caseins in Mature Human and Bovine Milks

Protein	Human[a] (g/liter)	Bovine[b] (g/liter)
Total	9.0	36.0
Total casein	2.7	29.5
α-S1	Not present	11.9
α-S2	Not present	3.1
β	2.3[c]	9.8
γ[d]	?	1.2
κ	0.4	3.5

[a]From Swaisgood, Chapter 4B, Table 1.
[b]Adapted from Kunz and Lonnerdal (1989a).
[c]Not precisely determined. Calculated by the authors based on κ-casein being 15% or less of total caseins.
[d]γ-Casein is a product of the proteolysis of the C-terminal of β-casein.

B. *Structure and Size Distribution*

1. *Human Milk*

The structure of casein micelles in human milk appears to be based on the association of highly phosphorylated β-casein which binds calcium, a low phosphorylated form, and glycosylated κ-casein (Chapter 5A and 7A). The interior of the micelle contains calcium phosphate. In human milk only 15% of the calcium is bound to casein and in bovine milk the amount is about 65% (Neville *et al.,* 1994). Submicelles may be grouped into spherical particles. Information on the micelles is given in Table V. Note that although there is less casein in human than in bovine milk, the numbers of micelles are about the same.

2. *Bovine Milk*

We mentioned earlier that the casein micelles in bovine milk occur as colloidal complexes of proteins and salts, primarily calcium (Farrell *et al.,* 1990). When calcium is removed submicelles are produced which contain four proteins: α-s1, α-s2, β, and, κ-caseins in ratios of about 4:1:4:1. These compounds have an average molecular weight of about 23,900 and are phosphorylated to various degrees. Farrell *et al.,* (1990, 1993) have postulated that these hydrophobically stabilized submicelles are incorporated into the micelle. According to Rollema (1992), the model for bovine casein micelles which best fits experimental data is the association of several subunits to form a large spherical micelle. The implication is that bonds between submicelles in the calcium phosphate phase and hydrophobic

TABLE V
Some Parameters of the Colloidal Casein Dispersion in Human and Bovine Milks[a]

Casein parameter	Unit	Human	Bovine
Concentration	g/dl	0.2–0.5	2.2–2.8
Types (α-s1:β:κ)	Ratio	ca. 0:7:3	ca. 1:0.8:0.3
Micellar Ca:P	Ratio	0.2:0.6	2.2:2.8
Number of micelles	Per milliliter	ca. 7×10^{15}	ca. 7×10^{15}
Average diameter of micelles			
d_n	nm	8–14 (43.0)[b]	21–24
d_v	nm	10–26 (44.9)	44–50
d_{vs}	nm	11–55 (46.9)	90–100
d_{vm}	nm	16–88 (49.9)	104–140
Average diameter of submicelles			
d_n	nm	6–8	10–11
d_v	nm	7–9	11–12
d_{vs}	nm	8–10	12–13
d_{vm}	nm	9–12	13–14

[a]Adapted from Rüegg and Blanc (1982).
[b]Carroll *et al.* (1989).

interactions between submicelles are responsible for the integrity of the casein micelles. The highly glycosylated ϰ-casein also has a structure-stabilizing role. The fact that casein micelles carry a strong negative charge, as indicated by their isoelectric point of pH 4.6, makes them strongly self-repelling at the normal pH of milk, 6.6–6.7. Table V contains the information on bovine casein micelles.

V Summary

Again, the extraordinary complexity of the physical organization of the components in human and bovine milks is obvious. Compartmentation influences the availability of the components as nutrient and as nonnutritive messages. The effect on bovine milk processing is also important. Obviously, more information is needed.

Acknowledgments

The preparation of the manuscript was supported in part by an NIH contract and by federal funds made available through provision of the Hatch Act, Scientific Contribution,

Storrs Agricultural Experiment Station, Storrs, Connecticut. We appreciate the advice of Dr. H. M. Farrell, Jr. on bovine casein micelles.

References

Bennett, R. (1993). The case for improving milk quality regulations: Milk somatic cell counts. *Dairy Food Environ. Sanit.* **13,** 278–282.

Brooker, B. E. (1980). The epithelial cells and cell fragments in human milk. *Cell. Tissue. Res.* **210,** 321–332.

Buchheim, W., Welsch, U., Huston, G. E., and Patton, S. (1988). Glycoprotein filament removal from human milk fat globules by heat treatment. *Pediatrics* **81,** 141–146.

Buescher, E. S., and Pickering, L. K. (1986). Polymorphonuclear leukocytes in human colostrum and milk. *In* "Human Milk in Infant Nutrition and Health" (R. P. Howell, F. H. Morriss, Jr., and L. K. Pickering, eds.), p. 160–173. Thomas, Springfield, IL.

Carroll, R. J., Basch, J. J., Phillips, J. G., and Farrell, H. M., Jr. (1985). Ultrastructural and biochemical investigations of mature human milk. *Food Microstruct.* **4,** 321–323.

Cyr, T. D., Lawrence, R. C., and Lovering, E. G. (1989). Determination of size distribution of fat globules in intravenous fat emulsions by photoncorrelation spectroscopy. *J. Assoc. Off. Anal. Chem.* **72,** 436–441.

Donnelly, W. J., McNeill, G. P., Buchheim, W., and McGann, F. C. A. (1984). A comprehensive study of the relationship between size and protein composition in natural bovine casein micelles. *Biochim. Biophys. Acta* **789,** 136–143.

Eigel, W. N., Butler, J. E., Ernstrom, C. A., Farrell, H. M., Jr., Harwalker, V. R., Jenness, R., and Whitney, R. M. L. (1984). Nomenclature of proteins of cow's milk. *J. Dairy Sci.* **67,** 1599–1631.

Farrell, H. M., Jr., Pessen, H., Brown, E. M., and Kamosinski, T. F. (1990). Structural insights into the bovine casein micelle: Small angle X-ray scattering studies and correlations with spectroscopy. *J. Dairy Sci.* **73,** 3592–3601.

Farrell, H. M., Jr., Brown, E. M., and Kumosinki, T. F. (1993). Three-dimensional molecular modeling of bovine caseins. *Food Struct.* **12,** 235–250.

Guidry, A. J. (1985). Mastitis and the immune system of the mammary gland. *In* "Lactation" (B. L. Larson, ed.), pp. 229–262. Iowa State University Press, Ames.

Hayward, A. R. (1983). The immunology of human milk. *In* "Lactation, Physiology, Nutrition, and Breastfeeding" (M. C. Neville and M. R. Neifert, eds.), pp. 258–261. Plenum Press, New York.

Heald, C. W. (1985). Milk collection. *In* "Lactation" (B. L. Larson, ed.), pp. 198–228. Iowa State Univ. Press, Ames.

Huston, G. E., and Patton, S. (1990). Factors related to the formation of cytoplasmic crescents on milk fat globules. *J. Dairy Sci.* **73,** 2061–2066.

Institute of Medicine (IOM). (1991). "Nutrition During Lactation," p. 137. National Academy of Science, Washington, DC.

Jensen, R. G., Ferris, A. M., Lammi-Keefe, C. J., and Henderson, R. A. (1990). Lipids of bovine and human milks: A comparison. *J. Dairy Sci.* **73,** 223–240.

Jensen, R. G., Lammi-Keefe, C. J., and Ferris, A. M. (1992). Effect of milk triacylglycerol structure on absorption and metabolism. *Inform* **3,** 560.

Jones, G. M., Person, R. E., Clabaugh, G. A., and Heald, C. W. (1984). Relationships between somatic cell counts and milk production. *J. Dairy Sci.* **67,** 1823–1831.

Kunz, C., and Lonnerdal, B. (1989a). Human milk protein: Separation of whey proteins and their analysis by polyacrylamide gradient gel electrophoresis, fast protein liquid chromatography (FPLC) gel filtration, and anion-exchange chromatography. *Am. J. Clin. Nutr.* **49,** 464–470.

Kunz, C., and Lonnerdal, B. (1989b). Casein micelles and casein subunits in human milk. *In* "Protein and Non-Protein Nitrogen in Human Milk" (S. Atkinson and B. Lonnerdal, eds.), pp. 10–27. CRC Press, Boca Raton, FL.

Lawrence, R. A. (1989). "Breastfeeding," 3rd Ed., pp. 120–126. C. V. Mosby, St. Louis, MO.

Lipkin, E., Shalom, M., Khabib, H., Soller, M., and Friedman, A. (1993). Milk as a source of deoxyribonucleic acid and as a substrate for the polymerase chain reaction. *J. Dairy Sci.* **76,** 2025–2032.

Mandyla, H., and Xanthou, M. (1986). Function of leukocytes in human milk. *In* "Human Lactation. 2. Maternal and Environmental Factors" (M. Hamosh and A. Goldman, eds.), pp. 533–540. Plenum Press, New York.

Mulder, H., and Walstra, P. (1984). "The Milk Fat Globule," Common. Agric. Bur., Farnham Royal, Bucks, England.

Neubauer, S. H., Ferris, A. M., Lammi-Keefe, C. J., Green, K. W., Hinckley, L. S., and Jensen, R. G. (1995). Composition and bacterial and somatic cell contents in the milk of women with and without insulin-dependent diabetes mellitus. Manuscript in preparation.

Neville, M. C., Keller, R. P., and Casey, C. (1994). Calcium partioning in human and bovine milk. *J. Dairy Sci.* **77,** 1964–1975.

Ogra, P. L., and Ogra, S. S. (1988). Cellular aspects of immunologic reactivity in human milk. In "Biology of Human Milk" (L. A. Hanson, ed.), pp. 171–184. Raven Press, New York.

Patton, S., and Huston, G. E. (1986). A method for the isolation of milk fat globules. *Lipids* **21,** 170–174.

Prentice, A., Prentice, A., and Lamb, W. (1985). Mastitis in rural Gambian mothers and the protection of the breast by milk antimicrobial factors. *Trans. R. Soc. Trop. Hygiene* **79,** 90–95.

Rollema, H. S. (1992). Casein association and micelle formation. *In* "Advanced Dairy Chemistry-1: Proteins" (P. F. Fox, ed.), pp. 111–140. Elsevier, New York.

Rüegg, M., and Blanc, B. (1981). The fat globule size distribution in human milk. *Biochim. Biophys. Acta* **666,** 7–14.

Rüegg, R., and Blanc, B. (1982). Structure and properties of the particulate constituents of human milk. A review. *Food Microstruct.* **3,** 25–47.

Senyk, G. F., Barbano, D. M., and Shipe, W. F. (1985). Proteolysis in milk associated with increasing somatic cell counts. *J. Dairy Sci.* **68,** 2189–2194.

Simonin, S., Rüegg, M., and Sideropoulis, D. (1984). Comparison of the fat globule size distribution of breast milk from mothers delivering term and preterm. *Am. J. Clin. Nutr.* **40,** 820–826.

Stewart, P. S., Puppiome, D. L., and Patton, S. (1972). The presence of microvilli and other membrane fragments in the non-fat phase of bovine milk. *Z. Zellforsch.* **123,** 161–167.

Timmen, H., and Patton, S. (1988). Milk fat globules: Fatty acid composition, size and in vivo regulation of fat liquidity. *Lipids* **23,** 685–689.

Verdi, R. J., and Barbano, D. M. (1991). Properties of proteases from milk somatic cells and blood leukocytes. *J. Dairy Sci.* **74,** 2077–2081.

Verdi, R. J., Dellavalle, M. E., Barbano, D. M., and Senyk, G. F. (1984). Relationship between milk somatic cell count, true protein, casein, nonprotein nitrogen, and tyrosine value. *J. Dairy Sci.* **67** (Suppl. 1), 70–71.

Verdi, R. J., Barbano, D. M., Dellavalle, M. E., and Senyk, G. F. (1987). Variability in true protein, casein, nonprotein nitrogen, and proteolysis in high and low somatic cell milks. *J. Dairy Sci.* **70,** 230–242.

Walstra, P., and Jenness, R. (1984). "Dairy Chemistry and Physics," Chap. 14. Wiley, New York.

C. Sampling and Storage of Human Milk

MARGARET C. NEVILLE

I. Introduction

The composition of human milk is influenced significantly by the stage of lactation, the presence or absence of mastitis, and the method of sampling. For some components the time of day, the nutritional state, and other conditions also affect milk composition. Although some general principles apply, for each previously unstudied milk component the sources and limits of variation must be determined. If the sample is to be stored, the method of storage that will best preserve the component must also be investigated. Although the focus of this article is on the human, the same principles apply in general to the milks of all species. One caveat is that an altogether representative sample of the milk of small animals, like mice and other small laboratory animals, may be difficult to obtain because of the problem of obtaining complete emptying of the mammary gland under laboratory conditions.

This chapter will deal with the sources of variation in milk composition and discuss some general principles for the sampling and storage of human milk. Because some of the variability in milk composition can be understood in terms of the physiology of milk secretion, this topic will be briefly summarized followed by a description of sampling methods that are most likely to provide a representative milk sample. The structure of milk as a fluid will then be outlined briefly to provide a basis for a practical discussion of methods of storage and handling of milk samples.

II. Mechanisms of Milk Secretion and Ejection

A. *Milk Ejection and the Anatomy of the Mammary Gland*

The mammary glands of all mammals consist of a series of ducts of epithelial origin coursing and branching through a connective tissue stroma and terminating distally in clusters of grape-like alveoli where milk secretion and storage take place (Gould, 1983). In many species, including rodents and dairy animals, the ducts terminate proximally in a cistern that

HANDBOOK OF MILK COMPOSITION

is connected to the exterior through a single canal terminating at the end of a teat. In other species, including humans, the ducts terminate directly through pinsized openings on the nipple (Figure 1).

Once lactation has commenced milk appears to be secreted continuously and stored in the alveoli in contact with the cells that secrete it (Neville *et al.*, 1983). The alveoli expand as their milk content increases, flattening the single layer of epithelial cells that forms a continuous lining of both ductules and alveoli. A network of basket-like myoepithelial cells is intimately associated with the basal surface of the mammary epithelial cells lining the ducts and alveoli (Figure 2). Milk ejection from the gland occurs when these cells contract in response to increased plasma levels of the posterior pituitary hormone, oxytocin. Oxytocin secretion is part of the "let-down" reflex engendered by stimulation of the nipples or, in some circumstances, by less direct stimuli such as the cry of the young. Without the let-down reflex it is not possible to extract milk stored within the alveoli, although some milk in the cistern and ducts can be expressed.

Once secreted, the nutrient composition of alveolar milk remains largely constant although the concentrations of lipid-soluble substances, such as steroid hormones and some drugs and toxins that pass freely through the mammary epithelium, vary with their blood concentrations (Walsh and Neville, 1992; Peterson and Bowes, 1983).

Unlike saliva the ionic composition of the milk is not altered as it passes through the ducts during milk ejection (Neville *et al.*, 1983). However, there is an important change in milk composition from the beginning to the end of the feed: milk extracted early in a feed or milking has a much lower fat content than milk extracted near the end of the feed (Hytten,

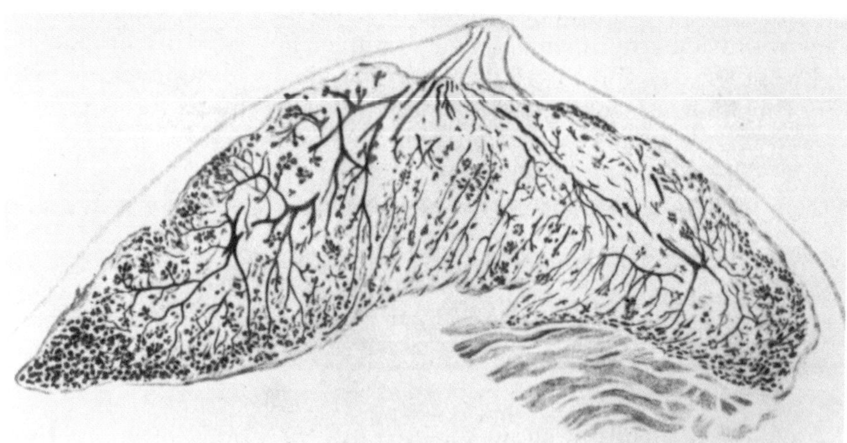

Figure 1 Anatomical structure of the human mammary gland showing ducts of epithelial origin coursing through a stroma consisting of adipose and connective tissue. Distally the ducts divide and terminate in clusters of alveoli; proximally the ducts terminate directly on the nipple. From Dabelow (1941).

1954; Woodward *et al.*, 1989). This change in fat content between fore- and hindmilk occurs in all species and presents one of the greatest challenges to the investigator attempting to extract a representative milk sample. Its origin is not entirely understood. Hytten (1954) noted that when milk was poured through a sponge early fractions had a lower fat content than later fractions and postulated that milk fat globules were retarded by the walls of the ducts as they passed toward the nipple. However, it is also possible that the mechanical effects of myoepithelial contraction loosen some forming milk fat globules from their cellular attachments during the let-down reflex so that later milk, presumably obtained from severely contracted

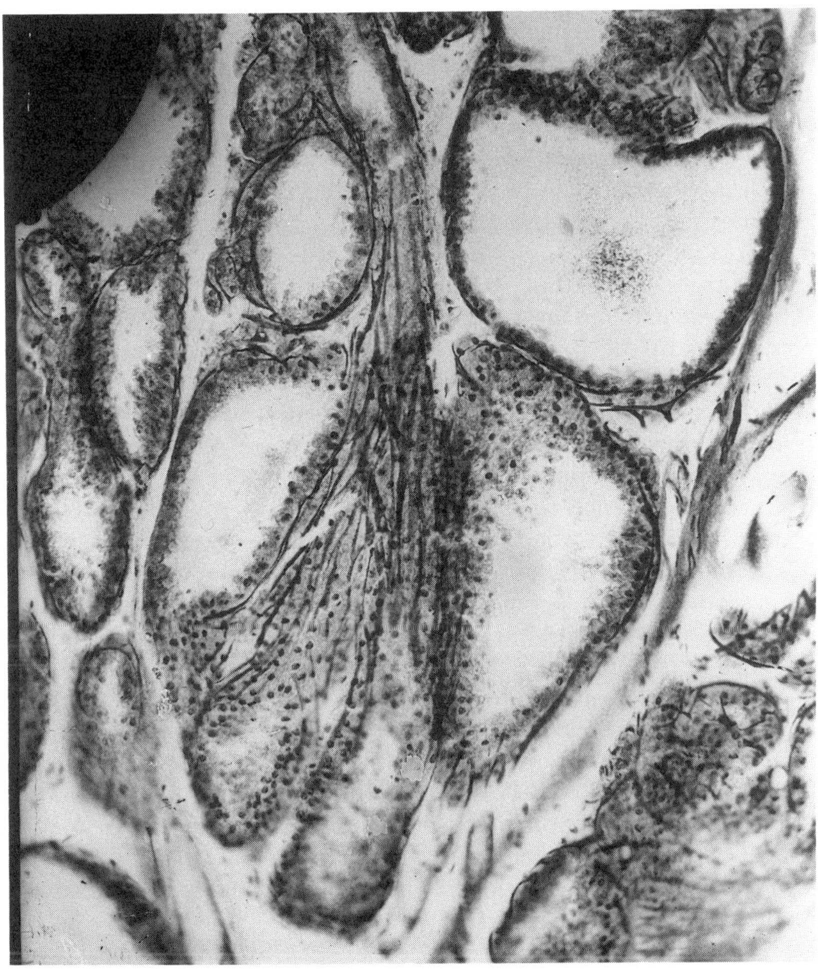

Figure 2 Myoepithelial cells surrounding mammary alveoli. Silver stain of ruminant mammary tissue. Note dark staining fibrillar cells running parallel to a duct and surrounding alveolar structures. This figure represents the original histological demonstration of the presence of myoepithelial cells in the mammary gland. From Richardson (1949).

alveoli, has a somewhat higher fat content than milk stored in expanded alveoli. Whatever the physiologic origin of the difference in fat content between fore- and hindmilk, when fat-soluble components of the mammary secretion are to be measured a sampling regimen that gives a milk sample that is as representative as possible must be instituted.

B. *Milk Secretion and the Anatomy of the Mammary Alveolar Cell*

A single layer of cuboidal mammary epithelial cells lines the smaller ducts and alveoli of the mammary gland. These cells, separated from the interstitial space by a basement membrane, are responsible for elaboration of all milk components, except the cellular components derived from the lymphoid system which presumably enter the milk space by "squeezing" between the mammary alveolar cells (Seelig and Beer, 1981). A number of distinct metabolic pathways are involved in milk formation (Figure 3). Quantitatively, the pathway responsible for the largest volume of milk is the exocrine pathway (pathway I in Figure 3) which is responsible for elaboration of most of the components of the aqueous fraction of milk. The basic mechanisms involved are no different from those of other secretory cells. That is, the proteins are synthesized on ribosomes and inserted across the membranes of the rough endoplasmic reticulum, transferred to the Golgi apparatus, sorted, and packaged into secretory vesicles for secretion by exocytosis (Mercier and Gaye, 1983). Secretion appears to be all or largely constitutive; that is, secretory vesicles are not retained within the mammary cell for any appreciable length of time (Ollivier and Denamur, 1975) although evidence for regulated secretion of a small fraction of mouse milk has been obtained by Burgoyne and co-workers (Turner *et al.*, 1992). The major milk proteins, casein, α-lactalbumin, and possibly transferrin (in rodents and lagomorphs) and lactoferrin (in humans), are secreted via pathway I (Siddiqui *et al.*, 1992). Within the terminal cisternae of the Golgi apparatus the enzyme galactosyl transferase acting jointly with α-lactalbumin promotes the synthesis of lactose from glucose and UDP-galactose (Kuhn, 1983). Because the membranes of the Golgi apparatus and secretory vesicles are impermeable to disaccharides, water enters these vesicles under the osmotic effect of the synthesized lactose. Kinases within the pathway phosphorylate casein (Bingham and Farrell, 1981). The presence of high calcium concentrations in the terminal Golgi (Neville and Watters, 1983) leads to formation of the casein micelle (Kumosinski and Farrell, 1991). Milk citrate and free phosphate are also secreted via this route (Linzell *et al.*, 1976; Neville and Peaker, 1979) which is probably also responsible for regulating the monovalent cation concentration of the milk by mechanisms which are not currently understood.

Milk lipids are secreted by a pathway (pathway II) unique to the mammary gland involving triglyceride synthesis from precursor fatty acids transported from the plasma or synthesized within the alveolar cell

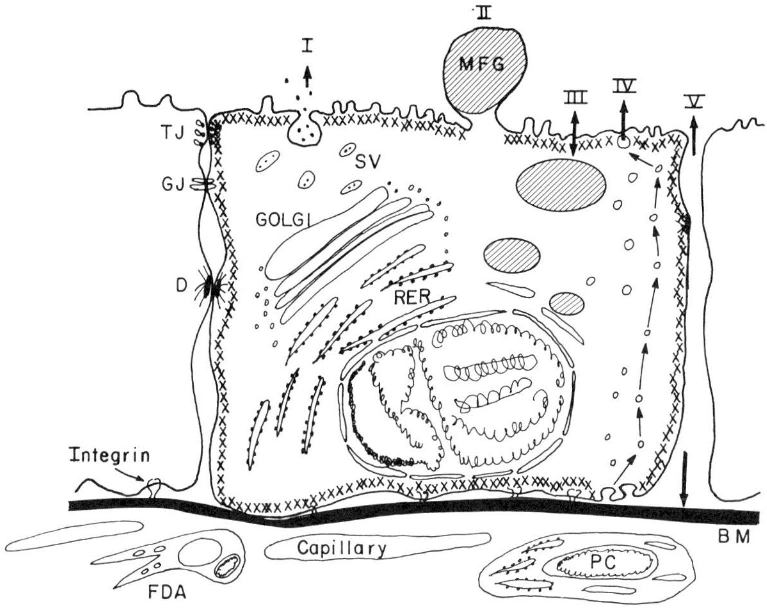

Figure 3 Diagrammatic represenıation of pathways for milk secretion. Pathway I, exocrine pathway showing rough endoplasmic reticulum (RER), Golgi vesicles, and secretory vesicles (SV) responsible for elaboration of the aqueous fraction of milk including proteins of epithelial origin such as casein, α-lactalbumin, and possibly lactoferrin and transferrin. Pathway II, pathway for secretion of the milk fat globule (MFG). Pathway III, apical membrane exchange pathway containing specific transport mechanisms for sodium, potassium, chloride, glucose, and probably bicarbonate. Pathway IV, transcytotic pathway for secretion of immunoglobulins and other extra-alveolar cell proteins including peptide hormones. Pathway V, paracellular pathway closed in lactation, open in pregnancy, after involution and during mastitis. The types of junctions that join neighboring cells include the tight junction (TJ), the gap junction (GJ), and a desmosome (D). Also shown are attachments to the basement membrane (BM) via integrins and, in the interstitial space, a fat-depleted adipocyte (FDA) and a plasma cell (PC) responsible for the elaboration of IgA.

(Dils, 1977). These triglycerides coalesce into larger and larger fat droplets that are pulled toward the apical surface of the cell eventually becoming enveloped in apical membrane and extruded from the cell as the milk fat globule (Mather, 1987). This pathway is dealt with in elegant detail in Chapter 2A.

The apical membrane of the cell itself (pathway III) is permeable to a limited number of ions and compounds including water, sodium, potassium, chloride, and glucose (Peaker, 1983). The concentrations of these compounds are characteristic of individual species but the mechanisms involved in the regulation of the monovalent ion concentrations are currently unclear. Of these substances, the concentration of glucose within the cell increases significantly after a carbohydrate meal and, therefore, milk glucose concentration, particularly in humans (Neville *et al.*, 1990), shows significant diurnal variation.

Certain compounds, most notably immunoglobulins (Solari and Kraehenbuhl, 1987), bind to a receptor at the basolateral membrane of the cell, are internalized in a coated pit, and are transcytosed to be secreted at the apical side of the cell. This transcytotic pathway (pathway IV) is probably responsible for the secretion of peptide hormones, such as prolactin and insulin, into milk; other plasma proteins may enter milk by the same pathway.

In full lactation in most species (the rabbit may be an exception; Peaker and Taylor, 1975) the junctional complexes between the cells are tightly closed and allow little or no traffic of solute directly between the interstitial space and the milk space through the pathway V, the *paracellular pathway*. As previously mentioned, the secretion of lymphoid cells into milk may be an exception; how these cells force their way across the epithelium is not yet understood. However, during pregnancy, after involution and with mastitis the paracellular pathway is open and small molecules from the interstitial space, most notably sodium, chloride, glucose, and phosphate, appear to pass freely into the milk space, while lactose, calcium, and potassium pass from the milk space into the plasma (see Chapter 3A; Neville *et al.*, 1991a). Under these conditions the concentrations of sodium and chloride are increased substantially in the milk, while those of lactose, potassium, and calcium are significantly reduced.

III. Methods for Obtaining a Representative Milk Sample

The outline of the mechanisms of milk secretion and ejection in the preceding sections suggests that milk composition can vary (1) within the feed due to changes in fat content (Hytten, 1954); (2) diurnally due to postprandial variation in the plasma concentration of nutrients, such as glucose (Neville *et al.*, 1990), amino acids (Donovan *et al.*, 1991), hormones, etc., whose plasma concentrations are reflected in the milk; (3) between breasts, if mastitis is present in one breast (Linzell and Peaker, 1972; Neville *et al.*, 1984); and (4) with duration of lactation due to variations in the permeability of the paracellular pathway or other secretory changes as lactation progresses (Neville *et al.*, 1991a; Allen *et al.*, 1991). To illustrate one dramatic effect of duration of lactation, Figure 4 shows the concentration of zinc as a function of time postpartum in women. The mechanism by which the zinc concentration is regulated is unknown.

Maternal nutritional status may alter milk composition particularly in high-producing dairy species or rodents, but also in humans. Recently, it has been shown that the fat content of the milk is reduced in women with low body fat (summarized in Allen *et al.*, 1991). Unknown factors may also contribute to variations in milk composition. In a carefully done recent study by Mock *et al.* (1992), biotin was found to vary widely among subjects, both diurnally and longitudinally (Figure 5). The results of this study

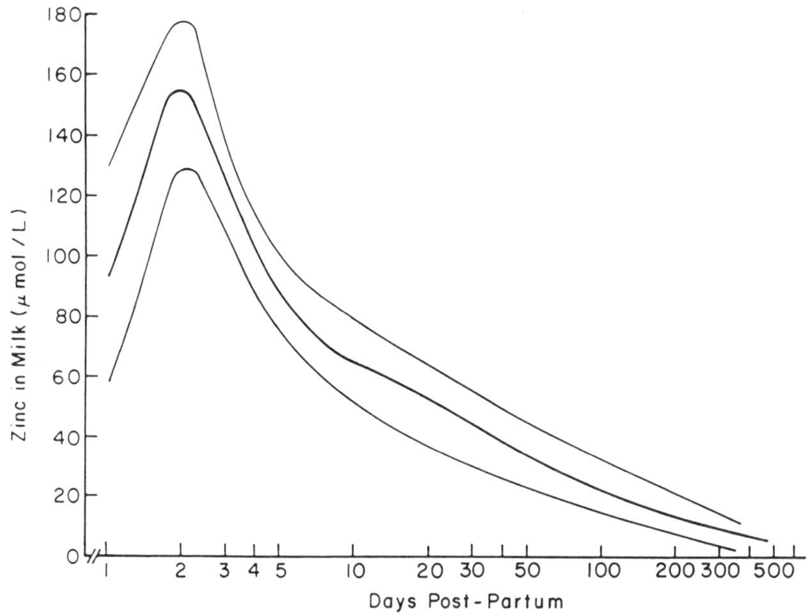

Figure 4 Changes in zinc concentration with duration of lactation in women. Drawn after Casey *et al.* (1989). The central line shows the mean value and the distance between the outer lines represents two standard deviations from the mean.

illustrate the point that sources of variation in the concentration of a milk component must be understood before any reliable population values are made available. Diurnal and longitudinal studies are necessary to reveal this variation. If a complete longitudinal survey is not possible, an effort should be made to collect milk samples at a specific time after birth and state that time explicitly in any publication.

The choice of sampling protocol depends on the substance to be measured and the nature of the population from whom milk samples are to be obtained. A number of acceptable protocols are described below. Because of the prevalence of mastitis and its effect on milk composition, it is opinion of the author that the sodium content of all milk samples should be routinely measured. This simple measurement is available in most hospital laboratories where the analyzer should be set with standards intended for urine. It can help rule out sporadic or chronic mastitis, a condition that has a profound effect on the concentration of many milk components. Conductivity measurements can be substituted for sodium concentrations if the latter are not available. Normal milk samples have a conductivity of 2.5 to 3.5 mmhos corresponding to an ionic strength of 24 to 32 m*M* (Allen and Neville, 1983). After the seventh day postpartum any sample with a sodium concentration above 20 m*M* or a conductivity above 6 mmhos should be considered mastitic unless involution is under way (Neville *et al.,* 1991a).

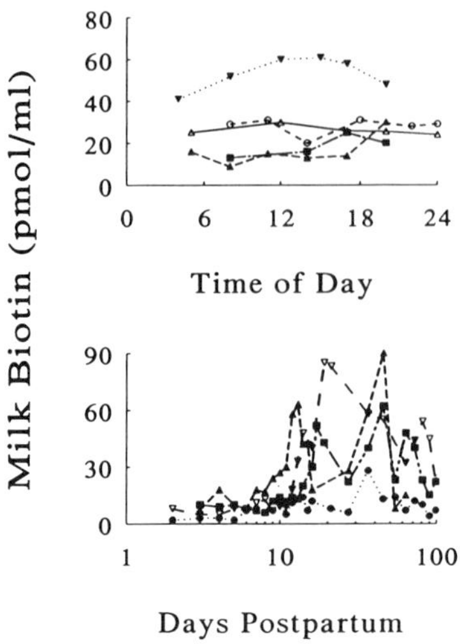

Figure 5 Changes in biotin concentration with time of day and duration of lactation in individual women, each represented by a different symbol. Redrawn from Mock *et al.* (1992).

A. Small Milk Samples

When only small samples can be taken, a mid-feed sample, taken 3 to 5 min after the onset of nursing on each breast, gives a reasonably representative sample, especially for population data (Prentice *et al.*, 1981; Allen *et al.*, 1991). However, the interruption of the feed is uncomfortable for many subjects so some investigators have obtained a small foremilk and a small hindmilk sample from each breast, analyzed these separately, and used a formula derived from the fat content of an entire emptying of the breast to determine the mean lipid content (Hartmann *et al.*, 1980). It is more satisfactory to use large milk samples whenever possible.

B. Pumped Samples from Alternate Breasts

The regimen first used by Butte and her colleagues (Butte *et al.*, 1984) probably gives the most satisfactory sample for estimates of daily nutrient transfer from mother to infant. Milk is obtained by a good electric breast pump (Egnell or Medela) from one breast while the infant nurses at the other, producing a satisfactory let-down even in a mother who does not let down in response to an electric pump. The pumped sample is thoroughly mixed, its volume is recorded, and an aliquot is taken and stored for later analysis. The remainder of the milk is fed to the infant by bottle if desired. The procedure is repeated at every feed for a 24-hr period alternating breasts for pumping and feeding. The aliquots are then combined in proportion to the volume pumped at the corresponding feed and the combined sample is stored for later analysis (see sample storage below). Samples for analysis by creamatocrit (Lucas *et al.*, 1978) may be drawn immediately from each pumped sample into capillary tubes for lipid analysis (see below). Some investigators, working under hospital conditions, have used the entire daily production from both breasts obtained by breast pump (Brown *et al.*, 1982). This method works particularly well for hospitalized populations. As long as a good let-down is achieved this method should give representative milk samples. Garza and Butte (1986) showed that abbreviated regimes gave higher intraindividual variability.

C. Single Large Pumped Sample

Many authors have used a single large pumped sample obtained 2 or more hours after a breast feed for lipid analysis (Ferris and Jensen, 1984). The total fat content of such a sample may be influenced by the amount of residual milk left by the infant at an earlier feed and gives a less accurate estimate of daily fat intake than the alternate pumped sample method described above. For dairy animals, usually milked out at each milking, this method is usually satisfactory. If residual milk is needed a small dose of oxytocin can be given iv or intranasally (Neville *et al.*, 1988).

For laboratory animals, such as rats, guinea pigs, and rabbits, oxytocin can be given up and various sorts of pumps used to extract the milk. We have used two methods to extract up to 1.5 ml of milk from lactating mice. In method I (Berga and Neville, 1985), the mouse is anesthetized and oxytocin in given by intracardiac injection. A small slit is then cut at the base of each teat and milk that exudes from the nipple is immediately drawn into a syringe. The mouse must be sacrificed at the end of this procedure. In a second procedure that extracts less milk but leaves the mouse intact, the mouse is anesthetized as above and oxytocin is given intraperitoneally. A Pasteur pipet fashioned into a 1-mm-diameter inverted bell is attached with plastic tubing to a vacuum jar in which a microfuge tube is nestled in a bed of ice and placed under the tubing from the milking device. The bell is moved up and down each nipple with a milking motion to extract the milk (Greenburg *et al.*, 1991).

IV. Sources of Changes in Composition During Storage

A. The "Structure" of Milk

Milk is a complex fluid consisting of several "compartments" or phases including a cellular component and the milk fat globules suspended in the aqueous fluid phase. These phases can be separated by centrifugation of fresh milk at *g* forces < 1000 for 15 to 30 min (Mather, 1987). The milk fat globules float to the surface and the cellular components form a loose pellet, both of which can be washed in saline and isolated for further analysis. Isolation of the milk fat globules, whose membrane coating prevents coalescence into a cake, is facilitated by addition of 5% sucrose to the milk and layering under distilled water prior to centrifugation (Patton and Huston, 1986). Centrifugation at high speed, $> 10{,}000g$, for periods exceeding 20 min creates shear forces that also result in breakage of the milk fat globule membrane. After such treatment the lipid forms a solid cake that can be easily removed. However, the milk fat globule membranes are broken and can be found in the cell pellet. The milk fat globules can then be lifted from the top. To separate the core triglycerides from the membranes surrounding them, milk fat globules from 10 ml of milk are churned in ice-cold deionized water for 5 min and transferred to an ultracentrifuge tube. The membranes can be pelleted by centrifuging at 78,000*g* for 75 min in the cold. Milk lipid forms a solid cake at the surface of the solution and can be removed for analysis.

The aqueous phase of milk is itself not a true solution but rather a suspension of aggregates of casein, calcium, and phosphate, with smaller amounts of many other components in a structure called a "micelle" (Kumosinski and Farrell, 1991). The casein micelles from human milk are illustrated in Figures 6A–6C. The casein micelle has a radius of 300–500 A and is a complex lattice of several thousand casein molecules with several thousand calcium ions bound primarily to phosphoserines (Kakalis *et al.*, 1990). This electron opaque particle serves as a convenient package for the transfer of large amounts of calcium and phosphate to the young and can be separated from most milks by centrifugation at 50,000*g* for 2 hr. Casein can also be precipitated by incubation at pH 4.0 or below or with the enzyme rennin. Rennin cuts a special type of casein, K-casein, that appears to have a surface location that stabilizes the micellar structure.

When skim milk is subjected to high-speed centrifugation a fluffy pellet forms above the solid pellet of casein micelles that contains a variety of membranous structures and some lipid (Figure 6D). If the milk has been previously frozen prior to removal of the lipid (see below), this "fluff" contains milk fat globule membranes that have been dissociated from the globule by the freezing process (Figure 6E).

B. *Effects of Freezing*

Freezing of milk samples can affect milk structure in a number of ways, the most important of which are destruction of the cells and breakage of the milk fat globule membranes so that the lipid is free to coalesce when the sample is centrifuged, even at low speed. The inner surface of the milk fat globule membranes is exposed by freezing and normally sequestered binding sites for ions, such as calcium and other milk components, may become available leading to redistribution of aqueous components (Neville *et al.*, 1994). Components of the aqueous solution with an affinity for the core milk fat may also redistribute as a result of freezing; the lipoprotein lipase of human milk has been shown to be particularly sensitive to this effect (Neville *et al.*, 1991b). Most of the nutrients in the aqueous fraction appear to be resistant to effects of freezing and thawing, particularly if the lipid is removed prior to freezing; however, repeated freezing and thawing of stored milk samples should be avoided.

C. Effects of Method of Extraction

In the mammary gland milk is equilibrated with the 5% CO_2 present in the plasma. When milk samples are extracted with a breast pump a vacuum is produced that extracts varying amounts of CO_2. This is a more serious problem when the milk has a fairly high bicarbonate concentration and therefore a fairly high pH as occurs in human milk. We found that pumped human milk had a pH of about 7.3 ± 0.07 with a pCO_2 below 20 mmHg (2.5%; Neville *et al.*, 1994; Allen and Neville, 1983). When the milk was equilibrated with 5% CO_2 the average pH was 7.18 ± 0.06, which is significantly lower ($p < 0.01$). This change in pH has the potential of altering the equilibria among the many ionic species present in milk. If the component under study has the potential to be sensitive to small changes in pH, the milk should be hand expressed and stored equilibrated with 5% CO_2. This effect is less of a problem in bovine milk which has a lower bicarbonate concentration and an initial pH closer to 6.0

V. Recommendations for Storage of Milk Samples

A. Choice of Storage Vessel

The storage vessel chosen should neither bind the milk component under study nor add additional material. For trace elements, vitamins, and other substances present in very low quantities, acid-washed glassware and pigment-free plastics should be used for collection, storage, and analysis of the sample (Casey *et al.*, 1985). Samples should be aliquoted for storage or analysis under assiduously clean conditions.

B. Handling of the Milk Sample

The milk sample should be chilled as soon as possible to refrigerator temperature to decrease bacterial growth and held at that temperature

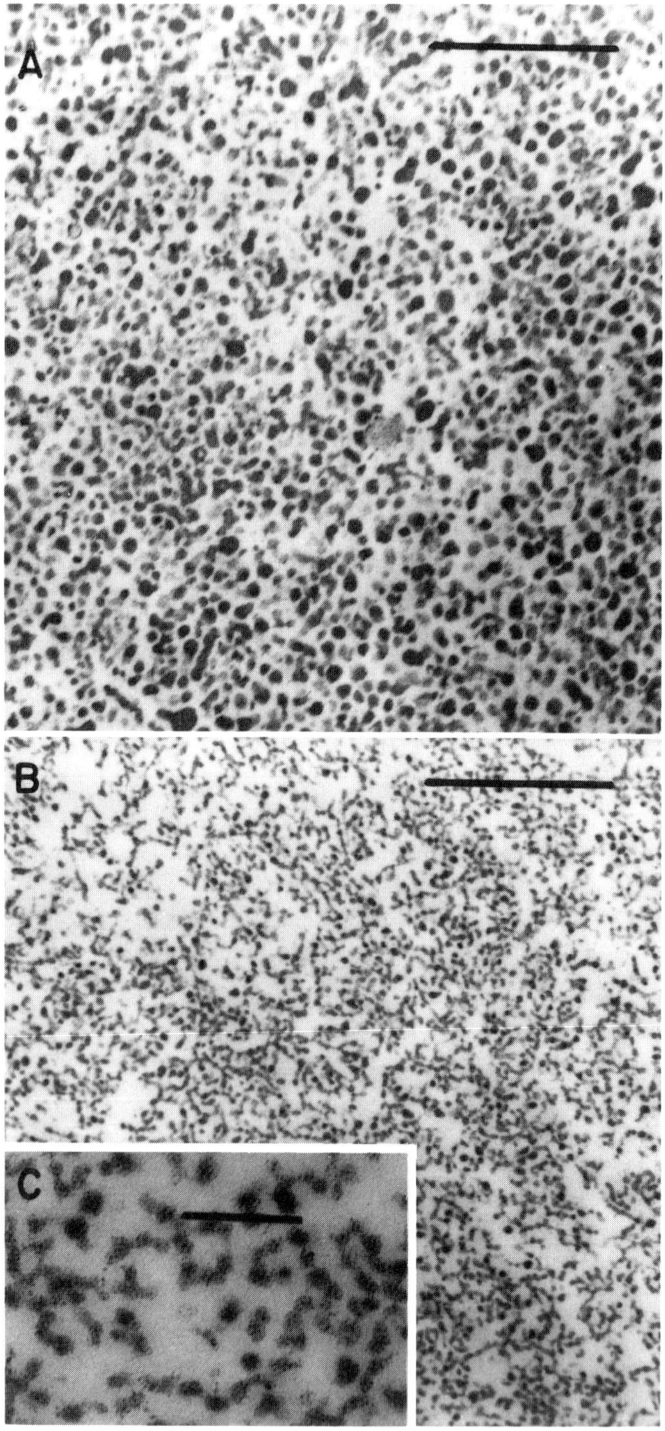

Figure 6 Electron micrographs of portions of the high-speed pellet from human milk centrifuged 2 hr at 50,000*g* after having been frozen. (A) The dense casein pellet from the bottom of the tube. (B) A lighter-colored casein pellet forming a second band above the dense casein pellet. (C) Higher magnification of the casein micelles in B. (D) Low-power view of the

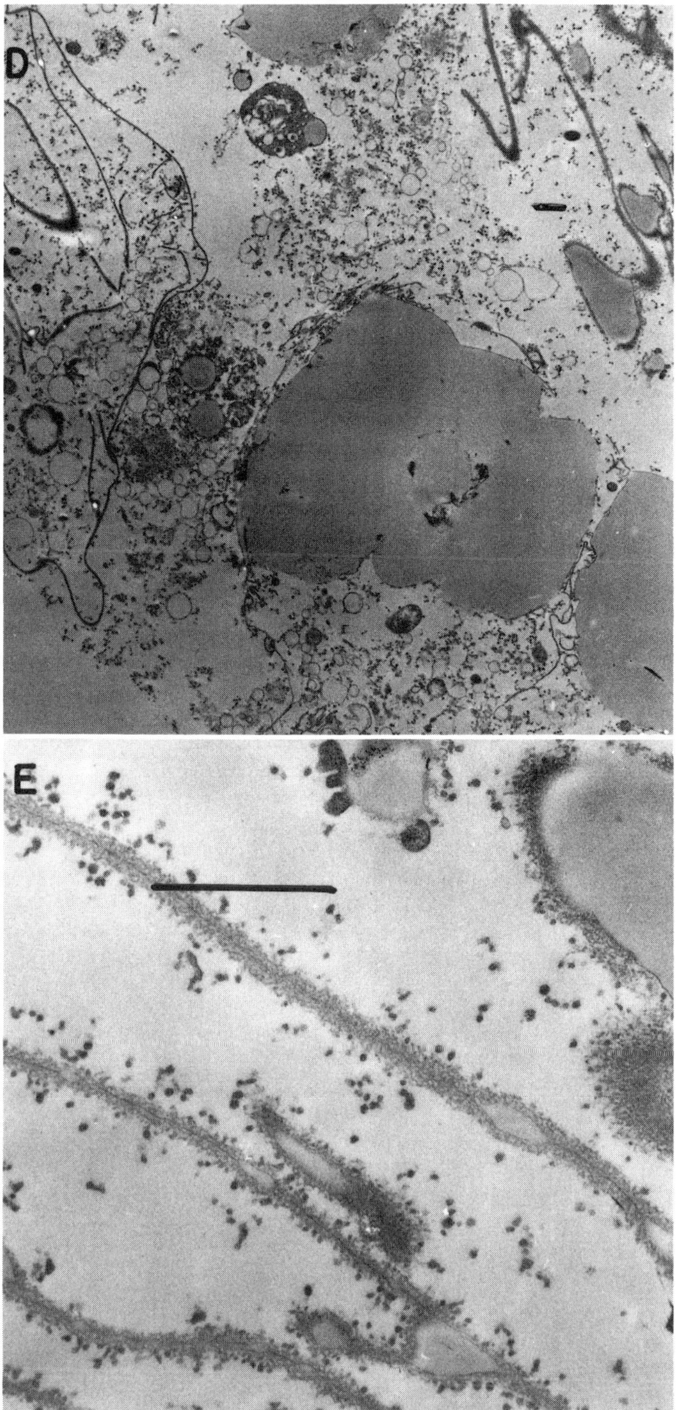

heterogeneous fluffy pellet that forms a third layer above the casein fractions. A variety of membranous structures as well as amorphous lipid droplets are visible. (E) Higher-power magnification of milk fat globule membranes from D. In all figures the bar represents 1 μm, except in C where it represents 0.2 μm. From Neville *et al.* (1994). Used by permission of *J. Dairy Sci.*

until the final sample is made up for long-term storage or the analysis is performed if freezing is not contemplated. Protease inhibitors and/or sodium azide may be added if these will not interfere with the intended analysis. It is our impression that most milk components are stable for several days at refrigerator temperature, but this must be determined for each milk component and is probably not true for cells. Separated lipid may be difficult to redistribute evenly if stored for too long a time period. If the sample is to be kept for long-term storage, it should be divided into aliquots of a size convenient for the analyses to be performed and stored frozen at −70°C. If a −70°C freezer is not available a −20°C freezer is often adequate for a few months storage if it has no automatic defrost mechanism. The temperature cycling in automatic defrost freezers causes changes in ice structure that can damage many milk components.

C. Special Considerations for Lipid Analyses

We have discussed previously the difficulty of obtaining milk samples with representative lipid contents. Storage of such samples also presents a problem because freezing breaks the emulsion between the milk fat globules and the aqueous fraction; often the lipid adheres to the sides of the container in a way that a representative milk sample cannot be recovered after freezing (Jensen, 1989). Two measures help alleviate this problem. Capillary tubes can be filled with milk immediately after expression and stored in the refrigerator for analysis of total fat by the "creamatocrit" method (Lucas, 1978). It is important that each technician who is using the creamatocrit method standardize her/his readings against lipid content obtained by a primary extraction method, such as the Folch (Jensen, 1989), using a fresh milk sample. Creamatocrit tubes can be conveniently filled in the home by trained mothers participating in lactation studies. If a gravimetric method is to be used the entire milk sample must be removed from the storage vessel with the organic solvent to be certain that some lipid has not been left behind. Clearly, this precaution is critical for any milk component associated with the milk lipid as well as for total energy analysis by bomb calorimetry.

VI. Summary

No guidelines for milk sampling and storage can be given that apply to all milk components. In this chapter possible sources of variation in milk composition and some techniques used to obtain and store milk samples have been outlined. However, it is important to evaluate within-feed, between-breast, diurnal, and longitudinal variation for any milk component under consideration and then devise a sampling scheme that will allow for the collection of representative samples. Likewise, the method of storage must be validated by comparing the concentration of the component in fresh samples with its concentration in samples frozen by whatever method seems appropriate. In general, it is also a good idea to analyze every milk sample for its sodium concentration, to rule out mastitis, and for lipid to allow correlation of the substance under study with the lipid content of that sample. When such precautions have not been taken, the limitations in the data must be recognized.

References

Allen, J. C., Keller, R. P., Neville, M. C., and Archer, P. (1991). Studies in human lactation: 6. Milk composition and daily secretion rates of macronutrients in the first year of lactation. *Am. J. Clin. Nutr.* **54,** 69–80.

Allen, J. C., and Neville, M. C. (1983). Ionized calcium in human milk determined with a calcium-selective electrode. *Clin. Chem.* **29,** 858–861.

Berga, S. E., and Neville, M. C. (1985). Sodium and potassium distribution in the lactating mouse mammary gland *in vivo. J. Physiol.* **361,** 219 230.

Bingham, E. S., and Farrell, H. M. (1974). Casein kinase from the Golgi apparatus of lactating mammary gland. *J. Biol. Chem.* **249,** 3647–3651.

Brown, K. H., Black, R. E., Robertson, A. D., Akhtar, N. A., Ahmed, G., and Becker, S. (1982). Clinical and field trials of human lactation: Methodological considerations. *Am. J. Clin. Nutr.* **35,** 747–756.

Butte, N. F., Garza, C., O'Brian Smith, E., and Nichols, B. L. (1984). Human milk intake and growth in exclusively breast-fed infants. *J. Pediatr.* **104,** 187–195.

Casey, C. E., Hambidge, K. M., and Neville, M. C. (1985). Studies in human lactation: Zinc, copper, manganese and chromium in human milk in the first month of lactation. *Am. J. Clin. Nutr.* **41,** 1193–1200.

Casey, C. E., Neville, M. C., and Hambidge, K. M. (1989). Studies in human lactation: Secretion of zinc, copper and manganese in human milk. *Am. J. Clin. Nutr.* **49,** 773–785.

Dabelow, A. (1941). Die postnatale Entwicklung der menschlichen Milchdruse und ihre Korrelationen. *Morphol. J.* **85,** 361–416.

Dils, T, Clark, S., and Knudsen, J. (1977). Comparative aspects of milk fat synthesis. *In* "Comparative Aspects of Lactation" (M. Peaker, ed.), pp. 43–55. Academic Press, New York.

Donovan, S. M., Freman, R. R., Dewey, K. G., and Lonnerdal, B. (1991). Postprandial changes in the content and composition of nonprotein nitrogen in human milk. *Am. J. Clin. Nutr.* **54,** 1017–1023.

Ferris, A. M., and Jensen, R. G. (1984). Lipids in human milk: A review. 1. Sampling, determination, and content. *J. Pediatr. Gastroenterol. Nutr.* **3,** 108–122.

Garza, C., and Butte, N. F. (1986). Concentration of human milk estimated from 24-hour pools and various abbreviated sampling schemes. *J.Ped. Gastroenterol.Nutr.***5**, 943–948.

Gould, S. F. (1983). Anatomy of the breast. *In* "Lactation: Physiology Nutrition and Breast-Feeding" (M. C. Neville and M. A. Neifert, eds.), pp. 23–47. Plenum Press, New York.

Greenberg, N. M., Anderson, J. W., Hsueh, A. J., Nishimori, K., Reeves, J. J., deAvila, D. M., Ward, D. N., and Rosen, J. M. (1991). Expression of biologically active heterodimeric bovine follicle-stimulating hormone in milk of transgenic mice. *Proc.Natl.Acad.Sci.USA* **88**, 8327–8331.

Hartmann, P. E., Kulski, J. K., Rattigan, S., and Saint, L. (1980). Lactation in Australian women. *Proc. Nutr. Soc. Aust.* **5,** 104–110.

Hytten, F. E. (1954). Clinical and chemical studies in human lactation. *Br. Med. J.* **2,** 175–182.

Jensen, R. G. (1989). "The Lipids of Human Milk." CRC Press, Boca Raton, FL.

Kakalis, L. T., Kumosinski, T. F., and Farrell, H. M., Jr. (1990). A multinuclear, high-resolution NMR study of bovine casein micelles and submicelles. *Biophys. Chem.* **38,** 87–98.

Kuhn, N. J. (1983). The biosynthesis of lactose. *In* "Biochemistry of Lactation" (T. B. Mepham, ed.), pp. 159–175. Elsevier, Amsterdam.

Kumosinski, T. F., and Farrell, H. M., Jr. (1991). Calcium-induced associations of the caseins: thermodynamic linkage of calcium binding to colloidal stability of casein micelles. *J. Protein Chem.* **10,** 3–16.

Linzell, J. L., Mepham, T. B., and Peaker, M. (1976). The secretion of citrate into milk. *J. Physiol.* **260,** 739–750.

Linzell, J. L., and Peaker, M. (1972). Day-to-day variations in milk composition in the goat and cow as a guide to the detection of subclinical mastitis. *Br. Vet. J.* **128,** 284–295.

Lucas, A., Gibbs, J. A. H., Lser, R. L. J., and Baum, J. P. (1978). Creamatocrit: Simple clinical technique for estimating fat concentration and energy value of human milk. *Br. Med. J.* **1,** 1018–1020.

Mather, I. H. (1987). Proteins of the milk-fat-globule membrane as markers of mammary epithelial cells and apical plasma membrane. *In* "The Mammary Gland: Development, Regulation and Function" (M. C. Neville and C. W. Daniel, eds.), pp. 268. Plenum Press, New York.

Mercier, J.-C., and Gaye, P. (1983). Milk protein synthesis. *In* "Biochemistry of Lactation" (T. B. Mepham, ed.), pp. 177–229. Elsevier, Amsterdam.

Mock, D. M., Mock, N. I., and Dankle, J. A. (1992). Secretory patterns of biotin in human milk. *J. Nutr.* **122,** 546–552.

Neville, M. C., Allen, J. C., and Watters, C. (1983). The mechanisms of milk secretion. *In* "Lactation: Physiology, Nutrition and Breast-feeding" (M. C. Neville and M. R. Neifert, eds.), pp. 49–104). Plenum Press, New York.

Neville, M. C., Keller, R. P., Casey, C., and Allen, J. C. (1994). Calcium partitioning in human and bovine milk. *J. Dairy Sci.* **77**, 1964–1975.

Neville, M. C., Keller, R. P., Seacat, J., Casey, C. E., Allen, J. C., and Archer, P. (1984). Studies on human lactation. I. Within-feed and between-breast variation in selected components of human milk. *Am. J. Clin. Nutr.* **40,** 635–646.

Neville, M. C., Keller, R. P., Lonnerdal, B., Atkinson, S., Wade, C. L., Butte, N., and Moser, P. B. (1985). Measurement of electrolyte and macromineral concentrations in human milk. *In* "Human Lactation: Milk Components and Methodologies" (R. G. Jensen and M. C. Neville, eds.), pp. 129–140. Plenum Press, New York.

Neville, M. C., Keller, R., Seacat, J., Lutes, V., Neifert, M., Casey, C., Allen, J. A., and Archer, P. (1988). Studies in human lactation: Milk volumes in lactating women during the onset of lactation and full lactation. *Am. J. Clin. Nutr.* **48,** 1375–1386.

Neville, M. C., Hay, W. W., Jr., and Fennessey, P. (1990). Physiological significance of the concentration of human milk glucose. *Protoplasma* **159,** 118–128.

Neville, M. C., Allen, J. C., Archer, P. G., Seacat, J., Casey, C., Sawicki, V., Oliva-Rasbach, J., and Neifert, M. (1991a). Studies in Human lactation: 7. Milk volume and nutrient composition during weaning and lactogenesis. *Am. J. Clin. Nutr.* **54,** 81–93.

Neville, M. C., Waxman, L. J., Jensen, D., and Eckel, R. H. (1991b). Lipoprotein lipase in human milk: Compartmentalization and effect of fasting, insulin, and glucose. *J. Lipid Res.* **32,** 251–257.

Neville, M. C., and Peaker, M. (1979). The secretion of calcium and phosphorus into milk. *J. Physiol.* **313,** 561–570.

Neville, M. C., and Watters, C. D. (1983). Secretion of calcium into milk: Review. *J. Dairy Sci.* **66,** 371–380.

Neville, M. C., and Walsh, C. (1995). Effect of xenobiotics on milk secretion and composition. *Am J. Clin. Nutr.* **61**, 5687–5694.

Ollivier-Bousquet, M., and Denamur, R. (1975). Effect de l'etat physiologique et du 3′5′ adenosine monophosphate cyclique sur le transit intracellulaire et l'ecretion des proteines du lait. Etude autoradiographique en microscopie electronique. *J. Microscopie et Biol. Cell.* **23,** 63–82.

Patton, S., and Huston, G. E. (1986). A method for isolation of milk fat globules. *Lipids* **21,** 170–174.

Peaker, M. (1983). Secretion of ions and water. *In* "Biochemistry of Lactation." (T. B. Mepham, ed.), pp. 285–307. Elsevier, New York.

Peaker, M., and Taylor, J. C. (1975). Milk secretion in the rabbit: changes during lactation and the mechanism of ion transport. *J. Physiol.* **253,** 527–545.

Peterson, R. G., and Bowes, W. A., Jr. (1983). Drugs, toxins and environmental agents inbreast milk. *In* "Lactation; Physiology, Nutrition and Breast-Feeding" (M. C. Neville and M. R. Neifert, eds.), pp. 367–403. Plenum Press, New York.

Prentice, A., Prentice, A. M., and Whitehead, R. G. (1981). Breast-milk fat concentrations of rural African women 2. Long-term variations within a community. *Br. J. Nutr.* **45,** 495–503.

Richardson, D. C. (1949). Contractile tissue in the mammary gland with special reference to the myoepithelium in the goat. *Proc. R. Soc. London Ser. B.* **136,** 30–45.

Seddiki, T., Delpal, S., and Ollivier-Bousquet, M. (1992). Endocytosis and intracellular transport of transferrin across the lactating rabbit mammary epithelial cell. *J. Histochem. Cytochem.* **40,** 1501–1510.

Seelig, L. L., Jr., and Beer, A. E. (1981). Transepithelial migration of leukocytes in the mammary gland of lactating rats. *Biol. Reprod.* **22,** 1157–1163.

Solari, R., and Kraehenbuhl, J.-P. (1987). Receptor-mediated transepithelial transport of polymeric immunoglobulins. *In* "The Mammary Gland" (M. C. Neville and C. W. Daniel, eds.), pp. 269–300. Plenum Press, New York.

Turner, M. D., Rennison, M. E., Handel, S. E., Wilde, C. J., and Burgoyne, R. D. (1992). Proteins are secreted by both constitutive and regulated secretory pathways in lactating mouse mammary epithelial cells. *J.Cell Biol.* **117**, 269–278.

Woodward, D. R., Rees, B., and Boon, J. A. (1989). Human milk fat content: Within-feed variation. *Early Hum. Dev.* **19,** 39–46.

D. Sampling and Storage of Bovine Milk

ROBERT G. JENSEN

I. Introduction

Milk and dairy products that are ready to be consumed will come from large pools so that genetic and environmental influences are minimized. Hundreds of gallons of milk are collected in bulk tanks on the farm and are gathered in tankers. The tankers transport the milk to plants, some of which can process up to 1,000,000 lbs of milk per day. One container of milk or a pound of butter could be representative of a large pool and a region.

II. Sampling

Bovine milk is sampled regularly on a huge scale for analyses of fat and protein content as a basis for payment to the producer. Standard procedures for sampling and storage must be used. These are described in "Official Methods of Analysis" (AOAC, 1990) and "Standard Methods for the Examination of Dairy Products" (APHA, 1993). These procedures are used to collect samples from individual cows and herds for the analyses of most components. However, as discussed by Neville in Chapter 2C, sampling and storage for some determinations, e.g., enzymes, may require special handling.

III. Storage

For long-term storage, a −70°C freezer is preferable, but a −20°C unit will suffice. Lipolytic activity is essentially inhibited at −70°C, but continues slowly at −20°C. Storage at both temperatures destabilizes the lipid globule emulsion, denatures some proteins and changes the nature of mineral complexes (Walstra and Jenness, 1984). Investigators should determine the procedure which is ideal for their requirements and dairy products. Some of the compounds in milk show seasonal changes, but this may be due to alterations in diet. An example is the lower content of carotenoids in winter compared to summer butter. This is caused by the unavailability of pastures during the winter. Again, the type of study being done will determine the sampling and storage protocols which should be used.

References

American Public Health Association (APHA). (1993). "Standard Methods for the Examination of Dairy Products," 16th ed. APHA, Washington, DC.

Association of Official Analytical Chemists. (AOAC). (1990). "Official Methods of Analysis of the AOAC," 15th ed., Vol. **2.** AOAC International, Arlington, VA.

Walstra, R., and Jenness, R. (1984). "Dairy Chemistry and Physics," pp. 316–318. Wiley–Interscience, New York.

E. The Physical Properties of Human and Bovine Milks

MARGARET C. NEVILLE
ROBERT G. JENSEN

I. Introduction

Milk is an extremely complex biological fluid with scores of nutrient chemicals contained in a fluid with characteristics of three physical phases: a dilute emulsion, a colloidal dispersion, and a solution. The emulsion can be broken by low-speed centrifugation and the milk separates into lipid and aqueous phases or compartments, each with a characteristic composition. With ultracentrifugation the casein micelles precipitate, bringing some other proteins, such as lysozyme (Neville, unpublished data) from human milk or lactoferrin from milks of animal species, with them. The supernatant remaining after this process has the characteristics of a true solution. The compartmentation is more thoroughly discussed in Chapters 2A, 2B, and 7A.

The physical properties of bovine milk have been thoroughly evaluated because of the importance of many of these parameters in processing and purity assessment. Values for few of these parameters are available for human and other animal milks as these are generally fed directly to the infant. Even for bovine milk, much of the information is available in technical publications and degree of variability and effects of physiological state are often not available. The physical properties are described below (Walstra and Jenness, 1984; Sherbon, 1988; National Dairy Council, 1993).

II. Electrical Conductivity

This is defined as a measure of the electrical resistance of the solution in reciprocal ohms (mhos). It is used to assess the total ionic content of milk. The greatest contributors to conductivity are the sodium, potassium, and chloride ions. Since the amounts of sodium and chloride increase with mastitis, measurements of conductivity in bovine milk are employed to screen for clinical cases of the disease. See Chapter 3F.

HANDBOOK OF MILK COMPOSITION

III. Freezing Point

The freezing point of milk is lower than that of pure water due to dissolved components. This property is measured to determine whether bovine milk has been diluted with water and is employed as a legal standard. As with osmolality, the freezing point is stable. The major contributors to the freezing point are lactose and chloride. Since the freezing point and osmolality are proportional and dependent upon the number of dissolved particles, they can be determined with the same instrument.

IV. Boiling Point

The boiling point in milk is higher than that of pure water again due to dissolved components. It is another of the colligative properties.

V. Osmolality or Osmotic Pressure

Osmolality is a measure of the total number of dissolved particles in a given volume of solution given in osmol/kg. Osmolality is one of the colligative properties (dependent on the number of dissolved particles, not their properties) of milk along with freezing and boiling points. It is measured in the instrument used to determine the freezing point. The osmotic pressure of milk is quite constant being equal to the osmotic pressure of blood. A result is that the variation in the dissolved substances in normal milk, primarily lactose, is small. The total concentration of dissolved materials is responsible for osmolality. Osmolality is proportional to the freezing point of milk (see Section III). As previously mentioned, osmolality remains constant in human and bovine milks because of the relationship between milk and blood. The osmolality of formulas is carefully controlled to resemble that in human milk (see Chapter 10D). The potential renal solute load is calculated from the contents of sodium, chloride, potassium, and protein (Fomon and Ziegler, 1993). Protein is included because it provides solutes from metabolism. Potential renal solute loads (mosmol/liter) are: human milk, 93; milk-based formula, 135; soy-protein based formula, 165; whole bovine milk, 308; and skim milk, 326. In addition to being low in iron, the potential renal solute load of the bovine milks is too high for them to be used as the sole food for young infants.

VI. pH

The pH of milk as generally measured outside the animal is higher than milk within the mammary gland due to loss of CO^2 to the ambient air (Allen *et al.*, 1983). See Chapter 7A for a full discussion of this principle.

TABLE I
The Physical Properties of Mature Human, Goat, and Bovine Milk[a]

Property	Human milk	Goat milk	Bovine milk
Electrical conductivity	0.0041		0.00465[b]
	(0.00150–0.00675)		(0.0042–0.0048)
Freezing point (°C)		−0.582	−0.552[c]
			−(0.512–0.550)
Boiling point (°C)			100.17
Osmolality (mosmol/μg)	290–299[d]		275[b]
pH	6.8[e]	6.37	6.62
	(6.57–6.85)	(6.33–6.52)	(6.22–6.77)
			(0.2065–0.2075)
Specific gravity	1.031	1.033	1.030
	(1.024–1.03)	(1.031–1.037)	(1.021–1.037)
Surface tension (dynes/cm²)		52	52.8
			(51.1–55)
Titratable acidity (percentage)			0.16 ± 0.02[b,c]
Specific heat (°C)			
0			0.920[c]
15			0.938
40			0.930
Coefficient of expansion (°C)			
10			0.9975[c]
15.6			0.9985
21.1			1.000
Viscosity (centipoise)			1.6314[c]

[a]All values from Macy *et al.* (1971) unless superscripted.
[b]From the National Dairy Council (1993).
[c]From Sherbon (1988).
[d]From Neubauer *et al.* (1993).
[e]From Allen *et al.* (1991).

While important for human milk, immediate determination of pH in bovine milk is never done except for research purposes. The processing which is done to bovine milk removes dissolved gases. Assays of pH and of titrable acidity (see below) are used to assure that lactic acid is being produced at the desired rate by added microorganisms during the preparation of cheeses and fermented milks, e.g., yogurt. The casein in milk forms into a gel or curd at pH 4.6.

VII. Specific Gravity

Specific gravity is the ratio of the mass of a solution or substance to the mass of a similar volume of water. This property is used to assess nonfat solids in milk and the addition of water to milk which lowers specific gravity. The dairy industry employs a special hydrometer, the lactometer, to determine specific gravity and total solids. Corrections are required for milk temperatures which differ from 20°C. A lactometer would be helpful for determining the specific gravity of human milk. None are available because the volume of milk required to float the lactometer, 150 to 300 ml, is never available for individual samples of human milk. The much smaller hydrometer used for specific gravity of urine can be employed for human milk.

VIII. Surface Tension

This is defined as the work required to increase the surface area of a solution and is usually expressed as erg/cm^2. This property is used to follow the changes in surface-active components during milk processing, to follow release of fatty acids during lipolysis, and as a measure of the foaming tendency of milk. Fatty acids and their salts and monoacylglycerols formed as a result of lipolysis are surface active and reduce surface tension. However, the method is not applied routinely for the assessment of lipolysis because the short-chain acids responsible for the flavor designated as hydrolytic rancidity are water soluble and do not affect surface tension. The interfacial tension between the fat-soluble surface and the aqueous medium, of considerable potential importance in emulsion stability and access by lipolytic enzymes, cannot be determined directly.

IX. Titratable Acidity

The amount of alkali required to bring the pH to neutrality (phenolphthalein) is titratable acidity. This property is used to determine bacterial growth during fermentations, such as during cheese making, as well as compliance with cleanliness standards. There is no lactic acid in fresh bovine milk. The titratable acidity is due mostly to the casein and phosphates. Lactic acid can be produced by bacterial contamination, although this is uncommon.

X. Specific Heat

Specific heat is the ratio between the amount of heat necessary to raise a given weight of a substance to a specified higher temperature and the amount of heat necessary to raise an equal weight of water to the same temperature. It is important in processing for determining the amount of heat or refrigeration necessary to change the temperature of milk.

XI. Coefficient of Expansion

This coefficient is the ratio of the increase in volume per unit increase in temperature to the increase in volume of water with the same temperature increase. It is used in the design of dairy equipment.

XII. Viscosity

This refers to the resistance to flow in centipoise units. It is used in the design of dairy processing equipment and to assess casein micellar aggregation.

Table I contains values for the physical parameters which describe bovine milk, with corresponding data for human and goat milks when available.

References

Allen, J. C., and Neville, M. C. (1983). Ionized calcium in human milk determined with a calcium-selective electrode. *Clin. Chem.* **29,** 858–861.

Allen, J. C., Keller, R. P., Archer, P., and Neville, M. C. (1991). Studies in human lactation: 6. Milk composition and dairy secretion rates of macronutrients in the first year of lactation. *Am. J. Clin. Nutr.* **54,** 69–80.

Fomon, S. J., and Ziegler, E. E. (1993). "Nutrition of Normal Infants" (S. J. Fomon, ed.), pp. 91–102, Mosby, St. Louis.

Macy, I. G., Kelly, H. J., and Sloan, R. E. (1971). "The Composition of Milks: A Compilation of the Comparative Composition and Properties of Human Cow and Goat; Milk, Colostrum, and Transitional Milk," pp. 1–70. National Academy of Sciences, National Research Council, Washington, DC.

National Dairy Council (1993). "Newer Knowledge of Milk and Other Fluid Dairy Products," pp. 1–52. National Dairy Council, Rosemont, IL.

Neubauer, S. H., Ferris, A. M., Chase, C. G., Fanelli, J., Thompson, C. A., Lammi-Keefe, C. J., Clark, R. M., Jensen, R. G., Bendel, R. B., and Green, K. W. (1993). Delayed lactogenesis in women with insulin-dependent diabetes. *Am. J. Clin. Nutr.* **58,** 54–60.

Sherbon, J. W. (1988). "Physical Properties of Milk," (N. P. Wong, ed.), 3rd Ed., pp. 409–460. Van Nostrand–Reinhold, New York.

Walstra, P., and Jenness, R. (1984). "Dairy Chemistry and Physics." Wiley, New York.

Determinants of Milk Volume and Composition

A. Lactogenesis in Women: A Cascade of Events Revealed by Milk Composition

MARGARET C. NEVILLE

I. Introduction

In the later stages of pregnancy the mammary glands of most mammals manufacture small quantities of a secretion product often called precolostrum. The composition of precolostrum differs considerably from true milk; in particular, it contains a high concentration of sodium, chloride, and protective proteins, such as lactoferrin and immunoglobulins (Fleet *et al.*, 1975), and low concentrations of such milk-specific substances as lactose and casein. Lactogenesis, defined here as the onset of copious milk secretion, occurs concomitantly with parturition in most species, particularly rodent and dairy animals. Lactogenesis occurs in two stages (Hartmann, 1973; Fleet *et al.*, 1975); the first signals the preparedness of the mammary glands for milk secretion and takes place sometime in later pregnancy. The second is the onset of copious milk secretion occurring around parturition. We shall be discussing the second stage in this article. For example, in rats, milk appears in the mammary ducts 4 hr prior to birth of the pups (Kuhn, 1977); in the cow, lactogenesis appears to coincide with parturition (Peaker and Linzell, 1975). In humans and guinea pigs, however, lactogenesis is delayed until the second or third day postpartum probably due to the slow postpartum fall in progesterone in these species (Neville, 1983).

Lactogenesis is perceived by most women as a more or less abrupt feeling of fullness or engorgement of the breasts occurring sometime between 40 and 72 hr postpartum (Arthur *et al.*, 1989) at which time the milk is said to "come in." However, a careful study of milk volumes focussing on the first 2 weeks after birth shows that there is a rapid increase in milk volume beginning about 36 hr postpartum and leveling off after 96 hr (Figure 1). The temporal sequence of the changes in human milk composition that accompany this volume increase allows us to draw some conclusions about the coordination of the physiologic mechanisms involved in lactogenesis. Such conclusions are difficult to obtain from studies in other mammals in which the process usually occurs more rapidly.

II. The Physiological Basis of Lactogenesis

The postembryonic development of the mammary gland begins at puberty with enlargement of the mammary fat pad and elongation of the mammary ducts under the direct or indirect influence of estrogen (Neville, 1983). With the onset of the estrus or menstrual cycle the cyclic appearance of progesterone stimulates limited alveolar development. Nonetheless, the full development of the gland requires exposure to the rich hormonal milieu of pregnancy. By midpregnancy in humans the mammary cells become competent to secrete milk and elaborate small amounts of precolostrum. Copious milk secretion is held in check by the high circulating levels of progesterone, elaborated by the placenta (Kuhn, 1977). Before describing the changes in milk volume and composition that take place during lactogenesis it is necessary to examine the prepartum secretion of the human breast and discuss the evidence that the junctions between the alveolar cells are leaky during pregnancy.

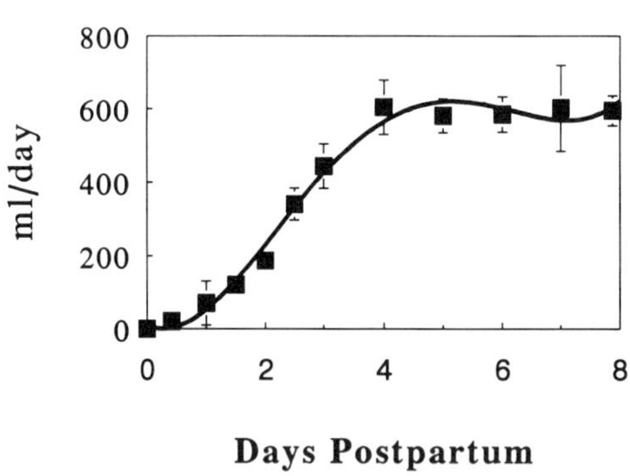

Figure 1 Milk volumes during the first week postpartum. Mean values from 12 multiparous Caucasian women who test-weighed their infants before and after every feed for the first 7 days postpartum. Redrawn from Neville *et al.* (1988).

III. The Composition of the Prepartum Mammary Secretion

A close examination of the composition of the prepartum secretion in women (Table I) and comparison with similar studies in goats provides evidence that the occluding junctions that join each mammary alveolar cell tightly to its neighbors are "leaky" during pregnancy, allowing fluid and solutes to flow between the milk space and the interstitial fluid of the mammary gland. Using the studies of Peaker and Linzell (Peaker, 1983) in the goat as a model, the lactose concentration of the milk of each of the nine individuals from whom prepartum samples were obtained was plotted as a function of the sodium, potassium, and chloride concentration in the same individual. Because we found no time dependence of the composition of prepartum milk the results from each individual were pooled. The results are shown in the lower graphs of Figure 2 along with similar values for goats taken from the work of Peaker and Linzell (Peaker, 1983). It can be seen that in both species the lactose concentration is directly correlated with the potassium concentration, whereas the sodium and

TABLE I
Composition of Prepartum Human Milk

Milk component	Units	Mean ± SD (*n*)
Mean days prepartum		20.21 ± 12.18 (11)
Lipid	%	2.07 ± 0.98 (11)
Lactose	m*M*	79.78 ± 21.68 (9)
Protein	g/dl	5.44 ± 1.71 (8)
Glucose	m*M*	0.35 ± 0.16 (8)
Sodium	m*M*	61.26 ± 25.82 (10)
Potassium	m*M*	18.30 ± 5.67 (10)
Chloride	m*M*	62.21 ± 17.44 (10)
Calcium	mg/dl	25.35 ± 8.48 (10)
Magnesium	mg/dl	5.64 ± 1.44 (10)
Citrate	m*M*	0.40 ± 0.17 (8)
Phosphate	mg/dl	2.32 ± 0.70 (9)
Ionized calcium	m*M*	3.25 ± 0.84 (6)
pH		6.83 ± 0.18 (6)
Urea	mg/dl	14.87 ± 2.40 (9)
Creatinine	mg/dl	1.47 ± 0.35 (9)

Note. Data from Allen *et al.* (1991). Small samples of mammary secretion were obtained three times in the prepartum period from each of 11 women. In some cases volumes were insufficient for all analyses.

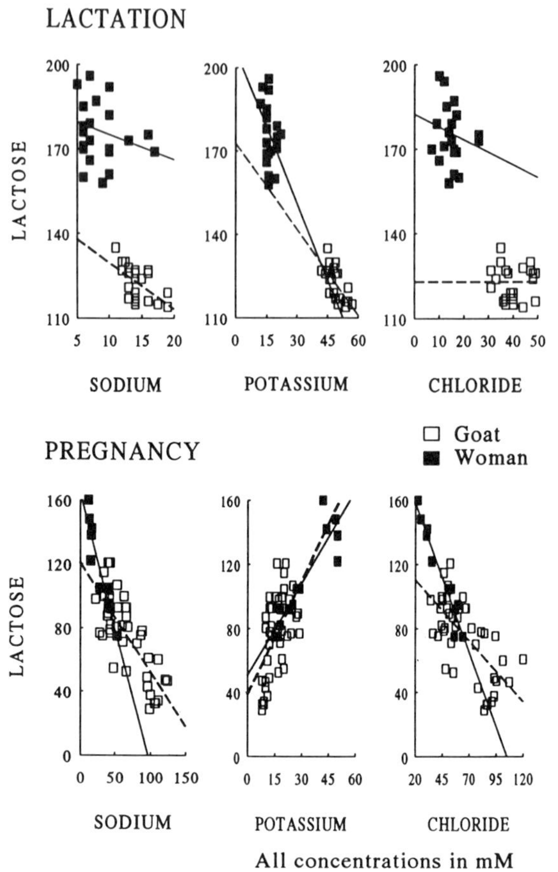

Figure 2 Correlation between lactose concentration and the concentrations of chloride, sodium, and potassium in antepartum and 3-month postpartum samples of mammary secretion from the same subjects as those in Figure 1 (closed symbols; Allen *et al.,* 1991) or from goats (open symbols; Peaker, 1983). For the pregnancy study each point represents the mean lactose concentration in the antepartum secretion of one individual plotted as a function of the mean monovalent ion concentration in the same samples. For the lactation study small (5–10 ml) midfeed samples were obtained at a morning feed from each breast of nine women at 3 months postpartum. Values for each milk sample are plotted separately. The lines are the best-fitting linear-regression lines. The slopes are significant ($p < 0.05$) for all lines except the relation between lactose and chloride in goats.

chloride concentrations are inversely related (see also Allen *et al.,* 1991). The most straightforward interpretation of these results is that during pregnancy lactose and potassium are secreted coordinately from the mammary alveolar cell. Plasma components, such as sodium and chloride, enter the milk down their concentration gradients through the paracellular pathway from the interstitial space bringing water with them. The higher the ratio of paracellular flux to cellular secretion, the higher the concentration of the interstitially derived components and the lower the concentrations of lactose and potassium.

In lactation, on the other hand, both sodium and potassium are inversely related to the lactose concentration during lactation in goats and humans, whereas chloride is poorly correlated (Figure 2, top). The inverse relation is expected because the conservation of osmotic equilibrium in the milk requires that an increase in osmotic pressure due to increased lactose concentration be balanced by a decrease in the concentration of other milk components, in particular, monovalent ions. The finding that sodium and potassium concentrations in milk vary coordinately during lactation is expected because both ions appear to be distributed according to their passive electrochemical gradients across the apical membrane of the mammary epithelial cells (Peaker, 1983; Berga and Neville, 1985). It is of interest that the monovalent ion concentrations in the prepartum secretions of the mammary glands of goats and humans appear to fall in the same range, whereas during lactation goat milk has much higher concentrations of all three monovalent cations shown in Figure 2 than human milk and a correspondingly lower lactose concentration.

It is difficult to obtain any other evidence in women that the paracellular pathway is indeed open in pregnancy. However, the evidence from experiments in goats provides strong support for this interpretation. For example, sucrose, placed in the udder, is lost from the milk space to the blood in pregnancy but not in lactation (Linzell and Peaker, 1974). In addition, there is a high potential difference, about 35 mV, between the blood and the milk space in lactation but not in pregnancy (Linzell and Peaker, 1974; Berga, 1984). This electrical potential could not be maintained across the mammary epithelium if the junctional complexes were leaky. The status of the paracellular pathway has been investigated morphologically in mice where freeze-fracture studies of the occluding junctions between mammary cells were consistent with leaky junctions during pregnancy and "tight" junctions in lactation (Pickett *et al.*, 1975).

During pregnancy the above model predicts that the concentration of substances that enter the lumen of the mammary gland through the paracellular pathway along with sodium should be directly related to the sodium concentration, whereas the concentrations of those substances that originate from the epithelial cells should be inversely related to the sodium concentration. To evaluate this hypothesis the concentrations of potassium, chloride, lactose, free phosphate, glucose, and ionized calcium were plotted as a function of the sodium concentration, this time for each of 45 prepartum milk samples obtained at different stages of pregnancy from the same women (Figure 3; Allen *et al.*, 1991). Lactose, potassium, and ionized calcium, all of which are found at higher concentration in the prepartum mammary secretion than in the blood, are negatively correlated with the sodium concentration. The concentrations of the other compounds are positively correlated, consistent with a principal site of origin from the plasma. A model showing the predominant pathways for the flux

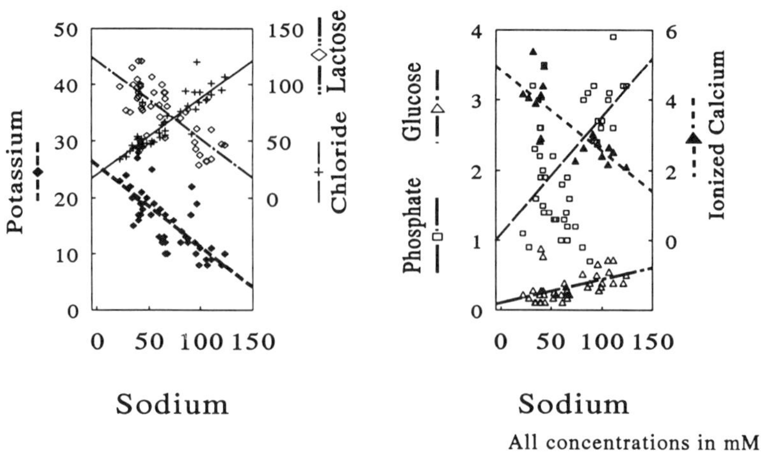

Figure 3 The relation between the concentrations of potassium, chloride, lactose, phosphate, glucose, and ionized calcium and the concentration of sodium in the prepartum mammary secretion from 10 individuals. Values for each milk sample were plotted individually. The data comprise varying numbers of samples from individual breasts at different times prepartum.

of the substances investigated in this study is shown in Figure 4 where the situations in both pregnancy and lactation are represented.

We can only speculate about the physiologic significance of leaky junctions during pregnancy. The most direct explanation is that they provide a pathway for resorption of secreted milk components under conditions in which secretion products are not being removed by suckling. Consistent with this idea, it has been shown that both lactose (Arthur *et al.*, 1991) and α-lactalbumin (Martin *et al.*, 1980) appear in the plasma during pregnancy. In any case the data presented in this section make it clear that many components of the mammary secretion during pregnancy originate directly from the plasma. In the next section the transition to lactation, a state in which this is not true, will be analyzed.

IV. The Implications of Changes in Milk Composition During Lactogenesis

The rate of milk secretion in the early postpartum period (Figure 1) started low, less than 100 ml/day up to 36 hr postpartum, then began to rise almost linearly to level off at about 600 ml/day at 96 hr postpartum (Neville *et al.*, 1988). The concomitant change in the citrate concentration is shown in Figure 5 replotted from Neville *et al.* (1991). When mean values are examined the change in milk volume and the change in the citrate concentration are almost precisely parallel as noted many years ago by

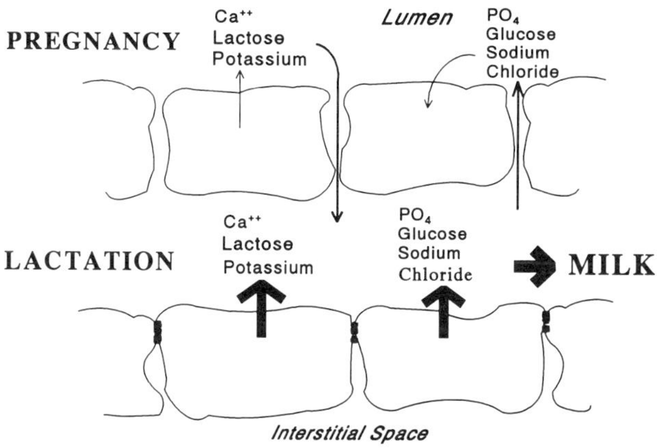

Figure 4 Model for the directions of the major fluxes of several macronutrients during pregnancy and lactation in women as predicted from the data in Figure 3.

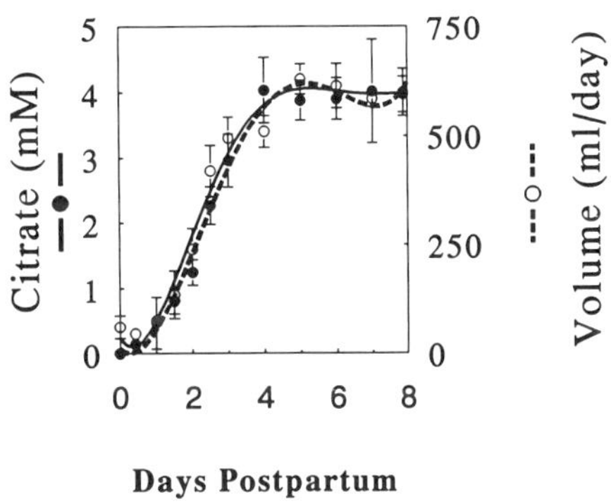

Figure 5 Changes in the concentration of citrate in human milk in the early postpartum period compared to the increase in milk volume. Data replotted from Neville *et al.* (1991).

Peaker and Linzell (1975). This parallelism led Peaker to refer to citrate as the "harbinger of lactogenesis." The increase in citrate concentration clearly parallels the metabolic activity of the mammary gland as it increases its production of milk lipid (Linzell *et al.,* 1976; Neville and Peaker, 1979).

In Figure 6 the temporal changes in the concentration of several other milk components are compared to the change in citrate concentration. After an initial rapid fall the phosphate concentration generally parallelled milk volume as did the glucose concentration. The changing concentrations of free phosphate may reflect the increased utilization of UDP-galactose for lactose synthesis with the subsequent generation of uridine monophosphate and phosphate in the Golgi compartment of the mammary alveolar cell (Neville, 1983). We have elsewhere provided evidence that the change in glucose concentration reflects an increase in glucose transport into the mammary alveolar cell across the basolateral cell membrane (Neville *et al.*, 1990). Casein (Patton *et al.*, 1986) as well as calcium and magnesium (Neville *et al.*, 1991) concentrations appear to increase coordinately with this metabolic sequence (data not shown).

On the other hand, not all concentration changes parallel the increase in milk synthesis rates. For example, as shown in Figure 6, the concentrations of lactose, sodium, chloride, and protein begin to change immediately after birth and achieve nearly stable values about 24 hr before peak milk volume is attained (note dotted line marking the point at which the lactose concentration stabilizes). These immediate changes likely reflect closure of the paracellular pathway with a corresponding decrease in the direct flux of interstitial constituents into the milk.

The composition changes during lactogenesis shown in Figure 6 can be explained by a simple two-step process in which junctional closure is followed by the onset of secretory activity. Unfortunately, other changes in milk composition fall less easily into this neat pattern. The abrupt postpartum fall in the protein content of the milk, for example, cannot be the result of closure of the junctions between the cells. Thus, the major

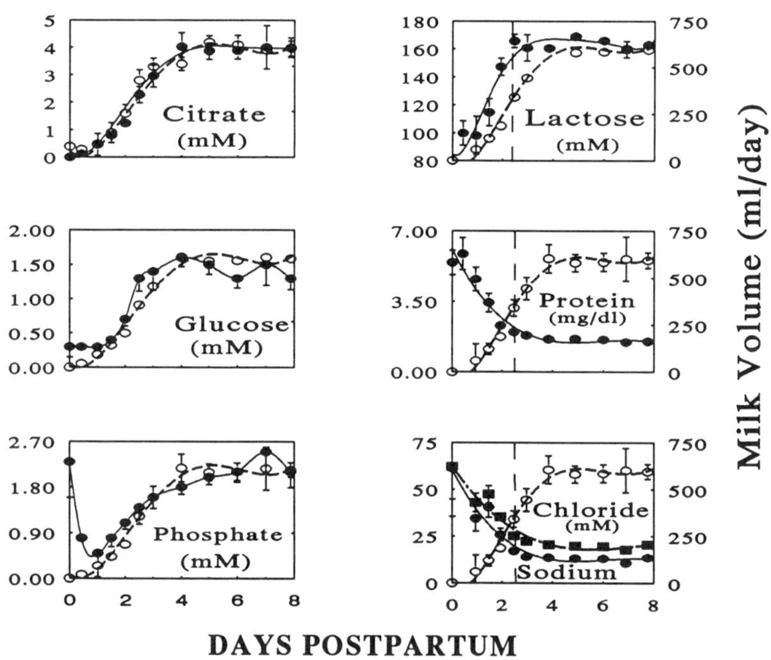

Figure 6 Changes in milk composition during the early postpartum period (closed symbols). The milk volumes during the corresponding period are plotted for comparison (open circles). Data replotted from Neville *et al.* (1991). The vertical lines in the graphs on the right hand side of the figure indicate the time at which the lactose concentration reaches a maximum. Note that this occurs about 24 hr before the milk volume reaches a constant value. Protein, chloride, and sodium have also largely stabilized by this time.

proteins in human colostrum are secretory IgA and lactoferrin. Secretory IgA reaches the milk via a specific transcytotic pathway that ferries dimeric IgA molecules from the interstitial space across the mammary cells themselves (Solari and Kraehenbuhl, 1987). Lactoferrin is actually synthesized in the mammary alveolar cells (Teng *et al.*, 1989; Schanbacher *et al.*, 1992). Quantitative data on the concentrations of these two proteins in human breast milk during the first week of lactation have been provided by Lewis-Jones and co-workers (1985). The concentrations of both fall in the early postpartum period (Figure 7, top) and are responsible for the decline in total protein concentration during this period. However, as shown in Figure 7 (bottom), the secretion rate of both proteins actually rises on the second or third day postpartum. sIgA secretion falls rapidly again on the third day but lactoferrin continues to be secreted at a more or less constant rate after the second day postpartum. Thus, the time courses of the changes in the secretion rates of these two proteins do not coincide with the other events taking place during the first week of lactation in women.

Examination of the concentration of cells in the mammary secretion (Figure 8) shows yet another pattern. The concentration of cells in milk is highest on Day 1 at about 3×10^6 cells/ml and falls by 50% on Day 3. However, there is substantial cellular secretion, particularly of polymorphonuclear leukocytes and macrophages, up to Day 10 after the cellular junctions have closed. This observation implies that the passage of cells into milk may involve more than passive transfer through open intercellular spaces.

It is likely that the gradual fall in progesterone during the postpartum period in women combined with maintained prolactin levels is responsible

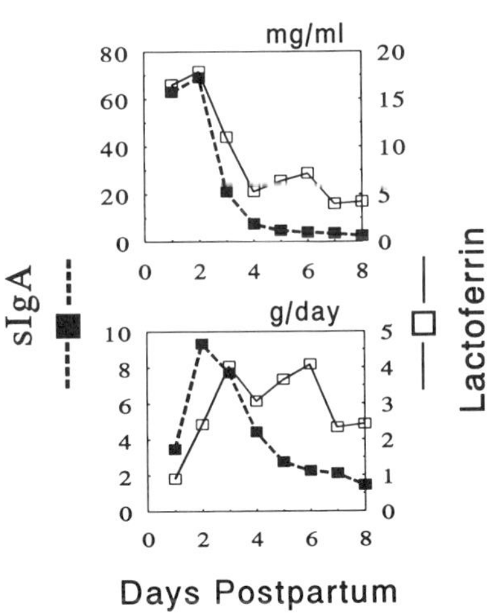

Figure 7 (Top) lactoferrin and secretory IgA concentrations in human milk during the first week postpartum. Data from Lewis-Jones *et al.* (1985). (Bottom) estimate of lactoferrin and secretory IgA secretion rates using data from Figure 6 multiplied by the mean volumes given in Figure 1.

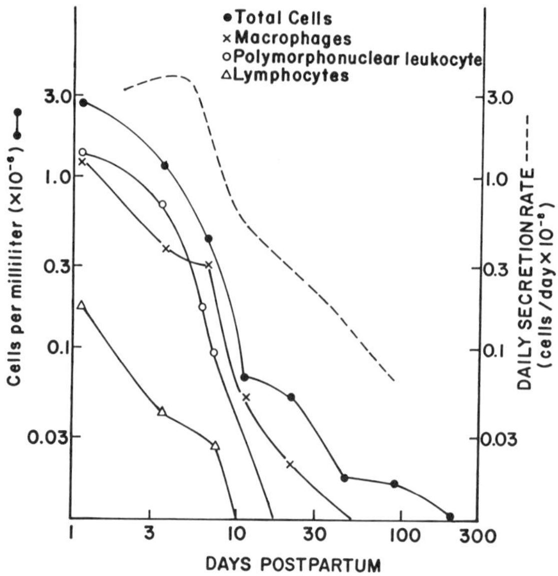

Figure 8 The concentration of leukocytes in milk. In addition to the leukocytes shown here epithelial cells are thought to be present at a concentration of about 10^4 cells per milliliter throughout lactation. The dotted line represents the total leukocyte secretion rate obtained by multiplying the total number of cells by the mean volume for the corresponding day. Data replotted from Ho *et al.* (1979). Used by permission of Plenum Press.

for the observed changes in mammary cell morphology and activity (Kuhn, 1977). However, the molecular mechanisms involved in both transduction of the secretory signal and initiation of the diverse components of lactogenesis remain almost totally unknown.

V. Summary and Conclusions

Several distinct metabolic and cellular functions are modified when the volume of the mammary secretion increases during lactogenesis. These include the permeability of the tight junctions, the rate of synthesis of lactose, lipids and nutrient proteins, the transport of glucose into the mammary alveolar cell, the transcytosis of sIgA, the movement of immune cells into the alveolar lumen, and the secretion of lactoferrin. The time course of some of these processes is diagrammed in Figure 9. The temporal sequence of these changes as they occur during lactogenesis suggests that they are either independently regulated or form a part of an orderly cascade of temporally separate events. In 1977, Nicholas Kuhn (Kuhn, 1977) predicted that "future studies will reveal a definite sequence of

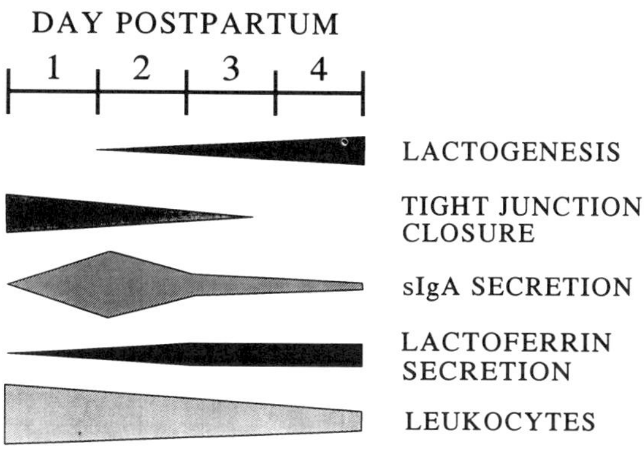

Figure 9 A summary model for the temporal sequence of changes in mammary gland function during lactogenesis in women.

biochemical responses, representing a 'fine structure' of the lactogenic response." The analysis of lactogenesis in women presented here allows us to see the framework of the lactogenic sequence; an understanding of the fine structure of the biological responses remains in the future.

References

Allen, J. C., Keller, R. P., Neville, M. C., and Archer, P. (1991). Studies in human lactation: 6. Milk composition and daily secretion rates of macronutrients in the first year of lactation. *Am. J. Clin. Nutr.* **54,** 69–80.

Arthur, P. G., Smith, M., and Hartman, P. (1989). Milk lactose, citrate and glucose as markers of lactogenesis in normal and diabetic women. *J. Ped. Gastroenterol. Nutr.* **90,** 488–496.

Arthur, P. G., Kent, J. C., Potter, J. M., and Hartmann, P. E. (1991). Lactose in blood in nonpregnant, pregnant and lactating women. *J. Ped. Gastroenterol. Nutr.* **13,** 254–259.

Berga, S. E. (1984). Electrical potentials and cell-to-cell dye movement in mouse mammary gland during lactation. *Am. J. Physiol.* **247,** C20–C25.

Berga, S. E., and Neville, M. C. (1985). Sodium and potassium distribution in the lactating mouse mammary gland *in vivo*. *J. Physiol.* **361,** 219–230.

Fleet, I. R., Goode, J. A., Hamon, M. H., Laurie, M. S., Linzell, J. L.,and Peaker, M. (1975). Secretory activity of goat mammary glands during pregnancy and the onset of lactation. *J. Physiol.* **251,** 763–773.

Hartmann, P. E. (1973). Changes in the composition and yield of the mammary secretion of cows during the initiation of lactation. *J. Endocrinol.* **59,** 231–247.

Ho, F. C. S., Wong, R. L. C., and Lawton, J. W. M. (1979). Human colostral and breast milk cells, a light and electron microscopic study. *Acta Paediatr. Scand.* **68,** 389–396.

Kuhn, N. J. (1977). Lactogenesis: The search for trigger mechanisms in different species. *In* "Comparative Aspects of Lactation" (M. Peaker, ed.), pp. 165–172. Academic Press, London.

Lewis-Jones, D. I., Lewis-Jones, M. S., Connolly, R. C., Lloyd, D. C., and West, C. R. (1985). Sequential changes in the antimicrobial protein concentrations in human milk during lactation and its relevance to banked human milk. *Pediatr. Res.* **19,** 561–565.

Linzell, J. L., Mepham, T. B., and Peaker, M. (1976). The secretion of citrate into milk. *J. Physiol.* **260,** 739–750.

Linzell, J. L., and Peaker, M. (1974). Changes in colostrum composition and in the permeability of the mammary epithelium at about the time of parturition in the goat. *J. Physiol.* **243,** 129–151.

Martin, R. H., Glass, M. R., Chapman, C., Wilson, G. D., and Woods, K. L. (1980). Human alpha-lactalbumin and hormonal factors in pregnancy and lactation. *Clin. Endocrinol.* **13,** 223–230.

Neville, M. C. (1983). Regulation of mammary development and lactation. *In* "Lactation: Physiology, Nutrition and Breast-feeding" (M. C. Neville and M. R. Neifert, eds.), pp. 103–140. Plenum Press, New York.

Neville, M. C., Keller, R., Seacat, J., Lutes, V., Neifert, M., Casey, C., Allen, J. A., and Archer, P. (1988). Studies in human lactation: Milk volumes in lactating women during the onset of lactation and full lactation. *Am. J. Clin. Nutr.* **48,** 1375–1386.

Neville, M. C., Hay, W. W., Jr., and Fennessey, P. (1990). Physiological significance of the concentration of human milk glucose. *Protoplasma* **159,** 118–128.

Neville, M. C., Allen, J. C., Archer, P. G., Seacat, J., Casey, C., Sawicki, V., Oliva-Rasbach, J., and Neifert, M. (1991). Studies in human lactation: 7. Milk volume and nutrient composition during weaning and lactogenesis. *Am. J. Clin. Nutr.* **54,** 81–93.

Neville, M. C., and Peaker, M. (1979). The secretion of calcium and phosphorus into milk. *J. Physiol.* **313,** 561–570.

Patton, S., Huston, G. E., Montgomery, P. A., and Josephson, R. V. (1986). Approaches to the study of colostrum—The onset of lactation. *In* "Human Lactation 2: Maternal and Environmental Factors" (M. Hamosh and A. S. Goldman, eds.), pp. 231–240. Plenum Press, New York.

Peaker, M. (1983). Secretion of ions and water. *In* "Biochemistry of Lactation" (T. B. Mepham, ed.), pp. 285–307. Elsevier, New York.

Peaker, M., and Linzell, J. L. (1975). Citrate in milk: Harbinger of lactogenesis. *Nature* **253,** 464–465.

Pickett, P. B., Pitelka, D. R., Hamamoto, S. T., and Misfeldt, D. S. (1975). Occluding junctions and cell behavior in primary cultures of normal and neoplastic mammary gland cells. *J. Cell Biol.* **66,** 316–332.

Schanbacher, F. L., Goodman, R. E., and Talhouk, R. S. (1992). Bovine mammary lactoferrin: Implications from mRNA sequence and regulation contrary to other milk proteins. *J. Dairy Sci.,* **76**, 3812–3831.

Solari, R., and Kraehenbuhl, J.-P. (1987). Receptor-mediated transepithelial transport of polymeric immunoglobulins. *In* "The Mammary Gland" (M. C. Neville and C. W. Daniel, eds.), pp. 269–300. Plenum Press, New York.

Teng, C. T., Pentecost, B. T., Chen, Y. H., Newbold, R. R., Eddy, E. M., and McLachlan, J. A. (1989). Lactotransferrin gene expression in the mouse uterus and mammary gland. *Endocrinology* **124,** 992–999.

B. Volume and Caloric Density of Human Milk

MARGARET C. NEVILLE

I. Introduction

A detailed knowledge of milk volume production and/or transfer to the offspring is important in assessing the metabolic impact of lactation on the mother, maternal factors that impact milk yield, and the role of mother's milk in nutrition of the young. In this discussion we will focus mainly on human lactation but the methodologies and principles are generally applicable to other mammals. One important difference is that, unlike women, the rate of milk production for those animal species that have received the most study, namely dairy animals and laboratory rodents, is limited by the amount of mammary tissue and the rate at which nutrients can be delivered to the mammary gland. Maximal production is stimulated by consistent milking in dairy species and by the often massive milk demands of a large litter in rodents. In these animals, the rate of milk production is, as expected, highly dependent on maternal factors including genetic heritage and nutrition. On the other hand, lactating women, and probably many other species in the wild, produce milk at a rate determined largely by infant demand (Prentice *et al.,* 1986; Dewey and Lonnerdal, 1986), a point we will return to several times in this discussion. The maximal human capacity for milk production may have been achieved by the wet nurses studied in the 1920s by Macy *et al.* (1930) who produced up to 3500 ml of milk per day, about four times the volume produced by women breast-feeding a single infant. Mothers of twins and triplets produced 2 or 3 liters of milk per day in a study by Hartmann and his colleagues (Saint *et al.,* 1986). It is especially important to keep this distinction in mind when maternal factors that may influence lactation capacity are under consideration.

It is also important to remember that partial rather than exclusive breast-feeding is the rule in most human societies after 3 to 6 months postpartum (Prentice *et al.,* 1986; Creed de Kanashiro *et al.,* 1990), yet most studies have focussed on volume transfer in the exclusively breast-feeding mother–infant dyad. In this chapter, following a discussion of methods of measurement of milk volume transfer, we will summarize the available data for the exclusively breast-feeding dyad followed by a shorter discussion of partial breast-feeding and weaning. A brief section on caloric density will complete the chapter.

HANDBOOK OF MILK COMPOSITION

II. Methods for Measurement of Milk Volume

In the measurement of milk volume it is important to distinguish between milk *yield* and milk *transfer to the infant.* In dairy animals milk yield, the amount of milk that is transferred from the mammary gland to the milk pail or milking machine, is the appropriate measure. In general, this quantity represents the productive capacity of the gland when it is milked empty at least twice a day. In humans, milk transfer to the infant is almost always the value sought. Because the breast is not usually emptied completely at a feed, this value is somewhat lower than the milk yield that can be obtained by expression of milk with a good electric breast pump several times during a 24-hr period (Neville *et al.,* 1987; Dewey *et al.,* 1991a).

Several methods are available for measurement of milk volume production. The first, expression of the contents of the breast by pump or manually, is, as mentioned above, the method almost universally used to obtain milk volume in dairy animals. Of the other available methods *test weighing* the young before and after each feed for a specified period of time is conceptually the simplest. *Isotope dilution,* virtually the only accurate method available for the measurement of milk production in small laboratory mammals, has recently been applied using stable isotopes to humans (Butte *et al.,* 1988). The *Doppler ultrasound human milk flowmeter* (Woolridge *et al.,* 1985), which uses ultrasound to measure milk velocity as it travels from mother to infant, has not been found to be particularly reliable for any but very short-term measurements of milk flow and will not be examined further here. A new method based on *topographical computer imaging* of the breast has recently become available for short-term measurement of milk production in women (Arthur *et al.,* 1989) and will be discussed briefly although it has not yet received extensive evaluation.

A. Extraction of Milk

Extraction of milk by pump, milking machine, or manually can achieve nearly complete emptying of the mammary gland if an adequate let-down is accomplished. Modern breeding practices have produced dairy cattle that let down to the milking machine, so that extraction efficiency is excellent. Nevertheless, to achieve complete extraction in goats, Linzell (1967) gave an iv dose of oxytocin just prior to milking. Linzell's technique, used to measure hourly milk yield, was adapted to women by Neville *et al.* (1988). Briefly, an electric breast pump with dual heads was used to extract milk from the breasts; after 10 min of pumping one drop of synthetic oxytocin was administered intranasally and pumping continued for a further 5 min. The volumes obtained during the first and second hours were higher than the average milk production due to extraction of residual milk (mean total excess 200 ± 25 ml). After the third pumping episode the mean hourly milk extraction in five women did not differ significantly from the milk volume measured by test weighing during a preceding 24-hr period. Such a pumping regimen may be useful when milk yield is to be assessed on a short-term basis.

Others have used milk extraction to measure 24-hr milk yield. Brown *et al.* (1982) extracted all milk by breast pump for 24 hr and compared extracted volumes with mean volumes obtained by 6 days of test weighing. Pumped volumes exceeded test-weigh volumes by 50 g/day or about 7%, again reflecting residual milk. Dewey *et al.* (1991a), using a breast pump to extract milk from alternate breasts at each feed over a 24-hr period, found residual volumes of about 110 ml/day. All investigators have, therefore, found that milk volume transfer is overestimated when measured by the extraction technique. The amount of the overestimate varies from 50 to 200 ml/day depending on the population and technique used.

B. Test Weighing

In most studies for which values are given in Table I, the infant was test weighed before and after every feed for at least 24 hr. This procedure is usually carried out by the mother in the home after a brief period of training, although trained workers have been used in the field or clinic in some studies, particularly in developing nations (Brown *et al,* 1982; Prentice *et al.*, 1983). Because the day-to-day coefficient of variation for milk transfer in a single mother–infant pair is quite high [15% in one study (Butte *et al.*, 1985), 8.9% in another (Dewey *et al.*, 1991a)], current recommendations suggest that test weighing be continued for 4 days when accurate measurements are needed for a single individual (Stuff *et al.*, 1986). However, 24-hr or even properly standardized 12-hr test weights appear to give equivalent population means (Prentice *et al.*, 1981; Creed de Kanashiro *et al.*, 1990). An integrating electronic balance provides the most accurate data, particularly if the infant is moving (Neville *et al.*, 1988). For older children it is possible to suspend a swing from the balance (Woolridge *et al.*, 1985). If the balance prints out the weight, errors in recording are minimized. Test weighing results in a systematic underestimation of the volume of milk produced because of insensible water loss through respiration and sweating during the feed. This loss from the infant amounted to 0.03 g/kg/min in one study (Hendrikson *et al.*, 1985) and 0.05 g/kg/min in another (Nommsen *et al.*, 1991). This number is multiplied by the total nursing time per 24 hours and the weight of the infant to obtain the correction which usually amounts to 3 to 5% of the measured daily intake.

It is possible to obtain satisfactory milk yields by weighing the mother rather than the infant before and after the feed if a sufficiently sensitive balance is available (Arthur *et al.*, 1987). However, the results obtained are less reliable than test weighing the infant because the amount of milk

TABLE I
Volume of Milk Transferred from Mother to Infant in the Exclusively Breast-Feeding Dyad

	Months postpartum							
Reference/Country	1	2	3	4	5	6	7	8
Neville *et al.* (1988)/ United States (Denver, CO)								
mean	668	694	734	711	838	820	848	818
SD	117	98	114	100	134	79	63	158
n	12	12	10	12	12	9	6	3
Pao *et al.* (1980)/United States (Ohio)								
mean	600		833					
SD	159							
n	11		2					
van Steenbergen *et al.* (1981)/Kenya								
mean	778**		619**					
SD	180		197					
n	7		13					
Hofvander *et al.* (1982)/ Uppsala, Sweden								
mean	656	773	776					
Range	360–860	575–985	600–937					
n	25	25	25					
Butte and Calloway, (1981)/United States (Navajo)								
mean	634							
SD	113							
n	10							
Butte *et al.* (1991a)/ United States (Texas)								
mean	738*	725	723*	755	741	818		
SD	157	129	114	113	103	166		
n	64	40	37	111	26	8		
Dewey *et al.* (1984)/ United States (California)								
mean							875	834
SD							142	99
n							8	8
Dewey and Lönnerdal (1983)/United States (California)								
mean	672	756	782	810	805	896		
SD	192	170	172	142	117	122		
n	16	19	16	13	11	11		

TABLE I *(continued)*

	Months postpartum							
Reference/Country	1	2	3	4	5	6	7	8
Salmenpera *et al.* (1985)/ Finland								
mean				790		800		
SD				140		120		
n				12		31		
Walgren (1944) Sweden (girls)								
mean	576**	704*	733*	747		740**		
Variance	80	98	113	19		16		
n	65	72	43	48		26		
Walgren (1944) Sweden (Boys)								
mean	645***	750***	798***	821***		817***		
Variance	97	107	113	122		133		
n	58	72	49	42		33		
Whitehead and Paul (1981) United Kingdom (girls)								
mean		677	742	775	814	838	854	786
SD		87	119	138	113			
n		20	17	14	6	1	1	1
Whitehead and Paul (1981) United Kingdom (boys)								
mean		791***	820	829	790	922		
SD		116	187	168	113			
n		27	23	18	5	1		
Stuff *et al.* (1986)/United States (Houston, TX)								
mean					735			
SD					85			
n					9			
Chandra (1981)/Canada								
mean				793	856	925**	872*	815
SD				71	99	112	126	97
n				33	31	28	27	24
van Raaij *et al.* (1991)/ The Netherlands								
mean	692	745						
SD	122	131						
n	16	40						
Nommsen *et al.* (1991)/ United States (California)								
mean			811*					
SD			133					
n			58					

TABLE I *(continued)*

Reference/Country	Months postpartum 1	2	3	4	5	6	7	8
Frigerio *et al.* (1991)/ The Gambia								
mean		738						
SD		47						
n		16						
Butte *et al.* (1992)/ Mexico (rural) (^2H_2O to mothers)								
mean				885**		869*		
SD				146		150		
n				15		15		
Goldberg *et al.* (1991)/ Cambridge, England ($H_2{}^{18}O$ to mother)								
mean	802**		792					
SD	179		177					
n	10		10					
Weighted mean (test weigh only)								
mean	657	735	767	773	802	827	809	819
SD	121	111	111	112	109	102	120	100
n	284	343	293	313	100	148	42	36

*$p < 0.05$ with respect to weighted mean.
**$p < 0.01$ with respect to weighted mean.
***Significantly different from girls, $p < 0.05$.

transferred compared to the mother's body weight is small and the correction for insensible weight loss is substantial. Milk volumes obtained using this procedure in Australia were usually considerably greater than the values given in Table I (Rattigan *et al.,* 1981); however, these early values were not corrected for insensible water loss. When the correction was made, milk volumes within the range of other reported studies were obtained (Arthur *et al.,* 1987).

The advantage of test weighing is that it is a reliable and relatively inexpensive way to measure milk transfer from mother to infant. The major disadvantage is that the technique requires a certain degree of education and dedication on the part of the mother or the presence in the home of a trained field worker. It can disrupt the feeding routine especially if frequent nighttime feeds are the norm.

C. Isotope Dilution

Isotope dilution was used first in animals and is the only technique available for small animals such as rats and mice. The mother is given a dose of tritiated water and the passage of isotope to the young is tracked by whole body analysis of pups at intervals (Thornburn *et al*, 1983). In humans it is not possible to use radioactive isotopes; fortunately, water can be obtained labeled with stable isotopes, namely, deuterium or ^{18}O. Deuterium oxide was first used by Coward and colleagues. In their earliest studies (Coward *et al.*, 1979, 1982a) the isotope was administered to the infant and the rate of dilution over a period of several days was determined by mass spectrometry. In this form the technique has been shown to be useful but only if a correction is made for exchange of water at the integumental and respiratory surfaces as well as intake of water from sources other than breast milk (Fjeld *et al.*, 1988; Butte *et al.*, 1991b). In later studies Coward *et al.* (1982b) and others (Butte *et al.*, 1988) administered the isotope to the mother and measured deuterium enrichment in the mother's milk and infant's urine to calculate milk transfer. It is necessary to know the infant's total body water when using this method; this parameter can best be estimated by administering a separate dose of ^2H_2O or $H_2{}^{18}O$ to the infant. Careful evaluation of isotope dilution against the test-weighing method (Butte *et al.*, 1988) in nine women gave a mean difference between the two methods of 2%, not statistically significant when metabolic water production was taken into account. Isotope dilution has the advantage of ease of sampling and requires little cooperation on the part of the mother. It gives mean milk intake over several days. It has the disadvantage of requiring sophisticated and expensive analytical techniques. Nevertheless, the technique is finding increasing use for studies in developing countries (Coward *et al.*, 1982b; Butte *et al.*, 1992; Goldberg *et al.*, 1991).

D. Topographical Computer Imaging

Hartmann and his colleagues (Arthur *et al*, 1989) have developed a technique in which moire patterns are projected onto the lactating breast. Video images obtained over a period of time are stored in the computer for later analysis. As the breast expands with increasing stored milk the moire patterns change in a way that can be related to the volume of milk produced. This promising technique has the potential of measuring short-term rates of milk synthesis between breast feeds and for that reason should be more completely evaluated. With the increasing availability of computers suitable for this type of task, the cost of this method may become quite reasonable, particularly in the clinical setting.

III. Milk Volumes in Exclusively Breast-Feeding Women

Table I is a compilation of worldwide values for milk transfer in exclusively breast-feeding women. When mean values, weighted for the number of subjects in each study, are calculated breast milk transfer is seen to increase gradually from about 650 g/day at 1 month to about 800 g/day at 6 months when it appears to level off (Figure 1). The variation between individuals is large with mean SDs about 100 g/day at all time points. Milk volumes from two underprivileged groups, Navajo Indians and women in The Gambia, did not differ from the overall mean in early lactation and milk volumes from Kenya at 1 month were actually significantly higher than the mean (Table I). The consistency between the various population groups suggest that nutritional and cultural factors have little effect on milk transfer in the exclusively breast-feeding dyad.

Milk volumes obtained by isotope dilution in Cambridge, England were significantly higher than the overall mean at 1 month postpartum but not significantly different thereafter from values obtained using the test-weighing method. The milk volumes measured in a group of undernourished Mesoamerindians using the isotope dilution technique were significantly higher at 4 and 6 months than the mean values in the table. This finding was attributed by the authors to the lower caloric density of the milk in this group (see below).

The range of milk volumes among individuals varies from about 500 to 1100 g/day (Figure 1). In a well-nourished population no maternal characteristics were found to correlate with breast milk transfer to the infant leading Dewey and her colleagues to propose that "infant demand is the main determinant of lactation performance" in affluent populations (Dewey *et al.,* 1991a). A similar study in malnourished populations has not been published, but Butte *et al.* (1992) proposed that milk production may be subject to maternal limitations on the basis of her study of Mesoamerindians in which milk transfer appeared to be insufficient to support optimal infant growth after 3 months postpartum. Infant factors related to milk production included infant weight (Neville *et al.,* 1988; Dewey *et al.,* 1991a) and total time nursing (Dewey *et al.,* 1991a). The difference between boys and girls disappears when their weight difference is taken into account. In general, although earlier authors suggested that infant morbidity would alter milk volume transfer (Prentice *et al.,* 1981), this has not proven to be the case. In a careful study by Brown and co-workers (Brown *et al,* 1990) diarrhea or fever were associated with no changes in the frequency of breast-feeding, total suckling time, or amount of breast milk consumed. Data supporting this conclusion have been obtained by Rowland *et al.* (1988) and Butte *et al.* (1992).

The caloric density of milk may be a factor in determining the volume of milk transferred to the infant. Butte *et al.* (1992) suggested that the low

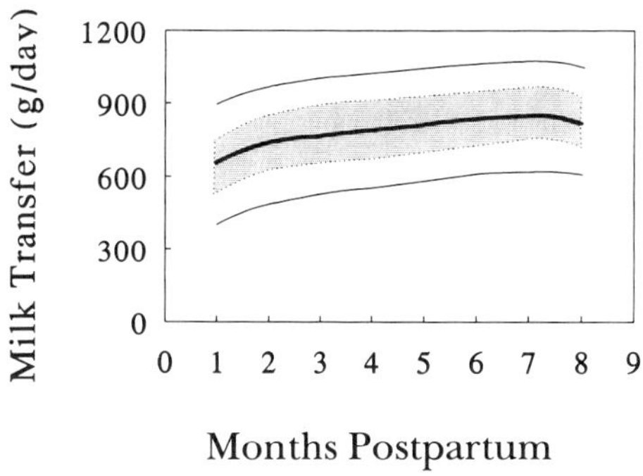

Figure 1 Milk transfer to exclusively breast-fed infants. Heavy line represents weighted mean from all studies using test weighing in Table I and shaded area represents one standard deviation from the mean. Area between light lines represents the 95% confidence interval.

caloric density of the milk (see below) was a factor in the high milk volumes noted in Mesoamerindians. Nommsen *et al.* (1991) found a significant inverse relationship between milk intake and milk energy density in a group of affluent American women. The effect of other composition variables on milk transfer has been incompletely investigated. Nommsen *et al.* (1991) found a significant ($p < 0.05$) inverse relation between milk volume and milk protein concentrations at 6 and 9 months postpartum in an affluent population. They also found a positive correlation with milk lactose concentration at the same stages. However, both these changes could have been the result of the inclusion of women in the early stages of weaning in their study population. In their study women were only considered to be weaning, and therefore excluded from the analysis if their milk volumes were below 200 ml per day. Neville *et al.* (1991) found an increase in protein concentration and a decrease in lactose concentration when milk volume fell below 400 ml/day in affluent American women. It would be of interest to know whether the inverse relation observed by Nommsen *et al.* (1991) persists when women transferring between 200 and 400 g of milk per day are excluded from statistical analysis. The correlation between milk volume and the rate of transfer of other milk components, such as calcium, magnesium, or other minerals, to the infant has not been investigated systematically although the relevant data are probably available in some studies. In one study (Allen *et al.*, 1991), for example, the *amounts* of lactose, magnesium, and ionized calcium transferred to a given infant on Day 21 of lactation were highly correlated with the amounts transferred at 6 months postpartum although the concentrations of these substances varied significantly through lactation. The number of subjects in this study (13) was not sufficient to make meaningful correlations between the rate of milk transfer and production of these nutrients.

IV. Breast Milk Volumes Transferred to Partially Breast-Fed Infants

It is difficult to give universal values of breast milk volumes in partially breast-fed infants because the pattern of supplementation differs from one cultural group to another. However, two circumstances are of particular interest. One is the infant who is breast-fed several times a day, receiving supplemental food on a meal-by-meal basis, but breast-feeds are not replaced by supplemental feedings or bottles. A small group of Western women who fed according to this pattern transferred about 600 g of milk per day to their infants at 1 year of age (Neville *et al.*, 1991). This pattern is prevalent in developing countries where breast-feeding of longer duration is the norm. The second pattern is the infant who is gradually weaned by replacing feeds with meals containing milk or formula and other foods. This is the pattern often noted during weaning in affluent societies.

A study by Creed de Kanoshira *et al.* (1990) in Peru provides an excellent example of the first pattern. The infants were supplemented from an early age (< 3 months) with a variety of foods including dairy products and, as time progressed, cereals. Breast milk consumed remained at a high level (> 550 ml per day) throughout the first year in 89% of the infants. The actual daily volumes of breast milk consumed are given in Table II. These volumes were about 93% of the volumes transferred to exclusively breast-fed infants at 2 months (Table I) and decreased slowly to about 75% at 6 months. By 10–12 months the infants were still receiving about 45% of their total energy from breast milk. Thus, these infants continued to obtain a substantial proportion of their nutrition from breast milk at 1 year and after. Data from The Gambia and Papua, New Guinea gave similar volume estimates. Note that the variability between individuals is about twice that seen in the exclusively breast-feeding group.

In contrast, Figure 2A shows the volumes of breast milk consumed by five American infants whose mothers deliberately began gradual weaning at about 6 months of age by substituting formula feeds for breast feeds. Milk volumes gradually decreased, reaching zero by 17 months postpartum or earlier. When milk transfer was plotted as a function of number of feeds per day for each mother–infant pair (Figure 2B), it became clear that milk volume was linearly related to the number of feeds, falling to zero at one feed per day. The data in Figure 2 also make it clear that on a cross-sectional basis little correlation between number of feds per day and milk volume would be expected since subjects producing over 600 g/day fed between 4 and 12 times per day, depending on the individual.

V. Caloric Density of Human Milk

The caloric density of human milk is best determined by bomb calorimetry (Garza *et al.*, 1985). However, many investigators calculate this quantity

TABLE II
Breast Milk Transfer in Partially Breast-Fed Infants

Reference	Milk volume ± SD (*n*)	Duration of lactation (Months)
Creed de Kanashiro *et al.* (1990)/ Peru (12-hr test weigh extrapolated to 24 hr)	685 ± 245 (128)	1.0–2.9
	690 ± 240 (121)	3.0–4.9
	655 ± 226 (108)	5.0–6.9
	624 ± 219 (103)	7.0–9.9
	565 ± 208 (89)	10.0–12.5
Prentice *et al.* (1986)/Keneba, The Gambia (24-hr test weigh)	582 ± 169 (10)	3–3.99
	643 ± 149 (17)	4–5.99
	607 ± 131 (16)	5–8.99
	594 ± 200 (16)	9–11.99
	633 ± 200 (15)	12–18
Prentice *et al.* (1986)/Cambridge, England (96-hr test weigh/4)	783 ± 176 (48)	3–3.99
	717 ± 207 (42)	4–5.99
	588 ± 206 (45)	5–5.99
	493 ± 216 (38)	6–7.99
	342 ± 228 (31)	7–8.99
	328 ± 292 (19)	9–10.99
Dewey *et al.* (1991b)/Davis, CA (4-day test weigh)	769 ± 171 (60)	6
	646 ± 217 (50)	9
	448 ± 251 (42)	12
Coward *et al.* (1982b)/The Gambia (^2H_2O to mother)	752 ± 36 (4)	0–4
	757 ± 44 (4)	5–9
	728 ± 170 (5)	> 9
Coward *et al.* (1982b)/Papua, New Guinea (^2H_2O to mother)	670 ± 184 (17)	0–4
	936 ± 183 (8)	5–9

from the proximate composition of the milk using the following factors in kcal/gm: 5.65 protein, 9.25 fat, 3.95 lactose, 5.65 nonprotein nitrogen. Other authors have used the factors 4, 9, and 4 kcal/gm for protein, fat, and lactose, respectively (Creed de Kanashiro *et al.*, 1990). Because lipids are the largest contributor to the caloric density, it is important that care be taken to obtain a representative milk sample. If this is not done, or the milk sample is stored in such a way that a representative aliquot is not assayed, erroneous values can result. A second caution is that some methods for lactose measurements, notably the automated enzyme procedure on the Yellow Springs Instrument Analyzer, probably do not include the oligosaccharides of human milk that comprise about 1% of the total weight of milk in normal circumstances and may contribute even more in conditions such as diabetes (Ferris, personal communication). Use of the standard factors in these circumstances may result in a small but systematic underestimation of the caloric content of the milk.

Representative values for caloric density are given in Table III. Note that the values for undernourished Mesoamerindians are about 100 cal/g

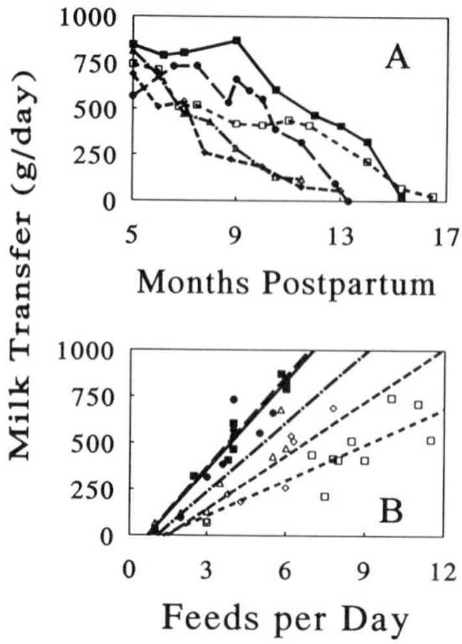

Figure 2 Milk volumes during gradual weaning. Milk transfer was measured by test weighing in five breast-feeding dyads who weaned gradually between 6 and 17 months lactation. (A) Milk volume transfer as a function of time postpartum. (B) Relation between the numbers of feeds per day and the milk volume. Each set of symbols represents an individual dyad. From Neville *et al.* (1991). Used by permission of *Am. J. Clin. Nutr.*

lower than values obtained by the same investigators for affluent American women. This difference is due to a difference in the fat content of the milk (Butte *et al.*, 1992), probably, as mentioned above, related to a lower body fat content. The values given by Dewey *et al.* (1991b) for a group of California women are about 10% higher than the values given by Butte *et al.* for the Texas women. Although this difference could be related to calculation from the proximate composition rather than actual analysis by bomb calorimetry, it is more likely related to the fact that the percentage fat obtained by the California group was systematically higher. Because of the systematic variation in the currently reported values, it is not possible at this time to give a single representative value for the caloric density of human milk.

VI. Conclusions

The mean volume of milk transfer in exclusively breast-feeding dyads can now be considered to be established for most, if not all, populations as about 770 g/day at 3 months postpartum. In exclusively breast-fed infants milk transfer varies significantly from one mother–infant pair to another with ranges from about 500 to 1200 g/day. The factors that govern milk consumption by breast-fed infants are imperfectly understood although most of the variability can be ascribed to infant factors. The weight of the

TABLE III
Caloric Density of Human Milk

Reference	Energy Density (cal/g)	Time postpartum (months)	Sampling method
Butte *et al.* (1985)/ Caucasians (U.S.A.) (bomb calorimetry)	680 ± 71 (37)	1	Expression of alternate breasts over 24 hr
	643 ± 83 (40)	2	
	625 ± 93 (37)	3	
	644 ± 102 (41)	4	
Butte *et al.* (1992)/ Mesoamerindians (Mexico) (bomb calorimetry)	560 ± 60 (15)	4	Pumped contents of one breast three times daily
	530 ± 70 (15)	6	
Nommsen *et al.* (1991)/ Caucasians (U.S.A.) (calculated from proximate composition)	697 ± 67 (67)	3	Expression of alternate breasts over 24 hr
	707 ± 92 (45)	6	
	709 ± 74 (28)	9	
	796 ± 110 (21)	12	

infant is the most significant variable but can account at most for 30% of the variation. Further, the rate of milk consumption per kilo decreases as the infant becomes older so age must also be taken into account. Differences in energy density of the milk produced by different women account for perhaps 10% of the variability. It is possible that milk volume transfer is regulated on the basis of milk components other than energy. The question boils down ultimately to determination of those factors that control infant appetite, an area that still requires substantial research.

References

Allen, J. C., Keller, R. P., Neville, M. C., and Archer, P. (1991). Studies in human lactation: 6. Milk composition and daily secretion rates of macronutrients in the first year of lactation. *Am. J. Clin. Nutr.* **54,** 69–80.

Arthur, P. G., Hartmann, P. E., and Smith, M. (1987). Measurement of the milk intake of breast-fed infants. *J. Ped. Gastroenterol. Nutr.* **6,** 758–768.

Arthur, P. G., Jones, T. J., Spruce, J., and Hartmann, P. E. (1989). Measuring short-term rates of milk synthesis in breast-feeding mothers. *Q. J. Exp. Physiol.* **74,** 419–428.

Brown, K. H., Black, R. E., Robertson, A. D., Akhtar, N. A., Ahmed, G., and Becker, S. (1982). Clinical and field trials of human lactation: Methodological considerations. *Am. J. Clin. Nutr.* **35,** 745–756.

Brown, K. H., Stallings, R. Y., Creed de Kanashiro, H., Lopez de Romaña, G., and Black, R. E. (1990). Effects of common illnesses on infants' energy intakes from breast milk and other foods during longitudinal community-based studies in Huascar (Lima), Peru. *Am. J. Clin. Nutr.* **52,** 1005–1013.

Butte, N. F., and Calloway, D. H. (1981). Evaluation of lactational performance of Navajo women. *Am. J. Clin. Nutr.* **34,** 2210–2215.

Butte, N. F., and Garza, C. (1985). Energy and protein intakes of exclusively brestfed infants during the first four months of life. *In* "Nutritional Needs and Assessment of Normal

Growth" (M. Gracey and F. Falkner, eds.), pp. 63–81. Nestle Nutrition, Vevey/Raven Press, New York.

Butte, N. F., Wong, W. W., Petterson, B. W., Garza, C., and Klein, P. D. (1988). Human-milk intake measured by administration of deuterium oxide to the mother: A comparison with the test-weighing technique. *Am. J. Clin. Nutr.* **47,** 815–821.

Butte, N. F., Wong, W. W., Garza, C., Stuff, J. E., O'Brian Smith, E., Klein, P. D., and Nichols, B. L. (1991a). Energy requirements of breast-fed infants. *J. Am. Colloq. Nutr.* **10,** 190–195.

Butte, N. F., Wong, W. W., Klein, P. D., and Garza, C. (1991b). Measurement of milk intake: Tracer-to-infant deuterium dilution method. *Br. J. Nutr.* **65,** 3–14.

Butte, N. F., Villalpando, S., Wong, W. W., Flores-Huerta, S., de Hernandez-Beltran, M., O'Brian, Smith, E., and Garza, C. (1992). Human milk intake and growth faltering of rural Mesoamerindian infants. *Am. J. Clin. Nutr.* **55,** 1109–1116.

Chandra, R. K. (1981). Breast feeding, growth and morbidity. *Nutr. Res.* **1,** 25–31.

Coward, W. A., Whitehead, R. G., Sawyer, M. B., Prentice, A. M., and Evans, J. (1979). New method for measuring milk intakes in breast-fed babies. *Lancet,* July 7, 13–14.

Coward, W. A., Cole, T. J., Gerber, H., Roberts, S. B., and Fleet, I. (1982a). Water turnover and the measurement of milk intake. *Pfluger's Archiv* **393,** 344–347.

Coward, W. A., Cole, T. J., Sawyer, M. B., Prentice, A. M., and Orr-Ewing, A. K. (1982b). Breast-milk intake measurement in mixed-feds infants by administration of deuterium oxide to their mothers. *Hum. Nutr. Clin. Nutr.* **36C,** 141–148.

Creed de Kanashiro, H., Brown, K. H., Lopez de Romaña, G., Lopez, T., and Black, R. E. (1990). Consumption of food and nutrients by infants in Huascar (Lima), Peru. *Am. J. Clin. Nutr.* **52,** 995–1004.

Dewey, K. G., and Lonnerdal, B. (1983). Milk and nutrient intake of breast-fed infants from 1 to 6 months: Relation to growth and fatness. *J. Ped. Gastroenterol. Nutr.* **2,** 497–506.

Dewey, K. G., and Lonnerdal, B. (1986). Infant self-regulation of breast milk intake. *Acta Paediatr. Scand.* **75,** 893–898.

Dewey, K. G., Finley, D. A., and Lonnerdal, B. (1984). Breast milk volume and composition during late lactation (7–20 months). *J. Ped. Gastroenterol. Nutr.* **3,** 713–720.

Dewey, K. G., Heinig, M. J., Nommsen, L. A., and Lonnerdal, B. (1991a). Maternal versus infant factors related to breast milk intake and residual milk volume: The DARLING study. *Pediatrics* **87,** 829–837.

Dewey, K. G., Heinig, M. J., Nommsen, L. A., and Lonnerdal, B. (1991b). Adequacy of energy intake among breast-fed infants in the DARLING study: Relationships to growth velocity, morbidity and activity levels. *J. Pediatr.* **119,** 538–547.

Fjeld, C. R., Brown, K. H., and Schoeller, D. A. (1988). Validation of the deuterium oxide method for measuring average daily milk intake in infants. *Am. J. Clin. Nutr.* **48,** 671–679.

Frigerio, C., Schutz, Y., Whitehead, R., and Jequier, E. (1991). A new procedure to assess the energy requirements of lactation in Gambian women. *Am. J. Clin. Nutr.* **54,** 526–533.

Garza, C., Butte, N. F., and Dewey, K. G. (1985). Determination of the energy content of human milk. *In* "Human Lactation 1: Milk Components and Methodologies" (R. G. Jensen and M. C. Neville, eds.), pp. 121–125. Plenum Press, New York.

Goldberg, G. R., Prentice, A. M., Coward, W. A., Daries, H. L., Murgatroyd, P. R., Sawyer, M. B., Ashford, J., and Black, A. E. (1991). Longitudinal assessment of the components of energy balance in well-nourished lactating women. *Am. J. Clin. Nutr.* **54,** 788–798.

Hendrikson, E. C., Seacat, J. M., and Neville, M. C. (1985). Insensible weight loss in children under one year of age. *Acta Paediatr. Scand.* **74**(5), 678–680.

Hofvander, Y., Hagman, U., Hillervik, C., and Sjolin, S. (1982). The amount of milk consumed by 1–3 months old breast- or bottle-fed infants. *Acta Paediatr. Scand.* **71,** 953–958.

Linzell, J. L. (1967). The effect of frequent milking and of oxytocin on the yield and composition of milk in fed and fasted goats. *J. Physiol.* **190,** 333–346.

Macy, I. G., Hunscher, H. A., Donelson, E., and Nims, B. (1930). Human milk flow. *Am. J. Dis. Child.* **6,** 492–515.

Neville, M. C., and Oliva-Rasbach, J. (1987). Is maternal milk production limiting for infant growth during the first year of life in breast-fed infants? *In* "Human Lactation: Effect of Human Milk on the Recipient Infant" (A. Goldman and S. A. Atkinson, eds.), pp. 123–133. Plenum Press, New York.

Neville, M. C., Keller, R., Seacat, J., Lutes, V., Neifert, M., Casey, C., Allen, J. A., and Archer, P. (1988). Studies in human lactation: Milk volumes in lactating women during the onset of lactation and full lactation. *Am. J. Clin. Ntur.* **48,** 1375–1386.

Neville, M. C., Allen, J. C., Archer, P. G., Seacat, J., Casey, C., Sawicki, V., Oliva-Rasbach, J., and Neifert, M. (1991). Studies in human lactation: 7. Milk volume and nutrient composition during weaning and lactogenesis. *Am. J. Clin. Nutr.* **54,** 81–93.

Nommsen, L. A., Lovelady, C. A., Heinig, M. U., Lönnerdal, B., and Dewey, K. G. (1991). Determinants of energy, protein, lactose concentrations in human milk during the first 12 mo of lactation: The DARLING study. *Am. J. Clin. Nutr.* **53,** 457–465.

Pao, E. M., Himes, J. M., and Roche, A. F. (1980). Milk intakes and feeding patterns of breast-fed infants. *J. Am. Diet. Assoc.* **77,** 540–545.

Prentice, A. M., Whitehead, R. G., Roberts, S. B., and Paul, A. A. (1981). Long-term energy balance in child-bearing Gambian women. *Am. J. Clin. Nutr.* **34,** 2790–2799.

Prentice, A. M., Roberts, S. B., Prentice A., Paul, A. A., Watkinson, M., Watkinson, A. A., and Whitehead, R. G. (1983). Dietary supplementation of lactating Gambian women. I. Effect on breast milk volume and quality. *Hum. Nutr. Clin. Nutr.* **37C,** 53–64.

Prentice, A. M., Paul, A., Prentice, A., Black, A., Cole, T., and Whitehead, R. (1986). Cross-cultural differences in lactational performance. *In* "Human Lactation: Maternal and Environmental Factors" (M. Hamosh and A. S. Goldman, eds.), pp. 13–50. Plenum Press, New York.

Rattigan, S., Ghisalberti, A. V., and Hartmann, P. E. (1981). Breast-milk production in Australian women. *Br. J. Nutr.* **45,** 243–249.

Rowland, M. G. M., Goh Rowland, S. G. I., and Cole, T. J. (1988). Impact of infection on the growth of children from 0 to 2 years in an urban West African community. *Am. J. Clin. Nutr.* **47,** 134–138.

Saint, L., Maggiore, P., and Hartmann, P. E. (1986). Yield and nutrient content of milk in eight women breast-feeding twins and one woman breast-feeding triplets. *Br. J. Nutr.* **56,** 49–58.

Salmenpera, L., Perheentupa, J., and Siimes, M. (1985). Exclusively breast-fed healthy infants grow slower than reference infants. *Pediatr. Res.* **19,** 307–312.

Stuff, J. E., Garza, C., Boutte, C., Fraley, J. K., Smith, E. O., Klien, E. R., and Nichols, B. L. (1986). Sources of variance in milk and caloric intakes in breast-fed infants: Implications for lactation study design and interpretation. *Am. J. Clin. Nutr.* **43,** 361–366.

Thornburn, C. C., and Bailey, C. J. (1983). Use of a τ- or hard β-emitting radioisotope to assess milk intake in suckling mouse pups. *J. Nutr.* **113,** 805–812.

van Raaij, J. M. A., Schonk, C. M., Vermaat-Miedema, S. H., Peek, M. E. M., and Hautvast, J. G. A. J. (1991). Energy cost of lactation and energy balances of well-nourished Dutch lactating women: Reappraisal of the extra energy requirement of lactation. *Am. J. Clin. Nutr.* **53,** 612–619.

van Steenbergen, W. M., Kusin, J. A., and Van Rens, M. M. (1981). Lactation performance of Akamba mothers, Kenya. Breast feeding behaviour, breast milk yield and composition. *J. Trop. Ped.* **27,** 155–161.

Walgren, A. (1944). Breast-milk consumption of healthy full-term infants. *Acta Paediatr. Scand.* **32,** 778–790.

Whitehead, R. G., and Paul, A. A. (1981). Infant growth and human milk requirements: A fresh approach. *Lancet* **2,** 161–163.

Woolridge, M. W., Butte, N., Dewey, K. G., Ferris, A. M., Garza, C., and Keller, R. P. (1985). Methods for the measurement of milk volume intake of the breast-fed infant. *In* "Human Lactation: Milk Components and Methodologies" (R. G. Jensen and M. C. Neville, eds.), pp. 5–20. Plenum Press, New York.

C. Volume and Caloric Density of Bovine Milk

ROBERT G. JENSEN

I. Volume

While the same principles discussed by Neville in Chapter 3A also apply to dairy cows, the emphasis is different. Dairy cows are selectively bred for production, from 10 to 40 or more kg/day. Artificial insemination and complete production records enable the breeder to select the most promising sires and dams. Cows that do not produce sufficient milk are sold (Touchstone, 1974). Volumes, or rather weights, of milk from each cow and each producer are recorded because these are used to evaluate the performance of the cow and as a parameter for payment. The volume of milk produced by cows has no meaning for the consumer because of pooling and packaging.

II. Caloric Density

When combusted completely, bovine milk contains 67 to 72 kcal/100 g (2.8–3.0 kJ/kg). The metabolically available energy is about 8.9, 4.1, and 4.0 kcal/g (37.0, 17.0, and 16.8 *M*J/g) for fat, protein, and lactose. When kcal/g for lactate, 3.6 (15 kJ), and citrate, 2.4 (10 kJ), are added, milk (4.3% fat) has the caloric density given above (Walstra and Jenness, 1984). Milk available to the consumer, with 3.34% fat, has 61 kcal/100 g (271 kJ) (NDC, 1993). The caloric density will fluctuate depending upon the fat content of the product. Skim milk contains 4 kcal/100 g (175 kJ) and light cream (19.3% fat) 195 kcal/100 g (818 kJ). Data for many dairy products can be found in the NDC (1993) publication.

References

National Dairy Council (NDC) (1993). "Newer Knowledge of Milk and other Fluid Dairy Products." National Dairy Council, Rosemont, IL.

Touchberry, R. W. (1974). Environmental and genetic factors in the development and maintenance of lactation. *In* "Lactation" (B. L. Larson and V. R. Smith, eds.), Vol. 3. Academic Press, New York.

Walstra, P., and Jenness, R. (1984). "Dairy Chemistry and Physics," pp. 158–159. Wiley, New York.

HANDBOOK OF MILK COMPOSITION

D. Regional Variations in the Composition of Human Milk

ANN PRENTICE

I. Summary

The composition of human milk varies between different parts of the world and between different women living in the same locality. The following five tables detail the information available to 1991 on the quality of milk from nonCaucasian mothers living outside of Europe and the United States. The tables cover the following main groups of milk components: Table I, lactose and fat; Table II, nitrogen and protein; Table III, specific proteins; Table IV, minerals and trace elements; and Table V, vitamins. The values for each constituent are grouped by country in three sections corresponding to the following areas of the world: (1) The Americas (2) Africa and Arabia, (3) Asia, Australasia, and Oceania. Within each country the mothers are identified by town/village of residence and by socioeconomic status. Only results from mothers who delivered their infants at term are included.

A close inspection of the tables reveals that the similarities in milk composition between women of varying geographic, ethnic, and socioeconomic backgrounds are more striking than the differences, particularly with regard to the major nutrients. Differences do occur, however, especially in the concentrations of certain proteins, minerals, and vitamins.

Many factors confound the interpretation of cross-cultural comparisons of breast milk composition. In the past the failure to fully comprehend the importance of these confounders has led to the erroneous impression that women living in underprivileged circumstances, particularly in the developing world, produce milk of inferior quality. Studies, such as those from Guatemala, Ethiopia, and The Gambia, which include parallel investigations in privileged mothers from the same country or from Europe, have demonstrated that, in the main, breast milk quality is conserved in mothers living in impoverished circumstances.

A. Stage of Lactation

The composition of milk changes dramatically during the first days after birth as the secretion changes from colostrum to milk. After the first 1 or 2 weeks the composition stabilizes and further changes are less marked and occur over a longer time frame. However, milk composition continues to alter throughout lactation. The concentration of many components declines during the first 3–6 months of lactation reaching a low plateau in late lactation. The concentration of some components, such as the antimicrobial protein lysozyme, increases during lactation. Milk composition is also modified by the weaning process. It is essential, therefore, in any comparison between groups of mothers, to take account of the stage of lactation at which milk samples were collected. This is particularly important when comparing mothers from traditional societies who lactate for 2 years or more with those mothers living in Europe and the United States who generally breast-feed for a much shorter period.

To facilitate direct comparisons between studies from different parts of the world, the data in the tables have been grouped into the following periods of lactation: colostrum (0–5 days), transitional (6–14 days), mature (0.5–6 months), and mature (> 6 months). Each data entry is accompanied by the stage of lactation as detailed by the investigators. Any published study which did not include a description of stage of lactation or which averaged the compositional results over more than one period has been omitted. In addition, the study design has been identified as either cross-sectional, in which only one sample per subject has been analyzed, or as one which contains a longitudinal element, in which results from the same subjects may appear in more than one period of lactation.

B. Within-Day Variations

The concentration of certain milk components, especially fat, varies substantially during a feed and throughout the day. The circadian and stage of feed variations are not consistent between societies and are related in part to the pattern and frequency of breast-feeds. For the comparison of such components between groups of mothers, particularly those with different breast-feeding behavior, it is essential that milk samples are collected which represent the average concentration over a full 24 hr. In practice this is very difficult to achieve and only a limited number of investigators have endeavored to address the problem. Energy determinations, which are heavily dependent on the concentration of fat, are further confounded by the choice of calculating metabolizable energy from constituent components or measuring gross energy by bomb calorimetry. The methods and calibrations used must be clearly detailed.

The inclusion of data on fat concentrations in the tables has been strictly confined to those studies which have attempted to obtain 24-hr information. Data on energy and total solids have been reviewed but no studies fulfilled all the criteria for inclusion.

C. Maternal Parity/Age

The composition of milk appears to be dependent on the parity and the age of the mother. Data from several studies suggest that primiparous/

young women have higher concentrations of several constituents, such as protein and fat, and that mothers of very high parity (>9) produce milk of reduced quality. This aspect of milk composition remains controversial and no attempt has been made in the tables to detail the data by maternal parity/age. However, it is important to keep in mind that the proportion of primiparous mothers in societies where women bear many children, such as in parts of Africa and Asia, will be considerably lower than that in other areas and that this may affect the average milk composition of that group.

D. Methodology

Differences in analytical methods, variations in the calibrators employed, and the lack of interlaboratory standardization have profound effects on the absolute concentrations of breast milk constituents reported in the literature. This ultimately influences any comparisons made between groups of mothers studied by different investigators. For this reason, the most definitive comparisons are those made by the same investigators employing the same methodology, and this type of study is to be encouraged.

Data have been included in the tables only if details of the methodology were given by the investigators. For components where the use of different methods may cause problems of interpretation, the analytical technique has also been tabulated.

Differences in methodology pose particular problems for certain components and these are outlined below.

1. Protein

Many investigators have estimated breast milk protein concentrations by measuring the nitrogen content and using a multiplication factor to calculate protein. Differences in the way these calculations were made affect the protein concentrations quoted in the published reports. In particular, a variety of factors were used, 6.25 or 6.38 being the most common. In addition, some authors corrected total nitrogen for nonprotein nitrogen before conversion, while others did not. For consistency, all protein values obtained by nitrogen analysis have been converted back to nitrogen before inclusion in Table II and referred to as "total nitrogen" if no correction for nonprotein nitrogen was made, or as "protein nitrogen" if it was. Any study which provided insufficient details for these calculations has been omitted.

Other methods have been used for the measurement of breast milk protein, particularly colorimetric methods. Many of these suffer from the problems of interference from lactose and of the different reactivities of individual protein species. The values from studies using these methods are detailed in Table II under Protein—Other Methods.

2. *Specific Proteins*

Concentrations of specific proteins are highly dependent on the analytical method used and in general comparisons between groups of mothers studied with different methods should be avoided. Many factors may be responsible for this, including losses due to milk preparation/storage, differing lower limits of detection, differences in the antisera used in immunological assays, and measurement of activity not concentration in microbiological assays. For secretory IgA and lysozyme use of serum IgA and hen egg-white lysozyme, respectively, as standards also contributes to the wide differences in values reported by different investigators. Data have been included in Table III only if details of the methods have been provided by the investigators. The analysis of human milk albumin, globulin, and casein using conventional precipitation assays has been shown to be unsatisfactory and no values obtained with these methods have been included in the table.

3. *Minerals*

In recent years the analysis of most minerals in breast milk has been conducted using atomic absorption spectrometry, or an equivalent method, with the inclusion of standardized reference materials. This facilitates the comparison of results from different laboratories. Any comparisons with values obtained with older, chemical methods must be undertaken with caution. Ashing and/or digestion of milk is a prerequisite for most of these methods. An exception to this is the analysis of phosphorus. The commonly used Fiske–Subbarrow colorimetric method can be performed on undigested samples. However, the results obtained are lower than those obtained on the same sample after digestion. Phosphorus assays in which preparation of the milk sample was not performed or for which no details were given have been identified in Table IV with footnote *g*.

E. Units

Published concentrations are expressed in a variety of forms. For consistency, all data have been expressed as weight (or activity) per unit volume of milk; the factors used to calculate these figures have been given where appropriate. Standard deviations have been given as the measure of variation, the values having been calculated from standard errors where necessary. The number of significant figures quoted in the tables are those used by the authors except in a few cases in which they have been abbreviated to three.

TABLE I
Lactose and Fat

Locality	SE class[a]	Reference	Design[b]	Colostrum (1–5 days)				Transitional (6–14 days)				Mature (0.5–6 months)				Mature (> 6 months)		
				Stage[c]	N	Mean(SD)	Range	Stage[c]	N	Mean(SD)	Range	Stage[c]	N	Mean(SD)	Range	Stage[c]	N	Mean(SD)
Lactose (g/dl)																		
The Americas																		
Brazil																		
Sao Paulo	UH	Carneiro and de Oliveira (1973)	L	1–5 d	3	5.8(1.0)	—	7 d	10	7.5(0.8)	—	28 d	10	6.8(1.2)	—			
								14 d	10	7.0(1.1)	—	56 d	8	6.2(1.0)	—			
												82 d	5	5.8(1.0)	—			
Sao Paulo	UL	Carneiro and de Oliveira (1973)	L	1–5 d	5	6.3(1.8)	—	7 d	9	6.9(0.8)	—	28 d	10	6.5(1.1)	—			
								14 d	9	6.6(0.8)	—	56 d	10	5.2(1.3)	—			
												82 d	8	5.4(1.3)	—			
												110 d	7	5.9(0.4)	—			
												138 d	7	5.6(1.0)	—			
												159 d	7	6.0(0.4)	—			
												186 d	5	5.5(0.6)	—			
Guatemala																		
Guatemala City	UH	WHO (1985)	X									1 m	32	7.70(0.68)	—			
												2 m	30	8.03(0.83)	—			
												3 m	28	8.04(0.49)	—			
Guatemala City	UL	WHO (1985)	X									1 m	27	7.67(1.14)	—	6 m	28	8.21(0.58)
												3 m	30	8.38(0.53)	—	9 m	24	8.37(0.53)
Santa Maria Cauque	RL	WHO (1985)	X									1 m	27	8.09(0.65)	—	6 m	28	8.45(0.57)
												3 m	26	8.29(0.77)	—	9 m	27	8.40(0.60)
																15 m	28	8.30(1.08)
																18 m	21	8.76(0.50)

TABLE I—*continued*

Locality	SE class[a]	Reference	Design[b]	Colostrum (1–5 days)				Transitional (6–14 days)				Mature (0.5–6 months)				Mature (> 6 months)		
				Stage[c]	*N*	Mean(SD)	Range	Stage[c]	*N*	Mean(SD)	Range	Stage[c]	*N*	Mean(SD)	Range	Stage[c]	*N*	Mean(SD)
United States (Navajo Indians)																		
Tuba City	RL	Butte and Calloway (1981)	X									19–62 d	23	6.1(0.6)	4.4–7.3			
Africa and Arabia																		
Ethiopia																		
Addis Ababa	UL	Lonnerdal *et al.* (1976)	X									0.5–1.5 m	3	7.42(0.51)	—	> 6.5 m	45	7.78(0.48)
												1.5–3.5 m	14	7.43(0.48)	—			
												3.5–6.5 m	26	7.49(0.31)	—			
Addis Ababa	UH	Lonnerdal *et al.* (1976)	X									0.5–1.5 m	15	6.60(0.63)	—			
												1.5–3.5 m	5	7.64(0.16)	—			
Ivory Coast																		
Kpouebo	RL	Lauber and Reinhardt (1979)	X									1 m	5	6.22(1.03)	—	12 m	23	7.00(0.97)
												6 m	13	7.15(0.72)	—	18 m	10	6.37(1.09)
Kenya																		
Machakos	RLn	Van Steenbergen *et al.* (1983)	X	3 d	39	6.2(0.4)	—											
Machakos	RLm	Van Steenbergen *et al.* (1983)	X	3 d	36	6.4(0.4)	—											
Nairobi	UM	Bwibo and Ondijo (1981) f	X									3 w–6 m	24	6.4(1.5)	3.3–9.7			

TABLE I—*continued*

Locality	SE class[a]	Reference	Design[b]	Colostrum (1–5 days)				Transitional (6–14 days)				Mature (0.5–6 months)				Mature (> 6 months)		
				Stage[c]	N	Mean(SD)	Range	Stage[c]	N	Mean(SD)	Range	Stage[c]	N	Mean(SD)	Range	Stage[c]	N	Mean(SD)
Nigeria																		
Ibadan	UL	Bassir (1958)	X					1 w	6	5.1(1.2)	2.9–5.6	3–26 w	12	6.3(1.4)	4.3–8.9	27–78 w	15	7.1(1.4)
Ibadan	XX	Naismith (1973)	X													7 m	12	7.67(—)
Sudan																		
Khartoum	XX	El Tom Ali and Zaki (1976)	X	1–3 d	7	4.2(—)	2.5–5.8											
The Gambia																		
Keneba	RL	Prentice *et al.* (1983)	X									0.5–2.9 m	11	7.79(0.27)	—	6.0–8.9 m	9	7.98(0.28)
												3.0–5.9 m	19	7.69(0.35)	—	9.0–11.9 m	10	7.86(0.32)
																12.0–14.9 m	12	8.02(0.28)
																15.0–17.9 m	15	7.91(0.47)
Keneba	RLs	Prentice *et al.* (1983)	X									0.5–2.9 m	13	7.77(0.52)	—	6.0–8.9 m	11	7.63(0.28)
												3.0–5.9 m	15	7.69(0.27)	—	9.0–11.9 m	6	7.82(0.37)
																12.0–14.9 m	13	7.62(0.58)
																15.0–17.9 m	19	7.70(0.33)
Zaire																		
Bakavu	UL	WHO (1985)	X									1 m	6	6.37(—)	—	9 m	6	6.33(—)
												2 m	6	6.56(—)	—	12 m	6	6.71(—)
												3 m	6	6.68(—)	—	18 m	4	6.81(—)
												4 m	6	6.78(—)	—			
												6 m	6	6.30(—)	—			
Kabare	RL	WHO (1985)	X									1 m	7	5.48(—)	—	9 m	7	6.29(—)
												2 m	7	6.02(—)	—	12 m	7	6.31(—)
												3 m	7	5.99(—)	—	15 m	7	6.15(—)
												4 m	7	6.07(—)	—	18 m	5	6.38(—)
												6 m	7	6.25(—)	—			

TABLE I—*continued*

Locality	SE class[a]	Reference	Design[b]	Colostrum (1–5 days)				Transitional (6–14 days)				Mature (0.5–6 months)				Mature (> 6 months)		
				Stage[c]	*N*	Mean(SD)	Range	Stage[c]	*N*	Mean(SD)	Range	Stage[c]	*N*	Mean(SD)	Range	Stage[c]	*N*	Mean(SD)
Kwango	RL	Holemans and Martin (1954); Holemans *et al.* (1954)	X	2–4 d	19	6.21(—)	—	6–9 d	18	6.21(—)	—	1–6 m	53	7.95(—)	—	7–12 m	62	8.21(—)
																13–18 m	40	7.32(—)
																19–24 m	27	7.71(—)
																25–36 m	12	8.23(—)
Yasa-Bonga	RLn	Barclay (1989) g	L					1 w	15	6.6(0.5)	—	1 m	15	6.9(0.6)	—	9 m	10	7.3(0.4)
												2 m	15	7.1(0.9)	—	12 m	10	7.5(0.4)
												4 m	15	7.0(0.8)	—	15 m	10	7.5(0.4)
												6 m	15	7.3(0.4)	—	18 m	10	7.0(1.0)
Yasa-Bonga	RLm	Barclay (1989) g	L					1 w	16	6.2(0.9)	—	1 m	16	7.4(0.8)	—	9 m	8	6.9(0.6)
												2 m	16	6.9(0.7)	—	12 m	8	7.2(0.3)
												4 m	16	7.1(0.6)	—	15 m	8	7.4(0.7)
												6 m	16	6.8(0.4)	—	18 m	8	7.4(0.3)
Asia, Australasia, and Oceania																		
Burma																		
Rangoon	UL	Khin-Muang-Naing *et al.* (1980)	X									1–3 m	29	7.19(0.24)	—	7–12 m	29	7.10(0.36)
												4–6 m	26	7.05(0.24)	—			
Central Pacific																		
Nauru	RL	Bray (1928)	X									8 w	q	7.64(—)	—			

TABLE I—*continued*

Locality	SE class[a]	Reference	Design[b]	Colostrum (1–5 days)				Transitional (6–14 days)				Mature (0.5–6 months)				Mature (> 6 months)		
				Stage[c]	*N*	Mean(SD)	Range	Stage[c]	*N*	Mean(SD)	Range	Stage[c]	*N*	Mean(SD)	Range	Stage[c]	*N*	Mean(SD)
China																		
Shanghai	UX	Huang *et al.*	L	1 d	11	3.7(–)	–	6 d	11	7.3(–)	–	16 d	11	7.5(–)	–			
		(1984) g		2 d	11	4.8(–)	–	7 d	11	7.4(–)	–	31 d	11	7.5(–)	–			
				3 d	11	5.8(–)	–	8 d	11	7.5(–)	–							
				4 d	11	6.6(–)	–											
				5 d	11	7.2(–)	–											
India																		
Baroda	UM	Karmarkar	X									1–3 m	45	7.14(0.40)	–	6–12 m	63	7.11(0.67)
		et al. (1959)										3–6 m	60	7.21(0.72)	–	> 12 m	24	7.24(1.29)
Baroda	UXa	Karmarkar and Ramakrishnan (1960)	X									3–4 m	15	7.20(0.35)	–			
Baroda	UXb	Karmarkar and Ramakrishnan (1960)	X									3–4 m	15	6.95(0.26)	–			
Baroda	UXc	Karmarkar and Ramakrishnan (1960)	X									3–4 m	17	7.18(0.24)	–			
Baroda	UXd	Karmarkar and Ramakrishnan (1960)	X									3–4 m	13	7.31(0.56)	–			
Coonoor	UL	Belavady	X									2–6 m	45	7.47(0.44)	–	7–12 m	29	7.54(0.41)
		(1959)														13–18 m	23	7.54(0.41)
																> 18 m	18	7.51(0.51)

TABLE I—*continued*

Locality	SE class[a]	Reference	Design[b]	Colostrum (1–5 days)				Transitional (6–14 days)				Mature (0.5–6 months)				Mature (> 6 months)		
				Stage[c]	*N*	Mean(SD)	Range	Stage[c]	*N*	Mean(SD)	Range	Stage[c]	*N*	Mean(SD)	Range	Stage[c]	*N*	Mean(SD)
Jaipur	XX	Mehta and Kala (1971)	L	0–2 d	84	6.7(—)	5.5–8.2					15 d	84	7.5(—)	6.4–10.0			
New Delhi	UH	Ashdir and Puri (1962)	L	3 d	10	6.33(0.45)	5.61–7.18	8 d	10	6.56(0.68)	5.59–7.78	18 d	10	7.03(0.71)	5.71–7.86			
Patna	UM	Sinha *et al.* (1959)	X									3 w–2 m	18	6.91(—)	—	8–9 m	12	6.99(—)
												3–4 m	16	7.05(—)	—			
Indonesia																		
Yogyakarta	UL	Boediman *et al.* (1979)	X													13–24 m	66	6.93(0.12)
																25–36 m	45	6.93(0.10)
Korea																		
Seoul	UH	Lee (1987)	L									15 d	16	6.52(0.49)	5.34–7.24			
												30 d	16	6.87(0.33)	6.16–7.40			
												60 d	12	7.01(0.59)	6.41–8.46			
												90 d	13	6.80(0.41)	5.72–7.40			
												120 d	12	6.86(0.67)	5.11–7.56			
												150 d	7	6.91(0.64)	6.04–7.52			
Pakistan																		
Karachi	UL	Lindblad and Rahimtoola (1974)	X									1.5–6 m	9	6.20(0.98)	4.51–7.31			
Papua New Guinea																		
Baiyer River	RL	Bailey (1965)	X													6–24 m	14	7.48(0.37)
Kalabu	RL	Bailey (1965)	X													6–12 m	19	7.78(0.44)

TABLE I—*continued*

Locality	SE class[a]	Reference	Design[b]	Colostrum (1–5 days)				Transitional (6–14 days)				Mature (0.5–6 months)				Mature (> 6 months)		
				Stage[c]	N	Mean(SD)	Range	Stage[c]	N	Mean(SD)	Range	Stage[c]	N	Mean(SD)	Range	Stage[c]	N	Mean(SD)
Wosera	RL	Bailey (1965)	X													6–12 m	16	7.59(0.44)
Philippines																		
Luzon	RL	WHO (1985)	X									1 m	27	6.02(0.54)	—	9 m	32	6.46(0.76)
												2 m	30	6.26(1.10)	—	12 m	26	6.56(0.66)
												3 m	28	6.56(0.68)	—	15 m	22	6.68(0.55)
												4 m	23	6.37(0.64)	—	18 m	17	6.60(0.76)
												6 m	29	6.22(0.60)	—			
Manila	UL	WHO (1985)	X									1 m	32	6.06(0.48)	—	9 m	31	6.46(0.67)
												2 m	23	5.83(0.79)	—	12 m	29	6.45(0.55)
												3 m	31	6.29(0.62)	—	15 m	18	6.71(0.68)
												4 m	27	6.39(0.74)	—			
												6 m	30	6.23(0.53)	—			
Manila	UH	WHO (1985)	X									1 m	34	5.96(1.17)	—	9 m	16	6.11(0.78)
												2 m	25	6.07(0.59)	—			
												3 m	20	5.86(1.03)	—			
												4 m	10	6.35(0.91)	—			
												6 m	16	6.25(0.55)	—			
Sri Lanka (Ceylon)																		
Colombo	XX	de Silva (1964)	X					7 d	36	6.65(—)	—	2 m	61	6.83(—)	—			
Taiwan																		
Taipei	UX	Lonnerdal *et* al. (1990)	L									1 m	q	7.0(—)	—			
												2 m	q	7.0(—)	—			
												3 m	q	7.0(—)	—			
												4 m	q	7.2(—)	—			

TABLE I—*continued*

Locality	SE class[a]	Reference	Design[b]	Colostrum (1–5 days)				Transitional (6–14 days)				Mature (0.5–6 months)				Mature (> 6 months)		
				Stage[c]	*N*	Mean(SD)	Range	Stage[c]	*N*	Mean(SD)	Range	Stage[c]	*N*	Mean(SD)	Range	Stage[c]	*N*	Mean(SD)
Vanatu (New Hebrides)																		
Port Vila	RL	Peters (1953)	X									2–5 m	18	5.05(0.47)	4.2–5.8	6–11 m	15	5.00(0.50)
																12–24 m	18	4.91(0.47)
Fat (24 hr values) (g/dl)																		
The Americas																		
Africa and Arabia																		
The Gambia																		
Keneba	RL	Prentice *et al.* (1981, (1983) f	X									0.5–2.9 m	11	3.91(0.83)	—	6.0–8.9 m	9	3.58(1.34)
												3.0–5.9 m	19	3.51(0.81)	—	9.0–11.9 m	10	3.48(0.82)
																12.0–14.9 m	12	3.58(0.72)
																15.0–17.9 m	15	3.52(0.95)
Keneba	RLs	Prentice *et al.* (1981, 1983) f	X									0.5–2.9 m	14	3.96(0.89)	—	6.0–8.9 m	11	3.84(0.85)
												3.0–5.9 m	15	3.73(0.67)	—	9.0–11.9 m	7	3.40(0.97)
																12.0–14.9 m	13	3.53(0.59)
																15.0–17.9 m	20	3.67(0.98)
Asia, Australasia, and Oceania																		
Burma																		
Rangoon	UL	Khin-Muang-Naing *et al.* (1980)	X									1–3 m	29	3.18(1.03)	—	7–12 m	29	3.55(1.05)
												4–6 m	27	3.47(1.49)	—			

TABLE I—*continued*

Locality	SE class[a]	Reference	Design[b]	Colostrum (1–5 days)				Transitional (6–14 days)				Mature (0.5–6 months)				Mature (> 6 months)		
				Stage[c]	*N*	Mean(SD)	Range	Stage[c]	*N*	Mean(SD)	Range	Stage[c]	*N*	Mean(SD)	Range	Stage[c]	*N*	Mean(SD)
Pakistan																		
Karachi	UL	Lindblad and Rahimtoola (1974)	X									1.5–6 m	9	2.73(1.21)	1.70–5.72			
Papua New Guinea																		
Biak Island	RL	Jansen *et al.* (1960)	X									2–5 m	3	3.4(—)	2.3–4.8	6–12 m	7	2.5(0.5)
																12–24 m	17	2.7(0.9)
Thailand																		
Chiang Mai	RL	Jackson *et al.* (1988)	X									3–4 w	6	3.57(0.79)	—	7–9 m	5	2.91(0.64)
												2–3 m	8	3.71(0.73)	—			

Note. f, figures calculated from information in text; g, figures taken from graphs; p, pooled sample; q, numbers of subjects not given; s, supplemented mothers; z, geometric mean (+1 geometric standard deviation).

[a]Socioeconomic class: RL, rural, poor; UH, urban, middle-high income; UL, urban, poor, and low-middle income; UM, urban, mixed socioeconomic class; UX, urban, socioeconomic class not stated; a, lowest to d, highest; quartiles of maternal intake; *m*, malnourished and undernourished mothers; *n*, good maternal nutritional status.

[b]Study design: L, longitudinal or semilongitudinal design; X, cross-sectional design.

[c]Stage of lactation: d, days postpartum; w, weeks postpartum; m, months postpartum.

TABLE II
Nitrogen and Protein

Locality	SE class[a]	Reference	Design[b]	Colostrum (1–5 days)				Transitional (6–14 days)				Mature (0.5–6 months)				Mature (> 6 months)		
				Stage[c]	*N*	Mean(SD)	Range	Stage[c]	*N*	Mean(SD)	Range	Stage[c]	*N*	Mean(SD)	Range	Stage[c]	*N*	Mean(SD)
Total Nitrogen (mg/dl) (Commonly used factors to convert nitrogen to protein: ×6.25 and ×6.38)																		
The Americas																		
Brazil																		
Brasilia	UH	Dorea *et al.* (1984) g	L									15 d	3	230(16)	—			
												30 d	3	200(17)	—			
												45 d	3	200(—)	—			
												60 d	3	160(36)	—			
												90 d	4	160(34)	—			
Brasilia	UL	Dorea *et al.* (1984) g	L									15 d	8	220(40)	—			
												30 d	7	190(32)	—			
												45 d	7	140(56)	—			
												60 d	6	180(22)	—			
												75 d	6	180(27)	—			
												90 d	6	210(71)	—			
												120 d	4	210(80)	—			
Guatemala																		
Guatemala City	UH	WHO (1985)	X									1 m	32	238(45)	—			
												2 m	30	201(38)	—			
												3 m	28	197(32)	—			
Guatemala City	UL	WHO (1985)	X									1 m	28	218(30)	—	6 m	28	182(24)
												3 m	30	185(35)	—	9 m	25	162(28)
Santa Maria Cauque	RL	WHO (1985)	X									1 m	27	195(26)	—	6 m	28	164(27)
												3 m	26	184(42)	—	9 m	27	162(21)
																15 m	28	165(43)
																18 m	21	153(23)

TABLE II—*continued*

Locality	SE class[a]	Reference	Design[b]	Colostrum (1–5 days)				Transitional (6–14 days)				Mature (0.5–6 months)				Mature (> 6 months)		
				Stage[c]	*N*	Mean(SD)	Range	Stage[c]	*N*	Mean(SD)	Range	Stage[c]	*N*	Mean(SD)	Range	Stage[c]	*N*	Mean(SD)
United States (Navajo Indians)																		
Tuba City	RL	Butte and Calloway (1981) f	X									19–62 d	23	254(—)	—			
Africa and Arabia																		
Ethiopia																		
Addis Ababa	UL	Lonnerdal *et al.* (1976)	X									0.5–1.5 m	3	250(17)	—	> 6.5 m	45	174(30)
												1.5–3.5 m	14	177(33)	—			
												3.5–6.5	26	169(30)	—			
	UH	Lonnerdal *et al.* (1976)	X									0.5–1.5 m	15	289(59)	—			
												1.5–3.5 m	5	197(29)	—			
Kenya																		
Machakos	RL	Van Steen-bergen *et al.* (1981) f	X	3 d	78	168(33)	—											
Machakos	RLn	Van Steen-bergen *et al.* (1983) f	X	3 d	39	188(31)	—											
Machakos	RLm	Van Steen-bergen *et al.* (1983) f	X	3 d	36	188(31)	—											
Sudan																		
Khartoum	XX	El Tom Ali and Zaki (1976)	X	1–3 d	7	533(—)	344–721											

TABLE II—*continued*

Locality	SE class[a]	Reference	Design[b]	Colostrum (1–5 days)				Transitional (6–14 days)				Mature (0.5–6 months)				Mature (> 6 months)		
				Stage[c]	*N*	Mean(SD)	Range	Stage[c]	*N*	Mean(SD)	Range	Stage[c]	*N*	Mean(SD)	Range	Stage[c]	*N*	Mean(SD)
The Gambia																		
Keneba	RL	Prentice *et al.* (1983)	X									0.5–2.9 m	11	201(31)	—	6.0–8.9 m	9	156(29)
												3.0–5.9 m	19	170(31)	—	9.0–11.9 m	10	157(17)
																12.0–14.9 m	11	159(21)
																15.0–17.9 m	14	161(43)
Keneba	RLs	Prentice *et al.* (1983)	X									0.5–2.9 m	14	210(63)	—	6.0–8.9 m	11	174(28)
												3.0–5.9 m	15	170(22)	—	9.0–11.9 m	7	167(29)
																12.0–14.9 m	13	166(16)
																15.0–17.9 m	18	181(35)
Keneba	RL	Kunz *et al.* (1990) f	X									0.5–2.9 m	8	151(—)	—	6.0–8.9 m	16	113(—)
												3.0–5.9 m	13	127(—)	—	9.0–11.9 m	11	113(—)
																12.0–14.9 m	11	110(—)
																15.0–17.9 m	19	112(—)
																18.0–20.9 m	18	120(—)
Zaire																		
Bakavu	UL	WHO (1985)	X									1 m	6	215(—)	—	9 m	6	197(—)
												2 m	6	197(—)	—	12 m	6	217(—)
												3 m	6	188(—)	—	18 m	4	175(—)
												4 m	6	223(—)	—			
												6 m	6	179(—)	—			
Kabare	RL	WHO (1985)	X									1 m	7	274(—)	—	9 m	7	217(—)
												2 m	7	246(—)	—	12 m	7	223(—)
												3 m	7	245(—)	—	15 m	7	254(—)
												4 m	7	212(—)	—	18 m	5	254(—)
												6 m	7	244(—)	—			

TABLE II—*continued*

Locality	SE class[a]	Reference	Design[b]	Colostrum (1–5 days)				Transitional (6–14 days)				Mature (0.5–6 months)				Mature (> 6 months)		
				Stage[c]	*N*	Mean(SD)	Range	Stage[c]	*N*	Mean(SD)	Range	Stage[c]	*N*	Mean(SD)	Range	Stage[c]	*N*	Mean(SD)
Kivu	R/U	Close *et al.* (1957)	X	1 or 2 d	15	493(271)	207–1021	6–12 d	35	292(59)	172–465	21–30 d	10	227(46)	146–308	181–210 d	33	166(29)
				3 d	20	313(42)	210–401					31–60 d	39	207(53)	108–346	211–240 d	29	165(28)
				4 or 5 d	19	302(53)	225–440					61–90 d	27	192(41)	138–333	241–270 d	22	180(33)
												91–120 d	37	175(35)	107–268	271–300 d	19	171(26)
												121–150 d	34	173(41)	93–266	301–330 d	15	160(21)
												151–180 d	30	160(35)	104–265	331–360 d	11	164(34)
																> 361 d	17	171(39)
Kwango	RL	Holemans and Martin (1954); Holemans *et al.* (1954)	X	2–4 d	19	273(–)	–	6–9 d	18	254(–)	–	1–6 m	53	201(–)	–	7–12 m	62	176(–)
																13–18 m	40	157(–)
																19–24 m	27	173(–)
																25–36 m	12	170(–)
Yasa-Bonga	RLn	Barclay (1989) g	L					1 w	15	251(–)	–	1 m	15	197(30)	–	9 m	10	144(20)
												2 m	15	171(–)	–	12 m	10	142(–)
												4 m	15	165(30)	–	15 m	10	152(20)
												6 m	15	149(30)	–	18 m	10	184(90)
Yasa-Bonga	RLm	Barclay (1989) g	L					1 w	16	256(–)	–	1 m	16	174(20)	–	9 m	8	133(20)
												2 m	16	184(–)	–	12 m	8	147(–)
												4 m	16	152(30)	–	15 m	8	144(20)
												6 m	16	144(20)	–	18 m	8	144(15)
Asia, Australasia, and Oceania																		
Burma																		
Rangoon	UL	Khin-Muang-Naing *et al.* (1980)	X									1–3 m	30	186(36)	–	7–12 m	29	166(27)
												4–6 m	26	180(39)				

TABLE II—*continued*

Locality	SE class[a]	Reference	Design[b]	Colostrum (1–5 days)				Transitional (6–14 days)				Mature (0.5–6 months)				Mature (> 6 months)		
				Stage[c]	*N*	Mean(SD)	Range	Stage[c]	*N*	Mean(SD)	Range	Stage[c]	*N*	Mean(SD)	Range	Stage[c]	*N*	Mean(SD)
China																		
Shanghai	UX	Huang *et al.* (1984) g	L	1 d	11	1380(–)	–	6 d	11	220(–)	–	16 d	11	175(–)	–			
				2 d	11	925(–)	–	7 d	11	205(–)	–	31 d	11	235(–)	–			
				3 d	11	520(–)	–	8 d	11	190(–)	–							
				4 d	11	315(–)	–											
				5 d	11	250(–)	–											
India																		
Baroda	UM	Karmarkar *et al.* (1959)	X									1–3 m	45	204(41)	–	6–12 m	63	190(38)
												3–6 m	60	190(44)	–	> 12 m	24	186(33)
Baroda	UXa	Karmarkar and Rama-krishnan (1960)	X									3–4 m	15	191(31)	–			
Baroda	UXb	Karmarkar and Rama-krishnan (1960)	X									3–4 m	18	218(40)	–			
Baroda	UXc	Karmarkar and Rama-krishnan (1960)	X									3–4 m	12	221(51)	–			
Baroda	UXd	Karmarkar and Rama-krishnan (1960)	X									3–4 m	15	231(47)	–			
Coonoor	UL	Belavady (1959)	X									2–6 m	14	168(39)	–	7–12 m	23	161(23)
																> 12 m	13	159(26)

TABLE II—*continued*

Locality	SE class[a]	Reference	Design[b]	Colostrum (1–5 days)				Transitional (6–14 days)				Mature (0.5–6 months)				Mature (> 6 months)		
				Stage[c]	N	Mean(SD)	Range	Stage[c]	N	Mean(SD)	Range	Stage[c]	N	Mean(SD)	Range	Stage[c]	N	Mean(SD)
Coonoor	UL	Belavady and Gopalan (1959)	X	1 or 2 d	18	1310(610)	—					2–6 m	45	166(38)	—	7–12 m	35	158(33)
																13–18 m	36	165(36)
																> 18 m	30	177(61)
Hyderabad	UX	Mohan *et* (1983)	L	1–5 d	19	310(40)	—	6–10 d	19	259(70)	—	16–20 d	19	243(45)				
								11–15 d	19	252(24)	—							
New Delhi	UH	Ashdir and Puti (1962)	L	3 d	10	718(130)	519–922	8 d	10	373(85)	165–596	18 d	10	255(63)	157–335			
New Delhi	XX	Khurana *et al.* (1970)	X	0–5 d	28	388(82)	132–678	6–15 d	30	304(96)	130–544	16–30 d	19	277(106)	151–576	7–9 m	23	143(64)
												1–3 m	45	233(123)	54–507	> 9 m	14	133(70)
												4–6 m	35	183(136)	69–419			
Patna	UM	Sinha *et al.* (1959)	X									3 w–2 m	18	199(—)	—	8 or 9 m	12	187(—)
												3–4 m	16	190(—)	—			
Varanasi	XX	Agarwal *et al.* (1975)	X	0–5 d	21	342(157)	142–659	6–15 d	17	240(110)	130–466							
Indonesia																		
Yogyakarta	UL	Boediman *et al.* (1979)	X													13–24 m	66	199(71)
																25–36 m	45	219(87)
Japan																		
Tokyo	UX	Nagasawa *et al.* (1972)	X	2–5 d	14	345(49)	—	6–10 d	15	321(52)	—							
Tokyo	UX	Nagasawa *et al.* (1973)	X									2 m	26	184(21)	—			
												3 m	8	165(23)	—			
												4 m	6	164(20)	—			

TABLE II—*continued*

Locality	SE class[a]	Reference	Design[b]	Colostrum (1–5 days)				Transitional (6–14 days)				Mature (0.5–6 months)				Mature (> 6 months)		
				Stage[c]	N	Mean(SD)	Range	Stage[c]	N	Mean(SD)	Range	Stage[c]	N	Mean(SD)	Range	Stage[c]	N	Mean(SD)
Korea																		
Seoul	UH	Lee (1987)	L									15 d	16	230(31)				
												30 d	16	199(41)				
												60 d	12	185(19)				
												90 d	13	172(41)				
												120 d	12	182(22)				
												150 d	9	176(39)				
Pakistan																		
Karachi	UL	Lindblad and Rahimtoola (1974)	X									1.5–6 m	9	185(31)	—			
Lahore	UL	Underwood *et al.* (1970)	L									6 w	133	226(36)	—	9 m	67	191(25)
												6 m	103	197(30)	—	12 m	37	187(23)
																18 m	27	213(73)
																24 m	7	205(29)
Papua New Guinea																		
Baiyer River	RL	Bailey (1965)	X													6–24 m	14	132(46)
Biak Island	RL	Jansen *et al.* (1960)	X									2–5 m	3	130(—)	110–160	6–12 m	7	140(10)
																12–24 m	17	130(20)
Kalabu	RL	Bailey (1965)	X													6–12 m	19	136(46)
Wosera	RL	Bailey (1965)	X													6–12 m	16	130(18)

TABLE II—*continued*

Locality	SE class[a]	Reference	Design[b]	Colostrum (1–5 days)				Transitional (6–14 days)				Mature (0.5–6 months)				Mature (> 6 months)		
				Stage[c]	*N*	Mean(SD)	Range	Stage[c]	*N*	Mean(SD)	Range	Stage[c]	*N*	Mean(SD)	Range	Stage[c]	*N*	Mean(SD)
Philippines																		
Luzon	RL	WHO (1985)	X									1 m	27	175(35)	—	9 m	32	143(24)
												2 m	30	156(25)	—	12 m	26	145(26)
												3 m	28	147(29)	—	15 m	22	145(24)
												4 m	23	152(27)	—	18 m	17	160(37)
												6 m	29	148(26)	—			
Manila	UL	WHO (1985)	X									1 m	32	191(32)	—	9 m	31	146(20)
												2 m	23	157(37)	—	12 m	29	149(26)
												3 m	31	159(29)	—	15 m	18	162(20)
												4 m	27	163(36)	—			
												6 m	30	156(27)	—			
Manila	UH	WHO (1985)	X									1 m	34	216(76)	—	9 m	16	169(74)
												2 m	25	184(53)	—			
												3 m	20	175(32)	—			
												4 m	10	163(37)	—			
												6 m	16	156(34)	—			
Taiwan																		
Taipei	UX	Lonnerdal *et al.* (1990) f	L									1 m	q	225(—)	—			
												2 m	q	187(—)	—			
												3 m	q	180(—)	—			
												4 m	q	172(—)	—			
Vanatu (New Hebrides)																		
Port Vila	RL	Peters (1953)	X									2–5 m	18	234(92)	170–268	6–11 m	15	218(85)
												6 m	103	197(30)	—	12–24 m	18	208(74)
																12 m	37	187(29)

TABLE II—*continued*

Locality	SE class[a]	Reference	Design[b]	Colostrum (1–5 days)				Transitional (6–14 days)				Mature (0.5–6 months)				Mature (> 6 months)		
				Stage[c]	*N*	Mean(SD)	Range	Stage[c]	*N*	Mean(SD)	Range	Stage[c]	*N*	Mean(SD)	Range	Stage[c]	*N*	Mean(SD)
Protein Nitrogen (mg/dl)																		
(Commonly used factors to convert N protein: ×6.25 and ×6.38)																		
The Americas																		
Argentina																		
Buenos Aires	UH	Ronayne de Ferrer *et al.* (1984)	X									20–30 d	11	190(43)				
Guatemala																		
Guatemala City	UH	WHO (1985)	X									1 m	32	182(43)	—			
												2 m	30	148(37)	—			
												3 m	28	152(30)	—			
Guatemala City	UL	WHO (1985)	X									1 m	28	167(28)	—	6 m	28	140(24)
												3 m	30	136(33)	—	9 m	25	121(27)
Santa Maria Cauque	RL	WHO (1985)	X									1 m	27	142(26)	—	6 m	28	124(27)
												3 m	26	135(42)	—	9 m	27	116(34)
																15 m	28	124(43)
United States (Navajo Indians)																18 m	21	116(27)
Tuba City	RL	Butte and Calloway (1981)	X									19–62 d	23	224(48)	160–368			
Africa and Arabia																		
Ethiopia																		
Addis Ababa	UL	Lonnerdal *et al.* (1976) f	X									0.5–1.5 m	3	207(—)	—	> 6.5 m	45	141(—)
												1.5–3.5 m	14	141(—)	—			
												3.5–6.5 m	26	135(—)	—			

TABLE II—continued

Locality	SE class[a]	Reference	Design[b]	Colostrum (1–5 days)				Transitional (6–14 days)				Mature (0.5–6 months)				Mature (> 6 months)		
				Stage[c]	*N*	Mean(SD)	Range	Stage[c]	*N*	Mean(SD)	Range	Stage[c]	*N*	Mean(SD)	Range	Stage[c]	*N*	Mean(SD)
Addis Ababa	UH	Lonnerdal *et al.* (1976) f	X									0.5–1.5 m	15	243(—)	—			
												1.5–3.5 m	5	156(—)	—			
Kenya																		
Machakos	RL	Van Steenbergen *et al.* (1981)	X	3 d	80	135(28)	—											
Nigeria																		
Ibadan	UL	Bassir (1958)	X					1 w	6	303(74)	207–396	3–26 w	12	192(88)	69–427	> 27 w	17	204(116)
Ibadan	XX	Naismith (1973)	X													7 m	12	192(—)
The Gambia																		
Keneba	RL	Kunz *et al.* (1990)	X									0.5–2.9 m	8	123(16)	—	6.0–8.9 m	16	83(18)
												3.0–5.9 m	13	98(26)	—	9.0–11.9 m	11	85(10)
																12.0–14.9 m	11	82(19)
																15.0–17.9 m	19	83(22)
																18.0–20.9 m	8	91(18)
Zaire																		
Bakavu	UL	WHO (1985) f	X									1 m	6	170(—)	—	9 m	6	165(—)
												2 m	6	156(—)	—	12 m	6	183(—)
												3 m	6	153(—)	—	18 m	4	146(—)
												4 m	6	186(—)	—			
												6 m	6	145(—)	—			
Kabare	RL	WHO (1985) f	X									1 m	7	220(—)	—	9 m	7	179(—)
												2 m	7	196(—)	—	12 m	7	186(—)
												3 m	7	202(—)	—	15 m	7	220(—)
												4 m	7	170(—)	—	18 m	5	215(—)
												6 m	7	200(—)	—			

TABLE II—*continued*

Locality	SE class[a]	Reference	Design[b]	Colostrum (1–5 days)				Transitional (6–14 days)				Mature (0.5–6 months)				Mature (> 6 months)		
				Stage[c]	*N*	Mean(SD)	Range	Stage[c]	*N*	Mean(SD)	Range	Stage[c]	*N*	Mean(SD)	Range	Stage[c]	*N*	Mean(SD)
Yasa-Bonga	RLn	Barclay (1989) g	L					1 w	15	200(30)	—	1 m	15	160(30)	—	9 m	10	120(10)
												2 m	15	140(20)	—	12 m	10	120(20)
												4 m	15	130(30)	—	15 m	10	120(—)
												6 m	15	120(—)	—	18 m	10	160(80)
Yasa-Bonga	RLm	Barclay (1989) g	L					1 w	16	210(80)	—	1 m	16	140(20)	—	9 m	8	110(20)
												2 m	16	150(80)	—	12 m	8	125(20)
												4 m	16	125(30)	—	15 m	8	120(—)
												6 m	16	120(—)	—	18 m	8	125(20)
Asia, Australasia, and Oceania																		
India																		
Coonoor	UL	Belavady (1959)	X									2–6 m	14	137(41)	—	7–12 m	23	128(22)
																> 12 m	13	131(25)
Japan																		
Tokyo	UX	Nagasawa *et al.* (1973)	X									2 m	26	140(—)	—			
												3 m	8	132(—)	—			
												4 m	6	120(—)	—			
Philippines																		
Luzon	RL	WHO (1985) f	X									1 m	27	134(35)	—	9 m	32	110(23)
												2 m	30	121(24)	—	12 m	26	118(28)
												3 m	28	113(31)	—	15 m	22	114(25)
												4 m	23	124(27)	—	18 m	17	134(36)
												6 m	29	118(28)	—			
Manila	UL	WHO (1985) f	X									1 m	32	144(29)	—	9 m	31	114(19)
												2 m	23	126(34)	—	12 m	29	125(25)
												3 m	31	123(29)	—	15 m	18	135(22)
												4 m	27	134(36)	—			
												6 m	30	127(26)	—			

TABLE II—*continued*

Locality	SE class[a]	Reference	Design[b]	Colostrum (1–5 days)				Transitional (6–14 days)				Mature (0.5–6 months)				Mature (> 6 months)		
				Stage[c]	N	Mean(SD)	Range	Stage[c]	N	Mean(SD)	Range	Stage[c]	N	Mean(SD)	Range	Stage[c]	N	Mean(SD)
Manila	UH	WHO (1985) f	X									1 m	34	168(73)	—	9 m	16	141(70)
												2 m	25	150(53)	—			
												3 m	20	141(30)	—			
												4 m	10	136(35)	—			
												6 m	16	121(36)	—			
Taiwan																		
Taipei	UX	Lonnerdal *et al.* (1990)	L									1 m	q	181(—)	—			
												2 m	q	150(—)	—			
												3 m	q	146(—)	—			
												4 m	q	141(—)	—			
Nonprotein nitrogen (mg/dl)																		
The Americas																		
Guatemala																		
Guatemala City	UH	WHO (1985)	X									1 m	32	55.9(14.5)	—			
												2 m	30	52.7(17.6)	—			
												3 m	28	45.3(7.5)	—			
Guatemala City	UL	WHO (1985)	X									1 m	28	50.4(11.1)	—	6 m	28	41.8(8.4)
												3 m	30	48.5(12.7)	—	9 m	25	41.7(11.6)
Santa Maria Cauque	RL	WHO (1985)	X									1 m	27	52.8(13.2)	—	6 m	28	40.2(8.9)
												3 m	26	48.7(12.6)	—	9 m	27	45.6(18.6)
																15 m	28	40.8(14.5)
																18 m	21	37.1(10.0)
United States (Navajo Indians)																		
Tuba City	RL	Butte and Calloway (1981)	X									19–62 d	23	29(10)	20–40			

TABLE II—*continued*

Locality	SE class[a]	Reference	Design[b]	Colostrum (1–5 days)				Transitional (6–14 days)				Mature (0.5–6 months)				Mature (> 6 months)		
				Stage[c]	*N*	Mean(SD)	Range	Stage[c]	*N*	Mean(SD)	Range	Stage[c]	*N*	Mean(SD)	Range	Stage[c]	*N*	Mean(SD)
Africa and Arabia																		
Ethiopia																		
Addis Ababa	UL	Lonnerdal *et al.* (1976)	X									0.5–1.5 m	3	43(5)	—	> 6.5 m	45	33(5)
												1.5–3.5 m	14	36(5)	—			
												3.5–6.5 m	26	34(5)	—			
Addis Ababa	UH	Lonnerdal *et al.* (1976)	X									0.5–1.5 m	15	46(8)	—			
												1.5–3.5 m	5	41(5)	—			
Kenya																		
Machakos	RL	Van Steen-bergen *et al.* (1981) f	X	3 d	78	33(—)	—											
The Gambia																		
Keneba	RL	Kunz *et al.* (1990)	X									0.5–2.9 m	8	28.1(4.4)	—	6.0–8.9 m	16	29.9(5.1)
												3.0–5.9 m	13	29.7(4.4)	—	9.0–11.9 m	11	28.2(3.7)
																12.0–14.9 m	11	28.7(3.9)
																15.0–17.9 m	19	28.6(5.0)
																18.0–20.9 m	8	28.6(5.0)
																21.0–23.9 m	3	29.2(0.9)
Zaire																		
Bakavu	UL	WHO (1985)	X									1 m	6	44.2(—)	—	9 m	6	32.7(—)
												2 m	6	40.4(—)	—	12 m	6	34.1(—)
												3 m	6	34.7(—)	—	18 m	4	29.8(—)
												4 m	6	37.0(—)	—			
												6 m	6	34.0(—)	—			

TABLE II—continued

Locality	SE class[a]	Reference	Design[b]	Colostrum (1–5 days)				Transitional (6–14 days)				Mature (0.5–6 months)				Mature (> 6 months)		
				Stage[c]	N	Mean(SD)	Range	Stage[c]	N	Mean(SD)	Range	Stage[c]	N	Mean(SD)	Range	Stage[c]	N	Mean(SD)
Kabare	RL	WHO (1985)	X									1 m	7	54.3(—)	—	9 m	7	38.1(—)
												2 m	7	50.1(—)	—	12 m	7	36.8(—)
												3 m	7	43.8(—)	—	15 m	7	34.2(—)
												4 m	7	41.6(—)	—	18 m	5	38.8(—)
												6 m	7	44.2(—)	—			
Yasa-Bonga	RLn	Barclay (1989) f,g	L					1 w	15	50(—)	—	1 m	15	37(—)	—	9 m	10	32(—)
												2 m	15	37(—)	—	12 m	10	28(—)
												4 m	15	35(—)	—	15 m	10	30(—)
												6 m	15	32(—)	—	18 m	10	28(—)
Yasa-Bonga	RLm	Barclay (1989) f,g	L					1 w	16	50(—)	—	1 m	16	37(—)	—	9 m	8	25(—)
												2 m	16	37(—)	—	12 m	8	28(—)
												4 m	16	35(—)	—	15 m	8	25(—)
												6 m	16	28(—)	—	18 m	8	28(—)
Asia, Australasia, and Oceania																		
India																		
Coonoor	UL	Belavady (1959)	X									2–6 m	14	31.5(6.4)	—	7–12 m	23	33.8(6.9)
																> 12 m	13	27.8(5.5)
Japan																		
Tokyo	UX	Nagasawa *et al.* (1973)	X									2 m	26	43.5(7.3)	—			
												3 m	8	33.4(15.1)	—			
												4 m	6	43.7(4.5)	—			

TABLE II—*continued*

Locality	SE class[a]	Reference	Design[b]	Colostrum (1–5 days)				Transitional (6–14 days)				Mature (0.5–6 months)				Mature (> 6 months)		
				Stage[c]	*N*	Mean(SD)	Range	Stage[c]	*N*	Mean(SD)	Range	Stage[c]	*N*	Mean(SD)	Range	Stage[c]	*N*	Mean(SD)
Philippines																		
Luzon	RL	WHO (1985)	X									1 m	27	41.0(10.9)	—	9 m	32	33.2(9.5)
												2 m	30	34.2(10.8)	—	12 m	26	26.2(4.9)
												3 m	28	33.3(11.7)	—	15 m	22	30.9(13.0)
												4 m	23	27.9(6.9)	—	18 m	17	25.3(6.4)
												6 m	29	29.9(6.8)	—			
Manila	UL	WHO (1985)	X									1 m	32	46.6(10.1)	—	9 m	31	31.9(7.6)
												2 m	23	31.3(8.0)	—	12 m	29	24.1(6.9)
												3 m	31	35.5(9.4)	—	15 m	18	26.8(8.3)
												4 m	27	28.7(8.9)	—			
												6 m	30	29.6(7.2)	—			
Manila	UH	WHO (1985)	X									1 m	34	48.1(11.1)	—	9 m	16	27.9(7.4)
												2 m	25	34.0(10.5)	—			
												3 m	20	34.1(6.5)	—			
												4 m	10	26.7(8.0)	—			
												6 m	16	35.2(14.3)	—			
Taiwan																		
Taipei	UX	Lonnerdal *et al.* (1990) f	L									1 m	q	44(–)	—			
												2 m	q	37(–)	—			
												3 m	q	34(–)	—			
												4 m	q	31(–)	—			

TABLE II—*continued*

Locality	SE class[a]	Reference	Design[b]	Colostrum (1–5 days)				Transitional (6–14 days)				Mature (0.5–6 months)				Mature (> 6 months)		
				Stage[c]	*N*	Mean(SD)	Range	Stage[c]	*N*	Mean(SD)	Range	Stage[c]	*N*	Mean(SD)	Range	Stage[c]	*N*	Mean(SD)
Amino nitrogen—Total (mg/dl)																		
The Americas																		
Africa and Arabia																		
Ethiopia																		
Addis Ababa	UM	Svanberg *et al.* (1977)	X									1–5 m	16	142(—)	—			
Zaire																		
Kivu	R/U	Close and Van de Walle (1957)	X	1–5 d	p	509(—)	—	6–12 d	p	309(—)	—	31–90 d	p	217(—)	—	211–270 d	p	179(—)
												91–150 d	p	174(—)	—	271–330 d	p	174(—)
																331–390 d	p	206(—)
																391–450 d	p	191(—)
Asia, Australasia, and Oceania																		
Amino nitrogen—Free (mg/dl)																		
The Americas																		
Africa and Arabia																		
Asia, Australasia, and Oceania																		
India																		
Coonoor	UL	Belavady (1959)	X									2–6 m	12	4.1(1.5)	—	7–12 m	23	4.2(1.4)
																> 12 m	13	3.1(1.3)
Varanasi	XX	Agarwal *et al.* (1975)	X	0–5 d	21	2.80(1.03)	1.04–5.46	6–15 d	17	3.50(1.12)	1.08–8.19							

TABLE II—*continued*

Locality	SE class[a]	Reference	Design[b]	Colostrum (1–5 days)				Transitional (6–14 days)				Mature (0.5–6 months)				Mature (> 6 months)		
				Stage[c]	*N*	Mean(SD)	Range	Stage[c]	*N*	Mean(SD)	Range	Stage[c]	*N*	Mean(SD)	Range	Stage[c]	*N*	Mean(SD)
Creatine nitrogen (mg/dl)																		
The Americas																		
Africa and Arabia																		
Asia, Australasia, and Oceania																		
India																		
Coonoor	UL	Belavady (1959)	X									2–6 m	8	4.5(2.3)	—	7–12 m > 12 m	16 10	3.7(1.9) 2.2(1.3)
Creatinine nitrogen (mg/dl)																		
The Americas																		
Africa and Arabia																		
Asia, Australasia, and Oceania																		
India																		
Coonoor	UL	Belavady (1959)	X									2–6 m	12	2.4(0.7)	—	7–12 m > 12 m	22 13	2.4(1.0) 1.9(0.7)
Glucosamine nitrogen (mg/dl)																		
The Americas																		

TABLE II—*continued*

Locality	SE class[a]	Reference	Design[b]	Colostrum (1–5 days)				Transitional (6–14 days)				Mature (0.5–6 months)				Mature (> 6 months)		
				Stage[c]	*N*	Mean(SD)	Range	Stage[c]	*N*	Mean(SD)	Range	Stage[c]	*N*	Mean(SD)	Range	Stage[c]	*N*	Mean(SD)
Africa and Arabia																		
Ethiopia																		
Addis Ababa	UM	Svanberg *et al.* (1977)	X									1–5 m	16	4(–)	–			
Asia, Australasia, and Oceania																		
Urea Nitrogen (mg/dl)																		
The Americas																		
Africa and Arabia																		
Ethiopia																		
Addis Ababa	UM	Svanberg *et al.* (1977)	X									1–5 m	16	16(–)	–			
Asia, Australasia, and Oceania																		
Total amino acids (g/dl)																		
The Americas																		
Africa and Arabia																		
Ethiopia																		
Addis Ababa	UM	Svanberg *et al.* (1977)	X									1–5 m	16	1.02(0.18)	–			

TABLE II—*continued*

Locality	SE class[a]	Reference	Design[b]	Colostrum (1–5 days)				Transitional (6–14 days)				Mature (0.5–6 months)				Mature (> 6 months)		
				Stage[c]	*N*	Mean(SD)	Range	Stage[c]	*N*	Mean(SD)	Range	Stage[c]	*N*	Mean(SD)	Range	Stage[c]	*N*	Mean(SD)
Ivory Coast																		
Kpouebo	RL	Lauber and Reinhardt (1979)	X									1 m	2	1.02(—)	—	12 m	8	0.96(—)
												6 m	8	0.78(—)	—	18 m	3	0.95(—)
Zaire																		
Kivu	R/U	Close and Van de Walle (1957)	X	1–5 d	p	2.93(—)	—	6–12 d	p	1.74(—)	—	31–90 d	p	1.25(—)	—	211–270 d	p	1.03(—)
												91–150 d	p	1.27(—)	—	271–330 d	p	1.02(—)
																331–390 d	p	1.18(—)
																391–450 d	p	1.10(—)
Asia, Australasia, and Oceania																		
Free amino acids (mg/dl)																		
The Americas																		
Africa and Arabia																		
Ethiopia																		
Addis Ababa	UM	Svanberg *et al.* (1977)	X									1–5 m	16	52.3(10.1)	—			
Asia, Australasia, and Oceania																		
Peptide amino acids (mg/dl)																		
The Americas																		

TABLE II—continued

Locality	SE class[a]	Reference	Design[b]	Colostrum (1–5 days)				Transitional (6–14 days)				Mature (0.5–6 months)				Mature (> 6 months)		
				Stage[c]	*N*	Mean(SD)	Range	Stage[c]	*N*	Mean(SD)	Range	Stage[c]	*N*	Mean(SD)	Range	Stage[c]	*N*	Mean(SD)
Africa and Arabia																		
Ethiopia																		
Addis Ababa	UM	Svanberg *et al.* (1977)	X									1–5 m	16	42.2(33.0)	—			
Asia, Autralasia, and Oceania																		
Protein amino acids (mg/dl)																		
The Americas																		
Africa and Arabia																		
Ethiopia																		
Addis Ababa	UM	Svanberg *et al.* (1977)	X									1–5 m	16	930(179)	—			
Asia, Australasia, and Oceania																		
Protein—Other methods (g/dl)[d]																		
The Americas																		
Brazil																		
Rio de Janeiro	UL	Donangelo *et al.* (1989)(1)	X	1–5 d	16	2.60(1.24)	1.47–5.12											
Rio de Janeiro	UL	Trugo *et al.* (1988)(1)	X	1–5 d	25	1.88(0.93)	—											

TABLE II—*continued*

Locality	SE class[a]	Reference	Design[b]	Colostrum (1–5 days)				Transitional (6–14 days)				Mature (0.5–6 months)				Mature (> 6 months)		
				Stage[c]	*N*	Mean(SD)	Range	Stage[c]	*N*	Mean(SD)	Range	Stage[c]	*N*	Mean(SD)	Range	Stage[c]	*N*	Mean(SD)
Sao Paulo	UH	Carniero and de Oliveira (1973)(3)	L	1–5 d	3	1.7(0.2)	—	7 d	10	1.4(0.2)	—	28 d	10	1.3(0.5)	—			
								14 d	10	1.2(0.2)	—	56 d	9	1.3(0.6)	—			
												82 d	5	1.0(0.4)	—			
												110 d	3	1.1(0.3)	—			
Sao Paulo	UL	Carniero and de Oliveira (1973)(1)	L	1–5 d	7	4.5(2.6)	—	7 d	10	1.5(0.3)	—	28 d	10	1.3(0.2)	—			
								14 d	9	1.3(0.2)	—	56 d	10	1.2(0.5)	—			
												82 d	8	1.2(0.3)	—			
												110 d	7	1.0(0.1)	—			
												138 d	7	1.0(0.1)	—			
												159 d	7	0.9(0.2)	—			
												186 d	5	1.2(0.2)	—			
Africa and Arabia																		
Ethiopia																		
Addis Ababa	UL	Duncan *et al.* (1983)(7)	X	1 d	12	9.27(5.54)	2.48–21.71											
Kenya																		
Nairobi	UM	Bwibo and Ondijo (1981)(7) f	X									3 w–6 m	26	0.79(0.15)	(0.48–1.10)			
Nigeria																		
Benin	UX	Omeme *et al.* (1981)(1)	X	1–3 d	q	1.6(0.6)	—					> 2 w	q	1.0(0.4)				
The Gambia																		
Keneba	RL	Prentice *et al.* (1989)(6)	X									1.5 m	5	1.19(0.27)	0.75–1.46	17 m	8	0.76(0.17)
												3 m	10	0.89(0.09)	0.75–1.06			

TABLE II—*continued*

Locality	SE class[a]	Reference	Design[b]	Colostrum (1–5 days)				Transitional (6–14 days)				Mature (0.5–6 months)				Mature (> 6 months)		
				Stage[c]	N	Mean(SD)	Range	Stage[c]	N	Mean(SD)	Range	Stage[c]	N	Mean(SD)	Range	Stage[c]	N	Mean(SD)
Asia, Australasia, and Oceania																		
Central Pacific																		
Nauru	RL	Bray (1928)(8)	X									8 w	q	1.06(—)	—			
India																		
Hyderabad	UL	Rao and Belavady (1981)(1)	X	1–5 d	3	1.42(0.12)	—	6–15 d	7	1.38(0.14)	—	16–30 d	8	1.15(0.27)	—	7–12 m	7	1.10(0.24)
												1–3 m	7	1.15(0.14)	—			
												4–6 m	12	1.00(0.21)	—			
Jaipur	XX	Mehta and Kali (1971)(5)	L	0–2 d	84	5.4(—)	2.5–10.8					15 d	84	1.5(—)	1.0–2.5			
New Delhi	UX	Mathur *et al.* (1990)(2)	X	1–3 d	10	3.1(0.5)	2.0–24.0											
Pakistan																		
Karachi	UL	Lindblad and Rahimtoola (1974)(4)	X									1.5–6 m	9	0.82(0.17)	—			
Taiwan																		
Tainan	UXn	Chang (1990)(2) g	L	3 d	50	2.9(2.3)	—	6 d	50	2.5(1.6)	—	21 d	50	1.82(0.78)	—			
				4 d	50	2.8(1.2)	—	7 d	50	2.3(1.6)	—	28 d	50	1.89(1.20)	—			
				5 d	50	2.4(1.0)	—	14 d	50	2.2(0.8)	—	42 d	50	1.55(0.28)	—			
												49 d	50	1.25(0.28)	—			
												56 d	50	1.20(0.21)	—			

TABLE II—*continued*

Locality	SE class[a]	Reference	Design[b]	Colostrum (1–5 days)				Transitional (6–14 days)				Mature (0.5–6 months)				Mature (> 6 months)		
				Stage[c]	*N*	Mean(SD)	Range	Stage[c]	*N*	Mean(SD)	Range	Stage[c]	*N*	Mean(SD)	Range	Stage[c]	*N*	Mean(SD)
Tainan	UXm	Chang	L	3 d	10	2.8(2.1)	—	6 d	10	2.4(0.6)	—	21 d	10	1.8(0.6)	—			
		(1990)(1) g		4 d	10	2.6(1.6)	—	7 d	10	2.1(0.6)	—	28 d	10	1.8(0.6)	—			
				5 d	10	2.2(1.5)	—	14 d	10	2.0(1.0)	—	42 d	10	1.4(0.2)	—			
												49 d	10	1.1(0.2)	—			
												56 d	10	1.2(0.2)	—			

Note. f, figures calculated from information in text; g, figures taken from graphs; p, pooled sample; q, numbers of subjects not given; s, supplemented mothers; z, geometric mean (+1 geometric standard deviation).

[a]Socioeconomic class; RL, rural, poor; UH, urban, middle-high income; UL, urban, poor, and low-middle income; UM, urban, mixed socioeconomic class; UX, urban, socioeconomic class not stated; a, lowest to d, highest; quartiles of maternal intake; *m*, malnourished and undernourished mothers; *n*, good maternal nutritional status.

[b]Study design: L, longitudinal or semilongitudinal design; X, cross-sectional design.

[c]Stage of lactation: d, days postpartum; w, weeks postpartum; m, months postpartum.

[d]Methods: 1, Lowry/Folin; 2, Biuret; 3, Ma and Zuazaga; 4, Amino acid analysis; 5, Formol titration; 6, BCA; 7, Reinhold; 8, not specified.

TABLE III
Specific Proteins

Locality	SE class[a]	Reference	Design[b]	Colostrum (1–5 days)				Transitional (6–14 days)				Mature (0.5–6 months)				Mature (> 6 months)		
				Stage[c]	*N*	Mean(SD)	Range	Stage[c]	*N*	Mean(SD)	Range	Stage[c]	*N*	Mean(SD)	Range	Stage[c]	*N*	Mean(SD)
Albumin (mg/dl)																		
The Americas																		
Colombia																		
Cali	UXn	Miranda *et al.* (1983)(6)[d] g	L	0–2 d	12	205(128)	—	2 w	12	75(42)	—	4 w	12	59(48)	—			
												8 w	12	39(21)	—			
Cali	UXm	Miranda *et al.* (1983)(6) g	L	0–2 d	11	82(46)	—	2 w	11	39(27)	—	4 w	11	35(20)	—			
												8 w	11	35(33)	—			
Africa and Arabia																		
Ethiopia																		
Addis Ababa	UL	Lonnerdal *et al.* (1976)(3)	X									0.5–1.5 m	3	43(2)	—	> 6.5 m	45	34(7)
												1.5–3.5 m	14	36(7)	—			
												3.5–6.5 m	26	36(8)	—			
Addis Ababa	UH	Lonnerdal *et al.* (1976)(3)	X									0.5–1.5 m	15	47(11)	—			
												1.5–3.5 m	5	38(6)	—			
Addis Ababa	UL	Duncan *et al.* (1983)(2)	X	1 d	12	184(136)	36–525											
Zaire																		
Yasa-Bonga	RLn	Barclay (1989)(2) g	L					1 w	15	31(12)	—	1 m	15	34(12)	—	9 m	10	25(8)
												2 m	15	33(12)	—	12 m	10	24(8)
												4 m	15	31(14)	—	15 m	10	24(6)
												6 m	15	29(13)	—	18 m	10	31(22)

TABLE III—*continued*

Locality	SE class[a]	Reference	Design[b]	Colostrum (1–5 days)				Transitional (6–14 days)				Mature (0.5–6 months)				Mature (> 6 months)		
				Stage[c]	*N*	Mean(SD)	Range	Stage[c]	*N*	Mean(SD)	Range	Stage[c]	*N*	Mean(SD)	Range	Stage[c]	*N*	Mean(SD)
Yasa-Bonga	RLm	Barclay (1989)(2) g	L					1 w	16	28(10)	—	1 m	16	22(5)	—	9 m	8	17(4)
												2 m	16	22(11)	—	12 m	8	19(17)
												4 m	16	17(7)	—	15 m	8	17(7)
												6 m	16	17(6)	—	18 m	8	17(6)
Asia, Australasia, and Oceania																		
Japan																		
Tokyo	UX	Nagasawa *et al.* (1973)(4)	X									2 m	26	73(–)	—			
												3 m	8	67(–)	—			
												4 m	6	68(–)	—			
Lactalbumin (mg/dl)																		
The Americas																		
Africa and Arabia																		
Ethiopia																		
Addis Ababa	UL	Lonnerdal *et al.* (1976)(2)	X									0.5–1.5 m	3	358(28)	—	> 6.5 m	45	258(37)
												1.5–3.5 m	14	276(29)	—			
												3.5–6.5 m	26	265(35)	—			
Addis Ababa	UH	Lonnerdal *et al.* (1976)(2)	X									0.5–1.5 m	15	372(35)	—			
												1.5–3.5 m	5	292(39)	—			
Asia, Australasia, and Oceania																		
Japan																		
Tokyo	UX	Nagasawa *et al.* (1973)(4)	X									2 m	26	226(–)	—			
												3 m	8	186(–)	—			
												4 m	6	223(–)	—			

TABLE III—*continued*

Locality	SE class[a]	Reference	Design[b]	Colostrum (1–5 days)				Transitional (6–14 days)				Mature (0.5–6 months)				Mature (> 6 months)		
				Stage[c]	N	Mean(SD)	Range	Stage[c]	N	Mean(SD)	Range	Stage[c]	N	Mean(SD)	Range	Stage[c]	N	Mean(SD)
Lactoferrin (mg/dl)																		
The Americas																		
Brazil																		
Rio de Janeiro	UL	Donangelo *et al.* (1989)(1)	X	1–5 d	14	686(408)	64–1638											
Africa and Arabia																		
Ethiopia																		
Addis Ababa	UL	Lonnerdal *et al.* (1976)(3)	X									0.5–1.5 m	3	264(6)	—	> 6.5 m	45	148(47)
												1.5–3.5 m	14	167(51)	—			
												3.5–6.5 m	26	172(67)	—			
Addis Ababa	UH	Lonnerdal *et al.* (1976)(3)	X									0.5–1.5 m	15	337(71)	—			
												1.5–3.5 m	5	189(51)	—			
Addis Ababa	UL	Duncan *et al.* (1983)(2)	X	1 d	12	905(491)	109–1830											
The Gambia																		
Keneba	RL	Prentice *et al.* (1984)(2) z	L									1.5–2.9 m	39	318(138)	—	6.0–8.9 m	47	232(91)
												3.0–5.9 m	43	256(95)	—	9.0–11.9 m	40	214(69)
																12.0–14.9 m	30	219(77)
																15.0–17.9 m	43	222(70)
																18.0–26.0 m	38	241(160)

TABLE III—*continued*

Locality	SE class[a]	Reference	Design[b]	Colostrum (1–5 days)				Transitional (6–14 days)				Mature (0.5–6 months)				Mature (> 6 months)		
				Stage[c]	*N*	Mean(SD)	Range	Stage[c]	*N*	Mean(SD)	Range	Stage[c]	*N*	Mean(SD)	Range	Stage[c]	*N*	Mean(SD)
Keneba	RL	Prentice *et al.* (1989)(1)	X									1.5 m	5	165(82)	105–305	17 m	8	89(27)
												3 m	10	133(34)	89–188			
Zaire																		
Kivu	UL	Hennart *et al.* (1991(1) g	X									3–6 m	10	69(38)	—	6–9 m	11	64(23)
																9–12 m	12	52(31)
																12–15 m	10	46(13)
																15–18 m	18	59(17)
Kivu	RL	Hennart *et al.* (1991)(1) g	X									3–6 m	11	66(46)	—	6–9 m	10	74(25)
																9–12 m	10	62(54)
																12–15 m	9	59(27)
																15–18 m	6	88(34)
Yasa-Bonga	RLn	Barclay (1989)(2) g	L					1 w	15	170(65)	—	1 m	15	123(50)	—	9 m	10	72(—)
												2 m	15	100(30)	—	12 m	10	65(20)
												4 m	15	90(30)	—	15 m	10	77(—)
												6 m	15	76(—)	—	18 m	10	125(133)
Yasa-Bonga	RLm	Barclay (1989)(2) g	L					1 w	16	195(155)	—	1 m	16	111(25)	—	9 m	8	68(—)
												2 m	16	145(175)	—	12 m	8	79(45)
												4 m	16	77(30)	—	15 m	8	73(—)
												6 m	16	75(—)	—	18 m	8	76(30)
Asia, Australasia, and Oceania																		
India																		
Hyderabad	UHn	Reddy *et al.* (1977)(2)	X	1–5 d	28	420(259)	—					1–6 m	17	250(268)	—			

TABLE III—*continued*

Locality	SE class[a]	Reference	Design[b]	Colostrum (1–5 days)				Transitional (6–14 days)				Mature (0.5–6 months)				Mature (> 6 months)		
				Stage[c]	*N*	Mean(SD)	Range	Stage[c]	*N*	Mean(SD)	Range	Stage[c]	*N*	Mean(SD)	Range	Stage[c]	*N*	Mean(SD)
Hyderabad	ULm	Reddy *et al.* (1977)(2)	X	1–5 d	19	520(301)	—					1–6 m	13	270(332)	—			
New Delhi	UX	Mathur *et al.* (1990)(2)	X	1–3 d	10	310(50)	190–400											
Japan																		
Tokyo	UX	Nagasawa *et al.* (1972)(3)	X	2–5 d	8	490(60)	400–630	6–10 d	10	450(80)	360–600	61–210 d	25	160(30)	50–210			
Lysozyme (mg/dl)																		
Using human lysozyme as standard																		
The Americas																		
Africa and Arabia																		
The Gambia																		
Keneba	RL	Prentice *et al.* (1984)(2)	L									1.5–2.9 m	39	4.1(2.2)	—	6.0–8.9 m	47	7.7(3.9)
		z										3.0–5.9 m	43	6.9(3.1)	—	9.0–11.9 m	40	7.7(4.1)
																12.0–14.9 m	30	10.9(5.2)
																15.0–17.9 m	43	13.1(12.0)
																18.0–26.0 m	38	19.5(21.9)
Zaire																		
Kivu	UL	Hennart *et al.* (1991)(1)	X									3–6 m	10	20(11)	—	6–9 m	11	19(17)
		g														9–12 m	12	28(26)
																12–15 m	10	37(28)
																15–18 m	18	40(38)

TABLE III—continued

Locality	SE class[a]	Reference	Design[b]	Colostrum (1–5 days)				Transitional (6–14 days)				Mature (0.5–6 months)				Mature (> 6 months)		
				Stage[c]	*N*	Mean(SD)	Range	Stage[c]	*N*	Mean(SD)	Range	Stage[c]	*N*	Mean(SD)	Range	Stage[c]	*N*	Mean(SD)
Kivu	RL	Hennart *et al.* (1991)(1) g	X									3–6 m	11	10(6)	—	6–9 m	10	13(4)
																9–12 m	10	10(5)
																12–15 m	9	19(5)
																15–18 m	6	27(15)
Asia, Australasia, and Oceania																		
India																		
New Delhi	UX	Mathur *et al.* (1990)(2)	X	1–3 d	10	3.4(0.8)	1.0–5.0											
Taiwan																		
Tainan	UXn	Chang (1990)(6) g	L	3 d	50	7.2(4.9)	—	6 d	50	4.3(2.1)	—	21 d	50	3.6(2.8)	—			
				4 d	50	5.9(3.5)	—	7 d	50	4.3(3.5)	—	28.d	50	4.4(3.5)	—			
				5 d	50	5.5(2.1)	—	14 d	50	4.1(2.8)	—	42 d	50	4.6(1.4)	—			
												49 d	50	5.0(1.4)	—			
												56 d	50	5.0(1.4)	—			
Tainan	UXm	Chang (1990)(6) g	L	3 d	10	3.9(0.9)	—	6 d	10	2.9(0.6)	—	21 d	10	2.8(0.6)	—			
				4 d	10	3.1(0.9)	—	7 d	10	2.1(0.6)	—	28 d	10	3.5(0.9)	—			
				5 d	10	2.8(0.6)	—	14 d	10	2.0(0.6)	—	42 d	10	2.6(0.9)	—			
												49 d	10	2.2(0.6)	—			
												56 d	10	2.1(0.9)	—			
Using hen egg white lysozyme or unspecified material as standard																		
The Americas																		
Colombia																		
Cali	UXn	Miranda *et al.* (1983)(5) g	0–2 d	12	112(17)	—	2 w	12	87(28)	—	4 w	12	86(28)	—				
											8 w	12	28(24)	—				

TABLE III—*continued*

Locality	SE class[a]	Reference	Design[b]	Colostrum (1–5 days)				Transitional (6–14 days)				Mature (0.5–6 months)				Mature (> 6 months)		
				Stage[c]	*N*	Mean(SD)	Range	Stage[c]	*N*	Mean(SD)	Range	Stage[c]	*N*	Mean(SD)	Range	Stage[c]	*N*	Mean(SD)
Cali	UXm	Miranda *et al.* (1983)(5) g	L	0–2 d	11	100(66)	—	2 w	11	73(46)	—	4 w 8 w	11 11	63(40) 54(53)	— —			
Africa and Arabia																		
Asia, Australasia, and Oceania																		
India																		
Hyderabad	XX	Rao and Belavady (1973)(5)	X	2–5 d	8	13.6(2.3)	—									> 12 m	10	75.5(41.8)
Hyderabad	UHn	Reddy *et al.* (1977)(5)	X	1–5 d	15	14.2(8.2)	—					1–6 mo	10	24.8(10.8)	—			
Hyderabad	ULm	Reddy *et al.* (1977)(5)	X	1–5 d	21	16.2(11.0)	—					1–6 mo	23	23.3(16.9)	—			
Secretory IgA (mg/dl)																		
Using 11s (secretory IgA) standard																		
The Americas																		
Colombia																		
Cali	UXn	Miranda *et al.* (1983)(2) g	L	0–2 d	12	270(113)	—	2 w	12	40(35)	—	4 w 8 w	12 12	25(52) 40(27)	— —			
Cali	UXm	Miranda *et al.* (1983)(2) g	L	0–2 d	11	205(108)	—	2 w	11	33(27)	—	4 w	11	45(27)	—			

TABLE III—*continued*

Locality	SE class[a]	Reference	Design[b]	Colostrum (1–5 days)				Transitional (6–14 days)				Mature (0.5–6 months)				Mature (> 6 months)		
				Stage[c]	*N*	Mean(SD)	Range	Stage[c]	*N*	Mean(SD)	Range	Stage[c]	*N*	Mean(SD)	Range	Stage[c]	*N*	Mean(SD)
Costa Rica																		
San Jose	UX	Yolken *et al.* (1978)(2)	X	1 d	32	287(—)	—											
Guatemala																		
Guatemala City	UL	Cruz *et al.* (1982)(1)	X									1 m 3 m	10 11	84(55) 60(21)	21–188 33–133			
Guatemala City	UH	Cruz *et al.* (1982)(1)	X									1 m 3 m	10 10	102(65) 58(24)	31–264 27–92			
Santa Maria Cauque	RL	Yolken *et al.* (1978)(2)	L	1 d	18	302(—)	—	1 w 2 w	19 15	195(—) 172(—)		3 w 4 w 5–12 w	17 16 10	157(—) 160(—) 158(—)		52–104 w	11	190(—)
Santa Maria Cauque	RL	Cruz *et al.* (1982)(1)	X									1 m 3 m 6 m	10 9 10	63(21) 41(13) 40(22)	41–109 19–65 20–87	9 m	10	43(15)
Santa Maria Cauque	RL	Cruz and Alveralo (1985)(1) g	L	5 d	20	122(28)	—	11 d	20	81(28)	—	12 w	20	58.5(28)				
Africa and Arabia																		
Ethiopia																		
Addis Ababa	UH	Cruz *et al.* (1982)(1)	L	3 d 5 d	5 5	560(654) 79(33)	139–1710 45–115					1 m	3	61(70)	34–111			
Addis Ababa	UL	Cruz *et al.* (1982)(1)	L	3 d 5 d	7 7	169(48) 72(27)	112–253 26–100					1 m	5	43(18)	21–72			

TABLE III—*continued*

Locality	SE class[a]	Reference	Design[b]	Colostrum (1–5 days)				Transitional (6–14 days)				Mature (0.5–6 months)				Mature (> 6 months)		
				Stage[c]	*N*	Mean(SD)	Range	Stage[c]	*N*	Mean(SD)	Range	Stage[c]	*N*	Mean(SD)	Range	Stage[c]	*N*	Mean(SD)
Addis Ababa	UL	Duncan *et al.* (1983)(2)	X	1 d	12	4654 (2165)	1167–9300											
The Gambia																		
Keneba	RL	Prentice *et al.* (1983)(1)	X									1.5 m	5	80(20)	58–106	17 m	8	67(23)
												3 m	10	74(19)	47–99			
Zaire																		
Kivu	UL	Hennart *et al.* (1989)(1) g	X									3–6 m	10	230(120)	—	6–9 m	11	177(86)
																9–12 m	12	204(194)
																12–15 m	10	201(66)
																15–18 m	18	183(89)
Kivu	RL	Hennart *et al.* (1991)(1) g	X									3–6 m	11	138(103)	—	6–9 m	10	183(73)
																9–12 m	10	138(51)
																12–15 m	9	146(69)
																15–18 m	6	206(115)
Yasa-Bonga	RLn	Barclay (1989)(2) g	L					1 w	15	165(—)	—	1 m	15	122(—)	—	9 m	10	79(22)
												2 m	15	94(—)	—	12 m	10	71(—)
												4 m	15	83(22)	—	15 m	10	75(—)
												6 m	15	79(25)	—	18 m	10	90(50)
Yasa-Bonga	RLm	Barclay (1989)(2) g	L					1 w	16	269(—)	—	1 m	16	128(—)	—	9 m	8	72(22)
												2 m	16	190(—)	—	12 m	8	75(—)
												4 m	16	70(22)	—	15 m	8	79(—)
												6 m	16	66(20)	—	18 m	8	75(50)

TABLE III—*continued*

Locality	SE class[a]	Reference	Design[b]	Colostrum (1–5 days)				Transitional (6–14 days)				Mature (0.5–6 months)				Mature (> 6 months)		
				Stage[c]	*N*	Mean(SD)	Range	Stage[c]	*N*	Mean(SD)	Range	Stage[c]	*N*	Mean(SD)	Range	Stage[c]	*N*	Mean(SD)
Asia, Australasia, and Oceania																		
China																		
Shanghai	UX	Huang *et al.*	L	1 d	11	972(—)	—	6 d	11	57(—)	—	16 d	11	25(—)				
		(1984)(2) g		2 d	11	670(—)	—	7 d	11	43(—)	—	31 d	11	22(—)				
				3 d	11	243(—)	—	8 d	11	44(—)	—							
				4 d	11	103(—)	—											
				5 d	11	65(—)	—											
India																		
Hyderabad	UHn	Reddy *et al.* (1977)(2)	X	1–5 d	17	336(154)	—					1–6 mo	12	120(27)	—			
Hyderabad	ULm	Reddy *et al.* (1977)(2)	X	1–5 d	10	374(133)	—					1–6 mo	10	118(51)	—			
New Delhi	UX	Mathur *et al.* (1990)(2)	X	1–3 d	10	530(130)	350–750											
Taiwan																		
Tainan	UXn	Chang	L	3 d	50	485(390)	—	6 d	50	155(190)	—	21 d	50	45(85)	—			
		(1990)(2) g		4 d	50	390(345)	—	7 d	50	140(130)	—	28 d	50	30(65)	—			
				5 d	50	260(345)	—	14 d	50	65(85)	—	42 d	50	45(65)	—			
												49 d	50	35(40)	—			
												56 d	50	15(40)	—			
Tainan	UXm	Chang	L	3 d	10	220(30)	—	6 d	10	90(60)	—	21 d	10	25(40)	—			
				4 d	10	165(40)	—	7 d	10	85(50)	—	28 d	10	10(20)	—			
				5 d	10	145(80)	—	14 d	10	35(80)	—	42 d	10	20(20)	—			
												49 d	10	20(20)	—			
												56 d	10	6(—)	—			

TABLE III—*continued*

Locality	SE class[a]	Reference	Design[b]	Colostrum (1–5 days)				Transitional (6–14 days)				Mature (0.5–6 months)				Mature (> 6 months)		
				Stage[c]	*N*	Mean(SD)	Range	Stage[c]	*N*	Mean(SD)	Range	Stage[c]	*N*	Mean(SD)	Range	Stage[c]	*N*	Mean(SD)
Using 7s (serum IgA) standard																		
The Americas																		
Argentina																		
Buenos Aires	UH	Ronayne de Ferrer *et al.* (1984)(2)	X									20–30 d	11	172(42)	—			
Guatemala																		
Santa Maria Cauque	RL	Wyatt *et al.* (1972)(2) g	L	0–3 d	34	333(—)	171–441	1 w	19	191(—)	128–267	3 w	17	157(—)	124–224			
								2 w	16	165(—)	20–193	4 w	15	146(—)	124–172			
												5–16 w	14	152(—)	104–178			
Africa and Arabia																		
The Gambia																		
Keneba	RL	Prentice *et al.* (1984)(2) z	L									1.5–2.9 m	39	45.6(19.7)	—	6.0–8.9 m	47	32.3(14.7)
												3.0–5.9 m	43	35.0(18.2)	—	9.0–11.9 m	40	26.6(11.0)
																12.0–14.9 m	30	32.3(14.9)
																15.0–17.9 m	43	34.3(14.7)
																18.0–26.0 m	38	38.1(25.4)
Asia, Australasia, and Oceania																		
IgG (mg/dl)																		
The Americas																		
Colombia																		
Cali	UXn	Miranda *et al.* (1983)(2) g	L	0–2 d	12	31(13)	—	2 w	12	16(13)	—	4 w	12	9.5(9)	—			
												8 w	12	7.5(—)	—			

TABLE III—*continued*

Locality	SE class[a]	Reference	Design[b]	Colostrum (1–5 days)				Transitional (6–14 days)				Mature (0.5–6 months)				Mature (> 6 months)		
				Stage[c]	N	Mean(SD)	Range	Stage[c]	N	Mean(SD)	Range	Stage[c]	N	Mean(SD)	Range	Stage[c]	N	Mean(SD)
Cali	UXm	Miranda *et al.* (1983)(2) g	L	0–2 d	11	9.5(13)	—	2 w	11	7.5(8)	—	4 w	11	5.5(—)	—			
												8 w	11	4.8(—)	—			
Africa and Arabia																		
Ethiopia																		
Addis Ababa	UL	Lonnerdal *et al.* (1976)(2)	X									0.5–1.5 m	3	4.5(1.2)	—	> 6.5 m	45	3.9(2.9)
												1.5–3.5 m	14	3.3(1.1)	—			
												3.5–6.5 m	26	4.8(2.4)	—			
Addis Ababa	UH	Lonnerdal *et al.* (1976)(2)	X									0.5–1.5 m	15	6.7(2.0)	—			
												1.5–3.5 m	5	4.6(1.2)	—			
Addis Ababa	UL	Duncan *et al.* (1983)(2)	X	1 d	12	47.5(—)	0–282											
The Gambia																		
Keneba	RL	Prentice *et al.* (1984)(2) z	L									1.5–2.9 m	39	6.5(3.2)	—	6.0–8.9 m	47	5.4(2.7)
												3.0–5.9 m	43	5.7(3.4)	—	9.0–11.9 m	40	4.5(1.5)
																12.0–14.9 m	30	5.9(3.8)
																15.0–17.9 m	43	5.6(2.1)
																18.0–26.0 m	38	6.1(3.2)
Zaire																		
Yasa-Bonga	RLn	Barclay (1989)(2) g	L					1 w	15	10(8)	—	1 m	15	10(6)	—	9 m	10	6(4)
												2 m	15	9(7)	—	12 m	10	6(4)
												4 m	15	8(5)	—	15 m	10	6(—)
												6 m	15	7(4)	—	18 m	10	6(2)

TABLE III—*continued*

Locality	SE class[a]	Reference	Design[b]	Colostrum (1–5 days)				Transitional (6–14 days)				Mature (0.5–6 months)				Mature (> 6 months)		
				Stage[c]	*N*	Mean(SD)	Range	Stage[c]	*N*	Mean(SD)	Range	Stage[c]	*N*	Mean(SD)	Range	Stage[c]	*N*	Mean(SD)
Yasa-Bonga	RLm	Barclay (1989)(2) g	L					1 w	16	12(5)	—	1 m	16	7(3)	—	9 m	8	6(2)
												2 m	16	8(7)	—	12 m	8	7(4)
												4 m	16	6(4)	—	15 m	8	5(—)
												6 m	16	6(3)	—	18 m	8	5(3)
Asia, Australasia, and Oceania																		
China																		
Shanghai	UX	Huang *et al.* (1984)(2) g	L	1 d	11	129(—)	—	6 d	11	6(—)	—	16 d	11	5(—)				
				2 d	11	80(—)	—	7 d	11	6(—)	—	31 d	11	13(—)				
				3 d	11	20(—)	—	8 d	11	6(—)	—							
				4 d	11	7(—)	—											
				5 d	11	6(—)	—											
India																		
Hyderabad	UHn	Reddy *et al.* (1977)(2)	X	1–5 d	17	5.9(6.5)	—											
Hyderabad	ULm	Reddy *et al.* (1977)(2)	X	1–5 d	10	5.3(7.3)	—											
New Delhi	UX	Mathur *et al.* (1990)(2)	X	1–3 d	10	26(7)	18–41											
IgM (mg/dl)																		
The Americas																		
Colombia																		
Cali	UXn	Miranda *et al.* (1983)(2)	L	0–2 d	12	155(107)	—											

TABLE III—*continued*

Locality	SE class[a]	Reference	Design[b]	Colostrum (1–5 days)				Transitional (6–14 days)				Mature (0.5–6 months)				Mature (> 6 months)		
				Stage[c]	*N*	Mean(SD)	Range	Stage[c]	*N*	Mean(SD)	Range	Stage[c]	*N*	Mean(SD)	Range	Stage[c]	*N*	Mean(SD)
Cali	UXm	Miranda *et al.* (1983)(2)	L	0–2 d	11	125(89)	—											
Guatemala																		
Santa Maria Cauque	RL	Wyatt *et al.* (1972)(2) g	L	0–3 d	8	36(—)	< 13–110											
Africa and Arabia																		
Ethiopia																		
Addis Ababa	UL	Lonnerdal *et al.* (1976)(2)	X									0.5–1.5 m	3	9.4(9.0)	—	> 6.5 m	45	4.0(4.9)
												1.5–3.5 m	14	2.9(1.3)	—			
												3.5–6.5 m	26	3.0(1.8)	—			
Addis Ababa	UH	Lonnerdal *et al.* (1976)(2)	X									0.5–1.5 m	15	4.5(2.5)	—			
												1.5–3.5 m	5	3.3(2.8)	—			
Addis Ababa	UL	Duncan *et al.* (1983)(2)	X	1 d	12	223(—)	0–840											
The Gambia																		
Keneba	RL	Prentice *et al.* (1984)(2) z	L									1.5–2.9 m	39	4.6(3.1)	—	6.0–8.9 m	47	2.8(1.7)
												3.0–5.9 m	43	3.5(2.1)	—	9.0–11.9 m	40	2.6(1.6)
																12.0–14.9 m	30	2.4(2.2)
																15.0–17.9 m	43	2.6(1.6)
																18.0–26.0 m	38	3.2(3.1)
Zaire																		
Yasa-Bonga	RLn	Barclay (1989)(2) g	L					1 w	15	13(—)	—	1 m	15	7(3)	—	9 m	10	4(3)
												2 m	15	5(3)	—	12 m	10	3(2)
												4 m	15	4(—)	—	15 m	10	4(3)
												6 m	15	4(3)	—	18 m	10	4(3)

TABLE III—*continued*

Locality	SE class[a]	Reference	Design[b]	Colostrum (1–5 days)				Transitional (6–14 days)				Mature (0.5–6 months)				Mature (> 6 months)		
				Stage[c]	*N*	Mean(SD)	Range	Stage[c]	*N*	Mean(SD)	Range	Stage[c]	*N*	Mean(SD)	Range	Stage[c]	*N*	Mean(SD)
Yasa-Bonga	RLm	Barclay (1989)(2) g	L					1 w	16	16(–)	—	1 m	16	6(4)	—	9 m	8	3(1)
												2 m	16	6(6)	—	12 m	8	3(2)
												4 m	16	4(–)	—	15 m	8	3(2)
												6 m	16	2(1)	—	18 m	8	3(3)
Asia, Australasia, and Oceania																		
China																		
Shanghai	UX	Huang *et al.* (1984)(2) g	L	1 d	1[illegible]	571(–)	—	6 d	11	10(–)	—	16 d	11	7(–)	—			
				3 d	1[illegible]	95(–)	—	7 d	11	8(–)	—	31 d	11	6(–)	—			
				4 d	1[illegible]	30(–)	—	8 d	11	8(–)	—							
				5 d	1[illegible]	15(–)	—											
India																		
Hyderabad	UHn	Reddy *et al.* (1977)(2)	X	1–5 d	17	17.1(17.7)	—											
Hyderabad	ULm	Reddy *et al.* (1977)(2)	X	1–5 d	10	15.3(7.9)	—											
New Delhi	UX	Mathur *et al.* (1990)(2)	X	1–3 d	10	120(170)	30–240											
Complement C3 (mg/dl)																		
The Americas																		
Colombia																		
Cali	UXn	Miranda *et al.* (1983)(2) g	L	0–2 d	12	81(35)	—	2 w	12	8(–)	—	4 w	12	5(–)	—			
												8 w	12	9(–)	—			
Cali	UXm	Miranda *et al.* (1983)(2) g	L	0–2 d	1[illegible]	65(40)	—	2 w	11	10(10)	—	4 w	11	7(6)	—			
												8 w	11	12(7)	—			

TABLE III—*continued*

Locality	SE class[a]	Reference	Design[b]	Colostrum (1–5 days)				Transitional (6–14 days)				Mature (0.5–6 months)				Mature (> 6 months)		
				Stage[c]	*N*	Mean(SD)	Range	Stage[c]	*N*	Mean(SD)	Range	Stage[c]	*N*	Mean(SD)	Range	Stage[c]	*N*	Mean(SD)
Africa and Arabia																		
Ethiopia																		
Addis Ababa	UL	Duncan *et al.* (1983)(2)	X	1 d	12	50.7(—)	13–189											
The Gambia																		
Keneba	RL	Prentice *et al.* (1984)(2)	L									1.5–2.9 m	39	1.9(1.5)	—	6.0–8.9 m	47	1.3(1.0)
												3.0–5.9 m	43	1.7(1.0)	—	9.0–11.9 m	40	1.0(0.6)
																12.0–14.9 m	30	1.0(1.1)
																15.0–17.9 m	43	1.0(0.8)
																18.0–26.0 m	38	1.1(1.3)
Asia, Australasia, and Oceania																		
Taiwan																		
Tainan	UXn	Chang *et al.* (1990)(2) g	L	3 d	50	86(71)	—	6 d	50	20(—)	—	21 d	50	14(—)	—			
				4 d	50	30(—)	—	7 d	50	19(—)	—	28 d	50	18(—)	—			
				5 d	50	23(—)	—	14 d	50	19 (—)	—	42 d	50	17(—)	—			
												49 d	50	9(—)	—			
												56 d	50	7(—)	—			
Tainan	UXm	Chang *et al.* (1990)(2) g	L	3 d	10	40(19)	—	6 d	10	10(—)	—	21 d	10	8(—)	—			
				4 d	10	13(22)	—	7 d	10	9(—)	—	28 d	10	8(—)	—			
				5 d	10	11(19)	—	14 d	10	7(—)	—	42 d	10	9(—)	—			
												49 d	10	4(—)	—			
												56 d	10	3(—)	—			

TABLE III—*ccntinued*

Locality	SE class[a]	Reference	Design[b]	Colostrum (1–5 days)				Transitional (6–14 days)				Mature (0.5–6 months)				Mature (> 6 months)		
				Stage[c]	*N*	Mean(SD)	Range	Stage[c]	*N*	Mean(SD)	Range	Stage[c]	*N*	Mean(SD)	Range	Stage[c]	*N*	Mean(SD)
Complement C4 (mg/dl)																		
The Americas																		
Colombia																		
Cali	UXn	Miranda *et al.* (1983)(2) g	L	0–2 d	12	21(16)	—	2 w	12	3(8)	—	4 w	12	4(—)	—			
												8 w	12	8(7)	—			
Cali	UXm	Miranda *et al.* (1983)(2) g	L	0–2 d	11	12(7)	—	2 w	11	5(—)	—	4 w	11	3(7)	—			
												8 w	11	2(7)	—			
Africa and Arabia																		
Ethiopia																		
Addis Ababa	UL	Duncan *et al.* (1983)(2)	X	1 d	12	28.1(—)	0–109											
The Gambia																		
Keneba	RL	Prentice *et al.* (1984)(2) z	L									1.5–2.9 m	39	2.4(3.0)	—	6.0–8.9 m	47	1.1(1.8)
												3.0–5.9 m	43	1.4(1.8)	—	9.0–11.9 m	40	0.7(1.1)
																12.0–14.9 m	30	1.1(1.4)
																15.0–17.9 m	43	1.1(1.3)
																18.0–26.0 m	38	1.3(2.5)
Asia, Australasia, and Oceania																		
Taiwan																		
Tainan	UXn	Chang *et al.* (1990)(2) g	L	3 d	50	14(16)	—	6 d	50	11(8)	—	21 d	50	6(3)	—			
				4 d	50	11(11)	—	7 d	50	10(13)	—	28 d	50	4(2)	—			
				5 d	50	10(8)	—	14 d	50	8(11)	—	42 d	50	4(2)	—			

TABLE III—*continued*

Locality	SE class[a]	Reference	Design[b]	Colostrum (1–5 days)				Transitional (6–14 days)				Mature (0.5–6 months)				Mature (> 6 months)		
				Stage[c]	N	Mean(SD)	Range	Stage[c]	N	Mean(SD)	Range	Stage[c]	N	Mean(SD)	Range	Stage[c]	N	Mean(SD)
Tainan	UXm	Chang *et al.* (1990)(2) g	L	3 d	10	5(1)	—	6 d	10	6(3)	—	21 d	10	4(1)	—			
				4 d	10	5(3)	—	7 d	10	6(4)	—	28 d	10	3(1)	—			
				5 d	10	4(4)	—	14 d	10	4(2)	—	42 d	10	2(—)	—			
α-1-Antitrypsin (mg/dl)																		
The Americas																		
Africa and Arabia																		
Nigeria																		
Benin	UX	Omeme *et al.* (1981)(1)	X	1–3 d	q	25(16)	—					> 2 w	q	11.5(2.5)	—			
Asia, Australasia, and Oceania																		
Amylase																		
The Americas																		
Africa and Arabia																		
The Gambia (IU/dl)																		
Keneba	RL	Dewit *et al.* (1990)(5) z	X									0.5–2.9 m	16	111(71)	—	6.0–8.9 m	17	79(64)
												3.0–5.9 m	17	104(95)	—	9.0–1.9 m	9	67(38)
																12.0–14.9 m	16	78(57)
																15.0–17.9 m	17	81(48)
Asia, Australasia, and Oceania																		

TABLE III—*continued*

Locality	SE class[a]	Reference	Design[b]	Colostrum (1–5 days)				Transitional (6–14 days)				Mature (0.5–6 months)				Mature (> 6 months)		
				Stage[c]	*N*	Mean(SD)	Range	Stage[c]	*N*	Mean(SD)	Range	Stage[c]	*N*	Mean(SD)	Range	Stage[c]	*N*	Mean(SD)
Bile salt-stimulated lipase																		
The Americas																		
Africa and Arabia																		
Ethiopia (μkat/dl)																		
Addis Ababa	UL	Hernell *et al.* (1977)(5)	X									0.3–3.5 m	9	64.0(22.9)	—			
Addis Ababa	UH	Hernel *et al.* (1977)(5)	X									0.3–3.5 m	10	87.1(16.4)	—			
Asia, Australasia, and Oceania																		
India (μmol FFA released/min/dl)																		
Hyderabad	UL	Rao and Belavady (1981)	X	1–5 d	3	254(44)	—	6–15 d	7	320(125)	—	16–30 d	8	411(181)	—	7–12 m	7	351(134)
												1–3 m	7	381(221)	—			
												4–6 m	12	407(163)	—			
Glutathione peroxidase activity																		
The Americas																		
Africa and Arabia																		
The Gambia (U/dl)																		
Keneba	RL	Funk *et al.* (1990)(5)	X									1–6 m	10	5.10(1.49)	—	13–19 m	10	4.08(1.61)
Asia, Australasia, and Oceania																		

TABLE III—*continued*

Locality	SE class[a]	Reference	Design[b]	Colostrum (1–5 days)				Transitional (6–14 days)				Mature (0.5–6 months)				Mature (> 6 months)		
				Stage[c]	*N*	Mean(SD)	Range	Stage[c]	*N*	Mean(SD)	Range	Stage[c]	*N*	Mean(SD)	Range	Stage[c]	*N*	Mean(SD)
Total peroxidase activity																		
The Americas																		
Africa and Arabia																		
The Gambia (U/dl)																		
Keneba	RL	Funk *et al.* (1990)(5)	X									1–6 m	10	12.6(3.0)	–	13–19 m	10	11.8(3.0)
Asia, Australasia, and Oceania																		

Note. f, figures calculated from information in text; g, figures taken from graphs; p, pooled sample; q, numbers of subjects not given; s, supplemented mothers; z, geometric mean (+1 geometric standard deviation).

[a]Socioeconomic class: RL, rural, poor; UH, urban, middle-high income; UL, urban, poor, and low-middle income; UM, urban, mixed socioeconomic class; UX, urban, socioeconomic class not stated; a, lowest to d, highest; quartiles of maternal intake; *m*, malnourished and undernourished mothers; *n*, good maternal nutritional status.

[b]Study design: L, longitudinal or semilongitudinal design; X, cross-sectional design.

[c]Stage of lactation: d, days postpartum; w, weeks postpartum; m, months postpartum.

[d]Methods: 1, enzyme-linked immunosorbent assay/radioimmunoassay; 2, single radial immunodiffusion; 3, immunoelectrophoresis; 4, electrophoresis; 5, enzyme activity; 6, other.

TABLE IV
Minerals and Trace Elements

Locality	SE class[a]	Reference	Design[b]	Colostrum (1–5 days)				Transitional (6–14 days)				Mature (0.5–6 months)				Mature (> 6 months)		
				Stage[c]	N	Mean(SD)	Range	Stage[c]	N	Mean(SD)	Range	Stage[c]	N	Mean(SD)	Range	Stage[c]	N	Mean(SD)
Ash (mg/dl)																		
The Americas																		
Brazil																		
Brasilia	UH	Dorea *et al.* (1984) g	L									15 d	3	260(16)	—			
												30 d	3	200(33)	—			
												45 d	3	210(26)	—			
												60 d	3	150(50)	—			
												90 d	4	170(29)	—			
Brasilia	UL	Dorea *et al.* (1984) g	L									15 d	8	210(31)	—			
												30 d	7	170(77)	—			
												45 d	7	180(24)	—			
												60 d	6	200(42)	—			
												75 d	6	160(51)	—			
												90 d	6	188(47)	—			
												120 d	5	189(20)	—			
Rio de Janeiro	UL	Trugo *et al.* (1988)	X	1–5 d	22	230(80)	—											
Africa and Arabia																		
Sudan																		
Khartoum	XX	El Tom Ali and Zaki (1976)	X	1–3 d	7	370(—)	110–640											

TABLE IV—*continued*

Locality	SE class[a]	Reference	Design[b]	Colostrum (1–5 days)				Transitional (6–14 days)				Mature (0.5–6 months)				Mature (> 6 months)		
				Stage[c]	N	Mean(SD)	Range	Stage[c]	N	Mean(SD)	Range	Stage[c]	N	Mean(SD)	Range	Stage[c]	N	Mean(SD)
Asia, Australasia, and Oceania																		
Central Pacific																		
Nauru	RL	Bray (1928)	X									8 w	q	160(—)	—			
India																		
Coonoor	UL	Belavady and Gopalan (1959)	X									2–6 m	41	178(58)	—	7–12 m	32	155(31)
																13–18 m	29	151(44)
																> 18 m	20	154(44)
Korea																		
Seoul	UH	Seol (1988)	L									15 d	16	240(10)	210–260			
												30 d	16	230(10)	200–240			
												60 d	12	220(10)	200–250			
												90 d	13	210(10)	200–230			
												120 d	12	210(20)	180–230			
												150 d	7	200(20)	180–220			
Vanatu (New Hebrides)																		
Port Vila	RL	Peters (1953)	X									2–5 m	18	188(13)	170–210	6–11 m	15	178(12)
																12–24 m	18	176(17)
Calcium (mg/dl)																		
The Americas																		
Brazil																		
Sao Paulo	UH	Carneiro and de Oliveira (1973)(2)[d]	L					7 d	7	23.8(4.0)	—	28 d	9	20.8(3.9)	—			
								14 d	6	22.2(6.3)	—	56 d	5	20.6(6.5)	—			
												82 d	6	23.6(6.7)	—			

TABLE IV—*continued*

Locality	SE class[a]	Reference	Design[b]	Colostrum (1–5 days)				Transitional (6–14 days)				Mature (0.5–6 months)				Mature (> 6 months)		
				Stage[c]	*N*	Mean(SD)	Range	Stage[c]	*N*	Mean(SD)	Range	Stage[c]	*N*	Mean(SD)	Range	Stage[c]	*N*	Mean(SD)
Sao Paulo	UL	Carneiro and de Oliveira (1973)(2)	L					7 d	9	20.8(2.6)	—	28 d	10	25.7(5.2)	—			
								14 d	8	25.3(4.9)	—	56 d	8	26.0(6.2)	—			
												82 d	8	27.3(5.8)	—			
												110 d	7	26.7(8.6)	—			
												138 d	7	26.5(8.4)	—			
												159 d	7	21.1(4.0)	—			
												186 d	5	20.3(4.8)	—			
Africa and Arabia																		
Egypt																		
Kalama	RL	Karra *et al.* (1988)(1) g	L									1 m	62	26.2(5.8)	—			
												2 m	61	26.7(6.9)	—			
												3 m	53	25.8(5.4)	—			
												4 m	55	25.6(5.5)	—			
												5 m	51	25.6(6.8)	—			
												6 m	51	23.7(4.9)	—			
Ethiopia																		
Addis Ababa	UL	Fransson *et al.* (1984)(1)	X	4 or 5 d	9	46.2(13.2)	—											
Addis Ababa	UH	Fransson *et al.* (1984)(1)	X	4 or 5 d	9	32.1(7.6)	—											
Kenya																		
Machakos	RL	Van Steenbergen *et al.* (1981)(1)	X	3 d	62	20.5(4.0)	—											
Nairobi	UM	Bwibo and Ondijo (1981)(1) f	X									3 w–6 m	21	24.5(4.0)	19–32			

TABLE IV—*continued*

Locality	SE class[a]	Reference	Design[b]	Colostrum (1–5 days)				Transitional (6–14 days)				Mature (0.5–6 months)				Mature (> 6 months)		
				Stage[c]	*N*	Mean(SD)	Range	Stage[c]	*N*	Mean(SD)	Range	Stage[c]	*N*	Mean(SD)	Range	Stage[c]	*N*	Mean(SD)
Nigeria																		
Lagos	UX	Bassir (1956)(2)	X									1 m	34	21.8(5.3)	13.0–32.0	7 m	7	19.1(2.2)
												2 m	19	21.3(4.6)	10.8–29.6	8 m	3	19.9(4.6)
												3 m	14	19.7(4.6)	14.0–26.0	9 m	3	18.7(0.8)
												4 m	10	21.3(3.4)	17.0–26.0	10 m	7	20.2(4.6)
												5 m	14	18.2(4.4)	10.0–24.4	11 m	6	17.3(3.4)
												6 m	5	17.4(1.2)	15.2–18.4	12 m	3	22.2(3.5)
Ibadan	UL	Bassir (1958)(2)	X					1 w	7	19.1(7.3)	14.3–33.4	3–26 w	10	19.5(5.6)	15.0–33.0	27–78 w	14	21.4(6.8)
Sudan																		
Khartoum	XX	El Tom Ali and Zaki (1976)(2)	X	1–3 d	7	39(—)	37–42											
The Gambia																		
Keneba	RL	Laskey *et al.* (1990)(1)	X									0.5–1.5 m	13	25.2(4.0)	—	6.0–8.9 m	17	20.0(4.6)
												1.5–2.9 m	20	24.5(4.5)	—	9.0–11.9 m	9	17.8(3.8)
												3.0–5.9 m	42	23.4(3.6)	—	12.0–14.9 m	16	18.3(3.7)
																15.0–17.9 m	15	18.7(3.7)
																18.0–26.0 m	12	17.4(3.2
Zaire																		
Yasa-Bonga	RL	Prentice (in preparation)(1)	L					10 d	12	25.2(6.3)	14.4–32	6 w	12	25.8(4.7)	18.7–32.5	9 m	12	19.8(3.7)
												9 w	12	26.0(5.9)	15.6–34.3	12 m	12	19.1(2.8)
												18 w	12	24.8(4.7)	17.1–33.6	15 m	12	17.5(2.7)
												6 m	12	22.0(3.6)	16.3–26.5	18 m	11	16.6(2.8)

TABLE IV—*continued*

Locality	SE class[a]	Reference	Design[b]	Colostrum (1–5 days)				Transitional (6–14 days)				Mature (0.5–6 months)				Mature (> 6 months)		
				Stage[c]	N	Mean(SD)	Range	Stage[c]	N	Mean(SD)	Range	Stage[c]	N	Mean(SD)	Range	Stage[c]	N	Mean(SD)
Asia, Australasia, and Oceania																		
India																		
Baroda	UXa	Karmarkar and Ramakrishnan (1960)(2)	X									3 or 4 m	17	20.5(5.8)	—			
Baroda	UXb	Karmarkar and Ramakrishnan (1960)(2)	X									3 or 4 m	13	22.3(4.5)	—			
Baroda	UXc	Karmarkar and Ramakrishnan (1960)(2)	X									3 or 4 m	14	20.3(5.5)	—			
Baroda	UXd	Karmarkar and Ramakrishnan (1960)(2)	X									3 or 4 m	15	20.2(4.8)	—			
Coonoor	UL	Belavady and Gopalan (1959)(2)	X									2–6 m	43	36.3(12.9)	—	7–12 m 13–18 m > 18 m	31 30 20	34.1(14.0) 29.8(9.4) 36.6(13.9)
Kanpur	UH	Singh (1984)(2)	X													> 6 m	6	19.0(2.2)
New Delhi	UH	Ashdir and Puri (1962)(2)	L	3 d	10	27.6(3.4)	21.1–30.9	8 d	10	23.3(3.3)	18.0–26.6	18 d	10	20.2(3.0)	17.1–25.0			

TABLE IV—*continued*

Locality	SE class[a]	Reference	Design[b]	Colostrum (1–5 days)				Transitional (6–14 days)				Mature (0.5–6 months)				Mature (> 6 months)		
				Stage[c]	*N*	Mean(SD)	Range	Stage[c]	*N*	Mean(SD)	Range	Stage[c]	*N*	Mean(SD)	Range	Stage[c]	*N*	Mean(SD)
Udaipur	UX	Gupta *et al.* (1984)(1)	L	2–5 d	50	25.6(4.5)	—	6–10 d	22	22.9(3.4)	—							
Korea																		
Seoul	UH	Seol (1988)(1)	L									15 d	16	29.5(1.2)	27.8–31.7			
												30 d	16	28.7(1.6)	26.2–31.6			
												60 d	12	28.2(1.4)	26.2–30.7			
												90 d	13	27.2(1.3)	25.2–29.7			
												120 d	12	25.8(1.4)	24.3–29.1			
												150 d	7	24.2(1.1)	22.6–25.5			
Nepal																		
Katmandu Valley	RL	Reynolds *et al.* (1986); Moser *et al.* (1988b)(1)	X									2–6 m	26	26.4(14.3)	21.0–33.0			
Pakistan																		
Karachi	UL	Lindblad and Rahim-toola (1974)(2)	X									1.5–6 m	9	28.4(7.4)	—			
Papua New Guinea																		
Biak Island	RL	Jansen *et al.* (1960)(2)	X									2–5 m	3	18.6(—)	16.9–19.8	6–12 m	7	21.7(3.1)
																12–24 m	17	19.5(3.5)
Taiwan																		
Taipei	UX	Lonnerdal *et al.* (1990)(1) g	L									1 m	q	29.0(—)	—			
												2 m	q	26.5(—)	—			
												3 m	q	27.5(—)	—			
												4 m	q	26.3(—)	—			

TABLE IV—*continued*

Locality	SE class[a]	Reference	Design[b]	Colostrum (1–5 days)				Transitional (6–14 days)				Mature (0.5–6 months)				Mature (> 6 months)		
				Stage[c]	*N*	Mean(SD)	Range	Stage[c]	*N*	Mean(SD)	Range	Stage[c]	*N*	Mean(SD)	Range	Stage[c]	*N*	Mean(SD)
Vanatu (New Hebrides)																		
Port Vila	RL	Peters (1953)(2)	X									2–5 m	18	27.4(4.2)	20.8–34.0	6–11 m 12–24 m	15 18	25.1(5.0) 24.9(5.1)
Chromium (μg/dl)																		
The Americas																		
Africa and Arabia																		
Asia, Australasia, and Oceania																		
Japan																		
Fukuyama	UX	Gunshin *et al.* (1985)(1)	X									19–178 d	11	0.56(0.45)	0.19–1.42	190–384 d	13	0.78(0.69)
Cobalt (μg/dl)																		
The Americas																		
Africa and Arabia																		
Asia, Australasia, and Oceania																		
Japan																		
Fukuyama	UX	Gunshin *et al.* (1985)(1)	X									19–178 d	8	0.16(0.12)	0.06–0.40	190–384 d	11	0.21(0.25)

TABLE IV—*continued*

Locality	SE class[a]	Reference	Design[b]	Colostrum (1–5 days)				Transitional (6–14 days)				Mature (0.5–6 months)				Mature (> 6 months)		
				Stage[c]	N	Mean(SD)	Range	Stage[c]	N	Mean(SD)	Range	Stage[c]	N	Mean(SD)	Range	Stage[c]	N	Mean(SD)
Copper (μg/dl)																		
The Americas																		
United States (Navajo Indians)																		
Tuba City	RL	Butte and Calloway (1981)(1)	X									19–62 d	23	30(20)	6–71			
Africa and Arabia																		
Ethiopia																		
Addis Ababa	UL	Fransson *et al.* (1984)(1)	X	4 or 5 d	9	37(20)	—											
Addis Ababa	UH	Fransson *et al.* (1984)(1)	X	4 or 5 d	9	17(4)	—											
Ivory Coast																		
Kpouebo	RL	Lauber and Reinhardt (1979)(1)	X									1 m 6 m	4 7	44(11) 14(7)	— —	12 m 18 m	8 6	13(4) 13(4)
Nigeria																		
Ibadan	UL	Atinmo and Omolulu (1982)(1)	L					8–14 d	20	31(19)	—	2 m	20	27(2)	—			
Asia, Australasia, and Oceania																		
India																		
Coonoor	UL	Belavady and Gopalan (1959)(2)	X									2–6 m	11	56(30)	—	13–18 m	14	26(10)

TABLE IV—continued

Locality	SE class[a]	Reference	Design[b]	Colostrum (1–5 days)				Transitional (6–14 days)				Mature (0.5–6 months)				Mature (> 6 months)		
				Stage[c]	N	Mean(SD)	Range	Stage[c]	N	Mean(SD)	Range	Stage[c]	N	Mean(SD)	Range	Stage[c]	N	Mean(SD)
Hyderabad	UM	Rajalakshmi and Srikantia (1980)(1)	X	2–5 d	76	46(17)	—	6–10 d	31	50(16)	—	1–3 m	77	29(9)	—	7–12 m	88	17(8)
												4–6 m	89	21(9)	—	> 13 m	23	16(7)
Hyderabad	RL	Rajalakshmi and Srikantia (1980)(1)	X									1–3 m	22	29(11)	—	7–12 m	73	17(9)
												4–6 m	41	21(12)	—	> 13 m	86	15(12)
Udaipur	XX	Gupta *et al.* (1984)(1)	L	2–5 d	50	11.4(4.3)	—	6–10 d	22	12.0(4.0)	—							
Japan																		
Kumamoto	XX	Higashi *et al.* (1982)(1)	L	1 d	65	45(23)	—	1 w	65	45(15)	—	1 m	65	44(10)	—			
												3 m	45	29(9)	—			
												5 m	35	22(8)	—			
Korea																		
Seoul	UH	Kim (1989)(1)	L									0.5 m	16	49.7(14.2)	—			
												1 m	16	45.0(10.1)	—			
												2 m	12	42.1(15.3)	—			
												3 m	13	35.6(10.5)	—			
												4 m	12	37.5(10.5)	—			
Nepal																		
Katmandu Valley	RL	Moser *et al.* (1988a)(1)	X									2–6 m	26	22.3(26.0)	—			
Taiwan																		
Taipei	UX	Lonnerdal *et al.* (1990)(1) g	L									1 m	q	49(—)	—			
												2 m	q	37(—)	—			
												3 m	q	40(—)	—			
												4 m	q	40(—)	—			

TABLE IV—*continued*

Locality	SE class[a]	Reference	Design[b]	Colostrum (1–5 days)				Transitional (6–14 days)				Mature (0.5–6 months)				Mature (> 6 months)		
				Stage[c]	*N*	Mean(SD)	Range	Stage[c]	*N*	Mean(SD)	Range	Stage[c]	*N*	Mean(SD)	Range	Stage[c]	*N*	Mean(SD)
Iron—Total (μg/dl)[e]																		
The Americas																		
Brazil																		
Rio de Janeiro	UL	Donangelo *et al.* (1989)(1)	X	1–5 d	17	123(74)	15–275											
Rio de Janeiro	UL	Trugo *et al.* (1988)(1)	X	1–5 d	22	104(62)	50–250											
United States (Navajo Indians)																		
Tuba City	RL	Butte and Calloway (1981)(1)	X									19–62 d	23	80(60)	1–222			
Africa and Arabia																		
Ethiopia																		
Addis Ababa	UL	Fransson *et al.* (1984)(1)	X	4 or 5 d	9	47(19)	—											
Addis Ababa	UH	Fransson *et al.* (1984)(1)	X	4 or 5 d	9	46(25)	—											
Ivory Coast																		
Kpouebo	RL	Lauber and Reinhardt (1979)(1)	X									1 m 6 m	4 7	90(20) 55(10)	— —	12 m 18 m	8 6	54(12) 67(37)
Niger																		
Diffa	RLn	Murray *et al.* (1978)(2)	X					2 w	24	121(180)	—							

TABLE IV—*continued*

Locality	SE class[a]	Reference	Design[b]	Colostrum (1–5 days)				Transitional (6–14 days)				Mature (0.5–6 months)				Mature (> 6 months)		
				Stage[c]	N	Mean(SD)	Range	Stage[c]	N	Mean(SD)	Range	Stage[c]	N	Mean(SD)	Range	Stage[c]	N	Mean(SD)
Diffa	RLd	Murray *et al.* (1978)(2)	X					2 w	31	113(233)	—							
Diffa	RLo	Murray *et al.* (1978)(2)	X					2 w	7	117(87)	—							
Nigeria																		
Ibadan	UL	Atinmo and Omololu (1982)(1)	L					8–14 d	20	49(20)	—	2 m	20	43(16)	—			
Sudan																		
Khartoum	XX	El Tom Ali and Zaki (1976)(2)	X	1–3 d	7	155(—)	79–213											
Asia, Australasia, and Oceania[f]																		
India																		
Baroda	UXa	Karmarkar and Ramakrishnan (1960)(2)	X									3 or 4 m	8	193(41)	—			
Baroda	UXb	Karmarkar and Ramakrishnan (1960)(2)	X									3 or 4 m	8	172(46)	—			
Baroda	UXc	Karmarkar and Ramakrishnan (1960)(2)	X									3 or 4 m	8	211(101)	—			

TABLE IV—*continued*

Locality	SE class[a]	Reference	Design[b]	Colostrum (1–5 days)				Transitional (6–14 days)				Mature (0.5–6 months)				Mature (> 6 months)		
				Stage[c]	*N*	Mean(SD)	Range	Stage[c]	*N*	Mean(SD)	Range	Stage[c]	*N*	Mean(SD)	Range	Stage[c]	*N*	Mean(SD)
Baroda	UXd	Karmarkar and Rama-krishnan (1960)(2)	X									3 or 4 m	8	189(49)	—			
Coonoor	UL	Belavady and Gopalan (1959)(2)	X									2–6 m	9	128(74)	—	13–18 m	11	115(50)
Japan																		
Fukuyama	UX	Gunshin *et al.* (1985)(1)	X									19–178 d	12	32(18)	10–60	190–384 d	13	32(15)
Korea																		
Seoul	UH	Kim (1989)(1)	L									0.5 m	16	210(60)	—			
												1 m	16	190(40)	—			
												2 m	12	220(70)	—			
												3 m	13	210(90)	—			
												4 m	12	190(80)	—			
Malaya																		
Kuala Lumpur	UXc	Loh and Sinnathuray (1971)(2)	X	2 d	14	82(38)	—											
Kuala Lumpur	UXi	Loh and Sinnathuray (1971)(2)	X	2 d	19	111(50)	—											
Kuala Lumpur	UXm	Loh and Sinnathuray (1971)(2)	X	2 d	12	116(47)	—											

TABLE IV—*continued*

Locality	SE class[a]	Reference	Design[b]	Colostrum (1–5 days)				Transitional (6–14 days)				Mature (0.5–6 months)				Mature (> 6 months)		
				Stage[c]	*N*	Mean(SD)	Range	Stage[c]	*N*	Mean(SD)	Range	Stage[c]	*N*	Mean(SD)	Range	Stage[c]	*N*	Mean(SD)
Taiwan																		
Taipei	UX	Lonnerdal *et al.* (1990)(1) g	L									1 m 2 m 3 m 4 m	q q q q	34(—) 34(—) 40(—) 34(—)	— — — —			
Iron—Whey fraction (μg/dl)[f]																		
The Americas																		
Brazil																		
Rio de Janeiro	UL	Donangelo *et al.* (1989)(1)	X	1–5 d	15	32(16)	10–170											
Rio de Janeiro	UL	Trugo *et al.* (1988)(1)	X	1–5 d	11	64(56)	10–160											
Africa and Arabia																		
Asia, Australasia, and Oceania																		
Malaya																		
Kuala Lumpur	UXc	Loh and Sinnathuray (1971)(2)	X	2 d	14	57(24)	—											
Kuala Lumpur	UXi	Loh and Sinnathuray (1971)(2)	X	2 d	19	61(36)	—											
Kuala Lumpur	UXm	Loh and Sinnathuray (1971)(2)	X	2 d	12	81(39)	—											

TABLE IV—*continued*

Locality	SE class[a]	Reference	Design[b]	Colostrum (1–5 days)				Transitional (6–14 days)				Mature (0.5–6 months)				Mature (> 6 months)		
				Stage[c]	*N*	Mean(SD)	Range	Stage[c]	*N*	Mean(SD)	Range	Stage[c]	*N*	Mean(SD)	Range	Stage[c]	*N*	Mean(SD)
Iron—Fat fraction (μg/dl)																		
The Americas																		
Brazil																		
Rio de Janeiro	UL	Trugo *et al.* (1988)(1)	X	1–5 d	11	29(26)	5–100											
Africa and Arabia																		
Asia, Australasia, and Oceania																		
Iron—Total binding capacity (μg/dl)[f]																		
The Americas																		
Africa and Arabia																		
Asia, Australasia, and Oceania																		
Malaya																		
Kuala Lumpur	UXc	Loh and Sinnathuray (1971)(2)	X	2 d	14	153(90)	—											
Kuala Lumpur	UXi	Loh and Sinnathuray (1971)(2)	X	2 d	19	180(84)	—											
Kuala Lumpur	UXm	Loh and Sinnathuray (1971)(2)	X	2 d	12	207(68)	—											

TABLE IV—continued

Locality	SE class[a]	Reference	Design[b]	Colostrum (1–5 days)				Transitional (6–14 days)				Mature (0.5–6 months)				Mature (> 6 months)		
				Stage[c]	*N*	Mean(SD)	Range	Stage[c]	*N*	Mean(SD)	Range	Stage[c]	*N*	Mean(SD)	Range	Stage[c]	*N*	Mean(SD)
Magnesium (mg/dl)																		
The Americas																		
Africa and Arabia																		
Egypt																		
Kalama	RL	Karra *et al.* (1988)(1) g	L									1 m	62	2.86(—)	—			
												2 m	61	3.16(0.35)	—			
												3 m	53	3.33(0.46)	—			
												4 m	54	3.41(0.52)	—			
												5 m	52	3.57(0.16)	—			
												6 m	51	3.40(—)	—			
Ethiopia																		
Addis Ababa	UL	Fransson *et al.* (1984)(1)	X	4 or 5 d	9	2.56(0.29)	—											
Addis Ababa	UH	Fransson *et al.* (1984)(1)	X	4 or 5 d	9	2.29(0.23)	—											
Asia, Australasia, and Oceania																		
India																		
Coonoor	UL	Belavady and Gopalan (1959)(2)	X									2–6 m	20	2.9(1.3)	—	13–18 m	37	2.4(1.0)
Hyderabad	UM	Rajalakshmi and Srikantia (1980)(1)	X	2–5 d	76	4.04(1.39)	—	6–10 d	31	3.41(9.8)	—	1–3 m	77	3.17(0.92)	—	7–12 m	88	3.08(1.00)
												4–6 m	89	3.07(1.28)	—	> 13 m	23	2.91(1.15)
Hyderabad	RL	Rajalakshmi and Srikantia (1980)(1)	X									1–3 m	22	3.58(2.33)	—	7–12 m	73	3.19(0.96)
												4–6 m	41	3.29(0.67)	—	> 13 m	86	2.89(0.51)

TABLE IV—*continued*

Locality	SE class[a]	Reference	Design[b]	Colostrum (1–5 days)				Transitional (6–14 days)				Mature (0.5–6 months)				Mature (> 6 months)		
				Stage[c]	*N*	Mean(SD)	Range	Stage[c]	*N*	Mean(SD)	Range	Stage[c]	*N*	Mean(SD)	Range	Stage[c]	*N*	Mean(SD)
Udaipur	XX	Gupta *et al.* (1984)(1)	L	2–5 d	50	4.41(0.92)	—	6–10 d	22	4.29(1.48)	—							
Bombay	UX	Raut and Viswanathan (1972)(1)	X	1–5 d	9	3.6(4.4)	2.2–8.7											
Korea																		
Seoul	UH	Seol (1990)(1)	L									15 d	16	2.42(0.01)	—			
												30 d	16	2.61(0.45)	—			
												60 d	16	2.97(0.51)	—			
												90 d	16	2.94(0.42)	—			
												120 d	16	2.98(0.44)	—			
												150 d	16	2.90(0.37)	—			
Nepal																		
Katmandu Valley	RL	Moser *et al.* (1988b)(1)	X									2–6 m	26	3.2(1.5)	—			
Taiwan																		
Taipei	UX	Lonnerdal *et al.* (1990)(1) g	L									1 m	q	2.25(—)	—			
												2 m	q	2.75(—)	—			
												3 m	q	3.00(—)	—			
												4 m	q	3.00(—)	—			

TABLE IV—*continued*

Locality	SE class[a]	Reference	Design[b]	Colostrum (1–5 days)				Transitional (6–14 days)				Mature (0.5–6 months)				Mature (> 6 months)		
				Stage[c]	*N*	Mean(SD)	Range	Stage[c]	*N*	Mean(SD)	Range	Stage[c]	*N*	Mean(SD)	Range	Stage[c]	*N*	Mean(SD)
Manganese (μg/dl)																		
The Americas																		
Africa and Arabia																		
Asia, Australasia, and Oceania																		
Japan																		
Fukuyama	UX	Gunshin *et al.* (1985)(1)	X									19–178 d	12	1.08(0.6)	0.39–2.12	190–384 d	13	0.83(0.62)
Molybdenum (μg/dl)																		
The Americas																		
Africa and Arabia																		
Asia, Australasia, and Oceania																		
Japan																		
Fukuyama	UX	Gunshin *et al.* (1985)(1)	X									19–178 d	12	2.0(1.6)	0.5–5.4	190–384 d	13	2.7(2.1)

TABLE IV—*continued*

Locality	SE class[a]	Reference	Design[b]	Colostrum (1–5 days)				Transitional (6–14 days)				Mature (0.5–6 months)				Mature (> 6 months)		
				Stage[c]	*N*	Mean(SD)	Range	Stage[c]	*N*	Mean(SD)	Range	Stage[c]	*N*	Mean(SD)	Range	Stage[c]	*N*	Mean(SD)
Nickel (μg/dl)																		
The Americas																		
Africa and Arabia																		
Asia, Australasia, and Oceania																		
Japan																		
Fukuyama	UX	Gunshin *et al.* (1985)(1)	X									19–178 d	9	0.35(0.25)	0.08–0.74	190–384 d	12	0.48(0.32)
Phosphorus (mg/dl)																		
The Americas																		
Brazil																		
Sao Paulo	UH	Carneiro and de Oliveria (1973)(2)[g]	L					7 d	6	18.0(5.9)	—	28 d	9	18.0(3.7)	—			
								14 d	7	16.6(4.6)	—	56 d	7	16.9(3.8)	—			
												82 d	4	13.7(0.7)	—			
Sao Paulo	UL	Carneiro and de Oliveria (1973)(2)[g]	L					7 d	8	16.8(5.9)	—	28 d	10	16.8(4.5)	—			
								14 d	8	16.7(4.3)	—	56 d	9	16.1(4.8)	—			
												82 d	8	16.7(5.1)	—			
												110 d	7	19.0(4.7)	—			
												138 d	7	16.7(2.2)	—			
												159 d	7	18.4(4.3)	—			
												186 d	6	20.9(4.1)	—			

TABLE IV—*continued*

Locality	SE class[a]	Reference	Design[b]	Colostrum (1–5 days)				Transitional (6–14 days)				Mature (0.5–6 months)				Mature (> 6 months)		
				Stage[c]	*N*	Mean(SD)	Range	Stage[c]	*N*	Mean(SD)	Range	Stage[c]	*N*	Mean(SD)	Range	Stage[c]	*N*	Mean(SD)
Africa and Arabia																		
Nigeria																		
Lagos	UX	Bassir (1956)(2)[g]	X									1 m	25	10.5(2.7)	6.0–17.7	7 m	7	7.9(1.5)
												2 m	18	8.6(1.9)	7.6–12.5	8 m	3	7.3(1.3)
												3 m	19	9.4(1.5)	6.9–12.4	9 m	3	7.1(0.4)
												4 m	10	9.8(2.7)	7.3–16.2	10 m	8	9.5(2.3)
												5 m	14	8.6(2.2)	5.9–11.7	11 m	6	9.5(3.1)
												6 m	7	9.0(2.0)	6.0–12.0	12 m	3	10.5(2.9)
Ibadan	UL	Bassir (1958)(2)[g]	X					1 w	6	14.6(2.6)	11.4–18.4	3–26 w	11	15.2(3.8)	9.2–22.0	27–78 w	16	14.2(4.7)
The Gambia																		
Keneba	RL	Laskey *et al.* (1991)(2)	X									0.5–2.9 m	29	16.5(2.7)	—	6.0–8.9 m	8	15.7(2.3)
												3.0–5.9 m	32	16.1(2.3)	—	9.0–26.0 m	30	15.8(2.2)
Zaire																		
Yasa-Bonga	RL	Prentice (1991)(2)	L					7–16 d	12	16.5(3.6)	10.0–22.8	6 w	12	14.8(2.4)	11.2–19.3	9 m	12	13.8(2.4)
												9 w	12	14.9(2.2)	10.4–17.5	12 m	12	13.9(1.8)
												18 w	12	14.6(2.1)	10.6–17.5	15 m	12	13.5(1.7)
												6 m	12	13.6(1.4)	10.7–15.1	18 m	11	13.6(1.8)
Asia, Australia, and Oceania																		
India																		
Baroda	UXa	Karmarkar and Ramakrishnan (1960)(2)[g]	X									3 or 4 m	14	11.4(4.1)				

TABLE IV—*continued*

Locality	SE class[a]	Reference	Design[b]	Colostrum (1–5 days)				Transitional (6–14 days)				Mature (0.5–6 months)				Mature (> 6 months)		
				Stage[c]	*N*	Mean(SD)	Range	Stage[c]	*N*	Mean(SD)	Range	Stage[c]	*N*	Mean(SD)	Range	Stage[c]	*N*	Mean(SD)
Baroda	UXb	Karmarkar and Ramakrishnan (1960)(2)[g]	X									3 or 4 m	14	9.9(4.4)				
Baroda	UXc	Karmarkar and Ramakrishnan (1960)(2)[g]	X									3 or 4 m	16	9.3(2.8)				
Baroda	UXd	Karmarkar and Ramakrishnan (1960)(2)[g]	X									3 or 4 m	15	11.8(5.0)				
Coonoor	UL	Belavady and Gopalan (1959)(2)	X									2–6 m	43	11.7(1.1)	—	7–12 m	31	12.0(2.1)
																13–18 m	30	12.6(2.7)
																> 18 m	20	11.1(1.9)
Korea																		
Seoul	UH	Seol (1988, (1990)(2)	L									15 d	16	17.3(2.0)	14.2–19.8			
												30 d	16	15.9(1.7)	13.6–18.0			
												60 d	12	14.9(0.9)	13.9–16.6			
												90 d	13	13.7(1.1)	11.1–15.0			
												120 d	12	13.4(1.2)	10.6–15.2			
												150 d	7	13.0(1.3)	11.2–14.8			
Taiwan																		
Taipei	UX	Lonnerdal *et al.* (1990)(2)[g] g	L									1 m	q	7.75(—)	—			
												2 m	q	6.75(—)	—			
												3 m	q	6.50(—)	—			
												4 m	q	6.50(—)	—			

TABLE IV—*continued*

Locality	SE class[a]	Reference	Design[b]	Colostrum (1–5 days)				Transitional (6–14 days)				Mature (0.5–6 months)				Mature (> 6 months)		
				Stage[c]	N	Mean(SD)	Range	Stage[c]	N	Mean(SD)	Range	Stage[c]	N	Mean(SD)	Range	Stage[c]	N	Mean(SD)
Vanatu (New Hebrides)																		
Port Vila	RL	Peters (1953)(2)[g]	X									2–5 m	18	15.9(1.7)	13–18	6–11 m	15	15.1(2.3)
Potassium (mg/dl)																		
The Americas																		
Africa and Arabia																		
Sudan																		
Khartoum	XX	El Tom Ali and Zaki (1976)(1)	X	1–3 d	7	28(—)	12–60											
The Gambia																		
Keneba	RL	Prentice (unpub lished)(1)	X									0.5–2.9 m	13	58.9(10.5)	—	6.0–8.9 m	9	52.7(9.0)
												3.0–5.9 m	6	53.4(4.3)	—	9.0–11.9 m	3	59.7(6.6)
																12.0–14.9 m	7	51.9(7.0)
																15.0–17.9 m	9	53.8(5.1)
																18.0–20.9 m	5	44.9(8.2)
Asia, Australasia, and Oceania																		
India																		
Coonoor	UL	Belavady and Gopalan (1959)(2)	X									2–6 m	15	37.3(8.6)	—	13–18 m	29	33.3(12.7)
Hyderabad	UX	Mohan *et al.* (1983)(1)	L	1–5 d	19	59.2(8.1)	—	6–10 d	19	66.5(15.2)	—	16–20 d	19	61.6(11.0)	—			
								11–15 d	19	65.4(21.1)	—							

TABLE IV—*continued*

Locality	SE class[a]	Reference	Design[b]	Colostrum (1–5 days)				Transitional (6–14 days)				Mature (0.5–6 months)				Mature (> 6 months)		
				Stage[c]	*N*	Mean(SD)	Range	Stage[c]	*N*	Mean(SD)	Range	Stage[c]	*N*	Mean(SD)	Range	Stage[c]	*N*	Mean(SD)
Korea																		
Seoul	UH	Seol (1989)(1)	L									0.5–5 m	16	47.2(—)	—			
Selenium (μg/dl)[h]																		
The Americas																		
Africa and Arabia																		
The Gambia																		
Keneba	RLw	Funk *et al.* (1990)	X									1–6 m	8	1.53(0.33)	—	13–19 m	13	1.75(0.34)
Keneba	RLd	Funk *et al.* (1990)	X									1–6 m	15	2.10(0.34)	—	13–19 m	19	1.94(0.35)
Asia, Australasia, and Oceania																		
Nepal																		
Katmandu Valley	RL	Reynolds *et al.* (1986); Moser *et al.* (1988a)	X									2–6 m	26	1.00(0.51)	—			

TABLE IV—*continued*

Locality	SE class[a]	Reference	Design[b]	Colostrum (1–5 days)				Transitional (6–14 days)				Mature (0.5–6 months)				Mature (> 6 months)		
				Stage[c]	*N*	Mean(SD)	Range	Stage[c]	*N*	Mean(SD)	Range	Stage[c]	*N*	Mean(SD)	Range	Stage[c]	*N*	Mean(SD)
Sodium (mg/dl)																		
The Americas																		
Africa and Arabia																		
Sudan																		
Khartoum	XX	El Tom Ali and Zaki (1976)(1)	X	1–3 d	7	16(—)	5–32											
The Gambia																		
Keneba	RL	Prentice (unpublished)(1)	X									0.5–2.9 m	13	16.1(5.3)	—	6.0–8.9 m	9	12.7(4.1)
												3.0–5.9 m	6	12.2(5.3)	—	9.0–11.9 m	3	11.7(3.9)
																12.0–14.9 m	7	13.1(5.1)
																15.0–17.9 m	9	16.6(7.6)
																18.0–20.9 m	5	15.0(6.2)
Asia, Australasia, and Oceania																		
India																		
Coonoor	UL	Belavady and Gopalan (1959)(2)	X									2–6 m	16	23.1(10.5)	—	13–18 m	26	21.5(15.1)
Hyderabad	UX	Mohan *et al.* (1983)(1)	L	1–5 d	19	27.4(5.6)	—	6–10 d	19	27.0(9.0)	—	16–20 d	19	24.4(5.1)	—			
								11–15 d	19	26.9(5.6)	—							
Korea																		
Seoul	UH	Seol (1989)(1)	L									0.5–5 m	16	19.8(—)				

TABLE IV—continued

Locality	SE class[a]	Reference	Design[b]	Colostrum (1–5 days)				Transitional (6–14 days)				Mature (0.5–6 months)				Mature (> 6 months)		
				Stage[c]	N	Mean(SD)	Range	Stage[c]	N	Mean(SD)	Range	Stage[c]	N	Mean(SD)	Range	Stage[c]	N	Mean(SD)
Zinc—Total (μg/dl)																		
The Americas																		
Brazil																		
Amazonas	UL	Shrimpton *et al.* (1985)(1) g	L									30 d 120 d	28 28	208(111) 133(63)	— —			
Amazonas	ULS	Shrimpton *et al.* (1985)(1) g	L									30 d 120 d	37 37	186(128) 142(73)	— —			
Amazonas	UL	Lehti (1989, 1990)(1)	X									0.5–1 m 1 or 2 m 2 or 3 m	37 59 41	220(90) 160(90) 140(90)	43–389 6–357 6–342			
Rio de Janeiro	UL	Donangelo *et al.* (1989)(1)	X	1–5 d	17	594(227)	138–925											
Rio de Janeiro	UL	Trugo *et al.* (1988)(1)	X	1–5 d	22	726(303)	130–1370											
United States (Navajo Indians)																		
Tuba City	RL	Butte and Calloway (1981)(1)	X									19–62 d	23	280(110)	70–460			

TABLE IV—*continued*

Locality	SE class[a]	Reference	Design[b]	Colostrum (1–5 days)				Transitional (6–14 days)				Mature (0.5–6 months)				Mature (> 6 months)		
				Stage[c]	*N*	Mean(SD)	Range	Stage[c]	*N*	Mean(SD)	Range	Stage[c]	*N*	Mean(SD)	Range	Stage[c]	*N*	Mean(SD)
Africa and Arabia																		
Egypt																		
Kalama	RL	Karra *et al.* (1988)(1) g	L									1 m	63	236(151)	—			
												2 m	62	147(87)	—			
												3 m	56	111(82)	—			
												4 m	60	96(62)	—			
												5 m	51	83(64)	—			
												6 m	52	78(72)	—			
Ethiopia																		
Addis Ababa	UL	Fransson *et al.* (1984)(1)	X	4 or 5 d	9	659(206)	—											
Addis Ababa	UH	Fransson *et al.* (1984)(1)	X	4 or 5 d	9	666(271)	—											
Ivory Coast																		
Kpouebo	RL	Lauber and Reinhardt (1979)(1)	X									1 m	4	350(90)	—	12 m	8	160(90)
												6 m	7	230(90)	—	18 m	6	150(130)
Nigeria																		
Ibadan	UL	Atinmo and Omololu (1982)(1)	L					8–14 d	20	549(73)	—	2 m	20	393(78)	—			
The Gambia																		
Jali	RL	Bates and Tsuchiya (1990)(1)	L									1 m	q	410(—)	—	6.1–9 m	q	145(—)
												2 m	q	340(—)	—	9.1–12 m	q	113(—)
												3 m	q	270(—)	—	12.1–15 m	q	126(—)
												4 m	q	166(—)	—	15.1–18 m	q	117(—)
												5 m	q	164(—)	—			
												6 m	q	209(—)	—			

TABLE IV—*continued*

Locality	SE class[a]	Reference	Design[b]	Colostrum (1–5 days)				Transitional (6–14 days)				Mature (0.5–6 months)				Mature (> 6 months)		
				Stage[c]	*N*	Mean(SD)	Range	Stage[c]	*N*	Mean(SD)	Range	Stage[c]	*N*	Mean(SD)	Range	Stage[c]	*N*	Mean(SD)
Asia, Australasia, and Oceania																		
India																		
Coonoor	UL	Belavady (1959)(2)	X									2–6 m	10	354(87)	—	13–18 m	12	330(162)
Hyderabad	UM	Rajalakshmi and Srikantia (1980)(1)	X	2–5 d	76	532(272)	—	6–10 d	31	472(153)	—	1–3 m	77	200(89)	—	7–12 m	88	112(49)
												4–6 m	89	133(52)	—	> 13 m	23	116(56)
Hyderabad	RL	Rajalakshmi	X									1–3 m	22	188(77)	—	7–12 m	73	103(57)
												4–6 m	41	132(72)	—	> 13 m	86	103(102)
Udaipur	XX	Gupta *et al.* (1984)(1)	L	2–5 d	50	222(96)	—	6–10 d	22	183(29)	—							
Japan																		
Kumamoto	XX	Higashi *et al.* (1982)(1)	L	1 d	65	1039(443)	—	1 w	65	456(301)	—	1 m	65	266(103)	—			
												3 m	45	114(67)	—			
												5 m	35	105(46)	—			
Nepal																		
Katmandu Valley	RL	Moser *et al.* (1988)(1)	X									2–6 m	26	110(50)	—			
Taiwan																		
Taipei	UX	Lonnerdal *et al.* (1990)(1) g	L									1 m	q	260(—)	—			
												2 m	q	205(—)	—			
												3 m	q	150(—)	—			
												4 m	q	120(—)	—			

TABLE IV—continued

Locality	SE class[a]	Reference	Design[b]	Colostrum (1–5 days)				Transitional (6–14 days)				Mature (0.5–6 months)				Mature (> 6 months)		
				Stage[c]	*N*	Mean(SD)	Range	Stage[c]	*N*	Mean(SD)	Range	Stage[c]	*N*	Mean(SD)	Range	Stage[c]	*N*	Mean(SD)
Zinc—Whey fraction (μg/dl)																		
The Americas																		
Brazil																		
Rio de Janeiro	UL	Donangelo *et al.* (1989)(1)	X	1–5 d	16	454(48)	141–759											
Rio de Janeiro	UL	Trugo *et al.* (1988)(1)	X	1–5 d	11	499(182)	110–770											
Africa and Arabia																		
Asia, Australasia, and Oceania																		

TABLE IV—*continued*

Locality	SE class[a]	Reference	Design[b]	Colostrum (1–5 days)				Transitional (6–14 days)				Mature (0.5–6 months)				Mature (> 6 months)		
				Stage[c]	*N*	Mean(SD)	Range	Stage[c]	*N*	Mean(SD)	Range	Stage[c]	*N*	Mean(SD)	Range	Stage[c]	*N*	Mean(SD)
Zinc—Fat fraction (μg/dl)																		
The Americas																		
Brazil																		
Rio de Janeiro	UL	Trugo *et al.* (1988)(1)	X	1–5 d	11	74(50)	20–180											
Africa and Arabia																		
Asia, Australasia, and Oceania																		

Note. f, figures calculated from information in text; g, figures taken from graphs; p, pooled sample; q, numbers of subjects not given; s, supplemented mothers; z, geometric mean (+1 geometric standard deviation).

[a]Socioeconomic class: RL, rural, poor; UH, urban, middle-high income; UL, urban, poor, and low-middle income; UM, urban, mixed socioeconomic class; UX, urban, socioeconomic class not stated; a, lowest to d, highest; quartiles of maternal intake; *m*, malnourished and undernourished mothers; *n*, good maternal nutritional status.

[b]Study design: L, longitudinal or semilongitudinal design; X, cross-sectional design.

[c]Stage of lactation: d, days postpartum; w, weeks postpartum; m, months postpartum.

[d]Methods: 1, atomic absorption/flame emission or equivalent; 2, other methods.

[e]Maternal iron status: n, normal; d, deficient; o, overload.

[f]c, Chinese; i, indian, m, malay.

[g]No prior ashing reported.

[h]w, Wet season; d, dry season.

TABLE V
Vitamins

Locality	SE class[a]	Reference	Design[b]	Colostrum (1–5 days)				Transitional (6–14 days)				Mature (0.5–6 months)				Mature (> 6 months)		
				Stage[c]	N	Mean(SD)	Range	Stage[c]	N	Mean(SD)	Range	Stage[c]	N	Mean(SD)	Range	Stage[c]	N	Mean(SD)
Vitamin A (μg retinol equivalents/dl)[d]																		
The Americas																		
Guatemala																		
Guatemala City	UH	WHO (1985)	X									1 m 3 m	30 28	79(34) 68(38)	— —			
Guatemala City	UL	WHO (1985)	X									1 m 3 m	27 29	77(36) 58(20)	— —	6 m 9 m	26 24	50(19) 47(23)
Santa Maria Cauque	RLe	WHO (1985)	X									1 m 3 m	28 27	51(27) 48(36)	— —	6 m 9 m 15 m	27 27 27	40(23) 38(17) 36(18)
United States (Navajo Indians)																		
Tuba City	RL	Butte and Calloway (1981)	X									19–62 d	23	32.9(15.7)	10.7–64.7			
Africa and Arabia																		
Ethiopia																		
Addis Ababa	UL	Gebre-Medhin *et al.* (1976)	X									0.5–1.5 m 1.5–3.5 m 3.5–6.5 m	3 14 25	29.0(9.5) 33.1(14.8) 28.1(15.0)	— — —	6.5–11.5 m 11.5–23.5 m	20 22	22.6(8.5) 21.2(7.2)
Addis Ababa	UH	Gebre-Medhin *et al.* (1976)	X					7–14 d	4	44.0(15.3)	—	0.5–1.5 m 1.5–3.5 m	14 5	36.2(7.7) 36.4(7.9)	— —			

TABLE V—*continued*

Locality	SE class[a]	Reference	Design[b]	Colostrum (1–5 days)				Transitional (6–14 days)				Mature (0.5–6 months)				Mature (> 6 months)		
				Stage[c]	N	Mean(SD)	Range	Stage[c]	N	Mean(SD)	Range	Stage[c]	N	Mean(SD)	Range	Stage[c]	N	Mean(SD)
Kenya																		
Machakos	RL	Van Steen-bergen *et al.* (1981)	X	3 d	59	34(26)	–											
Nigeria																		
Ibadan	XX	Naismith (1973)	X													7 m	12	112(–)
The Gambia																		
Manduar	RL	Villard and	L									3 w	18	118(–)	–			
		Bates (1987)										4 w	18	77(–)	–			
		g										5 w	18	75(–)	–			
												6 w	18	50(–)	–			
												7 w	18	54(–)	–			
												8 w	18	66(–)	–			
												9 w	18	63(–)	–			
												10 w	18	66(–)	–			
												11 w	18	73(–)	–			
												12 w	18	59(–)	–			
												13 w	18	63(–)	–			
												14 w	18	63(–)	–			

TABLE V—*continued*

Locality	SE class[a]	Reference	Design[b]	Colostrum (1–5 days)				Transitional (6–14 days)				Mature (0.5–6 months)				Mature (> 6 months)		
				Stage[c]	*N*	Mean(SD)	Range	Stage[c]	*N*	Mean(SD)	Range	Stage[c]	*N*	Mean(SD)	Range	Stage[c]	*N*	Mean(SD)
Keneba	RLs	Villard and Bates (1987) g	L									3 w	37	101(—)	—			
												4 w	37	100(—)	—			
												5 w	37	104(—)	—			
												6 w	37	83(—)	—			
												7 w	37	97(—)	—			
												8 w	37	85(—)	—			
												9 w	37	110(—)	—			
												10 w	37	80(—)	—			
												12 w	37	75(—)	—			
												13 w	37	97(—)	—			
												14 w	37	68(—)	—			
Asia, Australasia, and Oceania																		
India																		
Coonoor	UL	Belavady and Gopalan (1959)	X	3–10 d	26	103(75)	—					2–6 m	36	24(13)	—	7–12 m	28	20(16)
																13–18 nm	20	20(10)
																> 18 m	17	20(8)
Hyderabad	UL	Venkatachalam *et al.* (1962)	X	3 d	q	207(—)	—											
Hyderabad	ULs	Venkatachalam *et al.* (1962)	X	3 d	12	501(703)	—											
New Delhi	UH	Ashdir and Puri (1962)	L	3 d	10	114(30)	87–182	8 d	10	60(24)	49–127	18 d	10	48(9)	36–72			
Indonesia																		
Yogyakarta	UL	Boediman *et al.* (1979)	X													13–24 m	66	16.3(9.5)
																25–36 m	45	13.0(15.0)

TABLE V—*continued*

Locality	SE class[a]	Reference	Design[b]	Colostrum (1–5 days)				Transitional (6–14 days)				Mature (0.5–6 months)				Mature (> 6 months)		
				Stage[c]	N	Mean(SD)	Range	Stage[c]	N	Mean(SD)	Range	Stage[c]	N	Mean(SD)	Range	Stage[c]	N	Mean(SD)
Pakistan																		
Karachi	UL	Lindblad and Rahim-toola (1974)	X									1.5–6 m	9	49(22)	17–98			
Philippines																		
Luzon	RL	WHO (1985)	X									1 m	27	34(19)	—	9 m	32	25(11)
												3 m	28	36(18)	—	15 m	20	28(16)
Manila	UL	WHO (1985)	X									1 m	32	45(21)	—	9 m	31	28(14)
												3 m	29	39(24)	—	15 m	18	32(13)
Manila	UH	WHO (1985)	X									1 m	33	60(28)	—	9 m	15	35(13)
												3 m	20	48(21)	—			
Sri Lanka (Ceylon)																		
Colombo	XX	de Silva (1964)	X					7 d	36	14.5(—)	—							
β-Carotene (μg/dl)																		
The Americas																		
United States (Navajo Indians)																		
Tuba City	RL	Butte and Calloway (1981)	X									19–62 d	23	19.7(6.3)	11.1–32.3			

TABLE V—*continued*

Locality	SE class[a]	Reference	Design[b]	Colostrum (1–5 days)				Transitional (6–14 days)				Mature (0.5–6 months)				Mature (> 6 months)		
				Stage[c]	*N*	Mean(SD)	Range	Stage[c]	*N*	Mean(SD)	Range	Stage[c]	*N*	Mean(SD)	Range	Stage[c]	*N*	Mean(SD)
Africa and Arabia																		
Ethiopia																		
Addis Ababa	UL	Gebre-Medhin *et al.* (1976)	X									0.5–1.5 m	3	25.3(12.8)	—	6.5–11.5 m	20	21.0(9.2)
												1.5–3.5 m	14	23.9(8.8)	—	11.5–23.5 m	22	18.8(7.2)
												3.5–6.5 m	25	25.6(12.3)	—			
Addis Ababa	UH	Gebre-Medhin *et al.* (1976)	X					7–14 d	4	42.8(31.8)	—	0.5–1.5 m	14	28.1(16.1)	—			
												1.5–3.5 m	5	26.2(12.0)	—			
Kenya																		
Machakos	RL	Van Steenbergen *et al.* (1981)	X	3 d	40	30(20)	—											
Asia, Australasia, and Oceania																		
Vitamin B1 (thiamin) (μg/dl)																		
The Americas																		
Africa and Arabia																		
Kenya																		
Machakos	RL	Van Steenbergen *et al.* (1981)	X	3 d	28	23(1)	—											
The Gambia																		
Keneba	RL	Prentice *et al.* (1983)	X									0.5–6 m	21	16(3)	—			

TABLE V—*continued*

Locality	SE class[a]	Reference	Design[b]	Colostrum (1–5 days)				Transitional (6–14 days)				Mature (0.5–6 months)				Mature (> 6 months)		
				Stage[c]	*N*	Mean(SD)	Range	Stage[c]	*N*	Mean(SD)	Range	Stage[c]	*N*	Mean(SD)	Range	Stage[c]	*N*	Mean(SD)
Keneba	RLs	Prentice *et al.* (1983)	X									0.5–6 m	23	22(3)	—			
Asia, Australasia, and Oceania																		
India																		
Baroda	UM	Deodhar and Rama-krishnan (1959)	X									1–3 m 3–6 m	46 59	11.5(4.8) 10.5(5.1)	— —	6–12 m > 12 m	61 25	12.3(3.9) 14.0(5.2)
Baroda	UL	Deodhar *et al.* (1964)	L									1–3 m	10	10.9(0.3)	—	9–12 m	5	12.2(0.4)
Baroda	ULs	Deodhar *et al.* (1964)	L													9–12 m	5	26.8(1.6)
Coonoor	UL	Belavady and Gopalan (1959)	X	3–10 d	34	5.6(5.5)	—					2–6 m	34	15.3(5.6)	—	7–12 m 13–18 m > 18 m	25 31 18	14.5(5.1) 16.5(5.6) 15.1(5.7)
Sri Lanka (Ceylon)																		
Colombo	XX	de Silva (1964)	X					7 d	36	21(—)	—							

TABLE V—*continued*

Locality	SE class[a]	Reference	Design[b]	Colostrum (1–5 days)				Transitional (6–14 days)				Mature (0.5–6 months)				Mature (> 6 months)		
				Stage[c]	*N*	Mean(SD)	Range	Stage[c]	*N*	Mean(SD)	Range	Stage[c]	*N*	Mean(SD)	Range	Stage[c]	*N*	Mean(SD)
Vitamin B2 (riboflavin) (μg/dl)																		
The Americas																		
Africa and Arabia																		
Kenya																		
Machakos	RL	Van Steenbergen *et al.* (1981)	X	3 d	28	14(5)	—											
The Gambia																		
Keneba	RL	Prentice *et al.* (1983)	X									0.5–6 m	21	21(5)	—			
Keneba	RLs	Prentice *et al.* (1983)	X									0.5–6 m	23	28(5)	—			
Asia, Australasia, and Oceania																		
India																		
Baroda	UM	Deodhar and Ramakrishnan (1959)	X									1–3 m 3–6 m	46 59	26.0(10.8) 25.3(10.3)	— —	6–12 m > 12 m	61 25	25.9(9.0) 26.5(8.7)
Baroda	UL	Deodhar *et al.* (1964)	L									1–3 m	10	20.0(1.5)	—	9–12 m	5	22.0(1.3)
Baroda	ULs	Deodhar *et al.* (1964)	L													9–12 m	5	74.0(6.5)
Coonoor	UL	Belavady and Gopalan (1959)	X	3–10 d	17	28.9(20.0)	—					2–6 m	35	19.0(9.4)	—	7–12 m 13–18 m > 18 m	29 30 15	18.1(10.8) 17.1(11.0) 11.5(8.1)

TABLE V—*continued*

Locality	SE class[a]	Reference	Design[b]	Colostrum (1–5 days)				Transitional (6–14 days)				Mature (0.5–6 months)				Mature (> 6 months)		
				Stage[c]	*N*	Mean(SD)	Range	Stage[c]	*N*	Mean(SD)	Range	Stage[c]	*N*	Mean(SD)	Range	Stage[c]	*N*	Mean(SD)
Hyderabad	UL	Bamji *et al.* (1986)	L	< 5 d	16	25.7(8.5)	—					1–6 m	23	23.0(7.8)	—	7–12 m 13–18 m > 18 m	26 17 18	21.0(7.9) 22.4(13.3) 27.0(10.7)
Sri Lanka (Ceylon)																		
Colombo	XX	de Silva (1964)	X					7 d	36	23(—)	—							
Niacin (nicotinic acid) (μg/dl)																		
The Americas																		
Africa and Arabia																		
The Gambia																		
Keneba	RL	Prentice *et al.* (1983)	X									0.5–6 m	21	113(50)	—			
Keneba	RLs	Prentice *et al.* (1983)	X									0.5–6 m	23	162(10)	—			
Asia, Australasia, and Oceania																		
India																		
Baroda	UL	Deodhar *et al.* (1964)	L									1–3 m	10	99.0(4.1)	—	9–12 m	5	102(6)
Baroda	ULs	Deodhar *et al.* (1964)	L													9–12 m	5	275(20)
Sri Lanka (Ceylon)																		
Colombo	XX	de Silva (1964)	X					7 d	36	87(—)	—							

TABLE V—*continued*

Locality	SE class[a]	Reference	Design[b]	Colostrum (1–5 days)				Transitional (6–14 days)				Mature (0.5–6 months)				Mature (> 6 months)		
				Stage[c]	*N*	Mean(SD)	Range	Stage[c]	*N*	Mean(SD)	Range	Stage[c]	*N*	Mean(SD)	Range	Stage[c]	*N*	Mean(SD)
Pantothenic acid (μg/dl)[e]																		
The Americas																		
Africa and Arabia																		
The Gambia																		
Keneba	RL	Prentice *et al.* (1983)	X									0.5–6 m	21	204(50)	—			
Keneba	RLs	Prentice *et al.* (1983)	X									0.5–6 m	23	227(48)	—			
Asia, Australasia, and Oceania																		
India																		
Baroda	UM	Deodhar and Ramakrishnan (1959)	X									1–3 m 3–6 m	46 59	159(52) 145(61)	— —	6–12 m > 12 m	61 25	145(44) 139(59)
Baroda	UL	Deodhar *et al.* (1964)	L									1–3 m	10	99(3.1)	—	9–12 m	5	103(3[illegible]
Baroda	ULs	Deodhar *et al.* (1964)	L													9–12 m	5	303(18)

TABLE V—*continued*

Locality	SE class[a]	Reference	Design[b]	Colostrum (1–5 days)				Transitional (6–14 days)				Mature (0.5–6 months)				Mature (> 6 months)		
				Stage[c]	*N*	Mean(SD)	Range	Stage[c]	*N*	Mean(SD)	Range	Stage[c]	*N*	Mean(SD)	Range	Stage[c]	*N*	Mean(SD)
Biotin (μg/dl)																		
The Americas																		
Africa and Arabia																		
The Gambia																		
Keneba	RL	Prentice *et al.* (1983)	X									0.5–6 m	21	897(316)	—			
Keneba	RLs	Prentice *et al.* (1983)	X									0.5–6 m	23	717(230)	—			
Asia, Australasia, and Oceania																		
India																		
Baroda	UL	Deodhar *et al.* (1964)	L									1–3 m	10	160(30)	—	9–12 m	5	160(20)
Baroda	ULs	Deodhar *et al.* (1964)	L													9–12 m	5	500(110)
Vitamin B6 (pyridoxine) (μg/dl)[f]																		
The Americas																		
Africa and Arabia																		
The Gambia																		
Keneba	RL	Prentice *et al.* (1983)	X									0.5–6 m	21	12(2)	—			
Keneba	RLs	Prentice *et al.* (1983)	X									0.5–6 m	23	10(2)	—			

TABLE V—*continued*

Locality	SE class[a]	Reference	Design[b]	Colostrum (1–5 days)				Transitional (6–14 days)				Mature (0.5–6 months)				Mature (> 6 months)		
				Stage[c]	*N*	Mean(SD)	Range	Stage[c]	*N*	Mean(SD)	Range	Stage[c]	*N*	Mean(SD)	Range	Stage[c]	*N*	Mean(SD)
Asia, Australasia, and Oceania																		
India																		
Baroda	UL	Deodhar *et al.* (1964)	L									1–3 m	10	7.9(0.6)	—	9–12 m	5	8.0(0.4)
Baroda	ULs	Deodhar *et al.* (1964)	L													9–12 m	5	15.8(0.7)
Hyderabad	UL	Bamji *et al.* (1986)	X	< 5 d	15	2.09(1.52)	—					1–6 m	23	6.06(1.92)	—	7–12 m	26	7.11(2.29)
																13–18 m	17	8.04(2.68)
																> 18 m	18	6.24(2.42)
Nepal																		
Katmandu Valley	RL	Reynolds *et al.* (1986) f	X									2 m	q	18.8(4.7)	—			
												3 m	q	17.6(13.2)	—			
												4 m	q	19.5(8.8)	—			
												5 m	q	17.6(5.5)	—			
												6 m	q	17.7(2.5)	—			
Vitamin B12 (ng/dl)[g]																		
The Americas																		
Brazil																		
Rio de Janeiro	UL	Donangelo *et al.* (1989) f	X	1–5 d	17	111(145)	12–540											
Rio de Janeiro	UL	Trugo *et al.* (1988) f	X	1–5 d	26	120(94)	14–407											

TABLE V—*continued*

Locality	SE class[a]	Reference	Design[b]	Colostrum (1–5 days)				Transitional (6–14 days)				Mature (0.5–6 months)				Mature (> 6 months)		
				Stage[c]	*N*	Mean(SD)	Range	Stage[c]	*N*	Mean(SD)	Range	Stage[c]	*N*	Mean(SD)	Range	Stage[c]	*N*	Mean(SD)
Africa and Arabia																		
The Gambia																		
Keneba	RL	Prentice *et al.* (1983)	X									0.5–6 m	21	16(18)	—			
Keneba	RLs	Prentice *et al.* (1983)	X									0.5–6 m	23	12(10)	—			
Asia, Australasia, and Oceania																		
Thailand																		
Bangkok	UX	Areekul *et al.* (1977)	X	2 d	8	74(57)	—	6–10 d	7	21(32)	—							
				3 d	11	44(32)	—											
				4 d	11	32(27)	—											
				5 d	8	26(29)	—											
India																		
Baroda	UM	Deodhar and Ramakrishnan (1959)	X									1–3 m	46	11.7(6.6)	—	6–12 m	61	9.1(6.4)
												3–6 m	59	9.3(3.4)	—	> 12 m	25	11.0(4.0)
Baroda	UL	Deodhar *et al.* (1964)	L									1–3 m	10	7.8(0.4)	—	9–12 m	5	7.7(0.4)
Baroda	ULs	Deodhar *et al.* (1964)	L													9–12 m	5	10.0(0.9)

TABLE V—*continued*

Locality	SE class[a]	Reference	Design[b]	Colostrum (1–5 days)				Transitional (6–14 days)				Mature (0.5–6 months)				Mature (> 6 months)		
				Stage[c]	*N*	Mean(SD)	Range	Stage[c]	*N*	Mean(SD)	Range	Stage[c]	*N*	Mean(SD)	Range	Stage[c]	*N*	Mean(SD)
Vitamin B12—Unsaturated binding capacity (μg/dl)[h]																		
The Americas																		
Brazil																		
Rio de Janeiro	UL	Donangelo *et al.* (1989) f	X	1–5 d	18	20.8(10.8)	3.9–37.7											
Rio de Janeiro	UL	Trugo *et al.* (1988) f	X	1–5 d	24	10.2(5.1)	3.8–18.9											
Africa and Arabia																		
Asia, Australasia, and Oceania																		
Thailand																		
Bangkok	UX	Areekul *et al.* (1977)	X	2 d	8	5.84(2.35)	—	6–10 d	7	6.96(1.54)	—							
				3 d	11	3.73(2.18)	—											
				4 d	11	4.22(1.76)	—											
				5 d	8	5.63(4.77)	—											
Vitamin B12—total binding capacity (μg/dl)																		
The Americas																		
Brazil																		
Rio de Janeiro	UL	Donangelo *et al.* (1989) f	X	1–5 d	17	20.9(—)	—											
Rio de Janeiro	UL	Trugo *et al.* (1988) f	X	1–5 d	24	10.3(—)	—											

TABLE V—*continued*

Locality	SE class[a]	Reference	Design[b]	Colostrum (1–5 days)				Transitional (6–14 days)				Mature (0.5–6 months)				Mature (> 6 months)		
				Stage[c]	*N*	Mean(SD)	Range	Stage[c]	*N*	Mean(SD)	Range	Stage[c]	*N*	Mean(SD)	Range	Stage[c]	*N*	Mean(SD)
Africa and Arabia																		
Asia, Australasia, and Oceania																		
Thailand																		
Bangkok	UX	Areekul *et al.* (1977)	X	2 d	8	5.92(2.34)	—	6–10 d	7	6.98(1.52)	—							
				3 d	11	3.78(2.19)	—											
				4 d	11	4.26(1.77)	—											
				5 d	8	5.66(4.78)	—											
Vitamin C (ascorbic acid) (mg/dl)																		
The Americas																		
Africa and Arabia																		
Kenya																		
Machakos	RL	Van Steenbergen *et al.* (1981)	X	3 d	61	6.0(2.1)	—											
Asia, Australasia, and Oceania																		
India																		
Baroda	UM	Deodhar and Ramakrishnan (1959)	X									1–3 m	46	2.59(2.57)	—	6–12 m	61	3.42(1.44)
												3–6 m	59	3.19(1.46)	—	> 12 m	25	3.15(1.28)
Baroda	UL	Deodhar *et al.* (1964)	L									1–3 m	10	2.4(0.1)	—	9–12 m	5	2.4(0.4)
Baroda	ULs	Deodhar *et al.* (1964)	L													9–12 m	5	6.1(0.7)

TABLE V—*continued*

Locality	SE class[a]	Reference	Design[b]	Colostrum (1–5 days)				Transitional (6–14 days)				Mature (0.5–6 months)				Mature (> 6 months)		
				Stage[c]	N	Mean(SD)	Range	Stage[c]	N	Mean(SD)	Range	Stage[c]	N	Mean(SD)	Range	Stage[c]	N	Mean(SD)
Coonoor	UL	Belavady and Gopalan (1959)	X	3–10 d	73	4.4(1.5)	—					2–6 m	45	1.9(1.3)	—	7–12 m 12–18 m > 18 m	36 37 30	2.5(1.2) 3.2(1.6) 2.9(1.9)
New Delhi	UH	Ashdir and Puri (1962)	L	3 d	10	5.42(3.04)	4.13–8.31	8 d	10	4.29(1.35)	2.58–6.61	18 d	10	4.05(1.27)	2.38–6.90			
Folic acid—Total (μg/dl)[i]																		
The Americas																		
Brazil																		
Amazonas	UL	Lehti (1989, 1990)	X									0.5–1 m 1 or 2 m 2 or 3 m	52 68 45	3.7(2.4) 3.8(1.7) 3.9(1.7)	0.5–13.8 1.2–8.3 1.2–8.6			
Rio de Janeiro	UL	Donangelo *et al.* (1989) f	X	1–5 d	17	1.01(0.75)	0.17–2.34											
Rio de Janeiro	UL	Trugo *et al.* (1988) f	X	1–5 d	25	1.01(0.71)	0.22–3.18											
United States (Navajo Indians)																		
Tuba City	RL	Butte and Calloway (1981)	X									19–62 d 3 m	23 7	5.64(2.39) 7.77(—)	3.40–13.58 —			

TABLE V—*continued*

Locality	SE class[a]	Reference	Design[b]	Colostrum (1–5 days)				Transitional (6–14 days)				Mature (0.5–6 months)				Mature (> 6 months)		
				Stage[c]	*N*	Mean(SD)	Range	Stage[c]	*N*	Mean(SD)	Range	Stage[c]	*N*	Mean(SD)	Range	Stage[c]	*N*	Mean(SD)
Africa and Arabia																		
The Gambia																		
Keneba	RL	Prentice *et al.* (1983)	X									0.5–6 m	21	3.82(1.33)	—			
Keneba	RLs	Prentice *et al.* (1983)	X									0.5–6 m	23	4.73(1.89)	—			
Asia, Australasia, and Oceania																		
India																		
Baroda	UL	Deodhar *et al.* (1964)	L									1–3 m	10	0.20(0.04)	—	9–12 m	5	0.21(0.04)
Baroda	ULs	Deodhar *et al.* (1964)	L													9–12 m	5	0.56(0.09)
Hyderabad	UL.	Ramasastri (1965)	X	< 5 d	14	4.4(1.8)	2.4–8.4	6–14 d	9	8.4(2.3)	4.5–12.1							
Japan																		
Sendai	XX	Tamura *et al.* (1980)	X									3–25 w	16	13.0(4.6)	—			
Sendai	XXs	Tamura *et al.* (1980)	X									3–25 w	16	13.7(4.1)	—			

TABLE V—*continued*

Locality	SE class[a]	Reference	Design[b]	Colostrum (1–5 days)				Transitional (6–14 days)				Mature (0.5–6 months)				Mature (> 6 months)		
				Stage[c]	N	Mean(SD)	Range	Stage[c]	N	Mean(SD)	Range	Stage[c]	N	Mean(SD)	Range	Stage[c]	N	Mean(SD)
Folic acid—Free (μg/dl)																		
The Americas																		
Brazil																		
Amazonas	UL	Lehti (1989, 1990)	X									0.5–1 m	52	2.7(2.3)	0.2–13.8			
												1 or 2 m	68	2.7(1.4)	0.7–6.8			
												2 or 3 m	45	2.7(1.5)	0.5–7.8			
Africa and Arabia																		
Asia, Australasia and Oceania																		
Folic acid—Unsaturated binding capacity (μg/dl)																		
The Americas																		
Brazil																		
Rio de Janeiro	UL	Donangelo *et al.* (1989) f	X	1–5 d	17	7.32(4.5)	0.71–15.3											
Rio de Janeiro	UL	Trugo *et al.* (1988) f	X	1–5 d	26	4.37(2.56)	1.28–12.0											
Africa and Arabia																		
Asia, Australasia, and Oceania																		

TABLE V—*continued*

Locality	SE class[a]	Reference	Design[b]	Colostrum (1–5 days)				Transitional (6–14 days)				Mature (0.5–6 months)				Mature (> 6 months)		
				Stage[c]	*N*	Mean(SD)	Range	Stage[c]	*N*	Mean(SD)	Range	Stage[c]	*N*	Mean(SD)	Range	Stage[c]	*N*	Mean(SD)
Folic acid—Percentage saturation																		
The Americas																		
Brazil																		
Rio de Janeiro	UL	Donangelo *et al.* (1989)	X	1–5 d	17	17.8(18.6)	3.2–72.9											
Rio de Janeiro	UL	Trugo *et al.* (1988)	X	1–5 d	26	20.5(13.4)	4.9–53.5											
Africa and Arabia																		
Asia, Australasia, and Oceania																		

Note. f, figures calculated from information in text; g, figures taken from graphs; p, pooled sample; q, numbers of subjects not given; s, supplemented mothers; z, geometric mean (+1 geometric standard deviation).

[a]Socioeconomic class: RL, rural, poor; UH, urban, middle-high income; UL, urban, poor, and low-middle income; UM, urban, mixed socioeconomic class; UX, urban, socioeconomic class not stated; a, lowest to d, highest; quartiles of maternal intake; *m*, malnourished and undernourished mothers; *n*, good maternal nutritional status.

[b]Study design: L, longitudinal or semilongitudinal design; X, cross-sectional design.

[c]Stage of lactation: d, days postpartum; w, weeks postpartum; m, months postpartum.

[d]IU/dl converted by dividing by 3.3.

[e]s, Supplemented mothers.

[f]n*M* converted using × 0.0206.

[g]n*M* converted to ng/dl using × 135.7.

[h]n*M* converted to μg/dl using × 0.1357.

[i]nmol/liter converted to μg/dl by × 0.0441.

References

Agarwal, K. N., Khurana, V., and Agarwal, D. K. (1975). Protein and free amino acids content of human milk. *Ind. Pediatr.* **12,** 415–417.

Areekul, S., Ounarom, K., and Doungbarn, J. (1977). Determination of vitamin B12 and vitamin B12 binding proteins in human and cow's milk. *Mod. Med. Asia.* **13,** 17–23.

Ashdhir, S., and Puri, B. (1962). Human milk studies—Chemical composition of human milk at three different stages. *Ind. J. Pediatr.* **29,** 99–109.

Atinmo, T., and Omololu, A. (1982). Trace element content of breastmilk from mothers of preterm infants in Nigeria. *Early Hum. Dev.* **6,** 309–313.

Bailey, K. V. (1965). Quantity and composition of breastmilk in some New Guinean populations. *J. Trop. Pediatr.* **11,** 35–49.

Bamji, M. S., Prema, K., Jacob, C. M., Ramalakshi, B. A., and Madhavapeddi, R. (1986). Relationship between maternal vitamins B2 and B6 status and the levels of these vitamins in milk at different stages of lactation. A study in a low-income group of Indian women. *Hum. Nutr. Clin. Nutr.* **40C,** 119–124.

Barclay, D. V. (1989). Influence de l'etat nutritionnel de la mere sur l'allaitement au sein et sur la croissance de l'enfant en milieu rural au Zaire. Doctoral Thesis, University of Nancy, France.

Bassir, O. (1956). Nutritional studies on breast milk of Nigerian women—Some biochemical features of breast milk of Lagos women during the first year of lactation. *J. Trop. Med. Hyg.* **59,** 139–144.

Bassir, O. (1958). Nutritional studies on breast milk of Nigerian women. *J. Trop. Pediatr.* **3,** 3–12.

Bates, C. J., and Tsuchiya, H. (1990). Zinc in breast milk during prolonged lactation: Comparison between the UK and The Gambia. *Eur. J. Clin. Nutr.* **44,** 61–69.

Belävady, B. (1959). Distribution of nitrogen in breast milk of Indian mothers. *Ind. J. Med. Res.* **47,** 217–221.

Belavady, B., and Gopalan, C. (1959). Chemical composition of human milk in poor Indian women. *Ind. J. Med. Res.* **47,** 234–245.

Boediman, D., Ismail, D., Iman, S., Ismangoen, and Ismadi, S. D. (1979). Composition of breast milk beyond one year. *Trop. Pediatr. Environ. Child. Health* **25,** 107–110.

Bray, G. W. (1928). Vitamin-B deficiency in infants: Its possibility, prevalence and prophylaxis. *Trans. R. Soc. Trop. Med. Hyg.* **22,** 9–42.

Butte, N. F. and Calloway, D. H. (1981). Evaluation of lactational performance of Navajo women. *Am. J. Clin. Nutr.* **34,** 2210–2215.

Bwibo, N. O., and Ondijo, S. O. (1981). Composition of human breast milk in Kenyan women. *East African Med. J.* **58,** 510–514.

Carneiro, T. A., and de Oliveira, J. E. D. (1973). Nutritional studies in human lactation in Brazil 1. Chemical composition of breast milk. *J. Trop. Pediatr. Environ. Child. Health* **19,** 384–387.

Chang, S-J. (1990). Antimicrobial proteins of maternal and cord sera and human milk in relation to maternal nutritional status. *Am. J. Clin. Nutr.* **51,** 183–187.

Close, J. A., and Van de Walle, A. (1957). Composition en matieres proteiques du lait de femme au Kivu (Congo Belge). In "Third Inter-African Nutrition Conference," pp. 811–821. Luanda, Angola.

Close, J. A., Van de Walle, A., and Robyns, E. (1957). La composition du lait de femme au Congo belge. I. L'azote total. *Ann. Soc. Belge Med. Trop.* **37,** 191–201.

Cruz, J. R., Carlsson, B., Garcia, B., Gebre-Medhin, M., Hofvander, Y., Urrutia, J. J., and Hanson, L. A. (1982). Studies on human milk. III. Secretory IgA quantity and antibody levels against Escherichia coli in colostrum and milk from underprivileged and privileged mothers. *Pediatr. Res.* **16,** 272–276.

Cruz, J. R., and Arevalo, C. (1985). Fluctuation of specific IgA antibodies in human milk. *Acta Paediatr. Scand.* **74,** 897–903.

Deodhar, A. D., and Ramakrishnan, C. V. (1959). Studies on human lactation. II. Effect of socio-economic status on the vitamin content of human milk. *Ind. J. Med. Res.* **47,** 352–355.

Deodhar, A. D., Rajalakshmi, R., and Ramakrishnan, C. V. (1964). Studies on human lactation. III. Effect of dietary supplementation on vitamin contents of breast milk. *Acta Paediatr.* **53,** 42–48.

de Silva, C. C. (1964). Common nutritional disorders of childhood in the tropics. *Adv. Pediatr.* **13,** 213–264.

Dewit, O., Dibba, B., and Prentice, A. (1990). Breast-milk amylase activity in English and Gambian mothers: Effects of prolonged lactation, maternal parity, and individual variations. *Pediatr. Res.* **28,** 502–506.

Donangelo, C. M., Trugo, N. M. F., Koury, J. C., Barreto Silva, M. I., Freitas, L. A., Feldheim, W., and Barth, C. (1989). Iron, zinc, folate, and vitamin B12 nutritional status and milk composition of low-income Brazilian mothers. *Eur. J. Clin. Nutr.* **43,** 253–266.

Dorea, J. G., Horner, M. R., Bezerra, V. L. V. A., and Campanate, M. L. (1984). Longitudinal study of major milk components from two different socioeconomic groups of mothers in Brazil. *Nutr. Rep. Int.* **29,** 699–709.

Duncan, M. E., Samson, R. R., McGrath, J., and McClelland, D. B. L. (1983). Humoral defence factors in the breast milk of Ethiopian women with leprosy and healthy controls. *Am. J. Clin. Nutr.* **38,** 921–928.

El Tom Ali, K., and Zaki, J. (1976). The chemical composition of human, cow's, goat's and vendor's milk in Khartoum North. *Sudan J. Food Sci. Technol.* **8,** 12–17.

Fransson, G-B., Gebre-Medhin, M., and Hambraeus, L. (1984). The human milk contents of iron, copper, zinc, calcium and magnesium in a population with a habitually high intake of iron. *Acta Paediatr. Scand.* **73,** 471–476.

Funk, M. A., Hamlin, L., Picciano, M. F., Prentice, A., and Milner, J. A. (1990). Milk selenium of rural African women, influence of maternal nutrition, parity, and length of lactation. *Am. J. Clin. Nutr.* **51,** 220–224.

Gebre-Medhin, M., Vahlquist, A., Hofvander, Y., Uppsall, L., and Vahlquist (1976). Breast milk composition in Ethiopian and Swedish mothers. I. Vitamin A and β-carotene. *Am. J. Clin. Nutr.* **29,** 441–451.

Gunshin, H., Yoshikawa, M., Doudou, T., and Kato, N. (1985). Trace elements in human milk, cow's milk and infant formula. *Agric. Biol. Chem.* **49,** 21–26.

Gupta, A. P., Bhandari, B., Gupta, A., and Goyal, S. (1984). Mineral content of breast milk from North Indian mothers giving birth preterm and at term—Implications for mineral nutrition of preterm infants. *J. Trop. Pediatr.* **30,** 286–288.

Hennart, P. F., Brasseur, D. J., Delogne-Desnoeck, J. B., Dramaix, M. M., and Robyn, C. E. (1991). Lysozyme, lactoferrin, and secretory immunoglobulin A content in breast milk: Influence of duration of lactation, nutrition status, prolactin status, and parity of the mother. *Am. J. Clin. Nutr.* **53,** 32–39.

Hernell, O., Gebre-Medhin, M., and Olivecrona, T. (1977). Breast milk composition in Ethiopian and Swedish mothers. IV. Milk lipases. *Am. J. Clin. Nutr.* **30,** 508–511.

Higashi, A., Ikeda, T., Uehara, I., and Matsuda, I. (1982). Zinc and copper contents in breast milk of Japanese women. *Tohoku J. Exp. Med.* **137,** 41–47.

Holemans, K., and Martin, H. (1954). Etude de l'allaitement maternal et des habitudes alimentaires du sevrage chez les indigenes du Kwango. *Ann. Soc. Belge Med. Trop.* **34,** 915–923.

Holemans, K., Lambrechts, A., and Martin, H. (1954). Etude qualitative et quantitative du lait des femmes indigenes du Kwango (Congo belge). *Rev. Med. de Liege* **9,** 714–723.

Huang, L., Mao, X., Shi, Z., Cheng, H., and Su, T. F. (1984). A longitudinal study of the nutritional and immunological composition of human milk. *Nutr. Res.* **4,** 977–980.

Jackson, D. A., Imong, S. M., Silprasert, A., Preunglumpoo, S., Leelapat, P., Yootabootr, Y., and Amatayakul, K. (1988). Estimation of 24 h breast-milk fat concentration and fat intake in rural northern Thailand. *Br. J. Nutr.* **59,** 365–371.

Jansen, A. A. J., Luyken, R., Malcolm, S. H., and Willems, J. J. L. (1960). Quantity and composition of breast milk in Biak Island (Neth. New Guinea). *Trop. Geogr. Med.* **2,** 138–144.

Karmarkar, M. G., Kapur, J., Deodhar, A. D., and Ramakrishnan, C. V. (1959). Studies on human lactation. I. Diet survey of lactating women in different socioeconomic groups and the effects of socio-economic status and stage of lactation on the proximate principles and essential amino acids of human milk. *Ind. J. Med. Res.* **47,** 344–351.

Karmarkar, M. G., and Ramakrishnan, C. V. (1960). Studies on human lactation. Relation between the dietary intake of lactating women and the chemical composition of milk with regard to principal and certain inorganic constituents. *Acta Paediatr.* **49,** 599–604.

Karra, M. V., Kirksey, A., Galal, O., Bassily, N. S., Harrison, G. G., and Jerome, N. W. (1988). Zinc, calcium, and magnesium concentrations in milk from American and Egyptian women throughout the first 6 months of lactation. *Am. J. Clin. Nutr.* **47,** 642–648.

Khin-Maung-Naing, Tin-Tin-Oo, Kywe-Thein, and Nwe-New-Hlaing. (1980). Study on lactation performance of Burmese mothers. *Am. J. Clin. Nutr.* **33,** 2665–2668.

Khurana, V., Agarwal, K. N., Gupta., and Nath, T. (1970). Estimation of total protein in breast milk of Indian lactating mothers. *Ind. Pediatr.* **7,** 156–158.

Kim, M-Y. (1989). "A Longitudinal Study on Iron and Copper Contents in Korean Breast Milk." MSc Thesis, Graduate School, Dankook University, Seoul, Korea.

Kunz, C., Prentice, A., Moriera, P., Dibba, B., Ceesay, S., and Lonnerdal, B. (1990). Breast milk whey proteins and casein of rural Gambian mothers. In "Proceedings of the Fifth International Conference on Lactation." Asilomar, CA.

Laskey, M. A., Prentice, A., Shaw, J., Zachou, T., Ceesay, S. M., Vasquez-Velasquez, L., and Fraser, D. R. (1990). Breast-milk calcium concentrations during prolonged lactation in British and rural Gambian mothers. *Acta Paediatr. Scand.* **79,** 507–512.

Laskey, M. A., Dibba, B., and Prentice, A. (1991). Low ratios of calcium to phosphorus in the breast-milk of rural Gambian mothers. *Acta Paediatr. Scand.* **80,** 250–251.

Lauber, E., and Reinhardt, M. (1979). Studies on the quality of breast milk during 23 months of lactation in a rural community of the Ivory Coast. *Am. J. Clin. Nutr.* **32,** 1159–1173.

Lee, J-S. (1987). "A Longitudinal Study of Volume and Composition in Korean Human Milk." Doctoral Thesis, Graduate School, Dankook University, Seoul, Korea.

Lehti, K. K. (1989). Iron, folic acid and zinc intakes and status of low socio-economic pregnant and lactating Amazonian women. *Eur. J. Clin. Nutr.* **43,** 505–513.

Lehti, K. K. (1990). Breast milk folic acid and zinc concentrations of lactating, low socioeconomic, Amazonian women and the effect of age and parity on the same two nutrients. *Eur. J. Clin. Nutr.* **44,** 675–680.

Lindblad, B. S., and Rahimtoola, R. J. (1974). A pilot study of the quality of human milk in a lower socio-economic group in Karachi, Pakistan. *Acta Paediatr. Scand.* **63,** 125–128.

Loh, T. T., and Sinnathuray, T. A. (1971). Haematological data and milk iron in Malaysian women. *Aust. N.Z. J. Obstet. Gynaecol.* **11,** 254–258.

Lonnerdal, B., Forsum, E., Gebre-Medhin, M., and Hambraeus, L. (1976). Breast milk composition in Ethiopian and Swedish mothers. II. Lactose, nitrogen, and protein contents. *Am. J. Clin. Nutr.* **29,** 1134–1141.

Lonnerdal, B., Chang, H., and Chen, C-L. (1990). Breast milk and formula intake of Taiwanese infants during the first seven months of life: Effects on growth and metabolic parameters. *In* "Breastfeeding, Nutrition, Infection and Infant Growth in Developed and Emerging Countries" (S. A. Atkinson, L. A. Hanson, and R. K. Chandra, eds.), pp. 249–282. ARTS Biomedical Publishers and Distributors, St John's, Newfoundland.

Mathur, N. B., Dwarkadas, A. M., Sharma, V. K., Saha, K., and Jain, N. (1990). Anti-infective factors in preterm human colostrum. *Acta Paediatr. Scand.* **79,** 1039–1044.

Mehta, J. B., and Kala, N. (1971). Composition of mother's milk at 3 different stages of lactation. In Proceedings of the 13th International Congress on Paediatrics, Vol. 2, pp. 225–229.

Miranda, R., Saravia, N. G., Ackerman, R., Murphy, N., Berman, S., and McMurray, D. N. (1983). Effect of maternal nutritional status on immunological substances in human colostrum and milk. *Am. J. Clin. Nutr.* **37,** 632–640.

Mohan, V. M., Kumar, C. S., and Karan, S. (1983). Composition of milk from mothers of pre-term and term infants: Total nitrogen and electrolyte content. *Ind. Pediatr.* **20,** 163–166.

Moser, P. B., Reynolds, R. D., Acharya, S., Howard, M. P., Andon, M. A., and Lewis, S. A. (1988a). Copper, iron, zinc and selenium dietary intake and status of Nepalese lactating women and their breast-fed infants. *Am. J. Clin. Nutr.* **47,** 729–734.

Moser, P. B., Reynolds, R. D., Acharya, S., Howard, M. P., and Andon, M. A. (1988b). Calcium and magnesium dietary intakes and plasma and milk concentrations of Nepalese lactating women. *Am. J. Clin. Nutr.* **47,** 735–739.

Murray, M. J., Murray, A. B., Murray, N. J., and Murray, M. B. (1978). The effect of iron status of Nigerien mothers on that of their infants at birth and 6 months, and on the concentration of Fe in breast milk. *Br. J. Nutr.* **39,** 627–630.

Nagasawa, T., Kiyosawa, I., and Kuwahara, K. (1972). Amounts of lactoferrin in human colostrum and milk. *J. Dairy Sci.* **55,** 1651–1659.

Nagasawa, T., Kiyosawa, I., Fukuwatari, Y., Kitayama, T., Uechi, M., and Hyodo, Y. (1973). α-Lactalbumin and serum albumin in human milk. *J. Dairy Sci.* **56,** 177–180.

Naismith, D. J. (1973). Kwashiorkor in western Nigeria: A study of traditional weaning foods, with particular reference to energy and linoleic acid. *Br. J. Nutr.* **30,** 567–575.

Omeme, J. A., Lantos, J. D., and Ihongbe, J. C. (1981). Alpha-1-antitrypsin in breast milk of healthy Nigerian mothers. *East African Med. J.* **58,** 56–59.

Peters, F. E. (1953). The chemical composition of New Hebridean human milk. *Br. J. Nutr.* **7,** 208–211.

Prentice, A., Prentice, A. M., and Whitehead, R. G. (1981). Breast-milk fat concentrations of rural African women. 1. Short-term variations within individuals. *Br. J. Nutr.* **45,** 483–494.

Prentice, A., Prentice, A. M., Cole, T. J., Paul, A. A., and Whitehead, R. G. (1984). Breast-milk antimicrobial factors of rural Gambian mothers. I. Influence of stage of lactation and maternal plane of nutrition. *Acta Paediatr. Scand.* **73,** 796–802.

Prentice, A., MacCarthy A., Stirling, D. M., Vasquez-Velasquez L., and Ceesay, S. (1989). Breast-milk IgA and lactoferrin survival in the gastrointestinal tract—A study in rural Gambian children. *Acta Paediatr. Scand.* **78,** 505–512.

Prentice, A., and Barclay, D. V. (1991). The breast-milk calcium and phosphorus concentrations of Zairean mothers. *Eur. J. Clin. Nutr.* **45,** 611–617.

Prentice, A. (1994). Sodium and potassium concentrations of rural Gambian mothers. Unpublished results.

Prentice, A. M., Roberts, S. B., Prentice, A., Paul, A. A., Watkinson, M., Watkinson, A. A., and Whitehead, R. G. (1983). Dietary supplementation of lactating Gambian women. 1. Effect on breast-milk volume and quality. *Hum. Nutr. Clin. Nutr.* **37C,** 53–64.

Rajalakshmi, K., and Srikantia, S. G. (1980). Copper, zinc, and magnesium content of breast milk of Indian women. *Am. J. Clin. Nutr.* **33,** 664–669.

Ramasastri, B. V. (1965). Folate activity in human milk. *Br. J. Nutr.* **19,** 581–586.

Rao, P. S., and Belavady, B. (1981). Lipids and bile salt stimulated lipase in human milk of Indian mothers. *Ind. J. Med. Res.* **73,** 363–368.

Rao, P. U., and Belavady, B. (1973). Protein fractions in human milk: Part II. Isolation and characterization of basic protein from human milk and the lytic activity of milk samples. *Ind. J. Biochem. Biophys.* **10,** 87–96.

Raut, S. J., and Viswanathan, R. (1972). Distribution of magnesium in body fluids. *Ind. J. Med. Res.* **60,** 1272–1277.

Reddy, V., Bhaskaram, C., Raghuramulu, N., and Jagadeesan, V. (1977). Antimicrobial factors in human milk. *Acta Paediatr. Scand.* **66,** 229–232.

Reynolds, R. D., Acharya, S., Leklem, J. E., and Moser, P.B. (1986). Effects of low maternal dietary intake of calcium, selenium and vitamin B6 upon breast milk composition in

Nepal. *In* "Human Lactation 2: Maternal and Environmental Factors" (M. Hamosh and A. S. Goldman, eds.), pp. 205–213. Plenum Press, New York.

Ronayne de Ferrer, P. A., Slobodianik, N. H., Lopez, N., Sambucetti, M. E., and Sanahuja, J. C. (1984). Immunoglobulin A level in human milk from mothers delivering preterm. *Am. J. Clin. Nutr.* **40,** 465–467.

Seol, M-Y. (1988). "A Longitudinal Study of Calcium and Phosphorus in Korean Human Milk." MSc Thesis, Graduate School, Dankook University, Seoul, Korea.

Seol, M-Y, Kim, E. S., and Han, Y. I. (1989). A longitudinal study on ash, calcium, phosphorus, magnesium, sodium and potassium content in Korean breast milk. *In* "Proceedings of the 14th International Congress of Nutrition." Seoul, Korea.

Seol, M-Y. (1990). A longitudinal study on calcium, phosphorus and magnesium contents of breast milk from lactating women in Seoul area. *Korean J. Nutr.* **23,** 115–123.

Shrimpton, R., Alencar, F. H., Vasconcelos, J. C., and Rocha, Y. R. (1985). Effect of maternal zinc supplementation on the growth and diarrhoeal status of breast fed infants. *Nutr. Res. Suppl.* **1,** 338–342.

Singh, R. K., Nakra, V. K., Pandey, H. N., and Arora, S. R. (1984). Studies on circadian periodicity of plasma, breast milk and urinary calcium in lactating Indian women. *Trop. Geographr. Med.* **36,** 345–349.

Sinha, K. P., Sara, A., and Sinha, A. S. (1959). Composition of human milk. *Ind. J. Physiol. Pharmacol.* **3,** 173–176.

Svanberg, U., Gebre-Medhin, M., Ljungqvist, B., and Olsson M. (1977). Breast-milk composition of Ethiopian and Swedish mothers. III Amino acids and other nitrogenous substances. *Am. J. Clin. Nutr.* **30,** 499–507.

Tamura, T., Yoshimura, Y., and Arakawa, T. (1980). Human milk folate and folate status in lactating mothers and their infants. *Am. J. Clin. Nutr.* **33,** 193–197.

Trugo, N. M. F., Donangelo, C. M., Koury, J. C., Barreto Silva, M. I., and Freitas, L. A. (1988). Concentration and distribution pattern of selected micronutrients in preterm and term milk from urban Brazilian mothers during early lactation. *Eur. J. Clin. Nutr.* **42,** 497–507.

Underwood, B. A., Hepner, R., and Abdullah, H. (1970). Protein, lipid and fatty acids of human milk from Pakistani women during prolonged periods of lactation. *Am. J. Clin. Nutr.* **23,** 400–407.

Van Steenbergen, W. M., Kusin, J. A., and Van Rens, M. M. (1981). Lactation performance of Akamba mothers, Kenya. Breast feeding behaviour, breast milk yield and composition. *J. Trop. Pediatr.* **27,** 155–161.

Van Steenbergen, W. M., Kusin, J. A., De With, C., Lacko, E., and Jansen, A. A. J. (1983). Lactation performance of mothers with contrasting nutritional status in rural Kenya. *Acta Paediatr. Scand.* **72,** 805–810.

Venkatachalam, P. S., Belavady, B., and Gopalan, C. (1962). Studies on vitamin A nutritional status of mothers and infants in poor communities of India. *Trop. Pediatr.* **61,** 262–268.

Villiard, L., and Bates, C. J. (1987). Effect of vitamin A supplementation on plasma and breast milk vitamin A levels in poorly nourished Gambian women. *Hum. Nutr. Clin. Nutr.* **41C,** 47–58.

World Health Organization (WHO) (1985). "The Quantity and Quality of Breast Milk." Report on the WHO collaborative study on breast-feeding. WHO, Geneva.

Wyatt, R. G., Garcia, B., Caceres, A., and Mata, L. J. (1972). Immunoglobulins and antibodies in colostrum and milk of Guatemalan mayan women. *Archivos Latinoamericanos de Nutricion* **22,** 629–644.

Yolken, R. H., Wyatt, R. G., Mata, L., Urrutia, J. J., Garcia, B., Chanock, R. M., and Kapikian, A. Z. (1978). Secretory antibody directed against rotavirus in human milk—Measurement by means of enzyme-linked immuosorbent assay. *J. Pediatr.* **93,** 916–921.

E. Effects of Gestational Stage at Delivery on Human Milk Components

STEPHANIE A. ATKINSON

I. Introduction

Since 1978, more than 40 published studies have reported on the comparative analysis of milk produced by mothers delivering prematurely or at term, often referred to as preterm milk and term milk, respectively. Observations from many of these studies suggest that preterm milk is unique in nutrient composition compared to term milk but agreement on these observations is not universal.

Determination of the biochemical composition of milk from mothers giving birth prematurely (preterm milk) by Atkinson *et al.* (1978, 1980a,b, 1986, 1987; Anderson *et al.*, 1981) and subsequently by others (Gross *et al.*, 1980, 1981a,b; Lemons *et al.*, 1981, 1983; Thomas *et al.*, 1981; Lepage *et al.*, 1984; Bitman *et al.*, 1983, 1986; Britton, 1986; Ronayne de Ferrer *et al.*, 1984; Mendelson *et al.*, 1982; Pamblanco *et al.*, 1986; Guerrini *et al.*, 1981; Chandra, 1982; Chan, 1982; Chappell *et al.*, 1985) demonstrated that preterm milk had greater concentrations of nitrogen, immune proteins, lipid, medium-chain fatty acids, energy, vitamins and some minerals (e.g., calcium and sodium), and trace elements when compared to term milk at similar lactational stages. Some studies also found differences in protein and lipid composition within preterm mothers in relation to the degree of prematurity (Lepage *et al.*, 1984; Bitman *et al.*, 1983) or whether the infants were small or appropriate for gestational age (Barros and Carneiro-Sampaio, 1984). Some studies describing preterm milk composition (Ehrenkranz *et al.*, 1984; Jones *et al.*, 1982; Moran *et al.*, 1983a; Schanler *et al.*, 1980; Thomas *et al.*, 1986) did not use a comparison group of term mothers' milk and thus it is difficult to evaluate these data on a comparative basis.

Evidence for the observed compositional differences in milk between term and preterm mothers was not universal in all studies (Ferlin, 1980; Sann *et al.*, 1981; Moran *et al.*, 1983b; Udipi *et al.*, 1985; Vaisman *et al.*, 1985). Such lack of agreement among investigators may be attributed to differences in sample collection methodology, the inclusion of wide ranges of gestational stage, and the greater degree of interindividual variability in milk composition in preterm compared to term milk. To aid in the interpretation of the data summarized in this chapter on the nutrient content of preterm milk, the gestational stage at birth of the mothers and

HANDBOOK OF MILK COMPOSITION

the method of milk collection for each of the studies cited are summarized in Table I.

TABLE I
Gestational Age of Subjects and Method of Milk Collection for Studies Cited in Tables II–V

Reference	Gestational Age at delivery (weeks)	Method of milk collection
Anderson *et al.* (1983)	28–36	24-hr collections
Anderson *et al.* (1981)	26–33	Complete 24-hr expressions of both breasts with breast cleaning with sterile deionized water
Atkinson *et al.* (1978)	28–33	As above
Atkinson *et al.* (1980)	28–33	As above
Atkinson *et al.* (1982)	28–33	As above
Atkinson *et al.* (1987)	24–32	As above with use of acid-washed collection bottles
Bitman *et al.* (1983)	26–30 31–36	Entire content of one breast at 0900–1000 hr feeding by mechanical pump
Britton (1986)	30–36	Contents of one breast between 0800 and 1200 hr and 2 hr after a breast feeding
Butte *et al.* (1984)	27.5 ± 3.0 (SD)	Entire contents of one breast collected between 0800 and 1200 hr into acid-washed bottle
Chan (1982)	33–37	Fore- and hindmilk at 0900 hr
Chandra (1982)	28–35	Morning samples by complete manual expression of both breasts after cleaning breasts with sterile deionized water
Chappell *et al.* (1985)	28–88	Complete early morning expression
Ehrenkranz *et al.* (1984)	26–33	Complete 24-hr expressions
Goldman *et al.* (1982)	32 ± 3.5 (SD)	Expressed entire content of one breast into acid-washed containers at 1 to 3 hr after feeding the infant
Gross *et al.* (1980)	28–36	Complete emptying of both breasts at one time in morning
Gross *et al.* (1981)	28–36	Complete emptying of both breasts at one time in morning
Lemons *et al.* (1982)	27–37	Complete 24-hr expressions
Lepage *et al.* (1984)	26–31 32–36	Complete 24-hr expressions
Mendelson *et al.* (1982)	26–33	Complete 24-hr expression of both breasts into bottles rinsed with deionized water
Moran *et al.* (1983a)	27–32	24-hr expressions from both breasts

TABLE I—*continued*

Reference	Gestational Age at delivery (weeks)	Method of milk collection
Pamblanco *et al.* (1986)	26–32 33–36	Complete expression of both breasts at second morning feed
Ronayne de Ferrer *et al.* (1984)	28–35	Morning sample by complete manual emptying of one breast
Sann *et al.* (1981)	26–35	Milk expressed four to six times daily, pooled and pasteurized
Schanler *et al.* (1980)	29.7 ± 1 (SD)	Aliquots from each feeding over 24 hr were pooled for analysis
Vaisman *et al.* (1985)	25–35	Complete 24-hr expression
Udipi *et al.* (1985)	27–35	24-hr composite of 5 ml of fore- and hindmilk collected at each feeding

II. The Nitrogen Composition of Preterm Milk

Atkinson *et al.* (1978, 1980) were the first to report the nitrogen composition of preterm milk (PTM). They found that, over the first month of lactation, the total nitrogen (TN) and true protein content of PTM was significantly greater than that of milk from full-term mothers (FTM). However, the pattern of decrease in TN with time was similar between groups (Atkinson *et al.*, 1978). These differences in nitrogen quantity between PTM and FTM have been confirmed by others (Gross *et al.*, 1980; Lemons *et al.*, 1982; Chandra, 1982; Lemons *et al.*, 1983; Butte *et al.*, 1984; Hibberd *et al.*, 1982); although some reports indicate that the difference is marginal especially in mothers delivering after 30 weeks of gestation (Lepage *et al.*, 1984; Britton, 1986; Anderson *et al.*, 1983). Table II summarizes the reported data for nitrogen and protein composition of preterm milk.

The characterization of the protein content of preterm milk has been accomplished to some extent. The immune proteins, lactoferrin, lysozyme (Donovan *et al.*, 1987; Goldman *et al.*, 1982), IgA (Donovan *et al.*, 1987; Goldman *et al.*, 1982; Chandra, 1982; Ronayne de Ferrer *et al.*, 1984; Gross *et al.*, 1981), and other immunoglobulins (Chandra, 1982), have been reported to be significantly higher in preterm compared to term milk during early lactation. Immunoglobulin A comprises >90% of antibody molecules in human milk and over 90% of this is present in the polymerized form SIgA. The concentrations of immunoglobulins in preterm milk are summarized in Table II.

TABLE II
Total Protein and Immune Proteins in Preterm Milk

Milk component	Reference	Stage of lactation (days)					
		Colostrum		Transitional		Mature	
		X	SD (LD)[a]	X	SD (LD)[a]	X	SD (LD)[a]
Total protein (g/liter)	Anderson *et al.* (1983)	20	7 (3)	16	3 (7)	13	3 (14)
	Anderson *et al.* (1981)	21	(4–6)	18	(7–10)	14	(22–28)
	Atkinson *et al.* (1978)	22	(1–3)	22	(7–10)	16	(22–28)
	Butte *et al.* (1984)	—	—	15	1 (14)	13	8 (28)
	Chandra (1982)	24	6 (3)	23	6 (7)	15	9 (28)
	Ehrenkranz *et al.* (1984)	21	11 (0–10)	13	5 (10–20)	16	7 (20–30)
	Gross *et al.* (1981)	32	1.5 (3)	24	8 (7)	18	4 (28)
	Lemons *et al.* (1982)	—	—	20	3 (7)	15	3 (28)
	Schnurr and Atkinson (unpublished)	30	7 (3–5)	24	4 (7–8)	15	1 (28–30)
	Sann *et al.* (1981)	22	8.9 (< 6)	78	6 (7–14)	15	5 (> 15)
IgA (mg/g protein)	Chandra (1982)	109	93 (3)	92	63 (7)	64	70 (28)
	Ehrenkranz *et al.* (1984)	139	59 (0–10)	211	59 (10–20)	216	116 (20–30)
	Gross *et al.* (1981)	256[b]	(3)	114[b]	(7)	—	—
IgG (mg/g protein)	Chandra (1982)	1.1	1.1 (3)	1.0	0.6 (7)	2.1	1.5 (28)
	Gross *et al.* (1981)	1.8–2.8	(3–28)	—	—	—	—
IgM (mg/g protein)	Gross *et al.* (1981)	6.1–11.7	(3–28)	—	—	—	—
	Chandra (1982)	3.4	4.0 (3)	2.7	3.6 (7)	4.3	0.8 (28)

[a]LD, lactation day.
[b]Geometric mean.

III. The Acid-Soluble Nitrogen Fraction of Preterm Milk

A summary of the acid-soluble (ASN) or nonprotein nitrogen composition reported for preterm milk is provided in Table III. A significant decrease in the absolute quantity of ASN was observed for PTM but not for FTM over the first month of lactation (Atkinson *et al.*, 1980). The absolute quantity of ASN in PTM was significantly greater than that in FTM in early lactation (Atkinson *et al.*, 1980; Lemons *et al.*, 1983). However, the proportion of the TN contributed by the ASN was similar in PTM and FTM in some reports (Atkinson *et al.*, 1980; Lemons *et al.*, 1983). The reported proportion of TN contributed by ASN in PTM was highly variable, ranging from a low value of 16.8% (Lemons *et al.*, 1983) to a high of 23% (Butte *et al.*, 1984) at 1 month of lactation. Furthermore, the proportion of the major components within the ASN fraction—urea N and free amino acid N—of PTM increased with progressing lactation even though the proportion of ASN/TN remained relatively constant (Atkinson *et al.*, 1986). The nutritional significance of the ASN fraction to the infant has not been defined.

TABLE III
The Acid-Soluble Nitrogen (ASN) Fraction of Preterm Milk

Milk component	Reference	Stage of lactation					
		Colostrum		Transitional		Mature	
		X	SD (LD)[a]	*X*	SD (LD)[a]	*X*	SD (LD)[a]
ASN/TN (%)	Atkinson *et al.* (1980)	17.7	— (3)	18.5	(7)	20.4	(28)
	Butte *et al.* (1984)	—	—	23.6	(14)	23.6	(28)
	Lepage *et al.* (1984)	—	—	11.2	(7)	16.8	(28)
	Schurr & Atkinson (unpublished)	18.6	(3–5)	21.7	(7–8)	23.2	(28–30)
	Schanler *et al.* (1980)	15.2	4 (1–3)	17.9	4 (7–10)	16.9	7.2 (22–28)
Urea N (mg/liter)[b]	Atkinson *et al.* (1980)	136	(3)	150	(7)	156	(28)
	Schurr & Atkinson (unpublished)	113	40 (3–5)	111	29 (7–8)	126	26 (28–30)

[a]LD, lactation day.
[b]Urea N represents about 30% of total ASN in human milk.

IV. Macrominerals and Electrolytes

The mineral and electrolyte composition reported for preterm milk is summarized in Table IV. In all studies cited, sodium and chloride declined significantly during the first month and in parallel in both PTM and FTM. Potassium in milk declines over the first month of lactation but not to the same extent as sodium. In FTM, lactose is inversely correlated with Na and Cl ($p < 0.01$) and positively correlated with K ($p < 0.05$), but none of these relationships held for PTM (Atkinson *et al.,* 1986).

The pattern of change of electrolytes with lactational stage in milk follows that demonstrated by Linzell and Peaker (1974) in the goat. During the stage of lactation around parturition paracellular transport of ions and lactose is thought to occur. In serial measurements by pumping for PTM, the interrelationships of ions and lactose were not significant, suggesting that transport mechanisms of ions are different from those of FTM (Atkinson *et al.,* 1986). Certainly, in PTM, the patterns of change in ions observed do not reflect the "leaky junction" concept as described for the parturient goat and purported as an explanation of preterm milk composition (Anderson, 1984). In general, the increase in the concentrations of lactose and K and the decrease in the concentrations of Na and Cl in mammary secretions during lactogenesis are consistent with the existence of a paracellular pathway across the mammary gland just prior to parturition (Kulski and Hartman, 1981). Why there would be differences in the transport pathways between PTM and FTM during lactogenesis requires further study. The concept of "paracellular" transport in mammary glands has not been well explored in humans. In fact, the observation that milk glucose was not increased in association with other changes indicative of mammary epithelium permeability, during the colostral phase in women (Kulski and Hartman, 1981; Prosser and Hartman, 1983), casts some doubt as to whether the observations of mammary permeability in the goat (Linzell and Peaker, 1974) should be directly extrapolated to women (Hartman and Prosser, 1984).

The pattern of change in milk calcium and phosphorus content in PTM is variable between studies. Over the first 4 weeks of lactation there is only a difference of about 1 mmol/liter between lactational stages (Table IV). Magnesium remains relatively constant (Table IV).

V. Trace Elements

Information on the trace element concentration of preterm milk is limited. The studies summarized in Table V showed that Zn, Cu, and Fe

TABLE IV
Macromineral Element Content of Preterm Milk

Milk component (mmol/liter)	Reference	Stage of lactation					
		Colostrum		Transitional		Mature	
		X	SD (LD)[a]	X	SD (LD)[a]	X	SD (LD)[a]
Na	Atkinson *et al.* (1980)	23	2.5 (3–5)	13.5	1.1 (13–17)	—	—
	Butte *et al.* (1984)	—	—	11.6	4.3 (14)	6.5	1.3 (42)
	Lemons *et al.* (1982)	—	—	17.2	8.4 (7)	9.6	2.8 (28)
	Sann *et al.* (1981)	31.8	16 (< 6)	15.6	7.0 (7–14)	8.5	4.7 (> 15)
	Schanler *et al.* (1980)	22.2	9 (1–3)	11.6	6.0 (7–10)	8.8	2.0 (22–28)
K	Atkinson *et al.* (1980)	18.5	0.7 (3–5)	16.5	0.3 (13–17)	—	—
	Lemons *et al.* (1982)	—	—	17.3	3.0 (7)	13.5	1.6 (28)
	Sann *et al.* (1981)	19.3	6.2 (< 6)	20	5.4 (7–14)	15	5 (> 15)
	Schanler *et al.* (1980)	15.1	4.5 (1–3)	13.5	2.2 (7–10)	12.5	3.2 (22–28)
Ca	Atkinson *et al.* (1980)	8.15	0.68 (3.5)	6.95	0.39 (13–17)	—	—
	Butte *et al.* (1984)	—	—	5.3	1.2 (14)	5.4	1.5 (42)
	Lemons *et al.* (1982)	—	—	7.3	1.8 (7)	7.1	1.1 (28)
	Schanler *et al.* (1980)	6.75	1.7 (1–3)	8.0	1.8 (7–10)	7.2	1.3 (22–28)
	Sann *et al.* (1981)	6.6	2.8 (< 6)	7.6	2.1 (7–14)	5.0	1.4 (> 15)
P	Butte *et al.* (1984)	—	—	4.8	1.2 (14)	4.1	0.3 (42)
	Lemons *et al.* (1982)	—	—	4.3	0.9 (7)	4.2	0.7 (28)
	Sann *et al.* (1981)	3.8	1.1 (< 6)	4.9	1.4 (7–14)	3.0	0.8 (> 15)
Mg	Atkinson *et al.* (1980)	1.6 ± 0.06 (over the first 28 days)					
	Butte *et al.* (1984)	—	—	1.5	0.3 (14)	1.7	0.4 (42)
	Lemons *et al.* (1982)	—	—	1.5	0.2 (7)	1.3	0.3 (28)
	Sann *et al.* (1981)	1.1	0.3 (< 6)	1.05	0.2 (7–4)	1.0	0.3 (> 15)

[a] LD, lactation day.

TABLE V
Trace Element Composition of Preterm Milk

Milk component (μmol/liter)	Reference	Stage of lactation					
		Colostrum		Transitional		Mature	
		X	SD (LD)[a]	*X*	SD (LD)[a]	*X*	SD (LD)[a]
Zinc	Butte *et al.* (1984)	—	—	63	15 (14)	41	16 (42)
	Mendelson *et al.* (1982)	82	22 (3–5)	73	24 (8–10)	60	17 (28–30)
	Moran *et al.* (1983a)	78	(7)	—	—	24	(49)
	Sann *et al.* (1981)	62	37 (< 6)	49	21 (7–14)	40	12 (> 15)
Copper	Mendelson *et al.* (1982)	13.0	3 (3–5)	12.2	2.8 (8–10)	9.9	2.2 (28–30)
	Moran *et al.* (1983a)	9.0	4 (7)	8.2	3.5 (14)	4.5	2.2 (35)
	Sann *et al.* (1981)	9.9	4.2 (< 6)	11.0	3.8 (7–14)	12.8	5.8 (> 15)
Iron	Mendelson *et al.* (1982)	19.8	6 (3–5)	17.6	4.8 (8–10)	16.1	4.1 (28–30)

[a]LD, lactation day.

concentrations in preterm milk decline significantly over the first 4 weeks or so of lactation, a pattern similar to that observed for term milk (Siimes *et al.*, 1979; Vuori and Kuitunen, 1979).

VI. Vitamins

Concentrations of retinol, tocopherol, carotene, vitamin C, vitamin D, vitamin B_6, and folacin have been reported for preterm milk (Table VI). In some studies the concentrations of retinol and α-tocopherol (Chappell *et al.*, 1985) and vitamin D_3 (Atkinson *et al.*, 1982) were higher in preterm compared to term milk.

VII. Physiological Basis of Preterm Milk Composition

In general, the pattern of changes in milk composition in term milk is also observed in preterm milk. What then could be the possible physiological basis for the reported differences in nutrient density between the two

TABLE VI
Vitamin Content of Preterm Milk

Milk component	Reference	Stage of lactation					
		Colostrum		Transitional		Mature	
		X	SD (LD)[a]	X	SD (LD)[a]	X	SD (LD)[a]
Vitamin A (retinol) (μg/liter)	Chappell *et al.* (1985)	1450	100 (2)	2050	100 (5–6)	1080	60 (37)
	Moran *et al.* (1983)	203	18 (7)	132	13 (14)	137	20 (35)
	Vaisman *et al.* (1985)	1390	360 (7)	1250	320 (14)	830	120 (35)
Carotene (μg/liter)	Chappell *et al.* (1985)	2000	120 (1)	1000	40 (4)	230	50 (37)
Vitamin D (ng/liter)							
Total D_2 + D_3	Atkinson *et al.* (1987)	—	—	266	122 (14)	270	195 (31)
Total 25-OHD_2 + 25-OHD_3[b]	Atkinson *et al.* (1987)	—	—	320	73 (14)	310	98 (31)
Vitamin E tocopherol (mg/liter)							
α-Tocopherol	Chappell *et al.* (1985)	11	2.5 (4)	15[d]	(7)	5[d]	(37)
γ-Tocopherol	Chappell *et al.* (1985)	1.5	0.4 (4)	—	—	—	—
Total tocopherol	Moran *et al.* (1983)	12	2.3 (7)	6	0.5 (14)	4	0.5 (35)
	Vaisman *et al.* (1985)	4	0.7 (7)	2.8	0.4 (14)	1.1	2 (35)
Vitamin C (mg/liter)	Moran *et al.* (1983)	39	7 (7)	42	5 (14)	38	8 (35)
	Udipi *et al.* (1985)	190	(14)	180	(28)	—	—
Folate[c]							
Vitamin B_6[c]							

[a] LD, lactation day.
[b] 25-OHD, 25-hydroxyvitamin D.
[c] Data on folate and vitamin B_6 in preterm milk are reported (Udipi *et al.*, 1985), but only in graphical form and thus quantitative values are not available.
[d] Estimated values from published figure.

sources of milk? There is little evidence to suggest that this milk secretion has a teleological basis making it uniquely suited to meet the nutritional needs of the premature infant.

Possible reasons for the observed differences in nutrient composition between preterm and term milk include lack of preparedness of the mammary gland at premature delivery to support "normal" lactation (Anderson, 1984), a significantly different hormonal profile between PT and FT mothers resulting in differences in milk composition (Anderson, 1984), the artificial method of expressing PTM vs normal breast-feeding (Lemons *et al.*, 1982), or lower milk volumes from PTM resulting in a greater nutrient density (Lucas and Hudson, 1984; Anderson *et al.*, 1983).

A number of possible alterations in physiological mechanisms related to the lactational process may also contribute to the production of preterm milk of different nutrient composition (Table VII). Milk composition might be affected by variations in breast stimulation due to incomplete emptying of the breast because of weak sucking of the small premature infant or use of a breast pump versus suckling. Alternatively, there is indirect evidence from lactating animals to support the hypothesis that differences in milk composition may be secondary to an altered or an interrupted hormonal status at parturition and/or altered development of hormone receptors within the mammary gland. Prepartum lactation is thought to be repressed by high placental luteal hormone levels which serve to inhibit proliferation of mammary prolactin (PRL) receptors (Djiane and Durand, 1977). Since women at 30 weeks of gestation have relatively lower circulating progesterone/estrogen levels compared to those at 38 to 40 weeks of gestation (Parker *et al.*, 1979; Buster *et al.*, 1979), proliferation of PRL receptors and magnitude of PRL release may be

TABLE VII
Perinatal Events Related to Preterm Birth Which May Affect the Lactational Process and Milk Composition

Delay in initiating pumping
Mother too ill/anxious
Lack of nursing support for teaching
Unavailability of appropriate pumps
Pattern of nipple stimulation effecting milk volume
Time of initiation of breast expression
Pumping frequency (one to six times/day) and duration
Type of breast pump
Amount of suckling
Anxiety related to babe's medical condition
Poor milk let down—
Low milk volumes—? Effect on composition
Perinatal drugs, e.g., glucocorticoids, oxytocin
Early return of menses
Degree of prematurity

greater at preterm parturition. This hormonal milieu would be conducive to higher rates of mammary synthesis of lipid (Anderson *et al.*, 1981; Chappell *et al.*, 1983; Lepage *et al.*, 1984), protein (Atkinson *et al.*, 1978; Gross *et al.*, 1980; Lemons *et al.*, 1982; Butte *et al.*, 1982), or medium-chain fatty acids (Bitman *et al.*, 1983; Chapell *et al.*, 1983) as has been reported. Hypothetically, rate of milk production may also be affected by the interference of high maternal anxiety states with oxytocin secretion and "let down" of milk. As well, decreased blood flow to the mammary gland could alter substrate availability or the hormonal milieu with subsequent alterations in milk composition. Neither of the latter explanations have been systematically investigated.

A further possible explanation for altered composition of preterm milk is based on the hypothesis that preterm mothers have an immature mammary system which permits paracellular leakage of serum proteins and ions through incompletely matured tight junctions or leaky junctions between apical membranes as has been described in the goat. During the prepartum period paracellular transport is thought to allow direct diffusion between the extracellular fluid and milk thus causing elevated concentrations of serum proteins (IgG) and sodium and chloride, and lowered lactose and potassium content compared to milk obtained postparturition in the animal (Linzell and Peaker, 1974). In women, indirect evidence does exist for the presence of paracellular transport of ions in pregnancy (Kulski and Hartman, 1981), in the mastitic breast (Conner, 1979), and during mammary gland involution at weaning (Garza *et al.*, 1983). In these situations mammary secretions comprise elevated concentrations of sodium and chloride and lower lactose and potassium than mature milk, but not altered protein content. The factors which regulate alveolar tight junction structure and thereby permeability via paracellular transport are not well defined. Among the possibilities suggested are frequent breast stimulation which may alter integrity of the mammary epithelium (Linzell and Peaker, 1974); or local chemical mediators such as prostaglandins (Maule-Walker and Peaker, 1980) or ionic calcium concentration (Neville and Peaker, 1981).

Unfortunately, variability in design methodology limits the comparison of results between many of the reported studies on preterm milk composition. Some of the major important variables which have not been controlled between studies include inclusion of wide ranges of gestational stage in preterm groups; collapsed time intervals for stage of lactation; and milk sampling methodology which may represent a random sample, a pool of fore- and hindmilk, a single breast expression, a complete feeding expression (both breasts), or a complete 24-hr expression of milk (note variations in milk collection methods of studies cited in Table I). Unless one of the two latter collection methods is employed, assessment of the lactational capacity by measurement of milk volume cannot be achieved. In some, but not all, studies milk volume has been measured. When milk volume produced by a mother is less than that which meets her infant's needs, one should consider whether indeed this is representative of "normal" lactation. For example, in studies by Anderson *et al.* (1983) and Gross *et al.* (1980), reported 24-hr milk volumes are interpolated to be as low as 16 ml—a volume that is difficult to accept as being reflective of "full" or normal lactation. Anderson *et al.* (1983) noted that the protein content of milk was negatively correlated with milk volume but this would be predicted since milk volume increases and protein decreases with progressive stage of lactation. Statistical analysis of differences in preterm and term milk composition adjusted for milk volume as a cofactor has not found milk volume to be a significantly associated variable when the milk produced is a reasonable amount (Atkinson *et al.*, 1978).

Many investigators have demonstrated a greater degree of interindividual variability in milk composition in preterm compared to term milk. Whether such exaggerated variability in milk composition between prematurely delivered mothers reflects true biological variability or is an artifact of the lactational process and milk sampling methodology indigenous to this population has not been examined in detail.

One final consideration of the reason for differences in PTM and FTM composition is the frequency and pattern of breast stimulation. Mothers giving birth prematurely generally have a different timing pattern than mothers delivering at term in the initiation of lactation and frequency of pumping during early lactation (Atkinson *et al.*, 1986). Accordingly, comparison of milk composition between PTM and FTM using postpartum day as the dependent variable will misrepresent the cumulative lactational experience of these two groups of mothers. Because of this, it is best to employ sequential pumping/suckling numbers as the dependent variable in doing comparative analysis of preterm versus full-term milk composition.

The pattern of breast or nipple stimulation during lactogenesis may impact an important effect on the postparturient differentiation of the mammary gland which in turn effects transport of components into early milk. Certainly in the goat, preparturient milking of one mammary gland evoked changes in milk composition which were not evident in the contralateral nonmilked gland (Linzell and Peaker, 1974). The change in composition in the milked gland was matched by decreases in the permeability to labeled sucrose, sodium, and chloride passing from blood to milk. Maule and colleagues (1980) have suggested that a local factor, such as a prostaglandin, operates in late pregnancy to keep the paracellular pathway open and that this factor can be removed by milking. Further studies designed to examine preterm milk composition and the physiology of lactation in mothers giving birth prematurely should consider the breast stimulation and pumping pattern of the mothers since these activities may considerably impact the composition of the milk produced.

We propose that there are a number of perinatal events associated with preterm birth (Table VII) which may indeed influence the lactational process and the endocrine events that regulate milk synthesis and secretion. Further studies are needed to properly evaluate the impact of such perinatal events on the lactational capacity of mothers giving birth prematurely.

VIII. Summary

Preterm milk is a recently described entity that may be different in composition, at least for some nutrients, from term milk. The basis of differences in concentration of nutrients in preterm milk may be one or a combination of altered maternal hormonal milieu in the parturient period, an "immature" morphology of the mammary gland, or artifact(s) of perinatal events associated with premature birth. While preterm milk may not meet all of the nutrient needs of the growing premature infant (Lucas, 1993), its use should be encouraged with appropriate supplementation.

References

Anderson, D. M., Williams, F. H., Merkatz, R. B., Shulman, P. K., Kerr, D. S., and Pittard, W. B., III (1983). Length of gestation and nutritional composition of human milk. *Am. J. Clin. Nutr.* **37,** 810–814.

Anderson, G. H., Atkinson, S. A., and Bryan, M. H. (1981). Energy and macronutrient content of human milk during early lactation for mothers giving birth prematurely and at term. *Am. J. Clin. Nutr.* **34,** 258.

Anderson, G. H. (1984). The effect of prematurity on milk composition and its physiological basis. *Fed. Proc.* **43,** 2438.

Atkinson, S. A., Bryan, M. H., and Anderson, G. H. (1978). Human milk: Difference in nitrogen concentration in milk from mothers of term and premature infants. *J. Pediatr.* **93,** 67.

Atkinson, S. A., Radde, I. C., Chance, G. W., Bryan, M. H., and Anderson, G. H. (1980a). Macro-mineral content of milk obtained during early lactation from mothers of premature infants. *Early Hum. Dev.* **4,** 5–14.

Atkinson, S. A., Anderson, G. H., and Bryan, M. H. (1980b). Human milk: Comparison of the nitrogen composition in milk from mothers of premature and full-term infants. *Am. J. Clin. Nutr.* **33,** 811–815.

Atkinson, S. A., Wade, C. L., Stanhope, R., and Fraser, D. (1986). Pattern of change in milk composition during lactogenesis in term and preterm mothers. *In* "Human Lactation II" (M. Hamosh and A. Goldman, eds.), pp. 121–130. Plenum Press, New York.

Atkinson, S. A., Reinhardt, T. A., and Hollis, B. W. (1987). Vitamin D activity in maternal plasma and milk in relation to gestational stage at delivery. *Nutr. Res.* **7,** 1005–1011.

Barros, M. D., and Carneiro-Sampaio, M. M. S. (1984). Milk composition of low birth weight infants' mothers. *Acta Peadiatr. Scand.* **73,** 693–694.

Bitman, J., Wood, D. L., Mehta, N. R., Hamosh, P., and Hamosh, M. (1983). Comparison of the lipid composition of breast milk from mothers of term and preterm infants. *Am. J. Clin. Nutr.* **38,** 300–312.

Bitman, J., Wood, D. L., Mehta, N. R., Hamosh, P., and Hamosh, M. (1986). Comparison of the cholesteryl ester composition of human milk from preterm and term mothers. *J. Pediatr. Gastroenterol. Nutr.* **5,** 780–786.

Britton, J. R. (1986). Milk protein quality in mothers delivering prematurely: Implications for infants in the intensive care unit nursery setting. *J. Pediatr. Gastroenterol. Nutr.* **5,** 116–121.

Buster, J. E., Chang, R. J., Preston, D. L., Slashaff, B. M., Cousins, L. M., Abraham, G. E., Hobel, C. S., and Marshall, J. R. (1979). Interrelationships of circulating maternal steroid concentrations in third trimester pregnancies. *J. Clin. Endocrinol. Metab.* **48,** 133.

Butte, N. F., Garza, C., Johnson, C. A., O'Brian-Smith, E., and Nichols, B. L. (1984). Longitudinal changes in milk composition of mothers delivering preterm and term infants. *Early Hum. Dev.* **9,** 153–162.

Chan, G. M. (1982). Human milk calcium and phosphate levels of mothers delivering term and preterm infants. *J. Pediatr. Gastroenterol. Nutr.* **1,** 201–205.

Chandra, R. K. (1982). Immunoglobulin and protein levels in breast milk produced by mothers of preterm infant. *Nutr. Res.* **2,** 27–30.

Chappell, J. E., Francis, T., and Clandinin, M. T. (1985). Vitamin A and E content of human milk at early stages of lactation. *Early Hum. Dev.* **11,** 157–167.

Conner, A. E. (1979). Elevated levels of sodium and chloride in milk from mastitic breast. *Pediatrics* **63,** 910.

Djiane, J., and Durand, P. (1977). Prolactin-progesterone antagonism in self regulation of prolactin receptors in the mammary gland during pregnancy and lactation. *Nature* **266,** 641.

Donovan, S. M., Atkinson, S. A., and Lonnerdal, B. (1987). Whey proteins found in feces of preterm infants receiving preterm human milk and infant formulas. *In* "Human Lactation 3: Effects of Human Milk Upon the Recipient Infant" (A. S. Goldman, S. A. Atkinson, and L. A. Hanson, eds.), p. 377. Plenum Press, New York.

Ehrenhranz, R. A., Ackerman, B. A., and Nelli, C. M. (1984). Total lipid content and fatty acid composition of preterm human milk. *J. Pediatr. Gastroenterol. Nutr.* **3,** 755–758.

Ferlin, M. L., Santoro, J. R., Jorge, S. M., and Goncalves, A. L. (1986). Total nitrogen and electrolyte levels in colostrum and transition human milk. *J. Perinat. Med.* **14,** 251–257.

Garza, C., Johnson, C. A., O'Brian Smith, E., and Nichols, B. L. (1983). Changes in the nutrient composition of human milk during gradual weaning. *Am. J. Clin. Nutr.* **37,** 61.

Goldman, A. S., Garza, C., Nichols, B., Johnson, C. A., O'Brian Smith, E., and Goldblum, R. M. (1982). Effects of prematurity on the immunologic system in human milk. *J. Pediatr.* **101,** 901–905.

Gross, S. J., David, R. J., Bauman, L., and Tomarelli, R. M. (1980). Nutritional composition of milk produced by mothers delivering preterm. *J. Pediatr.* **96,** 641–644.

Gross, S. J., Buckley, R. H., Wakil, S. A., McAllister, D. C., David, R. J., and Faix, R. G. (1981a). Elevated IgA concentration in milk produced by mothers delivered of preterm infants. *J. Pediatr.* **99,** 389–393.

Gross, S. J., Geller, J., and Tomarelli, R. M. (1981b). Composition of breast milk from mothers of preterm infants. *Pediatrics* **68,** 490–493.

Guerrini, P., Bosi, G., Chierici, R., and Fabbri, A. (1981). Human milk: Relationship of fat content with gestational age. *Early Hum. Dev.* **5,** 187–194.

Hartman, P. E., and Prosser, C. G. (1984). Physiological basis of longitudinal changes in human milk yield and composition. *Fed. Proc.* **43,** 2448.

Hibberd, C. M., Brooke, O. G., Carter, N. D., Haug, M., and Harzer, G. (1982). Variation in the composition of breast milk during the first 5 weeks of lactation: Implications for the feeding of preterm infants. *Arch. Dis. Child.* **57,** 658–662.

Jones, J. B., Mehta, N. R., and Hamosh, M. (1982). α-Amylase in preterm human milk. *J. Pediatr. Gastroenterol. Nutr.* **1,** 43–48.

Kulski, J. K., and Hartman, P. E. (1981). Changes in milk composition during the initiation of lactation. *Aust. J. Exp. Biol. Med. Sci.* **59,** 101.

Lemons, J. A., Reyman, D., and Moye, L. (1983). Amino acid composition of preterm and term breast milk during early lactation. *Early Hum. Dev.* **8,** 323–329.

Lemons, J. A., Moye, L., Hall, D., and Simmons, M. (1982). Differences in the composition of preterm and term milk during early lactation. *Pediatr. Res.* **16,** 113–117.

Lepage, G., Collet, S., Bougle, D., Kien, L. C., Lepage, D., Daillaire, L., Darling, P., and Roy, C. (1984). The composition of preterm milk in relation to the degree of prematurity. *Am. J. Clin. Nutr.* **40,** 1042–1049.

Linzell, J. L., and Peaker, M. (1974). Changes in colostrum composition and in permeability of mammary epithelium at about the time of parturition in the goat. *J. Physiol. (London)* **243,** 129.

Lucas, A., and Hudson, G. (1984). Preterm milk as a source of protein for low birthweight infants. *Arch. Dis. Child.* **59,** 831–836.

Lucas, A. (1993). Enteral nutrition. *In* "Nutritional Needs of the Preterm Infant" (R. C. Tsang, A. Lucas, R. Uauy, and S. Zlotkin, eds.), pp. 209–223. Williams & Wilkins, New York.

Maule-Walker, F. M., and Peaker, M. (1980). Local production of prostaglandins in relation to mammary function at the onset of lactation in the goat. *J. Physiol.* **309,** 65.

Mendelson, R. A., Anderson, G. H., and Bryan, M. H. (1982). Zinc, copper and iron content of milk from mothers of preterm and full-term infants. *Early Hum. Dev.* **6,** 145–151.

Moran, J. R., Vaughan, R., Stroop, S., Coy, S., Johnston, H., and Greene, H. L. (1983a). Concentrations and total daily output of micronutrients in breast milk of mothers delivering preterm: A longitudinal study. *J. Pediatr. Gastroenterol. Nutr.* **2,** 629–634.

Moran, J. R., Courtney, M. E., Orth, D. N., Vaughan, R., Coy, S., Mount, C. D., Sherrell, B. J., and Greene, H. L. (1983b). Epidermal growth factor in human milk: Daily production and diurnal variation during early lactation in mothers delivering at term and at premature gestation. *J. Pediatr.* **103,** 402–405.

Neville, M. C., and Peaker, M. (1981). Ionized calcium in milk and integrity of the mammary epithelium in the goat. *J. Physiol.* **313,** 561.

Pamblanco, M., Ten, A., and Comin, J. (1986). Proteins in preterm and term milk from mothers delivering appropriate or small-for-gestational age infants. *Early Hum. Dev.* **14,** 267–272.

Parker, C. R., Everett, R. B., Querk, J. G., Whalley, P. J., and Gant, N. F. (1979). Hormone production during pregnancy in the primigravida patient I plasma levels of progesterone and 5-alpha-pregnone-β-3,20 dione throughout pregnancy of normal women and women who developed pregnancy-induced hypertension. *Am. J. Obstet. Gynecol.* **135,** 778.

Prosser, C. G., and Hartman, P. E. (1983). Saliva and milk composition during the menstrual cycle of women. *Aust. J. Exp. Biol. Med. Sci.* **61,** 265.

Ronayne de Ferrer, P. A., Slobodianik, N. H., Lopen, N., Sambucetti, M. E., and Sanahuja, J. C. (1984). Immunoglobulin A level in human milk from mothers delivering preterm. *Am. J. Clin. Nutr.* **40,** 465–467.

Sann, L., Bienvenu, F., Lahet, C., Bienvenu, J., and Bethenod, M. (1981). Comparison of the composition of breast milk from mothers of term and preterm infants. *Acta Paediatr. Scand.* **70,** 115–116.

Schanler, R. J., and Oh, W. (1980). Composition of breast milk obtained from mothers of premature infants as compared to breast milk obtained from donors. *J. Pediatr.* **96,** 679–681.

Siimes, M. A., Vuori, E., and Kuitunen, P. (1979). Breast milk iron—A declining concentration during the course of lactation. *Acta Pediatr. Scand.* **68,** 29–31.

Thomas, M. R., Chan, G. M., and Book, L. S. (1986). Comparison of macronutrient concentration of preterm human milk between two milk expression techniques and two techniques for quantitation of energy. *J. Pediatr. Gastroenterol. Nutr.* **5,** 597–601.

Thomas, M. F., Pearsons, M. H., DemKowiciz, M., Chan, J. M., and Lewis, C. G. (1981). Vitamin A and vitamin E concentration of the milk from mothers of preterm infants and milk of mothers of full term infants. *Acta Vitamino. Enzymol.* **3,** 135–144.

Udipi, S. A., Kirksey, A., West, K., and Giacoia, G. (1985). Vitamin B_6, vitamin C and folacin levels in milk from mothers of term and preterm infants during the neonatal period. *Am. J. Clin. Nutr.* **42,** 522–530.

Vaisman, N., Mogilner, B. M., and Sklan, D. (1985). Vitamin A and E content of preterm and term milk. *Nutr. Res.* **5,** 931–935.
Vuori, E., and Kuitunen, P. (1979). The concentration of copper and zinc in milk. *Acta Pediatr. Scand.* **68,** 33–68.

F. Miscellaneous Factors Affecting Composition and Volume of Human and Bovine Milks

ROBERT G. JENSEN

I. Introduction

Several of the many factors which influence the composition of human and bovine milks have been discussed in other chapters. In this section I will summarize and briefly describe those that were omitted. In general, the amounts and composition of most components respond to time and individuality. Once beyond the colostral stage, the amounts of constituents remain remarkably constant. Major changes thereafter may be caused by differences in diet and infections, mostly mastitis in dairy cattle. These variations occur primarily in the fatty acid profiles. Regional changes are usually due to alterations in the diet and again occur mostly in the composition of the fatty acids (see Chapter 6A). The factors associated with changes in the composition of human milk are listed in Table I and for bovine milk in Table II. These effects are important in human milk since with the exception of processing–banking, the milk is consumed on an individual mother infant basis. With bovine milk, the only important effects are diseases, season, and processing. This is because of the pooling, standardization, and selective breeding that are done in the dairy industry.

II. Human Milk

A. *During a Nursing*

The lipid content consistently rises during a nursing (WHO, 1985; Neville *et al.,* 1984; Chapter 3A). The amounts of sodium, chloride, calcium, magnesium, inorganic phosphate, lactose, glucose, urea (N), creatinine, zinc, and copper did not vary significantly (Neville *et al.,* 1984) (see

HANDBOOK OF MILK COMPOSITION

TABLE I
Factors Associated with Changes in the Composition of Human Milk[a]

A. During a nursing or feed
B. Time postpartum or stage of lactation
C. Diurnal or circadian rhythm
D. Between breasts
E. Gestational age at birth; preterm vs term
F. Diet. Region
G. Mother's weight
H. Infections, metabolic disorders
I. Medication
J. Mother's menstrual cycle or pregnancy
K. Parity
L. Season. Related to diet
M. Age of mother
N. Infant's birth weight
O. Processing–banking
P. Individuality
Q. Summary

[a]Adapted from Table I, Chapter 6A, and WHO (1985).

TABLE II
Factors Associated with Changes in the Composition of Bovine Milk[a]

A. During a nursing or feed
B. Age postpartum, stage of lactation
C. Diurnal rhythm
D. Between treatments
E. Breed
F. Diet. Region
G. Diseases and effects on consumers
H. Medication
I. Cow's pregnancy
J. Parity
K. Season. Related to diet
L. Age
M. Individuality
N. Processing

[a]Adapted from Table I, Chapter 6A.

Table III). Lactose showed a slight insignificant increase which vanished when the volume was corrected for the increase in fat. Neville *et al.* observed that a sample of milk taken at mid-feed had the same mean composition as the pooled, pumped contents of one breast.

B. Time Postpartum or Stage of Lactation

The changes in composition that occur as lactation progresses are shown in Tables IV–VI. Neville *et al.* (1991) were searching for indications of lactogenesis (Table IV). See Chapter 3A for discussion. Data on the composition further into lactation are presented in Tables V and VI. The amounts of milk, lactose, and fat increase and protein decreases as lactation progresses (Saint *et al.*, 1984) and then level off (Dewey and Lonnerdal, 1983).

C. Diurnal Rythm

The lipids have shown diurnal changes. See Chapter 6A and Table VII. These may be related to the interval between nursings, the degree of

TABLE III
Changes in Milk Composition during a Nursing[a]

Component	Foremilk	Midmilk	Hindmilk	p Value
Lipid (%)	2.1	3.0	4.1	0.001
Sodium (mmol)	7.5	8.0	8.0	ns
Potassium (mmol)	12.9	13.0	13.0	ns
Chloride (mmol)	11.7	12.0	12.8	ns
Urea N (mg/dl)	18.2	17.8	17.9	ns
Zinc (μg/ml)	1.5	1.3	1.3	ns
Copper (μg/ml)	0.25	0.28	0.25	ns
Creatinine (mg/dl)	2.6	2.6	2.4	ns
Inorganic phosphate (mmol)	5.1	5.5	5.3	ns
Calcium (mmol)	7.3	7.5	7.1	ns
Magnesium (mmol)	1.81	1.85	1.84	ns
Protein (g/dl)	1.42	1.44	1.44	ns
Glucose (mmol)	1.38	1.00	1.20	ns
Lactose (mmol)	2.08	206	1.96	0.005
Lactose, fat free (mmol)	210[b]	200	208	ns

[a]Adapted from Neville *et al.* (1984). Data were taken from figures and are approximate.
[b]Adjusted for volume of fat.

TABLE IV
Changes in Several Milk Components During Lactogenesis in Humans[a]

Component (mmol/liter)	Hours postpartum				
	21	48	60	96	120
Volume (ml/day)	—	180	350	560	540
Citrate	0.4	1.5	2.4	3.6	4.5
Glucose	0.38	0.64	1.28	1.50	1.40
Free phosphate	0.5	1.2	1.4	1.6	1.8
Magnesium	1.8	1.4	1.4	1.4	1.4
Lactose	100	140	160	160	160
Potassium	13.8	15.0	18.0	18.0	18.0
Sodium	34	25	16	14	14
Chloride	44	35	25	20	20
Calcium	4.0	6.0	6.6	7.6	8.0

[a]Adapted from Neville *et al.* (1991). Data taken from figures and are approximate.

TABLE V
Yield and Composition of Human Colostrum and Milk from Australian Donors 1 to 28 Days Postpartum[a]

Component	Days postpartum						
	1	2	3	4	5	14	28
Yield (g/24 hr)	50	190	400	625	700	1100	1250
Lactose (g/liter)	20	25	31	32	33	35	35
Fat (g/liter)	12	15	20	25	24	23	29
Protein (g/liter)	32	17	12	11	11	8	9

[a]Adapted from Saint *et al.* (1984).

emptying of the breast allowing carryover of the high-fat hindmilk into the next nursing, and the interval of sampling (Jensen *et al.*, 1995). Daly *et al.* (1993b) found that 41–95% of the variance in fat was related to the degree of breast emptying and suggested that this may explain the circadian rythmn in fat content of milk. However, their assumption is based on determinations of fat content in fore- and hindmilk samples only.

D. Between Breasts

Sporadic, inconsistent differences in the composition (components in Table III) of the milk from the right and left breasts have been observed

TABLE VI
Composition of Breast Milk Samples from California Donors 1 to 6 Months Postpartum[a]

Component	Month 1	2	3	4	5	6
Breast milk intake, mean (ml/day ± SD)	673 ± 192	756 ± 170	782 ± 182	810 ± 142	805 ± 113	896 ± 122
n	16	19	16	13	11	11
Energy (kcal/dl ± SD)	78.1 ± 10.0	75.3 ± 9.2	73.6 ± 14.8	78.7 ± 17.3	74.7 ± 14.8	74.8 ± 18.3
Protein (g/dl ± SD)	1.44 ± 0.20	1.33 ± 0.16	1.32 ± 0.16	1.30 ± 0.24	1.25 ± 0.17	1.27 ± 0.36
Fat (g/dl ± SD)	4.92 ± 1.05	4.58 ± 0.97	4.58 ± 1.65	4.62 ± 1.86	4.36 ± 1.67	4.30 ± 1.96
Fat intake[b] (g/day)	33.1	34.6	35.8	37.4	35.1	38.5
Lactose (g/dl ± SD)	7.05 ± 0.56	7.21 ± 0.62	7.13 ± 0.79	7.61 ± 0.40	7.62 ± 0.33	7.75 ± 0.27
n	13	16	18	16	14	18

[a]Adapted from Dewey and Lonnerdal (1983).
[b]Calculated by the author.

(Neville *et al.*, 1984). Mastitis may have contributed to these differences. See Chapter 3A for recommendations on sampling. As noted by Daly *et al.* (1993a,b), the degree of breast emptying is also important.

E. Gestational Age

See Chapter 3E

F. Diet

1. *Introduction*

It is difficult to evaluate the effect of maternal diet on the quantity and composition of human milk. The diet can be inadequate in general, resulting in malnourishment of the mother, or specific nutrients can be low in content or lacking altogether (Lonnerdal, 1986). The effects of malnourishment have been studied in regions where the food supply is inadequate or a specific nutrient, e.g., Se in soil, may be lacking. The effects of different nutrients have been determined by investigating the diets and milks in various regions (see Chapter 3D) or by supplementing maternal diets with different nutrients, e.g., fish oil concentrates, to

TABLE VII
Lipid and Cholesterol Contents and Volumes of Human Milk Produced during 24 hr[a]

Period (hr)	Lipid content (%)	Volume milk (ml/breast)	Lipid (g/vol)	Cholesterol contents: mmol/liter milk (mg/dl)	Cholesterol contents: mmol/100 g fat (mg/100 g fat)
1 (0600–1000)	2.93 ± 0.32[b,c]	86.8 ± 8.0[b,d]	2.54	0.36 ± 0.01[b,e] (14.0 ± 2.2)	1.24 ± 0.19[b] (478 ± 75)
2 (1000–1400)	3.89 ± 0.28	45.0 ± 7.4	1.75	0.42 ± 0.04 (16.2 ± 1.7)	1.1 ± 0.11 (416 ± 44)
3 (1400–1800)	3.87 ± 0.31	44.0 ± 7.7	1.94	0.56 ± 0.05 (21.7 ± 2.1)	1.45 ± 0.14 (561 ± 54)
4 (1800–2200)	4.37 ± 0.40	45.1 ± 9.8	1.97	0.57 ± 0.07 (22.0 ± 2.8)	1.30 ± 0.17 (503 ± 64)
5 (2200–0600)	2.86 ± 0.36	49.9 ± 9.0	1.43	0.33 ± 0.06 (12.9 ± 2.2)	1.17 ± 0.20 (451 ± 77)
Averages	3.56[f]	—	—	0.45 (17.4)	1.25 (484)
Totals	—	270.8	9.63	—	—

[a]Jensen *et al.* (1995).
[b]Least-square means ± SEM.
[c]Significant differences (p = 0.05) in fat content between periods 1 and 2–5; 2 and 5; 4 and 5.
[d]Significant differences (p = 0.05) in volume between periods 1 and 2–5.
[e]Significant differences (p = 0.05) in mg cholesterol/dl milk between periods 1 and 3 and 4; 2 and 3; 3 and 5; 4 and 5.
[f]Total lipid, 9.63 g/270.8 ml = 3.56% average lipid content.

increase the contents of the omega-3 polyunsaturated fatty acids in milk (see Chapter 6A).

A well-designed study should include: (a) a sufficient number of subjects, (b) estimation of the volume of milk (see Chapter 3A), (c) proper storage of the milk sample until it can be analyzed, (d) elimination of temporal effects, (e) assessment of the effects of diseases (mastitis) and parasitic infestations on the nutritional status of the mother and the volume and composition of the milk, (f) awareness that interactions may occur as a result of deficiencies (Lonnerdal, 1986; WHO, 1985), and (g) use of appropriate analytical methods (IOM, 1991). Most studies have not included b, e, and f from above. Volume is important because the amounts of nutrients delivered to the infant must be known to assess their effects. Volume can decrease, but the quantity of a nutrient will increase to compensate.

Mastitis alters the composition of milk increasing the sodium and chloride contents (Neville *et al.*, 1984) and, if severe, destroys secretory tissue thus reducing the volume of milk in later lactations (Prentice *et al.*,

1985) (see Section G). I do not know of any papers in which the extent of parasitic infections, endemic in Third World mothers with accompanying diversion of nutrients, was assessed before and after medication. On occasion it is possible to determine the effects of "alternative" diets on specific nutrients. An example is low vitamin B12 contents in milks from mothers on certain vegan diets.

Data on nutrition during lactation on the volume and composition of milk from women in the United States (IOM, 1991) which are applicable to affluent countries and to developing nations are outlined below (WHO, 1985).

2. *Volume*

The influences of nutrition on milk volume in the United States, other developed countries, and developing countries (IOM, 1991) are summarized below:

(a) The average daily amount of milk produced is 750 to 800 ml/day in women who consume different diets and whose nutritional status varies.

(b) The potential production of milk appears to be greater than the quantity consumed by the infant. This was confirmed by Daly *et al.* (1993a,b).

(c) Factors other than nutrition affect milk volume. See Table I and this section. Daly *et al.* (1992, 1993a) found that breast emptying was 76 + 1–20% of the total volume. They and others have suggested that infants are self-regulating in their intake of breast milk.

(d) Milk volume is not related to maternal nutritional status in affluent countries, but may be in less-developed countries. Severe malnutrition can stop the flow of milk (Sosa *et al.*, 1976). There was only one subject in this study.

(e) Maternal energy intake is not strongly related to average milk volumes from lactating women since the quantities differ between industrialized and developing nations regardless of major differences in nutrient and energy intakes.

(f) Dietary supplementation of lactating women in regions where malnutrition occurs has little or no effect on milk volume, but may benefit the mother more than the infant except when milk composition is affected (see below).

(g) Weight is usually lost during lactation, i.e., in the United States up to 2 kg or 4.5 lb/month, with no apparent effects on milk production.

(h) Regular exercise does not affect milk volume. The manual labor done by lactating women in The Gambia, for example, did not adversely influence milk volume (Prentice *et al.*, 1986).

(i) There are few investigations on the maternal intake of specific nutrients and their contents in milk, but there may be a relationship between protein intake and milk volume. See below for effects on composition.

(j) Fluid intake during lactation should be adequate, but additional amounts above thirst levels do not influence milk volume.

Again, I emphasize that the volume of milk consumed by the infant must be determined so that the actual amounts of nutrients delivered can be ascertained. Volumes consumed are best measured by test weighing of the infant (IOM, 1991). The more recent procedure of computerized breast measurement can also be used (Daly *et al.*, 1992). In evaluating this summary, I believe it is useful to consider the statements by Rasmussen (1992). She noted that randomized intervention among undernourished women shows that enhanced maternal diets during lactation increase milk intake and alleviate the growth deficit of the infants. While an adequate supply of nutrients is needed for milk biosynthesis, milk production increases only with adequate infant demand. This is a "catch 22" situation. Milk production will increase in response to the infant's demand, but the infant's demand will increase only if the infant grows as a result of sufficient milk. Nevertheless, maternal supplementation might break the cycle improving lactation performance and the infant's appetite. The infant regulates the production of milk, but as suggested by Daly *et al.* (1993a,b) and others, milk may contain appetite inhibitors and possibly stimulants. The overall regulatory mechanism is likely to be very complex.

3. *Composition*

It is useful at this point to reiterate the absolute necessity for proper sampling, storage, and analyses of the milk for the component(s) being studied. Proven procedures for all portions of this process are available (IDM, 1991). My admonition is not intended to discourage the development of new and improved analytical procedures, but they should always be compared to accepted methods. This is particularly applicable to lipids which are the most variable component in milk (see Chapter 6A).

The influences of maternal diet and the nutrient or nutrient class are summarized in Table VIII adapted from "Nutrition During Lactation" (IOM, 1991).

The IOM (1991) confirmed the value of human milk as the source of nutrients for protective substances and other useful messages for infants. They concluded that there is considerable evidence regarding the ability of women to produce milk containing adequate protein, lipid, carbohydrate, and most minerals regardless of the adequacy of their nutrient supply. If diets are inadequate, the quantities of vitamins, particularly B6, B12, A, and D, may be low. The amounts of macronutrients, most minerals, and folate are maintained in milk at the expense of maternal reserves. The amounts and types of fatty acids are influenced by diet (see Chapter 6A). However, maternal diet has no influence on milk cholesterol and phospholipids.

I postulate that there are two maternal set points in response to adequacy of maternal nutrient intake. Below set point I, dietary

TABLE VIII
Possible Effects of Maternal Intake on the Composition of Human and Nutrients for Which Clinical Deficiency Is Noticeable in Infants[a]

Nutrients or nutrient class	Effect on milk composition	Noticeable nutritional deficiency in infants
Macronutrients		
Proteins	+	Unknown[b]
Lipids	+	Unknown
Lactose	−	Unknown
Minerals		
Calcium	−	Unknown
Phosphorous	−	Unknown
Magnesium	−	Unknown
Sodium	−	Unknown
Chlorine	−	Unknown
Iron	−	Yes[d]
Copper	−	Yes
Zinc	±	Yes
Manganese	+	Yes
Selenium	+	Yes
Iodine	+	Yes
Fluoride	+	Yes
Vitamins		
Ascorbic acid	+	Yes
Thiamin	+	Yes
Riboflavin	+	Unknown
Niacin	+	Unknown
Pantothenic acid	+	Unknown
Pyridoxine	+	Yes
Biotin	+	Yes
Folate	+	Yes
Vitamin B12	+	Yes
Vitamin A	+	Yes
Vitamin D	+	Yes
Vitamin E	+	Yes
Vitamin K	+	Yes

[a]Adapted from IOM, p. 7 (1991).
[b]Data not available to classify as no.
[c]Types of fatty acids altered, but not the quantities of lipid or cholesterol.
[d]Not associated with maternal intake.
[e]Maternal intake not related to the amount in milk (Pietschnig *et al.*, 1993) or to infant's status.

inadequacy requires that the mother use her own reserves to maintain lactation. Below set point II, the mother is malnourished to the point that she cannot sustain lactation. This will vary with individuals and nutrients.

4. *Developing Countries*

The conclusions above also apply to milk from lactating women in developing countries with some exceptions. These women, who may consume diets that are relatively low in energy, protein, B12, and iron, do much more physical work and have repeated lactations. Yet the volume and composition of their milks is adequate for the nutriture of their infants. See Table IX in which the composition of milks from malnourished women in The Gambia and the United Kingdom are compared (Prentice *et al.*, 1986). The quantities differ, but the growth patterns of the infants in the two groups were similar. The ability of the mammary gland to maintain lactation performance and the composition of milk in undernourished women is remarkable.

G. Mother's Weight

This has two aspects, weight loss and gain, both obviously due to adequacy of the diet. Useful results have come from the group who have investigated breast-feeding in The Gambia. These women experience yearly cycles of food shortage (the rainy season, July–September) and adequate supplies (the dry season, February–May) (Prentice, 1980). During the wet season, lactating women consumed 1200–1300 kcal/day and lost weight, 1 kg/month with almost complete depletion of fat stores. Milk production decreased 40% and quality deteriorated. Some of the data are presented in Table X. The variations were termed seasonal but are due to food adequacy. The women gained weight and deposited fat stores with an intake of 1600–1700 kcal/day during the dry season. Prentice (1981) later noted a significant relationship ($p < 0.001$) between triceps skinfold thickness that was not related to the decrease due to parity, 4+. Michaelsen *et al.* (1990) detected a significant increase in fat content, 3.9 to 4.21%, as the body mass index (BMI; kg/m/2) increased from < 21 to > 27. The subjects were Danish mothers. Protein and lactose contents did not change much. Thus, as observed by Hachey (1994), the fat content of milk will respond to adequacy of the diet. The biological relationship between the depletion of body fat stores accompanied by inadequate intake seems clear. The association of BMI and fat content in milks from well-nourished Danish mothers is more difficult to explain, but may be a result of the relatively large number of samples tested (2554 from 244 subjects). A compensatory change in volume may also have occurred so that the total amount of fat delivered to the infant in 24 hr was about the same (Hachey, 1994).

TABLE IX
Comparison of Milk Composition from Malnourished (Gambia) and Well-Nourished (United Kingdom) Mothers[a]

Component	Gambia	United Kingdom	Gambia as % United Kingdom
Proximates[b]			
Volume intake (ml)	577[c]	763[d]	74
Energy (kcal/dl)	70	70	100
Protein (g/dl)	1.32	1.34	99
Fat (g/dl)	4.2	4.2	100
Lactose (g/dl)	7.7	7.4	104
Vitamins[e]			
Thiamin (μg/dl)	16	16	100
Riboflavin (μg/dl)	21	31	68
Pyridoxine (μg/dl)	12	6	200
Niacin (μg/dl)	113	230	49
Cobalamin (ng/dl)	16	10	160
Folic acid (μg/dl)	3.8	5.2	73
Biotin (ng/dl)	900	760	118
Pantothenic acid (μg/dl)	200	260	76
Ascorbic acid (mg/dl)	3.4	3.8	88
Immunoproteins[b]			
Lactoferrin (mg/dl)	318	216	147
IgA (arb. mg/dl)	46	37	123
IgG (mg/dl)	6.5	2.8	236
IgM (mg/dl)	4.6	2.4	194
C3 (mg/dl)	1.9	1.2	158
C4 (mg/dl)	2.4	0.9	259
Lysozyme (mg/dl)	4.1	4.0	101
Total immunoproteins (mg/dl)	382.9	264.3	145

[a]Adapted from Prentice *et al.* (1986).
[b]All values from 0 to 3 months postpartum.
[c]Milk intakes by test weighing, 0–9 months. Average calculated by author.
[d]Milk intakes by test weighing 2–9 months. Average calculated by author.
[e]United Kingdom values from 0 to 3 months; Gambian values from 0 to 18 months postpartum.

Michaelsen *et al.* (1994) did note a relationship over time in the fat content of milk and amount of weight gained postpartum by the mothers. The fat contents of the milk and amount of weight gains at 5 months were 3.0%, low (< 11.2 kg); 4.2%, medium; and 6.0%; high (> 16.8 kg).

TABLE X
Seasonal Variations in Composition of Milk from Gambian Mothers[a]

	Dry season[b] (adequate food)	Wet season[b] (inadequate food)
Mean stage of lactation (days)	149	132
Energy (kcal/dl)	69 ± 2	65 ± 0.2
Protein (gl/dl)	1.10 ± 0.05	0.98 ± 0.07
Fat (g/dl)	4.36 ± 0.25	3.45 ± 0.22*
Lactose (g/dl)	6.35 ± 0.09	7.17 ± 0.09**
Maternal caloric intake (kcal/day)	1600–1700	1200–1300
No. of subjects	22	21

[a]Adapted from Prentice (1980). Values are means ± SEM.
[b]Dry season, February–May. Wet season, July–September.
*$p < 0.02$.
**$p < 0.001$.

As a result of the obsession of many women with "leanness" the effects of maternal dieting and physical activity on lactation have been investigated and the results summarized by Dewey and McCrory (1994). Neither of these markedly affects the composition of milk. An exception is a small increase in the lactic acid contents of milks from women who exercise, but this is unlikely to affect consumption. Women who are already lean may have weight loss if energy intake drops below 1800 kcal/day. If intake drops to 1200 kcal then the changes observed in Gambian milks could occur.

H. Infections, Metabolic Disorders

1. Introduction

This area has not received much attention with the exception of diabetes. We have few data on mastitis, which can alter the composition and flavor of milk, stop lactation, and in the subclinical form, is probably more prevalent than we realize. The effects of diseases on lactation in general are discussed by Lawrence (1989) and on milk lipids in Chapter 6A. Hamosh and Bitman (1992) reviewed the effects of diabetes, cystic fibrosis, hypobetalipoproteinemia, type 1 hyperlpoproteinemia, breast milk jaundice, and ectopic eczema. Of these I will discuss diabetes and include mastitis.

2. Mastitis

Mastitis is the result of an infection in the breast causing tenderness and redness. Fever may occur. Prevention by stringent sanitation and treatment of the disease with appropriate medications are important to protect the mother from a potentially serious infection which could also adversely affect performance during the current and successive lactations. Mastitis reduces the volume and lactose and increases the sodium and chloride contents of the milk (Prentice *et al.* 1985). Serious mastitis can also destroy secretory tissue in the gland in the dairy cow (Kitchen, 1981) and possibly in the human.

Prentice *et al.* (1985) found that mastitis was common in rural Gambian mothers. In addition to the changes in lactose and sodium mentioned above, the secretory immunoproteins, IgA, lysozyme, and lactoferrin, increased. The normal milk from mothers with mastitis had lower concentrations of IgA, C3, and lactoferrin than controls. Prentice *et al.* (1985) suggested that the former group may have been predisposed to mastitis. Neubauer *et al.* (1995) found in a study of U.S. women that those with no mastitis had higher levels of protein and lower quantities of lactose than those who were afflicted. Conductivity was also greater in women with mastitis probably as a result of the increased sodium and chloride contents observed by others.

3. *Diabetes*

It would be expected that diabetes would affect lactation because of the deficiency in insulin and associated problems. Until recently, women with insulin-dependent diabetes (IDDM) were advised not to nurse their infants (Ferris *et al.,* 1993; Ferris and Reece, 1994). While lactogenesis is delayed in women with IDDM, probably because of poor metabolic control, differences in milk composition should not prevent them from breast-feeding their infants (Neubauer *et al.,* 1993). Data on the composition of milks from women with IDDM and appropriate controls are shown in Table XI. The statistically significant differences between the groups were few, occurring only with lactose, total nitrogen, milk and formula intakes of the infants, and prolactin contents. However, the lower milk lactose and higher total nitrogen in IDDM women indicated delayed lactogenesis but were within accepted ranges. Arthur *et al.* (1994) confirmed the delay of lactogenesis II (the onset of prolific milk flow) in mothers with IDDM. They determined the quantities of lactose metabolites in milk finding that there was a delay in the increase in mammary gland concentrations of glucose in IDDM mothers. They concluded that glucose availability may regulate lactose synthesis at lactogenesis II. Electrical conductivity was measured to assess changes in milk anions and cations, while osmolality was determined to see if the milks differed in lactose and electrolyte contents. As mentioned under Mastitis above, conductivity is utilized to evaluate damage to mammary gland tissue, resulting from mastitis, in bovine milk.

Comprehensive data on the fatty acid profiles and total lipid contents of milk from women with and without IDDM not available when Chapter 6A was written have been published (Jackson *et al.,* 1994). IDDM had little effect on the total lipid content after lactation was established. With the

TABLE XI
The Composition and Infant's Intake of Milk from Women with and without Insulin-Dependent Diabetes Mellitus (IDDM)[a]

Component and group	Days postpartum					
	2	3	7	14	42	84
Lactose (mmol/liter)						
IDDM[b]	95.87 ± 7.13 (6)	159.87 ± 4.18 (17)	163.34 ± 3.02 (29)	178.01 ± 3.44 (24)	185.58 ± 3.83(20)	188.16 ± 4.05 (18)
Control	130.81 ± 4.58 (14)	163.01 ± 3.36 (24)	175.84 ± 2.97 (29)	185.40 ± 3.38 (24)	192.49 ± 3.56 (22)	201.65 ± 4.23 (16)
Reference	142.80 ± 4.74 (11)	182.08 ± 4.74 (11)	182.08 ± 4.74 (11)	187.89 ± 4.74 (11)	202.00 ± 4.74 (11)	205.60 ± 4.74 (11)
Total nitrogen (g/liter)						
IDDM	7.78 ± 0.47 (3)[c]	4.78 ± 0.23 (13)[c]	3.09 ± 0.15 (25)	2.59 ± 0.16 (24)	2.24 ± 0.18 (20)	2.06 ± 0.19 (18)
Control	5.46 ± 0.29 (8)	3.49 ± 0.17 (22)	3.02 ± 0.15 (26)	2.49 ± 0.16 (24)	2.16 ± 0.17 (21)	1.89 ± 0.21 (14)
Reference	5.41 ± 0.25 (9)	3.38 ± 0.25 (9)	2.94 ± 0.22 (11)	2.49 ± 0.22 (11)	2.10 ± 0.22 (11)	1.82 ± 0.22 (11)
Conductivity (Ω)[d]						
IDDM		0.28 ± 0.02 (5)	0.31 ± 0.01 (17)	0.29 ± 0.01 (15)	0.25 ± 0.01 (14)	0.25 ± 0.02 (11)
Control		0.32 ± 0.02 (7)	0.29 ± 0.01 (20)	0.25 ± 0.01 (19)	0.20 ± 0.01 (19)	0.21 ± 0.01 (16)
Reference		0.34 ± 0.02 (8)	0.27 ± 0.01 (11)	0.25 ± 0.01 (11)	0.21 ± 0.01 (10)	0.20 ± 0.01 (11)
Osmolality (mOsmol/kg)						
IDDM		276.24 ± 9.09 (5)	297.98 ± 4.34 (16)	292.09 ± 4.55 (15)	279.02 ± 4.86 (14)	286.89 ± 5.54 (11)
Control		287.16 ± 7.04 (7)	290.19 ± 3.92 (20)	286.64 ± 4.22 (18)	290.75 ± 3.96 (20)	298.09 ± 4.52 (16)
Reference		293.15 ± 6.01 (8)	295.77 ± 4.95 (11)	296.95 ± 4.95 (11)	292.79 ± 5.27 (10)	290.45 ± 4.95 (11)

TABLE XI—*continued*

Component and group	Days postpartum					
	2	3	7	14	42	84
Milk intake of infants (g/day)[e]						
IDDM[f]			309.62 ± 32.37 (15)	426.05 ± 32.28 (15)	575.29 ± 33.98 (14)	530.93 ± 39.47 (11)
Control[f]			455.46 ± 29.71 (17)	504.80 ± 35.69 (13)	535.35 ± 36.04 (13)	511.95 ± 39.81 (11)
Reference			518.50 ± 38.20 (10)	592.17 ± 40.93 (9)	654.43 ± 38.20 (10)	673.72 ± 38.20 (10)
Milk and formula intake of infants (g/day)[e]						
IDDM[f]			329.89 ± 27.63 (15)[c]	447.10 ± 27.55 (15)[g]	624.70 ± 29.01 (14)	580.72 ± 33.69 (11)
Control[f]			456.52 ± 25.36 (17)	516.49 ± 30.46 (13)	542.44 ± 30.76 (13)	588.59 ± 33.98 (11)
Reference			528.12 ± 32.60 (10)	595.36 ± 34.93 (9)	654.43 ± 32.60 (10)	684.33 ± 32.60 (10)
Prolactin (μg/liter)[g]						
IDDM	66.3 ± 4.1 (6)[i]	65.8 ± 2.4 (17)[j]	62.0 ± 1.8 (29)[j]	48.7 ± 2.0 (24)	39.7 ± 2.2 (21)	33.8 ± 2.4 (17)
Control	66.7 ± 3.1 (11)	75.2 ± 2.0 (23)[j]	67.6 ± 1.7 (22)[j]	51.8 ± 2.1 (16)	41.2 ± 2.1 (16)	35.2 ± 2.5 (16)
Reference	91.9 ± 3.6 (7)[i]	76.5 ± 3.1 (9)[j]	63.3 ± 2.8 (11)[j]	52.6 ± 2.8 (11)	41.9 ± 2.9 (11)	35.6 ± 2.8 (11)

[a]Adapted from Neubauer *et al.* (1993). Data are least-square means ± SEM. When group names have superscripts, group least-square means collapsed over time are statistically different.
[b,c]Significantly different from both control and reference: $p < 0.01$; $p < 0.05$.
[d]Data not available at 2 days postpartum.
[e]Data not available at 2 or 3 days postpartum.
[f]Significantly different from reference: $p < 0.01$; $p < 0.05$.
[g]From Ostrom and Ferris (1993). Data are least-square means ± SEM; n in parentheses. Values with different lettered superscripts are statistically different.
[i]$p < 0.0001$.
[j]$p < 0.05$.

exception of long-chain polyunsaturated fatty acids (LC-PUFA), the other classes of fatty acids did not differ in content between IDDM and normal milks. The amounts of LC-PUFAs (20 and 22 C) were significantly lower in IDDM than in control milks from Days 14 to 84 postpartum. The reductions in 10:0, 12:0, and 14:0, and increases in 18:1 and LC-PUFA reported by Bitman *et al.* (1989) were not observed. van Beusekom *et al.* (1993) found that the macronutrient and fatty acid composition of milks from mothers with tight control were normal. Many of these women were treated with subcutaneous insulin infusion. In addition to fatty acids, they determined fat, protein, lactose, cholesterol, glucose, and myoinositol.

I. Medications

The effects of medications are discussed in Chapter 11A.

J. Mother's Menstrual Cycle or Pregnancy

Hartmann and Prosser (1982, 1984) observed two acute changes occurred in milk composition; the first 5 or 6 days before, and the second 6 or 7 days after, ovulation. At these times there was an increase in Na and Cl concentrations from (mean $\pm$ SE of mean) 4.6 and 11.1 ± 0.2 m*M* to 10.1 and 22.0 ± 0.9 m*M*, respectively. The concentrations of lactose and K decreased from 7.8 ± 0.2 g/dl and 13.6 ± 0.4 m*M* to 6.0 ± 0.2 g/dl and 10.2 ± 0.5 m*M*, respectively. The concentrations of these compounds remained relatively constant during lactational ammenorhea, anovulatory menstrual cycles, and for women taking oral contraceptives. The possibility that the changes were caused by mastitis can be eliminated by the abruptness of their occurrence and the absence of other indications of infection.

Hartmann and Prosser (1984) discounted earlier reports of differences between milks from pregnant and nonpregnant women as confounded by sampling inconsistencies. They found some changes, but these did not affect the suckling activity of the infant.

K. Parity

Prentice *et al.* (1981) found that parity has a major influence in fat content in milk from Gambian mothers. They observed a 25% decrease in the mean fat content between primiparous mothers and those of parity 4 or greater. Similar effects were found for nitrogen total energy and volume, but not for lactose, which remained constant (Prentice, 1986). Antimicrobial proteins showed a similar pattern except that they increased at parities

10 + . Prentice *et al.* (1989) observed that the fatty acid profiles of milks from Gambian mothers reflected the fatty acids in the diet except for parities 10 + . In these women, the amount of fatty acids synthesized endogenously was much lower (11.4%) than that in parity 1 (19.3%). *De novo* synthesis may have been impaired in these women. Prentice *et al.* (1981) also found a significant relationship ($p < 0.001$) between the fat content of milk and triceps skinfold thickness, but this was not related to parity. Later, the effects of parity on the daily production and composition of milk from mothers in the village of Keneba, Gambia were published (Prentice, 1986). The concentrations of protein, fat, lactose, and calculated energy were lower in parities 3–10 than in 1 or 2. The same trend was seen in antimicrobial proteins, except in parity 10 + when they increased.

L. Season

Seasonal variations in composition are caused by cyclical changes in the availability of food, at least in The Gambia (Paul and Muller, 1980; Prentice, 1980; Prentice *et al.*, 1981). There are no data on seasonal affects on milks from Western mothers probably because nutritional status is largely unchanged.

M. Age

About 20% of total births or 600,000 infants are born to teenage mothers in the United States (Lipsman *et al.*, 1985). Since the diets of teenage girls are often suboptimal, the burdens of pregnancy and lactation could cause problems for mother and infant. In a study of California teenagers, Lipsman *et al.* (1985) found as averages for 1 to 6 months: 1.43 g/dl protein, 5.62 g/dl fat, 6.76 g/dl lactose, 0.33 μg/ml iron, 0.28 μg/ml copper, 218 μg/ml calcium, 26.4 μg/ml magnesium, 111 μg/ml sodium, and 832 μg/ml potassium. The values for lactose, calcium, magnesium, sodium, and potassium were significantly lower in teenagers than in adults. These differences may have been a result of differences in timing intervals. The subjects were adequately nourished as were 88% of their infants. The nutritional status of an undernourished teenaged Gambian mother and her infant would be expected to be poor.

At the other end of the age spectrum, a 65 year old Nigerian woman was found who had been breast-feeding her grandchildren (Gindler *et al.*, 1985). Her milk and milks from 23 privileged Nigerian contained the following (g/dl): 5.22 and 7.08 lactose, 2.61 and 1.65 fat, and 1.59 and 0.74 protein. By Western standards the milk fat contents in both groups and protein content in milk from the 65 year old women are low.

N. *Infant's Birth Weight*

Michaelsen *et al.* (1990) detected a U-shaped curve in the relationship between the fat content of milk and the infant's birth weight. The fat content when the infant's birth weight was < 3.1 kg was 3.72% and 4.04% when it was > 4.0. The amounts of fat between were 3.18–3.43%. The cause of the difference is not apparent.

O. *Processing–Banking*

1. Introduction

Milk is stored or banked for future use primarily for the feeding of preterm infants (Lawrence, 1989; Jensen and Jensen, 1992). The mother's own milk may be stored in a home freezer and fed from a bottle if she is unable to breast-feed the infant regularly. The goal of banking and home storage is to preserve the protective, digestive, inductive, and nutrient carrier properties of milk. This is achieved by selection of donors and careful sanitation during expression of milk. Refrigerated or frozen storage and pasteurization followed by refrigerated storage are also used. The purposes of these processes are to prevent the entry of, delay the growth of, or destroy the microorganisms which are found in human milk. To rephrase, the purpose is to prolong the "shelf-life" of milk. The banking of human milk was reviewed by Garza *et al.* (1986) and Lawrence (1989). Guidelines for the establishment and operation of a human milk bank are available from the HMBNA (1994).

2. Refrigerated and Frozen Storage

Storage temperatures in use for raw milk are: fresh (4°C, refrigerator), frozen (−20°C, freezer), and deep frozen (−70°C) (see Table XII). The purposes of low-temperature storage are to retard microbial growth and delay some changes in the physiochemical character of milk. The maximum recommended lengths of storage for refrigerated milk and frozen milk (−20°C) are 72 hr and 12 months (HMBNA, 1994). Milk can be successfully stored at −70°C for longer than 12 months, but these freezers are not generally available. The effects of refrigerated and frozen storage are given in Table XII. Storage of milk overnight at 4°C will result in formation of a cream layer containing about 20% and skim milk with about 1.0% fat. The layers must be mixed before feeding the infant. This can be done by gentle inversion of the container several times. The numbers of microorganisms actually decrease and those are usually innocuous skin types. Sosa and Barness (1987) identified four types in milk and 71% of these were *Staphylococcus epidermis*. The total counts were low; mean, 10,000 colony-forming units (CFU)/ml; range, 1000–140,000. Interestingly, they

TABLE XII
Results of Various Treatments on Human Milk[a]

Treatment	Results
Refrigerated storage	
4°C, 72 hr	Creaming, decrease in bacterial growth, possible lipolysis
−20°C, 12 months	Lipolysis, possible demulsification and protein denaturation when thawed
−70°C, indefinite	Possible demulsification and protein denaturation when thawed
Pasteurization	
56°C, 30 min	Inactivation of enzymes and antimicrobial proteins, partial loss of some vitamins, destruction of microorganisms
62.5°C, 30 min	Inactivation of enzymes and antimicrobial proteins, partial loss of some vitamins, destruction of microorganisms
70°C, 15 sec	Inactivation of enzymes and antimicrobial proteins, partial loss of some vitamins, destruction of microorganisms
Microwave treatment	Decrease in IgA and lysozyme, substantial increase in coliforms
Sonication	Homogenization of milk fat globules
Selection	Selection of high protein milks, use of high-fat hindmilk
Supplementation	Addition of nutrients for preterm infants
Processing	Treatment of milk to isolate fats, proteins, etc. Fractions then added to milk
Manipulation of mother's diet	Change the fatty acid profile

[a]Adapted from Jensen and Jensen (1992).

observed a decrease in colony counts throughout a 5-day refrigeration period. Lin *et al.* (1988) obtained a mean of 15,000 CFU/ml on individual samples and of 290,000 CFU/ml on pooled samples. Most of the isolates from the individual samples were *S. epidermidis* (82%) and pooled samples (39%). Because of the diversity of microorganisms they identified, Lin *et al.* (1988) believe that all pooled human milk should be pasteurized. HMBANA (1994) requires that pooled milk must have less than 10,000 CFU/ml of normal skin flora before it can be dispensed raw.

There are two lipases in human milk, bile salt-stimulated (BSSL) and serum lipoprotein lipases; very little lipolysis should occur at 4°C because BSSL has not been activated and lipoprotein lipase activity is low. However, when in storage at −20°C for 1 month (Berkow *et al.*, 1984) or 3 months, lipolysis occurs (Friend *et al.*, 1983a). This will continue and soaps will be formed which the infant might find distasteful. Both lipases remained active after storage at −20 or −70°C and were not affected by freeze-thaw cycles (Berkow *et al.*, 1984). Protease and lysozyme were stable, but lactoperoxidase activity was lost (Friend *et al.*, 1983a). The vitamins are also unaffected (Friend *et al.*, 1983a; Lawrence, 1989). Frozen storage did not influence absorption of sodium, phosphorous, and calcium (Lawrence, 1989).

Storage of milk samples at −20°C does not maintain milk lipid composition, i.e., the triacylglycerols decrease and the free fatty acids increase (Berkow *et al.*, 1984). If milk samples are to be analyzed for lipid classes, they must be extracted immediately, pasteurized, or stored at −70°C where lipolysis does not occur. Freeze-thaw treatments cause destabilization of the fat globule emulsion and of proteins in bovine milk (Huung and Kuksis, 1967) and possibly in human milk (Garza *et al.*, 1986), but systematic studies are not available.

Lawrence (1989) reported that various antimicrobial proteins were not affected by storage at −20°C for 3 months, while the lymphocyte count was reduced by shorter periods. Storage at 4°C for 48 hr decreased macrophages and neutrophils, but not lymphocytes. These cells provide protection to the infant against infections (see Chapter 9A).

3. Pasteurization

The effects are listed in Table XII. The sole purpose of pasteurization is to destroy pathogenic microorganisms. The temperatures for holder and HTST pasteurization of bovine milk were selected because the relatively heat-resistant microorganisms which cause tuberculosis and Q fever are destroyed. Most other microorganisms are also destroyed, including HIV, and the shelf life of milk is extended. Pasteurization at 56 or 62.5°C for 30 min essentially eliminated added cell-free HIV-I and HIV-I-infected cell preparations by a minimum of five and six orders of magnitude, respectively (Orloff *et al.*, 1993). In addition, milk contains components that inactivate HIV-I, but are not lethal for cells used to replicate the virus.

The HMBANA (1994) recommends that human milk be heated at 56°C for 30 min. Holder pasteurization of bovine milk is done at 62.5°C for 30 min or with high-temperature short-time (HTST) processing at 70°C for 15 sec (Garza *et al.*, 1986). The lower temperature of 56°C is employed for human milk because there is less destruction of some of the immunologic components of milk. HTST treatment of human milk has minimal effect on these compounds, but equipment for small volumes is not available. However, all of these treatments inactivated BSSL and milk lipoprotein lipase (Pan *et al.*, 1983). The action of BSSL on human milk in the small intestine assists in the hydrolysis and absorption of milk fat. Heating at 62.5°C for 30 min significantly reduced the activities of lactoperoxidase and protease but not of lysozyme (Friend *et al.*, 1983b). Although not reported, most of the other enzymes will probably be deactivated. However, BSSL is the only human milk enzyme of major importance to the infant. When human milk was pasteurized, fat absorption in preterm infants was reduced by one-third (Williamson *et al.*, 1978). BSSL would be inactivated and lipolysis of milk fat in the small intestine reduced. Absorption of nitrogen, calcium, phosphorous, and sodium was not changed. See Chapter 5C for more information.

The effects of holder pasteurization on various antimicrobial factors in milk are summarized in Table XIII. I mentioned that HTST pasteurization has been investigated (Goldblum *et al.*, 1984), but the equipment needed for small volumes of milk is not available. Goldblum *et al.* injected their milk into a stream of sterile, distilled water pumped through the HTST pasteurizer.

Pasteurization did not reduce the quantities of biotin, niacin, and pantothenic acid (Friend *et al.*, 1983b). Nor were pyridoxine, folic acid, and ascorbic acid affected by HTST pasteurization, a treatment that as applied, was not equivalent to holder treatment (Goldblum *et al.*, 1984). Based upon experience with bovine milk, some losses of thiamin and ascorbic acid would be expected (see Chapter 8B). Riboflavin is photodegradable and milk should be protected from light. Vitamin A resists pasteurization but is photodegradable (Jensen, 1989). Vitamins D and E are heat resistant. There are apparently no reports on vitamin K.

4. *Microwave Treatment*

Frozen milk can be thawed by microwaving for 50 sec, but 30.5% of the IgA was destroyed and the number of bacteria substantially reduced

TABLE XIII
Effects of Pasteurization on Antimicrobial Factors in Human Milk[a]

Factor	56°C, 30 min[b]	62.5°C, 30 min[b]
Secretory IgA	Stable	0–30
IgM, IgG	—	IgM, 100; IgG, 33
B.bifidum factor (oligosaccharide)	—	Stable
Complement 1–9	100	—
Lactoferrin	—	67
Lactoperoxidase	—	50[c]
Lipase, BSSL, Lipoprotein[d]	100	100
Lysozyme	—	Stable
Lipids	—	Stable[e]
Lactose and oligosaccharides	—	Stable
Glycoproteins	—	Stable
Milk cells[f]	—	100

[a]Adapted from WHO (1985).
[b]Holder pasteurization in bottles.
[c]Friend *et al.* (1983b).
[d]Adapted from Pan *et al.* (1983).
[e]Lawrence (1989).

(Sigman *et al.*, 1989). Quan *et al.* (1992) found that microwaving at 72 to 98°C decreased the activities of lysozyme (96%), total IgA (98%), IgA directed against *Escherichia coli* antigen groups 01 (75%), 04 (88%), and 06 (83%). Treatment at low temperatures (20 to 53°C) did not affect total IgA and specific *E. coli* serotypes 01 and 04, but did decrease lysozyme (19%) and serotype 06 (33%). Astonishingly, the *E. coli* growth 3.5 hr after treatment was 5.2 times greater than the control at low microwave temperatures and 18 times greater at high temperatures. The increase was apparently due to the loss of the anti-infective factors. Microwaving should not be used to thaw frozen milk for feeding.

5. *Sonication*

Sonication has been employed to reduce the size of fat globules in milk from about 4.0 to 1.2 μm (Martinez *et al.*, 1987). This treatment reduced creaming and associated loss of fat during tube feeding from 16.8–47.4 to < 3.0%. Hamosh (1988) noted that some of the changes in milk compartmentation that occur as a result of homogenization could be detrimental. Among these in bovine milk are activation of lipoprotein lipase with rapid production of free fatty acids. Hamosh (1988) recommended caution in the application of sonication to human milk to prevent fat loss during continuous nasogastric feeding.

6. *Selection*

Michaelsen *et al.* (1990), using infrared spectrophotometry to determine the fat, protein, and lactose contents in milk at a large Danish milk bank, selected milks with high protein and energy contents of 1.2 g and 72.5 kcal/dl, respectively. These milks could be used for preterm infants of low birth weight. The infrared instrument, widely employed in the diary industry, is expensive (about $30,000) and must be calibrated for human milk. However, only 6 ml of milk is required and the results are available immediately. Routine analysis of the macronutrients in milk for use in neonatal units is recommended and desirable, but the purchase of an infrared analyzer cannot be justified. Polberger and Lonnerdal (1993) evaluated existing methods for analyses of protein, fat, and carbohydrate for this purpose. They modified the methods to minimize cost and time. They recommended these procedures: protein, Lowry or Bio-Rad; lipids, total lipids (phosphovanillin); and lactose, orcinol. These tests can be used in all laboratories associated with neonatal units caring for proterm infants to establish the need for and type of fortification.

Valentine *et al.* (1994) employed hindmilk to improve the weight gain of low birth weight babies. The foremilk they collected contained 2.86% fat and 62.9 kcal/dl; the hindmilk, 4.78% fat and 74.0 kcal/dl. Feeding the hindmilk (fortified) significantly increased the weight gain of the infants.

7. *Supplementation*

Supplementation of human milk has been done with commercial formulas and with addition of human cream obtained by centrifugation of human milk (Lucas *et al.*, 1980). Results from some of the many studies on supplementation or fortification were reviewed by Lucas (1993).

8. *Processing*

Lucas *et al.* (1980) and Schanler *et al.* (1985) described protocols for the processing of human milk to produce altered products suitable for the needs of preterm infants. These procedures require large volumes of milk and are beyond the capabilities of most neonatal units.

9. *Manipulation of Maternal Diet*

The contents of various polyunsaturated fatty acids can be increased in milk by inclusion of appropriate fats and oils in the maternal diet (see Chapter 6A). The contents of 10:0, 12:0, and 14:0 increase when the maternal diet is high in carbohydrate. Silber *et al.* (1988) investigated this change as a means of producing milk fat which might be more readily absorbed. They were able to increase the contents of 10:0, 12:0, and 14:0 by the use of a diet containing 5% fat, 15% protein, and 80% carbohydrate. The response was observed in women who delivered preterm and term and was highly variable. It may be controlled by total energy balance as well as by individual endocrine responses.

10. *Lyophilization*

Milk can be preserved with little loss of the compounds which have been determined. These are fatty acids, some enzymes, and some B vitamins (Friend *et al.*, 1983b). Lyophilization is not used in the United States and I did not list it in Table XII. However, a large milk bank in France lyophilizes 40,000 liters of milk per year (L. D. W. Arnold, personal communication).

P. Individuality

The effects of individuality are shown in Table XIV (Michaelsen *et al.*, 1990). As expected, the fat content had the greatest range: 1.84 to 8.9%. The table does not contain volumes which often show compensatory changes to offset the differences in fat content.

TABLE XIV
Percentile Distribution of Protein, Fat, Carbohydrate, and Energy Content in Human Milks[a]

	Percentile							
	2.5	10	25	50	75	90	97.5	No.
Protein (g/liter)	6.3	6.9	7.6	8.6	9.7	11.4	14.3	2553
Fat (g/liter)	18.4	23.8	29.4	36.1	43.4	54.6	89.0	2554
Carbohydrate (g/liter)	64.2	68.4	70.6	72.4	73.8	75.2	76.5	2554
Energy (kcal/liter)[b]	500	557	606	668	737	840	1155	2553

[a]Adapted from Michaelsen *et al.* (1990).
[b]Energy content has been calculated as combustible energy using the following caloric equivalents (in kcal/g): 5.65 protein, 9.25 fat, and 3.95 carbohydrate.

III. Bovine Milk

A. Introduction

Factors A–D, J, L, and M in Table II are applicable but not relevant because of exclusion of colostrum and mastitic milks, the pooling of milk, and the standardization of fat content. The effects of these factors have been reviewed by Jenness (1989). One of the major goals of dairy farmers has been and is to increase production by selective breeding, optimal nutrition, and control of animal diseases, primarily mastitis. They have been remarkably successful in that production per cow is approaching 30,000 lb/year. I provide information on factors E–I and N (Table II).

B. Breed

While of no importance in the United Stages, milk from the Channel Island breeds, Guernseys and Jerseys, is utilized elsewhere because of the higher fat contents of their milks (5.2 compared to 3.5% for Holsteins) (Jenness, 1988). The trend in the United States has been to utilize cows of the larger breeds, Holsteins, Brown Swiss, etc., because of the lower fat content and greater production.

C. Diet

The composition of the diet and the form in which fed affect composition particularly of milk fat (see Chapter 6B) (Sutton, 1989; and Palmquist *et al.*,

1993). The seasonal effect is due to changes in diet. One of the factors which alter the fats in the diet is biohydrogenation in the rumen. This reduces the amount of PUFA and is responsible for the presence of *trans* and positional isomers of fatty acids in milk fat. High-fat and/or low-roughage diets can reduce the fat content of milk. Diet has only small effects on the protein content (Sutton, 1988; Depeters and Cant, 1992) and virtually none on lactose. Regional and seasonal variations can usually be attributed to changes in diet (see Section G).

D. Diseases and Effects on Consumers

1. Mastitis

Mastitis is a serious problem in the dairy industry because production is decreased and the milk from cows treated for mastitis with antibiotics must be excluded from the market (see Chapters 2B and 11B). Mastitis alters the composition of milk as shown in Table XV (Harmon, 1994). The somatic cell counts are associated with increased incidence of mastitic infection and decreased milk production as shown by increases in sodium and chloride, which are diagnostic constituents (see Chapters 2B and 11B). Kitchen (1981) has written a comprehensive review on the effects of mastitis on the composition of milk. White *et al.* (1994) have refuted the

TABLE XV
Changes in Milk Constituents Associated with Elevated SCC[a]

Component	Normal milk (%)	Milk with high SCC (%)	Percentage of normal
SNF	8.9	8.8	99
Fat	3.5	3.2	91
Lactose	4.9	4.4	90
Total protein	3.61	3.56	99
Total casein	2.8	2.3	82
Whey protein	0.8	1.3	162
Serum albumin	0.02	0.07	350
Lactoferrin	0.02	0.10	500
Immunoglobulins	0.10	0.60	600
Sodium	0.057	0.105	184
Chloride	0.091	0.147	161
Potassium	0.173	0.157	91
Calcium	0.12	0.04	33

[a]Adapted from Harmon (1994). SCC, somatic cell count.

claims that treatment of cows with recombinant bovine somatotropin increases the incidence in the cows (see Section E).

2. *Effects on Consumers*

a. Introduction. The consumption of bovine milk is recommended for inclusion in healthful diets. The contributions of dairy foods to healthful diets have been reviewed by McBean (1994). My discussion is based on her review. Dairy foods are a major source of calcium, vitamin B12, phosphorous, vitamin D (fortified), riboflavin, magnesium, and vitamin A (fortified). Nevertheless, consumption has been associated as a causative or preventive factor with allergy, anemia, cancer, coronary heart disease, type 1 diabetes mellitus, hypertension, kidney stones, osteoporosis, and lactose intolerance.

b. Allergies. About 1 to 3% of infants have allergies to bovine milk proteins. These infants were fed milk earlier than recommended, less than 1 year, and some had a family history of allergies.

c. Anemia. Milk is deficient in iron and if fed in excess or too early, < 12 months, milk can contribute to iron deficiency in infants.

d. Cancer. A high intake of total fat has been related to increased risk of some cancers. However, there is no evidence that consumption of specific foods is involved. Furthermore, dairy foods contain conjugated linoleic acids which are anticarcinogenic in test animals (see Chapter 6A).

e. Coronary heart disease (CHD). The cholesterol content of milk is low (15 mg/dl). The content of hypercholesterolemic fatty acids is about 40% of the total (see Chapter 6B). Individuals with a family history of CHD are advised to control their diets accordingly. Otherwise, moderate consumption of dairy products is recommended because of the nutritional advantages.

f. Diabetes mellitus. There may be a link between early introduction of bovine milk consumption and increased risk of insulin-dependent diabetes mellitus. These preliminary findings have not been confirmed.

g. Hypertension. Adequate intakes of calcium, potassium, and magnesium are recommended as an approach to prevent and treat hypertension. Dairy foods are an important source of these nutrients.

h. Kidney stones. Traditional advice suggested that intake of calcium-rich foods contributed to the formation of kidney stones. The advice has been challenged because low calcium intakes may increase the risk of kidney stones perhaps by increasing urinary oxalate excretion.

i. Osteoporosis. This disease is caused by progressive loss of calcium from bone until they are porous. It is most often seen in elderly women and is the usual cause of hip bone fractures. Prevention requires daily consumption of adequate amounts of calcium and vitamin D throughout the life span. Dairy foods provide both and are the most readily available and absorbable sources of calcium.

j. Lactose intolerance. This is common in many adults throughout the world and is caused by a deficiency of intestinal lactase. The resulting gastrointestinal distress can be alleviated by limiting consumption of small quantities of regular milk; hard cheeses which contain almost no lactose, yogurt with active cultures, and products reduced in lactose can be consumed. Preparations of the enzyme lactase may be employed.

E. Medications

The only drug of importance to discuss here is recombinant methionyl bovine somtotropin (sometribove, BST). Other drugs, primarily the antibiotics used to treat mastitis, are almost always eliminated by disposal of the milk for required periods of time (see Chapter 11B). I previously mentioned that administration of BST to dairy cows does not increase the incidence of mastitis in dairy cows (White *et al.*, 1994). Fifteen full lactation trials (914 cows) in Europe and the United States and 70 short-term studies (2697) cows in eight countries were investigated. While production was increased from, on average, 1.9 to 6.2 kg/day/cow, the incidence of mastitis did not change.

F. Cow's Pregnancy

In order to continue production of milk, the cow must deliver a calf regularly. This is initiated by artificial insemination with semen from a bull whose progeny have a record of good production.

G. Season

Barbano (1990) reported the results of a massive investigation on seasonal and regional variations in milk composition throughout the United States. Fat, protein, nonprotein nitrogen, total solids, casein, lactose (anhydrous), and ash contents were determined in samples from 50 cheese plants in 19 states during January to December 1994. The milk received at these plants represented 10% of the U.S. supply. The purpose was to assess the impact of the variations on cheese yields. Summaries of the data can be seen in Table XVI. Regional and seasonal differences occurred. Both can be

TABLE XVI
Regional and Seasonal Changes in the Composition of Milk[a]

Component	Region		Season	
	Mean (%)	Range (%)	Mean (%)	Range (%)
Fat	3.61	3.39–3.72	3.61	3.40–3.78
Crude protein	3.27	3.22–3.29	3.27	3.13–3.38
True protein	3.11	3.06–3.17	3.11	2.97–3.20
Casein	2.56	2.51–2.61	2.56	2.48–2.65
Solids not fat	8.68	8.62–8.80	8.68	8.55–8.77
Lactose	4.54	4.51–4.61	4.54	4.47–4.62
Ash	0.72	0.71–0.72	0.72	0.69–0.75

[a]Adapted from Barbano (1990).

related to feedstuffs available, while the variation may have been caused by changes in ambient temperatures as all components were slightly lower in the summer in all regions. Heat is known to affect milk composition (Jenness, 1988; DePeters and Cant, 1992).

H. Processing

1. Introduction

The steps used in processing milk and the results are listed in Table XVII (Morr and Richter, 1988). My discussion is based on their review. I will not discuss the more complex processes used to make cheeses, ice cream, etc. In general, processing, with the exception of homogenization, is done to prevent the entry of, limit the growth of, or destroy microorganisms in milk. Shelf life is also extended. Milk is an excellent growth medium for microorganisms and they must be controlled. Colostrum and milk from cows treated for mastitis with antibiotics must be eliminated. The trend toward uniformity begins.

2. Agitation, Mixing, and Cooling

Milk is obtained from cows with vacuum milkers and conveyed into a refrigerated tank (bulk tank) equipped with an agitator. Therein, it is cooled rapidly to $< 5°C$ and held for 12 to 72 hr before transportation to the dairy plant in a large tanker. Pooling of milk from individual cows, from successive milkings, and several farms occurs. Agitation caused by air leaks produces foaming in the pipes from milker to tank can activate lipoprotein lipase. Excessive lipolytic activity while milk is stored in the bulk

TABLE XVII
The Processing of Bovine Milk[a]

Location and Process	Effects
Farm	
Agitation, mixing, cooling	Pooling, prevention of bacterial growth, crystallization of fat, clustering of globules, changes in casein micelles, possible activation of lipase
Farm to dairy plant	Agitation, pooling, cooling
Dairy plant	
Clarification	Centrifuged removal of somatic and bacterial cells, other sediment
Separation and standardization	Preparation of skim milk and cream, mixing of these for lower fat milk
Pasteurization (72°C for 15 sec)	Destruction of microorganisms, denaturation of enzymes and some proteins, partial loss of some vitamins, heated flavor
Vacuum removal of off flavors	Improvement of milk flavor
Homogenization	Reduction in size of fat globules, increase in globule numbers and surface area, absorption of casein onto globule surface, delayed creaming
Packaging and distribution	Conveyance of product to consumer
Sanitation	Removal of milk deposits and bacteria and destruction thereof

[a]Adapted from Morr and Richter (1988).

tank will cause hydrolytic rancidity: the accumulation of short-chain fatty acids, particularly butyric (4:0), to the point where the flavor and odor of the milk causes it to be unacceptable.

Changes that occur in and on the fat globule are crystallization of milk fat, absorption of whey proteins onto fat globules, increased clustering and creaming of globules, and partial removal of the fat-soluble membrane from the globules.

Changes occurring in casein micelles are release of proteolytic enzymes, some proteolysis, partial disaggregation of casein micelles, release of β-casein from micelles, and increased solvation of micelles with release of phosphorous and calcium.

3. Clarification

This process is done to centrifugally remove somatic cells, bacteria, and sediment. Some casein micelles are also removed. The clarifier slime contains lipase and has been used as a source of the enzyme.

4. *Separation and Standardization*

Centrifugal force is employed to separate milk into cream and skim milk. Low-fat milks are prepared by mixing these products as required. This is now done by direct continuous standardization. The fat content is monitored so as to maintain it as close to the legal minimum as possible. In the United States, this is 3.25% fat for whole milk.

5. *Pasteurization*

The primary purpose of pasteurization is the destruction of all pathogenic microorganisms and most others. The process is so effective that milk is almost never a cause of foodborne illness. The few that occur are usually the result of postpasteurization contamination. Shelf life is increased to 10 to 14 days. The most widely used method for pasteurization is HTST at a minimum in the United States of 72°C for 15 sec. The process is continuous and slightly higher temperatures are applied. Ultra-high temperature treatments at 138 to 150°C for 2 to 6 sec which produce an essentially sterile product are employed for coffee creamers, etc.

Most milk enzymes are destroyed by pasteurization. The loss of alkaline phosphatase is used to monitor pasteurization and determine if the heat treatment is adequate. The cooked flavor is caused by formation of sulfhydryl groups from disulfide bridges in α-lactalbumin. These groups also act as antioxidants. Some loss of ascorbic acid and thiamine occurs as a result of heating (see Chapter 8B). Ascorbic acid, riboflavin, and vitamin A are photodegradable. Milk should be packaged in opaque containers to protect these vitamins.

6. *Vacuum Removal of Off Flavors*

Many volatile, water-soluble flavor compounds are removed by this process. The compounds originate from weeds and feeds consumed by the cow.

7. *Homogenization*

The purpose is to delay creaming by reducing the size of the globules. Pasteurized milk at >60°C is forced through a small orifice by a high-pressure pump. The average diameter (volume/surface area) is decreased from 2.5–4.6 to 0.2–0.7 μm, the surface about 10-fold, and the numbers from 1.5×10^{10} to 1×10^{12}–10^{14} (see Chapter 2B). Casein is absorbed onto the globule surface from which the original membrane has been displaced. There is an increased susceptibility to light-induced flavor.

8. Packaging and Distribution

If the container is not opaque some riboflavin will be destroyed and its decomposition promotes the oxidation of ascorbic acid (see Chapter 8B). Some vitamin A will also be destroyed (Chapter 8E).

9. Sanitation

The purposes are to destroy microorganisms on surfaces of equipment which will contact milk and to remove deposits of milk proteins, etc. These are usually accomplished by pumping appropriate solutions of cleaners, sequestrants, detergents, and disinfectants through the equipment followed by a rinse. Some equipment, e.g., milking machines, may be cleaned by hand.

I. Summary

Although many factors influence the volume and composition of milk from individual cows and herds, these are almost eliminated by the processes described above. Exclusion of abnormal milks, pooling, and standardization result in milks of high uniform composition throughout the year and regions in the United States.

Acknowledgments

The preparation of the manuscript was supported in part by a NIH contract and federal funds made available through provision of the Hatch Act, Scientific Contribution 1602, Storrs Agricultural Experiment Station, University of Connecticut, Storrs, CT.

References

Arthur, P. G., Kent, J. C., and Hartmann, P. E. (1994). Metabolites of lactose synthesis in milk from diabetic and nondiabetic women during lactogenesis II. *J. Pediatr. Gastroenterol. Nutr.* **19,** 100–108.

Barbano, D. M. (1990). Seasonal and regional variation in milk composition in the U.S. *In* "1990 Cornell Nutrition Conference," pp. 96–105. Department of Animal Sciences, Cornell, University, Ithaca, NY.

Berkow, S. E., Freed, L. M., Hamosh, M., Bitman, J., Wood, D. L., Happ, B., and Hamosh, P. (1984). Lipases and lipids in human milk: Effect of freeze-thawing and storage. *Pediatr. Res.* **18,** 1257–1262.

Bitman, J., Hamosh, M., Hamosh, P., Lutes, V., Neville, M. C., Seacat, J., and Wood, D. L. (1990). Milk composition and volume during onset of lactation in a diabetic mother. *Am. J. Clin. Nutr.* **50,** 1364–1369.

Daly, S. E. J., Kent, J. C., Huynh, D. Q., Owens, R. A., Alexander, B. F., and Hartmann, P. E. (1992). The determination of short term breast volume changes and the rate of synthesis of human milk using computerized breast measurement. *Exp. Physiol.* **77,** 79–87.

Daly, S. E. J., Owens, R. A., and Hartmann, P. E. (1993a). The short-term synthesis and infant-regulated removal of milk in lactating women. *Exp. Physiol.* **78,** 209–220.

Daly, S. E. J., DiRosso, A., Owens, R. A., and Hartmann, P. E. (1993b). Degree of breast emptying explains changes in the fat content, but not fatty acid composition of human milk. *Exp. Physiol.* **78,** 741–745.

DePeters, E. J., and Cant, J. P. (1992). Nutritional factors influencing the nitrogen composition of bovine milk: A review. *J. Dairy Sci.* **75,** 2043–2070.

Dewey, K. G., and Lonnerdal, B. (1983). Milk and nutrient intake of breastfed infants from 1 to 6 months: Relation to growth and fatness. *J. Pediatr. Gastroenterol. Nutr.* **2,** 497–506.

Dewey, K. G., and McCrory, M. E. (1994). Effects of dieting and physical on pregnancy and lactation. *Am. J. Clin. Nutr.* **59**(Suppl), 446S–453S.

Ferris, A. M., and Reece, E. A. (1994). Nutritional consequences of chronic maternal conditions during pregnancy and lactation: Lupus and diabetes. *Am. J. Clin. Nutr.* **59**(Suppl), 465S–473S.

Ferris, A. M., Neubauer, S. H., Bendel, R. B., Green, K. W., Ingardia, C. J., and Reece, A. E. (1993). Perinatal lactation protocol and outcome in mothers with and without insulin-dependent diabetes mellitus. *Am. J. Clin. Nutr.* **58,** 43–48.

Friend, B. A., Shahani, K., Long, C. A., and Vaughn, C. A. (1983a). The effect of processing and storage on key enzymes, B vitamins, and lipids of mature human milk. Evaluation of fresh samples and effects of freezing and frozen storage. *Pediatr. Res.* **17,** 61–64.

Friend, B. A., Shahani, K. M., Long, C. A., and Hgel, E. N. (1983b). Evaluation of freeze-drying, pasteurization, high-temperature heating and storage on selected enzymes, B vitamins and lipids of human milk. *J. Food Prot.* **46,** 330–334.

Garza, C., Hopkinson, J., and Schanler, R. J. (1986). Human milk banking. *In* "Human Milk in Infant Nutrition and Health" (R. R. Howell, F. M. Morriss, Jr., and L. K. Pickering, eds.), Ch. 13. Thomas, Springfield, IL.

Gindler, J., Mwanko, M. U., Omene, J. A., and Roberts, I. M. (1985). The quality of milk in a 65 year-old Nigerian woman. *Nutr. Res.* **5,** 1209–1213.

Goldblum, R. M., Dill, C. W., Albrecht, T. B., Alford, E. S., Garza, C., and Goldman, A. S. (1984). Rapid high-temperature treatment of human milk. *J. Pediatr.* **104,** 380–385.

Hachey, D. L. (1994). Benefits and risks of modifying maternal fat intake in pregnancy and lactation. *Am. J. Clin. Nutr.* **59** (Suppl), 454S–464S.

Hamosh, M. (1988). Ultrasonic homogenization of expressed human milk to prevent fat loss during tube feeding. *J. Pediatr. Gastroenterol. Nutr.* **7,** 307–308.

Hamosh, M., and Bitman, J. (1992). Human milk in disease: Lipid composition. *Lipids* **27,** 848–857.

Harmon, R. J. (1994). Physiology of mastitis and factors affecting somatic cell counts. *J. Dairy Sci.* **77,** 2103–2112.

Hartmann, P. E., and Prosser, C. G. (1982). Acute changes in the composition of milk during the menstrual cycle in lactating women. *J. Physiol.* **324,** 21–30.

Hartmann, P. E., and Prosser, C. G. (1984). Physiological basis of longitudinal changes in human milk yield and composition. *Fed. Proc.* **43,** 2448–2453.

Human Milk Banking Association of North America (HMBANA) (1994). "Guidelines for the Establishment and Operation of a Donor Milk Bank" (L. D. W. Arnold and M. R. Tully, eds.). HMBANA, West Hartford, CT.

Institute of Medicine (IOM) (1991). "Nutrition During Lactation." Subcommittee on Nutrition During Lactation, Committee on Nutritional Status During Pregnancy and Lactation. Food and Nutrition Board, Institute of Medicine, National Academy of Sciences, 309P, National Academic Press, Washington, DC.

Jackson, M. B., Lammi-Keefe, C. J., Jensen, R. G., Couch, S. C., and Ferris, A. M. (1994). Total lipid and fatty acid composition of milk from women with and without insulin-dependent diabetes. *Am. J. Clin. Nutr.* **60,** 353–361.

Jenness, R. (1988). Composition of milk. *In* "Fundamentals of Dairy Chemistry" (N. P. Wong, ed.), 3rd Ed., pp. 1–38. Van Nostrand–Reinhold, New York.

Jensen, R. G., Lammi-Keefe, C. J., Ferris, A. M., Jackson, M. J., Couch, S. C., Capacchione, C. M., Ahn, H. S., and Murtaugh, M. (1995). Human milk total lipid and cholesterol are dependent on interval sampling during 24 hours. *J. Pediatr. Gastroenterol. Nutr.* **20,** 91–94.

Jensen, R. G. (1989). "The Lipids of Human Milk," pp. 167–180. CRC Press, Boca Raton, FL.

Jensen, R. G., and Jensen, G. L. (1992). Specialty lipids for infant nutrition. I. Milks and formulas. *J. Pediatr. Gastroenterol. Nutr.* **15,** 232–245.

Kitchen, B. J. (1981). Review of the progress of dairy science: Bovine mastitis: Milk compositional changes and related diagnostic tests. *J. Dairy Res.* **48,** 167–188.

Lawrence, R. A. (1989). "Breastfeeding," 3rd Ed. C. V. Mosby, St. Louis, MO.

Lin, F. J., Barnhart, H. M., Bailey, J. S., Cox, N. A., and Eitenmiller, R. R. (1988). Bacteriological profiles of human milk from individual donors and pooled samples from a commercial milk bank. *J. Food Prot.* **51,** 467–470.

Lipsman, S., Dewey, K. G., and Lonnerdal, B. (1985). Breast-feeding among teenage mothers: Milk composition, infant growth, and maternal dietary intake. *J. Pediatr. Gastroenterol. Nutr.* **4,** 426–434.

Lonnerdal, B. (1986). Effects of maternal dietary intake on human milk composition. *J. Nutr.* **116,** 499–513.

Lucas, A. (1993). Enteral nutrition. *In* "Nutritional Needs of the Preterm Infant" (R. C. Tsang, A. Lucas, R. Uauy, and S. Zlotkin, eds.), pp. 212–213. Williams and Watkins, Baltimore.

Lucas, A., Lucas, P. J., Chavin, S. I., Lyster, R. I. J., and Baum, J. D. (1980). A human milk formula. *Early Hum. Dev.* **4,** 15–21.

Martinez, F. E., Desai, I. D., Davidson, A. G. F., Nakai, S., and Radcliffe, A. (1987). Ultrasonic homogenization of expressed human milk to prevent fat loss during tube feeding. *J. Pediatr. Gastroenterol. Nutr.* **6,** 593–597.

May, J. T. (1988). Microbial contaminants and antimicrobial properties of human milk. *Microbiol. Sci.* **5,** 42–46.

McBean, L. D. (1994). Contributions of dairy foods to healthful diets. *Dairy Council Digest* **65**(5), 25–30.

Michaelsen, K. F., Skafte, L., Badsberg, J. H., and Jorgensen, M. (1990). Variation in macronutrients in human bank milk: Influencing factors and implications for human milk banking. *J. Pediatr. Gastroenterol. Nutr.* **11,** 229–239.

Michaelsen, K. F., Larsen, P. S., Thomson, B. L., and Samuelson, G. (1994). The Copenhagen cohort study on infant nutrition and growth: Breast-milk intake, human milk macronutrient content, and influencing factors. *Am. J. Clin. Nutr.* **59,** 600–611.

Morr, C. V., and Richter, R. L. (1988). Chemistry of processing. *In* "Fundamentals of Dairy Chemistry" (N. P. Weng, ed.), 3rd Ed., pp. 739–766. Van Nostrand–Reinhold, New York.

Neubauer, S. H., Ferris, A. M., Chase, C. G., Faneklli, J., Thompson, C. A., Lammi-Keefe, Clark, R. M., Jensen, R. G., Bendel, R. B., and Green, K. W. (1993). Delayed lactogenesis in women with insulin-dependent diabetes mellitus. *Am. J. Clin. Nutr.* **58,** 54–60.

Neubauer, S. H., Ferris, A. M., Lammi-Keefe, C. J., Green, K. W., Hinckley, L. S., and Jensen, R. G. (1995). Composition and bacterial and somatic cell contents in the milk of lactating women with and without insulin-dependent diabetes. Manuscript in preparation.

Neville, M. C., Keller, R. P., Seacat, J., Casey, C. E., Allen, J. C., and Archer, P. (1984). Studies on human lactation. 1. Within-feed and between-breast variation in selected components of human milk. *Am. J. Clin. Nutr.* **40,** 635–640.

Neville, M. C., Allen, J. C., Archer, P. C., Casey, C. E., Seacat, J., Keller, R. P., Lutes, V., Rasbach, J., and Neifert, M. (1991). Studies in human lactation: Milk volume and nutrient composition during weaning and lactogenesis. *Am. J. Clin. Nutr.* **54,** 81–92.

Orloff, S. L., Wallingford, J. C., and McDougal, J. S. (1993). Inactivation of human immunodeficiency virus type 1 in human milk: Effects of intrinsic factors in human milk and of pasteurization. *J. Hum. Lact.* **9,** 13–17.

Palmquist, D. L., Beaulieu, A. D., and Barbano, D. M. (1993). Feed and animal factors influencing milk fat composition. *J. Dairy Sci.* **76,** 1753–1771.

Paul, A. A., and Muller, E. M. (1980). *In* "Maternal Nutrition during Pregnancy and Lactation" (H. Aebi and R. G. Whitehead, eds.), pp. 105–116. Hans Hubor, Bern.

Pietschnig, B., Haschke, F., Vanura, H., Sheaker, M., Veitl, V., Kellner, S., and Schuster, E. (1993). Vitamin K in breast milk: No influence of maternal dietary intake. *Eur. J. Chin. Nutr.* **47,** 209–215.

Polberger, S., and Lonnerdal, B. (1993). Simple and rapid macronutrient analysis of human milk for individualized fortification. Basis for improved nutritional management of very-low birth-weight infants? *J. Pediatr. Gastroenterol. Nutr.* **17,** 283–290.

Prentice, A. M. (1980). Variations in maternal dietary intake, birth-weight, and breast-milk output in the Gambia. *In* "Maternal Nutrition during Pregnancy and Lactation" (H. Aeb and R. G. Whitehead, eds.), pp. 167–183. Hans Huber, Bern.

Prentice, A. (1986). The effect of maternal parity on lactational performance in a rural African community. *In* "Human Lactation 2. Maternal and Environmental Factors" (M. Hamosh and A. S. Goldman, eds.), pp. 165–186. Plenum Press, New York.

Prentice, A., Prentice, A. M., and Whitehead, R. G. (1981). Breast-milk fat concentrations of rural African women. *Br. J. Nutr.* **45,** 483–494.

Prentice, A., Prentice, A. M., and Lamb, W. H. (1985). Mastitis in rural Gambian mothers and the protection of the breast by milk antimicrobial factors. *Trans. R. Soc. Trop. Med. Hyg.* **70,** 90–95.

Prentice, A., Paull, A., Prentice, A., Black, A., Cole, T., and Whitehead, R. (1986). Cross-cultural differences in lactational performance. *In* "Human Lactation 2: Maternal and Environmental Factors" (M. Hamosh and A. S. Goldman, eds.), pp. 13–44. Plenum Press, New York.

Prentice, A. M., Landing, J. A. M., Drury, P. J., Dewit, O., and Crawford, M. A. (1989). Breast-milk fatty acids of rural Gambian mothers: Effects of diet and maternal parity. *J. Pediatr. Gastroenterol. Nutr.* **8,** 486–490.

Quan, R., Yang, C., Rubinstein, S., Lewiston, N. J., Sunshine, P., Stevenson, D. K., and Kerner, J. A., Jr. (1992). Effects of microwave radiation on antiinfective factors in human milk. *Pediatr.* **89,** 667–669.

Rasmussen, K. M. (1992). The influence of maternal nutrition on lactation. *Annu. Rev. Nutr.* **12,** 103–117.

Saint, L., Smith, M., and Hartmann, P. E. (1984). The yield and nutrient content of colostrum and milk of women giving birth to one month postpartum. *Br. J. Nutr.* **32,** 87–95.

Schanler, R. J., Garza, C., and Nichols, B. L. (1985). Fortified mother's milk for very low birth weight infants: Results of growth and balance studies. *J. Pediatr.* **107,** 437–445.

Sigman, M., Burke, K. J., Swarner, O. W., and Sharlik, G. W. (1989). Effects of microwaving human milk: Changes in IgA content and bacterial counts. *J. Am. Diet. Assoc.* **89,** 690–692.

Silber, G. H., Hachey, D. L., Schanler, R. J., and Garza, C. (1988). Manipulation of maternal diet to alter fatty acid composition of human milk intended for premature infants. *Am. J. Clin. Nutr.* **47,** 810–814.

Sosa, B., and Barness, L. (1987). Bacterial growth in refrigerated human milk. *Am. J. Dis. Child.* **41,** 111–112.

Sosa, R., Klaus, M., and Urrutia, J. J. (1976). Feed the nursing mother, thereby the infant. *J. Pediatr.* **88,** 668–670.

Sutton, J. D. (1989). Altering milk composition by feeding. *J. Dairy Sci.* **72,** 2801–2814.

Valentine, C. J., Hurst, N. M., and Schanler, R. M. (1994). Hindmilk improves weight gain in low-birth-weight infants fed human milk. *J. Pediatr. Gastroenterol. Nutr.* **18,** 474–477.

Van Beusekom, C. M., Zeegers, T. A., Martini, I. A., Herman, V. J. R., Visser, G. H. A., Van Doormal, J. J., and Muskiet, F. A. J. (1993). Milk of patients with tightly controlled insulin-dependent diabetes mellitus has normal macronutrient and fatty acid composition. *Am. J. Clin. Nutr.* **57,** 938–943.

White, T. C., Madsen, K. S., Hintz, R. L., Sorbert, R. H., Jr., Collier, R. J., Hard, D. L., Hartwill, G. F., *et al.* (1994). Clinical mastitis in cows treated with sometribove (recombinant bovine somatotropin) and its relationship to milk yield. *J. Dairy Sci.* **77,** 2249–2260.

Wiliamson, S., Finucane, E., Ellis, H., and Gamsu, H. R. (1978). Effect of heat treatment of human milk on absorption of nitrogen, fat, sodium, calcium, and phosphorous by preterm infants. *Arch. Dis. Child.* **53,** 555–563.

World Health Organization (WHO) (1985). "The Quantity and Quality of Breast Milk," Chap. 2. WHO, Geneva.

World Health Organization (WHO) (1989). "Infant Feeding: The Physiological Basis" (J. Akre, ed.), Suppl. to Bulletin 67, pp. 33–34. WHO, Geneva.

CHAPTER 4

Carbohydrates in Milks: Analysis, Quantities, and Significance

DAVID S. NEWBURG
SUZANNE H. NEUBAUER

I. Introduction

Lactose is readily purified by fractional crystallization, and was the first constituent of milk to be studied. Other carbohydrates of milk have been neglected until recently, because these glycoconjugates (oligosaccharides, glycolipids, glycoproteins, glycosaminoglycans, mucins, etc.) are complex, difficult to isolate, and have not yet been definitively measured. Along with a lack of good methodology that is specific for the glycoconjugates of milk, a great deal of heterogeneity exists among mothers, and, until recently, these carbohydrate fractions of milk were thought to be biologically irrelevant. However, their biological functions are now an active area of research.

The principal carbohydrate in most milks is lactose. Its concentration represents a balance between the high nutrient requirements of the infant and the constraints of carbohydrate concentration in milk due to osmolarity. Most milks contain small amounts of glucose and galactose, the biosynthetic precursors of lactose. The oligosaccharide content varies greatly among species, and within human populations oligosaccharides also manifest great heterogeneity both qualitatively and quantitatively.

In this chapter we compile, summarize, and analyze the most recent relevant information available on the carbohydrate content of milk.

II. Analytical Measurement of Carbohydrates in Milk

A. Lactose

1. Nonspecific Methods

Of the methods used for the analysis of lactose in milk, one of the least accurate and precise remains widely used, especially for the study of nonhuman milk, i.e., the dry weight minus the fat, protein, and ash weight of a milk sample. Based on the assumptions that the remaining solids are all carbohydrate and that the carbohydrate is all lactose, many report this weight as the lactose content of the milk sample. The earliest methods for the direct quantitative analysis of lactose in milk are based on reactions with the reducing end of the sugar molecule. These reactions are specific for carbohydrates and related compounds (Shaffer and Hartmann, 1920), but because they do not discriminate among sugars, analysis of a specific sugar requires its prior separation from other reducing constituents of milk. Aside from being inconvenient and tedious, these procedures allow the lactose values to be altered by residual nonlactose milk components, as well as by residual reagents used in the isolation of the sugars. Of these methods, the phenol–sulfuric acid reaction (Dubois *et al.,* 1956) is least sensitive to extraneous reagents and has been used extensively for the analysis of carbohydrate in milk and dairy products (Lawrence, 1968), usually with the explicit assumption that lactose is the only major carbohydrate present. The automated procedure based on the ferricyanide reaction on milk dialysate (Conetta *et al.,* 1970) suffers from the same defect. This problem of inaccurately high lactose values, especially for samples containing appreciable levels of nonlactose carbohydrate, was partially addressed by optimizing the reaction of methylamine–sodium sulfite so that it formed a chromophore with lactose, but not with glucose or galactose. In a complementary approach, ammonium molybdate was reacted with the sample under conditions which allowed formation of chromophores with glucose and galactose, but gave no color with lactose (Nickerson *et al.,* 1976). This procedure, however, did not address the contribution of oligosaccharides to the lactose values.

2. Specific Methods

Another class of analytical technique is based on the enzymatic hydrolysis of lactose. The level of free glucose before and after hydrolysis by β-galactosidase is measured, and the difference represents the quantity of lactose in the original sample (Kuhn and Lowenstein, 1967). The glucose is then quantitated enzymatically by the glucose oxidase method of Bergmeyer and Bernt (1963). An automated technique compares the quantity of sugars able to react with *p*-hydroxybenzoic acid hydrazide before and after hydrolysis by specific glycosidases. The difference in optical absorbance reflects the amount of hydrolyzable substrate in the original food sample (Hudson *et al.*, 1976). In an assay using a similar approach, the free glucose is converted to glucose-6-phosphate, and the oxidation of the glucose-6-phosphate to gluconate-6-phosphate is coupled to the reduction of nicotinamide adenine dinucleotide phosphate ($NADP^+$). The change in absorbance of the nucleotide is proportional to the concentration of lactose in the sample (Bahl, 1972). In a related scheme, β-galactosidase and glucose oxidase are immobilized on the surface of a sensor covered with a dialysis membrane; the diffusion of lactose from the sample into the sensor results in hydrogen peroxide, which is detected by a platinum electrode (Pilloton *et al.*, 1987). The concentration of lactose in the sample is related to the current measured at the electrode. This procedure is convenient for large numbers of analyses. These methods assume that lactose is the only constituent of milk that will release glucose when incubated with the β-glucosidase; the amount of lactose could be overestimated to the extent that glucose is released from oligosaccharides by the β-galactosidase. Although the relative contribution from oligosaccharides has not been directly tested to date, the presence of a lactose moiety at the reducing end of most known milk oligosaccharides makes possible the release of measurable glucose by the galactosidase. This source of error is minimized by the use of galactose dehydrogenase along with the β-galactosidase, in which the reduction of nicotinamide adenine dinucleotide (NAD^+) is coupled to the oxidation of galactose, both in the test tube (Berner, 1970; Wallenfels and Kurz, 1962) and in a sensor surrounded by immobilized enzyme (Yellow Springs Instrument Co., Scientific Division, Yellow Springs, OH 45387).

Another method for the quantitation of lactose in milk employs infrared (IR) analysis and is useful for quantitating macromolecules, but poor for quantitating lactose, as the oligosaccharides in milk are included in the results of IR analysis of lactose (Michaelsen *et al.*, 1988). A method for the analysis of glucose and other sugars in urine, plasma, and erythrocytes by the direct trimethylsilylation of the dry sample followed by a simple solvent partition cleanup step and by gas chromatography (GC) (Jansen *et al.*, 1986) could be quite useful in the analysis of sugars in milk. However, when van Beusekom *et al.* (1993) adapted this method for the analysis of lactose in milk, provision was not made for the poor solubility of lactose in the step that uses 80% methanol for deproteination. Most of the 80–90 mg of lactose present in a 1.5-ml sample of milk precipitates upon the addition of the 6 ml of methanol proscribed for deproteination and is lost to analysis. This brings into question the validity of the relatively low values of lactose reported by this group (van Beusekom *et al.*, 1993) and the lack of differences in milk lactose between the groups of women studied. Other methods of deproteination, perhaps including an ultrafiltration step, should be employed, and both the amount of interference by other milk components and the lactose recovery should be determined when this potentially useful method of lactose analysis is performed. Another potentially useful but underutilized approach toward the quantitation of human milk lactose is high-performance liquid chromatography (HPLC). Used in the analysis of lactose in human milk (Butte and Calloway, 1981), this approach resulted in slightly lower values than those produced by other procedures. Both HPLC and GC have the potential to be developed into relatively rapid and straightforward methods that are less confounded by the other types of carbohydrate found in human milk.

The lactose values available in the literature and summarized in this chapter were obtained by several methods; thus, these mean values should be considered definitive unless contradicted in the future by values determined by more advanced chromatographic techniques.

B. Monosaccharides

1. Nonspecific Methods

The early methods for sugar analysis all involved reaction with the reducing end of the sugar, and their specificity depended on separation of the sugars before analysis. One popular strategy was the reduction of cupric salt to cuprous oxide followed by measurement of the cuprous oxide by titration with iodine (Shaffer and Hartmann, 1920) or by quantitative conversion of cupric ion into a chromophore (Folin and Wu, 1920). The cuprous ion was susceptible to reoxidation, introducing error if a delay occurred between certain steps in the procedure. This problem was eliminated, however, by the use of barium salts for the precipitation step and the introduction of a different copper reagent (Somogyi, 1945). Another commonly employed strategy was the formation of chromophores through reaction with the reducing end of the sugar, as in the phenol–sulfuric acid reaction (Dubois *et al.*, 1956), the orcinol reaction (Svennerholm, 1956), or conjugation with *o*-toluidine (Hultman, 1959). These methods did not distinguish among the reducing sugars; thus, specificity was possible only with the quantitative and absolute separation of glucose, galactose, lactose, or oligosaccharides prior to development of the chromophore. In reality, such separations were seldom employed; in many cases, analyses using these techniques are performed on milk or other fluids without prior resolution of the sugars, and the results are expressed with the assumption that all chromophore development is due to the major sugar in the fluid (in the case of milk, lactose).

2. Specific Methods

The more recent methods permit greater specificity. For example, the oxidation of glucose by glucose oxidase produces hydrogen peroxide which, in the presence of peroxidase, oxidizes a dye. The amount of dye oxidized is proportional to the concentration of glucose in the original sample (Bergmeyer and Bernt, 1989). In a similar approach, glucose oxidase is immobilized onto an electrode sensitive to hydrogen peroxide; the amount of current produced in the electrode is proportional to the concentration of the glucose in the test solution (Yellow Springs Instrument Co.). Glucose can also be determined by the enzymatic conversion of glucose to glucose-6-phosphate followed by the oxidation of glucose-6-phosphate in the presence of $NADP^+$ by glucose-6-phosphate dehydrogenase to produce stoichiometric amounts of reduced nicotinamide adenine dinucleotide phosphate (NADPH) (Bahl, 1971). Thus, the amount of NADPH produced is proportional to the original concentration of glucose in the solution. Similar strategies are employed for the analysis of galactose through the use of analogous, galactose-specific enzymes (Kurz and Wallenfels, 1974).

The simultaneous quantitation of glucose, galactose, and lactose could be accomplished through modern chromatographic procedures. The amount of glucose in biological fluids has been determined by GC (Jansen *et al.*, 1986), and HPLC columns suitable for the resolution of other carbohydrates are now available.

C. Oligosaccharides

The oligosaccharide fraction of milk is usually obtained by first preparing the carbohydrate fraction and then by separating the oligosaccharides from the lactose. When the cream and the protein fractions are removed from a milk sample, the resulting fluid contains primarily lactose and oligosaccharides. Cream is generally removed by centrifugation in the cold (e.g., 4000*g*, 1 hr, 4°C). Protein is removed by ethanol precipitation (68% EtOH, 10°C, 18 hr) (Kobata *et al.*, 1978), acetone precipitation (50% acetone, 4°C, 16 hr) (Egge *et al.*, 1982), or ultrafiltration (Newburg, unpublished data). The oligosaccharides are separated from the lactose and other milk constituents by molecular sizing (Kobata *et al.*, 1978), fractional crystallization (Egge *et al.*, 1982), or passage over a charcoal column (Newburg *et al.*, 1990). Further isolation of individual oligosaccharides is accomplished through fractional crystallization, or, more commonly, through a combination of chromatographic techniques, including preparative thin-layer chromatography, molecular sizing, affinity chromatography, and HPLC with a variety of column and solvent systems.

D. Metanalysis

When compiling data on carbohydrate levels in milk across various studies in which sample size, analytical techniques, and experimental design are radically different, the statistic of choice is metanalysis. We employed two types of metanalysis to obtain mean values across studies. In both of these analyses we assume that each study is measuring the same quantity and that there is a common variance across studies.

1. Method 1

In the first method, the studies with the largest numbers of subjects are given the most weight independent of the variance and the mean of the study:

$$\overline{X}_{w1} = \frac{\sum_{i=1}^{k} n_i \overline{X}_i}{\sum_{i=1}^{k} n_i}, \tag{1}$$

where k is the number of studies, n_i is the sample size in study i, $\overline{X}_i$ is the mean in study i, and $\overline{X}_{w1}$ is the weighted mean (Weighted mean 1; Tables I–V) for the sample size of each study.

2. Method 2

In the second method of metanalysis, the studies with the largest numbers of subjects and the least variance are given the most weight independent of the mean of the study:

$$\overline{X}_{w2} = \frac{\sum_{i=1}^{k} n_i (\text{var}_i)^{-1} \overline{X}_i}{\sum_{i=1}^{k} n_i (\text{var}_i)^{-1}}, \tag{2}$$

where k is the number of studies, n_i is the sample size in study i, var_i is the variance in study i, $\overline{X}_i$ is the mean in study i, and $\overline{X}_{w2}$ is the weighted mean (weighted mean 2; Tables I–V).

The standard deviation across studies was pooled so that the contribution of each study was in proportion to its sample size:

$$SD^2\ pool = \frac{\sum_{i=1}^{k} (n_i - 1)\ \text{SD}_i^2}{\sum_{i=1}^{k} (n_i - 1)} \tag{3}$$

$$\text{SD pool} = \sqrt{\text{SD}^2 \text{ pool}}, \tag{4}$$

where k is the number of studies, n_i is the sample size in study i, SD_i is the standard deviation in study i, and SD^2 pool is the square of the standard deviation across the studies. The square root of SD^2 pool is the standard deviation (SD pool) across the studies.

3. Comparison of Methods

Our first method is the most simple form of metanalysis in which we pool the mean values of the studies, weighing only for the sample size. This approach allows the reviewer to make the fewest judgments, as each subject makes an equal contribution to the grand mean, regardless of which study she appeared in. This model assumes equal analytical precision across studies, equally valid sampling technique, etc.; the study with the largest sample size most influences the mean, regardless of its strengths or weaknesses.

The second method of metanalysis addresses the quality of the studies by using the variability within each study as a gauge of the precision of its methods and sampling techniques. The weakness of this approach is that studies of homogeneous populations may be given too much weight, thus allowing small, nonrepresentative subpopulations to have disproportionate weight toward the overall mean value. Also, unusually low error terms in a study could reflect methodological or calculation errors, rather than superior technique.

4. Alternative Approach to Metanalysis

A more complex model would give additional weight to studies whose deviation from the mean is closest to the average deviation shown across all studies, to those whose means are closest to the means across studies, and to those which are technically strong in the opinion of the reviewer. Although this approach would minimize the influence of technically flawed studies, it would allow the introduction of technical evaluations that could lead to inadvertent expression of personal bias by the author. This approach might also minimize the influence of newer studies that may be technically superior, but deviate from the older mean values. We chose not to utilize this approach, but rather to use agreement between the first two methods as a measure of our confidence in the weighted mean value.

5. Comparison of Results

The extent to which the results of the two methods of metanalysis agree indicates the extent to which the mean value obtained is independent of the assumptions employed in the metanalysis. Conversely, deviation of the two results indicates a defect in one or more of the assumptions or restrictions employed and lowers confidence in the weighted mean.

III. Human Milk Lactose

Levels of lactose are considered to be the most consistent of the macronutrients in human milk; 75% of the total variance is contributed by the heterogeneity of individuals (Butte *et al.*, 1988). The previously accepted value of 194 mmol/liter may slightly overestimate the actual lactose, as not all analytical methods distinguish oligosaccharides from lactose. Tables I and II illustrate this overestimation of lactose when it is analyzed by nonspecific methods (202 ± 9 mmol/liter) vs lactose-specific methods (185 ± 14 mmol/liter). Human drip milk has a lactose concentration similar to that of expressed milk (Gibbs *et al.*, 1977; Lucas *et al.*, 1978).

During pregnancy a paracellular pathway exists allowing plasma components to leak around the extracellular spaces of the mammary alveolar cells. Antepartum lactose, diluted by these plasma components (Allen *et al.*, 1991), ranges from 60 to 80 mmol/liter. Lactose marks the onset of lactogenesis (Saint *et al.*, 1984). Its levels increase significantly until approximately 4 or 5 days postpartum (Casey *et al.*, 1986; Kulski and Hartmann, 1983; Neville *et al.*, 1991; Viverge *et al.*, 1986). Lactose levels correlate positively with milk yield and negatively with whey protein content as lactation is established (Kulski and Hartmann, 1981) and the paracellular pathway is closed. The slight decrease in lactose within a feed (Hytten, 1954; Macy *et al.*, 1931; Watson *et al.*, 1982) is attributed to the increase in fat content that occurs (Neville *et al.*, 1984). Values are similar between breasts (Arthur *et al.*, 1991; Neville *et al.*, 1984) and at various times of the day (Hall, 1979; Hytten, 1954; Lammi-Keefe *et al.*, 1989). Lactose decreases with involution of the mammary gland upon weaning (Dewey *et al.*, 1984; Hartmann and Kulski, 1978; Neville *et al.*, 1991; Prosser *et al.*, 1984) and during the ovulatory menstrual cycle (Hartmann and Prosser, 1982); lactose is negatively correlated with sodium and chloride in milk, reflecting increases in the permeability of the mammary epithelium (Hartmann and Kulski, 1978; Hartmann and Prosser, 1982).

A. Nutritional Status

Lactose in milk of women in developing countries is not different from that of other populations. Low levels of milk lactose (139 ± 2 mmol/liter) were found in milk from women of New Hebrides (Peters, 1953), but this was likely due to technical problems. Milk composition of poorly nourished women has been reviewed (Jelliffe and Jelliffe, 1978; WHO, 1985). A study found milk lactose levels of a 65 year old, malnourished Nigerian woman were within normal range, though somewhat higher than those of the well-nourished control group (Gindler *et al.*, 1985). The same and other researchers found significantly elevated lactose levels in poorly nourished women (Gindler *et al.*, 1987; Ojofeitimi *et al.*, 1983; Van Steenbergen *et al.*,

1983), but in a study that used similar criteria to define "malnourished," Khin-Maung-Naing and co-workers (1980) found no such difference. Milk lactose of marginally nourished women, studied in a metabolic ward, was not correlated with maternal weight, arm circumference, or triceps skinfold measurements (Brown *et al.,* 1986). Lactose content of milk from Navajo women was lower than levels normally reported; this probably resulted from the use of a lactose-specific analytical method, HPLC, rather than from suboptimal nutrition (Butte and Calloway, 1981) or genetic factors. Race did not affect lactose concentration (Prinsloo *et al.,* 1970).

B. Effect of Diet

In some circumstances, lactose concentration has been altered by dietary manipulation, but such effects may be related to the mother's nutritional status. Reduction of the caloric intake of well-nourished exclusive breastfeeders did not influence milk lactose concentration (Strode *et al.,* 1986), nor did protein supplementation of poorly nourished women influence lactose concentration (Deb and Cama, 1962). However, a high-energy, balanced supplement given to nursing mothers in Keneba, The Gambia, whose dietary intakes were below recommended levels, resulted in a significant (7.6%) decrease in lactose concentration (Prentice *et al.,* 1983). Increasing carbohydrate from 35 to 65% of the diet, with consequent lowering of fat from 50 to 15%, resulted in significantly lower lactose (Harzer *et al.,* 1984). No effect of a vegetarian diet on milk lactose was found (Dagnelie *et al.,* 1992; Finley *et al.,* 1985). Lactose decreased significantly in fasting Gambian women, perhaps as a result of formation of a paracellular pathway (Prentice *et al.,* 1984).

C. Medications

Toaff and colleagues (1969) found no effect of estrogen or progestagen on breast milk lactose at Day 5 postpartum in nursing women; similarly, oral contraceptives have not been shown to affect lactose content (Abdel Kader *et al.,* 1976; Lönnerdal *et al.,* 1980; Ramadan *et al.,* 1972; Sammour *et al.,* 1973). Mean lactose at 2–5 days postpartum was significantly less in bromocriptine-treated women than in a placebo group (Kulski *et al.,* 1978). Use of oxytocin nasal spray by women who delivered prematurely did not affect lactose concentration at 3 or 5 days postpartum (Ruis *et al.,* 1981).

D. Preterm

Even when statistically adjusted for milk volume and stage of lactation (Gross *et al.,* 1981), lactose concentration in milk of women delivering

TABLE I
Lactose Content of Human Milk by Nonspecific Carbohydrate Methods (mmol/Liter)

	Days of lactation						
n^a	Antepartum	1	2	3	4–14	15+	Reference
10				171 ± 9^b			Gross *et al.* (1980)
13					185 ± 20		
23						199 ± 17	
1				180	174 ± 18		Hytten (1954)
4				183 ± 4	188 ± 6		Ruis *et al.* (1981)
15					165 ± 16		Lönnerdal *et al.* (1976a)
35						210 ± 13	
92						210 ± 15	Dewey and Lönnerdal (1983)
5						191 ± 41	Watson *et al.* (1982)
6						200 ± 4	Lönnerdal *et al.* (1976b)
313						189	Macy and Kelly (1961)
1743						201 ± 7	Michaelsen *et al.* (1990)
46						159 ± 3	Toaff *et al.* (1969)
20						187 ± 6	Sammour *et al.* (1973)
20						185 ± 7	Ramadan *et al.* (1972)
152						206 ± 6	Nommsen *et al.* (1991)
8						205 ± 8	Lovelady *et al.* (1990)
215						230 ± 9	Finley *et al.* (1985)
70						212 ± 9	Dewey *et al.* (1984)

TABLE I—*continued*

n[a]	Days of lactation						Reference
	Antepartum	1	2	3	4–14	15+	
10						188 ± 1	Lönnerdal *et al.* (1984a)
8						183 ± 31	Lönnerdal *et al.* (1980)
Weighted mean 1[c]				175 ± 8	176 ± 17	202 ± 9	
Weighted mean 2[d]				179 ± 8	181 ± 17	196 ± 9	
Weighted mean 1[c] (g/dl)[e]				6.3 ± 0.2	6.3 ± 0.5	7. 3 ± 0.4	

[a] n, No. of subjects.
[b] Mean ± SD.
[c] Weighted mean 1 is the mean of the mean of each study weighted for the number of samples analyzed per value.
[d] Weighted mean 2 is the mean of the mean of each study weighted for the number of samples analyzed per value and the variance.
[e] g/dl is calculated from the molarity using the molecular weight for lactose monohydrate, 360.3 g/mol.

TABLE II
Lactose Content of Human Milk Analyzed by Lactose-Specific Methods (mmol/Liter)

n^a	Days of lactation						Reference
	Antepartum	1	2	3	4–14	15+	
13	60						Kulski and Hartmann (1981)
11		53	119		150		
2	76	97	153	160			Kulski and Hartmann (1983)
3					164		
52	65 ± 14[b]						Kulski *et al.* (1981b)
77					153 ± 18		
47						183 ± 21	
13	80 ± 22					182 ± 8	Allen *et al.* (1991)
19		75	145	163	166		Kulski *et al.* (1981a)
38		87	125	162	170		Arthur *et al.* (1989)
5		108	139	172	184		Saint *et al.* (1984)
10		118		153	176	180	Harzer *et al.* (1986)
20			137 ± 18				Neubauer *et al.* (1993)
35				163 ± 11			
75					182 ± 16		
60						199 ± 18	
13			155	160			Neville *et al.* (1991)
9				172 ± 25			Anderson *et al.* (1983)
13					186 ± 15		
6					181 ± 12		Lemons *et al.* (1982)
11						194 ± 21	
17					152 ± 18		Anderson *et al.* (1981)

TABLE II—*continued*

n^a	Days of lactation						Reference
	Antepartum	1	2	3	4–14	15+	
5						181 ± 15	
12					173 ± 18	191 ± 19	Ferris *et al.* (1988)
8					160 ± 15		van Beusekom *et al.* (1993)
5						177 ± 18	
23						169 ± 17	Butte and Calloway (1981)
1						156	Verheul *et al.* (1986)
6						212	Lammi-Keefe *et al.* (1989)
5						189 ± 5	Butte *et al.* (1988)
155						183 ± 7	Butte *et al.* (1984)
6						199 ± 8	Neville *et al.* (1984)
46					165 ± 16	183 ± 20	Coppa *et al.* (1993)
15						185 ± 8	Villalpando *et al.* (1992)
10						186 ± 8	Dagnelie *et al.* (1992)
Weighted mean 1[c]	67 ± 16	85	134 ± 18	162 ± 18	168 ± 17	185 ± 14	
Weighted mean 2[d]	66 ± 16	85	134 ± 18	215 ± 18	169 ± 17	184 ± 14	
Weighted mean 1[c] (g/dl)[e]	2.4 ± 0.6	3.1	4.8 ± 0.6	5.8 ± 0.6	6.1 ± 0.6	6.7 ± 0.5	

[a] n, No. of subjects.
[b] Mean ± SD.
[c] Weighted mean 1 is the mean of the mean of each study weighted for the number of samples analyzed per value.
[d] Weighted mean 2 is the mean of the mean of each study weighted for the number of samples analyzed per value and the variance.
[e] g/dl is calculated from the molarity using the molecular weight for lactose monohydrate, 360.3 g/mol.

preterm was significantly lower than that in milk of women delivering full-term (Anderson *et al.*, 1981; Gross *et al.*, 1980); other investigations have found no such differences (Anderson *et al.*, 1983; Lemons *et al.*, 1982). The varied ranges of preterm gestational ages in each study may explain these contradictory findings. A lower gestational age corresponds to a less-developed mammary gland; lactose concentration should increase with maturation of the mammary gland (Anderson, 1984). The gestational age in studies in which significant differences between term and preterm milk were found ranged from 27 to 33.7 weeks, while in studies not finding any differences, the gestational age ranged from 27 to 36 weeks. Preterm milk does not vary within a feed (Thomas *et al.*, 1986) or at various times of the day (Gross *et al.*, 1981; Thomas *et al.*, 1986). Only one report from a developing country found significantly less lactose in preterm milk (Dawodu *et al.*, 1990); other reports on women with similar gestational ages found no differences (Jitta *et al.*, 1986; Kumbhat *et al.*, 1985).

E. Milk of Women with Insulin-Dependent Diabetes Mellitus (IDDM) (Table III)

In a longitudinal study, the lactose concentration of milk from women with IDDM was significantly lower than that of milk from women without IDDM (Neubauer *et al.*, 1993). Another study found a delay of 28 hr before milk lactose reached concentrations comparable to those seen in women without IDDM (Arthur *et al.*, 1989); this delay in lactogenesis was correlated with poor metabolic control (Neubauer *et al.*, 1993). No differences between lactose concentrations in milk from women with IDDM and from control subjects were found at 3–5 days postpartum (van Beusekom *et al.*, 1993) or in mature milk (Butte *et al.*, 1987; van Beusekom *et al.*, 1993). Milk lactose of women with IDDM may reach values comparable to those of women without IDDM by Day 4 postpartum (Arthur *et al.*, 1989), so differences will not be detected unless sampling is done in the immediate postpartum period.

F. Other Factors Influencing Milk Lactose

In lactating women, vigorous exercise (Lovelady *et al.*, 1990), parity (Prentice, 1985), delivery by cesarean section (Kulski *et al.*, 1981a), and lactation to 34 months postpartum (Boediman *et al.*, 1979) did not affect breast milk lactose, nor did heating the milk to 86°C for 8 min (Legge and Richards, 1978). Lactose in milk of low-income adolescents (mean age, 17.5 years) was significantly lower than that in milk of well-educated adults (range: 21–36 years) (Lipsman *et al.*, 1985). However, infant growth was satisfactory, indicating that the lower lactose may not be clinically relevant. Lactose in milk of a nonpuerperal woman, who induced lactation by manual

TABLE III
Lactose Content of Human Milk from Women with Insulin-Dependent Diabetes Mellitus (mmol/Liter)

n[a]	Days of lactation: Antepartum	1	2	3	4–14	15+	Reference
6			96 ± 17[b]				Neubauer *et al.* (1993)
17				160 ± 16			
53					170 ± 16		
33						187 ± 18	
6		50	50	125	175		Arthur *et al.* (1989)
5						194 ± 28[c]	Tolstoi (1935)
5						182 ± 5	Butte *et al.* (1987)
Weighted mean 1[d]		50	73 ± 17	151 ± 16	170 ± 16	187 ± 18	
Weighted mean 2[e]		50	73 ± 17	151 ± 16	170 ± 16	185 ± 18	
Weighted mean 1[d] (g/dl)[f]		1.8	2.6 ± 0.6	5.4 ± 0.6	6.1 ± 0.5	6.7 ± 0.6	

[a] n, No. of subjects.
[b] Mean ± SD.
[c] Analyzed by nonspecific method.
[d] Weighted mean 1 is the mean of the mean of each study weighted for the number of samples analyzed per value.
[e] Weighted mean 2 is the mean of the mean of each study weighted for the number of samples analyzed per value and the variance.
[f] g/dl is calculated from the molarity using the molecular weight for lactose monohydrate, 360.3 g/mol.

hyperstimulation to the breasts, was not different from normal values (Kulski *et al.,* 1981b). Milk lactose of mothers feeding twins (204 ± 13 mmol/liter) and of one mother feeding triplets (234 ± 20 mmol/liter) was generally higher than normal values (Saint *et al.,* 1986).

G. Other Breast Secretions

Breast fluid of a man with galactorrhea associated with hyperprolactinemia had slightly less lactose but was comparable to that of women during established lactation (Kulski *et al.,* 1981c).

IV. Human Milk Glucose

Glucose in mature human milk is 1.5 ± 0.4 mmol/liter (Table IV), with no significant differences between breasts (Arthur *et al.,* 1989; Kulski and Hartmann, 1983). It is generally accepted that glucose concentration varies greatly within individuals (Arthur *et al.,* 1991; Butte *et al.,* 1987). Diurnal variation has been reported by some (Arthur *et al.,* 1991), but not others (Lammi-Keefe *et al.,* 1989; Viverge *et al.,* 1986). Neville *et al.* (1984) found that glucose decreases in the course of a feed, consistent with a decrease in the milk's aqueous phase as lipid increases.

The antepartum milk glucose level, 0.3 ± 0.2 mmol/liter (Table IV), is negatively correlated with the lactose level (Allen *et al.,* 1991). In antepartum milk and early colostrum, passage of glucose from the extracellular fluid into the milk occurs via the paracellular pathway. During lactogenesis, milk glucose increases in parallel with lactose (Kulski and Hartmann, 1981; Kulski *et al.,* 1981a), reflecting the increased glucose transport capacity of the basolateral membrane of the mammary alveolar cells as volume increases (Neville *et al.,* 1990). Milk glucose in lactogenesis is also correlated with milk insulin and milk thyroid-stimulating hormone (TSH) levels (Kulski and Hartmann, 1983), as well as volume (Ereman *et al.,* 1988; Neville *et al.,* 1991). In established lactation, milk glucose is not correlated with milk lactose, indicating that glucose concentration is not rate limiting for lactose synthesis (Arthur *et al.,* 1991). During weaning, milk glucose decreases significantly; the paracellular pathway opens when milk volume falls below 0.4 liters per day (Hartmann and Kulski, 1978; Neville *et al.,* 1991) and as the number of glucose transporters in the mammary alveolar cell basolateral membrane decreases (Neville *et al.,* 1990). This decrease in glucose levels correlates with the number of feedings per 24 hours (Prosser *et al.,* 1984).

Glucose concentrations in milk and blood are not correlated at the time of pumping (Ratzmann *et al.,* 1988), and peak milk glucose lags 60–80 min behind peak whole blood glucose (Jovanovic-Peterson *et al.,* 1989). Glucose clamp studies in fully lactating women do not support a role for insulin in mammary gland transport (Neville *et al.,* 1990).

Decreases in milk glucose noted before and after ovulation may result from hormones controlling menstruation (Prosser and Hartmann, 1983). The milk glucose concentration of women with IDDM (Table V) is generally higher than that of women without IDDM (Butte *et al.,* 1987; Jovanovic-Peterson *et al.,* 1989; Neubauer *et al.,* 1987); however, in some cases in which the women were in tight metabolic control, milk glucose was similar to that of women without IDDM (Ratzmann *et al.,* 1988; van Beusekom *et al.,* 1993). Milk glucose significantly decreases with mastitis (Conner, 1979; Neubauer *et al.,* 1990).

Heating to 86°C for 8 min did not affect milk glucose (Legge and Richards, 1978). In one subject with galactosemia, glucose concentration was within the normal range (1.44 mmol/liter) (Forbes *et al.,* 1988).

V. Human Milk Galactose

Few measurements of galactose in human milk are available. Pooled samples at 7–12 days postpartum had 15 ± 2 mmol/liter (2.7 g/liter) and were not affected by heating to 86°C (Legge and Richards, 1978). One woman with galactosemia had milk galactose of 0.83 mmol/liter (Forbes *et al.,* 1988). Milk galactose from the mastitic breast of one woman was 32.3 mmol/liter (Conner, 1979).

VI. Human Milk Oligosaccharides (Table VI)

In addition to lactose, the carbohydrates of human milk include nucleotide sugars, glycolipids, glycoproteins, and oligosaccharides. A nonlactose carbohydrate fraction of human milk was prepared and studied in 1933 by Polonovsky and Lespagnol, who named this fraction gynolactose. It consisted mainly of oligosaccharides and was a major fraction of human milk.

A. Concentration

In 1960, Montreuil and Mullet analyzed dialysate of defatted human milk by classical specific colorimetric sugar assays. They calculated that oligosaccharides were present in milk at 12–13 g/liter and in colostrum at 22–24 g/liter. Viverge *et al.* (1990a) isolated three oligosaccharide fractions from milk dialysate whose combined weights represented 13–18 g/liter of dialysate; the concentration varied with the mother's genetic ability to

TABLE IV
Glucose Content of Human Milk (mmol/Liter)

	Days of lactation						
n[a]	Antepartum	1	2	3	4–14	15+	Reference
2	0.3				2.2		Kulski and Hartmann (1981)
2	0.4	0.6	1.1				Kulski and Hartmann (1983)
3					1.5		
13	0.3 ± 0.2[b]						Allen *et al.* (1991)
13		0.3	0.7	1.4	1.4		Neville *et al.* (1991)
38		0.05	0.2	0.6	0.9		Arthur *et al.* (1989)
19		0.3			1.3		Kulski *et al.* (1981a)
8					2.2 ± 1.0		van Beusekom *et al.* (1993)
5						2.6 ± 0.5	
6						1.7	Lammi-Keefe *et al.* (1989)
6						1.5 ± 0.2	Neville *et al.* (1984)
11						1.0 ± 0.3	Jovanovic-Peterson *et al.* (1989)
42						1.7 ± 0.4	Butte *et al.* (1987)
11						0.7 ± 0.5	Ratzmann *et al.* (1988)
114						1.4 ± 0.4	Neubauer *et al.* (1990)
16						1.6 ± 0.4	Neubauer *et al.* (1987)
5						1.6 ± 0.4	Arthur *et al.* (1991)
3						1.9 ± 0.3	Faulkner *et al.* (1981)
Weighted mean 1[c]	0.3 ± 0.2	0.2	0.4	0.8	1.2 ± 1.0	1.5 ± 0.4	
Weighted mean 2[d]	0.3 ± 0.2	0.2	0.4	0.8	1.2 ± 1.0	1.5 ± 0.4	

TABLE IV—*continued*

n^a	Days of lactation						Reference
	Antepartum	1	2	3	4–14	15+	
Weighted mean 1[c] (g/liter)[e]	0.06 ± 0.03	0.03	0.06	0.16	0.22 ± 0.19	0.26 ± 0.07	

[a]*n*, No. of subjects.
[b]Mean ± SD.
[c]Weighted mean 1 is the mean of the mean of each study weighted for the number of samples analyzed per value.
[d]Weighted mean 2 is the mean of the mean of each study weighted for the number of samples analyzed per value and the variance.
[e]g/liter is calculated from the molarity using the molecular weight for glucose, 180.1 g/mol.

TABLE V
Glucose Content of Human Milk from Women with Insulin-Dependent Diabetes Mellitus (mmol/Liter)

	Days of lactation						
n^a	Antepartum	1	2	3	4–14	15+	Reference
6		0.1	0.1	0.9	1.0		Arthur *et al.* (1989)
7					5.6 ± 2.0[b]		van Beusekom *et al.* (1993)
6						2.5 ± 1.2	
7						1.3 ± 0.5	Jovanovic-Peterson *et al.* (1989)
5						4.0 ± 0.8	Butte *et al.* (1987)
11						0.7 ± 0.5	Ratzmann *et al.* (1988)
9						3.4 ± 1.8	Neubauer *et al.* (1987)
Weighted mean 1[c]		0.1	0.1	0.9	3.5 ± 2.0	2.2 ± 1.1	
Weighted mean 2[d]		0.1	0.1	0.9	3.5 ± 2.0	1.4 ± 1.1	
Weighted mean 1[c] (g/liter)[e]		0.02	0.02	0.16	0.63 ± 0.36	0.4 ± 0.2	

[a] *n*, No. of subjects.
[b] Mean ± SD.
[c] Weighted mean 1 is the mean of the mean of each study weighted for the number of samples analyzed per value.
[d] Weighted mean 2 is the mean of the mean of each study weighted for the number of samples analyzed per value and the variance.
[e] g/liter is calculated from the molarity using the molecular weight for glucose, 180.1 g/mol.

TABLE VI
Oligosaccharides in Human Milk

I. Lactose (Levene and Sobotka, 1926)
Gal β(1→4) Glc

2′-Fucosyllactose (Kuhn *et al.* 1956a) (Not found in milk of nonsecretors)
Fuc α(1→2) Gal β(1→4) Glc

3-Fucosyllactose (Montreuil, 1956)

```
Gal β(1→4)↘
           Glc
Fuc α(1→3)↗
```

Lactodifucotetraose (Kuhn and Gauhe, 1958) (Not found in milk of nonsecretors)

```
Fuc α(1→2) Gal β(1→4)↘
                      Glc
           Fuc α(1→3)↗
```

3′-Sialyllactose (Kuhn and Brossmer, 1959)
NANA α(2→3) Gal β(1→4) Glc

6′-Sialyllactose (Kuhn, 1959)
NANA α(2→6) Gal β(1→4) Glc

6′-Galactosyllactose (Yamashita and Kobata, 1974)
Gal β(1→6) Gal β(1→4) Glc

***N*-Acetylneuramin lactose sulfate** (Sturman *et al.*, 1985)
NANA α(2→3) Gal-6-SO_4 β(1→4) Glc

Monofucosylmonosialyllactose (Grönberg *et al.*, 1989)

```
NANA α(2→3) Gal β(1→4)↘
                       Glc
            Fuc α(1→3)↗
```

II. Lacto-*N*-tetraose (Kuhn and Baer, 1956)
Gal β(1→3) GlcNAc β(1→3) Gal β(1→4) Glc

Lacto-*N*-fucopentaose I (Kuhn *et al.*, 1956b) (Not found in milk of nonsecretors)
Fuc α(1→2) Gal β(1→3) GlcNAc β(1→3) Gal β(1→4) Glc

Lacto-*N*-fucopentaose II (Kuhn *et al.*, 1958) [Not found in milk of Le^{a-b-} (Lewis ab negative)]

```
Gal β(1→3)↘
           GlcNAc β(1→3) Gal β(1→4) Glc
Fuc α(1→4)↗
```

Lacto-*N*-fucopentaose V (Ginsburg *et al.*, 1976)

```
Gal β(1→3) GlcNAc β(1→3) Gal β(1→4)↘
                                    Glc
                         Fuc α(1→3)↗
```

Lacto-*N*-difucohexaose I (Kuhn and Gauhe, 1958) (Not found in milk of nonsecretors; not found in milk of Le^{a-b-})

```
Fuc α(1→2) Gal β(1→3)↘
                      GlcNAc β(1→3) Gal β(1→4) Glc
           Fuc α(1→4)↗
```

Lacto-*N*-difucohexaose II (Kuhn and Gauhe, 1960) (Not found in milk of Le^{a-b-})

```
Gal β(1→3)↘
           GlcNAc β(1→3) Gal β(1→4)↘
Fuc α(1→4)↗                         Glc
                         Fuc α(1→3)↗
```

TABLE VI—*continued*

A-heptasaccharide (Strecker and Montreuil, 1973)
GalNAc α(1→3)↘
Gal β(1→3)↘
Fuc α(1→2)↗ GlcNAc β(1→3) Gal β(1→4) Glc
Fuc α(1→4)↗

Sialyllacto-*N*-tetraose (a) (Kuhn and Gauhe, 1965)
NANA α(2→3) Gal β(1→3) GlcNAc β(1→3) Gal β(1→4) Glc

Sialyllacto-*N*-tetraose (b) (Kuhn and Gauhe, 1965)
NANA α(2→6)↘
GlcNAc β(1→3) Gal β(1→4) Glc
Gal β(1→3)↗

Disialyllacto-*N*-tetraose (Grimmonprez and Montreuil, 1968)
NANA α(2→6)↘
GlcNAc β(1→3) Gal β(1→4) Glc
NANA α(2→3) Gal β(1→3)↗

Monosialylmonofucosyllacto-*N*-tetraose (Wieruszeski *et al.*, 1985)
NANA α(2→3) Gal β(1→3)↘
GlcNAc β(1→3) Gal β(1→4) Glc
Fuc α(1→4)↗

Monofucosylmonosialyllacto-*N*-tetraose (Wieruszeski *et al.*, 1985)
NANA α(2→6)↘
GlcNAc β(1→3) Gal β(1→4) Glc
Fuc α(1→2) Gal β(1→3)↗

Monofucosyldisialyllacto-*N*-tetraose (Kitagawa *et al.*, 1991)
NANA α(2→6)↘
NANA α(2→3) Gal β(1→3) GlcNAc β(1→3) Gal β(1→4) Glc
Fuc α(1→4)↗

A-hexasaccharide (Sabharwal *et al.*, 1984)
GalNAc α(1→3)↘
Gal β(1→3) GlcNAc β(1→3) Gal β(1→4) Glc
Fuc α(1→2)↗

III. Lacto-*N*-neotetraose (Kuhn and Gauhe, 1962)
Gal β(1→4) GlcNAc β(1→3) Gal β(1→4) Glc

Lacto-*N*-fucopentaose III (Kobata and Ginsburg, 1969)
Gal β(1→4)↘
GlcNAc β(1→3) Gal β(1→4) Glc
Fuc α(1→3)↗

Sialyllacto-*N*-neotetraose (c) (Kuhn and Gauhe, 1965)
NANA α(2→6) Gal β(1→4) GlcNAc β(1→3) Gal β(1→4) Glc

Monofucosylmonosialyllacto-*N*-neotetraose (Smith *et al.*, 1987)
NANA α(2→6) Gal β(1→4) GlcNAc β(1→3) Gal β(1→4)↘
Glc
Fuc α(1→3)↗

IV. Lacto-*N*-Hexaose (Kobata and Ginsburg, 1972a)
Gal β(1→4) GlcNAc β(1→6)↘
Gal β(1→4) Glc
Gal β(1→3) GlcNAc β(1→3)↗

TABLE VI—*continued*

Monofucosyllacto-*N*-hexaose I (Yamashita *et al.*, 1977a)
Gal β(1→4) GlcNAc β(1→6)↘
Gal β(1→4) Glc
Fuc α(1→2) Gal β(1→3) GlcNAc β(1→3)↗

Monofucosyllacto-*N*-hexaose II (Dua *et al.*, 1985)
Fuc α(1→3)↘
GlcNAc β(1→6)↘
Gal β(1→4)↗ Gal β(1→4) Glc
Gal β(1→3) GlcNAc β(1→3)↗

Difucosyllacto-*N*-hexaose (Dua *et al.*, 1985)
Fuc α(1→3)↘
GlcNAc β(1→6)↘
Gal β(1→4)↗
Gal β(1→4) Glc
Fuc α(1→4)↘
GlcNAc β(1→3)↗
Gal β(1→3)↗

Difucosyllacto-*N*-hexaose (a) (Yamashita *et al.*, 1977a)
Fuc α(1→3)↘
GlcNAc β(1→6)↘
Gal β(1→4)↗ Gal β(1→4) Glc
Fuc α(1→2) Gal β(1→3) GlcNAc β(1→3)↗

Trifucosyllacto-*N*-hexaose (Sabharwal *et al.*, 1988a)
Gal β(1→4)↘
GlcNAc β(1→6)↘
Fuc α(1→3)↗
Gal β(1→4) Glc
Fuc α(1→4)↘
GlcNAc β(1→3)↗
Fuc α(1→2)Gal β(1→3)↗

Monosialyllacto-*N*-hexaose (Kobata and Ginsburg, 1972a)
NANA α(2→6) Gal β(1→4) GlcNAc β(1→6)↘
Gal β(1→4) Glc
Gal β(1→3) GlcNAc β(1→3)↗

Disialyllacto-*N*-hexaose I (Kitagawa *et al.*, 1991)
NANA α(2→3) Gal β(1→3) GlcNAc β(1→3)↘
Gal β(1→4) Glc
NANA α(2→6) Gal β(1→4) GlcNAc β(1→6)↗

Disialyllacto-*N*-hexaose II (Kitagawa *et al.*, 1991)
NANA α(2→6)↘
GlcNAc β(1→3)↘
NANA α(2→3) Gal β(1→3)↗ Gal β(1→4) Glc
Gal β(1→4) GlcNAc β(1→6)↗

Monofucosylmonosialyllacto-*N*-hexaose (Yamashita *et al.*, 1977a)
NANA α(2→6) Gal β(1→4) GlcNAc β(1→6)↘
Gal β(1→4) Glc
Fuc α(1→2) Gal β(1→3) GlcNAc β(1→3)↗

Monofucosyldisialyllacto-*N*-hexaose I (Yamashita *et al.*, 1976a)
Fuc α(1→3) Gal β(1→4) GlcNAc β(1→6)↘
NANA α(2→3) Gal β(1→3)↘ Gal β(1→4) Glc
GlcNAc β(1→3)↗
NANA α(2→6)↗

TABLE VI—*continued*

Monofucosyldisialyllacto-*N*-hexaose II (Kitagawa *et al.*, 1991)

NANA α(2→6)↘
GlcNAc β(1→3)↘
NANA α(2→3) Gal β(1→3)↗
Gal β(1→4) Glc
Gal β(1→4)↘
GlcNAc β(1→6)↗
Fuc α(1→3)↗

Monofucosyldisialyllacto-*N*-hexaose III (Yamashita *et al.*, 1976a)

Fuc α(1→3)↘
GlcNAc β(1→6)↘
Gal β(1→4)↗
Gal β(1→4) Glc
NANA α(2→3) Gal β(1→3)↘
GlcNAc β(1→3)↗
NANA α(2→6)↗

Monofucosylmonosialyllacto-*N*-hexaose I (Grönberg *et al.*, 1992)

NANA α(2→6)↘
GlcNAc β(1→3)↘
Gal β(1→3)↗
Gal β(1→4) Glc
Gal β(1→4)↘
GlcNAc β(1→6)↗
Fuc α(1→3)↗

Monofucosylmonosialyllacto-*N*-hexaose II (Grönberg *et al.*, 1992)

NANA α(2→3) Gal β(1→3) GlcNAc β(1→3)↘
Gal β(1→4)↘ Gal β(1→4) Glc
GlcNAc β(1→6)↗
Fuc α(1→3)↗

Monofucosylmonosialyllacto-*N*-hexaose III (Grönberg *et al.*, 1992)

Fuc α(1→4)↘
GlcNAc β(1→3)↘
Gal β(1→3)↗ Gal β(1→4) Glc
NANA α(2→6) Gal β(1→4) GlcNAc β(1→6)↗

Difucosylmonosialyllacto-*N*-hexaose (Grönberg *et al.*, 1992)

Fuc α(1→4)↘
GlcNAc β(1→3)↘
Fuc α(1→2) Gal β(1→3)↗ Gal β(1→4) Glc
NANA α(2→6) Gal β(1→4) GlcNAc β(1→6)↗

V. Lacto-*N*-neohexaose (Kobata and Ginsburg, 1972b)

Gal β(1→4) GlcNAc β(1→6)↘
Gal β(1→4) Glc
Gal β(1→4) GlcNAc β(1→3)↗

Monofucosyllacto-*N*-neohexaose (Kobata and Ginsburg, 1972b)

⌈ Gal β(1→4) GlcNAc β(1→6))↘
Fuc α(1→ Gal β(1→4) Glc
⌊ Gal β(1→4) GlcNAc β(1→3))↗

TABLE VI—*continued*

Difucosyllacto-*N*-neohexaose (Haeuw-Fievre *et al.*, 1993)
Fuc α(1→3)↘
GlcNAc β(1→6)↘
Gal β(1→4)↗
Gal β(1→4) Glc
Gal β(1→4)↘
GlcNAc β(1→3)↗
Fuc α(1→3)↗

Monosialyllacto-*N*-neohexaose I (Kobata and Ginsburg, 1972b)
NANA α(2→6) Gal β(1→4) GlcNAc β(1→6)↘
Gal β(1→4) Glc
Gal β(1→4) GlcNAc β(1→3)↗

Monosialyllacto-*N*-neohexaose II (Grönberg *et al.*, 1989)
Gal β(1→4) GlcNAc β(1→6)↘
Gal β(1→4) Glc
NANA α(2→6) Gal β(1→4) GlcNAc β(1→3)↗

Disialyllacto-*N*-neohexaose (Grönberg *et al.*, 1992)
NANA α(2→6) Gal β(1→4) GlcNAc β(1→6)↘
Gal β(1→4) Glc
NANA α(2→6) Gal β(1→4) GlcNAc β(1→3)↗

Monofucosylmonosialyllacto-*N*-neohexaose (Grönberg *et al.*, 1989)
Fuc α(1→3)↘
GlcNAc β(1→6)↘
Gal β(1→4)↗
Gal β(1→4) Glc
NANA α(2→6) Gal β(1→4) GlcNAc β(1→3)↗

Monofucosylmonosialyllacto-*N*-neohexaose (Kobata and Ginsburg, 1972b)
┌ NANA α(2→6) Gal β(1→4) GlcNAc β(1→6)↘
Fuc α(1→
Gal β(1→4) Glc
└ Gal β(1→4) GlcNAc β(1→3)↗

Difucosylmonosialyllacto-*N*-neohexaose (Grönberg *et al.*, 1989)
Fuc α(1→2) Gal β(1→4)↘
GlcNAc β(1→6)↘
Fuc α(1→3)↗
Gal β(1→4) Glc
NANA α(2→6) Gal β(1→4) GlcNAc β(1→3)↗

Monofucosyldisialolacto-*N*-neohexaose (Yamashita *et al.*, 1976a)
NANA α(2→6/3) Gal β(1→4) GlcNAc β(1→6)↘
Fuc α(1→3)↘
Gal β(1→4) Glc
GlcNAc β(1→3)↗
NANA α(2→3/6)Gal β(1→4)↗

VI. para-Lacto-*N*-hexaose
Gal β(1→3) GlcNAc β(1→3) Gal β(1→4) GlcNAc β(1→3) Gal β(1→4) Glc

Fucosyl-*para*-lacto-*N*-hexaose (Sabharwal *et al.*, 1988b)
Gal β(1→3) GlcNAc β(1→3) Gal β(1→4)↘
GlcNAc β(1→3) Gal β(1→4) Glc
Fuc α(1→3)↗

Difucosyl-*para*-lacto-*N*-hexaose (Yamashita *et al.*, 1977b)
Gal β(1→3)↘
GlcNAc β(1→3) Gal β(1→4)↘
Fuc α(1→4)↗
GlcNAc β(1→3) Gal β(1→4) Glc
Fuc α(1→3)↗

TABLE VI—*continued*

Trifucosyl-*para*-lacto-*N*-hexaose (Strecker *et al.*, 1988)

Fuc α(1→4)↘
GlcNAc β(1→3) Gal β(1→4)↘
Fuc α(1→2) Gal β(1→3)↗
GlcNAc β(1→3) Gal β(1→4)Glc
Fuc α(1→3)↗

VII. *para*-Lacto-*N*-neohexaose

Gal β(1→4) GlcNAc β(1→3) Gal β(1→4) GlcNAc β(1→3) Gal β(1→4) Glc

Difucosyl-*para*-lacto-*N*-neohexaose (Yamashita *et al.*, 1977b)

Fuc α(1→3)↘
GlcNAc β(1→3) Gal β(1→4)↘
Gal β(1→4)↗
GlcNAc β(1→3) Gal β(1→4) Glc
Fuc α(1→3)↗

VIII. Lacto-*N*-octaose

Gal β(1→4) GlcNAc β(1→3) Gal β(1→4) GlcNAc β(1→6)↘
Gal β(1→4) Glc
Gal β(1→3) GlcNAc β(1→3)↗

Monofucosyllacto-*N*-octaose (Yamashita *et al.*, 1976b)

Fuc α(1→3)↘
GlcNAc β(1→6)↘
Gal β(1→4) GlcNAc β(1→3) Gal β(1→4)↗
Gal β(1→4) Glc
Gal β(1→3) GlcNAc β(1→3)↗

Difucosyllacto-*N*-octaose I (Tachibana *et al.*, 1978)

Fuc α(1→3)↘
Fuc α(1→3)↘
GlcNAc β(1→6)↘
GlcNAc β(1→3) Gal β(1→4)↗
Gal β(1→4) Glc
Gal β(1→4)↗
Gal β(1→3) GlcNAc β(1→3)↗

Difucosyllacto-*N*-octaose II (Tachibana *et al.*, 1978)

Fuc α(1→3)↘
GlcNAc β(1→6)↘
Gal β(1→4) GlcNAc β(1→3) Gal β(1→4)↗
Gal β(1→4) Glc
Gal β(1→3) GlcNAc β(1→3)↗
Fuc α(1→4)↗

Trifucosyllacto-*N*-octaose (Tachibana *et al.*, 1978)

Fuc α(1→3)↘
Fuc α(1→3)↘
GlcNAc β(1→6)↘
GlcNAc β(1→3) Gal β(1→4)↗
Gal β(1→4) Glc
Gal β(1→4)↗
Gal β(1→3) GlcNAc β(1→3)↗
Fuc α(1→4)↗

Monofucosylmonosialyllacto-*N*-octaose (sialyl Le[a]) (Kitagawa, 1993)

Gal β(1→4) GlcNAc β(1→3) Gal β(1→4) GlcNAc β(1→6)↘
Gal β(1→4) Glc
NANA α(2→3) Gal β(1→3) GlcNAc β(1→3)↗
Fuc α(1→4)↗

IX. Lacto-*N*-neooctaose

Gal β(1→3) GlcNAc β(1→3) Gal β(1→4) GlcNAc β(1→6)↘
Gal β(1→4) Glc
Gal β(1→4) GlcNAc β(1→3)↗

TABLE VI—*continued*

Monofucosyllacto-*N*-neooctaose (Yamashita *et al.*, 1976b)

```
                                  Fuc α(1→3)↘
                                              GlcNAc β(1→6)↘
Gal β(1→3) GlcNAc β(1→3) Gal β(1→4)↗                         Gal β(1→4) Glc
                                  Gal β(1→4) GlcNAc β(1→3)↗
```

Difucosyllacto-*N*-neooctaose I (Tachibana *et al.*, 1978)

```
                                  Fuc α(1→3)↘
Fuc α(1→4)↘                                   GlcNAc β(1→6)↘
           GlcNAc β(1→3) Gal β(1→4)↗                         Gal β(1→4) Glc
Gal β(1→3)↗                        Gal β(1→4) GlcNAc β(1→3)↗
```

Difucosyllacto-*N*-neooctaose II (Tachibana *et al.*, 1978)

```
                                  Fuc α(1→3)↘
                                              GlcNAc β(1→6)↘
Gal β(1→3) GlcNAc β(1→3) Gal β(1→4)↗
                                  Gal β(1→4)↘                 Gal β(1→4) Glc
                                              GlcNAc β(1→3)↗
                                  Fuc α(1→3)↗
```

Trifucosyllacto-*N*-neooctaose (Tachibana *et al.*, 1978)

```
                                  Fuc α(1→3)↘
Fuc α(1→4)↘                                   GlcNAc β(1→6)↘
           GlcNAc β(1→3) Gal β(1→4)↗
Gal β(1→3)↗                       Gal β(1→4)↘                 Gal β(1→4) Glc
                                              GlcNAc β(1→3)↗
                                  Fuc α(1→3)↗
```

X. *iso*-Lacto-*N*-octaose

```
Gal β(1→3) GlcNAc β(1→3) Gal β(1→4) GlcNAc β(1→6)↘
                                                   Gal β(1→4) Glc
                         Gal β(1→3) GlcNAc β(1→3)↗
```

Trifucosyl-*iso*-lacto-*N*-octaose (Strecker *et al.*, 1992)

```
                                                 Fuc α(1→3)↘
                                                             GlcNAc β(1→6)↘
Fuc α(1→2) Gal β(1→3) GlcNAc β(1→3) Gal β(1→4)↗                            Gal β(1→4) Glc
                                Fuc α(1→2) Gal β(1→3) GlcNAc β(1→3)↗
```

Tetrafucosyl-*iso*-lacto-*N*-octaose (Haeuw-Fievre *et al.*, 1993)

```
                                                 Fuc α(1→3)↘
                                                             GlcNAc β(1→6)↘
Fuc α(1→2) Gal β(1→3) GlcNAc β(1→3) Gal β(1→4)↗
                                                                           Gal β(1→4) Glc
                                      Fuc α(1→2) Gal β(1→3)↘
                                                             GlcNAc β(1→3)↗
                                                 Fuc α(1→4)↗
```

Pentafucosyl-*iso*-lacto-*N*-octaose (Haeuw-Fievre *et al.*, 1993)

```
                                                 Fuc α(1→3)↘
                      Fuc α(1→4)↘                            GlcNAc β(1→6)↘
                                 GlcNAc β(1→3) Gal β(1→4)↗
Fuc α(1→2) Gal β(1→3)↗                                                     Gal β(1→4) Glc
                                      Fuc α(1→2) Gal β(1→3)↘
                                                             GlcNAc β(1→3)↗
                                                 Fuc α(1→4)↗
```

TABLE VI—*continued*

Trifucosylmonosialyl-*iso*-lacto-*N*-octaose (sialyl Le[a]) (Kitagawa, 1993)

Fuc α(1→3)↘
GlcNAc β(1→6)↘
Fuc α(1→2) Gal β(1→3) GlcNAc β(1→3) Gal β(1→4)↗
Gal β(1→4) Glc
NANA α(2→3) Gal β(1→3)↘
GlcNAc β(1→3)↗
Fuc α(1→4)↗

XI. *para*-Lacto-*N*-octaose

Gal β(1→3) GlcNAc β(1→3) Gal β(1→4) GlcNAc β(1→3) Gal β(1→4) GlcNAc β(1→3) Gal β(1→4) Glc

Tetrafucosyl-*para*-lacto-*N*-octaose (Haeuw-Fievre *et al.*, 1993)

Fuc α(1→2) Gal β(1→3)↘
GlcNAc β(1→3) Gal β(1→4)↘
Fuc α(1→4)↗
GlcNAc β(1→3) Gal β(1→4)↘
Fuc α(1→3)↗
GlcNAc β(1→3) Gal β(1→4) Glc
Fuc α(1→3)↗

XII. Lacto-*N*-decaose

Gal β(1→4) GlcNAc β(1→6)↘
Gal β(1→4) GlcNAc β(1→6)↘
Gal β(1→3) GlcNAc β(1→3)↗
Gal β(1→4) Glc
Gal β(1→3) GlcNAc β(1→3)↗

XIII. Deviant Structures

A. Lactose-containing (Grönberg *et al.*, 1992)

NANA α(2→3) Gal β(1→3)↘
Gal β(1→4)↘
Gal β(1→4) Glc
GlcNAc β(1→6)↗
Fuc α(1→3)↗

A-pentasaccharide (Lundblad and Svensson, 1973; Strecker and Montreuil, 1973)

GalNAc α(1→3)↘
Gal β(1→4)↘
Fuc α(1→2)↗
Glc
Fuc α(1→3)↗

B. Nonlactose structures (Kitagawa *et al.*, 1990)

Fuc α(1→4)↘
GlcNAc
NANA α(2→3) Gal β(1→3)↗
Fuc α(1→4)↘
GlcNAc β(1→3)Gal
NANA α(2→3) Gal β(1→3)↗

Note. NANA, *N*-acetylneuraminic acid; Gal, galactose; Glc, glucose; Fuc, fucose; GlcNAc, *N*-acetylglucosamine; GalNAc, *N*-acetylgalactosamine.

synthesize specific fucosyl linkages. More recently, Coppa *et al.* (1993), using an HPLC separation with a refractive index detector, found that the oligosaccharide fraction was 21 g/liter on Day 4 of lactation and fell to 13 g/liter by Day 120.

The difference between our average lactose value for nonspecific methods (73 g/liter) and specific methods (67 g/liter) is 6 g/liter. Attributing this difference to the oligosaccharide content of the milk, and using Jenness' assumption (Jenness, 1979) that the effective "average size" of the oligosaccharide fraction is the same as that of a tetrasaccharide whose reducing power is equal to half that of lactose, we calculate that mature milk contains approximately 12 g/liter of oligosaccharides. These values are in reasonable agreement with the aforementioned values in the literature; thus, the oligosaccharide fraction of human milk is the third largest solid component, after fat and lactose.

However, just as it was incorrect 50 years ago to assume that all of the carbohydrate of human milk is lactose, it would be as incorrect now to assume that all of the nonlactose carbohydrate of human milk is oligosaccharide. The methods described above could overestimate the oligosaccharide content if other carbohydrate-containing materials copurify with the oligosaccharide fraction. Once routine methods for separating and quantifying the individual oligosaccharides become widely available, a more accurate value for the oligosaccharide fraction can be derived by totaling the individual oligosaccharide content of milk.

B. Qualitative Characteristics

The qualitative characterization of the oligosaccharides of human milk is already well under way. For example, a combination of chromatographic techniques and fast atom bombardment mass spectrometry has revealed evidence for 101 neutral components of the human milk oligosaccharides (Egge *et al.,* 1983). Approximately 80 neutral and acidic (sialic acid-containing) oligosaccharides have been isolated and identified to date; the structures are given in Table VI. Some of these structures are incompletely defined, and others undergo periodic revision. For example, some of the structures presented as derivatives of lacto-*N*-octaose and lacto-*N*-neooctaose may actually have the *iso*-lacto-*N*-octaose core structure (Haeuw-Fievre *et al.,* 1993). This list grows by several compounds each year, and the structures being defined become larger and more complex. Furthermore, the rule that all milk oligosaccharides contain a terminal lactose as part of their structure, although true for the vast majority of milk oligosaccharides, is no longer universal, as seen in the last portion of the table.

C. Biological Activity

Some of these oligosaccharides are thought to be biologically active. As they appear to be synthesized by some of the same glycosyltransferases that participate in the synthesis of glycoprotein and glycolipid cell surface components, it is reasonable to postulate that some of these oligosaccharides can act as analogs or homologs to host cell surface receptors for pathogens. Infants, whose stomachs are less acidified than those of adults and whose immune systems are not mature, may need additional protection from enteric pathogens; a major milk fraction composed of water-soluble cell surface analogs that can inhibit enteropathogen binding to host cell receptors could serve such a protective function.

This concept is supported by several findings. Andersson *et al.* (1986) reported that specific oligosaccharides can inhibit binding of *Streptococcus pneumoniae* and *Hemophilus influenzae* to their receptors. Similarly, Cravioto and co-workers (1991) described an oligosaccharide that inhibits adherence of enteropathogenic *Escherichia coli* to their receptors. We have found that a fucosylated oligosaccharide inhibits binding of invasive strains of *Campylobacter jejuni* to its host cell (Ruiz-Palacios *et al.,* 1992), and another fucosylated oligosaccharide inhibits the toxicity of stable toxin of *E. coli in vivo* (Newburg *et al.,* 1990). Milk oligosaccharides also contain human blood group antigens, such as Lewis a, Lewis b, Lewis x, A, B, O, and I. The oligosaccharide content of milk of women from different blood group types has distinct patterns (Viverge *et al.,* 1985, 1986, 1990b). Lactating mothers may differ genetically in their ability to produce protective oligosaccharides and thus may influence their breast-fed infants' susceptibilities to enteric disease. This hypothesis is currently under active investigation.

VII. Lactose In Nonhuman Milk (Table VII)

Lactose, the principal sugar of human and most other terrestrial eutherian milks, is, for the most part, unique to this fluid, although small amounts have been found in other sources, including plants (Kuhn and Löw, 1949; Venkataraman and Reithel, 1958). Although lactose was discovered and isolated in 1633 (Bartoletti, 1633), many fundamental questions regarding this sugar remain unanswered, including its actual concentration in the milks of animals. Unfortunately, much of the older literature on the lactose content of milk relied on analytical procedures that measure total carbohydrate (see Section II); the total carbohydrate measured was often assumed to be lactose, an assumption which proved to be invalid for many species. Most contemporary reports on milk composition now give the total carbohydrate content, and identify this measure as such, or use analytical methods that specifically measure lactose, i.e., GC, HPLC, and enzymatic analysis. In animals whose milk carbohydrate is primarily lactose (e.g., rat), the new values obtained with specific procedures agree with the older values obtained with the classical methods; in animals whose milk contains little lactose but relatively large amounts of oligosaccharides (e.g., pinnipeds, cetaceans, and bears), the discrepancy between old and new values is significant. Nonetheless, we have included many of these old values in Table VII because they are the only data available and, when compared within individuals, within species, and across species, they broadly indicate the variations in lactose concentrations seen over the period of lactation, among individuals of a species, and in entire species. For example, in nonhuman primates the lactose concentration of milk tends to increase as lactation is established over the course of a day or two, then the lactose levels tend to be quite stable over most of the remainder of lactation. In other animals (e.g., rats, ferrets, and camels) lactose concentration increases over a relatively prolonged period. Most species show a strong inverse correlation between the concentration of protein and the concentration of lactose in milk. Primates and some ruminants have the highest lactose concentrations (ranging from approximately 50 to 70 g/liter); rodent milk has intermediate levels of lactose (30–50 g/liter), whereas the milk of pinnipeds, cetaceans, bear, and marsupials contains the lowest levels of lactose. Although animals with low lactose levels often have high levels of other oligosaccharides, human milk contains high levels of both lactose and nonlactose oligosaccharides. A systematic investigation of the concentrations of milk oligosaccharides across the species would help to elucidate the structural relationships among the various forms of carbohydrates in milk. The phylogeny of the milk oligosaccharides may also prove to be interesting, in view of the increasing number of biological roles now being attributed to milk oligosaccharides.

VIII. Other Carbohydrates in Nonhuman Milk

A. Glucose (Table VIII)

The glucose concentration in milk from terrestrial animals ranges from 0.1 to 0.8 mmol/liter, lower than that found in human milk. Diurnal variation in milk glucose from rats has been reported, though these results are not consistent concerning the time of day when glucose is highest (Faulkner *et al.,* 1981; Grigor *et al.,* 1989).

TABLE VII
Lactose Content of the Milks of Various Species

Mammalian species and taxonomic position	n[a]	Days postpartum	Carbohydrate content[b] (g/liter)	Measured lactose[c] (mmol/liter)	Reference
Class Mammalia					
Subclass Prototheria					
Order Montremata					
Family Ornithorhynchidae					
Ornithorhynchus anatinus (duck-billed platypus)	12	Mature	32.8 ± 8.8[d]	Trace	Messer *et al.* (1983)
Family Tachyglossidae					
Tachyglossus aculeatus (echidna)	2	Mature	9		Jenness (1974)
Subclass Theria					
Infraclass Metatheria					
Order Marsupialia					
Family Didelphidae					
Didelphis virginiana (Virginia opossum)	1	Mature	41		Jenness (1974)
Monodelphis domestica (American marsupial)	12	1	30		Green *et al.* (1991)
		20	70		
		30	90		
		50	120		
		70	10		

TABLE VII—*continued*

Mammalian species and taxonomic position	n^a	Days postpartum	Carbohydrate content[b] (g/liter)	Measured lactose[c] (mmol/liter)	Reference
Family Dasyuridae					
Dasyurus viverrinus (eastern quoll)	21	0–28	30–50		Green *et al.* (1987)
	16,21	56	74		Green *et al.* (1987); Messer *et al.* (1987)
		119	52		
		154	< 20		
Family Peramelidae					
Isoodon macrourus (northern brown bandicoot)	6	0–10	20–40		Merchant and Libke (1988)
		11–55	50–80		
		55–60	10		
Family Petauridae					
Pseudocheirus peregrinus (common ringtail possum)	28	35	100		
		98	130		
(Common ringtail possum)/captive		126	50		Munks *et al.* (1991)
Family Phalangeridae					
Trichosurus vulpecula (brushtailed possum)	> 50	Mature	32		Jenness (1974)
	4	0–4	35 ± 3		Cowan (1989)
	3	5–10	45 ± 4		
	6	11–15	45 ± 2		
	5	16–20	49 ± 4		

TABLE VII—*continued*

Mammalian species and taxonomic position	n^a	Days postpartum	Carbohydrate content[b] (g/liter)	Measured lactose[c] (mmol/liter)	Reference
Family Macropodidae					
Setonix brachyurus (quokka)	4	Mature	34		Jenness (1974)
Mageleia rufa (red kangaroo)	1	Mature	67		Jenness (1974)
Macropus robustus (wallaroo)	3	Mature	13		Jenness (1974)
Macropus rufogriseus banksianus (rednecked wallaby)	na	Mature	45		Jenness (1974)
	3	30	70		Merchant *et al.* (1989)
	6	70	90		
	8	130	100		
	16	170	110		
	8	230	110		
	8	270	40		
	2	310	10		
Subfamily Potoroinae					
Potorous tridactylus (potoroo)	16	35	90	Trace	Crowley *et al.* (1988)
		105	150		
		175	20		
Infraclass Eutheria					
Order Insectivora					
Family Erinaceidae					
Erinaceus europaeus (hedgehog)	na	Mature	20		Jenness (1974)

TABLE VII—*continued*

Mammalian species and taxonomic position	n^a	Days postpartum	Carbohydrate content[b] (g/liter)	Measured lactose[c] (mmol/liter)	Reference
Family Soricidae					
Neomys fodiens (water shrew)	na	Mature	1		Jenness (1974)
Blarina brevicauda (short-tailed shrew)	na	Mature	32		Jenness (1974)
Suncus murinus (musk shrew)	2	Mature	8		Dryden and Anderson (1978)
Order Chiroptera					
Family Vespertilionidae					
Myotis thysanodes (fringed myotis)	1	Mature	34		Jenness (1974)
Myotis lucifugus	4	Early	32 ± 3		Kunz *et al.* (1983)
	6	10 Days after early	32 ± 4		
	3	22 Days after early	35 ± 3		
Eptesicus fuscus	4		25 ± 5		
Family Phyllostomatidae					
Leptonycteris sanborni (longnose bat)	na	Mature	54		Jenness (1974)
Family Molossidae					
Tadarida brasiliensis (Mexican freetail bat)	na	Mature	37		Jenness (1974)
Order Primates					
Family Tupaiidae					
Lyonogale tana	11		< 25		D'Souza and Martin (1974)
Tupaia belangeri	16		< 25		
Tupaia minor	16		< 25		

TABLE VII—*continued*

Mammalian species and taxonomic position	n[a]	Days postpartum	Carbohydrate content[b] (g/liter)	Measured lactose[c] (mmol/liter)	Reference
Family Lemuridae					
Lemur catta (ring-tailed lemur)	2	7–14	74		Buss *et al.* (1976)
		161	74		
		162	56		
Lemur fulvus	1	90	65		
Lemur macaco	2	1, 184	58, 53		
Hybrid lemur	1	88	76		
Family Lorisidae					
Subfamily Lorisinae					
Nycticebus coucang (slow loris)	1	Mature	62		Jenness (1974)
Subfamily Galaginae					
Galago crassicaudatus (Pangani thick-tailed galago)	5	Mature	45		Jenness (1974)
	2	0–1	36		Pilson and Cooper (1967)
	1	49–62	49		
	1	63–85	52		

TABLE VII—*continued*

Mammalian species and taxonomic position	n^a	Days postpartum	Carbohydrate content[b] (g/liter)	Measured lactose[c] (mmol/liter)	Reference
Family Cebidae					
Saimiri sciureus (squirrel monkey)	13	Mature	63		Jenness (1974)
Family Callithricidae					
Saguinus oedipus (cotton-headed tamarin)	3	Mature	58		Jenness (1974)
Callithrix jacchus (common marmoset)	2	47, 75	81		Turton *et al.* (1978)
	1	14	69		
Family Cercopithecidae					
Cercopithecus sabeus (green monkey)	1	Mature	102		Jenness (1974)
Cercopithecus talapoin (talapoin monkey)	5	Mature	72		Jenness (1974)
	1	1	56		Buss and Cooper (1970)
	1	2	72		
	1	17–19	64		
	1	23–25	62		
	1	24, 25	76		
	1	37–39	73		
	1	178–192	80		

TABLE VII—*continued*

Mammalian species and taxonomic position	n^a	Days postpartum	Carbohydrate content[b] (g/liter)	Measured lactose[c] (mmol/liter)	Reference
Cercocebus sp.	10	Mature	70		Jenness (1974)
Macaca mulatta (rhesus monkey)	1	Mature	70		Jenness (1974)
	6	0–5	78		Lönnerdal *et al.* (1984b)
	8	6–15	81		
	18	16–35	79		
	25	36 +	79		
Papio Papio,					
Papio anubis,					
Papio cynocephalus (baboon)	18	Mature	73		Jenness (1974)
	2	0–5	68		Buss (1968)
	2	6–11	74		
	7	12–35	77		
	21	36–279	73		
Family Pongidae					
Pongo pygmaeus (orangutan)	1	Mature	60		Jenness (1974)
Pan satyrus (chimpanzee)	na	Mature	70		Jenness (1974)
Order Lagomorpha					
Oryctolagus cuniculus (domestic rabbit)	12	Mature	21		Jenness (1974)
	na	Mature	18		Widdowson (1984)

TABLE VII—*continued*

Mammalian species and taxonomic position	n[a]	Days postpartum	Carbohydrate content[b] (g/liter)	Measured lactose[c] (mmol/liter)	Reference
New Zealand White and a black crossbred	3	7, 13, 19		59	Holt and Jenness (1984)
Lepus timidus (varying hare)	na	Mature	9		Jenness (1974)
Lepus townsendii (whitetail jackrabbit)	5	Mature	17		Jenness (1974)
Sylvilagus floridanus (eastern cottontail)	2	Mature	10		Jenness (1974)
	4	Mature	27 ± 5		Anderson *et al.* (1975)
Order Rodentia					
Suborder Hystricomorpha					
Family Echimyidae					
Thrichomys apereoides	16	7	46		Myerson-McCormick *et al.* (1990)
	16	14	43		
	16	21	26		
	16	28	29		
Family Capromyidae					
Myocastor coypus (nutria or coypu)	9	Mature	6		Jenness (1974)
Family Dasyproctidae					
Myoprocta pratti (acouchi)	2	Mature	18		Jenness (1974)
Family Hystricidae or Erethizontidae (porcupine)	na	Mature	18		Jenness (1974)

TABLE VII—*continued*

Mammalian species and taxonomic position	n[a]	Days postpartum	Carbohydrate content[b] (g/liter)	Measured lactose[c] (mmol/liter)	Reference
Family Chinchillidae					
Chinchilla chinchilla chinchilla (formerly *C. brevicaudata*)	20	0–1	16		Volcani *et al.* (1973)
		3–9	17		
Lagostomus maximus (plains viscacha)	8		18 ± 0.5		Goode *et al.* (1981)
Family Caviidae					
Cavia porcellus (guinea pig)	na	Mature	30		Jenness (1974)
	5	2–8	50		Mepham and Beck (1973)
		13–14	37		
	10	1	58		Anderson and Chavis (1986)
		2–5	48		
		13	33		
		21	5		
Suborder Sciuromorpha					
Family Sciuridae					
Sciurus carolinensis (eastern gray squirrel)	2	Mature	37		Jenness (1974)
Family Castoridae					
Castor fiber (beaver)	7	Mature	26		Jenness (1974)
Suborder Myomorpha					
Family Cricetidae					
Peromyscus eremicus (cactus mouse)	2	Mature	17		Jenness (1974)
Peromyscus californicus (California mouse)	1	Mature	15		Jenness (1974)

TABLE VII—*continued*

Mammalian species and taxonomic position	n^a	Days postpartum	Carbohydrate content[b] (g/liter)	Measured lactose[c] (mmol/liter)	Reference
Peromyscus leucopus (white-footed mouse)	2	Mature	18		Jenness (1974)
Peromyscus polionotus (oldfield mouse)	1	Mature	20		Jenness (1974)
Peromyscus crinitus (canyon mouse)	2	Mature	20		Jenness (1974)
Peromyscus floridanus (Florida mouse)	1	Mature	15		Jenness (1974)
Peromyscus melanophrys	1	Mature	24		Jenness (1974)
Peromyscus maniculatus bairdii (deer mouse)	2	Mature	17		Jenness (1974)
Peromyscus maniculatus gracilis (deer mouse)	1	Mature	20		Jenness (1974)
Mesocricetus auratus (golden hamster)	6	Mature	49		Jenness (1974)
Family Muridae					
Rattus norvegicus (Norway rat)	8	Mature	26		Jenness (1974)
Sprague–Dawley	24	19	(40)[e]	112	Holt and Jenness (1984)
	112	0–4	25		Keen *et al.* (1981)
		5–9	30		
		10–14	37		
		15–19	29		
		20–24	25		
		25–28	12		
	4	14	34 ± 4	(94 ± 11)[e]	Warman and Rasmussen (1983)
	10	9	28 ± 6	(78 ± 16)[e]	Kornbrust *et al.* (1986)
		20	28 ± 20	(78 ± 56)[e]	
	na	Mature	38		Widdowson (1984)

TABLE VII—*continued*

Mammalian species and taxonomic position	n[a]	Days postpartum	Carbohydrate content[b] (g/liter)	Measured lactose[c] (mmol/liter)	Reference
Wistar albino	38	0–5	25 ± 1	(70 ± 3)[e]	Nicholas and Hartmann (1991)
	5	10	32 ± 3	(89 ± 8)[e]	
	8	15–20	42 ± 3	(117 ± 8)[e]	
	11	Mature	42 ± 9		Treadway and Lederman (1986)
	6	0–9	(24)[e]	68	Chalk and Bailey (1979)
		14	(33)[e]	92	
		17–22	(40)[e]	110	
Wistar	5–8	10 hr	34 ± 5	(94 ± 14)[e]	Grigor *et al.* (1989)
		22	38 ± 2	(105 ± 6)[e]	
Mus musculus domesticus (house mouse)	5	Mature	30		Jenness (1974)
CBA/H/Orl	1	6,9	15,15	(42)[e]	Ragueneau (1987)
NZB/Orl	1	6,9	16,19	(44,53)[e]	
XLII/Orl	1	6,9	13,14	(36,39)[e]	
Balb/c/By	1	6,9	13,13	(36)[e]	
C57BL/6/By	1	6,9	17,22	(44,61)[e]	
SHN	9	12	24	(67)[e]	Nagasawa *et al.* (1989)
SLN	12	12	21	(58)[e]	
C3H/He	15	12	17	(47)[e]	
GR/A	11	12	21	(58)[e]	
Balb/c	4	1–5	23 ± 10		Baverstock *et al.* (1976)
	4	6–10	26 ± 10		
	2	11–15	19 ± 4		

TABLE VII—*continued*

Mammalian species and taxonomic position	n^a	Days postpartum	Carbohydrate content[b] (g/liter)	Measured lactose[c] (mmol/liter)	Reference
Notomys alexis	9	1–7	26 ± 9		Baverstock *et al.* (1976)
	9	8–14	26 ± 9		
	10	15–21	23 ± 6		
	4	22–28	23 ± 4		
Notomys cervinus	5	8–14	23 ± 4		
	5	15–21	28 ± 4		
Notomys mitchellii	2	1–7	26 ± 13		
	3	8–14	27 ± 10		
Notomys pseudomys australis (Australian hopping mouse)	5	1–6	34 ± 9		
	7	7–12	36 ± 3		
	2	13–18	29 ± 1		
Phloeomys cumingi (cloud rat)	1	Mature	55		Jenness (1974)
Order Carnivora					
Suborder Fissipedia					
Family Canidae					
Canis familiaris (domestic dog)	4	Mature	31		Jenness (1974)
Beagle	7	0–10	42 ± 1		Lönnerdal *et al.* (1981)
	17	11–20	45 ± 2		
	20	21–30	48 ± 1		
	25	31–40	42 ± 2		

TABLE VII—*continued*

Mammalian species and taxonomic position	*n*[a]	Days postpartum	Carbohydrate content[b] (g/liter)	Measured lactose[c] (mmol/liter)	Reference
	5	41–50	44 ± 4		
	5	7–9	35 ± 1		Oftedal (1984)
	5	15–16	36 ± 1		
	5	22–23	40 ± 2		
	5	29–30	41 ± 1		
	5	36–37	38 ± 2		
	na	Mature	38		Widdowson (1984)
Canis lupus (wolf)	na	Mature	34		Jenness (1974)
Canis latrans (coyote)	na	Mature	30		Jenness (1974)
Canis aureus (jackal)	na	Mature	30		Jenness (1974)
Alopex lagopus (arctic fox)	na	Mature	54		Jenness (1974)
Vulpes vulpes (red fox)	5	Mature	46		Jenness (1974)
Nyctereutes procyonides (raccoon dog)	30	Mature	66		Jenness (1974)
Lycaon pictus (African hunting dog)	na	Mature	35		Jenness (1974)
Family Ursidae					
Ursus americanus (black bear)/wild	5	Mature	4		Jenness (1974)
Ursus arctos arctos (brown bear)/wild	na	Mature	40		Jenness (1974)
Ursus arctos horribilis (grizzly bear)/wild	7	Mature	6		Jenness (1974)
Ursus arctos horribilis (grizzly bear)/zoo	2	Mature	24		Jenness (1974)
Ursus arctos yesoensis (yezo brown bear)/zoo	2	Mature	21		Jenness (1974)

TABLE VII—*continued*

Mammalian species and taxonomic position	n[a]	Days postpartum	Carbohydrate content[b] (g/liter)	Measured lactose[c] (mmol/liter)	Reference
Ursus arctos middendorffi (kodiak bear)/zoo	1	Mature	24		Jenness (1974)
Thalarctos maritimus (polar bear)/wild	7	Mature	3		Jenness (1974)
Family Procyonidae					
Procyon lotor (raccoon)	1	Mature	48		Jenness (1974)
Subfamily Ailurinae					
Ailuropoda melanoleuca (giant panda)	1	250	3		Hudson *et al.* (1984)
Nasua nasua (coati)	2	Mature	64		Jenness (1974)
Family Mustelidae					
Mustela putorius (ferret)	2	Mature	38		Jenness (1974)
	4	5	35		Schoknecht *et al.* (1985)
	4	11	36		
	4	19	40		
	4	25	38		
	4	33	34		
	4	39	25		
Mustela vison (mink)	na	Mature	20		Jenness (1974)
Conepatus mesoleucus (hognose skunk)	1	Mature	27		Jenness (1974)
Lutra sp. (otter)	na	Mature	1		Jenness (1974)
Enhydra lutris (sea otter)	5	Mature	1.1	No lactose	Jenness *et al.* (1981)
Taxidea taxus (badger)	1	Mature	35		Jenness (1974)

TABLE VII—*continued*

Mammalian species and taxonomic position	n[a]	Days postpartum	Carbohydrate content[b] (g/liter)	Measured lactose[c] (mmol/liter)	Reference
Family Felidae					
Felis lynx (lynx)	na	Mature	45		Jenness (1974)
Felis catus (domestic cat)	4	Mature	48		Jenness (1974)
	16	0–2	36 ± 4		Keen *et al.* (1982)
	16	3–7	37 ± 7		
	15	8–14	36 ± 7		
	14	15–21	38 ± 7		
	11	22–28	34 ± 6		
	11	29–35	39 ± 5		
	9	36–42	41 ± 7		
	6	43 +	43 ± 6		
	na	Mature	37		Widdowson (1984)
Felis concolor (puma)	na	Mature	39		Jenness (1974)
Panthera pardus (leopard)	na	Mature	42		Jenness (1974)
Panthera leo (lion)	1	Mature	34		Jenness (1974)
Acinonyx jubatus (cheetah)	na	Mature	35		Jenness (1974)
Suborder Pinnipedia					
Family Otariidae					
Callorhinus ursinus (northern fur seal)	5	Mature	1		Jenness (1974)
	na	na	24	Trace	Dosako *et al.* (1983)

TABLE VII—*continued*

Mammalian species and taxonomic position	n[a]	Days postpartum	Carbohydrate content[b] (g/liter)	Measured lactose[c] (mmol/liter)	Reference
Arctocephalus tropicalis gazella (Southern fur seal)	11	0–26	0.4	0.0	Peaker and Goode (1978)
Zalophus californianus (California sea lion)	1	Mature	0.0		Jenness (1974)
Family Phocidae					
Pagophilus groenlandicus (harp seal)	1	Mature	9		Jenness (1974)
Halichoerus grypus (gray seal)	1	Mature	26		Jenness (1974)
	56	0–25	7		Baker (1990)
Lobodon carcinophagus (crabeater seal)	4	na	0.2 ± 0.07 (lactose) 15 ± 3 (hexose)	(0.5 ± 0.2)[e]	Messer *et al.* (1988)
Leptonychotes weddelli (Weddell's seal)	8	Mature	1.0		Jenness (1974)
Cystophora cristata (hooded seal)	1	Mature	0.0		Jenness (1974)
Mirounga angustirostris (northern elephant seal)	7	Mature	7		Jenness (1974)
	20	Mature	< 2.5		Riedman and Ortiz (1979)
Order Proboscidea					
Elephas maximus (Indian elephant)	7	Mature	47		Jenness (1974)
Loxodonta africana (African elephant)	30	Mature	37		Jenness (1974)
Order Sirenia					
Family Trichechidae					
Trichechus manatus latirostris (Florida manatee)	2	210, 730	6 (hexose)	Trace	Pervaiz and Brew (1986)

TABLE VII—*continued*

Mammalian species and taxonomic position	n[a]	Days postpartum	Carbohydrate content[b] (g/liter)	Measured lactose[c] (mmol/liter)	Reference
Order Tubulidentata					
Orycteropus afer (aardvark)	1	Mature	46		Jenness (1974)
(Aardvark)/captive	1	3–8		19.4	White *et al.* (1985)
		9–14		17.0	
		15–20		13.7	
		21–26		12.7	
		27–32		13.4	
Order Perissodactyla					
Family Equidae					
Equus asinus (donkey)	1	Mature	74		Jenness (1974)
	na		61		Widdowson (1984)
Equus caballus (horse)	231	Mature	62		Jenness (1974)
Arabian	1	122	(80)[e]	222	Holt and Jenness (1984)
	5	10–11	68		Oftedal *et al.* (1983)
		17	69		
		24–25	68		
		31–33	68		
		38–40	69		
		45–47	69		
		52–54	71		
	na	Mature	69		Widdowson (1984)

TABLE VII—*continued*

Mammalian species and taxonomic position	*n*[a]	Days postpartum	Carbohydrate content[b] (g/liter)	Measured lactose[c] (mmol/liter)	Reference
Equus caballus (pony)	22	15–82	66		Pagan and Hintz (1986)
Equus przewalski (Przewalski's horse)	1	Mature	61		Jenness (1974)
Equus grevyi (Grevyi's zebra)	1	Mature	58		Jenness (1974)
Equus burchelli (common zebra)	1	Mature	83		Jenness (1974)
Family Rhinocerotidae					
Diceros bicornis (black rhinoceros)	na	Mature	61		Jenness (1974)
Diceros simus (white rhinoceros)	2	Mature	67		Jenness (1974)
Order Artiodactyla					
Suborder Suiformes					
Family Suidae					
Sus scrofa (pig)	4	Mature	55		Jenness (1974)
Landrace	2	21	(60)[e]	167	Holt and Jenness (1984)
German Landrace	25	0–6 hr	32		Klobasa *et al.* (1987)
		12–24 hr	44		
		2–3	50		
		5–28	56		
	na	Mature	50		Widdowson (1984)
Meishan	10	0	20 ± 5		Zou *et al.* (1992)
		1	32 ± 5		
		7	44 ± 4		
		21	51 ± 4		

–TABLE VII—*continued*

Mammalian species and taxonomic position	n^a	Days postpartum	Carbohydrate content[b] (g/liter)	Measured lactose[c] (mmol/liter)	Reference
Yorkshire	9	0	30 ± 5		
		1	38 ± 5		
		7	47 ± 5		
		21	50 ± 5		
Family Tayassuidae					
Tayassu tajocu (peccary)	2	Mature	65		Jenness (1974)
Family Hippopotamidae					
Hippopotamus amphibius (hippopotamus)	1	Mature	43		Jenness (1974)
Suborder Tylopoda					
Family Camelidae					
Lama glama (llama)	1	Mature	60		Jenness (1974)
Camelus bactrianus (Bactrian camel)	4	Mature	51		Jenness (1974)
Camelus dromedarius (dromedary)	15	Mature	50		Jenness (1974)
	4	0	28 ± 5		Yagil and Etzion (1980)
		1	38 ± 6		
		2–21	49 ± 4		
		Dehydration	29 ± 2		
	55 +	Mature	44 ± 0.9		Sawaya *et al.* (1984)
Majaheim	81	Mature	42 ± 1		Elamin and Wilcox (1992)

TABLE VII—*continued*

Mammalian species and taxonomic position	n[a]	Days postpartum	Carbohydrate content[b] (g/liter)	Measured lactose[c] (mmol/liter)	Reference
Suborder Ruminantia					
Family Giraffidae					
Okapia johnstoni (okapi)	1	Mature	51		Jenness (1974)
Giraffa camelopardalis (giraffe)	na	Mature	34		Jenness (1974)
	na	Mature	49		Widdowson (1984)
Family Cervidae					
Dama dama (fallow deer)	1	Mature	61		Jenness (1974)
Cervus nippon (sika deer)	na	Mature	34		Jenness (1974)
Cervus elaphus (red deer)	na	Mature	26		Jenness (1974)
(Red deer)/captive	5	3–30	44		Arman *et al.* (1974)
		31–100	44		
		101–261	45		
Odocoileus virginianus (whitetail deer)	2	Mature	46		Jenness (1974)
Odocoileus hemionus (mule deer)	1	Mature	54		Jenness (1974)
Alces alces (moose)	15	Mature	30		Jenness (1974)
Rangifer tarandus (reindeer)	4	Mature	28		Jenness (1974)
	7	21–25	34		Luick *et al.* (1974)
		37–45	33		
		62–70	33		
		87–93	27		
		116–140	28		

TABLE VII—*continued*

Mammalian species and taxonomic position	n[a]	Days postpartum	Carbohydrate content[b] (g/liter)	Measured lactose[c] (mmol/liter)	Reference
Rangifer arcticus (caribou)	3	Mature	37		Jenness (1974)
Capreolus capreolus (roe deer)	na	Mature	39		Jenness (1974)
Family Antilocapridae					
Antilocapra americana (pronghorn)	1	Mature	40		Jenness (1974)
Family Bovidae					
Subfamily Antelopinae					
Litocranius walleri (gerenuk)	1	Mature	40		Jenness (1974)
Gazella granti (Grant's gazelle)	na	Mature	28		Jenness (1974)
Gazella gazella (mountain gazelle)	na	Mature	33		Jenness (1974)
Gazella thomsoni (Thomson's gazelle)	na	Mature	27		Jenness (1974)
Aepyceros melampus (impala)	na	Mature	24		Jenness (1974)
Antidorcas marsupialis (springbok)	9	1–140	40		Spála and Váhala (1989)
Subfamily Bovinae					
Tragelaphus streptsiceros (greater kudu)	1	Mature	46		Jenness (1974)
Taurotragus oryx (eland)	51	Mature	39		Jenness (1974)
Bos taurus (cow)	na	Mature	48		Jenness (1974)
	na	Mature	46		Widdowson (1984)
	3806	na	48		Macy *et al.* (1953)
Friesian	1	Bulk	46		Williams *et al.* (1976)
Holstein	26	Bulk	49 ± 6		Cerbulis and Farrell (1975)
Jersey	25	Bulk	50 ± 3		

TABLE VII—*continued*

Mammalian species and taxonomic position	n^a	Days postpartum	Carbohydrate content[b] (g/liter)	Measured lactose[c] (mmol/liter)	Reference
Guernsey	24	Bulk	47 ± 3		
Ayrshire	25	Bulk	47 ± 3		
Brown Swiss	33	Bulk	52 ± 5		
Milking Shorthorn	18	Bulk	48 ± 3		
Bos indicus (zebu)	130	Mature	49		Jenness (1974)
Bos grunniens (yak)	na	Mature	46		Jenness (1974)
Bubalus bubalis (water buffalo)	na	Mature	48		Jenness (1974)
Bison bison (American buffalo)	1	Mature	51		Jenness (1974)
Subfamily Caprinae					
Oreamnos americanus (mountain goat)	1	Mature	28		Jenness (1974)
Ovibos moschatus (muskox)	1	Mature	41		Jenness (1974)
Hemitragus jemlahicus (tahr)	20	Mature	33		Jenness (1974)
Capra hircus (goat)	721	Mature	47		Macy *et al.* (1953)
	2662	Mature	41		Jenness (1974)
	na	Mature	47		Widdowson (1984)
	12	3	(50 ± 4)[e]	139 ± 10	Linzell and Peaker (1974)
Saanen	6	42	(44)[e]	123	Holt and Jenness (1984)
British Alpine	6	1–126	43.8 ± 3		Devendra (1972)
Anglo-Nubian	5	1–126	40.5 ± 6		
Red Sokoto	2	Early	47 ± 2		Mba *et al.* (1975)
		Mid	48 ± 1		

TABLE VII—*continued*

Mammalian species and taxonomic position	n[a]	Days postpartum	Carbohydrate content[b] (g/liter)	Measured lactose[c] (mmol/liter)	Reference
West African Dwarf	4	Early	57 ± 1		
		Mid	52 ± 2		
		Late	58 ± 1		
Saanen	3	Early	44.3 ± 1		
		Mid	45.6 ± 2		
		Late	44.3 ± 0.3		
Norwegian	70	0–90	48		Brendehaug and Abrahamsen (1986)
		120–240	43		
West African Dwarf	6	1–4	49		Akinsoyinu *et al.* (1977)
		14–126	63		
Ovis aries (sheep)	8	Mature	48		Jenness (1974)
Finn–Dorset	2	28,42	(50)[e]	147	Holt and Jenness (1984)
Suffolk × Clun Forest	6	2	55		Williams et al. (1976)
		21	47		
		49	44		
Dorset	41	14	53		Wohlt *et al.* (1984)
		28	55		
		42	56		
		56	52		
Ovis canadensis (bighorn sheep)	2	Mature	34		Jenness (1974)

TABLE VII—*continued*

Mammalian species and taxonomic position	n[a]	Days postpartum	Carbohydrate content[b] (g/liter)	Measured lactose[c] (mmol/liter)	Reference
Order Cetacea					
Family Monodontidae					
Delphinapterus leucas (beluga)	1	Mature	7		Jenness (1974)
Family Ziphiidae					
Mesoplodon stejnegeri (Stejneger's beaked whale)	1	20–40	0.0	(0.0)[e]	Ullrey *et al.* (1984)
Family Delphinidae					
Tursiops truncatus (Atlantic bottlenose dolphin)	1	Mature	11		Jenness (1974)
	4	168–210		61	Pervaiz and Brew (1986)
Stenella plagiodon (spotted dolphin)	1	Mature	6		Jenness (1974)
Stenella graffmani (spotted porpoise)	8	Mature	11		Jenness (1974)
Stenella microps (spinner porpoise)	1	Mature	10		Jenness (1974)
Family Phocoenidae					
Phocoena phocoena (Atlantic harbor porpoise)	1	Mature	13		Jenness (1974)
Suborder Mysticeti					
Family Balaenopteridae					
Balaenoptera musculus (blue whale)	2	Mature	13		Jenness (1974)

TABLE VII—*continued*

Mammalian species and taxonomic position	n[a]	Days postpartum	Carbohydrate content[b] (g/liter)	Measured lactose[c] (mmol/liter)	Reference
Balaenoptera physalus (finback whale)	2	Mature	3		Jenness (1974)
Megaptera novaeangliae (humpback whale)	8	Mature	11		Jenness (1974)

[a]For all Jenness (1974) references, n is the no. of samples; for all others, n is the no. of independent animals.
[b]Measured as total hexose.
[c]Analyzed by lactose-specific method.
[d]Mean ± SD
[e]Calculated using the molecular weight for lactose monohydrate, 360.3 g/mol.

TABLE VIII
Glucose Content of Animal Milk (mmol/liter)

Animal	n^a	Days of lactation						Weaning	Reference
		Antepartum	1	2	3	4–14	15+		
Cow									
Friesian	2						0.1		Faulkner *et al.* (1981)
Jersey	2						0.1		
Friesian	12						0.8 ± 0.3^b		Lück and Botha (1982);
Sp.	188						0.22 ± 0.03		Marschke and Kitchen (1984)
Goat									
British Saanen	5	0.03	0.2	0.08		0.12	0.12	0.025	Faulkner *et al.* (1982)
British Saanen	2						0.17		Faulkner *et al.* (1981)
Toggenburg	2						0.22		
Windsor	2						0.17		
Saanen × Windsor	2						0.24		
Sp.	6						0.16 ± 0.02		Faulkner (1985)
Rabbit									
New Zealand White	3						0.35		Faulkner *et al.* (1981)
Rat									
Wistar	5–8					0.8 ± 0.2			Grigor *et al.* (1989)
Sp.	13					0.2 ± 0.01			Faulkner *et al.* (1981)
	12					0.13 ± 0.01			

TABLE VIII—*continued*

Animal	n^a	Days of lactation						Weaning	Reference
		Antepartum	1	2	3	4–14	15+		
Sheep									
Merino	2						0.24		Faulkner *et al.* (1981)
Finn	2						0.06		
Clun	2						0.15		
Dalesbred	1						0.03		

[a] n, No. of subjects.
[b] Mean ± SD.

B. Galactose (Table IX)

Galactose levels in milk of goats, sheep, and cows are similar, and there do not appear to be species differences (Faulkner *et al.,* 1981). In goats, galactose concentrations are higher prepartum (0.45–0.50 mmol/liter) and at term (0.40 mmol/liter) than at mid-lactation (0.10 mmol/liter) or at cessation of lactation (0.03 mmol/liter) (Faulkner *et al.,* 1982).

C. Oligosaccharides (Table X)

The concentration of oligosaccharides varies widely across the milks of different species. For example, the milk of the pinnipeds contains far less carbohydrate than that of terrestrial animals (Oftedal *et al.,* 1987); however, much of this carbohydrate is in the form of oligosaccharides. The milk of both Otariidae (eared seals) and Phocidae (earless seals), for instance, contains only traces of lactose (Messer *et al.,* 1988); the carbohydrate is mainly in the form of oligosaccharides which contain galactose, *N*-acetyglucosamine, fucose, glucose, and sialic acid (*N*-acetylneuraminic acid, NANA) (Messer *et al.,* 1988). Manatees (Pervaiz and Brew, 1986) and some cetaceans have similar patterns, with a few exceptions [e.g., milk of the bottlenose dolphin (Pervaiz and Brew, 1986) has a relatively high carbohydrate content, mostly in the form of lactose]. Because of their oligosaccharide content, the low-lactose milks may still contain appreciable galactose, e.g., 7.8 mmol/liter for the manatee (Pervaiz and Brew, 1986) and 47 mmol/liter for the crabeater seal (Messer *et al.,* 1988).

The milks of marsupials (Green *et al.,* 1987; Messer *et al.,* 1987) and monotremes (egg-laying mammals) (Messer *et al.,* 1983; Messer and Kerry, 1973) are also low in lactose and contain appreciable levels of oligosaccharides. In marsupials, the oligosaccharide levels increase over the course of lactation from a low of 20–40 g/liter to a maximum of 80–100 g/liter. This occurs at 50 days for a northern brown bandicoot (Merchant and Libke, 1988) and American marsupial (Green *et al.,* 1991), 120 days for a common brushtail possum (Cowan, 1989) and potoroo (Crowley *et al.,* 1988), and 230 days for a rednecked wallaby (Merchant *et al.,* 1989). As the offspring leave the pouch, the carbohydrate content abruptly falls to 10–20 g/liter. The common ringtail possum shows a similar pattern, except that as the offspring leave the pouch (at 100 days) there is a significant shift to lactose synthesis (Munks *et al.,* 1991) which is maintained at approximately 50 g/liter until weaning at 200 days. This finding suggests an interesting parallel between a mature marsupial milk and milk of the eutherian (placental) species. Marsupial oligosaccharides seem to be rich in galactose, e.g., 239 mmol/liter for the eastern quoll (Messer *et al.,* 1987).

The structures of bovine colostrum oligosaccharides (Parkkinen and Finne, 1987) and of monotreme and marsupial oligosaccharides (Bradbury

TABLE IX
Galactose Content of Animal Milk (mmol/liter)

Animal	n^a	Days of lactation						Weaning	Reference
		Antepartum	1	2	3	4–14	15+		
Cow									
Friesian	1						0.19		Faulkner *et al.* (1981)
Jersey	1						0.26		
Goat									
British Saanen	5	0.45	0.4	0.3	0.15	0.1	0.1	0.03	Faulkner *et al.* (1982)
British Saanen	1						0.03		Faulkner *et al.* (1981)
Toggenburg	1						0.00		
Windsor	1						0.01		
Saanen × Windsor	1						0.03		
Sheep									
Merino	1						0.06		Faulkner *et al.* (1981)
Finn	1						0.07		
Clun	1						0.05		

[a] n, No. of subjects.

TABLE X
Animal Milk Oligosaccharides

Oligosaccharide		Reference
Bovine colostrum		
Sialyllactose		Kuhn and Gauhe (1965)
NANA α(2→3) Gal β(1→4) Glc	150 μmol/liter	
NGNA α(2→3) Gal β(1→4) Glc	2 μmol/liter	
NANA α(2→6) Gal β(1→4) Glc	30 μmol/liter	
Sialyllactosamine		
NANA α(2→6) Gal β(1→4) GlcNAc	70 μmol/liter	
Sialylgalactosyllactose		
NANA α(2→3) Gal β(1→3) Gal β(1→4) Glc	3 μmol/liter	
Disialyllactose		
NANA α(2→8) NANA α(2→3) Gal β(1→4) Glc	30 μmol/liter	
Sialyllactose phosphaste		
NANA α(2→6) Gal β(1→4) GlcNAc-1-PO_4	3 μmol/liter	
NANA α(2→6) Gal β(1→4) GlcNAc-6-PO_4	1 μmol/liter	
N-acetylgalactosyllactose		Urashima *et al.* (1991)
GalNAc α(1→3) Gal β(1→4) Glc		
Galactosyllactose[a]		
GalNAc α(1→3) Gal β(1→4) Glc		
Lacto-*N*-novotetraose[b]		
Gal β(1→4) GlcNAc β(1→6)↘ Gal β(1→4) Glc; Gal β(1→3)↗		

TABLE X—*continued*

Oligosaccharide		Reference
Goat		
GlcNAc β(1→6) Gal β(1→4) Glc Gal β(1→4) GlcNAc β(1→6) Gal β(1→4) Glc		Chaturvedi and Sharma (1988); Chaturvedi and Sharma (1990)
Gal β(1→4)↘ GlcNAc β(1→3) Gal β(1→4) Glc Gal β(1→3)↗		
Gal β(1→4)↘ GlcNAc β(1→6) Gal β(1→6) Glc Fuc α(1→3)↗		
Gal β(1→3) GlcNAc β(1→6) Gal β(1→4)↘ Glc Fuc α(1→3)↗		
Rat		
N-acetylneuramin lactose sulfate		Sturman *et al.* (1985)
NANA α(2→3) Gal-6-SO_4 β(1→4) Glc		
Dog, Monkey[c]		
Fucosyllactose		Grollman *et al.* (1965)
Fuc α(1→2) Gal β(1→4) Glc	0.2–1.0 mg/liter	
Monotremes (platypus, echidna)		
Fucosyllactose		Jenkins *et al.* (1984)
Fuc α(1→2) Gal β(1→4) Glc	2.9 g/liter (echidna)	
3,2′-Difucosyllactose		
Fuc α(1→3)↘ Glc Fuc α(1→2)Gal β(1→4)↗	1.3 g/liter (echidna) 10 g/liter (platypus)	

TABLE X—*continued*

Oligosaccharide	Reference
Marsupials	
Galactosyllactose, polygalactosyllactose	Collins *et al.* (1981); Messer *et al.* (1980)
[Gal β(1→3)]$_{1-5}$ Gal β(1→4) Glc	
Lacto-*N*-novotetraose	Messer *et al.* (1982)
GlcNAc β(1→6)↘ Gal β(1→4) Glc; Gal β(1→3)↗	
Gal β(1→4) GlcNAc β(1→6)↘ Gal β(1→4) Glc; Gal β(1→3)↗	
Gal β(1→4) GlcNAc β(1→6)↘ Gal β(1→4) Glc; Gal β(1→3) Gal β(1→3)↗	

Note. NANA, *N*-acetylneuraminic acid; NGNA, *N*-glycolylneuraminic acid; Gal, galactose; Glc, glucose; Fuc, fucose; GlcNAc, *N*-acetylglucosamine.

[a]Also found in bovine milk.

[b]Also found in horse colostrum; marsupial milk.

[c]Not found in cow, sheep, pig, rabbit, guinea pig, rat, and mouse.

et al., 1983; Collins *et al.*, 1981; Messer *et al.*, 1982, 1980) have been well defined; only a few other nonhuman milk oligosaccharides have been defined.

IX. Summary

The milks of most eutherian terrestrial species contain lactose as the principal carbohydrate. The lactose levels increase as lactation is established and tight junctions form in the mammary epithelium, limiting the movement of materials by the paracellular pathway. Lactose concentrations in mature human milk are quite stable showing little or no change in response to a variety of environmental or dietary challenges and little variation across genetically distinct groups. Conditions which weaken the mammary epithelial tight junctions (e.g., infection, weaning, ovulation) result in lower milk lactose levels. Human milk contains appreciable amounts of oligosaccharide and somewhat less lactose than had been previously reported. Common methods of lactose measurement fail to distinguish oligosaccharides from lactose and thus yielded false high values for measured lactose. Our best estimate for normal mature human milk lactose levels is 185 mmol/liter (6.7 g/dl). Oligosaccharide levels are approximately 12–14 g/liter (1.2–1.4 g/dl), making this the third largest solid component of human milk. In most eutherian species levels of glucose and galactose are quite low.

X. Speculation on Functions of Lactose

Small contributions to the survival of offspring by a milk component would be expected to result in strong genetic pressure toward the inclusion of this material in milk. Lactose is found in the milk of most mammals as the major source of carbohydrate; like many of the constituents of milk, it could serve several functions simultaneously. The first consideration is osmolarity: most carbohydrate is transported through mammalian fluids in the form of monosaccharide, mainly glucose, whereas the milk of humans and many other mammals contains appreciable carbohydrate in the form of disaccharide and oligosaccharides. A disaccharide has half the osmolarity as two equivalent monosaccharides, and thus would be less likely to cause postprandial osmotic stress in infants receiving large amounts of dietary calories in the form of carbohydrates. The digestion of disaccharides into monosaccharides is accompanied by the almost simultaneous absorption of the monosaccharides across the microvillus membrane, thus maintaining low osmolarity in the lumen of the intestine.

A second consideration is the unique structure of lactose, i.e., galactose $\beta(1\rightarrow4)$ glucose. Most carbohydrate macronutrients are glucans containing glucose $\alpha(1 \rightarrow 4)$ glucose. As lactose is found almost exclusively in milk, the presence of this particular linkage may preclude its digestion by large numbers of microbes commonly found in the environment that, in the presence of simple glucose polymers, might more readily infect either the lactating breast or the infant's gastrointestinal tract. Furthermore, the presence of large amounts of lactose in conjunction with traces of other specific carbohydrates may favor colonization of the infant intestine by organisms more able to split lactose. This could result in a symbiosis in which favorable microflora are established which compete with and exclude many potential pathogens.

The galactose, once digested and absorbed by the infant, can be converted to glucose by epimerization and used for energy. However, we speculate that preformed galactose per se may be of value to the infant. Milks which are low in lactose and other carbohydrates (such as those of cetaceans and pinnipeds) and marsupial milk (which is initially quite low in lactose) contain appreciable galactose, mostly derived from oligosaccharides. Most terrestrial eutherian milk contains a significant amount of galactose in the form of lactose. The presence of galactose in milk could be related to some unique requirement common to all young, growing mammals. For example, most young mammals are undergoing a period of rapid brain development during the nursing period, and myelination, which requires large amounts of galactosylceramide (galactocerebrosides) and other galactolipids, is a major component of this growth. It has been assumed that the liver is capable of providing all of the galactose required for synthesis of galactolipids through the enzymatic isomerization of glucose. However, many such adult liver functions are underdeveloped in the young. One hypothesis on the role of milk galactose is that it may ensure that galactose levels in the infant do not become limiting to galactosylceramide (galactocerebroside) production, thereby limiting optimal myelination and brain development. Brain growth and development during this period of life is known to be vulnerable to many types of nutritional deprivation, including deprivation of nutrients which are not essential in the diet of the adult (Newburg and Fillios, 1979, 1982; Newburg *et al.*, 1975).

Thus, we speculate that the galactose of milk could play a unique role in providing the requirements of the rapidly developing infant brain. This hypothesis would be strengthened should galactose prove to be a universal component of milk, but even if this proves to be true, definitive proof of such a requirement must be sought experimentally. A requirement for dietary galactose in young mammals, if real, could have important ramifications in the formulation of artificial diets for infants.

Acknowledgment

Supported in part by NIH Grant HD 13021.

References

Abdel Kader, M. M., Abdel Aziz, M. T., Bahgat, R., Hefnawi, F., Fawzi, G., and Badraoui, M. H. H. (1976). Effect of some progestational steroids on lactation in Egyptian women. II. Chemical composition of milk during the first year of lactation. *J. Biosocial Sci.* **8,** 49–51.

Akinsoyinu, A. O., Mba, A. U., and Olubajo, F. O. (1977). Studies on milk yield and composition of the West African dwarf goat in Nigeria. *J. Dairy Res.* **44,** 57–62.

Allen, J. C., Keller, R. P., Archer, P., and Neville, M. C. (1991). Studies in human lactation: Milk composition and daily secretion rates of macronutrients in the first year of lactation. *Am. J. Clin. Nutr.* **54,** 69–80.

Anderson, D. M., Williams, F. H., Merkatz, R. B., Schulman, P. K., Kerr, D. S., and Pittard, W. B., III (1983). Length of gestation and nutritional composition of human milk. *Am. J. Clin. Nutr.* **37,** 810–814.

Anderson, G. H. (1984). The effect of prematurity on milk composition and its physiological basis. *Fed. Proc.* **43,** 2438–2442.

Anderson, G. H., Atkinson, S. A., and Bryan, M. H. (1981). Energy and macronutrient content of human milk during early lactation from mothers giving birth prematurely and at term. *Am. J. Clin. Nutr.* **34,** 258–265.

Anderson, R. R., and Chavis, D. D. (1986). Changes in macroingredients of guinea pig milk through lactation. *J. Dairy Sci.* **69,** 2268–2277.

Anderson, R. R., Sadler, K. C., Knauer, M. W., Wippler, J. P., and Marshall, R. T. (1975). Composition of cottontail rabbit milk from stomachs of young and directly from gland. *J. Dairy Sci.* **58,** 1449–1452.

Andersson, B., Porras, O., Hanson, L. A., Lagergård, T.,and Svanborg-Edén, C. (1986). Inhibition of attachment of *Streptococcus pneumoniae* and *Haemophilus influenzae* by human milk and receptor oligosaccharides. *J. Infect. Dis.* **153,** 232–237.

Arman, P., Kay, R. N. B., Goodall, E. D., and Sharman, G. A. M. (1974). The composition and yield of milk from captive red deer (*Cervus elaphus* L.). *J. Reprod. Fertil.* **37,** 67–84.

Arthur, P. G., Smith, M., and Hartmann, P. E. (1989). Milk lactose, citrate, and glucose as markers of lactogenesis in normal and diabetic women. *J. Pediatr. Gastroenterol. Nutr.* **9,** 488–496.

Arthur, P. G., Kent, J. C., and Hartmann, P. E. (1991). Metabolites of lactose synthesis in milk from women during established lactation. *J. Pediatr. Gastroenterol. Nutr.* **13,** 260–266.

Bahl, R. K. (1971). An enzymatic method for the determination of skimmed milk powder in raw sausages. *Analyst* **96,** 88–92.

Bahl, R. K. (1972). An enzymatic method for the determination of lactose in milk including human milk. *Analyst* **97,** 559–561.

Baker, J. R. (1990). Grey seal (*Halichoerus grypus*) milk composition and its variation over lactation. *Br. Vet. J.* **146,** 233–238.

Bartoletti, F. (1633). "Methodus in Dyspnoeam; seu, de Respirationibus." N. Tebaldini, Bononiae (Bologna).

Baverstock, P. R., Spencer, L., and Pollard, C. (1976). Water balance of small lactating rodents. II. Concentration and composition of milk of females on *ad libitum* and restricted water intakes. *Comp. Biochem. Physiol. A* **53,** 47–52.

Bergmeyer, H.-U., and Bernt, E. (1963). Determination with glucose oxidase and peroxidase. *In* "Methods of Enzymatic Analysis" (H. U. Bergmeyer, ed.), Vol. 1, pp. 123–130. Academic Press, New York.

Bergmeyer, H. U., and Bernt, E. (1989). Determination with glucose oxidase and peroxidase. *In* "Methods of Enzymatic Analysis" (H. U. Bergmeyer, ed.), Vol. 3, pp. 1205–1215. Academic Press, New York.

Berner, G. (1970). Zucherabbau wahrend der Camenbert-Reifung. Enzymatische Bestimaung von D-Lactose, D-Glucose und D-Galactose. *Milchwissenschaft* **25,** 275–280.

Boediman, D., Ismail, D., Iman, S., Ismangoen, and Ismadi, S. D. (1979). Composition of breast milk beyond one year. *Trop. Pediatr. Environ. Child Health* **25,** 107–110.

Bradbury, J. H., Collins, J. G., Jenkins, G. A., Trifonoff, E., and Messer, M. (1983). ^{13}C-n.m.r. study of the structures of two branched oligosaccharides from marsupial milk. *Carbohydr. Res.* **122,** 327–331.

Brendehaug, J., and Abrahamsen, R. K. (1986). Chemical composition of milk from a herd of Norwegian goats. *J. Dairy Res.* **53,** 211–221.

Brown, K. H., Akhtar, N. A., Robertson, A. D., and Ahmed, M. G. (1986). Lactational capacity of marginally nourished mothers: Relationships between maternal nutritional status and quantity and proximate composition of milk. *Pediatrics* **78,** 909–919.

Buss, D. H. (1968). Gross composition and variation of the components of baboon milk during natural lactation. *J. Nutr.* **96,** 421–426.

Buss, D. H., and Cooper, R. W. (1970). Composition of milk from talapoin monkeys. *Folia Primatol.* **13,** 196–206.

Buss, D. H., Cooper, R. W., and Wallen, K. (1976). Composition of lemur milk. *Folia Primatol.* **26,** 301–305.

Butte, N. F., and Calloway, D. H. (1981). Evaluation of lactational performance of Navajo women. *Am. J. Clin. Nutr.* **34,** 2210–2215.

Butte, N. F., Garza, C., Smith, E. O., and Nichols, B. L. (1984). Human milk intake and growth in exclusively breast-fed infants. *J. Pediatr.* **104,** 187–195.

Butte, N. F., Garza, C., Burr, R., Goldman, A. S., Kennedy, K., and Kitzmiller, J. L. (1987). Milk composition of insulin-dependent diabetic women. *J. Pediatr. Gastroenterol. Nutr.* **6,** 936–941.

Butte, N. F., Garza, C., and Smith, E. O. (1988). Variability of macronutrient concentrations in human milk. *Eur. J. Clin. Nutr.* **42,** 345–349.

Casey, C. E., Neifert, M. R., Seacat, J. M., and Neville, M. C. (1986). Nutrient intake by breast-fed infants during the first five days after birth. *Am. J. Dis. Child.* **140,** 933–936.

Cerbulis, J., and Farrell, H. M., Jr. (1975). Composition of milks of dairy cattle. I. Protein, lactose, and fat contents and distribution of protein fraction. *J. Dairy Sci.* **58,** 817–827.

Chalk, P. A., and Bailey, E. (1979). Changes in the yield, and carbohydrate, lipid and protein content of milk during lactation in the rat. *J. Dev. Physiol.* **1,** 61–79.

Chaturvedi, P., and Sharma, C. B. (1988). Goat milk oligosaccharides: Purification and characterization by high performance liquid chromatography and high-field proton n.m.r. spectroscopy. *Biochim. Biophys. Acta* **967,** 115–121.

Chaturvedi, P., and Sharma, C. B. (1990). Purification, by high-performance liquid chromatography, and characterization, by high-field ^{1}H-n.m.r. spectroscopy, of two fucose-containing pentasaccharides of goat's milk. *Carbohydr. Res.* **203,** 91–101.

Collins, J. G., Bradbury, J. H., Trifonoff, E., and Messer, M. (1981). Structures of four new oligosaccharides from marsupial milk, determined mainly by ^{13}C-n.m.r. spectroscopy. *Carbohydr. Res.* **92,** 136–140.

Conetta, A., Stookey, L., and Zehnder, H. (1970). An automated system for the determination of milkfat, protein, and lactose in milk. *Adv. Autom. Anal.* **2,** 81–85.

Conner, A. E. (1979). Elevated levels of sodium and chloride in milk from mastitic breast. *Pediatrics* **63,** 910–911.

Coppa, G. V., Gabrielli, O., Pierani, P., Catassi, C., Carlucci, A., and Giorgi, P. L. (1993). Changes in carbohydrate composition in human milk over 4 months of lactation. *Pediatrics* **91,** 637–641.

Cowan, P. E. (1989). Changes in milk composition during lactation in the common brushtail possum, *Trichosurus vulpecula* (Marsupialia:Phalangeridae). *Reprod. Fertil. Dev.* **1,** 325–335.

Cravioto, A., Tello, A., Villafán, H., Ruiz, J., del Vedovo, S., and Neeser, J.-R. (1991). Inhibition of localized adhesion of enteropathogenic *Escherichia coli* to HEp-2 cells by immunoglobulin and oligosaccharide fractions of human colostrum and breast milk. *J. Infect. Dis.* **163,** 1247–1255.

Crowley, H. M., Woodward, D. R., and Rose, R. W. (1988). Changes in milk composition during lactation in the potoroo, *Potorous tridactylus* (Marsupialia:Potoroniae). *Aust. J. Biol. Sci.* **41,** 289–296.

Dagnelie, P. C., van Staveren, W. A., Roos, A. H., Tuinstra, L. G. M., and Burema, J. (1992). Nutrients and contaminants in human milk from mothers on macrobiotic and omnivorous diets. *Eur. J. Clin. Nutr.* **46,** 355–366.

Dawodu, A. H., Osibanjo, O., and Damole, I. O. (1990). Nutrient composition of milk produced by mothers of preterm infants in Nigeria. *East Afr. Med. J.* **67,** 873–877.

Deb, A. K., and Cama, H. R. (1962). Studies on human lactation. Dietary nitrogen utilization during lactation, and distribution of nitrogen in mother's milk. *Br. J. Nutr.* **16,** 65–73.

Devendra, C. (1972). The composition of milk of British Alpine and Anglo-Nubian goats imported into Trinidad. *J. Dairy Res.* **39,** 381–385.

Dewey, K. G., and Lönnerdal, B. (1983). Milk and nutrient intake of breast-fed infants from 1 to 6 months: Relation to growth and fatness. *J. Pediatr. Gastroenterol. Nutr.* **2,** 497–506.

Dewey, K. G., Finley, D. A., and Lönnerdal, B. (1984). Breast milk volume and composition during late lactation (7–20 months). *J. Pediatr. Gastroenterol. Nutr.* **3,** 713–720.

Dosako, S., Taneya, S., Kimura, T., Ohmori, T., Daikoku, H., Suzuki, N., Sawa, J., Kano, K., and Katayama, S. (1983). Milk of northern fur seal: Composition, especially carbohydrate and protein. *J. Dairy Sci.* **66,** 2076–2083.

Dryden, G. L., and Anderson, R. R. (1978). Milk composition and its relation to growth rate in the musk shrew, *Suncus murinus. Comp. Biochem. Physiol. A* **60,** 213–216.

D'Souza, F., and Martin, R. D. (1974). Maternal behaviour and the effect of stress in tree shrews. *Nature* **251,** 309–311.

Dua, V. K., Goso, K., Dube, V. E., and Bush, C. A. (1985). Characterization of lacto-N-hexaose and two fucosylated derivatives from human milk by high-performance liquid chromatography and proton NMR spectroscopy. *J. Chromatogr.* **328,** 259–269.

Dubois, M., Gilles, K. A., Hamilton, J. K., Rebers, P. A., and Smith, F. (1956). Colorimetric method for determination of sugars and related substances. *Anal. Chem.* **28,** 350–356.

Egge, H., Dell, A., von Nicholai, H, and Peter-Katalinic, J. (1982). Fucose containing oligosaccharides from human milk. *Hoppe–Seyler's Z. Physiol. Chem.* **363,** 1026.

Egge, H., Dell, A., and von Nicolai, H. (1983). Fucose containing oligosaccharides from human milk. I. Separation and identification of new constituents. *Arch. Biochem. Biophys.* **224,** 235–253.

Elamin, F. M., and Wilcox, C. J. (1992). Milk composition of Majaheim camels. *J. Dairy Sci.* **75,** 3155–3157.

Ereman, R., Lönnerdal, B., and Dewey, K. G. (1988). Relationship of breast milk composition to milk volume during initiation and termination of lactation. *FASEB J.* **2,** A647.

Faulkner, A. (1985). Glucose availability and lactose synthesis in the goat. *Biochem. Soc. Trans.* **13,** 496–497.

Faulkner, A., Chaiyabutr, N., Peaker, M., Carrick, D. T., and Kuhn, N. J. (1981). Metabolic significance of milk glucose. *J. Dairy Res.* **48,** 51–56.

Faulkner, A., Blatchford, D. R., White, J. M., and Peaker, M. (1982). Changes in the concentrations of metabolites in milk at the onset and cessation of lactation in the goat. *J. Dairy Res.* **49,** 399–405.

Ferris, A. M., Dotts, M. A., Clark, R. M., Ezrin, M., and Jensen, R. G. (1988). Macronutrients in human milk at 2, 12, and 16 weeks postpartum. *J. Am. Diet. Assoc.* **88,** 694–697.

Finley, D. A., Lönnerdal, B., Dewey, K. G., and Grivetti, L. E. (1985). Inorganic constituents of breast milk from vegetarian and nonvegetarian women: Relationships with each other and with organic constituents. *J. Nutr.* **115,** 772–781.

Folin, O., and Wu, H. (1920). A system of blood analysis. I. A simplified and improved method for determination of sugar. *J. Biol. Chem.* **41,** 367–373.

Forbes, G. B., Barton, L. D., Nicholas, D. L., and Cook, D. A. (1988). Composition of milk produced by a mother with galactosemia. *J. Pediatr.* **113,** 90–91.

Gibbs, J. H., Fisher, C., Bhattacharya, S., Goddard, P., and Baum, J. D. (1977). Drip breast milk: Its composition, collection and pasteurization. *Early Hum. Dev.* **1/3,** 227–245.

Gindler, J., Nwankwo, M. U., Omene, J. A., Glew, R. H., and Roberts, I. M. (1985). The quality of milk in a 65 year-old Nigerian woman. *Nutr. Res.* **5,** 1209–1213.

Gindler, J., Nwankwo, M. U., Omene, J. A., Roberts, I. M., LaRocca, G. M., and Glew, R. H. (1987). Breast milk composition and bile salt-stimulated lipase in well-nourished and under-nourished Nigerian mothers. *Eur. J. Pediatr.* **146,** 184–186.

Ginsburg, V., Zopf, D. A., Yamashita, K., and Kobata, A. (1976). Oligosaccharides of human milk. Isolation of a new pentasaccharide, lacto-*N*-fucopentaose V. *Arch. Biochem. Biophys.* **175,** 565–568.

Goode, J. A., Peaker, M., and Weir, B. J. (1981). Milk composition in the plains viscacha (*Lagostomus maximus*). *J. Reprod. Fertil.* **62,** 563–566.

Green, B., Merchant, J., and Newgrain, K. (1987). Milk composition in the eastern quoll, *Dasyurus viverrinus* (Marsupialia:Dasyuridae). *Aust. J. Biol. Sci.* **40,** 379–387.

Green, B., VandeBerg, J. L., and Newgrain, K. (1991). Milk composition in an American marsupial (*Monodelphis domestica*). *Comp. Biochem. Physiol. B* **99,** 663–665.

Grigor, M. R., Carrington, J. M., Arthur, P. G., and Hartmann, P. E. (1989). Lack of correlation between milk glucose concentrations and rates of milk production in the rat. *J. Dairy Res.* **56,** 37–43.

Grimmonprez, L., and Montreuil, J. (1968). Etude physico-chemicque de six nouveaux oligosaccharides isolés du lait de femme. *Bull. Soc. Chim. Biol.* **50,** 843–855.

Grollman, A. P., Hall, C. W., and Ginsburg, V. (1965). Biosynthesis of fucosyllactose and other oligosaccharides found in milk. *J. Biol. Chem.* **240,** 975–981.

Grönberg, G., Lipniunas, P., Lundgren, T., Erlansson, K., Lindh, F., and Nilsson, B. (1989). Isolation of monosialyated oligosaccharides from human milk and structural analysis of three new compounds. *Carbohydr. Res.* **191,** 261–278.

Grönberg, G., Lipniunas, P., Lundgren, T., Lindh, F., and Nilsson, B. (1992). Structural analysis of five new monosialylated oligosaccharides from human milk. *Arch. Biochem. Biophys.* **296,** 597–610.

Gross, S. J., David, R. J., Bauman, L., and Tomarelli, R. M. (1980). Nutritional composition of milk produced by mothers delivering preterm. *J. Pediatr.* **96,** 641–644.

Gross, S. J., Geller, J., and Tomarelli, R. M. (1981). Composition of breast milk from mothers of preterm infants. *Pediatrics* **68,** 490–493.

Haeuw-Fievre, S., Wieruszeski, J.-M., Plancke, Y., Michalski, J.-C., Montreuil, J., and Strecker, G. (1993). Primary structure of human milk octa-, dodeca- and tridecasaccharides determined by a combination of ^{1}H-NMR spectroscopy and fast-atom-bombardment mass spectrometry. Evidence for new core structure, the *para*-lacto-*N*-octaose. *Eur. J. Biochem.* **215,** 361–371.

Hall, B. (1979). Uniformity of human milk. *Am. J. Clin. Nutr.* **32,** 304–312.

Hartmann, P. E., and Kulski, J. K. (1978). Changes in the composition of the mammary secretion of women after abrupt termination of breast feeding. *J. Physiol. (London)* **275,** 1–11.

Hartmann, P. E., and Prosser, C. G. (1982). Acute changes in the composition of milk during the ovulatory menstrual cycle in lactating women. *J. Physiol. (London)* **324,** 21–30.

Harzer, G., Dieterich, I., and Haug, M. (1984). Effects of diet on the composition of human milk. *Ann. Nutr. Metab.* **28,** 231–239.

Harzer, G., Haug, M., and Bindels, J. G. (1986). Biochemistry of human milk in early lactation. *Z. Ernährungswiss,* **25,** 77–90.

Holt, C., and Jenness, R. (1984). Interrelationships of constituents and partition of salts in milk samples from eight species. *Comp. Biochem. Physiol. A* **77,** 275–282.

Hudson, G. J., John, P. M. V., Bailey, B. S., and Southgate, D. A. T. (1976). The automated determination of carbohydrate. Development of a method for available carbohydrates and its application to foodstuffs. *J. Sci. Food Agric.* **27,** 681–687.

Hudson, G. J., Bailey, P. A., John, P. M. V., Monsalve, L., Garcia del Campo, A.-L., Taylor, D. C., and Kay, J. D. S. (1984). Composition of milk from *Ailuropoda melanoleuca,* the giant panda. *Vet. Rec.* **115,** 252.

Hultman, E. (1959). Rapid specific method for determination of aldosaccharides in body fluids. *Nature* **183,** 108–109.

Hytten, F. E. (1954). Clinical and chemical studies in human lactation. *Br. Med. J.* **1,** 175–182.

Jansen, G., Muskiet, F. A. J., Schierbeek, H., Berger, R., and Slik, W. v. d. (1986). Capillary gas chromatographic profiling of urinary, plasma and erythrocyte sugars and polyols as their trimethylsilyl derivatives, preceded by a simple and rapid prepurification method. *Clin. Chim. Acta* **157,** 277–294.

Jelliffe, D. B., and Jelliffe, E. F. P. (1978). The volume and composition of human milk in poorly nourished communities. A review. *Am. J. Clin. Nutr.* **31,** 492–515.

Jenkins, G. A., Bradbury, J. H., Messer, M., and Trifonoff, E. (1984). Determination of the structures of fucosyl-lactose and difucosyl-lactose from the milk of monotremes, using ^{13}C-n.m.r. spectroscopy. *Carbohydr. Res.* **126,** 157–161.

Jenness, R. (1974). The composition of milk. *In* "Lactation: A Comprehensive Treatise" (B. L. Larson and V. R. Smith, eds.), Vol. III, pp. 3–107. Academic Press, New York.

Jenness, R. (1979). The composition of human milk. *Semin. Perinatol.* **3,** 225–239.

Jenness, R., Williams, T. D., and Mullin, R. J. (1981). Composition of milk of the sea otter (*Enhydra lutris*). *Comp. Biochem. Physiol. A* **70,** 375–379.

Jitta, J. N. S., Musoke, R. N., Bwibo, N. O., and Kioni, J. (1986). Composition of early human milk of Kenyan mothers of preterm and term infants. *East Afr. Med. J.* **63,** 693–698.

Jovanovic-Peterson, L., Fuhrmann, K., Hedden, K., Walker, L., and Peterson, C. M. (1989). Maternal milk and plasma glucose and insulin levels; Studies in normal and diabetic subjects. *J. Am. Coll. Nutr.* **8,** 125–131.

Keen, C. L., Lönnerdal, B., Clegg, M., and Hurley, L. S. (1981). Developmental changes in composition of rat milk: Trace elements, minerals, protein, carbohydrate and fat. *J. Nutr.* **111,** 226–230.

Keen, C. L., Lönnerdal, B., Clegg, M. S., Hurley, L. S., Morris, J. G., Rogers, Q. R., and Rucker, R. B. (1982). Developmental changes in composition of cats' milk: Trace elements, minerals, protein, carbohydrate and fat. *J. Nutr.* **112,** 1763–1769.

Khin-Maung-Naing, Tin-Tin-Oo, Kywe-Thein, and Nwe-New-Hlaing (1980). Study on lactation performance of Burmese mothers. *Am. J. Clin. Nutr.* **33,** 2665–2668.

Kitagawa, H., Nakada, H., Numata, Y., Kurosaka, A., Fukui, S., Funakoshi, I., Kawasaki, T., Shimada, I., Inagaki, F., and Yamashina, I. (1990). Occurrence of tetra- and pentasaccharides with the sialyl-Le^a structure in human milk. *J. Biol. Chem.* **265,** 4859–4862.

Kitagawa, H., Takaoka, M., Nakada, H., Fukui, S., Funakoshi, I., Kawasaki, T., Tate, S.-I., Inagaki, F., and Yamashina, I. (1991). Isolation and structural studies of human milk oligosaccharides that are reactive with a monoclonal antibody MSW 113. *J. Biochem.* **110,** 598–604.

Kitagawa, H., Nakada, H., Fukui, S., Funakoshi, I., Kawasaki, T., Yamashina, I., Tate, S-I., and Inagaki, F. (1993). Novel oligosaccharides with the sialyl-Le^a structure in human milk. *J. Biochem.* **114,** 504–508.

Klobasa, F., Werhahn, E., and Butler, J. E. (1987). Composition of sow milk during lactation. *J. Anim. Sci.* **64,** 1458–1466.

Kobata, A. (1972). Isolation of oligosaccharides from human milk. *In* "Methods in Enzymology" (V. Ginsburg, ed.), Vol. 28, pp. 262–271. Academic Press, New York.

Kobata, A., and Ginsburg, V. (1969). Oligosaccharides of human milk. II. Isolation and characterization of a new pentasaccharide, lacto-*N*-fucopentaose III. *J. Biol. Chem.* **244,** 5496–5502.

Kobata, A., and Ginsburg, V. (1972a). Oligosaccharides of human milk. I. Isolation and characterization of a new hexasaccharide, lacto-*N*-hexaose. *J. Biol. Chem.* **247,** 1525–1529.

Kobata, A., and Ginsburg, V. (1972b). Oligosaccharides of human milk. IV. Isolation and characterization of a new hexasaccharide, lacto-*N*-neohexaose. *Arch. Biochem. Biohys.* **150,** 273–281.

Kobata, A., Yamashita, K., and Tachibana, Y. (1978). Oligosaccharides from human milk. *In* "Methods in Enzymology" (V. Ginsburg, ed.), Vol. 50, pp. 216–220. Academic Press, New York.

Kornbrust, D., Gillis, B., Collins, B., Goehl, T., Gupta, B., and Schwetz, B. (1986). Effects of 1,1-dichloro-2,2-BIS[*p*-chlorophenyl]ethylene (DDE) on lactation in rats. *J. Toxicol. Environ. Health* **17,** 23–36.

Kuhn, R. (1959). Biochemie der Rezeptoren und Resistenzfaktoren. Von der Widerstandsfähigkeit der Lebewesen gegen Einwirkungen der Umwelt. *Naturwissenschaften* **46,** 43–50.

Kuhn, R., and Baer, H. H. (1956). Die Konstitution der Lacto-*N*-tetraose. *Chem. Ber.* **89,** 504–511.

Kuhn, R., and Brossmer, R. (1959). Über das durch Viren der Influenza-Gruppe spaltbare Trisaccharid der Milch. *Chem. Ber.* **92,** 1667–1671.

Kuhn, R., and Gauhe, A. (1958). Über die lacto-difuco-tetraose der Frauenmilch. *Justus Liebigs Ann. Chem.* **611,** 249–253.

Kuhn, R., and Gauhe, A. (1960). Über ein kristallisiertes, Le^a-aktives Hexasaccharid aus Frauenmilch. *Chem. Ber.* **93,** 647–651.

Kuhn, R., and Gauhe, A. (1962). Die Konstitution der Lacto-*N*-neotetraose. *Chem. Ber.* **95,** 518–522.

Kuhn, R., and Gauhe, A. (1965). Bestimmung der Bindungsstelle von sialinsäureresten in Oligosacchariden mit Hilfe von Perjodat. *Chem. Ber.* **98,** 395–413.

Kuhn, R., and Löw, I. (1949). Über ein Vorkommen von Milchzucker im Pflanzenreich. (The occurrence of lactose in the plant kingdom.) *Chem. Ber.* **82,** 479–481.

Kuhn, N. J., and Lowenstein, J. M. (1967). Lactogenesis in the rat. Changes in metabolic parameters at parturition. *Biochem. J.* **105,** 995–1002.

Kuhn, R., Baer, H. H., and Gauhe, A. (1956a). Kristallisierte fucosido-lactose. *Chem. Ber.* **89,** 2513.

Kuhn, R., Baer, H. H., and Gauhe, A. (1956b). Kristallisation und Konstitutionsermittlung der Lacto-*N*-fucopentaose I. *Chem. Ber.* **89,** 2514–2523.

Kuhn, R., Baer, H. H., and Gauhe, A. (1958). Die Konstitution der lacto-*N*-fucopentaose II. *Chem. Ber.* **91,** 364.

Kulski, J. K., and Hartmann, P. E. (1981). Changes in human milk composition during the initiation of lactation. *Aust. J. Exp. Biol. Med. Sci.* **59,** 101–114.

Kulski, J. K., and Hartmann, P. E. (1983). Milk insulin, GH and TSH: Relationship to changes in milk lactose, glucose and protein during lactogenesis in women. *Endocrinol. Exp.* **17,** 317–326.

Kulski, J. K., Hartmann, P. E., Martin, J. D., and Smith, M. (1978). Effects of bromocriptine mesylate on the composition of the mammary secretion in non-breast-feeding women. *Obstet. Gynecol.* **52,** 38–42.

Kulski, J. K., Smith, M., and Hartmann, P. E. (1981a). Normal and caesarian section delivery and the initiation of lactation in women. *Aust. J. Exp. Biol. Med. Sci.* **59,** 405–412.

Kulski, J. K., Hartmann, P. E., Saint, W. J., Giles, P. F., and Gutteridge, D. H. (1981b). Changes in the milk composition of nonpuerperal women. *Am. J. Obstet. Gynecol.* **139,** 597–604.

Kulski, J. K., Hartmann, P. E., and Gutteridge, D. H. (1981c). Composition of breast fluid of a man with galactorrhea and hyperprolactinaemia. *J. Clin. Endocrinol. Metab.* **52,** 581–582.

Kumbhat, M. M., Khanna, S. A., Bijur, A. M., and Jadhav, C. S. (1985). Breast milk composition in relation to gestation. *Indian Pediatr.* **22,** 229–233.

Kunz, T. H., Stack, M. H., and Jenness, R. (1983). A comparison of milk composition in *Myotis lucifugus* and *Eptesicus fuscus* (Chiroptera: Vespertilionidae). *Biol. Reprod.* **28,** 229–234.

Kurz, G., and Wallenfels, K. (1974). D-Galactose: UV-assay with galactose dehydrogenase. *In* "Methods of Enzymatic Analysis" (H. U. Bergmeyer, ed.), Vol. 3, pp. 1180–1184, 1279–1282. Academic Press, New York.

Lammi-Keefe, C. J., Ferris, A. M., and Jensen, R. G. (1989). Changes in several components in human milk every four hours from 0600 to 2200. *J. Pediatr. Gastroenterol. Nutr.* **11,** 83–88.

Lawrence, A. J. (1968). The determination of lactose in milk products. *Aust. J. Dairy Technol.* June, p. 103.

Legge, M., and Richards, K. C. (1978). Biochemical alterations in human breast milk after heating. *Aust. Paediatr. J.* **14,** 87–90.

Lemons, J. A., Moye, L., Hall, D., and Simmons, M. (1982). Differences in the composition of preterm and term human milk during early lactation. *Pediatr. Res.* **16,** 113–117.

Levene, P. A., and Sobotka, H. (1926). Lactone formation of lacto- and maltobionic acids and its bearing on the structure of lactose and maltose. *J. Biol. Chem.* **71,** 471–475.

Linzell, J. L., and Peaker, M. (1974). Changes in colostrum composition and in the permeability of the mammary epithelium at about the time of parturition in the goat. *J. Physiol. (London)* **243,** 129–151.

Lipsman, S., Dewey, K. G., and Lönnerdal, B. (1985). Breast-feeding among teenage mothers: Milk composition, infant growth, and maternal dietary intake. *J. Pediatr. Gastroenterol. Nutr.* **4,** 426–434.

Lönnerdal, B., Forsum, E., and Hambraeus, L. (1976a). A longitudinal study of the protein, nitrogen, and lactose contents of human milk from Swedish well-nourished mothers. *Am. J. Clin. Nutr.* **29,** 1127–1133.

Lönnerdal, B., Forsum, E., and Hambraeus, L. (1976b). The protein content of human milk. I. A transversal study of Swedish normal material. *Nutr. Rep. Int.* **13,** 125–134.

Lönnerdal, B., Forsum, E., and Hambraeus, L. (1980). Effect of oral contraceptives on composition and volume of breast milk. *Am. J. Clin. Nutr.* **33,** 816–824.

Lönnerdal, B., Keen, C. L., Hurley, L. S., and Fisher, G. L. (1981). Developmental changes in the composition of Beagle dog milk. *Am. J. Vet. Res.* **42,** 662–666.

Lönnerdal, B., Smith, C., and Keen, C. L. (1984a). Analysis of breast milk: Current methodologies and future needs. *J. Pediatr. Gastroenterol. Nutr.* **3,** 290–295.

Lönnerdal, B., Keen, C. L., Glazier, C. E., and Anderson, J. (1984b). A longitudinal study of rhesus monkey *(Macaca mulatta)* milk composition: Trace elements, minerals, protein, carbohydrate, and fat. *Pediatr. Res.* **18,** 911–914.

Lovelady, C. A., Lönnerdal, B., and Dewey, K. G. (1990). Lactation performance of exercising women. *Am. J. Clin. Nutr.* **52,** 103–109.

Lucas, A., Gibbs, J. A. H., and Baum, J. D. (1978). The biology of human drip breast milk. *Early Hum. Dev.* **2/4,** 351–361.

Lück, H., and Botha, W. C. (1982). Glucose content of milk as influenced by the stage of lactation, milk yield, energy intake and somatic cell count. *South Afr. J Dairy Technol.* **14,** 111–114.

Luick, J. R., White, R. G., Gau, A. M., and Jenness, R. (1974). Compositional changes in the milk secreted by grazing reindeer. I. Gross composition and ash. *J. Dairy Sci.* **57,** 1325–1333.

Lundblad, A., and Svensson, S. (1973). Letters: The structure of a urinary difucosyl pentasaccharide, characteristic of secretors with the blood-group A gene. *Carbohydr. Res.* **30,** 187–189.

Macy, I. G., and Kelly, H. J. (1961). Human milk and cow's milk in infant nutrition. *In* "Milk: The Mammary Gland and Its Secretion" (S. Kon and A. Cowie, ed.), Vol. II, pp. 265–304. Academic Press, New York.

Macy, I. G., Nims, B., Brown, M., and Hunscher, H. A. (1931). Human milk studies. VII. Chemical analysis of milk representative of the entire first and last halves of the nursing period. *Am. J. Dis. Child.* **42,** 569–589.

Macy, I. G., Kelly, H. J., and Sloan, R. E. (1953). "The Composition of Milks: A Compilation of the Comparative Composition and Properties of Human, Cow, and Goat Milk, Colostrum, and Transitional Milk." National Academy of Sciences–National Research Council, Publication 254, Washington, DC.

Marschke, R. J., and Kitchen, B. J. (1984). Glucose levels in normal and mastitic milk. *J. Dairy Res.* **51,** 233–237.

Mba, A. U., Boyo, B. S., and Oyenuga, V. A. (1975). Studies on the milk composition of West African dwarf, Red Sokoto and Saanen goats at different stages of lactation. I. Total solids, butterfat, solids-not-fat, protein, lactose and energy contents of milk. *J. Dairy Res.* **42,** 217–226.

Mepham, T. B., and Beck, N. F. G. (1973). Variation in the yield and composition of milk throughout lactation in the guinea pig (*Cavia porcellus*). *Comp. Biochem. Physiol. A* **45,** 273–281.

Merchant, J. C., and Libke, J. A. (1988). Milk composition in the northern brown bandicoot: *Isoodon macrourus* (Peramelidae, Marsupialia). *Aust. J. Biol. Sci.* **41,** 495–505.

Merchant, J., Green, B., Messer, M., and Newgrain, K. (1989). Milk composition in the red-necked wallaby, *Macropus rufogriseus banksianus* (Marsupialia). *Comp. Biochem. Physiol. A* **93,** 483–488.

Messer, M., and Kerry, K. R. (1973). Milk carbohydrates of the echidna and the platypus. *Science* **180,** 201–203.

Messer, M., Trifonoff, E., Stern, W., Collins, J. G., and Bradbury, J. H. (1980). Structure of a marsupial-milk trisaccharide. *Carbohydr. Res.* **83,** 327–334.

Messer, M., Trifonoff, E., Collins, J. G., and Bradbury, J. H. (1982). Structure of a branched tetrasaccharide from marsupial milk. *Carbohydr. Res.* **102,** 316–320.

Messer, M., Gadiel, P. A., Ralston, G. B., and Griffiths, M. (1983). Carbohydrates of the milk of the platypus. *Aust. J. Biol. Sci.* **36,** 129–137.

Messer, M., FitzGerald, P. A., Merchant, J. C., and Green, B. (1987). Changes in milk carbohydrates during lactation in the eastern quoll, *Dasyurus viverrinus* (Marsupialia). *Comp. Biochem. Physiol. B* **88,** 1083–1086.

Messer, M., Crisp, E. A., and Newgrain, K. (1988). Studies on the carbohydrate content of milk of the crabeater seal (*Lobodon carcinophagus*). *Comp. Biochem. Physiol. B* **90,** 367–370.

Meyerson-McCormick, R., Cranford, J. A., and Akers, R. M. (1990). Milk yield and composition in the punare (*Thrichomys apereoides*). *Comp. Biochem. Physiol. A* **96,** 211–214.

Michaelsen, K. F., Pedersen, S. B., Skafte, L., Jæger, P., and Peitersen, B. (1988). Infrared analysis for determining macronutrients in human milk. *J. Pediatr. Gastroenterol. Nutr.* **7,** 229–235.

Michaelsen, K. F., Skafte, L., Badsberg, J. H., and Jørgensen, M. (1990). Variation in macronutrients in human bank milk: Influencing factors and implications for human milk banking. *J. Pediatr. Gastroenterol. Nutr.* **11,** 229–239.

Montreuil, J. (1956). Structure de deux triholosides isolés du lait de femme. *C. R. Acad. Sci.* **242,** 192–193.

Montreuil, J., and Mullet, S. (1960). Étude des variations des constituants glucidiques du lait de femme au cours de la lactation. *Bull. Soc. Chim. Biol.* **42,** 365–377.

Munks, A., Green, B., Newgrain, K., and Messer, M. (1991). Milk composition in the common ringtail possum, *Psuedocheirus peregrinus* (Petauridae:Marsupialia). *Aust. J. Zool.* **39,** 403–416.

Nagasawa, H., Naito, T., and Kataoka, K. (1989). Relationship between milk composition and pup's growth in mice. *Proc. Soc. Exp. Biol. Med.* **191,** 78–81.

Neubauer, S. H., Ferris, A. M., Chase, C. G., Murtaugh, M., Lammi-Keefe, C., and Jensen, R. G. (1987). Lactose and glucose content of human milk from insulin-dependent diabetic women. *Fed. Proc.* **46A,** 439.

Neubauer, S. H., Ferris, A. M., and Hinckley, L. (1990). The effect of mastitis on breast milk composition in insulin-dependent diabetic and non-diabetic women. *FASEB J.* **4,** A915.

Neubauer, S. H., Ferris, A. M., Chase, C. G., Fanelli, J., Thompson, C. A., Lammi-Keefe, C. J., Clark, R. M., Jensen, R. G., Bendel, R. B., and Green, K. W. (1993). Delayed lactogenesis in women with insulin-dependent diabetes mellitus. *Am. J. Clin. Nutr.* **58,** 54–60.

Neville, M. C., Keller, R. P., Seacat, J., Casey, C. E., Allen, J. C., and Archer, P. (1984). Studies on human lactation. I. Within-feed and between-breast variation in selected components of human milk. *Am. J. Clin. Nutr.* **40,** 635–646.

Neville, M. C., Hay, W. W., Jr., and Fennessey, P. (1990). Physiological significance of the concentration of human milk glucose. *Protoplasma* **159,** 118–128.

Neville, M. C., Allen, J. C., Archer, P. C., Casey, C. E., Seacat, J., Keller, R. P., Lutes, V., Rasbach, J., and Neifert, M. (1991). Studies in human lactation: Milk volume and nutrient composition during weaning and lactogenesis. *Am. J. Clin. Nutr.* **54,** 81–92.

Newburg, D. S., and Fillios, L. C. (1979). A requirement for dietary asparagine in pregnant rats. *J. Nutr.* **109,** 2190–2197.

Newburg, D. S., and Fillios, L. C. (1982). Brain development in neonatal rats nursing asparagine-deprived dams. *Dev. Neurosci.* **5,** 332–344.

Newburg, D. S., Frankel, D. L., and Fillios, L. C. (1975). An asparagine requirement for rats fed the dietary combination of aspartic acid, glutamine and glutamic acid. *J. Nutr.* **105,** 356–363.

Newburg, D. S., Pickering, L. K., McCluer, R. H., and Cleary, T. G. (1990). Fucosylated oligosaccharides of human milk protect suckling mice from heat-stabile enterotoxin of *Escherichia coli. J. Infect. Dis.* **162,** 1075–1080.

Nicholas, K. R., and Hartmann, P. E. (1991). Milk secretion in the rat: Progressive changes in milk composition during lactation and weaning and the effect of diet. *Comp. Biochem. Physiol. A* **98,** 535–542.

Nickerson, T. A., Vujicic, I. F., and Lin, A. Y. (1976). Colorimetric estimation of lactose and its hydrolytic products. *J. Dairy Sci.* **59,** 386–390.

Nommsen, L. A., Lovelady, C. A., Heinig, M. J, Lönnerdal, B., and Dewey, K. G. (1991). Determinants of energy, protein, lipid, and lactose concentrations in human milk during the first 12 mo of lactation: The DARLING study. *Am. J. Clin. Nutr.* **53,** 457–465.

Oftedal, O. T. (1984). Lactation in the dog: Milk composition and intake by puppies. *J. Nutr.* **114,** 803–812.

Oftedal, O. T., Hintz, H. F., and Schryver, H. F. (1983). Lactation in the horse: Milk composition and intake by foals. *J. Nutr.* **113,** 2196–2206.

Oftedal, O. T., Boness, D. J., and Tedman, R. A. (1987). The behaviour, physiology and anatomy of lactation in the pinnipedia. *In* "Current Mammalogy" (H. H. Genoways, ed.), Vol. 1, pp. 175–246. Plenum, New York.

Ojofeitimi, E. O., Ogundiwin, J. O., and Saliu, O. Y. (1983). Protein and lactose contents of breast milk from mothers of malnourished children. *East Afr. Med. J.* **60,** 783–788.

Pagan, J. D., and Hintz, H. F. (1986). Composition of milk from pony mares fed various levels of digestible energy. *Cornell Vet.* **76,** 139–148.

Parkkinen, J., and Finne, J. (1987). Isolation of sialyl oligosaccharides and sialyl oligosaccharide phosphates from bovine colostrum and human urine. *In* "Methods in Enzymology" (V. Ginsburg, ed.), Vol. 138, pp. 289–300. Academic Press, New York.

Peaker, M., and Goode, J. A. (1978). The milk of the fur-seal, *Arctocephalus tropicalis gazella;* In particular the composition of the aqueous phase. *J. Zool.* **185,** 469–476.

Pervaiz, S., and Brew, K. (1986). Composition of the milks of the bottlenose dolphin (*Tursiops truncatus*) and the Florida manatee (*Trichechus manatus latirostris*). *Comp. Biochem. Physiol. A* **84,** 357–360.

Peters, F. E. (1953). The chemical composition of New Hebridean human milk. *Br. J. Nutr.* **7,** 208–211.

Pilloton, R., Mascini, M., Casella, I. G., Festa, M. R., and Bottari, E. (1987). Lactose determination in raw milk with a two-enzyme based electrochemical sensor. *Anal. Lett.* **20,** 1803–1814.

Pilson, M. E. Q., and Cooper, R. W. (1967). Composition of milk from *Galago crassicaudatus. Folia Primatol.* **5,** 88–91.

Polonovsky, M., and Lespagnol, A. (1933). Nouvelles acquisitions sur les composés glucidiques du lait de femme. *Bull. Soc. Chim. Biol.* **15,** 320–349.

Prentice, A. M., Roberts, S. B., Prentice, A., Paul, A. A., Watkinson, M., Watkinson, A. A., and Whitehead, R. G. (1983). Dietary supplementation of lactating Gambian women. I. Effect on breast-milk volume and quality. *Hum. Nutr. Clin. Nutr.* **37c,** 53–64.

Prentice, A. M., Lamb, W. H., Prentice, A., and Coward, W. A. (1984). The effect of water abstention on milk synthesis in lactating women. *Clin. Sci.* **66,** 291–298.

Prentice, A. (1985). The influence of maternal parity on breast-milk composition. *In* "Composition and Physiological Properties of Human Milk" (J. Schaub, ed.), pp. 309–322. Elsevier, Amsterdam.

Prinsloo, J. G., Wittmann, W., Strydom, E. S. P., deVilliers, D. B., Wehmeyer, A. S., Laubscher, N. F., and Botha, M. A. (1970). Composition of breast milk from Bantu and white women on the fifth postpartum day. *S. Afr. Med. J.* **44,** 738–739.

Prosser, C. G., and Hartmann, P. E. (1983). Saliva and breast milk composition during the menstrual cycle of women. *Aust. J. Exp. Biol. Med. Sci.* **61,** 265–275.

Prosser, C. G., Saint, L., and Hartmann, P. E. (1984). Mammary gland function during gradual weaning and early gestation in women. *Aust. J. Exp. Biol. Med. Sci.* **62,** 215–228.

Ragueneau, S. (1987). Early development in mice. IV: Quantity and gross composition of milk in five inbred strains. *Physiol. Behav.* **40,** 431–435.

Ramadan, M. A., Salah, M. M., Eid, S. Z., and Sammour, M. B. (1972). The effect of the oral contraceptive ovosiston on the composition of human milk. *J. Reprod. Med.* **9,** 81–83.

Ratzmann, K. P., Steindel, E., Hildebrandt, R., and Kohlhoff, R. (1988). Is there a relationship between metabolic control and glucose concentration in breast milk of Type 1 (insulin-dependent) diabetic mothers? *Exp. Clin. Endocrinol.* **92,** 32–36.

Riedman, M., and Ortiz, C. L. (1979). Changes in milk composition during lactation in the northern elephant seal. *Physiol. Zool.* **52,** 240–249.

Ruis, H., Rolland, R., Doesburg, W., Broeders, G., and Corbey, R. (1981). Oxytocin enhances onset of lactation among mothers delivering prematurely. *Br. Med. J.* **283,** 340–342.

Ruiz-Palacios, G. M., Cervantes, L. E., Newburg, D. S., Lopez-Vidal, Y., and Calva, J. J. (1992). In vitro models for studying *Campylobacter* infections. *In* "*Campylobacter jejuni.* Current-Status and Future Trends" (I. Nachamkin, M. J. Blaser, and L. S. Tomkins, eds), pp. 176–183. American Society for Microbiology, Washington, D.C.

Sabharwal, H., Nilsson, B., Chester, M. A., Sjöblad, S., and Lundblad, A. (1984). Blood group specific oligosaccharides from faeces of a blood group A breast-fed infant. *Mol. Immunol.* **21,** 1105–1112.

Sabharwal, H., Nilsson, B., Chester, M. A., Lindh, F., Grönberg, G., Sjöblad, S., and Lundblad, A. (1988a). Oligosaccharides from feces of a blood-group B, breast-fed infant. *Carbohydr. Res.* **178,** 145–154.

Sabharwal, H., Nilsson, B., Grönberg, G., Chester, M. A., Dakour, J., Sjöblad, S., and Lundblad, A. (1988b). Oligosaccharides from feces of preterm infants fed on breast milk. *Arch. Biochem. Biophys.* **265,** 390–406.

Saint, L., Smith, M., and Hartmann, P. E. (1984). The yield and nutrient content of colostrum and milk of women from birth to 1 month post-partum. *Br. J. Nutr.* **52,** 87–95.

Saint, L., Maggiore, P., and Hartmann, P. E. (1986). Yield and nutrient content of milk in eight women breast-feeding twins and one woman breast-feeding triplets. *Br. J. Nutr.* **56,** 49–58.

Sammour, M. B., Ramadan, M. E. A., and Salah, M. (1973). Effect of chlormadinone on the composition of human milk. *Fertil. Steril.* **24,** 301–304.

Sawaya, W. N., Khalil, J. K., Al-Shalhat, A., and Al-Mohammad, H. (1984). Chemical composition and nutritional quality of camel milk. *J. Food Sci.* **49,** 744–747.

Schoknecht, P. A., Cranford, J. A., and Akers, R. M. (1985). Variability in milk composition of the domestic ferret (*Mustela putorius*). *Comp. Biochem. Physiol. A* **81,** 589–591.

Shaffer, P. A., and Hartmann, A. F. (1920). The iodometric determination of copper and its use in sugar analysis. II. Methods for the determination of reducing sugars in blood, urine, milk, and other solutions. *J. Biol. Chem.* **45,** 365–389.

Smith, D. F., Prieto, P. A., McCrumb, D. K., and Wang, W.-C. (1987). A novel sialylfucopentaose in human milk. Presence of this oligosaccharide is not dependent on expression of the secretor or Lewis fucosyltransferases. *J. Biol. Chem.* **262,** 12040–12047.

Somogyi, M. (1945). Determination of blood sugar. *J. Biol. Chem.* **160,** 69–73.

Spála, P., and Váhala, J. (1989). Hand rearing and milk composition of springbok (*Antidorcas marsupialis*). *Zoologische Garten* **59,** 27–33.

Strecker, G., and Montreuil, J. (1973). Isolement et étude de la structure de 16 oligosaccharides isolés de l'urine humaine. *C. R. Acad. Sci.* **277,** 1393–1396.

Strecker, G., Wieruszeski, J.-M., Michalski, J.-C., and Montreuil, J. (1988). Structure of a new nonasaccharide isolated from human milk: VI^{2}Fuc, V^{4}Fuc, III^{3}Fuc-*p*-lacto-*N*-hexaose. *Glycoconjugate J.* **5,** 385–396.

Strecker, G., Fièvre, S., Wieruszeski, J.-M., Michalski, J.-C., and Montreuil, J. (1992). Primary structure of four human milk octa-, nona-, and undeca-saccharides established by ^{1}H- and ^{13}C-nuclear magnetic resonance spectroscopy. *Carbohydr. Res.* **226,** 1–14.

Strode, M. A., Dewey, K. G., and Lönnerdal, B. (1986). Effects of short-term caloric restriction on lactational performance of well-nourished women. *Acta Paediatr. Scand.* **75,** 222–229.

Sturman, J. A., Lin, Y. Y., Higuchi, T., and Fellman, J. H. (1985). N-acetylneuramin lactose sulfate: A newly identified nutrient in milk. *Pediatr. Res.* **19,** 216–219.

Svennerholm, L. (1956). The quantitative estimation of cerebrosides in nervous tissue. *J. Neurochem.* **1,** 42–53.

Tachibana, Y., Yamashita, K., and Kobata, A. (1978). Oligosaccharides of human milk: Structural studies of di- and trifucosyl derivatives of lacto-*N*-octaose and lacto-*N*-neooctaose. *Arch. Biochem. Biophys.* **188,** 83–89.

Thomas, M. R., Chan, G. M., and Book, L. S. (1986). Comparison of macronutrient concentration of preterm human milk between two milk expression techniques and two techniques for quantitation of energy. *J. Pediatr. Gastroenterol. Nutr.* **5,** 597–601.

Toaff, R., Ashkenazi, H., Schwartz, A., and Herzberg, M. (1969). Effects of oestrogen and progestagen on the composition of human milk. *J. Reprod. Fertil.* **19,** 475–482.

Tolstoi, E. (1935). The relationship of the blood glucose to the concentration of lactose in the milk of lactating diabetic women. *J. Clin. Invest.* **14,** 863–866.

Treadway, J. L., and Lederman, S. A. (1986). The effects of exercise on milk yield, milk composition, and offspring growth in rats. *Am. J. Clin. Nutr.* **44,** 481–488.

Turton, J. A., Ford, D. J., Bleby, J., Hall, B. M., and Whiting, R. (1978). Composition of the milk of the common marmoset (*Callithrix jacchus*) and milk substitutes used in hand-rearing programmes, with special reference to fatty acids. *Folia Primatol.* **29,** 64–79.

Ullrey, D. E., Schwartz, C. C., Whetter, P. A., Rao, T. R., Euber, J. R., Cheng, S. G., and Brunner, J. R. (1984). Blue–green color and composition of Stejneger's beaked whale (*Mesoplodon stejnegeri*) milk. *Comp. Biochem. Physiol. B* **79,** 349–352.

Urashima, T., Saito, T., Ohmisya, K., and Shimazaki, K. (1991). Structural determination of three neutral oligosaccharides in bovine (Holstein–Friesian) colostrum, including the novel trisaccharide; GalNAcα1-^{3}Galβ1-^{4}Glc. *Biochim. Biophys. Acta* **1073,** 225–229.

van Beusekom, C. M., Zeegers, T. A., Martini, I. A., Velvis, H. J. R., Visser, G. H. A., van Doormaal, J. J., and Muskiet, F. A. J. (1993). Milk of patients with tightly controlled insulin-dependent diabetes mellitus has normal macronutrient and fatty acid composition. *Am. J. Clin. Nutr.* **57,** 938–943.

van Steenbergen, W. M., Kusin, J. A., De With, C., Lacko, E., and Jansen, A. A. J. (1983). Lactation performance of mothers with contrasting nutritional status in rural Kenya. *Acta Paediatr. Scand.* **72,** 805–810.

Venkataraman, R., and Reithel, F. J. (1958). Carbohydrates of the Sapotaceae. I. The origin of lactose in *A. sapota*. *Arch. Biochem. Biophys.* **75,** 443–452.

Verheul, F. E. A. M., v. d. Bosch, M. J. A., Cornelissen, P. J. H. C., and Waelkens, J. J. J. (1986). Simplified and rapid methods for the determination of protein, fat and lactose in human milk and the energy intake by the breast-fed infant. *J. Clin. Chem.-Clin. Biochem.* **24,** 341–346.

Villalpando, S. F., Butte, N. F., Wong, W. W., Flores-Huerta, S., de Jesus Hernandez-Beltran, M., Smith, E. O., and Garza, C. (1992). Lactation performance of rural Mesoamerindians. *Eur. J. Clin. Nutr.* **46,** 337–348.

Viverge, D., Grimmonprez, L., Cassanas, G., Bardet, L., Bonnet, H., and Solère, M. (1985). Variations of lactose and oligosaccharides in milk from women of blood types secretor

A or H, secretor Lewis, and secretor H/nonsecretor Lewis during the course of lactation. *Ann. Nutr. Metab.* **29,** 1–11.

Viverge, D., Grimmonprez, L., Cassanas, G., Bardet, L., and Solère, M. (1986). Diurnal variations and within the feed in lactose and oligosaccharides of human milk. *Ann. Nutr. Metab.* **30,** 196–209.

Viverge, D., Grimmonprez, L., Cassanas, G., Bardet, L., and Solere, M. (1990a). Variations in oligosaccharides and lactose in human milk during the first week of lactation. *J. Pediatr. Gastroenterol. Nutr.* **11,** 361–364.

Viverge, D., Grimmonprez, L., Cassanas, G., Bardet, L., and Solère, M. (1990b). Discriminant carbohydrate components of human milk according to donor secretor types. *J. Pediatr. Gastroenterol. Nutr.* **11,** 365–370.

Volcani, R., Zisling, R., Sklan, D., and Nitzan, Z. (1973). The composition of chinchilla milk. *Br. J. Nutr.* **29,** 121–125.

Wallenfels, K., and Kurz, G. (1962). On the specificity of galactose dehydrogenase from *Pseudomonas saccharophila* and its use as an analytical aid. *Biochem Z.* **335,** 559–573.

Warman, N. L., and Rasmussen, K. M. (1983). Effects of malnutrition during the reproductive cycle on nutritional status and lactational performance of rat dams. *Nutr. Res.* **3,** 527–545.

Watson, M. A., Alford, E. S., Dill, C. W., Richter, R. L., and Garza, C. (1982). Compositional changes during sequential sampling of human milk. *Nutr. Rep. Int.* **26,** 1105–1111.

White, J. M., Williams, G., Samour, J. H., Drury, P. J. J., and Cheeseman, P. (1985). The composition of milk from captive aardvark (*Orycteropus afer*). *Zoo Biol.* **4,** 245–251.

Widdowson, E. M. (1984). Milk and the newborn animal. *Proc. Nutr. Soc.* **43,** 87–100.

Wieruszeski, J. M., Chekkor, A., Bouquelet, S., Montreuil, J., and Strecker, G. (1985). Structure of two new oligosaccharides isolated from human milk: Sialylated lacto-*N*-fucopentaoses I and II. *Carbohydr. Res.* **137,** 127–138.

Williams, A. P., Bishop, D. R., Cockburn, J. E., and Scott, K. J. (1976). Composition of ewe's milk. *J. Dairy Res.* **43,** 325–329.

Wohlt, J. E., Foy, W. L, Jr., Kniffen, D. M., and Trout, J. R. (1984). Milk yield by Dorset ewes as affected by sibling status, sex and age of lamb, and measurement. *J. Dairy Sci.* **67,** 802–807.

World Health Organization (WHO) (1985). "WHO Collaborative Study on Breast-Feeding. The Quantity and Quality of Breast Milk." WHO, Geneva.

Yagil, R., and Etzion, Z. (1980). Effect of drought condition on the quality of camel milk. *J. Dairy Res.* **47,** 159–166.

Yamashita, K., and Kobata, A. (1974). Oligosaccharides of human milk. V. Isolation and characterization of a new trisaccharides, 6′-galactosyllactose. *Arch. Biochem. Biophys.* **161,** 164–170.

Yamashita, K., Tachibana, Y., and Kobata, A. (1976a). Oligosaccharides of human milk. Isolation and characterization of three new disialylfucosyl hexasaccharides. *Arch. Biochem. Biophys.* **174,** 582–591.

Yamashita, K., Tachibana, Y., and Kobata, A. (1976b). Oligosaccharides of human milk: Isolation and characterization of two new nonasaccharides, monofucosyllacto-*N*-octaose and monofucosyllacto-*N*-neoctaose. *Biochemistry* **15,** 3950–3955.

Yamashita, K., Tachibana, Y., and Kobata, A. (1977a). Oligosaccharides of human milk: Structures of three lacto-*N*-hexaose derivatives with H-haptenic structure. *Arch. Biochem. Biophys.* **182,** 546–555.

Yamashita, K., Tachibana, Y., and Kobata, A. (1977b). Oligosaccharides of human milk. Structural studies of two new octasaccharides, difucosyl derivatives of *para*-lacto-*N*-hexaose and *para*-lacto-*N*-neohexaose. *J. Biol. Chem.* **252,** 5408–5411.

Zou, S., McLaren, D. G., and Hurley, W. L. (1992). Pig colostrum and milk composition: Comparisons between Chinese Meishan and US breeds. *Livest. Prod. Sci.* **30,** 115–127.

CHAPTER 5

Nitrogenous Components of Milk

A. Human Milk Proteins

BO LÖNNERDAL
STEPHANIE ATKINSON

I. Introduction

The protein concentration of human milk is high during early lactation, the *colostrum* period; then it gradually declines to a relatively low level of 0.8–1.0% in mature milk (Table I), particularly when compared to most other species (Hambraeus, 1977). It should be recognized, however, that the milk volume produced during early lactation is very low; thus, protein intake by the breast-fed infant is usually lower during the first weeks of life than later in life. The high protein concentration of colostrum is largely due to very high concentrations of secretory IgA and lactoferrin. In contrast, some milk proteins, like β-casein, are absent or present in very low concentrations during early lactation. It is apparent that milk protein gene expression is regulated by hormones; thus, the rapid changes in circulating hormones that accompany late pregnancy and early lactation will affect the concentrations of different milk proteins (Rosen *et al.*, 1986). Therefore, milk proteins that are synthesized by the mammary gland can be expected to be more affected by time postpartum (length of lactation) than proteins in milk that originate from serum and are likely passively transferred into milk (see below).

The nitrogen concentration of human milk largely follows the same developmental pattern as that shown for protein. The reason for this is that the concentration of nonprotein nitrogen (NPN) is relatively constant during lactation. Most NPN consists of free amino acids, urea, uric acid,

TABLE I
Concentrations of Nitrogen and Proteins in Human Milk[a]

	Time postpartum (months)			
g/Liter	0–0.5	0.5–1.5	1.5–3.5	3.5–6.5
Total nitrogen	3.05 ± 0.59	1.93 ± 0.24	1.61 ± 0.21	1.48 ± 0.17
NPN	0.53 ± 0.09	0.46 ± 0.03	0.41 ± 0.04	0.38 ± 0.07
True protein[b]	15.80 ± 4.2	9.2 ± 1.8	7.5 ± 1.6	6.9 ± 1.2
α-Lactalbumin	3.62 ± 0.59	3.26 ± 0.47	2.78 ± 0.49	2.68 ± 0.59
Lactoferrin	3.53 ± 0.54	1.94 ± 0.38	1.65 ± 0.29	1.39 ± 0.26
Serum albumin	0.39 ± 0.06	0.41 ± 0.07	0.39 ± 0.04	0.38 ± 0.04
SIgA	2.0 ± 2.5	1.0 ± 0.3	—	0.5 ± 0.1
IgM	0.12 ± 0.03	0.2	—	0.2
IgG	0.34 ± 0.01	0.05 ± 0.03	—	0.03

[a]Data are means ± SD. Adapted from Lönnerdal *et al.* (1976) and Goldman and Goldblum (1989).
[b]True protein = (total nitrogen − NPN) × 6.25.

nucleotides, etc. (Atkinson *et al.*, 1989); since most of these compounds are likely to originate from serum or are part of the mammary gland pool of metabolites necessary for milk synthesis, there is no pronounced effect of duration of lactation on their concentrations (see Chapter 5B).

The longitudinal changes in protein concentration of human milk shown in Table I are for women delivering at term. A very interesting finding was that the protein concentration of breast milk from women delivering prematurely is considerably higher than that of milk from women delivering at term (Atkinson *et al.*, 1980) (see Chapter 10A). However, although this is correct for each time point postpartum, the longitudinal pattern with high initial concentrations and a subsequent exponential decrease is similar for "preterm" and term human milk. These high protein concentrations and the longitudinal changes are likely a result of different hormonal stimuli brought about by the shortened gestation period. Little is known, however, about the precise role of different hormones in affecting milk protein synthesis. It is noteworthy, though, that these high protein concentrations are likely to benefit the prematurely born infant with its rapid catch-up growth and high protein requirement. Thus, when considering the aspect of human milk banking, milk from mothers delivering prematurely would be more appropriate for premature infants than early milk from mothers delivering at term or, particularly, mature milk. This, in itself, should also work as an incentive for women delivering premature infants to attempt to produce milk for their own infants and therefore help to establish breast-feeding.

The proteins in human milk have very different origin and, for the sake of clarity, it is usually operationally easier to separate them into various classes. When milk is synthesized by the mammary gland, synthesized proteins from the Golgi are mixed with cytosolic proteins, which are partially cell products and partially derived from serum. These proteins, together with some cells, are mixed with fat globules surrounded by the apical membrane. Thus, in milk there will be cells, with their protein constituents, milk fat globule membrane proteins, milk proteins, and serum proteins. The milk proteins have classically been divided into caseins and whey proteins; this separation has usually been achieved by precipitation/sedimentation procedures so that casein is defined as the proteins that can be found in the pellet after centrifugation. As serum proteins largely are soluble under these conditions, they will be found in the milk protein class of whey proteins. Quantitatively, caseins of human milk comprise some 10–50% of total protein, with a pronounced change during lactation, while whey proteins constitute 50–90% of total protein (Kunz and Lönnerdal, 1992). Milk fat globule membrane proteins (presented in detail in Chapter 9A) and protein derived from cells present in milk contribute a very small part of milk protein, or about 1–3% of total protein (Lönnerdal *et al.*, 1987).

II. Caseins

A. Micelles

The aggregates of casein protein subunits, calcium phosphate and some other ionic constituents, in the form of submicelles or micelles give milk its characteristic white appearance. Milk from most species have caseins as the major class of proteins; as mentioned above, this is not the case for human milk. In fact, colostrum and "preterm" milk do not contain or are very low in casein; with increasing time of lactation, however, casein will constitute a larger part of human milk protein (Kunz and Lönnerdal, 1992). Based on their behavior during electrophoresis, casein subunits were early classified into α, β, and ϰ-casein (Rowland, 1938; Jenness, 1985). Although there are some reports on γ-casein, this is not a true casein subunit, but rather a fragment resulting from degradation of β-casein (see below).

Human milk casein micelles are considerably smaller in size than cow milk micelles; human casein micelles are about 30–75 nm in diameter, while bovine micelles are as large as 600 nm (Calapaj, 1968). It is not known whether differences in casein subunit composition (and the inorganic constituents) and, consequently, the electrostatic forces within the micelle or a higher permeability of the lacteal ducts of the bovine mammary gland are responsible for these differences in micellar size. Human milk contains only β-casein and ϰ-casein, while cow milk contains α-casein in two different forms, α_{S1} and α_{S2} (Jenness, 1985). Micelle formation is likely dependent on both hydrophobic interactions and electrostatic binding. The latter type of binding occurs via bridges formed between charged parts of the casein subunits, calcium and phosphate. As β-caseins mostly are phosphorylated and κ-casein is a glycoprotein with charged sialic acid residues, it is believed that these post-translational modifications of the proteins are important for micelle formation. Studies on micelle formation *in vitro* (Azuma *et al.*, 1985) have shown that human casein micelles are formed at lower calcium concentration (5 m*M*) than bovine micelles (15 m*M*). This may be necessary as the calcium concentration of human milk (7 m*M*) is considerably lower than that of cow milk (30 m*M*). As human β-casein occurs in different phosphorylated forms, the capacity of these different forms to assemble micelles has also been explored *in vitro*. The highly phosphorylated form of β-casein was found to precipitate in the presence of calcium, while β-caseins with a low degree of phosphorylation stayed in solution (Azuma *et al.*, 1985). It has therefore been proposed that human casein micelles are built up by both highly phosphorylated β-casein-binding calcium and β-casein with a low degree of phosphorylation and κ-casein that aid in the stabilization of the micelle. However, considering the very low concentration of κ-casein in human milk with less than 15% of total casein being κ-casein (Kunz and Lönnerdal, 1992), it is unlikely that this casein subunit plays a major role in casein micelle stabilization.

The casein micelle does not only consist of the protein subunits. Within the micelle, calcium, phosphate, and to some extent other ions like magnesium and citrate form an insoluble aggregate usually referred to as colloidal calcium phosphate (CCP). The role of CCP in micelle aggregation and its structure have been studied to some extent for bovine milk, while there is limited information available for human milk. When dialyzing bovine casein micelles against EDTA or a calcium-free buffer, the micelles dissociate and form submicelles or free casein subunits (Schmidt, 1982). It is believed that the calcium phosphate in bovine micelles largely consists of tricalcium phosphate $Ca_3(PO_4)_2$, although some controversy still exists (Chaplin, 1984). The role of CCP in human casein micelle aggregation is likely to be less pronounced; while 65% of bovine milk calcium is found in CCP, only 6% of calcium in human milk is bound to casein (Fransson and Lönnerdal, 1983).

B. β-*Casein*

Human milk β-casein has a molecular weight of about 24 kDa. Its amino acid sequence was determined by biochemical methods (Greenberg *et al.*, 1984) and it was found to consist of 212 amino acids. The composition is shown in Table II. Several potential sites for phosphorylation were found and careful analysis of variants of human β-casein showed that up to five of these sites actually are phosphorylated *in vivo*. Phosphorylation at these serine and threonine residues appears to occur in a stepwise fashion, with the first phosphorylation occurring at Ser 9 or Ser 10. All the amino acid residues that are phosphorylated are located at the N-terminal of β-casein: Thr 3, Ser 6, Ser 8, Ser 9, and Ser 10. The unphosphorylated, mono- and

TABLE II
Amino Acid Composition (Residues) of Human Caseins

	β-Casein[a]	κ-Casein[b]
Asp	11	12
Thr	9	17
Ser	10	7
Glu	39	14
Pro	41	27
Gly	3	3
Ala	7	13
Val	19	12
Met	3	1
Cys	0	1
Ile	13	10
Leu	25	4
Tyr	7	10
Phe	5	3
Lys	11	5
His	5	3
Arg	3	7
Trp	1	1

[a]Adapted from Greenberg *et al.* (1984).
[b]Adapted from Yamauchi *et al.* (1981).

pentaphosphate forms appear to be less abundant than especially the di- and tetraphosphorylated forms. However, there are differences among milk samples from different donors and this does not appear to be true for all milk samples (Kunz and Lönnerdal, 1992). Further knowledge about human mammary gland casein kinases and regulation of their activity is needed to better understand the presence of these different forms in human milk.

The gene for human β-casein has been cloned and sequenced (Lönnerdal *et al.,* 1990). The amino acid sequence deduced from the nucleotide sequence differed from the published amino acid sequence at some points and, particularly, the protein was found to contain only 209 residues. Whether these discrepancies are due to genetic variants of β-casein, as has been suggested in some earlier studies (Azuma *et al.,* 1981), or to the inherent difficulties in the classical amino acid sequencing methods is not yet known. The genes for β-casein have been cloned and sequenced in several species like rat, mouse, cow, goat, and sheep (Bonsing and Mackinlay, 1987). Although some species differences are found, there are several conserved sequences among species, particularly at the N-terminal end where the sites of phosphorylation are located. Recombinant human β-casein has been produced in *Escherichia coli* and *Saccharomyces cerevisiae* (Hansson *et al.*, 1983), which will allow further studies on structure and biological function.

As mentioned above, there are reports of "γ-caseins" in human milk. However, recent studies have shown that these molecules are fragments of β-casein formed by proteolysis. Human milk contains several proteases (Borulf *et al.*, 1987) and it is therefore possible that some limited proteolysis of β-casein can occur during storage.

C. ϰ-Casein

Purification and characterization of human milk ϰ-casein has been proven to be a difficult task. This glycosylated protein occurs in human milk at a very low concentration and it is also sensitive to proteolysis. Further, as a large part of ϰ-casein consists of carbohydrate, the protein stains very poorly with conventional stains like Coomassie blue (Kunz and Lönnerdal, 1990a). Since most carbohydrate stains also are weak, detection of ϰ-casein in human milk and during purification is very difficult. Human ϰ-casein has a molecular weight of about 37 kDa, of which about 19 kDa is carbohydrate (Brignon *et al.*, 1985). Classical amino acid sequence analysis resulted in 158 residues (Yamauchi *et al.*, 1981). The composition is shown in Table II. The gene for human ϰ-casein was recently cloned and sequenced and the amino acid sequence deduced contained 162 residues (Bergström *et al.*, 1992). Comparisons between species demonstrate a large degree of homology.

The very high degree of glycosylation makes human ϰ-casein unique in comparison to ϰ-caseins from other species. It was found early (Johansson and Svennerholm, 1956) that whole human casein contained considerably more carbohydrate than did bovine casein (4 vs 0.8%). We now know that human ϰ-casein contains 40–60% of carbohydrate, while bovine ϰ-casein contains only 10% carbohydrate (Azuma *et al.*, 1984). Hexose and hexosamine were found to be particularly high in human casein but also sialic acid (van Halbeek *et al.*, 1985). Although the number of different carbohydrates in human ϰ-casein is low, namely galactose, *N*-acetylgalactosamine, *N*-acetylglucosamine, neuraminic acid, and fucose, the several glycosylation sites and the elaborate branched structures of the glycans provide numerous possibilities for structural variants of ϰ-casein. To date, however, only O-glycans have been described; no N-linked glycans have been detected. The complexity of the microheterogeneity of human ϰ-casein glycans is illustrated in several studies; in one study nine different saccharides were isolated and their structures determined (van Halbeek *et al.*, 1985). Further glycan structures are expected to be characterized; it is known that human milk contains a multitude of oligosaccharides, demonstrating the presence of several enzymes involved in oligosaccharide synthesis within the mammary gland.

Human κ-casein is easily cleaved by proteases at a sensitive peptide bond between Ile 105 and Met 106. This proteolytic cleavage leads to the formation of an N-terminal peptide called *para*-κ-casein and a C-terminal fragment which constitutes the casein glycopeptide. This cleavage causes destabilization of the casein micelle; the N-terminal fragment is insoluble and precipitates, while the carbohydrate-rich peptide that contains about one-third of the amino acids is soluble. Functional importance of the protease-sensitive peptide bond is suggested by the fact that the casein glycopeptide amino acid sequence varies considerably among species, but the particular region at and around this bond is highly conserved (Mercier and Chobert, 1976). The neighboring amino acids in the region of amino acid residues 97–116 contribute to the lability of the peptide bond at 105 and 106, as has been shown in studies on chymosin attack of synthetic peptide substrates (Raymond *et al.*, 1973).

D. Physiological Significance of Human Casein

In general, casein is considered to be an easily digested protein that will provide amino acids, calcium, and phosphorus to the newborn. While this may also be true for human casein, the relative contribution of amino acids, calcium, and phosphate from casein in human milk is relatively small compared to the total amount supplied. It is quite possible, however, that human casein subunits may have other physiological functions. Although studies on the biological activity of human milk caseins have been limited due to difficulties in preparing sufficient quantities of highly purified proteins, several areas of research strongly suggest that caseins, or rather their digestive fragments, can exhibit various activities.

Proteolytic degradation of human β-casein leads to the formation of N-terminal fragments containing the phosphorylated amino acid residues described previously. These so-called casein phosphopeptides (or CPPs) have been shown to keep calcium in soluble form, but also to facilitate calcium uptake by intestinal cells (Naito *et al.*, 1972; Sato *et al.*, 1986). Thus, formation of CPPs may be an "in-built" mechanism to assure adequate calcium absorption in the newborn. It is also possible that some effect is executed on trace element absorption as a fraction of iron, zinc, copper, and manganese in human milk and is bound to casein (Fransson and Lönnerdal, 1983). Although formation of CPPs from bovine casein has been shown *in vitro* in experimental animals, it is known that bovine casein is less easily digested in human infants and may therefore not exert a similar role, at least not at the cellular uptake phase.

Other peptides resulting from proteolysis of human β-casein have been shown to have opioid activity in that they both show affinity to opiate receptors and exhibit opiate-like effects (Brantl, 1984). These so-called casomorphins have been produced *in vitro* and *in vivo* and studies on isolated cells and in experimental animals have demonstrated their activity. It should be noted, however, that similar, but not identical, casomorphins have been isolated from cow milk. Also worth noting is that several peptides with antiopioid activity have been described (Yoshikawa *et al.*, 1986). To what extent these different types of peptides are produced in the infant and their relative physiological activity are important areas to explore. Besides the obvious potential effects on sleeping patterns and behavior via electrophysiological effects on the central nervous system (Reymann *et al.*, 1985), these peptides have been shown to modulate insulin and somatostatin activity and to affect pancreatic polypeptide release (Schusdziarra *et al.*, 1983a,b,c). In addition to these peptides with opioid or opioid agonist activity, a peptide that inhibits the activity of angiotensin I-converting enzyme (Maruyama *et al.*, 1985) and a peptide with immunostimulatory activity have been reported (Berthou *et al.*, 1987). Again, studies on the formation and activity of these peptides *in vivo* are needed to better evaluate their physiological significance.

III. Whey Proteins

A. *Origin and Function of Whey Proteins*

The proteins in human milk whey, i.e., the proteins remaining soluble after precipitation of caseins, are very diverse. It is possible to separate them into various groups depending on origin (milk proteins, serum proteins) or function (enzymes, binding proteins, immunoglobulins); however, there are no stringent borderlines between these categories and proteins may belong to several groups (e.g., alkaline phosphatase—a serum protein, most likely also a mammary-derived protein—an enzyme, and a zinc-binding protein). In this review, no attempt has been made to subdivide the whey proteins into classes. It should also be noted that the immunoglobulins, which are part of the whey protein fraction, are presented in Chapter 9A. Also, the enzymes in human milk, which are whey proteins, are presented in detail in Chapter 5C. The amino acid composition of several of the major whey proteins is shown in Table III.

B. *α-Lactalbumin*

One milk protein that appears to be present in milk from all species investigated to date is α-lactalbumin. In human milk, α-lactalbumin is a major protein and constitutes 10–20% of total protein. It has a molecular weight of 14.1 kDa and consists of a single polypeptide chain of 123 amino

TABLE III
Amino Acid Composition (Residues) of Human Whey Proteins[a]

	α-Lact-albumin	Lacto-ferrin	Serum albumin	Vitamin B_{12}-BP	FBP (sol.)	FBP (part.)	Vitamin D-BP
Asp	17	71	53	49	21	38	46
Thr	6	31	28	23	8	14	30
Ser	7	50	24	33	13	24	44
Glu	15	70	82	45	24	45	70
Pro	2	35	24	9	12	20	32 or 33
Gly	6	56	12	25	10	18	24
Ala	6	63	62	23	15	25	30
Val	2	49	41	26	7	12	26 or 27
Met	2	6	6	8	4	8	8
Cys	8	32	35	8	nd	nd	8
Ile	12	16	8	22	5	13	10
Leu	14	61	61	37	9	21	56
Tyr	4	20	18	17	7	13	4
Phe	4	31	31	13	14	22	18
Lys	12	46	59	19	8	24	44
His	2	9	16	3	14	18	2
Arg	1	46	24	8	12	16	12
Trp	3	11	1	7	nd	nd	nd

[a]See text for references.

acids (Phillips and Jenness, 1971). The gene for human α-lactalbumin has been cloned and sequenced and the deduced amino acid sequence agreed well with the previously published sequence obtained by amino acid sequencing (Hall *et al.*, 1987). Human α-lactalbumin is not glycosylated and contains no phosphate groups, but it does bind calcium in a 1:1 molar ratio (Lönnerdal and Glazier, 1985). The binding of calcium to α-lactalbumin dramatically changes its Stokes' radius and the protein becomes much more compact. As all α-lactalbumin appears to contain calcium, it is likely that the protein performs its physiological function in this state. It is unlikely, however, that α-lactalbumin has any significant role in calcium transport or absorption as only about 1% of human milk calcium is associated to this protein (Lönnerdal and Glazier, 1985).

α-Lactalbumin has been shown to be part of lactose synthase (EC 2.4.1.22), the enzyme responsible for lactose synthesis in the mammary gland. Lactose synthase consists of two proteins, α-lactalbumin and galactosyltransferase, which together catalyze the binding of glucose to UDP-galactose (Brew and Hill, 1975). The specific function of α-lactalbumin is to modify the catalytic site of galactosyltransferase and promote the binding of glucose to the enzyme part of the lactose synthase complex. Although glucose normally is a poor substrate for galactosyltransferase, α-lactalbumin markedly reduces the K_m of the enzyme for glucose by a factor of 1000, thereby allowing lactose synthesis to proceed at physiological concentrations of glucose (Richardson and Brew, 1980). The concentration of α-lactalbumin is considerably higher than that of galactosyltransferase and it is therefore unlikely that the concentration of α-lactalbumin has any regulatory effect on lactose synthesis. In fact, lactose synthase activity must be very high as human milk is almost saturated with lactose.

Human α-lactalbumin has a very high nutritional value (protein quality) and its amino acid composition appears to be very similar to the estimated amino acid requirement of newborns (Forsum, 1973). It has therefore been suggested that higher levels of α-lactalbumin (bovine) should be used in milk-based infant formulas in order to achieve an amino acid pattern of formula-fed infants that is more similar to that of breast-fed infants (Forsum and Lönnerdal, 1980; Heine *et al.*, 1991). However, bovine α-lactalbumin is not as well digested as human α-lactalbumin (Jakobsson *et al.*, 1982) and therefore the results may not be entirely as expected from amino acid composition data alone. Finally, it should be noted that the term "lactalbumin" previously was used for whey protein; with increased knowledge of whey proteins and the advent of biochemical separation techniques, the protein first isolated was named "α-lactalbumin." Thus, the term lactalbumin does not have any relevance any longer and should not be used (in order to avoid confusion).

C. *Lactoferrin*

As implied by its name, lactoferrin was the first iron-binding protein described in milk (Johansson, 1960). Although lactoferrin is similar in size to transferrin and binds two ferric ions together with carbonate or bicarbonate, it is a different gene product and it behaves completely different in biological systems. Lactoferrin has a molecular weight of 80 kDa and consists of 703 amino acids. Its amino acid sequence has been determined by biochemical methods (Metz-Boutigue *et al.*, 1984) and, following the cloning and sequencing of its gene, also by translating the nucleotide sequence (Powell and Ogden, 1990; Rey *et al.*, 1990).

The three-dimensional structure of lactoferrin has been determined at 2.8 Å resolution (Anderson *et al.*, 1989). The protein consists of two separate globular lobes that are connected via an α-helix. Each lobe binds one iron atom and the lobes have similar folding. The iron atoms are coordinated to four ligands: one histidine, one aspartate, and two tyrosines. Although the term "lactotransferrin" occasionally is used for lactoferrin, it should be recognized that despite similar molecular weight and iron-binding capacity, lactoferrin and transferrin are distinct molecules; their antibodies do not cross-react with each other, the affinity of lactoferrin for iron is much stronger ($K_d \sim 10^{-30}$ *M*) than that of transferrin, and their glycans are of different structure and composition. The structure of the carbohydrate side chains of human lactoferrin has been determined in detail and the side chains consist of an *N*-acetyllactosamine-type glycan with sialic acid, fucose, and galactose as terminals (Spik *et al.*, 1982).

Several physiological functions have been proposed for human milk lactoferrin. The earliest function suggested is related to the high iron-binding capacity of lactoferrin in human milk. As there is much more lactoferrin than iron in human milk (on a molar basis), only a small fraction (3–5%) of the iron-binding capacity of lactoferrin is utilized (Fransson and Lönnerdal, 1980). With the very high affinity of lactoferrin for iron, a bacteriostatic effect may be exerted by lactoferrin through withholding iron from iron-requiring pathogens. *In vitro* studies have supported such an effect of lactoferrin (Bullen *et al.*, 1972; Arnold *et al.*, 1977); however, there is little direct *in vivo* support for this bacteriostatic effect. In fact, studies on the fecal flora of infants fed formula without or with supplemental bovine lactoferrin do not show any significant effect (Balmer *et al.*, 1989). It is possible, however, that human lactoferrin may be more efficient than bovine lactoferrin. More recently, a bactericidal effect of lactoferrin has been suggested (Tomita *et al.*, 1991). A smaller fragment of lactoferrin, which does not contain iron, has been shown to kill certain bacteria at physiologically relevant concentrations. Whether these peptides are formed *in vivo,* though, remains to be documented. Since lactoferrin binds iron in human milk, a role for lactoferrin in iron absorption was proposed (Cox *et al.*, 1979). Again, there are several reports that do not show such an effect (Fairweather-Tait *et al.*, 1987; Schulz-Lell *et al.*, 1991). Even if carefully controlled studies in young infants still are lacking, the finding of a receptor in the small intestine of the infant that is specific for human lactoferrin (Kawakami and Lönnerdal, 1991) supports some physiological role of human milk lactoferrin. Other proposed roles for lactoferrin that need further study are related to lactoferrin as a growth factor (Nichols *et al.*, 1987) or an immunomodulatory factor (Birgens *et al.*, 1983). The presence of intact lactoferrin in the stool of breast-fed infants (Davidson and Lönnerdal, 1987) would also be in agreement with a biological function of lactoferrin in the gut.

D. Bile Salt-Stimulated Lipase

Lipid digestion in breast-fed infants has been documented to be very efficient (Fredrikzon *et al.*, 1978) (see Chapter 5C). It has been shown that part of this high digestibility is due to the presence of lipase in human milk at high concentration, which is stimulated by bile salts (Hernell and Bläckberg, 1983). This enzyme will aid in the formation of absorbable monoglycerides and also in the utilization of long-chain polyunsaturated fatty acids (Hernell *et al.*, 1993). Human milk bile salt-stimulated lipase has a molecular weight of 90 kDa and the protein has been cloned and sequenced (Nilsson *et al.,* 1990). It is a glycoprotein with several tandem repeats and it contains a significant part of O-linked glycans.

E. Lysozyme

Lysozyme is present in human milk at a concentration much higher than that in milk from other species (Chandan *et al.,* 1968) (see Chapter 5C). Its molecular weight is about 15 kDa and it is considered to be synthesized by the mammary gland, although there is no direct evidence for this. The human milk form of lysozyme appears to be identical to lysozyme from saliva, pancreatic juice, and leukocytes, as judged by immunochemical studies and by N-terminal analysis (Parry *et al.,* 1960). Lysozyme can catalyze the hydrolysis of specific bonds between *N*-acetylglucosamine and *N*-acetylmuramic acid in the cell walls of bacteria. Thus, lysozyme can initiate lysis of most gram-positive, but also some gram-negative bacteria. Lysozyme has therefore been suggested to contribute to the bacteriostatic properties of human milk; however, *in vivo* support for this hypothesis is lacking.

F. Serum Albumin

The major serum protein, serum albumin, is also found in human milk. The concentration in milk, however, is about 0.2–0.6 g/liter (Lönnerdal *et al.,* 1976), which is considerably lower than the serum concentration, which is normally 35–50 g/liter. The potential physiological function of serum albumin in human milk has not received much attention, but as this protein can bind many ligands, such as fatty acids, calcium, trace elements, hormones, drugs, etc., it is possible that it may act as a passive carrier of several ligands.

G. Folate-Binding Protein

A specific binding protein for folate has been shown to be present in milk (Ghitis, 1967). This protein appears similar to folate-binding proteins that have been found in serum and tissues from several species. In human milk, folate-binding protein is found in the whey (Waxman and Schreiber, 1975), but it has also been found in a membrane-bound form (Antony *et al.,* 1982). The whey form has a molecular weight of 25–27 kDa and it is glycosylated to about 22% of its molecular weight. It appears that this carbohydrate composition is variable, which may explain the occurrence of multiple forms. The membrane-bound form has a higher molecular weight, with estimates of 160 kDa or higher.

It has been suggested that milk folate-binding proteins may facilitate the uptake of folate in the gut. Studies in newborn goats have shown that the folate-binding protein may survive low gastric pH and limited proteolysis and appears in intact form in the small intestine (Salter and Mowlem, 1983). By using rat intestinal cells, Colman *et al.* (1981) found that uptake of folate was higher from folate-binding protein than from the free form. Other investigators have obtained different results (Said *et al.*, 1986) and hypothesize that folate-binding protein may slow down the release and uptake of folate in the proximal intestine and allow more gradual release and absorption of folate which may increase tissue utilization. The latter hypothesis would be in agreement with observations that protein-bound folate is less available to folate-requiring bacteria than free folate (Ford, 1974).

An alternative function for folate-binding protein in milk is that it would serve as a "trap" to ensure transfer of folylpolyglutamates into milk (Ford, 1974). It has been shown that human milk folate concentration is correlated to the concentration of folate-binding protein in milk (Selhub *et al.*, 1984).

H. Vitamin B_{12}-Binding Protein (Haptocorrin)

Vitamin B_{12} (cobalamin) has been shown to be bound to a specific binding protein in human milk. This protein is glycosylated to about 33% of its weight and it has a molecular weight of about 102 kDa (Sandberg *et al.*, 1981). It is of the R type and should, according to recent nomenclature, be classified as a haptocorrin. Transcobalamin II is also present in human milk, but its concentration is considerably lower than that of haptocorrin. The concentration of haptocorrin in human milk is considerably higher than that of cobalamin, which means that it has a high binding capacity for vitamin B_{12} (Samson and McClelland, 1980). This, in turn, has led some researchers to suggest that haptocorrin may have a bacteriostatic function by withholding cobalamin from vitamin B_{12}-requiring bacteria (Gullberg, 1973). It has been shown *in vitro* that cobalamin bound to haptocorrin is not taken up by intestinal pathogens, while free vitamin B_{12} is.

It has been proposed that haptocorrin may facilitate the absorption of cobalamin in the newborn. In newborn piglets, it was shown that suckled animals retained a significantly higher portion of haptocorrin-bound cobalamin than artificially reared piglets. This difference was considerable at 7 days of age, while at 15 days of age, no significant difference was found (Trugo *et al.*, 1985a,b). Thus, in the newborn, when the production of intrinsic factor is low, milk haptocorrin may assist in the uptake of cobalamin in the intestine. Such a scenario is supported by the fact that *in vitro* proteolysis of cobalamin-saturated haptocorrin is considerably lower than that from unsaturated haptocorrin. Binding studies in brush border membrane vesicles have shown saturation kinetics and it is possible that haptocorrin can facilitate cobalamin absorption by a receptor-mediated process. Thus, haptocorrin could potentially prevent acquisition of cobalamin by intestinal bacteria and therefore limit their growth, while at the same time it facilitates the uptake of cobalamin by the small intestine.

I. Vitamin D-Binding Protein

Human milk has been shown to contain a vitamin D-binding protein which appears identical to that reported to be present in plasma (Hollis *et al.*, 1986). The concentration of this protein in milk, however, seems to be only 3% of the concentration in plasma. It has a molecular weight of 59 kDa and it consists of one polypeptide chain. A single binding site appears to recognize both vitamin D_2 and D_3, as well as their hydroxylated analogues (Haddad and Walgate, 1986). Little is known yet about the mechanisms regulating the influx of vitamin D and its metabolites from serum to milk and the potential involvement of vitamin D-binding protein.

J. Thyroxine-Binding Protein

A thyroid hormone binding protein has been reported to be present in human milk (Oberkotter *et al.*, 1983). This protein appears similar to the thyroid-binding globulin which is present in serum. This protein, which binds thyroxine at one strong and one weak site, is present in human milk at a concentration of about 0.3 mg/liter. The potential role of this protein in mammary transfer of thyroxine remains to be studied.

K. Corticosteroid-Binding Protein

Human milk has been reported to contain a corticosteroid-binding protein (Payne *et al.*, 1976). Concentrations appear higher in colostrum than in mature milk and the protein has a molecular weight of about 93 kDa. Cortisol and progesterone bind tightly to the protein and the properties of the protein appear similar to the corticosteroid-binding globulin of serum. It has been suggested that the protein may play a role in regulating free and bound progesterone/cortisol in the mammary gland (Payne *et al.*, 1976); however, this has not been explored further.

References

Anderson, B. F., Baker, H. M., Norris, G. E., Rice, D. W., and Baker, E. N. (1989). Structure of human lactoferrin: Crystallographic structure analysis and refinement at 2.8 Å resolution. *J. Mol. Biol.* **209,** 711–734.

Antony, A. C., Utley, C. S., Marcell, P. D., and Kolhouse, J. F. (1982). Isolation, characterization, and comparison of the solubilized particulate and soluble folate binding proteins from human milk. *J. Biol. Chem.* **257,** 10081–10089.

Arnold, R. R., Cole, M. F., and McGhee, J. R. (1977). A bactericidal effect for human milk lactoferrin. *Science* **197,** 263–265.

Atkinson, S. A., Schnurr, C. M., Donovan, S. M., and Lönnerdal, B. (1989). The non-protein nitrogen components in human milk: Biochemistry and potential functional role. *In* "Protein and Non-Protein Nitrogen in Human Milk" (S. A. Atkinson and B. Lönnerdal, eds.), pp. 117–133. CRC Press, Boca Raton, FL.

Atkinson, S. A., Anderson, G. H., and Bryan, M. H. (1980). Human milk: Comparison of the nitrogen composition in milk from mothers of premature and full term infants. *Am. J. Clin. Nutr.* **33,** 811–816.

Azuma, N., Kobayashi, H., and Yamauchi, K. (1981). Variation in casein from individual human milk samples. *Agric. Biol. Chem.* **45,** 1007–1009.

Azuma, N., Kaminogawa, S., and Yamauchi, K. (1984). Properties of glycomacropeptide and para-κ-casein derived from human κ-casein and comparison of human and bovine κ-caseins as to susceptibility to chymosin and pepsin. *Agric. Biol. Chem.* **48,** 2025–2031.

Azuma, N., Kaminogawa, S., and Yamauchi, K. (1985). Reconstitution of human casein micelle and its properties. *Agric. Biol. Chem.* **49,** 2655–2660.

Balmer, S. E., Scott, P. H., and Wharton, B. A. (1989). Diet and faecal flora in the newborn: Lactoferrin. *Arch. Dis. Child.* **64,** 1685–1690.

Bergström, S., Hansson, L., Hernell, O., Lönnerdal, B., Nilsson, A. K., and Strömquist, M. (1992). Cloning and sequencing of human κ-casein cDNA. *DNA Seq. J. DNA Seq. Map.* **3,** 245–246.

Berthou, J., Migliore-Samour, D., Lifchitz, A., Delettre, J., Floch, F., and Jollès, P. (1987). Immunostimulating properties and three-dimensional structure of two tripeptides from human and cow caseins. *FEBS Lett.* **218,** 55–58.

Birgens, H. S., Hansen, N. E., Karle, H., and Kristensen, L. O. (1983). Receptor binding of lactoferrin by human monocytes. *Br. J. Haematol.* **54,** 383–391.

Bonsing, J., and Mackinlay, A. G. (1987). Recent studies on nucleotide sequences encoding the caseins. *J. Dairy Res.* **54,** 447–461.

Borulf, S., Lindberg, T., and Mansson, M. (1987). Immunoreactive anionic trypsin and anionic elastase in human milk. *Acta Paediatr. Scand.* **76,** 11–15.

Brantl, V. (1984). Novel opioid peptides derived from human β-casein. *Eur. J. Pharmacol.* **106,** 213–214.

Brew, K., and Hill, R. L. (1975). Lactose biosynthesis. *Rev. Physiol. Biochem. Pharmacol.* **72,** 105–157.

Brignon, G., Chtourou, A., and Ribadeau-Dumas, B. (1985). Preparation and amino acid sequence of human κ-casein. *FEBS Lett.* **188,** 48–54.

Bullen, J. J., Rogers, H. J., and Leigh, L. (1972). Iron-binding proteins in milk and resistance to Escherichia coli infection in infants. *Br. Med. J.,* 69–75.

Calapaj, G. G. (1968). An electron microscope study of the ultrastructure of bovine and human casein micelles in fresh and acidified milk. *J. Dairy Res.* **35,** 1–6.

Chandan, R. C., Parry, R. M., and Shahani, K. M. (1968). Lysozyme, lipase, and ribonuclease in milk from various species. *J. Dairy Sci.* **51,** 606–607.

Chaplin, L. C. (1984). Studies on micellar calcium phosphate: Composition and apparent solubility product in milk over a wide pH range. *J. Dairy Res.* **51,** 251–335.

Colman, N., Hettiarachchy, N., and Herbert, V. (1981). Detection of a milk factor that facilitates folate uptake by intestinal cells. *Science* **211,** 1427–1429.

Cox, T. M., Mazurier, J., Spik, G., Montreuil, J., and Peters, T. J. (1979). Iron binding proteins and influx of iron across the duodenal brush border. Evidence for specific lactotransferrin receptors in the human intestine. *Biochim. Biophys. Acta* **588,** 120–128.

Davidson, L. A., and Lönnerdal, B. (1987). Persistence of human milk proteins in the breast-fed infant. *Acta Paediatr. Scand.* **76,** 733–740.

Fairweather-Tait, S. J., Balmer, S. E., Scott, P. H., and Ninski, M. J. (1987). Lactoferrin and iron absorption in newborn infants. *Pediatr. Res.* **22,** 651–654.

Ford, J. E. (1974). Some observations on the possible nutritional significance of vitamin B_{12}- and folate-binding proteins in milk. *Br. J. Nutr.* **31,** 243–257.

Forsum, E. (1973). Nutritional evaluation of whey protein concentrates and their fractions. *J. Dairy Sci.* **57,** 665–670.

Fransson, G.-B., and Lönnerdal, B. (1980). Iron in human milk. *J. Pediatr.* **96,** 380–384.

Fransson, G.-B., and Lönnerdal, B. (1983). Distribution of trace elements and minerals in human milk and cow's milk. *Pediatr. Res.* **17,** 912–915.

Fredrikzon, B., Hernell, O., Bläckberg, L., and Olivecrona, T. (1978). Bile salt-stimulated lipase in human milk: Evidence of activity *in vivo* and of a role in the digestion of milk retinol esters. *Pediatr. Res.* **12,** 1048–1052.

Ghitis, J. (1967). The folate binding protein in milk. *Am. J. Clin. Nutr.* **20,** 1–5.

Goldman, A. S., and Goldblum, R. M. (1989). Immunoglobulins in human milk. *In* "Protein and Non-Protein Nitrogen in Human Milk" (S. A. Atkinson, and B. Lönnerdal, eds.), pp. 43–51. CRC Press, Boca Raton, FL.

Greenberg, R., Groves, M. L., and Dower, H. J. (1984). Human β-casein. Amino acid sequence and identification of phosphorylation sites. *J. Biol. Chem.* **259,** 5132–5128.

Gullberg, R. (1973). Possible influence of vitamin B12-binding protein in milk on the intestinal flora in breast-fed infants. *Scand. J. Gastroenterol.* **8,** 497–503.

Haddad, J. G., and Walgate, J. (1976). 25-Hydroxyvitamin D transport in human plasma. *J. Biol. Chem.* **251,** 4803–4809.

Hall, L., Emery, D. C., Davies, M. S., Parker, D., and Craig, R. K. (1987). Organization and sequence of the human α-lactalbumin gene. *Biochem. J.* **242,** 735–742.

Hambraeus, L. (1977). Proprietary milk versus human breast milk in infant feeding. *Pediatr. Clin. North Am.* **24,** 17–36.

Hansson, L., Bergström, S., Hernell, O., Lönnerdal, B., Nilsson, A. K., and Strömquist, M. (1993). Expression of human milk β-casein in Escherichia coli and Saccharomyces cerevisae: Comparison of recombinant protein with native isoforms. *Protein Expression Purification* **4,** 373–381.

Heine, W. E., Klein, P. D., and Reeds, P. J. (1991). The importance of α-lactalbumin in infant nutrition. *J. Nutr.* **121,** 277–283.

Hernell, O., and Bläckberg, L. (1983). Digestion of human milk lipids: Physiologic significance of sn-2monoacylglycerol hydrolysis by bile salt-stimulated lipase. *Pediatr. Res.* **16,** 882–885.

Hernell, O., Bläckberg, L., Chen, Q., Sternby, B., and Nilsson, Å. (1993). Does the bile salt-stimulated lipase of human milk have a role in the use of the milk long-chain polyunsaturated fatty acids? *J. Pediatr. Gastroentrol. Nutr.* **16,** 426–431.

Hollis, B. W., Pittard, W. B., III, and Reinhardt, T. A. (1986). Relationships among vitamin D, 25-hydroxyvitamin D, and vitamin D binding protein concentrations in the plasma and milk of human subjects. *J. Clin. Endocrinol. Metab.* **62,** 41–44.

Johansson, B. G. (1960). Isolation of an iron-containing red protein from human milk. *Acta Chem. Scand.* **14,** 510–512.

Jakobsson, I., Lindberg, T., and Benediktsson, B. (1982). In vitro digestion of cow's milk protein by duodenal juice from infants with various gastrointestinal disorders. *J. Pediatr. Gastroenterol. Nutr.* **1,** 183–191.

Jenness, R. (1985). Biochemical and nutritional aspects of milk and colostrum. *In* "Lactation" (B. L. Larson, ed.), pp. 78–94. University of Iowa Press, Ames.

Johansson, B., and Svennerholm, L. (1956). The content of carbohydrates in casein from different species. *Acta Physiol. Scand.* **37,** 324–329.

Kawakami, H., and Lönnerdal, B. (1991). Isolation and function of a receptor for human lactoferrin in human fetal intestinal brush-border membranes. *Am. J. Physiol.* **261,** G841–G846.

Kunz, C., and Lönnerdal, B. (1990a). Human milk proteins: Analysis of casein and casein subunits by anion-exchange chromatography, gel electrophoresis, and specific staining methods. *Am. J. Clin. Nutr.* **51,** 37–46.

Kunz, C., and Lönnerdal, B. (1990b). Casein and casein subunits in preterm milk, colostrum and mature human milk. *J. Pediatr. Gastroenterol. Nutr.* **10,** 454–461.

Kunz, C., and Lönnerdal, B. (1992). Re-evaluation of the whey protein/casein ratio of human milk. *Acta Paediatr.* **81,** 107–112.

Lönnerdal, B., and Glazier, C. (1985). Calcium-binding by α-lactalbumin in human milk. *J. Nutr.* **115,** 1209–1216.

Lönnerdal, B., Forsum, E., and Hambraeus, L. (1976). A longitudinal study of the protein, nitrogen, and lactose contents of human milk from Swedish well-nourished mothers. *Am. J. Clin. Nutr.* **29,** 1127–1133.

Lönnerdal, B., Keen, C. L., and Hurley, L. S. (1985). Manganese binding proteins in human and cow's milk. *Am. J. Clin. Nutr.* **41,** 550–559.

Lönnerdal, B., Woodhouse, L. R., and Glazier, C. (1987). Compartmentalization and quantitation of protein in human milk. *J. Nutr.* **117,** 1385–1395.

Lönnerdal, B., Bergström, S., Andersson, Y., Hjalmarsson, K., Sundquist, A. K., and Hernell, O. (1990). Cloning and sequencing of a cDNA encoding human milk β-casein. *FEBS Lett.* **269,** 153–156.

Maruyama, S., Nakagomi, K., Tomizuka, N., and Suzuki, H. (1985). Angiotensin-I-converting enzyme inhibitor derived from an enzymatic hydrolysate of casein. II. Isolation and bradykinin-potentiating activity of the uterus and the ileum of rats. *Agric. Biol. Chem.* **49,** 1405–1409.

Mercier, J.-C., and Chobert, J.-M. (1976). Comparative study of the amino acid sequence of the caseinomacropeptides from seven species. *FEBS Lett.* **72,** 208–214.

Metz-Boutigue, M.-H., Jolles, J., Mazurier, J., Schoentgen, F., Legrand, D., Spik, G., Montreuil, J., and Jolles, P. (1984). Human lactotransferrin: Amino acid sequence and structure comparisons with other transferrins. *Eur. J. Biochem.* **145,** 659–666.

Naito, H., Kawakami, A., and Imamura, T. (1972). In vivo formation of phosphopeptide with calcium-binding property in the small intestinal tract of the rat fed on casein. *Agric. Biol. Chem.* **36,** 409–415.

Nichols, B. L., McKee, K. S., Henry, J. F., and Nichols, V. N. (1987). Human lactoferrin stimulates thymidine incorporation into DNA of rat crypt cells. *Pediatr. Res.* **21,** 563–567.

Nilsson, J., Bläckberg, L., Carlsson, P., Enerbäck, S., Hernell, O., and Bjursell, G. (1990). cDNA cloning of human milk bile salt-stimulated lipase and evidence for its identity to pancreatic carboxylic ester hydrolase. *Eur. J. Biochem.* **192,** 543–550.

Oberkotter, L., Tenore, A., Pasquarillo, P., and Zavod, W. (1983). Thyroxine-binding proteins in human breast milk similar to serum thyroxine-binding globulin. *J. Clin. Endocrinol. Metab.* **57,** 1133–1139.

Parry, R. M., Chandan, R. C., and Shahani, K. M. (1960). Isolation and characterization of human milk lysozyme. *Arch. Biochem. Biophys.* **103,** 59–65.

Payne, D. W., Peng, L.-H., Pearlman, W. H, and Talbert, L. M. (1976). Corticosteroid-binding proteins in human colostrum and milk and rat milk. *J. Biol. Chem.* **251,** 5272–5279.

Phillips, N. I., and Jenness, R. (1971). Isolation and properties of human α-lactalbumin. *Biochim. Biophys. Acta* **229,** 407–410.

Powell, M. J., and Ogden, J. E. (1990). Nucleotide sequence of human lactoferrin cDNA. *Nucleic Acids Res.* **18,** 4013.

Raymond, M. N., Bvicas, E., and Mercier, J.-C. (1973). Synthesis of new oligopeptide substrates of chymosin (rennin) and kinetic parameters of their hydrolysis. *Neth. Milk Dairy J.* **27,** 298–302.

Rey, M. W., Woloshuk, S. L., deBoer, H. A., and Pieper, F. R. (1990). Complete nucleotide sequence of human mammary gland lactoferrin. *Nucleic Acids Res.* **18,** 5288.

Richardson, R. M., and Brew, K. (1980). Lactose synthase: An investigation of the interaction site of α-lactalbumin for galactosyltransferase by differential kinetic labeling. *J. Biol. Chem.* **255,** 3377–3385.

Rosen, J. M., Rodgers, J. R., Couch, C. H., Bisbee, C. A., David-Inouye, Y., Campbell, S. M., and Yu-Lee, L.-Y. (1986). Multihormonal regulation of milk protein gene expression. *Annu. N.Y. Acad. Sci.* **478,** 63–76.

Rowland, S. J. (1938). The protein distribution in normal and abnormal milk. *J. Dairy Res.* **9,** 47–57.

Said, H. M., Horne, D. W., and Wagner, C. (1986). Effect of human milk folate binding protein on folate intestinal transport. *Arch. Biochem. Biophys.* **251,** 114–120.

Salter, D. N., and Mowlem, A. (1983). Neonatal role of milk folate-binding protein: Studies on the course of digestion of goat's milk folate binder in the 6-d old kid. *Br. J. Nutr.* **50,** 589–596.

Samson, R. R., and McClelland, D. B. L. (1980). Vitamin B_{12} in human colostrum and milk. *Acta Paediatr. Scand.* **69,** 93–99.

Sandberg, D. P., Begley, J. A., and Hall, C. A. (1981). The content, binding, and forms of vitamin B_{12} in milk. *Am. J. Clin. Nutr.* **34,** 1717–1724.

Sato, R., Noguchi, T., and Naito, H. (1986). Casein phosphopeptide (CPP) enhances calcium absorption from the ligated segment of rat small intestine. *J. Nutr. Sci. Vitaminol.* **32,** 67–76.

Schmidt, D. G. (1982). Associations of caseins and casein micelle structure. *In* "Developments in Dairy Chemistry" P. F. Fox, ed., Vol. 1, pp. 61–86. Applied Science, London.

Schulz-Lell, G., Dörner, K., Oldigs, H. D., Sievers, E., and Schaub, J. (1991). Iron availability from an infant formula supplemented with bovine lactoferrin. *Acta Paediatr. Scand.* **80,** 155–158.

Schusdziarra, V., Holland, A., Schick, R., de la Fuente, A., Klier, M., Maier, V., Brantl, V., and Pfeiffer, E. F. (1983a). Modulation of post-prandial insulin released by ingested opiate-like substances in dogs. *Diabetologia* **24,** 113–116.

Schusdziarra, V., Schick, R., de la Fuente, A., Holland, A., Brantl, V., and Pfeiffer, E. F. (1983b). Effect of casomorphins on somatostatin release in dogs. *Endocrinology* **112,** 1948–1951.

Schusdziarra, V., Schick, R., Holland, A., de la Fuente, A., Specht, J., Maier, V., Brantl, V., and Pfeiffer, E. F. (1983c). Effect of opiate-active substances on pancreatic polypeptide levels in dogs. *Peptides* **4,** 205–210.

Selhub, J., Arnold, R., Smith, A., and Picciano, M. F. (1984). Milk folate binding protein (FBP): A secretory protein for folate. *Nutr. Res.* **4,** 181–187.

Spik, G., Strecker, G., Fournet, B., Bouquelet, S., and Montreuil, J. (1982). Primary structure of the glycans from human lactotransferrin. *Eur. J. Biochem.* **121,** 413–419.

Tomita, M., Bellamy, W., Takase, M., Yamauchi, K., Wakabayashi, H., and Kawase, K. (1991). Potent antibacterial peptides generated by pepsin digestion of bovine lactoferrin. *J. Dairy Sci.* **74,** 4137–4142.

Trugo, N. M. F., Ford, J. E., and Sansom, B. F. (1985a). Vitamin B_{12} absorption in the neonatal piglet. 1. Studies in vivo on the influence of the vitamin B_{12}-binding protein from sow's milk on the absorption of vitamin B_{12} and related compounds. *Br. J. Nutr.* **54,** 245–256.

Trugo, N. M. F., and Newport, M. J. (1985b). Vitamin B_{12} absorption in the neonatal piglet. 2. Resistance of the vitamin B_{12}-binding protein in sows' milk to proteolysis in vivo. *Br. J. Nutr.* **54,** 257–268.

van Halbeek, H., Vliegenhart, J. F. G., Fiat, A.-M., and Jolles, P. (1985). Isolation and structural characterization of the smaller size oligosaccharides from desialylated human κ-casein. Establishment of a novel type of core for a mucin-type carbohydrate chain. *FEBS Lett.* **187,** 81–88.

Waxman, S., and Schreiber, C. (1975). The purification and characterization of the low molecular weight human folate binding protein using affinity chromatography. *Biochemistry* **14,** 5423–5428.

Yamauchi, K., Azuma, N., Kobayashi, H., and Kaminogawa, S. (1981). Isolation and properties of human κ-casein. *J. Biochem.* **90,** 1005–1012.

Yoshikawa, M., Tani, F., Yoshimura, T., and Chiba, H. (1986). Opioid peptides from milk proteins. *Agric. Biol. Chem.* **50,** 2419–2421.

B. Nonprotein Nitrogen Fractions of Human Milk

STEPHANIE A. ATKINSON
BO LÖNNERDAL

I. Acid-Soluble Nitrogen Fraction

The nonprotein nitrogen (NPN) fraction of milk comprises 20 to 25% of the total nitrogen (TN) in human milk (Denis *et al.*, 1919; Erickson *et al.*, 1934; Courtney and Brown, 1930; Lonnerdal *et al.*, 1976a; Hambraeus *et al.*, 1978; Atkinson *et al.*, 1980) compared to only 3 to 5% in cow's milk (Hambraeus, 1977). Classically, the NPN fraction of human milk has been identified as the acid-soluble nitrogen remaining in the supernatant following protein precipitation, usually with trichloroacetic acid (TCA) (Lonnerdal *et al.*, 1976a; Atkinson, 1985), or the dialyzable nitrogen following dialysis of whole milk, using membranes with molecular weight cutoffs of about 12 kDa (Atkinson, 1985). However, since many large-molecular-weight glycoproteins—up to 40 kDa and containing up to 70% carbohydrate by weight—are soluble in the TCA used to precipitate proteins by standard techniques (Bezkorovainy and Nichols, 1976b), this fraction should more accurately be referred to as the acid-soluble nitrogen (ASN).

Interest in the ASN fraction of human milk stems from studies in the early to mid-1900s (Denis *et al.*, 1919; Erickson *et al.*, 1933, 1934; Macy, 1949; Courtney and Brown, 1930). At the time, quantitation of total ASN and the partitioning of ASN into nitrogen derived from urea, uric acid, creatine, creatinine α-amino N, and sometimes ammonia were reported (Table I). However, several studies demonstrated that the total recovery of ASN, as analyzed from the component parts, was incomplete, leaving up to 59% of the total analyzed ASN unidentified (Atkinson *et al.*, 1980; Svanberg *et al.*, 1977; Shahani and Shomer, 1951).

A comparison of more recent studies of ASN constituents in milk with earlier reports demonstrates a surprisingly good agreement (Table I) despite somewhat varying analytical techniques. Most of the variables that must be controlled in milk collection, sample preparation, and analysis when measuring the N components of human milk were recognized by the pioneer investigators in this area (Erickson *et al.*, 1933; Shahani and Shomer, 1951). Thus, there are more consistencies than not in comparing values from early to more recent studies.

HANDBOOK OF MILK COMPOSITION

TABLE I
The Nonprotein Nitrogen Constituents of Term Human Milk

Component	Reference[a]	Stage of lactation					
		Colostrum		Transitional		Mature	
		X	SD	*X*	SD	*X*	SD
Total NPN (mg/liter)	Denis *et al.* (1919)	—	—	—	—	284	45
	Macy (1949)	—	—	480	—	324	57
	Svanberg *et al.* (1977)	—	—	—	—	500	—
	Atkinson *et al.* (1980)	491	—	478	—	422	14
	Donovan and Lonnerdal (1985)	—	—	—	—	400	—
	Atkinson and Schnurr (1993)	586	90	740	198	597	68
% NPN/total N	Denis *et al.* (1919)					17.1	
	Macy (1949)					19.3	
	Svanberg *et al.* (1977)					30.0	
	Atkinson *et al.* (1980)	17.8	—	18.3	—	18.0	
	Donovan and Lonnerdal (1985)					24.0	
	Atkinson and Schnurr (1993)	16.9	—	26.9	—	28.7	
Urea N (μg/liter)	Denis *et al.* (1919)	—	—	—	—	123	21
	Macy (1949)	—	—	111	—	180	24
	Svanberg *et al.* (1977)	—	—	—	—	250	—
	Atkinson *et al.* (1980)	118	—	119	—	121	6
	Atkinson and Schnurr (1993)	118	30	133	45	152	25
	Harzer *et al.* (1984)	147	—	—	—	151	33

TABLE I—continued

Component	Reference[a]	Stage of lactation					
		Colostrum		Transitional		Mature	
		X	SD	*X*	SD	*X*	SD
Amino N (mg/liter)	Denis *et al.* (1919)	—	—	—	—	59	20
	Macy (1949)	—	—	44	—	50	14
	Svanberg *et al.* (1977)					130	—
	Atkinson *et al.* (1980)	37	—	36	—	36	5
	Atkinson and Schnurr (1993)	66	9	49	6	51	9
	Harzer *et al.* (1984)	43	—	49	—	51	20
Creatine N (mg/liter)	Denis *et al.* (1919)					13	2
	Macy (1949)					11	2
	Svanberg *et al.* (1977)					35	—
	Atkinson *et al.* (1980)					< 6.9	—
	Donovan and Lonnerdal (1985)					7.2	—
Creatine N (mg/liter)	Denis *et al.* (1919)					23	5
	Macy (1949)					11	7
					37	—	
Uric acid N (mg/liter)	Donovan and Lonnerdal (1985)					10	—
	Denis *et al.* (1919)					27	8
	Macy (1949)					22	5
	Svanberg *et al.* (1977)					5	
	Atkinson *et al.* (1980)	4.6	—	4.5	—	4.3	0.2

TABLE I—*continued*

		Stage of lactation					
		Colostrum		Transitional		Mature	
Component	Reference[a]	*X*	SD	*X*	SD	*X*	SD
Uric acid N (mg/liter)	Donovan and Lonnerdal (1985)	—	—	—	—	2.4	—
	Atkinson and Schnurr (1993)	5.7	3.9	5.8	0.4	5.2	0.6
Ammonia N (mg/liter)	Svanberg *et al.* (1977)	—	—	—	—	2	—
	Atkinson *et al.* (1980)	2.6	—	2.4	—	2.1	0.3
	Donovan and Lonnerdal (1985)	—	—	—	—	1.6	—
	Atkinson and Schnurr (1993)	10.3	3.2	8.7	0.5	10.6	5.1
Glucosamine N (mg/liter)	Svanberg *et al.* (1977)	—	—	—	—	47	—
	Donovan and Lonnerdal (1985)	—	—	—	—	16	—
	Atkinson and Schnurr (1993)	133	41	142	49	88	22
	Harzer *et al.* (1984)	270	—	—	—	190	—
	Carlson (1985a)	230	—	—	—	150	—
Sialic acid N (*N*-acetylneuraminic acid) (mg/liter)	Atkinson and Schnurr (1993)	60	14	60	9	38	7
	Harzer *et al.* (1984)	56	4	—	—	7	1
	Carlson (1985a)	63	5	39	4	19	3
Unidentified NPN (%)	Denis *et al.* (1919)					14	—
	Macy (1949)					16	—
	Atkinson *et al.* (1980)					59	—
	Donovan and Lonnerdal (1985)					27	—

TABLE I — *continued*

Component	Reference[a]	Stage of lactation					
		Colostrum		Transitional		Mature	
		X	SD	X	SD	X	SD
Carnitine	Atkinson and Schnurr (1993)	1.0	0.8	1.8	0.7	1.0	0.06
	Curry and Warshaw (1978)	1.0	—	—	—	0.7	—

[a]Definition of lactation stage (days postpartum)

Reference	Colostrum	Transitional	Mature
Denis *et al.* (1919)	—	—	Not defined
Macy (1949)	0–5	6–10	15–15 Months
Atkinson *et al.* (1980)	3	8	28
Donovan and Lonnerdal (1985)		Not reported	
Atkinson and Schnurr (1993)	3–5	7 or 8	28–30
Harzer *et al.* (1984)	1–3	5–15	22–36
Carlson (1985a)	0–14	14–28	28–42
Svanberg *et al.* (1977)	—	—	> 30 d
Curry and Warshaw (1978)		Not reported	

II. Components of Acid-Soluble Nitrogen Fraction

Characterization of the ASN has led to the quantification of more than 10 components: peptides (Svanberg *et al.*, 1977); urea, uric acid, and ammonia (Denis *et al.*, 1919; Erikson *et al.*, 1934; Shahani and Sommer, 1951; Atkinson *et al.*, 1980; Svanberg *et al.*, 1977; Forsum and Lonnerdal, 1980; Neville *et al.*, 1984); free amino acids (Atkinson *et al.*, 1980; Lemons *et al.*, 1983; Harzer *et al.*, 1984; Rassin *et al.*, 1978); creatine and creatinine (Denis *et al.*, 1919; Erickson *et al.*, 1934; Atkinson *et al.*, 1980; Neville *et al.*, 1984); nucleic acids and nucleotides (Janas and Picciano, 1982; Kobata *et al.*, 1962; Skala *et al.*, 1982); polyamines (Sanguansermsri *et al.*, 1974; Brosnan and Hu, 1985); carnitine (Borum, 1986; Sandor *et al.*, 1982); choline (Macy, 1949; Zeisel *et al.*, 1986); amino alcohols of phospholipids (Zeisel *et al.*, 1986); aminosugars (Harzer *et al.*, 1984; Bezkorovainy, 1976a, 1979); low-molecular-weight peptide hormones (reviewed by Koldovsky and Thornbury, 1989); and other biologically active compounds such as growth factors (reviewed by Kidwell and Saloman, 1989). The biological significance to the infant fed human milk of many of the low-molecular-weight N components identified in milk is not well established.

A. Urea

The urea content of milk comprises a surprisingly large proportion of the ASN (30 to 50%) and is known to vary with stage of lactation, as can be seen in Table II.

The increasing proportion of ASN represented by urea-N observed as lactation progresses from colostral to mature milk is partly due to the greater quantity of urea N and partly due to the decreasing TN content of milk with increasing lactation (from 3.2 g N/liter in colostrum to 1.7 g N/dl in mature milk) (Atkinson *et al.*, 1980; Donovan and Lonnerdal, 1985).

Several investigators have shown a correlation between blood and milk urea levels (Denis *et al.*, 1919; Erickson *et al.*, 1934; Svanberg *et al.*, 1977). If the origin of milk urea was solely due to passive diffusion from the maternal blood, a constant level of urea N would be expected irrespective of lactational stage. However, urea levels increased with increasing lactational stage, casting some doubt upon the postulated origin of urea in human milk as a diffused substrate from the maternal blood.

B. Peptides

The peptide fraction of human milk accounts for about 60 mg N/liter, equivalent to 3 to 5% of the total amino acids (Svanberg *et al.*, 1977). A recent study (Atkinson, 1991) showed that total peptide N represented 13% of ASN fraction in colostrum and about 8% of ASN in milk after 1

TABLE II
Urea N Content of Human Milk at Different Lactational Stages

Reference	Stage of lactation[a] (mg N/liter)					
	Colostrum		Transitional		Mature	
	X	SD	X	SD	X	SD
Donovan *et al.* (1986)	100	26	—	—	217	20
Atkinson *et al.* (1980)					120	9
Harzer *et al.* (1984)					152	—
Erickson *et al.* (1934)	118	—	119	—	180	24
Atkinson and Schnurr (1993)	118	30	133	45	152	25
% NPN						
Donovan *et al.* (1986)	16	—			50	—
Atkinson *et al.* (1980)					22	—
Harzer *et al.* (1984)					30	—
Erickson *et al.* (1934)	24	—	25	—	36	—
Atkinson and Schnurr (1993)	18	8.9	16.7	5.5	25	2

[a]Definition of lactation stage (days)[b]

Reference	Colostrum	Transitional	Mature
Donovan *et al.* (1986)	1–3		> 28
Atkinson *et al.* (1980)	3	8	28
Harzer *et al.* (1984)			25–36
Erickson *et al.* (1934)			> 30
Atkinson and Schnurr (1993)	3–5	7 or 8	28–30

[b]Refers to days postpartum that milk was collected.

month of lactation. The pattern of decline in peptide N over the first month of lactation (Table I) parallels that of total amino N. The molecular weights of the peptides in human milk may be up to 14,000 kDa (Svanberg *et al.*, 1977; Atkinson, 1985). Presumably, this fraction would include growth factors, such as epidermal growth factors and insulin-like growth factors; and peptide hormones such as thyroid hormones and insulin. These types of peptide compounds are detailed in several reviews (Carlson, 1985b).

Because of the wide variety of peptides present in human milk, quantitative recovery for the purpose of determining the amino acid and N contribution of these peptides to ASN is very difficult.

C. *Free Amino Acids*

Free amino acids (FAA) in human milk have been well characterized (Armstrong and Yates, 1963; Atkinson *et al.*, 1980; Svanberg *et al.*, 1977;

Harzer *et al.*, 1984; Lemons *et al.*, 1983; Rassin *et al.*, 1978; Wurtman and Fernstrom, 1979; Ghadimi and Pecora, 1963; Atkinson *et al.*, 1980). A typical pattern of free amino acids in milk collected serially over the first month of lactation is shown in Table III. For most amino acids, the absolute concentration decreased significantly with progressing lactational stage (Table III). Ghadimi and Pecora (1963) also reported that the concentration of most FAA was higher in colostrum than in transitional or mature milk. Changes in milk FAA with lactational stage have been examined in other previous studies (Atkinson *et al.*, 1980; Rassin *et al.*, 1978; Harzer *et al.*, 1984; Wurtman and Fernstrom, 1979) but because the data were not necessarily longitudinal and were of dissimilar lactational intervals, comparison of these studies is difficult. One striking similarity between studies is with respect to glutamate which consistently appears to increase in concentration with progressing lactation (Table III and Atkinson *et al.*, 1980; Wurtman and Fernstrom, 1979; Harzer *et al.*, 1984; Armstrong and Yates, 1963).

Free amino acids represent 3 to 5% of the total amino acids (Svanberg *et al.*, 1977; Ghadimi and Pecora, 1963; Lemons *et al.*, 1983) or 18 to 24% of the N in the ASN fraction (Harzer *et al.*, 1984; Lemons *et al.*, 1983; Carlson, 1985). Quantitatively, glutamate/glutamine and the nonprotein sulfonic amino acid taurine are the most predominant in the FAA fraction. The physiological significance of taurine in infant nutrition has been extensively reviewed because it has been implicated as a "conditionally essential" amino acid for the neonate (Raiha, 1980; Gaull *et al.*, 1972, 1977; Sturman *et al.*, 1976). The significance of the disproportionate amount of free glutamic acid (about 18% of the total glutamate in milk) relative to other FAA is open to conjecture. Since glutamate actually increases in concentration with progressing lactation (Atkinson *et al.*, 1980; Harzer *et al.*, 1984), while total N is decreasing, it is tempting to speculate on a nutritional role for this FAA. Levels of FAA in milk seem to reflect maternal protein intake, with high dietary protein intakes resulting in greater quantities of milk FAA (Wurtman and Fernstrom, 1979; Forsum and Lonnerdal, 1980; Lindblad and Rahimtoola, 1974). When lactating women consumed diets low in lysine and methionine (Wurtman and Fernstrom, 1979) or tryptophan and lysine (Lindblad and Rahimtoola, 1974), the milk levels of these amino acids in the free form were also low.

D. *Aminosugars:* **N**-*acetylneuraminic Acid,* **N**-*acetylglucosamine and Galactosamine*

Early reports indicated that glucosamine nitrogen (gluN) represented 2.0% of total milk nitrogen (Svanberg *et al.*, 1977) or 9.4% of the ASN (Hambraeus, 1977) in mature milk.

In the early 1950s a growth-promoting factor for *Lactobacillus bifidus* vs pennsylvanicus that had up to 100 times greater activity in human milk

TABLE III
Free Amino Acid Composition(μmol/Liter) of Term Human Milk at Varying Postpartum Stages over the First Month of Lactation[a]

	Postpartum day			
	3–5	7 or 8	14 or 15	28–30
Essential AA				
LEU[b]	306 ± 225^{c,*}	72 ± 20	55 ± 11^{+}	43 ± 8^{+}
THR[b]	143 ± 48*	73 ± 18^{+}	94 ± 13^{+}	97 ± 11^{+}
VAL[b]	192 ± 66*	90 ± 10^{+}	91 ± 9^{+}	77 ± 3^{+}
LYS[b]	275 ± 81*	47 ± 8^{+}	54 ± 7^{+}	35 ± 7^{+}
ILE[b]	76 ± 51*	21 ± 6^{+}	19 ± 6^{+}	12 ± 3^{+}
PHE[b]	31 ± 16*	34 ± 13	18 ± 3^{+}	16 ± 2^{+}
HIS	48 ± 16	38 ± 4	37 ± 10	41 ± 6
MET[b]	46 ± 24*	12 ± 2^{+}	13 ± 0.4^{+}	12 ± 2^{+}
1/2 CYS	40 ± 9	50 ± 6	52 ± 5	60 ± 7
TRP	19 ± 5	26 ± 0.1	—	23 ± 2
Nonessential AA				
GLU + GLN[b]	852 ± 240*	1236 ± 99	1393 ± 193	1656 ± 229^{+}
ASP + ASN[b]	104 ± 38*	56 ± 10^{+}	52 ± 12^{+}	55 ± 9^{+}
SER[b]	130 ± 52*	85 ± 22	7 8 ± 12^{+}	101 ± 12^{+}
PRO[b]	215 ± 140*	57 ± 16 $^{+}$	52 ± 13^{+}	29 ± 6^{+}
ALA	224 ± 77	218 ± 20	189 ± 52	247 ± 29
GLY[b]	59 ± 19	79 ± 12	102 ± 19	94 ± 4
ARG[b]	128 ± 47*	31 ± 3^{+}	27 ± 2^{+}	17 ± 2^{+}
TYR[b]	83 ± 33*	21 ± 6^{+}	17 ± 3^{+}	12 ± 3^{+}
Others				
Taurine	602 ± 47	571 ± 67	606 ± 65	574 ± 54
Phosphoserine[b]	176 ± 27*	91 om3^{+}	88 ± 6^{+}	71 + 5^{+}
Ethanolamine	77 ± 12	126 ± 17	120 ± 11	100 ± 12
PEA	131 ± 22	114 ± 5	110 ± 6	119 ± 0.7
Ornithine[b]	21 ± 7*	22 ± 6	6 ± 2^{+}	5 ± 2^{+}

[a]Atkinson and Schnurr (1993).
[b]Indicates a significant decline in amino acid concentration with increasing lactational stage, $p < 0.05$. Means with different symbols are different by Tukey's studentized range test ($p < 0.05$).
[c]Mean ± SEM ($n = 6$ milk samples at each lactational stage collected serially from six mothers).

than in cow's milk was discovered (Gyorgy, 1953). This was subsequently identified as aminosugar-containing oligosaccharides and glycoproteins (reviewed by Bezkorovainy, 1977). Gyorgy *et al.* (1954) reported that 40–60% of the total "Bifidus factor" activity was contained in the nondialyzable compartment, suggesting that Bifidus factor constituted several compounds of varying molecular size (Gyorgy *et al.*, 1954; Gauhe *et al.*, 1954). Although the total activity of Bifidus factor in colostrum was greater than that for mature milk, the distribution of the factor between the nondialyzable and dialyzable fractions was quite similar (Gyorgy *et al.*, 1954). The growth-stimulating activity of the Bifidus factor was linked to the presence of the aminosugar *N*-acetyl-D-glucosamine (gluNAC) in milk oligosaccharides (Gauhe *et al.*, 1954). Higher molecular weight oligosaccharides which also contained the aminosugar *N*-acetylneuraminic acid (NANA or sialic aid) were poor stimulators of *L. bifidus* growth unless NANA was first removed (Gyorgy *et al.*, 1974). Since N-containing oligosaccharides are soluble in acid, they are not precipitated with TCA and therefore may potentially contribute a significant amount of N from hexosamine N sources to the ASN fraction of milk (see Chapter 8A).

In addition to the N-containing oligosaccharides, glycoproteins exist in human milk which consist of aminosugar-containing oligosaccharide side chains covalently linked to the peptide backbone. They include secretory IgA, κ-casein, and lactoferrin. All glycoproteins contain the aminosugars gluN and galactosamine (galN), with their relative distribution varying from 50:1 to 2:1. Because of this large carbohydrate content, the majority of glycoproteins are soluble in TCA and, hence, contribute additional N sources from the aminosugars to the ASN fraction of milk.

Knowledge of the amino sugar content and distribution in human milk is limited partly due to the absence of standardized methods of analysis. The aminosugars gluN and galN are usually measured posthydrolysis using one of the many modifications of the Elson–Morgan colorimetric assays (Morgan and Elson, 1934). Although these methods allow quantitation of microgram quantities of aminosugars, interference from side reactions (neutral sugars or amino acids + chromogen) can result in an overestimation of aminosugar content in hydrolyzed milk samples (Marshall and Neuberger, 1972). The determination of NANA also involves a colorimetric method, posthydrolysis, which is sensitive and specific for NANA (Warren, 1959; Aminoff, 1969). Because the assay includes a final extraction of NANA in butanol or cyclohexanone, it avoids some of the interference problems present in the Elson–Morgan analysis. Methods using high-performance liquid chromatography (HPLC) analysis have been described for both postcolumn and precolumn derivitization of aminosugars from hydrolysate, and nanamolar amounts can be quantitated with accuracy and precision (Honda, 1984).

The NANA or sialic acid content of human milk at various lactational intervals has been described by Carlson (1985a), who noted an exponential decay in NANA content with increasing lactational stage. This dramatic fall was particularly noteworthy in the ASN fraction obtained after protein precipitation with 5% TCA. The NANA content of ASN dropped from 1138 ± 86 mg/liter milk at 0–2 weeks lactation to 135 ± 16 at 10–28 weeks, where it plateaued (Carlson, 1985a). NANA nitrogen represented 11% of the ASN during the initial postpartum period but dropped rapidly to contribute only 1.5% of the ASN in mature milk (Table I). Using HPLC methods, Atkinson and Schnurr (unpublished) conducted serial measures of NANA in human milk. While the absolute amount of NANA declined significantly over the first month of lactation, the percentage ASN as NANA was maintained between 6 and 9% (Table I).

The few reports of gluN concentrations in early milk indicate a dramatic fall within the first few weeks of lactation (Table I) similar to the pattern described for NANA. Reported values for glucosamine using modern assay techniques are fairly consistent: 350 and 240 mg *N*-acetylglucosamine/liter of milk at 5 and 36 days of lactation, respectively (Harzer *et al.*, 1985); 340 and 226 mg/liter at 3 and 28 days of lactation, respectively (Atkinson *et al.*, 1993); and 267 and 192 mg/liter at 0–14 and 14–28 days, respectively (Carlson, 1985a). The contribution of gluN to the ASN fraction is summarized in Table I. All of the NANA, gluN, and galN in human milk appear to be bound to milk oligosaccharides or glycopeptides since free NANA gluN and galN were not identified in the analysis of the free amino acid fraction of milk (Carlson, 1985; Atkinson and Schnurr, 1993). It is likely that a large amount of the ASN not quantitatively recovered in earlier studies was due to failure to measure the amino sugar N in this fraction.

E. Amino Alcohols

Analysis of human milk for the physiological amino acids yields quantitation of amino alcohols such as phosphoethanolamine (PEA), phosphoserine (PS), and phosphoglyceroethanolamine (Atkinson *et al.*, 1980; Harzer *et al.*, 1984). Additionally, choline—found as phosphatidylcholine (PC), sphingomyelin, and unesterified choline—has been quantitated in milk (Zeisel *et al.*, 1986). Together, the N derived from the amino alcohols may contribute significantly to the ASN of milk. Estimates ranging from 6 to 20 mg N/liter are derived from the amino alcohols PEA and PC alone (Carlson, 1985b). Unesterified choline may contribute an additional 3 to 9 mg N/liter milk, depending on the stage of lactation (see Chapter 6A).

The origin of amino alcohols in mammary secretions is likely via simple diffusion from maternal plasma. With the exception of choline, systematic studies which characterize the amino alcohols of milk as to the diurnal or within-feed variability, changes with lactational stage, or compartmentalization in milk have not been reported. It is presumed these alcohols are primarily situated in lipid membranes in the phospholipid fraction. The one report on amino alcohol levels during the first month of lactation showed that free PEA increased and PS decreased significantly over the first 5 weeks of lactation (Harzer *et al.,* 1984).

F. Carnitine

Carnitine is a quaternary amine which is functionally essential for the transport of long-chain fatty acids into the matrix of the mitochondrion for β-oxidation (Borum, 1986). Carnitine is found in the milk of all species tested and is not subject to variations in maternal dietary or urinary carnitine, (Sandor *et al.,* 1982) or time of day (Snyder and Mitchell, 1983). In human milk, 81% of total carnitine is present in the free form. With progressing lactation, carnitine content of human milk increased, peaking (98.2 μmol/liter) at 2 weeks of lactation and then decreased to 62.3 μmol/liter by 118 days (Snyder and Mitchell, 1983). Over the first month we observed no significant changes in milk carnitine (Table I; 0.1–0.13 mmol/liter = 1–1.8 mg N/liter) (Atkinson and Schnurr, 1993). A decrease in milk carnitine content during the second month of lactation was also observed by other investigators (Sandor *et al.,* 1982).

G. Nucleic Acids, Nucleotides, and Polyamines

Nucleic acids in human milk have been found in concentrations ranging from 100 to 5600 mg RNA/liter milk and from 10 to 120 mg DNA/liter milk (Brosnan and Hu, 1985; Janas and Picciano, 1982). Lactational stage and socioeconomic status appear to be important variables in nucleic acid content of milk; for instance, a Thai population consuming relatively low-protein diets had higher levels of DNA and RNA in milk than their European counterparts (Sanguansermsri *et al.,* 1974). The origin of nucleic acids in milk is unknown. Because milk cells may be disrupted during milk processing, there is every likelihood that these compounds derive from the cellular material. The contribution of N derived from nucleic acids might range from 105 mg N/liter during early lactation to 19 mg N/liter in pooled mature human milk (Carlson, 1985). This represents approximately 20 to 6% of the ASN, respectively.

Quantitative analyses of many monophosphate and disphosphate nucleotides (AMP, CMP, IMP, GMP, UMP, UDP, ADP, and GDP), but not triphosphate nucleotides, have been reported for human milk (Janas and Picciano, 1982; Johke, 1963; Kobata *et al.,* 1962). Cyclic nucleotides (cAMP and cGMP) have also been quantitated (Janas and Picciano, 1982; Skala *et al.,* 1981; Johke, 1963). Estimation of the N derived from all nucleotides in milk would yield only about 3 mg N/liter. Polyamines have been characterized—putrescine, cadaverine, spermidine, and spermine—and quantitated in human milk (Sanguansermsri *et al.,* 1974). However, the N contributed by these compounds is minute—0.05 to 0.2 mg N/liter of milk

(Carlson, 1985). The synthesis of polyamines appears to be an active process in the mammary gland throughout lactation (Brosnan and Hu, 1985). Thus, changes in pattern and amounts of polyamines at various lactational stages (Sanguansermsri *et al.*, 1974) are inconsequential in terms of the ASN content of the milk.

H. Uric Acid and Ammonia

The creatinine, creatine, uric acid, and ammonia contribution to the acid-soluble N fraction of mature milk has been well documented (Table I) (Lonnerdal *et al.*, 1976; Atkinson *et al.*, 1980; Harzer *et al.* 1984; Donovan *et al.*, 1986b). The total N levels of these components represent a very minor proportion of the ASN in human milk and a negligible contribution to the TN received by the infant. Creatinine usually represents less than 20 mg N/liter (Neville *et al.*, 1984; Atkinson *et al.*, 1980). Recent investigations of uric acid N in milk reported values of less than 8 mg/liter (Table I). Within the first month of lactation, ammonia contributes $<2\%$ ASN (Atkinson *et al.*, 1980; Donovan and Lönnerdal, 1985).

III. Factors Affecting Milk Acid-Soluble Nitrogen Composition

The origin of many of the ASN components of human milk is thought to arise from the filtration of metabolic breakdown products directly from the maternal plasma, and/or derived from normal or pathologic metabolism within the mammary gland itself (Denis *et al.*, 1919; Macy, 1949; Wurtman and Fernstrom, 1979). Denis *et al.* (1919) demonstrated a close parallel between blood and milk levels of ASN, particularly for urea. Maternal fever has been associated with an elevated milk ASN content especially in the concentrations of creatinine, creatine, urea, and free amino acids (Erickson *et al.*, 1934). Increased urea and free amino acid levels in milk have also been observed in women receiving a high- versus low-protein diet (20 versus 8% of energy intake) (Forsum and Lönnerdal, 1980) and free amino acid content has been noted to parallel protein quantity and quality of maternal diet (Wurtman and Fernstrom, 1979; Lindblad and Rahimtoola, 1974). Moreover, fasting plasma urea levels were closely correlated to milk urea levels irrespective of diet (Svanberg *et al.*, 1977; Forsum and Lönnerdal, 1980).

Postprandial rises in milk ASN components have recently been reported. Total NPN, urea, glutamate, and taurine concentrations increased at least 1.5-fold in some mothers within 15–45 min following consumption of a high-protein meal (Forsum and Lönnerdal, 1980). Perhaps extremes of dietary protein intake or catabolic stress must be invoked before significant changes in transfer of serum metabolic by-products into milk occur (Donovan *et al.*, 1986).

IV. Quantitative Recovery of Components in the Acid-Soluble Fraction of Milk

It is generally accepted that the ASN content of mature human milk represents approximately 20–25% of the TN (Table I; Erickson *et al.*, 1934; Lönnerdal *et al.*, 1976a; Atkinson *et al.*, 1980; Hambraeus *et al.*, 1978). Both lower (Lemons *et al.*, 1983) and higher (Hibberd *et al.*, 1982) amounts have been reported; however, this may result from indirect calculation of ASN content rather than direct measurement of N in the acid soluble/dialyzed fraction of delipidated milk. The ASN content of human milk was shown to decline with progressing stage of lactation (Hibberd *et al.*, 1982; Lönnerdal *et al.*, 1976b; Atkinson *et al.*, 1980). The absolute amount of ASN (mg N/liter) decreased proportionately with the decrease in TN, from approximately 500–600 mg N/liter in Week 1 (13–21% of TN) to approximately 400 mg N/liter in Week 4 (18–30% of TN) (Atkinson *et al.*, 1980; Lönnerdal *et al.*, 1976b; Hambraeus *et al.*, 1978; Chavalittamrong *et al.*, 1981; Hibberd *et al.*, 1982). It then remained relatively constant for the remainder of lactation (Macy, 1949; Lonnerdal, 1976b; Chavalittamrong *et al.*, 1981). Hibberd *et al.* (1982) have demonstrated a much more dramatic drop in ASN during the first week of lactation from levels of 1000 ± 700 mg N/liter on Day 1 to 500 ± 300 mg N/liter on Day 7, where it remained for the duration of the study (36 days) despite a continuing fall in the protein N and therefore the TN levels. These results emphasize the large amount of variability in milk N levels both between mothers and at different stages of lactation.

Previous attempts to fully characterize and recover all of the ASN from human milk have fallen short of their goal. The early work of Denis *et al.* (1919) and Macy (1949) reported recoveries of up to 86% of the ASN from small-molecular-weight components leaving 14% of this fraction of milk N unidentified.

Svanberg *et al.* (1977) accounted for almost 100% recovery of ASN as nitrogen from FAA, gluN, urea, uric acid, creatine, creatinine, and ammonia. However, the methodological techniques available in previous studies likely led to gross overestimations of the N contributions from creatinine + creatine + NH3 + uric acid, which were reported to comprise almost 25% of the total ASN (Table I). This is in sharp contrast to the less than 5% of the ASN reported recently using more accurate and specific enzyme assays (Donovan and Lönnerdal, 1985; Atkinson *et al.*, 1980; Atkinson and Schnurr, 1993). Consequently, the earlier studies do not accurately depict the true composition of the ASN fraction of human milk.

Atkinson *et al.* (1980) (Figure 1) reported an ASN recovery of 32% at Day 5 postpartum, rising to a 45% recovery at Day 28, from the contributions of FAA, urea, creatinine, uric acid, and ammonia sources in pooled milk samples. In the most recent study of the characterization of the NPN fraction of human milk (Atkinson and Schnurr, 1993), the additional quantitation of peptide N and aminosugar N accounted for the significantly greater recovery of ASN (Figure 1), since absolute quantities and proportions of the other ASN components measured were similar to those found by Atkinson *et al.* (1980). This demonstrates the important contribution of the aminosugars, NANA and gluN, to the total ASN, which, together, represent the greatest proportion of ASN in early lactation.

At 1 month of lactation, 12–14% of the TN in human milk still remains unaccounted for. In the ASN, approximately 30% of N is still of unknown origin (Figure 1). Nitrogenous components that are known to be present in the ASN fraction of human milk but which are not consistently all analyzed in the literature (e.g., nucleotides, nucleic acids, polyamines, creatinine, creatine, choline) represent approximately 50 mg N/liter (Carlson, 1985b), the equivalent of 8% of the ASN in mature milk. Incomplete recovery of nitrogen in the ASN fraction can also occur in amino acids which are vulnerable to acid hydrolysis conditions which were not quantitated (e.g., tryptophan 1/2 cystein, tyrosine, phenyalamine, arginine). Similarly, the aminosugar gluN may account for an even greater quantity of the ASN than reported, since it is particularly sensitive to destruction in the acidic conditions needed for its liberation from the glycoprotein bond. The methods employed in a recent study resulted in a 74.6% recovery of gluN, based upon gluN recovery from an amino acid hydrolyzate standard. If the amount of gluN in full-term milk and preterm milk is "corrected" to 100% recovery, an additional 2–6% of ASN is accounted for. Hence, the N contributions from the above sources (full recovery of gluN, uncharacterized amino acids of peptide origin, and other uncharacterized N-containing compounds) would account for an additional 11 to 16% of ASN, leaving approximately 20% of ASN still uncharacterized.

V. Summary

The partition of TN of human milk into protein and nonprotein fractions has been described since the very early work of Denis *et al.* (1919). Despite advances in the characterization of the component parts of the N fraction of human milk, we have yet to achieve complete quantitative recovery of the TN or of the ASN fraction obtained after protein precipitation with TCA. Part of the difficulty is that no single laboratory has analyzed *all* of the components of ASN in milk on the same samples. Thus, differences in sampling methodologies and quantitative analysis contribute to inaccurate estimation of ASN and the patterns of change with lactation.

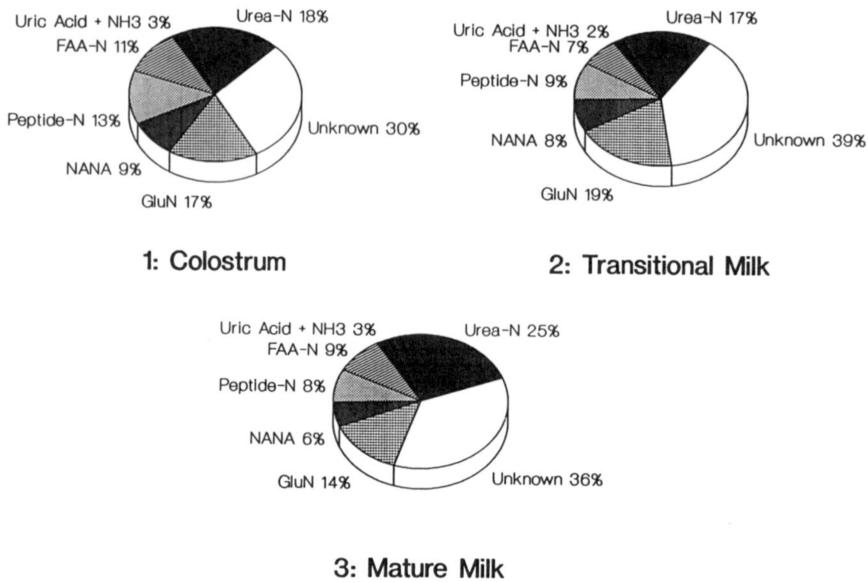

Figure 1 The relative contributions of various nitrogenous components comprising the nonprotein acid-soluble fraction of human milk in (1) colostrum (Days 3–5), (2) transitional milk (Days 7–10), and (3) mature milk (Days 28–30) (Atkinson and Schnurr, 1993).

Complete characterization and quantitation of the amino N and non-amino N in human milk has important ramifications for at least two reasons. First, knowledge of the utilizable (nutritionally available) N in human milk is essential as a basis for estimation of the recommended nutrient intakes of nitrogen (protein) for infants if human milk composition is to be used as a "gold standard." To achieve this we must ascertain the availability of N from components in milk, such as urea, creatinine, and aminosugars, to contribute to the total body pool of utilizable nitrogen.

It is evident that the nitrogen fraction of human milk is composed of a heterogeneous mixture of N-containing substances that are known to be influenced by such maternal variables as gestational interval and lactational stage. A significant proportion of the N fraction of human milk—the so-called nonprotein or acid-soluble nitrogen—is of nonamino origin and represents 20 to 25% of the TN. Contribution of the component parts of the ASN to either the nutritional value of human milk and/or the special biological value of human milk for the human infant have, for the most part, yet to be clarified.

References

Aminoff, D. (1961). Methods for the quantitative estimation of *N*-acetylneuraminic acid and their application to hydrolysates of sialomucoids. *Biochem. J.* **81,** 384–392.

Armstrong, M. D, and Yates, K. N. (1963). Free amino acids in milk. *Proc. Soc. Exp. Biol. Med.* **113,** 660–683.

Atkinson, S. A. (1985). Characterization of the non-protein nitrogen fractions of human milk: Methodological considerations. *In* "Human Lactation, Milk Components and Methodologies" (R. G. Jensen and M. Neville, eds.), pp. 39–44. Plenum Press, New York.

Atkinson, S. A. (1991). Protein metabolism: Nitrogenous sources for the newborn. *In* "Fetal and Neonatal Medicine: Physiology and Pathophysiology" (R. A. Polin and W. W. Fox, eds.), Chap. 42, pp. 443–449. Saunders, Philadelphia.

Atkinson, S. A., Schnurr, C., Donovan, S. M, and Lonnerdal, B. (1989). The non-protein nitrogen components in human milk: Biochemistry and potential functional roles. *In* "Protein and Non-Protein Nitrogen in Human Milk" (S. A. Atkinson and B. Lonnerdal, eds.), pp. 117–172. CRC Press, Boca Raton, FL.

Atkinson, S. A., Anderson, G. H., and Bryan, M. H. (1980). Human Milk: Comparison of the nitrogen composition in milk from mothers of premature and full term infants. *Am. J. Clin. Nutr.* **33,** 811–815.

Atkinson, S. A., and Schnurr, C. (1993). Unpublished results.

Bezkorovainy, A., and Nichols, J. H. (1976a). Glycoproteins from mature human milk whey. *Pediatr. Res.* **10,** 1–5.

Bezkorovainy, A., Nichols, J. H., and Sly, D. A. (1976b). Proteose-peptone fraction of human and bovine milk. *Int. J. Biochem.* **7,** 639–645.

Bezkorovainy, A. (1977). Human milk and colostrum proteins: A review. *J. Dairy Sci.* **60,** 1022–1037.

Bezkorovainy, A., Grohyan D., and Nichols, J. H. (1979). Isolation of glycopolypeptide fraction with *Lactobacillus bifidus* subspecies pennsylvanicus growth-promoting activity from whole humans milk casein. *Am. J. Clin. Nutr.* **32,** 1428–1432.

Borum, P. R. (1986). Human milk carnitine. *In* "Human Lactation 2: Maternal and Environmental Factors" (M. Hamosh and A. S. Goldman, eds.), pp. 335–337. Plenum Press, New York.

Brosnan, M. E., and Hu, Y-W. (1985). Synthesis and function of polyamines in mammary gland during lactation. *In* "Recent Progress in Polyamine Research" (L. Selmeci, M. E. Brosnan, and N. Seiler, eds.), pp. 169–180.

Carlson, S. E. (1985a). N-acetylneuraminic acid concentrations in human milk oligosaccharides and glycoproteins during lactation. *Am. J. Clin. Nutr.* **41,** 720–726.

Carlson, S. E. (1985b). Human milk nonprotein nitrogen: Occurrence and possible functions. *In* "Advances in Pediatrics" (L.A. Barness, ed.), Vol. 32, pp. 43–70.

Chavalittamrong, B., Suanpan, S., Boonvisut, S., *et al.* (1981). Protein and amino acids of breast milk from Thai mothers. *Am. J. Clin. Nutr.* **34,** 1126–1130.

Courtney, A. M., and Brown, A. (1930). The protein and non-protein fractions of some samples of woman's milk. *Arch. Dis. Child.* **5,** 36–39.

Curry, E., and Warshaw, J. B. (1978). Higher serum carnitine levels and ketogenesis in breast-fed as compared to formula-fed infants. *Pediatr. Res.* **12,** 504. [Abstract]

Denis, W., Talbot, F. B., and Minot, A. S. (1919). Non-protein nitrogenous constituents of human milk. *J. Biol. Chem.* **39,** 47–51.

Donovan, S. M., and Lonnerdal, B. (1985). Non-protein nitrogen constituents in mature human milk and colostrum. *Fed. Proc.* **44,** 1676. [Abstract]

Donovan, S. M., Dewey, K. G., and Lonnerdal, B. (1986). Postprandial changes in the non-protein nitrogen content and composition of mature human milk. *Fed. Proc.* **45,** 816. [Abstract]

Erickson, B. N., Stoner, N., and Macy, I. G. (1933). Human milk studies: *XIV*. A critique of the determinations of nitrogenous constituents. *J. Biol. Chem.* **103,** 235–247.

Erickson, B. N., Gulick, M., Hunscher, H. A., and Macy, I. G. (1934). Human milk studies. XV. The nonprotein nitrogen constituents. *J. Biol. Chem.* **106,** 145–159.

Forsum, E., and Lönnerdal, B. (1980). Effect of protein intake on protein and nitrogen composition of breast milk. *Am. J. Clin. Nutr.* **33,** 1809–1813.

Gauhe, A., Gyorgy, P., Hoover, J. R. E., Kuhn, R., Rose, C. S., Ruelius, H. W., and Zilliaen, F. (1954). Bifidus factor IV. Preparations obtained from human milk. *Arch. Biochem. Biophys.* **48,** 214–224.

Gaull, G. E., Sturnam, J. A., and Raiha, N. C. R. (1972). Development of mammalian sulfur metabolism: Absence of cystathionase in human fetal tissues. *Pediatr. Res.* **6,** 527–538.

Gaull, G. E., Rassin, D. K., Raiha, N. C. R., and Heinonen, K. (1977). Milk protein quantity and quality in low-birth-weight infants. III. Effects on sulfur amino acids in plasma and urine. *J. Pediatr.* **90,** 348–355.

Ghadimi, H., and Pecora, P. (1963). Free amino acids of different kinds of milk. *Am. J. Clin. Nutr.* **13,** 75–81.

Gyorgy, P. (1953). A hitherto unrecognized biochemical difference between human milk and cow's milk. *Pediatrics* **11,** 98–108.

Gyorgy, P., Hoover, J. R. E., Kuhn, R. and Rose, C. S. (1954). Bifidus Factor III. The rate of dialysis. *Arch. Biochem. Biophys.* **48,** 209–213.

Hambraeus, L., Lonnerdal, B., Forsum, E., and Gebre-Medhin, M. (1978). Nitrogen and protein components of human milk. *Acta. Paediatr. Scand.* **67,** 561–565.

Hambraeus, L. (1977). Proprietary milk versus human breast milk in infant feeding. A critical appraisal from the nutritional point of view. *Pediatr. Clin. North Am.* **24,** 17–36.

Harzer, G., Franzke, V., and Bindels, J. G. (1984). Human milk nonprotein nitrogen components: Changing patterns of free amino acids and urea in the course of early lactation. *Am. J. Clin. Nutr.* **40,** 303–309.

Hibberd, C. M., Brooke, O. G., *et al.* (1982). Variation in the composition of breast milk during the first 5 weeks of lactation: Implications for the feeding of preterm infants. *Arch. Dis. Child.* **57,** 658–662.

Honda, S. (1984). High-performance liquid chromatography of mono- and oligosaccharides. *Anal. Biochem.* **140,** 1–47.

Janas, L. M., and Picciano, M. F. (1982). The nucleotide profile of human milk. *Pediatr. Res.* **16,** 659–662.

Johke, T. (1963). Acid-soluble nucleotides of colostrum, milk and mammary gland. *J. Biochem.* **54,** 388–397.

Kidwell, W. R., and Salomon D. S. (1989). Growth factors in human milk: Sources and potential physiological roles. *In* "Protein and Non-Protein Nitrogen in Human Milk" (S. A. Atkinson and B. Lonnerdal, eds.), pp. 77–92. CRC Press, Boca Raton, FL.

Kobata, A. (1963). The acid soluble nucleotides of human milk. *J. Biochem. (Tokyo)* **53,** 167–175.

Koldovsky, O., and Thornberry, W. (1989). Peptide hormones and hormone-like substances in milk. *In* "Protein and Non-Protein Nitrogen in Human Milk" (S. A. Atkinson and B. Lonnerdal, eds.), pp. 53–66. CRC Press, Boca Raton, FL.

Lemons, J. A., Reyman, D., and Moye, L. (1983). Amino acid composition of preterm and term breast milk during early lactation. *Early Hum. Dev.* **8,** 323–329.

Lindblad, B. S., and Rahimtoola, R. J. (1974). A pilot study of the quality of human milk in a lower socioeconomic group in Kanachi, Pakistan. *Acta. Paediatr. Scand.* **63,** 125–131.

Lönnerdal, B., Forsum, E., and Hambraeus, I. (1976a). A longitudinal study of the protein, nitrogen and lactose contents of human milk from Swedish well-nourished mothers. *Am. J. Clin. Nutr.* **2,** 1127–1133.

Lönnerdal, B., Forsum, E., and Hambraeus, L. (1976b). The protein content of human milk I. A transversal study of Swedish normal material. *Nutr. Rep. Int.* **13,** 125–134.

Macy, I. G. (1949). Composition of human colostrum and milk. *Am. J. Dis. Child.* **78,** 589–603.

Marshall, R. D., and Neuberger A. (1972). Qualitative and quantitative analysis of the component sugars. *In* "Glycoproteins. Their Composition, Structure and Function Part A" (A. Gottschalk, ed.), pp. 270–274. Elsvier, New York.

Morgan, W. T. J., and Elson, L. A. (1934). A colorimetric method for the determination of N-acetylglucosamine and N-acetylchondrosamine. *Biochem. J.* **28,** 988–995.

Neville, M. C., Keller, R. P., Seacat, J., Casey, C. E., Allen, J. C., and Archer, P. (1984). Studies on human lactation within feed and between breast variation in selected components of human milk. *Am. J. Clin. Nutr.* **40,** 635–646.

Raiha, N. C. R. (1980). Protein in the nutrition of the preterm infant. Biochemical and nutritional considerations. *Adv. Nutr. Res.* **3,** 173–206.

Rassin, D. K., Sturman, J. A., and Gaull, G. E. (1978). Taurine and other free amino acids in milk of man and other mammals. *Early Hum. Dev.* **2,** 1–13.

Sandor, A., Pecsuvac, K., Kerner, J., and Alkonyi T. (1982). On carnitine content of the human breast milk. *Pediatr. Res.* **16,** 89–91.

Sanguansermsri, J., Gyorgy, P., and Zilliken, F. (1974). Polyamines in human and cow's milk. *Am. J. Clin. Nutr.* **27,** 859–865.

Shahani, K. M., and Sommer, K. H. (1951). The protein and nonprotein nitrogen fractions in milk. Methods of analysis. *J. Dairy Sci.* **34,** 1003–1009.

Skala, J. P., Koldovsky, O., and Hahn, P. (1981). Cyclic nucleotidase in breast milk. *Am. J. Clin. Nutr.* **34,** 343–350.

Snyder, E. A., and Mitchell, M. E. (1983). The effect of dietary carnitine on excretion of urinary and milk carnitine in adult and lactating women. *In* "Clinical Aspects of Human Carnitine Deficiency" (P. R. Borum, ed.), pp. 43–44. Pergamon Press, Elmsford, NY.

Svanberg, U., Gebre-Medhin, M., Ljungqvist, B.,and Olsson, M. (1977). Breast milk composition in Ethiopian and Swedish mothers. III. Amino acids and other nitrogenous substances. *Am. J. Clin. Nutr.* **30,** 499–507.

Warren, L. (1959). The thiobarbituric acid assay of sialic acids. *J. Biol. Chem.* **234,** 1971–1975.

Wurtman, J. J., and Funstrom, J. D. (1979). Free amino acids, protein, and fat contents of breast milk from Guatemalan mothers consuming a corn-based diet. *Early Hum. Dev.* **3,** 67–77.

Zeisel, S. H., Char, D., and Sheard, N. F. (1986). Choline, phosphatidylcholine and sphingomyelin in human and bovine milk and infant formulas. *J. Nutr.* **116,** 50–58.

C. Enzymes in Human Milk

MARGIT HAMOSH

I. Introduction

Human milk, like the milk of other species, contains numerous enzymes. Although this topic has been reviewed (Jenness, 1979; Shahani *et al.*, 1980; Hamosh *et al.*, 1985; Hamosh, 1986), the first two publications provide little information about the physiological significance of these enzymes. Shahani *et al.* (1980) compared the activity level of several enzymes in human and bovine milk, drawing attention to the great differences in the activity levels of numerous enzymes between the two species. There have been several additional publications within the last 6 or 7 years that have emphasized specific aspects of human milk enzymes, such as their function in the mammary gland, in milk, and in the infant (Hamosh, 1988), their role in nutrient digestion, gastrointestinal function, and nutrient delivery to the infant (Hamosh, 1989), and their protective function against infective agents (Isaacs and Thormar, 1991; Hamosh, 1991) Some reviews have emphasized specific enzymes in milk (Hamosh, 1981; Hernell *et al.*, 1989). This chapter incorporates the information presented by the author in earlier reviews on this topic (Hamosh *et al.*, 1985; Hamosh, 1986) and provides an update on recent publications in this field.

Because human milk provides the only way to investigate the physiology of the lactating human mammary gland [short of obtaining biopsy specimens from healthy nursing mothers (Blackberg *et al.*, 1987)], I also discuss enzymes that are present in milk but function only in the mammary gland.

It seems that the best way to approach a discussion of human milk enzymes is to arbitrarily divide the enzymes into three groups: (a) those that function in the mammary gland, (b) enzymes that might function in the infant, and (c) enzymes present in milk whose function is unknown. Milk enzymes that act in the infant would have to remain active during passage through the infant's digestive system.

This review does not aim to list or discuss the function of all enzymes in human milk; rather, specific enzymes are selected for discussion of their physiological role as components of human milk (Table I).

II. Milk Enzymes Active Mainly in the Mammary Gland

Although the physiology of lactation has been studied in experimental animals (*in vivo* and *in vitro,* in tissue explants or cell cultures), very little

HANDBOOK OF MILK COMPOSITION

TABLE I
General Functions of Enzymes in Human Milk

Function	Enzyme
Protection against protozoa bacteria and viruses	Lysozyme Peroxidase Lipase (LPL, MDL)[a]
Digestion	Amylase Lipase (MDL)
Repair	Sulfhydryl oxidase (SOX)
Transport (metal carrier)	Glutathione peroxidase Alkaline phosphatase Xanthine oxidase
Biosynthesis of milk components	Phosphoglucomutase Lactose synthetase Fatty acid synthetase Thioesterase

[a]LPL, lipoprotein lipase; MDL, milk digestive lipase.

is known about the physiology of lactation in humans. Prepartum mammary secretions and postpartum colostrum and milk can be used as "windows" through which one might obtain information on the function of the human mammary gland in the perinatal period (Table II).

The presence of enzymes in postpartum milk, as opposed to their absence in prepartum secretions, might indicate changes in gene expression associated with lactogenesis. An example is lipoprotein lipase, the enzyme that controls the uptake of lipoprotein fatty acids from the circulation into the mammary gland. This enzyme plays a key role in the

TABLE II
Milk Enzymes That Function in the Mammary Gland

Function	Enzyme
Phosphoglucomutase (PGM)	Galactose synthesis
Galactosyltransferase	Lactose synthesis
Lipoprotein lipase	Regulates transfer of triglyceride, cholesterol, and phospholipid from blood to milk
Antiproteases	Protection of mammary gland from proteolysis (leucocytes, lysosomes)?
γ-Glutamyltransferase	Endo and exocytosis of proteins?
Xanthine oxidase	Secretion of milk fat droplets?
Fatty acid synthetase	Lipogenesis
Thioesterase	Lipogenesis

delivery of long-chain fatty acids (Hamosh *et al.*, 1970), phospholipids, and cholesterol (Zinder *et al.*, 1976) to the lactating mammary gland. Its absence from prepartum mammary secretions (Hamosh, 1986) suggests low activity in the mammary gland, an assumption confirmed by low concentrations of fat (1 g/dl) in these secretions (Bitman *et al.*, 1986). A sharp rise in enzyme activity after birth is paralleled by an increase in milk fat concentrations to 3 or 4 g/dl (Bitman *et al.*, 1983, 1986). A second lipase, the bile salt-stimulated lipase of human milk, is present in early prepartum secretions (Hamosh, 1986). This enzyme is known to function in the intestine of the newborn, where it hydrolyzes dietary fat in the presence of bile salts (Frederikzon *et al.*, 1978; Alemi *et al.*, 1981). It remains to be determined whether this enzyme might also function in the mammary gland before and after parturition, possibly in the intracellular metabolism of fat. The question is, how does the enzyme, which has an obligatory dependence on bile salts (Freudenberg, 1953; Hernell, 1975), act in their absence? Although milk contains bile salts (Forsyth *et al.*, 1983), the concentration is several orders of magnitude lower than in the intestine. It could, however, be that (a) bile salt concentrations are higher in the mammary gland than in milk, or (b) that specific compartmentalization of bile salts and lipase might affect their interaction in the cell. Higher protein and enzyme concentrations in precolostrum and colostrum, compared to mature milk, might be the result of incomplete tight junctions and of the small volumes secreted before the second or third day postpartum. This is true for amylase (Hamosh, 1986; Jones *et al.*, 1982), lysozyme, lactoferrin, and other proteins, such as IgA (Lewis-Jones and Reynolds, 1983, 1985). With respect to these two lipases, there is a marked difference not only in the timing of their appearance in milk (Hamosh, 1986), but also in the change in levels of activity in milk during weaning (Freed *et al.*, 1989a). Thus, lipoprotein lipase activity is absent before delivery and is low in colostrum, increasing in the early period of lactation and decreasing during weaning. Indeed, during the weaning period there is a relationship between milk volume and lipoprotein lipase activity, the latter disappearing from milk when the volume decreases under 100 ml/day. Milk digestive lipase activity, however, seems to be independent of stage of lactation, being present in prepartum secretions as early as 2 months before term delivery and remaining constant in milk during weaning, irrespective of milk volume. It seems, therefore, that some milk enzymes might be constitutive in the mammary gland.

Mammary secretory cells, present in human milk throughout lactation, can also be used to learn about the function of the human mammary gland during lactation. For example, recent studies on lipogenesis in the lactating human mammary gland have used secretory cells isolated from human milk (Thompson and Smith, 1985). The data show that the human mammary gland contains the two enzymes necessary for lipogenesis—fatty acid synthetase and thioesterase II—and that it synthesizes the same type of fatty acids (i.e., C8–C16) as do other mammalian species (Smith and Abraham, 1975). It is possible that the lipogenic system is adaptive and can be repressed by maternal diets of high fat content or stimulated by high-carbohydrate low-fat diets (Insull *et al.*, 1958).

Other cells in milk or subcellular organelles might contain enzymes. Thus, macrophages in milk might contain lipoprotein lipase, as has been shown for macrophages of other tissues (Chait *et al.*, 1982). However, the higher number of macrophages in colostrum than in mature milk and the lower activity of lipoprotein lipase in colostrum than in mature milk suggest that this might not be the case. Milk fat globules, which are surrounded by the mammary secretory cell membrane, contain enzymes that are membrane bound such as xanthine oxidase and alkaline phosphatase. Since human milk fat globules also contain crescents of cytoplasm (Carroll *et al.*, 1985), the latter might be the reservoir of Golgi and endoplasmic reticulum enzymes, such as galactosyl transferase and glucose-6-phosphatase, found in milk.

The activity level of certain enzymes in milk has also been used as an indicator of the efficiency of pasteurization of human milk by different techniques (Rees, 1987). In general, the enzymes in human milk seem to have a more highly organized tertiary structure than the same enzymes from other sources. This results in greater hydrophobicity of human milk enzymes, possibly accounting for the remarkable resistance of many enzymes to proteolysis and denaturation in the infant's gastrointestinal system (Hamosh *et al.*, 1985; Hamosh, 1986). Indeed, there are also differences in the rate of disulfide bond formation (i.e., the acquisition of native, biologically active structure by the regeneration of disulfide bonds of denatured, reduced polypeptides) (Perraudin *et al.*, 1983), which might explain the function of the potent sulfhydryl oxidase of human milk (Isaacs *et al.*, 1984). It was shown that this oxidation proceeds slower with a human milk enzyme than with the identical enzyme (lysozyme) of hen egg white, suggesting a high-energy barrier, which would constitute a limiting step (Perraudin *et al.*, 1983). It is therefore possible that many milk proteins (enzymes included) might depend on enzyme-catalyzed oxidation of reduced sulfhydryl bonds (Claire *et al.*, 1981). There is evidence that suggests differences in isozymes, degree of glycosylation, and enzyme pattern between identical enzymes in the lactating mammary gland and milk and those in other tissues (Hamosh *et al.*, 1985; Hamosh, 1986).

A. Phosphoglucomutase (EC 2.7.5.1)

Phosphoglucomutase (PGM) catalyzes the production of glucose-1-phosphate, the first intermediary in the pathway of synthesis of the galactose moiety of lactose. Phosphoglucomutase (α-D-glucose-1,6-diphosphate:α-D-glucose-1-phosphate phosphotransferase) is the product of three loci: PGM_1, PGM_2, and PGM_3. In most tissues, PGM_1 isozymes account for 85 to 95% of total PGM activity, PGM_2 for 2–5%, and PGM_3 for 1 or 2%. In erythrocytes, PGM_1 and PGM_2 are found in equal amounts, whereas PGM_3 is absent. The PGM patterns in human milk are different and independent from those in the erythrocytes and can be explained on the basis of a distinct PGM_4 locus (Cantu and Ibarra, 1982). This was the first report of a distinct gene for a widely distributed protein being functionally restricted to the lactating mammary gland, since no evidence of its activity has been found in other tissues previously studied.

The antigenic relationship of the human isozymes has recently been examined using antirabbit muscle PGM polyclonal antibodies (Drago *et al.*, 1991). The conclusions of this study are that PGM1 shares no major antigenic determinants with PGM2 and PGM3 isozymes and is therefore structurally distinct, although all three isozymes are single-chain monomers of approx 60 kDa (McAlpine *et al.*, 1970; Whitehouse *et al.*, 1991). There is no expression of PGM1, PGM2, and PGM3 in human milk. The unique human milk PGM4 isozyme, however, showed similar cross-reactivity with PGM1, suggesting close structural similarity. The authors indicate that this antigenic similarity and shared catalytic properties suggest a recent gene duplication with preservation of considerable sequence homology. They also point out that, in the absence of data for unambiguous inherited variation, the data are also consistent with PGM4 being the product of post-transcriptional or post-translational processing of PGM1 gene products.

B. *Galactosyl Transferase (EC 2.4.1.74)*

Galactosyl transferases catalzye the synthesis of the heteropolysaccharide moieties of complex carbohydrates. One of the best known galactosyl transferases is UDP-galactose: *N*-acetylglucosamine galactosyl transferase or A-protein of the lactose synthetase system. This enzyme has been purified from various animal and human body fluids, the amino acid and carbohydrate content of the bovine milk enzyme are known, and recently information about its structure in human milk has been published. Galactosyl transferases from human amniotic and ascites fluids have similar isoelectric focusing patterns, whereas the milk enzyme is less negatively charged (Gerber *et al.*, 1979). The study suggests the milk enzyme contains less sialic acid, possibly as the result of the neuraminidase activity in human milk (Schauer *et al.*, 1976) and that the electrophoretic difference between the enzyme in milk and in other body fluids is of postribosomal rather than genetic origin. Galactosyl transferases are found in the Golgi membranes of many tissues. The binding of the regulatory protein α-lactalbumin to galactosyl transferase increases the latter's affinity for glucose (from a K_m for glucose of 1 M to 1 mM), thus enabling lactose synthesis at physiological glucose concentrations (Brew and Hill, 1975).

Studies of the enzyme prepared from human milk show that it is composed of a 14-kDa polypeptide containing the active site, a polypeptide backbone which is involved in the regulation of enzyme activity by α-lactalbumin, and a third part responsible for insertion into the membrane (Plancke *et al.*, 1987). Recent studies have also investigated the structure of the mucin-type sugar chains of galactosyl transferase purified from human milk (Amano *et al.*, 1991) as well as the specificity of the human milk enzyme for IgA (McGuire *et al.*, 1989).

Glycosyl transferases glycosylate proteins in the endoplasmic reticulum and Golgi system. The reaction seems to be specific to the nature of the protein recipient. For example, among the immunoglobulins acted on by galactosyl transferase in human milk only IgA and IgG serve as substrates, whereas IgM does not (McGuire *et al.*, 1989). Furthermore, among the IgA subclasses, the human milk enzyme seems to be specific for secreted IgA but not for plasma or myeloma IgA. The enzyme purified from human milk has a molecular weight of 64 kDa and a sixfold higher specific activity in colostrum than in milk. Earlier studies on glycosyl transferases in mouse and human milk have shown these enzymes to act on the incomplete peptide chain and to be present in milk in association with the milk fat globule. Because of activity on incomplete peptide chains, it has been questioned whether these enzymes might have any function in milk (Hernell *et al.*, 1989).

C. *Lipoprotein Lipase (EC 3.1.1.3)*

Another enzyme without function in milk is lipoprotein lipase (LPL). This enzyme has, however, a major role in providing the long-chain fatty acids that constitute the major component of milk fat. LPL regulates the uptake of circulating triglyceride fatty acids, cholesterol, and phospholipids by the lactating mammary gland (Hamosh *et al.*, 1970; Scow *et al.*, 1975; Hamosh, 1981). LPL has a central role in providing the lipid constituents of milk. In the mammary gland, LPL is located both in the endothelium (its site of activity) and in the alveolar cells, the site of its synthesis.

Recent studies in the mouse provide evidence that the cellular origin of LPL in the lactating mammary gland might be mammary adipocytes (Jensen *et al.*, 1991), and not, as previously reported (Clegg, 1981) and recently demonstrated in the guinea pig, the mammary epithelial cells (Camps *et al.*, 1990). The evidence for synthesis of LPL by cells other than the secretory epithelial cells of the mammary gland is twofold: absence of extracellular and intracellular LPL in two types of primary mouse mammary epithelial cell cultures and localization of LPL protein and mRNA to interstitial cells located between epithelial structures. The authors postulate that these interstitial cells are "regressed, lipid-depleted mammary adipocytes" (Jensen *et al.*, 1991). In support of such a source is also the origin of LPL in the lung, where the enzyme is synthesized in fibroblasts that accumulate fat and become the "lipid interstitial cells" of the lung, especially prominent in the young (Maksvytis *et al.*, 1985). The marked variation of LPL activity levels in the milk of different species (Hamosh and Scow, 1972; Hamosh and Hamosh, 1983) could be due to differences in the cellular source within the lactating mammary gland and/or to differences in the mechanism of milk secretion (Hamosh and Scow, 1972). Leakage of LPL into milk from ruptured mammary cells damaged in the process of milk expression is not the mechanism of its release into milk (Mehta *et al.*, 1982) and, in light of recent studies (Jensen *et al.*, 1991), neither is "missorting in the intracellular transport system" (Hernell *et al.*, 1989), i.e., transport from mammary secretory cells into milk rather than to the capillary endothelium, its site of action. In the human, the absence of LPL from prepartum mammary secretions (Hamosh, 1986) suggests low activity in the mammary gland, an assumption suggested by low concentrations of fat (1 g/dl) in these secretions (Bitman *et al.*, 1986). A sharp rise in enzyme activity after birth is paralleled by an increase in milk fat concentrations to 3 or 4 g/dl (Bitman *et al.*, 1983, 1986). The transfer of long-chain fatty acids from maternal blood to colostrum (Spear *et al.*, 1992a,b), indicates that LPL activity increases rapidly in the mammary gland after parturition.

Studies of two patients with familial LPL deficiency (type I hyperlipoproteinemia) show that LPL is absent from milk throughout lactation, suggesting that it is also absent from the mammary gland (Berger *et al.*, 1983; Myher *et al.*, 1984; Steiner *et al.*, 1985). The authors suggest that a common or closely related genetic locus might be implicated in the normal synthesis of LPL in different tissues as shown in the cld/cld mouse (Olivecrona *et al.*, 1985). These studies of LPL-deficient patients further highlight the key role of LPL in the control of milk fat content and composition. Thus, milk fat concentration was significantly lower in the patients when compared to normal lactating women; furthermore, its composition also differed, the milk containing higher amounts of lauric (C12:0) and myristic (C14:0) acids and considerably less oleic (C18:1) and especially linoleic acid (C18:2). The higher concentration of fatty acids synthesized in the breast tissue is probably due to the restricted fat intake, as well as to much restricted entry of long-chain fatty acids into the mammary epithelial cells (only nonesterified fatty acids would reach these cells from the circulation). Since long-chain fatty acids or their derivatives inhibit fatty acid synthetase (Bloch and Vance, 1977), their reduced uptake results in enhanced fatty acid synthesis in the mammary gland.

Recent studies show that contrary to earlier reports, most of LPL in human milk is in the skim milk, a shift to the lipid fraction occurring after freezing (Neville *et al.*, 1991; Hamosh *et al.*, 1991). The distribution of LPL in human milk is, thus, similar to this distribution in bovine milk (Korn, 1962). LPL stability is greater in milk than in blood and tissues (Neville *et al.*, 1991; Hamosh *et al.*, 1991; Hamosh and Hamosh, 1983) and is probably enhanced during freezing and thawing by protection of the enzyme by its association with milk lipid.

LPL has no known function in milk or in the newborn but has been implicated in hydrolysis of milk triglycerides and in the release of free fatty acids in human (Castberg and Hernell, 1975) and bovine milk (Jensen and Pitas, 1976) during storage. The increase in milk free fatty acid concentrations may be associated with the anti-infective activity of human milk

(Gillin *et al.*, 1985; Hernell and Blackberg, 1985), a phenomenon previously thought to depend only on bile salt-stimulated lipase activity (Gillin *et al.*, 1983).

D. Fatty Acid Synthetase and Thioesterase

Fatty acid synthetase and thioesterase catalyze *de novo* synthesis of fatty acids and have recently been described for the first time in the secretory cells of the human mammary gland (Thompson and Smith, 1985). Indirect evidence suggests that these enzymes may be regulated by the cellular concentration of long-chain fatty acids (see above).

Recent studies that have evaluated indirectly the activity of fatty acid-synthesizing enzymes in the human mammary gland by quantitating concomittantly milk and serum fatty acids (Spear *et al.*, 1922a,b) show that in the human, as in other species, parturition, irrespective of length of pregnancy, is the trigger for the synthesis of medium-chain fatty acids. Whether this is dependent upon hormonal changes associated with delivery and onset of lactation or with removal (Martyn *et al.*, 1981) of milk from the mammary gland (by suckling or pumping), or a combination of both, remains to be determined. As in other species (Bitman *et al.*, 1985), the rate of synthesis of medium-chain fatty acids increases with length of lactation (Bitman *et al.*, 1983; Harzer *et al.*, 1983).

E. γ-Glutamyl Transferase (EC 2.3.2.2)

γ-Glutamyl transferase activity is high in human colostrum, and although activity decreases thereafter, considerable amounts of enzyme are present in transitional and mature milk (Brinkley *et al.*, 1975; Patil and Rangnekar, 1982). The enzyme catalyzes the transfer of the γ-glutamyl group, the receptors differing according to the source of γ-glutamyl transferase (the renal enzyme utilizes different amino acid and peptide receptors than the milk enzyme). In addition to kidney, appreciable amounts of γ-glutamyl transferase are present in liver, pancreas, prostate, and breast cyst fluid. It has been suggested that the enzyme is localized in the Golgi apparatus and that it plays a role in the endo- and/or exocytotic transport of proteins. Indeed, the human and bovine milk enzyme is also associated with the membrane fraction of skim milk (Isaacs, 1985) or with the milk fat globules (Kitchen, 1974). The high levels of γ-glutamyl transferase in the serum of newborn infants have been attributed to absorption of the intact enzyme from breast milk (Patil and Rangnekar, 1982).

During storage at 4 or −20°C there is a 30% loss of total enzyme activity without change in compartmentalization between skim milk and skim milk membranes (fluff layer) (Isaacs, 1985). Storage led to an increase in enzyme activity in the fluff layer. The authors suggest that the latter could be due to an increase in the production of the fluff layer or to an altered membrane structure. Studies in the rat show that activity in the mammary gland changes during pregnancy, lactation, and involution (Pocius *et al.*, 1980). The timing and magnitude of these changes and the involvement of prolactin suggest a role in mammary gland function, possibly in the utilization of glutathione (Pocius *et al.*, 1980).

F. Xanthine Oxidase (EC 1.2.3.2)

Xanthine oxidase is a major component of the milk fat globule membrane in bovine milk (Jenness, 1979; Shahani *et al.*, 1980). It has been suggested that this enzyme has a function in the secretion of milk fat droplets, possibly by changing the fluidity of the plasma membrane by peroxidizing membrane-associated lipids (Mather and Keenan, 1983). The very low activity of xanthine oxidase in human milk suggests major differences in the composition of the milk fat globule membrane and possibly in the mechanism of secretion of fat globules into the milk of these two species. In addition to its function in the mammary gland, the enzyme might also act as a metal carrier thereby also having a function in the newborn.

Xanthine oxidase has specific binding sites for iron (eight atoms per molecule), which are important for its enzymatic activity, and for molybdenum (two atoms per molecule) (Lonnerdal *et al.*, 1981; Rumball and Baker, 1985).

Thirty-three percent of the iron in human milk is bound to xanthine oxidase (Fransson and Lonnerdal, 1983). Xanthine oxidase has been purified from human milk and polyclonal antibodies against the purified enzyme (a dimer of mw 244 kDa) have been prepared in rabbits (Graham *et al.*, 1989). A recently purified xanthine oxidase preparation from human milk has been characterized (Abadeh *et al.*, 1992) and the characteristics of the purified enzyme have been compared to those of xanthine oxidase purified from bovine milk. Xanthine oxidase can be purified from human milk at yields comparable to those from bovine milk. The enzyme in human milk is a homodimer with a total M_r of 290,000 with a slightly different ultraviolet/visible absorption spectrum. Activity levels differed among batches of milk and were between 3 and 46 mU/mg protein, two or three orders of magnitude lower than in bovine milk. Further studies showed that human milk probably contains mainly (98%) demolybdo (26%) and desulpho forms of the enzyme. Less than 2% of the human enzyme was active with xanthine (1–6% of activity in bovine milk), whereas activity involving NADH was of the same order of magnitude as that in bovine milk. Similar predominance of "inactive" xanthine oxidase has been described in human heart, consistent with detection of the enzyme by immunohistochemistry but not by enzyme assay (Harrison *et al.*, 1991).

G. Glutathione Peroxidase (EC 1.11.1.9)

There is a strong positive correlation between milk selenium concentration and milk glutathione peroxidase activity (Mannan and Picciano, 1987), and studies indicate that a large portion of the selenium in human milk is present as part of this enzyme (Milner *et al.*, 1987). The same association between selenium and glutathione peroxidase has also been found in cow and goat milk (Debski *et al.*, 1986). Furthermore, geographical variations in milk selenium concentration parallel similar variations in glutathione peroxidase activity of milk. Thus, Mannan and Picciano (1989) reported a mean selenium concentration and glutathione peroxidase activity of 16.8 μg/liter and 77.1 units/liter, respectively, for Illinois subjects, more than twice that of milk of New Zealand women (7.6 μg/liter and 31 units/liter, respectively) (Williams, 1983). Longitudinal studies during the first 16 weeks of lactation in eight women show that human milk selenium content and glutathione peroxidase activity are directly affected by maternal selenium nutrition (Mannan and Picciano, 1987). Both selenium concentrations and glutathione peroxidase activity are significantly higher in hindmilk than in foremilk.

It is interesting that in rural Gambian women milk selenium concentration is affected by maternal nutrition, and during late lactation, by parity, whereas glutathione peroxidase (a selenium-containing enzyme) is not affected by stage of lactation or parity (Funk *et al.*, 1990). Milk of vegetarian women contains higher glutathione peroxidase activity and selenium, although selenium intake was not higher in this population group. The high glutathione peroxidase activity in milk from vegetarians was associated with selenoproteins in the 90–100 kDa. There was a significant correlation between linoleic acid content and milk glutathione peroxidase activity (Ellis *et al.*, 1990; Bitman *et al.*, 1983). Glutathione peroxidase activity and selenium concentrations were also investigated in the milk of mothers of premature infants during the first 6 weeks of lactation. Enzyme activity in "preterm" milk differed from that of "term" milk and was higher in the former at established lactation (21 days postpartum) (Ellis *et al.*, 1990). The activity of glutathione peroxidase in milk of mothers of very premature (26–30 weeks gestation) and premature (31–37 weeks gestation) infants paralleled previously noted changes in long-chain polyunsaturated fatty acid content in human milk with the progression of lactation (Ellis *et al.*, 1990;). It is, therefore, possible that glutathione peroxidase activity is related to the structure and function of these fatty acids in milk. It is also possible that, in human milk, the enzyme has a similar function to that of xanthine oxidase in bovine milk in modulating the fluidity of the plasma membrane-associated lipids. The enzyme could also maintain the integrity of milk by neutralizing the damaging action of oxidants that could be produced by the activity of other milk enzymes such as superoxide dismutase (Willinger *et al.*, 1990) and the sulfhydryl oxidase (Isaacs *et al.*, 1984). Indeed, about half of the peroxidase activity in human milk was found to be associated with selenium-dependent glutathione peroxidase activity (Milner *et al.*, 1987).

Recent studies, in which plasma glutathione peroxidase was purified and partially sequenced, show that 90% of the milk enzyme is immunologically identical to the plasma enzyme (Avissar *et al.*, 1991). This study, which investigated only the milk of two women, reports that 3.6 and 14.3% of selenium was associated with glutathione peroxidase in milk. The molecular weight of the enzyme purified ($\times 4500$) from human milk was shown to be 92 kDa; the native enzyme consists of four identical subunits of 23 kDa (Bhattacharyia *et al.*, 1988). This study shows that glutathione peroxidase provides about 22% of total milk selenium, but only 0.025% of total protein.

III. Milk Enzymes Without Well-Defined Function

There are a number of enzymes without well-defined functions in the mammary gland, in milk, or in the infant. Following are a few examples.

A. Lactate Dehydrogenase (EC 1.1.1.27)

Similar to many enzymes in milk, lactate dehydrogenase (LD) activity is highest in colostrum and decreases as a function of lactation. Recent studies suggest that in addition to changes in enzyme concentration, there is also a change in isozyme pattern from an LD-S maximum in colostrum to LD-1 maximum in transitional milk (Patil and Rangnekar, 1983).

The LD molecule consists of four subunits of two different types, designated H (heart muscle) and M (skeletal muscle). Five different combinations of these subunits are possible, corresponding to LD-1 to LD-5. Cardiac muscle is richest in LD-1 and liver in LD-5. The patterns for colostrum and transitional milk differ from that of maternal serum for which LD-2 and LD-3 are the main isozymes. The change in the isozyme spectrum of the milk enzyme during the early stages of lactation is especially interesting in view of the fact that the LDH isozyme pattern of each organ is considered to be unique.

It is possible that this change in isoforms could be related to a change in the proportion of white blood cells in milk. It has been suggested that the pattern of changes in the activity of lactic dehydrogenase, malic dehydrogenase, and glucose-6-phosphate dehydrogenase of human colostrum and milk is close to that of these enzymes in the newborns' blood and that the former affect carbohydrate metabolism in the newborn (Nabukhotnyi *et al.*, 1986).

B. Plasminogen Activator (EC 3.4.31.21)

Although the enzyme was first reported in milk in 1953, it has only recently been characterized (Yamamoto *et al.*, 1980; Okamoto *et al.*, 1981). The purified enzyme has a molecular weight of 86,000 and was shown to be antigenically different from urokinase, a well-characterized plasminogen activator isolated from urine. Inhibition of activity by diisopropylfluorophosphate indicates that serine is at the active site, as in urokinase.

Research on all aspects of human milk has recently also led to an increased interest in the enzyme content of milk. In many instances, the description of enzymatic activity was followed by purification of the enzymes and by elucidation of possible function. I am listing a few recent reports of enzyme activities in human milk; in many cases function in lactation and in the newborn is as yet unknown.

C. DNase II

DNase II has recently been purified and described at high concentrations in secretory fluids, such as human milk, saliva, and semen, and in leukocytes (Yasuda *et al.*, 1992). This DNase is different from DNase I which is found in tissues, such as pancreas, liver, and kidney, and has been well-characterized biochemically and genetically. Several isosymes of DNase II have been described in urine and are thought to represent variations in the extent of sialilation. The enzyme has not been detected in serum or erythrocytes. Activity levels in milk and saliva are comparable.

D. RNase

RNase activity has been found to be associated with the lactoferrin of human milk (Furmansky *et al.*, 1989). This type of lactoferrin (an 80-kDa iron-binding glycoprotein present in high concentrations in human milk) does not differ from the iron-binding form and is similar in M_r, pI, amino-terminal amino acid sequence, partial proteolytic peptide pattern, and reactivity with monoclonal antibodies or polyclonal sera (Furmansky *et al.*, 1989). The authors suggest that the structural similarity and enzymatic difference could be related to the diverse functions of milk lactoferrin. It has been suggested that this RNase activity interferes with the detection of retrovirus like RNAs in human milk (Hemavathy and Das, 1985).

E. RNase II (EC 3.1.27.5)

RNase II, an enzyme present in bovine milk and colostrum, was found to also be present in human milk but at concentrations amounting to only 1% of those in cow's milk (Meyer *et al.*, 1987). 5-Nucleotidase (EC 3.1.3.5) activity has also been described in human milk (Chuang, 1987).

F. Lipoamidase

Lipoamidase is an enzyme that cleaves the bond between lipoic acid (6,8-dithiooctanoic acid) and ϵ-amino groups of lysine residues. Lipoic aid is a coenzyme of 2-oxoacid dehydrogenases and the function of lipoamidase is probably in the salvage pathway for lipoic acid (Beckman-Gullers *et al.*, 1990). The enzyme is present in bacteria, yeast, and mammalian tissues, with highest activities being reported in liver and kidney. Lipoamidase was recently reported to have high activity in human serum and milk, with activity being three times higher than that of serum (Hayakawa and Oizumi, 1988). The enzyme was purified from milk and was shown to be a glycoprotein of mw 135 kDa, probably consisting of a single polypeptide chain. Activity is completely lost by heating human milk for 5 min at 60°C. Differences were, however, reported between the serum and milk enzymes (Beckman-Gullers *et al.*, 1990). It was suggested that the serum and milk activities might be isoenzymes, the serum enzymes being cystein proteases.

G. Biotinidase (EC 3.5.1.12)

Biotinidase, another enzyme with a function similar to that of lipoamidase, i.e., the salvage of biotin, a cofactor of carboxylases, that is bound to lysine residues of those enzymes (Wolf *et al.*, 1985), was recently described in human milk (Oizumi and Hayakawa, 1989). Comparison between the biotinidase purified from human serum and milk showed that both are thiol enzymes; however, differences were reported in molecular weight (76 and 68 kDa for serum and milk enzymes, respectively) and structure. The authors indicate that the latter suggests differences at or near the active site as well as differences in the mechanism of secretion of the serum and milk enzymes. Activity was found in all milk specimens tested, was higher in colostrum than in mature milk, and was relatively unchanged at various times in lactation (Oizumi and Hayakawa, 1988). Milk biotinidase activity is lower than that of the serum enzyme. Taurine and glutathione were shown to enhance the activity of the milk enzyme (Oizumi *et al.*, 1989). The authors suggest the enzyme might be stable in milk and speculate that it might function in the infant in the absorption of biotin.

H. N-Acetylglucosaminyl Transferase

N-Acetylglucosaminyl transferase was recently reported to be present in human but not in bovine colostrum (Hosomi and Takeya, 1989). Human colostrum contained sufficient enzyme (15–25% of the level in human serum) to permit an investigation of its properties, which were found to be similar to the serum enzyme. The authors suggest that the absence of this enzyme from cow's colostrum and its presence in human colostrum correspond to the presence of oligosaccharides containing lacto-*N*-triose II structures in human colostrum.

I. Phosphatidylinositolglycan-Specific Phospholipase D

Phosphatidylinositolglycan-specific phospholipase D was recently reported to be present in mammalian milk and cerebrospinal fluid but at lower concentrations than that in serum (Hinemo *et al.*, 1991). The enzymes have been characterized in bovine and human sera. No specific function has been attributed to the milk enzymes.

J. Prosaposin

Prosaposin, a sphingolipid hydrolase activator precursor, was isolated from human milk and characterized recently (Hinemo *et al.*, 1989; Kondo *et al.*, 1991). Prosaposins are precursors of saposins, small heat-stable glycoproteins required for the hydrolysis of sphingolipids by specific lysohydrolases. It is suggested that their presence in secretory fluids (in addition to cell lysosomes) indicates functions, such as that of sphingolipid transporter (in milk and pancreatic juice), or by analogy to the apoproteins of serum lipoproteins, a regulatory function for sphingolipid digesting enzymes in the digestive tract. Based on studies in a single lactating woman, the authors show that activity increases rapidly in colostrum with fluctuation during later stages of lactation. They suggest that there might be a relationship with the relatively high concentration of sphingolipids in colostrum.

A comparative analysis of enzyme activities in human colostrum, milk, or serum shows different activity ratios between colostrum and serum, as well as different patterns of change during the transition from colostrum to mature milk, for several enzymes in the colostrum and milk of 14 women (Walentin *et al.*, 1988).

Quantitation of the pattern of change of milk proteins during 9 months of lactation indicates major changes during the initiation of lactation and relatively little change thereafter (Montgomery *et al.*, 1987). The authors suggest that the proteins of human milk (in skim milk and fat globules) are not coordinated to appear simultaneously in the colostral secretion.

IV. Milk Enzymes Important in Neonatal Development

A. Proteolytic Enzymes and Antiproteases in Human Milk

Human milk contains both proteolytic enzymes and protease-inhibiting activity: the net proteolytic activity will therefore depend on the quantitative interaction between the two proteins. These and some other milk enzymes important in human development are listed in Table III.

1. Proteolytic Enzymes

Earlier studies have reported the presence of caseinolytic activity and elastase-like activity. Evidence that plasmin cleaves human β-casein (Greenberg and Groves, 1984) suggests that plasmin activity might be present in human milk. A plasminogen activator has been described in colostrum and early milk (Astrup and Sterndorff, 1953). The origin of milk plasminogen activator is unclear; mammary tissue (Marshall *et al.*, 1986) and milk

TABLE III
Milk Enzymes with Functions in the Infant

Function	Enzyme
Proteases[a]	Hydrolysis of milk proteins?
Antiproteases	Protect bioactive proteins, (enyzmes, immunoglobulins) from hydrolysis in milk and in the intestine of the newborn
α-Amylase	Facilitates digestion of polysaccharides (in milk, formula, and Beikost) by the infant
Milk digestive lipase	Hydrolysis of fat in the intestine of the newborn; bactericidal activity
Sulfhydryl oxidase	Catalyzes oxidation of SH groups: possible role in maintaining structure and function of proteins containing disulfide bonds
PAF-AH[b]	Protection against necrotizing enterocolitis
Lysozyme	Bactericidal
Peroxidase	Bactericidal; present in leukocytes
Glutathione peroxidase	Selenium delivery to the infant
β-Glucuronidase	Breast-milk jaundice?

[a]It is not known whether the proteolytic enzymes of milk are active because of possible interaction with milk antiproteases.
[b]Platelet activating factor acetylhydrolase.

macrophages (LeDeist *et al.*, 1986) have been suggested as the source. Human milk contains little plasminogen activator (Karycka-Danl *et al.*, 1983), with activity being highest during early lactation (first 2 weeks) (Okamoto *et al.*, 1980, 1951). It was suggested that the activator could lead to the production of plasmin in the mammary gland, which in turn could facilitate the flow of milk through narrow ducts (Astrup and Sterndorff, 1953; LeDeist *et al.*, 1986). Plasminogen activator association with membrane permeability (Strickland and Beers, 1976) led to the suggestion that it might affect intestinal permeability in the newborn. It has been suggested that two types of plasminogen activators are present in human milk and that there are differences in their concentrations during lactation (Horie and Okamoto, 1987).

a. Caseinolytic activity. Casein is a major protein in bovine milk, but amounts to only 20% of the total protein of human milk. It forms micelles (Carroll *et al.*, 1985; Ruegg and Blanc, 1982) that are smaller than the casein micelles of bovine milk. The primary structure of human β-casein has recently been determined (Greenberg *et al.*, 1984). A number of smaller peptides, such as γ-casein (Greenberg and Groves, 1984) and galactothermin (Schade and Reinhart, 1970), are probably the products of endogenous human milk proteolytic activity (Greenberg, 1986), as was previously reported for the origin of γ-casein in bovine milk. Small peptides (three to eight amino acids) derived from casein, such as the casomorphines (Brantl, 1984; Yoshikawa *et al.*, 1984), have specific physiologic activity. The function of casomorphines has been reviewed recently (Tashemacher, 1987; Hamosh *et al.*, 1989). The wide-ranging effect of these opioid agonists has been investigated in several species in newborns and adults. In the human, it was suggested that these peptides might be associated with the sleeping pattern of newborns and with postpartum psychosis in some women (Lindstrom *et al.*, 1984).

b. Trypsin (EC 3.4.21.11). Trypsin and elastase activities (Monti *et al.*, 1986; Borulf *et al.*, 1987) have been the latest enzymes added to the list of proteases of human milk. Trypsin, purified by adsorption chromatography, has a mw of 24 kDa. Concentration in milk ranges between 2.9 and 5.6 μg/liter and does not seem to vary during the first month of lactation in a small number of women studied. In contrast to serum and duodenal juice, the trypsin in human milk was found to be anionic trypsin with only traces of cationic trypsin. The two trypsinogens have different isoelectric points and different clearance rates, the rate of clearance of anionic trypsinogen being 10 to 20% of that of the cationic form (Brodrick *et al.*, 1980). It was suggested that the anionic form of the enzyme might be preferentially transported across the mammary epithelial cells. Anionic trypsin, inactive in most milk specimens, was found to be complexed with IgA and not with the protein inhibitors of human colostrum and milk (α-1-antitrypsin and α-1-antichymotrypsin).

To be active in protein digestion by the infant, trypsin must be in its free form in the intestine. The link between trypsin and IgA is split by pepsin (Vojtek and Gjessing, 1971); furthermore, anionic trypsin may play a more important role than cationic trypsin in the activation of proteolytic zymogens in the duodenum (Vojtek and Gjessing, 1971). The possible digestive importance of proteolytic milk enzymes in neonatal protein digestion is suggested by the observation that infants with enteropeptidase deficiency thrive reasonably well when fed human milk (Antonowicz, 1987).

c. Elastase (EC 3.4.21.4). Anionic elastase present in milk has been shown not to originate in leukocytes (Borulf *et al.*, 1987), as was previously thought.

2. *Antiproteases*

The main protease inhibitors in human milk are α-1-antichymotrypsin and α-1-antitrypsin (Lindberg, 1979). Trace amounts of other antiproteases, such as inter-α-trypsin inhibitor, α-2-antiplasmin, α-2-macroglobulin, antithrombin-III, and antileucoprotease, are also present. As previously reported for bronchial lavage (Teguer, 1978) and uterine secretions (Casslen and Ohlsson, 1981), inactive forms of both α-1-antitrypsin and α-1-antichymotrypsin were present. The α-1-antichymotrypsin precipitation pattern differed from that in serum but was identical in milk specimens with and without chymotrypsin-inhibiting activity. The different migration pattern could be the result of limited proteolysis, which under some conditions leads to inactivation or to only changes in electrophoretic mobility (Lindberg *et al.*, 1982). Very high concentration of α-1-antichymotrypsin in colostrum suggests that it may be synthesized in the mammary gland, since no other biological fluids contain such high levels, with the possible exception of seminal plasma.

The physiological function of protease inhibitors is not clear at present. They may protect the mammary gland from local proteolytic activity by leukocytic and lysosomal proteases during different stages of differentiation and lactogenesis or during pathologic conditions such as mastitis, they may prevent the proteolytic breakdown of other enzymes and proteins in milk (and may thus be important in milk banking), and they may affect the absorption of milk proteins (immunoglobulins) in the newborn. Furthermore, the presence of antiproteases would facilitate the delivery of compensatory digestive enzymes (lipase and α-amylase) in active form from milk to the infant. It has been suggested that the antitryptic and antichymotryptic activity of human milk may prevent the absorption of endogenous and bacterial proteases in infants and thereby contribute to the passive protection of extraintestinal organs such as the liver (Udall *et al.*, 1984). The high concentration of antiproteases in colostrum coincides with the period of greatest transfer of nonimmunoglobulin protein from the intestine to the systemic circulation of the newborn (Udall *et al.*, 1985a).

Follow-up studies of infants of mothers with (35 infants) and without (18 infants) milk protease-inhibiting activity failed to show any difference (such as nutritional problems or increased susceptibility to infection) between the two groups (Lindberg *et al.*, 1982). Although this single study of full-term infants does not show a direct beneficial effect of inhibition of the proteolytic activity in milk, a large body of information supports the concept that the newborn of many species benefits from the transfer of macromolecules from maternal milk. In the human, who acquires maternal antibodies mainly *in utero* (Ogra, 1979), breast milk immunoglobulins are also transported across the intestine into the circulation of the newborn infant (Ogra *et al.*, 1977; Iyengar and Salvaroj, 1972, Udall *et al.*, 1981). This transfer is even more important in species that acquire maternal immunoglobulins only postnatally from colostrum and milk (cow, sheep, horse, and pig) or both prenatally and postnatally (rats and mice) (Ogra *et al.*, 1977; Iyengar and Salvaroj, 1972; Krahenbuhl and Campiche, 1969). Protease inhibitors in porcine colostrum (Carlson *et al.*, 1980; Westrom, 1982) have been shown to increase the efficiency of absorption of undergraded colostrum proteins in the intestine of the newborn pig, probably by inhibiting gastrointestinal proteolysis. The delivery of growth factors and hormones (Koldovsky and Thornburg, 1987) in active form from human milk to the newborn infant may also depend on the antiprotease activity of human milk. Estimates of the trypsin inhibitory activities in human milk suggest that 90% of the pancreatic trypsin secreted by the infant in the first 50 min after feeding could be inactivated if all the milk α-1-antitrypsin was biologically active when entering the intestine (Udall *et al.*, 1985b). Although in the adult α-1-antitrypsin is denatured in the stomach, it might remain active in the newborn because of the buffering capacity of human milk. α-1-Antitrypsin remains functionally and immunologically intact in the intestine (Florent *et al.*, 1981). Severe liver disease is less than expected in α-1-antitrypsin-deficient infants fed breast milk (Udall *et al.*, 1985a; Sveger, 1985).

Collagenase inhibitory activity is present in milk collected at 7 to 11 weeks lactation (Waxler *et al.*, 1985). This report of data from a single woman indicates that the collagenase inhibitory activity is associated with a 72-kDa protein (similar to that of α-1-antichymotrypsin and slightly bigger than α-1-antitrypsin, 70 and 50 kDa, respectively), that activity is highest in the earlier milk samples, and that it decreases during prolonged frozen storage (−70°C) of the milk. The authors postulate that invasive processes initiated by bacterial enzymes may be restrained by local enzyme inhibitors, such as anticollagenase and antiproteases, in host tissues or secretions. The presence of such inhibitors in milk may protect both the lactating mammary gland and the infant from bacterial infections and contribute to the stability of milk.

B. Enzymes that Digest Carbohydrate

1. *Amylase (EC 3.2.1.1)*

Amylase was detected in human milk in the last century (Bechamp, 1883); however, its properties and possible functions in the newborn have been investigated only recently. The enzyme digests polysaccharides that are not present in milk, such as starch and glycogen, by hydrolyzing the 1,4-glucan bonds. Amylase might therefore be more important to the newborn after initiation of starch supplements or when formula (which contains oligosaccharides hydrolyzed by amylase) is fed to partially breast-fed infants. At the time of supplementation (which is advised after 4–6 months of exclusive breast-feeding; National Academy of Sciences, 1991), the infant is still deficient in endogenously produced amylase. The latter, secreted from salivary glands and pancreas, does not reach adequate levels until 2 years after birth (Zoppi *et al.*, 1972; Hadorn *et al.*, 1968; Lebenthal, 1980). Indeed, in the newborn, amylase in the duodenum amounts to only 0.2–0.5% of the adult level. Another group of infants and toddlers that might benefit from milk amylase are those with pancreatic insufficiency caused by diseases such as cystic fibrosis (Lindberg and Skude, 1982) or malnutrition (Barbezat and Hansen, 1968; Damis *et al.*, 1970; Watson *et al.*, 1977; Collares and Brasil, 1979; Sauniere and Sarles, 1988). Milk amylase is identical to the salivary isozyme (Jones *et al.*, 1982; Lindberg and Skude, 1982; Fridhandler *et al.*, 1974). Studies of amylase in milk of mothers of preterm and term infants have shown similar levels of activity in both groups (Jones *et al.*, 1982; Hegardt *et al.*, 1984). Activity varies among women but is constant in individual women, beyond the initial phase of lactation, when activity is higher. Amylase activity is present in milk even after prolonged lactation of up to 27 months, maintaining a plateau at 6–27 months (Dewit *et al.*, 1990). This study has also shown that activity is unchanged during feeds or at different times during the day. Amylase is stable during storage at −20 and −70°C for months (Hamosh, 1986) and recent studies show there is no loss of activity even during storage of milk at higher temperatures (15, 25, and 38°C) for 24 hr (Ellis and Hamosh, 1992).

Human milk α-amylase has a broad pH optimum range of 4.5 to 7.5 and loses little activity during incubation for 2 hr at pH 3.0 and above (Jones *et al.*, 1982). Thus, hydrolysis of polysaccharide could start in the stomach (the postprandial pH of milk- or formula-fed infants is in the range of 5.0–6.0) (Armand *et al.*, 1993). The milk enzyme is relatively stable to peptic degradation (Heitlinger *et al.*, 1983) and remains active in the newborn's intestine (Hodge *et al.*, 1983; Lindberg and Skude, 1982). Starch and glucose polymers (from two to seven glucose molecules) protect salivary amylase from inactivation at low pH, whereas lactose, sucrose, and glucose had no effect. Starch also protects salivary amylase at low pH in the presence of pepsin and maltotriose and was shown to protect the enzyme at pH below 3 (Rosenblum *et al.*, 1988). This protection from inactivation in the stomach is similar to the protection of lingual lipase at low pH by lipids (Fink *et al.*, 1984). The level of α-amylase activity is 10 to 60 times higher in milk than in normal human serum (Friedhandler *et al.*, 1974; Heyndrickz, 1962). Little or no α-amylase activity is detected in fresh milk from cows, sheep, goats, or swine. The absence or very low activity of amylase in cow milk compared to that in human milk was previously reported (Jones *et al.*, 1982; Hamosh, 1989). High-parity (above 10 children) women produce milk with only half the amylase activity of primiparous women (Dewit *et al.*, 1990). The authors suggest that breast-feeding can provide infants with a continuing supply of amylase, which might be even more important in malnourished infants and toddlers during the first 2 years of life. Furthermore, supplementation with starch might be better tolerated in breast-fed infants because of high intestinal levels of amylase provided by human milk, as reported from Egypt (Hanafy *et al.*, 1971). A direct antibacterial effect of amylase has also been described. Thus, *Neisseria gonorrheae* is inhibited by salivary amylase (Mollersh *et al.*, 1979). Because milk amylase is of the same isozyme group as salivary amylase, it might likewise inhibit the growth of certain microorganisms. Some characteristics of the enzyme are listed in Table IV.

C. *Enzymes Active in Fat Digestion in the Infant*

1. *Milk Digestive Lipase (EC 3.1.1.3)*

Lipase activity was one of the first enzymatic activities described in human milk (Marfan, 1901). It is now well established that the milk of many species contains lipoprotein lipase (see Section II, C) and that the milk of some primates (Frendenberg, 1966), including the human (Frendenberg, 1953), and of carnivores (Freed *et al.*, 1986; Ellis and Hamosh, 1992) contains an additional lipase that is bile salt-dependent (Hernell and Olivecrona, 1974). Because this lipase can act in the digestion of milk lipid in the newborn, it has generated great interest and has become the most extensively studied enzyme in human milk. The potential importance of this enzyme stems from the fact that endogenous lipid digestive function is not well developed at birth, the newborn being deficient in pancreatic lipase and in bile salts which are necessary for the digestion and solubilization of fat during the digestive process (Watkins, 1974; Hamosh, 1979, 1990b). There is indirect evidence that this digestive lipase of milk improves fat absorption in the newborn (Williamson *et al.*, 1978; Alemi *et al.*, 1981; Wang *et al.*, 1989) and a greater body of evidence gathered from *in vitro* studies suggests that the enzyme remains active in the infant's gastrointestinal tract and therefore, might indeed contribute significantly to fat digestion (Hamosh, 1982; Hernell *et al.*, 1989; Hamosh, 1989). The

TABLE IV
Characteristics of Milk Enzymes Active in Infant Digestion

Characteristic	Enzyme: Amylase	Enzyme: BSSL[a]
Maternal factors		
High parity (≥10)	Low activity	?
Malnutrition	?	Decrease in activity
Diurnal and within feed activity	Constant	Constant
Pattern of secretion		
Prepartum	?	Present
Postpartum		
Presence in preterm (PT) and term (T) milk	Equal activity PT and T	Equal activity PT and T
Pattern through lactation	Colostrum greater than milk	Colostrum lower than milk
Weaning	?	Activity constant independent of milk volume
Distribution in milk	Aqueous phase	Aqueous phase
Effect of milk storage		
(−20, −70°C)	Stable	Stable
(15 to 38°C)	Stable (at least 24 hr)	Stable (at least 24 hr)
Stability to low pH (passage through stomach)	pH > 3.0	pH > 3.0
pH optimum	6.5–7.5	7.4–8.5
Enzyme characteristics	Salivary amylase isozyme	Identical with pancreatic carboxyl ester hydrolase
Evidence of activity in infant's intestine	Yes	Yes
Presence in milk of other species	?	Primates and carnivores

[a]BSSL, bile-salt stimulated or milk digestive lipase.

compensatory role of this enzyme might be especially important in premature infants, whose endogenous digestive system is more immature than that of full-term infants. The reader is advised that the milk digestive lipase has been discussed by various research groups under different names such as "bile salt stimulated lipase," "bile salt activated lipase," "bile salt dependent lipase," and "milk digestive lipase." Because of activity on lipid substrates as well as water soluble esters, the enzyme is also known as "bile salt stimulated esterase."

Great progress has been made recently in our understanding of this enzyme's origin, structure, enzymology, organ and species distribution, and possible function. This review does not permit a discussion of all these interesting aspects. I will, therefore, summarize only topics relevant to the newborn recipients' physiology. The enzyme is identical to carboxyl ester hydrolase, a pancreatic enzyme of wide species distribution. In the human and in carnivore species it is also expressed in the mammary gland (Blackberg *et al.*, 1987; Ellis and Hamosh, 1992). Enzyme activity varies in human milk, being lower in colostrum than in mature milk (Mehta *et al.*, 1982; Freed *et al.*, 1989b). As mentioned in the introduction to this chapter, contrary to other enzymes in milk and especially to the other milk lipase (LPL), milk digestive lipase is present in early prepartum secretions (> 2 months before term delivery) (Hamosh, 1986) and in milk expressed during weaning (Freed *et al.*, 1989a). There is no relationship between the milk volume secreted during these different stages of lactation and the level of digestive lipase activity, suggesting, as discussed above, that this lipase might be a constituent enzyme of the mammary gland. This suggestion is also supported by the high concentration of this lipase in milk (Hernell *et al.*, 1989; Ellis and Hamosh, 1992). Although activity varies among women, it seems to remain constant within each woman (Mehta *et al.*, 1982; Freed *et al.*, 1989b), a characteristic shared with the other milk digestive enzyme, amylase (Dewit *et al.*, 1990). Similar activity levels are present in the milk of women who deliver prematurely or at term (Mehta *et al.*, 1982; Freed *et al.*, 1989a), although one report that compared enzyme activity in these groups only in the initial colostrum stage indicates higher lipase activity in the preterm group than in the term group, irrespective of whether or not the latter included appropriate or small for gestatorial age infants (Pamblanco *et al.*, 1987). Although one earlier study (Gebre-Medhin *et al.*, 1976) reported similar activity levels in milk of well-nourished and undernourished women, two recent studies indicate that the milk of malnourished women has lower digestive lipase activity levels (Ginder *et al.*, 1987; Dupuy *et al.*, 1991) which decrease by 80–90% during the first 4 months of lactation (Dupuy *et al.*, 1991), contrary to the constant activity levels in well-nourished women (Dupuy *et al.*, 1991), even after prolonged lactation (Freed *et al.*, 1989b). This aspect is worrisome because it could adversely affect infants in undernourished areas or during periods of malnutrition. The effect would be not only the inability of mother's milk to provide sufficient digestive lipase especially needed because of the malnutrition-induced decrease in pancreatic digestive function (see Section IV, B, 1 for references), but could also affect the infant's resistance to infection. The latter effect is related to the production of free fatty acids and monoglycerides, products of fat digestion which have anti-infective properties (Kabara, 1980). This antiprotozoan, antibacterial, and antiviral effect is enhanced by the contribution of milk lipases (Gillin *et al.*, 1983, 1985) to the newborn's endogenous lipases, mainly gastric lipase (Canas-Rodriguez and Smith, 1966; Hamosh, 1991). The specific role of milk digestive lipase or lipoprotein lipase in this process is still the subject of debate (Hernell and Blackberg, 1985; Isaacs and Thormar, 1991; Gilin *et al.*, 1991).

The digestive lipase of human (Blackberg and Hernell, 1981; Wang and Jackson, 1983) and ferret (Ellis and Hamosh, 1992) milk has been purified and characterized. The enzyme in human milk has recently been cloned (Nilsson *et al.,* 1990). The existence of two variants of the cDNA for human milk digestive lipase (Baba *et al.,* 1991), as well as the existence of two active forms of the enzyme with molecular masses of 97 and 120 kDa, have been reported (Swan *et al.,* 1992). Thus, some women produce two forms of this enzyme in approximately equal amounts (Swan *et al.,* 1992). The human milk lipase mRNA encodes a 748-residue protein, including a 23-residue signal peptide (Hui and Kissel, 1990). Whether differences in activity levels associated with handling of certain milk specimens (Hall and Muller, 1983; Hamosh, 1982) are the result of different forms of the enzyme (Swan *et al.,* 1992) or of interaction between lactoferrin and lipase (Erlanson-Albertsson *et al.,* 1985) remains to be established.

Differences in molecular weight and levels of activity of milk digestive lipase among human, dog, cat, and ferret have been reported. The lipase in the milk of carnivore species, such as the ferret and cat, is slightly smaller (mw 90 and 75–80 kDa, respectively) (Ellis and Hamosh, 1992; Hernell *et al.,* 1989) than the enzyme in human milk (125 kDa). Activity levels are lower, about equal, and 20-fold higher in cat, dog (Freed *et al.,* 1986), and ferret (Ellis and Hamosh, 1992) milk than in human milk. Although the reason for the presence of this enzyme in the milk of certain species and its absence from milk of other species, such as the cow, donkey, and rabbit, (Freed *et al.,* 1986), remains to be investigated, so far all the species with milk digestive lipase secrete milk containing more than 90% long-chain fatty acids. It was suggested (Hamosh, 1989) that the short- and medium-chain fatty acids in milk of species without digestive lipase can probably be adequately hydrolyzed by the newborn's endogenous lingual and gastric lipases (Hamosh, 1990a). The presence of only long-chain triglyceride in milk, however, necessitates the action of pancreatic and/or milk digestive lipase.

In the human, enzyme characteristics are identical in milk from mothers of preterm and full-term infants (Freed *et al.,* 1987). Furthermore, activity levels are constant and do not change as a function of diurnal or in-feed variation (Freed *et al.,* 1986). Enzyme activity is also remarkably stable during prolonged storage (1 or 2 years) at either −20 or −70°C (Hamosh *et al.,* 1985); furthermore, the lipase is also stable at 15, 25, and 38°C for at least 24 hr (Ellis and Hamosh, 1992). Thus, banked human milk stored frozen maintains its fat-digesting capacity for long periods of time as does the milk that the working woman or the mother of a sick infant might store even at suboptimal conditions for short time periods. Some characteristics of the enzyme are listed in Table IV.

Although indirect evidence in the human suggested a function in the newborn as early as 1978, only recently has this aspect been studied in greater detail. Based on *in vitro* studies that simulate the gastrointestinal environment of the newborn, it is clear that the milk lipase, similar to pancreatic lipase (Cohen *et al.*, 1970; Plucinsky *et al.*, 1979), is unable to penetrate into milk fat globules (Hamosh *et al.*, 1987; Kirk *et al.*, 1991; Hernell *et al.*, 1991). Thus, the initiation of milk fat digestion by gastric lipase is a prerequisite for the subsequent digestion of the fat in the intestine by the combined action of milk and pancreatic lipases (Kirk, *et al.*, 1991; Hernell *et al.*, 1991). The lack of positional or fatty acid specificity of the milk lipase indicates that it is able to hydrolyze completely milk triglycerides. This is an important aspect of this enzyme's function because neither gastric lipase nor pancreatic lipase completely hydrolyze triglycerides; the former produces mainly diglycerides, and the latter monoglycerides (Hamosh, 1990b). It is of great physiological importance that the milk lipase hydrolyzes diglyceride (the product of gastric lipolysis) at higher rates than triglyceride (Wang *et al.*, 1988), whereas monoglyceride (the product of intestinal lipolysis by pancreatic lipase) hydrolysis does not require the presence of bile salts (Hernell and Blackberg, 1982). The lipolysis product of milk lipase, free fatty acids, is more readily absorbed than monoglyceride (Morgan and Borgstrom, 1969) at the low bile salt concentration present in the newborn (Watkins, 1974). Indeed, fat absorption in breast-fed contrary to formula-fed infants is not correlated to bile salt levels (Signer *et al.*, 1974). The low substrate specificity of milk lipase is probably the reason for hydrolysis of retinyl palmitate (Fredrikzon *et al.*, 1978; O'Connor *et al.*, 1988). The high extent of intragastric lipolysis, hydrolysis of 30–60% of milk fat (Iverson *et al.*, 1991; Armand *et al.*, 1993), indicates that the combined action of gastric lipase and milk digestive lipase, hydrolysis of 20–40% of milk fat (Hall and Muller, 1982), could accomplish the process of milk fat digestion in the presence of very little or even in the absence of pancreatic lipase.

Contrary to earlier suggestions of an association of breast milk jaundice with increased levels of free fatty acids produced as a result of higher activity of milk lipases (Constantopoulos *et al.*, 1980), no increase in free fatty acids (Forsyth *et al.*, 1990, Hamosh, and Bitman, 1992) or lipase activity–digestive lipase (Hamosh, 1990a; Forsyth *et al.*, 1990) or lipoprotein lipase (Hamosh, 1990)—was found in the milk of women whose infants developed breast milk jaundice, even when milk bile salt levels were higher (Forsyth *et al.*, 1990) than those in the milk of mothers of healthy infants (Forsyth *et al.*, 1983).

D. Enzymes with Diverse Functions

1. Sulfhydryl Oxidase

Sulfhydryl oxidase is another enzyme present in human and milk of other species (Hamosh *et al.*, 1985; Hamosh, 1986; Isaacs *et al.*, 1984) and may function in both milk and the gastrointestinal system of the newborn. The enzyme is present in colostrum as well as in mature milk. Sulfhydryl oxidase catalyzes the oxidation of sulfhydryl groups using O_2 as oxidant and producing equimolar quantities of H_2O_2 and the corresponding disulfide. Substrate specificity: the enzyme is acting on both small thiol compounds and protein, and might be essential to initiate or maintain the activity of proteins whose structure and function depend on intact disulfide bonds (Clare *et al.,* 1981). Differences in the rate of disulfide bond formation (i.e., the acquisition of native, biologically active structure by the regeneration of disulfide bonds of denatured, reduced polypeptides) between proteins of milk or of other origin (Perraudin *et al.,* 1983) might explain the function of the potent sulfhydryl oxydase of human milk. Slower rate of oxidation of milk proteins suggests that milk proteins (enzymes included) might depend on enzyme-catalyzed oxidation of reduced sulfhydryl bonds. The enzyme might, therefore, maintain the structural and functional integrity of milk proteins, enzymes, and immunoglobulins. Reports that this enzyme is stable at low pH suggest that it might retain activity during passage through the stomach and might function in the intestine of the newborn where it would be instrumental in the uptake of macromolecules by altering the physical state of the intestinal mucus diffusion barrier (Isaacs *et al.,* 1985). The enzyme is present in the skim milk membranes of human milk and is stable during storage at −20°C (Isaacs, 1986). Immunofluorescent labeling studies have shown that sulfhydryl oxidase is closely associated with membranes such as the plasma membranes of lactating rat and bovine mammary tissue (Clare *et al.,* 1984). The authors suggest that similar distribution of xanthine oxidase indicates that sulfhydryl oxidase may determine xanthine oxidizing activity *in vivo.* Sulfhydryl oxidase might be a potentially important enzyme with functions in the lactating mammary gland, in milk, and in the gastrointestinal tract of the newborn; its exact physiological function, however, remains to be established.

2. *β-Glucuronidase (EC 3.2.1.31)*

Breast-fed infants have a higher incidence of jaundice than formula-fed infants (Osborne *et al.,* 1984). Breast milk jaundice, due to prolonged nonconjugatged hyperbilirubinemia, was attributed initially to inhibition of hepatic UDP-gluronyl transferase by the steroid pregnane-3-γ-20-β-diol (Arias *et al.,* 1964) and, subsequently, to high levels of free fatty acids in jaundice-inducing milks (Bevan and Holton, 1972). However, no association was found between milk lipase levels and free fatty acid concentrations (Constantopoulos *et al.,* 1980). An association between high β-glucuronidase activity in milk and breast milk jaundice (Gourley and Arend, 1986) could not be confirmed by other investigators (Hamosh, 1990b; Wilson *et al.,* 1992). It was, however, recently reported that milk β-glucuronidase activity of diabetic women is higher than that of healthy women, and it is suggested that this might be the reason for hyperbilirubinemia in breast-fed infants of diabetic mothers (Siroti *et al.,* 1992). Glucuronidase cleaves glucuronic acid from bilirubin glucuronide, liberating unconjugated bilirubin, which is more easily absorbed from the intestine than from the conjugate. While unable to demonstrate a relationship between milk β-glucuronidase activity and serum bilirubin concentration, Alonso *et al.* (1991) confirmed and extended the studies of Gartner *et al.* (1983) that indicate that enhanced intestinal absorption of bilirubin contributes to the jaundice associated with breast-feeding.

3. Lysozyme (EC 3.2.1.17)

Lysozyme catalyzes the hydrolysis of the (1–4) linkage between *N*-acetylglucosamine and *N*-acetylmuramic acid in bacterial cell walls. The enzyme lyses mostly gram-positive and a few gram-negative bacteria; it is a major component of the human milk whey fraction and has been shown to play a role in the antibacterial activity of human milk. The lysozyme of human milk is composed of 130 amino acid residues and has a molecular weight of 14.4 kDa. Although its sequence exhibits considerable homology with the lysozyme of chicken egg white (Bezkorovainy, 1977), recent studies show a marked difference in the tertiary structure of the two proteins, resulting in greater organization and hydrophobicity of the human milk protein (Dubois *et al.*, 1982). Lysozyme has been crystallized from horse milk (Zeng *et al.*, 1990).

Crystalographic studies of equine lysozyme show that the conformation of the calcium binding loop is similar to α-lactalbumin (Tsuge *et al.*, 1992). Comparable backbone atomic displacements, and homologous tertiary structures (Tsuge *et al.*, 1992), indicate that these two proteins might be derived from a common ancestor molecule (Dayhoff, 1976). Several recent studies have investigated the structure of monotreme (Teahan *et al.*, 1991), avian, and mammalian lysozymes (Zhao *et al.*, 1991). Immunochemical identity of human milk and saliva lysozymes (Wang and Kloer, 1984) and amino acid sequence identity of human milk and leucocyte lysozyme (Jolles and Jolles, 1972) have been established. A relationship between lactation performance in the early postpartum period and milk and blood lysozyme concentration was recently reported (Sofronov *et al.*, 1991).

High concentrations of lysozyme are present in human milk throughout lactation (Butte *et al.*, 1984; Goldman *et al.*, 1982, 1983), whereas concentrations are several orders of magnitude lower in bovine milk. A comparison of changes in the concentrations of several protective factors in milk during lactation shows that, whereas secretory IgA and lactoferrin decrease after the early period, lysozyme is higher during 6 month–2 years of lactation than during the first month postpartum (Goldman and Goldblum, 1989). Conflicting data about the effect of maternal malnutrition on lysozyme concentrations have recently been reported. In a group of Zairean women, studied during 18 months of lactation, malnutrition led to a decrease in lactoferrin, no change in IgA, and a steady increase in milk lysozyme concentrations (Hennart *et al.*, 1991), whereas in women studied in Taiwan there was a decrease in lysozyme, IgA, and complement components C3 and C4 during the first 2 weeks of lactation. A subsequent increase in these specific proteins was attributed to improved maternal nutrition during lactation (Chang, 1990). In the later study total milk protein concentration was not affected by malnutrition, indicating a specific effect on host defense proteins secreted into milk. No effect of malnutrition on host defense proteins, including lysozyme, was previously reported in Indian women (Reddy and Strikantia, 1978), whereas a malnutrition-associated decrease of these components in colostrum and milk of Columbian women was reported by Miranda *et al.* (1983). However, in the latter study, while IgA, IgG, and C4 were markedly decreased, lysozyme and C3 concentrations in milk were not affected by malnutrition.

Less these 1% of milk lysozyme and IgA ingested by breast-fed infants is excreted (Eschenbury *et al.*, 1990). These authors have also investigated the resistance of lysozyme and IgA under conditions simulating the gastrointestinal tract of the infant and report that lysozyme, contrary to IgA, is resistant to peptic digestion but susceptible to tryptic digestion.

A bacteriostatic effect in the infant and possibly in milk (Reiter, 1985) has been suggested for lysozyme. Reports also suggest that lysozyme binds to bacterial lipopolysaccharide. This interaction results in reduction of the endotoxic effect of the lipopolysaccharide as well as in a dose-dependent inhibition of the enzymatic activity of lysozyme (Ohno and Morrison, 1989).

Lysozyme and α-lactalbumin of human milk seem to be derived from a common ancestor molecule on the basis of identical amino acids in 49 positions (Dayhoff, 1976).

Although not related to the topic of enzymes in human milk, it is worth noting that another antimicrobial agent in human milk, lactoferrin, was recently shown to be highly resistant to proteolysis both in its iron-free (apolactoferrin) and iron-containing forms (Brines and Brock, 1983; Samson *et al.*, 1980). Because native milk lactoferrin is largely free of iron, it can withhold iron from, and thus retard, the *in vitro* growth of microorganisms (Bullen *et al.*, 1972). The marked susceptibility of bovine milk lactoferrin to proteolysis led investigators to suggest that the unusual resistance of human apolactoferrin to proteolysis may reflect an evolutionary development designed to permit its survival in the intestine of the infant (Brines and Brock, 1983).

4. *Peroxidase (EC 1.11.1.17)*

Peroxidase in human milk was earlier considered to be lactoperoxidase (Gothefors and Marklund, 1975), but later studies indicated that the activity in milk is derived from milk leukocytes and is thus a myeloperoxidase (Moldoveanu *et al.*, 1982). The distinction between a true secretory peroxidase (lactoperoxidase) and a peroxidase derived from leukocytes (myeloperoxidase) is important because the two enzymes have different structures and catalyze the oxidation of thiocyanate ion products with bacteriostatic activity; however, only myeloperoxidase catalyzes the oxidation of the chloride ion; the products of the latter reaction only have bactericidal activity. The controversy about the nature of peroxidase activity in human milk continues (Pruitt *et al.*, 1991); the conflicting reports in the literature (Moldoveanu *et al.*, 1984; Hashinoda and Yamada, 1986; Langbakk and Flatmark, 1984, 1989) have been ascribed to qualitative and quantitative differences in peroxidase content of various sources as well as to sensitivity and specificity of the techniques used. The different peroxidases (myelo or lactoperoxidase) catalyze the oxidation of thiocyanate with the formation of bactostatic products. Pruitt *et al.* (1991) suggest that, in colostrum, the peroxidase system has an antibacterial function and that these enzymes might also protect the mammary gland from the accumulation of toxic levels of hydrogen peroxide. They also report that human milk and colostrum contain variable amounts of both peroxidase systems. Activity is present only in early milk, and decreases to very low levels in mature milk. Bovine milk and human saliva have potent lactoperoxidase activity. It remains to be established whether the transient peroxidase systems in human colostrum and early milk have a physiological function in milk or the infant.

5. *Alkaline Phosphatase (EC 3.1.3.1)*

Alkaline phosphatase is located on the luminal surface of the epithelial cells of the ducts and acinar glands. The enzyme is released into milk as part of the plasma membrane during the formation of milk fat globules. The high levels of the enzyme in colostrum and intermediate milk may be due to the sudden activation of the milk secretory mechanism. The enzyme was purified from bovine milk about 10 years ago and has been characterized in human milk (Hamilton *et al.*, 1979; Worth *et al.*, 1981). The conclusions reached by two groups of investigators differ as to the nature of the enzyme in milk. Whereas one group (Hamilton *et al.*, 1979) suggests that functional, antigenic, and structural analysis indicate that the milk enzyme is the same protein species as that of adult human liver, the data reported by the second group (Worth *et al.*, 1981) suggest that the milk enzyme is a mixture of isozymes similar to bone and liver alkaline phosphatase.

Alkaline phosphatase is a metal-carrying enzyme (Table I) (Rumball and Baker, 1985); it contains four atoms of zinc per molecule, two essential for its enzymatic activity and two fulfilling a structural role. In addition to zinc it also contains two magnesium atoms. The enzyme is heat stable and has a molecular weight of 160 kDa (Chuang, 1987). While the molecular size is similar to that of placental alkaline phosphatase and both enzymes are sialilated, enzyme characteristics are different. The milk enzyme has a higher surface charge and is not inhibited by L-phenylalanine and L-homoarginine.

Much remains to be learned about the function of many milk enzymes. It is important to know their origin, mechanism of secretion into milk, compartmentalization among the various milk fractions, as well as whether their activity changes as a function of length of pregnancy and lactation. Another important topic is to examine the function of some of these enzymes in the infant and their interaction with the infant's endogenous enzymes as well as whether there is an interaction between infant development and changes in their level in milk. Also of importance is to design studies in which the *in vivo* function of milk enzyme can be examined rather than the extrapolation from *in vitro* studies that simulate *in vivo* conditions.

6. *Platelet-Activating Factor Acetylhydrolase*

Recent reports suggest that this enzyme may have an important function in the prevention of necrotizing enterocolitis (NEC), an often fatal disease in premature infants. Platelet-activating factor (PAF) is a potent ulcerogen of the gastrointestinal tract (Snyder, 1990) and its administration into the descending aorta of experimental animals was shown to cause NEC within hours after injection (Gonsalez-Crussi and Hsueh, 1990). Recent studies show a protective effect of platelet-activating factor acetylhydrolase (PAF-AH) which hydrolyzes PAF to produce an inactive form against the development of NEC (Furukawa *et al.,* 1993a). The enzyme is present in serum and its level of activity is inversely related to experimental NEC development (Furukawa *et al.,* 1993b). Recent studies (Furukawa *et al.,* 1993a) report the presence of PAF-AH in the milk of several species (rat, pig, goat) including the human. The enzyme in human milk is the plasma-type isozyme. Furukawa *et al.* (1993b) report that the enzyme is secreted by milk macrophages and is resistant to low pH and proteolysis, and suggest that it may act in the intestine of the newborn to hydrolyze PAF and thereby prevent the development of NEC. The authors (Furukawa *et al.,* 1993a) emphasize that the reported lower incidence of NEC in breast-fed than in formula-fed infants (Lucas and Cole, 1990; DeCurtis *et al.,* 1987) could be associated with the presence of PAF-AH in human milk. It is interesting that among the species studied the only one devoid of milk PAF-AH was the bovine; thus, cow's milk cannot substitute for human milk (Park *et al.,* 1993). Destruction of PAF-AH by heat treatment and its dependence upon the presence of intact cells in milk would explain the lower protection provided by previously pasteurized (Narayanan *et al.,* 1984) or frozen milk or by milk after removal of the cell fraction (Pitt *et al.,* 1977).

References

Abadeh, S., Killacky, J., Benboubetra, M., and Harrison, R. (1992). Purification and partial characterization of xanthine oxidase from human milk. *Biochim. Biophys. Acta* **1117,** 25–32.

Abouakil, N., Rogalska, E., Bonicel, J., and Lombardo, D. (1988). Purification of pancreatic carboxyl ester hydrolase by immuno affinity and its application to the human bile-salt-stimulated lipase. *Biochim. Biophys. Acta* **961,** 299–308.

Alemi, B., Hamosh, M., Scanlon, J. W., and Hamosh, P. (1981). Fat digestion in low birth weight infants: Effect of addition of human milk to low birth weight formula. *Pediatrics* **68,** 484–488.

Alonso, E. M., Whitington, P. F., Whitington, S. H., Riverd, W. A., and Given, G. (1991). Enterohepatic circulation of nonconjugated bilirubin in rats fed with human milk. *J. Pediatr.* **118,** 425–430.

Antonowicz, I. (1979). The role of enteropeptidase in the digestion of protein and its development in human fetal small intestine. *In* "Development of Mammalian Àbsorptive Processes" (J. T. Harris, ed.), pp. 169–187. Ciba Foundation Series 70. Elsevier–North Holland, Amsterdam.

Arias, I. M., Gartner, L. M., Seifter, S., and Furman, M. (1964). Prolonged neonatal unconjugated hyperbilirubinemia associated with breast feeding and a steroid, pregnane-3 (alpha), 20(beta)-diol in maternal milk that inhibits glucuronid formation in vitro. *J. Clin. Invest.* **43,** 2037–2047.

Armand, M., Hamosh, M., Mehta, N. R., Angelus, P. A., and Philpott, J. R. (1993). Gastric lipolysis in premature infants, effect of diet: Human milk or formula. *FASEB J.* **7,** A207.

Armano, J., Straehl, P., Berger, R. G., Kochibe, N., and Kobata, A. (1991). Structures of mucin-type sugar chains of the galactosyl transferase purified from human milk. *J. Biol. Chem.* **266,** 11461–11477.

Astrup, T., and Sterndorff, I. (1953). A fibrinolytic system in human milk. *Proc. Soc. Exp. Biol. Med.* **84,** 605–608.

Avissar, N., Slemmon, J. R., Palmer, I. S., and Cohen, H. J. (1991). Partial sequence of human plasma glutathione peroxidase and immunologic identification of milk glutathione peroxidase as the plasma enzyme. *J. Nutr.* **121,** 1243–1249.

Baba, T., Downs, D., Jackson, K. W., Tang, J., and Wang, C-S. (1991). Structure of human milk bile salt activated lipase. *Biochemistry* **30,** 500–512.

Backman-Gullers, B., Hannestad, U., Nilsson, L., and Sorbo, B. (1990). Studies on lipoamidase: Characterization of the enzyme in human serum and breast milk. *Clin. Chim. Acta* **191,** 49–60.

Barbezat, G., and Hansen, J. D. L. (1968). The exocrine pancreas and protein-calorie malnutrition. *Pediatrics* **42,** 77–92.

Bechamp, A. (1883). Sur la zymase du lait de femme. *CR. Acad. Sci.* **96,** 1508–1510.

Berger, G. M. B., Spark, A., Baillie, P. M., Huskisson, J., Stockwell, G., and Van der Merwe, E. (1983). Absence of serum stimulated lipase activity and altered lipid content in milk from a patient with type I hyperlipoproteinemia. *Pediatr. Res.* **17,** 835–839.

Bernback, S., Blackberg, L., and Hernell, O. (1990). The complete digestion of human milk triacylglycerol in vitro requires gastric lipase, pancreatic colipase-dependent lipase and bile salt stimulated lipase. *J. Clin. Invest.* **85,** 1221–1226.

Bevan, B. R., and Holton, J. B. (1972). Inhibition of bilirubin conjugation in rat liver slices by free fatty acids, with relevance to the problem of breast-milk jaundice. *Clin. Chim. Acta.* **41,** 101–107.

Bezkorovainy, A. (1977). Human milk and colostrum proteins: a review. *J. Dairy Sci.* **60,** 1023–1027.

Bhattacharya, I. D., Picciano, M. F., and Milner, J. A. (1988). Characteristics of human milk glutathione peroxidase. *Biol. Trace Elem. Res.* **18,** 59–70.

Bitman, J., Freed, M., Wood, D. L., Hamosh, P., and Hamosh, M. (1986). Lipid composition of prepartum human mammary secretion and postpartum milk. *J. Pediatr. Gastroenterol. Nutr.* **5,** 608–615.

Bitman, J., Wood, D. L., Liao, T. H., Fink, C. S., Hamosh, P., and Hamosh, M. (1985). Gastric lipolysis of milk lipids in suckling rats. *Biochim. Biophys. Acta* **834,** 58–64.

Bitman, J., Wood, D. L., Hamosh, M., Hamosh, P., and Mehta, N. R. (1983). Comparison of the lipid composition of breast milk from mothers of term and preterm infants. *Am. J. Clin. Nutr.* **38,** 300–312.

Blackberg, L., and Hernell O. (1981). The bile salt stimulated lipase in human milk. Purification and characterization. *Eur. J. Biochem.* **116,** 221–225.

Blackberg, L., Angquist, K-A., and Hernell, O. (1987). Bile salt stimulated lipase in human milk: Evidence for its synthesis in the lactating mammary gland. *FEBS Lett.* **217,** 37–42.

Bloch, K., and Vance, D. (1977). Control mechanisms in the synthesis of saturated fatty acids. *Annu. Rev. Biochem.* **46,** 263–280.

Borulf, S., Lindberg, T., and Mansson, M. (1987). Immunoreactive anionic trypsin and elastase activity in human milk. *Acta Paediatr. Scand.* **76,** 11–15.

Brantl, V. (1984). Novel opioid peptides derived from human β-casomorphines. *Eur. J. Pharmocol.* **106,** 213–218.

Brew, K., and Hill, R. L. K. (1975). Lactose biosynthesis. *Rev. Physiol. Biochem. Pharmacol.* **72,** 103–158.

Brincs, R. D., and Brock, J. H. (1983). The effect of trypsin and chymotrypsin and the in vitro antimicrobial and iron binding properties of lactoferrin in human milk and bovine colostrum. *Biochim. Biophys. Acta* **759,** 229–235.

Brinkley, R., Weismann, M. L., Groth, D. P., and Powell, R. W. (1975). Glutamyl transferase: A secretory enzyme. *FEBS Lett.* **51,** 168–170.

Brodrick, J. W., Largman, C., Geokas, M. C., O'Rourke, M., and Roy, S. B. (1980). Clearance of circulating anionic and cationic pancreatic trypsinogens in the rat. *Am. J. Physiol.* **239,** G511–G515.

Bullen, J. J., Rodgers, H. J., and Leigh, L. (1972). Iron binding proteins in milk and resistance to Escherichia coli infections in infants. *Br. Med. J.* **1,** 69–75.

Camps, L., Reina, M., Llobera, M., Vilaro, S., and Olivecrona, T. (1990). Lipoprotein lipase: Cellular origin and functional distribution. *Am. J. Physiol.* **258,** C673–C681.

Canas-Rodriguez, A., and Smith, H. W. (1966). The identification of the antimicrobial factors of the stomach contents of suckling rabbits. *Biochem. J.* **100,** 79–82.

Cantu, I. M., and Ibarra, B. (1982). Phosphoglucomutase: Evidence of a new locus expressed in human milk. *Science* **216,** 639–640.

Carlsson, L. C. T., Westrom, B. R., and Karlsson, B. W. (1980). Intestinal absorption of proteins by the neonatal piglet fed on sow's colostrum with either natural or experimentally eliminated trypsin-inhibiting activity. *Biol. Neonate.* **38,** 309–320.

Carroll, R. J., Basch, J. J., Phillips, J. G., and Farrell, H. M. (1985). Ultrastructural and biochemical investigation of mature human milk. *Food Microstruct.* **4,** 323–331.

Castberg, H. B., and Hernell, O. (1975). Role of serum stimulated lipase in lipolysis in human milk. *Milchwissenschaft* **30,** 721–724.

Chait, A., Iverius, P. H., and Brunzell, J. D. (1982). Lipoprotein lipase secretion by monocyte derived macrophages. *J. Clin. Invest.* **69,** 490–493.

Chang, S. J. (1990). Antimicrobial proteins of maternal and cord sera and human milk in relation to maternal nutritional status. *Am. J. Clin. Nutr.* **51,** 183–184.

Chuang, N. N. (1987). Loss of sialic acid from 5′-nucleotidase in human milk. *Clin. Chim. Acta* **169,** 337–339.

Clegg, R. A. (1981). Lipoprotein lipase: Localization on plasma membrane fragments from lactating rat mammary tissue. *Biochim. Biophys. Acta* **664,** 397–408.

Cohen, M., Morgan, R. G. H., and Hofmann, A. F. (1971). Lipolytic activity of human gastric and duodenal juice against medium- and long-chain triglyceride. *Gastroenterology* **60,** 1–15.

Collares, E. F., and Brasil, M. R. (1979). Amilase salivar no kwashiorkor. III. Estudio evolutivo da concentraçao e secreçao em quatro pacientes apresentando emplicaçoes durante a recuperaçao nutritionel. *Arq. Gastroenterol.* **16,** 34–38.

Constantopoulos, A., Messaritakis, J., and Matsaniotis, N. (1980). Breast-milk jaundice: The role of lipoprotein lipase and the free fatty acids. *Eur. J. Pediatr.* **134,** 35–38.

Danus, O. V., Urbina, A. M., Valenzuela, I., and Solimano, G. C. (1970). The effect of refeeding on pancreatic exocrine function in marasmic infants. *J. Pediatr.* **77,** 334–337.

Dayhoff, M. D. (1976). "Atlas of Protein Sequence and Structure," Vol. 5, Suppl. 2. National Biomedical Research Foundation, Washington, DC.

Debski, B., Picciano, M. F., and Milner, J. A. (1986). Selenium content and distribution of human, cow and goat milk. *J. Nutr.* **117,** 1091–1097.

DeCurtis, M., Paone, C., Vetrano, G., Romano, G., Paludetto, R., and Ciccimarra, F. (1987). A case control study of necrotizing enterocolitis occurring over 8 years in a neonatal intensive care unit. *Eur. J. Pediatr.* **146,** 398–400.

Drago, G. A., Hopkinson, D. A., Westwood, S. A., and Whitehouse, D. B. (1991). Antigenic analysis of the major human phosphoglucomutase isozymes: PGM1, PGM2, PGM3 and PGM4. *Ann. Hum. Genet.* **55,** 263–271.

Dubois, T., Ghuillard, R., Prieels, J. B., and Perraudin, J. P. (1982). Comparison between the folding of reduced hen-egg white lysozyme and the reduced human milk lysozyme. *Biochemistry* **21,** 6515–6523.

Dupuy, P., Sauniere, J. F., Vis, H. L., Leclaire, M., and Lombardo, D. (1991). Change in bile salt dependent lipase in human breast milk during extended lactation. *Lipids* **26,** 134–138.

Ellis, L., and Hamosh, M. (1991). Human milk: Stability of digestive enzymes in expressed milk. *In* "Human Lactation V; Mechanisms Regulating Lactation and Infant Nutrient Utilization" (M. F. Picciano and B. Lonnerdal, eds.), pp. 389–393. Wiley, New York.

Ellis, L. A., and Hamosh, M. (1992). Bile salt stimulated lipase: Comparative studies in ferret milk and lactating mammary gland. *Lipids* **27,** 917–922.

Ellis, L., Picciano, M. F., Smith, A. M., Hamosh, M., and Mehta, N. R. (1990). The impact of gestational length on human selenium concentration and glutathione peroxidase activity. *Pediatr. Res.* **27,** 32–35.

Erlanson-Albertsson, C., Sternby, B., and Johannesson, V. (1985). The interaction between human pancreatic carboxylester hydrolase (bile salt stimulated lipase of human milk) and lactoferrin. *Biochim. Biophys. Acta* **839,** 282–287.

Eschenburg, G., Heine, W., and Peters, E. (1990). Fecal sIgA and lysozyme excretion in breast feeding and formula feeding. *Kinderartzl. Prax.* **58,** 255–260.

Florent, C., L'Hirondel, J., Desmazures, C., Aynes, C., and Bernier, J. (1981). Intestinal clearance of α-1-antitrypsin—A sensitive method for the detection of protein losing enteropathy. *Gastroenterology* **81,** 777–780.

Forsyth, J. S., Donnet, L., and Ross, P. E. (1990). A study of the relationship between bile salts, bile salt stimulated lipase, and free fatty acids in breast milk: Normal infants and those with breast milk jaundice. *J. Pediatr. Gastroenterol. Nutr.* **11,** 205–210.

Forsyth, J. S., Ross, P. E., and Bouchier, I. A. D. (1983). Bile salts in breast milk. *Eur. J. Pediatr.* **140,** 126–127.

Fransson, G. B., and Lonnerdal, B. (1983). Distribution of trace elements and minerals in human and cow's milk. *Pediatr. Res.* **17,** 912–915.

Fredrikzon, B., Hernell, O., Blackberg, L., and Olivecrona, T. (1978). Bile salt stimulated lipase in human milk: Evidence of activity in vivo and of a role in the digestion of milk retinol esters. *Pediatr. Res.* **12,** 1048–1052.

Freed, L. M., Berkow, S. E., Hamosh, P., York, C. M., Mehta, N. R., and Hamosh, M. (1989a). Lipase in human milk: Effect of gestational age and length of lactation on enzyme activity. *J. Am. Coll. Nutr.* **8,** 143–150.

Freed, L. M., Neville, M. C., and Hamosh, M. (1989b). Lipase activities in human milk during weaning. *Pediatr. Res.* **25,** 290A. [Abstract.]

Freed, L. M., York, C. M., Hamosh, P., Mehta, N. R., and Hamosh, M. (1987). Bile salt stimulated lipase of human milk: Characteristics of the enzyme in the milk of mothers of premature and full-term infants. *J. Pediatr. Gastroenterol. Nutr.* **6,** 598–604.

Freed, L. M., Neville, M. C., Hamosh, P., and Hamosh, M. (1986a). Diurnal and within-feed variations in lipase activity and triglyceride content of human milk. *J. Pediatr. Gastroenterol. Nutr.* **5,** 938–942.

Freed, L. M., York, C. M., Hamosh, M., Mehta, N. R., Sturman, J. A., Oftedal, O. T., and Hamosh, P. (1986b). Bile salt stimulated lipase: The enzyme is present in non primate milk. *In* "Human Lactation Vol. II: Maternal–Environmental Factors" (M. Hamosh and A. S. Goldman, eds.), pp. 595–602. Plenum Press, New York.

Freed, L. M., York, C. M., Hamosh, M., Sturman, J. A., and Hamosh, P. (1986c). Bile salt stimulated lipase in nonprimate milk: Longitudinal variation and lipase characteristics in cat and dog milk. *Biochim. Biophys. Acta* **878,** 209–215.

Freudenberg, E. (1953). "Die Frauenmilkch-lipase." Karger, Basel.
Freudenberg, E. (1966). A lipase in the milk of the gorilla. *Experientia* **22,** 317.
Fridhandler, L., Berk, J. E., Montgomery, K. A., and Wand, D. (1974). Column chromatographic studies of isoamylase in human serum, urine, and milk. *Clin. Chem.* **20,** 547–555.
Funk, M. A., Hamlin, L., Picciano, M. F., Prentice, A., and Milner, J. A. (1990). Milk selenium of rural African women: Influence of maternal nutrition, parity and length of lactation. *Am. J. Clin. Nutr.* **51,** 220–224.
Furmansky, P., Li, Z-P., Fortuna, M. B., Swamy, C. V. B., and Das, R. (1989). Multiple molecular forms of human lactoferrin. Identification of a class of lactoferrins that posses ribonuclease activity and lack iron binding capacity. *J. Exp. Med.* **170,** 415–429.
Furukawa, M., Lee, E. L., and Johnston, J. M. (1993a). Platelet-activating factor-induced ischemic bowel necrosis: The effect of platelet-activating factor acetylhydrolase. *Pediatr. Res.* **34,** 237–241.
Furukawa, M., Narahara, H., and Johnston, J. M. (1993b). The presence of platelet-activating factor acetylhydrolase in milk. *J. Lipid Res.* **34,** 1603–1609.
Gartner, L. M., Lee, K., and Moscioni, A. D. (1983). Effect of milk feeding on intestinal bilirubin absorption in the rat. *J. Pediatr.* **103,** 464–471.
Gerber, A. C., Kozdrowski, I., Wyss, S. R., and Berger, F. G. (1979). The charge heterogeneity of soluble human galactosyltransferases isolated from milk, amniotic fluid and malignant ascites. *Eur. J. Biochem.* **93,** 453–460.
Gillin, F. D., Cooper, R. W., Reiner, D. S., and Das, S. (1991). Secretory defenses against Giardia lamblia. *Adv. Exp. Med. Biol.* **310,** 227–233.
Gillin, F. D., Reiner, D. S., and Gault, M. J. (1985). Cholate-dependent killing of giardia lamblia by human milk. *Infect. Immun.* **47,** 619–622.
Gillin, F. D., Reiner, D. S, and Wang, C. S. (1983). Human milk kills parasitic intestinal protozoa. *Science* **221,** 1290–1292.
Ginder, J., Nwankwo, M. U., Omene, J. A., Roberts, I. M., La Rocca, G. M., and Glew, R. H. (1987). Breast milk composition and bile salt stimulated lipase in well nourished and undernourished Nigerian mothers. *Eur. J. Pediatr.* **146,** 184–186.
Goldman, A. S., and Goldblum, R. M. (1989). Immunologic system in human milk: Characteristics and effects. *In* "Textbook of Gastroenterology and Nutrition in Early Infancy" (E. Lebenthal, ed.), 2nd Ed., pp. 135–142. Raven Press, New York.
Goldman, A. S., Garza, C., Nichols, B. L., and Goldblum, R. M. (1982). Immunologic factors in human milk during the first year of lactation. *J. Pediatr.* **100,** 563–567.
Goldman, A. S., Goldblum, R. M., and Garza, C. (1983). Immunologic components in human milk during the second year of lactation. *Acta Paediatr. Scand.* **72,** 461–462.
Gonzalez-Crussi, F., and Hsueh, W. (1983). Experimental model of ischemic bowel necrosis. The role of platelet-activating factor and endotoxin. *Am. J. Pathol.* **112,** 127–135.
Gothefors, L., and Marklund, S. (1975). Lactoperoxidase activity in human milk and in saliva of newborn children. *Infect. Immun.* **11,** 1210–1215.
Gourley, G. R., and Arend, R. A. (1986). β-Glucuronidase and hyperbilirubinemia in breast- and formula-fed neonates. *Lancet* **1,** 644–646.
Graham, K., Fleming, J. E., Young, R., and Bensch, K. G. (1989). Preparation of antibodies against xanthine oxidase from human milk. *Int. J. Biochem.* **21,** 715–722.
Greenberg, R. The relationship of maternal factors to composition and structure of casein and whey proteins. *In* "Human Lactation: Maternal–Environmental Factors" (M. Hamosh and A. S. Goldman, eds.), pp. 187–194. Plenum Press, New York.
Greenberg, R., and Groves, M. L. (1984). Plasmin cleaves human β-casein. *Biochem. Biophys. Res. Commun.* **125,** 463–468.
Greenberg, R., Groves, M. L., and Dower, H. J. (1984). Human β-casein–amino acid sequence and identification of phosphorylation sites. *J. Biol. Chem.* **259,** 5132–5138.
Hadorn, B., Zoppi, G., Schmerling, D. H., Prader, A., McIntyre, I., and Anderson, C. M. (1968). Quantitative assessment of exocrine pancreatic function in infants and children. *J. Pediatr.* **73,** 39–50.

Hall, B., and Muller, D. P. R. (1982). Studies on the bile salt stimulated lipolytic activity in human milk using whole milk as a source of both substrate and enzyme. I. Nutritional implications. *Pediatr. Res.* **16,** 251–255.

Hall, B., and Muller, D. P. R. (1983). Studies on bile salt stimulated lipolytic activity in human milk. II. Demonstration of two groups of milk with different activities. *Pediatr. Res.* **17,** 716–720.

Hamilton, T. A., Gornicki, S. Z., and Sussman, H. H. (1979). Alkaline phosphatase from human milk. *Biochem. J.* **177,** 197–201.

Hamosh, M. (1979). Review, fat digestion in the newborn: role of lingual lipase and preduodenal digestion. *Pediatr. Res.* **13,** 615–622.

Hamosh, M. (1981). Physiological role of milk lipases. *In* "Textbook of Gastroenterology and Nutrition in Infancy" (E. Lebenthal, ed.), pp. 473–482. Raven Press, New York.

Hamosh, M. (1982). Lingual and breast milk lipases. *Adv. Pediatr.* **29,** 33–67.

Hamosh, M. (1986). Enzymes in human milk. *In* "Human Milk in Infant Nutrition and Health" (R. R. Howell and F. H. Morriss, eds.), pp. 66–97. Thomas, Springfield, IL.

Hamosh, M. (1988). Enzymes in milk: Their function in the mammary gland, in milk, and in the infant. *In* "Biology of Human Milk" (L. A. Hanson, ed.), pp. 45–58. Mosby, St. Louis, MO.

Hamosh, M. (1989). Enzymes in human milk: Their role in nutrient digestion, gastrointestinal function and nutrition in infancy. *In* "Textbook of Gastroenterology and Nutrition in Infancy" (E. Lebenthal, ed.), 2nd Ed., pp. 121–134. Raven Press, New York.

Hamosh, M. (1990a). Breast milk jaundice. *J. Pediatr. Gastroenterol. Nutr.* **11,** 145–149.

Hamosh, M. (1990b). "Lingual and Gastric Lipases: Their Role in Fat Digestion." CRC Press, Boca Raton, FL.

Hamosh, M. (1991). Free fatty acids and monoglycerides: Antiinfective agents produced during the digestion of milk fat by the newborn. *Adv. Exp. Med. Biol.* **310,** 151–158.

Hamosh, M., and Scow, R. O. (1971). Lipoprotein lipase activity in guinea pig and rat milk. *Biochim. Biophys. Acta* **231,** 282–289.

Hamosh, M., and Bitman, J. (1992). Human milk in disease: Lipid composition. *Lipids* **27,** 848–857.

Hamosh, M., and Hamosh, P. (1983). Lipoprotein lipase: Its physiological and clinical significance. *Mol. Aspects Med.* **6,** 199–289.

Hamosh, M., Clary, T. R., Chernick, S. S., and Scow, R. O. (1970). Lipoprotein lipase of adipose and mammary tissue and plasma triglyceride in pregnant and lactating rats. *Biochim. Biophys. Acta* **210,** 473–482.

Hamosh, M., Freed, L. M., Jones, J. B., Berkow, S. E., Bitman, J., Mehta, N. R., Happ, B., and Hamosh, P. (1985). Enzymes in human milk. *In* "Milk Components and Methodologies" (R. G. Jensen and M. C. Neville, eds.), pp. 251–266. Plenum Press, New York.

Hamosh, M., Freed, L. M., Fink, C. S., Mehta, N. R., and Hamosh, P. (1987). Development of Lipid Metabolism. *In* "New Aspects of Nutrition in Infancy and Prematurity" (M. Xanthou, ed.), pp. 67–78. Elsevier, Amsterdam.

Hamosh, M., Hong, M. H., and Hamosh, P. (1989). β-Casomorphins: Milk–β-casein derived opioid peptides. *In* "Textbook of Gastroentorology and Nutrition in Infancy" (E. Lebenthal, ed.), 2nd Ed., pp. 143–150. Raven Press, New York.

Hamosh, M., Mao, J., Ellis, L., and Hamosh, P. (1991). Lipoprotein lipase: Enzyme stability is greater in milk than in blood and tissues. *FASEB J* **5,** A1288. [Abstract]

Hanafy, M. M., El-Khateeb, S., Guirgis, F. K., and El-Lozy, M. (1971). Diastase in human milk. *Alexandria Med. J.* **17,** 299–305.

Harrison, R., Abadeh, S., and Benboubetra, M. (1991). Purification of xanthine oxidase from human milk. *In* "Purine and Pyrimidine Metabolism in Man" (R. A. Harkness, ed.), Vol. VII, Part A, pp. 335–338. Plenum Press, New York.

Harzer, G., Haug, M., Dieterich, I., and Gentner, P. R. (1983). Changing patterns of human milk lipids in the course of the lactation and during the day. *Am. J. Clin. Nutr.* **37,** 612–621.

Hashinaka, K., and Yamada, M. (1986). Identification of myeloperoxidase in human colostrum. *Arch. Biochem. Biophys.* **247,** 91–96.

Hayakawa, K., and Oizymi, J. (1988). Isolation and characterization of human breast milk lipoamidase. *Biochim. Biophys. Acta* **957,** 345–351.

Hegardt, P., Lindberg, T., Borjesson, J., and Skude, G. (1984). Amylase in human milk from mothers of preterm and term infants. *J. Pediatr. Gastroentrol. Nutr.* **3,** 563–566.

Heitlinger, L. A., Lee, P. C., Dillon, W. P., and Lebenthal, E. (1983). Mammary amylase: A possible alternate pathway of carbohydrate digestion in infancy. *Pediatr. Res.* **17,** 15–18.

Hemavathj, R., and Das, R. (1985). Purification and characterization of a high molecular weight human milk ribonuclease: Its potential use for diagnosis and prognosis of human mammary neoplasia. *In* "Retroviruses and Human Neoplasia" (R. C. Gallo, D. Stehelin, and O. E. Vernier, eds.), pp. 401–408. Humana Press, Clifton, N.J.

Hennart, P. F., Brasseur, D. J., and Delogne-Desnoeck, J. B. (1991). Lysozyme, lactoferrin, and secretory immunoglobulin A content in breast milk: Influence of duration of lactation, nutrition status, prolactin status, and parity of mother. *Am. J. Clin. Nutr.* **53,** 32–39.

Hernell, O., Gebre-Medhin, M. and Olivecrona, T. (1977). Breast milk composition in Ethiopian and Swedish mothers. IV. Milk lipases. *Am. J. Clin. Nutr.* **30,** 508–511.

Hernell, O. (1985). Human milk lipases. III. Physiological implications of bile salt stimulated lipase. *Eur. J. Clin. Invest.* **2,** 267–274.

Hernell, O., and Blackberg, L. (1982). Digestion of human milk lipids: Physiological significance of sn-2 monoacylglycerol hydrolysis by bile salt-stimulated lipase. *Pediatr. Res.* **16,** 882–885.

Hernell, O., and Blackberg, L. (1985). Lipolysis in human milk: Causes and consequences. *In* "Composition and Physiological Properties of Human Milk" (J. Schaub, ed.), pp. 165–188. Elsevier, Amsterdam.

Hernell, O., and Olivecrona, T. (1974a). Human milk lipases. II. Bile salt stimulated lipase. *Biochim. Biophys. Acta.* **369,** 234–244.

Hernell, O., and Olivecrona, T. (1974b). Human milk lipases. I. Bile salt stimulated lipase. *J. Lipid Res.* **15,** 367–372.

Hernell, O., Blackberg, L., and Lindberg, T. (1989). Human milk enzymes with emphasis on the lipases. *In* "Textbook of Gastroenterology and Nutrition in Infancy" (E. Lebenthal, ed.), 2nd Ed., pp. 209–217. Raven Press, New York.

Hernell, O., Ward, H., Blackberg, L., and Pereira, E. A. (1986). Killing of Giardia lamblia by human milk lipases: An effect mediated by lipolysis of milk lipids. *J. Infect. Dis.* **153,** 715–720.

Heyndrckz, G. V. (1962). Investigations on the enzymes in human milk. *Ann. Pediatr.* **198,** 356–359.

Hineno, T., Sano, A., Kondoh, K., Neno, S., Kakimoto, Y., and Yoshida, K. (1991). Secretion of sphingolipid hydrolase activator precursor, prosaposin. *Biochem. Biophys. Res. Commun.* **176,** 668–674.

Hodge, C., Lebenthal, E., Lee, P. C., and Topper, W. (1983). Amylase in the saliva and in the gastric aspirates of premature infants: Its potential role in glucose polymer hydrolysis. *Pediatr. Res.* **17,** 998–1001.

Horie, N., Okamoto, U., and Wijngaards, G. (1987). Demonstration of two types of plasminogen activator in human milk by SDS–PAGE copolymerizing fibrinogen and plasminogen: Different variations of the activators in puerperium. *Thromb. Res.* **45,** 703–707.

Hosoni, O., and Takeya, A. (1989). The relationship between the (β 1–3) *N*-acetylglucosaminyl transferase and presence of oligosaccharides containing lacto-N-triose II structure in bovine and human milk. *Nippon Juigaku Zassi* **51,** 1–6.

Hui, D. Y., and Kissel, J. A. (1990). Sequence identity between human pancreatic cholesterol esterase and bile salt-stimulated milk lipase. *FEBS Lett.* **276,** 131–134.

Institute of Medicine. (1991). "Nutrition during Lactation." National Academy Press, Washington, DC.

Insull, W., Jr., Hirsch, F., James, T., and Aherns, E. H., Jr. (1958). The fatty acids of human milk. II. Alterations produced by manipulation of caloric balance and exchange of dietary fats. *J. Clin. Invest.* **38,** 443–450.

Isaacs, C. E. (1985). Milk enzyme function: Effects of compartmentation and storage conditions on sulfhydryl oxidase and y-glutamyl transpeptidase. *In* "Human Lactation—Milk Components and Methodologies" (R. G. Jensen and M. C. Neville, eds.), pp. 277–282. Plenum Press, New York.

Isaacs, C. E., and Thormar, H. (1991). The role of milk derived antimicrobial lipids as antiviral and antibacterial agents. *Adv. Exp. Med. Biol.* **310,** 159–165.

Iverson, S. J., Kirk, C. L., Hamosh, M., and Newsome, J. (1991). Milk lipid digestion in the neonatal dog: The combined actions of gastric and bile salt stimulated lipases. *Biochim. Biophys. Acta* **1083,** 109–119.

Iyengar, L., and Selvaraj, R. J. (1972). Intestinal absorption of immunoglobulins by newborn infants. *Arch. Dis. Child.* **47,** 411–414.

Jenness, R. (1979). The composition of human milk. *Semin. Perinatol.* **31,** 225–239.

Jensen, D. R., Bessesen, D. H., Etienne, J., Eckel, R. H., and Neville, M. C. (1991). Distribution and source of lipoprotein lipase in mouse mammary gland. *J. Lipid Res.* **32,** 733–742.

Jensen, R. G., and Pitas, R. E. (1976). Milk lipoprotein lipase. A review. *J. Dairy Sci.* **59,** 1203–1214.

Jolles, J., and Jolles, P. (1972). Comparison between human and bird lysozymes. Previously observed deletion. *FEBS Lett.* **22,** 31–33.

Jones, J. B., Mehta, N. R., and Hamosh, M. (1982). α-Amylase in preterm human milk. *J. Pediatr. Gastroenterol. Nutr.* **1,** 43–48.

Kabara, J. J. (1980). Lipids as host resistance factors of human milk. *Nutr. Rev.* **38,** 65–73.

Kirk, C. L., and Hamosh, M. (1991). Initial lipolysis by gastric lipase is essential for the hydrolysis of milk or formula fat by milk lipase. *FASEB J.* **5,** A1288. [Abstract]

Kitchen, B. J. (1974). A comparison of the properties of membranes isolated from bovine milk and cream. *Biochim. Biophys. Acta* **356,** 257–269.

Koldovsky, O., and Thornburg, W. (1987). Hormones in milk. *J. Pediatr. Gastroenterol. Nutr.* **6,** 172–196.

Kondoh, K., Hineno, T., Sano, A., and Kakimoto, Y. (1991). Isolation and characterization of prosaposin from human milk. *Biochem. Biophys. Res. Commun.* **181,** 286–292.

Korn, E. D. (1962). The lipoprotein lipase of cow's milk. *J. Lipid Res.* **3,** 246–250.

Korycka-Danl, M., Ribadeau, D. B., Chene, N., and Martal, J. (1983). Plasmin activity in milk. *J. Dairy Sci.* **66,** 704–711.

Kraehenbuhl, J. P., and Campiche, M. A. (1969). Early stages of intestinal absorption of specific antibodies in the newborn. *J. Cell. Biol.* **42,** 345–365.

Langbakk, B., and Flatmark, T. (1984). Demonstration and partial purification of lactoperoxidase from human colostrum. *FEBS Lett.* **174,** 300–303.

Langbakk, B., and Flatmark, T. (1989). Lactoferrin from human colostrum. *Biochem. J.* **259,** 627–631.

LeDeist, F., De Saint-Basile, G., Angeles-Cano, E., and Griscelli, C. (1986). Prostaglandin E2 and plasminogen activators in human milk and their secretion by milk macrophages. *Am. J. Reprod. Immunol. Microbiol.* **11,** 6–10.

Lewis-Jones, D. I., and Reynolds, G. J. (1983). Suggested role for precolostrum in preterm and sick newborn infants. *Acta Paediatr. Scand.* **72,** 13–17.

Lewis-Jones, D. I., Lewis-Jones, M. S., Connolly, R. C., Lloyd, D. C., and West, C. R. (1985). Sequential changes in the antimicrobial protein concentrations in human milk during lactation and its relevance to banked human milk. *Pediatr. Res.* **19,** 562–565.

Lindberg, T. (1979). Protease inhibitors in human milk. *Pediatr. Res.* **13,** 969–972.

Lindberg, T., Ohlsson, K., and Westrom, B. (1982). Protease inhibitors and their relation to protease activity in human milk. *Pediatr. Res.* **16,** 479–483.

Lindstrom, L. H., Nyberg, F., Terenius, L., Bauer, K., Besev, G., Gunne, L. M., Lyrenas, S., Willdeck-Lund, G., and Lindberg, B. (1984). CSF and plasma–casomorphine-like opioid peptides in postpartum psychosis. *Am. J. Psychiatry* **141,** 1059–1066.

Lonnerdal, B., Keen, C. L., and Hurley, L. S. (1981). Iron, copper, zinc, and manganese in milk. *Annu. Rev. Nutr.* **1,** 149–174.

Lucas, A., and Cole, T. J. (1990). Breast milk and neonatal necrotizing enterocolitis. *Lancet* **336,** 1519–1523.

Maksvytis, H. J., Niles, R. M., Simanovsky, L., Minassian, M. M., Richardson, L. L., Hamosh, M., Hamosh, P., and Brody, J. S. (1984). In vitro characteristics of the lipid-filled interstitial cell associated with postnatal lung growth: Evidence for fibroblast heterogeneity. *J. Cell. Physiol.* **118,** 113–123.

Mannan, S., and Picciano, M. F. (1987). Influence of maternal selenium status on human milk selenium concentration and glutathione peroxidase activity. *Am. J. Clin. Nutr.* **46,** 95–100.

Marfan, A-B. (1901). Allaitment naturel et allaitment artificiel. *Presse Med.* **9,** 13–19.

Marshall, J. M., Rees, M. C. P., and Cederholm-Williams, S. A. (1986). Identification of t-PA as the major active plasminogen activator in human milk. *Thromb. Haemostat.* **55,** 279–281.

Martyn, P., and Hansen, I. A. (1981). Initiation of lipogenic enzyme activities in rat mammary glands. *Biochem. J.* **198,** 187–192.

Mather, I. H., and Keenan, T. W. (1983). Function of endomembranes and the cell surface in the secretion of organic milk constituents. *In* "Biochemistry of Lactation" (T. B. Mepham, ed.), pp. 231–283. Elsevier, Amsterdam.

McAlpine, P. J., Hopkinson, D. A., and Harris, H. (1970). Molecular size estimates of human phosphoglucomutase isozymes by gel filtration chromatography. *Ann. Hum. Genet.* **34,** 177–185.

McGuire, E. J., Kerlin, R., Cebra, J. J., and Roth, S. (1989). Human milk galactosyl transferase is specific for secreted, but not plasma IgA. *J. Immunol.* **143,** 2933–2938.

Mehta, N. R., Jones, J. B., and Hamosh, M. (1982). Lipases in human milk: Ontogeny and physiologic significance. *J. Pediatr. Gastroentorol. Nutr.* **1,** 317–326.

Meyer, D. H., Kumin, A. S., Maddalena, J., and Meyer, W. L. (1987). Ribonuclease activity and isoenyzmes in raw and prossessed cow's milk and infant formulas. *J. Dairy Sci.* **70,** 1797–1803.

Milner, J. A., Sherman, L., and Picciano, M. F. (1987). Distribution of selenium in human milk. *Am. J. Clin. Nutr.* **45,** 617–624.

Miranda, R., Saravia, N. G., Ackerman, R., Murphy, N., Berman, S., and McMurry, D. N. (1983). Effect of maternal nutritional status on immunologic substances in human colostrum and milk. *Am. J. Clin. Nutr.* **37,** 632–640.

Moldoveanu, Z., Tenovuo, J., Mestecky, J., and Pruitt, K. M. (1982). Human milk peroxidase is derived from milk leukocytes. *Biochim. Biophys. Acta* **718,** 103–108.

Montgomery, P. A., Patton, S., Huston, G. E., and Josephson, R. V. (1987). Gel electrophoretic analysis of proteins in human milk and colostrum. *Comp. Biochem. Physiol.* **86,** 635–639.

Monti, J. C., Mermoud, A. F., and Jolles, P. (1986). Trypsin in human milk. *Experientia* **42,** 39–41.

Morgan, R. G. H., and Borgstrom, B. (1969). The mechanism of fat absorption in the bile fistula rat. *Q. J. Exp. Physiol.* **54,** 228–243.

Nabukhotnyi, T. K., Markevich, V. E., Pavliuk, V. P., and Tkachenko, I. P. (1988). Effect of human milk enzymes on carbohydrate metabolism during adaptation of newborn infants in the early neonatal period. *Vopr. Pitan.* **3,** 34–38.

Narayanan, I., Prakash, K., Murthy, N. S., and Gurjal, V. V. (1984). Randomized controlled trial of effect of raw and holder pasteurised human milk and of formula supplements on incidence of neonatal infection. *Lancet* **2,** 1111–1113.

Neville, M. C., Waxman, L. J., Jensen, D., and Eckel, R. H. (1991). Lipoprotein lipase in human milk: Compartmentalization and effect of fasting, insulin and glucose. *J. Lipid Res.* **32,** 251–257.

Nilsson, J., Blackberg, L., Carlsson, P., Enerback, S., Hernell, O., and Bjursell, G. (1990). cDNA cloning of human milk bile-salt-stimulated lipase and evidence for its identity to pancreatic carboxylic ester hydrolase. *Eur. J. Biochem.* **192,** 543–550.

O'Connor, J., Butler, P. A., and Yaghi, B. M. (1988). The effect of bile salt stimulated lipase on the interconversion of retinylpalmitate and retinol. *N.Z. Med. J.* **101,** 583–584.

Ogra, P. L. (1979). Ontogeny of the local immune system. *Pediatrics* **64,** 765–774.

Ogra, S. S., Weintraub, D., and Ogra, P. L. (1977). Immunological aspects of human colostrum and milk. Fate and absorption of cellular and soluble components in the gastrointestinal tract of the newborn. *J. Immunol.* **119,** 245–248.

Ohno, N., and Morrison, D. C. (1989). Lipopolysaccharide interaction with lysozyme. Binding of lipopolysaccharide to lysozyme and inhibition of lysozyme enzymatic activity. *J. Biol. Chem.* **264,** 4434–4441.

Oizumi, J., and Hayakawa, K. (1988). Biotinidase in human breast milk. *Am. J. Clin. Nutr.* **48,** 295–297.

Oizumi, J., Hirano, M., Hayakawa, K., Daimatsu, T., and Zaima, K. (1989). The significance of breast milk biotinidase. *Acta Paediatr. Jpn.* **31,** 424–427.

Okamoto, U., Horie, N., Nagamotsu, Y., and Yamamoto, J. (1980). Distribution of plasminogen activator in human milk samples collected at various days after delivery. *Acta Haematol. Jpn.* **43,** 743–746.

Okamoto, U., Horie, N., Nagamotsu, Y., and Yamamoto, J. (1981). Plasminogen activator in human early milk: Its partial purification and characterization. *Thromb. Haemost.* **45,** 121–126.

Olivecrona, T., Chernick, S. S., Bengtsson-Olivecrona, G., Paterniti, J. R., Jr., Brown, V. W., and Scow, R. O. (1985). Combined lipase deficiency (cld/cld) in mice. Demonstration that an inactive form of lipoprotein lipase is synthesized. *J. Biol. Chem.* **260,** 2552–2557.

Osborn, L. M., Reiff, M. I., and Bolus, R. (1984). Jaundice in the full-term neonate. *Pediatrics* **73,** 520–525.

Pamblanco, M., Ten, A., and Conin, J. (1987). Bile salt stimulated lipase activity in human colostrum from mothers of infants of different gestational age and birth weight. *Acta Paediatr. Scand.* **76,** 328–331.

Park, D. A., Bulkley, G. B., and Granger, D. N. (1983). Role of oxygen-derived free radicals in digestive tract disease. *Surgery* **94,** 415–422.

Patil, K. P., and Rangnekar, N. R. (1982). Glutamyltransferase activity in human milk. *Clin. Chem.* **28,** 1724–1725.

Patil, K. P., and Rangnekar, N. R. (1983). Lactate dehydrogenase and its isoenzymes in human milk—A preliminary study. *Clin. Chem.* **29,** 1568–1570.

Perraudin, J. P., Guillard, R., Prieels, J. P., Torchia, T., and Dubois, T. (1983). Disulfide content of reduced hen egg white and human milk lysozymes during the folding process. *FEBS Lett.* **153,** 349–352.

Pitt, J., Barlow, B., and Heird, W. C. (1977). Protection against experimental necrotizing enterocolitis by maternal milk. I. Role of milk leukocytes. *Pediatr. Res.* **11,** 906–909.

Plancke, Y., Delpouve, B., and Montreuil, J. (1987). A solid phase assay for galactosyl transferases. Evidence of an active site of less than 14 kDa. *Biosci. Rep.* **7,** 721–729.

Plucinski, T. M., Hamosh, M., and Hamosh, P. (1979). Fat digestion in the rat: role of lingual lipase. *Am. J. Physiol.* **237,** E541–E547.

Pocius, P. A., Baumrucher, C. R., McNamara, J. P., and Bauman, D. E. (1980). α-Glutamyl transpeptidase in rat mammary tissue. *Biochem. J.* **188,** 565–568.

Pruitt, K. M., Rahemtulla, F., Mansson-Rahemtulla, B., Baldone, D. C., and Laven, G. T. (1991). Peroxidases in human milk. *Adv. Exp. Med. Biol.* **310,** 137–144.

Reddy, V., and Strikantia, S. C. (1978). Interaction of nutrition and immune response. *Indian J. Med. Res.* **68,** 48–57.

Rees, E. (1987). Human breast milk: Laboratory detection of enzymes, and their use as markers for control of the pasteurization process. *Med. Lab. Sci.* **44,** 345–349.

Reiter, B. (1985). Interaction between immunoglobulins and innate factors such as lysozyme, lactoferrin, lactoperoxidase. *In* "Composition and Physiological Properties of Human Milk" (J. Schaub, ed.), pp. 271–284. Elsevier, Amsterdam.

Ruegg, M., and Blanc, B. (1982). Structure and properties of the particulate constituents of human milk. A review. *Food Microstruct.* **1,** 25–40.

Rumball, S. V., and Baker, E. N. (1985). Trace elements in human milk: Structural aspects. *In* "Human Lactation: Milk Components and Methodologies" (R. J. Jensen and M. C. Neville, eds.), pp. 237–242. Plenum Press, New York.

Sauniere, J. F., and Sarles, H. (1988). Exocrine pancreatic function and protein-calorie malnutrition in Dakar and Abidjan (West Africa) silent pancreatic insufficiency. *Am. J. Clin. Nutr.* **48,** 1233–1238.

Schade, A. L., and Reinhart, R. W. (1970). Galactothermin, a reversibly heat-precipitable protein of human milk at neutral pH. *Biochem. J.* **118,** 181–186.

Schauer, R., Veh, R. W., and Wembler, M. (1976). Demonstration of a neuraminidase activity in human blood serum and human milk using a modified, radioactively labeled alpha-1-glycoprotein as substrate. *Hope-Seylers Z. Physiol. Chem.* **357,** 559–566.

Scow, R. O., Blanchette-Machie, E. J., Mendelson, C., Hamosh, M., and Zinder, O. (1975). Incorporation of dietary fatty acids into milk triglyderides: Mechanism and regulation. *Mod. Probl. Pediatr.* **14,** 31–44.

Shahani, K. M., Kwan, A. J., and Friend, B. A. (1980). Role and significance of enzymes in human milk. *Am. J. Clin. Nutr.* **33,** 1861–1868.

Signer, E., Murphy, G. M., Edkins, S., and Andersson, C. M. (1974). Role of bile salts in fat malabsorption of premature infants. *Arch. Dis. Child.* **49,** 174–180.

Sirota, L., Ferrera, M., Lerer, N., and Dulitzky, F. (1992). Beta glucuronidase and hyperbilirubinemia in breastfed infants of diabetic mothers. *Arch. Dis. Child.* **67,** 120–121.

Smith, S., and Abraham, S. (1975). The composition and synthesis of milk fat. *Adv. Lipid Res.* **13,** 195–239.

Snyder, F. (1990). Platelet-activating factor and related acetylated lipids as potent biologically active cellular mediators. *Am. J. Physiol.* **259,** C697–C708.

Sofronov, O. V., Bakuleva, L. P., Nesterova, A. A., and Babaian, S. S. (1991). Dynamics of lysozyme levels in the blood serum and milk of puerperae with various functional activities of the breasts. *Akush. Ginekol.* **5,** 42–45.

Spear, M. L., Bitman, J., Hamosh, M., Wood, D. L., Gavula, D., and Hamosh, P. (1992a). Human mammary gland function at the onset of lactation: Medium chain fatty acid synthesis. *Lipids* **27,** 908–911.

Spear, M. L., Hamosh, M., Bitman, J., Spear, M. L., and Wood, D. L. (1992b). Milk and blood fatty acid composition during two lactations in the same woman. *Am. J. Clin. Nutr.* **56,** 65–70.

Strickland, S., and Beers, W. H. (1976). Studies on the role of plasminogen activator in ovulation. *J. Biol. Chem.* **251,** 5694–5702.

Sveger, T. (1985). Breast-feeding, α-1-antitrypsin deficiency, and liver disease. *JAMA* **254,** 3036.

Swan, J. S., Hoffman, M. M., Lord, M. K., and Poechman, J. L. (1992). Two forms of human milk bile salt-stimulated lipase. *Biochem. J.* **283,** 119–122.

Teahan, C. G., McKenzie, H. A., Shaw, D. C., and Griffiths, M. (1991). The isolation and amino acid sequences of enchidna (Tochyoglossus oculeatus) milk lysozymes I and II. *Biochem. Int.* **24,** 85–95.

Teschemacher, H. (1987). β-Casomorphins: Do they have physiological significance? *In* "Human Lactation. Vol. 3: The Effects of Human Milk on the Recipient Infant." (A. S. Goldman, S. A. Atkinson, and L. A. Hanson, eds.), pp. 213–225. Plenum Press, New York.

Thompson, B. J., and Smith, S. (1985). Biosynthesis of fatty acids by lactating human breast epithelial cells: An evaluation of the contribution to the overall composition of human milk. *Pediatr. Res.* **19,** 139–143.

Tsuge, H., Ago, H., Noma, M., Nitta, K., Sugai, S., and Miyano, M. (1992). Crystallographic studies of a calcium binding lysozyme form equine milk at 2.5A resolution. *J. Biochem.* **111,** 141–143.

Udall, J. N., Dixon, M., Newman, A. P., Wright, J. A., Brent, J., and Bloch, K. J. (1985a). Liver disease in a-1-antitrypsin deficiency: A retrospective analysis of the influence of early breast vs. bottle-feeding. *JAMA* **253,** 2679–2682.

Udall, J. N., Dixon, M., Newman, A. P., Wright, J. A., Brent, J., and Block, K. J. (1985b). Liver disease in antitrypsin deficiency: A retrospective analysis of the influence of early breast vs. bottle feeding. *JAMA* **253,** 2679–2682.

Udall, J. N., Pan, K., Fritze, L., Kleinman, R., and Walker, W. A. (1981). Development of gastrointestinal mucosal barrier. The effect of age on intestinal permeability to macromolecules. *Pediatr. Res.* **15,** 241–244.

Voytek, P., and Gjessing, E. (1971). Studies of an anionic trypsinogen and its active enzyme from porcine pancreas. *J. Biol. Chem.* **246,** 508–516.

Walentin, S., Levay, G., Koranyi, L., and Endroezi, E. (1988). Comparative analysis of enzyme activity in human colostrum, milk and serum. *Clin. Biochem.* **21,** 131–133.

Wang, C-S., and Johnson, K. (1983). Purification of human bile salt activated lipase. *Anal. Biochem.* **133,** 457–461.

Wang, C-S., and Kloer, H-U. (1984). Purification of human lysozyme from milk and pancreatic juice. *Anal. Biochem.* **139,** 224–227.

Wang, C-S., Martindale, M. E., King, M. M., and Tang, J. (1987). Bile-salt-activated lipase: Effect on kitten growth rate. *Am. J. Clin. Nutr.* **49,** 459–463.

Wang, C. S., Hartsuck, J. A., and Downs, D. (1988). Kinetics of acylglycerol sequential hydrolysis by human milk bile salt activated lipase and effect of taurocholate as fatty acid acceptor. *Biochemistry* **27,** 4834–4840.

Watson, R. R., Tye, I. G., McMurray, D. N., and Reves, M. A. (1977). Pancreatic and salivary amylase activity in undernourished Colombian children. *Am. J. Clin. Nutr.* **30,** 599–604.

Westrom, B. R., Svendsen, J., and Karlsson, B. W. (1982). Protease inhibitor levels in porcine mammary secretions. *Biol. Neonate* **42,** 185–194.

Whitehouse, D. B., Putt, W., Lovegrove, J. U., Morrison, K., Hollyoake, M., Fox, M., Hopkinson, D. A., and Edwards, Y. H. (1992). Phosphoglucomutase 1: Complete human and rabbit mRNA sequences and direct mapping of this highly polymorphic marker of human chromosome 1. *Proc. Natl. Acad. Sci. USA* **89,** 411–415.

Williams, M. M. F. (1983). Selenium and glutathione peroxidase in mature human milk. *Proc. Univ. Otago. Med. Sch.* **61,** 20–21.

Willinger, L., Rosenfold, W., Koo, H. C., and Conception, L. (1990). Suproxide dismutase concentration in human breast milk. *Pediatr. Res.* **27,** 120A. [Abstract]

Wilson, D. C., Afrosiaby, M., and Reid, M. M. (1992). Breast milk beta-glucuronidase and exaggerated jaundice in the early neonatal period. *Biol. Neonate* **61,** 232–234.

Wolf, B., Heard, G. S., Secor McVoy, J. R., and Grier, R. E. (1985). Biotinidase deficiency. *Ann. N.Y. Acad. Sci.* **447,** 252–262.

Worth, G. K., Retallack, R. W., Gutteridge, D. H., Jeffries, M., Kent, J., and Smith, M. (1981). Serum and milk alkaline phosphatase in human milk. *Clin. Chim. Acta* **115,** 171–177.

Yamamoto, J., Horie, N., and Okamoto, U. (1980). Enzymatic properties of milk activator in relation to synthetic chromogenic substrates. *Thromb. Res.* **18,** 263–266.

Yasuda, T., Nadano, D., Awazu, S., and Kishi, K. (1992). Human urine deoxyribonuclease II (DNase II) isoenzymes: A novel immunoaffinity purification, biochemical multiplicity, genetic heterogeneity and broad distribution among tissue and body fluids. *Biochim. Biophys. Acta* **1119,** 185–193.

Yoshikawa, M., Yoshimura, T., and Chiba, H. (1984). Opioid peptides from human β-casein. *Agric. Biol. Chem.* **48,** 3185–3190.

Zeng, J., Rao, K. R., Brew, K., and Fenna, R. (1990). Crystallization of calcium-binding lysozyme from horse milk. *J. Biol. Chem.* **265,** 14886–14887.

Zhao, Y. J., Sulkowski, E., and Porath, J. (1991). Surface topography of histodine residues in lysozymes. *Eur. J. Biochem.* **202,** 1115–1119.

Zinder, O., Mendelson, C. R., Blanchette-Mackie, E. J., and Scow, R. O. (1976). Lipoprotein lipase and uptake of chylomicron triacylglycerol and cholesterol by perfused rat mammary tissue. *Biochim. Biophys. Acta* **431,** 526–537.

Zoppi, G., Andreotti, G., Pajno-Ferrara, F., Njai, D. M., and Gaburro, D. (1972). Exocrine pancreas function in premature and full-term neonates. *Pediatr. Res.* **16,** 880–884.

D. Hormones and Growth Factors in Human Milk

OTAKAR KOLDOVSKÝ
VLADIMÍR ŠTRBÁK

I. Introduction

The presence of hormones and hormone-related substances (for sake of brevity, in this review they will be known as "hormones") in milk was described more than 50 years ago (Yaida, 1929; Heim 1931a,b). Currently, this work is attracting a considerable number of researchers. Several reviews (Richardson and Mattarella, 1977; Koldovský, 1980, 1983, 1989a,b; Koldovský and Thornburg, 1987; Koldovský *et al.* 1987, 1988, 1990, 1992, 1993; Štrbák, 1985) and published symposia (Renner and Sawatzki, 1993; Štrbák, 1983, 1986, 1991) on this topic have clearly shown that the significance of these studies is not only theoretical but also potentially practical when designing infant nutrition formulas. Progress in this area has been possible because of the development of new methods for hormone determinations, although many methodological problems remain unsolved. Because milk is heterogenous, different substances can interfere with various determinations: thus, many results are affected, especially quantitatively. The estimated content of the thyroid and steroid hormones in milk strongly depends on the method used. Their concentration in human colostrum and milk was overestimated in the pioneer investigations who used less specific methods than those available later (for review see Pearlman, 1983; Štrbák, 1985). For practical reasons, only recent and reliable data are listed when available. In other cases old data (before 1980) are presented with the understanding of reader's reserved consideration.

II. Explanation of Data

Data are summarized in two tables: Table I—"Nonpeptide Hormones in Human Milk," and Table II—"Hormonally Active Peptides in Human Milk." Quoted reviews discuss important aspects of hormones in milk, specifically (a) their presence in various species; (b) factors influencing their

HANDBOOK OF MILK COMPOSITION

TABLE I
Nonpeptide Hormones in Human Milk

Hormone	ng/ml	References
Thyroid		
Thyroxine	1–4	Moller *et al.* (1983)
	0.3–2.0	Mallol *et al.* (1982)
	12.0	Oberkotter and Tenore (1983)
	1.16–2.4	Slebodzinski *et al.* (1986)
	0.8–2.3	Bohles *et al.* (1993)
Triiodothyronine	0.02–0.40	Slebodzinski *et al.* (1986)
	0.05–0.10	Bohles *et al.* (1993)
Reverse triiodothyronine	0.008–0.15	Bohles *et al.* (1993)
Adrenal gland		
Cortisol	0.2–32.0 (5–10)[a]	Kulski and Hartmann (1981)
	3.7	Alexandrová and Macho (1983)
Sexual		
Progesterone	10–40	Kulski *et al.* (1977)
Pregnane-3(α)20(β)-diol[b]	0–450	Munch (1954) Krauer-Mayer *et al.* (1968) Severi *et al.* (1970)
Estrogens	15–840 (15–60)[a]	Sas *et al.* (1969)
Contraceptives	Biol. sign. quantities	Nilsson *et al.* (1977a,b) Nilsson *et al.* (1978) Saxena *et al.* (1977)

[a]Ratio of values in colostrum/values in mature milk.
[b]Steroid possibly implicated in the etiology of the neonatal jaundice.

presence and concentration in milks and various preparations from bovine milk, especially infant nutrition formula; (c) the absorption of milk-borne hormones from the immature gastrointestinal tract; and (d) their effect after orogastric administration on the gastrointestinal tract and beyond in developing mammals.

TABLE II
Hormonally Active Peptides in Human Milk

Peptide	Concentration	Ratio (colostrum/ mature milk)	References
Erythropoietin	Bioassay	?	Bielecki *et al.* (1972)
Growth factors			
EGF	3–107 ng/ml	2/10	Beardmore *et al.* (1983); Connolly and Rose (1988); Corps *et al.* (1987, 1988); Hirata and Orth (1979); Jansson *et al.* (1985); Jaspar and Franchimont (1985); Moran *et al.* (1983); Petrides *et al.* (1985); Read *et al.* (1984, 1985); Yagi *et al.* (1986)
Insulin	0–80 μU/ml	3/10	Ballard *et al.* (1982); Čevreska *et al.* (1975); Jovanovic-Peterson *et al.* (1989); Kulski and Hartmann (1983); Nowak (1989); Read *et al.* (1984, 1985); Slebodzinski *et al.* (1986)
IGF-I	1.3–7 ng/ml	2/3	Baxter *et al.* (1984); Corps *et al.* (1988); Nagashima et al. (1990); Suikkari (1989)
NGF	Present		Wright *et al.* (1983)
TGFα	0–8.4 ng/ml	1	Connolly and Rose (1988); Okada *et al.* (1991)
Other GFs	Present	?	Corps *et al.* (1987); Kidwell *et al.* (1985); Noda *et al.* (1984); Shing and Klagsbrun (1984); Sinha and Yunis (1983)
Gastrointestinal regulatory peptides			
Gastrin	10–30 pg/ml	2/3	Berseth *et al.* (1990); Widström *et al.* (1988)
GIP	33–59 ng/ml	1	Berseth *et al.* (1990)
GRP	55–31 pg/ml 60–430 pg/ml	2/3	Berseth *et al.* (1990); Ekman *et al.* (1985); Takeyama *et al.* (1991); Widström *et al.* (1988)
Neurotensin	7–15 pg/ml	2/3	Berseth *et al.* (1990); Ekman *et al.* (1985); Werner *et al.* (1982)
PHM	3–32 pg/ml	5/10	Berseth *et al.* (1990)
PYY	15–30 pg/ml	2/3	Berseth *et al.* (1990)

TABLE II—*continued*

Peptide	Concentration	Ratio (colostrum/ mature milk)	References
Somatostatin	23–113 pg/ml	1	Koch *et al.* (1991); Werner *et al.* (1985, 1988); Widström *et al.* (1988)
VIP	7–13 pg/ml 67–161 pg/ml	1	Berseth *et al.* (1990); Werner *et al.* (1986)
Hypothal–hypohyseal hormones			
GnRH	0.1–4.0 ng/ml	?	Amarant *et al.* (1982); Sack *et al.* (1978); Sarda and Nair (1981)
GRF	23 ± 7 pg/ml	1.5	Werner *et al.* (1986)
	152–430 pg/ml		Werner *et al.* (1988)
Growth hormone	5–30 μU/ml	?	Kulski and Hartmann (1983)
Prolactin	20–71 ng/ml	2/5	Adampoulos and Kapolla (1983); Gala *et al.* (1975); Gala and Van DeWalle (1977); Healy *et al.* (1980); Tyson *et al.* (1972); Werner *et al.* (1982); Yuen (1986)
TRH	0.025–1.5 ng/ml	?	Amarant *et al.* (1982); Sack *et al.* (1978)
TSH	2.7–5.0 μU/ml	3	Tenore *et al.* (1981); Kulski *et al.* (1983)
Thyroid–parathyroid group			
Calcitonin-like	0–5 ng/ml	5	Arver *et al.* (1984); Bucht *et al.* (1983, 1986); Bucht and Sjöberg (1987); Werner *et al.* (1982)
Parathyroid hormone	15 pg/ml	?	Budayr *et al.* (1989)
Parathyroid hormone-related peptide	30–50 ng/ml	?	Budayr *et al.* (1989); Ratcliffe *et al.* (1990); Thurston *et al.* (1990)

References

Adamopoulos, D. A., and Kapolla, N. (1983). Prolactin concentration in milk and blood of patients with galactorrhoea. *Acta Endocrinologica* **261,** 5–7.

Alexandrová, M., and Macho, L. (1983). Glucocorticoids in human, cow and rat milk. *Endocrinol. Exp.* **17,** 183–189.

Amarant, T., Fridkin, M., and Koch, Y. (1982). Luteinizing hormone-releasing hormone and thyrotropin-releasing hormone in human and bovine milk. *Eur. J. Biochem.* **127,** 647–650.

Arver, S., Bucht, E., and Sjöberg, H. E. (1984). Calcitonin-like immunoreactivity in human milk, longitudinal alterations and divalent cations. *Acta Physiol. Scand.* **122,** 461–464.

Ballard, F. J., Nield, M. K., Francis, G. L., Dahlenburg, G. W., and Wallace, J. C. (1982). The relationship between the insulin content and inhibitory effects of bovine colostrum on protein breakdown in cultured cells. *J. Cell. Physiol.* **110,** 249–254.

Baxter, R. C., Zaltsman, Z., and Turtle, J. R. (1984). Immunoreactive somatomedin-insulin-like growth factor I and its binding protein in human milk. *J. Clin. Endocrinol. Metab.* **58,** 955–959.

Beardmore, J. M., Lewis-Jones, D. I., and Richards, R. C. (1983). Urogastrone and lactose concentrations in precolostrum, colostrum and milk. *Pediatr. Res.* **17,** 825–828.

Berseth, C. L., Michener, S. R., Nordyke, C. K., and Go, V. L. W. (1990). Postpartum changes in pattern of gastrointestinal regulatory peptides in human milk. *Am. J. Clin. Nutr.* **51,** 985–990.

Bielecki, M., Przala, F., and Lazewska, M. (1972). Poziom erytropoetyny w mleku kobiecym (in Polish) (Level of erythropoietin in human milk). *Acta Physiol. Pol.* **23,** 497–502.

Bohles, H., Aschenbrenner, M., Roth, M., Loewenich, V., Ball, F., and Usadel, K. H. (1993). Development of thyroid gland volume during the first 3 months of life in breast-fed versus iodine-supplemented and iodine-free formula-fed infants. *Clin. Invest.* **71,** 13–20.

Bucht, E., and Sjöberg, H. E. (1987). Evidence for precursors of calcitonin/PDN 21 in human milk. *Regul. Peptides* **19,** 65–71.

Bucht, E., Carlquist, M., Lindgren, E., Hedlund, B., and Torring, O. (1989). Parathyroid hormone like protein(s) in human milk. *J. Bone Mineral Res.* **4,** 606.

Bucht, E., Telenius-Berg, M., Lundell, G., and Sjöberg, H. E. (1986). Immunoextracted calcitonin in milk and plasma from totally thyrodectomized women. Evidence of monomeric calcitonin in plasma during pregnancy and lactation. *Acta Endocrinologica* **113,** 529–535.

Bucht, E., Arver, S., Sjöberg, H. E., and Low, H. (1983). Heterogeneity of immunoreactive calcitonin in human milk. *Acta Endocrinol.* **103,** 572–576.

Budayr, A. A., Halloran, B. P., King, J. C., Diep, D., Nissenson, R. A., and Strewler, G. J. (1989). High levels of a parathyroid hormone-like protein in milk. *Proc. Natl. Acad. Sci. USA* **86,** 7183–7185.

Carpenter, G. (1980). Epidermal growth factor is a major growth-promoting agent in human milk. *Science* **210,** 198–199.

Čevreska, S., Kovacev, V. P., Stankovski, M., and Kalamaras, E. (1975). The presence of immunologically reactive insulin in milk of women, during the first week of lactation and its relation to changes in plasma insulin concentration. *God. Zb. Med. Fak. Skopje.* **21,** 35–41.

Connolly, J. M., and Rose, D. P. (1988). Epidermal growth factor-like proteins in breast fluid and human milk. *Life Sci.* **42,** 1751–1756.

Corps, A. N., Blakeley, D. M., Carr, J., Rees, L. H., and Brown, K. D. (1987). Synergistic stimulation of Swiss mouse 3T3 fibroblasts by epidermal growth factor and other factors in human mammary secretions. *J. Endocrinol.* **112,** 151–159.

Corps, A. N., Brown, K. D., Rees, L. H., Carr, J., and Prosser, C. G. (1988). The insulin-like growth factor I content in human milk increases between early and full lactation. *J. Clin. Endocrinol. Metab.* **67,** 25–29.

Ekman, R., Ivarsson, S., and Jansson, L. (1985). Bombesin, neurotensin and pro-γ-melanotropin immunoreactants in human milk. *Regul. Peptides* **10,** 99–105.

Gala, R. R., and Van DeWalle, C. (1977). Prolactin heterogeneity in the serum and milk during lactation. *Life Sci.* **21,** 99–104.

Gala, R. R., Singhakowinta, A., and Brennan, M. J. (1975). Studies on prolactin in human serum, urine and milk. *Hormone Res.* **6,** 310–320.

Healy, D. L., Rattigan, S., Hartmann, P. E., Herington, A. C., and Burger, H. G. (1980). Prolactin in human milk: Correlation with lactose, total protein, and α-lactalbumin levels. *Am. J. Physiol.* **238,** E83–E86.

Heim, K. (1931a). Hormonale wirkungen der frauenmilch. *Klin. Wochenschr.* **10,** 357.

Heim, K. (1931b). Brustdruse und hypophysenvorderlappen. *Klin. Wochenschr.* **10,** 1598.

Hirata, Y., and Orth, D. N. (1979). Epidermal growth factor (urogastrone) in human fluids: Size heterogeneity. *J Clin. Endocrinol. Metab.* **48,** 673–679.

Jansson, L., Karlson, F. A., and Westermark, B. (1985). Mitogenic activity and epidermal growth factor content in human milk. *Acta Pediatr. Scand.* **74,** 250–253.

Jaspar, J. M., and Franchimont, P. (1985). Radioimmunoassay of human epidermal growth factor in human breast cyst fluid. *Eur. J. Cancer Clin. Oncol.* **21,** 1343–1348.

Jovanovic-Peterson, L., Fuhrmann, K., Hedden, K., Walker, L., and Peterson, C. M. (1989). Maternal milk and plasma glucose and insulin levels: Studies in normal and diabetic subjects. *J. Am. Coll. Nutr.* **8,** 125–131.

Kidwell, W. R., Bano, M., Burdette, K., Lococzy, I., and Salomon, L. (1985). Mammary derived growth factors in human milk. *In* "Human Lactation: Milk Components and Methodologies" (R. G. Jensen and M. C. Neville, eds.), pp. 209–219.

Koch, Y., Werner, H., and Fridkin, M. (1991). Hypothalamic hormones in milk. *Endocrinol. Exp.* **25,** 128–133.

Kohno, Y., Shiraki, K., and Mura, T. (1991). The effect of human milk on DNA synthesis of neonatal rat hepatocytes in primary culture. *Pediatr. Res.* **29,** 251–255.

Koldovský, O. (1980). Hormones in milk. *Life Sci.* **26,** 1833–1836.

Koldovský, O. (1989a). Hormones in milk: Their possible physiological significance for the neonate. *In* "Textbook of Gastroenterology and Nutrition in Infancy, 2nd Edition" (E. Lebenthal, ed.), pp. 97–119. Raven Press, New York.

Koldovský, O. (1989b). Critical review: Search for role of milk-borne biologically active peptides for the suckling. *J. Nutr.* **119,** 1543–1551.

Koldovský, O. (1993). Biologically active peptides in milk. *In* "New Perspectives in Infant Nutrition, Symposium Antwerpen 1992" (B. Renner and G. Sawatzki, eds.), pp. 123–130. Georg Thieme Verlag, Stuttgart/New York.

Koldovský, O., and Thornburg, W. (1987). Hormones in milk: A review. *J. Pediatr. Gastroenterol. Nutr.* **6,** 172–196.

Koldovský, O., Bedrick, A., Pollack, P., Rao, R. K., and Thornburg, W. (1987). Hormones in milk: Their presence and possible physiological difference. *In* "Human Lactation 3: The Effects of Human Milk on the Recipient Infant" (A. S. Goldman, S. A. Atkinson, and L. A. Hanson, eds.), pp. 183–196. Plenum Press, New York.

Koldovský, O., Bedrick, A., Pollack, P., Rao, R. K., and Thornburg, W. (1988). The possible physiological role of hormones and hormone-related substances present in milk. *In* "Biology of Human Milk" (L. A. Hanson, ed.), pp. 123–139. Raven Press, New York.

Koldovský, O., Bedrick, A., Pollack, P., Rao, R. K., and Thornburg, W. (1990). Hormones and modulating factors in human milk. *In* "Proceedings of the Third International Symposium on Nutrition and Gastroenterological Problems in Childhood" (O. Olivi and F. Balli, eds.), pp. 151–160. Editrice CSH Publishers, Milan, Italy.

Koldovský, O., Philipps, A., Rao, R. K., and Schaudies, P. (1992). Possible role of milk-borne peptide growth factors for the breast-fed infant. *In* "Regulatory Gut Peptides in Paediatric Gastroenterology and Nutrition; Frontiers of Gastrointestinal Research" (P. Heinz-Erian, J. Deutsch, and G. Granditsch, eds.), pp. 150–169. Karger, Basel/New York.

Koldovský, O., Kong, W., Rao, R. K., and Schaudies, P. (1993). Milk-borne peptide growth factors in human and bovine milk. *In* "Immunophysiology of the Gut" (W. Allan Walker, ed.), pp. 269–293. Academic Press, New York.

Krauer-Mayer, B., Keller, M., and Hottinger, A. (1968). Prolonged icterus in the newborn induced by mother's milk. *Helv. paed. Acta* **1,** 68–76.

Kulski, J. K., Smith, M., and Hartmann, P. E. (1977). Perinatal concentrations of progesterone, lactose and α-lactalbumin in the mammary secretion of women. *J. Endocrinol.* **74,** 509–510.

Kulski, J. K., and Hartmann, P. E. (1981). Changes in the concentration of cortisol in milk during different stages of human lactation. *Austr. J. Exp. Biol. Med. Sci.* **59,** 769–778.

Kulski, J. K., and Hartmann, P. E. (1983). Milk insulin, GH and TSH: Relationship to changes in milk lactose, glucose and protein during lactogenesis in women. *Endocrinol. Exp.* **17,** 317–326.

Mallol, J., Obregon, M. J., and Escobar, G. M. (1982). Analytical artifacts in radioimmunoassay of L-thyroxine in human milk. *Clin. Chem.* **28,** 1277–1282.

Moller, B., Bjorkhem, L., Falk, O., and Larsson, A. (1983). Identification of thyroxine in human breast milk by gas chromatography–mass spectrometry. *J. Clin. Endocrinol. Metab.* **30,** 30–34.

Moran, J. R., Courtney, M. E., Orth, D. N., *et al.* (1983). Epidermal growth factor in human milk: Daily production and diurnal variation during early lactation in mothers delivering at term and at premature gestation. *J. Pediatr.* **103,** 402–405.

Munch, U. (1954). Milk contains secretions of natural androgens and estrogens. *Milchwissenschaft* **9,** 150–153.

Nagashima, K., Itoh, K., and Kuroume, T. (1990). Levels of insulin-like growth factor in full and preterm human milk in comparison to levels in cow's milk and in milk formulas. *Biol. Neonate* **58,** 343–346.

Nilsson, S., Nygren, K-G., and Johansson, E. D. B. (1977a). Megestrol acetate concentrations in plasma and milk during administration of an oral contraceptive containing 4 mg megestrolacetate to nursing women. *Contraception* **16,** 615–624.

Nilsson, S., Nygren, K-G., and Johansson, E. D. B. (1977b). D-Norgesterol concentrations in maternal plasma, milk, and child plasma during administration of oral contraceptives to nursing women. *Am. J. Obstet. Gynecol.* **129,** 178–184.

Nilsson, S., Nygren, K-G., and Johansson, E. D. B. (1978). Ethinyl estradiol in human milk and plasma after oral administration. *Contraception* **17,** 131–139.

Noda, K., Umeda, M., and Ono, T. (1984). Transforming growth factor activity in human colostrum. *Gann* **75,** 109–112.

Nowak, J. (1989). Changes of insulin concentration in colostrum and milk of women, cows and sows. *Acta Physiol. Polonica* **40,** 349–355.

Oberkotter, L. V., and Tenore, A. (1983). Separation and radioimmunoassay of T3 and T4 in human breast milk. *Hormone Res.* **17,** 11–18.

Okada, M., Ohmura, E., Kamiya, Y., Murakami, H., Onoda, N., Iwashita, M., Wakai, K., Tsushima, T., and Shizume, K. (1991). Transforming growth factor (TGF)-alpha in human milk. *Life Sci.* **48,** 1151–1156.

Pearlman, W. H. (1983). Glucocorticoids in milk—A review. *Endocrinol. Exp.* **17,** 165–174.

Petrides, P. E., Hosang, M., Shooter, E., Esch, F. S., and Bohlen, P. (1985). Isolation and characterization of epidermal growth factor from human milk. *FEBS Lett.* **187,** 89–95.

Ratcliffe, W. A., Green, E., Emly, J., *et al.* (1990). Identification and partial characterization of parathyroid hormone-related protein in human and bovine milk. *J. Endocrinol.* **127,** 167–178.

Read, L. C., Upton, F. M., Francis, G. L., Wallace, J. C., Dahlenberg, G. W., and Ballard, F. J. (1984). Changes in the growth-promoting activity of human milk during lactation. *Pediatr. Res.* **18,** 133–139.

Read, L. C., Francis, G. L., Wallace, J. C., and Ballard, F. J. (1985). Growth factor concentrations and growth-promoting activity in human milk following premature birth. *J. Dev. Physiol.* **7,** 135–145.

Read, L. C., Summer, L., Gale, S. M., George-Nascimento, C., Ballard, F. J., and Wallace, J. C. (1986). Properties of synthetic-gene recombinant human epidermal growth factor:

Comparison with the natural growth factor from human urine and milk. *J. Endocrinol.* **109,** 245–250.

Renner, B., and Sawatzki, G. (eds.) (1993). "New Perspectives in Infant Nutrition, Symposium Antwerpen 1992." pp. 3–181. Georg Thieme Verlag, Stuttgart; New York.

Richardson, T., and Mattarella, N. (1977). Hormonal substances in human milk, cow's milk and dairy products. *J. Food Protein* **40,** 57–74.

Sack, J., Frucht, H., and Amado, O., and Lunenfeld, B. (1978). Thyroxine and thyrotropin-releasing hormone in human milk. *Israel J. Med. Sci.* **14,** 408–409.

Sarda, A. K., and Nair, R. M. G. (1981). Elevated levels of LRH in human milk. *J. Clin. Endocrinol. Metab.* **52,** 826–828.

Sas, M., Viski, S., and Gellen, J. (1969). The steroid content of mother's milk. *Arch. Gynak.* **207,** 452–459.

Saxena, B. N., Shrimankar, K., and Grudzinskas, J. G. (1977). Levels of contraceptive steroids in breast milk and plasma of lactating women. *Contraception* **16,** 605–613.

Severi, F., Rondini, G., Zaverio, S., and Vegni, M. (1970). Prolonged neonatal hyperbilirubinemia and pregnane-3(alpha),20(beta)-diol in maternal milk. *Helv. Paediatr. Acta* **25,** 517–521.

Shing, Y. W., and Klagsbrun, M. (1984). Human and bovine milk contain different sets of growth factors. *Endocrinology* **115,** 273–282.

Sinha, S. K., and Yunis, A. A. (1983). Isolation of colony stimulating factor from human milk. *Biochem. Biophys. Res. Commun.* **14,** 797–803.

Slebodzinski, A. B., Nowak, J., Gawecka, H., and Sechman, A. (1986). Thyroid hormones and insulin in milk: A comparative study. *Endocrinol. Exp.* **20,** 247–255.

Štrbák, V. (guest ed.) (1983). Hormones in milk: Their role in sucklings and lactating mothers (international symposium). *Endocrinol. Exp.* **17,** 163–370.

Štrbák, V. (1985). "The Role of Maternal Milk in Endocrine Regulation of Sucklings," pp. 1–108. VEDA, Bratislava.

Štrbák, V. (guest ed.) (1986). Hormones in milk: Their physiological role (international symposium). *Endocrinol. Exp.* **20,** 97–334.

Štrbák, V. (guest ed.) (1991). Hormones and bioactive substances in milk. (international symposium). *Endocrine Reg.* **25,** 3–143.

Suikkari, A-M. (1989). Insulin-like growth factor (IGF-I) and its low molecular weight binding protein in human milk. *Eur. J. Obstet. Gynecol. Reprod. Biol.* **30,** 19–25.

Takeyama, M., Kondo, K., Takayma, F., Kondo, R., Murata, H., and Miyakawa, H. (1991). High concentration of a gastrin releasing peptideside-like immunoreactive substance in pregnant human milk. *Biochem. Biophys. Res. Commun.* **176,** 931–937.

Tapper, D., Klagsbrun, M., and Neumann, J. (1979). The identification and clinical implications of human breast milk mitogen. *J. Pediatr. Surg.* **14,** 803–808.

Tenore, A., Oberkotter, L. V., Koldovsky, O., Parks, J. S., and Vandenberg, C. M. (1981). Thyrotropin in human breast milk. *Hormone Res.* **14,** 193–200.

Thurston, A. W., Cole, J. A., Hillman, L. S., *et al.* (1990). Purification and properties of parathyroid hormone-related peptide isolated from milk. *Endocrinology* **126,** 1183–1190.

Tyson, J. E., Hwang, P., Guyda, H., and Friesen, H. G. (1972). Studies of prolactin secretion in human pregnancy. *Am. J. Gastroenterol.* **113,** 14–20.

Werner, H., Amarant, T., Millar, R. P., Fridkin, M., and Koch, Y. (1985). Immunoreactive and biologically active somatostatin in human and sheep milk. *Eur. J. Biochem.* **148,** 353–357.

Werner, H., Katz, P., Fridkin, M., Koch, Y., and Levine, S. (1988). Growth hormone releasing factor and somatostatin concentrations in the milk of lactating women. *Eur. J. Pediatr.* **147,** 252–256.

Werner, H., Koch, Y., Fridkin, M., Fahrenkrug, J, and Gozes, I. (1986). High levels of vasoactive intestinal peptide in human milk. *Biochem. Biophys. Res. Commun.* **135,** 1084–1089.

Werner, S., Widström, A-M., Wahlberg, V., Eneroth, P., and Winberg, J. (1982). Immunoreactive calcitonin in maternal milk and serum in relation to prolactin and neurotensin. *Early Hum. Dev.* **6,** 77–82.

Widström, A-M., Marchini, G. Matthiesen, A-S., Werner, S., Winberg, J., and Uvnas-Moberg, K. (1988). Non-nutritive sucking in tube-fed preterm infants: Effects on gastric motility and gastric contents of somatostatin. *J. Pediatr. Gastroenterol. Nutr.* **7,** 517–523.
Wright, C. E., and Gaull, G. E. (1983). Nerve growth factor is present in human milk. *Pediatr. Res.* **17,** 347A.
Yagi, H., Suzuki, S., Noji, T., Nagashima, K., and Kuroume, T. (1986). Epidermal growth factor in cow's milk and milk formula. *Acta Pediatr. Scand.* **75,** 233–235.
Yaida, N. (1929). Ovarial hormone in blood of pregnant women, of pregnant animals; Ovarial hormone in urine of pregnant women; Ovarial hormone in milk of pregnant animals. *Trans. Jpn. Pathol. Soc.* **19,** 39–101.
Yuen, B. H. (1986). Immunoreactive prolactin in breast milk and plasma of women with hyper-prolactinemia, galactorrhea and menstrual dysfunction. *Int. J. Fertil.* **31,** 67–70.
Zwiebel, J. A., Bano, M., Nexø, E., Salomon, D. S., and Kidwell, W. R. (1986). Partial purification of transforming growth factors from human milk. *Cancer Res.* **46,** 933–939.

E. Nucleotides and Related Compounds in Human and Bovine Milks

ANGEL GIL
RICARDO UAUY

I. Introduction

Nucleotides play important roles in major physiologic and biochemical functions. They act as precursors for nucleic acid synthesis (DNA and RNA) and are also fundamental to intermediary metabolism. Adenosine triphosphate (ATP), an adenine nucleotide, is the major molecule responsible for the transfer of chemical energy from energy-yielding reactions to energy-requiring processes. Other adenine nucleotides, such as nicotinamide adenine dinucleotide (NAD), nicotinamide adenine dinucleotide phosphate, flavine adenine dinucleotide, and coenzyme A, are key in the synthesis of lipids, carbohydrates, and proteins. They also are essential for the transfer of reducing equivalents in cellular oxidative processes. Nucleotides and nucleic acids are constantly being formed and degraded, especially in tissues with rapid turnover rates such as skin, intestinal mucosa, white and red blood cells, and the immune system. Growing organs also require constant formation of new DNA and have a rapid turnover of RNA (Lehninger, 1982; McGillivray, 1983).

A. Nomenclature

Nucleotides contain three characteristic components: a nitrogenous base, a pentose, and one or more phosphate groups. The nitrogenous bases are derivatives of two parent heterocyclic compounds, purine and pyrimidine. The major pyrimidine bases are cytosine, thymine, and uracil. Adenine and guanine are the two major purines found in living organisms (Lehninger, 1982; McGillivray, 1983).

Nucleosides and nucleotides arise from substitutions on the ring structures of the parent bases. A nucleoside is formed from an oxygen–nitrogen glycosidic linkage of a pentose to a nitrogenous base. The pentose can be either D-ribose as in ribonucleic acid (RNA) or 2-deoxyribose as in deoxyribonucleic acid (DNA). A nucleotide is a phosphate ester of a nucleoside. The most common site of esterification is carbon No. 5 of the pentose, this is referred to as a 5′ nucleotide. The letters A (adenosine), G (guanosine), C (cytidine), I (inosine), T (thymidine), and U (uridine) designate the appropriate nucleoside. The prefix d is added if the sugar in the nucleoside is 2-deoxyribose. The capital letters MP, DP, or TP indicate the mono-, di-, and triphosphate esters (Lehninger, 1982; McGillivray, 1983). For example, the abbreviation ATP represents adenosine triphosphate. References for all abbreviated nucleotide forms used in this text can be found in Table I.

TABLE I
Major Nitrogenous Bases, Nucleosides, and Nucleotides, and Common Abbreviations

Nitrogenous bases	Nucleosides (N) (N base + pentose)		Nucleotides (N base + pentose + phosphate groups)	Abbreviations
Adenine	Adenosine	(A)	Adenosine mono , di , and triphosphate	AMP,[a] ADP, ATP
Cytosine	Cytidine	(C)	Cytidine mono-, di-, and triphosphate	CMP, CDP, CTP
Guanine	Guanosine	(G)	Guanosine mono-, di-, and triphosphate	GMP, GDP, GTP
Hypoxanthine	Inosine	(I)	Inosine mono-, di-, and triphosphate	IMP, IDP, ITF
Uracil	Uridine	(U)	Uridine mono-, di-, and triphosphate	UMP, UDP, UTP
Thymine	Thymidine	(T)	Thymidine mono-, di-, and triphosphate	TMP, TDP, TTP

[a]Other derived compounds: cAMP, 3′–5′ cyclic AMP.

B. Nucleotide and Nitrogenous Base Metabolism

Purine bases can be formed *de novo* from amino acid precursors or reutilized after being liberated from nucleic acid catabolism via the salvage pathway. The *de novo* synthesis of inosine, guanosine, and adenosine depends on the availability of the precursors glutamine, aspartic acid, glycine, formate, and CO_2. These bases can be linked to a ribose and phosphorylated to reconstitute the nucleotides via the salvage pathway. There is substantial evidence to suggest that this pathway is regulated by the availability of free purine and pyrimidine bases. Under fed conditions, 90% or more of the purine bases are recycled by this route. The oxidative catabolism of the purine nucleotides forms uric acid by the action of the enzyme xanthine oxidase. The compound xanthine is an intermediate product in this process. Mammals further oxidize urate via uricase to form allantoin except for primates, including humans. Excess formation of uric acid in the human leads to gout, a common disorder of purine metabolism. The hereditary deficiency of the enzyme responsible for the salvage of purine bases, hypoxanthine guanine phosphoribosyl transferase, also is associated with high uric acid in the pathologic condition known as the Lesch–Nyhan syndrome (Lehninger, 1982; McGillivray, 1983; Uauy, 1989; Nyhan, 1987). Purine nucleotides are hydrolytically degraded to nucleosides by nucleotidases. Adenosine must be deaminated before the formation of the free base hypoxanthine. Following dephosphorylation, the bases hypoxanthine and guanine are oxidized and deaminated, respectively, to yield xanthine (Lehninger, 1982; McGillivray, 1983; Uauy, 1989; Nyhan, 1987).

The synthesis of the pyrimidine bases requires the contribution of glutamine and CO_2 to form carbamoyl phosphate and aspartate. This pathway yields orotate that can be further metabolized to form uridine, cytidine, and thymidine specifically required for DNA synthesis (Lehninger, 1982; McGillivray, 1983; Uauy, 1989; Nyhan, 1987). The catabolism of pyrimidine nucleotides forms nucleosides that may be reutilized via the salvage pathway; free pyrimidine bases, however, are not salvaged efficiently by mammalian cells (Lehninger, 1982; McGillivray, 1983; Uauy, 1989; Nyhan, 1987). Alternatively, the carbon skeletons are oxidized to CO_2, and the nitrogen groups are cleaved by successive hydrolysis. Catabolism of pyrimidine occurs mainly in the liver. The only tracer for the excretion of the pyrimidine bases is α-aminoisobutyrate, which is slowly metabolized and can be quantified in the urine. Unlike uric acid, the purine end product, pyrimidine catabolism products are highly water soluble; alanine and aminoisobutyric acid are derived from cytosine–uracil and thymidine catabolism, respectively. Many of the initial steps of pyrimidine degradation are simply a reversal of the latter part of the synthetic pathway (Lehninger, 1982; McGillivray, 1983; Uauy, 1989; Nyhan, 1987).

There is evidence to indicate that some tissues possess limited capacity for the *de novo* synthesis of purine bases. These tissues require exogenously supplied bases that can be utilized via the salvage pathway rather than forming bases by *de novo* synthesis. The intestinal mucosa, bonc marrow hematopoietic cells including leukocytes and red cells, preferentially utilize preformed purine and pyrimidine bases. For these cells, the exogenous supply can be important for optimal function. Liver cells have especially active *de novo* synthesis of purine and pyrimidine bases. The liver not only produces enough for its own supply but it exports bases to other tissues with high requirements because of rapid cell turnover rates. In the absence of dietary nucleotides, the *de novo* pathway for purine and pyrimidine synthesis is activated; conversely, the addition of these bases to the diet suppresses *de novo* synthesis and activates the salvage pathway (McGillivray, 1983; Savaiano and Clifford, 1981; Rudolph *et al.*, 1984; Roux, 1973; Leleiko *et al.*, 1983).

II. Analytical Methodology

Multiple methods have been used to separate and quantitate purine and pyrimidine bases, nucleosides, nucleotides, and nucleic acids in milk. These compounds have usually been determined on protein-free milk filtrates obtained by dialysis or after protein precipitation using trichloroacetic acid (TCA) or perchloric acid (PCA).

A. Preparation of Protein-Free Milk Extracts

Deutsch and Mattsson (1960) conducted dialysis of milk for 48 hr at 3–5°C against 10 vol of glass-distilled water. Extraction with TCA was then performed by adding solid TCA to a precooled milk sample to a final TCA concentration of 10%. The precipitated proteins were reextracted four times. The filtrates were pooled and ether extracted to remove TCA. Rashid used a similar procedure to obtain protein-free milk filtrates (Rashid, 1973). Extraction with PCA was carried out by adding 70% PCA to the precooled milk sample to a final concentration of 4%. The filtrate was neutralized with potassium hydroxide at 0°C in order to remove PCA (Deutsch and Mattsson, 1960).

Denamur *et al.* (1959) used four extractions with 10% TCA and elimination of TCA by ether extraction. Johnke and Goto (1962) obtained milk protein-free filtrates by adding 2.5 vol of 10% TCA to precooled milk (2°C). The mixture was stirred with a glass rod and the protein pellet was centrifuged and reextracted with additional TCA. The acid of the combined supernatants was then removed by ether extraction. Kobata *et al.* (1962) added PCA to milk during 10 min stirring. The precipitate was centrifuged and again extracted with a quarter volume of cold 0.2 *M* PCA. The supernatants were combined and neutralized with KOH and potassium perchlorate was removed.

Gil and Sanchez-Medina (1981a) obtained protein-free milk extracts by mixing precooled milk with ice-cold 1 *M* PCA. After centrifuging the protein pellet was discarded and the supernatant was filtered and neutralized with KOH. Insoluble potassium perchlorate was allowed to precipitate and then removed by centrifugation.

Some authors have used a treatment with charcoal adsorption to purify the milk extracts prior to chromatography (Deutsch and Mattsson, 1960; Rashid, 1973; Kobata *et al.*, 1962). Dialysates or extracts are first adjusted to pH 2 passed through an activated charcoal or a purifying Norit-A column (Kobata *et al.*, 1962). The eluate is concentrated under reduced pressure to about 1/10 vol (Deutsch and Mattsson, 1960; Rashid, 1973; Kobata *et al.*, 1962).

B. Ion-Exchange Chromatography

The protein-free extracts or the charcoal eluates, brought to neutrality, are adsorbed on Dowex-1 columns in the formate form (Deutsch and Mattsson, 1960; Denamur *et al.*, 1959; Johnke and Gold, 1962; Kobata *et al.*, 1962; Gil and Sanchez-Medina, 1981a). After adsorption the columns are washed with water and eluted with increasing concentrations of formic acid and sodium formate (Bergkvist and Deutsch, 1954; Hurlbert *et al.*, 1954). Fractions are collected and measured at 254–280 nm and the formic acid and sodium formate are removed from the eluate by charcoal treatment (Deutsch and Mattsson, 1960; Kobata *et al.*, 1962), extraction with ether, and desalting with a cation-exchange resin (Denamur *et al.*, 1959) or freeze-drying (Deutsch and Mattsson, 1960; Gil and Sanchez-Medina, 1981a). Typical procedures used for the identification of the nucleotides in each peak obtained by the anion-exchange chromatography include rechromatography on a second system, in which the eluents are ammonium formate buffers, using paper chromatography, thin-layer chromatography, and paper electrophoretic identifications with reference standards using three or more different solvent systems (Deutsch and Mattsson, 1960; Denamur *et al.*, 1959; Gil and Sanchez-Medina, 1981a; Bergkvist and Deutsch, 1954; Hurlbert *et al.*, 1954).

C. Paper Chromatography

The most common mixtures of solvents used for identification of nucleotides using paper chromatography are (1) 95% ethanol–1 *M* ammonium acetate (pH 7.5) (75:30), (2) *n*-propanol–concentrated ammonia (d = 0.880)–water (60:30:10), (3) isopropanol–saturated ammonium sulfate solution–water (2:79:19), (4) isobutyric acid–0.5 *N* ammonia (10:6), (5) *n*-butanol–water (86:14), and (6) 0.2 *M* sodium phosphate (pH 6.8)–

ammonium sulfate–*n*-propanol (100:60:2); locations of the spots are made under ultraviolet light (Deutsch and Mattsson, 1960; Rashid, 1973; Denamur *et al.,* 1959; Johnke and Goto, 1962; Kobata *et al.,* 1962; Gil and Sanchez-Medina, 1981a). The base moieties can be further identified after purine and pyrimidine nucleotide hydrolysis in 12 *N* PCA at 100°C for 1 hr (Deutsch and Mattsson, 1960; Rashid, 1973; Denamur *et al.,* 1959; Johnke and Goto, 1962; Kobata *et al.,* 1962) or using 1*N* hydrochloric acid at 100°C for 1 hr (Deutsch and Mattsson, 1960; Johnke and Goto, 1962). These compounds have been chromatographed using more than three solvent systems as previously cited or special solvent systems: (1) *iso*-propanolhydrochloric acid (d = 1.18)–water (65:17:18), (2) *n*-butanol–glacial acetic acid–water (4:1:5), and (3) *n*-butanol saturated with ammonia and water (Deutsch and Mattsson, 1960; Johnke and Goto, 1962).

The carbohydrate moieties of nucleotide sugars have been usually identified after mild acid hydrolysis (Johnke and Goto, 1962; Kobata *et al.,* 1962; Gil and Sanchez-Medina, 1981a). After ion exchange removal of anions and cations, samples are concentrated under vacuum and chromatographed with specific solvent systems (Deutsch and Mattsson, 1960; Denamur *et al.,* 1959; Gil and Sanchez-Medina, 1981a). The sugars are identified by spraying papers with aniline hydrogen phthalate, naphthoresorcinol, ninhydrin, and Elson–Morgan reagent (Partridge, 1955; Pontis, 1955). All paper chromatogram spots can be estimated by the method of Markham and Smith (1949) based on the ultraviolet absorption spectra of the eluates. The following extinction molar coefficients have been used at pH 2 and 260 nm to estimate nucleotide content: adenosine nucleotide, 14,200; guanosine nucleotide, 11,600; uridine nucleotide, 11,800; cytidine nucleotide, 6800; and orotate, 4200 (Johnke and Goto, 1962; Gil and Sanchez-Medina, 1981a). Identification of nucleotides by paper electrophoresis has also been carried out on filter paper using a variety of buffer systems; acetate buffer and phosphate buffer have been the most commonly used (Partridge, 1955).

D. Enzymatic Methods

Gil and Sanchez-Medina (1981a) developed enzymatic methods to measure adenosine, cytidine, guanosine, and uridine 5′-monophosphates and total adenine, cytidine, guanine, and uridine nucleotides in milk. Uridine diphosphate glucose, uridine disphosphate galactose, and orotate were also measured by specific enzymatic methods. Total 5′ nucleotides were determined on neutralized protein-free milk extracts by a modification of Keppler method (Keppler, 1974). Protein-free milk extracts were desalted with Dowex-50 (H+) and concentrated by freeze-drying. Nucleotide pyrophosphates, dinucleotides, nucleoside diphosphate sugars, and related compounds present in milk extracts were hydrolyzed by snake venom phosphodiesterase and the nucleoside 5′-monophosphates were specifically determined. AMP was determined with myokinase, pyruvate kinase, and lactate dehydrogenase. The addition of nucleoside monophosphate kinase allowed the determination of CMP + UMP. The specific determination of UMP was carried out with nucleoside monophosphate kinase, nucleoside diphosphate kinase, UDP-glucose pyrophosphorylase, and UDP-glucose dehydrogenase. Determination of nucleoside 5′ monophosphates in milk was carried out as described above for total 5′ nucleotides but without prior hydrolysis. Determination of UDP-glucose and UDP-galactose in protein-free milk extracts was performed by a modification of the method of Keppler and Decker (Keppler and Decker, 1974). Specific determination of UDP-galactose as achieved by conversion to UDP-glucose by uridylyl transferase with NAD and UDP-glucose dehydrogenase. The orotic acid in milk was determined according to the method of Moellering (Moellering, 1974). In this method the orotate is converted to orotidine 5′-monophosphate in the presence of 1-phosphorybosil-5′-pyrophosphate by the action of the orotidine 5′-phosphate pyrophosphorylase and then to uridine 5′ monophosphate by orotine 5′-phosphate decarboxylase. The decrease in extinction at 295 nm was measured. Richardson *et al.* (1980) determined the levels and location of adenosine 5′-triphosphate in bovine milk. ATP was measured enzymatically in protein-free milk extracts obtained by treatment of the milk with TCA and luciferase using a scintillation counter to quantify the light emitted from the luciferase–ATP reaction.

E. High-Performance Liquid Chromatography (HPLC) Methods

The development of stable micron-particle chemically bonded column packings has allowed scientists to use this technique to separate and quantitate nucleotides, nucleosides, and bases in biological fluids and tissue extracts (Harwick and Brown, 1975; Brown, 1983a,b; Wynants and Van Belle, 1985; Kojima *et al.*, 1985). Nucleotides have traditionally been separated by anion-exchange HPLC (Harwick and Brown, 1975) although recently reverse-phase systems with ion-pairing agents have been employed (Brown, 1983a,b; Wynants and Van Belle, 1985; Kojima *et al.*, 1985).

Janas and Picciano determined the content of nucleoside 5′-monophosphates and nucleoside 5′-diphosphates in human milk by anion exchange HPLC (Janas and Picciano, 1989). Protein-free milk extracts were obtained by milk with PCA and successively filtering, neutralizing, centrifuging, and freeze-drying the samples. The material was injected into a HPLC system and nucleotide analyses were performed according to the method of Harwick and Brown (1975).

Schlimme *et al.* (1986a) have developed a nonline multidimensional high-performance affinity chromatography (HPAC) reverse-phase liquid chromatography (RPLC) method for the group selective separation and quantification of ribonucleosides in physiological fluids. The ribonucleoside profile of human and cow's milk has been established using this technique (Schlimme *et al.*, 1986b). Milk was adjusted to pH 4.6 with formic acid and deproteinized by centrifugal ultrafiltration in an anion, micropartition system. An aliquot of the ultrafiltrate was applied to a HPAC column [30-mm length × 4-mm inside diameter (i.d.)], filled by an upward slurrypacking technique with a laboratory-prepared boronic acid-substituted silica, and washed with ammonium phosphate buffer. Ribonucleosides are selectively retarded on the HPAC column and the sample matrix discharged. After this clean-up step the HPAC column is connected in series to a second column (250-mm length × 5-mm i.d.) packed with Lichrosorb RP-18, 7u, and ribonucleosides are eluted under acidic conditions using a multistep gradient of ammonium formate and methanol. Eighteen nucleosides are efficiently separated with this method. Recently, reverse-phase systems with or without ion-pairing agents have been increasingly employed for the separation and quantitation of nucleotides, nucleosides, purine and pyrimidine bases, and related compounds in one single run (Harwick and Brown, 1975; Brown, 1983a; Wynants and Van Belle, 1985; Niculesen-Duvaz and Voiculetz, 1989; Stocchi, 1988).

We are currently determining nucleotides, nucleosides, and bases in human milk, cow's milk, and adapted milk formulas using gradient reversed-phase ion-pair conditions. Basically, protein-free milk extracts obtained with perchloric acid are freeze-dried and reconstituted to 1/5–1/10 of the original milk volume. An aliquot of the milk extract, usually 10–20 μl, is injected in a HPLC system and eluted isocratically using 0.1 *M* potassium phosphate containing tetrabutyl ammonium phosphate buffer. Quantitation is achieved by comparison with authentic standards, using caffeine as internal standard. Determination is carried out using a uv detector at 254 nm wavelength.

F. Determination of Nucleic Acids

Sanguansermsri *et al.* (1974) determined DNA and RNA in human and cow's milk. Essentially, skimmed milk was dialyzed against running tap water for 72 hr and the nondialyzable fraction was extracted with PCA. DNA and RNA were determined in the clear acid extract obtained after centrifugation and filtration by an improved diphenylamine method and by the orcinol method, respectively. Hutjens *et al.* (1979) have determined the DNA content of somatic cells in cow's milk using membrane filter separation and DNA determination with diphenylamine. Likewise, Bremel *et al.* (1977, 1980) have developed a method to estimate the somatic cells in milk samples through the estimation of DNA with indol by a membrane-filter DNA procedure. Molina *et al.* (1980) have compared the accuracy and precision of both methods compared to the standard microscopic method for the estimation of somatic cells in nonrefrigerated and refrigerated cow's milk.

Milk is mixed with a Triton X-100–EDTA mixture to create a dispersion of fat globules and separate casein micelles into smaller aggregates. The mixture is then filtered through 3 μ pore size and the filters containing the somatic cells are treated either with an acid indol solution or with TCA solution followed by the addition of diphenylamine. The intensity of the developed color relates directly to the DNA content in the milk sample as measured by spectrophotometry at 490 nm for the DNA–indol reaction and at 600 nm for the DNA diphenylamine.

We have recently determined DNA and RNA in human and cow's milk by the following method. A volume of skimmed milk is centrifuged at 12,000*g* at 4°C for 20 min. The pellet is washed with cold saline and resuspended in sodium phosphate buffer, 2 *M* sodium chloride, and 2 × 10.3 *M* EDTA. The suspension is homogenized in a Potter–Evelheim for 5 min. Aliquots of the supernatant and cell pellets are used for DNA and RNA determination. DNA is determined by a spectrofluorimetric assay using bis-benzimide as fluorogenic agent (Labarca and Paigen, 1980) and RNA by the orcinol reaction (Renee *et al.,* 1985).

G. Critical Comparison of Analytical Methods

We will analyze the advantages and disadvantages of the various methods to separate and quantitate nucleotides and related compounds.

The obtainment of protein-free milk extracts previous to separation and quantitation of nucleotides can be done with PCA or TCA. Treatments with PCA have the advantage that this compound does not absorb ultraviolet light by itself, although dialysis or extraction with PCA instead of TCA results in milk filtrates containing larger amounts of other ultraviolet-absorbing material (Hurlbert *et al.,* 1954). Ion-exchange chromatography of TCA and PCA extracts and dialysates gives comparable results. Ion-exchange chromatography permits the best separation of most nucleotides; a few, i.e., UDP-sugars, are poorly separated, even after repeated chromatography using specific ammonium formate gradients (Deutsch and Mattsson, 1960; Denamur *et al.,* 1959; Johnke and Goto, 1962; Kobata *et al.,* 1962; Gil and Sanchez-Medina, 1981a). Recovery for pure solutions of nucleoside phosphates has been shown to be 94% or better with coefficients of variation about 3% with this technique. However, for an accurate quantitation of nucleotides in milk it is absolutely necessary to start from a large volume, usually from 100 to 1000 ml. The technique is time consuming and exposure to light and air during the analysis can lead to decomposition of some compounds. Paper chromatography and thin-layer chromatography enable faster analysis; however, high resolution of closely related compounds is not achieved and quantitation procedures do not permit reliable determination of small variations in concentrations (Brown,

1983a). In addition, this methodology is not practical for the analysis of a large number of samples.

The enzymatic methods developed for quantitation of nucleotides in milk provide approximately the same degree of accuracy yet are simpler and more reliable than the conventional methods using ion exchange column chromatography. Most information on the nucleotide profile of milk is expressed as CMP, AMP, GMP, and UMP, but only nucleoside 5′-monophosphates, UDP-glucose, UDP-galactose, and orotate can be quantitated separately (Gil and Sanchez-Medina, 1981a). The use of enzymatic assays has made it possible to determine the nucleotide contents of a larger number of milk samples at the same time, thus permitting the measurement of variations in the nucleotide content in milk of diverse mammal species during lactation (Gil and Sanchez-Medina, 1981b, 1982a).

Nucleotides have been separated and quantitated in human milk by anion exchange HPLC (Janas and Picciano, 1989). The method allows the identification and quantification of nine nucleotides but fails to separate and quantitate nucleotide sugars which are known to be present in human milk in relatively high amounts, mainly UDP-glucose, UDP-galactose, UDP-*N*-acetylhexosamines, UDP-glucuronate, GDP-sugars, and CDP-choline (Kobata *et al.*, 1962; Gil and Sanchez-Medina, 1982a). Moreover, the detection of nucleotides normally present in small amounts in marketed whole cow's milk and cow's milk-based formula is difficult using this method (Gil and Sanchez-Medina, 1982b). High concentrations of buffer are needed to elute nucleotides by HPLC in the ion-exchange creating problems for the sensitivity of the system because of the impurity of buffer salts, and increasing the risk of crystallization and obstruction of the system (Wynants and Van Belle, 1985).

Reverse-phase HPLC is currently the most commonly used technique for the separation and quantitation of bases, nucleosides, and nucleotides in biological tissues and fluids; the combination of chemically bonded packings with proper mobile phases has resulted in extremely efficient systems of separation of purine and pyrimidine derivatives (Brown, 1983a,b; Wynants and Van Belle, 1985; Kojima *et al.*, 1985; Schlimme *et al.*, 1986; Niculesen-Duvaz and Voiculetz, 1989; Stocchi, 1988). One of the advantages of RPLC is its high reproducability, especially when reverse-phase C18 columns are used. Nucleotides, nucleosides, and bases are successfully separated by RPLC. The operation is easy and reequilibration after gradient elution is rapid. Applying ion pairing allows for a better separation of nucleotides in complex mixtures of purine and pyrimidine derivatives. The disturbed equilibrium in the column needs long reequilibration and requires isocratical operation to obtain reproducible retention times.

Milk sample preparation for RPLC can be done under similar conditions to those necessary for other techniques. The removal of proteinaceous material from the sample is necessary to stop enzymatic reactions, mainly phosphatase activity, and to prevent irreversible adsorption of the proteins to the packing, thus plugging the column. Residual proteins in milk acid extracts may also be removed using ultrafiltration membranes with a 10,000 molecular weight cutoff. The use of sample precolumn purification is a simpler method of sample preparation. It can be used in combination with the analytical column. Borate affinity columns have proved to be useful for the selective analysis of ribonucleosides in milk and other physiological fluids eliminating the interference of bases and nucleotides (Schlimme *et al.*, 1986; Janas and Picciano, 1989).

The final identification of nucleotides, nucleosides, and bases in HPLC milk analysis is a challenging step, since retention times and cochromatography with the reference compounds only provide tentative identification of peaks. Positive identity of peaks in a multicomponent system, such as milk, requires the determination of absorbance ratios at several wavelengths. The assessment of refraction index and characterization by chemical, spectroscopic, chromatographic, and enzymatic methods are often needed (Brown, 1983b).

DNA and RNA have been determined in a small number of samples of human milk from European and Thai mothers, the latter from a low socioeconomic group, and in a few samples of cow's milk at different stages of lactation (Sanguanermsri *et al.*, 1974). The methodology used cannot discriminate whether DNA and RNA came from cells or were in solution in the milk samples (San Lin and Sehheide, 1969). The diphenylamine method to quantitate DNA used in this study gives poor reproducibility. Filter membrane techniques may enable the determination of cell-associated DNA (Bremel *et al.*, 1977; Bremel, 1980; Molina *et al.*, 1980; Labarca and Paigen, 1980), but analytical methodology, based in colorimetric determination of acid hydrolysates of DNA with either indol or diphenylamine, provide highly variable results. We prefer to measure DNA by the more specific spectrofluorometric method which uses bisbenzimide as fluorogenic reagent on homogenates of milk cells (Labarca and Paigen, 1980).

III. Composition of Nucleotides and Related Compounds in Milk

A. Human Milk and Cow's Milk

Nonprotein nitrogen accounts for 18–30% of the total nitrogen in human milk and only for about 5% in cow's milk (Carlson, 1985; Macy, 1949; Atkinson *et al.*, 1980; Forsum and Lonnerdal, 1980; Hambraeus, 1984; Sanchez-Pozo *et al.*, 1987). Nucleosides, acid-soluble nucleotides, and nucleic acids are present in substantial quantities in human and cow's milk (Deutsh and Mattsson, 1960; Rashid, 1973; Denamur *et al.*, 1959; Johnke and Goto, 1962; Kobata *et al.*, 1962; Gil and Sanchez-Medina, 1981a; Janas and Picciano, 1989; Schlimme *et al.*, 1986; Sanguanermsri *et al.*, 1974; Hutjens *et al.*, 1979; Bremel *et al.*, 1977; Bremel, 1980; Molina *et al.*, 1980; Gil and Sanchez-Medina, 1981b, 1982a).

1. *Nucleosides*

Schlimme *et al.* (1986) have analyzed the content of nucleosides in human and cow's milk by an on-line two-column reversed-phase HPLC system. Raw cow's milk was found to contain at least 10 nucleosides: cytidine, uridine, *N*-1-methyladenosine, inosine, *N*-3-methyluridine, adenosine, *N*-1-methylinosine, *N*-2-methylguanosine, *N*-2-dimethylguanosine, and *N*-6-carbamoyl-threonyladenosine (*t*-Ado). Cytidine uridine and adenosine are quantitatively the most important; lower concentrations were found for inosine, adenosine, and *t*-Ado. Other nucleosides were found in trace amounts. Recombined skim milk powder and ultra high temperature processed cow's milk exhibit a similar nucleoside profile although adenosine was reduced compared to raw cow's milk. The content of cytidine and uridine, but not of adenosine, was inversely related to the number of somatic cells. The content of nucleoside from milk of midlactation was significantly higher than that of late gestation.

Human milk showed a similar qualitative profile to that of cow's milk; however, 5-amino-imidazole-4-carboxamide-*N*-ribofuranoside (AICAR) was also found and *N*-2-methyl- and *N*-2,2-dimethylguanosine were present only in trace amounts. Cytidine, AICAR, uridine, and adenosine were the major nucleosides in human milk found in this study. The origin of nucleosides in milk is not clear, but it could be assumed that it is at least in part derived from nucleic acids degradation since methylated bases are found in all genetic material. Table II summarizes the average concentrations of major nucleosides found in cow's and human milk.

2. *Acid-Soluble Nucleotides*

In 1958, Denamur *et al.* (1958) reported for the first time the presence of acid-soluble nucleotide in ewe's milk. One year later, these authors described the nucleotide pattern of cow's, goat's, sheep's and sow's milks (Denamur *et al.*, 1959). Simultaneously, Deutsch and Mattson (1959) reported their findings on the purine and pyrimidine derivatives in cow's milk. Later on, these authors and Johnke and Goto described the variations of the acid-soluble nucleotide profiles from cow's, ewe's, and goat's milk at different stages of lactation (Deutsch and Mattsson, 1960); Johnke and Goto, 1962).

During the sixties Kobata *et al.* (1962, 1963, 1966) determined the quantitative and qualitative differences of nucleotides in human and cow's milk. In the early 1980s, Gil and Sanchez-Medina reported the qualitative and quantitative changes occurring in human, cow's, goat's, and sheep's

TABLE II
Average Levels of Major Nucleosides Found in Cow's and Human Milk

Milk type	Cytidine	Uridine	Adenosine	*N*-1-methyladenosine
Cow (midlactation)	0.48 ± 0.06	2.18 ± 0.37	0.09 ± 0.02	0.35 ± 0.10
Cow (late gestation)	0.22 ± 0.02	0.32 ± 0.07	0.07 ± 0.02	0.31 ± 0.05
UHT processed	1.9	1.74	0.06	0.3
Recombined skimmed milk powder	1.11	2.29	0.01	0.5
Human				
$n = 1$	0.42	0.68	0.05	0.2
$n = 10$	1.42	3.73	0.01	0.1

Note. Modified from Schlimme *et al*. (1986b). Results are expressed in μmol/dl as mean values ± SEM. Authors did not specify days of lactation and did not show dispersion values for ultra high temperature (UHT) processed cow's milk, recombined skim milk powder, or human milk.

milks during lactation determined not only by ion-exchange chromatography procedures but also using new enzymatic methods (Markham and Smith, 1949; Gil and Sanchez-Medina, 1981b, 1982a). Simultaneously, Janas and Picciano (1982) reported their findings on the nucleotide profile of human milk using ion-exchange HPLC. Skala *et al*. (1981) determined the content of cyclic nucleotides in breast milk.

Denamur *et al*. (1959) found in cow's milk at 30 days lactation high amounts of orotic acid and an unidentified orotate derivative, relative small levels of CMP, AMP, GMP, ADP, and very small amounts of UDP-*N*-acetylhexosamines and other UDP-sugar derivatives (11). Deutsch and Mattson (1960) established that cow's colostrum and cow's milk at different stages of lactation exhibited different nucleotide patterns. Colostrum of first day had three- or fourfold higher levels of nucleotides than dairy milk. Cow's colostrum has very high levels of pyrimidine derivatives, mainly UDP-glucose, UDP-galactose, UMP, UDP, and UDP-*n*-acetylhexosamines, and relative low levels of AMP, CMP, and guanosine derivatives. Uridine and guanosine derivatives sharply decrease from cow colostrum to milk of late gestation. Dairy milk has small amounts of CMP and high levels of orotic acid. Similar results were obtained by Johnke and Goto (1962), except that AMP and a derivative were found in cow's milk at higher levels throughout lactation.

Kobata *et al*. (1962) reported similar results for cow's colostrum of 2 days and cow's milk of 120 days of lactation. They observed the presence of 3′–5′-cyclic AMP both in colostrum and milk but they did not find significant amounts of AMP in the latter. Likewise, they showed that UDP-hexoses were the major components of nucleotides in cow's colostrum.

Gil and Sanchez-Medina (1981a) agreed with most published results and they reported the presence of CDP-choline in colostrum at 2, 27, and 78 hr, as well as in milk of 5, 15, and 21 days, and 2 and 6 months. Early colostrum (2 hr) presented similar levels of nucleotides to those of late lactation milk except that the orotate content was lower than that in mature human milk. Uridine and guanosine derivatives were present at very high levels in colostrum and decreased promptly; they were absent by 2 months of lactation. However, adenosine derivatives and cytidine derivatives decreased their levels gradually over time. On the contrary, orotate levels increased for the same period reaching average values of 36 μmol/dl at 2 months of lactation. Human milk has the same nucleotide pattern as that of human colostrum and it is relatively similar to that of cow's colostrum (Kobata *et al.*, 1962). It contains at least 12 acid-soluble nucleotides, mainly CMP, AMP, UMP, UDP-glucose, UDP-galactose, UDP-*N*-acetylhexosamines, and UDP- and GDP-mannose (Gil and Sanchez-Medina, 1981a). Kobata has also isolated and identified two UDP derivatives, namely UDP-*N*-acetylglucosaminegalactose and UDP-*N*-acetylglucosamine-D-galactose-L-fucose, present in small amounts both in human colostrum and milk (Kobata, 1963, 1966).

Janas and Picciano (1989) have separated and quantitated five nucleoside 5′-monophosphates (CMP, UMP, AMP, IMP, and GMP) and four nucleoside 5′-diphosphates. They found that mean concentrations of UMP, GMP, UDP, CDP, ADP, and GDP did not significantly change with the stage of lactation. However, CMP and AMP decreased and IMP increased their levels during the first 3 months of lactation. None of these studies has been able to detect orotate in human milk samples.

Richardson *et al.* (1980) demonstrated the presence of small amounts of ATP bound to calcium phosphate–citrate–caseinate micelles in bovine milk; 0.13 to 0.31 μmoles ATP/liter (mean, 0.23) was found in nine milk samples studied. Tables III–VII summarize reference values for acid-soluble nucleotide determined in human and cow's milk according to the type of methodology and author.

3. Nucleic Acids

Human milk contains substantial amounts of DNA and RNA (Sanguansermsri *et al.,* 1974). DNA content in milk from European mothers ranged from 8 to 22 mg/liter and from 44 to 117 mg/liter for Thai mothers. RNA content was 5- to 20-fold higher than DNA ranging between 111 to 400 and from 227 to 587 mg/liter in the two groups. DNA and RNA contents of European human and cow's milk paralleled each other. However, RNA content in human milk was much higher than that in cow's milk,

TABLE III

Mean Values for Purine Derivatives of Cow's Milk at Different Stages of Lactation as Determined by Different Authors Using Anion-Exchange Chromatography

	Stage of lactation (days)									
	Deutsch and Mattsson (1960)			Kobata *et al.* (1962)		Gil and Sanchez-Medina (1981a)				
	1	3	7	2	120	1	3	15	60	180
AMP	5.9	8.2	1.5	0.57	—	5.38	8.03	2.91	2.35	1.58
cAMP	—	—	—	1.45	0.40	—	—	—	—	—
ADP	—	1.8	1.6	—	—	—	—	—	—	—
Unidentified A	0.4	0.5	0.7	—	—	0.96	2.15	2.91	2.35	1.58
Total A derivatives	6.3	10.5	3.8	2.32	0.40	6.34	10.18	—	—	—
GMP	—	—	—	0.38	—	—	—	—	—	—
GDP	7.4	7.8	—	3.30	—	—	—	—	—	—
GDP-fucose	—	—	—	3.00	—	6.74	4.13	—	—	—
GDP-mannose	4.0	2.4	—	1.87	—	—	—	—	—	—
Unidentified G	—	1.2	0.2	—	—	—	—	—	—	—
Total G derivatives	11.4	11.4	0.2	8.55	—	6.74	4.13	—	—	—
Total Purines	17.7	21.9	4.0	10.87	0.40	13.08	14.31	2.91	2.35	1.58

Note. Results are expressed in μmol/dl. —, not detected. Data from Denamur *et al.* (1959) and Johnke and Goto (1962) are not represented since they expressed their results as percentages of total extinction in protein milk filtrates.

TABLE IV
Mean Values for Pyrimidine Derivatives of Cow's Milk at Different Stages of Lactation as Determined by Different Authors Using Anion-Exchange Chromatography

	Stage of lactation (days)									
	Deutsch and Mattsson (1960)			Kobata *et al.* (1962)		Gil and Sanchez-Medina (1981a)				
	1	3	7	2	120	1	3	15	60	180
CMP	7.0	14.2	5.6	5.92	0.92	3.68	8.68	4.90	3.22	1.23
CDP-choline	0.8	0.5	—	Trace	Trace	1.34	1.01	0.50	1.20	1.59
Total C derivatives	7.8	14.7	5.6	5.92	0.92	5.02	9.69	5.40	4.42	2.92
Uracil	2.0	—	5.1	—	—	—	—	—	—	—
Uridine	6.2	11.8	—	—	—	—	—	—	—	—
UMP	68.0	64.0	1.6	3.10	—	39.49	9.33	2.66	—	—
UDP	45.8	72.3	1.3	—	—	—	—	—	—	—
UDP-hexoses	113.0	16.1	—	63.50	—	90.73	48.91	1.90	—	—
UDP-*N*-Ac-Hexosamines	10.1	7.9	—	—	—					
UDP-glucuronate	6.7	5.8	—	4.72	—	13.47	5.53	—	—	—
Unidentified U	2.0	0.6	—	—	—	—	—	—	—	—
Total U derivatives	253.8	178.5	8.0	71.32	—	143.69	63.77	4.56	—	—
Orotic acid	—	6.9	27.2	10.84	39.60	7.37	16.67	26.38	26.84	41.94
Total pyrimidines	261.6	230.2	57.6	88.08	40.52	156.08	90.13	26.34	31.26	44.86

Note. Results are expressed in μmol/dl. —, Not detected. Data from Denamur *et al.* (1959) and Johnke and Goto (1962) are not presented since they expressed their results as percentages of total extinction.

TABLE V
Mean Values for Purine Derivatives of Human Milk at Different Stages of Lactation as Determined by Different Authors Using Anion-Exchange Chromatography

	Stage of lactation (days)							
	Kobata *et al.* (1962)		Gil and Sanchez-Medina (1982b)					
	2–4	30–100	2	3	6	15	30	180
AMP	0.39	0.25	3.34	2.41	2.24	2.60	2.02	1.51
cAMP	0.14	0.07	—	—	—	—	—	—
Unidentified A	—	—	0.69	—	0.38	0.41	0.64	0.53
Total A derivatives	0.53	0.32	4.03	2.41	2.62	3.01	2.66	2.04
GMP	0.19	0.16	0.33	0.36	0.50	—	0.32	—
GDP	Trace	Trace	—	—	—	—	—	—
GDP-mannose	0.11	0.06	0.53	0.97	0.54	0.46	0.44	0.44
Total G derivatives	0.30	0.22	0.86	1.33	1.04	0.47	0.76	0.44
Total purines	0.83	0.54	4.89	3.74	3.66	3.48	3.42	2.48

Notes: Results are expressed in μmol/dl.—, Not detected.

TABLE VI
Mean Values for Pyrimidine Derivatives of Human Milk at Different Stages of Lactation as Determined by Different Authors Using Anion-Exchange Chromatography

	Stage of lactation (days)							
	Kobata *et al.* (1962)		Gil and Sanchez-Medina (1982b)					
	2–4	30–100	2	3	6	15	30	180
CMP	3.81	3.34	5.51	3.45	3.10	2.64	1.87	1.83
CDP-choline	Trace	Trace	—	—	0.35	—	0.40	0.73
Total C derivatives	3.81	3.34	5.51	3.45	3.45	2.64	2.27	1.56
UMP	0.45	0.36	1.77	1.32	1.49	0.70	1.29	0.93
UDP	0.32	0.25	1.41	0.68	0.53	0.40	0.76	0.65
UDP-glucuronate	0.32	0.25	1.41	0.68	0.53	0.40	0.76	0.65
UDP-hexoses	0.18	0.15	0.45	3.57	1.31	3.14	0.82	1.01
UDP-*N*-Ac-hexosamines	0.82	0.68	0.45	3.57	2.32	3.14	1.96	2.20
Total U derivatives	1.77	1.44	3.63	5.57	5.65	4.24	4.83	4.79
Total pyrimidines	5.58	4.78	9.14	9.02	9.10	6.88	7.13	7.35

Note. Results are expressedin μmol/dl.—, Not detected.

TABLE VII
Acid-Soluble Nucleotides of Human Milk as Determined by Different Authors Using Anion-Exchange Chromatography HPLC and Enzymatic Methods

	Method			
	HPLC[a]		Enzymatic[b]	
	Mean	Range	Mean	Range
CMP	1.58 ± 0.06	0.5–14.05	2.92 ± 0.36	1.94–4.71
CDP	1.31 ± 0.12	n.d.–4.18	n.m.	n.m.
Total C derivatives	n.m.	n.m.	3.06 ± 0.83	2.16–4.87
UMP	0.61 ± 0.07	0.24–4.24	1.21 ± 0.08	1.05–1.57
UDP	0.49 ± 0.04	Trace–1.64	n.m.	n.m.
Total U derivatives	n.m.	n.m.	5.40 ± 0.36	4.93–7.47
AMP	0.56 ± 0.04	0.17–2.40	2.71 ± 0.20	1.97–3.29
ADP	0.18 ± 0.05	n.d.–1.28	n.m.	n.m.
Total A derivatives	n.m.	n.m.	2.88 ± 0.21	2.14–3.52
IMP	0.72 ± 0.05	0.17–2.03	n.m.	n.m.
GMP	0.42 ± 0.03	0.28–1.11	0.24 ± 0.03	10.16–0.32
GDP	0.24 ± 0.02	n.d.–1.35	n.m.	n.m.
Total G derivatives	n.m.	n.m.	0.81 ± 0.10	0.61–1.31

Note. Results are expressed in μmol/dl. n.m., Not measured. UDP and GDP sugars were not quantitated in Janas and Picciano (1989). Nucleoside 5′-monophosphates, nucleoside 5′-diphosphates, and nucleoside 5′-diphosphate sugars were quantitated as a whole and expressed in nucleoside 5′-monophosphate equivalents in Gil and Sanchez-Medina (1982b). For mean values data from 2, 4, 8, and 12 weeks of lactation for Janas and Picciano (1989) and from 3, 5, 10, and 20 days, and 1–3 months for Gil and Sanchez-Medina (1982b) were considered.
[a]Janas and Picciano (1989).
[b]Gil and Sanchez-Medina (1982b)

especially after 6 weeks of lactation. DNA content from somatic cells in fresh cow's milk has been reported to range from 5 to 20 mg/liter; a highly significant correlation was found between DNA content and number of cells counted by direct microscopy (Hutjens *et al.*, 1979; Bremel *et al.*, 1977; Bremel, 1980; Molina *et al.*, 1980). Our own data on dairy milk are in agreement with those previous data although our range (5–120 mg/dl) was wider. Our results also indicate that the number of cells found was related to nucleic acid content (samples with the highest levels were obtained from animals with higher cell count because of chronic mastitis). Table VIII shows a comprehensive estimate of nucleosides, nucleotides, and nucleic acid content of human and cow's milk.

TABLE VIII
Best Estimates for Nucleosides, Nucleotides, and Nucleic Acids Content of Human and Cow's Milk

	Cow		Human	
	μmol/dl	μg N/dl	μmol/dl	μg N/dl
Pyridine nucleosides (μmol/dl)	0.27–3.95	10–140	1.10–5.15	40–180
Purine nucleosides (μmol/dl)	0.08–0.80	6–60	0.11–0.25	8–18
Total nucleosides (μmol/dl)	0.35–4.75	16–200	1.21–5.40	48–198
Pyridine nucleotides (μmol/dl)	0.92–4.42	30–155	4.78–9.14	170–320
Purine nucleotides (μmol/dl)	0.40–2.35	28–165	0.54–4.89	38–340
RNA (mg/dl)[a]	8–19	480–1130	11–60	650–3570
DNA (mg/dl)[a]	11–39	650–2300	0.8–12	50–710
DNA (mg/dl)[b]				
Cell pellet	2.4–7.3	403–1226	—	—
Cell-free milk	5.7–6.9	958–1159	—	—
Total nucleic acids (mg/dl)	19–58	1130–3430	11.8–72	700–3280

Note. For conversion of pyridine derivatives into N equivalents an average factor of 35 μg N/μmol was used. For purine derivatives the factor was 70 μg N/μmol. To convert nucleic acids into N equivalents we considered 16.8 mg N per 100 mg of nucleic acid. Data for nucleotides are from Gil and Sanchez-Medina (1981a, 1982b).
[a]Data from Sanguansermsri *et al.* (1974).
[b]Unpublished results obtained by Angel Gil, Ph.D.

B. *Interspecies Comparison of Milk Nucleotide Composition*

Sheep's goat's, and sow's colostrum show qualitatively similar nucleotide patterns and are comparable to those of human and cow's colostrum (Deutsch and Mattsson, 1960; Denamur *et al.,* 1959; Johnke and Goto, 1962; Gil and Sanchez-Medina, 1981b). However sheep's colostrum has 3- to 10-fold higher nucleotide content than cow's colostrum and twice that of goat's colostrum at the same stage of lactation (Gil and Sanchez-Medina, 1981b). UDP-hexoses, UDP-hexosamines, and UDP, as well as GDP-fucose and GDP-mannose, are the main nucleotides in colostrum of these species. UDP derivatives decrease markedly from colostrum to mature milk; however, GDP derivatives and adenosine derivatives decrease gradually with advancing lactation. Sheep's, goat's and sow's milks have high levels of

UDP and GDP derivatives in contrast to cow's milk. Table IX shows a summary of nucleotide composition for sheep's, goat's, and sow's colostrum and milk.

C. *Nucleotides in Milk Formulas*

Currently available adapted milk formulas manufactured worldwide contain only trace amounts of purine and pyrimidine derivatives since cow's milk is a poor source of them. Furthermore, dairy milk, which is the base for infant milk formulas, is centrifuged before thermal treatment and cells that contain high amounts of DNA and RNA are discarded.

Based on Kobata's studies on nucleotide milk composition (Kobata *et al.*, 1962), Ziro *et al.* recommended adding nucleotides to infant milk formulas in a Japanese patent (Forsum and Lonnerdal, 1980). They suggested the addition to 1 liter of unprocessed cow milk about 10–20 mg of CMP, 0.2–0.4 mg of GMP, 1.2–1.4 mg of UMP, 0.4–0.6 mg GDP, and 1–3 mg of UDP glucuronate. Other nucleotide fractions could also be added by taking the difference between cow's milk and human milk in their nucleotides to common baby foods such as juice, fruits, vegetables, liver, chicken, meats, and custard. A Spanish company (Uniasa) has been adding nucleotides to their infant formula since 1983 in amounts which provide infants with 2.2–2.84 mg per 100 kcal. More recently, a United States company (Wyeth–Ayerst Labs) has added 3.8–4.2 mg/100 kcal nucleotides to their infant milk formulas. Table X summarizes the free nucleotide content of supplemented and regular formula compared to the range found in human milk. Infants taking routine cow's milk formulas receive a minimal supply of exogenous dietary nucleotides until solids are introduced in their diet. Furthermore, this supply is not fully characterized since there is insufficient data about purine and pyrimidine derivatives in infant formulas.

IV. Significance of Dietary Nucleotides in Infant Nutrition

Recent studies suggest that dietary nucleotides (purines and pyrimidine bases) may be semi-essential for newborn animals. Rapidly growing tissues, such as the intestinal epithelium and lymphoid cells, have an increased demand for purine and pyrimidine bases. Nucleic acids, nucleotides, and their related metabolic products are present in human milk in relatively large amounts. Their nutritional significance for the human infant has been the subject of recent studies and much interest (McGillivray, 1983; Quan *et al.*, 1990). The addition of nucleotides to infant formula, a practice initiated in Japan and currently being implemented in some European

TABLE IX
Acid-Soluble Nucleotides of Sheep's, Goat's and Sow's Colostrum and Milk

	Stage of lactation (days)					
	Sheep		Goat		Sow	
	2	60	1	60	1	22
Orotate	3.51 ± 0.27	4.09 ± 0.26	7.65 ± 0.37	11.87 ± 0.53	nd	nd
CMP	32.75 ± 2.79	8.74 ± 0.98	6.45 ± 0.68	5.40 ± 0.30	13.50 ± 1.21	6.18 ± 0.74
UMP	113.30 ± 7.01	25.09 ± 2.60	53.77 ± 0.76	14.48 ± 1.24	103.25 ± 12.37	12.31 ± 1.03
AMP	29.73 ± 0.80	9.35 ± 0.55	4.70 ± 0.41	6.73 ± 0.55	9.04 ± 0.73	5.54 ± 0.68
GMP	3.46 ± 0.43	nd	nd	nd	7.65 ± 0.65	3.79 ± 0.47
Total C derivatives	34.20 ± 2.93	7.96 ± 0.46	6.52 ± 0.57	5.53 ± 0.25	13.50 ± 1.21	6.18 ± 0.74
Total U derivatives	951.99 ± 55.81	160.89 ± 6.27	215.11 ± 6.97	73.09 ± 4.54	257.31 ± 10.42	120.09 ± 9.62
Total A derivatives	31.39 ± 2.08	10.57 ± 0.57	5.26 ± 0.39	6.76 ± 0.48	12.17 ± 1.34	5.84 ± 0.72
Total G derivatives	50.13 ± 2.14	28.85 ± 1.95	34.14 ± 1.19	22.74 ± 1.23	28.81 ± 2.68	17.43 ± 1.43

Note. Results are mean ± SEM expressed in μmol/dl. Nucleotides were estimated by enzymatic methods. C derivatives were mainly CMP and CDP-choline; U derivatives were mainly UDP-*N*-acetylhexosamines, UDP-hexoses, UDP-glucuronate, UDP, and UMP; A derivatives were mainly AMP and 3′–5′cAMP; and G derivatives were mainly GMP, GDP-mannose, and GDP-fucose. Data from Gil and Sanchez-Medina (1981a).

TABLE X
Nucleotide Content in Routine Formula and in Nucleotide Supplemented Milk Formulas

Formula	CMP	UMP	AMP	GMP	IMP
Puleva-1, Pre-Natur[a]	1.00	0.84	0.64	0.51	0.28
Nieda, Edacid-L[a]	1.00	0.68	0.66	0.30	0.20
SMA[b]	2.10	0.75	0.60	0.31	0.30
Enfamil[b]	0.40	0.03	0.07	0.04	—
Similac[b]	0.27	0.01	0.01	0.01	—
Breast milk, range[c]	0.99–2.25	2.28–3.46	1.06–1.75	0.32–0.68	0–4.57

Note. Results are expressed in mg/100 Kcal.
[a]Results kindly provided by UNIASA Research Department (Dr. Angel Gil).
[b]Results kindly provided by Wyeth–Ayerst Nutritional Division (Dr. Eric Lien).
[c]Based on nucleotidase and their derivatives as summarized in Tables 4–6. Puleva, Pre-Natur, Nieda, Edacid-L, and Edacid-V are registered trademarks of Uniasa, Granada, Spain. SMA is a registered trademark of Wyeth–Ayerst, Philadelphia. Enfamil is a registered trademark of Squibb–Bristol Myers–Mead Johnson, Evansville, In. Similac is a registered trademark of Ross Laboratories, Columbus, OH.

countries, based on studies that suggest potential benefits to immunity, iron absorption, intestinal flora, lipid metabolism, and gut growth and development.

The possibility of a role for exogenous nucleotides in the modulation of normal immune function has been suggested by experimental studies. The effects of dietary nucleotides have been examined in a newborn mouse heart allograft model. Prior to transplant, BALB/c mice were maintained on standard rodent chow, a nucleotide-free diet, or a nucleotide-free diet supplemented with 0.25% yeast RNA as a source of nucleotides. Allograft survival was significantly prolonged in the nucleotide-free group compared to both chow and RNA-supplemented groups (Van Buren *et al.*, 1983a). Using the same study diets, the investigators examined the acute lymphoproliferative response to alloantigen. Animals receiving nucleotide-free diets had significant suppression of the proliferative response (Van Buren *et al.*, 1985). Similarly delayed cutaneous hypersensitivity upon challenge with purified protein derivative or sheep red blood cells was diminished in the nucleotide-free diet group compared to the chow or RNA-supplemented groups (Van Buren *et al.*, 1982b).

Another experiment examined the effects of a nucleotide-free diet on the immune response of mouse syngeneic bone marrow chimeras. Onset of acute graft versus host disease was delayed in the nucleotide-free group (Kulkarni *et al.*, 1984). Additionally, the *in vitro* proliferative response of spleen cells to phytohemagglutinin was significantly reduced in this group. These experiments suggested that the T lymphocyte is the target of dietary nucleotide deprivation. To determine the mechanisms responsible for this phenomenon, the phenotypic characteristics of lymphocyte subpopulations were studied in mice maintained on a nucleotide-free diet and compared with chow-fed mice or mice receiving nucleotide-free diets repleted with RNA, adenine, or uracil. Restriction or nucleotides affected T-lymphocyte phenotypes and T cell function. The production of interleukin-2, a lymphokine vital for T-lymphocyte proliferation, was suppressed in irradiated splenic lymphocytes following concanavalin-A stimulation in the nucleotide-deficient group. These data suggest that helper/inducer T lymphocytes require exogenous nucleotides (Van Buren *et al.,* 1985). The same diets were used to determine the influence of dietary nucleotide deprivation on resistance to infection in mice. Nucleotide restriction increased mortality from *Staphylococcal sepsis* and adversely affected host resistance to *Candida* (Fenslow *et al.,* 1988). Addition of uracil to the nucleotide-free diet in both of the above experiments resulted in resistance similar to chow- or RNA-supplement-fed groups indicating that uracil may play a key role in resistance to infection. Phagocytic function was also assessed in the mice subjected to a *Staphylococcus aureus* challenge. Macrophages from mice on the nucleotide-free diet demonstrated diminished phagocytic activity as measured by uptake of radiolabeled bacteria (Kulkarni *et al.,* 1986).

The relationship of nucleotides to immunity has also been studied by Carver *et al.* (1990). The addition of nucleotides to a nucleotide-free formula-based diet fed to mice resulted in increased phagocytosis of macrophages along with increased natural killer (NK) cell activity of spleen cells. These authors also recently reported a controlled double-blind study demonstrating that 13 infants fed nucleotide-supplemented formula had similar NK activity to that of 9 breast-fed infants and significantly higher than 15 receiving nonsupplemented formula (Carver *et al.,* 1989). The animal studies and the preliminary human data are supportive, yet it is too early to conclude that all "healthy" infants would benefit from nucleotide supplementation of formula by increased resistance to infections.

Nucleotides may also affect the gastrointestinal microenvironment of infants since it is known to vary with diet. The gastrointestinal tract of a breast-fed infant has a predominance of bifidobacteria, which have been suggested to protect against gastroenteritis associated with enteropathogenic bacteria (Braun, 1981). *In vitro* experiments have revealed that the addition of nucleotides to bifidobacteria in minimal culture media enhanced their growth (Gil *et al.,* 1986). Infants fed nucleotide-supplemented formula had higher percentages of fecal bifidobacteria and lower percentages of gram-negative enterobacteria than formula-fed infants (Gil *et al.,* 1986). Thus, it is possible that dietary nucleotides may favor the development of a fecal flora similar to that of breast-fed infants.

In vitro and *in vivo* experiments show that *de novo* synthesis of purines is limited or inactive in gut epithelial cells, as measured by labeled glycine incorporation into mucosal nucleic acids (Savaiano and Clifford, 1981; Rudolph *et al.,* 1984). Enzymes responsible for the purine salvage pathway are high in the small intestine; the activity of the key enzyme for the *de novo* synthesis of purines, phosphoribosyl pyrophosphate amido transferase, is increased by a nucleotide-free diet. These data indicate that dietary nucleotides may play a role in determining the intestinal nucleotide pool (Rudolph *et al.,* 1984; Leleiko *et al.,* 1987). Since the intestine incorporates proportionately greater amounts of dietary nucleotides, it would be logical that the gut would be the most affected by dietary nucleotide supply. In the upper jejunum of weanling rats fed nucleotide-free diets there was less protein and DNA, and villi were shorter. Maltase activity was significantly lower through all portions of the intestine, but most significantly in the proximal portion (Uauy *et al.,* 1990). In a chronic diarrhea experimental model nucleotide supplementation led to increased maltase throughout the intestine (Nunez *et al.,* 1990a). In a scanning electron microscopy study on mice nucleosides supplementation induced increased villus height and greater surface area and a decrease in intraepithelial lymphocytes (Bueno *et al.,* 1994). Further studies must be done to explore the importance of these findings and their relevance to infant nutrition.

V. Summary

Human milk has a specific content of free nucleotides which differs from cow's milk. It has an abundance of CMP, AMP, and UMP, whereas orotate is low. It has been estimated that human milk or nucleotide-supplemented formula will provide 2–4% of daily nitrogenous base needs, whereas regular formula would provide < 1% (Table II). Total nucleotide content of human milk, if nucleic acid content is included, greatly exceeds the levels reached in nucleotide-supplemented formula currently available. Breast milk could provide 20 to 25% of daily needs assuming that nucleosides from nuclear-derived nucleic acids can be digested, absorbed, and salvaged. Further studies will be required to examine the fate of nuclear nucleic acids on human milk. This may lead to a better definition of how to supplement and at what level (Quan *et al.,* 1990).

To date, it appears that dietary nucleotides do have significant biological effects and that benefits for formula-fed human infants are possible. But even if clear benefits are demonstrated in animal experiments, further long-term human studies that demonstrate what levels of dietary nucleotides are effective and well tolerated are needed. Current levels of dietary-free nucleotide supplementation mimicking human milk content are safe; more information is required before higher supplementation levels are evaluated.

To fully justify the supplementation of formula with dietary nucleotides based on an enhanced immune response and other biologic effects, it should be conclusively demonstrated that an improved *in vitro* immunity translates into decreased morbidity. At present, there is insufficient knowledge on the functional effects of dietary nucleotides during the early months of life in the human infant. Additional research needs to be done to clarify pending questions as of the true significance of dietary nucleotides for healthy and sick infants.

References

Atkinson, S. A., Anderson, G. H., and Bryan, M. H. (1980). Human milk: Comparison of the nitrogen composition of milk from mothers of premature and full term infants. *Am. J. Clin. Nutr.* **33,** 811–815.

Bergkvist, R., and Deutsch, A. (1954). Ion-exchange chromatography of nucleoside polyphosphates. *Acta Chem. Scand.* **8,** 1877–1885.

Braun, O. H. (1981). Effect of consumption of human milk and other formulas on intestinal bacterial flora in infants. *In* "Textbook of Gastroenterology and Nutrition in Infancy" (E. Lebenthal, ed.), Vol. 1, pp. 247–253. Raven Press, New York.

Bremel, R. D. (1980). Membrane filter-deoxyribonucleic acid method of somatic cell counting: Collaborative study. *J. Assoc. Anal. Chem.* **63,** 211–218.

Bremel, R. D., Schultz, L. H., Gablet, R., and Peters, J. (1977). Estimating somatic cells in milk samples by the membrane-filter-NA procedure. *J. Food Protection* **40,** 32–38.

Brown, P. R. (1983a). Current high-performance liquid chromatographic methodology in analysis of nucleotides, nucleosides, and their bases. I. *Cancer Invest.* **1,** 439–454.

Brown, P. R. (1983b). Current high-performance liquid chromatographic methodology in analysis of nucleotides, nucleosides, and their bases. II. *Cancer Invest.* **1,** 527–536.

Bueno, J., Torres, M., Almendros, A., Carmona, R., Núnez, M. C. Rios, A., and Gil, A. (1994). Effect of dietary nucleotides on small intestinal repair after diarrhea: reistological and ultrastructural changes. *Gut* **35,** 926–933.

Carlson, S. E. (1985). Human milk non-protein nitrogen: Occurrence and possible functions. *In* "Yearbook of Pediatrics," pp. 43–70.

Carver, J. D., Pimentel, B., and Barness, L. A. (1989). Nucleotide effects in formula fed infants. Specialty poster presentation, APS–SPR Annual Meeting. *Pediatr. Res.* **25,** 286A.

Carver, T. D., Cox, W. I., and Barness, L. A. (1990). Dietary nucleotide effect upon murine NK killer cell activity and macrophage activation. *J. Parenter. Enterol. Nutr.* **14,** 18–22.

Denamur, R., Fauconneau, G., and Gruntz, G. (1958). Les variations de la composition nucleotidique du lait de brebis: Influence de l'activite secretoire de la glande mammaire. *C. R. Acad. Sci. (Paris)* **246,** 652–655.

Denamur, R., Fauconneau, G., and Guntz, G. (1959). Les nucleotides acido solubles des laits de brebis, vache, chevre et truie. *Rev. Esp. Fiiiol.* **15,** 301–310.

Deutsch, A., and Mattsson, S. (1959). Purine and pyrimidine derivatives in cow's milk. *XV Int. Dairy Congr.* **3,** 1700.

Deutsch, A., and Mattsson, S. (1960). Acid-soluble nucleotides in cow's milk and colostrum. *In* "Milk and Dairy Research," Report 63, pp. 1–18. Alnarp, Sweden.

Fanslow, W. C., Kulkarni, A. D., Van Buren, C. T., and Rudolph, F. B. (1988). Effect of nucleotide restriction and supplementation on resistance to experimental murine candidiasis. *J. Parenter. Enterol. Nutr.* **12,** 49–52.

Forsum, E., and Lönnerdal, B. (1980). Effects of protein intake on protein and nitrogen composition of breast milk. *Am. J. Clin. Nutr.* **33,** 1809–1813.

Gil, A., and Sanchez-Medina, F. (1981a). Acid-soluble nucleotides of cow's, goat's and sheep's milks at different stages of lactation. *J. Dairy Res.* **48,** 3544.

Gil, A., and Sanchez-Medina, F. (1981b). The determination of acid-soluble nucleotides in milk by improved enzymic methods: A comparison with the ion-exchange column chromatography procedure. *J. Sci. Food Agric.* **32,** 1123–1131.

Gil, A., and Sanchez-Medina, F. (1982a). Effects of thermal industrial processing on acid soluble nucleotides of milk. *J. Dairy Res.* **49,** 295–300.

Gil, A., and Sanchez-Medina, F. (1982b). Acid soluble nucleotides of human milk at different stages of lactation. *J. Dairy Res.* **4,** 301–307.

Gil, A., Corral, E., Martinez, A., and Molina, J. A. (1986). Effects of dietary nucleotides on the microbial pattern of feces of at term newborn infants. *J. Clin. Nutr. Gastroenterol.* **1,** 34–38.

Gillen, W. K., and Meyer, A. (1965). An improved diphenylamine method for the estimation of desoxyribonucleic acid. *Nature* **206,** 93.

Hambraeus, L. (1984). Human milk composition. *Nutr. Abst. Rev.* **54,** 219–233.

Hartwick, R. A., and Brown, P. R. (1975). The performance of microparticle chemically-bonded anion-exchange resins in the analysis of nucleotides. *J. Chromatogr.* **112,** 650–655.

Hurlbert, R., Schmitz, H., Brumm, A., and Potter, V. (1954). Nucleotide metabolism II. Chromatographic separation of acid-soluble nucleotides. *J. Biol. Chem.* **209,** 23–39.

Hutjens, M. F., Schultz, L. H., Ward, G. E., and Yamdagni, S. (1979). Estimation of somatic cells in milk using membrane filter separation and DNA determination with diphenylamine. *J. Milk Food Technol.* **33,** 227–231.

Janas, L. M., and Picciano, M. F. (1989). The nucleotide profile of human milk. *J. Pediatr. Res.* 659–662.

Johnke, T., and Goto, T. (1962). Acid-soluble nucleotides in cow's and goat's milk. *J. Dairy Sci.* **45,** 735–741.

Keppler, D. (1974). Determination of total 5′-nucleotides and nucleoside 5′-monophosphates. *In* "Methods of Enzymatic Analysis" (H. U. Bergmeyer, ed.), 2nd Ed., Vol. 4, pp. 2088–2096. Academic Press, New York.

Keppler, D., and Decker K. (1974). Uridine-5′-diphosphogalactose. *In* "Methods of Enzymatic Analysis" (H. U. Bergmeyer, ed.), 2nd Ed., Vol. 4, pp. 2221–2224. Academic Press, New York.

Kobata, A. (1963). The acid-soluble nucleotides of milk. II. Isolation and identification of two novel uridine nucleotide oligosaccharide conjugates from human milk and colostrum. *J. Biochem.* **53,** 167–175.

Kobata, A. (1966). The acid-soluble nucleotides of milk. IV. The chemical structure of UDP-X3. *J. Biochem.* **59,** 63–67.

Kobata, A., Ziro, S, and Kida, M. (1962). The acid-soluble nucleotides of milk. I. Quantitative and qualitative differences of nucleotides constituent in human and cow's milk. *J. Biochem.* **51,** 277–287.

Kojima, T., Nishima, T., Kitamura, M., Katatami, N., and Nishioka, K. (1985). Reversed-phase liquid chromatography determination of purine compounds in serum applied to studies of hypouricemia. *Clin. Chem.* **32,** 287–290.

Kulkarni, S. C., Bhateley, D. C., Zander, A. R., Van Buren, C. T., Rudolph, F. B., Kulkarni, A. D., and Dicke, K. A. (1984). T-cell impairment in mouse radiation chimeras by nucleotide-free diet (NFD). *Exp. Hematol.* **12,** 694–699.

Kulkarni, A. D., Fanslow, W. C., Drath, D. B., Rudolph, F. B., and Van Buren, C. T. (1986). Influence of dietary nucleotide restriction on bacterial sepsis and phagocytic cell function in mice. *Arch. Surg.* **121,** 169–172.

Labarca, C., and Paigen, K. (1980). A simple rapid and sensitive DNA assay procedure. *Anal. Biochem.* **102,** 344–352.

Lehninger, A. L. (1982). "Principles of Biochemistry." Worth, New York.

Leleiko, N. S., Bronstein, A. D., Baliga, S., and Munro, N. H. (1983). *De novo* purine nucleotide synthesis in the rat small and large intestines. Effect of dietary protein and purines. *J. Pediatr. Gastroenterol. Nutr.* **2,** 313–319.

Leleiko, N. L., Martin, B. A., Walsh, M., Kazlow, P., Rabinowitz, S., and Sterling, K. (1987). Tissue-specific gene expression results from a purine- and pyrimidine-free diet and 6-mercaptopurine in the rat small intestine and colon. *Gastroenterology* **93,** 1014–1120.

Macy, I. G. (1949). Composition of human colostrum and milk. *Am. J. Dis. Child* **78,** 589–603.

Markham, R., and Smith, J. D. (1949). Chromatographic studies of nucleic acids 1. A technique for the identification and estimation of purine and pyrimidine bases, nucleosides and related substances. *Biochem. J.* **45,** 294–298.

McGillivray, R. W. (1983). "Biochemistry: A Functional Approach." Saunders, Philadelphia.

Moellering, H. (1974). Orotate. *In* "Methods of Enzymatic Analysis" (H. U. Bergmeyer, ed.), 2nd Ed., Vol. 4, pp. 1959–1963. Academic Press, New York.

Molina, H., Heimlich, W., Horzella, M., Selive, S., Tronco, V., and Fuchslocher, E. (1984). Estimacion de celulas somaticas en leche por los metodos membrana-filtro DNA con difenilamina y membrana filtro DNA con indol. *Int. Dairy Fed. Bull. 177* **19,** 187–195.

Niculesen-Duvaz, I., and Voiculetz, N. (1989). Simultaneous analysis of bases, nucleosides and nucleotides by HPLC. *Chromatogram* **10,** 7–9.

Nunez, M. C., Ayudarte, M. V., Morales, D., Suarez, M., and Gil, A. (1990). Effect of dietary nucleotides on intestinal repair in rats with experimental chronic diarrhea. *J. Parenter. Enterol. Nutr.* **14.**

Nyhan, W. L. (1987). Inborn errors or purine metabolism. *In* "Diagnostic Recognition of Genetic Disease" (W. L. Nyhan and N. A. Sakati, eds.), pp. 1–40. Lea and Febiger, Philadelphia.

Partridge, S. M. (1955). Filter-paper partition chromatograph of sugars. *J. Biol. Chem.* **216,** 195–200.

Pontis, H. G. (1955). Uridine diphosphate acetyl hexosamine in liver. *J. Biol. Chem.* **216,** 195–200.

Quan, R., Barness, L. A., and Uauy, R. (1990). Do infants need nucleotide supplemented formula for optimal nutrition. *J. Pediatr. Gastroenterol. Nutr.,* in press.

Rashid, R. (1973). Application of active charcoal for the purification of purine and pyrimidine derivatives in milk for thin layer chromatography analysis. *Die Nahrung* **5,** 553–557.

Renee, A. R., Griffiths, J. M., and Wilkinson, M. L. (1985). "Basic Biochemical Methods." Wiley, New York.

Richardson, T., McGann, T., and Kearney, R. D. (1980). Levels and location of adenosine-5'triphosphate in bovine milk. *J. Dairy Res.* **47,** 91–96.

Roux, J. M. (1973). Nucleotide supply of the developing animal: Role of the so-called "Salvage Pathways." *Enzyme* **15,** 362–377.

Rudolph, F. B., Kulkarni, A. D., Schandel, V. B., and Van Buren, C. T. (1984). Involvement of dietary nucleotides in T-lymphocyte Function. *Exp. Med. Biol.* **165,** 175–178.

Sanchez-Pozo, A., Lopez, J., Izquierdo, A., Martinez-Valverde, A., and Gil, A. (1987). Protein composition of human milk in relation to mothers' weight and socioeconomic status. *Hum. Nutr.* **41C,** 115–125.

Sanguansermsri, J., Gyorgy, P., and Zilliken, F. (1974). Polyamines in human and cow's milk. *Am. J. Clin. Nutr.* **27,** 859–865.

San Lin, R. I., and Sehheide, O. A. (1969). Microestimation of RNA by the cupric ion catalyzed orcinol reaction. *Anal. Biochem.* **27,** 473–483.

Savaiano, D. A., and Clifford, A. J. (1981). Adenine. The precursor of nucleic acids in intestinal cells unable to synthesize purines *de novo. J. Nutr.* **111,** 1816–1822.

Schlimme, E., Boos, K. S., Hagemeier, E., Kemper, K., Meyer, U., Hobler, H., Schnelle, T., and Weise, M. (1986a). Direct clean-up and analysis of ribonucleosides in physiological fluids. *J. Chromatogr.* **378,** 349–360.

Schlimme, E., Boos, K. S., Frister, H., Pabst, K., Raezke, K. P., and Wilmers, B. (1986b). Gruppenselektive Hochleistungsflussigchromatographie von Ribonucleosiden in Milch. *Milchwissenschaft* **41,** 757–762.

Skala, J. P., Koldovsky, O., and Hahn, P. (1981). Cyclic nucleotides in breast milk. *Am. J. Clin. Nutr.* **34,** 343–346.

Stocchi, V. (1988). New RPLC columns improve resolution of important red blood cell nucleotides. *Biotext* **1,** 7–9.

Uauy, R. (1989). Dietary nucleotides and requirements in early life. *In* "Textbook of Gastroenterology and Nutrition in Infancy" (E. Lebenthal, ed.), 2nd Ed., pp. 265–280. Raven Press, New York.

Uauy, R., Stringel, G., Thomas, R., and Quan, R. (1990). Effect of dietary nucleosides on growth and maturation of the developing gut in the rat. *J. Pediatr. Gastroenterol. Nutr.* **10,** 497–503.

Van Buren, C. T., Kulkarni, A. D., Schandle, V. B., and Rudolph, F. B. (1983a). The influence of dietary nucleotides on cell mediated immunity. *Transplantation* **36,** 350–352.
Van Buren, C. T., Kulkarni, A. D., and Rudolph, F. B. (1983b). Nucleotide deprivation retards cutaneous hypersensitivity (DCH). *JPEN* **6,** 582.
Van Buren, C. T., Kulkarni, A. D., Fanslow, W. C., and Rudolph, F. B. (1985). Dietary nucleotides: A requirement for helper/inducer T lymphocytes. *Transplantation* **40,** 694–697.
Wynants, J., and Van Belle, H. (1985). Single run high-performance liquid chromatograph of nucleotides, nucleosides, and major purine bases and its application to different tissue extracts. *Anal. Biochem.* **144,** 258–266.

F. Protein and Amino Acid Composition of Bovine Milk

HAROLD E. SWAISGOOD

I. Introduction

Because milk is an excellent source of nutrients and since the proteins of milk are rather easily obtained, they have been the subject of many biochemical investigations. Thus, the primary structures of the major milk proteins have been determined by chemical sequencing and sequencing of complementary or genomic DNA (Swaisgood, 1982, 1993). There are six major gene products of the mammary epithelial cell; viz, α_{s1}-, α_{s2}-, β-, and κ-caseins, β-lactoglobulin, and α-lactalbumin (Swaisgood, 1982, 1993; Mepham *et al.,* 1993). Due to the inherent presence of blood plasmin and plasminogen in milk, a number of small (proteose–peptones) and large (γ-caseins) peptides are present in varying concentrations in milk as a result of post-translational proteolysis.

In addition to these proteins, there are numerous minor proteins and enzymes occurring in milk that are either derived from the epithelial cell or from the blood (Larson, 1993; Andrews, 1993; Walstra and Jenness, 1984). However, with respect to amino acid nutrition, the amounts of these proteins in normal milk are too small to be important. Enzymes are discussed in Chapter 5H, globule membrane proteins in Chapter 9A, and immuno proteins in Chapter 9B.

II. Protein Composition

The protein composition of mature herd milk is listed in Table I. The concentrations given represent approximate averages of values averaged

TABLE I
Protein Composition of Mature Bovine Herd Milk[a]

Protein	g/kg[b]	g/liter[b]
Total protein	35.1	36
Total casein	28.6	29.5
Whey protein	6.1	6.3
α_{s1}-Casein	11.5	11.9
α_{s2}-Casein	3.0	3.1
β-Casein	9.5	9.8
κ-Casein	3.4	3.5
γ-Casein	1.2	1.2
α-Lactalbumin	1.2	1.2
β-Lactoglobulin	3.1	3.2
Serum albumin	0.4	0.4
Immunoglobulin	0.8	0.8
Proteose–peptones	1.0	1.0

[a]Values averaged from those given by Walstra and Jenness (1984) and Swaisgood (1993). Colostrum excluded.
[b]Assuming a density of 1.03 g/ml of milk (Walstra and Jenness, 1984).

from the literature and reviewed by Swaisgood (1982, 1993) and Walstra and Jenness (1984). Proteose–peptones are small peptides largely derived from the N-terminus of β-casein and from a fat globule membrane glycoprotein; while the γ-caseins are derived from the C-terminus of β-casein (Swaisgood, 1993). The protein content of milk exhibits very little variability with a concentration range of approximately 30 to 35 g/kg.

III. Amino Acid Composition

Since the primary structures of the major milk proteins are known, very accurate amino acid compositions are available (Swaisgood, 1993). Compositions are listed in Table II for common genetic variants of each of the six gene products and for the largest γ-casein [β-CN A^2-1P (f 29–209)], i.e., the C-terminal 181 residues of β-casein A^2-5P. Other genetic variants differ in composition from the one listed in one to five residues (except for α_{s1}-CN A and α_{s2}-CN D), usually in only one or two. Thus, genetic variation is of little significance to amino acid nutrition.

The amino acid composition of milk is given in Table III. Experimental values are taken from the data of Williams *et al.* (1976). The calculated values are based on the protein composition given in Table I and the amino acid compositions listed in Table II. Hence, these values consider only

TABLE II
Chemical Composition of the Commonly Occurring Milk Proteins[a]

Acid	α_{s1}-CN[b] B-8P	α_{s2}-CN A-11P	κ-CN B-1P	β-CN A^2-5P	β-Lacto-globulin A	α-Lact-albumin B	β-CN A^2-1P (f29–209)
Asp	7	4	3	4	11	9	4
Asn	8	14	8	5	5	12	3
Thr	5	15	14	9	8	7	8
Ser	8	6	12	11	7	7	10
SerP	8	11	1	5	0	0	1
Glu	25	24	12	19	16	8	12
Gln	14	16	14	20	9	5	20
Pro	17	10	20	35	8	2	34
Gly	9	2	2	5	3	6	4
Ala	9	8	15	5	14	3	5
1/2 Cys	0	2	2	0	5	8	0
Val	11	14	11	19	10	6	17
Met	5	4	2	6	4	1	6
Ile	11	11	13	10	10	8	7
Leu	17	13	8	22	22	13	19
Tyr	10	12	9	4	4	4	4
Phe	8	6	4	9	4	4	9
Trp	2	2	1	1	2	4	1
Lys	14	24	9	11	15	12	10
His	5	3	3	5	2	3	5
Arg	6	6	5	4	3	1	2
Pyr or Glu	0	0	1	0	0	0	0
Total residues	199	207	169	209	162	123	181

[a]Based on their primary structures.
[b]In casein nomenclature, the letter indicates the genetic variant and the number represents the number of phosphorylated residues.

the six major milk proteins and the γ-caseins. However, they account for 94% of the total protein. For the latter, the composition of the largest C-terminal peptide was used in the calculations. The values were obtained using the equation:

$$C_n(\text{g/kg}) = 1/0.0036\ (\text{kg protein/liter milk}) \sum_{i,\ \text{all proteins}} [n(\text{residues/mol})\ M_r/M_i]c_i\ (g/liter),$$

where C_n is the concentration of amino acid n in milk, n is the number of residues of the amino acid in protein i, M_r is the molecular weight of amino

TABLE III
Amino Acid Composition of Milk

Amino acid	Experimental[a] (g/kg protein)	Calculated[b] g/kg Protein	Calculated[b] g/liter milk
Essential amino acids			
Thr	46	43	1.56
Val	66	66	2.36
Met	26	28	1.02
Cys	8	7	0.26
Ile	59	59	2.13
Leu	97	98	3.51
Phe	49	49	1.77
Lys	81	83	2.98
His	27	27	0.99
Arg	35	34	1.22
Trp	17	15	0.53
Nonessential amino acids			
Asp	79 (Asp + Asn)	34	1.21
Asn		42	1.49
Ser	56	63	2.26
Glu	219 (Glu + Gln)	121	4.37
Gln		94	3.38
Pro	99	100	3.59
Gly	20	18	0.63
Ala	34	33	1.19
Tyr	51	56	2.00

[a]Experimental values are taken from Williams *et al.* (1976).
[b]Calculated from the data listed in Tables I and II as described in the text.

acid n, M_i is the molecular weight of protein i and c_i is the protein concentration in milk. The excellent agreement between calculated and experimental values should be noted.

References

Andrews, A. T. (1993). Indigenous enzymes in milk: General introduction. *In* "Advanced Dairy Chemistry" (P. F. Fox, ed.), Vol. 1, pp. 285–292. Elsevier, London.

Larson, B. L. (1993). Immunoglobulins of the mammary secretions. *In* "Advanced Dairy Chemistry" (P. F. Fox, ed.), Vol. 1, pp. 231–254. Elsevier, London.

Mepham, T. B., Gaye, P., Martin, P., and Mercier, J. C. (1993). Biosynthesis of milk protein. *In* "Advanced Dairy Chemistry" (P. F. Fox, ed.), Vol. 1, pp. 491–543. Elsevier, London.

Swaisgood, H. E. (1982). Chemistry of milk protein. *In* "Developments in Dairy Chemistry" (P. F. Fox, ed.), Vol. 1, pp. 1–59. Applied Science, Barking, Essex.
Swaisgood, H. E. (1993). Chemistry of the caseins. *In* "Advanced Dairy Chemistry" (P. F. Fox, ed.), Vol. 1, pp. 63–110. Elsevier, London.
Walstra, P., and Jenness, R. (1984). *In* "Dairy Chemistry and Physics." Wiley, New York.
Williams, A. P., Bishop, D. R., Cockburn, J. E., and Scott, K. J. (1976). Composition of ewe's milk. *J. Dairy Res.* **43,** 325–329.

G. Nonprotein Nitrogen Compounds in Bovine Milk

BRENDA ALSTON-MILLS

I. Nitrogen Content in Milk

In milk, the three N-containing fractions are casein (78.5%), whey (16.5%), and nonprotein nitrogen (NPN, at 5 or 6%) (Roland, 1938; Cerbulis, 1975). Several factors affect the N content of milk. Within the United States, regional differences have been observed in the composition of milk fat, protein and solids-non-fat (Barbano, 1990; Bruhn, 1985). Breed differences affect milk total protein and casein from Holstein cows when compared to Jersey cows (Cerbulis, 1975). NPN is less variable, but with broad ranges within a breed. The NPN fraction is approximately 4.9% pf total N in milk across all breeds. However, the in-breed variation is 2.8–10.6% (Cerbulis, 1975). Seasonal differences affect protein content and NPN; generally, high environmental temperatures decrease the total milk protein. This observation was studied in depth and the findings were that protein concentration was higher in winter than in summer (Bruhn, 1977; Feagan, 1979). Conversely, NPN appears to be highest in summer and lowest in winter (Verdi *et al.,* 1987). Variations in feed constituents can affect milk N as a function of net energy intake. Energy intake from grain or roughage is positively correlated to milk protein concentration (Emery, 1978; Sporndly, 1989). There is general agreement that dietary protein has little effect on milk protein content.

Total protein and casein are highest in first lactation. Total casein N precipitously decreases after calving with the lowest concentration at 5–10 weeks followed by a gradual increase through the end of lactation. NPN follows a similar trend (Ng-Kwai-Hang *et al.,* 1985).

II. Milk NPN

Limited knowledge exists regarding the NPN fraction of total milk N concentration (Wolfschoon-Pombo and Klostermeyer, 1981). The NPN concentration is approximately 25–30 mg/100 ml of milk containing 5 or 6% of total N in cow's milk (Johnson, 1974). Coincidentally, compounds of the NPN fraction are similar to those found in cow urine suggesting that NPN compounds could be end products of N metabolism in the cow. Therefore, NPN in milks may be derived from blood (Jenness, 1989).

Although urea N contributes as much as 35–48% of the NPN (Kuzdzal-Savoie *et al.*, 1980; Wolfschoon-Pombo and Klostermeyer, 1981), a number of other constituents are reported in Table I. Orotic acid, for example, is unique in ruminant milks, little is found in other milks (Table II).

TABLE I
Major NPN Compounds in Bovine Milk[a]

	Nitrogen (mg/liter milk)		
	Recent Analyses[b]		
Compound	Mean	SD	Range in literature
Total NPN	296.4	37.7	229–308
Urea N	142.1	32.6	84–134
Creatine N	25.5	6.4	6–20
Creatinine N	12.1	6.8	2–9
Uric acid N	7.8	3.3	5–8
Orotic acid N	14.6	5.9	12–13
Hippuric acid N	4.4	1.2	4
Peptide N	32.0	14.9	—
Ammonia N	8.8	6.1	3–14
α-Amino acid N	44.3	8.2	39–51
Total	588.0		

Note. The amount of urea N is dependent on breed, stage of lactation, season, and diet. It has been suggested that dietary changes are the most influential variables in considering variations in milk urea N (Kaufmann, 1982). Additional nitrogenous compounds that have been detected are reported in Table II. Information on nucleotides and related compounds is in Chapter 5B.

[a]Wolfschoon-Pombo and Klostermeyer (1981).

[b]273 samples, each representing a single milking.

TABLE II
Some Nitrogenous Substances in Bovine Milk

Compound	Concn (mg/liter)	Reference
Amines		
1-Propylamine	3–15	Cole *et al.* (1961)
1-Hexylamine	5–24	Cole *et al.* (1961)
Ethanolamine	0.5–8.5	Armstrong and Yates (1963); Rassin *et al.* (1978)
Choline	43–285[a]	Hartman and Dryden (1974)
Putrescine, Cadaverine	0.003–0.021	Sanguansermsri *et al.* (1974)
Spermidine	0.009–0.028	Sanguansermsri *et al.* (1974)
Spermine	0.006–0.017	Sanguansermsri *et al.* (1974)
Amino acid derivatives		
N-methylglycine	+	Schwartz and Pallansch (1962a)
Histamine	0.03–0.05	Wrenn *et al.* (1963)
Salicyluric acid	0.016	Booth *et al.* (1962)
Phenylacetyl glutamine	> 0.01	Schwartz and Pallansch (1962b)
Kynurenine	0.023	Parks *et al.* (1967)
Indoxylsulfuric acid	0.124	Spinelli (1946)
Taurine	1–7	Armstrong and Yates (1963); Rassin *et al.* (1978)
Other compounds		
Carnitine	10–7	Erfle *et al.* (1970); Snoswell and Linzell (1975)
Acetyl carnitine	2–12	Erfle *et al.* (1970)
Morphine	0.0002–0.0005	Hazum *et al.* (1981)
N-acetylneuraminic acid (NANA)	120–270[b]	de Koning and Wijnand (1965); Kiermeier and Freisfeld (1965); Morrissey (1973)
N-acetylglucosamine	11	Hoff (1963)

[a]Total. About 25 mg/liter is in phospholipids. Adapted from Jenness (1989).
[b]Total. About 30 mg/liter is free dialyable NANA.

References

Armstrong, M. D., and Yates, K. N. (1963). Free amino acids in milk. *Proc. Soc. Exp. Biol. Med.* **113,** 680–683.

Barbano, D. M. (1990). Seasonal and regional variation in milk composition in the U.S. *In* "Proceedings of the Cornell Nutrition Conference on Feed Manufacturing, Rochester, New York." Cornell University, October 23–25.

Booth, A. N., Robbins, D. J., and Dunkley, W. L. (1962). Occurrence of salicyluric acid in milk. *Nature (London)* **194,** 290–291.

Bruhn, J. C. (1985). Suitability of California milk for cheese manufacture. *In* "Proceedings of the Annual Dairy Cattle Day, 24th," p. 1. Davis.

Bruhn, J. C., and Franke, A. A. (1977). Monthly variations in gross composition of California herd milks. *J. Dairy Sci.* **60,** 696–700.

Cerbulis, J., and Farrell, H. M., Jr. (1975). Composition of milks of dairy cattle. I. Protein, lactose, and fat contents and distribution of protein fraction. *J. Dairy Sci.* **58,** 817–827.

Cole, D. D., Harper, W. J., and Hankinson, C. L. (1961). Observations on ammonia and volatile amines in milk. *J. Dairy Sci.* **44,** 171–173.

deKoning, P. J., and Wijand, H. P. (1965). The effect of sugars and heat treatment on the determination of N-acetyl neuraminic acid in milk. *Neth. Milk Dairy J.* **19,** 73–81.

Emery, R. S. (1978). Feeding for increased milk protein. *J. Dairy Sci.* **61,** 825–828.

Erfle, J. D., Fisheer, L. J., and Sauer, F. (1970). Carnitine and acetyl-carnitine in the milk of normal and ketotic cows. *J. Dairy Sci.* **53,** 486–492.

Feagan, J. T. (1979). Factors affecting protein composition of milk and their significance to dairy processing. *Aust. J. Dairy Tech.* **34,** 77–81.

Hartman, A. M., and Dryden, L. P. (1974). The vitamins in milk and milk products. *In* "Fundamentals of Dairy Chemistry" (B. H. Webb, A. H. Johnson, and J. A. Alford, eds.), 2nd Ed. AVI, Westport, CT.

Hazum, E., Sabatka, J. J., Chang, K. J., Brent, D. A., Findlay, J. W. A., and Cuatrecasas, P. (1981). Morphine in cow and human milk: Could dietary morphine constitute a ligand for specific morphine (μ) receptors? *Science* **213,** 1010–1012.

Hoff, J. E. (1963). Determination of N-acetylglucosamine-1-phosphate and N-acetylglucosamine in milk. *J. Dairy Sci.* **46,** 573–574.

Jenness, R. (1989). Composition of milk. *In* "Fundamentals of Dairy Chemistry, Third edition" (N. P. Wong, ed.), pp. 1–38. Van Nostrand–Reinhold Company, New York.

Johnson, A. H. (1974). The composition of milk. *In* "Fundamentals of Dairy Chemistry" (B. H. Webb, A. H. Johnson, and J. A. Alford, eds.), 2nd Ed. AVI, Westport, CT.

Kaufman, W. (1982). Variation in composition of the raw material milk with special regard to the urea content. *Milchwissenschaft* **37,** 6–9.

Kiermeier, F., and Freisfeld, I. (1965). Neuraminic acid content of cow's milk. *Z. Lebensmitt. Untersuch. Forsch.* **128,** 207–217.

Kuzdzal-Savoie, S., Manson, W., and Moore, J. H. (1980). The constituents in cow's milk. *Int. Dairy Fed. Doc. (Brussels, Belgium)* **125,** 4.

Morrissey, P. A. (1973). The N-acetyl neuraminic acid content of the milk of various species. *J. Dairy Res.* **40,** 421–425.

Ng-Kavai-Hang, K., Hays, J. F., Maxely, J. E., and Monardes, H. G. (1985). Percentages of protein and non-protein nitrogen with varying fat and somatic cells in bovine milk. *J. Dairy Sci.* **68,** 1257–1262.

Parks, O. W., Schwartz, D. P., Nelson K., and Allen C. (1967). Evidence for kynurenine in milk. *J. Dairy Sci.* **50,** 10–11.

Rassin, D. K., Sturman, J. A., and Gaull, G. E. (1978). Taurine and other free amino acids in milk of man and other mammals. *Early Hum. Dev.* **2,** 1–13.

Rowland, S. J. (1938). The determination of the nitrogen distribution in milk. *J. Dairy Res.* **9,** 42–46.

Sanguansermsri, J., Gyorgy, P., and Zilliken, F. (1974). Polyamines in human and cow's milk. *Am. J. Clin. Nutr.* **27,** 859–865.

Schwartz, D. P., and Pallansch, M. J. (1962a). Identification of some nitrogenous constituents of cow's milk by ion exchange and paper chromatrography. *J. Agric. Food Chem.* **10,** 86–89.

Schwartz, D. P., and Pallansch, M. J. (1962b). Occurrence of phenylacetyl-glutamine in cow's milk. *Nature (London)* **194,** 186.

Snoswell, A. M., and Linzell, J. L. (1975). Carnitine secretion into milk of ruminants. *J. Dairy Res.* **42,** 371–380.
Spinelli, F. (1946). Indican in cow and goat milk. *Boll. Soc. Ital. Sper.* **21,** 210–211.
Sporndly, E. (1989). Effects of diet on milk composition and yield of dairy cows with special emphasis on milk protein content. *Swed. J. Agric. Res.* **19,** 99–106.
Verdi, R. J., Barbano, D. M., Dellavalle, M. E., and Senyk, G. F. (1987). Variability in true protein, casein, non-protein-nitrogen, and proteolysis in high and low somatic cell milks. *J. Dairy Sci.* **70,** 230–242.
Wolfschoon-Pombo, A., and Kostermeyer, H. (1981). The NPN-fraction of cow milk I. Amount and composition. *Milchwissenschaft* **36,** 598–600.
Wrenn, T. R., Bitman, J., Cecil, H., and Gilliam, D. R. (1963). Histamine concentration in blood, milk and urine of dairy cattle. *J. Dairy Sci.* **46,** 1243–1245.

H. Enzymes Indigenous to Bovine Milk

HAROLD E. SWAISGOOD

I. Introduction

Enzymes in milk occur in various states: (1) as unassociated forms in solution; (2) associated or an integral part of membrane fractions, such as the fat globule membrane or skim milk membrane vesicles, both of which are derived from the plasma membrane of the secretory cell; (3) associated with casein micelles; and (4) as part of microsomal particles. Enzymes that have been reported to be present in milk are listed in Table I. For more information and citations of the original work, the reader should consult the reviews from which the table was compiled (Shahani *et al.,* 1973; Kitchen, 1985; Farkye, 1992).

Some of the minor enzymes, such as aldolase, lactate dehydrogenase, arylsulfatase, catalase, and *N*-acetyl-β-D-glucosaminidase, are associated with somatic cells and thus their presence is related to disease of the mammary gland, particularly mastitis. Enzymes associated with membrane fractions will occur in both cream and skim milk. It should also be noted that skim milk membrane vesicles pass into the whey fraction upon casein curd formation. Thus, many enzymes are in whey although they do not represent unassociated forms "free in solution."

Enzymes of known or potential technological significance include plasmin, lipoprotein lipase, alkaline phosphatase, lactoperoxidase, sulfhydryl oxidase, *N*-acetyl-β-D-glucosaminidase, catalase, xanthine oxidase, superoxide dismutase, γ-glutamyltransferase, and lactose synthase.

HANDBOOK OF MILK COMPOSITION

TABLE I
Bovine Milk Enzymes[a]

Enzyme type	EC No.[b]	Enzyme type	EC No.[b]
Oxidoreductases		**Hydrolases**	
Amine oxidase	1.4.3.6	α-Amylase	3.2.1.1
Catalase	1.11.1.6	α-Fucosidase	3.2.1.51
Dihydrolipoamide dehydrogenase	1.8.1.4	β-Amylase	3.2.1.2
Glucose-6-phosphate dehydrogenase	1.1.1.49	β-Glucosidase	3.2.2.21
Glutathione peroxidase	1.1.1.9	Acid phosphatase (phosphoprotein phosphatase)	3.1.3.2
Isocitrate dehydrogenase	1.1.1.42	Alkaline phosphatase	3.1.3.1
L-Iditol dehydrogenase	1.1.1.14	Arylesterase	3.1.1.2
Lactate dehydrogenase	1.1.1.27	Arylsulfatase	3.3.6.1
Lactoperoxidase	1.11.1.7	ATPase	3.6.1.3
Malate dehydrogenase	1.1.1.37	Carboxylesterase	3.1.1.1
Malic enzyme	1.1.1.40	Cholinesterase	3.1.1.8
NADH dehydrogenase	1.6.99.3	5′-Nucleotidase	3.1.3.5
Phosphoglucuronate dehydrogenase (decarboxylating)	1.1.1.44	Glucose-6-phosphatase	3.1.3.9
Sulfhydryl oxidase	1.8.3	Inorganic pyrophosphatase	3.6.1.1
Superoxide dismutase	1.15.1.1	Leucine aminopeptidase	3.4.11.1
Xanthine oxidase	1.2.3.2	Lipoprotein lipase	3.1.1.3
		Lysozyme	3.2.1.17
Transferases		*N*-acetyl-β-glucosaminidase	3.2.1.30
Alanine aminotransferase	2.6.1.2	Plasmin	3.4.21.7
Aspartate aminotransferase	2.6.1.1	Ribonuclease	3.1.27.5
CMP-*N*-acetyllactosaminide-α-2,3-sialytransferase	2.4.99.6		

TABLE I—*continued*

Enzyme type	EC No.[b]	Enzyme type	EC No.[b]
γ-Glutamyltransferase	2.3.2.2	**Lyases**	
Lactose synthase	2.4.1.22	Aldolase	4.1.2.13
A protein: UDP galactosyltransferase		Carbonic anhydrase	4.2.1.1
B protein: α-lactalbumin			
N-acetyllactosamine synthase	2.4.1.90	**Isomerases**	
RNA-directed DNA polymerase	2.7.7.49	Glucose-6-phosphate isomerase	5.3.1.9
Thiamin–phosphate pyrophosphorylase	2.5.1.3		
Thiosulfate sulfurtransferase (rhodanase)	2.8.1.1	**Ligases**	
UDP-galactosyltransferase	2.4.1.38	Acetyl-CoA carboxylase	6.4.1.2

[a]Compiled from Shahani *et al.* (1973), Kitchen (1985), and Farkye (1992).
[b]Enzyme commission or EC No.

II. Enzymes of Technological Significance

Most of the plasmin, and its precursor plasminogen, is associated with casein micelles. Limited proteolysis of β-casein by this enzyme is responsible for the presence in milk of large polypeptides derived from this protein, known as the γ-caseins. Activity of this enzyme is also important to cheese ripening and the stability of casein micelles in various products such as ultra-high temperature pasteurized (UHT) milk. Lipoprotein lipase also is largely associated (80%) with casein micelles. Beneficial effects of its activity include the possible aid in initial digestion and absorption of milk lipids in the intestinal tract and flavor development in certain cheeses from raw milk. However, lipolytic activity also causes a hydrolytic rancid flavor. Normally the substrate is not accessible to the enzyme; however, rapid cooling can dissociate the enzyme from micelles allowing it to attach to fat globules resulting in "spontaneous lipolysis" and rough mechanical treatment of unpasteurized milk can disrupt the fat globule membrane allowing interaction with casein micelles and its associated lipase.

Many of the other technologically important enzymes are largely associated with membrane fractions. Alkaline phosphatase is of commercial importance because of its widespread use as an indicator of pasteurization efficiency. The enzyme's heat-stability profile closely follows that necessary for adequate pasteurization. Sulfhydryl oxidase catalyzes the oxidation of thiols and the formation of disulfide bonds in proteins and peptides. Treatment of UHT milk with the enzyme has been shown to eliminate cooked flavor. Also, because enzyme-catalyzed oxidation of thiols does not produce active oxygen species as does autoxidation, sulfhydryl oxidase-treated UHT milk may have longer flavor stability due to reduced lipid oxidation. Both xanthine oxidase and catalase have been implicated in oxidative flavor formation by virtue of the production of superoxide by the former and the heme iron content of the latter.

Catalase is most likely associated with membranes of somatic cells in milk. Several other important enzymes are also associated with somatic cell membranes. Xanthine oxidase contains all of the molybdenum in milk. A hypothesis, that the enzyme was associated with the initiation of atheromata in blood vessels of persons who consumed pasteurized, homogenized milk, has been discredited. *N*-acetyl-β-D-glucosaminidase is associated with cells in fresh milk but readily dissociates during various treatments or cold storage. Its activity is used as a diagnostic test for mastitis. Some superoxide dismutase may be associated with somatic cells because the activity increases with increasing cell counts. Superoxide dismutase activity has a protective effect on lipid oxidation as would be expected due to elimination of the superoxide anion.

Two membrane-associated enzymes of little importance to milk products, but of great importance to milk synthesis, are lactose synthase and γ-glutamyltransferase. As the name implies, the former is responsible for synthesis of lactose. γ-Glutamyltransferase appears to be involved in transport of amino acids into the mammary gland.

Recently, commercial interest has been expressed in the use of lactoperoxidase, activated by addition of thiocyanate, as an antibacterial agent. For example, the activity of this enzyme has been used to prevent microbial deterioration of nonrefrigerated unpasteurized milk during collection and storage in developing countries. Its availability through large-scale isolation from whey has also stimulated interest in other applications such as in dental products.

References

Farkye, N. Y. (1992). Other enzymes. *In* "Advanced Dairy Chemistry. 1. Proteins" (P. F. Fox, ed.), pp. 339–367. Elsevier, London.

Kitchen, B. J. (1985). Indigenous milk enzymes. *In* "Developments in Dairy Chemistry 3. Lactose and Minor Constituents" (P. F. Fox, ed.), pp. 239–279. Elsevier, London.

Shahani, K. M., Harper, W. J., Jensen, R. G., Parry, R. M., Jr., and Zittle, C. A. (1973). Enzyme in bovine milk: A review. *J. Dairy Sci.* **56,** 531–543.

I. Hormones and Growth Factors in Bovine Milk

W. M. CAMPANA
C. R. BAUMRUCKER

I. Introduction

Mammalian milk is unique in that it is one of the few substances naturally designed to sustain the newborn. Because of its ability to support growth and development, the composition of milk has been rigorously analyzed. The finding of hormones in milk was described as early as 1929 (Yaida, 1929). Reviews of hormones in milk during the late 1970s and early 1980s surveyed the known endocrine factors. At that time, milk hormones were thought to originate entirely from circulating endocrine hormones. However, studies in mammary cell biology indicate that bioactive substances in milk originate from mammary tissue and are capable of regulating mammary cell proliferation and perhaps differentiation through autocrine/paracrine action (Sporn and Roberts, 1985). Additionally, growth factors, such as EGF in rat milk, have been shown to stimulate intestinal growth suggesting a functional role in the neonate.

HANDBOOK OF MILK COMPOSITION

Bovine colostrum and milk are rich sources of various peptides which possess biological activity (Table I). Colostrum contains the highest concentrations of hormones/growth factors (Baumrucker and Blum, 1993). Bovine milk has been of significant interest since it is widely consumed and used for infant formulas. Over the years, improved methodology has enabled scientists to more accurately determine the concentration of these substances in milk. Also, with the advent of recombinant protein technology in the dairy industry (i.e., bovine somatotropin), public and political consciousness has been raised concerning other bioactive factors in commercially available milk. To date many growth factors have been identified in milk (Grosvenor *et al.*, 1993). This review will focus on growth factors and hormones found in bovine milk but will not address other biologically important peptides in milk such as immunoglobulins, allergins, opiates, enzymes, casomorphines, and cyclic nucleotides. However, excellent reviews recently published are available which thoroughly examine other hormones and bioactive substances present in mammalian milk (Koldovsky, 1989; Britton and Kastin, 1991; West, 1989).

It has not been established exactly how bioactive hormones are transported into milk. Many of the factors found in bovine milk exceed the

TABLE I
Bioactive Substances in Bovine Milk

Steroid hormones	**Hypothalamic hormones**
5-α Androstane-3,17-dione	Lutenizing hormone-releasing hormone
Corticosterone	Gonadotropin hormone-releasing hormone
Estradiol	Somatostatin
Estriol	Thyrotropin-releasing hormones
Estrone	
Progesterone	**Pituitary hormones**
Vitamin D	Growth hormone
	Prolactin
	Gastrointestinal hormones
Thyroid and parathyroid hormones	Bombesin
Parathyroid hormone-related peptide	Gastrin
Thyroxin (T3 and T4)	Gastrin-releasing hormone
	Neurotensin
Growth factors	**Others**
IGFs	PGFα
IGF-binding proteins	Transferrin
MDGI	
TGF-β	

concentration in maternal plasma [i.e., estrogen, gonadotropin-releasing hormone (GnRH), somatostatin, parathyroid hormone-related peptide (PTHrP), prolactin (PRL), insulin, and insulin-like growth factor (IGF-I)]. Some factors are rapidly transported (unchanged in structure and activity) into milk from the maternal circulation. Also, some peptide hormones are synthesized by the mammary gland and post-translationally modified. Some of the peptides secreted into milk may be proteolytically cleaved, rendering them biologically active or inactive. Interestingly, hormones, such as thyroid hormones, relaxin, PTHrP, estrogen, GnRH, prolactin, insulin-like growth factors, epidermal growth factor, and nonhormonal bioactive substances (lactoferrin, transferrin, casomorphines), appear to be synthesized by mammary tissue of various species and transported from maternal circulation.

Determining the absolute amount of a specific hormone in milk continues to be challenging. Because milk is a complex substance, inaccuracy arises from using inefficient techniques that do not adequately separate the desired hormone from interfering substances. Thus, quantification of these hormones is limited by the sensitivity and specificity of the assay. Another factor influencing accurate quantification is the separation of the fat layer from the rest of the milk. For example, steroid hormones are found at higher concentration in milk fat because they are fat-soluble compounds. To ensure accurate assessment of hormones/growth factors, milk samples should be taken from complete milkings since fat content differs at the beginning and end of milking. Unfortunately, differences in methodology between published results limit absolute amounts of hormones in milk.

Three nonexclusive lines of thought have been proposed to explain the function of bioactive substances in milk. The first is that many, if not all, bioactive substances which appear in milk are a result of disposal by the mammary gland. This concept suggests that although a mechanism exists to transport or synthesize and secrete bioactive substances, it does not have a specific function for neonatal development. Unless function is ascribed to these bioactive components found in milk, this idea remains viable. The second idea is that many of these bioactive agents in milk are part of maternal mammary cell regulation. Data clearly show that infusion of bioactive agents into the mammary gland cistern via the teat canal (which serves as a milk reservoir) has dramatic effects upon mammary function. Additionally, many studies have shown endocrine receptors exist and function on the apical side of mammary epithelial cells. Finally, another line of thinking suggests importance for the neonate. Primary emphasis has focused upon the newborn gastrointestinal tract and has clearly shown acute effects by these bioactive agents in newborn and suckling models. Long-term effects of systemic regulation have been implied as well. However, the neonatal concept will need further investigation to define overall effects of development and health before actual functions can be assigned.

II. Hormones (Table II)

A. Adrenal Gland Hormones

In the mid 1950s corticosteroids (includes glucocorticoids and mineral corticoids) were identified in cow's milk (Ratsimamanga *et al.*, 1956, 1961; Rappi and Rossi, 1955). Interestingly, in bovine milk, corticosterone levels are higher than cortisol levels, whereas, in plasma, cortisol levels are higher than corticosterone levels suggesting enhanced mammary generation of C19 steroids.

The concentration of glucocorticoids in milk is lower than that in plasmas (Gwazdauskas *et al.*, 1977) but can be increased temporarily in cows by systemic (Gwazdauskas *et al.*, 1977) or intramammary (Paape *et al.*, 1975) administration of hydrocortisone. Paape *et al.* (1975) measured corticosteroid values in milk before intramammary infusion of 1600 mg of hydrocortisone. Milk corticosteroids went from 3.7 ng/ml to >400 ng/ml after 4 hr. Milk corticosteroid values were unaffected by milking. Gwazdaukas *et al.* (1977) measured total glucocorticoids in milk and plasma using a competitive protein-binding assay. Values for glucocorticoids are similar (0.7 to 1.4 ng/ml) for both whole milk and skim milk. Even after injection of hydrocortisone or ACTH (Gwazdauskas *et al.*, 1977), the concentration of corticosteroids in milk is the same in bovine whole milk and skim milk suggesting glucocorticoids are not exclusively associated with fat. However, as expected, stressed cows exhibit elevated glucocorticoids in milk (Holdsworth *et al.*, 1983; Gwazdauskas *et al.*, 1977). The range in corticosteroid levels in milk may be attributed to the physiological condition of the animals and methods of extraction and analysis.

B. Gonadal Hormones

There is considerable interest in estrogen and progesterone found in farm animal milk because these compounds can be used as indicators of reproductive status. Specifically, estrogen predicts the functional state of the Graffian follicle and progesterone indicates the development of the corpus luteum (Koldovsky, 1989). Although there is little information available on the androgen content in milk, bovine milk contains 5-α-androstane-3,17-dione (Darling *et al.*, 1974) and testosterone (Hoffman and Rattenberger, 1977). 5-α-Androstane-3,17-dione does not change during pregnancy, after delivery, or during the estrous cycle (Darling *et al.*, 1974). During pregnancy, mean levels in milk ($\pm$SEM) range from 0.9 ± 0.4 to 03.9 ± 0.9 ng/ml. 5-α-Androstane-3,17-dione may be involved in the development of milk secretion as it was not detected in colostrum of cows. Testosterone levels in milk range from 45 to 71 pg/ml (Hoffman and Rattenberger, 1977).

TABLE II
Amounts of Hormones in Bovine Milk

Hormone	Reference	Amount	Stage of lactation	Methods
5-α-Androstane-3,17-dione	Darling *et al.* (1974)	0.0–7.2 ng/ml	Days 0–57	Gas chromatography
Estradiol 17-β	Wolford and Argoudelis (1979)	4–14 pg/ml	Commercial milk	RIA
		361 ± 30 ng/ml	Days 0 to 2	RIA
	Erb *et al.* (1977b)	13 ± 1 ng/ml	Days 3 to 25	RIA
Estradiol 17-α	Erb *et al.* (1977b)	474 ± 152 ng/ml	Days 0 to 2	RIA
		160 ± 14 ng/ml	Days 3 to 25	RIA
Estriol	Wolford and Argoudelis (1979)	9–31 pg/ml	Commercial milk	RIA
Estrone	Wolford and Argoudelis (1979)	34–55 pg/ml	Commercial milk	RIA
		1032 ± 264 ng/ml	Days 0 to 2	RIA
	Erb *et al.* (1977b)	28 ± 3 ng/ml	Days 3 to 25	RIA
		1341 ± 880 pg/ml	Day 0	RIA
	Kesler *et al.* (1976)	241 ± 32 pg/ml	Day 5	RIA
Estrogen (total)	Erb *et al.* (1977b)	1867 ± 438 ng/ml	Days 0 to 2	RIA
		201 ± 15 ng/ml	Days 3 to 25	RIA
Glucocorticoids	Gwazdauskas *et al.* (1977)	0.7–1.4 ng/ml	Whole milk	Competitive protein binding assay
		0.7–1.3 ng/ml	Skim milk	
GnRH	Baram *et al.* (1977)	0.1–3 ng/ml		RIA
Growth hormone	Torkelson (1987)	< 1 ng/ml	Mature milk	RIA
Insulin	Ballard *et al.* (1982)	0.67–5.7 n*M*	Colostrum	RIA
	Malven *et al.* (1987)	37.1 ± 14 ng/ml	Prepartum	RIA
		6.2 ± 2.1 ng/ml	Postpartum	RIA
		5.5 ± 0.6 ng/ml	Days 4 to 6	RIA

TABLE II—*continued*

Hormone	Reference	Amount	Stage of lactation	Methods
IGF-I	Vega *et al.* (1991)	2949 ± 1158 ng/ml	2 weeks prepartum	Acid ext./RIA
		5 ± 2 ng/ml	Day 49	Acid ext./RIA
IGF-II	Vega *et al.* (1991)	1825 ± 608 ng/ml	2 weeks prepartum	Acid ext./RIA
		1 ± 0.1 ng/ml	Day 49	Acid ext./RIA
LHRH	Amarant *et al.* (1982)	3.9–11.8 ng/ml	Mature milk	RIA
PGFα	Mann (1975); Hansel (1976)	0.2–0.4 ng/ml	Mature milk	RIA
Progesterone	Ginther *et al.* (1974)	23 ng/ml	Day 60	RIA
		18.4 ng/ml	Day 210	RIA
	Darling *et al.* (1974)	10 ng/ml	Day 50	Gas chromatography
Prolactin	Kacsóh *et al.* (1991a,b, 1993)	500–800 ng/ml	Colostrum	RIA
		6–8 ng/ml	Late milk	RIA
PTHrP	Goff *et al.* (1991)	56 ± 12 ng/ml	Colostrum	RIA
		77 ± 19 ng/ml	1 Month	RIA
		106 ± 11 ng/ml	5 Months	RIA
		168 ± 17 ng/ml	9 Months	RIA
Somatostatin	Takeyama *et al.* (1990)	20 pmol/liter	Pre-and postpartum	Enzyme immunoassay
Testosterone	Hansel (1976)	45–71 pg/ml	Active corpus luteum	RIA
TRH	Amarant *et al.* (1982)	16–34 ng/ml	Mature milk	RIA

1. Estrogen

Analysis of estrogen (includes estrone and estradiol) in bovine colostrum and milk has been done by bioassay (Monk *et al.*, 1975), colorimetry (Chicchini, 1965), spectroflourimetry (Ittrich and Mbohb, 1960), gas chromatography (Darling *et al.*, 1974), and radioimmunoassay (RIA) (Kesler *et al.*, 1976). Absolute estrogen levels in milk have been difficult to evaluate and compare due to different methodologies of measurement. McGariggle and Lachelin (1983) discovered that spectroflourimetric detection was not appropriately sensitive and lead to falsely elevated values. The development of estrogen analysis by RIA improved sensitivity and demonstrated that conjugated estrogens comprise more than 90% of total human plasma and milk. The concentration of estrogen exceeds 1 ng/ml in prepartum secretions and colostrum (Monk *et al.*, 1975). These concentrations have been correlated with total estrogen in blood plasma and urine before and after parturition. The estimated excretion of estrogen through milk represents < 1% of the total excreted during the estrous cycle and decreases as lactation proceeds. Estrogen is found in several commercial dairy products including nonfat dry milk, butter, whey, and dry curd cottage cheese (Wolford and Argoudelis, 1979). Not surprisingly, estrogen is found in high concentration in whole compared to skim milk. Wolford *et al.* (1979) measured free natural estrogens in raw and commercial whole milk by RIA. They reported concentration ranges of estrone, estradiol-17-β and estriol to be 34–55, 4–14, and 9–31 pg/ml, respectively.

Induced parturition by dexamethasone and estradiol injections into prepartum cows does not change concentrations of estrone or estradiol in milk (Kesler *et al.*, 1976). However, elevated levels of estrogen occur in milk when cows are injected 3 days postpartum. Estrogen concentrations in milk change during the estrus cycle, during gestation, and following insemination (Koldovsky, 1989). Estrogen concentrations are higher in bovine mammary secretions compared to blood during the week before parturition and decrease within several days postpartum (Monk *et al.*, 1975; Erb *et al.*, 1977a). Evidence indicates that at parturition the mammary gland synthesizes and secretes estrogen (Prandi and Gaiani, 1984; Peaker and Taylor, 1990) and that it may account for the large increase in estradiol-17-β in the blood (Maule Walker and Peaker, 1978; Maule Walker *et al.*, 1983). Changes in estrogen concentration in bovine milk may also be the result of increased mammary steroid aromatase activity (Peaker and Taylor, 1990).

2. Progesterone

Progesterone has been measured in milk of many species. It is found in many dairy products with the highest concentration (130–300 ng/g) found in butter (Erb *et al.*, 1977a; Ginther *et al.*, 1974). Progesterone levels are influenced by total fat content in milk; thus, collecting homogenous samples of milk is important for measuring consistent values (Cowie and Swinburne, 1977). Milk concentration of progesterone is 2.4 times that of blood. The concentration of progesterone in skim milk is similar to that of blood but is approximately one-fifth the amount in whole milk.

Progesterone levels decrease in plasma and milk during the final 2 days of gestation (Kulski *et al.*, 1977). This depletion of progesterone allows milk secretion to begin. Ginther *et al.* (1974) determined the amounts of progesterone in bovine milk by RIA over the course of lactation. Progesterone levels were 2.3 μg/100 ml at Day 60 of lactation and fell to 1.84 μg/100 ml 150 days later. Darling *et al.* (1974) reported 1 μg/100 ml of milk on Day 50 of pregnancy which decreased to 1 μg/100 ml after 50 days of pregnancy. However, the Darling group used gas chromatography to determine progesterone levels so differences are likely due to methods used. The ratio of milk progesterone to plasma progesterone changes with the reproductive cycle. During late lactation, the ratio is one or less due to low extraction from the circulation (Holdsworth *et al.*, 1983); there is no indication that the mammary gland can synthesize this hormone. However, when cows are pregnant the ratio increases to 3–5 (Erb *et al.*, 1977a; Koontz, 1984).

Progesterone metabolites are present in milk but in smaller quantities than progesterone as determined by the gas chromatography (Darling *et al.*, 1974). These metabolites (pregnanediols, pregnanolones, and pregnanediones) are measured as 5α- and 5β-pregnane-3,20-diones. In milk, the 5α are present in greater quantity than the 5β steroids.

C. Vitamin D

Vitamin D is found in bovine milk although the amount is variable depending upon the housing and seasonal lighting conditions under which the animals are exposed (Henry, 1942). Vitamin D associates with the fat layer; thus, homogenous samples of milk are critical when comparing Vitamin D values. Large dietary increases (14-fold) of vitamin D only double vitamin D in milk (40–80 IU/liter). Reeve *et al.* (1982) determined that cholecalciferol is the major form of vitamin D (281 ng/ml) followed by 25-OH-CC (145 ng/ml), 1,25-$(OH)_2$CC (27 ng/ml), and 24,25$(OH)_2$CC (5 ng/ml). See Chapter 8C for additional information.

D. Brain Gut Hormones

1. *GnRH*

GnRH has been identified in bovine milk (Amarant *et al.*, 1982), but is not detectable in infant formula (Nair *et al.*, 1987). GnRH in milk exceeds the concentration in plasma five- to sevenfold (Sarda and Nair, 1981; Nair *et al.*, 1987). There is evidence that GnRH is synthesized in the mammary glands of rats (Smith and Ojeda, 1986). Similar evidence has not been validated in the bovine mammary gland.

When measured by RIA, GnRH concentration in bovine milk is 0.1–3 ng/100 ml. GnRH may influence the secretion of gonadotropic hormones in neonates (Baram *et al.*, 1977).

2. *Somatostatin (SS)*

Biological and immunoreactive SS has been identified in milk of many species (Holst *et al.*, 1990; Rao *et al.*, 1990) including bovine (Werner *et al.*, 1988). Unlike maternal serum, milk has SS 14 and not the amino terminal-extended SS-28-like material. SS-14 is the originally identified peptide which is predominantly found in the brain, whereas SS-28 is predominantly found in the gut. A sensitive and specific enzyme immunoassay for SS was developed by using β-D-galactosidase-labeled antigen. Bovine prepartum secretions contained 20 pmol/liter and the level was unchanged after delivery (Takeyama *et al.*, 1990).

3. *Thyrotropin-Releasing Hormone (TRH) and Lutenizing Hormone-Releasing Hormone (LHRH)*

TRH is present in bovine milk and exceeds the concentration in maternal serum (Amarant *et al.*, 1982). Initially it was thought that TRH was synthesized in bovine mammary tissue; however, Strbak (1991) found that the TRH gene is not expressed. When quantified by RIA (after acid–methanol extraction) milk extracts contain 16–34 ng/ml TRH and 3.9–11.8 ng/ml LHRH. Both hormones (when analyzed by HPLC) co-migrated with their corresponding hypothalamic-derived marker hormones. Bovine milk LHRH was shown to be equal in bioactivity compared to synthetic LHRH (Amarant *et al.*, 1982).

Gut hormones have been found in bovine milk and include bombesin (Jahnke and Lazarus, 1984) and neurotensin (Wood *et al.*, 1988; Thurston *et al.*, 1990). Both hormones in milk exceed the concentration in blood. Gastrin (Scanff *et al.*, 1992) and gastrin-releasing peptide are found in cow's milk and are highest in bovine prepartum secretions. Levels of these peptides decline in concentration 1 week after delivery to about 10% the amount in prepartum secretions (Takeyama *et al.*, 1990).

E. Other Hormones

1. *PTHrP*

High concentrations of immunoreactive and bioactive PTHrP are found in bovine milk (Law *et al.*, 1991; Ratcliffe *et al.*, 1990), while lowered levels of PTHrP are found in maternal serum (Budayr *et al.*, 1989b; Law *et al.*, 1991). Several research groups have quantified the amounts of PTHrP in bovine milk. Law *et al.* (1991) analyzed PTHrP by RIA and reported 59.2 ± 18.5 μg/liter in milk samples pooled from various stages of lactation. PTHrP levels vary with the breed of cows. Jersey milk PTHrP was shown to be significantly higher than that of Fresians (52.6 ± 5.4 compared to 41 ± 4.8 μg/liter). Correspondingly, the concentration of calcium was higher in Jersey milk. Generally, the levels of PTHrP in milk correlate positively with total milk calcium (Law *et al.*, 1991) and suggest a role in mammary calcium transport from blood to milk. Budayr *et al.* (1989a) reported bovine PTHrP concentration in milk to be 40 to 75 ng eq of PLP amide/ml as determined by RIA. Bioactivity, determined by cyclic AMP production by ROS 17/2.8 cells, correlated similarly with immunoreactivity.

Stage of lactation affects PTHrP concentrations in milk. During early lactation, PTHrP is low, but increases as lactation proceeds (Law *et al.*, 1991). Goff *et al.* (1991), using RIA, showed that Jersey colostrum was 56 ± 12 ng/ml. Milk concentrations from 1, 2, 3, 5, 7, and 9 months of lactation were 77, 59, 57, 106, 119, and 168 ng/ml, respectively. At all stages, PTHrP was bioactive.

2. *Thyroid Gland Hormones*

Thyroxine (T_4) and triodothroxine (T_3) are found in domestic animals (Koldovsky, 1989). Levels of thryoxine in bovine milk are very low (Mann, 1969). The determination and quantification of thyroid hormones in milk is subject to problems in methodology (Koldovsky, 1989).

F. Growth Factors

Klagsbrun showed that bovine mammary secretions (Klagsbrun, 1980; Steimer *et al.*, 1981; Shing and Klagsbrun, 1984) contained "factors" which stimulated the growth of cells in culture. Growth factors in bovine milk have also been shown to alter the differentiation of cells (Sporn and Roberts, 1988).

1. *IGFs*

IGFs are part of the insulin family of protein hormones consisting of insulin, IGF-I, IGF-II, and relaxin. IGFs are ubiquitous and act as mediators of growth, development, and differentiation (Lowe, 1991). IGFs interact with several receptors designated type I and type II IGF receptors and the insulin receptor (Rechler, 1987). IGFs also bind IGF-binding proteins (IGFBP). To date, six IGFBPs have been sequenced and cloned (Ballard *et al.*, 1989, 1990). Bovine secretions contain both IGFs, some IGFBPs, and IGF receptors (Baumrucker, 1994). Of particular interest in milk is the stability of IGFs in heat and acid treatment (Lowe, 1991). These characteristics contribute to the survivability of IGFs in commercial milk products and to their potential bioactivity in the gastrointestinal tract of the consumer.

Malven *et al.* (1987) described IGFs in bovine prepartum secretions. Bovine milk concentrations of both forms of IGFs declined rapidly after parturition. IGF-II concentrations were approximately double those of IGF-I during the course of the experimental period. Campbell and Baumrucker (1989) quantified the temporal patterns of IGF-I in bovine colostrum and milk. Multiparous cows had higher IGF-I concentration (306 μg/liter) at parturition than primaparous cows (147 μg/liter). By Day 2 of lactation, milk IGF-I concentrations were 30 to 50% of initial values. By Day 56 of lactation, milk IGF-I concentrations were 34 μg/liter for combined parity groups. The decline of milk IGF-I after parturition cannot be solely attributed to a dilution effect because rates of change (IGF-I vs milk volume) are not coincident (Campbell and Baumrucker, 1989).

IGF-I in bovine colostrum and milk is principally associated with a 45-kDa IGFBP (Pyke and Baumrucker, 1988; Vega *et al.*, 1991). Skaar *et al.* (1991) reported that both blood and milk secretion patterns of IGFBP change with the reproductive and lactational stage of the cow. In mammary secretions, four specific IGF-binding protein bands were observed by Western blot, two of which have been positively identified as IGFBP-3 and IGFBP-2 (Skaar and Baumrucker, 1992). Vega *et al.* (1991) described the substantial increase in concentrations of IGFs and IGFBP that occur in bovine prepartum mammary secretions. IGFBP profile changes in milk are more dramatic and distinct from those observed in blood (Skaar *et al.*, 1991).

Milk shows a sizable decrease in IGFBP-3 after parturition with an increase in the occurrence of a 20-kDa IGFBP (Skaar *et al.*, 1991, 1993). Unlike the milk of pigs and humans, bovine milk does not contain the acid-labile subunit capable of binding IGFBP-3 (Binoux and Hossenlopp, 1992).

2. *Insulin*

Insulin concentrations in bovine colostrum range from 0.67 to 5.7 n*M* and are 100 times higher than blood levels (Ballard *et al.*, 1982) Malven *et al.* (1987) reported insulin concentrations in prepartum secretions to be higher than those after parturition (37.1 ± 14 vs 5.5 ± 0.6 ng/ml, respectively).

3. *Transforming Growth Factor (TGF-*α and TGF-β)

TGF-β is important for the regulation of cell proliferation and differentiation (Massague, 1987) and is found in bovine milk (Cox and Bürk, 1991; Jin *et al.*, 1991). Transcription of the TGF-β1, -2, and -3 isoforms has been demonstrated in the nonlactating and lactating bovine mammary gland by *in situ* hybridization (Maier *et al.*, 1991). TGF-α mRNA has also been identified in mammary glands of cows, albeit there are no reports indicating the presence of TGF-α in bovine milk.

4. *Growth Inhibitors in Milk*

Interest in growth inhibitors of mammary tissue proliferation has heightened due to their potential use in breast cancer treatment. Inhibitory polypeptides controlling growth, differentiation, and regression of mammary epithelial cells has been postulated (Grosse and Langen, 1993). A 13-kDa polypeptide termed mammary-derived growth inhibitor (MDGI) has been purified from bovine mammary tissue and from milk fat globule membranes (Brandt *et al.*, 1987) that reversibly inhibits the proliferation of several normal and transformed mammary epithelial cells (Grosse and Langen, 1993). MDGI appears to function in synchronizing the cell cycle by transient inhibition of mammary cell growth (Politis *et al.*, 1992).

5. *Other Growth Factors*

Tissue plasminogen activator (tPA) may be a potential mammary trophic factor and involved in tissue remodeling (Turner and Huynh, 1991). Tissue PA catalyzes the conversion of plasminogen to the active enzyme plasmin. Both plasmin and plasminogen have been identified in bovine milk, although a principle source may be from white blood cells present in milk (Politis *et al.*, 1991). Enhanced production of tPA by GPK cell line in the presence of cow milk has also been reported (Electricwala *et al.*, 1992).

G. Pituitary Hormones

1. *Growth Hormone (GH)*

Growth hormone (bovine somatotropin; bST) has been of significant interest and extensively researched due to its dramatic impact on dairy cows. Recently, the FDA has approved treatment of dairy cows with GH to increase milk volume. Torkelson *et al.* (1987) developed a highly sensitive RIA for bovine milk bST with a lower detection limit of 0.3 ng/ml. Bulk milk from lactating cows that never received recombinant bST was obtained from 120 different farms. Bovine ST was shown to be less than 1 ng/ml. Animals that were injected with recombinant bST (600 mg every 2 weeks) showed no increase in milk bST concentrations over control animals. The safety of commercially available milk products containing growth hormone has been reviewed previously (Daughaday and Barbano, 1990; Juskevich and Guyer, 1990).

2. *PRL*

Prolactin in bovine milk was first suggested by the observation of Geschicker and Lewis (1936) when colostrum demonstrated lactogenic activity in the pigeon crop assay. Prolactin in bovine milk has been detected by RIA (Malven and McMurtry, 1974). New insight in PRL variants suggested that milk PRL exhibits molecular heterogeneity with varying patterns of immunoreactivity and biological activity (Baumrucker, 1994). Milkborne PRL survives passage through the neonatal gastrointestinal tract and functions in neuroendocrine regulation (Grosvenor *et al.,* 1992; Kacsóh *et al.,* 1991b).

Recent methodology has allowed the comparison of bioactivity (NB2 lymphoma proliferation assay) and immunoreactivity (RIA) of PRL. The amount of immunoreactive PRL in bovine colostrum is 500–800 ng/ml and has a bioassay/immunoassay ratio (B/I) of 0.2 to 0.3. Late bovine milk contains 6 to 8 ng/ml and has a B/I ratio of 0.3 to 0.6 (Kacsóh *et al.,* 1991a,b, 1993). The bioactivity of bovine milk PRL is greatest in early milk and decreases as lactation proceeds. The B/I ratio is less than 1 in early bovine milk vs rat and human milk which show a ratio greater than 1. Differential B/I ratios have been speculated to be related to PRL heterogeneity that includes phosphorylation and glycosylation. Recent studies (Kurtz *et al.,* 1993) indicate that mammary epithelial cells transcribe the PRL gene and are capable of post-translational modifications.

H. Prostaglandins (PG)

Prostaglandins of the E and F series (includes PG, PG2, PGα, and PGFα) have been identified in bovine milk (Simmons *et al.,* 1979; Mann, 1975). The mammary gland is capable of PG_2 synthesis (Bennet *et al.,* 1977). Macrophages found in milk also synthesize and secrete prostaglandins (Blau *et al.,* 1986; LeDiest *et al.,* 1986). Thus, the source(s) of milk PG is uncertain. Mann (1975) and Hansel (1976) measured 0.2 to 0.4 ng/ml of PGα in bovine milk by RIA. Simmons *et al.* (1979) demonstrated that injecting PGα (30 mg) into cows results in a 10-fold increase of PGα in the plasma and 3-fold increase of PGα into milk (Simmons *et al.,* 1979). Early *in vitro* work by Mann (1975) showed that native and added PG were stable during incubation at 37°C for 6 hr in bovine milk.

III. Summary

It is not surprising that many of the hormones in blood and mammary tissue are present in colostrum and milk. First, circulating proteins may be transferred to milk via active transport or "leaky" junctions of the mammary epithelial cells. Second, epithelial cells from the alveolus are sloughed into milk which contribute to milk constituents. Third, mammary tissue can locally synthesize hormones/growth factors which are secreted into milk. All of these events, occurring in combination or individually, contribute to the finding of hormones in milk.

The fact that some hormones/growth factors appear in greater concentration in milk than blood suggests a specific role for these substances in mammary and neonatal function. However, lower concentrations in milk do not lessen their importance; they suggest that a specific mechanism of hormonal entry does not exist or has not yet been detected. Some hormones/growth factors appear in colostrum and milk in distinct or multiple forms when compared to maternal serum. It is speculated that each different form has distinct target tissue and a unique function. These observations, complemented with high concentrations in milk, strengthen the evidence for synthesis and post-translational processing by the mammary gland.

Finally, regardless of modes of entry or amounts of hormones/growth factors in milk, evidence supports a role for bioactive substances in the development and endocrine/metabolic functions in the neonate. Further investigations of the specific functions of milkborne hormones will be beneficial for understanding these relationships.

References

Amarant, T., Fridkin, M., and Koch, Y. (1982). Luteinizing hormone-releasing hormone and thyrotropin-releasing hormone in human and bovine milk. *Eur. J. Biochem.* **127,** 647–650.

Ballard, F. J., Nield, M. K., Francis, G. L., Dahlenburg, G. W., and Wallace, J. C. (1982). The relationship between the insulin content and inhibitory effects of bovine colostrum on protein breakdown in cultured cells. *J. Cell. Physiol.* **110,** 249–254.

Ballard, J., Baxter, R., Binoux, M., Clemmons, D., Drop, S., Hall, K., Hintz, R., Rechler, M., Rutanen, E., and Schwander, J. (1989). On the nomenclature of the IGF binding proteins. *Acta Endocrinol.* **121,** 751–752.

Ballard, F. J., Baxter, R. C., Binoux, M., Clemmons, D. R., Drop, S. L. S., Hall, K., Hintz, R. L., Rechler, M. M., Rutanen, E. M., and Schwander, J. C. (1990). Report on the nomenclature of the IGF binding proteins. *J. Clin. Endocrinol. Metab.* **70,** 817.

Baram, T., Koch, Y., Hazum, E., and Fridkin, M. (1977). Gonadotropin-releasing hormone in milk. *Science* **198,** 300–302.

Baumrucker, C. R., Campana, W. M., Gibson, C. A., and Kerr, D. E. (1994). Insulin-like growth factors (IGFs) and IGF binding proteins in milk; Sources and functions. *Endocr. Regul.* **27,** 152–172.

Baumrucker, C. R., and Blum, J. W. (1993). Effects of dietary recombinant human insulin-like growth factor I on concentrations of hormones and growth factors in the blood of newborn calves. *J. Endocrinol.* **140,** 15–21.

Bennet, A., Charlier, E., McDonald, A. M., Simpson, J. S., Stamford, I. F., and Zebro, T. (1977). Prostaglandins and breast cancer. *Lancet* **2,** 624–626.

Binoux, M., and Hossenlopp, P. (1992). Insulin-like growth factor (IGF) and IGF binding proteins: Comparison of human serum and lymph. *J. Clin. Endocrinol. Metab.* **67,** 509–514.

Blau, H., Passwell, J., Levanon, M., Davidson, J., Kohen, D., and Ramot, B. (1986). Studies on human milk macrophages: Effect of activation of phagocytosis and secretion of prostaglandin E2 and lysozyme. *Pediatr. Res.* **11,** 6–10.

Brandt, R., Pepperle, M., Otto, A., Kraft, R., Beohmer, F. D., and Grosse, R. (1987). A 13-kilodalton protein purified from milk fat globule membranes is closely related to a mammary-derived growth inhibitor. *Biochem. J.* **27,** 1420–1425.

Britton, J. R., and Kastin, A. J. (1991). Biologically active polypeptides in milk. *Am. J. Med. Sci.* **301,** 124–132.

Budayr, A. A., Halloran, B. P., King, J. C., Diep, D., Nissenson, R. A., and Strewler, G. J. (1989a). High levels of a parathyroid hormone-like protein in milk. *Proc. Natl. Acad. Sci. USA* **86,** 7183–7185.

Budayr, A. A., Halloran, B. P., King, J. C., Diep, D., Nissenson, R. A., and Strewler, G. J. (1989b). High levels of a parathyroid hormone-like protein in milk. *Proc. Natl. Acad. Sci. USA* **86,** 7183–7185.

Campbell, P. G., and Baumrucker, C. R. (1989). Insulin-like growth factor-I and its association with binding proteins in bovine milk. *J. Endocrinol.* **120,** 21–29.

Chicchini, U. (1965). Behavior of oestrogens in blood and milk during pregnancy in Friesians. *Atti. Soc. Ital. Sci. Vet.* **23,** 360–367.

Cowie, A. T., and Swinburne, J. K. (1977). Hormones, drugs, metals and pesticides in milk: A guide to the literature. *Dairy Sci. Abstr.* **39,** 391–402.

Cox, D. A., and Bürk, R. R. (1991). Isolation and characterisation of milk growth factor, a transforming-growth-factor-b2-related polypeptide, from bovine milk. *Eur. J. Biochem.* **197,** 353–358.

Darling, J. A. B., Laing, A. H., and Harkness, R. A. (1974). A survey of the steroids in cow's milk. *J. Endocrinol.* **62,** 291–297.

Daughaday, W. H., and Barbano, D. M. (1990). Bovine somatotropin supplementation of dairy cows. Is the milk safe. *JAMA* **264,** 1003–1005.

Electricwala, A., Clark, S., Griffiths, J. B., and Atkinson, T. (1992). Tissue plasminogen activator production from cells cultured in milk extract. *Fibrinolysis* **6,** 51–55.

Erb, R. E., Chew, B. P., and Keller, H. F. (1977a). Relative concentration of estrogen and progesterone in milk and blood, and excretion of estrogen in urine. *J. Anim. Sci.* **46,** 617–266.

Erb, R. E., Chew, B. P., Keller, H. F., and Malven, P. V. (1977b). Effect of hormonal treatments prior to lactation on hormones in blood plasma, milk, and urine during early lactation. *J. Dairy Sci.* **60,** 557–565.

Geschickter, C. F., and Lewis, D. (1936). Lactogenic substance in the human breast: Its use in experimental stimulation of mammary secretion and its assay in cases of cystic disease. *Arch. Surg.* **32,** 598–617.

Ginther, O. J., Nuti, L., Wentworth, B. C., and Tyler, W. J. (1974). Progesterone concentration in milk and blood during pregnancy in cows. *Proc. Soc. Exp. Biol. Med.* **146,** 354–357.

Goff, J. P., Reinhardt, T. A., Lee, S., and Hollis, B. W. (1991). Parathyroid hormone-related peptide content of bovine milk and calf blood assessed by radioimmunoassay and bioassay. *Endocrinology* **129,** 2815–2819.

Grosse, R., and Langen, P. (1993). Mammary derived growth inhibitor. *In* "Handbook of Experimental Pharmacology" (B. V. R. Born, P. Cuatrecasas, H. Herken, and A. Schwartz, eds.), in press. Springer-Verlag, Heidelberg.

Grosvenor, C. E., Toth, B. E., and Kacsóh, B. (1992). Importance of milk prolactin (PRL) in the ontogeny of lactotrophe function in the rat. *In* "Mechanisms Regulating Lactation and Infant Nutrient Utilization" (M. F. Picciano and B. Lonnerdal, eds.), pp. 309–377. Wiley–Liss, New York.

Grosvenor, C. E., Picciano, M. F., and Baumrucker, C. R. (1993). Hormones and growth factors in milk. *Endocr. Rev.* **14,** 710–728.

Gwazdauskas, F. C., Pappe, M. J., and McGilliard, M. L. (1977). Milk and plasma glucocorticoid alterations after injections of hydrocortisone and adrenocorticotropin. *Proc. Soc. Exp. Biol. Med.* **154,** 543–545.

Hansel, W. (1976). Concentrations and activities of prostoglandins of the F series in bovine tissue, blood, and milk. *J. Dairy Sci.* **59,** 1353–1365.

Henry, K. M., and Kon, S. K. (1942). The vitamin D content of English butter fat throughout the year. *Biochem. J.* **36,** 456–459.

Hoffman, B., and Rattenberger, E. (1977). Testosterone concentrations in tissue from veal calves, bulls and heifers and in milk samples. *J. Anim. Sci.* **46,** 635–641.

Holdsworth, R. J., Heap, R. B., Goode, J., Peaker, M., and Walters, D. E. (1983). Mammary uptake and metabolism of progesterone in goats and its effect on milk progesterone concentrations during the oestrous cycle and early pregnancy. *J. Endocrinol.* **98,** 263–270.

Holst, N., Jenssen, T. G., and Burhol, P. G. (1990). A characterization of immunoreactive somatostatin in human milk. *J. Pediatr. Gastroenterol. Nutr.* **10,** 47–52.

Ittrich, G. T., and Mbohb, M. (1960). Eine methode zur chemischen bestimmung von ostrogenen hormonen in Blut, milch und colostrum. *Hoppe.-Seyler. Z. Physiol. Chem.* **320,** 103–110.

Jahnke, G. D., and Lazarus, L. H. (1984). A bombesin immunoreactive peptide in milk. *Proc. Natl. Acad. Sci. USA* **81,** 578–582.

Jin, Y., Cox, D. A., Knecht, R., Raschdorf, F., and Cerletti, N. (1991). Separation, purification, and sequence identification of TGF-b1 and TGF-b2 from bovine milk. *J. Protein Chem.* **10,** 565–575.

Juskevich, J. C., and Guyer, C. G. (1990). Bovine growth hormone: Human food safety evaluation. *Science* **249,** 875–884.

Kacsóh, B., Toth, B. E., Avery, L. M., Deaver, D. R., Baumrucker, C. R., and Grosvenor, C. E. (1991a). Biological and immunological activities of glycosylated and molecular weight variants of bovine prolactin in colostrum and milk. *J. Anim. Sci.* **69** (Supp.1), 456. [Abstract]

Kacsóh, B., Toth, B. E., Avery, L. M., Yamamuro, Y., and Grosvenor, C. E. (1991b). Molecular heterogeneity of prolactin in lactating rats and their pups: Biological and immunological activities in the pituitary gland, serum, and milk. *Endocr. Regul.* **25,** 98–111.

Kacsóh, B., Toth, B. E., Veress, Z., Avery, L. M., and Grosvenor, C. E. (1993). Bioactive and immunoreactive variants of prolactin in milk and serum of lactating rats and their pups. *J. Endocrinol.* **138,** 243–257.

Kesler, D. J., Peterson, R. C., Erb, R. E., and Callahan, C. J. (1976). Concentrations of hormones in blood and milk during and after induction of parturition in beef cattle with dexamethasone and estradiol-17b. *J. Anim. Sci.* **42,** 918–926.

Klagsbrun, M. (1980). Bovine colostrum supports the serum-free proliferation of epithelial cells but not of fibroblasts in long-term culture. *J. Cell Biol.* **84,** 808–814.

Koldovsky, O. (1989). Hormones in milk: Their possible physiological significance for the neonate. *In* "Textbook of Gastroenterology and Nutrition in Infancy" (E. Ledbenthal, ed.), pp. 97–119. Raven Press, New York.

Koontz, J. W. (1984). Role of the insulin receptor in mediating the insulin stimulated growth response in reuber h-35 cells. *Mol. Cell. Biochem.* **58,** 139–146.

Kulski, J. K., Smith, M., and Hartmann, P. E. (1977). Perinatal concentrations of progesterone, lactose and a-lactalbumin in the mammary secretion of women. *J. Endocrinol.* **74,** 509–510.

Kurtz, A., Bristol, L. A., Tóth, B. E., Lazar-Wesley, E., Takács, L., and Kacsóh, B. (1993). Mammary epithelial cells of lactating rats express prolactin messenger ribonucleic acid. *Biol. Reprod.* **48,** 1095–1103.

Law, F. M. K., Moate, P. J., Leaver, D. D., Diefenbach-Jagger, H., Grill, V., Ho, P. W. M., and Martin, T. J. (1991). Parathyroid hormone-related protein in milk and its correlation with bovine milk calcium. *J. Endocrinol.* **128,** 21–26.

LeDiest, F., de Saint-Basile, B., Angelos, C., and Griscelli, C. (1986). Prostaglandin E2 and plasminogen activator in human milk and their secretion by milk macrophages. *Am. J. Reprod. Immunol. Microbiol.* **11,** 6–10.

Lowe, W. L. J. (1991). Biological actions of the insulin-like growth factors. *In* "Insulin-Like Growth Factors: Molecular and Cellular Aspects" (D. LeRoith, ed.), pp. 49–85. CRC Press, Boca Raton, FL.

Maier, R., Schmid, P., and Cox, D., Bilbe, G., and McMaster, G. K. (1991). Localization of transforming growth factor-b1, -b2, and -b3 gene expression in bovine mammary gland. *Mol. Cell. Endocrinol.* **82,** 191–198.

Malven, P. V., Head, H. H., Collier, R. J., and Buonomo, F. C. (1987). Periparturient changes in secretion and mammary uptake of insulin and in concentrations of insulin and insulin-like growth factors in milk of dairy cows. *J. Dairy Sci.* **70,** 2254–2265.

Malven, P. V., and McMurtry, J. P. (1974). Measurement of prolactin in milk by radioimmunoassay. *J. Dairy Sci.* **57,** 411–415.

Mann, E. B. B. (1969). Butanol-extractable iodine in human and bovine colostrum and milk. *Clin. Chem.* **15,** 1141–1146.

Mann, J. G. (1975). The excretion of prostaglandin F2a in milk of cows. *Prostaglandins* **9,** 463–474.

Massague, J. (1987). The TGF-beta family of growth and differentiation factors. *Cell* **49,** 437–438.

Maule Walker, F. M., Davis, A. J., and Fleet, I. R. (1983). Endocrine activity of the mammary gland: Oestrogen and prostaglandin secretion by the cow and sheep mammary glands during lactogenesis. *Br. Vet. J.* **139,** 171–177.

Maule Walker, F. M., and Peaker, M. (1978). Production of oestradiol-17b by the goat mammary gland during late pregnancy in relation to lactogenesis. *J. Physiol. London* **284,** 71P–72P.

McGarrigle, H. H. G., and Lachelin, G. C. L. (1983). Oestrone, oestradiol and oestriol glucosiduronates and sulphates in human puerperal plasma milk. *J. Steroid Biochem.* **18,** 607–611.

Monk, E. L., Erb, R. E., and Mollet, T. A. (1975). Relationships between immunoreactive estrone and estradiol in milk, blood, and urine of dairy cows. *J. Dairy Sci.* **58,** 34–40.

Nair, R. M., Somasundaran, M., Katikaneni, L. D., and Purohit, D. M. (1987). Studies on LHRH and physiological fluid amino acids in human colostrum and milk. *Endocrinol. Exp.* **21,** 23–30.

Paape, M., Schultze, W. D., and Smith, J. W. (1975). Milk and plasma corticoid levels following hydrocortisone. *J. Dairy Sci.* **58,** 1244–1251.

Peaker, M., and Taylor, E. (1990). Oestrogen production by the goat mammary gland: Transient aromatase activity during late pregnancy. *J. Endocrinol.* **125,** R1–R2.

Politis, I., Zhao, X., McBride, B. W., Burton, J. H., and Turner, J. D. (1991). Plasminogen activator production by bovine milk macrophages and blood monocytes. *Am. J. Vet. Res.* **52,** 1208–1213.

Politis, I., Zavizion, B., Grosse, R., Turner, J. D., and Gorewit, R. C. (1992). Mammary-derived growth inhibitor production by bovine mammary epithelial cells. *J. Dairy Sci.* **75,** 181. [Abstract]

Prandi, A., and Gaiani, R. (1984). Studio in vitro ed in vivo dell'attivata'steroidogenetica dell mammella della capra nelle ultimae della gravidanza. *Atti. Soc. Ital. Sci. Vet.* **38,** 194–197.

Pyke, S. N., and Baumrucker, C. R. (1988). Analysis of insulin-like growth factor (IGF) binding proteins in bovine serum, colostrum and milk. *Endocrine Soc. 70th Annu. Meeting* **152.** [Abstract]

Rao, R. K., Koldovsky, O., and Davis, T. P. (1990). Inhibition of intestinal degradation of somatostatin by rat milk. *Am. J. Physiol.* **258,** G426–G431.

Rappi, G., and Rossi, R. (1955). Ricerca degli ormoni steroidi nel latte umano (alcuni rilievi sulla reazione di Porter e Silber). *Riv. Clin. Pediat.* **56,** 267–280.

Ratcliffe, W. A., Green, E., Emly, J, Norbury, S., Lindsay, M., Heath, D. A., and Ratcliffe, J. G. (1990). Identification and partial characterization of parathyroid hormone-related protein in human and bovine milk. *J. Endocrinol.* **127,** 167–176.

Ratsimamanga, A. R., Nigeon-Dureuil, M., and Rabinowicz, M. (1956). Presence d'hormone-type cortinique-das le lait de la vache gravide. *C. R. Soc. Biol.* **150,** 2179–2182.

Ratsimamanga, A., Mouton, M., and Bein, M. (1961). Mise en evidence d'hormones du type cortinique dans le lait. *Ann. Radiol.* **196,** 9–17.

Rechler, M. M. (1987). Receptors for insulin like growth factors. *Poly Peptide Hormone Receptors* **4,** 227–297.

Reeve, L. E., Jorgensen, N. A., and DeLuca, H. F. (1982). Vitamin D compounds in cows' milk. *J. Nutr.* **112,** 667–672.

Sarda, A., and Nair, R. M. G. (1981). Elevated levels of LRH in human milk. *J. Clin. Endocrinol. Metab.* **52,** 826–828.

Scanff, P., Yvon, M., Thirouin, S., and Pélissier, J.-P. (1992). Characterization and kinetics of gastric emptying of peptides derived from milk proteins in the preruminant calf. *J. Dairy Res.* **59,** 437–447.

Shing, Y. W., and Klagsbrun, M. (1984). Human and bovine milk contain different sets of growth factors. *Endocrinology* **115,** 273–282.

Simmons, K. R., Moses, S. C., and Perkins, P. L. (1979). Prostaglandin in milk, days open, and estrus detection in dairy cows treated with prostaglandin F2a. *J. Dairy Sci.* **62,** 1443–1448.

Skaar, T. C., Vega, J. R., Pyke, S. N., Baumrucker, C. R. (1991). Changes in insulin-like growth factor-binding proteins in bovine mammary secretions associated with pregnancy and parturition. *J. Endocrinol.* **131,** 127–133.

Skaar, T. C., Baumrucker, C. R., Deaver, D. R., and Blum, J. W. (1993). Diet effects and ontogeny of alterations of circulating insulin-like growth factor binding proteins in newborn dairy calves. *J. Anim. Sci.* **72,** 421–427.

Skaar, T. C., and Baumrucker, C. R. (1992). Purification of bovine mammary insulin-like growth factor binding protein 3 (IGFBP-3). *J. Dairy Sci.* **75** (Suppl. 1). [Abstract]

Smith, S. S., and Ojeda, S. R. (1986). Presence of luteinizing hormone releasing hormone (LHRH) in milk. *Endocrinol. Exp.* **20,** 147–153.

Sporn, M. B., and Roberts, A. B. (1985). Introduction: Autocrine, paracrine and endocrine mechanisms of growth control. *Cancer Surveys* **4,** 627–632.

Sporn, M. B., and Roberts, A. B. (1988). Peptide growth factors are multifunctional. *Nature* **322,** 217–219.

Steimer, K. S., Packard, R., Holden, D., and Klagsbrun, M. (1981). The serum-free growth of cultured cells in bovine colostrum and in milk obtained later in the lactation period. *J. Cell. Physiol.* **109,** 223–234.

Strbak, V., Giraud, P., Resetkova, E., Ouafik, L., Dutour, A., Oliver, C., Povazanova, K., and Randuskova, A. (1991). Thyroliberin (TRH) and TRH free acid (TRH-OH) present in milk do not originate from local synthesis in mammary gland. *Endocr. Regul.* **25,** 134–138.

Takeyama, M., Yanaga, N., Yarimizu, K., Ono, J., Takaki, R., Fujii, N., and Yajima, H. (1990). Enzyme immunoassay of somatostatin (SS)-like immunoreactive substance in bovine milk. *Chem. Pharm. Bull.* **38,** 456–459.

Thurston, A. W., Cole, J. A., Hillman, L. S., Im, J. H., Thorne, P. K., Krause, W. J., Jones, J. R., Eber, S. L., and Forte, L. R. (1990). Purification and properties of parathyroid hormone-related peptide isolated from milk. *Endocrinology* **126,** 1183–1190.

Torkelson, A. R. (1987). Radioimmunoassay of somatotropin in milk from cows administered recombinant bovine somatotropin. *Proc. Am. Dairy Sci. Assoc.* **70,** (Suppl. 1), 146.

Turner, J. D., and Huynh, H. T. (1991). Role of tissue remodeling in mammary epithelial cell proliferation and morphogenesis. *J. Dairy Sci.* **74,** 2801–2807.

Vega, J. R., Gibson, C. A., Skaar, T. C., Hadsell, D. L., and Baumrucker, C. R. (1991). Insulin-like growth factor (IGF)-I and -II and IGF binding proteins in serum and mammary secretions during the dry period and early lactation in dairy cows. *J. Anim. Sci.* **69,** 2538–2547.

Werner, H., Katz, P., Fridkin, M., Koch, Y., and Levine, S. (1988). Growth hormone releasing factor and somatostatin concentrations in the milk of lactating women. *Eur. J. Pediatr.* **147,** 252–256.

West, D. W. (1989). The origin, transport and function of hormones and growth factors in milk. *Exp. Clin. Endocrinol.* **8,** 145–156.

Wolford, S. T., and Argoudelis, C. J. (1979). Measurement of estrogen in cow's milk, human milk, and dietary products. *J. Dairy Sci.* **62,** 1458–1463.

Wood, J. G., Hoang, H. D., Bussjaeger, J. L., and Solomon, T. E. (1988). Neurotensin stimulates growth of small intestine in rats. *Am. J. Physiol.* **255,** G813–G817.

Yaida, N. (1929). Ovarial hormone in blood of pregnant women, of pregnant animals: Ovarial hormone in urine of pregnant women: Ovarial hormone in milk of pregnant animals. *Trans. Jpn. Pathol. Soc.* **189,** 93–101.

CHAPTER 6

Milk Lipids

A. Human Milk Lipids

ROBERT G. JENSEN
JOEL BITMAN
SUSAN E. CARLSON
SARAH C. COUCH
MARGIT HAMOSH
DAVID S. NEWBURG

I. Introduction

A. Definitions and Nomenclature

According to Gurr and Harwood (1991), lipids are a chemically heterogeneous group of substances sharing the property of insolubility in water but solubility in the nonpolar solvents such as the ethers and chloroform. There are many classes of lipids and literally thousands of subclasses.

The following shorthand notation for fatty acids is widely employed by lipidologists and will be used in this chapter. For example, palmitic acid is 16:0; oleic acid, 18:1; and linoleic acid, 18:2. The figure to the left of the colon is the number of carbons, and the figure to the right, the number of double bonds. The location of the double bond is usually given by the number of carbons from the carboxyl group: 9-18:1, 9,12-18:2, etc. However, with polyunsaturated fatty acids (PUFA) another notation is valuable, locating the double bonds from the terminal methyl; 9,12-18:2 becomes 18:2n6. The letter n is preferable to omega, which is prevalent in older literature. Geometric isomers of 18:1 are designated by *cis* (*c*) or *trans* (*t*) 18:1*c* or 18:1*t*. Linoleic acid is 9,12-18:2*cc*. With the exception of ruminant milk and partially hydrogenated fats, the configuration of the double bonds in all dietary fats is *cis*.

The location of fatty acids on triacylglycerols (TG) will be identified by stereospecific numbering or sn. With this nomenclature, if the hydroxyl or substituent group on glycerol or a TG is drawn to the left, the group above is numbered sn-1 and the one below, sn-3. L-α-Phosphatidylcholine (lecithin) becomes sn-3 phosphatidylcholine. Note that if different fatty acids occupy sn positions 1 and 3 (the identity of the fatty acids at sn-2 do not matter), the TG is an optical isomer or enantiomer.

Another convenience is used for acylglycerols; 1-oleoyl-2-palmitoyl-3-stearoyl sn-glycerol becomes sn-18:1–16:0–18:0, with the sn-1 group starting at the left. If the TG is a racemic mixture, it is rac 18:1–16.0–18.0. If the enantiomeric composition is unknown, the prefix is X. Enantiomeric diacylglycerol (DG) and monoacylglycerols (MG) are similarly identified.

B. *The Nature of Lipids in Milk*

The lipids (3–5%) occur as globules emulsified in the aqueous phase (87%) of milk. The globules contain nonpolar or core lipids such as TG, cholesteryl esters, retinol esters, etc., (Jensen, 1989a,b). They are coated with bipolar materials, phospholipids, proteins, cholesterol, enzymes, etc. into a loose layer called the milk lipid globule membrane (MLGM). The MLGM prevents the globules from coalescing and acts as an emulsion stabilizer. The diameter of the globules ranges from 1 to 10 μm, with most of the globules at 1 μm, but those of 4 μm account for most of the weight. The globules present a large surface area (4.6 m^2/dl) to the lipolytic enzymes encountered during their passage through the digestive tract facilitating lipolysis of milk TG and absorption of the digestion products. See Chapter 2A for more information.

II. Collection, Preparation, and Storage of Samples

These have been described in Chapter 2C, but a few additional remarks are necessary. Care must be taken to obtain a sample which represents the total lipid content of the milks being studied. The factors which could influence the lipid content, time postpartum, time of day, beginning or end of nursing, gestational age, etc., should be identified (Jensen, 1989a). The phases, colostrum, transitional, and mature, should be replaced by hours, days, and weeks. These suggestions also apply to the lipid classes and fatty acids even though they may not be affected. Maternal dietary information, even if only regional, should be provided for all samples.

III. Determinations of Lipid Content

These have been described by Jensen (1989a,b) and Jensen *et al.* (1992). Briefly, these are the creamatocrit, solvent extraction, enzymatic, gas-liquid chromatographic (GLC), and infrared. In the creamatocrit procedure, about 75 μl of milk is drawn into a capillary tube which is sealed at one and spun for 15 min at 1200*g* in a hematocrit or other centrifuge. The length of the cream layer is measured and calculated as a percentage of the total length of the milk column. This figure is converted to percentage fat using standard curves obtained by one of the other methods in which the fat content is determined. The method is useful in the field or when large numbers of samples must be analyzed. Solvent extraction employs dichloromethane or chloroform and methanol (modified Folch, dry column) or hexane-ethyl ether-ethyl alcohol–ammonia (Roese-Gottlieb or Mojonnier). The extracted lipids are usually measured gravimetrically, although densitometry has been employed. If the extracted samples must be stored, it should be at −70°C. In the enzymatic procedure, TG glycerol is determined. With the GLC method, an internal standard is used to calculate the weight of each acid and this is converted to TG. The infrared procedure is used in the dairy industry to analyze large numbers of samples for fat, lactose, and protein. However, the instruments are expensive and not likely to be available for the analysis of human milk.

IV. Factors Affecting Total Lipid Content

The factors that are known or believed to alter the fat content of human milk are listed in Table I. The direction of the effects to the extent that they are known is also presented. Data will be given under Section V, E.

Investigators must define their goals and identify their subjects so as to control these influences. The effect on volume of milk must also be considered.

V. Lipid Classes

A. Introduction

The data obtained mostly by Bitman *et al.* (1983) on milk from mothers of term and preterm infants are given in Table II. These data have been corrected for the products of lipolysis, DG, MG, and free fatty acids (FFA). The lipid content increased, and the phospholipid and cholesterol contents

TABLE I
Factors Associated with Changes in the Total Lipid Content of Human Milk[a]

During a nursing or feed (increases)
Age postpartum, stage of lactation (increases)
Diurnal rhythm (variable)
Between breasts (occurs)
Gestational age at birth; preterm vs term (occurs)
Diet (may occur)
Infections, metabolic disorders (usually decreases)
Medication (?)
Mother's menstrual cycle or pregnancy (?)
Parity (decreases)
Season (? related to diet)
Age (?)
Miscellaneous
Individuality (adiposity increases)

[a]Adapted from Jensen (1989a,b).

decreased as lactation progressed. In general, fat contents range from 3.5 to 4.5% (Jensen, 1989a,b). The milk lipid content of milks from mothers of preterm babies decreased with the advance of lactation. All milk samples will contain small amounts of some of the products of lipolysis seen under the column, immediate extraction. These have no nutritional significance unless the milk has been "banked" for storage at −20°C for several months. Milk lipases are still active at this temperature and an undesirable flavor may develop. The relative amounts of TG will decrease and those of DG and MG will increase.

Cholesteryl esters were not determined in this study, but later, Bitman *et al.* (1986) found that they decreased from about 5 mg/dl at 3 days postpartum to 1 mg/dl at Day 21 and thereafter. There are traces of hydrocarbons (squalene), glyceryl ethers, and other miscellaneous lipids in human milk (Jensen, 1989a). In the amounts present, these are of little nutritional significance.

B. *Triacylglycerols*

1. *Introduction*

Triacylglycerols account for 98+% of the lipids in milk. The composition of TG is usually given as the kinds and amounts of fatty acids present. However, structure must be considered. Structure means the distribution of fatty acids within and among the TG molecules and

TABLE II
Lipid Class Composition of Human Milk During Lactation[a]

Lipid class	Percentage of total lipids at Lactation Day					Immediate extraction
	3	7	21	42	84	
Total lipid, % in milk	2.04 ± 1.32[b]	2.89 ± 0.31	3.45 ± 0.37	3.19 ± 0.43	4.87 ± 0.62	
Phospholipid	1.1	0.8	0.8	0.6	0.6	0.81
Monoacylglycerol						ND
Free fatty acids						0.08
Cholesterol (mg/dl)[c]	1.3 (34.5)	0.7 (20.2)	0.5 (17.3)	0.5 (17.3)	0.4 (19.5)	0.34
1,2-Diacylglycerol						0.01
1,3-Diacylglycerol						ND
Triacylglycerol	97.6	98.5	98.7	98.9	99.0	98.76
Cholesterol esters (mg)[d]						
N	39	41	25	18	8	6

[a]Adapted from Bitman *et al.* (1983).
[b]Mean ± SEM.
[c]The total cholesterol content ranges from 10 to 20 mg/dl after 21 days in most milks.
[d]Not reported, but in Bitman *et al.* (1986) was 5 mg/dl at 3 days and 1 mg/dl at 21 days and thereafter.

ultimately identification of the individual molecular species. Structure is important because TGs are exposed as such to lipolytic enzymes and not as fatty acids. Structure is also one of the key factors controlling the products formed by gastric lipase in the stomach and by pancreatic and bile salt-stimulated lipases in the small intestine. These in turn help control intestinal absorption. The 12:0 and 18:2 and their MG are potent microbicides and will help control infections in the stomach and small intestine (Jensen *et al.*, 1992).

2. *Structure*

Human milk contains about seven fatty acids in amounts >1%. If these were randomly distributed among the three positions of glycerol, 343 (or 7 to the third power) TGs could result. These are numbers, not quantities, that would be determined by the amounts of individual fatty acids. However, the distribution of fatty acids in human milk TG is not random.

The identities of the fatty acids in sn positions 1–3 have been determined. Some of these data are given in Table III. Note that the distribution is unique with most of the 16:0 at sn-2, 12:0 at sn-3, 18:0 at sn-1, and 18:1 and 18:2 at sn-1 and -3. By combining their data, Breckenridge *et al.* (1967, 1969) and Kuksis and Breckenridge (1968) surmised that the TGs listed in Table IV were present in major quantities in human milk TGs. The

TABLE III
Positional Distribution of Major Fatty Acids in Triacylglycerols from Human Milk[a]

Fatty acid (mole %)	sn Position[b]		
	1	2	3
8:0			
10:0	0.2	0.2	1.8
12:0	1.3	2.1	6.1
14:0	3.2	7.3	7.1
16:0	16.1	58.2	6.2
16:1	3.6	4.7	7.3
18:0	15.0	3.3	2.0
18:1	46.1	12.7	49.7
18:2	11.0	7.3	14.7
18:3	0.4	0.6	1.6
20:1	1.5	0.7	0.5
20:4	Trace	0.9	0.3

[a]Adapted from Breckenridge *et al.* (1967, 1969) and Kuksis and Breckenridge (1968).
[b]sn, stereospecific numbering.

TABLE IV
Major Triacylglycerols in Human Milk Lipids[a]

Trisaturates	
sn[b]-1-18:0–16:0–14:0	
sn-1-18:0–16:0–16:0	
sn-1-18:0–16:0–18.0	
Monoenes	
sn-1-18:1–16:0–12:0[c]	sn-1-12:0–16:0–18:1
sn-1-18:1–16:0–14:0	sn-1-14:0–16:0–18:1
sn-1-18:1–16:0–16:0	sn-1-16:0–16:0–18:1
sn-1-18:1–16:0–18:0	sn-1-18:0–16:0–18:1
Dienes	
sn-1-18:0–18:1–18:1	sn-1-18:1–16:0–18:1
Trienes	
sn-1-18:1–18:1–18:1	sn-1-18:1–16:0–18:2
Tetraenes	
sn-1-18:1–18:1–18:2	sn-1-16:0–18:2–18:2

[a]Adapted from Breckenridge *et al.* (1967, 1969); Kuksis and Breckenridge (1968).
[b]sn, stereospecific numbering.
[c]Fatty acids in these columns distributed at random.

monoenoic TGs in the left column are enantiomers of those to the right. More recent data from Dotson *et al.* (1992) are shown in Table V. These were obtained by HPLC separation of the TG and determination of the fatty acids in the fractions by GLC. Milks were analyzed at 4, 6, and 8 weeks postpartum. The samples contained more than 27 peaks. By combining

TABLE V
Major Triacylglycerols in Human Milk[a]

Triacylglycerol	Area (%)	TG	Area (%)
14:0–14:0–12:0[b]	2.62	16:0–12:0–18:1	4.84
18:2–18:2–18:1	3.27	12:0–16:0–16:0	14.69
18:1–14:0–14:0	4.70	12:0–18:0–18:1	6.21
16:0–14:0–14:0	4.47	18:1–18:1–18:1	3.17
18:2–18:2–16:0	8.84	16:0–16:0–18:1	11.96
18:2–16:0–12:0	6.07	18:0–18:0–18:1	7.38
12.0–18.1–18.1	1.99	18:0–18:0–18:1	3.44

[a]Adapted from Dotson *et al.* (1992).
[b]14:0–14:0–12:0 represents the isomeric TGs 14:0–14:0–12:0, 12:0–14:0–14:0, and 14:0–12:0–14:0.

these data with those in Table IV, we can surmise that in this study, the major enantiomeric TG in human milk were similar to those in Table IV. The value of these data is in the profiles of lipolysis products, FFA, and sn-1,2 DG produced in the stomach during the window of 0 to 45 min therein (Jensen *et al.*, 1992). The FFA have microbicidal effects and are needed to facilitate digestion by pancreatic lipase in the small intestine. The sn-1,2 DG are second messengers affecting events in the small intestine. These DG will be lipolyzed in the small intestine to 2-MG and FFA.

C. Phospholipids (PL) and Sphingolipids

1. Introduction

Total amounts and recent data on the quantities of total PL in human milk are presented in Table VI. A wide range of amounts of total PL (6 to 200 mg/dl) have been reported (Jensen, 1989a,b). The sources of variation are the decrease in PL content as lactation progresses and the considerable variation in earlier analyses done by oxidation of total lipid followed by colorimetry of organic P. Recently, PL have been separated by TLC and quantitated and these values are about 25 mg/dl or 0.6%/100 g total lipid in milks containing 4% total lipid. Sphingomyelin is not a glycerophospholipid, but is isolated with the PLs and is traditionally reported as a member of this class.

TABLE VI
Total Phospholipid Contents of Human Milks

Reference	Method	Total lipid (%)	mg/dl
Harzer *et al.* (1983); 36 weeks	TLC	0.55	37
Bitman *et al.* (1984); mature	Sep-pak, TLC	0.4–0.5 0.77 + 0.23[a]	15–20 17.1 + 10.9[a]
Yonekubo *et al.* (1987)	Oxidation, colorimetry	0.57	21.0 + 3.13[a]
van Beusekom *et al.* (1990); mature	Sum of fatty acids		
Dominica ($n = 6$)		0.97 (1.88–6.81)[b]	37.2 + 19.4[a] (6.2–37.2)[b]
Belize ($n = 6$)		0.39 (2.20–6.49)[b]	16.3 + 10.8[a] (8.5–58.8)[b]

[a]Mean + SD.
[b]Range.

Some of the variations in Table VI are probably due to the different methods employed. van Beusekom *et al.* (1990) suggested that the large difference they observed between the PL in milks from women in Belize and Dominica was due to the much larger intake of dietary carbohydrate of the Dominican subjects (70 en%) compared to those from Belize (55 en%). The high carbohydrate diet increases the amounts of 6:0–14:0 in the milk. Some of these are incorporated into the milk PL which are derived from membranes in the secreting cell and this change could presumably affect the amounts and accretion of PL onto the milk fat globule. Bitman *et al.* (1983) attributed the decrease in total PL to the increase in fat content and globule diameter that occurs postpartum. This means that a fixed amount of PL would result in thinner globule membranes. See Chapter 2A for information on the milk lipid globule membrane.

2. *Classes*

The kinds and quantities of phospholipids are shown in Table VII. The separations by Harzer *et al.* (1983) and Bitman *et al.* (1984) were done with TLC and those of Hundrieser and Clark (1988) and van Beusekom *et al.* (1990) with HPLC. The procedure of the latter authors did not resolve, as shown in Table VII, sphingomyelin (Sph), plasmalogen phosphatidylcholine (pl PC), and lysophosphosphatidylcholine (LPC). After hydrolysis with acetic acid to produce the 2-acyllyso analogs, there was a large increase in pl phosphatidylethanolamine (PE) and a moderate increase in pl PC. These results indicate the presence of substantial amounts, at least 13.3% of plasmalogens, which have not been previously reported. These data are from Dominican milks.

Hundrieser and Clark (1988) did not report the resolution of the plasmalogens. The amounts of Sph found by van Beusekom *et al.* (1990) were much greater than the others, but included pl PC and LPC. Hallgren *et al.* (1974) detected unsubstituted glyceryl ethers (0.18% of total PL) and 2-methoxy-substituted glyceryl ethers (trace) in mature (8–90 day) human milk.

3. *Complex lipids other than phospholipids*

Milk contains sphingomyelins, neutral glycosylceramides, and acidic glycosphingolipids or gangliosides. Sphingomyelin is often classified and reported as a phospholipid but it is also a sphingolipid. Like phosphatidylcholine (lecithin), it contains a choline moiety and a phosphodiester linkage, but in contrast to phosphatidylcholine which is a glycerolipid, the backbone (lipid moiety) of sphingomyelin is a ceramide. Neutral glycolipids are composed of ceramides bound to one or more glycosyl units. Monoglycosylceramides are termed cerebrosides. Gangliosides are glycosphingolipids that contain sialic acid (*N*-acetylneuraminic acid, NANA) as part of their carbohydrate moiety and are therefore acidic. These compounds are

TABLE VII
Phospholipids (% of Total Lipids) in Human Milk

Phospholipid	Reference			
	Harzer *et al.* (1983)	Bitman *et al.* (1984)	van Beusekom *et al.* (1990)	Hundrieser and Clark (1988)
Phosphatidylcholine (PC)	24.9	28.4	13.06 ± 5.48[a] (4.74–19.87)[b]	33.2 ± 5.5
Phosphatidylethanolamine (PE)	27.7	19.3	7.35 ± 3.31 (3.44–11.93)	23.8 ± 3.3
Phosphatidylserine (PS)	9.3	8.8	8.65 ± 4.21 (2.95–15.75)	3.7 ± 1.5
Phosphatidylinositol (PI)	5.4	6.1	13.06 ± 5.48 (4.74–19.87)	5.3 ± 3.0
Sphingomyelin (Sph)	32.4	37.5	54.75 ± 13.73[c] (40.76–75.01)	29.0 ± 6.4
Lysophosphatidylcholine (LPC)	—	—	Present	Trace
Lysophosphatidylethanolamine (LPE)	—	—	Present	5.0 ± 3.1
Plasmalogens (pl)	—	—	13.31 ± 8.06[d] (31.6–66.98)	—

[a]Mean ± SD, mol%.
[b]Range.
[c]Sph major component plus traces of pl PC and LPC.
[d]Mostly pl PE plus trace of pl PC.

concentrated in the lipid globule membrane which is derived from membranes in the secreting, mammary gland cell (see Chapter 9B). The kinds and amounts are listed in Table VIII.

Bouhours and Bouhours (1981) isolated sphingomyelin from the human milk fat globule membrane, characterized it chemically, and determined its concentrations in whole milk and its distribution in the cream and skim components of milk, using a single undefined milk sample (Table

TABLE VIII
Sphingolipids in Human Milk

Constituent	Concentration in milk	Reference
Sphingomyelin	110 μmol/liter (μ*M*)[a]	Bouhours and Bouhours (1981)
	100–200 μ*M*	Zeisel *et al.* (1986)
Gangliosides	15–20 mg/liter[b]	Grimmonprez and Montreuil (1977)
	178 μg/liter[c]	Bouhours and Bouhours (1979)
GM1	12 μg/liter	Laegreid *et al.* (1986)
GM2	250 μg/liter	Laegreid *et al.* (1986)
GM3	8.1 mg/liter	Laegreid *et al.* (1986)
Days 1–6	1.6 ± 0.1 μ*M*[c]	Takamizawa *et al.* (1986)
Days 8–40	3.0 ± 0.5 μ*M*[c]	Takamizawa *et al.* (1986)
Days 120–390	8.6 ± 0.9 μ*M*[c]	Takamizawa *et al.* (1986)
GD3	2.7 mg/liter	Laegreid *et al.* (1986)
Days 1–6	6.7 ± 0.5 μ*M*[c]	Takamizawa *et al.* (1986)
Days 8–40	3.7 ± 0.3 μ*M*[c]	Takamizawa *et al.* (1986)
Days 120–390	0.8 ± 0.2 μ*M*[c]	Takamizawa *et al.* (1986)
Neutral glycolipids	20–25 mg/liter[b]	Grimmonprez and Montreuil (1977)
glucosylceramide	32 μg/liter[c]	Bouhours and Bouhours (1979)
	383 n*M*[c]	Newburg and Chaturvedi (1992)
galactosylceramide	235 μg/liter[c]	Bouhours and Bouhours (1979)
	3183 n*M*[c]	Newburg and Chaturvedi (1992)
lactosylceramide	133 μg/liter[c]	Bouhours and Bouhours (1979)
	1032 n*M*[c]	Newburg and Chaturvedi (1992)
GB_3 (globotriaosylceramide)	123 n*M*[c]	Newburg *et al.* (1992)
GB_4 (globoside)	91 n*M*[c]	Newburg *et al.* (1992)

[a]Single analysis.
[b]Crude fraction.
[c]Value derived from data presented in paper.

VIII). Their values agree well with those of Zeisel *et al.* (1986) who determined the sphingomyelin content in the milk of several donors. Among these donors, the sphingomyelin content was elevated by 20% in hindmilk compared with fore- and midmilk, was stable through 90 days of lactation, and did not exhibit consistent diurnal variation.

The identity of the acyl moieties attached to sphingosine by a peptide linkage was reported by Bouhours and Bouhours (1981). The major fatty acid was 18:1 (61.8%) and most of the remaining fatty acids were 20:0, 22:0, 24:0, and 24:1. The monohexosyl ceramides isolated by Bouhours and Bouhours (1979) contained long-chain hydroxylated and nonhydroxylated fatty acids, mostly 24:0.

Grimmonprez and Montreuil (1977), using classical partition techniques, demonstrated the presence of gangliosides in human milk, but did not isolate and quantitate individual species; thus, this measurement of the concentration of the ganglioside fraction (Table VIII) surely included much nonganglioside material. The amount of gangliosides calculated from the report by Bouhours and Bouhours (1979) is likewise derived from the weight of a crude fraction of gangliosides (which includes large neutral glycolipids), but from isolated milk lipid globule membrane; this value is thus expected to be low due to losses in the isolation of the membrane and because of the presence of glycolipids in other membranous compartments in human milk. Laegreid *et al.* (1986) extracted gangliosides from the creams of 10 mothers who were in the second to tenth month of lactation. The gangliosides were separated and quantitated by HPTLC densitometry (and HPTLC immunodensitometry for GMI). These values agree well with those of mature milk measured by Takamizawa *et al.* (1986) by TLC densitometry (Table VIII). The Tokyo laboratory measured gangliosides from Days 2 to 390 of lactation and found a reciprocal relationship between milk GD3, the predominant ganglioside early in lactation, and milk GM3, which predominates late in lactation. The ganglioside GM1 is known to bind to cholera toxin, the labile toxin of *Escherichia coli,* and a similar toxin from *Campylobacter jejuni.* Its presence in human milk may have a significant role in protection of the infant, although this has not been proven *in vivo.* (Laegrid *et al.,* 1986; Kolsto-Otnaess, 1989). Keenan and Patton (Chapter 2A) postulate that gangliosides promote fusion of microdroplets of lipid in the globules.

The presence of the neutral glycolipids from the lower (organic) phase of a Folch distribution (8:4:3, chloroform:methanol:water) of human milk was reported by Grimmonprez and Montreuil (1977). Again, the total fraction was determined gravimetrically, and the value undoubtedly includes materials that are nonglycosphingolipids and/or species which have not yet been recognized. The neutral glycolipid levels of Bouhours and Bouhours (1979) were determined by classical gravimetric, TLC, and GLC techniques on extracts of human milk lipid globule membrane. These values are not corrected for the neutral glycolipids which are found in other membranous components of human milk. Recently, whole milk samples obtained from women representing various stages of lactation, parity, etc. were extracted directly into a Folch distribution, and the lower-phase neutral glycolipids were perbenzoylated, separated by gradient HPLC, and quantitated by uv absorbance (Newburg and Chaturvedi, 1992). In addition to the expected cerebrosides and lactosylceramide (Table VIII), low levels of Gb_3 and Gb_4 were found consistently in human milk. Newburg *et al.* (1992) demonstrated that these globo-series glycolipids from human milk can bind to Shiga toxin at levels that could be relevant to the protection of infants from diarrhea by human milk. Gb3 also binds to Shiga-like toxin produced by some enterohemorrhagic *E. coli;* Gb4 binds to a variant of the Shiga-like toxin. The clinical relevance of such binding toward protection of pediatric patients has not yet been demonstrated. These compounds contained hydroxylated and nonhydroxylated fatty acids.

D. Sterols

Earlier research has established that the sterol content of human milk ranges from 10 to 20 mg/dl with cholesterol as the major component (Jensen, 1989a,b; Jensen *et al.,* 1992). Most of the cholesterol is located in the milk lipid globule membrane and the amount is not affected by diet or by maternal plasma levels. Cholesterol has been determined by colorimetry (*O*-phthalaldehyde), GLC, and HPLC. Kallio *et al.* (1989) analyzed milk for cholesterol, its precursors, and other sterols by GLC and mass spectrometry. They obtained the data in Table IX and some of the data in Table X.

TABLE IX
Amounts (μg/dl) of Cholesterol and Its Precursors and Triacylglycerol in Human Milk at 2, 6, and 9 Months of Lactation[a]

Compound	Months of lactation		
	2 (n = 88)	6 (n = 28)	9 (n = 6)
Squalene	386	493	452
Lanosterol	94	98	115
Dimethylsterol	45	62	62
Methostenol	48	72	88
Lathosterol	43	89	112
Desmosterol	1,509	1,351	1,140
Cholesterol	15,800	18,000	18,900
Triacylglycerol (g/dl)	2.96	3.89	4.25

[a]Adapted from Kallio *et al.* (1989).

TABLE X
Other Sterols Detected in Human Milk[a]

Sterol	μg/dl	Reference
7-dehydrocholesterol	tr	Bracco *et al.* (1972)
Stigma- and campesterol	100	Haug and Harzer (1984)
7-ketocholesterol	10	Haug and Harzer (1984)
Sitosterol	200	Haug and Harzer (1984)
Cholestanol	TR	Kallio *et al.* (1989)
β-Lathosterol	TR	Kallio *et al.* (1989)
Vitamin D metabolites	See Chapter 8D	
Steroid hormones	See Chapter 5D	

[a]Sterols other than those listed in Table IX.

The authors concluded that in the mammary gland, cholesterol is synthesized from lanosterol via preservation of the side-chain double bond. However, the amount of cholesterol synthesized in the gland is unknown.

Lammi-Keefe *et al.* (1990) detected a diurnal pattern in the cholesterol content of milk ranging from 8.75 mg/dl at 0600 hr to 11.2 mg/dl at 2200 hr confirming earlier results that the pattern exists. Clark and Hundrieser (1989) found that the mean total cholesterol content of 25 milk samples was 13.5 mg/dl and was significantly correlated with the lipid content. The cholesteryl ester content was about 20% of this value. The composition of the component fatty acids is provided in this paper. Boersma *et al.* (1991) found cholesterol (mg/dl) in milks from women in St. Lucia as follows: 0 to 4 days postpartum, 36.0; 5 to 9 days, 19.7; and 10 to 30 days, 19.0.

E. Fatty Acids and Related Compounds

1. Introduction

Some data are listed in Table XI (Tomarelli, 1988; see also Jensen, 1989a,b; Jensen *et al.*, 1992) and were obtained by GLC instruments equipped with packed columns. These columns, while reliable for the major fatty acids, are incapable of the resolution attainable with wide-bore capillary columns of suitable length (at least 30 m, preferably longer), coated with polar stationary phases and utilizing temperature programming. The recently developed stationary phases, e.g., SP-2340 and SP-2560, will separate *trans* isomers and PUFA and should be used. Details of operation are in the papers quoted below. In addition, we urge investigators to determine the lipid content of milk so that the actual amounts of fatty acids conveyed to the infant can be calculated and the effects of any intervention be seen. Data should be presented as weight % (g/100 g fatty

TABLE XI
Fatty Acids of Human Milk Lipids[a]

Saturates			Monounsaturates			Polyunsaturates		
Fatty acid	wt%	mg/dl[b]	Fatty acid	wt%	mg/dl	Fatty acid	wt%	mg/dl
							n6 Series	
4:0	0.19	7	*c*-14:1n5[c]	0.41	15	18:2*cc*	10.85	391
6:0	0.15	5	*t*-14:n5[c]	0.07	3	18:2*tt*	0.46	17
8:0	0.46	17	15:1	0.11	4	18:2*ct*	0.69	25
10.0	1.03	37	*c*-16:1n7	3.29	118			
			t-16:1n7	0.36	13	18:3	0.25	9
12:0	4.40	158	17:1	0.37	13			
13:0	0.06	2				20:2	0.27	10
i-14:0[d]	0.04	1	*c*-18:1n9	31.30	1127	20:3	0.32	12
14:0	6.27	226	*t*-18:1n9	2.67	96	20.4	0.46	17
a-15:0[d]	0.21	8				22:2	0.11	4
15:0	0.43	15	20:1n9	0.67	24	22:4	0.09	3
i-16:0	0.17	6				22:5	0.09	3
16:0	22.00	792	22:1n9	0.08	3	Total n6	13.59	489
a-17:0	0.23	8	24:1n9	0.12	4		n3 Series	
17:0	0.58	21	Total monounsaturates	39.45	1420	18:3	1.03	37
i-18:0	0.11					20:4	0.09	3
18:0	8.06	290				20:4	0.09	3
						20.05	0.12	4

TABLE XI—*continued*

Saturates			Monounsaturates			Polyunsaturates		
Fatty acid	wt%	mg/dl[b]	Fatty acid	wt%	mg/dl	Fatty acid	wt%	mg/dl
20:0	0.44	16						
21:0	0.13	5				22:5	0.19	7
22:0	0.12	4				22:6	0.25	9
24:0	0.25	9				Total n3	1.68	60
						Total PUFA	15.27	550
Total saturates	45.33	1632				Total n6/n3	8.10	

[a]Modified from Tomarelli (1988). Data are normalized compilations from 15 papers.
[b]Calculated by the authors based on 3.6% fat in milk.
[c]*c* is *cis*; *t*, *trans*.
[d]*i* is *iso*; *a*, *anteiso*.

acid) and gravimetric reports (wt of fatty acid/dl of milk). We calculated the data on mg/dl in Table XI. The fat content can be determined gravimetrically by weighing a tared flask containing the sample after removal of the extraction solvents or by GLC with an internal standard (Clark and Hundrieser, 1990).

Fatty acids were converted to methyl esters prior to analysis by GLC to increase volatility and efficiency of separation. Methanolysis of the extracted lipid is usually done by reactions with sodium hydroxide–methanol or boron trifluoride–methanol. Lepage and Roy (1984) observed slightly better recovery of human milk fatty acids with a direct transesterification procedure they developed. If cost of extraction solvents is a limiting factor than direct transesterification may be the best method since none are used in this analysis. However, hexane should be substituted for the benzene used in the original method (Sukhija and Palmquist, 1988). Benzene is too toxic for general laboratory use. Many investigators add antioxidants to the solvents to prevent possible loss of PUFA by oxidation (van der Steege *et al.,* 1987).

When analyzing human milk lipids for fatty acids, we recommend that investigators (a) use columns which will resolve *trans* isomers and PUFA; (b) determine and report the total lipid content (it is desirable to present the data as wt% and g or mg of fatty acid/dl); and (c) if possible, obtain information on maternal diets. Analysts may want to investigate the bracketing procedure used by van der Steege and co-workers (1987). Internal standards were employed to quantitate the fatty acids; 5:0–15:0 for milk 6:0–14:0 and 14:1 and 17:0 for the long-chain acids. While the method improved the quantitation of 6:0–14:0, milk odd-chain fatty acids cannot be determined. Data on within-series and series-to-series precision and biological variation are given in this paper.

2. *Factors affecting composition*

Most of the factors affecting fat content in Table I do not alter the fatty acid composition of milk, but those which change the fat content will influence the amounts of acids. Exceptions are time postpartum (Tables XIIa–XIVb), gestational age (Table XV), parity, diseases, individuality, and diet (Jensen 1989a,b; Jensen *et al.* 1992). Diet has the greatest effect. Influences attributed to region are undoubtedly due to diet. We present data from selected papers obtained with GLC capillary columns, when possible, on the effects of some of the factors listed above. The data on time postpartum are in Tables XIIa–XIVb. In Tables XIIIa and XIVa Harzer *et al.* (1983) observed *trans* isomers of 14:1, 16:1, 17:1, and 18:2, but not 18:1, which should have been present in amounts of 3 or 4%. These isomers are derived mostly from partially hydrogenated food fats and the primary isomer is elaidic acid (18:1*t*n9). In general, the fatty acids, 6:0–12:0, which were synthesized in the mammary gland increased palmitic acid (16:0) decreased, while the changes in 18:0 were variable. These

TABLE XIIa
Effect of Time Postpartum on Saturated Fatty Acids (wt%) in Human Milk Lipids

	Day of lactation								
	Harzer *et al.* (1983)[a]			Jackson *et al.* (1993)[b]					
	3	8	36	2	3	7	14	42	84
n	17	17	17	6	11	11	11	11	10
Lipid (%)	2.8	3.5	3.6	1.17 ± 0.43	2.31 ± 0.30	3.48 ± 0.30	3.81 ± 0.30	3.95 ± 0.30	3.93 ± 0.32
Fatty acid									
10:0	0.26	0.61	0.86	0.06 ± 0.09	0.24 ± 0.06	0.62 ± 0.06	0.84 ± 0.06	0.59 ± 0.06	0.67 ± 0.67
11:0	TR	TR	TR	—	—	—	—	—	—
12:0	2.25	3.87	5.47	1.20 ± 0.47	1.94 ± 0.33	3.65 ± 0.33	4.41 ± 0.33	2.80 ± 0.33	3.18 ± 0.35
13:0	TR	TR	0.03	—	—	—	—	—	—
14:0	5.84	5.91	7.20	4.10 ± 0.48	4.72 ± 0.34	5.41 ± 0.34	5.51 ± 0.34	4.49 ± 0.34	4.73 ± 0.36
15:0	—	—	—	0.46 ± 0.06	0.55 ± 0.04	0.61 ± 0.04	0.61 ± 0.04	0.74 ± 0.04	0.58 ± 0.04
16:0	26.01	24.79	23.10	27.7 ± 0.75	22.38 ± 0.53	20.06 ± 0.53	20.38 ± 0.53	21.29 ± 0.53	20.53 ± 0.56
17:0	0.79	0.68	—	—	—	—	—	—	—
18:0	8.30	8.55	8.37	8.01 ± 0.36	7.46 ± 0.25	7.33 ± 0.25	7.67 ± 0.25	8.10 ± 0.25	7.23 ± 0.27
22:0	TR	TR	TR	—	—	—	—	—	—
24:0	0.29	0.15	0.20	—	—	—	—	—	—
Total	43.34	44.56	45.23	36.54	38.24	37.73	39.42	37.99	36.91

[a]Adapted from Harzer *et al.* (1983). German mothers. Packed GLC column.
[b]Adapted from Jackson *et al.* (1993). Least-square means ± SEM. Data from Connecticut reference women. Capillary GLC column.

TABLE XIIb
Effects of Time Postpartum on the Saturated Fatty Acids (wt%, Means ± SD) in Human Milk Lipids

	Age Postpartum (days)				
	Gibson and Kneebone (1981)[a]		Boersma *et al.* (1991)[b]		
	3–5	40–45	0–4	5–9	10–30
n	59	61	13	11	12
Lipid (%)	—	—	1.0 ± 0.8	2.7 ± 1.2[c]	4.3 ± 1.5[d,e]
Fatty acid					
6:0	0.08 ± 0.03		0.03 ± 0.07	0.04 ± 0.02[c]	0.07 ± 0.02[e]
8:0	0.04 ± 0.03	0.13 ± 0.06	0.06 ± 0.04	0.22 ± 0.7[f]	0.37 ± 0.09[f,g]
10:0	0.40 ± 0.23	1.11 ± 0.31	0.73 ± 0.44	1.65 ± 0.47[f]	2.39 ± 0.39[f,g]
12:0	2.41 ± 1.01	4.07 ± 1.40	3.76 ± 2.20	8.66 ± 2.97[e]	12.32 ± 3.16[f,g]
13:0	0.03 ± 0.04	0.09 ± 0.02	—	—	—
14:0	0.02 ± 0.01	0.04 ± 0.02	—	—	—
14:0	5.09 ± 1.10	5.63 ± 1.45	6.52 ± 1.90	11.11 ± 3.62[e]	11.78 ± 4.07[d,f]
15:0	0.14 ± 0.04	0.21 ± 0.03	—	—	—
15:0	0.44 ± 0.08	0.50 ± 0.14	—	—	—
16:0	0.14 ± 0.03	0.10 ± 0.04	—	—	—
16:0	24.47 ± 1.70	22.44 ± 1.82	28.43 ± 1.68	26.12 ± 1.54[f]	23.61 ± 2.65[d,e]
17:0	0.44 ± 0.07	0.45 ± 0.10	—	—	—
17:0	0.64 ± 0.08	0.66 ± 0.12	—	—	—
18:0	8.24 ± 1.27	9.20 ± 1.43	8.66 ± 0.93	6.60 ± 0.77[f]	5.83 ± 0.07[f,h]
20:0	0.71 ± 0.12	0.75 ± 0.16	0.54 ± 0.18	0.28 ± 0.06[e]	0.24 ± 0.06[f]
22:0	0.24 ± 0.15	0.06 ± 0.24	0.43 ± 0.25	0.14 ± 0.03[e]	0.12 ± 0.04[e,g]
24:0	0.31 ± 0.17	0.08 ± 0.03	0.43 ± 0.27	0.14 ± 0.03[e]	0.06 ± 0.02[g,i]
Total	43.84	45.56	49.6 ± 5.2	55.2 ± 6.0	56.8 ± 7.7[i]

[a]Adapted from Gibson and Kneebone (1981). Australian donors. Prior separation of fatty acid classes by $AgNO_3$ thin-layer chromatography. GLC analyses of classes with a packed column.

[b]Adapted from Boersma *et al.* (1991). St. Lucian (Windward Islands, Caribbean) donors. Capillary GLC column.

[c,e,f,i]Significantly different from 0 to 4 days; [c]$P = 0.05$; [e]$P = 0.005$; [f]$P = 0.001$; [i]$P = 0.01$.

[d,g,h,j]Significantly different from 5 to 9 days; [d]$P = 0.01$; [g]$P = 0.001$; [h]$P = 0.05$; [i]$P = 0.005$.

TABLE XIIIa
Effect of Time Postpartum on Monounsaturated Fatty Acids (wt%) in Human Milk Lipids

Fatty acid	Day of lactation								
	Harzer *et al.* (1983)[a]			Jackson *et al.* (1993)[b]					
	3	8	36	2	3	7	14	42	84
14:1*t*n5[c]	0.11	0.11	0.11						
14:1n5	0.53	0.57	0.49						
16:1*t*n7	0.57	0.45	0.48						
16:1n7	3.55	3.92	3.90	2.98 ± 0.21	3.21 ± 0.15	3.51 ± 0.15	3.09 ± 0.15	3.58 ± 0.15	3.27 ± 0.16
17:1*t*n7	TR	TR	TR	—	—	—	—	—	—
17:1n7	0.19	0.22	0.16	—	—	—	—	—	—
18:1*t*n9	—	—	—	—	—	—	—	—	—
18:1n9	37.76	36.36	35.02	35.57 ± 1.32	38.43 ± 0.93	36.32 ± 0.93	36.35 ± 0.03	30.11 ± 0.64	37.43 ± 0.98
20.1n9	1.37	0.88	0.85	2.04 ± 0.12	1.96 ± 0.08	1.97 ± 0.08	1.86 ± 0.08	1.86 ± 0.08	1.77 ± 0.09
22:1n9	0.20	0.19	TR	—	—	—	—	—	—
24.1n9	0.36	0.18	0.22	0.63 ± 0.07	0.40 ± 0.05	0.23 ± 0.05	0.14 ± 0.05	0.10 ± 0.05	0.09 ± 0.05
Total	44.64	42.88	41.23	43.32 ± 1.35	43.99 ± 0.96	42.03 ± 0.96	41.44 ± 0.96	43.29 ± 0.96	42.56 ± 1.01

[a]Adapted from Harzer *et al.* (1983). Packed GLC column. See Table XIIa for lipid content and *n*.
[b]Adapted from Jackson *et al.* (1993). Least-square means ± SEM. Data from reference women.
[c]*t* is *trans*.

TABLE XIIIb
Effects of Time Postpartum on the Monounsaturated Fatty Acids (wt%, Means ± SD) in Human Milk Lipids

	Time Postpartum (days)				
	Gibson and Kneebone (1981)[a]		Boersma *et al.* (1991)[b]		
	3–7	40–45	0–4	5–9	10–30
n	89	61	13	11	12
Lipid (%)	—	—	1.0 ± 0.8	2.7 ± 1.2[c]	4.3 ± 1.5[d,e]
Fatty acid					
14:1n5	0.30 ± 0.08	0.43 ± 0.13	0.11 ± 0.03	0.22 ± 0.05[i]	0.29 ± 0.09[b,h]
16:1n7	3.92 ± 0.61	3.79 ± 0.65	2.08 ± 0.45	2.97 ± 0.82[c]	3.55 ± 0.89[e]
17:1	0.51 ± 0.06	—	—	—	—
18:1n7	—	—	4.77 ± 0.53	4.42 ± 0.60[c]	3.64 ± 0.80[d,e]
18:1n9	37.18 ± 2.47	35.00 ± 2.31	26.78 ± 3.36	23.66 ± 3.59[c]	22.63 ± 4.03[c,h]
20:1n9	1.17 ± 0.22	0.60 ± 0.08	1.15 ± 0.17	0.64 ± 0.08[f]	0.42 ± 0.11[f,g]
Unknown	0.24 ± 0.04	0.26 ± 0.03	—	—	—
22:1	0.08 ± 0.03	0.06 ± 0.04	—	—	—
24:1n9	0.24 ± 0.11	0.03 ± 0.01	0.41 ± 0.29	0.11 ± 0.02[c]	0.04 ± 0.02[c,g]
Total	43.64	40.68	35.3 ± 3.5	33.3 ± 1.7[c]	30.6 ± 4.7[h,i]

[a–i]See Table XIIb footnotes.

TABLE XIVa
Effect of Time Postpartum on the Polyunsaturated Fatty Acids (wt%) in Human Milk Lipids

Fatty acid	Day of lactation								
	Harzer *et al.* (1983)[a]			Jackson *et al.* (1993)[b]					
	3	8	36	2	3	7	14	42	84
n6 Series									
18:2*tt*[c]	TR	0.26	—						
18:2	10.30	10.76	11.78	14.57 ± 1.02	14.48 ± 0.72	15.22 ± 0.72	16.30 ± 0.73	16.38 ± 0.72	18.37 ± 0.76
18:3	0.70	TR	0.85	—	—	—	—	—	—
20:2	0.65	0.44	0.35	1.09 ± 0.06	0.85 ± 0.04	0.57 ± 0.04	0.42 ± 0.04	0.31 ± 0.04	0.31 ± 0.05
20:3	0.35	0.34	0.30	0.75 ± 0.06	0.57 ± 0.05	0.48 ± 0.05	0.45 ± 0.05	0.38 ± 0.05	0.35 ± 0.05
20:4	0.55	0.50	0.39	1.15 ± 0.07	0.87 ± 0.05	0.79 ± 0.05	0.71 ± 0.05	0.56 ± 0.05	0.56 ± 0.05
22.2	0.17	0.16	0.20	—	—	—	—	—	—
22:4	0.15	0.09	0.05	0.69 ± 0.06	0.49 ± 0.04	0.30 ± 0.04	0.19 ± 0.04	0.15 ± 0.04	0.17 ± 0.04
22:5	0.10	0.06	TR	—	—	—	—	—	—
Total	12.97	12.61	13.97	19.34 ± 0.25	17.26 ± 0.74	17.35 ± 0.74	18.08 ± 0.74	17.78 ± 0.74	19.77 ± 0.78

TABLE XIVa—*continued*

Fatty acid	Day of lactation								
	Harzer *et al.* (1983)[a]			Jackson *et al.* (1993)[b]					
	3	8	36	2	3	7	14	42	84
n3 Series									
18:3	0.70	0.69	0.71	0.31 ± 0.07	0.26 ± 0.05	0.23 ± 0.05	0.28 ± 0.05	0.31 ± 0.05	0.25 ± 0.05
20:5	0.43	0.22	0.05	0.16 ± 0.02	0.15 ± 0.01	0.11 ± 0.01	0.08 ± 0.01	0.06 ± 0.01	0.06 ± 0.01
22:5	0.13	0.09	0.05	0.29 ± 0.03	0.25 ± 0.02	0.14 ± 0.02	0.12 ± 0.02	0.12 ± 0.02	0.10 ± 0.02
22:6	0.21	0.26	0.16	0.31 ± 0.04	0.30 ± 0.03	0.27 ± 0.04	0.21 ± 0.03	0.16 ± 0.03	0.10 ± 0.03
Total	1.28	1.26	0.97	1.08 ± 0.11	0.95 ± 0.08	0.75 ± 0.08	0.69 ± 0.08	0.64 ± 0.08	0.54 ± 0.08

[a]Adapted from Harzer *et al.* (1983). Packed GLC column. See Table XIIa for lipid content and *n*.
[b]Adapted from Jackson *et al.* (1993). Least-square means ± SEM. Data from reference women. Capillary GLC column.
[c]*tt* is *trans*, *trans*.

TABLE XIVb
Effects of Time Postpartum on the Polyunsaturated Fatty Acids (wt%, Means ± SD) in Human Milk Lipids

	Time Postpartum (days)				
	Gibson and Kneebone (1981)[a]		Boersma *et al.* (1991)[b]		
	3–7	40–45	0–4	5–9	10–30
n	59	61	13	11	12
Lipid (%)	—	—	1.0 ± 0.08	2.7 ± 1.2	4.3 ± 1.5[d,e]
Fatty acid					
n9 Series					
20:3	—	—	0.04 ± 0.04	0.06 ± 0.01	0.05 ± 0.01
n6 Series					
18:2	7.82 ± 2.01	10.75 ± 4.22	8.84 ± 2.10	8.61 ± 1.81	9.57 ± 3.02
18:3	0.34 ± 0.05	0.35 ± 0.05	0.00 ± 0.01	0.06 ± 0.05[e]	0.09 ± 0.04[f]
20:2	0.65 ± 0.24	0.24 ± 0.11	0.80 ± 0.23	0.52 ± 0.09[c]	0.31 ± 0.06[f,g]
20:3	0.49 ± 0.18	0.31 ± 0.09	0.81 ± 0.42	0.54 ± 0.11	0.42 ± 0.08[c,g]
20:4	0.71 ± 0.18	0.40 ± 0.10	1.60 ± 0.75	0.84 ± 0.14[c]	0.58 ± 0.14[i,j]
22:4	0.40 ± 0.24	0.10 ± 0.06	0.77 ± 0.35	0.31 ± 0.10[i]	0.15 ± 0.04[f,j]
22:5	—	—	0.22 ± 0.11	0.13 ± 0.04[c]	0.07 ± 0.08[g,i]
Total	10.41	12.15	13.0 ± 2.9	11.0 ± 2.8	11.2 ± 3.2
n3 Series					
18:3	0.41 ± 0.09	0.59 ± 0.16	0.45 ± 0.11	0.58 ± 0.07[f]	0.52 ± 0.20[c]
20:5	0.43 ± 0.20	0.16 ± 0.07	0.03 ± 0.05	0.07 ± 0.03	0.07 ± 0.03
22:5	0.35 ± 0.17	0.21 ± 0.05	0.44 ± 0.22	0.24 ± 0.08[c]	0.16 ± 0.05[e,h]
22:6	0.64 ± 0.27	0.32 ± 0.17	1.10 ± 0.53	0.88 ± 0.31	0.56 ± 0.76[d]
Total	1.83	1.28	2.0 ± 0.8	1.8 ± 0.4	1.4 ± 0.4[g]
Grand total	12.14	13.43	15.1 ± 3.3	12.8 ± 1.7	12.7 ± 3.3

[a–j] See Table XIIb.

changes are relative, because if one acid actually increases, then the rest must decrease. Oleic acid (18:1) decreased. The changes observed in the PUFA may have been controlled by maternal diet instead of maturation of the mammary gland. The influence of gestational age is presented in Table XV, in which the changes observed as age increased were somewhat similar to those seen for time postpartum. It is doubtful if these occur in anticipation of the infant's perceived needs, since premature birth is an abnormal event.

Fatty acid profiles of milks from women on Western diets are depicted in Tables XVI–XVIII, those consuming non-Western diets in Tables XIX–XXIV, sea mammal flesh or fish oil in Tables XXV and XXVI, and

TABLE XV
Effects of Gestational Age on the Fatty Acids (wt%, Means ± SD) in Human Milk Lipids at Day 42 Postpartum[a]

	Gestational age (weeks)		
	26–30	31–36	37–40
n	18	28	6
Lipid (%)	2.4	2.6	3.8
Fatty acid			
Saturates			
10:0	1.37 ± 0.17	1.27 ± 0.18	0.97 ± 0.28
12:0	7.47 ± 0.72	6.55 ± 0.77	4.46 ± 1.17
14:0	8.41 ± 0.83	7.55 ± 0.89	5.68 ± 1.36
15:0	0.23 ± 0.04	0.27 ± 0.05	0.31 ± 0.07
16:0	20.13 ± 1.40	23.16 ± 1.49	22.20 ± 2.28
17:0	0.34 ± 0.22	0.60 ± 0.24	0.49 ± 0.36
18:0	7.24 ± 1.13	7.25 ± 1.21	7.68 ± 1.85
20:0	0.17 ± 0.07	0.09 ± 0.08	0.32 ± 0.11
21:0	0.05 ± 0.07	0.07 ± 0.08	0.17 ± 0.12
Total	45.41	46.81	42.28
Monounsaturates			
16:1	2.56 ± 0.24	2.92 ± 0.26	3.83 ± 0.39
18:1	33.41 ± 1.67	33.74 ± 1.79	35.51 ± 2.73
Total	35.97	36.66	39.34
Polyunsaturates			
18:2	15.75 ± 1.22	13.83 ± 1.30	15.58 ± 1.99
18:3	0.76 ± 0.13	0.76 ± 0.14	1.03 ± 0.21
20:2	0.35 ± 0.13	0.33 ± 0.13	0.18 ± 0.20
20:3	0.51 ± 0.09	0.43 ± 0.10	0.53 ± 0.15
20:4	0.55 ± 0.18	0.58 ± 0.19	0.60 ± 0.29
20:5	0.04 ± 0.05	0.00	0.00
22:4	0.13 ± 0.10	0.24 ± 0.11	0.07 ± 0.16
22:5n6	0.11 ± 0.05	0.04 ± 0.05	0.03 ± 0.08
22:5n3	0.42 ± 0.09	0.12 ± 0.10	0.11 ± 0.15
22:6	0.24 ± 0.09	0.21 ± 0.09	0.23 ± 0.14
Total	18.86	16.54	18.36

[a]Adapted from Bitman *et al.* (1983). Maryland donors. Packed GLC column.

TABLE XVI
The Saturated Fatty Acids (wt%) in Milk Lipids from Mothers Consuming Western Diets

	Reference				
	Koletzko *et al.* (1988)[a]	Spear *et al.* (1992)[b]	Dotson *et al.* (1992)[c]		Putnam *et al.* (1982)[d]
n	15	1	30		9
Lipid (%)	—	4.12	—		
Fatty acid				CV	
6:0	—	0.10	—	—	—
8:0		0.32	0.02	0.04	0.03
10:0	0.71	1.93	0.06	0.05	1.4
12:0	4.41	7.54	4.34	0.11	6.2
13:0	0.05	—	0.05	0.14	—
14:0	6.73	1.98	4.65	0.07	7.6
15:0	0.46	0.31	0.34	0.09	—
16:0	21.83	21.66	19.25	0.03	20.5 ± 0.70
17:0	0.57	0.31	0.43	0.07	—
18:0	8.15	7.41	7.97	0.06	9.0 ± 0.46
19:0	0.04	—	0.10	0.07	—
20:0	0.22	0.16	0.21	0.10	0.3 ± 0.02
21:0	0.35	—	—	—	—
22:0	0.09	—	0.12	0.06	—
23:0	—	—	0.03	0.02	—
24:0	—	—	0.22	0.14	0.5 ± 0.01
26:0	—	—	0.03	0.31	—
Total	43.61	46.72	37.82	—	38.6 ± 0.72

[a]Adapted from Koletzko *et al.* (1988). German donors, 3 or 4 weeks postpartum. Capillary GLC column.
[b]Adapted from Spear *et al.* (1992). Maryland donor, 8 weeks postpartum. Capillary GLC column.
[c]Adapted from Dotson *et al.* (1992). Illinois donors, 4 to 8 weeks postpartum. Capillary GLC column. Means; CV, coefficient of variation.
[d]Adapted from Putnam *et al.* (1982). Floridian donors, 8 weeks postpartum. Packed and capillary GLC columns. Means ± SEM.

TABLE XVII
The Monounsaturated Fatty Acids (wt%) in Milk Lipids from Mothers Consuming Western Diets

	Reference				
	Koletzko *et al.* (1988)[a]	Spear *et al.* (1992)[b]	Dotson *et al.* (1992)[c]		Putnam *et al.* (1982)[d]
n	15	1	30	9	
Lipid (%)	—	1.12	—	—	
Fatty acid					
				CV	
13:1	—	—	0.03	0.10	—
c-14:1n5	0.29	0.22	0.31	0.20	—
t-14:n5	0.19	—	—	—	—
15:1	—	—	0.09	0.09	—
c-16:1n7	2.68	1.77	2.58	0.07	—
t-16:1n7	0.46	—	—	—	—
17:1	0.32	0.21	0.33	0.07	—
c-18:1n7	—	—	—	—	—
c-18:1n9	34.31	30.97	33.23	0.04	37.6 (0.75)
t-18:1n9	3.12	—	4.72	0.07	—
19:1n9	—	—	0.11	0.08	—
20:1n11	—	—	0.17	0.20	—
20:1n9	—	0.56	0.38	0.11	0.9 (0.7)
21:1n9	—	—	0.01	0.29	—
22:1n9	0.08	—	0.07	0.05	0.0 (0.02)
24:1n9	—	—	0.03	0.07	—
Total	42.38	—	42.06	—	38.5

[a–d]See Table XVI.

TABLE XVIII
The Polyunsaturated Fatty Acid (wt%) in Milk Lipids from Mothers Consuming Western Diets

	Reference			
	Koletzko *et al.* (1988)[a]	Spear *et al.* (1992)[b]	Dotson *et al.* (1992)[c]	Putnam *et al.* (1982)[d]
n	15		30	9
Lipid (%)				
Fatty acid				
n6 Series				
18:2cc	10.76	16.80	15.55	15.8
18:2tt	0.14	—	—	—
18:2ct	0.14	—	—	—
18:2tc	0.07	—	—	—
18:3	0.16	0.14	0.18	—
20:2	0.34	0.27	0.38	0.4
20:3	0.26	0.30	0.46	0.4
20:4	0.36	0.51	0.53	0.6
22:2	0.11	—	0.05	—
22:4	0.08	0.09	0.06	0.2
22:5	—	—	—	0.1
Total	12.26	18.11	17.21	17.4
n3 Series				
18:3	0.81	1.05	1.11	0.8
20:3	0.06	—	0.03	—
20:4	—	0.06	—	—
20:5	0.04	0.06	0.07	0.1
22:3	—	—	0.13	—
22:5	0.17	0.09	—	0.1
22:6	0.22	0.10	0.16	0.1
Total	1.38	1.44	1.5	1.1
Total PUFA	13.64	19.55	19.22	17.4

[a–d]See Table XVI.

TABLE XIX
The Fatty Acids (wt%) in Human Milk Lipids from Mothers Consuming Non-Western Diets

	References					
	Kneebone *et al.* (1985)[a]			Prentice *et al.* (1989)[b]	van Westhuyzen *et al.* (1988)[c]	
	Chinese	Malaysian	Indian	Gambian	South African	
					Urban	Rural
n	15	26	10	23	12	18
Lipid (%)	—	—	—	—	—	—
Fatty acid						
6:0	—	—	—	—	—	—
8:0	—	—	—	—	0.1 ± 0.1[d]	0.1 ± 0.2
10:0	0.84 ± 0.45[d]	0.90 ± 0.44	1.22 ± 0.68	0.92 ± 0.10[e]	0.5 ± 0.9	1.3 ± 0.2
12:0	5.28 ± 2.66	8.86 ± 3.72	8.44 ± 4.10	6.99 ± 0.59	2.6 ± 4.8	7.4 ± 6.7
14:0	6.53 ± 2.88	10.05 ± 3.02	8.90 ± 4.05	8.80 ± 0.85	12.7 ± 2.4	15.9 ± 7.6
15:0	0.17 ± 0.03	0.16 ± 0.05	0.22 ± 0.05	—	—	—
16:0	22.00 ± 3.13	26.86 ± 2.71	25.77 ± 3.28	14.1 ± 0.50	23.7 ± 4.3	21.8 ± 4.9
17:0	0.30 ± 0.07	0.27 ± 0.09	0.36 ± 0.07	—	—	—
18:0	5.16 ± 0.43	4.09 ± 0.67	4.98 ± 0.71	3.94 ± 0.11	7.7 ± 3.6	6.8 ± 2.8
20:0	0.28 ± 0.05	0.19 ± 0.05	0.26 ± 0.08	0.47 ± 0.06	0.1 ± 0.1	0.1 ± 0.1
22:0	0.09 ± 0.02	0.09 ± 0.11	0.06 ± 0.02	0.03 ± 0.04	TR	0.1 ± 0.3
24:0	0.16 ± 0.06	0.16 ± 0.08	0.19 ± 0.07	—	TR	0.1 ± 0.2
Total	40.87 ± 5.49	52.72 ± 5.36	50.49 ± 8.52	34.9	47.6 ± 4.8	53.9 ± 7.8

TABLE XIX—*continued*

	References					
	Kneebone *et al.* (1985)[a]			Prentice *et al.* (1989)[b]	van Westhuyzen *et al.* (1988)[c]	
	Chinese	Malaysian	Indian	Gambian	South African	
					Urban	Rural
	Diet descriptions					
	Meat, fish, polyunsaturated oil	Fish, mixed oils	Meat, fish, saturated oil	Peanuts, 75% of dietary fat. Fat 16 en%	Animal fats: protein, 10 to 12 en%; fat, 15 to 25 en%; CHO, 65 to 75 en%	Vegetable fats: protein, 8 to 10 en%; fat, 10 to 15 en%; CHO, 70 to 80 en%
Energy (kcal/day)	1745	1243	1386			
Protein (% total energy)	13	12	13			
Fat (% total energy)	30	33	39			
Carbohydrate (% total energy)	47	45	48			

[a]Adapted from Kneebone *et al.* (1985). Penangese donors. Time postpartum not given. Packed GLC column.
[b]Adapted from Prentice *et al.* (1989). Donors from Keneba, The Gambia. Time postpartum, 2 to 21 months. Includes parity through 10*t*. Packed GLC columns.
[c]Adapted from van Westhuyzen *et al.* (1988). South African donors, 3 to 10 months postpartum. Capillary GLC column.
[d]Means ± SD.

TABLE XX
The Monounsaturated Fatty Acids (wt%) in Human Milk Lipids from Mothers Consuming Non-Western Diets

	References					
	Kneebone *et al.* (1985)[a]			Prentice *et al.* (1989)[b]	van Westhuyzen *et al.* (1988)[c]	
	Chinese	Malaysian	Indian	Gambian	South African	
					Urban	Rural
n	15	26	10	23	12	18
Lipid (%)	—	—	—	—	—	—
Fatty acid						
Monounsaturates						
14:1	0.18 ± 0.07[d]	0.33 ± 0.10	0.26 ± 0.07	0.23 ± 0.04[d]	0.5 ± 0.4	0.6 ± 0.8[e]
16:1	3.03 ± 0.74	4.17 ± 1.15	3.39 ± 0.74	0.66 ± 0.13	4.3 ± 1.6	3.1 ± 1.3
17:1	0.17 ± 0.04	0.15 ± 0.06	0.20 ± 0.07	—	—	—
18:1	33.82 ± 4.30	30.82 ± 4.07	30.66 ± 5.14	47.0 ± 1.5	28.5 ± 3.1	23.9 ± 6.0
20:1	0.66 ± 0.31	0.51 ± 0.20	0.63 ± 0.26	0.83 ± 0.05	0.36 ± 0.1	0.1 ± 0.1
22:1	0.18 ± 0.08	0.20 ± 0.10	0.18 ± 0.10	0.22 ± 0.04	TR	TR
24:1	0.13 ± 0.11	0.09 ± 0.07	0.07 ± 0.04	0.05 ± 0.01	—	—
Total	38.14 ± 4.86	36.25 ± 4.79	35.34 ± 5.23	48.8 ± 1.5	33.6 ± 4.4	27 ± 5.8

[a,b,c]See Table XIX for dietary information.
[d]Means ± SD.
[e]Means ± SE.

TABLE XXI
The Polyunsaturated Fatty Acids (wt%) in Human Milk Lipids from Mothers Consuming Non-Western Diets

	References					
	Kneebone *et al.* (1985)[a]			Prentice *et al.* (1989)[b]	van Westhuyzen *et al.* (1988)[c]	
	Chinese	Malaysian	Indian	Gambian	South African	
					Urban	Rural
n	15	26	10	23	12	18
Lipid (%)	—	—	—	—	—	—
Fatty acids						
Polyunsaturates						
n6 Series						
18:2	11.96 ± 4.20[d]	8.84 ± 4.20	10.71 ± 4.66	13.0 ± 0.3[d]	16.2 ± 5.7[e]	15.3 ± 39[e]
18:3	TR	TR	TR	—	0.1 ± 0.3	0.3 ± 0.1
20:2	0.71 ± 0.44	0.29 ± 0.16	0.39 ± 0.22	0.83 ± 0.09	0.4 ± 0.1	0.3 ± 0.2
20:3	0.51 ± 0.14	0.27 ± 0.14	0.40 ± 0.16	0.21 ± 0.01	0.4 ± 0.2	0.8 ± 0.8
20:4	0.64 ± 0.17	0.47 ± 0.20	0.57 ± 0.17	0.31 ± 0.03	0.6 ± 0.2	1.0 ± 1.4
22:4	—	—	—	0.08 ± 0.01	0.1 ± 0.1	TR
22:5	—	—	—	0.30 ± 0.03	—	—
Total	19.16 ± 4.30	10.03 ± 1.21	12.37 ± 4.73	14.9 ± 0.80	17.9	17.8

TABLE XXI—*continued*

	References					
	Kneebone *et al.* (1985)[a]			Prentice *et al.* (1989)[b]	van Westhuyzen *et al.* (1988)[c]	
	Chinese	Malaysian	Indian	Gambian	South African	
					Urban	Rural
n3 Series						
18:3	0.38 ± 0.19	0.30 ± 0.14	0.33 ± 0.20	0.84 ± 0.20	0.4 ± 0.5	0.1 ± 0.1
20:5	—	—	—	—	0.1 ± 0.4	0.1 ± 0.2
22:5	0.21 ± 0.10	0.21 ± 0.11	0.19 ± 0.12	0.20 ± 0.05	0.2 ± 0.2	0.1 ± 0.2
22:6	0.71 ± 0.14	0.90 ± 0.29	0.90 ± 0.36	0.39 ± 0.28	0.2 ± 0.2	0.1 ± 0.2
Total	13.1 ± 0.30	1.41 ± 0.42	1.42 ± 0.40	1.4 ± 0.3	0.9	0.5

[a,b,c]See Table XIX for dietary information.
[d]Means ± SD.
[e]Means ± SE.

TABLE XXII
Saturated Fatty Acids (wt%, Means ± SD) in Human Milk Lipids from Women Consuming Non-Western Diets

	References						
	Muskiet *et al.* (1987)[a]			Muskiet *et al.* (1988)[b]		Borschel *et al.* (1986)[c]	
	Tanzanian	Curacao	Surinam	Dominica	Belize	Egypt	U.S.A.
n	11	47	20	6	6	22	21
Lipid (%)	—	3.17	2.81	4.60	2.86	—	—
Fatty acid							
6:0	0.18 ± 0.04	0.07 ± 0.02	0.13 ± 0.04	0.09	0.11	—	—
8:0	0.55 ± 0.10	0.36 ± 0.08	0.59 ± 0.20	0.56	0.33	—	—
10:0	2.75 ± 0.52	2.40 ± 0.56	3.08 ± 0.64	3.42	2.07	1.6 ± 0.9	1.0 ± 1.1
12:0	16.51 ± 5.12	10.71 ± 3.12	13.12 ± 2.89	16.34	9.47	9.1 ± 3.2	6.5 ± 2.3
14:0	14.72 ± 5.77	10.47 ± 3.52	10.80 ± 2.78	14.95	9.57	9.1 ± 3.8	6.2 ± 1.7
16:0	19.98 ± 3.48	19.77 ± 3.67	20.85 ± 2.34	21.59	22.85	22.4 ± 3.2	20.9 ± 4.4
18:0	3.65 ± 1.06	6.23 ± 0.94	4.51 ± 0.04	5.18	7.64	5.5 ± 2.2	7.3 ± 1.9
20:0	0.12 ± 0.02	0.27 ± 0.07	0.18 ± 0.05	0.24	0.19	—	—
22:0	0.06 ± 0.03	0.14 ± 0.04	0.11 ± 0.03	0.08	0.08	—	—
24:0	0.06 ± 0.02	0.14 ± 0.06	0.11 ± 0.03	0.08	0.07	—	—
Total	58.55	51.00	53.48	62.53	52.38	47.7	58.2

TABLE XXII—*continued*

	References						
	Muskiet *et al.* (1987)[a]			Muskiet *et al.* (1988)[b]		Borschel *et al.* (1986)[c]	
	Tanzanian	Curacao	Surinam	Dominica	Belize	Egypt	U.S.A.
	Dietary information						
Tanzania	High CHO, low fat and protein						
Curacao	Approaching Western types						
Surinam	High CHO, low fat and protein						
Dominica	High CHO, low fat and protein. Fish						
Belize	Approaching Western types						
Egypt	CHO, 70 en%; fat, 18–20 en%, unhydrogenated cottonseed oil, predominant fat						

[a]Adapted from Muskiet *et al.* (1987). Tanzanian donors, Bantus Curacao, African, Surinam, African, Asian, Indonesian. Mature more than 10 days postpartum milks. Capillary GLC column.
[b]Adapted from Muskiet *et al.* (1988). Capillary GLC column. See Tables XIIa–XIVb for St. Lucian data.
[c]Adapted from Borschel *et al.* (1986). Donors from Kalama, Egypt, and U.S.A. university community. Packed GLC column.

TABLE XXIII
Unsaturated Fatty Acids (wt%, Means ± SD) in Human Milk Lipids from Women Consuming Non-Western Diets

	References						
	Muskiet *et al.* (1987)[a]			Muskiet *et al.* (1988)[b]		Borschel *et al.* (1986)[c]	
	Tanzanian	Curacao	Surinam	Dominica	Belize	Egypt	U.S.A.
n	11	47	20	6	6	22	21
Lipid (%)							
Fatty acid							
Monounsaturates							
14:1n5	0.19 ± 0.07	0.20 ± 0.09	0.22 ± 0.08	0.22	0.25	—	—
16:1n7	0.27 ± 0.96	1.97 ± 0.66	3.23 ± 0.89	2.65	2.70	3.1 ± 0.09	2.9 ± 1.4
18:1n7	2.31 ± 1.35	3.83 ± 1.14	2.38 ± 0.48	2.59	3.30	—	—
18:1n9	19.42 ± 5.16	23.40 ± 3.50	25.16 ± 3.06	17.93	28.09	24.6 ± 5.7	36.7 ± 4.8
20:1n9	0.20 ± 0.07	0.47 ± 0.11	0.40 ± 0.13	0.47	0.47	—	—
24:1n9	0.02 ± 0.02	0.12 ± 0.06	0.08 ± 0.04	0.04	0.06	—	—
Total	24.41 ± 6.7	29.99 ± 4.17	31.46 ± 3.80	23.93	34.83	27.7	39.6
Polyunsaturates							
n6 Series							
18:2	13.88 ± 5.23	16.06 ± 3.60[d]	12.02 ± 3.74	10.14	10.27	23.8 ± 5.1	17.2 ± 6.2
18:3	0.11 ± 0.05	0.08 ± 0.05	0.10 ± 0.04	0.07	0.08	—	—
20:2	0.29 ± 0.10	0.58 ± 0.14	0.32 ± 0.08	0.30	0.34	—	—
20:3	0.40 ± 0.10	0.50 ± 0.12	0.41 ± 0.09	0.39	0.38	—	—

TABLE XXIII—*continued*

	References						
	Muskiet *et al.* (1987)[a]			Muskiet *et al.* (1988)[b]		Borschel *et al.* (1986)[c]	
	Tanzanian	Curacao	Surinam	Dominica	Belize	Egypt	U.S.A.
20:4	0.60 ± 0.02	0.09 ± 0.04	0.09 ± 0.03	0.45	0.51	0.2 ± 0.3	0.1 ± 0.2
22:4	0.13 ± 0.02	0.22 ± 0.10	0.14 ± 0.04	0.11	0.16	—	—
22:5	0.07 ± 0.02	0.43 ± 0.14	0.08 ± 0.03	0.03	0.05	—	—
Total	15.48	18.26	13.63	11.49	11.79	24.0	17.3
n3 Series							
18:3	9.98 ± 0.44	—	0.70 ± 0.39	0.73	0.62	0.7 ± 0.5	0.1 ± 0.2
20:5	0.19 ± 0.38	0.05 ± 0.03	0.06 ± 0.05	0.05	0.00	—	—
22:5	0.15 ± 0.07	0.21 ± 0.06	0.17 ± 0.07	0.11	0.13	—	—
22:6	0.27 ± 0.11	0.43 ± 0.14	0.41 ± 0.18	1.15	0.25	—	—
Total	1.59	0.69[e]	1.34	2.60	1.81	0.7	0.1
20:3n9	0.06 ± 0.02	0.09 ± 0.04	0.09 ± 0.03	0.04	0.06	—	—

[a,b,c]See Table XXI.
[d]Includes 18:3n3.
[e]Does not include 18:3n3.

TABLE XXIV
Saturated, Monounsaturated, and *trans*-Isomeric Fatty Acids (wt%) in Milk Lipids from Women ($n = 10$) Consuming Non-Western (Nigerian) Diets[a]

Fatty acid	Median Range	Fatty acid	Median Range
Saturated		cis-monounsaturates	
6:0	0.01(0.00–0.05)	14:1n5	0.08(0.05–2.40)
8:0	nd	15:1n5	0.05(0.00–2.28)
10:0	0.54(0.00–1.14)	16:1n7	0.91(0.64–2.19)
11:0	0.06(0.03–0.23)	17:1n7	0.13(0.00–1.58)
12:0	8.34(1.05–11.87)	18:1n9	18.52(9.44–25.30)
13:0	0.15(0.03–0.94)	18:1n7	0.95(0.77–2.32)
14:0	9.57(4.38–21.90)	20:1n9	0.34(0.12–0.69)
14:0*i*[b]	0.00(0.00–0.25)	22:1n9	0.75(0.12–2.06)
15:0	0.54(0.16–2.34)	24:1n9	0.59(0.31–1.25)
15:0*ai*[b]	0.04(0.00–1.18)	Total	22.82(14.76–29.30)
16:0	23.35(16.09–30.42)		
16:0*i*	0.00(0.00–5.08)	*Trans*-isomers	
17:0*ai*	0.44(0.20–1.41)	14:1	0.04(0.03–1.04)
18:0	10.15(6.86–14.76)	16:1	0.27(0.08–3.91)
19:0*ai*	0.09(0.06–0.44)	18:1	0.86(0.52–4.94)
20:0	0.42(0.26–0.58)	18:2	0.12(0.06–0.39)
20:0*i*	0.00(0.00–0.23)	Total	1.20(0.79–10.29)
22:0	0.41(0.19–0.50)		
24:0	0.39(0.17–0.58)		
Total saturates	54.07(38.42–71.74)		

Dietary information

Low animal and total fat, high CHO and fiber. Fresh and dried sea fish readily available.

Polyunsaturated fatty acids (wt%) in milk lipids from women ($n = 10$) consuming non-western (Nigerian) diets

n6 Series		n3 Series	
18:2n6	11.06(5.40–13.78)	18:3n3	1.41(0.64–5.45)
18:3n6	0.12(0.01–0.35)	18:4n3	0.09(0.04–0.32)
20:2n6	0.26(0.19–2.04)	20:3n3	0.27(0.05–1.09)
20:3n6	0.49(0.39–0.98)	20:4n3	0.14(0.00–0.28)
20:4n6	0.82(0.38–1.48)	20:5n3	0.48(0.17–1.57)
22:2n6	0.14(0.05–0.91)	22:3n3	0.21(0.03–2.02)
22:4n6	0.09(0.05–0.16)	22:5n3	0.39(0.12–0.72)

TABLE XXIV—*continued*

Fatty acid	Median Range	Fatty acid	Median Range
22:5n6	0.09(0.05–0.59)	22:6n3	0.93(0.70–2.16)
Total n6	12.52(7.47–16.64)	Total n3	4.63(2.16–7.97)
		Other:	
		20:3n9	0.43(0.22–0.72)

[a]Adapted from Koletzko *et al.* (1991). Women from Udo, Bendel State, Nigeria; 3 to 5 months postpartum. Capillary GLC column.
[b]*i, iso; ai, anteiso.*

TABLE XXV
Effects of Maternal Dietary Omega-3 Fatty Acids on the Saturated Fatty Acids (wt%) in Human Milk Lipids

	References			
	Innis and Kuhnlein (1988)[a]		Henderson *et al.* (1992)[b]	
	Inuit	Vancouver	Day 1	Day 21
n	5	12	5	5
Lipid (%)	2.8 ± 0.2[c]	3.1 ± 0.3[c]	2.74 ± 1.06[d]	2.10 ± 0.83[d]
Fatty acid				
10:0	1.2 ± 2.2	1.2 ± 0.2	—	—
12:0	6.2 ± 1.0	5.2 ± 0.7	4.8 ± 1.9	5.4 ± 2.0
14:0	5.7 ± 1.0	0.7 ± 0.5	5.3 ± 1.5	6.5 ± 1.6
16:0	18.0 ± 0.4	22.1 ± 2.7	20.6 ± 2.3	21.28 ± 3.0
18:0	7.1 ± 0.5	8.2 ± 0.8	6.5 ± 1.6	0.8 ± 1.5
Total	38.2	43.4	37.2 ± 3.6	39.9 ± 3.8

Dietary Information

Inuit: Marine mammal flesh Vancouver: mixed Western	Day 1: Baseline, mixed Western Day 21: Supplementation of some women with 1080 mg 20:5n3 and 720 mg 22:6n3 per day for 21 days.

[a]Adapted from Innis and Kuhnlein (1988). Inuit, Broughton Island, Canada donors, 4 to 28 months postpartum. Vancouver donors, 2 to 4 months postpartum. Capillary GLC column.
[b]Adapted from Henderson *et al.* (1992). Connecticut donors, 2 to 5 weeks postpartum. Capillary GLC column.
[c]Means ± SE.
[d]Means ± SD.

TABLE XXVI
Effects of Maternal Dietary Omega-3 Fatty Acids on the Unsaturated Fatty Acids (wt%) in Human Milk Lipids

	References			
	Innis and Kuhnlein (1988)[a]		Henderson *et al.* (1992)[b]	
	Inuit	Vancouver	Day 1	Day 21
n	5	12	5	5
Lipid (%)	2.8 ± 0.2[c]	3.1 ± 0.3[c]	2.74 ± 1.06[d]	2.10 ± 0.83[d]
Fatty acid				
Monounsaturates				
16:1	5.0 ± 0.8	3.3 ± 0.6	2.9 ± 0.7	2.8 ± 0.7
18:1	38.1 ± 2.4	36.3 ± 2.7	38.6 ± 3.1	35.0 ± 2.5
20:1	1.4 ± 0.3	0.7 ± 0.3	0.67 ± 0.20	0.73 ± 0.11
22:1	0.2 ± 0.4	0.2 ± 0.4	0.20 ± 0.06	0.18 ± 0.05
Total	44.7	40.3	42.4 ± 2.9	38.6 ± 2.7
Polyunsaturates				
n6 Series				
18:2	11.5 ± 0.7	12.7 ± 1.8	13.0 ± 1.7	12.5 ± 3.0
18:3	—	—	0.15 ± 0.07	0.15 ± 0.12
20:2	0.2 ± 0.0	0.4 ± 0.1	0.39 ± 0.12	0.23 ± 0.05
20:3	—	—	0.55 ± 0.16	0.45 ± 0.26
20:4	0.6 ± 0.0	0.7 ± 0.0	0.67 ± 0.11	0.52 ± 0.10
22:4	—	—	0.24 ± 0.16	0.21 ± 0.15
22:5	0.2 ± 0.0	0.2 ± 0.4	—	—
Total	12.5	14.0	15.0 ± 1.5	14.1 ± 2.9
n3 Series				
18:3	0.5 ± 0.2	0.6 ± 0.2	0.77 ± 0.12	0.76 ± 0.23
20:5	1.1 ± 0.3	0.2 ± 0.2	0.88 ± 0.04	0.50 ± 0.12
22:5	0.8 ± 0.2	0.4 ± 0.1	0.14 ± 0.05	0.34 ± 0.08
22:6	1.4 ± 0.4	0.4 ± 0.1	0.37 ± 0.26	0.70 ± 0.12
Total	3.8	1.6	1.4 ± 0.4	2.3 ± 0.5

[a,b]See Table XXV.
[c]Means SE.
[d]Means SD.

vegetarian diets in Tables XXVII and XXVIII. These data confirm earlier observations that maternal diets high in carbohydrates increase the synthesis of 6:0–12:0 in the mammary gland and that fatty acids (PUFA) which cannot be synthesized or are converted to a limited extent by the mother respond to changes in the diet, e.g., 18:2n6 and 18:3n3. The essentiality of 18:2n6 has been established. Linoleic acid (18:3n3) and its elongation–desaturation products also appear to be essential for infants since they are involved in the maturation of brain and nervous tissues and in the visual process (Koletzko *et al.*, 1991).

TABLE XXVII
Effects of Maternal Vegetarian Diets on the Saturated Fatty Acids (wt%) in Human Milk Lipids

	References				
	Specker *et al.* (1987)[a]		Sanders and Reddy (1992)[b]		
	Vegetarian	Control	Vegan	Vegetarian	Omnivore
n	12	7	19	5	21
Lipid (%)					
Fatty acid					
6:0	—	—	—	—	—
8:0	0.16 ± 0.03[c]	0.22 ± 0.01[c]	—	—	—
10:0	1.56 ± 0.13	1.57 ± 0.09	1.8 ± 0.40[c]	1.3 ± 0.51[c]	0.4 ± 0.23[c]
12:0	7.07 ± 0.78	5.47 ± 0.66	6.6 ± 0.54	3.2 ± 0.49	1.7 ± 0.35
14:0	8.16 ± 1.00	6.54 ± 0.73	6.9 ± 0.58	5.2 ± 0.50	4.5 ± 0.35
16:0	15.31 ± 0.73	20.48 ± 0.64	18.1 ± 1.34	21.2 ± 1.07	25.1 ± 0.78
18:0	4.48 ± 0.37	8.14 ± 0.55	4.9 ± 0.36	7.4 ± 0.35	9.7 ± 0.68
20:0	0.54 ± 0.02	0.57 ± 0.03	—	—	—
Total	37.28	42.99			

Dietary information

Specker *et al.* (1987)	Sanders and Reddy (1992)
Vegetarian: whole cereal grains, 50–60%; soup, 5%; vegetables, 20–25%; beans and sea vegetables, 5–10%; macrobiotic diet for a mean of 81 months. No meat or dairy products. Occasional seafood, nuts/fruit Control: Typical U.S. diet	Vegans: no foods of animal origin Vegetarians: Exclude meat/fish Omnivores: typical Western diet

[a]Adapted from Specker *et al.* (1987). New England donors: vegetarians, 3–13 months postpartum. Controls, 1–5 months. Capillary GLC columns.
[b]Adapted from Sanders and Reddy (1992). British donors, 6 weeks postpartum. Packed GLC columns.
[c]Means ± SEM.

TABLE XXVIII
Effect of Maternal Vegetarian Diets on the Unsaturated Fatty Acids (wt%) in Human Milk Lipids

	References				
	Specker *et al.* (1987)[a]		Sanders and Reddy (1992)[b]		
	Vegetarian	Control	Vegan	Vegetarian	Omnivore
n	12	7	19	5	21
Lipid (%)					
Fatty acid					
Monounsaturates					
16:1	1.66 ± 0.14^c	3.35 ± 0.28^c	4.9 ± 0.24^c	2.9 ± 0.37^c	3.4 ± 0.35^c
18:1	26.89 ± 1.47	34.7 ± 0.86	32.2 ± 1.06	35.3 ± 1.94	38.7 ± 1.27
Total	28.55	38.06	37.10	38.2	42.1
Polyunsaturates					
n6 Series					
18:2	28.82 ± 1.39	14.47 ± 1.98	23.8 ± 1.40	19.5 ± 3.62	10.9 ± 0.96
20:2	0.72 ± 0.03	0.50 ± 0.03	—	—	—
20:3	0.62 ± 0.03	0.56 ± 0.03	0.44 ± 0.03	0.42 ± 0.07	0.40 ± 0.08
20:4	0.68 ± 0.03	0.68 ± 0.03	0.32 ± 0.02	0.38 ± 0.05	0.35 ± 0.03
Total	30.84	16.21	31.4	27.5	18.4
n3 Series					
18:3	2.76 ± 0.16	1.85 ± 0.16	1.36 ± 0.18	1.25 ± 0.22	0.49 ± 0.06
22:6	0.22 ± 0.08	0.27 ± 0.08	0.14 ± 0.06	0.30 ± 0.05	0.36 ± 0.07
Total	3.05	2.12	1.50	1.55	0.86

[a,b]See Table XXVII.
[c]Means ± SEM.

3. *Summarized data*

We have consolidated and normalized the data from mature milks from women on Western diets and non-Western diets and calculated the weights based on the average lipid contents (Tables XXIX–XXXI). Ranges are also given. We have not provided fatty acid data on the influences of parity and disease. See Prentice *et al.* (1989) and Hamosh *et al.* (1992) for more information. Additional discussion is available in Jensen (1989a,b) and Jensen *et al.* (1992) as well as data on prostaglandins, alkyl ethers, and the many fatty acids found in human milk in trace amounts. We have summarized in Table XXXII the 184 different fatty acids that have been found in human milk lipids. For those who are interested, there are certainly many more because of the large differences in dietary fatty acids consumed around the world.

TABLE XXIX
Saturated Fatty Acids in Human Milk Lipids

	Western diets[a]			Non-Western diets[b]		
	wt%	Range	g/dl	wt%	Range	g/dl
Lipid		3.60–4.30	3.9540		2.10–4.60	2.9630
Fatty acid						
6:0	0.07	0.07–0.10	0.0028	0.15	0.07–0.18	0.0043
8:0	0.17	0.02–0.37	0.0067	0.37	0.10–0.59	0.0110
10:0	1.01	0.06–2.39	0.0399	1.63	0.50–3.42	0.0483
12:0	4.94	1.70–12.32	0.1951	8.12	2.40–16.51	0.2406
13:0	0.06	0.03–0.09	0.0024	—	—	—
14:0	5.63	1.98–11.78	0.2224	9.59	5.30–15.90	0.2842
15:0	0.44	0.31–0.74	0.0174	0.18	0.16–0.22	0.0542
16:0	20.33	19.25–25.10	0.8030	21.46	14.10–25.77	0.6359
17:0	0.54	0.31–1.11	0.0213	0.31	0.27–0.36	0.0092
18:0	7.54	5.83–9.70	0.2978	5.61	0.80–8.20	0.1663
20:0	0.32	0.16–0.75	0.0126	0.22	0.10–0.47	0.0065
22:0	0.09	0.06–0.12	0.0036	0.11	0.06–0.30	0.0033
24:0	0.19	0.07–0.50	0.0075	0.12	0.06–0.16	0.0035
Total	41.33	37.82–46.72	1.6326	47.88	34.90–62.53	1.4186

[a]From Tables XVI–XVIII.
[b]From Tables XIX–XXIV.

Acknowledgments

Some of the research reported herein was supported in part by federal funds made available through the provision of the Hatch Act and NIH Contracts N01-HD6-2917 and HD13021. Scientific Contribution 1531, Storrs Agricultural Experiment Station, The University of Connecticut, Storrs, CT 06269-4017.

References

Bitman, J., Wood, D. L. Hamosh, M., Hamosh, P., and Mehta, N. R. (1983). Comparison of the lipid composition of breast milk from mothers of term and preterm infants. *Am. J. Clin. Nutr.* **38,** 300–312.

Bitman, J., Wood, D. L., Mehta, N. R., Hamosh, P., and Hamosh, M. (1984). Comparison of the phospholipid composition of breast milk from mothers of term and preterm infants during lactation. *Am. J. Clin. Nutr.* **40,** 1103–1119.

Bitman, J., Wood, D. L., Mehta, N. R., Hamosh, P., and Hamosh, M. (1986). Comparison of the cholesteryl ester composition of human milk from preterm and term infants. *J. Pediatr. Gastroenterol. Nutr.* **5,** 780–785.

Bitman, J., Hamosh, M., Wood, D. L., Freed, L. M., and Hamosh, P. (1987). Lipid composition of milk from mothers with cystic fibrosis. *Pediatrics* **80,** 927–932.

TABLE XXX
Monounsaturated Fatty Acids in Human Milk Lipids

	Western diets[a]			Non-Western diets[b]		
	wt%	Range	g/dl	wt%	Range	g/dl
Lipid		3.60–4.30	3.9540		2.10–4.60	2.9630
Fatty acid						
14:1tn5	0.14	0.11–0.19	0.0055	—	—	—
14:1n5	0.31	0.22–0.49	0.0122	0.68	0.19–5.00	0.0202
16:1tn7	0.44	0.46–0.48	0.0174	—	—	—
17:1n7	0.24	0.16–0.32	0.0095	0.17	0.15–0.20	0.0051
18:1n7	3.37	3.20–3.78	0.1331	2.89	2.31–3.83	0.0854
18:1tn9	3.63	3.12–4.72	0.1434	—	—	—
18:1n9	30.96	22.63–38.70	1.2230	30.50	17.93–47.00	0.9027
20:1n9	0.74	0.38–1.86	0.0292	0.51	0.10–0.83	0.0151
22:1n9	0.07	0.06–0.08	0.0028	0.20	0.18–0.22	0.0057
24:1n9	0.07	0.03–0.22	0.0028	0.07	0.02–0.13	0.0022
Total	42.97	30.60–43.29	1.6973	35.02	23.93–48.80	1.0365

[a]From Tables XVI–XVIII.
[b]From Tables XIX–XXIV.

Boersma, E. R., Offringa, P. J., Muskiet, F. A. J., Chase, M. W., and Simmons, I. J. (1991). Vitamin E, lipid fractions, and fatty acid composition of colostrum, transitional milk, and mature milk: An international collaborative study. *Am. J. Clin. Nutr.* **53,** 1197–1204.

Borschel, M. W., Elkin, R. G., Kirksey, A., Story, J. A., Galal, O., Harrison, G. G., and Jerome, N. W. (1986). Fatty acid composition of mature human milk of Egyptian and American women. *Am. J. Clin. Nutr.* **44,** 330–335.

Bracco, U., Hidalgo, J., and Bohren, H. (1973). Lipid composition of the fat globule membrane of bovine and human milks. *J. Dairy Sci.* **55,** 165–172.

Breckenridge, W. C., and Kuksis, A. (1967). Molecular weight distribution of milk fat triglycerides from seven species. *J. Lipid Res.* **8,** 473–478.

Breckenridge, W. C., Marai, L., and Kuksis, A. (1969). Triglyceride structure of human milk fat. *Can. J. Biochem.* **47,** 761–769.

Bouhours, J-F., and Bouhours, D. (1979). Galactosylceramide is the major cerebroside of human milk fat globule membrane. *Biochem. Biophys. Res. Commun.* **88,** 1217–1222.

Bouhours, J-F., and Bouhours, D. (1981). Ceramide structure of sphingomyelin from human milk fat globule membranes. *Lipids* **16,** 726–731.

Clark, R. M., Ferris, A. M., Brown, P. B., Hundrieser, K. E., and Jensen, R. G. (1982). Changes in the lipids of human milk from 2 to 16 weeks postpartum. *J. Pediatr. Gastroenterol. Nutr.* **1,** 311–315.

Clark, R. M., Fey, M. B., Jensen, R. G., and Hill, D. W. (1984). Desmosterol in human milk. *Lipids* **18,** 264–266.

Clark, R. M., and Hundrieser, K. E. (1989). Changes in cholesteryl esters with total milk lipid. *J. Pediatr. Gastroenterol. Nutr.* **9,** 347–350.

Clark, R. M., and Hundrieser, K. E. (1990). Gas chromatographic procedure for measuring total lipid in breast milk. *J. Pediatr. Gastroenterol. Nutr.* **10,** 271–272.

TABLE XXXI
Polyunsaturated Fatty Acids in Human Milk Lipids

	Western diets[a]			Non-Western diets[b]		
	wt%	Range	g/dl	wt%	Range	g/dl
n6 Series						
18:2	12.55	9.57–16.80	0.4957	13.78	8.84–23.80	0.4079
18:3	0.37	0.09–1.03	0.0146	0.14	0.07–0.30	0.0041
20:2	0.31	0.18–0.50	0.0122	0.37	0.20–0.83	0.0108
20:3	0.36	0.26–0.56	0.0142	0.41	0.21–0.55	0.0122
20:4	0.47	0.36–0.68	0.0186	0.48	0.09–0.70	0.0142
22:2	0.11	0.05–0.20	0.0043	—	—	—
22:4	0.10	0.05–0.20	0.0039	0.17	0.08–0.24	0.0052
22:5	0.07	0.03–0.10	0.0028	0.17	0.03–0.43	0.0050
Total	14.34	11.20–18.42	0.5664	15.52	10.03–24.00	0.4599
n3 Series						
18:3	0.69	0.31–1.85	0.0273	0.52	0.10–0.98	0.0153
20:3	0.05	0.03–0.06	0.0020	—	—	—
20:5	0.07	0.00–0.16	0.0028	0.24	0.05–1.10	0.0072
22:3	0.12	0.11–0.13	0.0047	—	—	—
22:5	0.12	0.05–0.21	0.0047	0.24	0.10–0.80	0.0070
22:6	0.23	0.10–0.56	0.0091	0.57	0.10–1.40	0.0169
Total	1.28	0.64–2.20	0.0506	1.57	0.10–3.80	0.0465

[a]From Tables XVI–XVIII.
[b]From Tables XIX–XXIV.

Dotson, K. D., Jerrel, J. P., Picciano, M. F., and Perkins, E. G. (1992). High performance liquid chromatography of human milk triacylglycerols and gas chromatography of component fatty acids. *Lipids,* 933–939.

Gibson, R. A., and Kneebone, G. M. (1981). Fatty acid composition of human colostrum and mature breast milk. *Am. J. Clin. Nutr.* **34,** 252–257.

Grimmonprez, L., and Montreuil, J. (1977). Etude des fractions glycanniques des glycosphingolipides to taux de la membrane des globules lipidiaues du lait de frimme. *Biochimie* **59,** 899–908.

Gurr, M. I., and Harwood, J. L. (1991). "Lipid Biochemistry," 4th Ed. Chapman & Hall, New York.

Hallgren, B., Niklasson, A., Stallberg, G., and Thorin, H. (1978). On the occurrence of 1-O-alkylglycerols and 1-O-(2-methoxyalkyl) glycerols in human colostrum, human milk, cow's milk, sheep's milk, human and bone marrow, red cells, blood plasma and a uterine carcinoma. *Acta Chem. Scand.* **B28,** 1029–1034.

Hamosh, M., and Bitman, J. (1992). Human milk in disease. Lipid composition. *Lipids* **27,** 848–857.

Harzer, G., Haug, M., Dieterich, I., and Gentner, P. (1983). Changing patterns of human milk lipids in the course of the lactation and during the day. *Am. J. Clin. Nutr.* **37,** 612–621.

TABLE XXXII
Fatty Acids in Human Milk Lipids[a]

Number	Type	Identity
	Saturates	
10	Normal, even	4:0–22:0
7	Normal, odd	11:0–23:0
49	Monobranched	10:0–18:0
5	Multibranched	12:0–13:0
	Monoenes	
63	*cis*	10:1–18:1, 20:1, 23:1–24:1–26:1
4	*Trans*	14:1, 16:1, 18:1, 20:1
22	Dienes	12:2–22:2, all even, *cis, cis; cis, trans; trans,* and positional isomers
	Polyenes	
7	Tri-	18:3, 20:3, 22:3, geometric acid positional isomers
3	Tetra-	20:4, 22:4
3	Penta-	20:5, 22:5
1	Hexa-	22:6
1	Cyclic hexane	11, terminal hexane
9	Hydroxy-	16:0, 18:0, 20:0, 22:0, 23:0, 24:0, 24:1, 25:0, 26:0
Total 184		

[a]Adapted from Jensen (1989a,b); Jensen *et al.* (1990).

Haug, M., and Harzer, G. (1984). Cholesterol and other sterols in human milk. *J. Pediatr. Gastroenterol. Nutr.* **3,** 816–817.

Henderson, R. A., Jensen, R. G., Lammi-Keefe, C. J., Ferris, A. M., and Dardick, K. R. (1992). Effect of fish oil in the fatty acid composition of human milk and maternal and infant erythrocytes. *Lipids* **27,** 863–869.

Hundrieser, K. E., and Clark, R. M. (1988). A method for separation and phospholipid classes in human milk. *J. Dairy Sci.* **71,** 61–67.

Hundrieser, K. E., Clark, R. M., and Brown, P. B. (1983). Distribution of *trans* octadecenoic acid in the major phospholipids of human milk. *J. Pediatr. Gastroenterol. Nutr.* **2,** 635–639.

Innis, S. M., and Kuhnlein, H. V. (1988). Long-chain n-3 fatty acids in breast milk of Inuit women consuming traditional foods. *Early Hum. Dev.* **18,** 185–189.

Jackson, M. B., Lammi-Keefe, C. J. Jensen, R. G., Couch, S. C., and Ferris, A. M. (1993). Total lipid and fatty acid composition of milk in women with and without insulin-dependent diabetes mellitus. Submitted for publication.

Jensen, R. G. (1989a). "The Lipids of Human Milk," pp. 7–42. CRC Press, Boca Raton, FL.

Jensen, R. G. (1989b). Lipids in human milk—Composition and fat soluble vitamins. *In* "Textbook of Gastroenterology in Infancy" (E. Lebenthal, ed.), 2nd Ed., pp. 57–208. Raven Press, New York.

Jensen, R. G., Ferris, A. M., and Jensen, R. G. (1992). Lipids in human milk and infant formulas. *Annu. Rev. Nutr.* **12,** 417–441.

Jensen, R. G., Ferris, A. M., Lammi-Keefe, C. J., and Henderson, R. A. (1990). Lipids of bovine and human milk: A comparison. *J. Dairy Sci.* **73,** 233–240.

Kallio, M. T. J., Siims, M. A., Perheentupa, J., Salmenpera, L., and Miettinen, T. A. (1989). Cholesterol and its precursors in human milk during prolonged exclusive breast-feeding. *Am. J. Clin. Nutr.* **50,** 782–785.

Kallio, M. T. J., Salmenpera, L., Siims, M. A., Perheentupa, J., and Miettinen, T. A. (1992). Exclusive breastfeeding and weaning: Effect on serum cholesterol and lipoprotein in concentrations in infants during the first year of life. *Pediatrics* **8a,** 663–666.

Kneebone, G. M., Kneebone, R., and Gibson, R. A. (1985). Fatty acid composition of breast milk from three racial groups from Penang, Malaysia. *Am. J. Clin. Nutr.* **41,** 765–769.

Koletzko, B., Mrotzek, M., and Bremer, H. J. (1988). Fatty acid composition of mature human milk in Germany. *Am. J. Clin. Nutr.* **47,** 954–959.

Koletzko, B., Mrotzek, M., and Bremer, H. J. (1991). Fatty acid composition of mature human milk. *Z. Ernaharungswiss–Evr. J. Nutr.* **30,** 289–297.

Kolsto-Otnaess, A. B. (1989). Nonimmunoglobulin components in human milk candidates for prophylaexis against infantile infections. *In* "Protein and Non-protein Nitrogen in Human Milk. (S. A. Atkinson and B. Lonnerdal, eds.), pp. 211–220. CRC Press, Boca Raton, FL.

Kuksis, A., and Breckenridge, W. C. (1968). Triglyceride composition of milk fats. *In* "Dairy Lipids and Lipid Metabolism" (M. F. Brink and D. Kritchevsky, eds.), pp. 28–98. Raven Press, New York.

Laegrid, A., Otnaess, A-BK., and Fugelsang, J. (1986). Human and bovine milks: Comparison of ganglioside compositions and enterotoxin inhibitory activity. *Pediatr. Res.* **20,** 416–420.

Lammi-Keefe, C. J., Ferris, A. M., and Jensen, R. G. (1990). Changes in human milk at 0600, 1000, 1400, 1800, and 2200 h. *J. Pediatr. Gastroenterol. Nutr.* **11,** 83–88.

Lepage, G., and Roy, C. C. (1984). Improved recovery of fatty acids through direct transesterification without prior extraction or purification. *J. Lipid Res.* **25,** 1391–1396.

Muskiet, F. A. J., Hutter, N. H., Martini, I. A., Jonxis, J. H. P., Offringa, P. J., and Boersma, E. R. (1987). Comparison of the fatty acid composition of human milk from mothers in Tanzania, Curacao and Surinam. *Hum. Nutr. Clin. Nutr.* **41C,** 149–159.

Muskiet, F. A. J., Offringa, P. J., and Boersma, E. R. (1988). Lipid content and fatty acid composition of human milk in relation to developing countries. *In* "A Holistic Approach to Perinatal Care and Prevention of Handicap" (E. R. Boersma, H. H. Huisjes, and H. M. C. Poortman, eds.), pp. 294–305. Vanderkamp, Groningen, The Netherlands.

Newburg, D. S., and Chaturvedi, P. (1992). Neutral glycolipids of human and bovine milk. *Lipids* **27,** 923–927.

Newburg, D. S., Ashkenuzl, S., and Cleary, T. G. (1992). Human milk contains the Shiga toxin and shiga-like toxin receptor, glycolipid Gb3. *J. Infect. Dis.* **166,** 832–836.

Prentice, A. M., Landing, M. A. J., Drury, P. J., Dewit, O., and Crawford, M. A. (1989). Breast-milk fatty acids of rural Gambian mothers: Effects of diet and maternal parity. *J. Pediatr. Gastroenterol. Nutr.* **8,** 486–490.

Putnam, J. C., Carlson, S. E., DeVoe, P. W., and Barness, L. W. (1982). The effect of variations in dietary fatty acids on the fatty acid composition of erythrocyte phosphatidylcholine and phosphatidylethanolamine in human infants. *Am. J. Clin. Nutr.* **41,** 619–623.

Reiser, R., O'Brien, B. C., Henderson, G. R., and Moore, G. R. (1979). Studies on a possible function cholesterol in milk. *Nutr. Rpts. Int.* **19,** 835–849.

Sanders, T. A. B., and Reddy, S. (1992). The influence of a vegetarian diet on the fatty acid composition of human milk and the essential fatty acid status of the infant. *J. Pediatr.* **120,** S71–77.

Spear, M. L., Hamosh, M., Bitman, J., Spear, M. L., and Wood, D. L. (1992). Milk and blood fatty acid composition during two lactations in the same woman. *Am. J. Clin. Nutr.* **56,** 65–70.

Specker, B. L., Wey, H. E., and Miller, D. (1987). Differences in fatty acid composition of human milk in vegetarian and nonvegetarian women: Long-term effect of diet. *J. Pediatr. Gastroenterol. Nutr.* **6,** 764–768.

Sukhija, P. S., and Palmquist, D. L. (1988). Rapid methods for the determination of total fatty acid content and composition of feedstuffs and feces. *J. Agric. Food Chem.* **16,** 1202–1206.

Takamizawa, K., Iwamori, M., Mutai, M., and Nagai, Y. (1986). Selective changes in gangliosides of human milk for the period of lactation. *Biochem. Biophys. Acta* **879,** 73–77.

Tomarelli, R. M. (1988). Suitable fat formulations for infant feeding. *In* "Dietary Fat Requirements in Health and Development" (J. Beare-Rodgeas, ed.), Chap. 1. American Oil Chemists' Society, Champaign, IL.

van der Steege, G., Muskiet, F. A. J., and Martini, I. (1987). Simultaneous quantification of total medium-and long-chain fatty acids in human milk by capillary gas chromatography with split injection. *J. Chromatogr.* **415,** 1–11.

van der Westhuyzen, van der Chetty, N., and Atkinson, P. A. (1988). Fatty acid composition of human milk from South African black mothers consuming a traditional maize diet. *Eur. J. Clin. Nutr.* **149,** 126–129.

van Beusekom, C. M., Martini, I. A., Rutgers, H. K., Boersma, E. R., and Muskiet, F. A. J. (1990). A carbohydrate-rich diet not only leads to incorporation of medium-chain fatty acids (6:0–14:0) in milk triglycerides but also in each milk phospholipid subclass. *Am. J. Clin. Nutr.* **52,** 326–334.

Yonekubo, K. A., Arima, H., and Yamamoto, Y. (1987). The composition of Japanese human (III)—Polyunsaturated fatty acid composition and sterol and phospholipid contents. *J. Child Health* **3,** 349–352.

Zeisel, S. H., Char, D., and Sheard, N. F. (1986). Choline, phosphatidylcholine and sphingomyelin in human and bovine milk and infant formulas. *J. Nutr.* **116,** 50–58.

B. Bovine Milk Lipids

ROBERT G. JENSEN
DAVID S. NEWBURG

I. Introduction

A. Definitions and Nomenclature

See Chapter 6A I,A.

B. The Nature of Lipids in Milk

See Chapter 6A I,B.

II. Collection, Preparation, and Storage of Samples

See the latest editions of "Standard Methods for the Analysis of Dairy Products" (APHA, 1993) and "Official Methods of Analysis" (AOAC, 1990) for detailed descriptions of these procedures. Additional information is presented in Chapter 2D.

However, most bovine milks are pooled and processed, while human milks are not (Jensen and Jensen, 1992). Most all bovine milk and its products are clarified (centrifugal removal of particulates), pasteurized, and homogenized (Jensen, 1992). Pasteurization apparently does not affect the lipid content and composition, although this has not been thoroughly investigated. Conversely, homogenization, a process which reduces the size of the lipid globules from about 3 to 0.8 μm and increases their number at least 100-fold and the surface area about 6 to 10 times, alters the globule membrane structure and composition. The globule surface is recoated largely, but not completely, with caseins. However, some semblance of the original globule membrane is retained (see Chapter 2A). Unfortunately, we have very few data on the composition of processed products and virtually none about the digestion of the lipids therein.

III. Determination of Lipid Content

See the reference under Section I, B and in Chapter 6A,III. Most of the methods employed for human milk were borrowed from the dairy industry.

IV. Factors Affecting Total Lipid Content

Most of the factors listed in Table I, in Chapter 6A also influence the lipid contents of milk from individual cows (Jenness, 1985). However, production and processing practices eliminate most of these. The current tendency is to select and breed for low-fat milks, e.g., Holsteins vs Guernseys. Colostral, late, and milks from mastitic or otherwise diseased cows and those treated with antibiotics are excluded and pooling occurs. Milk and dairy products have legal minimal lipid contents (3.25% for whole milk) and the contents are held very close to these standards, usually 3.34% for whole milk.

The processor can adjust the lipid content of milk to any amount desired lower than the original quantity by controlled separation (centrifugation).

V. Lipid Classes

A. Introduction

The average composition for milk lipids is given in Table I. See Christie (1983), Jensen and Clark (1988), and Jensen *et al.* (1991). We show the data reported by Bitman and Wood (1990). These data, obtained by

TABLE I
Lipids in Milk

Lipid class	% of total lipid (g/100 g)	
	Bitman and Wood (1990)	National Dairy Council (1993)
Phospholipid	1.11[a]	0.20–100[a]
Cholesterol	0.46	0.419[b]
Triacylglycerol	95.80	97 or 98
1,2-Diacylglycerol	2.25	0.28–0.59
Free fatty acids	0.28	0.10–0.44
Monoacylglycerol	0.08	0.16–0.38
Cholesteryl ester	0.02	—
Hydrocarbons[c]	TR	TR

[a]Includes sphingomyelin.
[b]Does not include cholesteryl esters.
[c]Includes squalene and carotenoids.

densitometric analysis of separated milk lipids on TLC plates, are the first reported in many years. Earlier results are compiled in the data from the National Dairy Council (NDC, 1993). The overwhelming mass of the lipids is triacylglycerol (TG) with much smaller quantities of sterols and phospholipids which are primarily associated with the membrane. The sterols are mostly cholesterol with about 10% of this in the ester form. Traces of hydrocarbons, carotenoids, retinyl esters, squalene, etc., are found in freshly drawn and extracted or processed milks. Only traces of free fatty acids (FFA), and di- (DG) and monoacylglycerol (MG) will be detected. The presence of large quantities of these and smaller amounts of TG is indicative of lipolysis as indicated by the data of Bitman and Wood (1990) in Table I. If raw milk is to be analyzed, it should be extracted immediately, frozen to −70°C, or pasteurized to prevent lipolytic action. Lipolysis will not change the total fatty acid composition, unless some of the volatile short-chain acids are lost, but will alter the relative amounts of FFA, TG, DG, and MG. In milks that have been processed the amounts of lipid classes will be similar to those of the NDC (1993) in Table I.

The large preponderance of TG makes it difficult to quantitate and/or resolve the other lipids. Bitman and Wood (1990) separated the polar and nonpolar lipids of milk with a Sep-Pak column. Bitman and Wood (1990) further resolved the nonpolar and polar lipids into classes by thin-layer chromatography (TLC) then quantitated them with densitometry. Christie *et al.* (1987) used the same method to separate the nonpolar and polar lipids, then resolved and quantitated the phospholipid classes by high-performance liquid chromatography (HPLC). Earlier investigators measured phospholipids by determination of organic P, a procedure which may have resulted in the lower quantities reported by the NDC (1993).

The easiest procedure for routine determination of cholesterol was developed by Bachman *et al.* (1976). With this technique, the milk is saponified directly, extracted with hexane, and the cholesterol in the extract determined by treatment with *o*-phthalaldehyde, and then spectrophotometry. Gas–liquid chromatography (GLC) should be employed if the investigator is searching for sterols other than cholesterol.

B. Triacylglycerols

1. Introduction

The composition of TGs is usually defined in terms of the kinds and amounts of fatty acids present and will be discussed later. Structure includes the distribution of fatty acids within the TG molecule and among the TG molecules, as well as the identification of the individual molecular species of TGs.

The structure of the TGs influences the action of lipolytic enzymes and, therefore, absorption (Jensen *et al.*, 1992) and flavor of cheeses.

Structure of milk TGs is responsible for the melting points, crystallization behavior, and rheological properties of milk fat as globules, and in butter and butter oil. The fatty acid composition and, hence, bovine milk TG structure, is not greatly affected directly by ordinary changes in diet because of the biohydrogenation and production of short-chain fatty acids in the rumen.

Bovine milk lipids contain about 12 fatty acids in amounts greater than 1% (Table II). Therefore, it would be theoretically possible to have 12×10^3 or 1728 TG species if all the acids were randomly distributed. The total theoretical possibilities are much greater, since bovine milk lipids contain at least 406 fatty acids. With 406 fatty acids the theoretical maximum is 406^3 or 66, 923, and 416 TGs. Since the distribution of fatty acids in milk TG is not random, the numbers of TG do not approach this figure, but several or many enantiomers are present. Investigations of structure can be divided into pre- and post-HPLC. For reviews see Christie (1983), Jensen and Clark (1988), and Jensen *et al.* (1991).

2. *Structure: Pre-HPLC*

These semi-identifications were achieved by use of chromatographic procedures other than HPLC and by enzymatic resolutions. Analyses done

TABLE II
Positional Distribution of Fatty Acids in the Triacylglycerols from Normal and Linoleic Acid-Rich Bovine Milk[a]

Fatty acids (mol%)	Bovine milk				Linoleic acid-rich bovine milk			
	TG	sn-1	sn-2	sn-3	TG	sn-1	sn-2	sn-3
4:0	11.8	—	—	35.4	10.8	—	—	32.3
6:0	4.6	—	0.9	12.9	3.8	—	0.6	10.6
8:0	1.9	1.4	0.7	3.6	1.5	1.6	0.7	2.3
10:0	3.7	1.9	3.0	6.2	2.7	2.4	3.0	2.7
12:0	3.9	4.9	6.2	0.6	3.4	3.3	4.8	2.0
14:0	11.2	9.7	17.5	6.4	7.5	8.3	12.4	1.7
15:0	2.1	2.0	2.9	1.4	1.0	1.2	1.4	0.4
16:0	23.9	34.0	32.3	5.4	15.7	22.1	23.3	1.7
16:1	2.6	2.8	3.6	1.4	0.8	0.8	1.2	0.5
17:0	0.8	1.3	1.0	0.1	0.4	0.6	0.4	0.1
18:0	7.0	10.3	9.5	1.2	10.4	14.3	11.1	5.7
18:1	24.0	30.0	18.9	23.1	26.9	32.3	24.4	24.1
18:2	2.5	1.7	3.5	2.3	15.3	13.1	16.8	16.0
18:3	TR	—	—	—	TR	—	—	—

[a]Adapted from Christie and Clapperton (1982).

over two decades and in several countries have produced remarkably similar results. Representative data are given in Table II (Christie and Clapperton, 1982). We selected these data because they were obtained by a nonselective method of producing the 1,2(2,3) DGs needed for analysis.

The data in the linoleic acid-rich bovine milk are from cows fed protected oils rich in 18:2. The oils were encapsulated in denatured casein. When fed to the cows, the capsules pass through the rumen unaffected by biohydrogenation. The casein is digested in the abomasum releasing the oil. An 18:2-rich milk results (Table II). The fatty acid profile can also be manipulated by altering the dietary regimens of the cattle.

Additional information is presented in Table III (Kuksis and Breckenridge, 1968). The data in Table III were obtained by analyses of the most volatile (2.5%) molecular distillate of butter oil from the original most volatile 10% cut with carbon numbers (total of acyl groups) ranging from C24 to C40.

Based on these and other data we can conclude that the distribution of fatty acids in milk TGs is asymmetrical. Enantiomers are present when there are major differences in the fatty acid composition of the sn-1 and sn-3 positions. Note in Table II that all of the 4:0, 83% of the 6:0, and 63% of the 8:0 are esterified to the sn-3 position. Also, note in Table III that many of the short-chain fatty acids are combined with two long-chain fatty acids in TGs.

Kuksis *et al.* (1973) summarized the results of their extensive analyses of milk TG structure as follows. There are three types of TGs. The first has acyl carbons totaling 48–54 composed of sn-1,2 DGs containing 18:0, 18:1, and 18:2. The 3-position is acylated with 12:0, 14:0, 16:0, or the acids above. In type 2, the carbon numbers are 36–46 and the sn-3 position acids are 4:0, 6:0, and 8:0. These TGs are enantiomers. The carbon numbers in type 3 are 26–34, the sn-1,2 TGs contain medium-chain fatty acids and the sn-3 position acids are short and medium chain. The TGs in types 1 and 3 that have different fatty acids in the sn-1 and -3 positions are enantiomers.

Myher *et al.* (1988) reanalyzed their most volatile 2.5% molecular distillate described above with capillary GLC and mass spectrometry. Over 1000 TGs were identified. The major TGs in this fraction are listed in Table IV. The sequence of the fatty acids in the table is not their real location in the TGs. When these data are coupled with the information in Tables II and III, several TGs can be identified. These are sn-18:0–14:0–6:0, 18:0–16:0–4:0, 18:1–18:1–4:0, etc. The data from Myher *et al.* (1988) vividly illustrate the complexity of bovine milk TGs and the difficulties entailed in their analysis. The fraction they analyzed represented a very small portion of milk TGs, yet over 100 were detected and their positional and stereoconfigurations were not determined.

Kallio *et al.* (1989) employed supercritical fluid GLC and electron impact mass spectrometry to determine the TGs in butter. The TGs were resolved in molecular weight fractions by GLC and the degree of

TABLE III
Estimates of Specific Triglyceride Types[a]

Saturates (16.9% total)			
C32(1.8% total)		C34(3.8% total)	
18,10,4	12	18,12,4 =	10
16,12,4	30	16,14,4 =	83
16,10,6	25	16,12,6 =	7
16,8,8	55		100%
14,14,4	24		
10,10,12	3		
	100%		
C35(0.6% total)		C36(6.1% total)	
17,14,4	21	18,14,4	18
17,12,6	9	16,16,4	77
15,16,4	51	16,14,6	3
15,14,6	19	16,12,8	2
	100%		100%
C37(0.6% total)		C38(4.0% total)	
18,15,4	48	18,16,4	70
16,15,6		16,16,6	17
16,17,4	52	16,14,8	12
14,17,6		16,12,10	1
	100%		100%

Monoenes (10.6% total)			
C38(6.8% total)		C37:1 (0.6% total)	
18:1,16,4	87	18:1,15,4	74
16:1,18,6	5	16:1,17,4	26
18:1,14,6	6	16:1,15,6	Trace
18:1,12,8	2		100%
	100%		

C40:1 (3.2% total)	
18:1,18,4	62
18:1,16,6	26
16:1,18,6	3
18:1,14,8	3
18:1,12,10	2
16:1,14,10	2
	100%

[a]Adapted from Kuksis and Breckenridge (1968).

TABLE IV
Major Triacylglycerol Types in a Volatile Molecular Distillate from Butter Oil[a]

Types	Mol%	Types	Mol%
18,14,16 + 10,14,14 + 10,12,16[b]	2.53	6,18:1,18 + 8, 16, 18:L	0.62
6,14,18 + 6,16,16 + 8, 16:1,14	5.37	6,18:1, 18:1	0.74
6,18:1, 14 + 6, 16:1, 16	1.48	6,18:1,18:2	0.29
4, 16, 18	5.40	10, 1, 18 + 12,16,16 + 14,14,16	1.23
4,16,18:1 + 4, 16:1, 18	9.65		
4,16:1,18:1	0.96	10,18:1, 16	1.32
10,14,16	1.74		
8,16,16 + 8,14,16	1.61	8,18:1,18:1	0.45
6,16,18 + 8,18:1,14 + 10, 18:1, 12	2.18	14,16,16	0.55
6,18:1, 16	2.90	12,16,18:1, + 14,14,18:1	0.66
4,18:1, 18	1.08	10,18:1, 18:1	0.18
4,18:1, 18:1	2.14	16,16,16	0.26
10,14,18 + 10,16,10	2.04	14,16,18:1	0.54
8,16,18 + 10,18:1,14	0.92	12,18:1,18:1	0.06
8,18:1,16	1.34	16,16,18	0.09
		16,16,18:1	0.33
		14,18:1,18:1	0.16

[a]Adapted from Myher *et al.* (1988). Fraction was the fourth most volatile 2.5% distillate from molecular redistillation of the original most volatile 10% weight from distillation of 777 lbs of butter oil.
[b]8,14,16 is 8:0–14:0–16:0. Sequence of fatty acids is not their real location.

unsaturation in each was identified by mass spectrometry. Their data can be seen in Table V. The presence of 16:0–16:0–16:0, 16:0–16:0–18:1, and 16:0–18:1–4:0 can be inferred although individual fatty acid distributions were not given. When the data in Table II are utilized, 16:0–18:1–4:0 becomes sn-16:0–18:1–4:0.

3. *Structure: Post-HPLC*

Barron *et al.* (1991) employed reversed-phase liquid chromatography to separate butterfat into 62 fractions. The fatty acid and TG (carbon number) compositions of the fractions were determined by GLC resulting in the tentative identification of 116 TGs. However, their assignments were based on random distributions of fatty acids. The asymmetry of distribution precludes randomness. Consequently, we have not provided the data of Barron *et al.* (1991).

Maniongui *et al.* (1991) made the same preparations and analyses as Barron *et al.* above (1991), but they employed different methods to

TABLE V
Distribution (wt%) of Butter Triacylglycerols According to Carbon Number and Unsaturation[a]

	No. of Double Bonds			
Carbon No.[b]	0	1	2	3
34	4.8	1.4	—	—
36	5.0	4.9	2.6	
38	4.6	6.9	2.9	3.1
40	2.0	4.6	3.1	1.2
42	1.5	2.4	2.1	1.2
44	1.0	2.8	2.9	1.0
46	1.3	2.1	2.2	1.0
48	1.6	2.2	2.2	1.0
50	2.6	3.4	2.7	0.8
52	2.7	5.7	1.9	0.4
54	2.2	1.4	0.3	—
Total	29.3	37.9	22.9	9.9

[a]Adapted from Kallio *et al.* (1989).
[b]Sum of carbons in fatty acids.

calculate the amounts and kinds of TGs. They identified 223 individual TGs. Some of their data are presented in Table VI (Gresti *et al.*, 1993). Positions of fatty acids in TG were not determined. We have given data only on those TGs which were present in amounts greater than 1 mol% as well as the quantities which would have been present if the distributions of fatty acids had been random. Inspection of the data in Table VI and of the remainder in the paper (Gresti *et al.*, 1993) reveals that almost none of the TGs were present in random amounts. A notable exception was 18:1–18:1–18:1 (Table VI). We have summarized their data in Table VII. Again, by combining data from Tables II and VI, we can expect sn-16:0–14:0–4:0, sn-14:0–16:0–4:0, sn-6:0–16:0–4:0, etc. to be present in the TG. With the exception of 18:1–18:1–18:1, all of the TGs in Table VI are enantiomers.

Kermasha *et al.* (1993) separated butterfat TG by preparative HPLC with a laser light-scattering detector. They recovered fractions containing 12:0, 14:0, and 16:0, fatty acids believed to be hypercholesterolemic. The fractions, 16.2% of total TGs, were analyzed stereospecifically to determine the location of the fatty acids in the TGs (see Table VIII). Again, the asymmetry of fatty acid distribution is obvious. All of the TGs in Table VIII are enantiomers. Also notable is the location of relatively large amounts of 14:0 and 16:0 at the sn-2 position. It has been suggested that

TABLE VI
Major Triacylglycerols in Bovine Milk Lipids[a]

Triacylglycerol species			Amounts (mol%)[b]		Triacylglycerol species			Amounts (mol%)[b]	
			Experimental	Random				Experimental	Random
4:0	14:0	16:0[c]	3.05	1.42	10:0	16:0	18:1	1.60	1.15
6:0	14:0	16:0	1.37	0.72	12:0	16:0	18:1	1.22	1.11
4:0	14:0	18:0	1.35	0.65	14:0	16:0	18:1	2.82	3.39
4:0	16:0	16:0	3.20	1.54	14:0	18:0	18:1	1.45	1.55
6:0	16:0	16:0	1.50	0.78	16:0	16:0	18:1	2.34	3.69
4:0	16:0	18:0	2.47	1.42	16:0	18:0	18:1	2.16	3.39
6:0	16:0	18:0	1.12	0.71	4:0	18:1	18:1	1.48	1.32
4:0	14:0	18:1	1.79	1.31	14:0	18:1	18:1	1.26	1.58
4:0	16:0	18:1	4.17	2.87	16:0	18:1	18:1	2.50	3.43
6:0	16:0	18:1	2.02	1.45	18:0	18:1	18:1	1.21	1.57
4:0	18:0	18:1	1.58	1.31	18:1	18:1	18:1	1.02	1.06
Totals								42.68	37.43

[a]Adapted from Gresti *et al.* (1993).
[b]Experimental (determined) or random (calculated) mol%.
[c]Position of the acyl chains within each TG not determined. However, in all butyroyl TG, 4:0 is at sn-3; in hexanoyl TG, > 90% of 6:0 is at sn-3; in octanoyl TG, about 63% is at sn-3; and in decanoyl TG, 56%, see Table II.

TABLE VII
Summary of TG in Bovine Milk as Determined by Reverse-Phase Liquid and Capillary Gas–Liquid Chromatography[a]

Quantitated 223 TG containing even numbered FA accounting for 80% of the total

Major TG were (mol%): 18-1-16:0–4:0, 4.2; 16:0–16:0–4:0, 3.2; and 16:0–14:0–4:0, 3.1. In these and all other butyroyl TG, 4:0 will be esterified to the sn-3 position. These TG are enantiomers

Twenty-two TG (1 > mol%), 42.68 mol% of total, contained at least two of the major FA: 14:0, 16:0, 18:0, and 18:1. In this group there were eight butyroyl diacylglycerols —19% of the total. See Table VI.

Thirty-six mol% of the TG contained 4:0 or 6:0 and two long-chain FA. All of the 4:0 and at least 90% of the 6:0 will be at sn-3. See Table II.

In the TG with 4:0 and 6:0, 14:0, 16:0, and 18:0 were equally distributed among the sn-1 and -2 positions. See Table II.

With the exception of 18:1–18:1–18:2 (1.02 mol%), there were no monoacid TG of 4:0–12:0 and very small quantities of 14:0, 16:0, and 18:0

8:0, 10:0, and 12:0 were located at sn-3; a decreasing amount of 18:1 was in all positions

There were no predominant TG and the amounts of almost all TG were nonrandom. These will change as the fatty acid profiles are altered, but the distributions will remain nonrandom

[a]Adapted from Gresti *et al.* (1993).

TABLE VIII
Positional Distribution of Fatty Acids in Butter Triacylglycerol Fractions[a]

	Triacylglycerol Fatty acids (sn position)			
Fraction	1	2	3	Percentage
9a	16:0	14:0	18:1	20
9b	18:1	14:0	16:0	20
9c	18:1	16:0	14:0	25
9d	16:0	18:1	14:0	35
11a	18:1	16:0	18:1	50
11b	18:1	18:1	16:0	50
12a	18:1	16:0	16:0	85
12b	16:0	18:1	16:0	15

[a]Adapted from Kermasha *et al.* (1993). Fractions representing 16.2% of total TGs were separated by HPLC.

the capability of dietary saturated fats to raise serum cholesterol depends on the amounts of saturated fatty acids in the sn-2 position of milk TGs.

Itabashi *et al.* (1993) determined the distribution of 2:0, 4:0, 6:0, and 10:0 in the fraction described in Table IV and the relevant text. They used chiral-phase HPLC of the derived DG. The quantities of these acids in the sn-3 position were (mol%): 100, 100, 100, 85, and 50. The other acids were usually long chain. Myher *et al.* (1993) further analyzed the most volatile 2.5% molecular distillate from butter oil with reversed-phase HPLC coupled with mass spectrometry and capillary column GLC. They identified and quantitated over 150 molecular species. Their results, too voluminous to present, show that much of the 4:0 and 6:0 was associated with two long-chain fatty acids. However, some of the TG contained two short-chain fatty acids. The complexity of the fractions, which represented only 0.25% of the sample, emphasizes the large number of TG likely to be present in milk fat.

4. *Significance of Triacylglycerol Structure*

We have described earlier the influence of TG structure in processing parameters. These are particularly applicable to butter in which spreadability is affected by the TG or, ultimately, the fatty acid composition of milk lipids (Parodi, 1981). The location of the flavor acids (4:0–10:0) in the primary positions of TGs makes them accessible to lipases. When free, the acids contribute to the flavor of cheeses.

Of equal or possibly more importance are the physiological and nutritional effects of TG structure. Milk TGs, when consumed by humans, are lipolyzed first in the stomach by gastric lipase. The lipase preferentially hydrolyzes sn-3 position fatty acids 4 to 1 compared to sn-1 and selectively releases the shorter acids (Jensen *et al.*, 1992). The result is that 4:0–10:0 pass through the stomach wall in decreasing quantities as the molecular weight increases, enter the portal vein, and are transported to the liver where they are oxidized. About 25–40% of the TG is digested in the stomach. Milk lipid globules are resistant to pancreatic lipolysis in the small intestine unless they are first exposed to gastric lipolysis. The digestate entering the small intestine will contain bioactive sn-1,2 DGs and very small quantities of 4:0–10:0. These aspects of milk TG digestion have not been investigated. The hypercholesterolemic potential of some milk TGs has been described (Kermasha *et al.*, 1993).

5. *Summary of Structure*

Bovine milk contains a unique assembly of TGs, possibly thousands, which are very difficult to identify. The TGs are characterized by the location of most of the 4:0–8:0 at the sn-3 position and 12:0, 14:0, and 16:0 at the sn-2 position. Structure affects the behavior of milk lipids during and after processing, the metabolism of milk lipids, and possibly their hypercholesterolemic potential. All of these factors present a challenge to the analyst, who may possibly need to identify a minor atherogenic TG.

C. Phospholipids (PL) and Glycosphingolipids

1. Introduction

Data on the contents of these and the other lipids are given in Table I (Bitman and Wood, 1990). The amounts of PL, similar to data reported earlier (Jensen and Clark, 1988), do not change much during time postpartum in individual cows. The small variations will probably not be observed in pooled, homogenized milks in which the amounts will range from 20 to 30 mg/dl.

2. Composition of PL

The composition of the PL determined by Bitman and Wood (1990) throughout lactation is presented in Table IXa. These data and those in Table IXb are probably the most reliable in the literature because extracted milk lipids were first separated into neutral and PL. Sphingomyelin is usually reported as a PL even though it is also a member of the sphingalipid class. The PL classes were then resolved by TLC.

Christie *et al.* (1987) reported on the separation of phospholipids and dairy products. They extracted the lipids from 10 ml of fresh milk with the Bligh–Dyer method, separated the total lipids into neutral and phospholipids with a Sep-Pak column as described by Bitman and Wood (1990), and resolved the total phospholipids into classes by HPLC using a mass detector. We believe that the first use of HPLC for this purpose is described by Christie *et al.* (1987). They also presented data on some of the phospholipids in ultra-high temperature (UHT) pasteurized pooled milks from April through October, the semiskimmed and skimmed milks therefrom, and in buttermilk from the preparation of butter. In UHT pasteurization the milk is heated to 138–150°C for 2 to 6 sec. They noted that almost no phospholipids were found in powdered whole milk and buttermilk. They attributed the loss to autoxidation of the polyunsaturated fatty acids in the phospholipids as a result of exposure to heat during processing. Earlier analyses were often done on spray-dried buttermilk powders, which, according to Christie *et al.* (1987), may not contain any PL because of autoxidation. The fresh, liquid product should be used and can be readily obtained by agitating cream containing 30–35% fat for a few minutes in a Waring Blendor. Buttermilk (not cultured) is the fluid product formed during the production of butter. Earlier data compiled by Jensen and Clark (1988) are presented in Table X. Milk lipids also contain small quantities (0.009%) of glycerol ethers in neutral lipids and PL (Ahrne *et al.*, 1979).

TABLE IXa
Phospholipid Composition of Cow's Milk during Lactation

Class[a]	Concentration at lactation day (mg/dl) 3	7	42	180	Pooled SE
SM	5.8^{2}	11.9^{1}	7.1^{2}	3.9^{3}	0.5
PC	5.8^{3,4}	8.9^{1}	6.7^{2,3}	4.5^{4}	0.6
PS	1.6^{2}	3.0^{1}	2.1^{2}	0.3^{3}	0.2
PI	0.8^{2}	1.6^{1}	1.3^{1}	1.5^{1}	0.1
PE	6.4^{2}	10.0^{1}	7.9^{2}	2.6^{3}	0.8
Total	20.4	35.4	25.1	12.8	2.0
	Percentage of phospholipids				
SM	28.7^{2}	34.1^{1}	28.7^{2}	31.4^{1,2}	1.1
PC	28.0^{2}	25.1^{2}	26.4^{2}	35.1^{1}	1.1
PS	8.1^{1}	8.4^{1}	8.5^{1}	1.9^{2}	0.4
PI	4.1^{4}	4.6^{3,4}	5.2^{2,3}	11.8^{1}	0.3
PE	31.0^{1}	27.8^{1}	31.1^{1}	19.8^{2}	1.6

[a]Bitman and Wood (1990).
[b]SM, sphingomyelin; PC, phosphatidyl choline; PS, phosphatidyl serine; PI, phosphatidyl inositol; and PE, phosphatidyl ethanolamine.
^{1,2,3,4}Means within a row with different superscripts differ ($P < 0.05$).

TABLE IXb
Lipid Class Content of Phospholipids and Glycolipids from Bovine Milk as Determined by High-Performance Liquid Chromatography[a]

Lipid class	mg/dl[b]
Ceramide monohexoside	1.13 ± 0.04
Ceramide dihexoside	0.65 ± 0.20
Phosphatidylethanolamine (PE)	7.78 ± 0.10
Phosphatidylinositol (PI)	1.42 ± 0.20
Phosphatidylserine (PS)	0.64 ± 0.07
Phosphatidylcholine (PC)	5.79 ± 0.05
Sphingomyelin (SPH)	5.37 ± 0.06

[a]Calculated from Christie *et al.* (1987). Lipids extracted from 10 ml fresh milk were separated into neutral and phospholipids and the phospholipids resolved by HPLC.
[b]Means and standard errors of six analyses.

TABLE X
Phospholipid and Sphingolipid Composition of Bovine Milk[a]

Phospholipid	Mol%
Phosphatidylcholine	34.5
Phosphatidylethanolamine	31.8
Phosphatidylserine	3.1
Phosphatidylinositol	4.7
Sphingomyelin	25.2
Lysophosphatidylcholine	Trace
Lysophosphatidylethanolamine	Trace
Total choline phospholipids	59.7
Plasmalogens	3
Diphosphatidylglyerol	Trace
Ceramides	Trace
Cerebrosides	Trace
Gangliosides	Trace

[a]Adapted from Jensen and Clark (1988).

The PL and sphingolipids (SL) contain relatively larger quantities of polyunsaturated fatty acids (PUFA) than the TG, but the amounts are so small that they will have little nutritional significance. The PL and SL bind cations, help stabilize the emulsion, and probably orient enzymes on the globule surface, but their effects in processed milks are unknown. Data on the fatty acids in the PL and SL are available in the references above. For information on the lysophospholipids, plasmalogens, diphosphatidylglycerol, and ceramides (Table IXb), see Jensen and Clark (1988) and the following section.

3. Glycosphingolipids

a. Neutral glycolipids. The glycolipid fraction of bovine milk was first reported to comprise 6% of the total "phospholipid" fraction (Hladik and Michalec, 1966). As the phospholipid fraction is approximately 1% of total lipids of bovine milk, the glycolipid fraction is found in bovine milk at approximately 20–24 mg/liter, assuming 3.3–4% butterfat. The principal constituents of this fraction are glucosylceramide and lactosylceramide (Morrison and Smith, 1964) (see also Table IXb). The amounts and structures of the compounds are listed in Table XI. In contrast, human milk contains a large amount of galactosylceramide (see Chapter 6A). Over 70% of the glycolipids of bovine milk are associated with the milk lipid globule membrane (MFLM). Thus, bovine MFLM contains glucosylceramide and lactosylceramide as its major constituents; both of these

TABLE XI
Glycosphingolipids of Bovine Milk

		mg/l	Reference
Neutral glycolipids			
1. Glucosylceramide	Glc β-1-Cer[a]	6.0	Morrison and Smith (1964)
2. Lactosylceramide	Gal β(1–4) Glc β-1-Cer[b]	15.0	Morrison and Smith (1964)
Gangliosides			
1. GM_1	Gal β(1–3) GalNAc β(1–4)↘ Gal β(1–4) Glc β-1-Cer[c,d]; NANA α(2–3)↗	0.0012	Laegreid *et al.* (1986)
2. GD_{1b}	Gal β(1–3) GalNAc β(1–4)↘ Gal β(1–4) Glc β-1-Cer; NANA α(2–8) NANA α(2–3)↗	1.2	Laegreid *et al.* (1986)
3. GM_2	GalNAc β(1–4)↘ Gal β(1–4) Glc β-1-Cer; NANA α(2–3)↗	0.7	Laegreid *et al.* (1986)
4. GD_2[e]	GalNAc β(1–4)↘ Gal β(1–4) Glc β-1-Cer; NANA α(2–8) NANA α(2–3)↗	Trace	Puente *et al.* (1992); Bushway and Keenan (1978)
5. GM_3	NANA α(2–3) Gal 3(1–4) Glc β-1-Cer	0.3	Laegreid *et al.* (1986)
6. GD_3	NANA α(2–8) NANA α(2–3) Gal β(1–4) Glc-1-Cer	Mature, 8.8 Colostrum, 12.5	Laegreid *et al.* (1986)
		Mature, 2.0	Puente *et al.* (1992)

TABLE XI—*continued*

		mg/l	Reference
7. GT_3	NANA α(2–8) NANA α(2–8) NANA α(2–3) Gal β(1–4) Glc β-1-Cer	28 mg/kg Buttermilk solids	Takamizawa *et al.* (1986)
8.	Gal β(1–4) GlcNAc β(1–6) ↘ Gal β(1–4) Glc β-1-Cer; NANA α(2–6) Gal β(1–4) GlcNAc β(1–3) ↗	78 mg/kg Buttermilk solids	Takamizawa *et al.* (1986)
9.	$(NANA)_2$ [Gal β(1–4) GlcNAc β(1–6) ↘ Gal β(1–4) Glc β-1-Cer; NANA α(2–6) Gal β(1–4) GlcNAc β(1–3) ↗]	8.5 mg/kg Buttermilk solids	Takamizawa *et al.* (1986)
10. 9-*O*-acetyl GD_3	9-*O*-Ac NANA α(2–8) NANA α(2–3) Gal β(1–4) Glc β-1-Cer[f]	22 mg/kg Buttermilk solids	Ren *et al.* (1992); Bonafede *et al.* (1989)
11. 7-*O*-acetyl GD_3	7-*O*-Ac NANA α(2–8) NANA α(2–3) Gal β(1–4) Glc β-1-Cer	1.2 mg/kg Buttermilk solids	Ren *et al.* (1992)
12. 7,9,di-*O*-acetyl GT_3	7,9 diAc NANA α(2–8) NANA α(2–8) NANA α(2–3) Gal β(1–4) Glc β-1-Cer	24 mg/kg Buttermilk solids	Ren *et al.* (1992)

[a]Glc, glucose; Cer, ceramide.
[b]Gal, galactose.
[c]GalNAc, *N*-acetylgalactosamine; GlcNAc, *N*-acetylglucosamine.
[d]NANA, *N*-acetylneuraminic acid.
[e]Found in mammary tissue, presumed in milk.
[f]Ac, acetyl.

glycolipids contain mainly nonhydroxy fatty acids, and lactosylceramide is present in higher concentrations than glucosylceramide. The fatty acids of the glucosylceramide associated with the MFLM tend to be 20–26 carbons long, while the glucosylceramide found in skim milk has fatty acids of 18 carbons or less (Kayser and Patton, 1970). Quantitative analysis by HPLC indicates that bovine milk contains mainly normal or nonhydroxylated fatty acid (NFA), glucosylceramide (8 μmol/liter; 6 mg/liter), and NFA lactosylceramide (17 μmol/liter; 15 mg/liter); bovine milk contains little globotriaosylceramide or globoside (Newburg and Chaturvedi, 1992).

b. Gangliosides. The milk gangliosides were initially studied in bovine milk. Keenan *et al.* (1972) showed that the apical membrane of bovine mammary secretory cells, the source of membrane for the MFLM, contains 10–25% of the total cellular gangliosides of these cells. This was followed by a report (Keenan, 1974) that the MFLM has the same ganglioside profile as the mammary gland, that 90% of milk gangliosides are found in the MFLM, and that the principal gangliosides are GM3, GM2, and GD3 (Table XI). The sialic acids of the bovine gangliosides include both *N*-acetylneuraminic acid (NANA) and *N*-glycolylneuraminic acid (Bushway and Keenan, 1978). Minor gangliosides in the mammary gland include GD1b, GD2, and GM1. Huang (1973) found that bovine buttermilk, which contains high concentrations of MFLM, is a rich source of gangliosides, even richer than the gray matter of the brain (10–20 mg/liter vs 0.5 mg/kg). The principal gangliosides are GM3 (20% of gangliosides) and GD3 (50% of gangliosides). Hauttecoeur *et al.* (1985) found that GD3 comprises 85% of the total gangliosides of buttermilk, and that it consists of two species, one having long-chain (C22–C25) fatty acids and an equimolar proportion of C16 and C18 sphingosine bases, and the other species consisting of mainly 16:0 and C18 sphingosine.

Takamizawa *et al.* (1986) reported that buttermilk contains 0.92 μmol of lipid-bound sialic acid per gram of dry weight (approx 72 mg sialic acid/kg), 80% of which is in the form of GM3, GD3, and GT3. In addition, 41 nmol/g is in the form of a ganglioside with a novel branched structure (Structure 7, Table XI); the trisialo derivative of this structure (Structure 8, Table XI) was also found. The 9-*O*-acetyl form of GD3 was found in buttermilk (Bonafede *et al.*, 1989). Other *O*-acetyl derivatives of gangliosides also have been found, including 7-*O*-acetyl GD3 (1.2 mg/kg), 9-*O*-acetyl GD3 (22 mg/kg), and 7,9 di-*O*-acetyl GT3 (24 mg/kg) (Ren *et al.*, 1992).

The variation in ganglioside levels during lactation was studied in six Spanish–Brown cows. Puente *et al.* (1992) found that colostrum (Days 1–3) contains 7.5 mg NANA/liter, transitional milk (Day 5) contains 2.3 mg NANA/liter, and mature milk (Days 30–180) contains 1.4 mg NANA/liter. GM3 increases from Days 1–5, while GD3 decreases during this period; GM3 decreases from Day 5 to the end of lactation, as GD3 increases. The GM3, GD3, and GT3 account for 80–90% of the gangliosides in bovine milk, with GD3 initially being the major single component (60–70%).

Laegreid *et al.* (1986) compared the cholera toxin inhibition by human milk (11 mg gangliosides/liter) with that of bovine milk and bovine milk-based formula (6 mg gangliosides/liter). Less than 1 ml of human milk inhibited 0.1 μg of cholera toxin *in vitro* (enzyme-linked immunosorbent assay) and *in vivo* (rabbit small bowel loops), while 5- to 10-fold the amount of bovine milk was needed to achieve comparable results. The amounts of individual gangliosides were thought to be related to this difference in biological activity. The gangliosides in human milk consist of 74% GM3, while only 3% of bovine milk gangliosides is GM3. Conversely, human milk contains a lower amount of GD3 (25%) than bovine milk (80% GD3). Although only trace amounts of GM1 are found in milk, human milk contains 10 times the concentration of bovine milk.

Other trace gangliosides of milk may bind to other toxins to which infants are exposed (Ochanda *et al.*, 1986); it would be of interest to see if human or bovine milk would inhibit such toxin binding to its host receptors.

D. Sterols

Milk contains 10 to 20 mg/dl of cholesterol or 308 to 606 mg/100 g fat in whole milk containing 3.3% fat (Jensen and Clark, 1988; Jensen, 1990; Jensen *et al.*, 1991; Bitman and Wood, 1990; NDC, 1993) (see Table I). The amount is positively correlated with the fat content of the dairy product (see Table XII for examples). Cholesterol is the major sterol and

TABLE XII
The Cholesterol Content of Various Dairy Products[a]

Identity of product	Fat (%)	Cholesterol (mg/100 g)	Identity of product	Fat (%)	Cholesterol (mg/100 g)
Skim milk	0.25	2	Blue	28.74	75
Whole milk	3.30	14	Brie	27.68	100
Half and half	11.50	37	Cheddar	33.14	105
Light cream	19.31	66	Cream	34.87	110
Medium cream	25.00	88	Mozzarella whole milk	21.60	78
Nonfat dry	0.77	20	Neufchatel	23.43	76
Cottage cheese creamed	4.51	15	Swiss	27.45	92
Cream cheese	34.87	110	Butter	81.11	219
Ice cream	10.77	45	Sherbert orange	1.98	7

[a]Posati and Orr (1976).

is located mostly in the milk lipid globule membrane (see Chapter 2A). About 10% of the cholesterol is esterified (see Table I, and for the fatty acid composition of the cholesteryl esters, see Wood and Bitman, 1985).

Cholesterol has been determined by colorimetry with *o*-phthalaldehyde (Bachman *et al.*, 1976), GLC (Tsui, 1989; IDF, 1992), and HPLC (Hurst *et al.*, 1983). Small quantities of many other sterols have been detected (Mincione *et al.*, 1977; Walstra and Jenness, 1984). The amounts of 7-dehydrocholesterol range from 0.7 to 4.0% of total sterols with less than 1% of phytosterols (IDF, 1992). Since cholesterol accounts for at least 95% of the sterols, the inexpensive and rapid method of Bachman *et al.* (1976) will provide satisfactory results. The other methods will separate most of the sterols which may be present. The manner in which milk is obtained, elimination of colostrum, and pooling eliminate the effects of time post-partum and diurnal rhythm seen in individual samples.

E. Fatty Acids and Related Compounds

1. Production

Milk lipids have attracted the interest of investigators for many years. The lipids are readily available, for example, in butter, but are exceptionally complex, both with respect to lipid classes and to component fatty acids. Furthermore, the latter have been difficult to analyze because the short-chain fatty acids present, being water-soluble and volatile, are easily lost, and because of the large number of fatty acids in general. The number was 400 in 1992 (Jensen, 1992) and is now 406.

The application of several chromatographic procedures to the separation and identification of milk lipids was mainly responsible for these achievements. The first paper on analysis of ruminant milk fatty acids was published by James and Martin (1956). By 1960, many laboratories were using GLC for routine analysis of fatty acids. A complete analysis of milk fatty acid composition now requires no more than 2 hr, including extraction of lipids.

Column and TLC came into use about the same time as GLC, with the latter widely accepted because of its speed, ease of use, versatility, resolving power, and, probably most important, ease of visualization. Thin-layer chromatography has been particularly useful in the separation and nondestructive recovery of lipid classes. Tentative identifications can be made by comparison to known compounds and purity can be checked. The methyl esters of fatty acids can be separated into saturates, monoenes, etc. with TLC using $AgNO_3$ impregnated film and the esters can then be analyzed by GLC. A more recent innovation, HPLC, is only now being applied to milk lipids and fatty acids.

As a result of many extensive efforts, a large amount of information is available on the composition of bovine milk lipids, although most of it is not recent. With the exception of flavors derived from milk lipids, there has been little research on milk lipids since 1970. Lipid composition has been reviewed by Jensen and Clark (1988), Jensen *et al.* (1991), and Jensen (1992). We will present summarized and recent data on the amounts of fatty acids in milk products and discuss methods for analysis.

The major reason for concern about the fatty acids in milk is that some of them are atherogenic (Jensen *et al.,* 1991; Jensen, 1992; Berner, 1993). These include myristic (14:0), palmitic (16:0), and, possibly, lauric (12:0) acids. Although widely regarded as a high-cholesterol food, whole milk contains only about 15 mg/dl and cannot be considered a major contributor to dietary cholesterol levels. However, the fat and cholesterol contents of dairy products increase roughly in parallel (see Table XII).

2. *Analysis of Milk Fatty Acids*

a. Extraction of lipids. Assuming that a representative sample is available, extraction of the milk lipids is the next step. Since most of the lipids are TG, the standard AOAC method (Roese–Gottlieb or Mojonnier) will suffice unless more than traces of FFA are present (AOAC, 1992). The latter procedure utilizes ammonium hydroxide and acidification of pH 2 is needed to ensure recovery of the FFA. A modified Folch procedure (Timmen and Dimick, 1972) or a column method (Maxwell *et al.,* 1986) are available to extract "all" of the lipids. Determination of total lipids, best done by weighing of the solvent-free extract, is an absolute requirement, even though milk and most dairy products contain expected amounts of fat. The fat content must be known in order to calculate the weight of fatty acids/dl or 100 g of milk or dairy product. The dietitian or nutritionist must have this information to properly advise individuals who should alter the amounts of fatty acids in their diets. The contents of fatty acids are usually reported as weight percentage or g/100 g of fatty acid, but must also be given as gravimetric data, mg or g/dl or 100 g of edible product. Analysis of bovine milk lipids has been reviewed by Christie (1987) and Jensen (1992). An HPLC method for the determination of free fatty acids has been described by Elliot *et al.* (1989).

Instructions for the gathering and preparation of samples were given previously in the references (see Chapter 2D). In general, milk purchased almost anywhere, at least in the United States, will be representative because of pooling. There are seasonal and regional effects which are caused by the availability of different feeds for dairy cattle. Butter is also a good index material because of the very large amounts of milk which are needed for its production and also because of pooling.

b. Analysis of fatty acids. GLC is the best method for the routine separation and tentative identification of common milk fatty acids, as well as the resolution of the less abundant and less common acids. Although 406 fatty acids are listed herein as being present in milk, not all of these were rigorously identified. For nutritional evaluation of milk and dairy products, the amounts of major fatty acids, *trans*, and PUFA acids are needed and these data have not been provided.

GLC analysis requires that the fatty acids be converted to a more volatile derivative such as the methyl ester. Unfortunately, the short-chain fatty acid esters cause problems. A major difficulty has not been the GLC separation of these esters, which is done with temperature programming, but in transferring them from the esterification mixture to the GLC instrument without loss of the volatile esters. A widely used procedure is a slight modification of the method developed by Christopherson and Glass (1969) which uses sodium methoxide for transesterification. Sukhija and Palmquist (1988) have adapted a sealed tube method for preparation of methyl esters of milk fat which eliminates losses during transfer. Butyl ester could also be prepared by this procedure. In order to decrease volatility, butyl esters have been used rather than methyl esters. See Jensen (1992) for more information.

The analyst can use a variety of GLC columns which will provide analyses previously unattainable. Wide-bore, wall-coated open tubular capillary columns of suitable length, at least 30 m, will provide excellent resolution of many minor fatty acids including *trans* isomers. Temperature programming has been employed to separate the peaks. Otherwise, the short esters may not be resolved, being retained in the solvent peak.

3. Milk Fatty Acids

Barbano (1990) published a fatty acid composition of reference milk lipids. The data, in Table XIII, are the means of samples from 50 cheese plants in 10 regions of the United States in February, May, August, and November of 1984. These are the most comprehensive data available, but were not done by the recent analytical methods described earlier. Other reference data can be seen in Table XIV representing analyses of butter using mostly butyl esters, but done with packed GLC columns. Data obtained with capillary columns, butyl esters, and temperature programming are presented in Table XV. These are almost the only data on the fatty acids in milk and butter that have been obtained with modern methods. There are no such data for the other dairy products.

The contents of *trans* isomers, mostly 11-18:1 in milk, must be considered. They are produced by ruminal biohydrogenation of polyunsaturated fatty acids and will always be present. The methods described above could not resolve these isomers. They have been analyzed by infrared spectrophotometry, polar capillary GLC columns, and other methods. A large number of isomers have been identified (Jensen and Clark, 1988) and data on these will be presented later. A wide range of contents has been reported, but an average value of 2.5% total *trans* fatty acids, mostly the

TABLE XIII
Fatty Acid Composition of a Reference Milk Fat[a]

Fatty acid	wt%
4:0	3.32
6:0	2.34
8:0	1.19
10:0	2.81
12:0	3.39
14:0	11.41
14:1	2.63
16:0	29.53
16:1	3.38
18:0	9.84
18:1	27.39
18:2	2.78

[a]Adapted from Barbano (1990) and Palmquist *et al.* (1993). Analyses done with packed GLC columns and methyl esters of fatty acids. Mean of measures from 50 cheese plants in 10 regions of the United States in February, May, August, and November.

18:1 isomers, is given by Renner (1983). Careful GLC analyses by Enig *et al.* (1983) with a 15-m capillary column obtained an average content of 3.32% [recalculated by the authors to include 4:0–10:0 not reported by Enig *et al.* (1983)] 18:1*t* from three samples of butter. The total *trans* fatty acids in butter (Table XV) is 1.97%. These separations were achieved with a capillary column. Wolff (1994) found 3.22% ± 0.44 SD total *trans*-18:1 in French autumn butters and 4.25% ± 0.47 in spring butters. The annual mean value for total *trans*-18:1 was 3.8% with about 2% for *trans*-11-18:1.

In most papers on milk fatty acids, the authors have not given the quantities of LC-PUFA beyond 18:3 because of instrumental and column limitations. Excellent resolution and sensitivity and instrument stability are now available, and dependable data on the contents of omega-3 and -6 LC-PUFA can be achieved. The acids are nutritionally important and their contents can and should be reported. We need to know what and how much LC-PUFA are present in milk and dairy products; even though the quantities may be low, so are the amounts needed.

4. *Factors Affecting Fatty Acid Composition*

Palmquist *et al.* (1993) reviewed these factors: animal, genetic and stage of lactation, feed, grain, amount and composition of dietary fat, dietary protein, and seasonal and regional effects. The influences of all these

TABLE XIV
Major Fatty Acids of 50 Butters Obtained in 1982 Determined as Butyl Esters for Major Acids and Methyl Esters for Minor Acids[a]

Saturates		Monoenes		Trienes		Branched	
Fatty acid	wt% (SEM)[b]	Fatty acid	wt% (SEM)	Fatty acid	wt% (SEM)	Fatty acid	wt%
4:0	4.84 ± 0.126	10:1	0.15	18:3	1.13 ± 0.037[c]	13:0*i*[d]	0.03
6:0	2.20 ± 0.030	12:1	0.06	20:3	0.10	14:0*a*	0.02
8:0	1.30 ± 0.016	13:1	0.03	22:3	0.07	15:0*i*	0.40
10:0	2.88 ± 0.033	14:1	0.40	**Total**	**1.30**	15:0*a*	0.44
11:0	0.20	16:1	1.70[e]			16:0*i*	0.40
12:0	3.33 ± 0.034	17:1	0.36	Other polyenes		17:0*i*	0.50
13:0	0.19	18:1	24.10 ± 0.206[f]	20:4	0.14	17:0*a*	0.52
14:0	10.76 ± 0.102	19:1	0.16	20:5	0.09	18:0*i*	0.16
15:0	1.48	20:1	0.32	22:4	0.03	19:0*i*	0.10
16:0	26.10 ± 0.265	21:1	0.04	22:5	0.04	20:0*i*	TR
17:0	0.60	22:1	0.06	22:6	0.01	22:0*i*	TR
18:0	10.76 ± 0.102	23:1	TR[g]	**Total**	**0.31**	Total	2.57
19:0	0.15	24:1	TR				
20:0	0.35	**Total**	**27.38**				
21:0	0.04					Multibranched	
22:0	0.20	Dienes				16:0	TR
23:0	0.12	18:2	2.37 ± 0.038			19:0	TR
24:0	0.14	20:2	0.04			20:0	TR
25:0	0.03	22:2	0.04				
26:0	0.06	Total	2.48				
Total	**65.83**						

[a]Adapted from Iverson and Sheppard (1986).
[b]SEM given for major acids.
[c]At least 90% omega-3.
[d]*i, iso; a, anteiso.*
[e]From Enig *et al.* (1983).
[f]Includes about 3% of total fatty acids as *trans* 18:1.
[g]TR, less than 0.01%.

factors except seasonal and regional effects are eliminated by the pooling of milk. Palmquist *et al.* (1993) published the data in Tables XVI and XVII showing the affects of various feeding regimens on milk fatty acids. Since the feeding regimens vary according to season (Table XVIII) and region (Table XIX) the latter trends are probably the result of differences in feeds.

TABLE XV
Fatty Acids in Bovine Milk Fat as Determined by Gas–Liquid Chromatography with Capillary Columns

Fatty acid	Milk[a] (wt%)	Butter[b] (wt% ± SD)	Fatty acid	Milk (wt%)	Butter (wt% ± SD)
		Saturated fatty acids			
4:0	4.5	5.31 ± 0.30	16:0	—	0.29 ± 0.01
6:0	2.3	2.81 ± 0.09	16:0	28.2	28.13 ± 0.37
8:0	1.3	1.56 ± 0.08	17:0	0.7	0.52 ± 0.01
10:0	2.7	3.14 ± 0.06	17:0	—	0.50 ± 0.01
11:0	0.3	—			
12:0	3.0	3.39 ± 0.06	17:0	0.6	0.57 ± 0.01
13:0	—	0.13 ± 0.01	18:0	—	0.09 ± 0.01
13:0	0.2	0.11 ± 0.00	18:0	12.6	10.62 ± 0.11
14:0	0.1	0.15 ± 0.00	19:0	—	0.14 ± 0.03
14:0	10.6	10.78 ± 0.17	20:0	0.2	0.20 ± 0.03
15:0	0.7	0.30 ± 0.01			
15:0	—	0.49 ± 0.01			
15:0	1.0	1.03 ± 0.01			
		Monounsaturated fatty acids			
10:1	—	0.31 ± 0.01	18:1*t*	1.7	—
12:1	—	0.09 ± 0.01	18:1n9	21.4	20.84 ± 0.79
14:1n5	0.9	0.90 ± 0.02	18:1n7	—	0.15 ± 0.02
15:1	0.3	—			
16:1*t*	—	0.27 ± 0.02	20:1n9	0.6	0.29 ± 0.06
16:1n7	1.8	1.38 ± 0.03	22:1	—	0.09 ± 0.05
17:1	0.4	0.28 ± 0.04			
		Polyunsaturated fatty acids			
18:2*t*	0.4	0.47 ± 0.04	20:2n6	—	0.03 ± 0.02
18:2n7	—	0.15 ± 0.02	20:3	—	0.10 ± 0.01
18:2n6	2.9	2.01 ± 0.14	20:4n6	0.2	0.14 ± 0.01
18:2n4	—	0.09 ± 0.03	20:4n3	—	0.11 ± 0.05
18:3n6	2.9	0.08 ± 0.02	20:5n3	—	0.08 ± 0.04
18:3n3	0.3	0.48 ± 0.05	22:6n3	—	0.09 ± 0.05
18:4n3	—	0.27 ± 0.04	Unknown	—	0.23 ± 0.26

[a]Personal communication, J. Sampugna, University of Maryland, (1993). Analyses done with butyl esters of 4:0–14:0 and methyl esters of 14:0 and up and temperature.
[b]Personal communication, S. J. Iverson and R. G. Ackman, Technical University of Nova Scotia (1993). Analyses done with butyl esters and temperature programming.

TABLE XVI
Fatty Acid Composition (wt%) of Milk Fat from Cows Fed Diets High in Cereals and Low in Forage[a]

Cereal Source	Milk fat (%)	Fatty acid											
		4:0	6:0	8:0	10:0	12:0	14:0	14:1	16:0	16:1	18:0	*trans* 18:1	*cis* 18:1	18:2
Barley	2.2	—	1.6	0.9	2.8	3.3	9.7	1.9	24.7	3.3	11.1	—	29.1	5.7
Hay	3.4	—	2.1	1.0	2.8	3.1	9.7	2.3	26.6	3.9	11.2	—	27.9	2.9
Corn	< 2	0.9	0.8	0.4	1.7	2.4	7.3	2.9	21.0	4.8	6.5	—	39.9	7.7
Control	4.1	3.4	2.2	1.0	3.9	4.3	10.5	1.7	23.0	2.6	12.1	—	26.0	1.2
Corn	1.8	0.6	0.9	0.5	1.6	2.9	9.3	3.2	25.8	3.6	3.4	11.1	16.7	7.6
Control	3.3	2.4	2.3	1.6	4.1	5.1	13.3	0.8	37.2	2.1	9.0	0.7	17.2	2.9
Corn	1.6	2.7	1.1	0.4	1.7	3.9	10.9	—	25.7	—	4.6	—	35.8	4.7
Control	3.6	3.7	2.5	1.6	3.8	5.5	15.1	—	28.0	—	9.7	—	22.1	1.3

[a]Adapted from Palmquist *et al.* (1993).

TABLE XVII

Effect of Dietary Fat Source and Level on Milk Fatty Acid Composition in a Feeding Trial[1]

Fatty acid	g/100 g of methyl esters								
	Source ($n = 12$)						Level (n = 36)		
	Basal	Animal–vegetable blend	Ca soap	Hydrogenated animal fat	Saturated fatty acids	Tallow	Low	High	*P*
4:0	3.34	3.66	3.81	3.79	3.62	3.49	3.54	3.69	NS[2]
6:0	2.70^{a}	2.40^{ab}	2.48^{ab}	2.53^{ab}	2.46^{ab}	2.34^{b}	2.48	2.48	NS
8:0	1.75^{x}	1.34^{y}	1.35^{y}	1.39^{y}	1.41^{y}	1.34^{y}	1.47	1.39	0.03
10:0	3.97^{x}	2.51^{y}	2.57^{y}	2.63^{y}	2.72^{y}	2.60^{y}	3.01	2.66	0.001
12:0	4.64^{x}	2.75^{y}	2.84^{y}	2.88^{y}	3.03^{y}	2.89^{y}	3.38	2.97	0.001
14:0	13.01^{x}	9.33^{y}	9.54^{y}	10.28^{y}	10.10^{y}	10.30^{y}	10.83	10.03	0.001
14:1	1.46^{a}	1.08^{b}	1.07^{b}	1.26^{ab}	1.26^{ab}	1.31^{ab}	1.33	1.15	0.001
15:0	1.28^{x}	0.87^{y}	0.84^{y}	1.07^{y}	1.06^{y}	1.04^{y}	1.05	1.01	0.01
16:0	$29.87^{b,y}$	$26.45^{c,z}$	$34.15^{a,x}$	$28.42^{bc,x}$	$32.67^{a,xy}$	$28.41^{bc,z}$	29.73	30.26	0.07
16:1	1.68^{y}	1.64^{y}	1.64^{y}	1.72^{y}	1.99^{x}	1.80^{xy}	1.78	1.71	NS
17:0	0.60^{y}	0.52^{y}	0.39^{z}	0.88^{x}	0.78^{x}	0.82^{x}	0.66	0.67	NS
18:0	$9.05^{b,yz}$	$11.50^{a,xy}$	$7.71^{c,z}$	$11.68^{a,x}$	$9.86^{b,xy}$	$10.43^{ab,xy}$	10.16	9.92	NS
18:1	$17.22^{c,z}$	$25.74^{a,x}$	$22.80^{ab,xy}$	$22.89^{ab,xy}$	$20.30^{b,yz}$	$23.26^{ab,xy}$	21.60	22.46	0.06
18:2	$2.24^{b,xy}$	$2.00^{bc,yz}$	$2.58^{a,x}$	$1.67^{c,z}$	$1.74^{c,z}$	$1.59^{c,z}$	2.01	1.92	0.05
18:3	$0.55^{c,z}$	$1.16^{a,x}$	$0.63^{c,yz}$	$0.72^{bc,yz}$	$0.62^{c,yz}$	$0.91^{b,xy}$	0.76	0.85	0.01

[1]Adapted from Palmquist *et al.* (1993).

[2]$P > 0.05$.

[a,b,c]Values with different superscripts differ ($P < 0.05$).

[x,y,z]Values with different superscripts differ ($P < 0.01$).

TABLE XVIII
Seasonal Variation in Fatty Acid Composition (wt%) of Milk Fat in the United States[a]

	Fatty acid											
Month	4	6	8	10	12	14	14:1	16	16:1	18	18:1	18:2
February	3.48	2.44	1.24	2.95	3.52	11.63	2.57	29.89	3.32	9.68	26.51	2.77
May	3.42	2.36	1.20	2.82	3.38	11.20	2.58	28.40	3.36	10.14	28.10	3.05
August	3.07	2.28	1.12	2.55	3.10	10.92	2.66	28.76	3.41	10.28	29.00	2.86
November	3.33	2.31	1.20	2.90	3.54	11.80	2.69	30.78	3.37	9.37	26.19	2.53

[a]Adapted from Barbano (1990) and Palmquist *et al.* (1993).

5. *Types of Fatty Acids*

a. Saturated and branched-chain fatty acids. Saturated even and odd *n*-chain acids from 2 to 28 carbons have been found in milk (Jensen, 1992). Many of the identifications were unequivocally confirmed by mass spectrometry. Many of these are present in small quantities, less than 1%, and are of little known importance. The amounts of major fatty acids are listed in Tables XIV and XV and the numbers of all in Table XX. Milk also contains a small quantity, up to 2.5% (Tables XIV and XV), of a variety of branched-chain fatty acids.

b. Monounsaturated fatty acids. Oleic acid (*cis*-9-18:1) accounts for most of these, about 97.5% of the geometric isomers (*cis* or *trans*), 85% of the positional isomers (delta-5-16), and 85% of the fatty acids (14:1–22:1) (Patton and Jensen, 1976; Jensen and Clark, 1988). In contrast, the double bond in the *trans*-18:1s was mostly at the 11 position (35.7%) but with approximately 10% each at 9, 10, 13, and 14. Recent data were obtained by Wolff (1994) who found 3.8% *trans*-18:1 in French butters. About 53% was the *trans*-11 isomer, with smaller quantities of 6–10 and 12–16 isomers. The number of these acids are given in Table XX.

c. Polyunsaturated fatty acids. Although dairy cattle consume relatively large amounts of PUFA, the amounts in milk are low because of ruminal biohydrogenation. The amounts in Tables XIII and XIV range from 2.37 to 2.9% 18:2, presumably n6 and all *cis*. Even though the amounts of this and other PUFA are small, efforts should be made to include them in GLC analyses of milk and dairy products. In a mixed diet, they contribute significant quantities and the total should be known. The importance of the PUFA are the roles they play as essential fatty acids, membrane components, and eicosanoid precursors.

As would be expected, many other PUFA and their isomers have been identified (Patton and Jensen, 1976; Jensen and Clark, 1988). These are listed in Table XX.

TABLE XIX
Regional Variation in Fatty Acid Composition (wt%) of Milk Fat in the United States[a]

	Fatty acid											
Region[b]	4	6	8	10	12	14	14:1	16	16:1	18	18:1	18:2
1	3.33	2.30	1.14	2.70	3.25	10.79	2.52	29.21	3.23	10.31	27.26	3.47
2	3.30	2.33	1.17	2.83	3.44	11.30	2.55	30.31	3.34	10.00	26.46	2.97
3	3.41	2.35	1.20	2.83	3.43	11.41	2.66	28.97	3.30	9.97	27.72	2.75
4	3.37	2.35	1.21	2.86	3.44	11.57	2.61	29.82	3.35	9.75	27.16	2.50
5	3.29	2.35	1.20	2.86	3.47	11.61	2.66	29.70	3.44	9.65	27.12	2.66
6	3.20	2.39	1.22	2.89	3.47	11.71	2.72	30.02	3.49	9.47	26.86	2.54
7	3.25	2.34	1.21	2.85	3.44	11.72	2.71	30.08	3.50	9.52	26.88	2.50
8	3.33	2.40	1.23	2.87	3.43	11.67	2.71	29.77	3.49	9.57	26.93	2.60
9	3.44	2.37	1.20	2.77	3.31	11.25	2.61	28.74	3.34	9.91	28.19	2.88
10	3.24	2.28	1.14	2.64	3.19	11.04	2.56	28.66	3.28	10.24	28.79	2.94

[a]Barbano (1990) and Palmquist *et al.* (1993).
[b]Regions: 1, far south; 2, western states; 3, NE, IA, KS, MO; 4, ND, SD, western MN; 5, eastern MN; 6, southern WI, IL, IN, lower MI, OH; 7, northeast WI, upper MI; 8, northwest WI; 9, southern NY, PA, VA, WV, MD; 10, northern NY, New England.

TABLE XX
Fatty Acid Composition of Bovine Milk Lipids as of March 1994[a]

Type	No.	Identity
Saturates		
Normal	27	14:2–28, even; 3–27, odd
Monobranched	39	11–24; 13–19; three or more positional isomers, 5–10; Me, Et
Multibranched	16	16–28
Monoenes		
cis	61	9–25, positional isomers of 12:1–14:1, 16:1–18:1, and 23:1–25:1
Trans	49	14; 16–25; positional isomers of 14:1–16:1, 18:1 and 23:1 to 25:1
Dienes	45	14–26, evens only; *cis, cis; cis, trans;* or *trans, cis* and *trans,* unconjugated and conjugated and positional isomers
Polyenes		
Tri-	10	18, 20, 22; geometric positional conjugated and unconjugated isomers
Tetra-	5	18, 20, 22; positional isomers
Penta-	2	20, 22
Hexa-	1	22
Keto (oxo)		
Saturated	44	6, 8–10, 12, 14, 15–20, 22, 24; positional isomers
Unsaturated	21	14, 16, 18; positional isomers of carbonyl and double bond
Hydroxy		
2-Position	16	14:0, 16:0–26:0, 16:1, 18:1, 21:1, 24:1, 25:1
4 and 5 Position	9	10:0–16:0, 14:1–6, and 12:1–9
Other positions	60	
Cyclic		
Hexyl	1	11; terminal cyclohexyl
Total	406	

[a]Corrected from Jensen (1992).

Recently, conjugated linoleic acid isomers, notably 9,11-18:2*ct,* with anticarcinogenic activity, have been found in bovine milk (2.8 mg/100 g fat) and a variety of cheeses and ground beef (Ha *et al.,* 1989), as well as Australian dairy products and human milk (Fogerty *et al.,* 1988). The 9,11 isomer is a potent antioxidant equivalent to BHT (Ha *et al.,* 1990). We have detected the isomer in human milk, infant formulas, and evaporated milk. The anticarcinogenic effects of consuming dairy products must also be considered.

d. Other acids. Many keto (oxo) and hydroxy fatty acids and others have been identified in milk fat (Jensen and Clark, 1988; and of oxo acids by Brechany and Christie, 1992, 1994). These are listed in Table XX with a corrected total of 406, more than any other fat.

e. Fatty acids and flavor of dairy products. The free volatile short-chain fatty acids, *n*- and branched-chain, contribute to the characteristic flavors of ripened cheeses (Seitz, 1990; Ha and Linday *et al.,* 1990). Unfortunately, 4:0, and to a lesser extent, 6:0–10:0, can produce an extremely unpleasant flavor in raw milk when the lipoprotein lipase therein is activated (IDF, 1991). This can result when excessive foaming or agitation of raw milk occurs. The 2-oxo and 4- and 5-hydroxy acids are precursors of methyl ketones and γ- or δ-lactones which contribute to flavor as do the aldehydes resulting from the oxidation of unsaturated fatty acids (Seitz, 1990).

f. Related compounds. Ahrne *et al.* (1979) identified these alkyl ethers in the glycerol ethers from neutral and phospholipids: 14, 14:1, 15:1, 16, 16:1, 17:1, 18, 18:1, 19:1, and 20:1.

VI. Summary

Bovine milk lipids are very complex containing over 400 fatty acids, probably thousands of TG, and many microlipids. Since they are readily available in butter, they have been extensively investigated, yet because of their complexity, much remains to be done. We believe that areas in which more research is needed are (1) determination of the fatty acids in market milk and dairy products with modern methods, (2) studies of compartmentation in homogenized milk, (3) determination of TG species with reference to atherogenesis, (4) identification and roles of microlipids, and (5) recognition that we still have much to learn about milk.

Acknowledgments

Some of the research reported herein was supported in part by federal funds made available through the provision of the Hatch Act and by NIH Contract N01-HD-2817 and Grant H-9414. Scientific Contribution 1539, Storrs Agricultural Experiment Station, The University of Connecticut, Storrs, CT 06269.

References

Ahrne, L., Bjorck, L., Raznikiewicz, T., and Claesson, O. (1979). Glycerol ethers in colostrum and milk from cow, goat, pig, and sheep. *J. Dairy Sci.* **63,** 741–745.

American Public Health Association (APHA) (1993). "Standard Methods for the Examination of Dairy Products," 16th Ed. APHA, Washington, DC.

Association of Official Analytical Chemists Inc. (AOAC) (1990). "Official Methods of Analyses of the Association of Official Analytical Chemists" (K. Hydrich, ed.), 15th ed., part 2.

Bachman, K. E., Lin, J-H, and Wilcox, C. J. (1976). Sensitive colorimetric determination of cholesterol in dairy products. *J. Assoc. Off. Anal. Chem.* **59,** 1146–1149.

Barbano, D. M. (1990). Seasonal and regional variation in milk composition in the U.S. *In* "Proceedings of the 1990 Cornell Nutrition Conference," pp. 96–105. Cornell University, Ithaca, NY.

Barron, L. J. R., Hierro, M. T. G., and Santa Maria, G. (1990). HPLC and GLC analysis of the triglyceride composition of bovine, ovine and caprine milk fat. *J. Dairy Res.* **57,** 517–526.

Berner, L. A. (1993). Defining the role of milk fat in balanced diets. *Adv. Food Nutr. Res.* **17,** 131–257.

Bitman, J., and Wood, D. L. (1990). Changes in milk phospholipids during lactation. *J. Dairy Sci.* **73,** 1208–1216.

Bonafede, D. M., Macala, L. J., Constantine-Paton, M., and Yu, R. K. (1989). Isolation and characterization of ganglioside 9-0-acetyl-GD3 from bovine buttermilk. *Lipids* **24,** 680–684.

Brechany, E. Y., and Christie, W. W. (1992). Identification of the saturated oxo fatty acids in cheese. *J. Dairy Res.* **59,** 57–64.

Brechany, Y., and Christie, W. W. (1994). Identification of the unsaturated fatty acids in cheese. *J. Dairy Res.* **61,** 111–115.

Bushway, A. A., and Keenan, T. W. (1978). Composition and synthesis of three higher ganglioside analogs in bovine mammary tissue. *Lipids* **13,** 59–65.

Christie, W. C. (1983). The composition and structure of milk lipids. *In* "Developments in Dairy Chemistry—2. Lipids" (P. F. Fox, ed.), pp. 1–36. Applied Sciences, New York.

Christie, W. W. (1987). The analysis of lipids with special reference to milk fat. *In* "Recent Advances in Chemistry and Technology of Fats and Oils" (R. J. Hamilton and A. Bhati, eds.), pp. 57–78. Applied Sciences, New York.

Christie, W. W., and Clapperton, J. L. (1982). Structures of the triglycerides of cows' milk, fortified milks (including infant formulae), and human milk. *J. Soc. Dairy Technol.* **35,** 22–24.

Christie, W. W., Noble, R. C., and Davies, C. (1987). Phospholipids in milk and dairy products. *J. Soc. Dairy Technol.* **40,** 10–12.

Christophersen, S. W., and Glass, R. L. (1969). Preparation of milk fat methyl esters by alcoholysis in an essentially nonalcoholic solution. *J. Dairy Sci.* **52,** 1289–1290.

Elliot, J. M., de Haan, B., and Parkin, K. L. (1989). An improved liquid chromatographic method for the quantitative determination of free fatty acids in milk products. *J. Dairy Sci.* **72,** 2478–2482.

Enig, M. L., Pallansch, L. A., Sampugna, J., and Keeney, M. (1983). Fatty acid composition of the fat in selected food items with emphasis on *trans* components. *J. Am. Oil Chem. Soc.* **60,** 1789–1795.

Fogerty, A. C., Ford, G. L., and Suoronus, D. (1988). Octadeca-9, 11-dienoic acid in foodstuffs and in the lipids of human blood and breast milk. *Nutr. Rep. Int.* **38,** 937–944.

Gresti, J., Bugaut, M., Maniongul, C., and Bezard, J. (1993). Composition of molecular species of triacylglycerols in bovine milk fat. *J. Dairy Sci.* **76,** 1850–1869.

Ha, J. K., and Lindsay, R. C. (1990). Method for the quantitative analysis of volatile free and total branched-chain fatty acids in cheese and milk fat. *J. Dairy Sci.* **73,** 1988–1999.

Ha, Y. L., Storkson, J., and Pariza, M. W. (1990). Inhibition of benzo (α) pyrene induced mouse forestomach neoplasia by conjugated dienoic derivatives of linoleic acid. *Cancer Res.* **50,** 1097–1101.

Ha, Y. L., Grimm, N. K., and Pariza, M. W. (1989). Newly recognized anticarcinogenic fatty acids: Identification and quantification in natural and processed cheese. *J. Agric. Food Chem.* **37,** 75–81.

Hauttecoeur, B., Sonnino, S., and Ghidoni, R. (1985). Characterization of two molecular species GD3 gangliosides from bovine buttermilk. *Biochim. Biophys. Acta* **833,** 303–307.

Hladik, J., and Michalec, C. (1966). Ceramide-monohexosides and ceramide-dihexosides in lipoproteins of the membranes of fat globules in bovine milk. *Acta Biologica Medica Germanica* **16,** 696–699.

Huang, R. T. C. (1973). Isolation and characterization of the gangliosides of buttermilk. *Biochim. Biophys. Acta* **306,** 82–84.

Hurst, W. J., Aleo, M. D., and Martin, R. A., Jr. (1983). High performance liquid chromatographic analysis of cholesterol in milk. *J. Dairy Sci.* **66,** 2192–2194.

International Dairy Federation (IDF) (1991). Determination of free fatty acids in milk and dairy products. *In* "Bulletin of the International Dairy Federation," Vol. 265. Brussels, Belgium.

International Dairy Federation (IDF) (1992). Milk fat and milk fat products: Determination of cholesterol content. *In* "Bulletin of the International Dairy Federation Standard," Vol. 159, pp. 92. Brussels, Belgium.

Itabashi, Y., Myher, J. J., and Kiksis, A. A. (1993). Determination of positional distribution of short-chain fatty acids in bovine milk fat on chiral columns. *J. Am. Oil Chem. Soc.* **70,** 1177–1181.

James, A. T., and Martin, A. J. P. (1956). Gas–liquid chromatography: The separation and identification of the methyl esters of saturated and unsaturated acids from formic to n-octadecanoic acid. *Biochem. J.* **63,** 144–152.

Jenness, R. (1985). Biochemical and nutritional aspects of milk and colostrum. *In* "Lactation" (B. L. Larson, ed.), pp. 164–197. Iowa State University Press, Ames.

Jensen, R. G. (1992). Fatty acids in milk and dairy products. *In* "Fatty Acids in Food and their Health Implications" (C. K. Chow, ed.), Chap. 5. Dekker, New York.

Jensen, R. G., and Clark, R. M. (1988). Lipid composition and properties. *In* "Fundamentals of Dairy Chemistry" (N. P. Wong, ed.), 3rd Ed., pp. 171–214. Van Nostrand–Reinhold, New York.

Jensen, R. G., Ferris, A. M., and Lammi-Keefe, C. J. (1991). The composition of milk fat. *J. Dairy Sci.* **74,** 3228–3243.

Jensen, R. G., Lammi-Keefe, C. J., and Ferris, A. M. (1992). Effect of milk triacylglycerol structure in absorption and metabolism. *Inform* **3,** 560. [Abstract]

Kallio, H P., Laakso, R., Huopalahti, R., Linko, R. R., and Oksman, P. (1989). Analysis of butterfat triacylglycerols by supercritical fluid chromatography/electron impact mass spectrometry. *Anal. Chem.* **61,** 698.

Kayser, S. G., and Patton, S. (1970). The function of very long chain fatty acids in membrane structure: Evidence from milk cerebrosides. *Biochem. Biophys. Res. Commun.* **41,** 1572–1578.

Keenan, T. W. (1974). Composition and synthesis of gangliosides in mammary gland and milk of the bovine. *Biochim. Biophys. Acta* **337,** 255–270.

Keenan, T. W., Huang, C. M., and Morre, D. J. (1972). Gangliosides: Nonspecific localization in the surface membranes of bovine mammary gland and liver. *Biochem. Biophys. Res. Commun.* **47,** 1277–1283.

Kermasha, S., Kubow, S., Safari, M., and Reid, A. (1993). Determination of the positional distribution of fatty acids in butterfat triacylglycerols. *J. Am. Oil Chem. Soc.* **70,** 169–173.

Kuksis, A., and Breckenridge, W. C. (1968). Triglyceride composition of milk fats. In "Dairy Lipids and Lipid Metabolism" (M. F. Brink and D. Kritchevsky, eds.), pp. 28–98. Avi, Westport, CT.

Kuksis, A., Marai, L., and Myher, J. J. (1973). Triglyceride structure of milk fats. *J. Am. Oil Chem. Soc.* **50,** 193–201.

Laegrid, A., Otnaess, A-B. K., and Fuglesang, J. (1986). Human and bovine milk: Comparison of ganglioside composition and enterotoxin-inhibitory activity. *Pediatr. Res.* **20,** 416–421.

Maniongui, C., Gresti, J., Bugant, M., and Gauthier, J. (1991). Determination of bovine butterfat triacylglycerols by reversed-phase liquid chromatography and gas chromatography. *J. Chromatogr.* **543,** 81–103.

Maxwell, R. J., Mondimore, D., and Tobias, J. (1986). Rapid method for the quantitative extraction and simultaneous class separation of milk lipids. *J. Dairy Sci.* **60,** 321–325.

Mincione, B., Spagna Musso, S., and de Francisco, G. (1977). Studies on milk from different species. 2. Sterol content in cows' milk. *Milchwissenschaft* **32,** 599–303.

Morrison, W. R., and Smith, L. M. (1964). Identification of ceramide monohexoside and ceramide dihexoside in bovine milk. *Biochim. Biophys. Acta* **84,** 759–761.

Myher, J. J., Kuksis, A., Marai, L., and Sandra, P. (1988). Identification of the more complex triacylglycerols in bovine milk fat by gas chromatography–mass spectrometry using polar capillary columns. *J. Chromatogr.* **452,** 93–118.

Myher, J. J., Kuksis, A., and Marai, L. (1993). Identification of the less common isologous short-chain triacylglycerols in the most volatile 2.5% molecular distillate of butter oil. *J. Am. Oil Chem. Soc.* **70,** 1183–1191.

National Dairy Council (NDC) (1993). "Newer Knowledge of Milk and Other Fluid Dairy Products." NDC, Rosemont, IL.

Newburg, D. S., and Chaturvedi, P. (1992). Neutral glycolipids of human and bovine milk. *Lipids* **27,** 923–927.

Ochanda, J. O., Syoto, B., Ohishi, I., Naiki, M., and Kubo, S. (1986). Binding of *Clostridium botulinum* neurotoxin to gangliosides. *J. Biochem.* **100,** 27–33.

Palmquist, D. L., Beaulieu, A. D., and Barbano, D. M. (1991). Feed and animal factors influencing milk fat composition. *J. Dairy Sci.* **76,** 1753–1771.

Parodi, P. W. (1981). Relationship between triglyceride structure and the softening point of milk fat. *J. Dairy Res.* **48,** 131–138.

Patton, S., and Jensen, R. G. (1976). "Biomedical Aspects of Lactation." Pergamon, New York.

Posati, L. P., and Orr, M. L. (1976). "Composition of Foods: Dairy and Egg Products: Raw–Processed–Prepared." Agriculture Handbook 8.1, ARS, USDA, Washington, DC.

Puente, R., Garcia-Pardo, L.-A., and Hueso, P. (1992). Gangliosides in bovine milk. Changes in content and distribution of individual ganglioside levels during lactation. *Biol. Chem. Hoppe-Seyler.* **373,** 283–288.

Ren, S., Scarsdale, J. N., Ariga, T., Zhang, Y., Klein, R. A., Hartmann, R., Kushi, Y., Egge, H., and Yu, R. K. (1992). O-acetylated gangliosides in bovine buttermilk. *J. Biol. Chem.* **267,** 12632–12683.

Renner, E. (1983). "Milk and Dairy Products in Human Nutrition," p. 14. W-GmbH, Volkswirtsscheftlicher Verlag, Munich.

Seitz, E. W. (1990). Microbial and enzyme-induced flavors in dairy foods. *J. Dairy Sci.* **73,** 3664–3691.

Sukhija, P. S., and Palmquist, D. L. (1988). Rapid method for determination of total fatty acid content and composition of feedstuffs and feces. *J. Agric. Food Chem.* **36,** 1202–1206.

Takamizawa, K., Iwamori, M., Mutai, M., and Nagai, Y. (1986). Gangliosides of bovine buttermilk. Isolation and characterization of a novel monosialoganglioside with a new branching structure. *J. Biol. Chem.* **261,** 5625–5630.

Timmen, H., and Dimick, P. S. (1972). Structure and synthesis of milk fat. X. Characterization of the major hydroxy compounds of milk lipids. *J. Dairy Sci.* **55,** 919–925.

Tsui, I. C. (1989). Rapid determination of total cholesterol in homogenized milk. *J. Assoc. Off. Anal. Chem.* **72,** 421–424.

Walstra, P., and Jenness, R. (1984). Table A.4. Various neutral lipids in milk fat. *In* "Dairy Chemistry and Physics," p. 399. Wiley, New York.

Wolff, R. L. (1994). Contributions of *trans*-18:1 acids from dairy fat to European diets. *J. Am. Oil Chem. Soc.* **71,** 277–283.

Wood, D. L., and Bitman, J. (1986). Cholesteryl esters of cow's milk. *J. Dairy Sci.* **69,** 2203–2208.

CHAPTER 7

Minerals, Ions, and Trace Elements in Milk

A. Ionic Interactions in Milk

MARGARET C. NEVILLE
PEIFANG ZHANG
JONATHAN C. ALLEN

I. Introduction

Knowledge of the structural and electrochemical compartmentalization of the ionic components of milk is important to the understanding of ion secretion, ion absorption from the gastrointestinal tract, and ionic effects on the properties of food products derived from milk. Certain ions, sodium, potassium, and chloride exist largely in the ionized state in the aqueous compartment of milk. The other, divalent cations of milk are distributed among the structural compartments and protein components in a highly specific manner. Most, including calcium, magnesium, and zinc, have measurable concentrations of free ion and are part of the complex electrochemical equilibrium depicted in Figure 1 for calcium and magnesium. Many of these ions also are bound with very high affinity to specific milk proteins. In this chapter we focus on both the ionic equilibria in the aqueous compartment of milk and the distribution of ions among its structural compartments. We will begin by considering the methodologies available for measurement of these parameters. It will then be necessary to consider milk pH and the major hydrogen ion buffers that regulate it. Monovalent ions will be briefly considered; then the compartmentalization and equilibria of the divalent cations and the anions to which they bind will be discussed at some length. In general, our discussion will be confined to

HANDBOOK OF MILK COMPOSITION

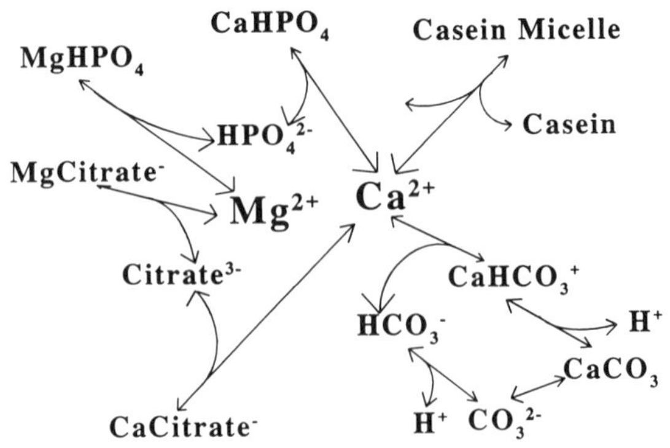

Figure 1 Principle interactions of calcium and magnesium in the aqueous compartment of milk.

human and bovine milk, the two species for which substantial data are available. Calcium and zinc will receive a good bit of attention because of their nutritional importance and the complexity of their interactions.

II. Methodologies

A. Analysis of Ionic Distribution in the Structural Compartments of Milk

The structural compartments of milk have been described in Chapter 2 as consisting of the milk fat, the cellular compartment, the aqueous compartment, casein, and the membrane or "fluff" compartment. These compartments are best separated by centrifugation as illustrated in Figure 2. A 20-min centrifugation at 6000*g* separates the cream and cellular fractions from fresh whole milk. The aqueous infranatant is carefully removed using a needle and syringe to avoid disturbing the loose cream layer. An aliquot is reserved for analysis and the remainder subjected to ultracentrifugation at 100,000*g* for 1 or 2 hr (4°C) to bring down the casein micelles which form a solid pellet in the bottom of the tube. A loose pellet, called the fluff, directly above the casein contains the so-called skim milk membranes (Huston and Patton, 1986). A thin lipid layer on the surface contains residual milk fat. If necessary, the infranatant can be centrifuged through a 10,000 MW cutoff filter to separate ionic species bound to macromolecules. If this step is used it is important to determine that the ionic species in question does not bind to the filter.

Although it is possible to analyze the cream and casein fractions directly for their ionic content, because these fractions are difficult to handle it is much simpler to analyze only the aqueous and fluff fractions

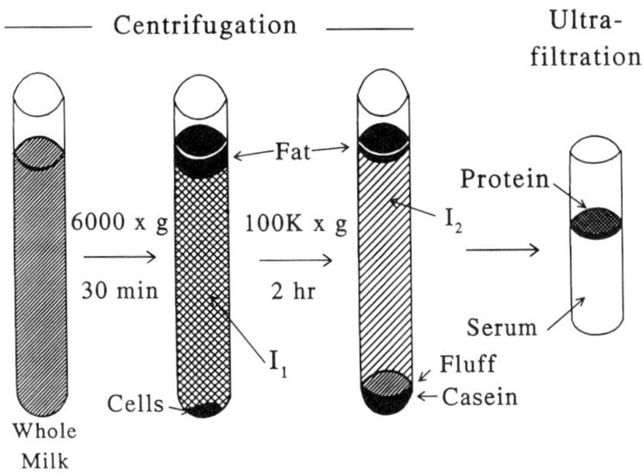

Figure 2 Protocol for separation of structural compartments of human milk. The first centrifugation is carried out at room temperature, the second at 4°C. To prevent protease action and bacterial growth, 0.01 μg of sodium azide and 50 μl of a solution of protease inhibitors (15 mg/ml of iodoacetamide and 15 mg/ml of phenylmethyl sulfonyl fluoride) are added to each milk sample. After measurement of pH, small aliquots are taken for analysis of total lipid by the creamatocrit (Lucas *et al.*, 1978) or other convenient method. Another aliquot is put aside for the ion concentration in whole milk, and the remainder is subjected to the centrifugation protocol shown. To determine ion binding to macromolecular components of the aqueous phase after ultracentrifugation (I_2), the fluid can be subjected to centrifugation through a 10,000 MW cutoff filter. The filtrate contains free ions and ions bound to small molecular weight components of milk.

using the following equations to calculate the ionic concentrations in the cream and casein fractions. The equations also correct for the presence of aqueous infranatant in the fluff fraction. The assumptions made in these calculations are (i) that the infranatant (I_1) composition after the first centrifugation reflects loss of the fat fraction, (ii) that the cellular fraction is too small to affect ionic distribution, and (iii) that after the high-speed centrifugation the composition of the infranatant (I_2) reflects the loss of the casein and fluff fractions. The amount of solute is referred to its concentration in whole milk. Solute associated with the fluff fraction is calculated directly from analysis of this fraction correcting for the presence of aqueous supernatant (I_2). The units of the quantities resulting from these calculations are mmol or grams of ion per liter of whole milk.

Cation associated with the fat fraction is

$$[S_f] = [W] - [I_1](1 - F), \qquad [1]$$

where $[S_f]$ is the cation associated with the fat, $[W]$ is the concentration in whole milk, $[I_1]$ is the cation concentration in the infranatant after the 6000g centrifugation, and F is the fraction of fat in the whole milk sample.

Cation associated with the casein and fluff fractions, S_p, is

$$[S_p] = ([I_1] - [I_2])(1 - F). \qquad [2]$$

Cation associated with the aqueous fraction, S_{aq}, is

$$[S_{aq}] = [W] - [S_f] - [S_p]. \qquad [3]$$

Cation associated with the fluff fraction, S_m, is

$$[S_m] = ([\mathrm{Fl}] - [I_2])(V), \qquad [4]$$

where V is the volume of the fluff fraction divided by the initial volume of the milk sample.

Cation associated with casein is

$$[S_c] = [S_p] - [S_m]. \qquad [5]$$

B. *Analysis of Electrochemical Equilibria in the Aqueous Compartment*

In order to determine the electrochemical interactions of any ionic species in milk it is necessary to know the ionic activity, the concentrations of all potential ligands, and the equilibrium constants for the relevant interactions. Table I gives the total concentration of the relevant ionic species in human and bovine milk and Table II gives the equilibrium constants and equations for the interactions of these species. Zinc is not included in Table I because its concentration in milk is highly dependent on the duration of

TABLE I
Major Ionic Constituents of Human and Bovine Milk[a]

Component	Units	Human milk	Bovine milk
Sodium	mmol/liter	6.3	24.2
Potassium	mmol/liter	13.9	34.7
Chloride	mmol/liter	11.6	30.2
Magnesium, total	mmol/liter	1.8	5.1
Calcium, total	mmol/liter	7.5	29.4
Calcium, ionized	mmol/liter	3.0	2.0[b]
Citrate	mmol/liter	2.6	9.2
Phosphate, free	mmol/liter	1.8	11.2
Bicarbonate	mmol/liter	6.0	4.8
Casein	g/liter	18.0	261
pH		6.8	6.7

[a]Values for composition of human milk at 90 days postpartum from Allen *et al.* (1991) except casein which is taken from Casey and Hambidge (1983). Values for bulk herd bovine milk from White and Davies (1958).

[b]Value obtained by calculation using a series of assumed equilibrium constants (Holt *et al.*, 1981). The murexide method gave values closer to 3.0 (Holt *et al.*, 1981).

TABLE II
Equations for Ionic Equilibria in Milk

	Equilibrium constant	Products/reactants
k_w	−13.357	$[H^+] \times [OH^-]/[HOH]$
k_c	−10.585	$([H^+] \times [HCO_3^-])/pCO_2$
k_3	−10.059	$([H^+] \times [CO_3^=])/[HCO_3^-]$
k_4	−3.029	$([\text{Citrate}^-] \times [H^+])/[\text{citrate}]$
k_5	−4.580	$([\text{Citrate}^{2-}] \times [H^+])/[\text{citrate}^-]$
k_6	−6.102	$([\text{Citrate}^{3-}] \times [H^+])/[\text{citrate}^{2-}]$
k_7	−2.131	$([H_2PO_4^-] \times [H^+])/[H_3PO_4]$
k_8	−7.255	$([HPO_4^{2-}] \times [H^+])/[H_2PO_4^-]$
k_9	−12.104	$([PO_4^{3-}] \times [H^+])/[HPO_4^{3-}]$
j_1	4.606	$[\text{Ca·citrate}^-]/(Ca^{2+}] \times [\text{citrate}^{3-}])$
j_2	2.536	$[\text{Ca·citrate}]/(Ca^{2+}] \times [\text{citrate}^{2-}])$
j_3	1.250	$[\text{Ca·citrate}^+]/(Ca^{2+}] \times [\text{citrate}^-])$
j_4	0.840	$[Ca{\cdot}H_2PO_4^+]/([Ca^{2+}] \times [H_2PO_4^-]$
j_5	2.401	$[Ca{\cdot}HPO_4]/([Ca^{2+}] \times [HPO_4^{2-}]$
L1	4.72	$[\text{Zn·citrate}^-]/(Zn^{2+}] \times [\text{citrate}^{3-}])$
L2	3.72	$[\text{Zn·citrate}]/(Zn^{2+}] \times [\text{citrate}^{2-}])$
L6	4.446	$[\text{Mg·citrate}^-]/(Mg^{2+}] \times [\text{citrate}^{2-}])$
L7	2.21	$[\text{Mg·citrate}]/(Mg^{2+}] \times [\text{citrate}^{2-}])$
L8	2.529	$[Mg{\cdot}HPO_4]/([Mg^{2+}] \times [HPO_4^{2-}]$

Note: Equilibrium constants are expressed as $\log_{10}$ of the constant for the reaction in the forward direction. All constants are corrected to the ionic strength of human milk (μ = 0.03 *M*). Subscripted letters represent the arbitrarily chosen designation for the equilibrium constant.

lactation (Casey *et al.*, 1989). The concentration of ionized species can be determined using either ion-selective electrodes or equilibrium dialysis. Highly specific ion-selective electrodes are available for hydrogen ion, sodium, potassium, and calcium. It has been necessary to use equilibrium dialysis to determine the concentration of free zinc in milk for reasons discussed below (Zhang and Allen, 1992).

Older methods for measuring Ca^{2+} and Mg^{2+} activity included resin equilibrium (Christianson *et al.*, 1954) and dye binding with murexide (Tessier and Rose, 1957). These methods were compared with an ion-selective electrode method and gave higher values for $[Ca^{2+}]$ in milk (Holt *et al.*, 1981). The low readings from the electrode were thought by the authors to be a matrix effect; some electrode systems may be more sensitive to matrix effects than others. Also, alterations of the calcium complex formation by either the resin or the murexide is a possibility.

If resins or dyes are to be used for measuring the activity of metals, the resin or dye should have an appropriate binding and adequate sensitivity.

Allen and Zhang (unpublished) attempted to use these methods to measure zinc activity ([Zn^{2+}]). However, the cation-exchange resins tested had such high zinc affinity that they disrupted the equilibrium in the milk and bound most of the zinc. Lower-affinity resins did not have a high enough affinity and selectivity to bind a measurable quantity of zinc in the presence of a 1000-fold excess of calcium. Similar problems occurred with metal-sensitive dyes. An equilibrium dialysis approach using the colloidal calcium phosphate of bovine casein as the zinc complexing agent proved successful. The zinc content of this material, placed inside a dialysis bag, is proportional to the zinc activity of the dialysate (McGann *et al.*, 1983). Zinc chelators must be used in the dialysate to achieve reliable and reproducible zinc activities in the nanomolar range and to diminish the effect of zinc contamination.

Milk samples should be taken and maintained under anaerobic conditions for measurement of pH and the activity of certain ions for reasons described under Section III. We have described an appropriate technique for human milk (Allen and Neville, 1983).

III. Hydrogen Ion Equilibria in Milk

The following reaction is at equilibrium in milk:

$$CO_2 + H_2O \rightleftharpoons H_2CO_3 \rightleftharpoons H^+ + HCO_3{}^- \rightleftharpoons 2H^+ + CO_3{}^=. \qquad [6]$$

In all mammals the body fluids, including milk, are equilibrated with 5% CO_2. When milk is removed from the breast or udder and exposed to the air, CO_2 is lost, the equation shifts to the left, and the pH rises. The decreased hydrogen ion also shifts the reaction, $H^+ + HCO_3 \rightarrow 2H^+ + CO_3{}^=$, to the right, increasing [$CO^={}_3$]. Because of the high [Ca^{2+}] in milk the increased [$CO^={}_3$] probably also results in formation of a $CaCO_3$ complex, explaining decreased [Ca^{2+}] as milk is stored and pCO_2 decreases.

The bicarbonate concentration in human milk can be calculated from the pH and pCO_2 to vary between about 4.3 and 6.0 mmol/liter (Allen *et al.*, 1991). Equilibrated with a pCO_2 of 40 mm Hg, the pH of human milk as present in the breast is 6.6 to 6.8. Measured values are often given as 7.2 or above because anaerobic precautions are not usually taken and considerable loss of CO_2 occurs particularly when milk is expressed with a breast pump. We have been unable to find values for the pH of bovine milk taken under conditions in which the bicarbonate–CO_2 equilibrium was maintained under *in vivo* conditions.

IV. Distribution of Monovalent Ions in Milk

Measurement of the total concentration of sodium and potassium in human milk by ion-selective electrodes or by flame photometry gave equal values (Neville *et al.*, 1984), implying that these ions are present as the free ionic species. A similar conclusion was reached by Holt *et al.* (1981) for bovine milk. The assumption is usually made that chloride is also completely ionized. At the ionic strength of milk, there may be a small amount of interaction of chloride with sodium, calcium, magnesium, and potassium, amounting to less than 5% of the total in bovine milk (Holt *et al.*, 1981). A similar conclusion applies to human milk with its even lower ionic strength. The concentration of these ions in milk ultrafiltrates (Holt and Jenness, 1984) is also consistent with the conclusion that monovalent ions are present in free solution in the aqueous compartment of milk only.

V. Distribution of Divalent Cations among the Structural Compartments of Milk

The centrifugation protocol illustrated in Figure 2 was used to determine the divalent cation distribution among the compartments of human milk. The results are shown in Table III. Calcium was found only in the aqueous compartment and associated with casein. Contrary to earlier reports from the literature (Neville *et al.*, 1985; Lonnerdal and Fransson, 1981), no calcium was associated with the fat fraction. This is expected because milk fat globules originate from a cellular compartment with a submicromolar calcium concentration. When human milk samples were frozen and thawed prior to analysis, 6.0 ± 2.1% of the calcium was associated with the lipid, suggesting that damage to the milk fat globule membrane during expression or storage may increase calcium binding either to free fatty acids or to milk fat globule membrane fractions (Rüegg and Blanc, 1982). Blake and Henning (1988) observed that negligible calcium was associated with washed cream in rat milk suggesting that the same principle may hold in other species.

TABLE III
Structural Compartmentalization of Divalent Ions in Human Milk[a]

Milk fraction	Calcium (mmol/liter)	Magnesium (mmol/liter)	Copper (μmol/liter)	Zinc (μmol/liter)
Whole milk	7.96 ± 0.46	1.59 ± 0.08	3.34 ± 0.50	26.78 ± 4.37
Aqueous	6.76 ± 0.35	1.38 ± 0.09	1.51 ± 0.22	12.05 ± 2.66
Fat	−0.02 ± 0.23	0.11 ± 0.09	0.89 ± 0.19	8.13 ± 1.94
High-speed pellet	1.15 ± 0.20	−0.07 ± 0.04	1.18 ± 0.24	7.26 ± 1.98
Membranes	0.08 ± 0.02	0.17 ± 0.05	0.16 ± 0.04	0.65 ± 0.09

[a]All concentrations expressed per liter of whole milk.

Although magnesium was found mainly in the aqueous compartment of human milk (Figure 3) significant amounts were also associated with the lipid and membranous compartments; none was associated with the casein pellet. Copper and zinc were more or less evenly distributed between the aqueous, lipid, and casein fractions. Another group found that casein micelles in human milk contained 14% of the total zinc, serum albumin bound 28%, 29% was found to be present in the aqueous compartment, and the remaining 29% was associated with the fat, possibly bound to the alkaline phosphatase in the milk fat globule membrane (Lonnerdal *et al.*, 1982; Hurley and Lonnerdal, 1982; Lonnerdal and Fransson, 1981).

Extrinsic labeling studies (Sandstrom *et al.*, 1983) demonstrated that zinc bioavailability was 28% from human milk, 25% from whey-adjusted cow's milk formula, 15% from cow's milk, and 10% from soy-based formula. It is reasonable to conclude that it is not only the amount of zinc, but also the compounds binding the element that affect the degree to which it is absorbed. In contrast to human milk, practically all the zinc in bovine milk was in the skim milk fraction (Blakesborough *et al.*, 1983). Casein micelles in bovine milk separated by ultracentrifugation (100,000*g* for 1 hr) contained about 90% of the total zinc; only 10% was associated with the soluble phase (Parkash and Jenness, 1966; Blakesborough *et al.*, 1983; Singh *et al.*, 1989b). About half of the soluble zinc was nondialyzable, indicating that it was tightly bound to protein (Singh *et al.*, 1989b). Thus, only about 5% of total zinc in cow's milk should be associated with small-molecular-weight ligands.

Soluble macromolecules, including nonmicellar casein, remain in the infranatant after high-speed centrifugation. In order to determine whether divalent ions are bound to these milk components, the infranatant can be forced through a 10,000 MW cutoff filter. Results of such an

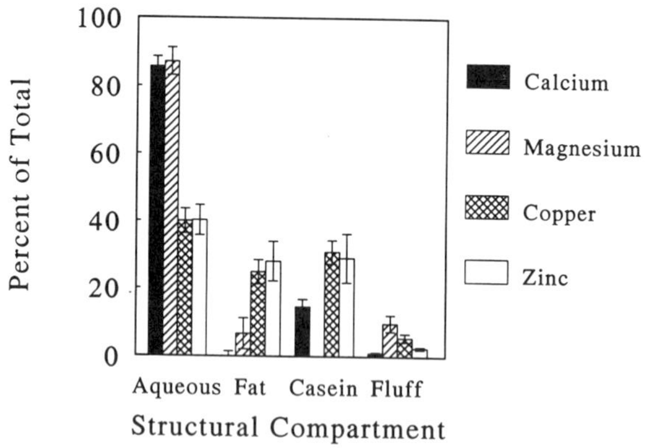

Figure 3 Distribution of divalent cations among the structural compartments of human milk. Milk samples were fractionated according to the protocol in Figure 2 and the whole milk, infranatant, and membrane fraction analyzed for calcium, magnesium, copper, and zinc by atomic adsorption spectroscopy as described previously (Casey *et al.*, 1989; Allen *et al.*, 1991).

experiment for the calcium of human milk suggested that approximately 13% of the calcium in the infranatant, or 1 mmol/liter, was retained in the macromolecular fraction (Neville *et al.*, 1994). Added to the 15% of calcium associated with casein, a total of 28% of the calcium in human milk is associated with macromolecules, and 72% is free or associated with small-molecular-weight components. In a very careful study Arnaud and Favier (1992) found that 20 ± 12% of the zinc in human was present in the ultrafiltrable fraction at all stages of lactation. These values are consistent with the earlier work of Lönnerdal *et al* (1982).

VI. Calcium and Zinc Binding to Casein

Figure 4 shows a comparison of the distribution of calcium among the structural compartments of human and bovine milk. About 65% of the calcium is associated with casein in bovine milk. The difference between bovine and human milk is due to a number of factors: (i) in bovine milk the concentration of casein is nine times that of human milk (Table I); (ii) in bovine milk some of the citrate is associated with the casein micelle (Farrell, 1988) so that some of the protein-associated calcium may be in the form of calcium citrate. In human milk only about 0.1 mmol/liter of citrate is not ultrafiltrable (Holt and Jenness, 1984). (iii) Finally, human casein is not fully phosphorylated (Groves and Gordon, 1970). The calcium binding capacity of human casein can be calculated to be only about 14 mol of calcium/mol of casein (Neville *et al.*, 1994); in most species the ratio is 20 mol of calcium/mol of casein (Jenness, 1979).

The casein micelles in bovine milk are composed of subunits linked together by colloidal calcium phosphate and hydrophobic bonding (Slattery, 1976; Schmidt, 1982). Parkash and Jenness (1966) reported that the zinc in bovine casein micelles is present in two forms, one of which is loosely bound and is readily removed by dialysis against dilute EDTA (< 2

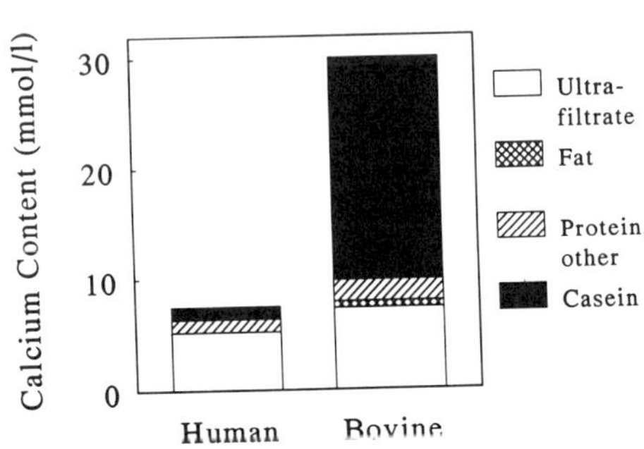

Figure 4 Comparison of structural compartmentation of calcium in human and bovine milk. Bovine milk values taken from the data of Griffin *et al.* (1988). Figure from Neville *et al.* (1994). Used by permission of the *Journal of Dairy Science*.

mm) or extraction with dithiazone in acetone. Singh and co-workers (Singh *et al.*, 1989b) found that 32% of the zinc in bovine skim milk was directly bound to caseins, while about 63% was associated with colloidal calcium phosphate.

Zinc binding by whole bovine and human casein and by purified bovine casein and whey proteins was investigated by equilibrium dialysis by Singh and co-workers (Singh *et al.*, 1989a). There were large differences in the estimated numbers of binding sites on the different caseins: 11, 8, and 2 atoms zinc/mol for bovine α_{S1}-, β-, and κ-casein, respectively. The zinc-binding capacities of the individual bovine caseins, i.e., α_{S1}- > β- > κ-casein, were in the same order as their phosphoserine contents which are 8 in α_{S1}-casein, 5 in β-casein, and 1 in κ-casein (Ribadeau *et al.*, 1972; Grosclaude *et al.*, 1973). Dephosphorylation of bovine whole casein markedly reduced its zinc-binding capacity. These results suggest that the phosphoserine groups of the casein are the primary binding sites for zinc (Harzer and Kauer, 1982; Singh *et al.*, 1989a). However, it appears that casein contains zinc-binding sites other than phosphoserine residues because the total molecules of zinc bound (~11) exceeded the number of phosphoserine residues (~8) and dephosphorylation of casein did not eliminate its zinc-binding capacity. Experiments in our laboratory with dialysis and chelators suggested that the zinc in colloidal calcium phosphate is in equilibrium with the free zinc in the dialysate or milk.

Casein from human milk had a slightly higher zinc-binding capacity (7 or 8 atoms Zn/mol protein) than whole bovine casein (506 atoms Zn/mol protein), but the apparent association constants were the same (Singh *et al.*, 1989b), indicating a similarity in the nature of zinc binding to the phosphoserine residues in casein from these two species. The above research shows that, on an equimolar basis, the zinc-binding capacity and affinity of whole human casein are similar to those of whole bovine casein. The large difference between the casein concentrations of bovine and human milk probably accounts for the higher proportion of zinc associated with casein micelles of cow's milk. Human milk contains little or no colloidal calcium phosphate (Jenness, 1973). With the exception of bovine serum albumin, which bound over 8 atoms Zn/mol, the bovine whey proteins, β-lactoglobulin, α-lactalbumin, and lactoferrin had little capacity for zinc binding (Singh *et al.*, 1989b).

VII. Divalent Cation Equilibria in the Aqueous Compartment of Milk

A. Calcium

Calcium is pumped into the saccules of the terminal Golgi apparatus and secretory vesicles of the mammary alveolar cell by an ATPase (Neville and

Watters, 1983). There it interacts with casein, citrate, phosphate, bicarbonate, and carbonate leading to the formation of casein micelles. Although equilibrium is attained, the calcium in all the fractions appears to be readily exchangeable because all fractions of milk equilibrated with ^{45}Ca within 4 hr (Neville and Keller, unpublished; see also Sandström *et al.*, 1983). The equations given in Table II can be solved numerically using any one of a number of modern computer programs. When the concentrations of the various ions present in human milk at 3 months lactation and the measured ionized calcium (3 m*M*) are used as independent variables, the only significant calcium salts are found to be calcium citrate (1 m*M*) and calcium phosphate (0.4 m*M;* Table IV).

It is instructive to examine the composition of human milk during lactogenesis. In humans there is a delay of about 2 days after birth before the onset of copious milk secretion takes place; a major volume increase takes place on Days 3 and 4 postpartum (Neville *et al.*, 1988; see also Chapter 3A). Figure 5 depicts the concentrations of the relevant components of milk over this period. Total calcium nearly doubled between Days 1 and 3 postpartum (see also Kent *et al.*, 1992), although ionized calcium actually fell slightly. Magnesium and pH remained relatively constant; changes in these parameters cannot account for the observed changes in total calcium. Citrate and phosphate rose in parallel with the calcium suggesting that the increase in these two anions was largely responsible for the increase in total calcium in early lactation. To determine whether this was indeed the case, we calculated the amounts of calcium bound to citrate

TABLE IV
Calculated Values for Major Ionic Forms of Calcium and Magnesium in Human and Bovine Milk

Ionic species	Human milk (mmol/liter)[a]	Bovine milk (mmol/liter)[b]
Calcium		
$[Ca^{2+}]$	3.0	2.0
$[CaCit^-]$	2.0	6.9
$[CaPO_4]$	0.4	0.6
Magnesium		
$[Mg^{2+}]$	0.94	0.8
$[MgCit^-]$	0.82	2.0
$[MgPO_4]$	0.03	0.3

[a]Values for human milk from Neville *et al.* (1994) and unpublished data from Neville and Allen and calculated using data from human milk at 3 months postpartum. Ionized calcium was measured on anaerobic samples using an ionized calcium electrode as described. All other values were calculated from the equations in Table II.

[b]Values for bovine milk taken from Holt *et al.* (1981). These may be subject to reevaluation because ionized calcium was derived by calculation and loss of CO_2 from the milk was not considered.

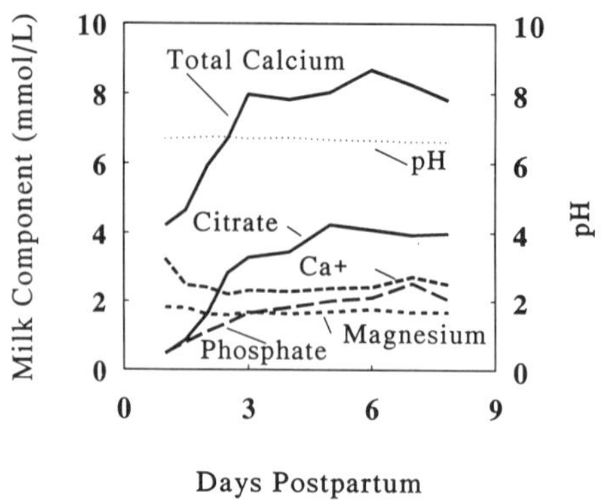

Figure 5 Changes in the concentration of total calcium and other milk components involved in calcium equilibria during lactogenesis. Data replotted from those of Neville *et al.* (1991). Figure from Neville *et al.* (1994). Used by permission of the *Journal of Dairy Science.*

and phosphate for each day. The results are shown in Figure 6. On Day 1 of lactation nearly all the calcium can be accounted for as ionized calcium. By Days 3 and 4 about 6 mmol/liter of the calcium is present as ionized calcium, calcium phosphate, and calcium citrate, and about 2 mmol/liter is presumed to be bound to casein. Unfortunately, there are no accurate measurements of the casein concentration in human milk during lactogenesis. However, Patton *et al.* (1986) showed clearly that casein is very low prior to day 2 after which it increases rapidly. We assume, therefore, that the shaded area largely represents calcium bound to casein.

In late lactation the concentration of calcium in human milk (Figure 7, left) declines significantly from nearly 7 mmol/liter at 150 days postpartum

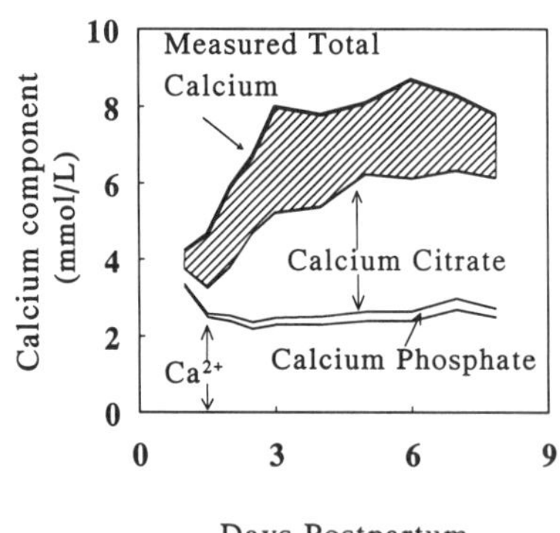

Figure 6 Calcium equilibria during lactogenesis. The lines are calculated using the measured milk composition shown in Figure 5 and the equations in Table II. The shaded area represents calcium not accounted for by the equilibrium calculations, presumably bound to protein. Figure from Neville *et al.* (1994). Used by permission of the *Journal of Dairy Science.*

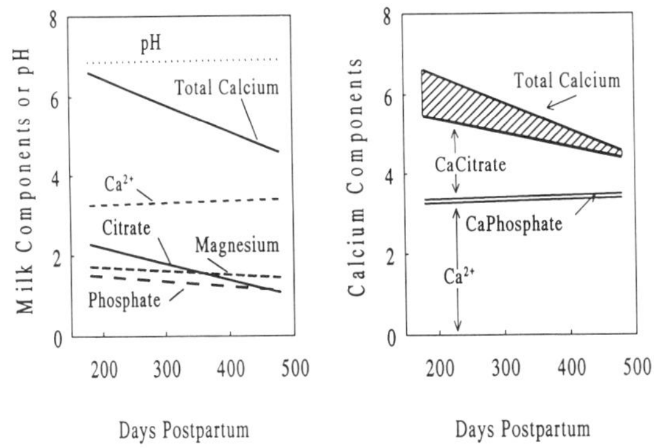

Figure 7 Human milk components and calcium equilibria in late lactation (all values in mmol/liter). Milk composition extrapolated from Neville *et al.* (1991); calculations carried out as those for Figure 6. All subjects were producing more than 400 ml of milk/day. Figure from Neville *et al.* (1994). Used by permission of the *Journal of Dairy Science.*

to about 4.5 mmol/liter at 450 days. The question of which components of milk are responsible for the decline is answered by solving the electrochemical equations of Table II. The results of this analysis are shown in Figure 7 (right). Clearly, the fall in calcium is not due to a change in ionized calcium, which actually increases slightly over this time. Rather, both citrate and the protein fraction represented by the shaded area decrease. These results, together with the observations during lactogenesis, suggest that the total concentration of calcium in human milk is not a regulated variable in the physiological sense, but varies as a function of the concentrations of citrate and caseins in the milk. The finding that the calcium in cow's milk also varies with the citrate concentration (Holt and Muir, 1979) suggests that this might be a general phenomenon.

B. Magnesium

As shown in Table III about 18% of the magnesium in human milk is associated with the lipid and membrane fractions. The remainder, in the aqueous fraction, is found to be divided among free magnesium, magnesium citrate, and magnesium phosphate (Table IV) when the equations of Table II are solved. Holt *et al.* (1981) obtained a similar result for bovine milk.

C. Zinc

Using several zinc chelators to prepare standard solutions of known $[Zn^{2+}]$, the $[Zn^{2+}]$ of bovine skim milk was found by equilibrium dialysis to be approximately 5×10^{-11} *M* (Zhang and Allen, 1992). Studies of the binding of zinc in milk ultrafiltrates have been very controversial. Although both critic and picolinic acids have been proposed as low-molecular-weight ligands, the concentration of the complexes depends on the concentrations of the ligands, the free zinc, and the concentrations of competing ligands. Using the value for [Zn^{2+}] obtained by Zhang and Allen (1992), the citrate concentration in Table I, and the association constants from Table II, the concentration of the zinc–citrate complex in bovine milk is only about 0.68 m*M*, or about 0.02% of the total.

Picolinic acid has been implicated in the zinc absorption process. Zinc picolinate was found to be efficacious when fed to children with disorders which responded to zinc therapy (Krieger, 1980). However, the picolinic acid concentration in human milk is very low, less than 3.7 μ*M* (Rebello *et al.,* 1982) as measured with high-performance liquid chromatography. This low concentration together with the low zinc activity would appear to rule out any important role for this compound as a zinc complexing ligand in human milk. Computer simulations used to determine the distribution of zinc among low-molecular-weight ligands, namely citrate, glutamate, and picolinate, in both human and bovine milk (May *et al.,* 1982) showed that at high concentrations of picolinate, this ligand may form a neutral complex, which could facilitate intestinal absorption of the metal, but at lower levels of picolinate, such as those found in milk, the concentration of zinc picolinate would be vanishingly small.

VIII. Summary and Conclusions

In considering the ionic interactions in milk it is important to know how the ion in question is compartmentalized in the structural compartments as well as the concentrations of all interacting species in the aqueous compartment. The activity of the free species should be measured using an ion-selective electrode, if available, or equilibrium dialysis under conditions in which the CO_2 content of the milk can be controlled. When this analysis is carried out for the calcium content of human milk it becomes clear that the ionized calcium in milk is more or less constant at about 3.0 mmol/liter. Physiologic variations in the total calcium are due primarily to changes in citrate and casein during the course of lactation. Recent results using equilibrium dialysis to determine the free zinc suggest that the majority of zinc, in bovine milk at least, is bound to high-molecular-weight species. Similar experiments remain to be carried out with human milk.

References

Allen, J. C., Keller, R. P., Neville, M. C., and Archer, P. (1991). Studies in human lactation: 6. Milk composition and daily secretion rates of macronutrients in the first year of lactation. *Am. J. Clin. Nutr.* **54,** 69–80.

Allen, J. C., and Neville, M. C. (1983). Ionized calcium in human milk determined with a calcium-selective electrode. *Clin. Chem.* **29,** 858–861.

Arnaud, J., and Favier, A. (1992). Determination of ultrafiltrable zinc in human milk by electrothermal atomic absorption spectrometry. *Analyst* **177,** 1593–1598.

Blake, H. H., and Henning, S. J. (1988). Absorption and transport of milk calcium by infant rats. *Am. J. Physiol.* **254,** B12–G19.

Blakesborough, P., Salter, D. N., and Gurr, W. I. (1983). Zinc binding in cow's milk and human milk. *Biochem. J.* **209,** 505–515.

Casey, C. E., Neville, M. C., and Hambidge, K. M. (1989). Studies in human lactation: Secretion of zinc, copper and manganese in human milk. *Am. J. Clin. Nutr.* **49,** 773–785.

Casey, C. E., and Hambidge, K. M. (1983). Nutritional aspects of human lactation. *In* "Lactation: Physiology, Nutrition and Breast-feeding" (M. C. Neville and M. A. Neifert, eds.), pp. 199–248. Plenum Press, New York.

Christianson, G., Jenness, R., and Coulter, S. T. (1954). Determination of ionized and calcium and magnesium in milk. *Anal. Chem.* **26,** 1923–1927.

Farrell, H. M. (1988). Physical equilibria: Proteins. *In* "Fundamentals of Dairy Chemistry" (N. P. Wong, ed.), pp. 461–509. Van Nostrand–Reinhold, New York.

Griffin, M. C. A., Lyster, R. L., and Price, J. C. (1988). The disaggregation of calcium-depleted casein micelles. *Eur. J. Biochem.* **174,** 339–343.

Grosclaude, G., Mahe, M. F., and Ribadeau-Dumas, B. (1973). Primary structures of bovine α_{S1}- and β-caseins. *Eur. J. Biochem.* **40,** 323–324.

Groves, M. L., and Gordon, W. G. (1970). The major component of human casein: A protein phosphorylated at different levels. *Arch. Biochem. Biophys.* **140,** 47–51.

Harzer, G., and Kauer, H. (1982). Binding of zinc to casein. *Am. J. Clin. Nutr.* **35,** 981–987.

Holt, C., Dalgleish, D. G., and Jenness, R. (1981). Calculation of the ion equilibria in milk diffusate and comparison with experiment. *Anal. Biochem.* **113,** 154–163.

Holt, C., and Jenness, R. (1984). Interrelationships of constituents and partition of salts in milk samples from eight species. *Comp. Biochem. Physiol. A* **77,** 275–282.

Holt, C., and Muir, D. D. (1979). Inorganic constituents of milk: I. Correlation of soluble calcium with citrate in bovine milk. *J. Dairy Res.* **46,** 433–439.

Hurley, L. S., and Lonnerdal, B. (1982). Zinc binding in human milk: Citrate versus picolinate. *Nutr. Rev.* **40,** 65–82.

Huston, G. E., and Patton, S. (1986). Membrane distribution in human milks as revealed by phospholipid and cholesterol analyses. *In* "Human Lactation: Milk Components and Methodologies" (R. G. Jensen and M. C. Neville, eds.), pp. 85–94. Plenum Press, New York.

Jenness, R. (1973). Caseins and caseinate micelles of various species. *Neth. Milk Dairy J.* **27,** 251–260.

Jenness, R. (1979). Comparative aspects of milk proteins. *J. Dairy Res.* **46,** 197–210.

Kent, J. C., Arthur P. G., Retallack, R. W., and Hartmann, P. E. (1992). Calcium, phosphate and citrate in human milk at initiation of lactation. *J. Dairy Res.* **59,** 161–167.

Krieger, I. (1980). Picolinic acid in the treatment of disorders requiring zinc supplementation. *Nutr. Rev.* **38,** 148–150.

Lonnerdal, B., Hoffman, B., and Hurley, L. S. (1982). Zinc and copper binding proteins in human milk. *Am. J. Clin. Nutr.* **36,** 1170–1175.

Lonnerdal, B., Keen, C. L., and Hurley, L. S. (1985). Zinc binding ligands and complexes in zinc metabolism. *Adv. Nutr. Res.* **6,** 139–167.

Lonnerdal, B., and Fransson, G. B. (1981). Distribution of copper, zinc, calcium and magnesium in human milk. *In* "Nutrition in Health and Disease and International Development: Symposia from the 12th International Congress of Nutrition" (A. E. Harper and G. K. Davis, eds.). A. R. Liss, New York.

Lucas, A., Gibbs, J. A. H., Lser, R. L. J., and Baum, J. P. (1978). Creamatocrit: Simple clinical technique for estimating fat concentration and energy value of human milk. *Br. Med. J.* **1,** 1018–1020.

May, P. M., Smith, G. L., and Williams, D. R. (1982). Computer calculation of zinc(II)-complex in milk. *J. Nutr.* **112,** 1990–1993.

McGann, T. C. A., Buchheim, W., Kearney, R. D., and Richardson, T. (1983). Composition and ultrastructure of calcium phosphate-citrate complexes in bovine milk systems. *Biochim. Biophys. Acta* **760,** 415–420.

Neville, M. C., Keller, R. P., Seacat, J., Casey, C. E., Allen, J. C., and Archer, P. (1984). Studies on human lactation. I. Within feed and between breast variation in selected components of human milk. *Am. J. Clin. Nutr.* **40,** 635–646.

Neville, M. C., Keller, R. P., Lonnerdal, B., Atkinson, S., Wade, C. L., Butte, N., *et al.* (1985). Measurement of electrolyte and macromineral concentrations in human milk. *In* "Human Lactation: Milk Components and Methodologies" (R. G. Jensen and M. C. Neville, eds.), pp. 129–140. Plenum Press, New York.

Neville, M. C., Keller, R. P., Seacat, J., Lutes, V., Neifert, M., Casey, C. E., *et al.* (1988). Studies in human lactation: Milk volumes in lactating women during the onset of lactation and full lactation. *Am. J. Clin. Nutr.* **48,** 1375–1386.

Neville, M. C., Allen, J. C., Archer, P. G., Seacat, J., Casey, C., Sawicki, V., *et al.* (1991). Studies in human lactation: 7. Milk volume and nutrient composition during weaning and lactogenesis. *Am. J. Clin. Nutr.* **54,** 81–93.

Neville, M. C., Keller, R. P., Casey, C. E., and Allen, J. C. (1994). Calcium partitioning in human and bovine milk. *J. Dairy Sci.,* in press.

Neville, M. C., and Watters, C. D. (1983). Secretion of calcium into milk: Review. *J. Dairy Sci.* **66,** 371–380.

Parkash, S., and Jenness, R. (1966). Status of zinc in cow's milk. *J. Dairy Sci.* **50,** 127–133.

Patton, S., Huston, G. E., Montgomery, P. A., and Josephson, R. V. (1986). Approaches to the study of colostrum—The onset of lactation. *In* "Human Lactation 2:Maternal and Environmental Factors" (M. Hamosh and A. S. Goldman, eds.), pp. 231–240. Plenum Press, New York.

Rebello, T., Lonnerdal, B., and Hurley, L. S. (1982). Picolinic acid in milk, pancreatic juice and intestine: Inadequate for role in zinc absorption. *Am. J. Clin. Nutr.* **35,** 1–5.

Ribadeau, D. B., Brignon, G., Grosclaude, F., and Mercier, J. C. (1972). Primary structure of bovine β-casein. Complete amino acid sequence. *Eur. J. Biochem.* **25,** 505–511.

Rüegg, M., and Blanc, B. (1982). Structure and properties of the particulate constituents of human milk. A review. *Food Microstruct.* **1,** 25–47.

Sandstrom, B., Keen, C. L., and Lonnerdal, B. (1983). An experimental model for studies of zinc bioavailability from milk and infant formulas using extrinsic labelling. *Am. J. Clin. Nutr.* **38,** 420–428.

Schmidt, D. G. (1982). Association of caseins and casein micelle structure. *In* "Developments in Dairy Chemistry 1. Proteins" (P. F. Fox, ed.), pp. 61–86. Applied Science Publishers, London.

Singh, H., Flynn, A., and Fox, P. F. (1989a). Binding of zinc to bovine and human milk proteins. *J. Dairy Res.* **56,** 235–248.

Singh, H., Flynn, A., and Fox, P. F. (1989b). Zinc binding in bovine milk. *J. Dairy Res.* **56,** 249–263.

Slattery, C. W. (1976). Review: Casein micelle structure: An examination of models. *J. Dairy Sci.* **59,** 1547–1549.

Tessier, H., and Rose, C. (1957). Calcium ion concentration in milk. *J. Dairy Sci.* **41,** 351–359.

White, J. C. D., and Davies, D. T. (1958). The relation between the chemical composition of milk and the stability of the caseinate complex—I. General introduction, description of samples, methods, and chemical composition of samples. *J. Dairy Res.* **30,** 171–189.

Zhang, P.-F., and Allen, J. C. (1992). Zinc activity measurement in bovine milk. *J. Dairy Sci.* **75** (Suppl), 106.

B. Major Minerals and Ionic Constituents of Human and Bovine Milks

STEPHANIE ATKINSON
BRENDA ALSTON-MILLS
BO LÖNNERDAL
MARGARET C. NEVILLE

I. Introduction

The major ionic constituents of milk consist of the monovalent ions sodium, potassium, and chloride and the divalent species calcium, magnesium, citrate, phosphate, and sulfate. After considering the secretion mechanisms for these milk constituents as well as the analytical methodology used to determine their concentrations, in this chapter we focus on effects of maternal physiologic and pathologic states in human and bovine milk. Interactions among these milk constituents as well as their partitioning among the physical compartments of milk have been considered in Chapter 7A. Tabular data gives results of studies carried out with modern technologies for the concentrations of all these constituents of human and bovine milk in units of mmol per liter. A consensus concentration, obtained by averaging the means from each laboratory at each time period, is given at the bottom of each table where the data are also translated into units of mg/liter.

II. Major Monovalent Ions: Sodium, Potassium, and Chloride

The monovalent ions, sodium, potassium, and chloride, are among the most prevalent minerals in milk collectively contributing 30 mosmol or one-tenth of the total osmolarity of human milk, 82 mosmol or one-fourth the osmolarity of bovine milk, and 196 mosmol or nearly two-thirds of the osmolarity of rabbit milk (Peaker, 1977). The sum of the monovalent ion concentrations is more or less inversely proportional to the lactose concentration (Figure 1). The mechanisms that regulate the monovalent ion concentrations in milk are only partially understood. In most species [the

HANDBOOK OF MILK COMPOSITION

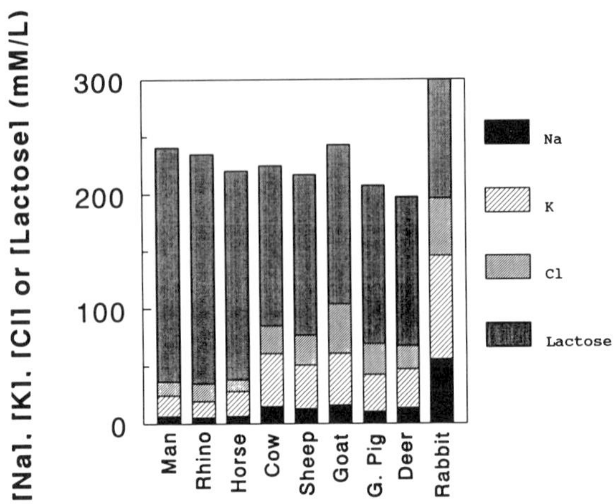

Figure 1 The concentrations of sodium, potassium, chloride, and lactose in the milks of several species. Drawn from data of Peaker (1977).

rabbit may be an exception (Peaker and Taylor, 1975)] the paracellular junctions between the mammary alveolar cells are tightly closed in full lactation and milk composition is strictly controlled by the secretory activity of the alveolar cell. Under these conditions the concentrations of the monovalent ions are regulated entirely by the secretion mechanisms in this cell.

A. Secretion Mechanism

Because we lack a good model system for the study of the molecular mechanisms of ion fluxes across the mammary epithelium, the scheme proposed by Linzell and Peaker more than 20 years ago (Linzell and Peaker, 1971b,c) remains largely unchallenged today (Figure 2). The major features of this scheme are (1) All mammary membranes are freely permeable to water; (2) the ducts have the same permeability properties as the mammary alveolar cells and milk composition is not changed as the milk travels from the alveoli to the infant; (3) ion concentrations are established in the Golgi and secretory vesicles which possess ion channels and pumps similar to those on the apical membrane of the cell. Firm evidence for this assertion is lacking although it is clear that elements of the apical membrane are derived from the secretory vesicles (see Chapter 2A); (4) ionic concentrations in the cytoplasm are maintained by ion pumps and exchangers in the basolateral membrane of the alveolar cells. Early immunocytochemical evidence documented the mostly basal localization of Na/K ATPase (Johnson and Wooding, 1978). More recent studies on a model

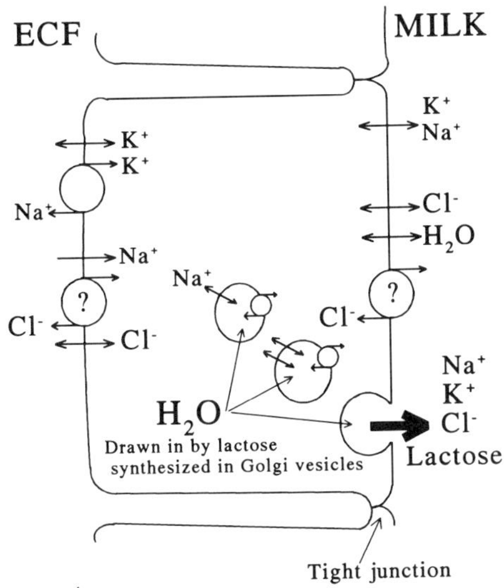

Figure 2 Hypothesis for the secretion of monovalent ions into milk. After Peaker (1977).

mammary membrane grown on filter supports (Sjaastad *et al.,* 1993) provide evidence for both a Na/HCO_3 cotransporter and the amiloride-sensitive Na/H exchanger on the basolateral membrane of cells grown in the presence of prolactin; (5) the apical membrane is permeable to sodium, potassium, and chloride. Experimental evidence for this statement includes the observation made in goats that NaCl or KCl solutions infused up the teat are absorbed into the blood stream (Linzell and Peaker, 1974) and that isotonic sucrose solutions infused up the teat draw ions into the milk space. In addition, measurement of the effect of changes in the Na, K, and Cl concentrations in the milk space of the goat mammary gland on the transepithelial potential difference provide evidence for the presence of a nonselective cation channel and a chloride channel in the apical membrane; and (6) there is evidence that sodium and chloride are at electrochemical equilibrium across the apical membrane of the alveolar cell. The ratios of the concentrations of sodium and potassium in the milk to their concentrations in the cytoplasm of the mammary alveolar cells are approximately equal and predicted by the potential across the apical membrane (Peaker, 1977; Berga and Neville, 1985).

As has been discussed elsewhere in this book, during pregnancy, involution, and with mastitis, the junctions between the alveolar cells are open (Peaker, 1977; Neville *et al.,* 1983) and the sodium and chloride enter the milk space drawing water with them. Lactose and potassium are also thought to move from the milk space to the blood. The net result is that the mammary secretion product under these conditions has much higher concentrations of sodium and chloride and lower concentrations of lactose and potassium. The presence of high sodium concentrations in human milk is diagnostic of either mastitis or low milk volume secretion (Morton, 1994).

B. Methodological Considerations

Milk samples collected for the determination of the concentrations of the major monovalent ions should be expressed into clean containers, refrigerated for short-term storage, and stored frozen at −20°C if they are to be maintained for a longer time. There does not appear to be either within-feed[1] (Neville *et al.*, 1984; Gillies and Niell, 1985; Gunther *et al.*, 1965) or diurnal variation (Gillies and Niell, 1985; Neville, unpublished data) in these milk components so that sampling time is not a concern. However, there is substantial longitudinal variation in monovalent cation concentrations in human milk (Allen *et al.*, 1991; Neville *et al.*, 1991; Gunther *et al.*, 1965) so the stage of lactation is important. Similarly, there are changes in sodium and chloride concentrations associated with weaning (Neville *et al.*, 1991), so that daily milk volumes or at least number of feeds per day should be noted. Samples from different breasts or different teats should be kept separate to rule out mastitis affecting only one gland or quarter.

Flame photometry and atomic emission spectroscopy on ashed or diluted samples are well-established methods for measurement of sodium and potassium (Neville *et al.*, 1985). If dilution and flame photometry are used ionization enhancement between sodium and potassium can be suppressed with 15 m*M* LiCl so that both ions can be determined on the same sample. Whole and defatted samples give similar results.

More recently, ion-selective electrodes have been found to be equally satisfactory for measurement of these ions (Neville *et al.*, 1984, 1985). One precaution necessary with the use of any of these methods is that standards should be chosen with care. When commercial ion-selective electrode systems in clinical laboratories are utilized for milk analysis, urine standards should be used. Plasma standards have sodium concentrations as much as 30-fold those of mature human milk and are, therefore, unsuitable. Occasionally, human milk samples will have sodium levels below the level of detection of standard ion-selective electrode systems. In this case, more sensitive analytical methods must be used for accurate results.

The standard method for determination of chloride in biological samples is potentiometric titration with silver (Cotlove, 1964); this method is satisfactory for use with undiluted milk samples and gave similar values as ion-selective electrodes used after ashing in a closed flask to prevent volatilization (Picciano *et al.*, 1981). Clinical laboratories utilize an auto-

[1]Very accurate measurements will reveal a small decrease in sodium and potassium between fore- and hindmilk (Neville *et al.*, 1984) that can largely be accounted for by the increase in fat content in hindmilk. The concentration of these ions in the aqueous fraction is, however, constant.

mated colorimetric method based on the displacement of thiocyanate from mercuric thiocyanate to react with ferric ions (Schoenfeld and Lewellen, 1964). This method gave satisfactory correlation with the results of potentiometric titration (Neville *et al.*, 1984, 1985). However, because it is a colorimetric method, milk samples must be carefully defatted prior to analysis to avoid light-scattering artifacts (Neville *et al.*, 1984).

C. Factors that Influence Sodium, Potassium, and Chloride Concentrations in Milk

Major changes in the concentrations of the major monovalent cations in milk are associated with conditions that promote opening of the tight junctions between epithelial cells. For example, colostrum has much higher levels of sodium and chloride than mature milk because the gland is undergoing the transition between pregnancy, when the junctions are open, to lactation, when they are closed. More minor changes are associated with duration of lactation and prematurity (see below). Dehydration as can occur, for example, with the fasting associated with Ramadan in Moslem countries, can lead to hyperosmolarity (Prentice *et al.*, 1983) and associated slight increases in monovalent ions. However, the magnitude of such changes is small. The levels of monovalent ions are species specific (Figure 1) and do not appear to be influenced by nutritional factors or systemic disease such as diabetes. Early reports that cystic fibrosis was associated with high milk sodium concentrations have been shown to be erroneous (Shiffman *et al.*, 1989).

The major pathological process that alters monovalent cation content is mastitis or localized inflammation of breast tissue. Inflammation does open the junctions between the cells and changes in sodium and chloride are large enough that mastitis can often be detected by measurement of the electrical conductivity of milk (Linzell and Peaker, 1971a). Figure 3 shows the major changes in the concentration of several milk components associated with a minor bout of mastitis in one breast between Days 7 and 11 postpartum in an exclusively breast-feeding woman. Both sodium and chloride increased by about 40 m*M*, the protein concentration increased on the third day, and lactose concentration fell as the sugar was diluted by water entering with NaCl through the more permeable junctional complexes between the cells.

D. Monovalent Cation Concentrations During Lactogenesis and Weaning in Women

Figure 4 shows the mean monovalent ion concentrations in the milks of 12 women who provided frequent milk samples during the first week postpartum (Neville *et al.*, 1991). The concentrations of lactose, glucose, and

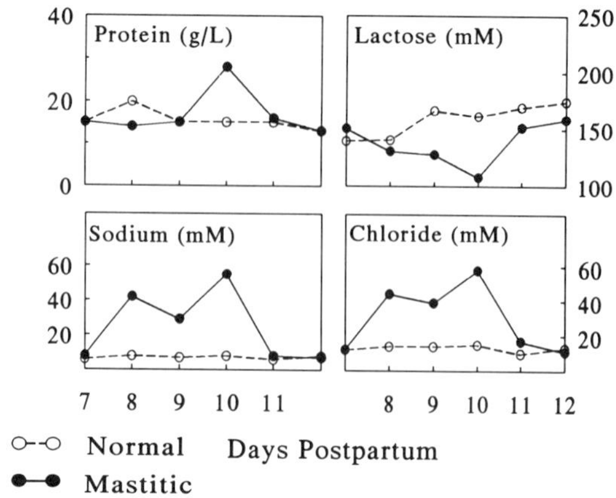

Figure 3 Changes in the concentrations of sodium, chloride, protein, and lactose associated with unilateral mastitis in an exclusively breast-feeding woman. Data replotted from Neville *et al.* (1984).

several divalent ions are also shown. In late pregnancy the sodium and chloride were about 80 and 60 m*M*, respectively, falling rapidly during the postpartum period to 13 m*M* and 19 m*M*, on Day 7. Potassium fell rapidly immediately after birth from 18 to 13.5 m*M* and then rose again to 19 m*M* on Day 7. Lactose rose rapidly during the same period. These changes in the concentration of sodium, chloride, lactose, and potassium occur mostly during the first 2 days postpartum as a result of the closure of the tight junctions between epithelial cells. They precede the major increase in milk volume that occurs between Days 2 and 4 postpartum (see Chapter 3B).

Of the 12 women in whom lactogenesis was studied, 5 were also followed during gradual weaning between 6 and 12 months postpartum (Neville *et al.*, 1991). As the milk volume fell below 300 ml/day sodium and chloride began to rise, reaching about 50 m*M* when the volume was less than 50 ml/day and the number of feeds two or fewer per day. The lactose concentration fell concomitantly with the rise in sodium and chloride. Interestingly, however, the potassium concentration rose from about 12 to 16 m*M* as the volume fell, possibly reflecting the time postpartum rather than changes in milk volume. These data illustrate the importance of knowing both the time postpartum and the rate of the milk production when evaluating monovalent ion concentrations in human milk.

E. *Effects of Prematurity on Monovalent Ion Concentrations in Human Milk*

Premature birth of the infant has significant consequences on the monovalent cation concentrations of human milk (Gross *et al.*, 1980; Lemons *et al.*, 1982; Atkinson *et al.*, 1980). Figure 5 depicts the observed sodium and

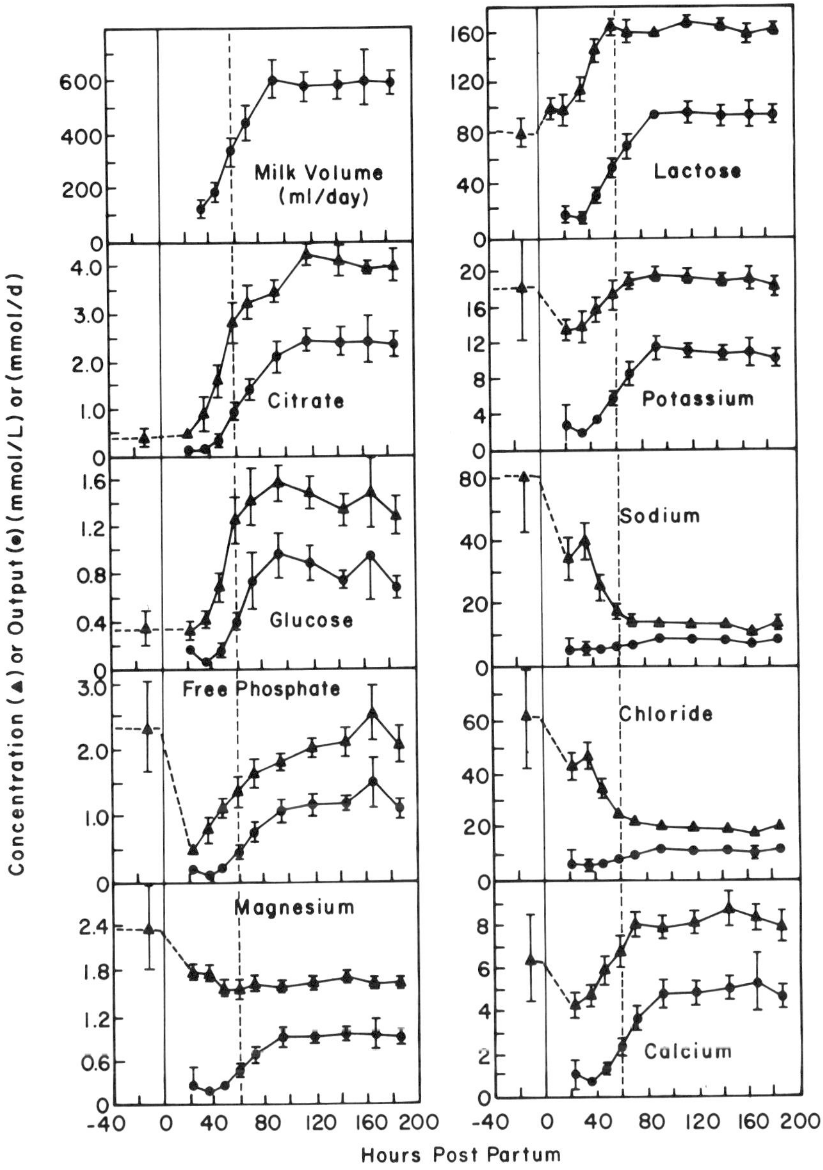

Figure 4 Changes in the concentrations of minerals and sugars during lactogenesis. Triangles represent the mean and standard deviation of the concentrations of various milk components for samples obtained at frequent intervals during the first week postpartum from 12 multiparous, exclusively breast-feeding Caucasian women in Denver, Colorado. Mean secretion rates are shown by the closed circles. Used by permission from Neville *et al.* (1991).

chloride concentrations in milks from mothers of preterm and term infants (Gross *et al.*, 1980; Lemons *et al.*, 1982). In the two studies shown the concentrations of sodium and chloride were about 3 to 5 m*M* lower in the milks from the preterm mothers (average gestational age, 31 to 33 weeks)

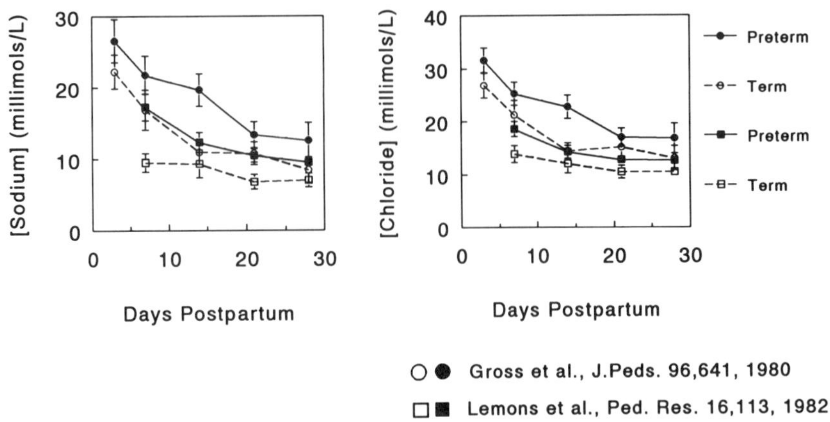

Figure 5 Sodium and chloride concentrations in the milks of mothers of preterm and term infants measured in two laboratories at the times indicated.

than in the milks from mothers of term infants (average gestational age, 40 weeks). Potassium concentrations were not significantly different between the two groups. After about 30 days postpartum the concentrations became similar in the two groups (Lemons *et al.,* 1982; Butte *et al.,* 1984). The increases in sodium and chloride concentration are accompanied by an increase in protein concentration suggesting that maturation of the secretory epithelium is delayed in mothers of preterm infants.

F. Mineral Content of Human and Bovine Milk

Tables I–III give representative values for the monovalent cation content of human milk at different stages of lactation. The analytical methods in each table represent several different methodologies and many different sampling protocols (Table IV). This being the case, the extent of the agreement between laboratories is remarkable. Table V gives similar but less extensive values for bovine milk. The main point to be made here is that bovine milk contains much higher concentrations of the three monovalent cations than human milk. Milk from mastitic cows, a common disorder, can have much higher sodium and chloride concentrations than shown in the table.

III. Divalent Ions: Calcium, Magnesium, Citrate, Phosphate, and Sulfate

The divalent ions, calcium, magnesium, citrate, phosphate, and sulfate, are the second most abundant mineral components of human milk, next

TABLE I
Sodium Concentration in Human Milk[a]

Reference	Subject No	Method[b]	Days postpartum						
			3	7–9	14	30	90	180	360
Neville *et al.* (1991); Allen *et al.* (1991); U.S.	12	c	17.1 ± 2.1	1.06 ± 1.2	10.4 ± 1.8	7.6 ± 2.0	6.0 ± 1.3	6.0 ± 1.3	6.0 ± 1.2
Gross *et al.* (1980); U.S.	12	a	22.3 ± 2.4	16.9 ± 2.8	11.0 ± 1.7	8.5 ± 1.8			
Atkinson *et al.* (1980); Canada	7	b	18.6 ± 1.3	13.9 ± 1.7		11.0 ± 2.0			
Hibberd *et al.* (1982); UK	10	a	21.4 ± 6.4	17.8 ± 13.6	10.9 ± 4.8	17.1 ± 15.8			
Hazebroek and Hofmann (1983); U.S.	269[c]	b	19.8 ± 10.2	15.7 ± 8.7	13.1 ± 6.0	8.5 ± 6.1	6.0 ± 3.1	6.0 ± 3.8	
Lemons *et al.* (1982); U.S.	7	b		9.5 ± 1.3	9.4 ± 1.9	7.0 ± 1.0			
Butte *et al.* (1984); U.S.	13	a			11.6 ± 3.3	8.0 ± 2.3	5.7 ± 1.8		
Dewey and Lonnerdal (1983); U.S.	18–20	b				9.9 ± 6.6	8.0 ± 6.0	5.8 ± 3.4	
Butte *et al.* (1987); U.S.	45	b				5.9 ± 1.4	4.7 ± 0.9		
Picciano *et al.* (1981); U.S.	26	b				6.6 ± 2.4	5.5 ± 2.0		
Dewey *et al.* (1984); U.S.	90	b						3.6 ± 1.0	3.3 ± 1.6, 7–11 months 4.8 ± 2.4, 12–20 months

TABLE I—*continued*

Reference	Days postpartum						
	3	7–9	14	30	90	180	360
Consensus (mmol/liter)	17.9 ± 4.5	14.1 ± 4.9	11.1 ± 3.3	9.0 ± 4.1	6.0 ± 2.5	5.4 ± 2.4	4.6 ± 1.4
Consensus (mg/liter)	411 ± 4.0	324 ± 113	255 ± 76	207 ± 94	138 ± 58	124 ± 55	106 ± 32

[a] All values in millimol per liter; mean ± SD, unless otherwise noted.
[b] Methods: a, atomic absorption on dry ashed samples; b, flame emission spectroscopy; c, ion-selective electrode.
[c] Subjects in entire study.

TABLE II
Potassium Concentration in Human Milk[a]

Reference	Subject No.	Method[b]	Days postpartum						
			3	7–9	14	30	90	180	360
Neville *et al.* (1991); Allen *et al.* (1991); U.S.	12	c	17.5 ± 2.0	19.2 ± 1.4	16.9 ± 1.8	15.4 ± 1.7	12.4 ± 1.4	12.7 ± 1.4	13.4 ± 2.5
Gross *et al.* (1980); U.S.	12	b	18.5 ± 2.0	16.5 ± 1.3	15.4 ± 1.8	15.0 ± 1.9	13.9 ± 1.6		
Atkinson *et al.* (1980); Canada	7	b	18.5 ± 0.9		16.5 ± 1.1	15.0 ± 2.5			
Hibberd *et al.* (1982); UK	10	a	18.0 ± 2.4	16.4 ± 3.2	15.2 ± 2.7	13.9 ± 2.3			
Lemons *et al.* (1982); U.S.		a		16.9 ± 2.8	14.6 ± 1.7	13.0 ± 1.7			
Dewey and Lonnerdal (1983); U.S.	18–20	b				13.5 ± 1.8	12.0 ± 2.1	11.0 ± 1.6	
Picciano *et al.* (1981); U.S.	26	b				11.9 ± 2.4	10.4 ± 2.1		
Dewey *et al.* (1984); U.S.	90	b						9.4 ± 1.8	9.0 ± 2.0, 7–11 months 8.8 ± 1.9, 12–20 months
Consensus (mmol/liter)			18.2 ± 2.1	17.2 ± 2.2	15.6 ± 1.8	13.9 ± 2.0	12.1 ± 1.8	11.0 ± 1.6	11.2 ± 2.2
Consensus (mg/liter)			712 ± 82	672 ± 86	610 ± 70	543 ± 78	473 ± 70	430 ± 63	437 ± 86

[a]All values in millimol per liter; mean ± SD, unless otherwise noted.
[b]Methods: a, atomic absorption on dry ashed samples; b, flame emission spectroscopy; c, ion-selective electrode.

TABLE III
Chloride Concentration in Human Milk[a]

Reference	Subject No.	Method[b]	Days postpartum 3	7–9	14	30	90	180	360
Neville *et al.* (1991); Allen *et al.* (1991); U.S.	12	c	25.4 ± 1.9	20.5 ± 1.9	20.5 ± 1.6	16.5 ± 0.7	13.9 ± 1.0	11.6 ± 0.8	13.0 ± 1.1
Gross *et al.* (1980); U.S.	12	a	26.9 ± 2.4	13.9 ± 1.6	12.1 ± 1.8	10.5 ± 0.9			
Atkinson *et al.* (1980); U.S.	7	b	23.0 ± 0.5	18.1 ± 1.8		12.0 ± 1.0			
Lemons *et al.* (1982); U.S.	7	c		21.3 ± 2.7	14.5 ± 1.5	13.1 ± 2.3			
Picciano *et al.* (1981); U.S.	26	b				12.0 ± 2.4	12.0 ± 2.6		
Consensus (mmol/liter)			25.1 ± 1.6	18.4 ± 2.0	15.7 ± 1.6	12.8 ± 1.5	12.9 ± 1.8	11.6 ± 0.8	13.0 ± 1.1
Consensus (mg/liter)			888 ± 56	651 ± 71	556 ± 57	453 ± 53	456 ± 64	411 ± 28	460 ± 39

[a] All values in millimol per liter; mean ± SD.
[b] Methods: a, amperometric titration; b, ashing followed by ion-specific electrode measurement; c, automated colorimetric procedure.

TABLE IV
Methods for Sampling Human Milk in the Quoted Studies

Reference	Method
Neville *et al.* (1991); Allen *et al.* (1991); U.S.	Longitudinal study of 12 nonsmoking multiparous Caucasian women. Midfeed samples at frequent intervals during the first 2 weeks postpartum, monthly samples to 1 year
Gross *et al.* (1980); U.S.	Complete emptying of both breasts at a morning feed
Atkinson *et al.* (1980); Canada	Longitudinal study of 10 Canadian mothers of full-term infants. Complete 24-hr expressions obtained by breast pump
Hibberd *et al.* (1982); UK	Longitudinal study of 10 European women during the first 5 weeks postpartum. Complete expression by breast pump over a 24-hr period with pooling of milk aliquots
Feeley *et al* (1983); U.S.	Longitudinal study of 102 middle-class American women during the first 6 weeks postpartum. Pooled fore-, mid-, and hindmilk samples from evening and morning feeds
Kirksey *et al.* (1979); U.S.	Mostly cross-sectional study of 52 Caucasian American women. Five to 10-ml expressed after let down at first morning feed
Dai and Tang (1994); China	Longitudinal study of nine Chinese women with full-term infants during the first month postpartum. Manual expression of milk sample between 8:00 and 10:00 AM
Lemons *et al.* (1982); U.S.	Seven mothers of term infants studied from 1 to 44 weeks postpartum. Complete 24-hr expressions of milk from both breasts obtained by electric breast pump
Prentice and Barclay (1991); Zaire	Longitudinal study of 12 mothers living in poor, rural Zaire. Complete expression of one breast not suckled overnight
Butte *et al.* (1984); U.S.	Thirteen American women studied longitudinally for 12 weeks postpartum. Entire contents of one breast expressed 2 hr after a morning feeding
Dewey and Lonnerdal (1983); U.S.	Longitudinal study of full and partially breast-feeding American women with volumes recorded between 1 and 6 months postpartum. Complete manual expression of all milk from one breast at second feed of the morning
Butte *et al.* (1987); U.S.	Longitudinal study of 45 normal, healthy American mother–infant pairs. Milk intake by test-weighing, 24-hr collection alternating breasts
Picciano *et al.* (1981); U.S.	Longitudinal study of 26 American women with volumes recorded between 1 and 3 months postpartum. Thirty-milliliter samples of foremilk expressed at morning, midday, and evening feeds and pooled

TABLE IV—*continued*

Reference	Method
Laskey *et al.* (1990); UK	Cross-sectional study of 72 British subjects partially breast-feeding after 3 months. Pooled samples from both breasts not controlled for time of day or feeding pattern
Laskey *et al.* (1990); Gambia	Cross-sectional study of 144 Gambian mothers 0.5 to 25 months postpartum. Samples pooled from both breasts
Karra *et al.* (1988); U.S.	Longitudinal study of educated middle-class American women. Ten-milliliter milk samples collected at each feed during a 24-hr period from 1 to 6 months postpartum
Karra *et al.* (1988); Egypt	Longitudinal study of marginally malnourished, low-income rural Egyptian women. Ten-milliliter samples collected monthly from 1 to 6 months postpartum by manual expression between 10:00 AM and 2:00 PM
Tanzer and Sunel (1991); Turkey	Longitudinal study for 26 weeks postpartum of 20 Turkish mothers living in poor socioeconomic circumstances. Milk samples expressed at middle and end of feed repeated three times daily
Moser *et al.* (1988); U.S.	Cross-sectional study of 26 lactating American women 1 to 6 months postpartum. Early morning milk sample (5–10 ml) collected by manual expression. Dietary intake of calcium and magnesium measured
Moser *et al.* (1988); Nepal	Cross-sectional study of 26 lactating Nepalese women 2 to 6 months postpartum. Early morning milk sample (5–10 ml) collected by manual expression. Dietary intake of calcium and magnesium measured
Greer *et al.* (1982); U.S.	Longitudinal study of 18 American women, 17 Caucasian, one Asian, in origin for 6 months postpartum. Pooled fore-, mid-, and hindmilk samples collected from first daylight feed from a single breast
Dewey *et al.* (1984); U.S.	Cross-sectional study of full and partially breast-feeding American women with volumes recorded between 4 and 20 months postpartum. Complete expression of all milk from one breast at second feeding of the morning
Karra *et al.* (1986); U.S.	Longitudinal study of 55 partially breast-feeding women starting at 7 months postpartum. Foremilk samples expressed between 7 and 10 AM monthly
McNally *et al.* (1991); Canada	Cross-sectional study of eight mothers. Complete 24-hr collection from both breasts by breast pump at 2–4 or 23–30 days postpartum

TABLE V
Monovalent Ion Concentrations in Bovine Milk[a]

	Concentration—millimol per liter (mg/100 g)		
Milk component	Mean	Range	SD
Sodium	25.2 (58)	20.4–33.4 (47–77)	4.3 (10)
Potassium	35 (140)	28.5–42.8 (113–171)	9.6 (35.8)
Chloride	28.9 (104)	25–35.3 (90–127)	8.1 (29)

[a]Source: White and Davies (1958).

to the monovalent ions. The interactions of calcium, magnesium, phosphate, and citrate with each other as well as the distribution of calcium and magnesium among the physical compartments of human and bovine milk have been discussed in Chapter 7A (Neville *et al.*, this volume, Chapter 7A).

In distinction to the monovalent ions, the total concentration of many of the divalent ions depends on the concentration of a particular binding entity. For example, when examined across species the concentrations of both phosphate and calcium are generally proportional to the concentration of casein (Jenness, 1979). Figure 6 shows this relation for calcium. The high concentration of this protein with its ability to bind large quantities of

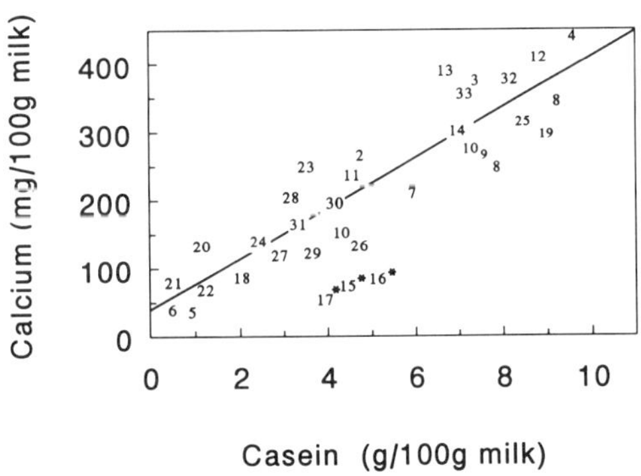

Figure 6 The relation between the calcium and casein concentrations in the milks of many species. 1, Long-tailed bat; 2, little brown bat; 3, tree-tailed bat; 4, rabbit; 5, baboon, 6, human; 7, hamster; 8, rat; 9, mouse; 10, guinea pig; 11, dog; 12, black bear; 13, grizzly bear; 14, polar bear; 15, fur seal; 16, elephant seal; 17, harp seal; 18, Indian elephant, 19, aardvark; 20, horse; 21, burro; 22, rhinoceros; 23, pig; 24, camel; 25, reindeer; 26, giraffe; 27, cow; 28, buffalo; 29, goat; 30, sheep; 31, pygmy sperm whale; 32, fin whale; 33, blue whale. Starred points were omitted from the regression. The best-fitting line had a slope of 8.8 mM/g casein and an intercept of 9.1 mM/kg milk. Redrawn from Jenness (1979).

calcium and phosphate assures that the milk provided to the young of rapidly growing species can satisfy their nutrient requirements for both bone and muscle accretion.

A. Mechanisms of Secretion

There is considerable evidence that calcium, phosphate, and citrate all enter milk via the exocytotic pathway (reviewed in Neville *et al.*, 1983). As Figure 7 shows, calcium is pumped from the cytoplasm into the Golgi via a calcium ATPase (Baumrucker and Keenan, 1975; Watters, 1984). Within the Golgi the ionized calcium concentration is thought to be about 3 m*M* (Neville *et al.*, 1994), a sufficiently high concentration so that almost all the citrate is completed to calcium and that casein micelle formation is complete. Citrate is made in the mitochondria from pyruvate and transported to the cytoplasm where it is available for lipid synthesis and for transport into the Golgi (Linzell *et al.*, 1976). Nothing is known about the mechanism of its transport across the Golgi membrane. Phosphate is formed in the Golgi vesicles as a by-product of lactose synthesis and presumably equilibrates across the Golgi membrane, although again the transport mechanisms are not understood. Although magnesium is believed to be secreted via the exocytotic pathway, because there is no convenient isotope of magnesium that allows study of its transport, there is no firm evidence supporting this belief. As these compounds enter the Golgi vesicles they interact with the casein which is undergoing phosphorylation in the same

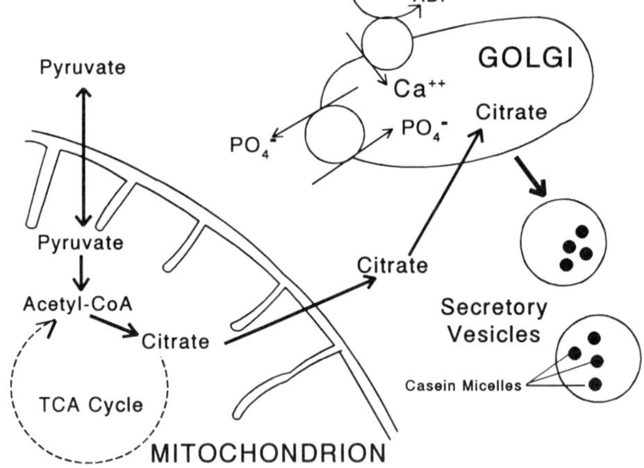

Figure 7 Pathways for the secretion of calcium, phosphate, and citrate into milk. The Golgi vesicles contain soluble casein as well as these ions. The casein condenses to micelles containing up to 30,000 casein molecules, each binding as many as 30 molecules of calcium and phosphate, and is packaged into secretory vesicles whose contents are secreted via exocytosis.

compartment. These interactions lead eventually to formation of the casein micelle, a complex package of protein, calcium, phosphate, and other ions that is large enough to be seen as a discreet particle in the electron microscope.

B. Methodological Considerations

Milk samples collected for the determination of the concentrations of the major divalent ions should be handled in a way similar to those for monovalent ion analysis; that is, expression into clean containers and storage at a temperature of −20°C or below. There does not appear to be either within-feed (Neville *et al.*, 1984; Gillies and Niell, 1985) or diurnal variation (Kirksey *et al.*, 1979; Gillies and Niell, 1985) in the calcium and magnesium contents of human milk so that sampling may be carried out at any time during the day and feed (see also Table IV). For citrate data neither within-feed nor diurnal variation are available. No diurnal variation was observed in the total sulfur content of milk (McNally *et al.*, 1991). Free phosphate showed no within-feed variation in one study (Neville *et al.*, 1984). There is, however, substantial longitudinal variation in the concentrations of most of these ions (Karra *et al.*, 1986; Laskey *et al.*, 1990; Figure 8, Tables VI–IX) so the duration of lactation is important to consider.

Traditionally, calcium and magnesium have been assayed by atomic absorption spectrometry after wet or dry ashing or simple dilution (Neville *et al.*, 1985). More recently, automated colorimetric procedures have been found to be satisfactory and much more rapid when a number of samples must be analyzed (Neville *et al.*, 1984; Laskey *et al.*, 1991). Samples for colorimetric procedures must be assiduously defatted to avoid light-scattering artifacts. Citrate has been measured by the method of

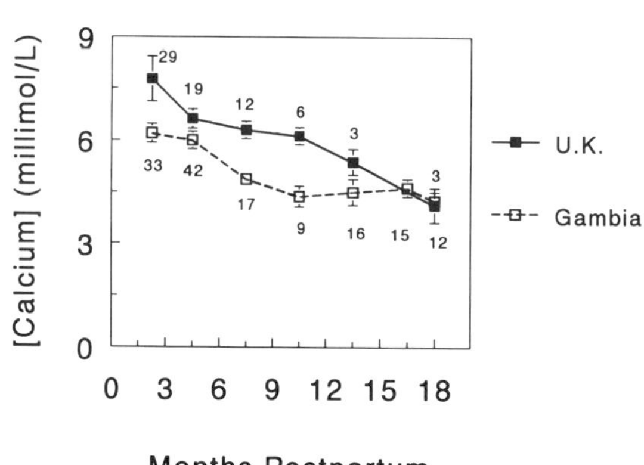

Figure 8 Calcium concentration in human milk from British and Gambian women. Replotted from Laskey and Prentice (1991).

TABLE VI
Calcium Concentration in Human Milk[a]

Reference	Subject No.	Method[b]	Days postpartum							
			3	7–9	14	30	90	180	360	440
Neville *et al.* (1991); Allen *et al.* (1991); U.S.	12	d	6.3 ± 3.5	8.0 ± 2.8	7.6 ± 0.3	7.7 ± 0.2	7.5 ± 0.2	6.3 ± 0.2	5.0 ± 0.22	
Gross *et al.* (1980); U.S.	12	a	5.4 ± 1.0	6.4 ± 2.8	6.5 ± 4.3	6.2 ± 4.5				
Atkinson *et al.* (1980); Canada	7	b	8.0 ± 0.4	7.1 ± 0.7	7.1 ± 0.7	6.8 ± 2.0				
Hibberd *et al.* (1982); UK	10	e	5.8 ± 1.5	5.3 ± 1.23	4.6 ± 1.1	4.3 ± 1.3				
Feeley *et al.* (1983); U.S.	102	a	6.6 ± 1.4		6.3 ± 1.6	6.5 ± 1.6				
Kirksey *et al.* (1979); U.S.	12	d	5.9 ± 1.3		5.5 ± 1.1			5.9 ± 0.9	4.4 ± 0.7	
Dai and Tang (1994); China	14	b	6.8 ± 3.0	6.6 ± 1.9	5.9 ± 1.0	5.9 ± 1.5				
Lemons *et al.* (1982); U.S.	7	e		7.3 ± 1.9	7.1 ± 3.3	6.7 ± 3.2				
Prentice and Barclay (1991); Zaire	12	f		6.3 ± 1.6		6.5 ± 0.4		5.5 ± 0.3	4.8 ± 0.2	4.5 ± 1.4
Butte *et al.* (1984); U.S.	13	a			6.4 ± 1.3	6.4 ± 1.3	6.5 ± 1.7			
Dewey and Lonnerdal (1983); U.S.	19	b				6.5 ± 1.1	6.8 ± 1.5	6.4 ± 1.1		
Butte *et al.* (1987a); U.S.	45	b				5.9 ± 1.4	4.7 ± 0.9			
Picciano *et al.* (1981); U.S.	26	b				7.2 ± 1.5	7.1 ± 1.3			

TABLE VI—*continued*

Reference	Subject No.	Method[b]	Days postpartum							
			3	7–9	14	30	90	180	360	440
Laskey *et al.* (1990); UK	7	f				7.8 ± 2.3	6.5 ± 1.2	6.3 ± 0.9	5.7 ± 0.6	
Laskey *et al.* (1990); Gambia	12	f				6.3 ± 0.9	6.1 ± 1.5	5.4 ± 0.7	4.4 ± 1.1	
Karra *et al.* (1988); U.S.	45	d				6.5 ± 0.5	6.6 ± 1.7	6.3 ± 0.8		
Karra *et al.* (1988); Egypt	50	d				6.5 ± 1.4	6.4 ± 1.1	6.0 ± 1.1		
Tanzer and Sunel (1991); Turkey	20	d				6.7 ± 0.3		5.9 ± 0.9		
Moser *et al.* (1988); U.S.	26	c					6.1 ± 0.1			
Moser *et al.* (1988); Nepal	26	c					6.6 ± 0.7			
Greer *et al.* (1982); U.S.	24	d					6.9 ± 0.9	6.2 ± 0.9		
Dewey *et al.* (1984); U.S.	90	b						5.6 ± 1.1	5.2 ± 1.3	4.4 ± 1.4
Karra *et al.* (1986); U.S.	24	d							5.3 ± 0.9	4.6 ± 0.3
Consensus (mmol/liter)			6.4 ± 1.7	6.7 ± 1.8	6.3 ± 1.6	6.5 ± 1.5	6.5 ± 1.0	6.0 ± 0.8	4.9 ± 0.8	4.4 ± 0.9
Consensus (mg/liter)			255 ± 68	268 ± 73	253 ± 65	259 ± 59	259 ± 39	238 ± 31	194 ± 30	176 ± 34

[a]All values in millimol per liter; mean ± SD, unless otherwise noted.
[b]Methods: a, atomic absorption on dry ashed samples; b, atomic absorption on wet ashed samples; c, atomic absorption on a combination of dry and wet ashed samples; d, atomic absorption on diluted samples; e, colorimetric; f, semiautomated dry ashing followed by automated colorimetric assay (Laskey *et al.*, 1991).

TABLE VII
Magnesium Concentration in Human Milk[a]

Reference	Subject No.	Method[b]	Days postpartum							
			3	7–9	14	30	90	180	360	440
Neville *et al.*, (1991); Allen *et al.*, (1991); U.S.	12	e	1.6 ± 0.1	1.6 ± 0.1	1.7 ± 0.1	1.6 ± 0.1	1.7 ± 0.2	1.6 ± 0.2	1.5 ± 0.1	
Gross *et al.* (1980); U.S.	12	a	1.0 ± 0.4	1.2 ± 0.2	1.1 ± 0.2	1.0 ± 0.2				
Atkinson *et al.* (1980); Canada	7	a	1.3 ± 0.2	1.2 ± 0.1	1.3 ± 0.2	1.3 ± 0.1				
Hibberd *et al.* (1982); UK	10	e	1.5 ± 0.3	1.6 ± 0.5	1.4 ± 0.5	1.4 ± 0.5				
Feeley *et al.* (1983); U.S.	102	a	2.2 ± 0.4		2.0 ± 0.5	2.0 ± 0.5				
Lemons *et al.* (1982); U.S.	12	e		1.3 ± 0.3	1.2 ± 0.3	1.2 ± 0.3				
Butte *et al.* (1984); U.S.	8	a			1.4 ± 0.3	1.3 ± 0.2	1.6 ± 0.4			
Dewey and Lonnerdal (1983); U.S.	274	b				1.1 ± 0.2	1.4 ± 0.2	1.4 ± 0.2		
Picciano *et al.* (1981); U.S.	26	b				1.2 ± 0.3	1.4 ± 0.2			
Karra *et al.* (1988); U.S.	80	d				1.2 ± 0.2	1.4 ± 0.3	1.5 ± 0.3		
Karra *et al.* (1988); Egypt	50	d				1.2 ± 0.2	1.4 ± 0.3	1.4 ± 0.4		
Tanzer and Sunel (1991); Turkey	20	d				1.7 ± 0.3		1.9 ± 0.3		
Greer *et al.* (1982); U.S.	18	e					1.2 ± 0.2	1.3 ± 0.3		
Moser *et al.* (1988); U.S.	20	c					1.4 ± 0.2			

TABLE VII—*continued*

Reference	Subject No.	Method[b]	Days postpartum							
			3	7–9	14	30	90	180	360	440
Moser *et al.* (1988); Nepal	20	c					1.3 ± 0.6			
Dewey *et al* (1984); U.S.	90	b						1.3 ± 0.2	1.3 ± 0.2	1.1 ± 0.2
Karra *et al.* (1986); U.S.	24	d							1.5 ± 0.4	1.3 ± 0.3
Consensus (mmol/liter)			1.5 ± 0.3	1.4 ± 0.2	1.4 ± 0.3	1.3 ± 0.3	1.4 ± 0.3	1.5 ± 0.3	1.3 ± 0.2	1.2 ± 0.3
Consensus (mg/liter)			35.7 ± 6.4	32.0 ± 5.8	33.4 ± 7.5	31.4 ± 5.9	33.1 ± 6.6	34.8 ± 6.6	31.4 ± 5.8	27.7 ± 7.2

[a] All values in millimol per liter; mean ± SD, unless otherwise noted.
[b] Methods: a, atomic absorption on dry ashed samples; b, atomic absorption on wet ashed samples; c, atomic absorption on a combination of dry and wet ashed samples; d, atomic absorption on diluted samples; e, colorimetric.

TABLE VIII
Phosphorus Concentration in Human Milk[a]

Reference	Subject No.	Days postpartum						
		3	7–9	14	30	90	180	360
Total Phosphate								
Gross *et al.* (1980); U.S	8	3.5 ± 0.9	4.9 ± 1.7	5.4 ± 0.6	5.1 ± 1.4			
Atkinson *et al.* (1980); Canada	7	3.9 ± 0.4	4.8 ± 0.3	5.0 ± 0.3	4.4 ± 0.2			
Feeley *et al.* (1983); U.S.	102	4.7 ± 1.2		4.6 ± 1.2	4.3 ± 1.2			
Lemons *et al.* (1982); U.S.	7		5.5 ± 0.7	4.9 ± 0.7	4.5 ± 0.8			
Prentice and Barclay (1991); U.S.	12		5.3 ± 1.1		4.8 ± 0.8		4.4 ± 0.4	4.5 ± 0.5
Butte *et al.* (1984); U.S.	13			5.8 ± 0.5	5.3 ± 0.8	4.4 ± 0.8		
Picciano *et al.* (1981); U.S.	26				5.0 ± 0.8	4.7 ± 0.8		
Greer *et al.* (1982); U.S.	18					4.6 ± 2.1	3.5 ± 0.6	
Consensus (mmol/liter)		4.0 ± 0.8	5.1 ± 0.9	5.1 ± 0.6	4.8 ± 0.8	4.6 ± 1.2	3.9 ± 0.5	4.5 ± 0.5
Consensus (mg/liter)		124 ± 25	158 ± 28	158 ± 19	142 ± 25	143 ± 37	121 ± 16	140 ± 16
Free Phosphate								
Neville *et al.*, (1991; Allen *et al.*, 1991); U.S.	12	1.6 ± 0.3	1.9 ± 0.3	1.9 ± 0.4	1.9 ± 0.4	1.8 ± 0.3	1.8 ± 0.3	1.7 ± 0.4

Note. Methods: Fiske Subbarow reaction. All samples ashed except free phosphate.
[a]All values in millimol per liter; mean ± SD, unless otherwise noted.

TABLE IX
Citrate Concentration in Human Milk[a]

Reference	Subject No.	Days postpartum						
		3	7–9	14	30	90	180	360
Neville *et al.* (1991); Allen *et al.* (1991); U.S.	12	3.0 ± 0.3	3.7 ± 0.3	3.5 ± 0.3	3.6 ± 0.4	2.6 ± 0.3	2.3 ± 0.3	1.3 ± 0.3

Note. Method: citrate lyase (Boehringer-Mannheim).
[a]All values in millimol per liter; mean ± SD.

Moellering and Gruber (Moellering and Gruber, 1966) in which the reduction of citrate by citrate lyase is coupled to the oxidation of NADH, measured spectrophotometrically on skim milk samples. A separate blank without citrate lyase must be run for each sample (Allen *et al.*, 1991). A satisfactory kit is available from Boehringer-Mannheim. Phosphate is measured by modification of the method of Fiske and Subbarow (1925) which involves interaction with molybdate in an acid medium to form a blue phosphomolybdate complex that can be measured colorimetrically or spectrophotometrically. If the sample is ashed prior to measurement the total phosphate of the milk is measured; use of the method on defatted milk samples allows measurement of inorganic phosphate alone (Neville *et al.*, 1985).

C. Factors that Influence Calcium, Magnesium, Phosphate, and Citrate Concentrations in Milk

The concentrations of the major divalent ions are species specific (Figure 6). In general, concentration changes have not been associated with systemic disease, such as diabetes (Butte *et al.*, 1987b) or cystic fibrosis, or with local disease such as mastitis. It should be cautioned, however, that citrate concentrations in milk are not often measured so it is not possible to make conclusive statements about this milk component. Dietary effects on calcium and magnesium have been claimed by some authors (reviewed in Lonnerdal (1986a,b), but are not consistently reported for any of these milk components in humans. There is a single report that pharmacological doses of magnesium sulfate increase the magnesium concentration in colostrum (Cruikshank *et al.*, 1982). There are dietary and seasonal effects on the calcium and citrate concentrations in bovine milk (Holt and Muir, 1979).

Longitudinal effects on the concentrations of calcium, magnesium, citrate, and phosphorus have been consistently reported in human milk (Figure 8, Tables VI–IX). Calcium increases markedly during the first few days postpartum as citrate increases and then falls gradually stabilizing near 6.6 mmol/liter for the first 3 months postpartum. After 3 months calcium falls gradually and continuously so that milk from women who have been lactating more than 1 year contains only about 4.4 mmol/liter calcium (Figures 8 and 9). As the analysis in Chapter 7A shows, this fall in total calcium is related largely to a fall in calcium citrate; ionized calcium changes minimally during this period (see also Figure 9). Magnesium falls gradually throughout lactation; this trend is real as shown in two studies that contained careful longitudinal analyses of the same subjects through the first year of lactation or longer (Karra *et al.*, 1986; Allen *et al.*, 1991). Both citrate and phosphate rise in parallel with the sharp increase in milk volume between 2 and 4 days postpartum (Figure 4; Neville *et al.*, 1991), then gradually decrease over the first year of lactation (Tables VIII and IX;

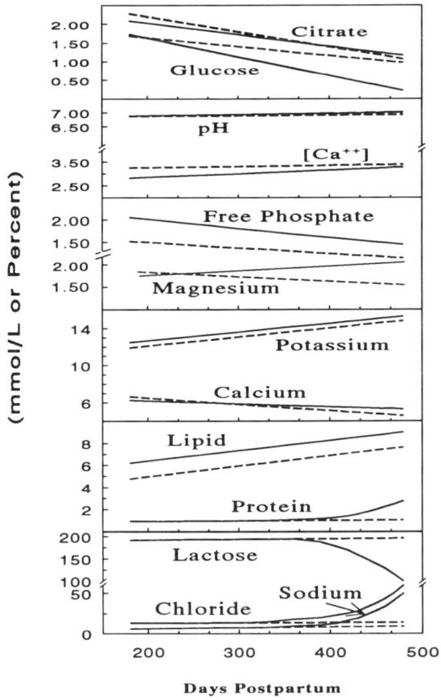

Figure 9 Schematic showing changes in milk composition with duration of lactation in weaning (solid lines) and nonweaning (dotted lines) women. Eleven subjects were studied between 6 and 18 months postpartum. Five women gradually weaned their infants by decreasing the number of feeds and, therefore, the milk volume output per day. Six women maintained six or more feeds per day and milk volumes above 400 ml/day for the duration of the study. The slopes of the glucose and magnesium data are significantly different between the two groups of women ($P < 0.05$).

Allen *et al.*, 1991). Free phosphate and total phosphorus appear to vary in parallel (Table VIII).

Recently, Laskey and Prentice reported that milk from women in Zaire and The Gambia contained significantly less calcium than milk from British women when analyses were performed by identical methods (Prentice and Barclay, 1991; Laskey *et al.*, 1990; Figure 8). The basis for this difference is not known; nutritional effects did not seem to be involved.

D. Divalent Ion Concentrations During Lactogenesis and Weaning in Women

Ionic changes during lactogenesis are shown in Figure 4. Phosphate and citrate concentrations increase in parallel with milk volume between Days 2 and 4 after birth. The calcium concentration increases in parallel with these two ions. The magnesium concentration drops sharply at birth then remains more or less constant over the period of lactogenesis. Effects of gradual weaning were distinguished from effects of lactation duration in a study by Neville and colleagues (Neville *et al.,* 1991). The data are summarized in Figure 9. The late lactation fall in calcium was arrested in women who gradually weaned their infants. In these same women the magnesium concentration actually increased in contrast to the fall observed in women who maintained six or more feeds and a milk volume output greater than 400 ml per day.

E. Effects of Prematurity on Divalent Ion Concentrations in Human Milk

Several studies have provided extensive longitudinal analysis of the effect of prematurity on the concentrations of calcium, magnesium, and phosphorus in human breast milk (Gross *et al.,* 1980); Lemons *et al.,* 1982; Butte *et al.,* 1984; Atkinson *et al.,* 1980). Three studies showed no effect of gestational age on the calcium concentration, whereas the calcium concentration was consistently lower in the milk from preterm mothers in one study (Butte *et al.,* 1984). In three studies for the first month postpartum phosphorus was higher in the milk of mothers delivering at term than in the milk of preterm mothers, the difference reaching significance only in the study by Lemons *et al.* (1982). For the first 2 weeks postpartum magnesium was higher in the milks of mothers of preterm infants in all studies; however, the difference reached significance only in the study of Lemons *et al.* (1982).

F. Mineral Content of Human and Bovine Milk

Tables VI–X give representative values for the calcium, magnesium, phosphorus, and sulfate concentrations in human milk at different stages of lactation. The analytical methods in Tables V–VII represent several methodologies and many different sampling protocols. Given these facts the extent of the agreement between laboratories is remarkable. The free phosphate, useful in understanding ionic interactions in milk, was determined in one laboratory (Table VIII). Longitudinal values for the citrate concentration in human milk (Table IX) are available from only one laboratory despite the importance of citrate to the binding of calcium (Neville *et al.,* this volume, Chapter 7A) and possibly zinc and iron. The limited data for sulfur and its distribution in human milk are summarized in Table X. Table XI gives similar but less extensive values for bovine milk. A major point is that bovine milk contains much higher concentrations of all of the substances described here except sulfur.

TABLE X
Sulfur Concentration in Human Milk[a]

	Days postpartum			
	3		30	
	Subject No.	Concentration	Subject No.	Concentration
McNally *et al.* (1991); Canada				
Total Sulfur	5	10.4 ± 4.0 (333 ± 128 mg/liter)	3	4.5 ± 0.8 (144 ± 26 mg/liter)
Ester Sulfate	5	0.28 ± 0.05	3	0.18 ± 0.22
Free Sulfate	5	0.066 ± 0.021	3	0.029 ± 0.006

Note. Method: oxidation, followed by acid digestion and precipitation with radioactive barium chloride.
[a]All values in millimol per liter; mean ± SD.

TABLE XI
Divalent Ion Concentrations in Bovine Milk

	Concentration—millimol per liter (mg/liter)		
Milk component	Mean	Range	SD
Calcium	29.5 (1180)	27.8–30 (1110–1200)	0.6 (25)
Magnesium	4.9 (120)	4.6–5.4 (110–130)	0.2 (6)
Phosphorus	30.1 (930)		— —
Inorganic phosphate	20.4	16.7–22.6	

Note. Source: White and Davies (1958).

References

Allen, J. C., Keller, R. P., Archer, P., and Neville, M. C. (1991). Studies in human lactation: 6. Milk composition and daily secretion rates of macronutrients in the first year of lactation. *Am. J. Clin. Nutr.* **54,** 69–80.

Atkinson, S. A., Radde, I. C., Chance, G. W., Bryan, M. H., and Anderson, G. H. (1980). Macromineral content of milk obtained during early lactation from mothers of premature infants. *Early Hum. Dev.* **4,** 5–14.

Baumrucker, C. R., and Keenan, T. W. (1975). Membranes of the mammary gland. X. ATP-dependent calcium accumulation by a Golgi apparatus-rich fraction from bovine mammary gland. *Exp. Cell. Res.* **90,** 253–260.

Berga, S. E., and Neville, M. C. (1985). Sodium and potassium distribution in the lactating mouse mammary gland *in vivo*. *J. Physiol.* **361,** 219–230.

Blatchford, D. R., and Peaker, M. (1988). Effect of ionic composition of milk on transepithelial potential in the goat mammary gland. *J. Physiol.* **402,** 533–541.

Butte, N. F., Garza, C., Johnson, C. A., O'Brian Smith, E., and Nichols, B. L. (1984). Longitudinal changes in milk composition of mothers delivering preterm and term infants. *Early Hum. Dev.* **9,** 153–162.

Butte, N. F., Garza, C., Smith, E. O., and Nichols, B. L. (1987a). Macro- and trace-mineral intakes of exclusively breast-fed infants. *Am. J. Clin. Nutr.* **45,** 42–48.

Butte, N. F., Garza, C., Burr, R., Goldman, A. S., Kennedy, K., and Kitzmiller, J. L. (1987b). Milk composition of insulin-dependent diabetic women. *J. Pediatr. Gastroenterol. Nutr.* **6,** 936–941.

Cotlove, E. (1964). Determination of chloride in biological materials. *Methods Biochem. Anal.* **12,** 277–391.

Cruikshank, D. P., Varner, M. W., and Pitkin, R. M. (1982). Breast milk magnesium and calcium concentrations following magnesium sulfate treatment. *Am. J. Obstet. Gynecol.* **143,** 685–688.

Dai, D., and Tang, Z. (1994). Consecutive study of nutritional composition of milk from mothers of preterm delivery. *Hua Hsi I Ko Ta Hsueh Hsueh Pao (J. West China Univ.)* **22,** 332–336.

Dewey, K. G., Finley, D. A., and Lonnerdal, B. (1984). Breast milk volume and composition during late lactation (7–20 months). *J. Pediatr. Gastroenterol. Nutr.* **3,** 713–720.

Dewey, K. G., and Lonnerdal, B. (1983). Milk and nutrient intake of breast-fed infants from 1 to 6 months: Relation to growth and fatness. *J. Pediatr. Gastroenterol. Nutr.* **2,** 497–506.

Feeley, R. M., Eitenmiller, R. R., Jones, J. B., and Barnhart, H. (1983). Calcium, phosphorus and magnesium contents of human milk during early lactation. *J. Pediatr. Gastroenterol. Nutr.* **2,** 262–267.

Fiske, C. H., and Subbarow, Y. (1925). The colorimetric determination of phosphorus. *J. Biol. Chem.* **66,** 375–400.

Gillies, M. E., and Niell, A. E. (1985). Variations in the mineral concentrations in breast milk during a single nursing, diurnally and on consecutive days. *Hum. Nutr. Appl. Nutr. A* **39,** 370–375.

Greer, F. R., Tsang, R. C., Levin, R. S., Searcy, J. E., Wu, R., and Steichen, J. J. (1982). Increasing serum calcium and magnesium concentrations in breast-fed infants: Longitudinal studies of minerals in human milk and in sera of nursing mothers and their infants. *J. Pediatr.* **100,** 59–64.

Gross, S. J., David, R. J., Bauman, L., and Tomarelli, R. M. (1980). Nutritional composition of milk produced by mothers delivering preterm. *J. Pediatr.* **96**(4), 641–644.

Gunther, M., Hawkins, D. F., and Whyley, G. A. (1965). Some observations on the sodium and potassium content of human milk. *J. Obstet. Gynecol.* **72,** 69–74.

Hazebroek, A., and Hofmann, A. (1983). Sodium content of breast milk in the first six months after delivery. *Acta Paediatr. Scand.* **6400,** 203–4099.

Hibberd, C. M., Brooke, O. G., Carter, N. D., Haug, M., and Harzer, G. (1982). Variations in the composition of breast milk during the first 5 weeks of lactation: implications for the feeding of preterm infants. *Arch. Dis. Child.* **57,** 658–662.

Holt, C., and Muir, D. D. (1979). Inorganic constituents of milk: I. Correlation of soluble calcium with citrate in bovine milk. *J. Dairy Res.* **46,** 433–439.

Jenness, R. (1979). Comparative aspects of milk proteins. *J. Dairy Res.* **46,** 197–210.

Johnson, M. P., and Wooding, F. B. P. (1978). Adenosine triphosphatase distribution in mammary tissue. *Histochem. J.* **10,** 171–183.

Karra, M. V., Udipi, S. A., Kirksey, A., and Roepke, J. L. B. (1986). Changes in specific nutrients in breast milk during extended lactation. *Am. J. Clin. Nutr.* **43,** 495–503.

Kirksey, A., Ernst, J. A., Poepke, J. L., and Tsai, T.-L. (1979). Influence of mineral intake and use of oral contraceptives before pregnancy on the mineral content of human colostrum and of more mature milk. *Am. J. Clin. Nutr.* **32,** 30–39.

Laskey, M. A., Prentice, A., Shaw, J., Zachou, T., Ceesay, S. M., Vasquez-Velasquez, L., and Fraser, D. R. (1990). Breast-milk calcium concentrations during prolonged lactation in British and rural Gambian mothers. *Acta Paediatr. Scand.* **79,** 507–512.

Laskey, M. A., Dibba, B., and Prentice, A. (1991). A semi-automated micromethod for the determination of calcium and phosphorus in human milk. *Ann. Clin. Biochem.* **28,** 49–54.

Lemons, J. A., Moye, L., Hall, D., and Simmons, M. (1982). Differences in the composition of preterm and term human milk during early lactation. *Pediatr. Res.* **16,** 113–117.

Linzell, J. L., Mepham, T. B., and Peaker, M. (1976). The secretion of citrate into milk. *J. Physiol.* **260,** 739–750.

Linzell, J. L., and Peaker, M. (1971a). Early detection of mastitis. *Vet. Rec.* **89,** 393–394.

Linzell, J. L., and Peaker, M. (1971b). Intracellular concentrations of sodium, potassium and chloride in the lactating mammary gland and their relation to the secretory mechanism. *J. Physiol.* **216,** 683–700.

Linzell, J. L., and Peaker, M. (1971c). Mechanism of milk secretion. *Physiol. Rev.* **51,** 564–597.

Linzell, J. L., and Peaker, M. (1974). Changes in colostrum composition and in the permeability of the mammary epithelium at about the time of parturition in the goat. *J. Physiol.* **243,** 129–151.

Lonnerdal, B. (1986a). Nutrition and lactation. *In* "Human Lactation 2: Maternal Factors" (M. Hamosh and A. Goldman, eds.), Plenum Press, New York.

Lonnerdal, B. (1986b). Effects of maternal dietary intake on human milk composition. *J. Nutr.* **116,** 499–513.

McNally, M. E., Atkinson, S. A., and Cole, D. E. (1991). Contribution of sulfate and sulfoesters to total sulfur intake in infants fed human milk. *J. Nutr.* **121,** 1250–1254.

Moellering, H., and Gruber, W. (1966). Determination of citrate with citrate lyase. *Anal. Biochem.* **17,** 369–376.

Morton, J. A. (1994). The clinical usefulness of breast milk sodium in the assessment of lactogenesis. *Pediatrics* **93,** 802–806.

Moser, P. B., Reynolds, R. D., Acharya, S., Howard, P., and Andon, M. B. (1988). Calcium and magnesium dietary intakes and plasma and milk concentrations of Nepalese lactating women. *Am. J. Clin. Nutr.* **47,** 735–739.

Neville, M. C., Allen, J. C., and Watters, C. (1983). The mechanisms of milk secretion. *In* "Lactation: Physiology, Nutrition and Breast-feeding" (M. C. Neville and M. R. Neifert, eds.), pp. 49–104. Plenum Press, New York.

Neville, M. C., Keller, R. P., Seacat, J., Casey, C. E., Allen, J. C., and Archer, P. (1984). Studies on human lactation. I. Within-feed and between-breast variation in selected components of human milk. *Am. J. Clin. Nutr.* **40,** 635–646.

Neville, M. C., Keller, R. P., Lonnerdal, B., Atkinson, S., Wade, C. L., Butte, N., *et al.* (1985). Measurement of electrolyte and macromineral concentrations in human milk. *In* "Human Lactation: Milk Components and Methodologies" (R. G. Jensen and M. C. Neville, eds.), pp. 129–140. Plenum Press, New York.

Neville, M. C., Allen, J. C., Archer, P., Seacat, J., Casey, C., Lutes, V., Rasbach, J., and Neifert, M. (1991). Studies in human lactation: Milk volume and nutrient composition during weaning and lactogenesis. *Am. J. Clin. Nutr.* **54,** 81–93.

Neville, M. C., Keller, R. P., Casey, C. E., and Allen, J. C. (1994). Calcium partitioning in human and bovine milk. *J. Dairy Sci.* **77,** 1964–1975.

Neville, M. C., and Watters, C. D. (1983). Secretion of calcium into milk: Review. *J. Dairy Sci.* **66,** 371–380.

Peaker, M. (1977). The aqueous phase of milk: Ion and water transport. *Symp. Zool. Soc. London* **41,** 113–134.

Peaker, M., and Taylor, J. C. (1975). Milk secretion in the rabbit: Changes during lactation and the mechanism of ion transport. *J. Physiol. (London)* **253,** 527–545.

Picciano, M. F., Calkins, E. J., Garrick, J. R., and Deering, R. H. (1981). Milk and mineral intakes of breastfed infants. *Acta Paediatr. Scand.* **70,** 189–194.
Prentice, A., and Barclay, D. V. (1991). Breast-milk calcium and phosphorus concentrations of mothers in rural Zaire. *Eur. J. Clin. Nutr.* **45,** 611–617.
Prentice, A. M., Prentice, A., Lamb, W. H., Lunn, P. G., and Austin, S. (1983). Metabolic consequences of fasting during Ramadan in pregnant and lactating women. *Hum. Nutr. Clin. Nutr. C* **37,** 283–294.
Schoenfeld, R. G., and Lewellen, C. (1964). A colorimetric method for determination of serum chloride. *Clin. Chem.* **10,** 533–539.
Shiffman, M. L., Seale, T. S., Flux, M., Rennert, O. R., and Swender, P. T. (1989). Breast-milk composition in women with cystic fibrosis: Report of two cases and a review of the literature. *Am. J. Clin. Nutr.* **49,** 612–617.
Sjaastad, M. D., Zettl, K. S., Parry, G., Firestone, G. L., and Machen, T. E. (1993). Hormonal regulation of the polarized function and distribution of Na/H exchange and $NaHCO_3$ cotransport in cultured mammary epithelial cells. *J. Cell Biol.* **122,** 589–600.
Tanzer, F., and Sunel, S. (1991). Calcium, magnesium and phosphorus concentrations in human milk and in sera of nursing mothers and their infants during 26 weeks of lactation. *Indian Pediatr.* **28,** 391–399.
Watters, C. (1984). A Ca2+-stimulated adenosine triphosphatase in Golgi-enriched membranes of lactating murine mammary tissue. *Biochem. J.* **224,** 39–45.
White, J. C. D., and Davies, D. T. (1958). The relation between the chemical composition of milk and the stability of the caseinate complex—I. General introduction, description of samples, methods, and chemical composition of samples. *J. Dairy Res.* **30,** 171–189.

C. Microminerals in Human and Animal Milks

CLARE E. CASEY
ANNE SMITH
PEIFANG ZHANG

I. Nutritional Aspects of Microminerals

A. Introduction

Microminerals, also referred to as trace minerals or trace elements, are defined in physiological terms as substances that comprise less than 0.01% of the body mass (Mertz, 1981). In practice, the term includes all elements except those making up the organic matrix (carbon, hydrogen, nitrogen, oxygen, and sulfur), and the bulk minerals of the body fluids and skeleton (calcium, magnesium, potassium, sodium, chlorine, and phosphorus). Iron is on the borderline between the macro- and microminerals and, because

HANDBOOK OF MILK COMPOSITION

of its long history in medicine and well-documented physiology, it is frequently treated separately. Here it is included with the microminerals (Nielsen, 1990).

Based on practical significance, the trace minerals may be divided into four categories: (1) essential: required in the diet of humans and other animal species—iron, zinc, copper, manganese, molybdenum, cobalt, selenium, iodine, fluorine; (2) possibly essential: apparently required in the diet of some animal species under strict experimental conditions but not yet considered proven essential for humans—chromium, nickel, silicon, tin, vanadium; (3) toxic: problems arise from excess rather than deficiency under free-living conditions in humans and animals—aluminum, arsenic, cadmium, lead, mercury; and (4) all other elements: currently considered adventitious contaminants of animals but may change category as improvements in analytical techniques and understanding of their biology produce evidence for a role in metabolism.

B. Trace Element Nutrition in Infants

Because infants typically receive their entire nutrition from a single type of food, human milk or formula, it is important to know how much of any particular trace element is required for adequate growth and development during this nutritionally demanding phase of life (Milner, 1990). Unfortunately, complete data of this sort is unavailable for any trace element, in part, because a number of factors influence trace element requirements including internal stores at birth, the rate of growth and the bioavailability—or the fraction of the element that is absorbed and utilized. Because absorption and assimilation are affected by many factors and because extrinsic labeling may not produce trace element distributions in milk characteristic of the intrinsic element (Davidson *et al.*, 1994), the last factor is very difficult to assess. However, in general, trace element bioavailability appears to be higher from human than from cow's milk. The relative difficulty and cost of trace element analysis also contribute to our state of ignorance about trace element requirements.

Specific clinical syndromes associated with deficiencies of zinc, copper, and iodine are well described (Casey and Walravens, 1988; Milner, 1990). Iron deficiency is not manifest prior to 6 months in human infants because of extensive iron stores present at birth (Cavell and Widdowson, 1964; Dallman, 1988). Although specific physiological and/or enzymatic functions can be ascribed to molybdenum and manganese, instances of nutritional deficiency of these elements have not been documented in human infants (Casey and Walravens, 1988). Of the large number of trace elements thought or suspected to be essential for infant growth information is available on bioavailability only for iron, zinc, copper, manganese, and selenium.

Recommended dietary intakes for essential trace elements for infants are given in Table I. It should be noted that these recommendations are for formula-fed infants and do not apply to fully breast-fed infants under 4–6 months of age. Of particular note are the high recommended intakes of iron and zinc largely because of the considerably lower bioavailability of these elements from formulas compared with human milk.

The concentration of most trace elements in human milk is little affected by maternal intakes or blood levels (Institute of Medicine, 1991). The exceptions are the "anionic" elements iodine, fluorine, and selenium. Excess intake of many elements, particularly those metabolized as anions, may be associated with a risk of toxicity to the nursing infant and supplementation of the lactating mother is generally not recommended. Fluorine levels in milk are only slightly affected by maternal supplementation and this is not thought to be associated with serious risk (Institute of Medicine, 1991; Ekstrand *et al.*, 1984).

C. Analytical Issues

As their name implies, many trace elements are present in milk at very low concentrations and assiduous attention to analytical methodology is of utmost importance. Most important are methods of sampling and storing the sample to avoid contamination, choice of an analytical method with high specificity and low detection limits, and the use of prepared reference materials for standardization of results. These issues have been extensively discussed in several recent reviews (Carl *et al.*, 1992; Versieck and Cornelis, 1989; Casey *et al.*, 1985b) and will be only briefly summarized here.

TABLE I
Recommended Dietary Intake of Trace Elements during the First Half of Infancy

Element	Recommended intake (mg/day)
Iron	6
Zinc	5
Iodine	0.4
Copper	0.4–0.6
Manganese	0.3–0.6
Molybdenum	0.015–0.03
Chromium	0.01–0.04
Selenium	0.01
Fluoride	0.1–0.5

Note. Data from Milner (1990). These recommendations are for formula-fed infants and do not apply to fully breast-fed infants under 4–6 months of age.

1. Sampling and Storage of Sample

Avoidance of contamination during sampling and storage requires attention to both collection vessels, which should be of inert materials such as cleaned, colorless polyethylene, and sample handling conditions. To avoid contamination from ambient air, samples should be handled in a laminar flow cabinet with HEPA air filter. Reagents and water of high purity should be utilized and checked regularly for contamination. Standard solutions and electrode materials should be carefully handled to prevent adsorption and desorption phenomena. Digestion vessels must be properly cleaned. Samples can be stored frozen at −20°C.

2. Mineralization and Concentration of Sample

Both wet and dry ashing procedures are available to concentrate the sample and rid it of unwanted organic matrix. Dry ashing has the advantage that reagents are not necessary and the sample is concentrated. However, ashing conditions must be carefully controlled to avoid volatilization of the minerals to be analyzed. For this reason wet ashing procedures utilizing a variety of strong acids are often preferred. The drawback to this method is the necessity for pure reagents and a limitation on the amount of sample that can be used. To preconcentrate samples for analysis a variety of methods may be used including extraction with organic solvents or solid phases, use of ion-exchange columns, coprecipitation, or stripping voltametry. The choice depends on the element(s) to be analyzed and the requirements of the instrumental technique used.

3. Measurement Techniques

By far the most commonly used methods involve atomic absorption spectrometry in which the sample is atomized by the high temperature of a flame or electric element and the atom cloud passed through light from a monochromatic source of the wavelength of the most intense absorption line of the element of interest. The amount of light absorbed is measured and is proportional to the number of atoms present (and hence, the concentration in the sample introduced to the light path). Description of the instrumentation available for this type of measurement is beyond the scope of this review. However, regardless of instrumentation, a number of types of interference must be minimized if accurate measurements are to be made: chemical interferences arise from the presence or formation of compounds of the element in question that are excessively volatile or refractory, physical interference results from differences in viscosity and surface tension between standards and samples, and ionization interference results when ionization of an element alters its spectral response. Special applications of atomic absorption are used for some elements: hydride generation techniques are used for selenium, arsenic, and antimony, and the cold-vapor technique is used for mercury. Although atomic absorption remains a technique of choice for many elements, recent developments in applications of mass spectrometry, using various atomization methods, particularly inductively coupled plasma, have made this technique more widely available and it is becoming increasingly popular for accurate routine determinations of many elements. Voltametric techniques take advantage of the oxidation/reduction potential of a given element; polarography is a voltametry technique using a dropping mercury electrode. Use of the method of standard additions for calculation of concentrations with these analytical techniques will usually make allowance for any interferences arising from the complex sample matrix of milk.

4. *Quality Assurance*

Both accuracy and precision of analyses must be demonstrated. Precision is usually checked by use of a stock of the same type of matrix as the study samples, which is stored and analyzed at frequent intervals in the same manner as the true samples. Accuracy is best checked by analysis at intervals of appropriate standard or certified reference materials. No reference materials made from human milk are currently available so that some other material with levels of the elements similar to those in milk and with a similar matrix (especially with regard to protein and calcium contents) should be chosen. Cow's milk-based reference materials are available from the European Community, the International Atomic Energy Agency and the U.S. National Bureau of Standards (addresses for these agencies are: Community Bureau of Reference, Commission of the European Communities, 200, rue de la Loi, B-1049 Brussels, Belgium; International Atomic Energy Agency, Analytical Quality Control Services, Laboratory Seibersdorf, P.O. Box 100, A-1400 Vienna, Austria; and National Institute of Standards and Technology, Office of Standard Reference Materials, Chemistry Building, Washington, DC 20234). For publication, the matrix analyzed should be clearly stated (e.g., whole, skim, dried milk); trace element concentrations are usually reported in terms of weight (μg or ng) per unit volume (ml or liters). In general, reporting levels in dried milk is not useful, except for formula sold in powder form.

II. Microminerals in Milks

A. Iron

In addition to its role as an oxygen carrier in the heme respiratory pigments, iron is active in a variety of metalloenzymes involved in redox reactions with oxygen, such as superoxide dismutase and cytochrome oxidase. Compared with calculated requirements for the growing infant, human milk is relatively low in iron. However, the full-term neonate is born with large physiological stores in the liver and hemoglobin, which are adequate to meet requirements for 3 or more months (Dallman, 1988; Siimes *et al.*, 1984). Nonetheless, breast milk iron is controlled within a narrow concentration range and is highly bioavailable (McMillan *et al.*, 1977), suggesting some dietary supply is important for the infant.

1. Methodology

Because iron is ubiquitous in the environment, and at levels often greater than those in milks, great care must be taken to avoid contamination at all stages of sample collection and analysis. The most common technique currently used for reliable iron analysis of human milk is atomic absorption spectrophotometry following either wet or dry ashing. Precision has been shown to be superior with wet ashing (Lonnerdal *et al.*, 1981a). The accuracy of methods should be evaluated using appropriate reference material, such as National Bureau of Standards Bovine Milk or IAEA A-11 Milk Powder (Dybczinski *et al.*, 1980), and the precision of the method should be established by repeated analysis of milk powder.

2. Human Milk

a. Stage of lactation. The iron concentration of human milk is highest during the first few days after birth and diminishes with the progression of lactation (Figure 1; Cavall and Widdowson, 1964; Feeley *et al.*, 1983; Mbofung *et al.*, 1984; Siimes *et al.*, 1979; Vaughan *et al.*, 1979). A 30% decrease in the iron concentration of human milk was reported during the first month of lactation (Feeley *et al.*, 1983). Reported mean iron concentrations of mature human milk range from 0.2 to 0.8 mg/liter (Picciano and Guthrie, 1976; Parr *et al.*, 1991; Siimes *et al.*, 1979). No significant decrease in iron content has been reported after the first 5 months of lactation (Mbofung *et al.*, 1984; Siimes *et al.*, 1979). Trugo *et al.* (1988) found no significant difference in Brazilian milk iron content due to prematurity (Table II).

Figure 1 gives values for iron in milk from 26 studies carried out in 20 different countries. Data included in the figure are from both cross-sectional and longitudinal studies in which time values were given to within 1 or 2 months. Picciano and Guthrie (1976) reported that concentrations of iron in individual milk samples, collected under various protocols from women at 6–12 weeks, ranged between 0.1 and 1.6 μg/ml with the variance about equally due to differences within and between subjects.

The combined data in Figure 1 suggest that the iron concentration rises in late lactation, which is also apparent from the individual studies of Lauber and Reinhardt (1979), Siimes *et al.* (1979), and Vaughan *et al.* (1979), in which samples were collected throughout lactation for 9 months or more. Garza and co-workers (1983) reported that iron levels increased nearly twofold as women gradually weaned their infants over a period of

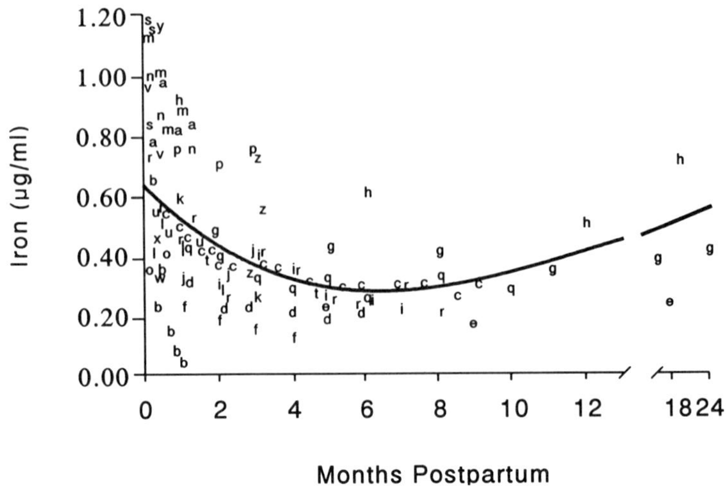

Figure 1 Concentration of iron in human milk. Data are means, with some medians, from 26 longitudinal and cross-sectional studies from 20 countries. a, Lemons *et al.* (1982) (U.S.A., full-term); b, Burguera *et al.* (1988) (Venezuela); c, Siimes *et al.* (1979) and Vuori (1979) (Finland, medians); d, Dewey and Lonnerdal (1983) (U.S.A.); e, Dewey *et al.* (1984) (U.S.A., full lactation); f, Butte *et al.* (1987) (U.S.A.); g, Vaughan *et al.* (1979) (U.S.A.); h, Lauber and Reinhardt (1979) (Ivory Coast); i, Lipman *et al.* (1985) (U.S.A., teenage mothers); j, Picciano *et al.* (1981) (U.S.A.); k, Saner and Yuzbasiyan (1984) (Turkey); l, Saner and Garibagaoglu (1988) (Turkey, full-term infants); m, Mendelson *et al.* (1982) (Canada, full-term infants); n, Feeley *et al.* (1983) (U.S.A.); o, Nassi *et al.* (1974) (Italy); p, Ruz *et al.* (1982) (Chile, well-nourished mothers); q, Mbofung *et al.* (1984) (Nigeria); r, Howell *et al.* (1986) (Houston, full-term infants); s, Loh and Sinnathuray (1971) (Malaysia, all groups); t, Vuori *et al.* (1980) (Finland); u, Atinmo and Omololu (1982) (Nigeria); v, Vega-Franco *et al.* (1987) (Mexico, +54 hr postpartum); w, Celada *et al.* (1982) (Spain); x, Fransson *et al.* (1984) (Sweden, Ethiopia, all groups); y, Murray *et al.* (1978) (Niger, normal); z, WHO/IAEA (1989) (Guatemala, Hungary, Nigeria, Philippines, Sweden, Zaire, medians, all mothers for each country). The solid line is the smoothed line drawn from the arithmetic averages of the mean values at each major time point.

12 weeks. None of these studies collected data in such a way as to determine whether this late rise was due to duration or an effect of declining milk volumes (Neville *et al.*, 1991). However, Dewey and co-workers (1984) found that iron levels were higher in women producing less than 300 ml/day compared with more than 500 ml/day after 7 months postpartum.

b. Dietary intake and maternal status. Geographic variation in human milk iron content apparently occurred in a multinational study on the determination of trace elements in human milk (Parr *et al.*, 1991). This study was coordinated by the IAEA in cooperation with the WHO and six countries from different geographical locations (Parr *et al.*, 1991). The results of the iron concentration in human milk from this and other studies are shown in Table II.

TABLE II
Iron Concentrations in Human Milk

	Colostrum	Mature (1–6 months)
	(μg/ml)	
Brazil		
Term	1.04 (40)[a]	0.90 (35)
Preterm	0.86 (33)	0.69 (26)
Trugo *et al.* (1988)		
Finland		
Siimes *et al.* (1979)		0.30 (12)
Guatemala		
Parr *et al.* (1991)		0.35 (13)
Hungary		
Parr *et al.* (1991)		0.37 (14)
Japan		
Gunshin *et al.* (1985)		0.32 (12)
Nigeria		
Murray *et al.* (1978)		
Parr *et al.* (1991)		0.52 (20)
Mbofung *et al.* (1984)	0.55 (21)	0.38 (15)
Philippines		
Parr *et al.* (1991)		0.72 (28)
Sweden		
Parr *et al.* (1991)		0.45 (17)
United States		
Feeley *et al.* (1983)	0.97 (37)	0.76 (29)
Vaughan *et al.* (1979)		0.46 (18)
Macy and Kelly (1961)		0.30 (12)
United Kingdom		
Dept HSS (1977)		0.76 (29)
Zaire		
Parr *et al.* (1991)		0.56 (22)

[a]Values in parentheses are μmol/liter.

Despite the reported geographic differences in human milk iron content, very little correlation has been found between the amount and distribution of iron in the maternal diet and the iron content of the mother's milk (Celada *et al.*, 1982; Vuori *et al.*, 1980; Karmarkar and Ramakrishnan, 1960; Murray *et al.*, 1978). Furthermore, no relationship has been established between the mother's iron status and the iron concentration of human milk (Dallman, 1986; Murray *et al.*, 1978; Siimes *et al.*,

1984). Neither iron supplementation of mothers with adequate iron status (Karmarkar and Ramakrishnan, 1960; Murray *et al.*, 1978; Vuori *et al.*, 1980) nor poor maternal iron status (Karmarkar and Ramakrishnan, 1960; Murray *et al.*, 1978) has been shown to significantly affect milk iron content. Fransson *et al.* (1985) found that there was a negative correlation between maternal hemoglobin levels and milk iron content when hemoglobin was below 120 g/liter; above this level there was no relationship (Celada *et al.*, 1982). Human milk from vegetarians and nonvegetarians has been shown to have similar amounts of iron (Finley *et al.*, 1985).

c. Distribution. Although the iron content of human milk is low, its bioavailability is very high (McMillan *et al.*, 1976; Saarinen and Siimes, 1977; Garry *et al.*, 1981). The bioavailability of human milk iron is most likely influenced by its distribution on the various milk fractions. About one-third of the iron in human milk is associated with the low-molecular-weight aqueous fraction, one-third with the milk fat, mainly the outer fat globule membrane, and of the remainder, about 10% is found with the casein fraction (Fransson and Lonnerdal, 1984, 1983, 1980). At least some of the 20–30% associated with the whey protein fraction may be bound to lactoferrin. This is a unique milk glycoprotein, MW 76 000, with a concentration in mature human milk of about 25 μM, that may have a bacteriostatic function. It has two binding sites for ferric iron, for which it has a high binding affinity, but will also bind other divalent minerals such as zinc and manganese. In human milk, lactoferrin is highly unsaturated, only about 3–5% of total iron-binding capacity being used (Lonnerdal *et al.*, 1985a). Levels of lactoferrin are high in early lactation (~3 g/liter) and decline during the first month by about one-half (Lonnerdal *et al.*, 1976). The lactoferrin content of milk is not related to the iron status of the mother over the normal range of dietary iron, but may be elevated in women with very high iron intakes and is lower in women who are generally malnourished (Houghton *et al.*, 1985; Lonnerdal *et al.*, 1976). Human milk also contains small amounts of transferrin and ferritin, the concentration of which is about half that in maternal serum (Fransson and Lonnerdal, 1984; van der Westhuyzen *et al.*, 1986).

3. *Animal Milks*

The concentration of iron in most domestic animal milks is in the range of 0.2–1.0 μg/ml, with slightly higher levels in colostrum. In companion animals, levels are generally an order of magnitude higher (Table III). The considerably higher concentrations of iron in milk from monotremes and marsupials, and those of the rat and mouse, compared with most eutherian species may be related to a lack of iron stores in the newborn of the former groups, with their very small body and liver sizes at birth (Kaldor and Ezekiel, 1962). The higher iron levels in these milks and in dolphin milk

TABLE III
Concentration of Iron in Milks from Various Species

Species	Colostrum	Mature milk[a]	Reference
	(μg/ml)		
Rhesus monkey	1.8	1.2	Lonnerdal *et al.* (1984b)
Cow	1–2	0.2–0.6	Anderson (1992); Kincaid and Cronrath (1992); Visser *et al.* (1991)
Buffalo	1.5	0.2–0.3	Lonnerdal *et al.* (1981a)
Goat	1–2	0.3–0.4	Lonnerdal *et al.* (1981a)
Sheep	0.5	0.4–0.6	Lonnerdal *et al.* (1981a)
Pig	1–2	1–3	Lonnerdal *et al.* (1981a)
Horse, domestic	1	0.3–0.8	Anderson (1992); Schryver *et al.* (1986a,b); Ullrey *et al.* (1974)
Zebra		2–4	Schryver *et al.* (1986a)
Dog			
Labrador	6	6.5	Anderson *et al.* (1991)
Beagle	13	7.6	Lonnerdal *et al.* (1981b)
Cat	4–6	3–4	Keen *et al.* (1982)
Rat	8–16	3–6	Anaokar and Garry (1981); Keen *et al.* (1981); Loh and Kaldor (1970); Kaldor and Ezekiel (1962)
Guinea pig	0.6	0.8	Anderson (1990)
Mouse[b]	20–30	15	Reis *et al.* (1991)
Rabbit		2–4	Tarvydas *et al.* (1968)
Echidna		33	Griffiths *et al.* (1984)
Platypus		21	Griffiths *et al.* (1984)
Quokka		12–24	Kaldor and Ezekiel (1962)
Dolphin		36	Peddemors *et al.* (1989)

[a]Monkey > 6 days; horse and zebra > 85 days; rat, guinea pig, dog, cat, mouse, and quokka > 20 days.
[b]μg/g curd.

are probably associated with a high casein content. Lonnerdal and coworkers (1982) reported that around 20% of iron was found in the fat fraction of a range of animal milks, with 30–60% being in the casein fraction. The higher percentage of iron associated with casein in cow's milk compared with human milk (25% vs. 10%) is probably due to the higher casein and lower lactoferrin (whey protein fraction) of bovine milk (Fransson and Lonnerdal, 1983). Iron supplementation does not appear to affect milk levels in cows, pigs, or goats (Lonnerdal *et al.*, 1982), but levels in rat milk may be altered by the iron intake or status of the dam (Anaokar and Garry, 1981).

B. Zinc

Zinc is an essential component of over 200 enzymes which may play both a catalytic and structural role (O'Dell, 1992; Anonymous, 1991; Hambidge *et al.*, 1986). It appears to have a critical role in gene expression: many DNA-binding proteins are zinc complexes. Zinc metalloproteins are also important in maintaining the integrity of cell membranes and extracellular matrix architecture (Waxman and Wasan, 1992).

Naturally occurring nutritional zinc deficiency has been widely reported in a number of species including humans. Depending on the degree of zinc depletion, deficiency in the young may cause growth delay, failure to thrive, anorexia, and when severe, diarrhea and typical skin lesions (Casey and Walravens, 1988; Hambidge *et al.*, 1986). Mild nutritional zinc deficiency occurs most readily in the young and zinc may be a limiting nutrient for growth in otherwise well-nourished human infants and children (Hambidge, 1985; Walravens *et al.*, 1983). Because of their faster growth rates, males and preterm infants have a higher requirement for zinc and are more vulnerable to deficiency (Krebs and Hambidge, 1986).

1. Human Milk

The concentration of zinc in human milk varies considerably with the stage of lactation and among individuals (Vaughan *et al.*, 1979; Karra *et al.*, 1988; Hurley and Lonnerdal, 1986; Casey *et al.*, 1989). Zinc concentration is greater in colostrum, then declines throughout lactation (Figure 2). Reported average zinc concentrations in colostrum range from 8 to 12 μg/ml and in mature milk from 1 to 3 μg/ml (Casey *et al.*, 1989). The most marked reduction in zinc level occurs during the first 2 weeks postpartum. This trend appears to be physiological and has been related to changes in distribution of zinc-bound proteins in the early stages of lactation (Suzuki *et al.*, 1991) and gradual involution of the mammary gland from decreased frequency of breast-feeding (Karra *et al.*, 1986).

The concentration of zinc in milk does not appear to be related directly to the nutritional status of the mother (Vaughan *et al.*, 1979; Vuori *et al.*, 1980; Moser-Veillon and Reynolds, 1983; Krebs *et al.*, 1985). Values reported from developed countries are similar to those reported from developing countries. The effect of mineral supplementation may depend on the zinc status of the lactating mother. When lactating women with normal serum zinc levels took a supplement of 50 to 150 mg zinc daily for a week, the milk zinc concentration was not increased, although serum zinc was significantly elevated (Moore *et al.*, 1984). In a different report, another group of women took a zinc supplement up to 27 mg per day; the

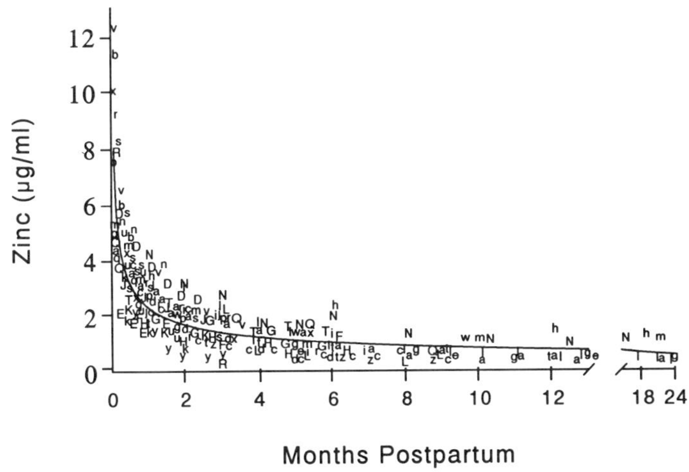

Figure 2 Concentrations of zinc in human milk. Data are means, with some medians, from 41 longitudinal and cross-sectional studies from 18 countries. The solid line is the smoothed line drawn from the arithmetic averages of the mean values at each major time point. a, Casey *et al.* (1989) (U.S.A.); b, Casey *et al.* (1985a) (U.S.A.); c, Vuori and Kuitunen (1979) (Finland, medians); d, Dewey and Lonnerdal (1983) (U.S.A.); e, Dewey *et al.* (1984) (U.S.A., full lactation); f, Butte *et al.* (1987) (U.S.A.); g, Vaughan *et al.* (1979) (U.S.A.); h, Lauber and Reinhardt (1979) (Ivory Coast); i, Lipsman *et al.* (1985) (U.S.A., teenage mothers); j, Picciano *et al.* (1981) (U.S.A.); k, Saner and Garibagaoglu (1988) (Turkey, full-term infants); l, Mbofung *et al.* (1984) (Nigeria); m, Rajalakshmi and Srikantia (1980) (India, cross-sectional); n, Atinmo and Omololu (1982) (Nigeria, term infants); o, Feeley *et al.* (1983) (U.S.A.); p, Ruz *et al.* (1982) (Chile, control group); q, Mendelson *et al.* (1982) (Canada, full-term infants); r, Hambidge (1976) (U.S.A.); s, Howell *et al.* (1986) (Houston, full-term infants); t, Simmer *et al.* (1990) (Bangladesh); u, Moran *et al.* (1983) (U.S.A.); v, Ohtake *et al.* (1981) (Japan); w, Belavady *et al.* (1978) (India); x, Higashi *et al.* (1982) (Japan); y, Dorea *et al.* (1985) (Brazil); z, Hibberd *et al.* (1982) (UK); A, Bhandari *et al.* (1985) (India); B, Nagra (1989) (Pakistan); D, Ehrenkranz *et al.* (1984) (U.S.A., preterm infants); E, Tkachenko (1970) (Ukraine); F, Moser and Reynolds (1983) (U.S.A.); G and H, Karra *et al.* (1988) (U.S.A., no supplement, 24 hr; Egypt); I, Karra *et al.* (1986) (U.S.A.); J, Butte *et al.* (1984) (U.S.A., full-term infants); K, Lehti (1990) (Brazil); L and N, Bates and Tsuchiya (1990) (England; The Gambia); Q, Moser-Veillon and Reynolds (1990) (U.S.A., 0 mg zinc); R, Nyazema *et al.* (1989) (Zimbabwe); T, Lamounier *et al.* (1989) (Brazil, medians); U, Krebs *et al.* (1985) and Krebs and Hambidge (1985) (U.S.A., unsupplemented); X, Berfenstam (1952) (Sweden).

rate of decline in milk zinc level over 1–9 months was significantly less than that in a similar unsupplemented group of mothers (Krebs *et al.*, 1985). No significant difference has been found between the milk from mothers delivering at term and prematurely in terms of the mean concentration or rate of change in the concentration of zinc (Butte *et al.*, 1984). Feeley *et al.* (1983) found no correlation between zinc levels in breast milk and age, number of gestations, or previous lactation history. The mechanism of the secretion of zinc into milk is unknown, but is probably under genetic regulation.

It is well known that zinc in human milk is more efficiently utilized by infants than zinc in cow's milk or milk formulas based on cow's milk (Evans and Johnson, 1977; Johnson and Evans, 1978; Hambidge *et al.*, 1979; Casey *et al.*, 1981; Sandstrom *et al.*, 1983; Lonnerdal *et al.*, 1984a, 1985b; Anonymous, 1986). The plasma zinc level was higher in breast-fed infants than in formula-fed infants even when the concentration of zinc in the formula was about three times that of breast milk (Sandstrom *et al.*, 1982; Hambidge *et al.*, 1979). In an extrinsic labeling study using 16 day old rat as an animal model, Sandstrom *et al.* (1982) demonstrated that the bioavailability of zinc was 28% from human milk, 24% from whey-adjusted cow's milk formula, 15% from cows' milk, and 10% from soy formula. Human milk has therapeutic value in treating infants suffering from a hereditary zinc-deficiency disease, acrodermatitis enteropathica (Moynahan, 1974), whereas cow's milk is not efficacious even though the concentration of zinc is higher in cow's milk than in human milk (Lonnerdal *et al.*, 1981a). In general, the symptoms of the disease arise when the afflicted infant is weaned from human milk to bovine milk. So it is not only the amount of zinc, but also the molecular localization of zinc in milk that affects the degree to which it is absorbed.

The distribution of zinc in human milk has been studied qualitatively and quantitatively and compared with that of cow's milk. Zinc is present in different chemical forms in the three major milk fractions: fat, casein, and whey. It is reported (Lonnerdal *et al.*, 1982; Hurley and Lonnerdal, 1982) that casein micelles in human milk contained 14% of the total zinc content and the whey fraction contained two major zinc-binding ligands, serum albumin and citrate, which bound 28 and 29% of the total zinc, respectively. Another 29% was associated with fat. Phosphoserine residues were demonstrated to be the primary zinc-binding sites in casein (Singh *et al.*, 1989). Alkaline phosphatase, bound to the fat globule membrane, was suggested to be the major zinc-binding protein of milk fat (Fransson and Lonnerdal, 1984). Since the zinc in cow's milk is bound primarily to casein and, to a smaller extent, to citrate (Lonnerdal *et al.*, 1981a, 1984a), and bovine casein is difficult to digest (Fomon and Filer, 1974), the superior bioavailability of zinc in human milk compared with cow's milk may be due, at least partially, to the less extensive association with casein micelles.

2. *Animal Milks*

Compared with the human data, there is relatively little information on zinc in milks from other species. Table IV gives values for the concentration of zinc in colostrum (2–5 days) and mature milk from 19 species. With few exceptions (e.g., the cat), levels in colostrum are higher than those in secretions from later lactation. It is apparent from the longitudinal studies carried out by Lonnerdal's group that the changes with time of zinc in milk from the dog and cat are not marked, whereas in the milk of the rat, mouse, and rhesus monkey zinc levels change with time in a manner similar to human milk (Reis *et al.*, 1991; Lonnerdal *et al.*, 1984b; Keen *et al.*, 1981,

TABLE IV
Concentrations of Zinc in Milks from Various Species

Species	Colostrum	Mature milk[a]	Reference
	(μg/ml)		
Human[b]	8–12	1–3	Casey *et al.* (1989); Fig. 2
Rhesus monkey	5	2	Lonnerdal *et al.* (1984b)
Cow	12–18	4	Benemariya *et al.* (1993); Anderson (1992); Kincaid and Cronrath (1992); Singh *et al.* (1989); Lonnerdal *et al.* (1986); Varo *et al.* (1980); Casey (1977)
Buffalo	1	0.2–0.3	Lonnerdal *et al.* (1982)
Goat	11–13	3–6	Benemariya *et al.* (1993); Lonnerdal *et al.* (1982)
Sheep	5–15	1–2	Lonnerdal *et al.* (1982)
Pig	11	4–6	Lonnerdal *et al.* (1982)
Horse, domestic	3–6	1–3	Anderson (1992); Schryver *et al.* (1986a,b); Ullrey *et al.* (1974)
Zebra	2	2–3	Schryver *et al.* (1986a)
Dog	8–10	7–8	Anderson *et al.* (1991); Lonnerdal *et al.* (1981b)
Cat	4–6	5–7	Keen *et al.* (1982)
Rat	14	5–10	Keen *et al.* (1981)
Guinea pig	5	4	Anderson (1990)
Mouse[c]	40	10–17	Ackland and Mercer (1992); Reis *et al.* (1991); Witsell *et al.* (1990)
Rabbit		2–4	Lonnerdal *et al.* (1982)
Echidna		15	Griffiths *et al.* (1984)
Platypus		19	Griffiths *et al.* (1984)
Dolphin		11	Honda and Taksukawa (1983)

[a]See Table I.
[b]Very time dependent.
[c]μg/g curd.

1982; Lonnerdal *et al.*, 1981b). In a study in which colostral samples were collected from cows on an 8 or 10-hr schedule, Kincaid and Cronrath (1992) found an average level of 5–8 μg/ml zinc both pre- and post- (-240 to 288 hr)-partum, but there was a "spike" up to 18 μg/ml in the first sampling after parturition (at 1 hr), possibly analogous to the spike in zinc levels seen in human colostrum at about 40 hr. They postulated that the elevation in colostral zinc levels may be caused by the increase in glucocorticoids at parturition, which causes an increased transfer of zinc from the blood to the mammary gland (Vaillancourt and Allen, 1990).

A mutation has been described in mice, lethal milk (lm), which causes zinc deficiency in pups nursed by lm dams (Piletz and Ganschow, 1978). The milk of the lm dams is relatively deficient in zinc and the genetic defect appears to result in a decrease (about 50% of normal) in the transport of zinc into the milk (Lee *et al.*, 1992).

Levels of zinc in the milk of cows and pigs are not generally related to dietary intakes but may be increased by very large supplements (grams per day). Lowered zinc concentrations have been found in the inflamed quarter of the udder of mastitic cows (Hambidge *et al.*, 1986). A similar observation was made with human milk: the zinc concentration of milk from the mastitic breast of a woman was about half that in the healthy breast (Casey, unpublished observations).

C. Copper

Copper is a constituent of many metalloenzymes, including cytochrome oxidase, superoxide dismutase, ceruloplasmin which functions to transport copper to the tissues and to release iron from stores, and enzymes involved in the synthesis of connective tissue, melanin and catecholamines (Casey and Walravens, 1988). Copper deficiency due to inadequate dietary intakes occurs in humans and other species, particularly the young. Interactions with excess molybdenum and sulfate may be important in the etiology of copper deficiency of pastoral animals. Generally, the symptoms of deficiency in the young include anemia, unresponsiveness to iron supplementation, defects in bone, cartilage, hair/wool, and pigmentation, and diarrhea (Davis and Mertz, 1987).

The newborn of many species, including humans, but not sheep, have copper concentrations in the liver which are higher than adult levels. In humans and rats, fetal liver copper levels may be up to 10 times adult values. This copper is tightly bound to intracellular metallothionein and may represent a way of protecting the fetus from the excess copper crossing the placenta from the elevated levels in the maternal circulation. Nevertheless, it also constitutes a store which is used up gradually during early postnatal life to meet the needs of the growing infant in conjunction with the highly bioavailable copper in milk. Neonates which lack liver copper stores (e.g., premature human infants, rat pups with toxic milk syndrome) are more vulnerable to nutritional copper deficiency (Mercer *et al.*, 1992; Casey and Walravens, 1988).

1. *Human Milk*

Figure 3 gives values for copper concentrations in human milk samples over 24 months of lactation. The data are taken from 35 studies in 15

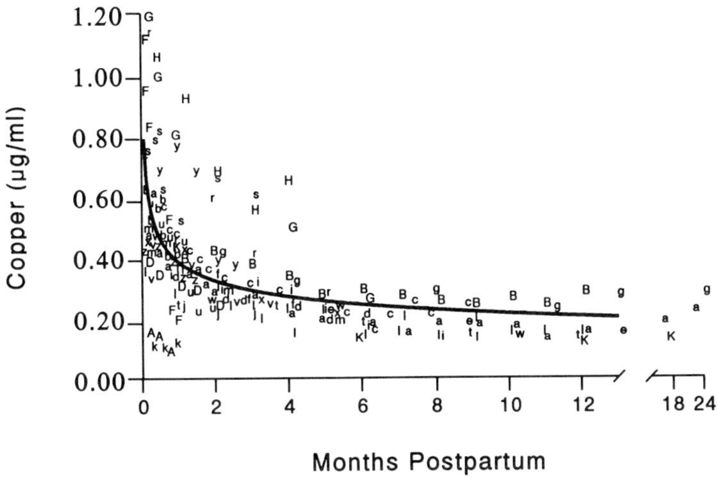

Figure 3 Concentrations of copper in human milk. Data are means, with some medians, from 34 longitudinal and cross-sectional studies from 15 countries. The solid line is the smoothed line drawn from the arithmetic averages of the mean values at each major time point. a–E same as for zinc (Fig. 2); F, Burguera *et al.* (1988) (Venezuela); G, Kleinbaum (1962) (West Germany); H, Dorner *et al.* (1989) (West Germany, foremilk); I, Salmenpera *et al.* (1986) (Finland, medians).

different countries, in which at least three different time periods were given and in which the information on duration of lactation was provided in days prior to 1 months and to within 1 month thereafter. Both cross-sectional and longitudinal studies are included.

Copper concentrations are generally higher in colostrum and fall throughout lactation, with the largest change being in the first month. The decline appears to be less marked than that of zinc and copper and zinc concentrations are not correlated. The initial decline in copper levels during the colostral phase was related to the fall in protein content (Casey *et al.*, 1989). Although in some studies copper levels appear to increase in late lactation (Casey *et al.*, 1989; Vuori and Kuitunen, 1979), the change was not significant in relation to either duration of lactation or declining volumes (Neville *et al.*, 1991; Dewey *et al.*, 1984).

The data presented in Figure 3 show a very wide range in the mean values from different studies up to about 3 months, and particularly in very early lactation. A relatively wide range is also seen between individual women in longitudinal studies (Burguera *et al.*, 1988; Saner and Garibagaoglu, 1988; Howell *et al.*, 1986; Casey *et al.*, 1985a; Feeley *et al.*, 1983; Hibberd *et al.*, 1982; Rajalakshmi and Srikantia, 1980). For example, Casey *et al.* (1989) found individual values to range between 0.16 and 1.34 µg/ml among 11 women in the first 5 days, but by 3 months the variation was down to 20%. Picciano and Guthrie (1976) found that, in mature milk from American mothers, the day-to-day variation in copper was accounted for largely (60–80%) by differences between women. It can be seen in Figure 3 that several studies stand out from the range that includes most of the data. These include, on the high side, the two studies from Germany (G, Kleinbaum, 1962; H, Dorner *et al.*, 1989) and the three from South America (p, Ruz *et al.*, 1982; y, Dorea *et al.*, 1985; F, Burguera *et al.*, 1988). On the low side are two studies from Asia (k, Saner and Yuzbasiyan, 1984; A, Bhandari *et al.*, 1985); other studies from Asia are within the "normal" range. However, it is not currently possible to say if these differences represent a true geographical variation or arise from analytical differences. During the WHO/IAEA (1989) collaborative study on minor and trace elements in human milk, an analytical coefficient of over 50% was reported for interlaboratory assays of a powdered milk reference material (Casey *et al.*, 1985). Copper concentrations in human milk are not affected by maternal dietary intake or nutritional status of copper (Vuori *et al.*, 1980; Vaughan *et al.*, 1979).

Nearly 80% of human milk copper is found in the whey fraction, with only 5–15% in the fat and the remainder in the casein precipitate. The major copper-binding protein in the whey appears to be serum albumin, and some of the mineral may also be associated with low-molecular-weight ligands such as citrate and free amino acids (Fransson and Lonnerdal, 1984, 1983; Lonnerdal *et al.*, 1982).

2. *Animal Milks*

Table V gives values for copper concentrations in milks from 16 animal species, all of which fall within a similar range. In all species, levels in colostrum are higher than later milk and there is a decline throughout lactation which is generally of smaller magnitude than that seen in zinc levels.

Copper levels in milk from cows and sheep are lowered by dietary deficiency and may be increased by oral or subcutaneous administration of supplementary copper (Davis and Mertz, 1987).

A much higher percentage, 30–80% of the copper in animal milks, is found bound to casein compared with human milk. With the exception of sow milk, in which 40% is found in the fat fraction, less than 10% of milk copper is associated with milk lipids, mainly in the outer milk fat globule membrane (Fransson and Lonnerdal, 1983; Lonnerdal *et al.*, 1982).

An autosomal recessive genetic mutation, toxic milk (tm), has been identified in mice which causes a marked hepatic accumulation of copper in the adult with levels of copper in the milk of lactating dams being one-quarter to one-half of those in normal mice (Rauch, 1983). Pups die in the second week from severe copper deficiency due to low liver stores and low intakes from maternal milk (Mercer *et al.*, 1992).

D. Manganese

Manganese metalloenzymes have a wide range of metabolic functions: mucopolysaccharide synthesis, gluconeogenesis, lipid metabolism, neuro-

TABLE V
Concentrations of Copper in Milks from Various Species

Species	Colostrum	Mature milk[a]	Reference
	(μg/ml)		
Human	0.5–0.8	0.2–0.4	Casey *et al.* (1989); Fig. 3
Rhesus monkey	3	1	Lönnerdal *et al.* (1984b)
Cow	0.2–0.4	0.05–0.2	Benemariya *et al.* (1993); Anderson (1992); Kincaid and Cronrath (1992); Lönnerdal *et al.* (1982)
Buffalo	0.3–0.4	0.2–0.3	Lönnerdal *et al.* (1982)
Goat	0.4–1.2	0.1–0.2	Benemariya *et al.* (1993); Lönnerdal *et al.* (1982)
Sheep	3–4	0.2–0.4	Kincaid and White (1988); Lönnerdal *et al.* (1982)
Pig	6	0.6–1	Lönnerdal *et al.* (1982)
Horse			
Domestic	0.6–1	0.2–0.4	Anderson (1992); Schryver *et al.* (1986b); Ullrey *et al.* (1974)
Przewalski	0.4	0.2	Schryver *et al.* (1986a)
Zebra	1	0.2–1	Schryver *et al.* (1986a)
Dog	1–2	1.3–2	Anderson *et al.* (1991); Lönnerdal *et al.* (1981b)
Cat	1–1.5	0.8–1.2	Keen *et al.* (1982)
Rat	9	1–2	Keen *et al.* (1981)
Guinea pig	1	0.4–0.6	Anderson (1990)
Mouse[b]	8	1–2	Ackland and Mercer (1992); Reis *et al.* (1991); Witsell *et al.* (1990)
Echidna		4	Griffiths *et al.* (1984)
Platypus		1	Griffiths *et al.* (1984)

[a]See Table I.
[b]μg/g curd.

transmitter synthesis, and, in mitochondrial, superoxide dismutase. Deficiencies of manganese have been produced experimentally in a number of species, but have only been found to occur naturally in pigs and poultry on some diets, and never in humans (Hurley and Keen 1987; Doisy, 1974). Fetal life and early infancy are the periods most vulnerable to manganese deficiency.

When flameless (graphite furnace) atomic absorption spectroscopy became available in the late 1970s, concentrations of manganese in biological fluids, including milk, were found to be orders of magnitude lower than previously reported (Versieck and Cornelis, 1989). Along with the improvements in analytical technology, there has been an increased appreciation of the ease with which contamination may occur during sample collection and preparation, leaving in question the reliability of reports of manganese values which do not include information on quality control or reference material.

Figure 4 gives data for manganese levels in human milk from six published studies in which samples were collected in a longitudinal fashion from the same mothers throughout the study period (Casey *et al.*, 1989; Saner and Garibagaoglu, 1988; Casey *et al.*, 1985a; Dorner *et al.*, 1985; Stastney *et al.*, 1984; Vuori, 1979). Average concentrations of manganese in human milk after 2–4 weeks postpartum are generally in the range of 3–6 ng/ml, with a between-subject variation of 40–60%. Casey *et al.* (1989) reported that median values were not different from means after the first month postpartum, with individual values generally falling in the range of 2–8 ng/ml.

Table VI gives values for average concentrations of manganese in colostrum and mature milk from different species. Nonhuman milks (Witsell *et al.*, 1990; Anderson, 1992; Keen *et al.*, 1981, 1982; Lonnerdal *et al.*, 1981a,b; Varo *et al.*, 1980; Casey, 1977) typically have levels one or two orders of magnitude higher than values reported for human milk in recent well-controlled studies (Casey *et al.*, 1989; WHO/IAEA, 1989; Saner and Garibagaoglu, 1988; Casey *et al.*, 1985a; Dorner *et al.*, 1985; Saner and Yuzbasiyan, 1984; Stastney *et al.*, 1984; Cumming *et al.*, 1983; Kosta *et al.*, 1983; Vuori, 1979). In the rat (Keen *et al.*, 1981), dog (Lonnerdal *et al.*, 1981b), and human (Vega-Franco *et al.*, 1987; Casey *et al.*, 1985a), levels of manganese in colostrum are slightly higher than those in mature milk and the changes with duration of lactation are not as marked as those seen for iron (Siimes *et al.*, 1979), copper, or zinc (Casey *et al.*, 1989).

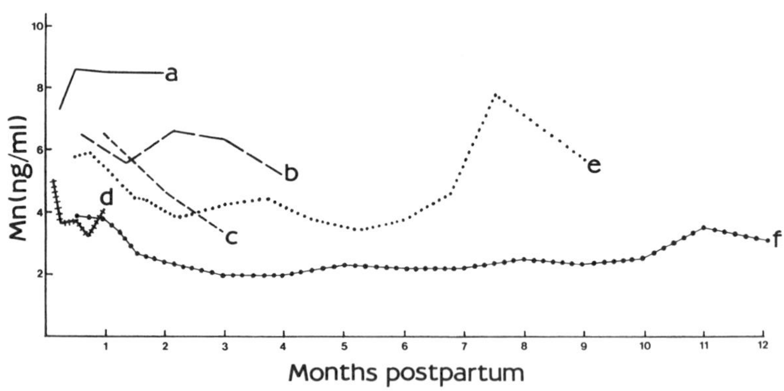

Figure 4 Manganese in human milk: longitudinal studies. f, Casey *et al.* (1989) (U.S.A., medians, late volumes < 400 ml/day); a, Saner and Garibagaoglu (1988) (Turkey, means, full-term infants); d, Casey *et al.* (1985a) (U.S.A., means); b, Dorner *et al.* (1989) (Germany, means); c, Stastny *et al.* (1984) (U.S.A., means); e, Vuori (1979) (Finland, medians).

TABLE VI
Concentrations of Manganese in Milks from Various Species

Species	Colostrum	Mature milk[a]	Reference
	(ng/ml)		
Human	5–12	3–6	Casey *et al.* (1989); WHO/IAEA (1989); Saner and Garibagaoglu (1988); Vega-Franco *et al.* (1987); Casey *et al.* (1985a); Dorner *et al.* (1989); Saner and Yuzbasiyan (1984); Stastney *et al.* (1984); Cumming *et al.* (1983); Kosta *et al.* (1983)
Cow		21	Anderson (1992)
Goat	100–160	200–500	Lonnerdal *et al.* (1981a)
Sheep	100–160	200–500	Lonnerdal *et al.* (1981a)
Horse		14	Anderson (1992)
Dog	160	140	Lonnerdal *et al.* (1981b)
Cat	140	300	Keen *et al.* (1982)
Rat	330	120	Keen *et al.* (1981)
Guinea pig	14	11–26	Anderson (1992, 1990)
Mouse[b]		50	Witsell *et al.* (1991)

[a]Human, 1–12 months; dog, > 20 days; cat, > 20 days; rat, 10–20 days; guinea pig, 20 days.
[b]ng/g curd.

Longitudinal studies in humans (Figure 4; Casey *et al.*, 1989; Vuori, 1979), cats (Keen *et al.*, 1982), the rat (Keen *et al.*, 1981), and guinea pig (Anderson, 1990) suggest that manganese levels increase in late lactation. This appears, however, at least in the human, to be related to declining milk volumes as the suckling is weaned: Casey and co-workers (1989) found that manganese concentrations in milk from women who were producing more than 400 ml/day remained at normal levels (3–5 ng/ml), whereas rapid and large increases in manganese levels, up to 35 ng/ml, were observed as volumes fell below this and particularly as milk production neared termination, as shown in Figure 5. Secretion of large amounts of manganese may be associated with an increase in levels of α-lactalbumin (Neville and Casey, unpublished observations). Manganese may be a cofactor, along with α-lactalbumin, for galctosyltransferase in the lactose synthetase complex (Witsell *et al.*, 1990). Milk volumes fall during weaning with a decline in the rate of lactose secretion (Neville *et al.*, 1991), and the late spike in manganese concentration may reflect these terminal events.

The level of manganese in cow's milk responds to changes in manganese intake (Hurley and Keen, 1987). Vuori *et al.* (1980) found that manganese in human milk was also related to maternal intake. It is not

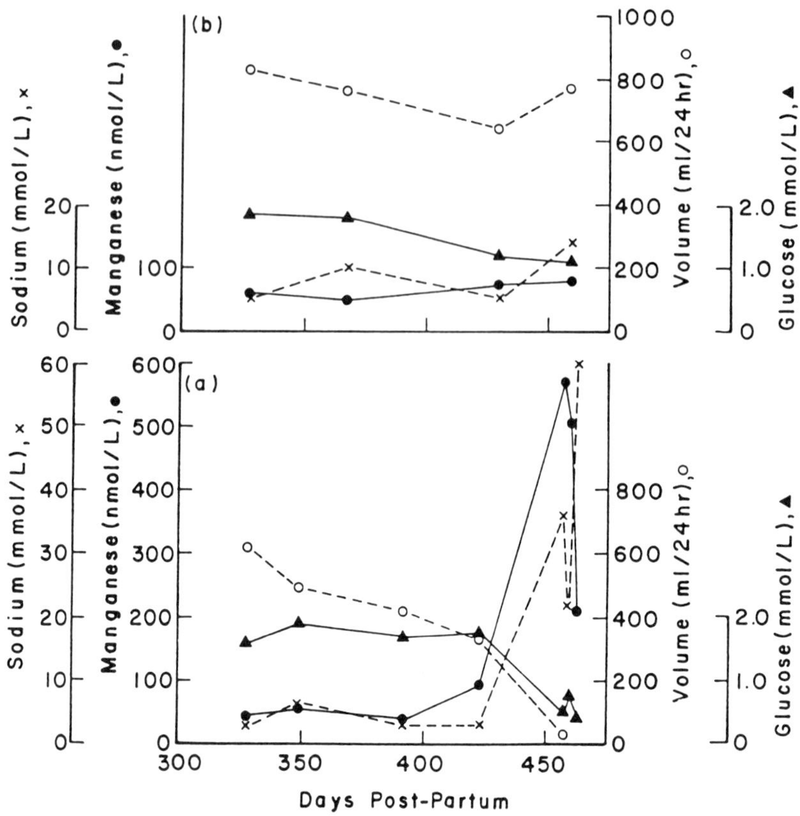

Figure 5 Manganese in human milk in late lactation. (a) Weaning subject. Concentrations of manganese in all samples collected from one breast between 327 days and cessation of lactation at 462 days are given with concentrations of glucose and sodium in the same milk samples and with 24-hr milk volumes. (b) Nonweaning subject (producing > 399 ml/day). Concentrations of manganese in all samples collected from one breast between 334 and 460 days postpartum are given with concentrations of glucose and sodium in the same milk samples and with 24-hr milk volumes. From Casey *et al.* (1989) with permission.

possible at present, however, to determine whether the small differences in mean concentrations of manganese in milk from different parts of the world (Figure 4) reflect maternal exposure or analytical variation.

Seventy percent of the manganese in human milk is found in the whey fraction, mostly bound to lactoferrin, 11% is in the casein fraction, and the remainder with the milk fat. This is in contrast to cow's milk, in which 67% of the total manganese is in the casein fraction and almost all the remainder associated with small-molecular-weight compounds is in the whey fraction (Lonnerdal *et al.*, 1985a). These differences in casein-bound mineral may explain, in part, the better absorption of manganese from human milk compared with cow's milk (Davidson *et al.*, 1989).

E. Selenium

Selenium concentrations in tissues and body fluids are directly affected by levels in the food chain and, hence, reflect the geochemical environment in which the animal lives. Worldwide, the full range of nutritional conditions from severe, lethal deficiency to severe, lethal toxicity occurs naturally in humans and animals in relation to the selenium in their environment (Levander, 1987, 1986). Selenium has two (known) important functions in biological systems: it is an integral component of the enzyme glutathione peroxidase (GSHPx), which acts in the cytosol to protect cell components from oxidant damage, and iodothyronine deiodinase, which catalyses the conversion of T_4 to T_3 in the peripheral tissues (Berry *et al.,* 1991). Most selenium in human milk is bound to protein (Milner *et al.,* 1987). A small fraction (9–17%) is associated with the selenoprotein, glutathione peroxidase (Avissar *et al.*, 1991; Milner *et al.*, 1987). The GSHPx in milk has been shown to be the plasma form of the enzyme (Avissar *et al.,* 1991).

1. *Methodology*

Information on human milk selenium concentration and variations with stage of lactation, geographical region, dietary intake, and maternal status have become well established over the past decade. The appearance of reliable selenium data has coincided with the availability of accurate techniques for trace element analysis and the availability of appropriate reference material, especially the IAEA A-11 milk powder (Dybczinski *et al.,* 1980). Reliable techniques currently used for selenium analysis of human milk include neutron activation analysis (Kosta *et al.,* 1983; Schramel *et al.,* 1988), gas chromatography with electron capture detection (McCarthy *et al.,* 1981), and hydride generation atomic absorption spectrometry (Verlinden *et al.,* 1980).

2. *Human Milk*

Geographic differences in the selenium content of mature milk have been closely linked to variations in dietary selenium intake (Table VII). Low values, 10 ng/ml or less, have been reported for mature human milk from New Zealand (Williams, 1983; Casey, 1988), Finland (Kumpulainen *et al.,* 1983b), Nepal (Moser *et al.,* 1988), Sweden (Walivaara *et al.,* 1986), the Kenshan region of China (Levander, 1987), and Belgium (Robberecht *et al.,* 1985), countries known to have low dietary selenium intakes for adults. In contrast, higher mean values, 16–28 ng/ml, have been reported for human milk from Germany (Lombeck *et al.,* 1978), the United States (Smith *et al.,* 1982), and Japan (Higashi *et al.,* 1983). The effect of dietary selenium intake on milk selenium content was demonstrated well in Finland where between 1976 and 1980 the importation and use of wheat

TABLE VII
Selenium Content in Human Milk from Different Geographical Areas

	Colostrum	Mature (1–6 months)
	(ng/ml)	
Belgium		
Robberecht *et al.* (1985)	14.8 (0.19)[a]	9.4 (0.12)
Finland		
1976		5.6–10.7 (0.07–0.14)
1980		10.0–11.8 (0.13–0.15)
Kumpulainen *et al.* (1984)		
Sweden		
1978	17.6 (0.22)	9.4 (0.12)
1983	14.2 (0.18)	11.9 (0.15)
Walivaara *et al.* (1986)		
New Zealand		
Dunedin (low environmental selenium) (Williams, 1983)		7.6 (0.10)
Other areas (Millar and Sheppard 1972)	22.6 (0.29)	9.8–13.3 (0.12–0.17)
Nepal		
Moser *et al.* (1988)		9.9 (0.13)
United States		
Smith *et al.* (1982)	41.2 (0.52)	15.3 (0.19)
Ellis *et al.* (1990)	32.4 (0.41)	21.3 (0.27)
South Dakota (high environmental selenium) (Shearer and Hadjimarkos, 1975)		17–60 (0.21–0.76)
Gambia		
Rainy		15.3 (0.19)
Dry		21.0 (0.26)
Funk *et al.* (1990)		
Germany		
Lombeck *et al.* (1978)	83 (1.05)	28.3 (0.36)
Schramel *et al.* (1988)	43 (0.54)	21 (0.26)
Japan		
Higashi *et al.* (1983)	80 (1.01)	18 (0.23)
China		
Keshan area (low environmental selenium)		3 (0.04)
(High environmental selenium)		283 (3.63)
Levander (1987)		

[a]Values in parentheses are μmol/liter.

higher in selenium increased the maternal dietary selenium intake and milk selenium concentrations (Kumpulainen *et al.*, 1984). A similar increase in milk levels was observed in Sweden between 1978 and 1980, with increased agricultural use of selenium supplements (Walivaara *et al.*, 1986). In rural Africa, where the selenium content of the diet varies with food availability, Funk *et al.* (1990) found that when maternal selenium intake is low, milk selenium concentrations are low and decrease with increasing parity. Whereas total selenium intake is a strong determinant of milk selenium content, the form of dietary selenium may also influence milk selenium concentration. The milk of vegetarian women in California was found to contain more selenium than that of nonvegetarians with similar selenium intakes, suggesting a variation in availability (Debski *et al.*, 1989).

The effect of selenium intake on milk concentration appears to be mediated through maternal selenium status. The results of Mannan and Picciano (1987) indicate a direct relationship between maternal plasma selenium concentration and their milk selenium concentrations. In addition, selenium supplementation of lactating women in Finland (Kumpulainen *et al.*, 1985) and the United States (McGuire *et al.*, 1993) has been shown to increase maternal serum selenium and their corresponding milk selenium concentrations. Other workers (Cumming *et al.*, 1992; Levander *et al.*, 1987; Higashi *et al.*, 1983) found no relationship between selenium in blood and milk in lactating women in Australia, the United States, and Japan.

The selenium concentration of human milk varies with the stage of lactation and with maternal selenium intake and status. The selenium concentration of colostrum is relatively high (41 ng/ml) compared to that of mature milk (Smith *et al.*, 1982). Selenium concentrations of mature milk stabilize by 1 month with mean values ranging from 10 to 30 ng/ml depending on maternal selenium intake and status. As lactation progresses no further declines have been reported except where maternal selenium intakes are low (Kumpulainen *et al.*, 1983b). The selenium composition of human milk from mothers of preterm infants parallels that of mothers of term infants with the selenium concentration of colostrum greater than that of mature milk (Ellis *et al.*, 1990).

Between 60 and 70% of human milk selenium is found in the whey fraction, and 5–10% in the lipid, mostly associated with the outer milk fat globule membrane, the remaining 20–35% precipitates with the casein pellet (van Dael *et al.*, 1988; Debski *et al.*, 1987; Milner *et al.*, 1987). Picciano's group (Debski *et al.*, 1987; Milner *et al.*, 1987) found that about 20–35% of the total selenium was associated with GSHPx. Levels of GSHPx activity in milk from American and Gambian women are in the range of 25–80 mU/ml (Ellis *et al.*, 1990; Funk *et al.*, 1990; Debski *et al.*, 1987; Mannan and Picciano, 1987; Milner *et al.*, 1987). In the Illinois women, there was a strong correlation between milk selenium and milk GSHPx (Debski *et al.*, 1989; Mannan and Picciano, 1987). Glutathione peroxidase activity in milk from New Zealand mothers was at the lower end of this range (30 mU/ml), despite a selenium concentration being half that in the American milk samples, suggesting GSHPx is protected in the face of low selenium status (Williams, 1983).

3. *Animal Milks*

Although selenium nutrition is of considerable practical importance in animal husbandry in many parts of the world (Levander, 1986), there are relatively few published data on concentrations of this essential trace mineral in milks of species other than human. Table VIII gives values for some dairy animals from different regions. As for human milk, levels of selenium in animal milks vary considerably in relation to maternal intakes. A wide range of selenium levels is found in milks from cows raised in different parts of the United States, in relation to selenium in the geochemical environment. Where plants are deficient in selenium, milk levels are 5–30 ng/ml; moderate levels in plants are associated with milk levels of 30–66 ng/ml; concentrations up to 1300 ng/ml were found in milk from cows living in seleniferous areas of South Dakota (Levander, 1986; Maus *et al.*, 1980). Benemariya *et al.*, (1993) found that selenium levels in Day 2 colostrum were about threefold higher than those in mature milk from both cows and goats. Concentrations of selenium in milk are readily increased by maternal supplementation.

Most of the selenium in cow's and goat's milk is found in the skim milk with only 2–10% being in the fat fraction. About 30% of the total selenium in goat milk is found in the whey, compared with over 70% in bovine milk, of which 80% is found with β-lactoglobulin (van Dael *et al.*, 1989; Debski *et al.*, 1987). Debski and co-workers (1987) reported that the amount of selenium associated with the casein pellet in goat's milk was double the 25% of human and cow's milks. Conversely, Yoshida *et al.* (1981) reported that selenium associated with the casein fraction of bovine milk was twice that in the whey fraction. The variations in selenium distribution reported by different groups may arise from different methods of preparation as the selenium associated with the casein fraction is readily removed during purification steps (Debski *et al.*, 1987).

F. Iodine

Iodine is an integral part of the thyroid hormones which have a role in the regulation of growth and metabolism. Levels of iodine in tissues and body fluids, like those of selenium, are affected by intakes of the element. Low soil levels of iodine are responsible for inadequate dietary intakes by humans and animals, and which are readily overcome by supplementation, usually of salt. Nonetheless, iodine-deficiency disorders, endemic goiter,

TABLE VIII
Selenium in Animal Milks

Species	Region	Selenium (ng/ml)	Reference
Cow	Illinois	10	Debski *et al.* (1987)
	South Dakota	50	Levander (1987)
	South Dakota (high selenium area)	160–1270	Levander (1987)
	Burundi	26	Benemariya *et al.* (1993)
	Finland (1976)	2	Varo *et al.* (1980)
	Germany (West)	24	Lombeck *et al.* (1978)
	Israel	73	Lavi and Alfassi (1990)
	Japan	22–28	Hojo (1982)
	New Zealand	5–7	Dolamore *et al.* (1992); Millar and Sheppard (1972)
Goat	Illinois	13	Debski *et al.* (1987)
	Burundi	23	Benemariya *et al.* (1993)
Sheep	Illinois	16	Debski *et al.* (1987)
Pig	U.S.A.	13–15	Levander (1987)

and endemic cretinism, remain major problems in public health and agriculture in large parts of the world (Delange and Burgi, 1989; Hetzel and Maberly, 1986; Matovinovic, 1983). The widespread use of iodophore disinfectants in the dairy and food industries is a significant source of iodine in human diets, particularly where milk and other dairy products are widely consumed, as in the United States and New Zealand (Park *et al.*, 1981).

Iodine is usually analyzed by colorimetric methods which measure the total iodine (Etling *et al.*, 1986). The iodine ion-specific electrode measures only ionic iodine, but Gushurst *et al.* (1984) found that this was 84% of the total iodine in milk, an acceptable level of error considering the ease and precision of the method.

Concentrations of iodine in milks vary widely according to geographical region and intakes from the diet, dietary supplements, and iodine-containing pharmaceuticals (Hetzel and Maberly, 1986; Postellon and Aronow, 1982). Table IX gives values for iodine in milks by species and country of study. Concentrations have been reported to be high in colostrum and decline with duration of lactation in cows, goats (Groppel *et al.*, 1985), and humans (Johnson *et al.*, 1990; Hetzel and Maberly, 1986). However, Etling *et al.* (1986) found lower levels in colostrum, with a slight increase with duration, in their Italian mothers, but provided no data to explain the discrepancy.

TABLE IX
Iodine Concentrations in Mature Milks from Different Geographical Regions

Species	Region	Iodine (ng/ml)	Reference
Human	Belgium	95	Delange *et al.* (1988)
	West Germany[a]	17	Heidmann *et al.* (1984)
	West Germany[b]	25	
	France	82	WHO/IAEA (1989)
	East Germany[a]	12	Delange and Burgi (1989)
	Hungary	64	WHO/IAEA (1989)
	Italy	59	Etling *et al.* (1986)
	Sicily[a]	27	Delange and Burgi (1989)
	Spain	77	Delange and Burgi (1989)
	Sweden	90	WHO/IAEA (1989)
	United Kingdom	70	DHSS (1977)
	Yugoslavia	88	Kosta *et al.* (1983)
	Philippines	57	WHO/IAEA (1989)
	Nigeria	62	WHO/IAEA (1989)
	Zaire[a]	15	Delange *et al.* (1988)
	Zaire[b]	146	Delange *et al.* (1988)
	Guatemala	60	WHO/IAEA (1989)
	California	142	Bruhn and Franke (1983)
	N. Carolina	178	Gushurst *et al.* (1984)
	New Zealand	49	Johnson *et al.* (1990)
Cow	Germany	98	Hetzel and Maberly (1986)
	New Zealand	219 70	Johnson *et al.* (1990) Sutcliff (1990)
Goat	East Germany[a]	6	Groppel *et al.* (1985)
	New Zealand	60	Sutcliffe (1990)
	New Zealand[b]	247	

[a]Endemic goiter area.
[b]Adequate iodine intake, from diet and/or supplement.

G. Fluorine

Ionic flouride is deposited in the bones and teeth in which it serves to strengthen the crystalline mineral structures, thus protecting tooth enamel from dental decay. The optimum level of intake for good dental health is 0.7–1.21 mg/liter in the water supply. When the drinking water contains > 2 mg/liter, mottling of teeth may occur. Chronic fluorosis occurs in both humans and animals when the naturally occurring levels of fluoride in soils and water supply are above 20 mg/liter. Chronic fluoride toxicity causes severe tooth damage and crippling bone and joint disorders (Krishnamachari, 1987).

With the general introduction of the fluoride ion-specific electrode in the late 1960s, concentrations of fluoride in milks were found to be an order of magnitude lower than previously estimated by colorimetric methods (Rao, 1984; Dirks *et al.,* 1974). There is still some uncertainty about what proportion of the total fluoride present in foods is measured by the ion-specific electrode, depending on sample pretreatment (Rao, 1984; Duff, 1981). The Association of Official Analytical Chemists recommends a method of acid diffusion from unashed samples, with extraction and quantification by ion-specific electrode. Microdiffusion methods for analysis of fluoride in foods and milk have been published (Taves, 1983; Dabeka and McKenzie, 1981; Dabeka *et al.,* 1979).

Dirks *et al.* (1974) reported that breast milk from mothers in a low fluorine area contained about 46 ng/ml fluoride as measured by gas–liquid chromatography and by ion-specific electrode. They reported that about 10% (4 ng/ml) of this was in the free ionic form. Adair and Wei (1978) also reported that about 10% of the total fluorine was present as the free ionic fluoride, at a higher level of 15 ng/ml, in milk from mothers living in a 1 ppm fluoridated water area. However, most recent reports on concentrations of fluoride in human milk give values for the total fluoride below 10–20 ng/ml, measured by ion-specific electrode with various methods of sample preparation. While methodological issues have not been completely resolved, it is now generally accepted that human milk from women living in areas with 1 ppm fluoride in the water supply contains 4–15 ng/ml fluoride (WHO/IAEA, 1989; Spak *et al.,* 1983; Esala *et al.,* 1982). When a large, acute oral dose of fluoride is given to the lactating woman, very little appears to be transferred to the milk (Ekstrand *et al.,* 1984, 1981). However, average concentrations in milk do reflect the level in the local environment. Finnish and Swedish mothers living in areas with 0.2 ppm fluoride in the water had 5–7 ng/ml in their milk compared with 7–11 ng/ml in milk in areas with 1–1.7 ppm fluoride in water. Women living in an area of Kenya with a high natural exposure to fluoride (9 ppm in the water supply) had considerably higher concentrations in their milk: 33 ng/ml on average (Opinya *et al.,* 1991). High levels have also been found in milk from Nigeria (25 ng/ml) and the Philippines (120 ng/ml), but no information was given on maternal exposure (WHO/IAEA, 1989).

There are few recent reports of fluoride in other milks. Dirks *et al.* (1974) found about 100 ng/ml, of which about 15% was in the free ionic form, in cows feeding on "normal" pasture in the Netherlands. Animals grazing on fluoride-contaminated pasture, with an intake to 10 times higher, had about 300 ng/ml fluoride in milk, with a much higher percentage (50%) in the free ionic form. Earlier studies from the United States have reported levels in cow's milk of 100–400 ng/ml, with concentrations up to 640 ng/ml in milk from cows from Michigan which showed signs of fluorosis (Krishnamachari, 1987). More recently, Taves (1983) gave a value of about 19 ng/ml in cow's milk with no difference between inorganic and total fluoride. Adair and Wei (1978) also reported 19 ng/ml ionic fluoride in cow's milk from Iowa, but this was only 16% of the total fluoride.

H. Essential and Possibly Essential Ultratrace Elements

1. Molybdenum

Deficiencies and toxicities of molybdenum are important in animal husbandry. Interactions with copper and sulfate modify the nutritional impact of a given level of molybdenum; an excessive intake of molybdenum may precipitate copper deficiency in animals even when copper intake is apparently adequate (Mertz, 1981). Disorders of molybdenum metabolism in humans are unlikely to be dietary in origin but several inborn errors of molybdoenzymes have been reported in which there is a defect in the synthesis of the molybdenum cofactor of sulfite and xanthine oxidases (Casey and Walravens, 1988).

Of the few published reports of molybdenum in milks (Table X), even fewer have given analytical quality control information. The concentration of molybdenum in mature (> 1 month postpartum) human milk from a number of countries is 1 or 2 ng/ml (WHO/IAEA, 1989; Bougle *et al.*, 1988; Casey and Neville, 1987), with approximately 40% associated with the milk fat (Casey, 1989). The molybdoenzyme xanthine oxidase is a major component of the milk fat globule membrane; Mather *et al.*, 1979). Levels in colostrum are 5- to 10-fold higher (10–20 ng/ml at 1–4 days) and show the same rapid fall as zinc levels during the first 1 or 2 weeks postpartum (Bougle *et al.*, 1988; Casey and Neville, 1987; Vega-Franco *et al.*, 1987). Studies from India (Dang *et al.*, 1984, 1983a; Krishnamachari, 1982), the Philippines (WHO/IAEA, 1989), and Finland (Varo *et al.*, 1980) have reported levels an order of magnitude higher, but it is not possible to say whether the differences are due to a true geographical variation or reflect analytical problems. Cow's milk from the United States (Missouri) contained 22 ng/ml (Anderson, 1992), and goat milk had 15 ng/ml under normal feeding conditions in (East) Germany (Anke *et al.*, 1985).

2. Cobalt

Higher animals cannot utilize free cobalt; it is required in the diet as preformed vitamin B_{12}, of which cobalt forms part of the active site. Vitamin B_{12} is synthesized only by algae and microorganisms, including those found in the rumen (Smith, 1987). No reports have distinguished between vitamin-bound and free cobalt in milks.

The best estimates of cobalt concentrations in human milk (Table X), from the few published analyses, are 0.1–2.0 ng/ml, with a mean of about

TABLE X
Essential and Possibly Essential UltraTrace Elements in Mature Milks

Element	Species	Value (ng/ml)	Reference
Molybdenum	Human	1–2[a]	WHO/IAEA (1989); Bougle *et al.* (1988); Casey and Neville (1987)
	Cow	12–22	Anderson (1992); Lavi and Alfassi (1990)
	Goat	15	Anke (1985)
	Horse	16	Anderson (1992)
	Guinea pig	26	Anderson (1992)
Cobalt	Human	0.1–0.2	WHO/IAEA (1989); Cumming *et al.* (1983); Clemente *et al.* (1982)
	Cow	0.4	Smith (1987)
	Goat	2–4	Anderson (1992)
Chromium	Human	0.2–0.4	Anderson *et al.* (1993); Casey and Hambidge (1984); Kumpulainen and Vuori (1980)
	Cow	5–15	Anderson *et al.* (1985); Varo *et al.* (1980)
Nickel	Human	0.5–2	Casey and Neville (1987)
	Cow	4–40	Lavi and Alfassi (1990); Varo *et al.* (1980); Casey (1977); Jaulmes and Hamelle (1971)
	Goat	37	Nielsen (1987)

[a]Colostrum, 10–20 ng/ml.

0.5 ng/ml (WHO/IAEA, 1989; Clemente *et al.*, 1982; Cumming *et al.*, 1982). Mature cow's milk contains about 0.4 ng/ml (Smith, 1987). Levels of cobalt are higher in colostrum than in mature milk in both species (Smith, 1987; Krishnamachari, 1982). There appears to be no effect of geographical region.

3. *Chromium*

Chromium is considered an essential trace mineral, playing a role in glucose metabolism and insulin functioning (Mertz, 1993), but the requirement has recently been questioned (Uustitupa *et al.*, 1992). The measurement of chromium in biological fluids and tissues has been particularly beset by analytical problems, arising both from technical difficulties with atomic absorption spectrophotometry and the practical problems in preventing contamination from the sampling environment (Kumpulainen *et al.*, 1983a). As these problems have been gradually appreciated and overcome, the accepted concentration of chromium in human serum has fallen 100-fold to <0.5 ng/ml (Anderson *et al.*, 1985). A similar decline has occurred in published values for human milk, and the accepted levels are now <1 ng/ml (Casey and Hambidge, 1984) (Table X).

Mean values for chromium in human milk are in the range of 0.2–0.4 ng/ml, with a coefficient of variation of about 40%; there is no effect of duration of lactation (Anderson *et al.*, 1993; Engelhardt *et al.*, 1990; Casey *et al.*, 1985; Casey and Hambidge, 1984; Kumpulainen and Vuori, 1980). Engelhardt and co-workers (1990) found that a bolus oral supplement of chromium 13 times the usual daily intake increased milk chromium twofold. However, dietary and serum levels do not affect the concentration of chromium in milk under normal conditions (Anderson *et al.*, 1993; Kumpulainen *et al.*, 1980).

4. Nickel

Nickel is regarded as a dietary requirement for animals but, to date, no nutritional problems have been reported under nonexperimental conditions (Nielsen, 1991). Some values for the concentration of nickel in milks from different species are given in Table X. Casey and Neville (1987) reported nickel levels in human milk collected from American women at 0–30 days postpartum, analyzed by graphite furnace atomic absorption spectrometry with good quality control. The average concentration was 1.2 ± 0.4 ng/ml with no effect of duration of lactation. The WHO/IAEA (1989) study reported levels a magnitude higher (averages from six countries, 11–16 ng/ml), but stated that they were unable to collaborate analytical quality control. Varo *et al.* (1980) found 4–10 ng/ml in milk from Finnish mothers.

Point-of-sale cow's milk from New Zealand contained 10 ng/ml (Casey, 1977), in the same range as European levels (4–40 ng/ml) (Lavi and Alfassi, 1990; Varo *et al.*, 1980; Jaulmes and Hamelle, 1971). Nielsen (1987) reported values from the German literature for nickel in some animal milks—goat, 37 ng/ml; pig, 230 ng/ml; and minipig, 110 ng/ml. Anke *et al.* (1985) reported 1100 ng/ml in rat milk. These values are high in relation to other species; no quality control information was given.

I. Toxic Elements

1. Aluminum

In healthy individuals, aluminum is not regarded as a significant toxicant. Normally the skin, lungs, and intestinal tract act as very effective barriers to aluminum uptake and very little of orally ingested mineral is absorbed. Where the gut does not provide a barrier, however (for example, in individuals on hemodialysis, in preterm infants with an immature gastrointestinal tract, and in patients on intravenous feeding), aluminum may accumulate in the tissues with severe consequences (Committee on Nutrition, 1986). Impaired renal function prevents the secretion of excess aluminum, further increasing the body burden. Aluminum is neurotoxic in humans and may also cause a fracturing osteomalacia (Klein *et al.*, 1989).

Levels of aluminum in blood and milk appear to reflect dietary intake/environmental exposure to some extent (Alfrey, 1986), but there is little published information on concentrations in milks. The lowest level, 3 ng/ml, has been reported from the Netherlands (Semmerkrot *et al.*, 1989) and levels in the United States are also low (4–14 ng/ml) (Koo *et al.*, 1988; McGraw *et al.*, 1986; Freudlich *et al.*, 1985; Sedman *et al.*, 1985), but Anderson (1993) reported a high level (125 ng/ml) in milk from mothers living in Missouri. Levels up to 30 ng/ml have been reported in milk from Italy and Australia (De Curtis *et al.*, 1989; Weintraub *et al.*, 1986). Concentrations in milk from Austrian mothers were twofold higher (74 ng/ml), probably from higher dietary intakes compared with other countries (Haschke *et al.*, 1989).

Aluminum in cow's milk from the United States (27 ng/ml) was considerably lower than the 95–100 ng/ml reported for Italy and Australia (Koo *et al.*, 1988; De Curtis *et al.*, 1988; Weintraub *et al.*, 1986). Anderson (1992) found levels in milk from cows grazing in Missouri comparable to these European levels, 98 ng/ml, and 123 ng/ml in mare's milk from the same area. Guinea pig milk may contain 80–450 ng/ml aluminum (Anderson, 1992, 1990).

2. *Arsenic*

Although arsenic is not generally a problem in human nutrition, chronic arsenosis may occur in areas exposed to industrial and mining operations using arsenic and in areas where high environmental levels occur naturally, usually due to active volcanism (Guha *et al.*, 1992; Anke, 1986). Evidence has been presented for rats and goats to suggest that arsenic is an essential nutrient, but it is not yet firmly established as such (Nielsen, 1991; Anke, 1986). The arsenic content of tissues and body fluids is markedly influenced by the level of intake (Fordyce *et al.*, 1924), but experiments in cows suggest there is a barrier to excessive mammary uptake, as milk concentrations were not increased by feeding diets containing 25 times the arsenic level in normal rations. Similar studies have not been done in humans, but the variation in arsenic concentrations in mother's milk from different parts of the world probably reflects chronic exposure to differing environmental levels (WHO/IAEA, 1989). Arsenic concentrations in milk from India (Dang *et al.*, 1983a,b), Yugoslavia (Kosta *et al.*, 1983), Guatemala, Hungary, Sweden, and Zaire (WHO/IAEA, 1989) are all in the range of 0.2–0.7 ng/ml. Higher levels have been found in Filipino women (19 ng/ml) (WHO/IAEA, 1989), Greek women (6 ng/ml) (Anke, 1986), and in mothers from northern Chile (5 ng/ml) which is known as a region of high natural arsenic (WHO/IAEA, 1989). Normal milk from cows and goats contains rather higher levels of arsenic (20–60 ng/ml) (Anke, 1986).

3. Cadmium

Cadmium levels in tissues are low at birth and increase with age in relation to exposure from the diet and cigarette smoking (Casey *et al.*, 1982). The ingested mineral is sequestered in the liver and kidney by metallothionein which has a higher affinity for cadmium than for zinc or copper. Accumulation of excessive levels causes renal damage and dysfunction (Chisholm, 1985).

Well-controlled, modern surveys of cadmium in human milk have generally found average levels to be < 1 ng/ml, with mean values in the range of 0.1–0.5 ng/ml, from many parts of the world with a "normal" environmental burden, including Canada (Dabeka *et al.*, 1988), New Zealand (Eynon *et al.*, 1985), the United Kingdom (Kovar *et al.*, 1984), Germany (Schrammel *et al.*, 1988; Muller, 1987; Radisch *et al.*, 1987), and Finland (Varo *et al.*, 1980). Where samples were collected from several regions within a country, no geographical variation was seen (Dabeka *et al.*, 1888; Schrammel *et al.*, 1988). The WHO/IAEA (1989) worldwide study found cadmium concentrations were low in most areas in accord with other reports, but levels of up to 20 ng/ml were seen in some samples from mothers living in urban areas of Nigeria, Guatemala, and the Philippines. Higher levels, up to 40 ng/ml, have also been reported from urban areas in Europe in earlier studies (Bates and Prentice, 1988), but it is not possible to say if these reflect methodology or are a true reflection of higher maternal exposure. Maternal smoking increased breast milk cadmium (Dabeka *et al.*, 1988; Eynon *et al.*, 1985) in a dose–response manner up to twofold for women smoking more than 20 cigarettes per day (Radisch *et al.*, 1987). One study from Germany (Schulte-Lobbert and Bohn, 1977) found that cadmium levels were higher in colostrum than in later milk, but this has not been confirmed by other workers (Eynon *et al.*, 1985). Almost all cadmium in human milk appears to be associated with the fraction eluting as metallothionein on high-performance liquid chromatography (Michalke and Schrammel, 1990).

The cadmium concentration of cow's milk is also low, generally < 5 ng/ml (Lavi and Alfassi, 1990; Varo *et al.*, 1980; Casey, 1977), and may vary regionally.

4. Lead

Exposure of infants and young children to even modest amounts of lead in the environment appears to have detrimental effects on intellectual and behavioral development. Soils and air, particularly in urban environments are readily contaminated from lead-based paints, sewage sludge, and leaded petroleum products (Rhein, 1991; Casey and Walravens, 1988; Chisholm, 1985). Lead is poorly absorbed by mammals and is concentrated in bone, so levels in milks are generally low and are not considered an important source of exposure.

Concentrations of lead in human milk of less than 10 ng/ml, with averages of about 2 ng/ml, have been reported for Canada (Dabeka *et al.*, 1988), the UK (Kovar *et al.*, 1984), Germany (Schrammel *et al.*, 1988), Sweden (Larsson *et al.*, 1981), Arizona (Rockway *et al.*, 1984), Guatemala, Zaire, and Nigeria (WHO/IAEA, 1989). Higher means, 10–30 ng/ml, and values up to 200 ng/ml in some urban areas, have been found in other parts of Europe (WHO/IAEA, 1989; Bates and Prentice, 1988), the Philippines (WHO/IAEA, 1989), Malaysia (Ong *et al.*, 1988), and Tennessee (Dillon *et al.*, 1974). It is not clear whether these differences are due to higher maternal exposures or to methodology; lead is not an easy element to analyze in milk, but most of the quoted studies were well controlled (WHO/IAEA, 1989; Camara Rica and Kirkbright, 1982).

Cow's milk from uncontaminated areas contains 20–50 ng/ml lead (MAFF, 1982; Casey, 1977). Cows given a high-lead silage had up to 140 ng/ml in their milk.

5. *Mercury*

Mercury occurs widely in the environment and has long been known to be neurotoxic on occupational exposure. There is an increasing concern about its toxic properties because of the widespread use in industry and agriculture. The alkyl derivatives, particularly methylmercury, are more toxic than other chemical forms and readily enter the food chain through marine and freshwater fish which take up methylmercury from microorganisms and sediments (Clarkson, 1987; Chisholm, 1985). Major, catastrophic outbreaks of methylmercury poisoning occurred in Iraq in 1960 and 1971/1972, through consumption of wheat seed dressed with methylmercury fungicide (Amin-Zaki *et al.*, 1976; Bakir *et al.*, 1973), and in Minamata, Japan, in the 1950s from fish contaminated with factory waste (Fujita *et al.*, 1977). In both Iraq and Japan, babies born to contaminated mothers also showed signs of mercury poisoning, but one infant who was only exposed postnatally through high levels in mother's milk was unaffected (Amin-Zaki *et al.*, 1976).

Concentrations of mercury in tissues and blood reflect exposure of the animal to mercury in the diet, particularly through fish intake, and to airborne sources (Clarkson, 1987). In women with a normal exposure to mercury, levels in breast milk are about 0.2–5 ng/ml (WHO/IAEA, 1989; Schrammel *et al.*, 1988; Bates and Prentice, 1986; Kosta *et al.*, 1983; Varo *et al.*, 1980). Where mothers had a high fish consumption, in Japan (Fujita and Takabatake, 1977) and Sweden (Skerfving, 1988), mercury levels in milk were not elevated (3 ng/ml). In the Swedish samples, about 20% of the total mercury was present as methylmercury. In healthy women from the contaminated Minamata area, breast milk concentrations of up to 60 ng/ml were found (Fujita and Takabatake, 1977). During the methylmercury poisoning episodes in Iraq in 1960, milk collected from lactating women was found to contain up to 150 ng/ml of methylmercury (Bakir *et al.*, 1973). In the second outbreak in 1971/1972, the total concentration of mercury in milk from one mother was found to be 100 ng/ml but declined to normal levels (2 ng/ml) over 9 months as maternal blood levels declined (Amin-Zaki *et al.*, 1976).

There is very little information on levels of mercury in milks from other species. Cow's milk from Scandinavia is generally low (< 1 ng/ml) (Skerfving, 1988; Varo *et al.*, 1980).

J. Other Elements

Understandably, analytical work has focused mainly on those microminerals which are known to be essential or for which problems of deficiency or toxicity have been identified in human and domestic animal populations. There are, therefore, few published studies of other elements in human or cow's milks and almost no data were found for milks of other species. Several reports have included elements additional to those covered above when multielement analytical technology has been used, such as X-ray fluorescence (Howell *et al.*, 1986) or neutron activation analysis (Cumming *et al.*, 1983; Kosta *et al.*, 1983; Clemente *et al.*, 1982; Iyengar *et al.*, 1982). The large, multinational WHO/IAEA (1989) study reported additional elements measured by single-element techniques. Values are presented in Table XI for concentrations of 12 nonessential trace elements in mature milks from various species. Most, but not all, of these studies presented adequate quality control information; the criteria used above to judge the accuracy of the analytical method have not been strictly applied and the values in Table XI are given as "best available." There are insufficient data to indicate whether concentrations of these elements vary with factors such as duration of lactation, geographical region, or dietary intake.

III. Radioisotopes

Under conditions of environmental contamination with radioactivity, milk is an important source of human exposure because, unlike most foodstuffs, it is "harvested" daily and may be consumed well within the half-life of even short-lived species of isotopes (Bouville *et al.*, 1990; Watson, 1986).

TABLE XI
Concentrations of Some Nonessential Trace Elements in "Mature" Milks

Element	Species	Concentration	Reference
Antimony	Human	0.2–3 ng/ml	WHO/IAEA (1989); Kosta *et al.* (1983); Clemente *et al.* (1982); Iyengar *et al.* (1982)
	Cow	< 10	MAFF (1985)
Barium	Human	0.15 μg/ml	Anderson (1992)
	Cow	0.2	Anderson (1992)
	Horse	0.08	Anderson (1992)
	Guinea pig	0.2	Anderson (1990)
Boron	Human	0.08–0.2 μg/ml	Anderson (1992); Nielsen (1986)
	Cow	0.2–1	Anderson (1992); Nielsen (1986)
	Horse	0.1	Anderson (1992)
	Guinea pig	0.6	Anderson (1990)
	Buffalo	1	Nielsen (1986)
Bromium	Human	1–8 μg/ml	Khalkhali and Parsa (1972); Howell *et al.* (1986)
	Cow	10	Nielsen (1986)
Caesium	Human	1–6 ng/ml	Cumming *et al.* (1983)
Lithium	Human	6 ng/ml	Anderson (1992)
	Cow	24	Anderson (1992)
	Horse	15	Anderson (1992)
	Guinea pig	34	Anderson (1992)
Rubidium	Human	0.6–0.8 μg/ml	Howell *et al.* (1986); Cumming *et al.* (1983); Clemente *et al.* (1982); Varo *et al.* (1980)
	Cow	2.4	Varo *et al.* (1980)
Scandium	Human	< 0.01 ng/ml	Clemente *et al.* (1982)
Silicon	Human	0.5 μg/ml	Anderson (1992)
	Cow	0.4–2	Anderson (1992); Carlisle (1986)
	Horse	0.2	Anderson (1992)
	Guinea pig	0.6	Anderson (1992)
Strontium	Human	0.06 μg/ml	Anderson (1992); Howell *et al.* (1986)
	Cow	0.4	Anderson (1992)
	Horse	0.4	Anderson (1992)
	Guinea pig	1	Anderson (1990)

TABLE XI—*continued*

Element	Species	Concentration	Reference
Sulfur	Human	0.12 mg/ml (5% free sulfate)	McNally *et al.* (1991)
	Cow	0.32	Varo *et al.* (1980)
Tin	Human	0.5–3 ng/ml	WHO/IAEA (1989)
	Cow	< 10	MAFF (1985)
Titanium	Human	0.25 μg/ml	Anderson (1992)
	Cow	0.1–0.3	Anderson (1992); Lavi and Alfassi (1990)
	Horse	0.15	Anderson (1992)
Vanadium	Human	0.1–0.5 ng/ml	Casey (unpublished); Kosta *et al.* (1983); WHO/IAEA (1989)
	Cow	0.1	Anderson (1992)
	Goat	0.4	Anke *et al.* (1985)

The milk transfer coefficient, Fm, is used to describe the fraction of the daily intake of radionuclide by the mother that is secreted into the milk.

$$Fm \text{ (day/liter)} = \frac{\text{Milk concentration (Bq/liter)}}{\text{Daily radionuclide intake (Bq/day)}}$$

The Fm term has been used to predict the transfer of radioactivity resulting from environmental contamination, particularly following the Chernobyl accident, through the dairy food chain to cow and sheep milk, but has not been used for human milk (Ward and Johnson, 1989).

The Fm may be affected by a number of factors of varying practical importance: (1) the physical–chemical form of the radionuclide: fallout cesium (^{134}Cs, ^{137}Cs) from Chernobyl had a lower Fm than that from atmospheric nuclear weapons testing. Uptake from wet (in rainfall) or dry deposition on pasture differs; (2) feed: type and source of forage, pasture, hay, etc.; (3) soil: access to plant roots and via direct ingestion by the animal, e.g., clay soils may bind divalent cations and make them unavailable to grazing cattle; (4) species: sheep and goats typically have Fm values an order of magnitude higher than cows, possibly from differences in metabolic rates or milk production rates; and (5) other routes of ingestion: inhalation and via drinking water, particularly for ^{131}I.

With the exception of some case reports on the secretion of radiopharmaceuticals in human milk (Lazarus and Edwards, 1988), and studies on cesium isotopes in cows post-Chernobyl, there has been very little systematic study of radioisotopes in milks. Radioactive isotopes of minerals may occur in milk from four sources, which will be considered separately. Table XII gives the main examples of each class, with physical half-life and the implications for breast-feeding.

A. Natural Background Radiation

The contribution of naturally occurring radionuclides to the average effective dose equivalent (to the human body) is 52% ^{238}U series, 17% ^{232}Th series, 15% ^{40}K, and the remaining 16% from cosmic rays and

TABLE XII
Radioisotopes Possibly Found in Human Milk

Isotope	$t_{1/2}$	Notes
Natural background		
^{238}U series	4×10^9 Years	
^{232}Th series	1.4×10^{10} Years	Total background exposure
^{40}K	1.2×10^9 Years	~1 mSv/year
Radiopharmaceuticals		
		Interruption of breast-feeding
^{32}P	14 Days	Discontinue
^{51}Cr	28 Days	4 hr
^{67}Ga	3 Days	2 Weeks
^{75}Se	118 Days	1 Week
^{99m}Tc	6 hr	4–16 hr, depending on carrier
^{111}In	3 Days	24 hr
^{125}I	60 Days	16 hr–10 days, depending on carrier
^{131}I	8 days	0–48 hr, depending on carrier
Weapons testing		Not currently a source of exposure
^{90}Sr	29 Years	
^{131}I	8 Days	
^{134}Cs	2 Years	
^{137}Cs	30 Years	
Nuclear accident		
^{131}I	8 Days	Recommended maintain breast-feeding for as long as possible: excretion in human milk 0.1xlevel in cow's milk
^{134}Cs	2 Years	
^{137}Cs	30 Years	

cosmogenic radiation (Gori *et al.*, 1988). In the absence of any information suggesting isotopic fractionation, the natural radiation in milks is likely to arise from the same sources but no data are available. Spencer *et al.* (1990) reported that commercial cow's milk consumed by individuals taking part in a balance study in Illinois contained about 0.64 Bq/liter from ^{234}U and ^{238}U.

B. Radiopharmaceuticals

A range of chemical compounds utilizing radioisotopes of chromium, gallium, indium, iodine, phosphorus, selenium, and technetium has been used medically for diagnosis and treatment of a number of disorders, particularly cancers and thyroid disease, in lactating women. Several recent reviews discuss such uses, giving information on physical and biological half-lives, kinetic data for the excretion of the isotope in breast milk, calculated maximum exposure of the suckling infant, and advice on the interruption or cessation of breast-feeding (Lazarus and Edwards, 1988; Mountford and Coakley, 1988; Ahlgren *et al.*, 1985). The fraction of administered isotope which will be secreted into the milk will depend on the rate of decay of radioactivity, the pharmacokinetics of the carrier compound, and its partitioning into milk. For example, ^{131}I levels in breast milk are eightfold those in plasma. Table XII shows the radioisotopes most widely used in clinical practice, with half-life data and recommendations for continuing breast-feeding.

C. Nuclear Weapons Testing

During the 1950s and early 1960s, some populations were exposed to fallout from atmospheric testing of nuclear weapons. The most significant source of radiation exposure for humans was from ^{131}I entering domestic cow's milk supplies (Bouville *et al.*, 1990), but secretion of ^{90}Sr and ^{137}Cs in cow and human milks were also of concern (Baker *et al.*, 1970; Straub and Murthy, 1965; Aarkrog, 1963). Since the Partial Test Ban Treaty of 1963, this has no longer been a source of radioactive contamination of milk.

D. Nuclear Industrial Accident

Subsequent to the accident at the nuclear power plant at Chernobyl, Ukraine, on April 26, 1986, studies have been published from various parts of Europe on the contamination of human and animal milks with fallout isotopes of iodine and caesium (Table XII). In the weeks immediately following the accident, milk from cows and sheep in Hungary contained 10–96 Bq/liter ^{137}Cs, depending on location and type of feed. By 1 year later, values were below 8 Bq/liter (Ward *et al.*, 1989). Transfer coefficients, Fm, from Chernobyl fallout were lower than those reported for worldwide fallout from weapons testing. Experimental studies, with animals fed forage harvested from contaminated pasture in Greece and Wales, showed that about 5% of ingested ^{137}Cs was excreted in the milk once equilibrium was reached by 11 or 12 days (Assimakopoulos *et al.*, 1989a; Mitchell *et al.*, 1989).

Austria was among the countries having the highest deposition of radioactive material. Haschke *et al.* (1988) found that ^{131}I activity peaked in the first week of May: levels in breast milk were about 40 Bq/liter compared with a peak of about 180 Bq/liter in cow's milk from the Vienna area. By late June, ^{131}I was at background level in both milks. Radiocesium peaked in cow's milk in the first weeks of June, with counts up to 300 Bq/liter; activity declined only slowly and plateaued at about 50 Bq/liter over the winter, when animals received contaminated fodder. ^{137}Cs in human milk remained below 40 Bq/liter. In human milk from Italy, ^{131}I remained below detection limits; ^{134}Cs and ^{137}Cs were measurable after July 1986 but remained at low levels (1–4 Bq/liter) (Gori *et al.*, 1988). Gattavecchia and colleagues (1989) followed ^{137}Cs levels in breast milk from Italian mothers over 1 year after the accident; activity peaked at about 4 Bq/liter by March 1987 then declined gradually. Conversely, Assimakopoulos *et al.* (1989b) found that colostrum from Greek mothers contained on average 16 Bq/liter radiocesium as late as April 1987.

References

Aarkog, A. (1963). Caesium-137 from fall-out in human milk. *Nature* **197,** 667–668.

Ackland, M. L., and Mercer, J. F. B. (1992). The murine mutation, lethal milk, results in production of zinc-deficient milk. *J. Nutr.* **122,** 1214–1218.

Adair, S. M., and Wei, S. H. Y. (1978). Supplemental fluoride recommendations for infants based on dietary fluoride intake. *Caries Res.* **23,** 76–82.

Ahlgren, L. Ivarsson, S., Johansson, L., Mattsson, S., and Nosslin, B. (1985). Excrtion of radionuclides in human breast milk after the administration of radiopharmaceuticals. *J. Nucl. Med.* **26,** 1085–1090.

Alfrey, A. C. (1986). Aluminum metabolism. *Kidney Int. Suppl.* **18,** S8–S11.

Amin-Zaki, L., Elhassani, S., Majeed, M. A., Clarkson, T. W., Doherty, R. A., Greenwood, M. R., and Giovanoli-Jakubczak, T. (1976). Perinatal methylmercury poisoning in Iraq. *Am. J. Dis. Child.* **130,** 1070–1076.

Anaokar, S. G., and Garry, P. J. (1981). Effects of maternal iron nutrition during lactation on milk iron and rat neonatal iron status. *Am. J. Clin. Nutr.* **34,** 1505–1512.

Anderson, R. A., Bryden, N. A., Patterson, K. Y., Veillon, C., Andon, M. B., and Moser-Veillon, P. B. (1993). Breast milk chromium and its association with chromium intake, chromium excretion and serum chromium. *Am. J. Clin. Nutr.* **57,** 519–523.

Anderson, R. A., Bryden, N. A., and Polansky, M. M. (1985). Serum chromium of human subjects: Effects of chromium supplementation and glucose. *Am. J. Clin. Nutr.* **41,** 571–577.

Anderson, R. R. (1992). Comparison of trace elements in milk of four species. *J. Dairy Sci.* **75,** 3050–3055.

Anderson, R. R. (1990). Trace elements in milk of guinea pigs during a 20-day lactation. *J. Dairy Sci.* **73,** 2327–2332.

Anderson, R. S., Carlos, G. M., Robinson, I. P., Booles, D., Burger, I. H., and Whyte, A. L. (1991). Zinc, copper, iron and calcium concentrations in bitch milk. *J. Nutr.* **121,** S81–S82.

Anke, M. (1986). Arsenic. *In* "Trace Elements in Human and Animal Nutrition" (W. Mertz, ed.), 5th ed., Vol. 2, pp. 347–372. Academic Press, Orlando.

Anke, M., Groppel, B., and Grun, M. (1985). Essentiality, toxicity, requirement and supply of molybdenum in humans and animals. *In* "Trace Elements in Man and Animals—TEMA 5" (C. F. Mills, I. Bremmer, and J. K. Chesters, eds.), pp. 154–157. Commonwealth Agricultural Bureau, Slough, UK.

Anonymous (1986). Zinc bioavailability of human and cow's milk. *Nutr. Rev.* **44,** 181–183.

Anonymous (1991). Importance of zinc for hormone binding and signal transduction: Limiting mechanisms in zinc deficiency? *Nutr. Rev.* **49,** 369–370.

Assimakopoulos, P. A., Ioannides, K. G., Pakou, A. A., and Mantzios, A. S. (1989a). A study of radiocaesium contamination and decontamination of sheep's milk. *Sci. Total Environ.* **85,** 279–285.

Assimakopoulos, P. A., Ioannides, K. G., Pakou, A. A., Lolis, D., Zikopoulos, K., and Dusias, B. (1989b). Radiocaesium levels measured in breast milk one year after the reactor accident at Chernobyl. *Health Phys.* **56,** 103–106.

Atinmo, T., and Omololu, A. (1982). Trace element content of breastmilk from mothers of preterm infants in Nigeria. *Early Hum. Dev.* **6,** 309–313.

Avissar, N., Slemmon, J. R., Palmer, I. S., and Cohen, H. J. (1991). Partial sequence of human plasma glutathione peroxidase and immunologic identification of milk glutathione peroxidase as the plasma enzyme. *J. Nutr.* **121,** 1243–1249.

Baker, B. E., Nelson, C. H., and Samuels, E. R. (1970). Strontium-90 and Caesium-137 in human and other milks collected in Alaska. *J. Dairy Sci.* **51,** 241–244.

Bakir, F., Damluji, S. F., Amin-Zaki, L., Murtadha, M., Khalidi, A., Al-Rawi, N. Y., Tikriti, S., Dhahir, H. I., Clarkson, T. W., Smith, J. C., and Doherty, R. A. (1973). Methylmercury poisoning in Iraq. *Science* **181,** 230–241.

Bates, C. J., and Tsuchiya, H. (1990). Zinc in breast milk during prolonged lactation: comparison between the UK and The Gambia. *Eur. J. Clin. Nutr.* **44,** 61–69.

Bates, C., and Prentice, A. (1988). Vitamins, minerals and essential trace elements. *In* "Drugs and Human Lactation" (P. N. Bennett, ed.), pp. 433–493. Elsevier, Amsterdam.

Belavady, B. (1978). Lipid and trace element composition of human milk. *Acta Paediatr. Scand.* **67,** 560–570.

Benemariya, H., Robberecht, H., and Deelstra, H. (1993). Zinc, copper, and selenium in milk and organs of cow and goat from Burundi, Africa. *Sci. Total Environ.* **128,** 83–98.

Berfenstam, R. (1952). Studies on blood zinc. *Acta Paediatr.* **41** (Suppl. 87), 1–92.

Berry, M. J., Banu, L., and Larsen, P. R. (1991). Type I iodothyronine deiodinase is a selenocysteine-containing enzyme. *Nature* **349,** 438–440.

Bhandari, B., Gupta, A. P., and Gupta, A. (1985). Breast milk minerals contents. *Indian Pediatr.* **22,** 23–26.

Bougle, D., Bureau, F., Foucault, P., Duhamel, J.-F., Muller, G., and Drosdowsky, M. (1988). Molybdenum content of term and preterm human milk during the first 2 months of lactation. *Am. J. Clin. Nutr.* **48,** 652–654.

Bouville, A., Dreicer, M., Beck, H. L., Hoecker, W. H., and Wachholz, B. W. (1990). Models of radioiodine transport to populations within the continental U.S. *Health Phys.* **59,** 659–668.

Bruhn, J. C., and Franke, A. A. (1983). Iodine in human milk. *J. Dairy Sci.* **66,** 1396–1398.

Burguera, M., Burguera, J. L., Garaboto, A. M., and Alarcon, O. M. (1988). Iron and copper content of human milk at early stage of lactation in Venezuelan women. *Trace Elem. Med.* **5,** 60–63.

Butte, N. F., Garza, C., O'Brian Smith, E., Wills, C., and Nichols, B. (1987). Macro- and trace-mineral intakes of exclusively breast-fed infants. *Am. J. Clin. Nutr.* **45,** 42–48.

Butte, N. F., Garza, C., Johnson, C. A., O'Brian Smith, E., and Nichols, B. L. (1984). Longitudinal changes in milk composition of mothers delivering preterm and term infants. *Early Hum. Dev.* **9,** 153–162.

Camara Rica, C., and Kirkbright, G. F. (1982). Determination of trace concentrations of lead and nickel in human milk by electrothermal atomization atomic absorption spectrophotometry and inductively-coupled plasma emission spectroscopy. *Sci. Total Environ.* **22,** 193–201.

Carl, M., and the members of the IDF/ISO/AOAC group E15. (1992). Trace elements in milk and milk products. *Bull. Int. Dairy Fed.* **278.**

Carlisle, E. M. (1986). Silicon. *In* "Trace Elements in Human and Animal Nutrition" (W. Mertz, ed.), 5th Ed., Vol. 2, pp. 373–390. Academic Press, Orlando.

Casey, C. E. (1989). Distribution of molybdenum in human milk. *In* "Nutrient Availability: Chemical and Biological Aspects." (D. A. T. Southgate, I. T. Johnson, and G. R. Fenwick, eds.), Special Publication 72, pp. 109–111. Royal Society of Chemistry, Cambridge.

Casey, C. E. (1988). Selenophilia. *Proc. Nutr. Soc.* **47,** 55–62.

Casey, C. E. (1977). The content of some trace elements in infant milk foods and supplements available in New Zealand. *N.Z. Med. J.* **85,** 275–278.

Casey, C. E., and Walravens, P. A. (1988). Trace elements. *In* "Nutrition during Infancy" (R. C. Tsang and B. L. Nichols, eds.), pp. 190–215. Mosby, St. Louis, MO.

Casey, C. E., and Neville, M. C. (1987). Studies in human lactation 3: Molybdenum and nickel in human milk during the first month of lactation. *Am. J. Clin. Nutr.* **45,** 921–926.

Casey, C. E., and Hambidge, K. M. (1984). Chromium in human milk from American mothers. *Br. J. Nutr.* **52,** 73–77.

Casey, C. E., Neville, M. C., and Hambidge, K. M. (1989). Studies in human lactation: Secretion of zinc, copper, and manganese in human milk. *Am. J. Clin. Nutr.* **49,** 773–785.

Casey, C. E., Hambidge, K. M., and Neville, M. C. (1985a). Studies in human lactation: Zinc, copper, manganese and chromium in human milk in the first month of lactation. *Am. J. Clin. Nutr.* **41,** 1193–1200.

Casey, C. E., Howell, R. R., Lonnerdal, B., Moser, P. B., Picciano, M. R., and Rumball, S. V. (1985b). Principles of trace element analysis and notes on some important elements. *In* "Human Lactation I: Milk Components and Methodologies" (R. G. Jensen and M. C. Neville, eds.), pp. 223–236. Plenum Press, New York.

Casey, C. E., Guthrie, B. E., and Robinson, M. F. (1982). Copper, manganese, zinc, and cadmium in tissues from New Zealanders. *Biol. Trace Elem. Res.* **4,** 105–15.

Casey, C. E., Walravens, P. A., and Hambidge, K. M. (1981). Availability of zinc: Loading tests with human milk, cow's milk and infant formulas. *Pediatrics* **68,** 394–396.

Cavell, P. A., and Widdowson, E. M. (1964). Intakes and excretions of iron, copper, and zinc in the neonatal period. *Arch. Dis. Child* **39,** 496–501.

Celada, A., Busset, R., Gutierrez, J., and Herreros, V. (1982). No correlation between iron concentration in breast milk and maternal stores. *Helv. Paediatr. Acta* **37,** 11–16.

Chisholm, J. J. (1985). Pediatric exposures to lead, arsenic, cadmium, and methyl mercury. *In* "Trace Elements in Nutrition of Children (R. K. Chandra, ed.), pp. 229–261. Nestle Nutrition, Vevey/Raven Press, New York.

Clarkson, T. W. (1987). Mercury. *In* "Trace Elements in Human and Animal Nutrition" (W. Mertz, ed.), 5th Ed., Vol. 1, pp. 417–428. Academic Press, Orlando.

Clemente, G. F., Ingrao, G., and Santaroni, G. P. (1982). The concentration of some trace elements in human milk from Italy. *Sci. Total Environ.* **24,** 255–265.

Committee on Nutrition (1986). Aluminum toxicity in infants and children. *Pediatrics* **78,** 1150–1154.

Cumming, F. J., Fardy, J. J., and Briggs, M. H. (1983). Trace elements in human milk *Obstet. Gynecol.* **62,** 506–508.

Dabeka, R. W., and McKenzie, A. D. (1981). Microdiffusion and fluoride-specific electrode determination of fluoride in infant foods: collaborative study. *J. Assoc. Off. Anal. Chem.* **64,** 1021–1026.

Dabeka, R. W., Karpinski, K. F. McKenzie, A. D., and Badjik, C. D. (1988). Survey of lead and cadmium in human milk and correlation of levels with environmental and food factors. *Sci. Total Environ.* **71,** 65–66.

Dabeka, R. W., McKenzie, A. D., and Conacher, H. B. S. (1979). Microdiffusion and fluoride-specific electrode determination of fluoride in foods. *J. Assoc. Off. Anal. Chem.* **62,** 1065–1069.

Dallman, P. R. (1988). Nutritional anemia of infancy: iron, folic acid, and vitamin B_{12}. *In* "Nutrition during Infancy" (R. C. Tsang and B. L. Nichols, eds.), pp. 216–235. Mosby, St. Louis, MO.

Dallman, P. R. (1986). Iron deficiency in the weanling: A nutritional problem on the way to resolution. *Acta Paediatr. Scand. Suppl.* **323,** 59–67.

Dang, H. S., Jaiswal, D. D., and Somasunderam, S. (1983a). Distribution of arsenic in human tissues and milk. *Sci. Total Environ.* **29,** 171–175.

Dang, H. S., Jaiswal, D. D., Wadhwani, C. N., Somasunderam, S., and Dacosta, H. (1983b). Infants with a congenital anomaly and the concentration of Mo, As, Mn, Zn and Cu in the mother's milk. *Sci. Total Environ.* **27,** 43–47.

Davidson, L., Cederblad, A., Lonnerdal, B., and Sandstrom, B. (1989). Manganese absorption from human milk, cow's milk, and infant formulas in humans. *Am. J. Dis. Child.* **143,** 823–827.

Davidson, L., Kastenmayer, P., Yuen, M., Lonnerdal, B., and Hurrell, R. F. (1994). Influence of lactoferrin on iron absorption from human milk in infants. *Pediatr. Res.* **35,** 117–124.

Davis, G. K., and Mertz, W. (1987). Copper. *In* "Trace Elements in Human and Animal Nutrition" (W. Mertz, ed.), vol. 1, pp. 301–364. Academic Press, San Diego.

Debski, B., Finley, D. A., Picciano, M. F., Lonnerdal, B., and Milner, J. (1989). Selenium content and glutathione peroxidase activity of milk from vegetarian and nonvegetarian women. *J. Nutr.* **119,** 215–220.

Debski, B., Picciano, M. F., and Milner, J. A. (1987). Selenium content and distribution of human, cow and goat milk. *J. Nutr.* **117,** 1091–1097.

De Curtis, M., Napolitano, E., Ciccimarra, F., Mellone, M. C., and Del Rio, A. (1989). Aluminum content in human milk and in infant formulas (letter). *Eur. J. Clin. Nutr.* **43,** 887.

Delange, F., and Burgi, H. (1989). Iodine deficiency disorders in Europe. *WHO Bull.* **67,** 317–325.

Delange, F., Bourdoux, P., Chanoine, J. P., and Ermans, A. M. (1988). Physiopathology of iodine nutrition during pregnancy, lactation, and early postnatal life. *In* "Vitamins and Minerals in Pregnancy and Lactation" (H. Berger, ed.), pp. 205–214. Nestle Nutrition Workshop Series, Vol. 16, Nestle, Vevey/Raven Press, New York.

Department of Health and Social Security (1977). "Report on Health and Social Subjects: The Composition of Mature Human Milk," pp. 47. Her Majesty's Stationery Office, London.

Dewey, K. G., and Lonnerdal, B. (1983). Milk and nutrient intake of breast-fed infants from 1 to 6 months: Relation to growth and fatness. *J. Pediatr. Gastroenterol. Nutr.* **2,** 497–506.

Dewey, K. G., Finley, D. A., and Lonnerdal, B. (1984). Breast milk volume and composition during late lactation (7–20 months). *J. Pediatr. Gastroenterol. Nutr.* **3,** 713–720.

Dillon, H. K., Wilson, D. J., and Schaffner, W. (1974). Lead concentrations in human milk. *Am. J. Dis. Child.* **128,** 491–492.

Dirks, O. B., Jongeling-Eijndhoven, M. P. A., Flissebaalje, T. D., and Gedalia, I. (1974). Total and free ionic fluoride in human and cow's milk as determined by gas–liquid chromatography and the fluoride electrode. *Caries Res.* **8,** 181–186.

Doisy, E. A. (1974). Effects of deficiency in manganese upon plasma levels of clotting proteins and cholesterol in man. *In* "Trace Element Metabolism in Animals, – 2" (W. G. Hoekstra *et al.,* eds.), pp. 668–669. University Park Press, Baltimore.

Dolamore, B. A., Brown, J., Darlow, B. A., George, P. M., Sluis, K. B., and Winterbourne, C. C. (1992). Selenium status of Christchurch infants and the effect of diet. *N.Z. Med. J.* **105,** 139–142.

Dorea, J. G., Horner, M. R., and Campanate, M. L. (1985). Lacteal zinc and copper in relation to volume, total ash and energy during the first three months of lactation in Brazilian women. *Acta Paediatr. Scand.* **74,** 891–896.

Dorner, K., Dziadzka, S., Hohn, A., Sievers, E., Oldigs, H.-D., Schulz-Lell, G., and Schaub, J. (1989). Longitudinal manganese and copper balances in young infants and preterm infants fed on breast-milk and adapted cow's milk formulas. *Br. J. Nutr.* **61,** 559–572.

Duff, E. J. (1981). Total and ionic fluoride in milk. *Caries Res.* **15,** 406–408.

Dybczinski, R., Veglia, A., and Suschny, O. (1980). Report on the intercomparison run A-11 for the determination of inorganic constituents of milk powder, IAEA/RL68, IAEA, Vienna, July 1980; and IAEA Information Sheet, Reference Material A-11 July.

Ehrenkranz, R. A., Ackerman, B. A., and Nelli, C. M. (1984). Calcium, phosphorus, zinc and copper content of preterm human milk. *Pediatr. Res.* **18,** 195A.

Ekstrand, J., Boreus, L. O., and de Chateau, P. (1981). No evidence of transfer of fluoride from plasma to breast milk. *Br. Med. J.* **283,** 761–762.

Ekstrand, J., Spak, C.-J., Falch, J., Afseth, J., and Ulverstad, H. (1984). Distribution of fluoride to human breast milk following high doses of fluoride. *Caries Res.* **18,** 93–95.

Ellis, L., Picciano, M. F., Smith, A. M., Hamosh, M., and Mehta, N. R. (1990). The impact of gestational length on human milk selenium concentration and glutathione peroxidase activity. *Pediatr. Res.* **27,** 32–35.

Engelhardt, S., Moser-Veillon, P B. Mangels, A. R., Patterson, K. Y., and Veillon, C. (1990). Appearance of an oral dose of chromium (^{53}Cr) in breast milk. *In* "Breastfeeding, Nutrition, Infection and Infant Growth in Developed and Emerging Countries" (S. A. Atkinson, L. A. Hanson, and R. K. Chandra, eds.), pp. 485–487. ARTS Biomedical Publishers, St. John's, Newfoundland.

Esala, S., Vuori, E., and Helle, A. (1982). Effect of maternal fluorine intake on breast milk fluorine content. *Br. J. Nutr.* **47,** 201–204.

Etling, N., Padovani, E., Fouque, F., and Tato, L. (1986). First-month variations in total iodine content of human breast milks. *Early Hum. Dev.* **13,** 81–85.

Evans, G. W., and Johnson, P. E. (1977). Determination of zinc availability in foods by the extrinsic label technique. *Am. J. Clin. Nutr.* **30,** 873–878.

Eynon, G. R., McKenzie-Parnell, J. M., Robinson, M. F., and Wilson, P. D. (1985). Cadmium in non-smoking New Zealand women immediately following child birth. *Proc. Uni. Otago Med. School* **63,** 38–40.

Feeley, R. M., Eitenmiller, R. R., Jones, J. B., and Barnhart, H. (1983). Copper, iron, and zinc contents of human milk at early stages of lactation. *Am. J. Clin. Nutr.* **37,** 443–448.

Finley, E. A., Lonnerdal, B., Dewey, K. G., and Grivetti, L. E. (1985). Inorganic constituents of breast milk from vegetarian and nonvegetarian women: Relationships with each other and with organic constituents. *J. Nutr.* **115,** 772–781.

Fomon, S. J., and Filer, L. J., Jr. (1974). Milks and formulas. In "Infant Nutrition" (S. J. Fomon, ed.), pp. 370–371. Saunders, Philadelphia.

Fordyce, J. A., Rosen, I., and Myers, C. N. (1924). Quantitative studies in syphilis from a clinical and biological point of view. X. Arsenic in human milk after intravenous injections of Salvarsan. *Am. J. Syph.* **8,** 65–73.

Fransson, G.-B., and Lonnerdal, B. (1984). Iron, copper, zinc, calcium, and magnesium in human milk fat. *Am. J. Clin. Nutr.* **39,** 185–189.

Fransson, G.-B., and Lonnerdal, B. (1983). Distribution of trace elements and minerals in human and cow's milk. *Pediatr. Res.* **17,** 912–915.

Fransson, G.-B., and Lonnerdal, B. (1980). Iron in human milk. *J. Pediatr.* **96,** 380–384.

Fransson, G.-B., Agarwal, K. N., Gebre-Medhin, M., and Hambraeus, L. (1985). Increased breast milk iron in severe maternal anemia: Physiological "trapping" or leakage? *Acta Paediatr. Scand.* **74,** 290–291.

Fransson, G.-B., Gebre-Medhin, M., and Hambraeus, L. (1984). The human milk contents of iron, copper, zinc, calcium and magnesium in a population with a habitually high intake of iron. *Acta Paediatr. Scand.* **73,** 471–476.

Freundlich, M., Zilleruelo, G., Abitbol, C., Strauss, J., Faugere, M.-C., and Malluche, H. H. (1985). Infant formula as a cause of aluminum toxicity in neonatal uraemia. *Lancet* **2,** 527–529.

Fujita, M., and Takabatake, E. (1977). Mercury levels in human maternal and neonatal blood, hair and milk. *Bull. Environ. Contam. Toxicol.* **18,** 205–209.

Funk, M. A., Hamlin, L., Picciano, M. F., Prentice, A., and Milner, J. A. (1990). Milk selenium of rural African women: Influence of maternal nutrition, parity, and length of lactation. *Am. J. Clin. Nutr.* **51,** 220–224.

Garry, P. J., Owen, G. M., Hooper, E. M., and Gilbert, B. A. (1981). Iron absorption from human milk and formula with and without iron supplementation. *Pediatr. Res.* **15,** 822–828.

Garza, C., Johnson, C. A., O'Brian Smith, E., and Nichols, B. (1983). Changes in the nutrient composition of human milk during gradual weaning. *Am. J. Clin. Nutr.* **37,** 61–65.

Gattavecchia, E., Ghini, S., Tonelli, D., Gori, G., Cama, G., and Guerresi, E. (1989). Caesium-137 levels in breast milk and placentae after fallout from the reactor accident at Chernobyl. *Health Phys.* **56,** 245–248.

Gori, G., Cama, G., Guerresi, E., Cocchi, G., Dalla Casa, P., Gattavecchia, E., Ghini, S., and Tonelli, D. (1988). Radioactivity in breast milk and placentas during the year after Chernobyl. *Am. J. Obstet. Gynecol.* **159,** 1232–1234.

Griffiths, M., Green, B., Leckie, R. M. C., Messer, M., and Newgrain, K. W. (1984). Constituents of platypus and echidna milk, with particular reference to the fatty acid complement of the triglycerides. *Aust. J. Biol. Sci.* **37,** 323–329.

Groppel, B., Anke, M., and Kronemann, H. (1985). Influence of iodine supply on reproduction and the iodine content of milk, blood, hair and several organs of ruminants. *In* "Trace Elements in Man and Animals—TEMA 5" (C. F. Mills, I. Bremmer, and J. K. Chesters, eds.), pp. 279–282. Commonwealth Agricultural Bureau, Slough, UK.

Guha Mazumder, D. N., Das Gupta, J., Chakraborty, A. K., Chatterjee, A., Das, D., and Chakraborti, D. (1992). Environmental pollution and chronic arsenosis in South Calcutta. *Bull. WHO* **90,** 481–485.

Gunshin, H., Yoshikawa, M., Doudou, T., and Kata, N. (1985). Trace elements in human milk, cow's milk, and infant formula. *Agric. Biol. Chem.* **49,** 21–26.

Gushurst, C. A., Mueller, J. A., Green, J. A., and Sedor, F. (1984). Breast milk iodide: Reassessment in the 1980s. *Pediatrics* **73,** 354–357.

Hambidge, K. M., (1976). The importance of trace elements in infant nutrition. *Curr. Med. Res. Opin.* **4** (Suppl. 1), 44–53.

Hambidge, K. M. (1985). Low serum zinc levels in Aboriginal children [letter]. *Lancet* **1,** 1333.

Hambidge, K. M., Casey, C. E., and Krebs, N. F. (1986). Zinc. *In* "Trace Elements in Human and Animal Nutrition" (W. Mertz, ed.), 5th Ed., Vol. 2, pp. 1–138. Academic Press, Orlando.

Hambidge, K. M., Walravens, P. A., Casey, C. E., Brown, R. M., and Bender, C. (1979). Plasma zinc concentrations of breast-fed infants. *J. Pediatrics* **94,** 607–608.

Haschke, F., Heil, M., Steffan, I., Camaya, Z., Pietschnig, B., Bock, A., and Huemer, C. (1989). Aluminum in European formulas and in breast milk. *In* "6th International Trace Element Symposium" (M. Anke, W. Baumann, H. Braunlich, Chr. Bruckner, B. Groppel, and M. Grun, eds.), pp. 1368–1373. Karl-Marx-Universitat, Leipzig.

Haschke, F., Pietschnig, B., and Karg, V. (1988). ^{131}I, ^{134}Cs, and ^{137}Cs in Austrian milk. *In* "Vitamins and Minerals in Pregnancy and Lactation" (H. Berger, ed.), pp. 351–352. Nestle Nutrition, Vevey/Raven Press, New York.

Heidemann, P. H., Stubbe, P., v. Reuss, K., Schurnbrand, P., Larson, A., and v. Petrykowski, W. (1984). Jodausscheidung und alimentare jodversorgung bei neugeborenen in jodmangelgebeiten der Bundesrepublik. *Dtsch. Med. Wschr.* **109,** 773–778.

Hetzel, B. S., and Maberly, G. F. (1986). Iodine. *In* "Trace Elements in Human and Animal Nutrition" (W. Mertz, ed.), 5th ed., Vol. 2, pp. 139–208. Academic Press, Orlando.

Hibberd, C. M., Brooke, O. G., Carter, N. D., Haug, M., and Harzer, G. (1982). Variation in the composition of breast milk during the first 5 weeks of lactation: Implications for the feeding of preterm infants. *Arch. Dis. Child.* **57,** 658–662.

Higashi, A., Tamari, H., Kuroki, Y., and Matsuda, I. (1983). Longitudinal changes in selenium content of breast milk. *Acta Paediatr. Scand.* **72,** 433–436.

Higashi, A., Ikeda, T., Uehara, I., and Matsuda, I. (1982). Zinc and copper contents in breast milk of Japanese women. *Tohoku J. Exp. Med.* **137,** 41–47.

Hojo, Y. (1982). Selenium concentration and glutathione peroxidase activity in cow's milk. *Biol. Trace Elem. Res.* **4,** 233–239.

Honda, K., and Tatsukawa, R. (1983). Distribution of cadmium and zinc in tissues and organs and their age-related changes in striped dolphins, Stenella coeruleoalba. *Arch. Environ. Contam. Toxicol.* **12,** 543–550.

Houghton, M. R., Gracey, M., Burke, V., Bottrell, C., and Spargo, R. M. (1985). Breast milk lactoferrin levels in relation to maternal nutritional status. *J. Pediatr. Gasterenterol. Nutr.* **4,** 230–233.

Howell, R. R., Palma, P. A., West, M. S., Caprioli, R. M., and Seifert, W. E. (1986). Trace elements in human milk: Differences over time and between population groups. *In* "Human Milk in Infant Nutrition and Health" (R. R. Howell, F. H. Morriss, and L. K. Pickering, eds.), pp. 28–50. Thomas, Springfield.

Hurley, L. S., and Lonnerdal, B. (1982). Zinc binding in human milk: Citrate versus picolinate. *Nutr. Rev.* **40,** 65–71.

Hurley, L. S., and Lonnerdal, B. (1986). Trace elements in human milk. *In* "Biology of Human Milk" (L. A. Hanson, ed.), Nestle Nutrition Workshop Series Vol. 15, pp. 75–94. Nestle, Vevey/Raven Press, New York.

Hurley, L. S., and Keen, C. L. (1987). Manganese. *In* "Trace Elements in Human and Animal Nutrition" (W. Mertz, ed.), 5th Ed., Vol. 1, pp. 185–224. Academic Press, Orlando.

Institute of Medicine (1991), "Nutrition in Lactation." National Academy Press. Washington, DC.

Iyengar, G. V., Kasperek, K., Feinendegen, L. E., Wang, Y. X., and Weese, H. (1982). Determination of Co, Cu, Fe, Hg, Mn, Sb, Se and Zn in milk samples. *Sci. Total. Environ.* **24,** 267–274.

Jaulmes, P., and Hamelle, G. (1971). Presence and levels of trace elements in the food and beverages of man. *Ann. Nutr. Aliment.* **25,** B133–B203.

Johnson, L. A., Ford, H. C., Doran, J., and Richardson, V. F. (1990). A survey of the iodide concentration of human milk. *N.Z. Med. J.* **103,** 393–394.

Johnson, P E., and Evans, G. W. (1978). Relative zinc availability in human breast milk, infant formulas, and cow's milk. *Am. J. Clin. Nutr.* **31,** 416–421.

Kaldor, I., and Ezekiel, E. (1962). Iron content of mammalian breast milk: Measurements in the rat and a marsupial. *Nature* **196,** 175.

Karmarkar, M. J. G., and Ramakrishnan, C. V. (1960). Studies on human lactation. Relation between the dietary intake of lactating women and the chemical composition of milk with regard to principal and certain inorganic constituents. *Acta Paediatr.* **49,** 599–604

Karra, M. V., Kirksey, A., Galal, O., Bassily, N. S., Harrison, G. G., and Jerome, N. (1988). Zinc, calcium, and magnesium concentrations in milk from American and Egyptian women throughout the first 6 months of lactation. *Am. J. Clin. Nutr.* **47,** 642–648.

Karra, M. V., Udipi, S. A., Kirksey, A., and Roepke, J. L. B. (1986). Changes in specific nutrients in breast milk during extended lactation. *Am. J. Clin. Nutr.* **43,** 495–503.

Keen, C. L., Lonnerdal, B., Clegg, M. S., Hurley, L. S., Morris, J. G. Rogers, Q. R., and Rucker, R. B. (1982). Developmental changes in composition of cats' milk: Trace elements, minerals, protein, carbohydrate and fat. *J. Nutr.* **112,** 1763–1769.

Keen, C. L., Lonnerdal, B., Clegg, M., and Hurley, L. S. (1981). Developmental changes in composition of rat milk: Trace elements, minerals, protein, carbohydrate and fat. *J. Nutr.* **111,** 226 230.

Khalkhali, Z., and Parsa, B. (1972). Measurement by non-destructive neutron activation analysis of bromine concentration in the secretions of nursing mothers. *In* "Nuclear Activation Techniques the Life Sciences," pp. 461–467. IAEA, Vienna.

Kincaid, R. L., and Cronrath, J. D. (1992). Zinc concentration and distribution in mammary secretions of peripartum cows. *J. Dairy Sci.* **75,** 481–484.

Kincaid, R. L., and White, C. L. (1988). The effects of ammonium tetramolybdate intake on tissue copper and molybdenum in pregnant ewes and lambs. *J. Anim. Sci.* **66,** 3252–3258.

Klein, G. L., Snodgrass, W. R., Griffin, M. P., Miller, N. L., and Alfrey, A. C. (1989). Hypocalcemia complicating desferoxamine therapy in an infant with parental nutrition-associated aluminum overload: Evidence for a role of aluminum in the bone disease of infants. *J. Pediatr. Gastroenterol. Nutr.* **9,** 400–403.

Kleinbaum, H. (1962). Uber den kupfergehalt der nahrungsmittel des kindes. *Zeitschr. Kinderheilk.* **86,** 655–666.

Koo, W. W. K., Kaplan, L. A., and Krug-Wispe, S. K. (1988). Aluminum contamination of infant formulas. *J. Paren. Enter. Nutr.* **12,** 170–173.

Kosta, L., Byrnne, A. R., and Dermelj, M. (1983). Trace elements in some human milk samples by radiochemical neutron activation analysis. *Sci. Total Environ.* **29,** 261–268.

Kovar, I. Z., Strehlow, C. D., Richmond, J., and Thompson, M. G. (1984). Perinatal lead and cadmium burden in a British urban population. *Arch Dis. Child.* **59,** 36–39.

Krebs, N. F., and Hambidge, K. M. (1986). Zinc requirements and intakes of breast-fed infants. *Am. J. Clin. Nutr.* **43,** 288–292.

Krebs, N. F., and Hambidge, K. M. (1985). Zinc supplementation during lactation: Effects on maternal zinc status and milk zinc concentrations. *In* "Trace Element Metabolism in Man and Animals—TEMA 5" (C. F. Mills, I. Bremmer, and J. K. Chesters, eds.), pp. 416–419. Commonwealth Agricultural Bureau, Slough, UK.

Krebs, N. F., Hambidge, K. M., Jacobs, M. A., and Oliva Rasbach, J. (1985). The effects of a dietary zinc supplement during lactation on longitudinal changes in maternal status and milk zinc concentrations. *Am. J. Clin. Nutr.* **41,** 560–570.

Krishnamachari, K. A. V. R. (1987). Fluorine. *In* "Trace Elements in Human and Animal Nutrition" (W. Mertz, ed.), 5th Ed., Vol. 1, pp. 365–416. Academic Press, Orlando.

Kumpulainen, J., and Vuori, E. (1980). Longitudinal study of chromium in human milk. *Am. J. Clin. Nutr.* **33,** 2299–2302.

Kumpulainen, J., Salmenpera, L., Siimes, M. A., Koivistoinen, P., and Perheentupa, J. (1985). Selenium status of exclusively breast-fed infants as influenced by maternal organic or inorganic selenium supplementation. *Am. J. Clin. Nutr.* **42,** 829–835.

Kumpulainen, J., Vuori, E., and Siimes, M. A. (1984). Effect of maternal dietary selenium intake on selenium levels in breast milk. *Int. J. Vit. Nutr. Res.* **54,** 252–255.

Kumpulainen, J., Lehto, J., Koivistoinen, P., Usitupa, M., and Vuori, E. (1983a). Determination of chromium in human milk, serum and urine by electrothermal atomic absorption spectrometry without preliminary ashing. *Sci. Total Environ.* **31,** 71–80.

Kumpulainen, J., Vuori, E., Kuitunen, P., Makinen, S., and Kara, R. (1983b). Longitudinal study on the dietary selenium intake of exclusively breast-fed infants and their mothers in Finland. *Int. J. Vit. Nutr. Res.* **53,** 420–426.

Kumpulainen, J., Vuori, E., Makinen, S., and Kara, R. (1980). Dietary chromium intake of lactating Finnish mothers: Effect on the Cr content of their breast milk. *Br. J. Nutr.* **44,** 257–263.

Lamounier, J. A., Danelluzzi, J. C., and Vannucchi, H. (1989). Zinc concentrations in human milk during lactation: A 6-month longitudinal study in southern Brazil. *J. Trop. Pediatr.* **35,** 31–34.

Larsson, B., Slorach, S. A., Hagman, U., and Hofvander, Y. (1981), WHO collaborative breast feeding study. II. Levels of lead and cadmium in Swedish human milk, 1978–1979. *Acta Paediatr. Scand.* **70,** 281–284.

Lauber, E., and Reinhardt, M. (1979). Studies on the quality of breast milk during 23 months of lactation in a rural community of the Ivory Coast. *Am. J. Clin. Nutr.* **32,** 1159–1173.

Lavi, N., and Alfassi, Z. B. (1990). Determination of trace amounts of cadmium, cobalt, chromium, iron, molybdenum, nickel, selenium, titanium, vanadium and zinc in blood and milk by neutron activation analysis. *Analyst* **115,** 817–822.

Lazarus, C. R., and Edwards, S. (1988). Radiopharmaceuticals. *In* "Drugs and Human Lactation" (P. N. Bennett, ed.), pp. 495–549. Elsevier, Amsterdam.

Lee, D.-Y., Shay, N. F., and Cousins, R. J. (1992). Altered zinc metabolism occurs in murine lethal milk syndrome. *J. Nutr.* **122,** 2233–2238.

Lehti, K. K. (1990). Breast milk folic acid and zinc concentrations of lactating, low socioeconomic, Amazonian women and the effect of age and parity on the same two nutrients. *Eur. J. Clin. Nutr.* **44,** 675–680.

Lemons, J. A., Moye, L., Hall, D., and Simmons, M. (1982). Differences in the composition of preterm and term milk during early lactation. *Pediatr. Res.* **16,** 113–117.

Levander, O. A. (1987). A global view of human selenium nutrition. *Annu. Rev. Nutr.* **7,** 227–250.

Levander, O. A. (1986). Selenium. *In* "Trace Elements in Human and Animal Nutrition" (W. Mertz, ed.), 5th Ed., Vol. 2, pp. 209–280. Academic Press, Orlando.

Levander, O. A., Moser, P. B., and Morris, V. C. (1987). Dietary selenium intake and selenium concentrations of plasma, erythrocytes, and breast milk in pregnant and postpartum lactating and nonlactating women. *Am. J. Clin. Nutr.* **46,** 694–698.

Lipsman, S., Dewey, K. G., and Lonnerdal, B. (1985). Breast-feeding among teenage mothers: Milk composition, infant growth, and maternal dietary intake. *J. Pediatr. Gastroenterol. Nutr.* **4,** 426–434.

Loh, T. T., and Sinnathuray, T. A. (1971). Haematological data and milk iron in Malaysian women. *Aust. N.Z. J. Obstet. Gynaecol.* **11,** 254–258.

Loh, T. T., and Kaldor, I. (1971). Intestinal iron absorption in suckling rats. *Biol. Neonate* **17,** 173–186.

Lombeck, I., Kasperek, K., Bonnerman, B., Feinendegen, L. E., and Bremer, H. J. (1978). Selenium content of human milk, cow's milk and cow's milk infant formulas. *Eur. J. Pediatr.* **129,** 139–145.

Lonnerdal, B., Keen, C. L., and Hurley, L. S. (1985a). Manganese binding proteins in human and cow's milk. *Am. J. Clin. Nutr.* **41,** 550–559.

Lonnerdal, B., Keen, C. L., Bell, J. G., and Hurley, L. S. (1985b). Zinc uptake and retention from chelates and milk fractions. *In* "Trace Elements in Man and Animals—TEMA 5" (C. F. Mills, I. Bremner, and J. K. Chesters, eds.), pp. 427–430. Commonwealth Agricultural Bureaux, Farnham Royal, England.

Lonnerdal, B., Cederblad, A., Davidsson, L., and Sandstrom, B. (1984a). The effect of individual components of soy formula and cow's milk formula on zinc bioavailability. *Am. J. Clin. Nutr.* **40,** 1064–1070.

Lonnerdal, B., Keen, C. L., Glazier, C. E., and Anderson, J. (1984b). A longitudinal study of rhesus monkey (Macaca mulatta) milk composition: Trace elements, minerals, protein, carbohydrate and fat. *Pediatr. Res.* **18,** 911–914.

Lonnerdal, B., Keen, C. L., and Hurley, L. S. (1982). Trace elements in milk from various species. *In* "Trace Element Metabolism in Man and Animals, TEMA-4" (J. McC. Howell, J. M. Hawthorne, and C. L. White, eds.), pp. 249–252. Netley, Griffin Press, Australia.

Lonnerdal, B., Keen, C. L., and Hurley, L. S. (1981a). Iron, copper, zinc, and manganese in milk. *Annu. Rev. Nutr.* **1,** 149–174.

Lonnerdal, B., Keen, C. L., Hurley, L. S., and Fisher, G. L. (1981b). Developmental changes in the composition of Beagle dog milk. *Am. J. Vet. Res.* **42,** 662–666.

Lonnerdal, B., Forsum, E., Gebre-Medhin, M., and Hambraeus, L. (1976). Breast milk composition in Ethiopian and Swedish mothers. II. Lactose, nitrogen and protein. *Am. J. Clin. Nutr.* **29,** 1134–1141.

Lonnerdal, B., and Hoffman, B. (1982). Zinc- and copper-binding proteins in human milk. *Am. J. Clin. Nutr.* **36,** 1170.

Macy, I. G., and Kelly, H. J. (1961). Human milk and cows' milk in infant nutrition. *In* "Milk: The Mammary Gland and Its Secretion" (S. K. Kon and A. T. Cowie, eds.), Vol. II, pp. 265–304. Academic Press, New York.

MAFF (1985). "Survey of Aluminum, Antimony, Chromium, Cobalt, Indium, Nickel, Thallium and Tin in Food." Food Surveillance Paper 15, HMSO, London.

Mannan, S., and Picciano, M. F. (1987). Influence of maternal selenium status on human milk selenium concentration and glutathione peroxidase activity. *Am. J. Clin. Nutr.* **46,** 95–100.

Mather, I. H., Weber, K., and Keenan, T. W. (1977). Membranes of mammary gland XII. Loosely associated proteins and compositional heterogeneity of bovine milk fat globule membrane. *J. Dairy Sci.* **60,** 394–402.

Matovinovic, J. (1983). Endemic goiter and cretinism at the dawn of the third millennium. *Annu. Rev. Nutr.* **3,** 341–412.

Maus, R. W., Martz, F. A., Belyea, R. L., and Weiss, M. F. (1980). Relationship of dietary selenium to selenium in plasma and milk from dairy cows. *J. Dairy Sci.* **63,** 532–537.

Mbofung, C. M. F., Atinmo, T., and Omololu, A. (1984). Mineral content of colostrum and mature milk of lactating Nigerian women as influenced by stage of lactation. *Nutr. Rep. Int.* **30,** 1137–1146.

McCarthy, T. P., Brodie, B., Milner, J. A., and Bevill, R. F. (1981). Improved method for selenium determination in biological samples by gas chromatography. *J. Chromatogr.* **225,** 9–16.

McGraw, M., Bishop, N., Jameson, R., Robinson, M. J., O'Hara, M., Hewitt, C. D., and Day, J. P. (1986). Aluminum content of milk formulae and intravenous fluids used in infants [letter]. *Lancet* **1,** 157.

McGuire, S. L., Burgert, S. L., Milner, J. A., Glass, L., Kummer, R., Deering, R., Boucek, R., and Picciano, M. F. (1993). Selenium status of infants is influenced by supplementation of formula or maternal diets. *Am. J. Clin. Nutr.* **58,** 643–648.

McMillan, J. A., Landaw, S. A., and Oski, F. A. (1976). Iron sufficiency in breast-fed infants and the availability of iron from human milk. *Pediatrics* **58,** 686–691.

McMillan, J. A., Oski, F. A., Lourie, G., Tomarelli, R. M., and Landaw, S. A. (1977). Iron absorption from human milk, simulated human milk, and proprietary formulas. *Pediatrics* **60,** 896–900.

McNally, M. E., Atkinson, S. A., and Cole, D. E. C. (1991). Contribution of sulfate and sulfoesters to total sulfur intake in infants fed human milk. *J. Nutr.* **121,** 1250–1254.

Mendelson, R. A., Anderson, G. H., and Bryan, M. H. (1982). Zinc, copper and iron content of milk from mothers of preterm and full-term infants. *Early Hum. Dev.* **6,** 145–151.

Mercer, J. F. B., Grimes, A., and Rauch, H. (1992). Hepatic metallothionein gene expression in toxic milk mice. *J. Nutr.* **122,** 1254–1259.

Mertz, W. (1993). Chromium in human nutrition: A review. *J. Nutr.* **123,** 626–633.

Mertz, W. (1981). The essential trace elements. *Science* **213,** 1332–1338.

Michalke, B., and Schramel, P. (1990). Protein fractionation and Cd-speciation in human breast milk by HPLC and voltammetry. *J. Trace Elem. Electrolytes Health Dis.* **4,** 163–167.

Millar, K. R., and Sheppard, A. D. (1972). a-Tocopheral and selenium levels in human and cow's milk. *N.Z. J. Sci.* **15,** 3–15.

Milner, J. A. (1990). Trace minerals in the nutrition of children. *J. Pediatr.* **117,** S147–S155.

Milner, J. A., Sherman, L., and Picciano, M. F. (1987). Distribution of selenium in human milk. *Am. J. Clin. Nutr.* **45,** 617–624.

Mitchell, N. G., Coughtrey, P. J., Beetham, C. J., Hughes, J. G., Clench, S. F., and Walters, B. (1989). Transfer of caesium from silage to cows milk: Observations and models. *Sci. Total Environ.* **85,** 307–316.

Moore, M. E. C., Moran, J. R., and Greene, H. L. (1984). Zinc supplementation of lactating women: Evidence for mammary control of zinc secretion. *J. Pediatr.* **105,** 600–602.

Moran, J. R., Vaughan, R., Stroop, S., Coy, S., Johnston, H., and Greene, H. L. (1983). Concentrations and total daily output of micronutrients in breast milk of mothers delivering preterm: A longitudinal study. *J. Pediatr. Gastroenterol. Nutr.* **2,** 629–634.

Moser, P. B., Reynolds, R. D., Acharya, S., Howard, M. P., Andon, M. B., and Lewis, S. A. (1988). Copper, iron, zinc, and selenium dietary intake and status of Nepalese lactating women and their breast-fed infants. *Am. J. Clin. Nutr.* **47,** 729–734.

Moser-Veillon, P. B., and Reynolds, R. D. (1990). A longitudinal study of pyridoxine and zinc supplementation of lactating women. *Am. J. Clin. Nutr.* **52,** 135–141.

Moser-Veillon, P. B., and Reynolds, R. D. (1983). Dietary zinc intake and zinc concentrations of plasma, erythrocytes, and breast milk in antepartum and postpartum lactating and nonlactating women: A longitudinal study. *Am. J. Clin. Nutr.* **38,** 101–108.

Mountford, P. J., and Coakley, A. J. (1989). A review of the secretion of radioactivity in human breast milk: Data, quantitative analysis and recommendations. *Nuclear Med. Commun.* **10,** 15–27.

Moynahan, E. J. (1974). Acrodermatitis enteropathica: A lethal inherited human zinc-deficiency disorder. *Lancet* **2**, 399–400.

Muller, C. (1987). Cadmium content of human milk. *Trace Elem. Med.* **4,** 4–7.

Murray, M. J., Murray, A. B., Murray, N. J., and Murray, M. B. (1978). The effect of iron status of Nigerian mothers on that of their infants at birth and 6 months, and on the concentration of Fe in breast milk. *Br. J. Nutr.* **39,** 627–630.

Nagra, S. A. (1989). Longitudinal study in biochemical composition of human milk during the first year of lactation. *J. Trop. Pediatr.* **35,** 126–128.

Nassi, L., Poggini, G., Vecchi, C., and Galvan, P. (1974). Considerazioni sul contentuto di zinco, di rame e di ferro nel colostro e nel latte umano. *Min. Pediatr.* **26,** 832–836.

Neville, M. C., Allen, J. C., Archer, P. C., Casey, C. E., Seacat, J., Keller, R. P., Lutes, V., Rasbach, J., and Neifert, M. (1991). Studies in human lactation: Milk volumes and nutrient composition during weaning and lactogenesis. *Am. J. Clin. Nutr.* **54,** 81–92.

Nielsen, F. (1991). Nutritional requirements for boron, silicon, vanadium, nickel, and arsenic: Current knowledge and speculation. *FASEB J.* **5,** 2661–2667.

Nielsen, F. H. (1990). New essential trace elements of the life sciences. *Biol. Trace Elem. Res.* **26–27,** 599–611.

Nielsen, F. H. (1987). Nickel. *In* "Trace Elements in Human and Animal Nutrition" (W. Mertz, ed.), 5th Ed., Vol. 1, pp. 245–274. Academic Press, Orlando.

Nielsen, F. H. (1986). Other elements: Sb, ba, B, Br, Cs, Ge, Rb, Ag, Sr, Sn, Ti, Zr, Be, Bi, Ga, Au, In, Nb, Sc, Te, Tl, W. *In* "Trace Elements in Human and Animal Nutrition" (W. Mertz, ed.), 5th ed., Vol. 2, pp. 415–464. Academic Press, Orlando.

Nyazema, N. Z., Mahomva, O., and Andifasi, W. (1989). The levels of zinc in breast milk of urban African women in Zimbabwe. *Afr. J. Med. Med. Sci.* **18,** 159–162.

O'Dell, B. (1992). Zinc plays both structural and catalytic roles in metalloproteins. *Nutr. Rev.* **50,** 48–50.

Ohtake, M., Chiba, R., Mochizuki, K., and Tada, K. (1981). Zinc and copper concentrations in human milk and in serum from exclusively-breast-fed infants in the first 3 months of life. *Tohoku J. Exp. Med.* **135,** 335–343.

Ong, C. N., Phoon, W. O., Law, H. Y., Tye, C. Y., and Lim, H. H. (1985). Concentrations of lead in maternal blood, cord blood, and breast milk. *Arch Dis. Child.* **60,** 756–759.

Opinya, G. N., Bwibo, N., Valderhaug, J., Birkeland, J. M., and Lokken, P. (1991). Intake of fluoride and excretion in mother's milk in a high fluoride (9 ppm) area in Kenya. *Eur. J. Clin. Nutr.* **45,** 37–41.

Park, Y. K., Harland, B. F., Vanderveen, J. E., Shank, F. R., and Prosky, L. (1981). Estimation of dietary iodine intake of Americans in recent years. *J. Am. Diet. Assoc.* **79,** 17–24.

Parr, R. M., DeMaeyer, E. M., Iyengar, V. G. Byrne, A. R., Kirkbright, G. F. Schoch, G., Niinisto, L., Pineda, O., Vis, H. L., Hofvander, U., and Omololu, A. (1991). Minor and trace elements in human milk from Guatemala, Hungary, Nigeria, Philippines, Sweden, and Zaire. *Biol. Trace Elem. Res.* **29,** 51–75.

Peddemors, V. M., de Muelenaere, H. J. H., and Devchand, K. (1989). Comparative milk composition of the bottlenosed dolphin (Tursiops truncatus), humpbacked dolphin (Sousa plumbea) and common dolphin (Delphinus delphis) from southern African waters. *Comp. Biochem. Physiol. A* **94,** 639–671.

Picciano, M. F., and Guthrie, H. A. (1976). Copper, iron, and zinc contents of mature human milk. *Am. J. Clin. Nutr.* **29,** 242–254.

Picciano, M. F., Calkins, E. F., Garrick, J. R., and Deering, R. H. (1981). Milk and mineral intakes of breastfed infants. *Acta Paediatr. Scand.* **70,** 189–194.

Piletz, J. E., and Ganschow, R. E. (1978). Zinc deficiency in murine milk underlies expression of the lethal milk (lm) mutation. *Science* **199,** 181–183.

Postellon, D. C., and Aronow, R. (1982). Iodine in mother's milk [letter]. *JAMA* **247,** 463.

Radisch, B., Luck, W., and Nau, H. (1987). Cadmium concentrations in milk and blood of smoking mothers. *Toxicol. Lett.* **36,** 147–152.

Rajalakshmi, K., and Srikantia, S. G. (1980). Copper, zinc, and magnesium content of breast milk of Indian women. *Am. J. Clin. Nutr.* **33,** 664–669.

Rao, G. S. (1984). Dietary intake and bioavailability of fluoride. *Annu. Rev. Nutr.* **4,** 115–136.

Rauch, H. (1983). Toxic milk, a new mutation affecting copper metabolism in the mouse. *J. Heredity* **74,** 141–144.

Reis, B. L., Keen, C. L., Lonnerdal, B., and Hurley, L. S. (1991). Longitudinal changes in the mineral composition of mouse milk and the relationship to zinc metabolism in the suckling neonate. *J. Nutr.* **121,** 687–699.

Rhein, R. (1991). US lowers lead limits (news). *Br. Med. J.* **303,** 943.

Robberecht, H., Roekens, E., van Caille-Bertrand, M., Deelstra, H., and Clara, R. (1985). Longitudinal study of the selenium content in human breast milk in Belgium. *Acta Paediatr. Scand.* **74,** 254–258.

Rockway, S. W., Weber, C. W., Lei, K. Y., and Kemberling, S. R. (1984). Lead concentrations of milk, blood, and hair in lactating women. *Int. Arch. Occup. Environ. Health* **53,** 181–187.

Ruz, M., Atalah, E., Bustos, P., Masson, L., Oliver, H., Hurtado, C., and Araya, J. (1982). Composicion quimica de leche materna. Influencia del estado nutricional de la nodriza. *Arch. Latinoamer. Nutr.* **32,** 697–712.

Saarinen, U. M., Siimes, M. A., Dallman, P. R. (1977). Iron absorption in infants: High bioavailability of breast milk iron as indicated by the extrinsic tag method of iron absorption and by the concentration of serum ferritin. *J. Pediatr.* **91,** 36–39.

Salmenpera, L., Pereentupa, J., Pakarinen, P., and Siimes, M. (1986). Cu nutrition in infants during prolonged exclusive breast-feeding: Low intake but rising serum concentrations of Cu and ceruloplasmin. *Am. J. Clin. Nutr.* **43,** 251–257.

Sandstrom, B., Keen, C. L., and Lonnerdal, B. (1982). An experimental model for studies of zinc bioavailability from milk and infant formulas using extrinsic labeling. *Am. J. Clin. Nutr.* **38,** 420–428.

Sandstrom, B., Cederblad, A., and Lonnerdal, B. (1983). Zinc absorption from human, cow's milk and infant formulas. *Am. J. Dis. Child.* **137,** 726–729.

Saner, G., and Garibagaoglu, M. (1988). Zinc, copper, iron and manganese levels in milk from mothers of preterm and term infants. *Nutr. Rep. Int.* **37,** 211–217.

Saner, G., and Yuzbasiyan, V. (1984). Diurnal and longitudinal variations in fat, energy and trace element content of human milk. *Nutr. Rep. Int.* **29,** 1181–1189.

Schramel, P., Hasse, S., and Ovcar-Pavlu, J. (1988). Selenium, cadmium, lead, and mercury concentrations in human breast milk, placenta, maternal blood, and the blood of the newborn. *Biol. Trace Elem. Res.* **15,** 111–124.

Schryver, H. F., Oftedal, O. T., Williams, J., Cymbaluk, N. F., Antczak, D., and Hintz, H. F. (1986a). A comparison of the mineral composition of milk of domestic and captive wild equids (Equus przewalski, E. zebra, E. burchelli, E. caballus, E. assinus). *Comp. Biochem. Physiol. A* **85,** 233–235.

Schryver, H. F., Oftedal, O. T., Williams, J., Soderholm, L. V., and Hintz, H. F. (1986b). Lactation in the horse: Composition of mare milk. *J. Nutr.* **116,** 2142–2147.

Schulte-Lobbert, F. J., and Bohn, G. (1977). Determination of cadmium in human milk during lactation. *Arch. Toxicol.* **37,** 155–157.

Sedman, A. B., Klein, G. L., Merritt, R. J., Miller, N. L., Weber, K. O., Gill, W. L., Anand, H., and Alfrey, A. C. (1985). Evidence of aluminum loading in infants receiving intravenous therapy. *N. Engl. J. Med.* **312,** 1337–1343.

Semmerkrot, B. A., Monnens, L. A. H., and Baadenhuysen, H. (1989). Levels of aluminum in infant formulae [letter]. *Lancet* **1,** 1024–1025.

Shearer, T. R., and Hadjimarkos, D. M. (1975). Geographic distribution of selenium in human milk. *Arch. Environ. Health* **30,** 230–233.

Siimes, M., Vuori, E., and Kuitunen, P. (1979). Breast milk iron—A declining concentration during the course of lactation. *Acta Paediatr. Scand.* **68,** 29–31.

Siimes, M. A., Salmenpera, L., and Perheentupa, J. (1984). Exclusive breast-feeding for 9 months: Risk of iron deficiency. *J. Pediatr.* **104,** 196–199.

Simmer, K., Ahmed, S., Carlsson, L., and Thompson, R. P. H. (1990). Breast milk zinc and copper concentration in Bangladesh. *Br. J. Nutr.* **63,** 91–96.

Singh, H., Flynn, A., and Fox, P. F. (1989). Binding of zinc to bovine and human milk proteins. *J. Dairy Res.* **56,** 235–248.

Skerfving, S. (1988). Mercury in women exposed to methylmercury through fish consumption, and in their newborn babies and breast milk. *Bull. Environ. Contam. Toxicol.* **41,** 475–482.

Smith, R. M. (1987). Cobalt. *In* "Trace Elements in Human and Animal Nutrition" (W. Mertz, ed.), 5th Ed., Vol. 1, pp. 143–184. Academic Press, Orlando.

Smith, A. M., Picciano, M. F., and Milner, J. A. (1982). Selenium intakes and status of human milk and formula fed infants. *Am. J. Clin. Nutr.* **35,** 521–526.

Spak, C. J., Hardell, L., and de Chateau, P. (1983). Fluoride in human milk. *Acta Paediatr. Scand.* **72,** 699–701.

Spencer, H., Osis, D., Fisenne, I. M., Perry, P. M., and Harley, N. H. (1990). Measured intake and excretion patterns of naturally occurring ^{234}U, ^{238}U, and calcium in humans. *Radiat. Res.* **124,** 90–95.

Stastney, D. S., Vogel, M. S., and Picciano, M. F. (1984). Manganese intake and serum manganese concentration of human milk-fed and formula-fed infants. *Am. J. Clin. Nutr.* **39,** 872–878.

Straub, C. P., and Murthy, G. K. (1965). A comparison of Sr^{90} component of human and cows' milk. *Pediatrics* **36,** 732–735.

Sutcliff, E. (1990). Iodine in New Zealand milk. *Food Technol N.Z.* **97,** 32–38.

Suzuki, K. T., Tamagawa, H., Hirano, S., Kobayashi, E., Takahashi, K., and Shimojo, N. (1991). Changes in element concentration and distribution in breast-milk fractions of a healthy lactating mother. *Biol. Trace Elem. Res.* **28,** 109–121.

Tarvydas, H., Jordan, S. M., and Morgan, E. H. (1968). Iron metabolism during lactation in the rabbit. *Br. J. Nutr.* **22,** 565–573.

Taves, D. R. (1983). Dietary intake of fluoride ashed (total fluoride) v. unashed (inorganic fluoride) analysis of individual foods. *Br. J. Nutr.* **49,** 295–301.

Tkachenko, S. K. (1970). Intake and balance of some trace elements in preterm infants. *Pediatriia* **49,** 10–14.

Trugo, N. M. F., Donangelo, C. M., Koury, J. C., Barreto-Silva, M. I., and Freitas, L. A. (1988). Concentration and distribution pattern of selected micronutrients in preterm and term milk from urban Brazilian mothers during early lactation. *Eur. J. Clin. Nutr.* **42,** 497–507.

Ullrey, D. E., Ely, W. T., and Covert, R. L. (1974). Iron, zinc and copper in mare's milk. *J. Anim. Sci.* **38,** 1276–1277.

Uusitupa, M. I. J., Mykkanen, L., Siitonen, O., Laakso, M., Sarlund, H., Kolehainen, P., Rasanen, I., Kumpulainen, J., and Pyorala, K. (1992). Chromium supplementation in impaired glucose tolerance of elderly: Effects on blood glucose, plasma insulin, C-peptide and lipid levels. *Br. J. Nutr.* **68,** 209–216.

Vaillancourt, S. J., and Allen, J. C. (1991). Glucocorticoid effects on zinc transport into colostrum and milk of lactating cows. *Biol. Trace Elem. Res.* **30,** 185–196.

van Dael, P., Deelstra, H., Vlaemynck, G., and van Renterghem, R. (1989). Separation of selenium in bovine whey. *In* "Nutrient Availability: Chemical and Biological Aspects" (D. A. T. Southgate, I. T. Johnson, and G. R. Fenwick, eds.), pp. 112–115. Royal Society of Chemistry, Cambridge.

van Dael, P., Deelstra, H., Vlaemynck, G., and van Renterghem, R. (1988). Distribution of selenium in cow's and human milk (abstract). *J. Trace. Elem. Electrolytes Health Dis.* **2,** 121.

van der Westhuyzen, J., van Tonder, S. V., and Fernandes-Costa, F. J. (1986). Ferritin in breast milk of black urban and rural mothers. *Internat. J. Vit. Nutr. Res.* **56,** 287–289.

Varo, P., Nuurtamo, M., Saari, E., and Koivistoinen, P. (1980). Mineral element composition of Finnish foods VIII. Dairy products, eggs and margarine. *Acta Agric. Scand. Suppl.* **22,** 115–126.

Vaughan, L. A., Weber, C. W., and Kemberling, S. R. (1979). Longitudinal changes in the mineral content of human milk. *Am. J. Clin. Nutr.* **32,** 2301–2306.

Vega-Franco, L., Batista-Primers, E., and Meza-Camacho, C. (1987). Manganeso, cobre, hierro y molibdeno en la secrecion temprana de calostro humano. *Bol. Med. Hosp. Infant. Mex.* **44,** 86–91.

Verlinden, M., Baart, J., and Deelstra, H. (1980). Optimization of the determination of selenium by atomic absorption spectrometry: Comparison of two hydride generation systems. *Talanta* **27,** 633–639.

Versieck, J., and Cornelis, R. (1989). "Trace Elements in Human Plasma or Serum." CRC Press, Boca Raton, FL.

Vuori, E. (1979). A longitudinal study of manganese in human milk. *Acta Paediatr. Scand.* **68,** 571–573.

Vuori, E., and Kuitunen, P. (1979). The concentrations of copper and zinc in human milk. *Acta Paediatr. Scand.* **68,** 33–37.

Vuori, E., Makinen, S. M., Kara, R., and Kuitunen, P. (1980). The effects of the dietary intakes of copper, iron, manganese, and zinc on the trace element content of human milk. *Am. J. Clin. Nutr.* **33,** 227–231.

Walivaara, R., Jansson, L., and Akesson, B. (1986). Selenium content of breast milk sampled in 1978 and 1983 in Sweden. *Acta Paediatr. Scand.* **75,** 236–239.

Walravens, P. A., Krebs, N. F., and Hambidge, K. M. (1983). Linear growth of low income preschool children receiving a zinc supplement. *Am. J. Clin. Nutr.* **38,** 195–201.

Ward, G. M., and Johnson, J. E. (1989). Assessment of milk transfer coefficients for use in prediction models of radioactivity transport. *Sci. Total Environ.* **85,** 287–294.

Ward, G. M., Keszthelyi, Z., Kanyar, B., Kralovanszky, U. P., and Johnson, J. E. (1989). Transfer of ^{137}Cs to milk and meat in Hungary from Chernobyl fallout with comparisons of worldwide fallout in the 1960s. *Health Phys.* **57,** 587–592.

Watson, W. S. (1986). Human ^{134}Cs/^{137}Cs levels in Scotland after Chernobyl. *Nature* **323,** 763–764.

Waxman, J., and Wasan, H. (1992). The architecture of cancer. *Br. Med. J.* **305,** 1306–1307.

Weintraub, R., Hams, G., Meerkin, M., and Rosenberg, A. R. (1986). High aluminum content of infant milk formulas. *Arch. Dis. Child.* **61,** 914–916.

Williams, M. M. F. (1983). Selenium and glutathione peroxidase in mature human milk. *Proc. Uni. Otago Med. School* **61,** 20–21.

Witsell, D. L., Casey, C. E., and Neville, M. C. (1990). Divalent cation activation of galactosyltransferase in native mammary golgi vesicles. *J. Biol. Chem.* **265,** 15731–15737.

World Health Organization/IAEA (WHO/IAEA) (1987). "Minor and Trace Elements in Breast Milk." WHO, Geneva.

Yoshida, M., Yasumoto, K., Iwami, K., and Tashiro, H. (1981). Distribution of selenium in bovine milk and selenium deficiency in rats fed casein-based diets, monitored by lipid peroxide levels and glutathione peroxidase activity. *Agric. Biol. Chem.* **45,** 1681–1688.

CHAPTER **8**

Vitamins in Milk

A. Water-Soluble Vitamins in Human Milk

MARY FRANCES PICCIANO

I. Introduction

The water-soluble vitamins in human milk consist of ascorbic acid (vitamin C), thiamin (vitamin B_1), riboflavin (vitamin B_2), niacin, pyridoxine (vitamin B_6), folate, pantothenate, biotin, and vitamin B_{12}. The water-soluble vitamins represent a diverse group of low-molecular-weight organic compounds that function in intermediary metabolism. They are grouped together not because of similarity in either structure or function, but on their physical characteristic of solubility in water. The water-soluble vitamins was the last group of essential nutrients to be discovered because they are present in minute amounts in plant and animal tissues and are required in human nutrition in microquantities (milligram quantities or lower). In general, quantities of the water-soluble vitamins found in human milk are several-fold greater than quantities in maternal plasma suggesting regulated transport, but mechanisms of secretion remain largely unexplored. In this chapter, analytical methodologies for the determinations of water-soluble vitamins are briefly considered followed by a discussion of physiological and environmental factors capable of influencing the quantities of these vitamins in human milk. Tabular data are results of studies using modern and/or appropriate methodologies from a variety of countries providing representative values.

II. Methodological Considerations

The collection, storage, and measurement of human milk water-soluble vitamins requires special consideration because many are unstable to light (i.e., riboflavin and folate), degrade during storage at −20°C (i.e., ascorbic acid, pyridoxine, and folate), and detection methods are often nonspecific, notably microbiological assays. As a general rule, samples obtained for water-soluble vitamin assay should be protected from light during collection, transported on ice to the laboratory, treated to maintain stability, divided among storage vessels, and stored at −70°C if not assayed immediately. Several of the water-soluble vitamins exist as bound species in human milk (i.e., folate, vitamin B_{12}, thiamin, and pantothenic acid) and extraction procedures must liberate the bound vitamin from its carrier protein to permit accurate detection. The methods most commonly employed for detection of water-soluble vitamins following suitable extraction are colorimetric (ascorbic acid), fluorometric, radiometric, and microbiological. Since many exist in multiforms in human milk, methodologies using high-performance liquid chromatography prior to detection provide the opportunity to furnish an accurate determination of the water-soluble vitamins and their active metabolites. The recent application of such methodology to the assay of riboflavin, for example, shows that human milk contains nearly twofold higher quantities than previously estimated using only partially discriminatory microbiological or fluorometric analyses (Roughead and McCormick, 1990). For a discussion of appropriate analytical schemes for the accurate and reliable detection of the water-soluble vitamins in biological samples, the reader is referred to "The Handbook of Vitamins" edited by Machlin (1991).

III. Factors That Influence the Water-Soluble Vitamin Concentrations in Human Milk

The composition of the water-soluble vitamins in human milk shows variation due to stage of lactation, maternal intake, and premature initiation of lactation from interrupted gestation. However, much of the variation in reported values for the water-soluble vitamins reflects analytical difficulties rather than variable patterns of secretion.

A. Stage of Lactation

Since the mammary gland cannot synthesize the water-soluble vitamins, their origin is maternal plasma, ultimately derived from the maternal diet. The mammary gland does actively transport and metabolize the vitamins as evidenced by generally higher concentrations of water-soluble vitamins in milk compared to maternal plasma and secretion profiles widely different than corresponding plasma vitamin profiles (Brown *et al.*, 1986). For most of the water-soluble vitamins, concentrations are lower in early secretions (1–5 days) compared to mature milk (> 1 month) with the possible exception of vitamin B_{12} (Table I).

TABLE I
Relative Contents of Human Milk Water-Soluble Vitamin Content as Influenced by Stage of Lactation, Maternal Intake, and Premature Delivery

Vitamin	Stage of Lactation[a]	Maternal intake[b]	Premature delivery[c]
Ascorbic acid	⇓	+	⇑
Thiamin	⇓	+	⇓
Riboflavin	⇓	+ +	=
Niacin	⇓	+	=
Folate	⇓	+	=
Vitamin B_6	⇓	+ +	⇓
Vitamin B_{12}	⇑	+	⇑
Pantothenic acid	⇓	+	⇑
Biotin	⇓	+	=

[a]The direction of arrow indicates whether reported values are higher (⇑) or lower (⇓) in early lactation (1–5 days) relative to later in lactation (> 1 month).
[b]A plus sign indicates that maternal intake of the vitamin can influence milk vitamin content but primarily in women deficient in the vitamin. Double plus signs indicate that level of maternal vitamin intake influences milk content even when vitamin status of the women is adequate.
[c]The direction of the arrow indicates whether reported values for the specific vitamin are higher (⇑) or lower (⇓) in milk from mothers delivering prematurely. An equal sign indicates that values are similar for milk from mothers of term and preterm infants.

B. Maternal Intake

Unquestionably, there is regulation of the quantities of water-soluble vitamins secreted in human milk. When appropriate methodology is applied to human milk samples, maternal supplementation affects milk water-soluble vitamin content in a linear fashion only when maternal stores are depleted or absent (Figure 1). In fact, there is evidence to support the maintenance of milk water-soluble vitamin secretion patterns during the development of maternal depletion (Salmenperä *et al.*, 1986) and preferential vitamin partitioning to the mammary gland in the face of frank maternal deficiency (Ghitis, 1966). In mothers judged to be nutritionally adequate, maternal supplementation (see Table I) in supraphysiological doses either has no effect (i.e., ascorbic acid, folate, riboflavin) or the effect on milk vitamin content is less marked (Styslinger and Kirksey, 1985), and when marked, it is often transient (West and Kirksey, 1976). The impact of maternal nutrition on milk water-soluble vitamin content also can be dependent on the stage of lactation. For example, in early lactation (< 20 days) low milk concentrations of vitamin B_6 cannot be altered by maternal supplementation (5 and 100 mg vitamin B_6, levels corresponding to 2.3

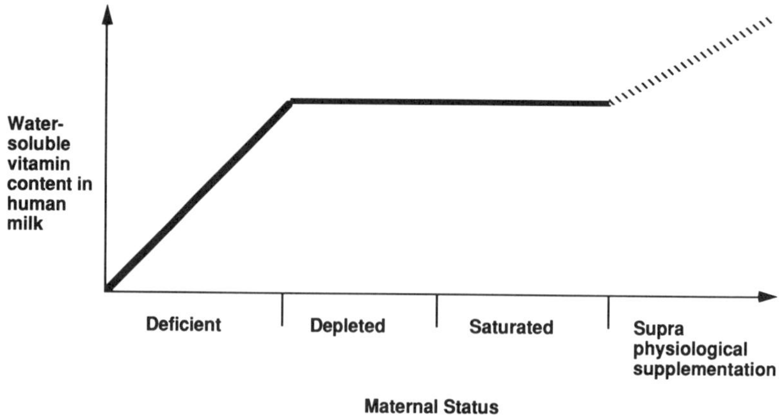

Figure 1 Schematic representation of the relationship between maternal water-soluble vitamin intakes and human milk contents

and 47.6 times recommended intakes, respectively), but after 20–22 days, such supplementation markedly increases the vitamin concentration in milk (Kirksey and Rahmanylar, 1988). For this reason, studies designed to assess the relationship between maternal intake and corresponding milk water-soluble vitamin contents must control for the possible influence of stage of lactation.

Although rarely encountered in industrialized countries, such as the United States, there are reported cases of deficiency in exclusively breast-fed infants for ascorbic acid (scurvy), thiamin (beri-beri), riboflavin, folate, vitamin B_6, and vitamin B_{12} (Fomon, 1993). In contrast, there are no reports of toxic symptoms in breast-fed infants from ingestion of milk with grossly elevated levels of the water-soluble vitamins due to maternal megadosing.

C. Premature Delivery

The water-soluble vitamin content of human milk can be influenced by premature interruption of gestation (Table I). Evidence exists to indicate that compared to mothers of term infants, concentrations of ascorbic acid (Udipi *et al.*, 1985), pantothenic acid (Song *et al.*, 1984), and vitamin B_{12} (Ford *et al.*, 1983) are higher, while concentrations of thiamin (Ford *et al.*, 1985) and vitamin B_6 (Udipi *et al.*, 1985) are lower in milk samples from mothers of preterm infants. For the other water-soluble vitamins, premature initiation of lactation is not reported to influence milk contents. The mechanisms underlying these differences are not defined but possibly reflect incomplete differentiation of the mammary epithelial cells, leaky junctions between epithelial cells, decreased blood flow to the mammary gland, and/or decreased milk volume resulting from prematurity (Anderson, 1984).

There is an amazing dearth of data on water-soluble vitamin contents of milk from mothers delivering prematurely. This probably stems from the fact that current feeding protocols for preterm infants are designed to achieve intrauterine growth and nutrient accretion rates that cannot be achieved with human milk feeding whether the milk is derived from mothers of term infants or the infant's own mother. Nutritional management of the preterm infant often involves total parenteral nutrition prior to enteral feeding that consists of either specially prepared preterm infant formulas or human milk with nutrition fortifiers (Tsang *et al.*, 1993). In all three cases, high levels of water-soluble vitamins are furnished and contributions from human milk assume little clinical relevance. Nonetheless, valuable insights into mammary regulation of water-soluble vitamin secretions could be obtained from further study particularly since no animal model for premature initiation of lactation exists. See Chapter 3E for more information on the effects of gestational age.

IV. Water-Soluble Vitamin Contents of Human Milk

Representative values for the water-soluble vitamin contents in mature human milk are presented in Table II. For reasons that are not always clear, there are quantitative differences for reported mean values of vitamins among investigators even when seemingly similar methodology was employed. This may be due to genetic differences among donors, to sample handling procedures that inadvertently destroy labile vitamins, or to interferences in analytical schemes that are reproducible and therefore not controlled.

A. Ascorbic Acid

The ascorbic acid content of human milk from well-nourished U.S. women can be expected to average approximately 100 mg/liter. Customary maternal intakes in the United States ($>$ 100 mg/day) exceed recommended intake of 95 mg/day for the first 6 months of lactation (IOM/NAS, 1991) and intakes above this level do not alter milk ascorbic acid content.

B. Thiamin

Human milk thiamin content increases with the progression of lactation and reported values are amazingly similar, approximately 200 μg/liter, despite the fact that widely different methods for analysis were employed.

TABLE II
The Water-Soluble Vitamin Contents of Human Milk

Reference	Country	Comment	Sample No.	Concentration
Ascorbic acid				
Department of Health and Social Security (1977)	UK	Pooled sample from 96 mothers from five different cities		38 mg/liter
Thomas *et al.* (1979, 1980)	U.S.A.	Maternal intake		
		130 mg/day	12	35 mg/liter
		174 mg/day	6	61 mg/liter
		215 mg/day	7	87 mg/liter
Sneed *et al.* (1981)	U.S.A.	Maternal intake		
		152 mg/day	7	61 mg/liter
		193 mg/day	9	72 mg/liter
Bates *et al.* (1982, 1983)	Gambia	Maternal intake		
		$<$ 10 mg/day	100	20 mg/liter
		34 mg/day	80	34 mg/liter
		100 mg/day	80	55 mg/liter
Salmenpera (1984)	Finland	Maternal intake		
		138 mg/day	200	45 mg/liter
Byerley and Kirksey (1985)	U.S.A.	Maternal intake		
		$<$ 100 mg/day	7	85 mg/liter
		$>$ 1000 mg/day	8	100 mg/liter
Anderson and Pitlard (1985)	U.S.A.	Maternal intake		
		$>$ 1500 mg/day	1	105 mg/liter
Thiamin				
Department of Health and Social Security (1977)	UK	Pooled sample from 96 mothers from five different cities		166 μg/liter

TABLE II—*continued*

Reference	Country	Comment	Sample No.	Concentration
Nail *et al.* (1980)	U.S.A.	Maternal intake		
		1.3 mg/day	5	220 μg/liter
		3.3 mg/day	7	238 μg/liter
Thomas *et al.* (1980)	U.S.A.	Maternal intake		
		1.5 mg/day	6	208 μg/liter
		3.3 mg/day	6	228 μg/liter
Ford *et al.* (1983)	UK	Dietary evaluation not performed	26	183 μg/liter
Prentice *et al.* (1983)	Gambia	Unsupplemented women	21	160 μg/liter
		Supplemented women (1.4 mg/day)	23	220 μg/liter
Riboflavin				
Rönnholm (1986)	U.S.A.	Samples collected at Weeks 1, 2, 6, and 12; supplementation with 2.5 to 5.0 mg/day had no effect on milk content	39	475 μg/liter
Roughead and McCormick (1990)	U.S.A.	Riboflavin intake was estimated at 1.1 to 2.9 mg/day for 4 mothers	5	580 μg/liter
Niacin				
Department of Health and Social Security (1977)	UK	Pooled sample from 96 mothers from five cities		2.3 mg/liter
Ford *et al.* (1983)	UK	Dietary evaluation not performed	24	1.8 mg/liter
Prentice *et al.* (1983)	Gambia	Nonsupplemented mothers	21	1.1 mg/liter
		Supplemented mothers (19 mg/day)	23	1.6 mg/liter
Vitamin B_6				
West and Kirksey (1976)	U.S.A.	Maternal intake		
		< 2.5 mg/day	6	0.13 mg/liter
		> 2.5 mg/day	5	0.31 mg/liter

TABLE II—*continued*

Reference	Country	Comment	Sample No.	Concentration
Prentice *et al.* (1983)	Gambia	No dietary intervention	21	0.12 mg/liter
Vanderslide *et al.* (1983)	U.S.A.	Maternal intake > 4 mg/day	7	0.31 mg/liter
Styslinger and Kirksey (1985)	U.S.A.	Maternal intake		
		2.0 mg/day	6	0.09 mg/liter
		4.4 mg/day	6	0.19 mg/liter
		11.3 mg/day	6	0.25 mg/liter
Karra *et al.* (1986)	U.S.A.	No dietary intervention	40	0.15 mg/liter
Bamji (1986)	India	No dietary intervention	73	0.07 mg/liter
Folate				
Tamura *et al.* (1980)	Japan	Samples collected between 3 and 29 weeks of lactation, no effect of supplementation	16	133 μg/liter
Smith *et al.* (1983)	U.S.A.	Samples collected at 4 and 8 weeks of lactation; maternal supplementation without an effect on milk folate	132	79 μg/liter
O'Connor *et al.* (1991)	U.S.A.	Samples independently analyzed in two laboratories	8	83 μg/liter
Vitamin B_{12}				
Samson *et al.* (1980)	UK	No dietary intervention	16	260 ng/liter
Sanberg *et al.* (1981)	U.S.A.	Supplementation without an effect	19	970 ng/liter
Sneed *et al.* (1981)	U.S.A.	Nonsupplemented mothers–estimated intake of 5.2 μg/day;	7	550 ng/liter
		Supplemented mothers–estimated intake of 11.8 μg/day	9	790 ng/liter
Ford *et al.* (1983)	UK	Dietary evaluation not performed	23	230 ng/liter
Prentice *et al.* (1983)	Gambia	No dietary intervention	16	160 ng/liter

TABLE II—*continued*

Reference	Country	Comment	Sample No.	Concentration
Pantothenic acid				
Department of Health and Social Security (1977)	UK	Pooled sample from 96 women from five cities		2.2 mg/liter
Ford *et al.* (1983)	UK	Dietary evaluation not performed	26	2.3 mg/liter
Prentice *et al.* (1983)	Gambia	No dietary intervention	21	2.0 mg/liter
Song *et al.* (1984)	U.S.A.	Positive effect of maternal supplementation suggested	43	2.5 mg/liter
Biotin				
Department of Health and Social Security (1977)	UK	Pooled samples from 96 women from five cities		7.6 μg/liter
Ford *et al.* (1983)	UK	Dietary evaluation not performed	26	5.3 μg/liter
Prentice *et al.* (1983)	Gambia	No dietary intervention	19	9.0 μg/liter
Mock *et al.* (1992a,b)	U.S.A.	Multiple samples collected from 7 and 3 women on self-selected diets		5.0 μg/liter

C. Riboflavin

Problems in methodology led early investigators to report values for human milk riboflavin activity that were approximately one-half actual values (Roughead and McCormick, 1990). A sizable fraction (38–62%) of human milk riboflavin activity is furnished by flavin adenine dinucleotide (FAD) which is underestimated using microbiological and fluorometric analyses unless FAD is converted to riboflavin before detection or values are corrected for internal fluorescence quenching, respectively. The application of appropriate methodology to human milk analyses indicates that typical values for riboflavin activity are 400–600 μg/liter.

D. Niacin

Average niacin content of human milk increases from 0.5 to 1.8–2.0 mg/liter based principally on microbiological analyses. Actual niacin values for human milk would be higher owing to the possible contribution from conversion of tryptophan. Modern chromatographic analyses have not been applied to human milk for determination of the relative distribution of niacin forms yielding vitamin activity.

E. Vitamin B_6

The average vitamin B_6 content of human milk is low in early milk and varies between 0.09 and 0.31 mg/liter in mature secretions. Human milk vitamin B_6 content responds to maternal intake over a wide range (< 2.5 to > 20 mg/day) during established lactation. High-pressure liquid chromatographic analysis indicates that vitamin B_6 in human milk exists in multiple forms: 81% as pyridoxal, 7% as pyridoxal-5-phosphate, 5% as pyridoxamine and as pyridoxal, and 2% as pyridoxamine-5′-phosphate. Supplementation of mothers with 2.5 or 15 mg of pyridoxine yields similar distribution patterns for the vitamers in milk (Hamaker *et al.*, 1985).

F. Folate

The folate content of human milk typically secreted by well-nourished women averages about 80–130 μg/liter. These values are substantially greater than those obtained by early and modern investigators owing to analytical problems. Folate in human milk is quantitatively bound to folate-binding proteins and present in multiple labile forms. Accurate analysis requires not only heat treatment to release folates from their binding proteins, but also use of an antioxidant as a preservative, enzymatic cleavage of polyglutamate forms, and application of analytical schemes capable of detecting all of the substituted ring species of the vitamins in samples (O'Connor *et al.*, 1991). Folate values typically increase with the progression of lactation even during established lactation. Folate levels in milk are maintained during the development of maternal folate depletion (Salmenperä *et al.*, 1986).

G. Vitamin B_{12}

Levels of vitamin B_{12} in human milk show wide variation which may reflect analytical difficulties rather than true biological variance. In well-nourished women, supplementation appears to be without an effect or to have a minimal effect on milk vitamin B_{12} content. Levels are reportedly low in samples from strict vegetarians by at least an order of magnitude below values presented in Table I and vitamin B_{12} deficiency in breast-fed infants is observed in some cases (Johnson and Roloff, 1982).

H. Pantothenic Acid

The pantothenic acid content of human milk averages approximately 2.0 to 2.5 mg/liter. A weak correlation between maternal intake and milk pantothenic acid content is observed ($r = 0.5$). Pantothenic acid analysis involves a two-stage assay: enzymatic cleavage of the bound vitamin from its carrier protein and microbiological or radiometric detection. Some commercial sources of enzyme can be contaminated with pantothenic acid and result in overestimations of amounts in human milk (Song *et al.*, 1984).

I. Biotin

Human milk is reported to contain between 5 and 9 μg/liter of biotin. Less than 5% of total human milk biotin is protein bound and concentrations in milk are 20 to 50 times greater than corresponding levels in maternal plasma (Mock *et al.*, 1992a,b).

V. Summary

Human milk contents of the water-soluble vitamins of well-nourished women and respective intakes of their exclusively breast-fed infants provide the primary knowledge base for estimates of infant water-soluble vitamin requirements and recommended levels of intakes and for the formulation of human milk substitutes. Our knowledge of contents and forms of the water-soluble vitamins secreted and factors capable of having an impact is far from complete. Many of the techniques used to assay the water-soluble vitamins in human milk are insensitive, nondiscriminatory, and inaccurate. Newer methods are available that can furnish accurate determination of the vitamins and their active metabolite and their application to human milk analyses are warranted. There is an amazing lack of data on milk levels secreted in advanced lactation (>3 months) even though human milk feeding is recommended for the entire first year of life. Investigations on mechanisms of water-soluble vitamin secretion are virtually nonexistent. These are areas where further research is not only warranted, it is necessary.

Acknowledgment

The preparation of the manuscript was supported in part by U.S.D.A. Grant.

References

Anderson, D. M., and Pittard, W. B. (1985). Vitamin E and C concentrations in human milk with maternal megadosing. A case report. *J. Am. Diet. Assoc.* **85,** 715–717.

Anderson, G. H. (1984). The effect of prematurity on milk composition and its physiological basis. *Fed. Proc.* **43,** 2438–2442.

Bamji, M. S., Prema, K., Jacob, C. M., Ramalakshmi, B. A., and Madhavapeddi, R. (1986). Relationship between maternal vitamins B_2 and B_6 status and the levels of these vitamins in milk at different stages of lactation. A study of a low-income group of Indian women. *Hum. Nutr. Clin. Nutr.* **40C,** 119–124.

Bates, C. J., Prentice, A. M., Prentice, A., Lamb, W. H., and Whitehead, R. G. (1983). The effect of vitamin C supplementation on lactating women in Keneba, a West African rural community. *Int. J. Vit. Nutr. Res.* **53,** 68–76.

Bates, C. J., Prentice, A. M., Prentice, A., Paul, A. A., and Whitehead, R. G. (1982). Seasonal variations in ascorbic acid status and breast milk ascorbic acid levels in rural Gambian women in relation to dietary intake. *Trans R. Soc. Trop. Med. Hyg.* **76,** 341–347.

Brown, C. M., Smith, A. M., and Picciano, M. F. (1986). Forms of human milk folacin and variation patterns. *J. Pediatr. Gastroenterol. Nutr.* **5,** 278–282.

Byerley, L. O., and Kirksey, A. (1985). Effects of different levels of vitamin C intake on the vitamin C concentration in human milk and the vitamin C intakes of breast-fed infants. *Am. J. Clin Nutr.* **41,** 665–671.

Department of Health Social Security (1977). The composition of mature human milk. *Rep. Health Soc. Subj.* **12.** HMSO, London

Fomon, S. J. (1993). "Nutrition of Normal Infants." Mosby, St. Louis, MO.

Ford, J. E., Zechalko, A., Murphy, J., and Brooke, O. G. (1983). Comparison of the B vitamin composition of milk from mothers of preterm and term babies. *Arch. Dis. Child.* **58,** 367–372.

Hamaker, B., Kirksey, A., Ekanayaki, A., and Borschel, M. (1985). Analysis of B-6 vitamins in human milk by reverse-phase liquid chromatography. *Am. J. Clin. Nutr.* **42,** 650–655.

Institute of Medicine (IOM)/National Academy of Sciences (NAS) (1991). "Nutrition During Lactation." National Academy Press, Washington, DC.

Johnson, P. R., Jr., and Roloff, J. S. (1982). Vitamin B_{12} deficiency in an infant strictly breast-fed by a mother with latent pernicious anemia. *J. Pediatr.* **100,** 917–919.

Karra, M. F., Udipi, S. A., Kirksey, A., and Roepke, J. L. B. (1986). Changes in specific nutrients in breast milk during extended lactation. *Am. J. Clin. Nutr.* **43,** 495–503.

Kirksey, A., and Rahmanifar, A. (1988). Vitamin and mineral composition of preterm human milk: Implications for the nutritional management of the preterm infant. *In* "Vitamins and Minerals in Pregnancy and Lactation" (H. Berger, ed.), pp. 301–329. Raven Press, New York.

Machlin, L. J. (1991). "Handbook of Vitamins," 2nd ed., revised and expanded. Dekker, New York.

Mock, D. M., Mock, N. I., and Dankle, J. A. (1992a). Secretory patterns of biotin in human milk. *J. Nutr.* **122,** 546–552.

Mock, D. M., Mock, N. I., and Langbehn, S. E. (1992b). Biotin in human milk: Methods, location and chemical form. *J. Nutr.* **122,** 535–545.

Nail, P. A., Thomas, M. R., and Eakin, R. (1980). The effect of thiamin and riboflavin supplementation on the level of those vitamins in human breast milk and urine. *Am. J. Clin. Nutr.* **33,** 198–204.

O'Connor, D. L., Tamura, T., and Picciano, M. F. (1991). Pteroylpolyflutamates in human milk. *Am. J. Clin. Nutr.* **53,** 930–934.

Prentice, A. M., Roberts, S. B., Prentice, A., Paul, A. A., Watkinson, M., Watkinson, A. A., and Whitehead, R. G. (1983). Dietary supplementation of lactating Gambian women. I. Effect on breast-milk volume and quality. *Hum. Nutr. Clin. Nutr.* **37C,** 53–64.

Ronnholm, K. A. R. (1986). Need for riboflavin supplementation in small prematures fed human milk. *Am. J. Clin. Nutr.* **43,** 1–6.

Roughead, Z. K., and McCormick, D. B. (1990). Flavin composition of human milk. *Am. J. Clin. Nutr.* **52,** 854–857.

Salmenperä, L. (1984). Vitamin C nutrition during prolonged lactation: Optimal in infants while marginal in some mothers. *Am. J. Clin. Nutr.* **40,** 1050–1056.

Salmenperä, L., Perheentupa, J., and Siimes, M. A. (1986). Folate nutrition is optimal in exclusively breast-fed infants but inadequate in some of their mothers and in formula-fed infants. *J. Pediatr. Gastroenterol. Nutr.* **5,** 283–289.

Samson, R. R., and McClelland, D. B. L. (1980). Vitamin B_{12} in human colostrum and milk. *Acta Paediatr. Scand.* **69,** 93–99.

Sandberg, D. P., Begley, J. A., and Hall, C. A. (1981). The content, binding, and forms of vitamin B_{12} in milk. *Am. J. Clin. Nutr.* **34,** 1717–1724.

Smith, A. M., Picciano, M. F., and Deering, R. H. (1983). Folate supplementation during lactation: Maternal folate status, human milk folate content and their relationship to infant folate status. *J. Pediatr. Gastroenterol. Nutr.* **2,** 622–628.

Sneed, S. M., Zane, C., and Thomas, M. R. (1981). The effects of ascorbic acid, vitamin B_6, vitamin B_{12}, and folic acid supplementation on the breast milk and maternal status of low socioeconomic lactating women. *Am. J. Clin. Nutr.* **34,** 1338–1346.

Song, W. O., Chan, G. M., Wyse, B. W., and Hansen, R. G. (1984). Effect of pantothenic acid status on the content of the vitamin in human milk. *Am. J. Clin. Nutr.* **40,** 317–324.

Styslinger, L., and Kirksey A. (1985). Effects of different levels of vitamin B_6 supplementation on vitamin B_6 concentrations in human milk and vitamin B_6 intakes of breast fed infants. *Am. J. Clin. Nutr.* **41,** 21–31.

Tamura, T., Yoshimura, Y., and Arakawa, T. (1980). Human milk folate and folate status in lactating mothers and their infants. *Am. J. Clin. Nutr.* **33,** 193–197.

Thomas, M. R., Kawamoto, J., Sneed, S. M., and Eakin R. (1979). The effects of vitamin C, vitamin B_6, and vitamin B_{12} supplementation on the breast milk and maternal status of well-nourished women. *Am. J. Clin. Nutr.* **32,** 1679–1685.

Thomas, M. R., Sneed, S. M., Wei, C., Nail, P. A., Wilson, M., and Sprinkle, E. E. (1980). The effects of vitamin C, vitamin B_6, vitamin B_{12}, folic acid, riboflavin, and thiamin on the breast milk and maternal status of well-nourished women at 6 months postpartum. *Am. J. Clin. Nutr.* **33,** 2151–2156.

Tsang, R. C., Lucas, A., Uauy, R., and Zlotkin, S. (1993). "Nutrition Needs of the Preterm Infant: Scientific Basis and Practical Guidelines." Caduceres Medical Publishers, Rawling, NY.
Udipi, S. A., Kirksey, A., West, K., and Giacola, G. (1985). Vitamin B_6, vitamin C and folacin levels in milk from mothers of term and preterm infants during the neonatal period. *Am. J. Clin. Nutr.* **42,** 522–530.
Vanderslice, J. T., Brownlee, S. G., Maire C. E., Reynolds, R. D., and Polansky M. (1983). Forms of vitamin B_6 in human milk. *Am. J. Clin. Nutr.* **37,** 867–871.
West, K. D., and Kirksey, A. (1976). Influence of vitamin B_6 intake on the content of the vitamin in human milk. *Am. J. Clin. Nutr.* **29,** 961–969.

B. Water-Soluble Vitamins in Bovine Milk

ROBERT G. JENSEN

I. Introduction

Milk and its products contain varying quantities of the B vitamins and ascorbic acid and are an excellent dietary source of some (Table I). Almost all of the analyses therein were done with microbiological assays, except for ascorbic acid and one for riboflavin, and most on pasteurized milk, but not on milk sold at retail outlets. Recently developed high-performance liquid chromatographic (HPLC) methods utilized in the analyses of some B vitamins and their various forms in human milk have apparently not been applied to bovine milk, again with the exception of riboflavin. See Chapter 8A for a discussion of the methods.

Table I also contains the latest RDAs for an adult male so that the reader can assess the contribution of pasteurized whole milk to the vitamin intake of the consumer. Low-fat and nonfat milks will contain about the same amounts of these vitamins as whole milk. The amounts of the vitamins in other dairy products are listed in USDA Handbook 8.1 (Posati and Orr, 1976). For information on the roles, etc., of these vitamins see Machlin (1991).

II. Forms and Stability

A. Thiamine (Vitamin B_1)

Thiamine occurs in the free form (50–70%), phosphorylated (18–45%), and protein bound (5–17%) (Crenin and Power, 1982; Renner *et al.*, 1989). Most of the thiamine in milk is produced by microorganisms in the rumen,

TABLE I
The Contents (μg/liter) of Water-Soluble Vitamins in Bovine Milks[a]

Vitamin	References							
	Posati and Orr (1976)	Cremin and Power (1982)	Scott *et al.* (1984)	Renner *et al.* (1989)	Fomon (1993)	Scott (1989)		Food and Nutrition Board RDAs (1989)[b]
Thiamine	350	450	460	370–460	388	400	400[c]	1,500
Riboflavin	1620	1,750	1,780	1,610–1,900	914[d]	1670	1830	1,700
Pyridoxine	420	500	610	400–600	554	600	650	2,000
Cobalamin	4	4	4	3–5	4	4	4	2
Niacin	900	900	710	710–930	1300	830	890	—
Niacin equivalents[e]	9870[c]	—	—	—	8500	—	—	19,000
Folic acid	50	55	60	50–60	60	57	53	200
Pantothenic acid	3140	3,500	3,600	3,130–3,600	3251	3400	3200	—
Biotin	—	35	20	20–36	47	20	23	—
Vitamin C total[f]	9000	—	—	—	30,000	—	—	—
Ascorbic acid	—	20,000	12,500	15,000	—	8000	8000	60,000

[a]Pasteurized whole milks unless otherwise noted.
[b]For males, 20–50 years.
[c]Skim milk.
[d]Contained (%); riboflavin, 60.5; flavin adenine nucleotide, 25.6; and the hydroxyethyl form, 11 (Roughead and McCormick, 1990).
[e]Includes preformed niacin and niacin derived from tryptophan; 50 mg = 1 mg niacin.
[f]Includes dehydroascorbic acid.

so the nutritional status of the cow has little influence. Nevertheless, seasonal (feed) effects have been observed (Scott, 1984). Pasteurization, either HTST (72°C, 15 sec) or UHT (138–150°C, 2–6 sec), decreases the thiamine content about 10% (Scott, 1989).

B. Riboflavin (Vitamin B_2)

Riboflavin was found to exist in several forms when the flavins in milk were analyzed by HPLC (Roughead and McCormick, 1990). They observed (see Table I) that pasteurized bulk milk contained 914 μg/liter of total flavins. The types of flavins were (%): riboflavin, 60.5; flavin adenine dinucleotide, 25.6; hyroxylethyl form, 11; and traces of three other derivatives. The hydroxyethyl derivative is a potential antivitamin, which illustrates the usefulness of the resolution attainable by HPLC analyses. Riboflavin is not affected by pasteurization but is photodegradable when exposed to sunlight in clear containers (Cremin and Power, 1982; Renner *et al.*, 1989). Destruction of riboflavin catalyzes the photochemical oxidation and loss of ascorbic acid. In a recent survey of milks in the United States for vitamin contents, it was found that the percentage of milks containing 81–120% of the stated label contents of riboflavin decreased as the fat content dropped (Tanner *et al.*, 1988). These contents were the percentages of the RDA contained in the product. It is unlikely that the water-soluble riboflavin is preferentially partitioned into cream during separation. A more plausible explanation is that antioxidants in milk fat, i.e., carotenoids and tocopherols, protect the vitamin from photodegradation.

C. Pyridoxine (Vitamin B_6)

The vitamin activity in raw bovine milk is partitioned into (%): pyriodoxal, 80; pyridoxamine, 20; and pyridoxamine phosphate, traces (Cremin and Power, 1982; Renner *et al.*, 1989). The vitamin is not affected by pasteurization treatments ($<$ 10% loss) but is photodegradable (Scott, 1989).

D. Cobalmin (Vitamin B_{12})

In bovine milk, the predominant form is hydroxycobalamin, with minor amounts of methyl- and adenosynlcobalamins (Cremin and Power, 1982). The contents are not reduced by pasteurization and refrigerated storage. In milk, the vitamin is associated with an R-binder glycoprotein. For absorption of the vitamin to occur in the small intestine, pancreatic digestion is required. The vitamin is heat stable, with less than 10% loss due to pasteurization (Scott, 1989).

E. Niacin

Most of the niacin activity in milk occurs as the niacinamide (Crenin and Power, 1982; Renner *et al.*, 1989). Tryptophan is a precursor of niacin; 60 mg of dietary tryptophan is equivalent to 1 mg of niacin in the body. This, plus the preformed niacin, are reported as niacin equivalents (see Table I). The vitamin is not affected by pasteurization, but is somewhat photolabile.

F. Folic acid

The major chemical form in bovine milk is 5-methyltetrahydrofolic acid (Cremin and Power, 1982). Folate is bound to a specific glycoprotein. About 40% occurs as the conjugated polyglutamate form (Renner *et al.*, 1989). The vitamin is affected little by pasteurization ($<10\%$) but is photolabile to some extent (Scott, 1989).

G. Pantothenic acid

The vitamin apparently occurs in the free form in milk. It resists pasteurization (Cremin and Power, 1982; Scott, 1989).

H. Biotin

The vitamin is apparently in the free form and is not affected by pasteurization (Cremin and Power, 1982; Scott, 1989).

I. Ascorbic acid (Vitamin C)

The contents in milk have been assayed with a redox method using an indophenol indicator. Better methods are available but have not been applied to milks. The vitamin is secreted as the L-ascorbate, but is rapidly oxidized to dehydroascorbate which remains biologically active (Cremin and Power, 1982). Ascorbate is heat labile ($<20\%$) (Scott, 1989). The oxidation of ascorbate is catalyzed by the photodegradation of riboflavin (See above). Milk is not a good source of the vitamin.

III. Summary

The influence of pasteurization on vitamin contents is known, but is not particularly relevant, since most milks are pasteurized. However, the effects of further handling, types of packaging, length of storage, and storage temperatures have not been systematically studied, particularly with modern analytical methods. Research on this area as well as most aspects of milk composition is and has been at a standstill for years. Additional analyses are needed.

Acknowledgments

The preparation of the manuscript was supported in part by an NIH contract and by federal funds made available through provision of the Hatch Act. Scientific Contribution, Storrs Agricultural Experiment Station, Storrs, Connecticut.

References

Cremin, F. M., and Power, P. (1982). Vitamins in bovine and human milks. *In* "Developments in Dairy Chemistry—3" (P. F. Fox, ed.), pp. 337–398. Elsevier, New York.

Fomon, S. D., and McCormick, D. B. (1993). B vitamins and choline. *In* "Nutrition of Normal Infants," pp. 366–394. Mosby, St. Louis, MO.

Food and Nutrition Board (1989). "Recommended Dietary Allowances (RDAs)." National Academy of Sciences, National Research Council, National Academy Press, Washington, DC.

Machlin, E. D. (1991). "Handbook of Vitamins." Dekker, New York.

Posati, L. P., and Orr, M. L. (1976). "Composition of Foods, Dairy and Egg Products, Raw–Processed–Prepared." Agriculture Handbook 8.1, ARS. USDA, Washington, D.C.

Renner, E., Schaafsma, G., and Scott, K. J. (1989). Micronutrients in milk. *In* "Micronutrients in Milk and Milk-Based Products" (E. Renner, ed.), pp. 1–70. Elsevier, New York.

Roughead, Z. K., and McCormick, D. B. (1990). Qualitative and quantitative assessment of flavins in cow's milk. *J. Nutr.* **120,** 382–388.

Scott, K. J., Bishop, D. R., Zechalko, A., Edwards-Webb, J. D., Jackson, P. A., and Scuffam, D. (1984). Nutrient content of liquid milk I. Vitamins A, D-3, C and of the B complex in pasteurized bulk liquid milk. *J. Dairy Res.* **51,** 37–50.

Scott, K. J. (1989). Micronutrients in milk products. *In* "Micronutrients in Milk and Milk-Based Products" (E. Renner, ed.), pp. 71–124. Elsevier, New York.

Tanner, J. T., Smith, J., Defibaugh, P., Augyal, G., Villalobos, M., Bueno, M. P., McGarrahan, E. T., Wehr, H. M., Muniz, J. F., Hollis, B. W., Koh, Y., Reich, P., and Simpson, K. L. (1988). Survey of vitamin content of fortified milk. *J. Assoc. Off. Anal. Chem.* **71,** 607–610.

C. Carotenoids, Retinoids, and Vitamin K in Human Milk

LOUISE M. CANFIELD
ANNA R. GIULIANO
ELLEN J. GRAVER

I. Introduction

We present here the current knowledge of the content of fat-soluble vitamins K and A and the carotenoids in human milk along with recommended procedures for sample collection, processing, and storage. Many important studies were conducted previously which are germane to current issues, particularly the early work on the relationship of the effects of maternal status on vitamin A content of human milk. This work has been reviewed previously (Lesher *et al.,* 1945; Kon and Mawson, 1950; Wallingford and Underwood, 1986) and is not included here. The work reviewed here is for the most part based on high-performance liquid chromatography (HPLC) methodology. Only studies which provided detailed methodological procedures were considered.

II. Retinoids

Several studies have shown an inverse correlation of risk of morbidity and mortality in children and vitamin A status, reemphasizing the importance of vitamin A in infant growth and development (De Sole *et al.,* 1987; Sommer *et al.,* 1986; Hussey and Klein, 1990; Rahmathalluh *et al.,* 1990). This is of particular concern in Third World countries where the majority of the vitamin A requirement is met by consumption of plant products (i.e., carotenoids) and the dietary supply of preformed vitamin A is limited. The efficacy of carotenoids in decreasing the risk of childhood mortality is therefore potentially of great significance. Although there are a number of early studies of the breast milk content of vitamin A and total carotenes using spectral techniques (Rodriguez and Irwin, 1972; Wallingford and Underwood, 1986), quantitative HPLC techniques (Patton *et al.,* 1990; Wollard, 1989; Giuliano *et al.,* 1992; Ross, 1986; Taylor, 1983; Chappell *et al.,* 1985) have only recently been applied. Some early data may have included carotenoids in "vitamin A" values and in most cases relied on bioassay or spectral assay of crude lipid extracts. However, despite the vast

HANDBOOK OF MILK COMPOSITION

improvement which HPLC technology has made in our ability to fractionate and quantitate retinoids, current estimates of 30 to 60 μg/dl (Woolard, 1989; Chappell *et al.*, 1985; Ollilainen *et al.*, 1989) agree well with earlier reported concentrations of retinol in mature human milk (Lesher *et al.*, 1945; Kon and Mawson, 1950; Rodriguez and Irwin, 1972; Wallingford and Underwood, 1986) (Table I).

In well-nourished mothers, free retinol is a minor component of total milk retinoids in mature milk; ≥95% of retinol is present as retinyl esters. This is the opposite ratio to that in plasma. Thus, the predominant chemical form of vitamin A ingested differs between the fetus and the newborn (Ross, 1986; Jensen, 1989; Wallingford and Underwood, 1985). This proportion does not appear to change diurnally or on loading with vitamin A; however, the retinol:retinyl ester ratio may rise to 30% in vitamin A-deficient mothers (Gebre-Medin *et al.*, 1976). The ratio of retinol to retinyl esters in human colostrum has not been reported.

At least 12 retinyl esters have been identified in mature human milk, with fatty acid chain lengths ranging from octanoate (C8) to stearate (C16) (Figure 1). The major esters appear to be retinyl palmitate and retinyl stearate in approximately equal amounts. However, although these two esters are the most *predominant,* together they contribute only about 60% of the retinoids in milk (Ross, 1986). For quantitation, therefore, differences in molecular weights and polarities of the various retinyl esters (and thus their solubility in organic solvents) must be considered. Human milk differs from bovine and caprine in the relative contribution of retinyl esters, reflecting species specificity.

Concentrations of retinol (Wallingford and Underwood, 1985) as well as retinyl esters (Rodriguez and Irwin, 1972; Chappell *et al.*, 1985; Jensen, 1989) decline rapidly over the first month after parturition in normal mothers and mothers delivering prematurely. In addition, this was confirmed in the 1971–1974 National Health and Nutrition Examination Survey (NHANES 1) as well as the Survey of Hispanic Americans in the southwest (Life Science Research Office, Federation of American Societies for Experimental Biology, 1985). In one study (Chappell *et al.*, 1985) retinyl esters reached a maximum of 200 μg/dl at Day 4 for mothers delivering at term and Day 6 for preterm samples. Concentrations declined to 62 and 108 μg/dl in term and preterm samples respectively by the end of the first month (Figure 2).

The relationship between body stores and concentrations of milk retinoids has not been studied and the mechanisms regulating retinoid storage, mobilization, and secretion from mammary cells are unknown. There does not appear to be a direct relationship between moderate increases in dietary intake of vitamin A in well-nourished mothers and milk retinol (Lesher *et al.*, 1945; Rodriguez and Irwin, 1972; Wallingford and Underwood, 1986). Data obtained in animal studies predict that liver stores are a better indicator of milk retinol concentrations than recent dietary intake (Davile *et al.*, 1985; Tomlinson *et al.*, 1974).

TABLE I
Mean Concentrations of Fat-Soluble Vitamins in Human Milk

Colostrum (μg/dl)	Mature (μg/dl)	Assay method	Reference
		Retinol	
	29.5[a]	UV/Vis	Hussein *et al.* (1987)
	18.8[a,b]	UV/Vis	Hussein *et al.* (1987)
200 (Day 5)	62.0	HPLC	Chappell *et al.* (1985)
	52.2[a]	HPLC	Ollilainen *et al.* (1989)
	31.2	HPLC	Woolard (1989)
	43.0	HPLC	Giuliano *et al.* (1992)
	57.6	TLC	Cumming and Briggs (1983)
	32.9[b]	UV/Vis	Butte and Calloway (1981)
	19.2[b]	UV/Vis	Thein (1979)
	40.3[a]	Fluorometric	Bates *et al.* (1985)
	112.0[a]	UV/Vis	Naismith (1973)
		Carotenoids[c]	
120	40.0	UV/Vis	Ostrea *et al.* (1986)
200 (Day 1)	23.0	HPLC	Chappell *et al.* (1985)
	65.2[a]	UV/Vis	Hussein *et al.* (1987)
	19.7	UV/Vis	Butte and Calloway (1981)
218	—	HPLC	Patton *et al.* (1990)
66[e]	—	HPLC	Patton *et al.* (1990)
	1.0[e]	HPLC	Giuliano *et al.* (1992)
		Vitamin K[d]	
0.52	0.92	HPLC	Fournier *et al.* (1987)
0.23	0.21	HPLC	Haroon *et al.* (1982)
0.18	0.12	HPLC	Von Kries *et al.* (1987)
0.34	0.28	HPLC	Canfield *et al.* (1991)

[a]Lipids saponified prior to extraction.
[b]Malnourished mothers.
[c]Total carotenoids unless otherwise indicated.
[d]Quantitated by HPLC.

Early data indicated that minimal maternal requirements are satisfied before vitamin A is furnished to milk. This would predict that marginally deficient mothers would have low milk retinol. Based on early estimates (Kon and Mawson, 1950) a lactating woman requires ≥1200 μg of retinol daily in order to furnish 750 μg for her own needs. In agreement with this hypothesis, there are reports (Gebre-Medin *et al.*, 1976; Butte and Calloway, 1981; Thein, 1979; Wallingford and Underwood, 1986) of lower than normal concentrations of retinol in milk from poorly nourished mothers

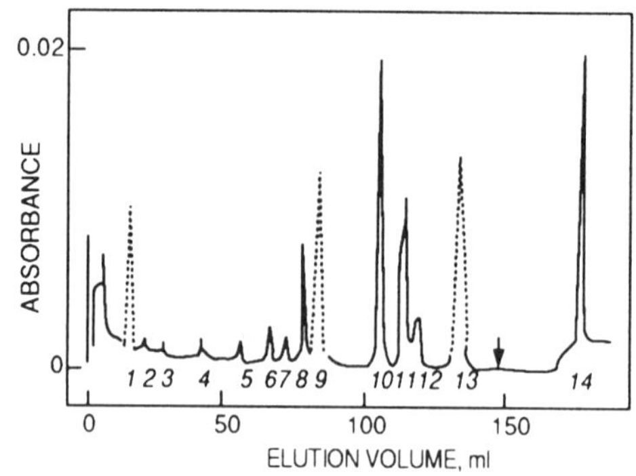

Figure 1 HPLC chromatogram of retinyl esters in human milk [reprinted from Ross (1986) with permission].

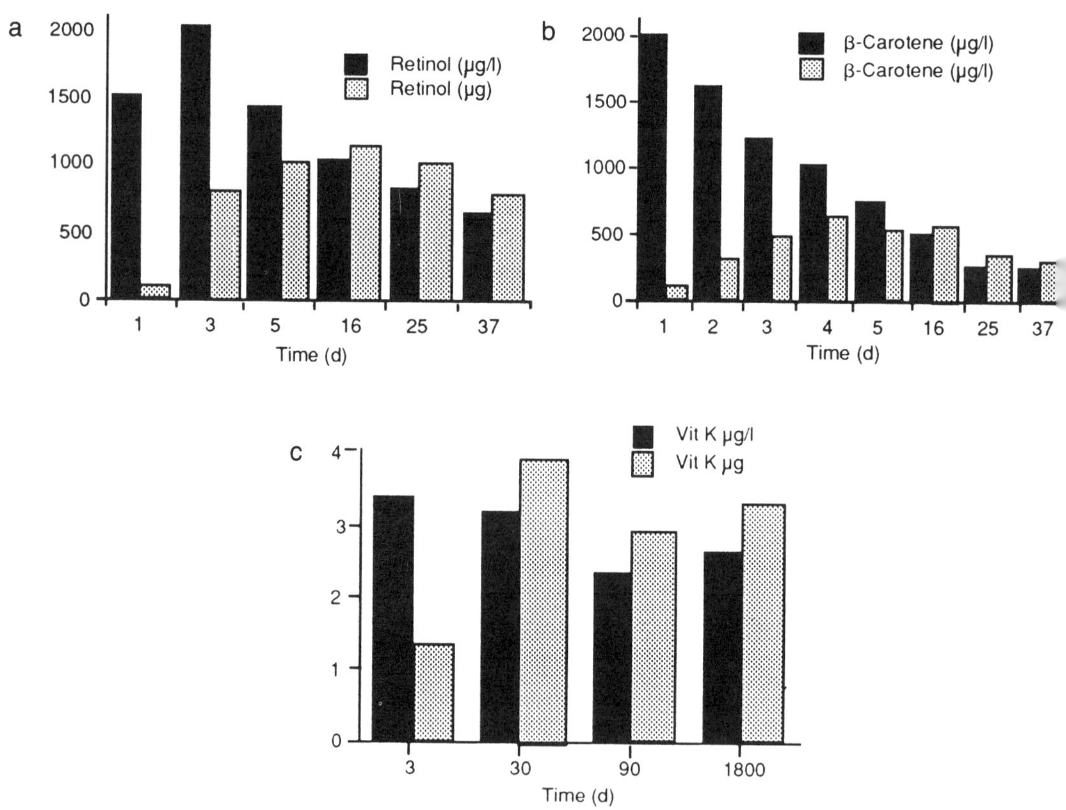

Figure 2 Effect of changing milk volumes over the lactation period on concentrations and total quantities of retinol (a), β-carotene (b), and vitamin K (c). Retinol and β-carotene were followed for 37 days and vitamin K was measured over the period of 1 to 6 months.

(see also Table I), although mean serum concentrations of retinol were in the normal range in these studies. In contrast, slightly higher retinol concentrations were reported in unsupplemented Gambian women compared to UK controls (Villard and Bates, 1987). Similarly, extremely high retinol concentrations of milk in Nigerian mothers were reported, although this may be the result of diets high in palm oil which is enriched in carotenoids and fatty acids (Naismith, 1973). Interestingly, in one study (Thein, 1979), lactating mothers had higher serum retinol than nonlactating controls, possibly indicating mobilization of retinol from stores to the mammary gland.

Unfortunately, at present, direct measurement is the only way to determine whether retinol is adequate in an individual mother's milk. Attempts to show a correlation between serum and milk retinol have produced mixed results. In one study (Butte and Calloway, 1981) a weak correlation between mature milk and serum retinol ($r = 0.317$) was observed, while others (Villard and Bates, 1987) saw no correlation over the lactation period. However, as mothers with serum vitamin A concentrations < 30 μg/dl more frequently had infants with respiratory and gastrointestinal infection (Hussein *et al.*, 1987), the mother's retinol status may be crucial to the well-being of the infant. According to WHO estimates (Underwood, 1984), the calculated *daily* intake of retinol in breast-fed infants in countries where serum retinol concentrations are commonly < 30 μg/dl ranges from 90 to 170 μg, significantly less than minimal requirements of 300 μg/day. Indeed, where concentrations of vitamin A in breast milk are ≤20 μg/dl, 1400 to 1600 ml of breast milk would be needed to meet the recommended intakes for vitamin A.

Following studies in the mid-1960s which established teratogenic effects of vitamin A (Kochhar, 1967), vitamin A supplementation of pregnant women has been discontinued. However, a number of studies were done prior to this time. When small daily doses of vitamin A were given during pregnancy, there was no effect on colostrum or mature milk vitamin A concentrations (Rodriguez and Irwin, 1972). However, vitamin A given just prior to parturition, either as one massive (240,000 to 600,000 IU) or smaller dose (30,000 IU) (Venkatachalem *et al.*, 1962), significantly increased the concentration of vitamin A in both colostrum and mature milk. In early studies (Lesher *et al.*, 1945; Rodriguez and Irwin, 1972; Wallingford and Underwood, 1986), vitamin A supplements given *after* parturition significantly increased the vitamin A concentration of milk if given in doses ≥15 mg/day. Modest daily doses of vitamin A (< 15 mg/day) did not affect breast milk concentration, regardless of the vitamin A content of the mother's diet (Wallingford and Underwood, 1985). In contrast, supplementation of retinol-low pregnant or lactating Gambian women with 650 μg/day vitamin A (Villard and Bates, 1987) significantly increased breast milk retinol levels. In a study with marginally vitamin A-deficient women in Indonesia, one massive dose (300,000 IU) given 2 weeks postpartum significantly increased the vitamin A concentration of breast milk (Stolzfus *et al.*, 1993) confirming the results from older studies. The increase in breast milk vitamin A was measurable through 8 months of lactation. The effect of dietary carotenoids on milk carotene and retinol concentrations has been variable (Wallingford and Underwood, 1985) and appears to be closely related to maternal diet as well as retinol status.

Mechanisms regulating the production of vitamin A in the human mammary gland are not understood. Despite the fundamental importance of these processes to child health, absorption and transport of vitamin A, or for that matter any of the fat-soluble vitamins, has received little attention. In preliminary studies, intestinal bioconversion of β-carotene to retinol appears higher in vitamin A-deficient compared to vitamin A-replete rats (Gronowska-Senger and Wolf, 1969). Early investigators (Lesher *et al.*, 1945; Rodriguez and Irwin, 1972) assumed that retinol transfer from blood to milk was accomplished by passive diffusion with associated lipid. More recently, in monkeys, it was shown that most (≥90%) of vitamin A in milk was derived from serum retinol-binding protein (RBP) (Vahlquist and Nilsson, 1979), and that the concentration of RBP in serum determines the amount of retinol delivered to milk. Additional study is needed to clarify these mechanisms.

III. Carotenoids

A. Colostrum

In the first study in which carotenoids were separated and quantitated in human colostrum, a mean concentration of total carotenoids of 218 μg/dl was reported with β-carotene accounting for about 30% (Patton *et al.*, 1990). The major carotenoids identified were lutein, cryptoxanthin, lycopene, and β-carotene (Figure 3). All studies to date have shown a sharp decrease in concentrations of carotenoids over the first month postpartum (Patton *et al.*, 1990; Chappell *et al.*, 1985; Jensen, 1989; Ostrea *et al.*, 1986). For example, Ostrea *et al.*, (1986) reported a 100% decrease in carotenoid concentrations from Days 1 to 5.

B. Mature Milk

As shown in Figure 3, within an individual, carotenoids in mature milk are qualitatively the same as those in colostrum. Concentrations of carotenoids in mature milk are shown in Table I. In early studies absorption at 452 nm was attributed to "carotene" and in some cases no distinction between total carotenoids and β-carotene was made. Later studies (Giuliano *et al.*, 1992) have shown that β-carotene accounts for only about one-tenth of total milk

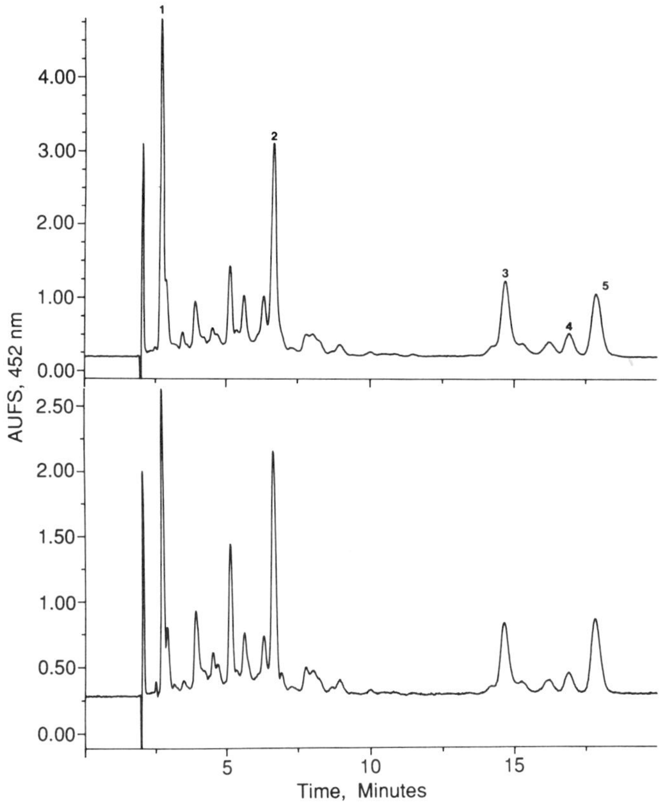

Figure 3 HPLC chromatograms of carotenoids in colostrum (top) and mature milk (bottom) of the same mother. Major carotenoids are (1) lutein, (2) cryptoxanthin, (3) lycopene, (4) α-carotene, (5) β-carotene.

carotenoids, and the total carotenoids recovered were lower than those reported earlier. These discrepancies do not appear to be related to differences in extraction methodology, but could reflect removal of compounds with competing absorbance at 452 nm or differences in the diets of the populations being sampled. In fact, β-carotene concentrations vary significantly both within and among individuals (Patton *et al.*, 1990; Jensen, 1989; Butte and Calloway, 1981; Hussein *et al.*, 1987; Ostrea *et al.*, 1986). The concentration of β-carotene may be determined by both the dietary intake and the serum β-carotene levels. The ratio of the major milk carotenoids, e.g., lycopene, β-cryptoxanthin, and β-carotene, to total milk carotenoids varies significantly among individuals. In early studies, the major carotenoids detected in milk were lycopene and lutein, carotenoids without provitamin A activity (Chandra *et al.*, 1951; Kon and Mawson, 1950). This has not been confirmed in well-nourished mothers (Giuliano *et al.*, 1992) and may have reflected bioconversion of available provitamin A carotenoids to vitamin A. To avoid collecting 24-hr breast milk samples, while accurately estimating the mean value, statistical methods for predicting 24-hr concentrations of milk lipids using two daily samples are recommended (Giuliano *et al.,* 1994; Jackson *et al.,* 1988).

The relationship between vitamin A status and carotenoid concentrations is not clear. Mothers with low or marginal serum retinol concentrations might be expected to have correspondingly low carotenoids; however, this has not been documented. Milk carotenoids in vitamin A-low Ethiopian mothers (Gebre-Medin *et al.,* 1976) were 23–36 μg/dl compared to 16–20 μg/dl in well-nourished Swedish mothers. Low carotenoid content of the Swedish diet cannot be ruled out as a factor; however, an average of 65 μg/dl was reported in mature milk of Egyptian mothers with marginal vitamin A status (Hussein *et al.,* 1987). It is of interest (Prentice *et al.,* 1986) that none of the water-soluble vitamins in the milk of Gambian women was lower than that of women in the UK. Apparently, the mothers' stores are mobilized to provide adequate production of vitamins in milk. Thus, the effect of β-carotene supplementation on milk β-carotene and concentrations deserves further investigation.

IV. Vitamin K

Due to increased risk of the solely breast-fed infant for hemorrhagic disease of the newborn (HDN), relatively more recent research has been done on vitamin K compared to the other fat-soluble vitamins in human milk. Recent reports are summarized in Table I and an HPLC chromatogram of vitamin K in human milk is shown in Figure 4. A detailed review is available (Canfield and Hopkinson, 1989). Although reported quantities differ, there is consensus that due to the risk of HDN in the newborn, breast milk should not be the sole source of vitamin K for newborns (Canfield and Hopkinson, 1989; Greer, 1992).

The amount of vitamin K in human milk is near the detection limit of HPLC methodology and therefore greater than usual precision in the method is required. In order to separate the vitamin from triglyceride in milk (approximately 4 g/liter), sample cleanup is required prior to HPLC. Due to the sensitivity required, detection by electrochemistry or fluorimetry is preferred over UV. As seen in Figure 4, UV detection cannot provide baseline resolution of vitamin K_1 due to the large number of coeluting impurities. In addition, when developing new methodology, verification of the vitamin by another technique, e.g., mass spectroscopy, is recommended. Due to the trace quantities present, small errors in recovery calculations can significantly prejudice the results. Statistical procedures to control for such variance have been presented (Canfield *et al.,* 1990).

As vitamin K is localized in the milk fat globule (Canfield *et al.,* 1990), sampling techniques significantly affect the amount of vitamin K which is

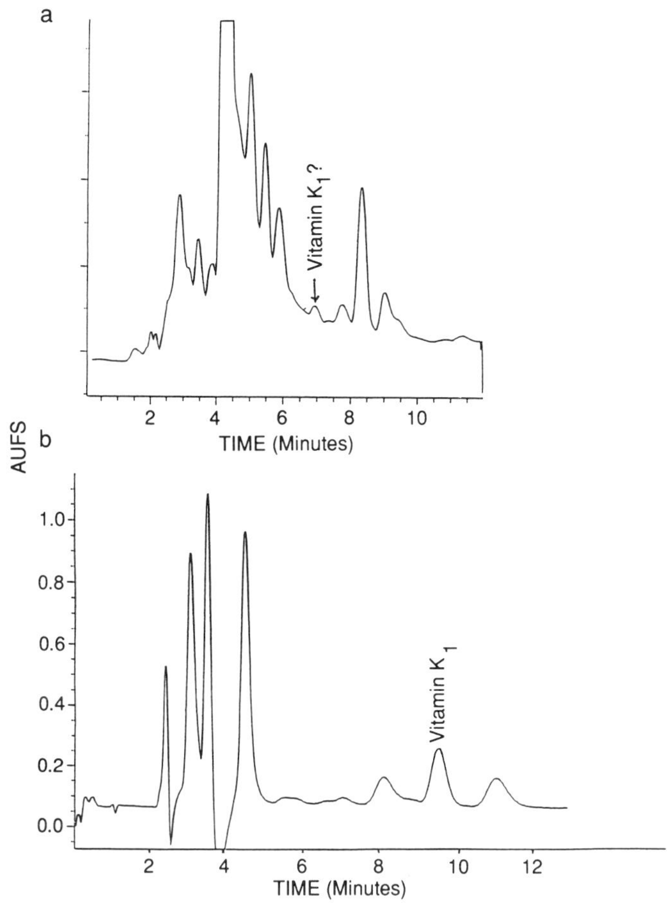

Figure 4 HPLC chromatogram of vitamin K in human milk detected using UV (a) and electrochemical (b) detection. The unidentified compounds did not coelute with standards for common menaquinones.

collected. Thus, in hindmilk (Fournier *et al.,* 1987), approximately twofold higher vitamin K concentrations were reported than those where full breast collections were made (Canfield *et al.,* 1991; Von Kries *et al.,* 1987; Haroon *et al.,* 1982). In addition, common bacteria produce vitamin K as menaquinones in significant quantity. As these vitamins, particularly MK_4, MK_5, MK_6, and MK_7, exhibit chromatographic properties similar to phylloquinone in commonly used systems, precautions must be taken to avoid bacterial contamination (Canfield *et al.*, 1990). It is generally agreed that vitamin K_1 (phylloquinone) is the major endogenous form of vitamin K in human milk.

In contrast to the 5- to 10-fold decrease in retinoid and carotenoid concentrations over the lactation period, vitamin K concentrations are only slightly higher, if at all, in colostrum than in mature milk. Three laboratories (Fournier *et al.*, 1987, Canfield *et al.*, 1991; Von Kries *et al.*, 1987) reported mean concentrations of 1.5- to 2-fold higher in colostrum compared to mature milk and one group (Haroon *et al.*, 1982) saw no significant difference over the lactation period. Thus, as milk volumes increase significantly over the first 30 days postpartum, so does the total amount of vitamin K available to the infant (Figure 2).

Recommended intakes of vitamin K for the infant are 5 μg/day for the first 6 months and 10 μg/day for the second 6 months (Lane and Hathaway, 1985). Assuming an intake of about 750 ml of milk per day (Subcommittee on Nutrition during Lactation, 1991) and using the data in Table I, daily vitamin K intake by the infant could range from a maximum of 6.9 μg/day to a minimum of 0.9 μg/day. Thus, vitamin K in human milk may not be sufficient to meet the needs of all newborns. In compliance with recommendations of the Committee on Nutrition of the American Academy of Pediatrics, vitamin K is administered parenterally to all newborn infants in this country as well as in most industrialized countries in the world (National Research Council, 1989; Lane and Hathaway, 1985). This prophylaxis appears to protect most newborns against HDN for 6 months; however, little is known about vitamin K absorption and utilization in infants.

Although a rapid and substantial increase in vitamin K levels in milk occurs in response to supplementation (Greer, 1992; Canfield *et al.*, 1991; Von Kries *et al.*, 1987), the response is extremely variable and the effects of diet on vitamin K levels in milk have not been systematically studied. It is therefore recommended that infants, particularly exclusively breast-fed infants, receive vitamin K supplements under the care of a physician.

V. Fat-Soluble Vitamins—Methodological Considerations

A. Relationship of Milk Volume to Concentration

Concentrations of fat-soluble vitamins, particularly vitamin A and carotene, in human milk decrease significantly during the first week of lactation (Figure 2). This observation has been interpreted to mean that fat-soluble vitamins are stored in the mammary gland prior to parturition and released in response to endocrine changes associated with parturition and lactogenesis (Chappell and Clandinin, 1984). However, milk volumes *increase* substantially (10- to 20-fold) over the first month postpartum (Subcommittee on Nutrition during Lactation, 1991); thus, consideration of concentrations alone can be misleading. The total *amount* (μg) of retinol and carotene available to the infant peaks during late colostrum and decreases by the first month to about one-half the maximum value, while the total amount of vitamin K increase over the same period (Figure 2). For these calculations, published milk volume data were used (Jensen, 1989; Subcommittee on Nutrition during Lactation, 1991); however, volumes differ significantly from mother to mother. Also, it should be noted that the volume consumed by the infant, typically 600 to 700 ml/day at 1 month (Jensen, 1989; Subcommittee on Nutrition during Lactation, 1991), is considerably less than the amount produced (1000–1250 ml/day).

B. Technical Variability

Reported concentrations of fat-soluble vitamins in milk reflect not only the true concentrations but also sampling procedures, handling and storage techniques, and analytical methodology. For example, alkaline hydrolysis is necessary to completely recover both carotenoids and retinoids from the milk matrix (Giuliano *et al.*, 1992; Taylor, 1983). Photolysis due to light exposure as well as losses due to adherence to plastic storage bottles, syringes, and tubing can be significant (Ross, 1986; Bates *et al.*, 1985). Bacterial contamination can lead to significant overestimation of vitamin K concentrations (Canfield *et al.*, 1990). Phthalates and other chemicals in plasticizers commonly interfere with uv analysis of fat-soluble vitamins and losses due to recovery can be substantial. Finally, large intra- and interindividual variations in fat-soluble vitamin concentrations in milk may indicate multiple sampling protocols (Giuliano *et al.*, 1994; Jackson *et al.*, 1988). Thus, methodological error should be carefully considered when quantitating fat-soluble vitamins in milk.

References

Bates, C. J., Liu, D. S., Fuller, N. J., and Lucas, A. (1985). Susceptibility of riboflavin and vitamin A in breast milk to photodegradation and its implications for the use of banked breast milk in infant feeding. *Acta Pediatr. Scand.* **74,** 40–44.

Butte, N. F., and Calloway, D. H. (1981). Evaluation of lactational performance of Navajo women. *Am. J. Clin. Nutr.* **34,** 2210–2215.

Canfield, L. M., and Hopkinson, J. M. (1989). State of the art: Vitamin K in human milk. *J. Pediatr. Gastroenterol. Nutr.* **8,** 430–441.

Canfield, L. M., Hopkinson, J. M., Lima, A. F., Martin, G. S., Sugimoto, K., Burr, J., Clark, L., and McGee, D. L. (1990). Quantitation of vitamin K in human milk. *Lipids* **25,** 406–411.

Canfield, L. M., Hopkinson, J. M., Lima, A. F., Silva, B., and Garza, C. (1991). Vitamin K in colostrum and mature milk over the lactation period—A cross-sectional study. *Am. J. Clin. Nutr.* **53,** 730–735.

Chappell, J. E., and Clandinin, M. J. (1984). Physiology of lactation: An integrative perspective *In* "Malnutrition: Determinants and Consequences" (Philip L. White and Nancy Selve, eds.), pp. 85–100. A. R. Liss, New York.

Chappell, J. E., Francis, T., and Clandinin, M. T. (1985). Vitamin A and E content of human milk at early stages of lactation. *Early Hum. Dev.* **11,** 157–167.

Cumming, F. J., and Briggs, M. H. (1983). Changes in plasma vitamin A in lactating and nonlactating oral contraceptive users. *Br. J. Obstet. Gynecol.* **90,** 73–77.

Davila, M. E., Norris, L., Cleary, M. P., and Ross, A. C. (1985). Vitamin A during lactation: Relationship of maternal diet to milk vitamin A content and to the vitamin A status of lactating rats and their pups. *J. Nutr.* **115,** 1033–1041.

De Sole, G., Belay, Y., and Zegeye B. (1987). Vitamin A deficiency in southern Ethiopia. *Am. J. Clin. Nutr.* **45,** 780–784.

Fournier, B., Sann, L., Guillaumont, M., and Leclercq, M. (1987). Variations of phylloquinone concentration in human milk at various stages of lactation and in cow's milk at various seasons. *Am. J. Clin. Nutr.* **45,** 551–558.

Gebre-Mehdin, M., Vahlquist, A, Hofvander, Y., Uppsall, L., and Vahlquist, B. (1976). Breast milk composition in Ethiopian and Swedish mothers. I. Vitamin A and β-carotene. *Am. J. Clin. Nutr.* **29,** 441–451.

Giuliano, A. R., Nielsen, E. M., and Canfield, L. M. (1992). Simultaneous quantitation and separation of carotenoids and retinol in human milk by high performance liquid chromatography (HPLC). *Methods Enzymol.* **213,** 391–399.

Giuliano, A. R., Matzener, M. B., and Canfield, L. M. (1993). Assessing variability in quantitation of carotenoids in human plasma: Variance component model. *Methods Enzymol.* **214,** 94–101.

Giuliano, A. R., Neilson, E. M., Matzner, M. B., and Canfield, L. M. (1994). Carotenoids of mature human milk: Inter/intraindividual variability. *J. Nutr. Biochem.,* in press.

Greer, F. (1992). Factors influencing vitamin K status of the newborn. *In* "Mechanisms Regulating Lactation and Infant Nutrient Utilization" (M. F. Picciano and B. Lonnerdal, eds.), pp. 259–271. Wiley–Liss, New York.

Gronowska-Senger, A., and Wolf, G. (1969). Effect of dietary protein on the enzyme from rat and human intestine which converts β-carotene to retinal. *J. Nutr.* **100,** 300–308.

Haroon, Y., Shearer, M. J., Rahim, S., Gunn, W. G., McEnery, G., and Barkhan, P. (1982). The content of phylloquinone (vitamin K_1) in human milk, cow's milk and infant formula foods determined by high-performance liquid chromatography. *J. Nutr.* **112,** 105–117.

Hussein, L., Drar, A., Horeyah, Al., and El Naggar, B. (1987). Lipid and retinol contents in the milk of Egyptian mothers with normal and sick infants. *Int. J. Vitamin Nutr. Res.* **57,** 3–11.

Hussey, G. D., and Klein, M. B. (1990). A randomized, controlled trial of vitamin A in children with severe measles. *N. Engl. J. Med.* **323,** 160–164.

Jackson. D. A., Imong, S. M., Silprasert, A., Preunglumpoo, S., Leelapat, P., Yootabootyr, Y., and Amatayakul, K. (1988). Estimation of 24 h breast milk fat and fat intake in rural northern Thailand. *Br. J. Nutr.* **59,** 365–371.

Jensen, R. G. (1989). Lipids in human milk—Composition and fat-soluble vitamins. *In* "Textbook of Gastroenterology and Nutrition in Infancy" (E. Lebenthal, ed.), 2nd ed., pp. 157–208. Raven Press, New York.

Kochhar, D. M. (1967). Teratogenic activity of retinoic acid. *Acta Pathol. Microbiol. Scand.* **70,** 398–404.

Kon, S. K., and Mawson, E. H. (1950). "Human Milk: Wartime Studies of Certain Vitamins and Other Constituents." Medical Research Council Special Report Series 269, H.M. Stationery Office, London.

Lane, P. A., and Hathaway W. E. (1985). Vitamin K in infancy. *J. Pediatr.* **106,** 351–359.

Lesher, M., Brody, J. K., Williams, H. K., and Macy, I. C. (1945). Human milk studies. Vitamin A and carotenoid contents of colostrum and mature human milk. *Am. J. Dis. Child.* **70,** 182–192.

Life Science Research Office, Federation American Societies for Experimental Biology (LSRO–FASEB) (1985). "Assessment of the Vitamin A Status of the U.S. Population Based on Data Collected in the Health and Nutrition Examination Surveys" (S. J. Pilch, ed.). LSRO–FASEB, Bethesda, MD.

Naismith, D. J. (1973). Kwashiorkor in western Nigeria: A study of traditional weaning foods with particular reference to energy and linoleic acid. *Br. J. Nutr.* **30,** 567–576.

Ollilainen, V., Heinonen, M., Linkola, E., Varo, P., and Koivistoinen, P. (1989). Carotenoids and retinoids in Finnish foods: Dairy products and eggs. *Dairy Sci.* **72,** 2257–2265.

Ostrea, E. M., Balun, J. E., Winkler, R., and Porter, T. (1986). Influence of breast-feeding on the restoration of the low serum concentration of vitamin E and β-carotene in the newborn infant. *Am. J. Obstet. Gynecol.* **154,** 1014–1017.

Patton, S., Canfield, L. M., Huston, G. E., Ferris, A. M., and Jensen, R. G. (1990). Carotenoids of human colostrum. *Lipids* **25,** 159–165.

Prentice, A., Paul, A., Prentice, A., *et al.,* (1986). Cross-cultural differences in lactational performance *In* "Human Lactation 2: Maternal and Environmental Factors" (M. Hamosh and A. S. Goldman, eds.), pp. 13–44. Plenum Press, New York.

Rahmathullah, L., Underwood, B. A., Thulasiraj, R. D., Milton, R. C., Ramaswamy, K., Rahmathullah, R, and Babu, G. (1990). Reduced mortality among children in southern India receiving a small weekly dose of vitamin A. *N. Engl. J. Med.* **323,** 929–935.

Rodriguez, M. S., and Irwin, M. I. (1972). A conspectus of research on vitamin A requirements in man. *J. Nutr.* **102,** 909–968.

Ross, A. C. (1986). Separation and quantitation of retinyl esters and retinol by high-performance liquid chromatography. *Methods Enzymol.* **123,** 68–74.

Sommer, A., Tarwotjo, I., Djunaedi, E., *et al.* (1986). Impact of vitamin A supplementation on childhood mortality: A randomized controlled community trial. *Lancet* **1,** 1169–1173.

Stolzfus, R. J, Hakimi, M., Miller, K. W., Rasmussen, K. M., Dawiesah, S., Habicht, J. P., and Dibley, M. J. (1993). High-dose vitamin A supplementation to breastfeeding Indonesian mothers: Effects on the vitamin A status of mother and infant. *J. Nutr.* **123,** 666–675.

Subcommittee on Nutrition during Lactation, Committee on Nutritional Status During Pregnancy & Lactation, Food & Nutrition Board, Institute of Medicine, National Academy of Science (1991). "Nutrition during Lactation," p. 82. National Academy Press, Washington, DC.

Taylor, R. F. (1983). Chromatography of carotenoids and retinoids. *Adv. Chromatogr.* **22,** 157–213.

Thein, M. (1979). Study on milk vitamin A, serum vitamin A and serum protein levels of lactating mothers of Bochess Village, rural Ethiopia. *East African Med. J.* **56,** 542–547.

Tomlinson, J. E., Mitchell, G. E., Jr., Bradley, N. W., Tucker, R. E., Boling, J. A., and Schelling, G. T. (1974). Transfer of vitamin A from bovine liver to milk. *J. Anim. Sci.* **39,** 813–817.

Underwood, B. A. (1984). Vitamin A in animal and human nutrition. *In* "The Retinoids" (M. B. Spron, A. B. Roberts, and D. S. Goodman, eds.), Vol. 1, pp. 263–274. Academic Press, New York.

Vahlquist, A., and Nilsson, S. (1979). Mechanisms for vitamin A transfer from blood to milk in Rhesus monkeys. *J. Nutr.* **109,** 1456–1463.

Venkatachalem, P. S., Belavady, B., and Gopalan, C. (1962). Studies on vitamin A nutrition status of mothers and infants in poor communities of India. *Ind. J. Pediatr.* **61,** 262–268.

Villard, I., and Bates, C. J. (1987). Effect of vitamin A supplementation on plasma and breast milk vitamin A levels in poorly nourished Gambian women. *Hum. Nutr. Clin. Nutr.* **41C,** 47–58.

Von Kries, R., Shearer, M., McCarthy, P. T., Haug, M., Harzer, G., and Gobel, U. (1987). Vitamin K_1 content of maternal milk; Influence of the state of lactation, lipid composition and vitamin K supplements given to the mother. *Pediatr. Res.* **22,** 513–517.

Wallingford, J. C., and Underwood, B. A. (1986). Vitamin A deficiency in pregnancy, lactation and the nursing child. *In* "Vitamin A Deficiency and Its Control" (J. C. Bauernfeind, ed.), pp. 101–153. Academic Press, Orlando, FL.

Woolard, D. (1989). The distribution of retinyl esters in milks and milk products. *J. Micronutr. Anal.* **5,** 35–52.

D. Vitamins D and E in Human Milk

CAROL J. LAMMI-KEEFE

I. Introduction

Presented here is our knowledge of the human milk content of vitamins D and E. The current evidence points to an inadequacy of vitamin D in human milk for the breast-fed infant, while vitamin E appears to generally be adequate for the full-term infant but not for the low birth weight premature infant. The issue of supplementing infants with vitamin D is unresolved and remains in question.

II. Vitamin D

A. Introduction

Vitamin D is a group of related fat-soluble compounds with antirachitic activity. Ergocalciferol (D_2) and cholecalciferol (D_3) are the most important members of this vitamin group. The most common source of vitamin D is the plant steroid, ergosterol. Vitamin D is synthesized in the skin upon photoradiation. Active metabolites in the plasma include 25-OH cholecalciferol (25-OH-D) and 1,25-$(OH)_2$ cholecalciferol [1,25-$(OH)_2$-D] (Miller and Norman, 1984).

B. Quantities in Milk

Reported levels of vitamin D in human milk (4 to 40 IU/liter; 0.1 to 1.0 μg/liter cholecalciferol) (Harris and Bunker, 1939; Polskin *et al.*, 1945; Macy and Kelly, 1961; Hollis *et al.*, 1981; Ala-Houhala *et al.*, 1988; Jensen *et al.*, 1992) (Table I) are below the minimum amount required to prevent rickets and ensure proper bone mineralization (100 IU; 2.5 μg cholecalciferol) and much lower than the RDA for infants from 0 to 0.5 years of 7.5 μg (300 IU) or from greater than 0.5 years of 10 μg (400 IU) (National Research Council, 1989) which is based on public health considerations (National Research Council, 1989; Fomon, 1986). The 400 IU is not toxic and prevents vitamin D deficiency. Clearly, if the infant requirement for vitamin D is 300–400 IU, based on these reports (Harris and Bunker,

HANDBOOK OF MILK COMPOSITION

TABLE I
Vitamins D and E in Human Milk

	Reference
Vitamin D (μg/liter)	
0.11[a]	Harris and Bunker (1939)
Traces - 1.07[a]	Polskin *et al.* (1945)
0.1[a,b]	Macy and Kelly (1961)
0.62[a,c]	Hollis *et al.* (1981)
0.31 (25-OHD)	
0.35[a,d] (Winter, foremilk)	Ala-Houhala *et al.* (1988)
3.1[a,d] (Summer, foremilk)	
0.33[e] (Preterm)	Atkinson *et al.* (1987)
0.36[e] (25-OHD) (preterm)	
0.33[e] (Term)	
0.25[e] (25-OHD) (term)	
Vitamin E (mg/dl)	
0.779 (0.310–0.340)[f] (2–7 days)	Kobayashi *et al.* (1975)
0.180 (0.083–0.310)[f] (30–39 days)	
1.0 ± 0.550[g] (1–4 days)	Jansson *et al.* (1981)
0.320 ± 0.180[g] (12 days–5 months)	
0.67[h] (2 weeks)	Lammi-Keefe *et al.* (1985)
0.40[h] (6 weeks)	
0.37[h] (12 weeks)	
0.37[h] (16 weeks)	
1.14[i] (3 days, term)	Haug *et al.* (1987)
0.28[i] (36 days, term)	
1.45[i] (3 days, preterm)	
0.29[i] (36 days, preterm)	
0.31[h] (4+ weeks)	Moffatt *et al.* (1987)
0.34[h] (4+ weeks)	Collins *et al.* (1989)
0.80[h] (10–30 days, St. Lucia)	Boersma *et al.* (1991)
0.50[h] (10–30 days, Dominica and Belize)	
2.2[h] (0–4 days, St. Lucia)	

[a]Antirachitic activity.
[b]Average value from summary compilation by the Food and Nutrition Board of the National Research Council (U.S.A.) (1953) (Table V).
[c]All metabolites.
[d]Calculated from vitamin D + 25-OHD.
[e]Recalculated, as described in text.
[f]α-, β-, γ-tocopherol.
[g]α-, β-, γ-, δ-tocopherol.
[h]α-Tocopherol.
[i]α-TE (α- and β- + γ-tocopherol).

1939; Polskin *et al.*, 1945; Macy and Kelly, 1961; Hollis *et al.*, 1981; Ala-Houhala, 1985; Jensen *et al.*, 1992), human milk is not ideally suited for providing the recommended amount of this vitamin to the breast-feeding infant. However, the majority of breast-feeding infants do not develop vitamin D-deficiency rickets. Two possibilities have surfaced to explain this phenomenon.

First, sunlight exposure of the light-skinned infant may protect against deficiency. While it is difficult to estimate "adequate" exposure, less than 1 hr per week of exposure of the extremities of infants may suffice (Fomon, 1986). Dark-skinned infants would not receive the same benefit from sunlight exposure. Indeed, Hayward *et al.* (1987) reported nutritional rickets in a black-skinned infant in San Diego which is known for its sunny climate. A combination of factors was suspected to have contributed to the deficiency symptoms, including the dark skin pigmentation, avoidance of sunlight, and unsupplemented breast-feeding. Those authors recommended unconditional vitamin D supplementation for all breast-fed infants.

Another possibility, that of a water-soluble vitamin D sulfate compound that was not being detected in the analyses of the lipid component, was explored between the mid-1960s and 1980, but was finally shown not to be a major source of vitamin D activity by Hollis *et al.* (1981). Those reports have been reviewed and summarized by Lammi-Keefe and Jensen (1984).

Until recently, with the publication of a method that currently allows quantitation of vitamin D and its metabolites (Hollis *et al.*, 1983; Hollis and Frank, 1986), data on the antirachitic activity of milk were determined with the rat line test. That test was a bioassay in which the deposition of new bone was measured. Hollis and colleagues (Hollis *et al.*, 1983; Hollis and Frank, 1986), using a liquid binding assay coupled with high-performance liquid chromatography (HPLC), have quantitated vitamins D_2 and D_3, 25-hydroxyvitamin D_2, and 25-hydroxyvitamin D_3 in human milk. These techniques have provided information regarding the levels of each of the antirachitic sterols in human milk and they have provided a means for examining the factors that can influence the levels of these compounds in milk.

The report from Atkinson *et al.* (1987) is a contemporary study of the vitamin D content of human milk. These investigators studied the milk from women delivering prematurely versus at term. Their findings placed the calculated vitamin D activity of human milk at the higher end of the reported range: 80 IU/liter for preterm milk and 60 IU/liter for term milk, which was not significantly different from the preterm milk. These values are considerably higher than many of the other reported values (Harris and Bunker, 1939; Polskin *et al.*, 1945, Macy and Kelly, 1961; Hollis *et al.*, 1981; Ala-Houhala, 1985; Jensen *et al.*, 1992), but still well below the estimated daily requirement of 300–400 IU for the infant. The difference in these reported values of Atkinson *et al.* (1987), compared to earlier reports, was explained on a technical basis. That is, the majority of vitamin D activity in milk is the hydroxylated metabolites of D_2 and D_3, which may not have been accurately estimated in milk in previous studies (Atkinson *et al.*, 1987). Greater than 90% of the antirachitic activity is D (D_2 and D_3) and 25-hydroxyvitamin D (25-OHD_3) (Reeve *et al.*, 1982). Currently, using the method published by Hollis (1983), it is possible to quantitate vitamin D_2 and D_3 and the hydroxylated metabolites in human milk. Employing that procedure, Atkinson *et al.* (1987) presented data for the D and hydroxylated forms of D and then calculated a total vitamin D activity, where for the hydroxylated forms, 1 μg = 200 IU, and for vitamin D, 1 μg = 40 IU. However, the activity of 25-OHD is approximately 1.5 times that of the vitamin D, or 1 μg 25-OHD is approximately 60 IU (National Research Council, 1989). A recalculation of their data gives values closer to 34 and 28 IU/liter, respectively, for preterm and term milks, values which more closely approximate some of the earlier values and which are less than 50% of the authors' values (Atkinson *et al.*, 1987) calculated D activities, which were 80 and 60 IU/liter, respectively. In μg/liter, recalculated levels of D and 25-OHD were approximately 0.33 and 0.36 for preterm and 0.33 and 0.25 for full-term human milk (Table I).

C. *Factors Affecting Milk Levels*

Evidence that season and supplementation can affect milk vitamin D levels is provided by reports deriving from studies conducted in Finland (Ala-Houhala, 1985; Ala-Houhala *et al.*, 1988). Particularly during periods of decreased sunlight exposure, maternal dietary intake may be a significant factor increasing 25-OHD in mothers during lactation. Foremilks ranged from 0.35 to 3.1 μg/dl, dependent on season (Table I). Based on that observation, maternal sunlight exposure can increase milk D to levels much higher than those generally reported; however, that elevated concentration would still fall short of the amount required by the infant. Supplementation of women with 25 μg/day (1000 IU) vitamin D slightly increased hindmilk levels of 25-OHD, the major D metabolite in milk; D levels were not affected. Twice that amount, 2000 IU/day, had a more pronounced effect on milk levels of 25-OHD, increasing that metabolite from 0.157 to 0.40 μg/liter (9.4 to 24 IU). Supplementation effects of vitamin D were more pronounced in hind- than foremilk, raising the issue of sampling protocol for studies in which milk vitamin D will be investigated. Also notable was the wide variation among individuals in response to supplementation. Supplementation of women in winter with 50 μg vitamin D should, in theory, have increased the winter milk levels to amounts comparable to those measured in unsupplemented women in September. This was not observed due to variable responses. Others (Kunz *et al.*, 1984) reported that lower levels of supplementation (400 IU/day) also increased 25-OHD.

The results of the studies conducted by Hollis *et al.* (1982, 1983) contrast with those described for the Finnish study. In these studies (Hollis *et al.*, 1982, 1983) supplementation with 2400 IU/day (60 μg) over 2 weeks did not change 25-OHD but did increase milk vitamin D.

Vegetarian diets which are low in dietary calcium resulted in increased serum 1,25-dihydroxyvitamin D in lactating women compared to lactating women consuming nonvegetarian diets (Specker *et al.*, 1987). The serum 25-OHD, reflecting vitamin D status, was lower in the vegetarian women. Presumably, the low calcium intake, or lower mineral availability due to higher dietary phytate in the vegetarian women, may have caused increased serum 1,25-dihydroxyvitamin D to increase intestinal calcium absorption. The mechanisms are unexplored. None of the breast-fed infants of either vegetarian or nonvegetarian women exhibited vitamin D-deficiency symptoms. In another report unsupplemented breast-fed infants in Madison, Wisconsin had no evidence of vitamin D deficiency during the first 6 months of life (Greer and Marshall, 1989).

D. Summary

The recommendations for infant and maternal intakes of vitamin D appear to be more than adequate (National Research Council, 1989). Clearly, the issue of supplementation of vitamin D for the breast-feeding infant has not been resolved. The large difference between reported levels in breast milk and the infant requirement warrants further investigation. Avoidance of sunlight exposure and dark skin pigmentation are risk factors for vitamin D deficiency in the unsupplemented breast-fed infant. The premature infant who may have a higher requirement than the term infant for vitamin D is also a candidate for supplementation during breast-feeding.

III. Vitamin E

A. Introduction

There are two groups of compounds in nature with vitamin E activity, but from the standpoint of biological activity the most important group is the tocopherols. This group is characterized by a long saturated side chain and contains four members, α, β, γ, and δ, which differ in biopotency (National Research Council, 1989). The other group, the tocotrienols, have an unsaturated side chain. The naturally occurring α-tocopherol is designated RRR-α-tocopherol (formerly, *d*-α-tocopherol). For purposes of estimating dietary vitamin E, RRR-α-tocopherol equivalents (α-TE) is used. One milligram of this isomer is 1 α-TE. The other isomers, β, γ, and δ, have relatively less biological activity than the α-tocopherol.

The biological function of vitamin E is as an antioxidant. This role is particularly important in protecting against free radical peroxidation of polyunsaturated fatty acids (PUFA) of cell membranes (Tappel, 1965; Lucy, 1972). Deficiency of vitamin E can lead to cell damage which may manifest itself with clinical symptoms of a neurological nature. The hemolytic anemia reported in vitamin E-deficient premature infants (Lucy, 1972; Hassan *et al.*, 1966; Oski and Barness, 1967; Ritchie *et al.*, 1968; Melhorn *et al.*, 1971) can be corrected or prevented with vitamin E.

It is estimated that 3 mg vitamin E per day should meet the needs of healthy term infants fed human milk or formula through the first 6 months of age (National Research Council, 1989). Infants older than 6 months have a proportionally higher requirement (4 mg) due to growth. These estimates are based on published human milk vitamin E data of Jansson *et al.* (1981), taking into account 750 ml estimated daily volume of milk ingestion and a coefficient of variation of 12.5% (National Research Council, 1989).

For the premature infant, in contrast to the term infant, human milk may not provide sufficient vitamin E (Dallman, 1974; Jansson *et al.*, 1978, 1981). These infants have relatively lower levels of vitamin E at birth due to limited placental transfer and have reduced intestinal absorption (Gross and Melhorn, 1972) coupled with relatively greater growth rates. Oral supplementation with vitamin E at 17 mg per day may be necessary for premature infants in the first 3 months (National Research Council, 1989). Whether a deficiency manifests itself clinically in a premature infant may be at least partly dependent on iron supplementation practices since iron can act as a prooxidant (Williams *et al.*, 1975). Alternatively, increased milk content of linoleic acid (18:2) could increase the vitamin E requirement (Dallman, 1974). Increased content of other PUFAs should presumably pose a similar risk for inadequacy.

B. Quantities in Milk

For this review the estimates of human milk vitamin E that have been made most recently by HPLC will be reported (Table I). With HPLC the isomers can be resolved and quantitation can be accomplished with speed, sensitivity, and precision. Most investigators in the field today rely upon this methodology for analysis. With the ability to distinguish the individual isomers comparisons across studies become more reliable. Previous reviews (Lammi-Keefe and Jensen, 1984a,b; Jensen, 1989) have summarized the earlier data that were collected by methodologies other than HPLC.

There is great interindividual variation in the human milk content of vitamin E. For individuals, the concentration of α-tocopherol (collected 4 weeks+ postpartum), determined by HPLC, ranged from less than 0.1 mg/dl (Kobayashi *et al.*, 1975) to 0.86 α-TE/dl (Haug *et al.*, 1987). For the purpose of comparison, if it is assumed that 10% of the activity reported by Haug *et al.*, as α-TE, is attributed to the γ- and β-isomers, then the range is <0.1 to approximately 0.8 mg/dl. Average α-tocopherol values range from less than 0.18 mg/dl (Haug *et al.*, 1987) to 0.34 mg/dl (Collins *et al.*, 1989) (Table I). In the data just cited values for milk collected at 4+ weeks were reported. The problem that is encountered in making comparisons across studies and in attempting to summarize the data is that many times concentrations of nutrients are dependent upon the stage of lactation. This is true for vitamin E. Additionally, it has become apparent that the classification of stages into colostral, transition, and mature also may not suffice. There is a need to be more precise. It is preferable to designate the stage of lactation by days postpartum. Lammi-Keefe and associates (1985) showed that the mean value for α-tocopherol at 2 weeks postpartum was 0.67 mg/dl which was significantly higher than the values of 0.40, 0.37, and 0.37 mg/dl at 6, 12, and 16 weeks postpartum. Vitamin E was correlated with lipid at 6, 12, and 16 weeks, but not at 2 weeks postpartum. From these data, between 2 and 6 weeks milk becomes "mature" with respect to vitamin E. This transition likely occurs very close to the 2 weeks that has traditionally been looked upon as the transition point between "transition" milk and mature milk, but there are probably slight differences in this timing between individuals. Based on that study (Lammi-Keefe *et al.*, 1985), the mature milk in the study of Jansson *et al.* (1981), which was collected from 12 days postpartum to the fifth month, likely included collections of milk from women who had higher milk vitamin E that could be explained by the stage/time in lactation that milk was collected. Boersma *et al.* (1991) reported 0.8 α-TE/dl in mature milk from women in St. Lucia. That value is considerably higher than the range noted above for mature milk collected from women in industrialized countries. The time postpartum that the milk in that study (Boersma *et al.*, 1991) was collected may partly explain the higher levels. Milk was collected between 10 and 30 days, and thus may have included less mature milks that would be expected to have higher vitamin E levels. The vitamin E in milk samples from several countries in the Caribbean were lower than the St. Lucian value: Curacao, 0.6 α-TE/dl; Dominica, 0.5 α-TE/dl; and Belize, 0.5 α-TE/dl (Boersma *et al.*, 1991). Do differences in dietary pattern account, at least in part, for the differences?

Milk collected 1 to 4 days postpartum ("colostral"), compared to milk collected at 4+ weeks postpartum, has a higher concentration of α-tocopherol: 1.0 ± 0.5 mg/dl, mean ± SD (Jansson *et al.*, 1981); 1.14 (0.63–4.21) α-TE, median and range (Haug *et al.*, 1987). The values of Haug *et al.* (1987) were reported as α-TE which included γ- and β-tocopherols that had median values at 1–3 days that were approximately one-tenth of the total activity. Thus, in the first week after parturition, when neonate vitamin E levels are low and absorption inefficient, the increased levels of this vitamin in human milk during that time are beneficial to the infant. In infants who were breast-fed, plasma tocopherol increased in the first week after parturition (Wright *et al.*, 1951). Milk collected between 0 and 4 days in Boersma *et al.*'s (1991) study in St. Lucia was higher in vitamin E than the values of Jansson *et al.* (1981) and Haug *et al.* (1987) by a factor of approximately two (2.2 α-TE/dl), raising the question of effect of maternal dietary intake on milk levels. While there is no direct evidence for a relationship between maternal dietary vitamin E intake and concentrations in human milk (Chappel *et al.*, 1985), high vitamin E intake (megadoses via supplementation) may increase the content of human milk vitamin E (Anderson and Pittard, 1985).

C. Factors Affecting Milk Levels

Tocopherol declines as lactation progresses into the second+ week (Kobayashi *et al.*, 1975; Jansson *et al.*, 1981; Haug *et al.*, 1987; Lammi-Keefe *et al.*, 1987) to the values cited above at 4+ weeks. Jansson *et al.* (1981) reported no differences between β- and γ-tocopherol levels at different lactation stages and Chappell *et al.* (1985) also demonstrated no effect of lactation stage on γ-tocopherol content. Haug *et al.* (1987) reported a decrease in the ratio of α- to β+γ-tocopherol in the first 2 weeks of lactation: 10:1 to 4:1. No further decline in the ratio was observed to 36 weeks. The dramatic decline in α-tocopherol in the first 2 weeks, coupled with no change in the other isomers, would explain the decrease in the ratio. δ-tocopherol was not detected by fluorescence and the methodology of Haug *et al.* (1987) did not resolve β from γ. Lammi-Keefe (1986) previously reported that β- and δ-tocopherols are seldom present in milk samples in amounts large enough to be quantitated with uv detection. Thus, in milk, α- and γ-tocopherol are generally the isomers of greatest interest.

Henderson *et al.* (1992) detected 0.18 mg/dl of α-tocopherol and 0.9 mg/dl of γ-tocopherol in milk collected at 2 weeks. That average vitamin E content is low compared to other published mean values, approximating the lowest published value that has been estimated on an individual basis. However, the γ-tocopherol is consistent with other published values: 0.9 mg/dl (Jansson *et al.*, 1981), 0.8 mg/dl (Harzer and Haug, 1985), and 0.66 mg/dl (Moffatt *et al.*, 1987). In the study of Henderson *et al.* (1992) the maternal plasma α-tocopherol levels were also low (range, 0.4–2.6 versus 7–15 μg/ml), while γ-tocopherol levels (1.1–5.6 μg/ml) were similar to the 0.7–5.5 μg/ml reported by others (Chow, 1975, Meydani *et al.*, 1989). It is tempting to speculate that the trend and levels observed for the milk reflect the pattern seen in the plasma, though such a relationship previously has not been demonstrated. Henderson *et al.* (1992) studied a small number of subjects.

Supplementation of mothers for 1 week with 50 mg vitamin E per day did not increase milk vitamin E (Haug *et al.*, 1987). Maternal supplementation with relatively large amounts of vitamin E increased the amount of vitamin E in milk (Anderson and Pittard, 1985). In that report the woman was consuming a rotational diet to control allergies and she had a 10-year history of megavitamin supplementation. Jansson *et al.* (1981) reported very small quantities (0.2 μmol/liter) of β-tocopherol in the milk from Swedish subjects.

Haug *et al.* (1987) demonstrated that the pattern of change over lactation showed no difference between term versus preterm milk. Additionally, those investigators provided evidence that milk for preterm infants contained tocopherol that approximated the levels in term milk. They reported milk vitamin E contents (mg α-TE/dl, medians and ranges) at Days 3 and 36 to be 1.45 (0.64–6.4) and 0.29 (0.17–0.48) for preterm and 1.14 (0.63–4.21) and 0.28 (0.19–0.86) for term. Based on those findings they concluded that, given the greater requirement of the preterm infant for vitamin E, even preterm human milk can probably not provide enough of this vitamin for these infants. These findings do not corroborate an earlier report from Chappell *et al.* (1985). Those investigators observed that milk from mothers of preterm infants contained more α-tocopherol out to 37 days. γ-Tocopherol was not, on the other hand, influenced by gestational age (Chappell *et al.*, 1985). The question of term versus preterm milk with respect to vitamin E content may deserve further attention. However, from the available evidence (Chappell *et al.*, 1985), given that there may be higher levels of vitamin E in preterm milk, these are still not adequate to meet the vitamin E requirement of preterm infants. The data for vitamin E during lactation and for terms versus preterm infants are summarized in Table I.

The relationship of milk vitamin E to the polyunsaturated fatty acids deserves further study. There is limited evidence for adequacy of vitamin E with respect to the human milk content of 18:2 (Lammi-Keefe *et al.*, 1985; Moffatt *et al.*, 1987; Boersma *et al.*, 1991) according to the current recommendation of 0.5 mg/g (American Academy of Pediatrics Committee on Nutrition, 1976). Are there dietary patterns that would result in ratios less than the recommended?

D. Relationship to Other Milk Lipids

The correlations between α-tocopherol and the milk lipids, triglyceride, cholesterol, and phospholipid, were examined in an effort to determine vitamin E's place of origin in the mammary gland. Harzer and Haug (1985) reported that vitamin E was correlated only with cholesterol in mature milk and not with triglyceride or phospholipid. They concluded that vitamin E is secreted, in part, as a constituent of the apical membrane of secretory cells and therefore is transferred to the globule during secretion. The existence of a second secretory pathway, which is a major pathway early in lactation, was postulated.

Collins *et al.* (1989) noted high correlations between α-tocopherol and triacylglycerol and choloesterol, but not phospholipid. Thus, these authors suggested that, at least part of the vitamin E is dependent on the secretion of triacylglycerol and cholesterol and independent of phospholipid secretion. Further, the apical membrane appears not to be the place of origin for vitamin E in human milk. If this were the case, a correlation between vitamin E and the membrane phospholipid would be expected.

Jensen (1989) has proposed that a pathway of vitamin E secretion into milk may involve a "rearrangement" of vitamin E and cholesterol on the globule surface after extrusion from the cell. This rearrangement is speculated to be in response to "interfacial forces." By that mechanism, vitamin E would correlate with globule surface area. This possibility has not been examined.

E. Summary

In summary, levels of vitamin E in human milk are adequate for the mature, but may not be for the premature, infant. The overall vitamin E status of the infant is dependent on interactions with PUFA, iron, and other antioxidants, such as selenium. Human milk contains α-tocopherol approximating 0.2 to 0.3 mg/dl. Small amounts of other isomers are also present, with γ-tocopherol the next highest concentration. Concentrations of α-tocopherol are highest during early lactation and decrease through approximately 2 weeks postpartum. The question of effect of maternal diet on milk vitamin E levels deserves further study.

References

Ala-Houhala, M. (1985). 25-Hydroxvitamin D levels during breast-feeding with or without maternal or infantile supplementation of vitamin D. *J. Pediatr. Gastroenterol. Nutr.* **4,** 220–226.

Ala-Houhala, M., Koskinen, T., Parviainen, M. T., and Visakorpi, J. K. (1988). 25-Hydroxyvitamin D and vitamin D in human milk: Effects of supplementation and season. *Am. J. Clin. Nutr.* **48,** 1057–1060.

American Academy of Pediatrics Committee on Nutrition (1976). Commentary on breast feeding and infant formula including proposed standards for formulas. *Pediatrics* **57,** 278–285.

Anderson, D. M., and Pittard, W. B., III (1985). Vitamin E and C concentrations in human milk with maternal megadosing: A case report. *J. Am. Diet. Assoc.* **43,** 925–930.

Atkinson, S. A., Reinhardt, T. A., and Hollis, B. W. (1987). Vitamin D activity in maternal plasma and milk in relation to gestational stage at delivery. *Nutr. Res.* **7,** 1005–1011.

Boersma, E. R., Offringa, P. J., Muskiet, F. A. J., Chase, W. M., and Simmons, I. J. (1991). Vitamin E, lipid fractions, and fatty acid composition of colostrum, transitional milk, and mature milk: An international comparative study. *Am. J. Clin. Nutr.* **53,** 1197–1204.

Chappell, J. E., Francis, T., and Clandinin, M. T. (1985). Vitamin A and E content of human milk at early stages of lactation. *Early Hum. Dev.* **11,** 157–167.

Chow, C. K. (1975). Distribution of tocopherols in human plasma and red blood cells. *Am. J. Clin. Nutr.* **28,** 756–760.

Collins, S. E., Jackson, M. B., Lammi-Keefe, C. J., and Jensen, R. G. (1989). The simultaneous separation and quantitation of human milk lipids. *Lipids* **24,** 746–749.

Dallman, P. R. (1974). Iron, vitamin E and folate in the preterm infant. *Pediatrics* **85,** 742–752.

Fomon, S. J. (1986). Breast-feeding and evolution. *J. Am. Diet. Assoc.* **3,** 317–318.

Greer, F. R., and Marshall, S. (1989). Bone mineral content, serum vitamin D metabolite concentrations, and ultraviolet B light exposure in infants fed human milk with and without vitamin D-2 supplements. *J. Pediatr.* **114,** 204–212.

Gross, S., and Melhorn, D. K. (1972). Vitamin E, red cell lipids and red cell stability in prematurity. *Ann. N.Y. Acad. Sci.* **203,** 141–162.

Harris, R. S., and Bunker, J. W. (1939). Vitamin D potency of human breast milk. *Am. J. Public Health* **29,** 744–747.

Harzer, G., and Haug, M. (1985). Correlation of human milk vitamin E with different lipids. *In* "Composition and Physiological Properties of Human Milk" (J. Schaub, ed.), pp. 247–258. Elsevier, Amsterdam.

Hassan, E., Hashim, S. A., van Italie, T. B., and Sebrell, W. H. (1966). Syndrome in premature infants associated with low vitamin E levels and high polyunsaturated fatty acid diet. *Am. J. Clin. Nutr.* **19,** 147–157.

Haug, M., Laubach, C., Burke, M., and Harzer, G. (1987). Vitamin E in human milk from mothers of preterm and term infants. *J. Pediatr. Gastroenterol. Nutr.* **6,** 605–609.

Hayward, I., Stein, M. T., and Gibson, M. I. (1987). Nutritional rickets in San Diego. *Am. J. Dis. Child* **141,** 1060–1062.

Henderson, R. A., Jensen, R. G., Lammi-Keefe, C. J., Ferris, A. M., and Dardick, K. R. (1992). Effect of maternal fish oil supplementation on human milk and infants. *Lipids* **27,** 863–869.

Hollis, B. W. (1983). Individual quantitation of vitamin D_2, vitamin D_3, 25-hydroxyvitamin D_2 and 25-hydroxyvitamin D_3 in human milk. *Anal. Biochem.* **131,** 211–219.

Hollis, B. W., and Frank, N. E. (1986). Quantitation of vitamin D_2, vitamin D_3, 25-hydroxyvitamin D_2 and 25-hydroxyvitamin D_3 in human milk. *In* "Methods in Enzymology," Vol. 123, pp. 167–177. Academic Press, New York.

Hollis, B. W., Roos, B. A., Draper, H. H., and Lambert, P. W. (1981). Occurrence of vitamin D sulfate in human milk whey. *J. Nutr.* **111,** 384–390.

Hollis, B. W., Greer, F. R., and Tsang, R. C. (1982). The effects of oral vitamin D supplementation and ultraviolet phototherapy on the antirachitic sterol content of human milk. *Calcif. Tissue Int.* **34** (suppl 1), s52, [Abstract].

Hollis, B. W., Lambert, P. W., and Horst, R. L. (1983). Factors affecting the antirachitic sterol content of native milk. *In* "Perinatal Calcium and Phosphorus Metabolism" (M. F. Holick, T. K. Gray, and C. S. Anast, eds.), pp. 157–182. Elsevier, Amsterdam

Jansson, L., Holmberg, L., Nilsson, B, and Johansson, B. (1978). Vitamin E requirement of preterm infants. *Acta Paediatr. Scand.* **67,** 459–463.

Jansson, L., Akesson, B., and Holmberg, L. (1981). Vitamin E and fatty acid composition of human milk. *Am. J. Clin. Nutr.* **34,** 8–13.

Jensen, R. G. (1989). "The Lipids of Human Milk." CRC Press, Boca Raton, Fl.

Jensen, R. G., Ferris, A. M., and Lammi-Keefe, C. J. (1992). Lipids in human milk and infant formulas. *In* "Annual Review of Nutrition" (R. E. Olson, ed.), Vol. 12, pp. 417–441. Annual Reviews, Palo Alto, CA.

Kobayashi, H., Kanno, C., Yamauchi, K., and Tsugo, T. (1975). Identification of α, β, γ, and δ-tocopherols and their contents in human milk. *Biochim. Biophys. Acta* **381,** 282–290.

Kunz, C., Niesen, M., von Lilienfeld-Toal, H., and Burmeister, W. (1984). Vitamin D, 25-hydroxyvitamin D and 1,25-hydroxyvitamin D in cow's milk, infant formulas and breast milk during different stages of lactation. *Int. J. Vitamin Nutr. Res.* **54,** 141–148.

Lammi-Keefe, C. J. (1986). Tocopherols in human milk: Analytical method using high-performance liquid chromatography. *J. Pediatr. Gastroenterol. Nutr.* **5,** 934–937.

Lammi-Keefe, C. J., and Jensen, R. G. (1984a). Lipids in human milk: A review. 2: Composition and fat-soluble vitamins, *J. Pediatr. Gastroenterol. Nutr.* **3,** 172–198.

Lammi-Keefe, C. J., and Jensen, R. G. (1984b). Fat-soluble vitamins in human milk. *Nutr. Rev.* **42,** 365–371.

Lammi-Keefe C. J., Jensen, R. G., Clark, R. M., and Ferris, A. M. (1985). Alpha tocopherol, total lipid and linoleic acid contents of human milk at 2, 6, 12, and 16 weeks. *In* "Composition and Physiological Properties of Human Milk" (J. Schaub, ed.), pp. 241–246. Elsevier, Amsterdam.

Lucy, J. A. (1972). Functional and structural aspects of biological membranes: A suggested structural role for vitamin E in the control of membrane permeability and stability. *Ann. N.Y. Acad. Sci.* **203,** 4–11.

Macy, I. G., and Kelly, H. J. (1961). Human milk and cow's milk in infant nutrition, *In* "Milk: The Mammary Gland and Its Secretions" (S. K. Kon and A. T. Cowie, eds.), pp. 265–304. Academic Press, New York.

Melhorn, D. K., Gross, S., and Childers, G. (1971). Vitamin E-dependent anemia in the premature infant. II. Relationship between gestational age and absorption of vitamin A. *J. Pediatr.* **79,** 581–588.

Meydani, M., Cohn, J. S., Macauley, J. B., McNamara, J. R., Blumberg, J. B., and Schaefer, E. J. (1989). Postprandial changes in the plasma concentration of α- and γ-tocopherol in human subjects fed a fat-rich meal supplemented with fat-soluble vitamins. *J. Nutr.* **119,** 1252–1258.

Miller, B. E., and Norman, A. W. (1984). Vitamin D. *In* "Handbook of Vitamins" (L. J. Machlin, ed.), Chap. 2. Dekker, New York.

Moffatt, P. A., Lammi-Keefe, C. J., Ferris, A. M., and Jensen, R. G. (1987). Alpha and gamma tocopherols in pooled mature human milk after storage. *J. Pediatr. Gastroenterol. Nutr.* **6,** 225–227.

National Research Council (1989). "Recommended Dietary Allowances," 10th Ed., National Academy Press, Washington, DC.

Oski, G. A., and Barness, L. A. (1967). Vitamin E deficiency: A previously unrecognized cause of hemolytic anemia in the premature infant. *J. Pediatr.* **70,** 211–220.

Polskin, L. J., Kramer, B., and Sobel, A. (1945). Secretion of vitamin D in milks of women fed fish liver oil. *J. Nutr.* **30,** 451–466.

Reeve, L. E., Chesney, R. W., and DeLuca, H. F. (1982). Vitamin D of human milk: Identification of biologically active forms. *Am. J. Clin. Nutr.* **36,** 122–126.

Ritchie, J. H., Fish, M. B., McMasters, V., and Grossman, M. (1968). Edema and hemolytic anemia in premature infants. *N. Engl. J. Med.* **279,** 1189–1190.

Specker, B. L., Tsang, R. C., Ho, M., and Miller, D. (1987). Effect of vegetarian diet on serum 1,25-dihydroxyvitamin D concentrations during lactation. *Obstet. Gynecol.* **70,** 870–874.

Tappel, A. L. (1965). Free-radical lipid peroxidation damage and its inhibition by vitamin E and selenium. *Fed. Proc.* **24,** 73, [Abstract].

Williams, M. L., Shott, R. J., O'Neal, P. L., and Oski, F. A. (1975). Role of dietary iron and fat in vitamin E deficiency anemia in infancy. *N. Engl. J. Med.* **292,** 887–890.

Wright, S. W., Filer, L. J., and Mason, K. E. (1951). Vitamin E blood levels in premature and full term infants. *Pediatrics* **7,** 386–393.

E. Fat-Soluble Vitamins in Bovine Milk

ROBERT G. JENSEN

I. Introduction

The methods of analysis and nutritional aspects for and of fat-soluble vitamins are discussed in Chapter 8C. The most recent and best method of analysis, high-performance liquid chromatography (HPLC), has not been used much for the analysis of retinoids, carotenoids, vitamin D, tocopherols, and vitamin K in bovine milk. This is in part because of pooling and of the fortification of various milks with 2000 IU of retinyl palmitate (vitamin A palmitate) and 400 IU of cholecalciferol (vitamin D_3). Milk contains very little vitamin D, tocopherols, and vitamin K, but is an excellent source of carotenoids and retinoids. The retinoids, carotenoids, and some of the vitamins D and K are found in the diet of the cow. Most of the vitamin D activity is produced by the action of sunlight on the skin of the cow. Some of the vitamin K may be produced in the rumen (Renner *et al.*, 1989; McBean and Speckman, 1988). The RDAs were taken from NRC (1989).

II. Carotenoids and Retinoids

The amounts and relevant information are shown in Table I. The older data from the USDA compilation of Posati and Orr (1976) were obtained mostly by spectrophotometric methods either of the compounds themselves or of a derivative. They are sensitive and precise, but do not resolve the complex mixture of compounds in the extracts from milks or tissues. HPLC is used for these separations (Olson, 1991). However, better resolution almost always increases complexity. In human milk, many carotenoids and retinoids of differing biopotency or retinol equivalents (RE) have been detected. Thus, the real RE of any food could be higher or lower than the published amounts. One microgram of all-*trans*-retinol = 1 RE, while 1 μg of β-carotene = 0.167 RE. 1 RE is equal to 3.33 of IU activity from retinol. Some carotenoids, e.g., zeaxanthin, have no activity. The question of availability is moot in many countries (United States, Canada, etc.) where vitamin A is added to milks. In the United States the addition of vitamin A, usually retinol palmitate at a level of 2000 IU per quart or 946 ml, is required in skim and low-fat milks shipped interstate because retinoids and carotenoids are reduced or removed when the fat content is

HANDBOOK OF MILK COMPOSITION

TABLE I
Retinoids and Carotenoids in Bovine Milk and Dairy Products

Product	Compound	Amount/dl or 100 g			Method	Reference
		RE[a]	IU	μg		
Milk, 3.3% fat	Vitamin A	31	126	31	Spectrophotometric	Posati and Orr (1976)
Butter	Vitamin A	754	3058	759	Spectrophotometric	Posati and Orr (1976)
Milk, non-Channel Island (NCI) breeds[b]						
Summer	Total retinol	—	—	61.9	HPLC	Scott *et al.* (1984)
	β-Carotene	—	—	31.5		
Winter	Total retinol	—	—	41.2	HPLC	Scott *et al.* (1984)
	β-Carotene	—	—	10.5		
	RE	43.0	—			
Channel Island (CI) breeds[b]						
Summer	Total retinol	—	—	64.9	HPLC	Scott *et al.* (1984)
	β-Carotene	—	—	114.3		
Winter	Total retinol	—	—	26.5	HPLC	Scott *et al.* (1984)
	β-Carotene	—	—	26.6		
Finnish milk, 3.9% fat	β-Carotene	—	—	16.7 (0.03)	HPLC	Ollilainen *et al.* (1989)
Fall, winter	Retinol	—	—	32.6 (0.3)		
	RE	37	—	—		
United States milk						
Summer, 3.9% fat	Retinol	46.6	—	46.6 (0.9)	HPLC	Zahar and Smith (1990)
Winter, 3.67% fat	Retinol	28.0	—	28.0 (0.17)		
New Zealand pasteurized	Retinyl ester		89 (99.6)[c]	26.7	HPLC	Woolard and Indyk (1989)
UHT milk, 2% fat	Retinyl ester		53 (99.6)[c]	15.9		
Bovine milk fat, anhydrous	Retinyl ester		3100 (97.8)[c]	930		

[a]RE, retinol equivalents + all-*trans*-retinol = 1 μg + 13-*cis*-retinol, μg × 0.75.
[b]Non-Channel Island breeds, Holsteins, etc. CI, Channel Island breeds, Guernseys, and Jerseys.
[c]Total ester (%).

reduced or eliminated. The restored amount, 2000 IU, is 601 RE of vitamin A activity from retinol (NDC, 1993). The RDA for an adult male is 1000 RE (NRC, 1989). Addition of vitamin A to whole milk is optional. A quart of whole milk will contain 310 to 413 RE based on data from HPLC analyses (see Table I for RE/dl).

The retinoids are photo and heat sensitive. Pasteurization reduces the content, but this is not important since most milk is pasteurized and some is fortified. Bruhn (1990) found that losses of added vitamin A in low-fat and nonfat milks range from 5.4 to 33% by the retail pull date, 10 days after filling.

There is a regional or seasonal effect due to the feeding regimen, with lower amounts of total retinol and β-carotene in winter compared to summer (Scott *et al.*, 1984). Breed has an influence, with more β-carotene in milk presumably from Guernseys.

Ollilainen *et al.* (1989) found trace amounts of lutein in milk and other products. They also detected these retinoids (μg/100 g) in whole milk: 13-*cis*, 2.1; 11-*cis*, TR; 9,11-di-*cis*,, TR; 9-*cis*,, 0; and all-*trans*,, 32.6. According to Woolard and Indyk (1989), 99.6% of the retinol in milk is esterified. The amounts of the most prevalent esters were (% of RE): 16:0, 36.7; 18:1, 20.3; 18:3, 8.5; 18:0, 8.4; and 18:2, 7.3. This profile is quite different from that of the fatty acids in that the amounts of 18:3 and 18:2 are much higher (see Chapter 6B).

Milk and dairy products can provide substantial amounts of retinoids and carotenoids in the diet. One quart of whole milk contains about 36–41% of the RDA for an adult male. The carotenoids, while a minor source of vitamin A, are potent antioxidants and may aid in the prevention or delay of onset of a host of problems, e.g., the formation of cataracts (Hankinson *et al.*, 1992).

III. Vitamin D

Bovine milk as secreted is deficient in vitamin D for the needs of humans, particularly for infants and children (McBean and Speckmann, 1988; Renner *et al.*, 1989). It contains about 47 to 105 IU/liter and the RDA is 400 IU or 10 μg of cholecalciferol or vitamin D_3 (NRC, 1989). Since milk is deficient, it and other dairy products are fortified with 400 IU of cholecalciferol per quart (NDC, 1993). Fortification of milk in North America and Europe eliminated rickets as a health problem in children (Fomon, 1993).

Earlier analyses of vitamin D or antirachitic activity in milk were done by measurement of increased bone growth in rats when milk was fed compared to standards (Norman and Miller, 1991). This assay is called the rat line test. With newer methods of extraction, ligand binding, and HPLC the analyst can separate, identify, and quantify vitamin D and its metabolites. The major metabolite, 25-OHD, is five times more active than vitamin D. It contributes most of the antirachitic activity in milk. See Table II for results. In Table II, the data of Leerbeck and Sondergaard (1988) were obtained by the rat line method and the remainder by the newer procedures. The "real" IUs of antirachitic activity in bovine milk are considerably higher than those obtained by older methods.

Earlier reports of water-soluble activity in milk attributed to vitamin D_3 sulfate have been disproven (Hollis *et al.*, 1981). The compound was found in negligible quantities and has very little antirachitic activity (Reeve *et al.*, 1981).

About one-third of the vitamin D was lost when low-fat milk was stored for 10 days at 4°C (Bruhn, 1990), presumably because of photoxidation. Care must be exercised when the milk processor adds the concentrate to milk. Tanner *et al.* (1988) and Holick *et al.* (1992) found that the actual amounts in 71% of the milk samples they analyzed were less than the quantities on the labels. Conversely, overfortification occurs. Jacobus *et al.*, (1992) observed that the cause of hypervitaminosis D in eight patients, all of whom drank milk, 118 to 710 ml daily from one dairy plant, was overfortification. When milks from this plant were analyzed the results suggested that fortification was sporadic with amounts of vitamin D_3 ranging from < 40 to 232, 565 IU/qt. While it is important that regulatory agencies analyze milk for vitamin D_3 and other materials regularly, proper training and supervision of plant employees is equally important. Analysis of milk for the vitamin is readily done by HPLC and routine testing is necessary and required.

IV. Tocopherols (Vitamin E)

The methods of analysis and nutritional roles are discussed in Chapter 8D (see also Renner *et al.*, 1989; McBean and Speckman, 1989; Fomon and Bell, 1993; Machlin, 1991; NDC, 1993).

Bovine milk contains relatively small quantities of tocopherols compared to, for example, human milk. Lehman *et al.* (1986) detected 30 and 10 μg each of the α- and γ-isomers in 100 g of whole milk containing 3.3% fat (see Table III). Again, HPLC provides much better resolution, precision, and sensitivity than the older colorimetric method. Results from the few published HPLC analyses of tocopherols in milk are shown in Table III. A quart of milk contains only about 0.3 mg of α-tocopherol equivalents, and the RDA for adult males is 10 mg. Milk is not consumed as a source of tocopherols. Tocopherols are antioxidants and their effectiveness in delaying the onset of oxidative rancidity in milk has been investigated (Hidiroglou, 1989).

In addition to α-tocopherol, Hidiroglou (1989) detected an unspecified amount of the γ-isomer, but it was equivalent to about 10% of the α form.

TABLE II
Vitamin D and Metabolites in Bovine Milk (Amounts / Liter)

D[a]		25-OHD		24,25-$(OH)_2$-D		25,26-$(OH)_2$-D		1,25-$(OH)_2$-D		Total	Reference
ng	IU	ng	IU	ng	IU	ng	IU	ng	IU	IU	
—	38	—	—	—	—	—	—	—	—	38	Leerbeck and Sondergaard (1980)
43.8 ± 8[b]	2[c]	372 ± 42	74	45 ± 6	9	21 ± 1.2	4	5.4 ± 1.2	2	91	Hollis *et al.* (1981)
281	11	145	29	27	5	—	—	4.9	2	47	Reeve (1982)
330	13	—	—	—	—	—	—	—	—	13	Scott *et al.* (1984)
50.4 ± 4.1[d]	2	497 ± 47	99	—	—	NR		9.7 ± 1.0	4	105	Kunz *et al.* (1984)

[a]D is D_2 or D_3. 25-OHD is 25-hydroxyvitamin D_3, 25,26-$(OH)_2$-D is 25,26-dihydroxyvitamin D_3, etc.
[b]Means ± SD.
[c]1 IU = 25 ng D_2 or D_3 = 5 ng 25-OHD, 24,25-$(OH)_2$-D, 25,26-$(OH)_2$-D = 1 ng 1,25-$(OH)_2$-D.
[d]Means ± SEM.

TABLE III
Tocopherol Contents of Bovine Milk

Product	Amount (μg/100 g or dl)	Reference
Milk, 3.3% fat		
α-Isomer	30	Lehman *et al.* (1986)
γ-Isomer	10	
Milk, Holsteins		
α-Isomer	28	Hidiroglou (1989)
γ-Isomer	Present	
Finnish milk pooled, pasteurized, 3.9% fat		Syvaoja *et al.* (1985)
Summer, α-isomer	70	
Winter, α-isomer	20	
Tocotrienol, α-isomer	TR	

Traces of β- and δ-isomers were detected. Lehman *et al.* (1986) found a 3 to 1 ratio of the α- and γ-isomers. Syvaoja (1985) found traces of γ- and δ-tocopherols.

V. Vitamin K

Vitamin K is difficult to analyze even with the aid of HPLC (see Chapter 8B). Because of this problem and because milk contains very little of the vitamin, few reports on the contents obtained by modern methods are available. These are collected in Table IV. More phylloquinone (K_1) was found in the milk from Jersey or Guernsey cows (8.7 μg/liter) compared to the Friesian breed (4.9 μg/liter) by Haroon *et al.* (1982). The milk from Friesian cows contains much less fat than milk from Guernseys or Jerseys. Fournier *et al.* (1987) found more K_1 (19.7 μg/liter) than Haroon *et al.* (1982) (4.9 or 8.7 μg/liter). Fournier *et al.* (1987) analyzed pooled samples collected over a year, as did Haroon *et al.* They also mentioned differences in methodology as a cause. However, when compared to the values (3 μg/liter) of Booth *et al.* (1993), they seem to be high. Although milk is an insignificant source of vitamin K_1, further analyses, as discussed by Booth *et al.* (1993), are indicated. The total contents of any nutrients consumed in a mixed diet are important.

TABLE IV
The Vitamin K_1 Contents of (μg/liter) Bovine Milk

Product	Amount (range)	Reference
Whole milk		Haroon *et al.* (1982)
Friesians	4.9 (3.6–18.9)	
Guernseys	8.7 (3.8–17.8)	
Jerseys		
Pooled whole milk	19.7 (7.46–37.65)	Fournier *et al.* (1987)
Whole milk, 3.5% fat	3.0	Booth *et al.* (1993)

Acknowledgments

The preparation of the manuscript was supported in part by an NIH contract and by federal funds made available through provision of the Hatch Act, Scientific Contribution, Storrs Agricultural Experiment Station, Storrs, Connecticut.

References

Booth, S. L., Sadowski, J. A., Welrauch, J. L., and Ferland, G. (1993). Vitamin K-1 (Phylloquinone) content of foods: A provisional table. *J. Food Comp. Anal.* **6.**

Bruhn, J. C. (1990). Vitamin A and D additions to lowfat and nonfat milks. *J. Dairy Sci.* **73**(Suppl. 1), 96. [Abstract D83]

Collins, E. D., and Norman, A. W. (1991). Vitamin D. *In* "Handbook of Vitamins" (L. J. Machlin, ed.), 2nd Ed., pp. 60–98. Dekker, New York.

Fomon, S. J., and Ziegler, E. E. (1993). Vitamin D. *In* "Nutrition of Normal Infants" (S. J. Fomon ed.), pp. 323–338. Mosby, St. Louis, MO.

Fournier, B., Sann, L., Guillaumont, M., and Leclercq, M. (1987). Variations of phylloquinone concentration in human milk at various stages of lactation and in cow's milk at various seasons. *Am. J. Clin. Nutr.* **45,** 551–558.

Hankinson, S. L., Stampfer, M. J., Seddon, J. M., Colditz, G. A., Rosner, B., Speizer, F. E., and Willet, W. C. (1992). Nutrient intake and cataract extraction in women: A prospective study. *Br. Med. J.* **305,** 335–339.

Haroon, Y., Shearer, M. J., Rahim, S., Gunn, W. G., McEnery, G., and Barkhan, P. (1982). The content of phylloquinone (Vitamin K-1) in human milk, cow's milk and infant formulas determined by high-performance liquid chromatography. *J. Nutr.* **112,** 1105–1117.

Hidiroglou, M. (1989). Mammary transfer of vitamin E in dairy cows. *J. Dairy Sci.* **72,** 1067–1071.

Holick, M. F., Shao, Q., Liu, W. W., and Chen, T. C. (1992). The vitamin D content of fortified milk and infant formula. *N. Engl. J. Med.* **326,** 1178–1181.

Hollis, B. W., Roos, B. A., Draper, H. H., and Lambert, P. W. (1981). Vitamin D and its metabolites in human and bovine milk. *J. Nutr.* **111,** 1240–1248.

Jacobus, C. H., Holick, M. F., Shao, Q., Chen, T. C., Holm, I. A., Kolodny, J. M., Fuleihan, G. E-H, and Seely, E. W. (1992). Hypervitaminosis D associated with drinking milk. *N. Engl. J. Med.* **326,** 1173–1177.

Kunz, C., Niesen, M., Lilienfield, H.V., and Burmeister, W. (1984). Vitamin D, 25-hydroxy-vitamin D and 1,25-dihydroxy-vitamin D in cow's milk, infant formulas, and breast milk during different stages of lactation. *Int. J. Vit. Nutr. Res.* **34,** 141–148.

Leerbeck, E., and Sondergaerd, H. (1980). The total content of vitamin D in human milk and cow's milk. *Br. J. Nutr.* **44,** 7–12.

McBean, L. D., and Speckmann, L. W. (1988). Nutritive value of dairy foods. *In* "Fundamentals of Dairy Chemistry" (N. P. Wong, R. Jenness, M. Keeney, and E. H. Marth, eds.), 3rd Ed., pp. 343–408. Van Nostrand–Reinhold, New York.

National Dairy Council (NDC) (1993). "Newer Knowledge of Milk and Other Fluid Dairy Products." NDC, Rosemont, IL.

National Research Council (NRC) (U.S.) (1989). Committee on dietary allowances. Fat-soluble vitamins. *In* "Recommended Dietary Allowances (RDA)," 10th ed., pp. 78–104. National Academy Press, Washington, DC.

Ollilainen, V., Heinonen, M., Linkola, E., Varo, P., and Koivistionein, P. C. (1989). Carotenoids and retinoids in Finnish foods: Dairy products and eggs. *J. Dairy Sci.* **72,** 2257–2265.

Olsen, J. A. (1991). Vitamin A. *In* "Handbook of Vitamins" (L. J. Machlin, ed.), 2nd Ed., pp. 1–58. Dekker, New York.

Posati, L. P., and Orr, M. L. (1976). "Composition of foods, Dairy and Egg Products, Raw–Processed–Prepared." Agriculture Handbook 8.1, ARS, USDA, Washington, DC.

Reeve, L. E., DeLuca, H. F., and Schnoes, H. K. (1981). Synthesis and biological activity of vitamin D-3 sulfate. *J. Biol. Chem.* **256,** 823–826.

Reeve, L. E., Jorgensen, N. E., and DeLuca, H. F. (1982). Vitamin D compounds in cow's milk. *J. Nutr.* **112,** 667–672.

Renner E., Schaafsma, G., and Scott, K. J. (1989). Micronutrients in milk, *In* "Micronutrients in Milk and Milk-Based Food Products" (E. Renner, ed.), pp. 36–38. Elsevier, New York.

Scott, K. J., Bishop, D. R., Zechalko, A., and Edwards-Webb, J. D. (1984). Nutrient content of liquid milk. I. Vitamins A, D-3, C and of the B complex in pasteurized bulk liquid milk. *J. Dairy Res.* **51,** 37–50.

Syvaoja, E. L., Piironen, V., Varo, P., Koivistionen, P., and Salminen, K. (1985). Tocopherols and tocotrienols in Finnish foods: Dairy products and eggs. *Milchwissenschaftt* **40,** 467–469.

Tanner, J. T., Smith, J., Defigbaugh, P., Angyal, G., Villalobos, M., Bueno, M. P., Wahr, B. W., Hollis, B. W., Koh, T., Rerch, P., and Simpson, K. L. (1988). Survey of vitamin content of fortified milk. *J. Assoc. Anal. Chem.* **71,** 607–610.

Woolard, D. C., and Indyk, H. (1989). The distribution of retinyl esters in milks and milk products. *J. Micronutr. Anal.* **5,** 35–52.

Zahar, M., and Smith, D. E. (1990). Vitamin A quantification in fluid dairy products: Rapid method for vitamin A extraction for high performance liquid chromatography. *J. Dairy Sci.* **73,** 3402–3407.

CHAPTER 9

Defense Agents in Milk

A. Defense Agents in Human Milk

ARMOND S. GOLDMAN
RANDALL M. GOLDBLUM

I. Introduction

A. Factors with Nonnutritive Functions in Human Milk

During the past three decades, there has been a growing realization that breast-feeding not only provides the nutritional requirements of the infant, but also supplies a host of defense factors for the protection of the recipient and/or the mammary gland. The study of this remarkable defense system in human milk has been difficult, however, because of its biochemical complexities, the small concentrations of certain very potent agents in human milk, the need to develop special methods to quantify certain factors because of their particular forms in human milk, the compartmentalization of some of the agents, and the dynamic effects of the length of lactation and other maternal factors upon the concentrations or functions of the components of the system. In this chapter, we summarize the current information concerning the molecular forms, concentrations in milk during the several phases of lactation, biological activities, fate in the recipient infant, and *in vivo* functions of the defense agents in human milk.

B. General Characteristics of Defense Agents in Human Milk

The defense agents in human milk, though biochemically diverse, share certain features: (1) there is often an inverse relationship between the

HANDBOOK OF MILK COMPOSITION

production of these factors in the mammary gland and their production by the infant during the same time frames of lactation and postnatal development. As lactation progresses, the concentrations of many of the factors in human milk fall. Concomitantly, the production at mucosal sites of those very factors rises in the developing infant; (2) most components of the immunologic system in human milk are produced throughout lactation and during gradual weaning; (3) the factors are usually common to other mucosal sites; (4) they are adapted to resist digestion in the gastrointestinal tract of the recipient infant; (5) they protect by noninflammatory mechanisms; and (6) the agents act synergistically with each other or with defense agents produced by the recipient. Representative examples of soluble defense agents are listed in Table I.

II. Types of Defense Agents in Human Milk

A. Direct-Acting Antimicrobial Agents

1. *Oligosaccharides–Glycoconjugates*

These agents include monosialogangliosides that are receptor analogues for heat-labile toxins produced by *Vibrio cholerae* and *Escherichiae coli* (Holmgren *et al.*, 1981), fucose-containing oligosaccharides that inhibit the hemagglutinin activity of the classical strain of *V. cholerae* (Holmgren *et al.*, 1983), fucosylated oligosaccharides that protect against heat-stable enterotoxin of *E. coli* (Newburg *et al.*, 1990), mannose-containing high-molecular-weight glycoproteins that block the binding of the El Tor strain of *V. cholerae* (Holmgren *et al.*, 1981), and glycoproteins and glycolipids that interfere with the binding of CFA/II fimbrae on enterotoxigenic *E. coli* (Holmgren *et al.*, 1987). The inhibition of toxin binding appears to be associated with acidic glycolipids that contain sialic acid (gangliosides). Although the total quantities of gangliosides in human and bovine milk are similar, the relative frequency of each type of ganglioside composition is distinct. Monosialo-ganglioside 3 predominates in human milk (about 74% of total gangliosides), but not in bovine milk (Laegreid *et al.*, 1986a,b), and the enterotoxin receptor ganglioside, GM1, as measured by a highly sensitive immunostaining method, is 10-fold greater in human than bovine milk (Laegreid *et al.*, 1986a). In that respect, GM1 inhibits the enterotoxins of *E. coli* and *V. cholerae* (Laegreid and Otnaess, 1987).

Oligosaccharides in human milk also interfere with the attachment of *Haemophilus influenzae* and *Streptococcous pneumoniae* (Andersson *et al.*, 1986). In particular, G1cNAc (β1–3)Gal-disaccharide subunits block the attachment of *S. pneumoniae* to respiratory epithelium. It is anticipated that carbohydrate side chains on a number of whey proteins in human milk will be found to function as receptor analogues.

TABLE I
Representative Soluble Defense Agents in Human Milk

	Representative function
Anti-infectious agent	
Oligosaccharides–glycoconjugates	Inhibit binding of bacterial pathogens and toxins to epithelium
Lactoferrin	Decrease multiplication of siderophilic bacteria/fungi by Fe^{3+} chelation
Lysozyme	Disrupts peptidoglycans of cell walls on susceptible bacteria
Secretory IgA	Antibodies inhibit adherence of pathogens to epithelium; neutralize toxins
Mucin	Inhibits rotavirus
Lipids	Disrupt enveloped viruses
Anti-inflammatory agents	
Uric acid, ascorbate, α-tocopherol, β-carotene	Antioxidants
Prostaglandins	Cytoprotective
Cortisol, lactoferrin, EGF	Epithelial growth factors
Platelet-activating factor–acetylhydrolase	Degrades PAF
Immunomodulators	
Interleukin-1β	Activates T cells/monocytes
Interleukin-6	Aids terminal differentiation of IgA-producing cells
Tumor necrosis factor-α	Upregulates production of secretory component.
	Activates T cells/monocytes

In addition, there is recent evidence that human milk interferes with the binding of human immunodeficiency virus envelope antigen gp120 to CD4 molecules on T cells (Newburg *et al.*, 1992). The physical characteristics of the inhibitor were consistent with mucins, sulfated proteins, glycoproteins, or glycoaminoglycans.

Some data from animal models suggest that the oligosaccharides and glycoconjugates in human milk protect *in vivo* (Cleary *et al.*, 1983; Otnaess *et al.*, 1982), but there is little information from human studies that pertains to this question (Glass *et al.*, 1983).

2. *Proteins*

Many whey proteins in human milk have direct antimicrobial properties. The principal ones are as follows.

a. Lactoferrin. Lactoferrin, a member of the transferrin family of iron-binding glycoproteins, is the dominant whey protein in human milk. Lactoferrin, a single-chain glycoprotein with 703 amino acids, has an M_r of 79 kDa and two globular lobes, both of which display a site that binds ferric iron (Anderson *et al.*, 1987). Except for a 20-kDa fragment of lactoferrin that immunologically cross-reacts with bovine β-lactoglobulin (Monti *et al.*, 1989), the vast majority of lactoferrin in human milk consists of intact molecules. Over 90% of the lactoferrin in human milk is in the form of apolactoferrin (i.e., does not contain ferric iron) (Fransson and Lonnerdal, 1980). This is highly advantageous, since apolactoferrin competes with siderophilic bacteria for ferric iron (Arnold *et al.*, 1977; Bullen *et al.*, 1972; Spik *et al.*, 1978; Stephens *et al.*, 1980; Stuart *et al.*, 1984) and thus disrupts the proliferation of those microbial pathogens. The epithelial growth-promoting activities of lactoferrin in human milk may also be advantageous to the defense of the recipient infant (Nichols *et al.*, 1987). The mean concentration of lactoferrin in human colostrum as measured by electro-immunodiffusion is between 5 and 6 mg/ml (Goldblum *et al.*, 1982). The concentration at 4 weeks falls to about 2 mg/ml (Goldblum *et al.*, 1981; Goldman *et al.*, 1982). Afterwards, the concentration of lactoferrin averages about 1 mg/ml (Goldman *et al.*, 1982).

In keeping with its resistance to proteolysis (Brines and Brock, 1983; Samson *et al.*, 1980; Spik and Montreuil, 1966), a number of groups have reported that the excretion of lactoferrin in the stools is higher in human milk- than in cow's milk-fed infants (Butte *et al.*, 1984; Davidson and Lonnerdal, 1985, 1987; Spik *et al.*, 1982). The total daily secretion of lactoferrin in human milk during the first 4 months of lactation has been investigated by a test-weighing procedure and immunologic assays. The approximate mean intake of milk lactoferrin per day in healthy full-term infants was reported to be 260 mg per kilogram per day at 1 month and 125 mg per kilogram per day by 4 months (Butte *et al.*, 1984). The amount of lactoferrin excreted in the stools of low birth weight infants fed human milk appears to be about 185 times that in infants fed a cow's milk formula (Schanler *et al.*, 1986). However, this estimate is probably too high because of the presence of immunoreactive fragments of that protein (Goldman *et al.*, 1990).

In addition, there is a significant increment in the urinary excretion of intact and fragmented lactoferrin as a result of human milk feedings (Goldblum *et al.*, 1989; Goldman *et al.*, 1990; Prentice, 1987). Recent stable isotope studies suggest that the increment in urinary lactoferrin and its fragments is principally from ingested human milk lactoferrin (Hutchens *et al.*, 1991).

b. Lysozyme. Lysozyme, a 15-kDa single-chain protein, is found in relatively high concentrations in external secretions including human milk (Chandan *et al.*, 1964; Jolles and Jolles, 1967; Goldblum *et al.*, 1981; Goldman *et al.*, 1982, 1983a,b; Peitersen *et al.*, 1975). Lysozyme lyses susceptible bacteria by hydrolyzing β-1,4 linkages between *N*-acetylmuramic acid and 2-acetylamino-2-deoxy-D-glucose residues in cell walls (Chipman and Sharon, 1969). The agent is relatively resistant to digestion by trypsin or denaturation due to acid. The mean concentration of lysozyme in colostrum as measured by electroimmunodiffusion is about 70 μg/ml (Goldblum *et al.*, 1981). The concentration drops to about 20 μg/ml at 1 month of lactation and then rises to a mean of 250 μg/ml by 6 months (Goldman *et al.*, 1982). The approximate mean intake of milk lysozyme per day in healthy full-term infants was reported to be 3 or 4 mg per kilogram per day at 1 month and 6 mg per kilogram per day by 4 months (Butte *et al.*, 1984).

Limited studies have been conducted concerning the fate of human milk lysozyme that is ingested by the infant. The amount of lysozyme excreted in the stools of low birth weight infants fed human milk is about eight times that found in infants fed a cow's milk formulation (Schanler *et al.*, 1986). There was no increment in the urinary excretion of this protein as a result of human milk feedings. Otherwise, the *in vivo* fate and functions of the agent remain to be determined.

c. Fibronectin. Significant amounts of fibronectin, a high-molecular-weight protein that facilitates the uptake of many types of particulates by mononuclear phagocytic cells, are present in human milk (mean concentrations in colostrum, 13.4 mg/liter) (Friss *et al.*, 1988). Structural analyses of fibronectin in human milk have not been reported. The *in vivo* effects of this broad spectrum opsonin in human milk are not known.

d. Complement components. Although there is evidence that the components of the classical and alternative pathways of complement are present in human milk, the concentrations of these components, except for C3, are exceptionally low (Ballow *et al.*, 1974; Nakajima *et al.*, 1977). The quantitation of these components has been limited to hemolytic assays. Additional studies with newer immunoassays have not been reported. It is also unclear whether the structures of these molecules are the same as those in human blood.

e. Immunoglobulins. The pattern of the concentrations of major immunoglobulin isotypes in human milk is quite different from those found in blood and interstitial fluids. The predominant immunoglobulin in human milk is IgA (Goldman and Goldblum, 1989). IgA is the dominant immunoglobulin in human milk, whereas IgG is the most common one in adult blood and interstitial fluids. The principal molecular forms of IgA in blood and milk are different. The main form in serum is a four-chain structure consisting of two identical heavy polypeptide chains (predominantly the α 1 isotype) and two identical light chains (either κ or λ) linked by disulfide bonds. In contrast to the IgA monomers, a polymeric form of IgA, secretory IgA, comprises over 95% of IgA in human milk (Goldman and Goldblum, 1989). This type of IgA consists of two identical IgA monomers united by a 15-kDa polypeptide called the joining or J chain and complexed to a 75-kDa glycopeptide designated as secretory component (Mostov and Blobel, 1982; Mostov *et al.*, 1984; Solari and Kraehenbuhl, 1984). This form of IgA is assembled when dimeric IgA binds to the first domain of polymeric immunoglobulin receptors (Bakos *et al.*, 1991) on the basolateral surface membranes of epithelial cells. Before the assembled molecule is secreted, the intracellular part of the original receptor is cleaved so that the final molecule consists of the ectoplasmic portion (secretory component) and dimeric IgA. Secretory IgA is more resistant to proteolytic enzymes and therefore is more able to persist in the intestinal tract than other immunoglobulins.

Specific antibodies in human milk arise from a triggering of enteromammary (Allardyce *et al.*, 1974; Goldblum *et al.*, 1975; Roux *et al.*, 1977; Weiz-Carrington *et al.*, 1978) and bronchomammary (Fishaut *et al.*, 1981) immune pathways. In the case of the enteromammary pathway, the responsible immunogen is taken up by M cells on the surface of Peyer's patches. The immunogen is recognized by B cells which display specific IgM antibodies to the immunogen. The immunoglobulin isotype of the B cells is switched to IgA, and those B cells are then launched into a migration pathway that begins in the intestinal lymphatics and ends in the lamina propia of the mammary gland. Those B cells undergo terminal differentiation to become plasma cells that produce dimeric IgA antibodies that are specific for the very immunogens that originally triggered the pathway.

The concentrations of IgM are much lower in human milk than in serum (Jatsyk *et al.*, 1985; Mata and Wyatt, 1971). IgM molecules in blood and milk display a pentameric structure. However, in contrast to serum IgM, some human milk IgM is complexed to secretory component (Brandtzeg, 1974). In the few studies that have been published, the antibody specificities of human milk IgM are similar to those of the major immunoglobulin isotype in human milk, secretory IgA.

IgG, the principal immunoglobulin in human serum, is present in modest amounts in human milk (Jatsyk *et al.*, 1985; Keller *et al.*, 1983; Mata and Wyatt, 1971; McClelland *et al.*, 1978; Ogra and Ogra, 1978; Peitersen *et al.*, 1975). Each IgG subclass has been detected in human milk, but the relative proportion of IgG4 is higher in human milk than serum (Keller *et al.*, 1983). It has therefore been suggested that IgG4 may be produced in or specifically transported to the mammary gland. An alternate explanation is that the increased proportion of IgG4 is due to a more efficient exclusion of other IgG subclasses from human milk.

The quantity of IgD in human milk is lower than that in serum, but the decrease is proportionally less than is reported for IgG and IgM (Keller *et al.*, 1984). IgE, the principal type of antibodies responsible for immediate hypersensitivity reactions, is essentially absent in human milk (Underdown *et al.*, 1976).

There has been considerable interest in specific antibodies in human milk. Depending upon the precise question, it may be necessary to determine the fine specificity, avidity, isotypes, and quantities of antibodies in milk. The determination of the fine specificity depends upon the use of highly specific antigens. Even then, immunologic cross-reactivity, particularly against common enteric microorganisms and food antigens, may occur. Because of the structure of secretory IgA, immunoassays that are designed to quantify IgA will also detect secretory IgA, but the resultant data may also reflect the presence of other molecular forms of IgA, such as monomeric or dimeric IgA, that are not complexed to secretory component. Secretory IgA antibodies may be distinguished from other types of IgA antibodies by using specific antibodies to secretory component in solid-phase immunoassays, although secretory IgM antibodies will also be detected.

Solid-phase immunoassays in which the capture antibody is directed against the α-chain of IgA and the antibody detector recognizes secretory component have been used to quantitate total secretory IgA in human milk (Goldblum *et al.*, 1981). The concentrations of secretory IgA in human milk were highest in colostrum (Goldblum *et al.*, 1981) and gradually declined to a plateau of about 1 mg/ml for the remainder of lactation (Goldman *et al.*, 1982). The approximate mean intake of secretory IgA per day in healthy full-term breast-fed infants was found to be 125 mg per kilogram per day at 1 month and ~75 mg per kilogram per day by 4 months (Butte *et al.*, 1984).

The fate of human milk secretory IgA fed to infants has been examined. In one study, the amount of secretory IgA excreted in the stools of low birth weight infants who were fed human milk was about 30 times that in infants fed a cow's milk formula (Schanler *et al.*, 1986). In addition, there was a significant increment in the urinary excretion of intact secretory IgA antibodies as a result of human milk feedings (Goldblum *et al.*, 1989). The origin of those antibodies in the urine of infants fed human milk is undetermined.

f. Mucins. Human milk mucins have recently been reported to be antimicrobial. Membrane mucins on human milk fat globules interfere with the binding of S-fimbriated *E. coli* (Schroten *et al.*, 1992) and human milk mucins defend against rotavirus (Yolken *et al.*, 1992), the most common cause of infectious enteritis in human infants (Kapikian *et al.*, 1981). The range of the antimicrobial effects of these compounds in human milk and their abilities to cooperate with other defense agents in milk are unclear.

B. Growth Promoters of Protective Microorganisms

In contrast to bovine milk, human milk contains a growth-promoting activity for *Lactobacillus bifidus* var. *Pennsylvania* (György *et al.*, 1974). It appears that this activity is responsible for the predominance of *Lactobacillus* in the bacterial flora of large intestines of breast-fed infants. Those bacteria produce acetic acid, which aids in suppressing the multiplication of enteropathogens. The bifidus growth-factor activity is due to *N*-containing oligosaccharides (György *et al.*, 1974) and glycoproteins and glycopeptides (Bezkorovainy *et al.*, 1979; Nichols *et al.*, 1975). The bifidus growth-promoter activity associated with caseins may reside in the oligosaccharide moiety of those molecules (Bezkorovainy and Topouzian, 1981).

C. Defense Agents Created from Partially Digested Substrates from Human Milk

Human milk may also protect by supplying defense agents from substrates that are partially digested in the recipient's alimentary tract. Fatty acids and monoglycerides produced from milk fats by bile salt-stimulated lipase or lipoprotein lipase in human milk or lingual/gastric lipase from the recipient are able to disrupt enveloped viruses (Issacs *et al.*, 1986; Stock and Francis, 1940; Thromar *et al.*, 1987; Welsh *et al.*, 1979; Welsh and May, 1979). These antiviral lipids may aid in preventing coronavirus infections of the intestinal tract (Resta *et al.*, 1985). They may also defend against intestinal parasites such as *Giardia lambdia* (Gillin *et al.*, 1983, 1985).

The second example of the generation of biologically active agents from enzymatic digestion of substrates in human milk is the production of β-casomorphins from ingested human casein (Brantl, 1985). These peptide fragments have not only opioid, but also immunostimulating effects (Berthou *et al.*, 1987; Parker *et al.*, 1984).

A 20-kDa fragment of lactoferrin has been described in human milk (Monti *et al.*, 1989), but its function is not known. Fragments of human lactoferrin have been demonstrated in the stools of human milk-fed infants, and the multiplicity of those fragments suggests that some apolactoferrin from human milk feedings is partially cleaved in the gastrointestinal tract of the recipient (Goldman *et al.*, 1990). Similar fragments of lactoferrin were demonstrated in the urine suggesting that they were from absorbed from the gastrointestinal tract and excreted into the urinary tract (Goldman *et al.*, 1990). The biologic activity of the fragments of lactoferrin in the excreta of the recipients is undetermined. It is undetermined whether one of those fragments is similar to the pepsin-derived fragment of lactoferrin, lactoferricidin-B, that is bactericidal (Bellamy *et al.*, 1993; Yamauchi *et al.*, 1993).

D. Leukocytes

Living white blood cells are present in human milk particularly during the first 3 or 4 months of lactation. The concentrations of these leukocytes are highest in the first 2–4 days of lactation and gradually decline during the next few months. Neutrophils and macrophages are the most prominent cells in human milk. It is necessary to employ special cytochemical stains or immunologic markers to distinguish the neutrophils from the macrophages in human milk because their morphology is altered by the large amount of lipid vesicles in their cytoplasm (Smith and Goldman, 1968; Smith *et al.,* 1971).

Although human milk neutrophils are phagocytic, they are unable to adhere to common substrata, move as rapidly as neutrophils from venous blood, or respond to chemotactic agents (Thorpe *et al.,* 1986). Those features may be due to prior activation in that recent flow cytometry studies demonstrate that neutrophils in human milk display changes in their surface phenotypes (increased expression of CD11b, decreased expression of leukocyte adhesion molecule-1) that are found on activated neutrophils (Keeney *et al.,* 1992).

Macrophages in milk also appear to be activated. This is suggested from their morphology (Smith *et al.,* 1971), their surface phenotypic features (Keeney *et al.,* 1992), and their enhanced motility *in vitro* (Mushtaha *et al.,* 1989; Özkaragoz *et al.,* 1988). Human milk macrophages have also been found to produce toxic oxygen radicals (Tsuda *et al.,* 1984) and express class II gene products of the major histocompatibility complex (Wirt *et al.,* 1992). The *in vivo* functions of these leukocytes are not established.

Lymphocytes are also found consistently in human milk. About 80% of them are T cells (Wirt *et al.,* 1992). There is controversy concerning the relative frequencies of the major subsets of T cells in human milk and some of the differences in the results from different studies may be due to methodologic variables. The proportions of $CD3^+CD4^+$ and $CD3^+CD8^+$ in unfractionated human milk leukocytes examined by immunofluorescent microscopy were similar to those in human blood (Keller *et al.,* 1986), whereas the proportions of CD^+CD8^+ in human milk leukocytes that were separated by density gradient centrifugation and examined by flow cytometry were higher than those in peripheral blood (Richie *et al.,* 1982). Recently, unfractionated human milk leukocytes were found by flow cytometry to have a higher relative frequency of $CD3^+CD8^+$ than those in peripheral blood (Wirt *et al.,* 1992). The cytotoxic responses of these cells are poor (Kohl *et al.,* 1978, 1980), but the cells are capable of generating certain lymphokines, including interferon-γ and monocyte chemotactic factor Emodi and Just, 1974; Keller *et al.,* 1981; Lawton *et al.,* 1979). In that respect, essentially all T cells in human milk bear the phenotypic marker of memory T cells (leukocyte common antigen isoform, CD45RO) (Bertotto *et al.,* 1990; Wirt *et al.,* 1992), and that type of T cell is the principal producer of interferon-γ (Berlotto *et al.,* 1990; Sanders *et al.,* 1988).

The *in vivo* role of T cells in human milk is uncertain, but it is of considerable interest that very small numbers of memory T cells are detected in blood in infancy (Hayward *et al.,* 1989). Thus, it may be possible that maternal memory T cells in milk compensate for the developmental delay in their production in the infant. There is evidence from experimental animal studies that milk lymphocytes enter tissues of the neonate (Head *et al.,* 1977; Jain *et al.,* 1989; Schnorr and Pearson, 1984; Weiler *et* al., 1983), but that has not been demonstrated in humans. In that regard, comparisons between the phenotypic expression of CD45RO on T cells in the blood of breast-fed and nonbreast-fed infants will be of interest. In addition, further studies are in order of the pattern of antigens to which these T cells respond (Parmeley *et al.,* 1976) and the repertoire of the T cell antigen receptors of those cells in human milk.

E. Anti-inflammatory Agents

One of the extraordinary features of the protection afforded by human milk is the virtual absence of clinical signs of inflammation during the gastrointestinal infections. This may be due in part to the more rapid elimination or neutralization of microbial pathogens in the lumen of the gastrointestinal tract by defense agents from human milk, but other features of human milk suggest that this is not the sole explanation. Phlogistic agents and the systems that give rise to them are poorly represented in human milk. In addition, human milk contains a host of anti-inflammatory agents (Goldman *et al.,* 1986; Goldman *et al.,* 1989b), some that double as antimicrobial factors (lactoferrin, secretory IgA, and lysozyme). The major classes of these anti-inflammatory agents in human milk include factors that promote the growth of epithelium, such as cortisol (Kulski *et al.,* 1981), epithelial growth factor (Carpenter, 1980), polyamines (Romain *et al.,* 1992), and lactoferrin (Nichols *et al.,* 1987), antioxidants (ascorbate-like compound, uric acid, β-carotene), and agents that inhibit nonoxidative inflammatory systems such as prostaglandins (Nen *et al.,* 1988) and platelet-activating factor acetylhydrolase (Furukawa *et al.,* 1992). Like the antimicrobial factors, these factors are well adapted to operate in the hostile environment of the alimentary tract.

The nature of the antioxidants in human milk has recently been investigated (Buescher and McIlheran, 1992). It was found that the peaks of antioxidant activity in colostrum were due to an ascorbate-like compound and uric acid. In addition, two other antioxidants in human milk, α-tocopherol (Chapell *et al.,* 1985; Ostrea *et al.,* 1986) and β-carotene (Ostrea *et al.,* 1986), are absorbed into the circulation where they may have systemic effects. In that regard, serum vitamin E concentrations rise in breast-fed infants from a mean of 0.3 mg/ml at birth to ~0.9 mg/ml on the fourth day of life. Otherwise, there is little information concerning the *in vivo* fate and functions of the anti-inflammatory agents in human milk.

F. Immunostimulating Agents

If human milk stimulates certain defense systems in the infant, one might predict that the effects might lead to long-lasting resistance. Supporting epidemiologic evidence for that premise has been mounting for several years. The incidence of juvenile diabetes mellitus (Mayer *et al.,* 1988) and Crohn's disease (Koletsko *et al.,* 1989) appears to be less among children who have been breast-fed during infancy. In addition, a recent retrospective analysis suggests that breast-feeding lessens the risk from lymphomas (Davis *et al.*, 1988). In each study, considerable reliance has been placed upon the abilities of mothers to recall the type and duration of feeding given to their offspring; yet, recall of events that transpired many years beforehand may be suspect. Undoubtedly, prospective studies of the possible long-term protective role of human milk will be required to further explore the possible long-term benefits of human milk.

A number of experimental observations also suggest that human milk provides active protection for the recipient infant. The production of IgA at mucosal sites may be enhanced by human milk (Goldblum *et al.*, 1989; Prentice, 1987; Schanler *et al.*, 1986; Stephens, 1986; Stephens *et al.*, 1984). Although in most of those studies it has been difficult to exclude the effect of passively transferred secretory IgA, in two investigations the excretion of IgA was increased in the urinary tract, a system removed from direct contact with human milk (Goldblum *et al.*, 1989; Prentice, 1987). Furthermore, in one report the urinary excretion of free secretory component was also remarkably increased in infants fed human milk (Goldblum *et al.*, 1989). The M_r of these proteins far exceeds the size of molecules filtered by glomeruli and neither secretory component nor secretory IgA are transported from the systemic circulation into external secretions by epithelial cells. It is, therefore, likely that human milk feedings stimulated the synthesis of secretory component by epithelial cells in the urinary tract and that this, in turn, enhanced the transport of secretory IgA into the urine of the infants. The components in human milk that may be responsible for such an enhancement are undetermined at this time.

The second type of evidence that the recipient infant's immune system is stimulated by breast-feeding is the increase in certain immune factors in the blood of breast-fed infants that cannot be accounted for by the levels of those factors in human milk. The response of breast-fed and nonbreast-fed infants to respiratory syncytial virus (RSV) infection was compared by measuring their serum interferon-α levels (Chiba *et al.*, 1987). The serum levels of interferon-α were strikingly increased in breast-fed infants in the first 2–4 weeks after RSV infection. Since there is little interferon-α in human milk, it seems likely that human milk is able to prime leukocytes in the host to produce that cytokine. In addition, there is evidence that the plasma concentrations of fibronectin are higher in breast-fed than nonbreast-fed infants (237 and 17/mg/liter, respectively) (Friss *et al.*, 1988). Since the amount of ingested fibronectin is not sufficient to account for the increment in plasma fibronectin that has been observed, it seems likely that human milk induces the synthesis of that opsonin in the infant.

The last piece of evidence comes from discoveries of immunomodulators in human milk. Human milk contains a high concentration of α-tocopherol, an agent which, in addition to its antioxidant effects, is known to stimulate the development of immunity (Bendich *et al.*, 1983, 1984, 1986; Tengerdy *et al.*, 1981). Several glycoproteins that orchestrate the development and functions of the immune system have been found in human milk. These agents, termed cytokines, require only minute quantities to be bioactive. Moreover, there are many interrelationships between those agents. The cytokines and their concentrations (mean values unless otherwise indicated) in early human milk are (1) interleukin-1β (IL-1β) (~1130 pg/ml) (Munoz *et al.*, 1990), (2) IL-6 (~150 pg/ml) (Saito *et al.*, 1991; Rudloff *et al.*, 1993), (3) IL-8 (~3680 pg/ml) (Palkowetz *et al.*, 1994), (4) IL-10 (~3300 pg/ml) (Garofalo *et al.*, 1995), (5) tumor necrosis factor-α (~620 pg/ml) (Mushtaha *et al.*, 1989; Rudloff *et al.*, 1992), and (6) transforming growth factor-β (~130 pg/ml) (Noda *et al.*, 1984; Palkowetz *et al.*, 1994), granulocyte colony-stimulating factor (45–1554 pg/ml) (Gilmore *et al.*, 1994), and macrophage colony-stimulating factor (Hara *et al.*, 1995). The effects of these agents in human milk on the recipient infant are as yet unknown, but it is likely that they will influence the development of the defenses of the respiratory and alimentary tract.

III. Coda

Because of the heterogeneity and complexity of the immunologic system in human milk, definitive investigations of the molecular, quantitative, and functional features of the components of the system have often been incomplete. Although considerable progress has been made toward defining many aspects of the defense system in human milk, further research will be required to identify the entire system, unravel the molecular biology and mechanisms of production and secretion, and determine the discrete role of each component of the system in the defense of the mammary gland or the recipient infant. In the process, it will continue to be important to anticipate that the ultimate role of these defense agents may depend upon a multiplicity of interactions with other factors in the system found in milk or with defense agents or cells produced by the breast-fed infant.

Acknowledgment

We thank Susan C. Kovacevich for her secretarial assistance in the preparation of the manuscript.

References

Allardyce, R. A., Shearman, D. J. C., McClelland, D. B. L., Marwick, K., Simpson, A. J., and Laidlaw, R. M. (1974). Appearance of specific colostrum antibodies after clinical infection with *Salmonella typhimiurim*. *Br. Med. J.* **3,** 307–309.

Anderson, B. F., Baker, H. M., Dodson, E. J., Norris, G. E., Rumball, S. V., Waters, J. M., and Baker, E. N. (1987). Structure of human lactoferrin at 3.1-Å resolution. *Proc. Natl. Acad. Sci. USA* **84,** 769–1773.

Andersson, B., Porras, O., Hanson, L.Å., Lagergard, T., and Svanborg-Eden, C. (1986). Inhibition of attachment of *Streptococcus pneumoniae* and *Haemophilus influenzae* by human milk and receptor oligosaccharides. *J. Infect. Dis.* **153,** 232–237.

Arnold, R. R., Cole, M. F., and McGhee, J. R. (1977). A bactericidal effect for human milk lactoferrin. *Science* **197,** 263–265.

Bakos, M-A., Kurosky, A., and Goldblum, R. M. (1991). Characterization of a critical binding site for human polymeric Ig on secretory component. *J. Immunol* **147,** 3419–3426.

Ballow, M., Fang, F., Good, R.A., and Day, N.K. (1974). Developmental aspects of complement components in the newborn. The presence of complement components and C3 proactivator (properdin factor B) in human colostrum. *Clin. Exp. Immunol.* **18,** 257–266.

Bellamy, W., Wakabayashi, H., Takase, M., Kawase, K., Shimamura, S., and Tomita, M. (1993). Killing of *Candida albicans* by lactoferricin B, a potent antimicrobial peptide derived from the N-terminal region of bovine lactoferrin. *Med. Microbiol. Immunol. (Berlin)* **182,** 97–1051.

Bendich, A., D'Apolito, P., Gabriel, E., and Machlin, L. J. (1983). Modulation of the immune system function of guinea pigs by dietary vitamins E and C following exposure to 100% O_2. *Fed. Proc.* **42,** 923.

Bendich, A., D'Apolito, P., Gabriel, E., and Machlin, L. J. (1984). Interaction of dietary vitamin C and vitamin E on guinea pig immune responses to mitogens. *J. Nutr.* **114,** 1588–1593.

Bendich, A., Gabriel, E., and Machlin, L. J. (1986). Dietary vitamin E requirement for optimum immune responses in the rat. *J. Nutr.* **116,** 675–681.

Berthou, J., Migliore-Samour, D., Lifchitz, A., Delettre, J., Floch, F., and Jolles, P. (1987). Immunostimulating properties and three-dimensional structure of two tripeptides from human and cow caseins. *FEBS Lett.* **218,** 55–58.

Bertotto, A., Gerli, R., Fabietti, G., Crupi, S., Arcangeli, C., Scalise, F., and Vaccaro, R. (1990). Human breast milk T cells display the phenotype and functional characteristics of memory T cells. *Eur. J. Immunol.* **20,** 1877–1880.

Bezkorovainy, A., Grohlich, D., and Nichols, J. H. (1979). Isolation of a glycopeptide fraction with *Lactobacillus bifidus* subspecies *Pennsylvanicus* growth-promoting activity from whole human milk casein. *Am. J. Clin. Nutr.* **32,** 1428–1432.

Bezkorovainy, A., and Topouzian, N. (1981). *Bifidobacterium bifidus* var. *Pennsylvanicus* growth promoting activity of human milk casein and its derivatives. *Int. J. Biochem.* **13,** 585–590.

Brandtzeg, P. (1974). Mucosal and glandular distribution of immunoglobulin components: Differential localization of free and bound SC in secretory epithelial cells. *J. Immunol.* **112,** 1553–1559.

Brantl, V. (1985). Novel opioid peptides derived from human β-casein: Human β-casomorphins. *Eur. J. Pharmacol.* **106,** 213–214

Brines, R. D., and Brock, J. H. (1983). The effect of trypsin and chymotrypsin on the *in vitro* antimicrobial and iron-binding properties of lactoferrin in human milk and bovine colostrum. *Biochem. Biophys. Acta* **759,** 229–235.

Buescher, S. E., and McIlheran, S. M. (1992). Colostral antioxidants: Separation and characterization of two activities in human colostrum. *J. Pediatr. Gastroenterol. Nutr.* **14,** 47–56.

Bullen, J. J., Rogers, H. J., and Leigh, L. (1972). Iron-binding proteins in milk and resistance of *Escherichia coli* infection in infants. *Br. Med. J.* **1,** 69–75.

Butte, N. F., Goldblum, R. M., Fehl, L. M., Loftin, K., Smith E. O., Garza, C., and Goldman, A. S. (1984). Daily ingestion of immunologic components in human milk during the first four months of life. *Acta Paediatr. Scand.* **73,** 296–301.

Carpenter, G. (1980). Epidermal growth factor is a major growth-promoting agent in human milk. *Science* **210,** 198–199.

Chandan, R. C., Shahani, K. M., and Holly, R. G. (1964). Lysozyme content of human milk. *Nature (London)* **204,** 76–77.

Chapell, J. E., Francis, T., and Clandinin, M. T. (1985). Vitamin A and E content of human milk at early stages of lactation. *Early Hum. Dev.* **11,** 157–167.

Chiba, Y., Minagawa, T., Mito, K., Nakane, A., Suga, K., Honjo, T., and Nako, T. (1987). Effect of breast feeding on responses of systemic interferon and virus-specific lymphocyte transformation in infants with respiratory syncytial virus infection. *J. Med. Virol.* **21,** 7–14.

Chipman, D. M., and Sharon, N. (1969). Mechanism of lysozyme action. *Science* **165,** 454–465.

Cleary, T. G., Chambers, J. P., and Pickering, L. K. (1983). Protection of suckling mice from the heat-stable enterotoxin of *Escherichae coli* by human milk. *J. Infect. Dis.* **148,** 1114–1119.

Cruz, J. R., Gil, L., Cano, F., Caceres, P., and Pareja, G. (1988). Breast milk anti-*Escherichia coli* heat-labile toxin IgA antibodies protect against toxin-induced infantile diarrhea. *Acta Paediatr. Scand.* **77,** 658–662.

Davidson, L. A., and Lonnerdal, B. (1985). Lactoferrin and secretory IgA in the feces of exclusively breast-fed infants. *Am. J. Clin. Nutr.* **41,** 852A.

Davidson, L. A., and Lonnerdal, B. (1987). The persistence of human milk proteins in the breast-fed infant. *Acta Paediatr. Scand.* **76,** 733–740.

Davis, M. K., Savitz, D. A., and Grauford, B. (1988). Infant feeding in childhood cancer. *Lancet* **2,** 365–368.

Emodi, G., and Just, M. (1974). Interferon production by lymphocytes in human milk. *Scand. J. Immunol.* **3,** 157–160.

Fishaut, M., Murphy, D. S., Neifert, M., McIntosh, K., and Ogra, P. L. (1981). Broncho-mammary axis in the immune response to respiratory syncytial virus. *J. Pediatr.* **99,** 186–191.

Fransson, G. B., and Lonnerdal, B. (1980). Iron in human milk. *J. Pediatr.* **96,** 380–384.

Friss, H. E., Rubin, L. G., Carsons, S., Baranowski, J., and Lipsitz, P. J. (1988). Plasma fibronectin concentrations in breast fed and formula fed neonates. *Arch. Dis. Child.* **63,** 528–532.

Furukawa, M., Narahara, H., and Johnston, J. M. (1992). Platelet-activating factor acetyl-hydrolase activity (PAF-AH) in human milk (HM). *In* "Fourth International Congress on PAF and Related Lipid Mediators." [Abstract]

Garofalo, R., Chheda, S., Mei, F., Palkowetz, K. H., Rudloff, H. E., Schmalstieg, F. C., and Goldman, A. S. (1995). Interleukin-10 in human milk. *Pediatr. Res.* **37,** 444–449.

Gillin, F. D., Reiner, D. S., and Gault, M. J. (1985). Cholate-dependent killing of Giardia Lamblia by human milk. *Infect. Immun.* **47,** 619–622.

Gillin, F. D., Reiner, D. S., and Wang, C-S. (1983). Human milk kills parasitic protozoa. *Science* **221,** 1290–1292.

Gilmore, H. S., McKelvey-Martin, V. J., Rutherford, S., Strain, J. J., Kell, M., and Miller, S. (1944). Human milk contains granulocyte-colony stimulating factor. *Eur. J. Clin. Nutr.* **48,** 222–224.

Glass, R. I., Svennerholm, A-M., Stoll, B. J., Khan, M. R., Belayet Hossain, K. M., Imdadul Huq, M., and Holmgren, J. (1983). Protection against cholera in breast-fed children by antibodies in breast milk. *N. Engl. J. Med.* **308,** 1389–1392.

Goldblum, R. M., Ahlstedt, S., Carlsson, B., Hanson, L. Å., Jodal, U., and Lindin-Janson, G. (1975). Antibody-forming cells in human colostrum after oral immunization. *Nature* **257,** 797–799.

Goldblum, R. M., Garza, C., Johnson, C. A., Harrist, R., Nichols, B. L., and Goldman, A. S. (1981). Human milk banking I. Effects of container upon immunologic factors in mature milk. *Nutr. Res.* **1,** 449–459.

Goldblum, R. M., Garza, C., Johnson, C. A., Nichols, B. L., and Goldman, A. S. (1982). Human milk banking. II. Relative stability of immunologic factors in stored colostrum. *Acta Paediatr. Scand.* **72,** 143–144.

Goldblum, R. M., Schanler, R. J., Garza, C., and Goldman, A. S. (1989). Human milk feeding enhances the urinary excretion of immunologic factors in low birth weight infants. *Pediatr. Res.* **25,** 184–188.

Goldman, A. S., Garza, C., Johnson, C. A., Nichols, B. L., and Goldblum, R. M. (1982). Immunologic factors in human milk during the first year of lactation. *J. Pediatr.* **100,** 563–567.

Goldman, A. S., Garza, C., Johnson, C. A., Nichols, B. L., and Goldblum, R. M. (1983a). Immunologic components in human milk during weaning. *Acta Paediatr. Scand.* **72,** 133–134.

Goldman, A. S., Goldblum, R. M., and Garza, C. (1983b). Immunologic components in human milk during the second year of lactation. *Acta Paediatr, Scand.* **72,** 461–462.

Goldman, A. S., Thorpe, L. W., Goldblum, R. M., and Hanson, L. Å. (1986). Anti-inflammatory properties of human milk. *Acta Paediatr. Scand.* **75,** 689–695.

Goldman, A. S., and Goldblum, R. M. (1989a). Immunoglobulins in human milk. *In* "Protein and Non-Protein Nitrogen in Human Milk" (S. A. Atkinson and B. Lonnerdal, eds.), pp. 43–51. CRC Press, Boca Raton, FL.

Goldman, A. S., Goldblum, R. M., and Hanson, L. Å. (1989b). Anti-inflammatory systems in human milk. *In* "Antioxidant Nutrients and the Immune Response" (A. Bendich, M. Phillips, and R. Tengerdy, eds.), pp. 69–76. Plenum Press, New York.

Goldman, A. S., Garza, C., Schanler, R. J., and Goldblum, R. M. (1990). Molecular forms of lactoferrin in stool and urine from infants fed human milk. *Pediatr. Res.* **27,** 252–255.

György, P., Jeanloz, R. W., Von Nicolai, H., and Zilliken, F. (1974). Undialyzable growth factors for *Lactobacillus bifidus* var. *Pennsylvanicus. Eur. J. Biochem.* **43,** 29–33.

Haneberg, B., and Finne, P. (1974). Lysozymes in feces from infants and children. *Acta Paediatr. Scand.* **63,** 588–594.

Hara, T., Irie, K., Saito, S., Ichijo, M., Yamada, M., Yanai, N., and Miyazaki, S. (1995). Identification of macrophage colony-stimulating factor in human milk and mammary gland epithelial cells. *Pediatr. Res.,* in press.

Hayward, A. R., Lee, J., and Beverley, P. C. L. (1989). Ontogeny of expression of UCHL1 antigen on TcR-1$^+$ (CD4/8) and TcRδ^+ T cells. *Eur. J. Immunol.* **19,** 771–773.

Head, J. R., Beer, A. A., and Billingham, R. E. (1977). Significance of the cellular component of the maternal immunologic endowment in milk. *Transplant Proc.* **9,** 1465–1471.

Hernell, O., Ward, H., Blackberg, L., and Pereira, M. E. (1987). Killing of Giardia Lamblia by human milk lipases: An effect mediated by lipolysis of milk lipids. *J. Infect. Dis.* **153,** 715–720.

Holmgren, J., Svennerholm, A-M., and Ahren, C. (1981). Nonimmunoglobulin fraction of human milk inhibits bacterial adhesion (hemagglutination) and enterotoxin binding of *Escherichiae coli* and *Vibrio cholerae. Infect. Immun.* **33,** 136–141.

Holmgren, J., Svennerholm, A-M., Lindblad, M. (1983). Receptor-like glycocompounds in human milk that inhibit classical and *E1 Tor Vibrio cholerae* cell adherence (hemagglutination). *Infect. Immun.* **39,** 147–154.

Holmgren, J., Svennerholm, A-M., Lindblad, M., and Strecker, G. (1987). Inhibition of bacterial adhesion and toxin binding by glycoconjugate and oligosaccharide receptor analogues in human milk. *In* "Human Lactation 3: The Effects of Human Milk on the Recipient Infant" (A. S. Goldman, S. A. Atkinson, and L. Å. Hanson, eds.), pp. 251–259. Plenum Pres, New York.

Hutchens, T. W., Henry, J. F., Yip, T-T., Hachey, D. L., Schanler, R. J., Motil, K. J., and Garza, C. (1991). Origin of intact lactoferrin and its DNA-binding fragments found in the urine of human milk-fed preterm infants. Evaluation of stable isotopic enrichment. *Pediatr. Res.* **29,** 243–250.

Issacs, C. E., Thormar, H., and Pessolano, T. (1986). Membrane-disruptive effect of human milk: Inactivation of enveloped viruses. *J. Infect. Dis.* **154,** 966–971.

Jain, L., Vidyasagar, D., Xanthou, M., Ghai, V., Shimada, S., and Blend, M. (1989). *In vivo* distribution of human milk leucocytes after ingestion by newborn baboons. *Arch. Dis. Child.* **64,** 930–933.

Jatsyk, G. V., Kuvaeva, I. B., and Gribakin, S. G. (1985). Immunologic protection of the neonatal gastrointestinal tract: The importance of breast-feeding. *Acta Paediatr. Scand.* **74,** 246–249.

Jolles, J., and Jolles, P. (1967). Human tear and human milk lysozymes. *Biochemistry* **6,** 411–417.

Kapikian, A. Z., Greenburg, H. B., Kalica, A. R., Wyatt, R. G., Kim, H. W., Brandt, C. D., Rodriguez, W. J., *et al.* (1981). New developments in viral gastroenteritis. *In* "Acute Enteric Infections in Children. New Prospects for Treatment and Prevention" (T. Holme, J. Holmgren, M. H. Merson, and R. Molby, eds.), pp. 9–57. Elsevier, Amsterdam.

Keeney, S. E., Schmalstieg, F. C., Palkowetz, K. H., Rudloff, H. E., Schmalstieg, F. C., Jr., and Goldman, A. S. (1992). Activated neutrophils in human milk. *Clin. Res.* **39,** 822A.

Keller, M. A., Faust, J., Rolewic, L. J., and Stewart, D. D. (1986). T cell subsets in human milk. *J. Pediatr. Gastroenterol. Nutr.* **5,** 439–443.

Keller, M. A., Heiner, D. C., Myers, A. S., and Reisinger, D. M. (1984). IgD—A mucosal immunoglobulin? *Pediatr. Res.* **18,** 258A.

Keller, M. A., Heiner, D. C., Kidd, R. M., and Myers, A. S. (1983). Local production of IgG4 in human colostrum. *J. Immunol.* **130,** 1654–1657.

Keller, M. A., Kidd, R. M., Bryson, Y. J., Turner, J. L., and Carter, J. (1981). Lymphokine production by human milk lymphocytes. *Infect. Immun.* **32,** 632–636.

Koch, G., Wiedermann, K., and Teschemacher, H. (1985). Opioid activities of human β-casomorphins. *Naunyn Schmiedebergs Arch. Pharmacol.* **331,** 351–354.

Kohl, S., Malloy, M. D., Pickering, L. K., Morriss, F., Adcock, E. W., and Waters, D. L. (1978). Human colostral cytotoxicity. I. Antibody-dependent cellular cytotoxicity against herpes simplex viral-infected cells mediated by colostral cells. *J. Clin. Lab. Immunol.* **1,** 221–224.

Kohl, S., Pickering, L. K., Cleary, T. G., Steinmetz, K. D., and Loo, L. S. (1980). Human colostral cytotoxicity. II. Relative defects in colostral leukocyte cytotoxicity and inhibition of peripheral blood leukocyte cytoxicity by colostrum. *J. Infect. Dis.* **142,** 884–891.

Koletzko, S., Sherman, P., Corey, M., Griffiths, A., and Smith, C. (1989). Role of infant feeding practices in development of Crohn's disease in childhood. *Br. Med. J.* **298,** 1617–1618.

Kulski, J. K., and Hartmann, P. E. (1981). Changes in the concentration of cortisol in milk during different stages of human lactation. *Aust. J. Exp. Biol. Med. Sci.* **59,** 769–780.

Laegreid, A., and Kolsto Otnaess, A.-B. (1987). Trace amounts of ganglioside GM1 in human milk inhibit enterotoxins from *Vibrio cholerae* and *Escherichia coli. Life Sci.* **40,** 55–62.

Laegreid, A., Kolsto Otnaess, A.-B., and Bryn, K. (1986a). Purification of human milk gangliosides by silica gel chromatography and analysis of trifluoroacetate derivatives by gas chromatography. *J. Chromatogr.* **377,** 59–67.

Laegreid, A., Kolsto Otnaess A.-B., and Fuglesang, J. (1986b). Human and bovine milk: Comparison of ganglioside composition and enterotoxin-inhibitory activity. *Pediatr. Res.* **20,** 416–421.

Lawton, J. W. M., Shortridge, K. F., Wong, R. L. C., and Ng, M. H. (1979). Interferon synthesis by human colostral leukocytes. *Arch. Dis. Child.* **54,** 127–130.

Mata, L. J., and Wyatt, R. G. (1971). Host resistance to infections. *Am. J. Clin. Nutr.* **24,** 976–986.

Mayer, E. J., Hamman, R. F., Gay, E. C., Lezotte, D. C., Savitz, D. A., and Klingensmith, G. J. (1988). Reduced risk of IDDM among breast fed children. The Colorado IDDM Registry. *Diabetes* **37,** 1625–1632.

McClelland, D. B. L., McGrath, J., and Samson, R. R. (1978). Antimicrobial factors in human milk. Studies of transfer to the infant during the early stages of lactation. *Acta Paediatr. Scand.* **57** (Suppl. 271), 1–20.

Monti, J. C., Mermoud, A.-F., and Jolles, P. (1989). Anti-bovine β-lactoglobulin antibodies react with a human lactoferrin fragment and bovine β-lactoglobulin present in human milk. *Experientia* **45,** 178–180.

Mostov, K. E., Kraehenbuhl, J. P., and Blobel, G. (1980). Receptor-mediated transcellular transport of immunoglobulin: Synthesis of secretory component as multiple and larger transmembrane forms. *Proc. Natl. Acad. Sci. USA* **77,** 7257–7261.

Mostov, K. E., and Blobel, G. A. (1982). A transmembrane precursor of secretory component. The receptor for transcellular transport of polymeric immunoglobulins. *J. Biol. Chem.* **257,** 11816–11821.

Mostov, K. E., Friedlander, M., and Blobel, G. (1984). The receptor for transepithelial transport of IgA and IgM contains multiple immunoglobulin-like domains. *Nature (London)* **308,** 37–43.

Munoz, C., Endres, S., van der Meer, J., Schlesinger, L., Arevalo, M., and Dinarello, C. (1990). Interleukin-1β in human colostrum. *Res. Immunol.* **141,** 501–513.

Mushtaha, A. A., Schmalstieg, F. C., Hughes, T. K., Rajaraman, S., Rudloff, H. E., and Goldman, A. S. (1989). Chemokinetic agents for monocytes in human milk: Possible role of tumor necrosis factor-α. *Pediatr. Res.* **25,** 629–633.

Nakajima, S., Baba, A. S., and Tamura, N. (1977). Complement system in human colostrum. *Int. Arch. Allergy Appl. Immunol.* **54,** 428–433.

Nen, J., Wu-Wang, C. Y., Measel, C. P., and Gimotty, P. (1988). Prostaglandin concentrations in human milk. *Am. J. Clin. Nutr.* **47,** 649–652.

Newburg, D. S., Hundreiser, K. E., and McCluer, R. H. (1990a). Novel glycolipids of human and bovine milk. *In* "Breastfeeding, Nutrition, Infection and Infant Growth in Developed and Emerging Countries" (S. A. Atkinson, L. Å. Hanson, and R. K. Chandra, eds.). ARTS Biomedical Publishers, St. Johns, Newfoundland, Canada.

Newburg, D. S., Pickering, L. K., McCluer, R. H., and Cleary, T. G. (1990b). Fucosylated oligosaccharides of human milk protect suckling mice from heat-stable enterotoxin of *Escherichia coli. J. Infect. Dis.* **162,** 1075–1080.

Newburg, D. S., Viscidi, R. P., Ruff, A., and Yolken, R. H. (1992). A human milk factor inhibits binding of human immunodeficiency virus to the CD4 receptor. *Pediatr. Res.* **31,** 22–28.

Nichols, B. L., McKee, K. S., Henry, J. F., and Putnam, M. (1987). Human lactoferrin stimulates thymidine incorporation into DNA of rat crypt cells. *Pediatr. Res.* **21,** 563–567.

Nichols, J. H., Bezkorovainy, A., and Paque, R. (1975). Isolation and characterization of several glycoproteins from human colostrum whey. *Biochem. Biophys. Acta* **412,** 99–108.

Noda, K., Ruegg, C., and Ono, T. (1984). Transforming growth factor activity in human colostrum. *Gann* **75,** 109–112.

Ogra, S. S., and Ogra, P. L. (1978). Immunologic aspects of human colostrum and milk I. Distribution characteristics and concentrations of immunoglobulins at different times after the onset of lactation. *J. Pediatr.* **92,** 546–549.

Ostrea, E. A., Jr., Balun, J. E., Winkler, R., and Porter, T. (1986). Influence of breast-feeding on the restoration of the low serum concentration of vitamin E and β-carotene in the newborn infant. *Am. J. Obstet. Gynecol.* **154,** 1014–1017.

Otnaess, A. B., Laegreid, A., and Ertresvag, K. (1983). Inhibition of enterotoxin from *Escherichia coli* and *Vibrio cholerae* by gangliosides from human milk. *Infect. Immunol.* **40,** 563–569.

Otnaess, A-B., and Svennerholm, A-M. (1982). Non-immunoglobulin fraction of human milk protects against enterotoxin-induced intestinal fluid secretion. *Infect. Immun.* **35,** 738–740.

Özkaragoz, F., Rudloff, H. E., Rajaraman, S., Mushtaha, A. A., Schmalstieg, F. C., and Goldman, A. S. (1988). The motility of human milk macrophages in collagen gels. *Pediatr. Res.* **23,** 449–452.

Palkowetz, K. H., Royer, C. L., Garofalo, R., Rudloff, H. E., Schmalstieg, F. C., and Goldman, A. S. (1994). Production of interleukin-6 and interleukin-8 by human mammary gland epithelial cells. *J. Reprod. Immunol.* **26,** 57–64.

Parker, F., Migliore-Samour, D., Floch, F., Zerial, A., Werner, G. H., Jolles, J., Casaretto, M., Zahn, H., and Jolles, P. (1984). Immunostimulating hexapeptide from human casein: Amino acid sequence, synthesis, and biological properties. *Eur. J. Biochem.* **145,** 677–682.

Parmely, M. J., Beer, A. E., and Billingham, R. E. (1976). *In vitro* studies on the T-lymphocyte population of human milk. *J. Exp. Med.* **144,** 358–370.

Peitersen, B., Bohn, L., and Anderson, H. (1975). Quantitative determination of immunoglobulins, lysozyme, and certain electrolytes during a 24-hour period, and in milk from the individual mammary gland. *Acta Paediatr. Scand.* **64,** 709–717.

Prentice, A. (1987). Breast feeding increases concentrations of IgA in infants' urine. *Arch. Dis. Child.* **62,** 792–795.

Prentice, A., Ewing, G., Roberts, S. B., Lucas, A., MacCarthy, A., Jarjou, L. M. A., and Whitehead, R. G. (1987). The nutritional role of breast milk IgA and lactoferrin. *Acta Paediatr. Scand.* **76,** 592–598.

Resta, S., Luby, J. P., Rosenfeld, C. R., and Siegel, J. D. (1985). Isolation and propagation of a human enteric coronavirus. *Science* **229,** 978–981.

Richie, E. R., Bass, R., Meistrich, M. L., and Dennison, D. K. (1982). Distribution of T lymphocyte subsets in human colostrum. *J. Immunol.* **129,** 1116–1119.

Romain, N., Dandrifosse, G., Jeusette, F., and Forget, P. (1992). Polyamine concentration in rat milk and food, human milk, and infant formula. *Pediatr. Res.* **32,** 58–63.

Roux, M. E., McWilliams, M., Phillips-Quagliata, J. M., Weisz-Carrington, P., and Lamm, M. E. (1977). Origin of IgA-secreting plasma cells in the mammary gland. *J. Exp. Med.* **146,** 1311–1322.

Rudloff, H. E., Schmalstieg, F. C., Jr., Mushtaha, A. A., Palkowetz, K. H., Liu, S. K., and Goldman, A. S. (1992). Tumor necrosis factor-α in human milk. *Pediatr. Res.* **31,** 29–33.

Rudloff, H. E., Schmalstieg, F. C., Jr., Palkowetz, K. H., Paskiewicz, E. J., and Goldman, A. S. (1993). Interleukin-6 (IL-6) in human milk (HM). *J. Reprod. Immunol.* **23,** 3–20.

Saito, S., Maruyama, M., Kato, Y., Moriyama, I., and Ichijo, M. (1991). Detection of IL-6 in human milk and its involvement in IgA production. *J. Reprod. Immunol.* **20,** 267–276.

Samson, R. R., Mirtle, C., and McClelland, D. B. L. (1980). The effect of digestive enzymes on the binding and bacteriostatic properties of lactoferrin and vitamin B12 binder in human milk. *Acta Paediatr. Scand.* **69,** 517–523.

Sanders, M. E., Makgoba, M. W., Sharrow, S. O., Stephany, D., Spinger, T. A., Young, H. A., and Shaw, S. (1988). Human memory T lymphocytes express increased levels of three cell adhesion molecules (LFA-3, CD2, and LFA-1) and have three other molecules (UCHL1, CDw29, and Pgp-1) and have enhanced INF-γ production. *J. Immunol.* **140,** 1401–1407.

Schanler, R. J., Goldblum, R. M., Garza, C., and Goldman, A. S. (1986). Enhanced fecal excretion of selected immune factors in very low birth weight infants fed fortified human milk. *Pediatr. Res.* **20,** 711–715.

Schnorr, K. L., and Pearson, L. D. (1984). Intestinal absorption of maternal leukocytes by newborn lambs. *J. Reprod. Immunol.* **6,** 329–337.

Schroten, J., Hanisch, F. G., Plogmann, R., Hacker, J., Uhlernbruch, G., Nobis-Bosch, R., and Wahn, V. (1992). Inhibition of adhesion of S-fimbriated *Escherichia coli* to buccal epithelial cells by human milk fat globule membrane components: A novel aspect of the protective function of mucins in the nonimmunoglobulin fraction. *Pediatr. Res.* **32,** 58–63.

Smith, C. W., and Goldman, A. S. (1968). The cells of human colostrum. I. *In vitro* studies of morphology and functions. *Pediatr. Res.* **2,** 103–109.

Smith, C. W., and Goldman, A. S. (1970). Interactions of lymphocytes and macrophages from human colostrum: Characteristics of the interacting lymphocyte. *Res. J. Reticuloendothelial Soc.* **8,** 91–104.

Smith, C. W., Goldman, A. S., and Yates, R. D. (1971). Interactions of lymphocytes and macrophages from human colostrum: Electron microscopic studies of the interacting lymphocyte. *Exp. Cell Res.* **69,** 409–415.

Soder, O. (1987). Isolation of interleukin-1 from human milk. *Int. Arch. Allergy Appl. Immunol.* **83,** 19–23.

Solari, R., and Kraehenbuhl, J. P. (1984). Biosynthesis of the IgA antibody receptor: A model for the transepithelial sorting of a membrane glycoprotein. *Cell* **36,** 61–71.

Spik, G., Brunet, B., Mazurier-Dehaine, C., Fontaine, G., and Montreuil, J. (1982). Characterization and properties of the human and bovine lactotransferrins extracted from the feces of newborn infants. *Acta Paediatr. Scand.* **71,** 979–985.

Spik, G., Cheron, A., Montreuil, J., and Dolby, J. M. (1978). Bacteriostasis of a milk-sensitive strain of *Escherichia coli* by immunoglobulins and iron-binding proteins in association. *Immunology* **35,** 663–671.

Spik, G., and Montreuil, J. (1966). Etudes comparatives de la structure de la tranferrine de la lactotransferrine humaines. Finger-printing des hydrolytes proteasiques des deux glycoproteides. *C. R. Seances Soc. Biol. Paris* **160,** 94–98.

Stephens, S., Dolby, J. M., Montreuil, J., and Spik, G. (1980). Differences in inhibition of the growth of commensal and enteropathogenic strains of *Escherichia coli* by lactoferrin and secretory immunoglobulin A isolated from human milk. *Immunology* **41,** 597–603.

Stephens, S., Kennedy, C. R., Lakhani, P. K., and Brenner, M. K. (1984). *In-vivo* immune responses of breast- and bottle-fed infants to tetanus toxoid antigen and to normal gut flora. *Acta Paediatr. Scand.* **73,** 426–432.

Stephens, S. (1986). Development of secretory immunity in breast fed and bottle fed infants. *Arch. Dis. Child.* **61,** 263–269.

Stock, C. C., and Francis, T., Jr. (1940). The inactivation of the virus of epidemic influenza by soaps. *J. Exp. Med.* **71,** 661–681.

Stuart, J., Norrel, S., and Harrington, J. P. (1984). Kinetic effect of human lactoferrin on the growth of *Escherichia coli. J. Biochem.* **16,** 1043–1047.

Tengerdy, R. P., Mathias, M. M., and Nockels, C. F. (1981). Vitamin E, immunity and disease resistance. *In* "Diet and Resistance to Disease" (M. Phillips and A. Baetz, eds.), pp. 27–42. Plenum Press, New York.

Teschemacher, H. (1987). β-Casomorphils: Do they have physiological significance? *In* "Human Lactation 3. The Effects of Human Milk on the Recipient Infant" (A. S. Goldman, S. A. Atkinson, and L. Å. Hanson, eds.), pp. 213–225. Plenum Press, New York.

Thorpe, L. W., Rudloff, H. E., Powell, L. C., and Goldman, A. S. (1986). Decreased response of human milk leukocytes to chemoattractant peptides. *Pediatr. Res.* **20,** 373–377.

Thromar, H., Isaacs, C. E., Brown, H. R., Barshatzky, M. R., and Pessolano, T. (1987). Inactivation of enveloped viruses and killing of cells by fatty acids and monoglycerides. *Am. Soc. Microbiol.* **32,** 27–31.

Tsuda, H., Takeshige, K., Shibata, Y., and Minakami, S. (1984). Oxygen metabolism of human colostral macrophages. *J. Biochem.* **95,** 1237–1245.

Underdown, B. J., Knight, A., and Papsin, F. R. (1976). The relative paucity of IgE in human milk. *J. Immunol.* **116,** 1435–1438.

Weiler, I. L., Hickler, W., and Spenger, R. (1983). Demonstration that milk cells invade the neonatal mouse. *Am. J. Reprod. Immunol.* **4,** 95–98.

Weiz-Carrington, P., Roux, M. E., McWilliams, M., Philips-Quaglita, J. M., and Lamm, M. E. (1978). Hormonal induction of the secretory immune system in the mammary gland. *Proc. Natl. Acad. Sci. USA* **75,** 2928–2932.

Welsh, J. K., and May, J. T. (1979). Anti-infective properties of breast milk. *J. Pediatr.* **94,** 1–9.

Welsh, J. K., Arsenakis, M., Coelen, R. J., and May, J. T. (1979). Effect of antiviral lipids, heat, and freezing on the activity of viruses in human milk. *J. Infect. Dis.* **140,** 332–338.

Wirt, D. P., Adkins, L. T., Palkowetz, K. H., Schmalstieg, F. C., and Goldman, A. S. (1992). Activated-memory T lymphocytes in human milk. *Cytometry* **13,** 282–290.

Yolken, R. H., Peterson, J. A., Vonderfecht, S. L., Fouts, E. T., Midthun, K., and Newburg, D. S. (1992). Human milk mucin inhibits rotavirus replication and prevents experimental gastroenteritis. *J. Clin. Invest.* **90,** 1984–1991.

Yamauchi, K., Tomita, M., Giehl, T. J., and Ellison, R. T. (1992). Antibacterial activity of lactoferrin and a pepsin-derived lactoferrin peptide fragment. *Infect. Immun.* **61,** 713–728.

B. Defense Agents in Bovine Milk

ROBERT G. JENSEN

I. Introduction

Raw bovine milk contains several antimicrobial agents which are beneficial to the calf, primarily in colostrum. However, these may be of little significance to the human consumer of milk and its products, since colostrum is generally excluded from the milk supply and the effectiveness of the systems is reduced or eliminated by clarification (centrifugal removal of cells, etc.), pasteurization, and homogenization. It is important to remember that bovine milk is intended to be consumed raw by the calf and should not be considered as an effective conveyer of antimicrobial systems to the human consumer. However, in locales where refrigeration and processing facilities are not available, one of the agents can be used to extend the shelf life of milk.

The antimicrobial agents are lysozyme, lactoferrin, lactoperoxidase, immunoglobulins, vitamin-binding proteins (IDF, 1991), and lipids. These systems are discussed below. Emphasis is on the human consumer although the agents contribute to the immunological status of the mammary gland and the calf.

II. Lysozyme

The enzyme is a 1,4-β-*N*-acetylmuramidase (EC 3.2.1.17), which hydrolyzes glycosidic bonds in Gram-positive bacterial cell walls. The amount in bovine milk is about 13 μg/dl (Banks and Tranter, 1986) and is inactivated to some extent by pasteurization. It seems unlikely that the enzyme is antimicrobial in processed milks and its effect is probably minimal in raw milk. See Chapters 5F and 2E for more discussion.

III. Lactoferrin

Lactoferrin is one of several proteins which binds iron. The iron is sequestered from microorganisms that require the cation, thus limiting their growth. However, according to Renner *et al.* (1989), the protein in

HANDBOOK OF MILK COMPOSITION

bovine milk is not bacteriostatic. Milk contains 2 to 20 mg/dl each of lactoferrin and transferrin. See Chapters 5F and 2E for more information.

IV. Lactoperoxidase

Lactoperoxidase (EC 1.11.1.7) catalyzes the conversion of H_2O_2 to H_2O. Milk contains about 3 mg/dl (Bjorck, 1991). When H_2O_2 (1 mg/dl) and thiocyanate (1 mg/dl) are added to raw milk, the SCN is oxidized by the enzyme–H_2O_2 complex producing bactericidal compounds which destroy Gram-negative bacteria (Renner *et al.*, 1989; Schmekel and Harnulv, 1986 IDF, 1991). The system is being used to improve the keeping quality of raw milk in developing countries. The enzyme is presumably inactivated by pasteurization at 72°C for 15 sec. See Chapter 5F for more information.

V. Immunoglobulins

Bovine milk contains the following immunoglobulins (mg/dl): IgG1, 59; IgG2, 20; IgA, 10; and IgM, 5 (IDF, 1991; Larson, 1992). Since these are bovine proteins they may not be effective in humans. They are inactivated by pasteurization. See Chapters 5F and 2E for more information.

VI. Vitamin-Binding Proteins

Milk contains proteins which bind vitamin B12, folate, and riboflavin (IDF, 1991; Fox and Flynn, 1992). It has been postulated but not proven that these proteins may inhibit the growth of microorganisms that require the vitamins for growth. The proteins are partially denatured by pasteurization. It is therefore unlikely that the proteins have any effect on bacterial growth in processed milk which has low bacterial contents and is refrigerated.

VII. Lipids

Gastric lipase can produce microbicidal free fatty acids and monoacylglycerols from milk triacylglycerols (TG) in the stomach as can pancreatic lipase in the small intestine (Hernell *et al.*, 1989). There are no studies on the identity of these compounds produced by the physiological digestion of milk TG. Milk contains 1.3 μg/liter of the ganglioside, GMI, which binds some enterotoxins (Laegrid *et al.*, 1986). The effectiveness of the compound in processed milk has not been studied,

VIII. Summary

Lactoperoxidase is the only one of these factors that appears to have any effect on milk for human consumption. This is because it can be used to extend the shelf life of milk in developing countries, thus increasing the amount of milk in locales where it is relatively scarce.

References

Banks, J. G., and Tranter, H. S. (1986). Lysozyme. *In* "Antimicrobial Systems in Milk," pp. 39–48. International Dairy Federation, 41, Square Vergote, 1040, Brussels, Belgium. Available in the United States from USNAC-IDF, 464 Central Ave., Northfield, IL 60093.

Bjorck, L. (1992). Lactoperoxidase. *In* "Proteins—Advanced Dairy Chemistry" (P. F. Fox, ed.), Vol. 1, pp. 332–336. Elsevier, New York.

Hernell, O., Blackberg, L., and Bernback, S. (1989). Milk lipases and in vivo lipolysis. *In* "Protein and Non-Protein Nitrogen in Human Milk" (S. A. Atkinson and B. Lonnerdal, eds.), pp. 221–236. CRC Press, Boca Raton, FL.

International Dairy Federation (IDF) (1991). Significance of the indigenous antimicrobial agents of milk to the dairy industry. *In* "Bulletin of the International Dairy Federation," No. 264, pp. 2–19.

Laegrid, A., Kolsto-Otnaess, A-B, and Fugelsan, J. (1986). Human and bovine milk: Comparison of ganglioside composition and enterotoxin-inhibitory activity. *Pediatr. Res.* **20,** 416–421.

Renner, E., Schaafsma, G, and Scott, K. J. (1989). Micronutrients in milk. *In* "Micronutrients in Milk and Milk-Based Food Products" (E. Renner, ed.), Lactoferrin, pp. 14–18. Lactoperoxidase, pp. 46–47. Elsevier, New York.

Schmekel, J., and Harnulv, G. (1986). Activation of the Lactoperoxidase as a Means of Saving Milk in Tropical Countries for Early Spoilage. *In* "Antimicrobial Systems in Milk" pp. 75–79. International Dairy Federation.

CHAPTER **10**

Comparative Analysis of Nonhuman Milks

A. Phylogenetic Variation in the Gross Composition of Milks

OLAV T. OFTEDAL
SARA J. IVERSON

I. Introduction

Our objective in preparing a compilation of data on the composition of mammalian milks is to assemble unbiased information that may allow an understanding of phylogenetic trends and an interpretation of the underlying evolutionary adaptations. For example, some primates and perissodactyls (horses and rhinoceroses) secrete dilute, low-fat milks, while bears, seals, and whales produce remarkably high-fat, energy-dense secretions. Marsupial milks are very high in carbohydrate, especially oligosaccharides, while the milks of sea lions and fur seals have only traces of carbohydrate and no lactose. These differences in milk secretion represent highly divergent patterns of nutrient transfer to the young, and presumably reflect adaptations in maternal rearing to environmental opportunities and physiological constraints.

Perhaps equally important is the fact that milk composition tables form the basis for the development of artificial formulas with which abandoned, orphaned, or injured neonates of a wide variety of wild and zoo animals are fed. If the data included in tables are misleading, as has so often been the case, the consequences may not only be faulty conclusions but also loss of life of neonates.

At the time of the last comprehensive review of mammalian milks (Oftedal, 1984a), there were relatively few careful or systematic studies of nondomestic species. Although data were located for 194 species, in only 55 cases were as many as 10 samples analyzed at all lactation stages. Much of the available information was from opportunistic situations in which effects of stage of lactation, compromised maternal or infant health, and sampling bias could not be explored. Fortunately, the past decade has seen a surge of interest in the energetics of mammalian reproduction, and with it an emphasis on more systematic studies of lactation. This is particularly true in studies of marsupials, primates, pinnipeds, and ungulates; the data base on rodents, carnivores, bats, and cetaceans remains meager, despite the widespread distribution and central role these organisms play in ecosystems and in evolutionary debates.

In this chapter our goal is to provide an update on the milks of a broad range of nonfarm species and to encourage investigators to evaluate carefully both the sampling and analytical methods that they select. Although investigators are now more aware of the potential bias introduced by sampling methods, inappropriate analytical methods continue to be used in otherwise exemplary studies. This may stem from a desire to use rapid methods that do not require much milk, coupled with inattention to the specificity of assays.

II. Factors Affecting Milk Composition Data

A. What Should Samples Represent?

Analyses of milks marketed for human use, whether as fluid milk, for cheese manufacture, or for production of other dairy products, must be focused on the milk as it enters trade or the production process; the degree to which such milk is representative of the fluid consumed by suckling young in a nonagricultural setting has little practical importance. However, from a biological point of view, it is the milk consumed by suckling young that needs investigation. This is especially true if the intent is to quantify the demands of lactation on mammalian mothers or to develop models for the development of substitute formulas. Thus, we define the optimal biological sample as that which most closely represents the milk ingested by suckling young under conditions of normal maternal rearing.

B. The Definition of Lactation Stages

The definition of an optimal sample, while necessary, is not sufficient for comparative analysis. It is widely recognized that the composition of milk changes over the course of lactation in most species of both placental and marsupial mammals (Oftedal 1984a; Green and Merchant, 1988). If stage of lactation is not controlled for in comparing one species to another, intraspecific (and intraanimal) variation may be confounded with interspecific differences, leading to erroneous or misleading conclusions. Unfortunately, this has been the case in most of the earlier milk compilations (e.g., Ben Shaul, 1962; Smith, 1970; Jenness and Sloan, 1970; Jenness, 1974).

The delineation of lactation stages is complicated by the fact that mammals are born at different stages of development, are suckled for differing lengths of time, and are weaned at different rates. We have chosen to use a nutritional, rather than developmental, definition of lactation stage. All mammalian young are completely dependent on mother's milk until they begin to feed on their own or are weaned. Although milk composition may change as lactation progresses, both the absolute requirements of the young for milk nutrients and the demands of lactation on maternal resources are likely to peak just before the young begin to eat significant amounts of solid (nonmilk) foods. We consider this period of maximal lactation performance to be the stage of midlactation. The period of changing milk composition prior to midlactation (including colostrum and transitional milk) is termed early lactation, and the period of declining yields and mixed feeding by the young is termed late lactation.

To be consistent with the earlier compilation, we operationally define the duration of midlactation by the stability of milk composition (Oftedal 1984a, p. 40) and use this stage as the basis for species comparison. However, it should be recognized that the demarcation of stages is particularly problematic in some mammalian groups. For example, kangaroos and many other marsupials give birth to very immature young and there is a long period of dependence during which the milk undergoes gradual change, before reaching plateau levels at about the time that the young emerge from the pouch to begin independent feeding. We have considered this entire period of changing composition to be early lactation. However, during the relatively brief period between pouch emergence and the onset of substantial solid food intake by the young, the milk may undergo even more marked changes characterized by a fall in carbohydrate, a replacement of oligosaccharides by mono- and disaccharides, and a rise in fat concentration (Green and Merchant, 1988). We have considered the plateau and the period immediately thereafter to be "midlactation" but the pattern of change makes delineation of this stage difficult. In mammals with prolonged and complex patterns of milk secretion, it might be of value to define more than three stages of lactation.

In the true seals the opposite is true as young are born very well developed and lactation lasts only a few days or weeks (Bonner, 1984; Oftedal *et al.*, 1987a). In most if not all species the stage of late lactation, as defined above, is absent. Mothers typically abandon their pups before the pups begin to feed on solid foods, and the milk, which may undergo relatively large changes in composition in the first half of lactation, remains relatively stable in the later part of lactation (Oftedal *et al.*, 1987a). Thus, in these species midlactation, as we define it, ends when the pup is weaned.

C. Avoidance of Sampling Bias

In many species of ungulates and in humans, a pronounced rise in fat concentration occurs as milk is being withdrawn from the mammary glands, whether by suckling young or by investigators (see Oftedal, 1984a). If it were possible to replicate the pattern and amounts of milk withdrawn by suckling young, the samples obtained would presumably be identical. Unfortunately, this is rarely the case. In practice, manual expression of mammary contents is usually incomplete, especially in species with a high proportion of stored milk in mammary alveoli and ducts (rather than sinuses or cisterns) (e.g., Cross, 1977). Exogenous oxytocin may facilitate milk collection via induction of the milk ejection response, but the response is transitory and followed by a refractory period (Denamur, 1965). In nonfarm animals, investigators rarely obtain as much milk as suckling young would.

If fat concentration rises during milk expression, incomplete milking should result in underestimation of the mean fat concentration in mammary milk. However, an added complication arises from the fact that the offspring may have removed much of the stored milk in a prior suckling session. Since residual milk remaining after suckling may contain elevated fat concentrations, milk samples collected shortly after suckling may overestimate fat levels (Erb *et al.*, 1977). One approach is to collect samples both before and after suckling in the hope that the average fat concentration of these two samples will more closely approximate the milk received by the young (Atwood and Hartmann, 1992).

Given these effects, it is not surprising that milk fat is usually the most variable constituent assayed, and the one most likely to differ among studies. Small differences in sampling protocols may have a substantial effect on fat concentration. We recommend the following steps to minimize sampling bias: (1) the young should be prevented from suckling prior to milk collection to avoid sampling residual milk. This may be done by physical separation, use of muzzles on the young, nipple covers, etc. The method of prevention should not be unduly stressful to the mother, since adrenalin exerts an inhibitory effect on the action of oxytocin (Denamur, 1965) and may thus interfere with milk collection. If possible, the period between suckling and milking should approximate the normal intersuckling interval. Prolonged separation should be avoided, as this may initiate mammary involution with consequent changes in milk composition (Lascelles and Lee, 1976). Although separation without stress may be difficult to achieve in the wild, it is sometimes possible to observe suckling events and to time milk collection accordingly; (2) the use of exogenous oxytocin is recommended to induce the milk ejection response, and thereby increase the amount of milk collected. However, repeated milking at frequent intervals should be avoided if oxytocin is used, since excessive oxytocin use may have unintended effects on milk composition (Linzell, 1967; Linzell *et al.*, 1975; Oftedal, 1984a); (3) it is preferable to express as much milk as possible from one or two glands, rather than taking partial samples from all glands. Although the total amount of milk collected may be reduced, the sample obtained should be more representative of mammary contents; and (4) the sampling method and amount of milk collected should be included in the published report as it assists other investigators in judging the degree to which mammary evacuation was achieved.

Unfortunately, investigators rarely provide enough detail on milk sampling protocols to allow a critical evaluation of the likelihood of sampling bias. Studies are also needed in species other than humans and ungulates to determine the extent to which fat levels change with mammary evacuation. For example, fat concentrations do not appear to change during mammary evacuation in several species of seals and sea lions (Oftedal *et al.*, 1987a, 1988; Iverson *et al.*, 1993) and the same may be true of rabbits, rats, goats, and the black rhinoceros (Glass, 1956; Gregory *et al.*, 1965; Parkash and Jenness, 1968; Cowie, 1969; Jaoen and Mens, 1981). Although some earlier studies indicated that fat concentration did not change during mammary evacuation in pigs, Atwood and Hartmann (1992) demonstrated a small (16%) but significant increase from fore- to hindmilk.

In view of the difficulty in replicating the milk removal pattern of suckling young, some investigators have allowed suckling to occur and then collected milk from the stomachs of the young, either by incision or by gastric intubation (see Oftedal, 1984a for references). Unfortunately such samples are contaminated by salivary, lingual, and gastric secretions and have likely undergone partial digestion. If gastric acid and proteases induce formation of milk curds, the aqueous portion (including whey proteins and sugars) may escape to the small intestine, leading to elevation of fat and/or decline in sugar concentration of the remaining material (e.g., Naismith *et al.*, 1969; Anderson *et al.*, 1975). In seals and dogs, the fat concentration of gastric "milk" is lower than that of mammary samples due to the diluting effect of secretory fluids, partial lipid hydrolysis, and the rapid passage of fat from the stomach (Oftedal *et al.*, 1987a; Iverson *et al.*, 1991, 1992). Thus, this procedure cannot be considered valid.

D. Problems Associated with Methods of Analysis

Although methods of analysis have been well established for cow's milk (e.g., AOAC, 1990), problems may arise when these methods are applied to milks of other species that are comprised of different constituents. Analytical procedures should be chosen carefully, with particular attention to the sensitivity of the method to the chemical and/or structural properties in the constituents being analyzed. Ideally, the analytical method should measure the aggregate amount of protein, fat, or carbohydrate without any dependence on the specific proteins, fats, or carbohydrates present in the sample. We will briefly review some of the common methods used to assay dry matter, fat, protein, and carbohydrate in the milks of nonfarm mammals.

1. Dry Matter

The standard method for determining the dry matter or total solids concentration in milk is to oven dry the sample to constant weight at a temperature near the boiling point of water (AOAC, 1990). However, it is important to recognize that oven-dried samples may not be suitable for subsequent analysis. At this temperature a nonenzymatic browning or Maillard reaction occurs, resulting in a condensation of sugars with amino acids (especially lysine) and the formation of relatively inert polymers. Amino acid analysis indicates that the extent of Maillard product formation is dependent on the carbohydrate content of the sample and, hence, varies among species (O. Oftedal, S. Crissey and O. Thomas, unpublished data). Some investigators have circumvented this limitation by freeze-drying samples. However, freeze-dried samples typically contain a small amount of residual water, causing a minor overestimation of dry matter. If this is important (as in isotope studies of milk intake), a correction factor can be determined by oven drying a subset of samples.

2. Fat

Various methods have been used to extract the lipid fraction in milks, usually employing one or more lipid solvents such as diethyl ether and petroleum ether (e.g., the Roese–Gottlieb method) or chloroform and methanol (e.g., the Folch method). The Roese–Gottlieb method involves additions of ammonium hydroxide and alcohol to disrupt the fat globule membrane. While suitable for fresh or well-preserved milk, this may cause an underestimation of total lipids if substantial hydrolysis of triacylglycerols has occurred, as in samples collected from stomachs of suckling young or in samples stored with preservatives but not with organic solvents and antioxidants (Iverson and Oftedal, Chapter 10B). The Roese–Gottlieb procedure can be adapted to samples of about 100 μl by a proportional reduction in the size of glassware and amounts of reagents, so long as enough milk fat is recovered for gravimetric measurement.

Some investigators have estimated fat by centrifugation of milk in hematocrit tubes and measurement of the cream layer (Fleet and Linzell, 1964). However, the data for each species must be calibrated against true extraction methods to account for species differences in the separation and packing of lipid droplets in the cream layer (Linzell and Fleet, 1969; Lucas *et al.*, 1978; Knight *et al.*, 1986; Atwood and Hartmann, 1992). For example, Linzell and Fleet (1969) reported that the factor required to convert percentage cream to true fat (g/100 g) ranged from 0.58 to 0.79 in 11 species.

For even smaller (30–60 mg) milk samples, Glass *et al.* (1968) describe a gas chromatographic method based on comparison to an added standard, but the accuracy of this method depends on the chromatographic procedures (see Iverson and Oftedal, Chapter 10B) as well as the degree to which such small samples are representative. The latter problem is particularly problematic with regard to samples which have been frozen and thawed, a process which may destabilize fat droplets and complicate efforts to obtain homogeneity. It may also be difficult to transfer small milk samples quantitatively without loss or fat or other constituents that adhere to the walls of the container or to the transfer pipetts.

In recent years, a number of studies have measured milk fat concentrations by spectrophotometric assays such as the methods of Stern and Shapiro (1953) and Zöllner and Kirsch (1962). As these procedures are based on chemical reactions, they are often sensitive to the chemical properties of the lipids assayed. For example, in the sulfuric acid–phosphoric acid–vanillin reaction employed by Zöllner and Kirsch (1962) saturated fatty acids give little if any color development, such that fatty acid composition will directly affect the measured fat concentration. It is critical that such methods are calibrated against fat extracted from milk of the species to be studied, not cow's milk. Unfortunately, this is rarely done, with the result that spectrophotometric assays have produced erroneous data.

3. *Protein*

Milk protein concentration is most commonly assayed by the Kjeldahl method, originally developed in 1883. In this method the amount of ammonia is measured after digestion of protein with concentrated acid, catalyst, and an additive to increase the boiling point of the solution. The potential errors, necessary cautions, and various modifications have been extensively reviewed by Bradstreet (1965). More recent applications often include use of semiautomated equipment (Jones, 1991). A micro adaptation called the Nessler procedure (Koch and McMeekin, 1924) will measure microgram quantities of protein, and is thus suited to small samples. However, this protocol needs careful standardization to ensure complete digestion of organic nitrogen and should only be used if results comparable to those of the Kjeldahl method can be demonstrated.

The total nitrogen (TN) value obtained by the Kjeldahl method is multiplied by a conversion factor to determine crude protein content (Jones, 1931). The conversion factor most commonly used for milks, 6.38, assumes that the average nitrogen concentration in the proteins being assayed is 15.7%, as in cow's milk. This may introduce minor error for species with different proportions of caseins and whey proteins, but until detailed and rigorous studies are conducted on the nitrogen content of the milk proteins of other species, the use of 6.38 remains an appropriate convention.

A more significant error arises from the inclusion of nitrogenous compounds other than protein in Kjeldahl analyses (Barbano *et al.*, 1991). In cow's milk, nonprotein nitrogen (NPN) accounts for about 6% of total nitrogen, but in some species, such as humans, mink, asses, pigs, and Indian rhinoceroses (*Rhinoceros unicornis*), NPN accounts for 15–17% of total nitrogen (Oftedal, 1984a). Thus, in these species multiplication of total nitrogen by 6.38 will overestimate true protein concentration by about one-sixth. In Table I Kjeldahl protein values which have been corrected for NPN are indicated by an asterisk.

Spectrophotometric methods that have been used to measure the protein concentrations in milks include the Folin–phenol or Lowry method (Lowry *et al.*, 1951) and dye-binding methods using amido black (Weidner and Jakobsen, 1966), acid orange 12 (Sherbon, 1967), and Coomassie brilliant blue (Bradford, 1976; Sedmak and Grossberg, 1977). As these methods are influenced by the amino acid composition and, in some instances, the tertiary structure of proteins, they are plagued by an unequal sensitivity to different proteins (Lowry *et al.*, 1951; Sedmak and Grossberg, 1977; Macart and Gerbaut, 1983), so that no single protein standard can adequately represent the various proteins found in milks. In the Biuret method (e.g., Gornall *et al.*, 1949), color development is proportional to the peptide linkage and, hence, does not vary greatly from protein to protein (Jenness and Patton, 1959), but this procedure has not been widely used for milks of nonfarm animals.

In a study of pig milk, Atwood and Hartmann (1992) demonstrated that the Bradford dye-binding method, which uses bovine serum albumin as the protein standard, underestimated crude protein concentration (TN × 6.37) by 31% and true protein concentration [(TN-NPN) × 6.37] by 23%. In the same study the Lowry method overestimated true protein by about 10% (Atwood and Hartmann, 1992). However, since the estimates provided by both methods were highly correlated to true protein concentration, these methods may yield accurate results if calibrated to the milk proteins in the species being studied. Unfortunately, studies on the milks of marsupials, rodents, and other species which have used the Lowry or dye-binding methods (Table I) have failed to calibrate the methods to species-specific milk proteins and may include unacceptable analytical error. The effect of lactation stage on milk protein concentration may also be misinterpreted by these methods if the changes in the relative proportions of caseins and various whey proteins over the course of lactation are large, as in marsupials (see below).

4. Sugars

The notion that lactose or "milk sugar" is the predominant carbohydrate in most milks has led to the widespread assumption that methods for

TABLE I
The Gross Composition of Mammalian Milks

Species	*N*	Lactation stage (days)	Dry matter (%)	Fat (%)[a]	Crude protein (%)[b]	Lactose and sugars (%)[c]	Ash (%)	Source of data
Monotremata								
Tachyglossidae								
Short-beaked echidna (*Tachyglossus aculeatus*)	15	37–99	48.9	31.0[E]	12.4[N]	2.3[A]	—	Griffiths *et al.* (1984)
Ornithorhynchidae								
Platypus (*Ornithorhynchus anatinus*)	10	Mature	39.1	22.2[E]	8.2[N]	3.7[A]	—	Griffiths *et al.* (1984)
Marsupialia								
Dasyuridae								
Eastern native quoll (*Dasyurus viverrinus*)	8–35	70–91	29.6	10.9[E]	7.3[N]	5.6[P]	—	Green *et al.* (1987)
Paramelidae								
Northern brown bandicoot (*Isoodon macrourus*)	8–10	30–37	26.0	10.0[E]	9.0[N]	6.9[P]	—	Merchant and Libke (1988)
Phalangeridae								
Brushtail possum (*Trichosurus vulpecula*)	20–23	100–120	24.0	4.4[E]	7.0[D]	11.0[P]	1.5	Cowan (1989); Gross and Bollinger (1959)
Petauridae								
Ringtail possum (*Pseudocheirus peregrinus*)	> 8	91–98	23.0	3.0[C*]	4.5[D]	12.5[P]	—	Munks *et al.* (1991)
Macropodidae								
Tasmanian bettong (*Bettongia gaimardi*)	3–6	84–91	25.0	4.0[C*]	11.0[D]	11.0[P]	—	Smolenski and Rose (1988)

TABLE I—*continued*

Species	N	Lactation stage (days)	Dry matter (%)	Fat (%)[a]	Crude protein (%)[b]	Lactose and sugars (%)[c]	Ash (%)	Source of data
Tammar wallaby (*Macropus eugenii*)	18	168–182	25.0	4.0[E]	6.0[N]	12.5[P]	—	Green *et al.* (1980, 1983); Messer and Green (1979)
Red kangaroo (*Macropus rufus*)	6	200–232	24.1	6.1[E]	7.2[N]	—	—	Lemon and Barker (1967)
Red-necked wallaby (*Macropus rufogriseus*)	8–39	226	25.0	7.2[E]	6.8[N]	10.9[P]	—	Merchant *et al.* (1989)
Long-nosed potoroo (*Potorous tridactylus*)	3–5	98–112	27.0	3.0[C*]	10.0[D]	14.0[P]	—	Smolenski and Rose (1988); Crowley *et al.* (1988)
Insectivora								
Soricidae								
White-toothed shrew (*Crocidura russula*)	3	8–12	51.0	30.0[E]	9.4[L]	3.0[A]	1.6	Mover *et al.* (1985)
Chiroptera								
Phyllostomatidae								
Jamaican fruit bat (*Artibeus jamaicensis*)	21	13–43	17.8	9.0[E]	3.6[N]	6.1[P]	—	O. Oftedal and L. Taft (unpublished data)
Vespertilionidae								
Little brown bat (*Myotis lucifugus*)	3	13–19+	27.1	15.8[E]	8.5[N]	4.0[P]	—	Kunz *et al.* (1995)
Cave bat (*Myotis velifer*)	3	20–32	25.4	19.9[E]	10.7[N]	4.4[P]	—	Kunz *et al.* (1995)

TABLE I—*continued*

Species	*N*	Lactation stage (days)	Dry matter (%)	Fat (%)[a]	Crude protein (%)[b]	Lactose and sugars (%)[c]	Ash (%)	Source of data
Molossidae								
Mexican free-tailed bat (*Tadarida brasiliensis*)	21	22–42	36.5	25.8[E]	7.7[N]	3.4[P]	—	Kunz *et al.* (1995)
Primates								
Lemuridae								
Brown lemur (*Eulemur fulvus*)	5	28–74	9.6	0.9[E]	1.3[N*]	8.5[P]	0.2	Tilden and Oftedal (1995)
Black lemur (*Eulemur macaco*)	7	30–82	10.1	1.1[E]	1.5[N*]	8.4[P]	0.3	Tilden and Oftedal (1995)
Red-bellied lemur (*Eulemur rubriventer*)	3	26–57	10.3	0.8[E]	1.1[N*]	8.9[P]	0.2	Tilden and Oftedal (1995)
Mongoose lemur (*Eulemur mongoz*)	4	45–81	9.8	0.7[E]	1.3[N*]	7.9[P]	0.2	Tilden and Oftedal (1995)
Ruffed lemur (*Varecia variegata*)	5	17–48	14.0	3.2[E]	4.2[N*]	7.7[P]	0.4	Tilden and Oftedal (1995)
Lorisidae								
Garnett's bushbaby (*Otolemur garnettii*)	14	14–73	18.5	7.3[E]	5.2[N*]	6.6[P]	0.6	Tilden and Oftedal (1995)
Thick-tailed bushbaby (*Otolemur crassicaudatus*)	8	19–60	18.6	8.0[E]	4.8[N*]	6.4[P]	0.6	Tilden and Oftedal (1995)
Slow loris (*Nycticebus coucang*)	4	18–90	16.3	7.0[E]	3.9[N*]	6.6[P]	0.7	Tilden and Oftedal (1995)
Callitrichidae								
Golden lion tamarin (*Leontopithecus rosalia*)	4	10–55	19.4	10.2[E]	3.0[N]	6.8[P]	—	O. Oftedal and M. Power (unpublished data)

TABLE I—*continued*

Species	*N*	Lactation stage (days)	Dry matter (%)	Fat (%)[a]	Crude protein (%)[b]	Lactose and sugars (%)[c]	Ash (%)	Source of data
Cebidae								
Red howler (*Alouatta seniculus*)	7	30–150	11.3	1.1^{E}	1.9^{N}	6.6^{P}	—	O. Oftedal, S. Crissey, and R. Rudran (unpublished data)
Mantled howler (*Alouatta palliata*)	7	30–150	11.7	1.6^{E}	2.2^{N}	6.7^{P}	—	O. Oftedal and K. Glander (unpublished data)
Cercopithecidae								
Talapoin monkey (*Cercopithecus talapoin*)	4	17–38	12.3	3.0^{E}	2.1^{N}	7.2^{R}	0.3	Buss and Cooper (1970)
Crab-eating macaque (*Macaca fascicularis*)	8	44–119	12.2	5.2^{E}	1.6^{N*}	—	0.4	Nishikawa *et al.* (1976)
Japanese macaque (*Macaca fuscata*)	7	35–56	14.0	4.2^{E}	1.6^{L}	6.2^{Z}	—	Ota *et al.* (1991)
Rhesus macaque (*Macaca mulatta*)	13–18	16–35	—	4.6^{S}	2.3^{L}	7.9^{Z}	—	Lönnerdal *et al.* (1984)
Baboons (*Papio anubis, Papio cynocephalus, Papio papio*)	24	21–63	14.0	4.5^{E}	1.5^{N}	7.8^{R}	0.3	Buss (1968); Roberts *et al.* (1985)
Carnivora								
Canidae								
Arctic fox (*Alopex lagopus*)	100?	Mid?	28.6	13.5^{U}	11.1^{U}	3.0^{U}	1.0	Dubrovskaya (1975)
Dog (domestic) (*Canis familiaris*)	25	7–37	22.7	9.5^{E}	7.5^{N*}	3.8^{P}	1.1	Oftedal (1984b); Rüsse (1961)
Raccoon dog (*Nyctereutes procyonoides*)	22	7–59	18.6	3.4^{E}	7.8^{N}	—	1.1	Iwata and Ishii (1946)

TABLE I—*continued*

Species	*N*	Lactation stage (days)	Dry matter (%)	Fat (%)[a]	Crude protein (%)[b]	Lactose and sugars (%)[c]	Ash (%)	Source of data
Red fox (*Vulpes vulpes*)	3	28–35	18.1	5.8[E]	6.7[N]	4.6[M]	0.9	Young and Grant (1931); Laxa (1930)
Ursidae								
Brown bear (*Ursus arctos*)	9	60–98	31.9	17.1[E]	9.2[N*]	2.2[P]	1.5	Jenness *et al.* (1972); Ando *et al.* (1979)
Black bear (*Ursus americanus*)	6	60–90	37.6	25.1[E]	7.0[N]	3.0[P]	—	Oftedal *et al.* (1993a)
Mustelidae								
Striped skunk (*Mephitis mephitis*)	15	20–48	30.6	13.8[E]	9.9[N*]	3.0[P]	—	Oftedal (1981)
Ferret (*Mustela putorius*)	18	11–25	—	9.7[C]	6.9[L]	3.8[R]	—	Schoknecht *et al.* (1985)
American mink (*Mustela vison*)	20	10–27	21.7	7.3[E]	5.6[N*]	4.5[P]	1.0	Oftedal (1981); Conant (1962)
Felidae								
Cat (domestic) (*Felis catus*)	15	6–38	—	10.8[O]	10.6[N]	3.7[R]	1.0	Folin *et al.* (1919)
African lion (*Panthera leo*)	6	45–90	26.8	8.7[E]	11.8[N]	3.2[P]	—	O. Oftedal, A. Pusey, and C. Packer (unpublished data)
Pinnipedia								
Phocidae								
Hooded seal (*Cystophora cristata*)	15	2–4	69.8	61.1[E]	4.9[N]	1.0[P]	—	Oftedal *et al.* (1988)
Grey seal (*Halichoerus grypus*)	13	8–15	71.1	59.8[E]	9.2[N]	—	—	Iverson *et al.* (1993)
Weddell seal (*Leptonychotes weddellii*)	7	10–43	66.2	53.6[E]	8.9[N]	0.02[Z]	—	Tedman (1980)

TABLE I—*continued*

Species	*N*	Lactation stage (days)	Dry matter (%)	Fat (%)[a]	Crude protein (%)[b]	Lactose and sugars (%)[c]	Ash (%)	Source of data
Northern elephant seal (*Mirounga angustirostris*)	20–24	20–28	65.8	51.9[E]	10.2[L,N]	< 0.025[R]	—	Riedman and Ortiz (1979); Kretzmann *et al.* (1993)
Southern elephant seal (*Mirounga leonina*)	5	11–26	—	46.9[C*]	7.4[N]	0.02[A]	—	Peaker and Goode (1978); M. Peaker (personal communication)
Harp seal (*Phoca groenlandica*)	8	10–13	65.7	53.5[E]	7.7[N]	0.8[P]	—	Oftedal *et al.* (1995)
Otariidae								
South American fur seal (*Arctocephalus australis*)	4	≈150	54.4	44.4[E]	9.7[N]	—	—	Ponce de Leon (1984)
Northern fur seal (*Callorhinus ursinus*)	5	30–120	63.3	50.7[E]	10.3[N]	0.1[R]	—	Ashworth *et al.* (1966); Dosako *et al.* (1983)
Australian sea lion (*Neophoca cinerea*)	20–38	14–125	37.6	25.4[E]	10.5[N]	—	0.9	Kretzmann *et al.* (1991)
California sea lion (*Zalophus californianus*)	9	≈3–60	41.0	31.7[E]	8.6[N*]	0.3[P]	—	Oftedal *et al.* (1987b)
Cetacea								
Delphidae								
Spotted dolphin (*Stenella attenuata*)	3	Mid–late?	—	22.5[E]	8.4[N*]	1.2[R]	—	Pilson and Waller (1970)
Bottlenose dolphin (*Tursiops truncatus*)	4	198–210	—	29.4[E]	12.2[D]	2.5[P]	—	Pervaiz and Brew (1986)

TABLE I—*continued*

Species	N	Lactation stage (days)	Dry matter (%)	Fat (%)[a]	Crude protein (%)[b]	Lactose and sugars (%)[c]	Ash (%)	Source of data
Balaenopteridae								
Minke whale (*Balaenoptera acutorostrata*)	12	Mid?	41.5	22.2[U]	14.6[U]	—	1.9	Best (1982)
Blue whale (*Balaenoptera musculus*)	4	≈210	55.0	40.9[E]	11.9[N]	1.3[R]	1.4	White (1953); Gregory *et al.* (1955)
Fin whale (*Balaenoptera physalus*)	7–9	≈210	46.5	33.2[E]	10.5[N]	2.3[R]	1.1	White (1953); Ohta *et al.* (1955)
Humpback whale (*Megaptera novaeangliae*)	8	≈300	48.4	33.0[U]	12.5[N]	—	1.6	Chittleborough (1958)
Proboscidea								
Elephantidae								
Asian elephant (*Elephas maximus*)	3	60–120	17.7	7.3[U]	4.5[N]	5.2[R]	0.6	Simon (1959)
African elephant (*Loxodonta africana*)	6	60–80	17.3	5.0[O]	4.0[N]	5.3[A]	0.7	McCullagh and Widdowson (1970)
Perissodactyla								
Equidae								
Ass (*Equus asinus*)	9	30–180	10.8	1.8[E]	1.7[N*]	5.9[P]	0.4	Oftedal and Jenness (1988)
Plains zebra (*Equus burchelli*)	5	90–240	11.3	2.2[E]	1.6[N*]	7.0[P]	0.4	Oftedal and Jenness (1988)
Przewalski horse (*Equus przewalskii*)	14	90–360	10.5	1.5[E]	1.6[N*]	6.7[P]	0.3	Oftedal and Jenness (1988)
Mountain zebra (*Equus zebra*)	7	90–360	10.0	1.0[E]	1.6[N*]	6.9[P]	0.3	Oftedal and Jenness (1988)

TABLE I—*continued*

Species	*N*	Lactation stage (days)	Dry matter (%)	Fat (%)[a]	Crude protein (%)[b]	Lactose and sugars (%)[c]	Ash (%)	Source of data
Tapiridae								
Baird's tapir (*Tapirus bairdii*)	4	15–31	13.3	1.9^{E}	4.6^{N}	5.3^{P}	0.7	O. Oftedal (unpublished data)
Brazilian tapir (*Tapirus terrestris*)	3	15–20	15.0	3.9^{E}	4.4^{N}	5.3^{R}	0.7	R. Jenness (personal communication)
Rhinocerotidae								
Black rhinoceros (*Diceros bicornis*)	11	30–330	8.8	0.2^{E}	1.4^{N*}	6.6^{R}	0.3	Gregory *et al.* (1965); Aschaffenburg *et al.* (1961)
Artiodactyla								
Tayassuidae								
Collared peccary (*Tayassu tajacu*)	4	21–48	16.2	4.2^{E}	5.1^{N}	6.2^{M}	—	Lochmiller *et al.* (1985)
Camelidae								
Bactrian camel (*Camelus bactrianus*)	30	23–91	15.2	4.3^{E}	4.3^{N*}	—	0.9	O. Oftedal, C. Wemmer, and J. Murtaugh (unpublished data)
Cervidae								
Moose (*Alces alces*)	15	Mid?	21.5	10.0^{U}	8.4^{U}	3.0^{U}	1.5	Ivanova (1965)
North American elk (*Cervus elaphus nelsoni*)	28	14–77	19.0	6.7^{E}	5.7^{N}	4.2^{P}	1.3	Robbins *et al.* (1981)
Red deer (*Cervus elaphus scoticus*)	6	3–31	21.1	8.5^{V}	7.1^{N*}	4.5^{P}	1.4	Arman *et al.* (1974)

TABLE I—*continued*

Species	N	Lactation stage (days)	Dry matter (%)	Fat (%)^{a}	Crude protein (%)^{b}	Lactose and sugars (%)^{c}	Ash (%)	Source of data
Mule deer (*Odocoileus hemionus*)	24	14–35	18.5	5.5^{E}	7.0^{N}	4.5^{P}	1.4	Carl and Robbins (1988); Mueller and Sadleir (1977)
White-tailed deer (*Odocoileus virginianus*)	4+	21–28	22.5	7.7^{U}	8.2^{U}	4.6^{U}	1.5	Compilation (see Oftedal, 1984a)
Reindeer (*Rangifer tarandus*)	6	21–30	26.3	10.9^{E}	9.5^{N}	3.4^{R}	1.3	Luhtala *et al.* (1968); Luick *et al.* (1974)
Giraffidae								
Giraffe (*Giraffa camelopardis*)	3	Mid	14.5	4.8^{V}	4.0^{N}	—	0.8	Hall-Martin *et al.* (1977)
Bovidae								
Gayal (*Eos frontalis*)	4–	11–50	20.0	7.0^{V}	6.3^{T}	5.2^{M}	—	Scheurmann *et al.* (1977)
Ibex (*Capra ibex*)	24	30–60	23.3	12.4^{U}	5.7^{U}	—	1.2	Maltz (1979)
Dorcas gazelle (*Gazella dorcas*)	16	30–60	24.1	8.8^{U}	8.8^{U}	—	1.1	Maltz (1979)
Tahr (*Hemitragus jemlahicus*)	9	60?	—	7.9^{E}	5.4^{N}	3.1^{P}	—	Rammell and Caughley (1964)
Sable antelope (*Hippotragus niger*)	6–8	≈30–107	17.9	$5.0^{E,U}$	6.2^{N}	$5.3^{P,U}$	0.9	O. Oftedal (unpublished data); Wilson and Hirst (1977)
Muskox (*Ovibus moschatus*)	6	≈100	28.5	14.3^{E}	8.7^{N*}	3.6^{R}	1.2	R. Jenness (personal communication)
Rocky mountain goat (*Oreamnos americanus*)	28	14–35	18.0	7.0^{E}	6.5^{N}	4.5^{P}	0.7	Carl and Robbins (1988)

TABLE I—*continued*

Species	*N*	Lactation stage (days)	Dry matter (%)	Fat (%)[a]	Crude protein (%)[b]	Lactose and sugars (%)[c]	Ash (%)	Source of data
Dall sheep (*Ovis dalli*)	4	21–42	22.9	9.5[E]	7.2[N]	5.3[R]	0.9	Cook *et al.* (1970)
Eland (*Taurotragus oryx*)	11	30–60	21.9	9.9[U]	6.3[U]	4.4[U]	1.1	Treus and Kravchenko (1968)
Rodentia								
Castoridae								
European beaver (*Castor fiber*)	14	10–50	34.1	19.0[V]	11.2[N*]	1.7[M]	1.1	Zurowski *et al.* (1974)
Muridae								
Golden hamster (*Mesocricetus auratus*)	6	Mid	22.6	4.9[U]	9.4[N]	4.9[P]	1.4	Jenness and Sloan (1970)
House mouse (*Mus musculus*)	5	9–10	40.8	27.0[C*]	12.5[N]	2.6[R]	—	Knight *et al.* (1986); Baverstock *et al.* (1976)
Spinifex hopping mouse (*Notomys alexis*)	3–12	8–14	29.3	15.0[E]	5.5[D]	2.6[R]	—	Baverstock *et al.* (1976)
Fawn-colored hopping mouse (*Notomys cervinus*)	3–7	8–14	30.2	10.3[E]	5.6[D]	2.3[R]	—	Baverstock *et al.* (1976)
Mitchell's hopping mouse (*Notomys mitchelli*)	2–4	8–14	33.3	7.5[E]	6.5[D]	2.7[R]	—	Baverstock *et al.* (1976)
Eastern native mouse (*Pseudomys australis*)	6–7	7–12	25.4	12.1[E]	6.4[D]	3.6[R]	—	Baverstock *et al.* (1976)
Brown or Norway rat (*Rattus norvegicus*)	3–18	8–17	22.1	8.8[E]	8.1[N*]	3.8[R]	1.2	Luckey *et al.* (1955); Glass (1956); Venkatachalam and Ramanathan (1964)

TABLE I—*continued*

Species	N	Lactation stage (days)	Dry matter (%)	Fat (%)[a]	Crude protein (%)[b]	Lactose and sugars (%)[c]	Ash (%)	Source of data
Caviidae								
Guinea pig (*Cavia porcellus*)	10	4–9	17.5	5.7[E]	6.3[N*]	4.8[P]	0.8	Oftedal (1981); Nelson *et al.* (1951)
Chinchillidae								
Chinchilla (*Chinchilla laniger*)	60	3–7	—	11.2[E]	7.3[N]	1.7[M]	1.0	Volcani *et al.* (1973)
Echimyidae								
Punare (*Thricomys apereoides*)	18–30	7–14	—	22.3[C]	11.0[D]	4.4[R]	—	Meyerson-McCormick *et al.* (1990)
Lagomorpha								
Leporidae								
European brown hare (*Lepus capensis*)	30	2–26	32.5	15.6[E]	10.0[N]	1.5[Z]	—	Lhuillery *et al.* (1984)
Rabbit (*Oryctolagus cuniculus*)	56	5–21	31.2	15.2[C*]	10.3[N]	1.8[M]	1.8	Cowie (1969); Coates *et al.* (1964)
Eastern cottontail rabbit (*Sylvilagus floridanus*)	4	12–15	35.2	14.4[E]	15.8[N]	2.7[M]	2.1	Anderson *et al.* (1975)

[a]Fat: [E]extraction with solvents, and gravimetric determination of fat, such as by Roese–Gottlieb and Folch methods (AOAC, 1990); [V]volumetric measurement of fat after separation of fat in concentrated acid, such as Babcock and Gerber methods (see Jenness and Patton, 1959); [C]measurement of cream layer in capillary tube after centrifugation, such as by methods of Fleet and Linzell (1964) and Ganguli *et al.* (1969); [C*]measurement of cream layer, as above, but procedure calibrated for species being studied using extraction or volumetric procedures; [S]spectrophotometric measurement of lipids, such as by methods of Stern and Shapiro (1953) and Zöllner and Kirsch (1962); [O]Other lipid methods; [U]uncertain methodology, as no description of analytical procedures available.

TABLE I—*continued*

[b]Protein: [N]total nitrogen multiplied by 6.38, as assayed by the Kjeldahl procedure and various modifications (including the Nessler procedure; Koch and McMeekin 1924); [N*]protein nitrogen (TN-NPN) multiplied by 6.38, as assayed by the Kjeldahl procedure and various modifications; [D]dye-binding methods, such as the procedures using amido black (Weidner and Jakobsen, 1966) and Coomassie brilliant blue (Bradford, 1976; Sedmark and Grossberg, 1977); [L]Lowry method (Lowry *et al.*, 1951); [T]Biuret method, such as Gornall *et al.* (1949); [U]uncertain methodology, as no description of analytical procedures available.

[c]Sugars: [P]phenol–sulfuric acid method (Marier and Boulet, 1959; Messer and Green, 1979); [R]reducing sugar methods such as copper precipitation method (Munson and Walker, 1906), copper titration method (Folin and Wu, 1919), picric acid method (Perry and Doan, 1950), and chloramine-T method (see Jenness and Patton, 1959); [A]anthrone method, such as Morris (1948); [Z]enzymatic methods specific for lactose (e.g., Bahr, 1972); [M]miscellaneous other methods; [U]uncertain methodology, as no description of analytical procedures available.

measuring lactose in cow's milk will accurately measure the sugar concentration of other milks. This will not be true if (1) the milks of other species contain sugars other than lactose, and (2) the analytical methods used have a differential response to different sugars.

Jenness *et al.* (1964) surveyed the milk sugars of more than 50 species of mammals based on paper chromatography of protein-free dialysates. Sugars that migrated with the mobility of tri- and oligosaccharides were chromatographically prominent in a broad range of mammals, including species in the Marsupialia, Insectivora, Rodentia, Edentata, Carnivora, Perissodactyla (Tapiridae only), and Artiodactyla (Suidae, Tayassuidae, Camelidae, and Cervidae only). Sugars larger than lactose appeared to be the primary sugar in marsupials, an insectivore (Western hedgehog, *Erinaceous europaeus*), grey and flying squirrels (*Sciurus carolinensis, Glaucomys volans*), and brown and black bears. Oligosaccharides are also predominant in at least some seals (Messer *et al.*, 1988). Thus, methods which are specific for lactose, such as enzymatic methods based on lactase (e.g., Bahl, 1972; Essig and Kleyn, 1983), omit important sugar constituents in many mammalian species and may greatly underestimate total sugar content.

A critical consideration in selecting a sugar assay for use on the milks of a broad array of species is that the various mono-, di-, and oligosaccharides that may be present are all measured with similar sensitivity. The phenol–sulfuric acid colorimetric method (Dubois *et al.*, 1956; Marier and Boulet, 1959) is relatively nonspecific among sugars and provides a measure of total sugar content. However, at the concentration of phenol specified by Marier and Boulet (1959), the absorbance readings per milligram galactose are only about 80% that of glucose (Dubois *et al.*, 1956; Messer and Green, 1979). The total sugar content of milks containing high proportions of galactose (as in the oligosaccharides of marsupial milks) may be somewhat underestimated. Messer and Green (1979) proposed a modification to equalize the absorbances of the two sugars in the analysis of marsupial milks. Although the phenol–sulfuric acid method fails to measure amino sugars (Montgomery, 1961), this is typically a small error as these are only minor constituents of milk oligosaccharides.

The anthrone procedure is another colorimetric method that responds to a wide variety of sugars, but the absorbance difference between glucose and galactose is even greater: 100 μg galactose produces the same color as 54 μg glucose (Morris, 1948). For harp seal milk, the sugar concentration estimated by the anthrone method was 54% that estimated by the phenol–sulfuric acid method (Stewart *et al.*, 1983). Messer *et al.* (1988) found that the ratio of galactose to glucose in crabeater seal (*Lobodon carcinophagus*) milk was 6:1, and if the same is true of harp seals, the reduced anthrone response to galactose may explain the discrepancy between the two methods. The anthrone method cannot be considered reliable for milks containing galactose-rich oligosaccharides.

Early sugar methods were based on the reducing power of sugars, usually assessed by titration or gravimetric measurement of reduced copper compounds (e.g., Munson and Walker, 1906; Folin and Wu, 1919; Shaffer and Somogyi, 1933; Somogyi, 1952). Subsequently, more simple methods of measuring reducing power have been developed and extensively used for milks, including the picric acid method (Perry and Doan, 1950) and the chloramine-T method (see Jenness and Patton, 1959). While these methods are suitable for milks that only contain lactose, other sugars differ in reducing power per milligram sugar. This is particularly a problem when oligosaccharides are present, as reducing methods based on lactose standards will underestimate total sugar concentration unless the oligosaccharides are hydrolyzed to monosaccharide constituents prior to measurement. For example, in early studies of marsupial milks (e.g., Bolliger and Pascoe, 1953; Lemon and Barker, 1967; Bergman and Housley, 1968) the measured sugar concentrations obtained by reducing methods without prior hydrolysis were only about 1–3% compared to more recent midlactation estimates of 6–14% by the phenol–sulfuric acid method, as modified by Messer and Green (1979) (Table I).

The extent to which the method of analysis influences sugar measurements may be further illustrated by a study (O. Oftedal, unpublished data) in which five samples of midlactation milk from each of four species [cow [homogenized whole milk], horse, dog, and black bear] were assayed in duplicate by the phenol–sulfuric acid method (P) (Marier and Boulet, 1959), the picric acid-reducing method (R) (Perry and Doan, 1950), and an enzymatic lactase procedure (Z) (Boehringer-Mannheim kit; Essig and Kleyn, 1983). The means and standard errors were:

Horse: $P = 6.73 \pm 0.206$; $R = 6.74 \pm 0.083$; $Z = 6.93 \pm 0.112$
Cow: $P = 4.98 \pm 0.049$; $R = 4.78 \pm 0.021$; $Z = 4.97 \pm 0.023$
Dog: $P = 3.67 \pm 0.218$; $R = 4.39 \pm 0.180$; $Z = 3.99 \pm 0.162$
Bear: $P = 2.91 \pm 0.228$; $R = 2.45 \pm 0.134$; $Z = 0.17 \pm 0.025$.

Although all three methods produced similar results in the cow and horse, in the dog the mean obtained by the picric acid method was significantly higher than that of the phenol–sulfuric acid method, and in the bear the mean of the enzymatic method was much lower than the means of either other procedure. Although the phenol–sulfuric acid procedure tends to be less precise (as indicated by the larger standard errors), it is least influenced by the presence of sugars other than lactose and thus is preferred for interspecific comparisons. About one-half of the sugar values included in Table I were measured by this method.

III. Phylogenetic Patterns in Milk Composition

A. *Species Selection*

The 100 species included in Table 1 were selected to fulfill criteria established by Oftedal (1984a): (1) sufficient information was given in the publication to assess that samples were collected in midlactation; (2) excessive (≥24 hr) separation of mother from young, that might induce mammary involution, did not occur (this criterion was not applied to fur seals, rabbits, and other species in which intersuckling intervals are normally ≥24 hr); (3) samples were collected directly from the mammary glands, not from neonatal stomachs, nor via vacuum systems that could cause evaporative water losses; and (4) at least three qualifying samples per species were located.

In general, farm animals were not included as they are treated elsewhere in this volume (Alston-Mills, Chapter 10C). However, there may be instances in which duplication is appropriate. For example, reindeer milk has been employed as human food in northern Europe and Asia for many years, and thus a considerable number of studies have examined the composition of reindeer milk (Luick *et al.*, 1974). However, as milk is not normally harvested in early and midlactation when the fawns depend almost entirely on milk, most studies emphasize late-lactation milk. We include reindeer in this chapter so midlactation data can be compared to other species of deer.

If several studies were available for a particular species, the study which appeared most reliable, especially in terms of analytical methodology, was selected for Table I. In some cases a secondary source was used to fill in missing data, such as dry matter or ash concentrations. The arrangement of orders and families in Table I follows Corbet and Hill (1980); taxonomic binomials are provided in the text only for species that are not included in Table I.

B. *Egg-Laying Mammals (Order Monotremata)*

The midlactation secretion of the platypus and short-beaked echidna is remarkably high in dry matter and fat (Table I); a few samples obtained shortly after hatching indicate that the milk is initially much more dilute (12–20% dry matter; Griffiths *et al.*, 1969, 1984). The apparent rise in dry matter and fat concentrations resembles what is seen in marsupials (below), although sugar concentrations are never as high. Free lactose is only a very minor component of the sugar fraction; the major sugars are difucosyllactose in the platypus and sialyllactose and fucosyllactose in the echidna (Messer and Kerry, 1973; Messer *et al.*, 1983; Griffiths *et al.*, 1984).

The high fat and energy density of echidna milk in midlactation may relate to the fact that the young suckle only once every few days and, thus, must receive large amounts of energy at each suckling (Griffiths, 1978; Griffiths *et al.*, 1984). Caseins comprise about half of the total protein in echidna milk (Teahan *et al.*, 1991) and presumably play a role in gastric curd formation and retention of fat in the stomach. The suckling frequency of the aquatic platypus is not known. High fat concentrations in other aquatic mammals are thought to relate to high energy requirements of the young and/or a need to deposit a layer of subcutaneous fat as insulation in a thermally demanding environment (Jenness and Sloan, 1970; Oftedal *et al.*, 1987b, 1988).

C. Marsupials (Order Marsupialia)

The milks of marsupials are unique in the extent to which milk composition changes over the course of lactation. In the early period after birth milks are typically quite dilute (often no more than 8–15% dry matter), but since fat, protein, and carbohydrate all increase up until the time of teat detachment and pouch emergence, dry matter concentrations are usually 23–30% at this lactation stage (Green, 1984; Green and Merchant, 1988; Munks *et al.,* 1991; Table I). Another remarkable feature of marsupial milks in early and midlactation is the preponderance of oligosaccharides that are comprised chiefly of galactose with lesser amounts of glucose, hexosamines, and sialic acid (Messer and Mossop, 1977; Messer and Green, 1979; Messer *et al.,* 1987; Crisp *et al.,* 1989). In late lactation the concentrations of oligosaccharides drop to low levels, being replaced by mono- and disaccharides, including, in some species, lactose. The amounts and proportions of various milk proteins also change during lactation, and some whey proteins appear only in late lactation (Green and Renfree, 1982; Nicholas *et al.,* 1987; Nicholas, 1988a; Nicholas *et al.,* 1989).

The large changes in milk composition during lactation complicate interspecific comparisons as there is no consistent or easily defined plateau. We have followed Munks *et al.* (1991) in considering the period around pouch emergence (or teat detachment in species in which young are not contained in a pouch) as the period of midlactation. This period precedes significant intake of solid foods. Unfortunately, milk composition changes during this period in some species with the result that the summary values presented in Table I are influenced by the range of days included. In many of these studies the measured protein concentrations are suspect as they were obtained by dye-binding methods using bovine serum albumin and were not calibrated to the changing mix of proteins in the species being studied.

Of the nine marsupial species included in Table I, seven appear to produce milks that are very similar at midlactation: 23–27% dry matter, 3–7% fat, 6–11% protein, and 11–14% carbohydrate. Among mammals, such high carbohydrate concentrations have only been found in marsupials and are only possible because of the lower osmotic effect (relative to mass) of oligosaccharides than mono- or disaccharides. These seven species represent three families (Phalangeridae, Petauridae, and Macropodidae) and are all herbivorous. The milks of the carnivorous quoll (family Dasyuridae) and omnivorous bandicoot (family Paramelidae) contain only about half the carbohydrate (6 or 7%) and are somewhat higher in fat (10 or 11%).

An unusual feature of reproduction in kangaroos and wallabies (family Macropodidae) is that a newborn offspring may attach to one nipple, while an older sibling continues to suckle at another nipple. The respective glands simultaneously produce milks that differ not only in the proportions of fat, protein, and carbohydrate, but also in the specific carbohydrates and proteins synthesized, suggesting local control over milk synthesis (Nicholas, 1988b).

D. Insectivores (Order Insectivora)

The order Insectivora is a diverse assemblage of mammals that includes both very small species with high metabolic rates (e.g., soricine shrews) and larger species that undergo periodic torpor (e.g., tenrecs). Unfortunately, only a few studies have examined milk composition and these data are of limited reliability (Oftedal, 1980). For example, Dryden and Anderson (1978) reported high values for dry matter (37.5%), fat (17.5%), and protein (10.7%) in the milk of musk shrews (*Suncus murinus*), but the samples were obtained by excising the stomachs of recently suckled young and may not be representative. However, Mover *et al.* (1985) also found very high concentration of these constituents in white-toothed shrews (Table I).

Blaxter (1961) predicted that gastrointestinal limitations should favor secretion of milks of high energy density in small mammals since the young will have high mass-specific metabolic rates without a corresponding increase in gastrointestinal capacity. However, mass and volume constraints may also influence the mother directly. If the peak milk energy output of 9 g white-toothed shrews is predicted by the equation developed by Oftedal (1984a) for mammals with litters, the predicted output (16 kJ/day) is equivalent to about 0.9 g milk (18 kJ/g) or 10% of body weight. Lower energy density of milk would require increased volume production which might strain mammary storage capacity, locomotor ability, and foraging success.

E. Bats (Order Chiroptera)

As flying mammals, bats might be expected to produce concentrated milks as a means of reducing weight and, hence, wing loading. Although this is one of the largest and most diverse mammalian orders, relatively few species have been studied. For example, only one sample (*Epomophorus wahlbergi;* Quicke *et al.,* 1984) has been analyzed for the ca. 170 species of Old World fruit bats (Family Pteropidae). Jenness and Studier (1976) observed considerable variation in the composition of samples analyzed for 8 species of New World bats, but as only one or two pooled samples were assayed per species and lactation stage was not known, these data are difficult to interpret. Kunz *et al.* (1983) found little difference in the milks of two insectivorous bats in the family Vespertilionidae, *Myotis lucifugus*

(13.5% fat, 7.4% protein, and 3.3% sugars, $n = 13$) and *Eptesicus fuscus* (16.4% fat, 6.2% protein, and 2.5% sugars, $n = 4$), but lactation stage could only be approximated.

More recent data indicate that the Mexican free-tailed bat produces milk that is high in dry matter and fat, and relatively low in sugars, compared to two species of *Myotis* (Table I). This may correlate to the longer foraging trips and, hence, longer durations and distances over which milk must be stored and transported, in the free-tailed bat (Kunz *et al.*, 1995). In contrast, the Jamaican fruit bat produces milk that is much lower in dry matter, fat, and protein, and higher in lactose, than the other bats that have been studied (Table I). This species forages primarily on fruit, whereas the other species catch insects on the wing.

F. Primates (Order Primates)

Data on 16 species are included in Table I, a considerable improvement over the 3 nonhuman primates that met the criteria in 1984 (Oftedal, 1984a). The notion that primates as a group produce milks characterized by low levels of protein and energy now appears overly simplistic. New data on the milks of four species of *Eulemur* and two species of *Allouata* are consistently low in dry matter (10–12%), fat (0.7–1.6%), and protein (1.1–2.2%), but the milks of the bushbabies and slow loris are not (Table I). Even among anthropoid primates, there is evidence that callitrichids, such as the golden lion tamarin (Table I) and the common marmoset, *Callithrix jacchus* (Turton *et al.*, 1978), produce more concentrated milks, and the same may be true of a small cebid, the squirrel monkey, *Saimiri sciureus* (Buss and Cooper, 1972). As more data become available, it will be possible to reevaluate the hypothesis put forth by Powers (1933) that the low protein:energy ratio of human and other primate milks is related to a slow rate of postnatal growth. It would be interesting to know if the milks of chimpanzees and other great apes resemble that of humans, but reliable midlactation data are not available.

G. The Carnivores (Order Carnivora)

Among the carnivores, the milk of the domestic dog has received most attention. Despite variation in sampling procedures, breeds sampled, and analytical methods, the mean values for midlactation milk were relatively similar in most of the 15 studies tabulated by Oftedal (1984b): 21–26% dry matter, 8–12% fat, 7–10% protein, and 3 or 4% sugar. However, Lönnerdal *et al.* (1981) obtained considerably lower values for fat measured by the sulfuric acid–phosphoric acid–vanillin reaction (Zöllner and Kirsch, 1962) and for protein measured by the binding of Coomassie brilliant blue dye (Sedmak and Grossberg, 1977). These rapid spectrophotometric methods appear to have been inaccurate with dog milk.

Most other carnivores that have been studied produce milks that resemble dog milk (Table I). However, the milks of bears contain higher dry matter and fat concentrations, and protein represents a smaller proportion of total dry matter. These attributes are thought to reflect the need of the lactating bear to conserve water and protein while fasting in the winter den (Oftedal *et al.*, 1993a; Oftedal, 1993). Jenness *et al.* (1972) reported that milk samples from zoo bears differed in composition from those from wild bears, but the two groups represented different stages of lactation. In the black bear, fat and protein concentrations increase markedly from early to late lactation; sugar concentration initially rises during the denned period and then falls to low levels (Oftedal *et al.*, 1993a). Lactose is only a minor component of the sugars in bear milk, but the other sugar constituents have not been specifically identified (Jenness *et al.*, 1972; Oftedal *et al.*, 1993a).

Given the abundance and prolific reproduction of domestic cats, it is surprising that so little is known about cat milk. Although the studies of Commaille (1866), Abderhalden (1898), Hurni and Montalta (1980), and Keen *et al.* (1982) indicate fat concentrations of only 3–5%, we believe the data of Folin *et al.* (1919), indicating about 11% fat, are more representative (Table I). Linzell and Fleet (1969) reported fat concentrations of 8.6–10.6% in cat milk, but gave no sampling details. African lion milk contains about 9% fat, as measured by the Roese–Gottlieb method (O. Oftedal, A. Pusey, and C. Packer, unpublished data). Ben Shaul (1962) listed values of 6–19% fat in milks of five felid species, but no information is provided on numbers of samples or sampling procedure, stage of lactation, or methods of analysis. Although it appears that felid milks typically contain about 8–11% fat, further research is needed on both the domestic cat and other species. Much of the literature on the milks of carnivores is of uncertain reliability (Gittleman and Oftedal, 1987).

H. Seals and Sea Lions (Order Pinnipedia)

A phylogenetic dichotomy in lactation pattern has been described among pinnipeds: the true seals (family Phocidae) lactate for relatively short periods (4–45 days) during which mothers fast or feed only a little, whereas the fur seals and sea lions (family Otariidae) lactate much longer (ca. 120–720 days) and undertake regular foraging trips to sea (Bonner, 1984; Oftedal *et al.*, 1987a; Costa, 1991; but see Boness *et al.*, 1995). Lactating phocids would be expected to conserve body water and protein and convert to a reliance on lipids during fasting (Oftedal, 1993). This may partly explain why phocid milks are so very high in fat, and low in water, by comparison to other mammals (Table I). The high fat concentrations may also be critical for the rapid deposition of body fat by pups, especially in species such as the hooded seal which lactates for only 4 days (Bowen *et al.*, 1985; Oftedal *et al.*, 1988).

The milks of fur seals and sea lions tend to have lower concentrations of dry matter and fat, at least in midlactation (Table I). Trillmich and Lechner (1986) concluded that the fat concentration in the milks of otariids is correlated to the species-specific duration of maternal foraging trips. Unfortunately, the analysis was confounded by data that may not be representative, due to sampling problems and inattention to lactation stage. However, even in the restricted data set of Table I there is evidence of the pattern proposed by Trillmich and Lechner (1986). The northern fur seal, which makes long foraging trips (6 or 7 days during lactation; Gentry and Holt, 1986), produces milk with higher dry matter and fat concentrations than is found in species with short trips (e.g., 1.5–2.5 days in the Australian and California sea lions; Ono *et al.*, 1987; Kretzmann *et al.*, 1993). Secretion of high-fat milk enables more milk energy to be accumulated per unit of mammary volume, which is presumably beneficial in species with finite storage capacity but long intervals between suckling.

It is remarkable that active secretion is maintained by the mammary glands of fur seals even though no suckling occurs during the period at sea, which is as long as 9–12 days in the Juan Fernandez fur seal, *Arctocephalus philippii* (J. M. Francis, D. J. Boness, and H. Ochoa, personal communication). Equally unusual is the fact that the milks of otariids appear to be virtually devoid of lactose and other carbohydrates (Pilson and Kelly, 1962; Oftedal *et al.*, 1987a), raising into question the mechanism by which the aqueous phase of the milk is secreted (Peaker, 1977). Little is known about the carbohydrates of phocid milks, although a recent study demonstrated that the carbohydrate in the milk of the crabeater seal (*L. carcinophagus*) is predominantly oligosaccharides containing galactose and hexosamines, with only a trace (0.02%) of lactose (Messer *et al.*, 1988).

I. Whales and Dolphins (Order Cetacea)

Most of the data on the milks of whales and dolphins is of limited value due to the lack of information on stage of lactation, coupled with the opportunistic nature of sample collection. The early literature on the milk of great whales is based on samples obtained by whaling vessels at a time when international convention prohibited the killing of whales with attendant calves. These samples represent milk secreted at about weaning (White, 1953; Chittleborough, 1958). Oftedal (1984a) was unable to find any published reports that represented three or more samples at midlactation. A paper on captive bottlenose dolphins of known lactation stage has since appeared (Pervaiz and Brew, 1986), but other recent reports are primarily of single samples from strandings or incidental capture (e.g., in fishing nets) (Jenness and Odell, 1978; Ullrey *et al.*, 1984; Peddemors *et al.*, 1989; O. Oftedal, unpublished data).

For comparative purposes, data from several species of the great (baleen) whales are included in Table I. These species resemble phocid seals in that mothers and calves feed little if at all during lactation, and lactation is brief (6 or 7 months in most species, which is surprisingly short given such large body size) (Lockyer, 1984; Oftedal, 1993). Thus, it is possible that many of these whales do not have a true stage of late lactation, as defined by declining yields and the progressive weaning of the young. By contrast, the toothed whales (Odontocetes) feed regularly during a long lactation (12–24 months in most species), and weaning is gradual (Perrin *et al.,* 1984).

These limited data suggest that dolphins and whales produce milks that are relatively high in dry matter (42–55%) and fat (22–41%) concentrations, although the effect of lactation stage remains uncertain. The values for spotted dolphins and minke whales only include samples from animals believed to be in midlactation based on the relatively large measured depth of the mammary tissue (≥3.0 and 10 cm, respectively) (Pilson and Waller, 1970; Best, 1982).

J. Elephants (Order Proboscidea)

The milks of Asian and African elephants are very similar to the milks of many ruminants which contain moderate concentrations of dry matter, fat, protein, and sugars (Table I). There is some evidence that dry matter, fat, and protein may increase over the 2 or more years that elephants lactate (Simon, 1959; McCullagh and Widdowson, 1970; Peters *et al.,* 1972).

K. Horses, Rhinos, and Tapirs (Order Perissodactyla)

All studied species of horses, zebras, and asses (genus *Equus,* family Equidae), including the domestic horse, produce very dilute milks containing only 10 or 11% dry matter and 1 or 2% fat (Oftedal *et al.,* 1983; Oftedal and Jenness, 1988). The milk of the black rhinoceros is even lower in dry matter and contains but a trace amount of fat (Table I). It is not clear why these species produce milk that is so low in dry matter and, hence, high in water. Perhaps zebra or rhino foals require high water intakes to compensate for the water lost during evaporative cooling in hot environments; unlike large adults, heat storage may not be a viable option (Schmidt-Nielsen and Schmidt-Nielsen, 1952). Certainly foals need to ingest relatively large volumes of these low-energy milks to meet their energy requirements (Oftedal, 1985). The milks of tapirs differ from other perissodactyls in having higher dry matter and protein concentrations, and somewhat lower sugar concentrations.

L. Ruminants and Related Species (Order Artiodactyla)

Most nondomestic ruminants produce milks that are relatively similar in composition, containing 18–24% dry matter, 5–10% fat, 5–8% protein, and 3–5% sugars (Table I). Thus, the milks of dairy cattle and goats are not typical of wild ruminants, being lower in dry matter, fat, and protein (see other chapters in this book). Other species with milks low in dry matter include camels and giraffes, both of which are of large size and inhabit hot climates. Perhaps the young of these species have relatively high water requirements, as suggested for zebras and rhinos (see above). However, dorcas gazelle are also adapted to hot climates but do not produce dilute milks (Maltz and Shkolnik, 1984). Insufficient data are available to determine if small ruminants, such as dik-dik, duikers, and muntjac, produce more concentrated milks than larger species. Taylor *et al.* (1990) reported rather high concentrations of dry matter (28%), fat (12%), and crude protein (10%) in milk of the blue duiker, *Cephalophus monticola,* but these means may not be representative of midlactation as they include samples collected near weaning. The relatively high dry matter, fat, and protein concentrations in reindeer milk are thought to relate to the need for rapid nutrient transfer to the young during the short season when food is of high quality in extreme northern latitudes (Luick *et al.*, 1974; White and Luick, 1984). The same may also apply to muskoxen and moose (Table I).

M. Rodents (Order Rodentia)

Although 11 rodent species are represented in Table I, this is a rather poor representation of the approximately 1700 species in this order. The rodent data are especially plagued by methodological problems. For example, the midlactation milk of the house mouse is reported to contain mean values of 29.3 to 61.3% dry matter, 13.1 to 41.6% fat, and 7.2 to 14.1% protein in six different studies (Meier *et al.*, 1965; Hanrahan and Eisen, 1970; Jenness and Sloan, 1970; Baverstock *et al.*, 1976; Knight *et al.*, 1986; König *et al.*, 1988). The careful study of Knight *et al.* (1986) was chosen as most representative; the dry matter and fat results of this study agree with those of Baverstock *et al.* (1976) and König *et al.* (1988) but the protein values are considerably higher (12.5 vs 7 or 8%). It appears that the Bradford dye-binding method used by Baverstock *et al.* (1976) and König *et al.* (1988) underestimated protein concentration; Knight *et al.* (1986) used a more appropriate Kjeldahl method. The protein in *Notomys* and *Pseudomys* milks was probably underestimated as well, which accounts in part for the large discrepancy between dry matter on the one hand and the sum of fat, protein, and sugars on the other (Table I). Dye binding was also used to measure protein in punare milk (Meyerson-McCormick *et al.*, 1990).

A lack of methodological standardization also characterizes research on the milk composition of the laboratory rat (Brown or Norway rat, *Rattus norvegicus*). In the past 15 years, fat has been measured by solvent extraction, the sulfuric acid–vanillin reaction of Zöllner and Kirsch (1962), a glycerol assay, and a turbidometric method, protein has been measured by Kjeldahl (TN × 6.38), the Lowry procedure, and the Bradford dye-binding method, and sugars have been measured by reducing sugar, enzymatic, and other assays (Chalk and Bailey, 1979; Keen *et al.*, 1981; Roberts and Coward, 1985; Treadway and Lederman, 1986; Grigor *et al.*, 1987; Nicholas and Hartmann, 1991). In these studies, the ranges for the mean values for fat, protein, and lactose (sugars) at midlactation (ca. 8–14 days postpartum) are 8–19, 8–15, and 2.8–4.3%, respectively. As none of the studies included direct comparisons of the different methods, it is not possible to determine the extent to which this variation is due to analytical inaccuracy. In the absence of a more recent definitive study, we have retained the values compiled by Oftedal (1984a) from Luckey *et al.* (1955), Glass (1956), and Venkatachalem and Ramanathan (1964).

The neonatal guinea pig is born with substantial stores of body fat, which it proceeds to catabolize during the lactation period (Widdowson and McCance, 1955). The prenatal store appears to relieve the necessity of much postnatal fat transfer from the mother, and the milk is correspondingly low in lipids (Oftedal, 1981; Anderson and Chavis, 1986). The contrast to the very high-fat milk of the punare is striking, but the adaptive significance of high fat in this species is not clear (Meyerson-McCormick *et al.*, 1990). The high fat concentration in European beaver milk would be expected based on its aquatic habits.

N. Rabbits and Hares (Order Lagomorpha)

Referring to data of Ben Shaul (1962) and Jenness and Sloan (1970), Martin (1984) suggested that mammals giving birth to altricial (undeveloped) young tended to produce more concentrated milks than mammals with precocial offspring. This is certainly not true of rabbits and hares. At birth domestic rabbits are altricial, hares are precocial, and cottontail rabbits are intermediate, but all three species produce milks of similar composition (Table I). The high but similar dry matter concentrations of these species may relate to the fact that in each species the young are suckled infrequently, about once per day (Zarrow *et al.*, 1965; Lhuillery *et al.*, 1984).

IV. Conclusion

Although the quality and quantity of data available on the milk composition of nonfarm animals have improved in the 10 years since the last comprehensive review (Oftedal, 1984a), considerable inaccuracy may remain in some of the 100 entries in Table I due to the small numbers of samples, difficulties in defining stage of lactation, biases introduced during sampling, and flawed analytical procedures. Researchers have shown a willingness to invest substantial amounts of time, effort, and resources to obtain series of milk samples from a wide variety of species. We hope that a similar commitment will be applied to the assessment and improvement of sampling and laboratory protocols so that analytical results need not be questioned. At the present time, oven-drying, solvent extraction, the Kjeldahl procedure (corrected for NPN), and a modified phenol–sulfuric acid procedure appear to be the most reliable methods for measuring total dry matter, fat, protein, and sugars, respectively, in the milks of diverse species. If alternative methods are used, it is important that they be compared to, or calibrated against, these standard methods using the milk of the species to be tested, not just cow's milk.

References

Abderhalden, E. (1898). Die Beziehungen der Wachsthums-geschwindigkeit des Saüglings zur Zusammensetzung der Milch beim Kaninchen, bei der Katz und beim Hunde. *Hoppe-Seyler's Z. Physiol. Chem.* **26,** 487–497.

Anderson, R. R., and Chavis, D. D. (1986). Changes in macroingredients of guinea pig milk through lactation. *J. Dairy Sci.* **69,** 2268–2277.

Anderson, R. R., Sadleir, K. C., Knaver, M. W., Wippler, J. P., and Marshall, R. T. (1975). Composition of cottontail rabbit milk from stomachs of young and directly from gland. *J. Dairy Sci.* **58,** 1449–1452.

Ando, K., Mori, M., Kato, I., Yusa, K., and Goda, K. (1979). General composition and chemical properties of the main components of Yeso brown bear (*Ursus arctos yesoensis*) milks. *J. Coll. Dairying, Natl. Sci.* **8,** 9–21.

Arman, P., Kay, R. N. B., Goodall, E. D., and Sharman, G. A. M. (1974). The composition and yield of milk from captive red deer (*Cervus elaphus* L.). *J. Reprod. Fertil.* **37,** 67–84.

Aschaffenburg, R., Gregory, M. E., Rowland, S. J., Thompson, S. Y., and Kon, V. M. (1961). The composition of the milk of the African black rhinoceros (*Diceros bicornis*). *Proc. Zool. Soc. London* **137,** 475–479.

Ashworth, U. S., Ramaiah, G. D., and Keys, M. C. (1966). Species difference in the composition of milk with special reference to the northern fur seal. *J. Dairy Sci.* **49,** 1206–1211.

Association of Official Analytical Chemists (AOAC) (1990). "Official Methods of Analyses of the Association of Official Analytical Chemists" (K. Helrich, ed.), 15th Ed. AOAC, Arlington, Va.

Atwood, C. S., and Hartmann, P. E. (1992). Collection of fore and hind milk from the sow and the changes in milk composition during suckling. *J. Dairy Res.* **59,** 287–298.

Bahl, R. K. (1972). An enzymic method for the determination of lactose in milk including human milk. *Analyst* **97,** 559–561.

Barbano, D. M., Lynch, J. M., and Fleming, J. R. (1991). Direct and indirect determination of true protein content of milk by Kjeldahl analysis: Collaborative study. *J. Assoc. Off. Anal. Chem.* **74,** 281–288.

Baverstock, P. R., Spencer, L., and Pollard, C. (1976). Water balance of small lactating rodents—II. Concentration and composition of milk of females on ad libitum and restricted water intakes. *Comp. Biochem. Physiol. A* **53,** 47–52.

Ben Shaul, D. M. (1962). The composition of the milk of wild animals. *Int. Zool. Yearbook* **4,** 333–342.

Bergman, H. C., and Housley, C. (1968). Chemical analyses of American opossum (*Didelphys virginiana*) milk. *Comp. Biochem. Physiol.* **25,** 213–218.

Best, P. B. (1982). Seasonal abundance, feeding, reproduction, age and growth in minke whales off Durban (with incidental observations from the Antarctic). *Rep. Int. Whal. Commission* **32,** 759–786.

Blaxter, K. L. (1961). Lactation and the growth of the young. *In* "Milk: The Mammary Gland and Its Secretion" (S. K. Kon and A. T. Cowie, eds.), pp. 305–361. Academic Press, New York.

Bolliger, A., and Pascoe, J. V. (1953). Composition of kangaroo milk (wallaroo, *Macropus robustus*). *Aust. J. Sci.* **15,** 215–217.

Boness, D. J., Bowen, W. D., and Oftedal, O. T. (1994). Evidence of a maternal foraging cycle resembling that of otariid seals in a small phocid, the harbor seal. *Behav. Ecol. Sociobiol.* **34,** 95–104.

Bonner, W. N. (1984). Lactation strategies in pinnipeds: Problems for a marine mammalian group. *Symp. Zool. Soc. London* **51,** 253–272.

Bowen, W. D., Oftedal, O. T., and Boness, D. J. (1985). Birth to weaning in four days: Remarkable growth in the hooded seal, *Cystophora cristata*. *Can. J. Zool.* **63,** 2841–2846.

Bradford, M. M. (1976). A rapid and sensitive method for the quantitation of microgram quantities of protein utilizing the principle of protein-dye binding. *Anal. Biochem.* **72,** 248–254.

Bradstreet, R. B. (1965). "The Kjeldahl Method for Organic Nitrogen." Academic Press, New York.

Buss, D. H. (1968). Gross composition and variation of the components of baboon milk during natural lactation. *J. Nutr.* **96,** 421–426.

Buss, D. H., and Cooper, R. W. (1970). Composition of milk from talapoin monkeys. *Folia Primat.* **13,** 196–206.

Buss, D. H., and Cooper, R. W. (1972). Composition of squirrel monkey milk. *Folia Primat.* **17,** 285–291.

Carl, G. R., and Robbins, C. T. (1988). The energetic cost of predator avoidance in neonatal ungulates: Hiding versus following. *Can. J. Zool.* **66,** 239–246.

Chalk, P. A., and Bailey, E. (1979). Changes in the yield, and carbohydrate, lipid and protein content of milk during lactation in the rat. *J. Dev. Physiol.* **1,** 61–79.

Chittleborough, R. G. (1958). The breeding cycle of the female humpback whale *Megapter nodosa*. *Aust. J. Marine Freshwater Res.* **9,** 1–18.

Coates, M. E., Gregory, M. E., and Thompson, S. Y. (1964). The composition of rabbit's milk. *Br. J. Nutr.* **18,** 583–586.

Commaille, M. A. (1966). Analyse du lait de chatte. *Compte Rend.* **63,** 692–693.

Conant, R. A. (1962). A milking technique and the composition of mink milk. *Am. J. Vet. Res.* **23,** 1104–1106.

Cook, H. W., Pearson, A. M., Simmons, N. M., and Baker, B. E. (1970). Dall sheep (*Ovis dalli dalli*) milk. I. Effects of stage of lactation on the composition of the milk. *Can. J. Zool.* **48,** 629–633.

Corbet, G. B., and Hill, J. E. (1980). "A World List of Mammalian Species," Cornell University Press, Ithaca, NY.

Costa, D. P. (1991). Reproductive and foraging energetics of pinnipeds: Implications for life history patterns. *In* "Behaviour of Pinnipeds" (D. Renouf, ed.), pp. 300–344. Chapman and Hall, London.

Cowan, P. E. (1989). Changes in milk composition during lactation in the common brushtail possum, *Trichosurus vulpecula* (Marsupialia: Phalangeridae). *Reprod. Fertil. Dev.* **1,** 325–335.

Cowie, A. T. (1969). Variation in the yield and composition of the milk during lactation in the rabbit and galactopoietic effect of prolactin. *J. Endocrinol.* **44,** 437–450.

Crisp, E. A., Cowan, P. E., and Messer, M. (1989). Changes in milk carbohydrates during lactation in the common brushtail possum, *Trichosurus vulpecula* (Marsupialia: Phalangeridae). *Reprod. Fertil. Dev.* **1,** 309–314.

Cross, B. A. (1977). Comparative physiology of milk removal. *Symp. Zool. Soc. London* **41,** 193–210.

Crowley, H. M., Woodward, D. R., and Rose, R. W. (1988). Changes in milk composition during lactation in the potoroo, *Potorous tridactylus* (Marsupialia:Potorinae). *Aust. J. Biol. Sci.* **41,** 189–296.

Denamur, R. (1965). The hypothalamo-neurohypophysial system and the milk ejection reflex. *Dairy Sci. Abstr.* **27,** 193–224, 263–280.

Dosako, S., Taneya, S., Kimura, T., Ohmori, T., Daikoku, H., Suzuki, N., Sawa, J., Kano, K., and Katayama, S. (1983). Milk of northern fur seal: Composition, especially carbohydrate and protein. *J. Dairy Sci.* **66,** 2076–2083.

Dryden, G. L., and Anderson, R. R. (1978). Milk composition and its relation to growth rate in the musk shrew, *Suncus murinus. Comp. Biochem. Physiol. A* **60,** 213–216.

Dubois, M., Gilles, K. A., Hamilton, J. K., Rebers, P. A., and Smith, F. (1956). Colorimetric method for determination of sugars and related substances. *Anal. Chem.* **28,** 350–356.

Dubrovskaya, R. M. (1975). Milk production in female blue polar foxes and its effect on the development of the young. *Dairy Sci. Abstr.* **29,** 115 [Abstract]

Erb, R. E., Sitarz, N. E., and Malven, P. V. (1977). Blood plasma and milk prolactin, and effects of sampling technique on composition of milk from ewes. *J. Dairy Sci.* **60,** 197–203.

Essig, A. M., and Kleyn, D. H. (1983). Determination of lactose in milk: Comparison of methods. *J. Assoc. Off. Anal. Chem.* **66,** 1514–1516.

Fleet, I. R., and Linzell, J. L. (1964). A rapid method of estimating fat in very small quantities of milk. *J. Physiol.* **175,** 15P.

Folin, O., and Wu, H. (1919). A system of blood analysis. *J. Biol. Chem.* **38,** 81–110.

Folin, O., Denis, W., and Minot, A. S. (1919). Lactose, fat and protein in milk of various animals. *J. Biol. Chem.* **37,** 349–352.

Ganguli, M., Smith, J., and Hanson, L. (1969). Indirect micromethod of milk fat determination. *J. Dairy Sci.* **52,** 126–127.

Gentry, R. L., and Holt, J. R. (1986). Attendance behavior of northern fur seals. *In* "Fur Seals: Maternal Strategies on Land and at Sea" (R. L. Gentry and G. L. Kooyman, eds.), pp. 41–60. Princeton University Press, Princeton, NJ.

Gittleman, J. L., and Oftedal, O. T. (1987). Comparative growth and lactation energetics in carnivores. *Symp. Zool. Soc. London* **57,** 41–77.

Glass, R. J. (1956). "Chemical, Physical and Biological Studies of Rat's Milk and Its Components." Ph.D. thesis, University of Minnesota.

Glass, R. L., Lohse, L. W., and Jenness, R. (1968). Chromatographic procedure for determination of fat content of small specimens of milk. *J. Dairy Sci.* **51,** 1847–1849.

Gornall, A. G., Bardawill, C. J., and David, M. M. (1949). Determination of serum proteins by means of the biuret reaction. *J. Biol. Chem.* **177,** 751–766.

Green, B. (1984). Composition of milk and energetics of growth in marsupials. *Symp. Zool. Soc. London* **51,** 369–387.

Green, B., and Merchant, J. C. (1988). The composition of marsupial milk. *In* "The Developing Marsupial. Models for Biomedical Research" (C. H. Tyndale-Biscoe and P. A. Janssens, eds.), pp. 41–54. Springer-Verlag, Berlin, Germany.

Green, B., Newgrain, K., and Merchant, J. (1980). Changes in milk composition during lactation in the tammar wallaby (*Macropus eugenii*). *Aust. J. Biol. Sci.* **33,** 35–42.

Green, B., Griffiths, M., and Leckie, R. M. C. (1983). Qualitative and quantitative changes in milk fat during lactation in the tammar wallaby (*Macropus eugenii*). *Aust. J. Biol. Sci.* **36,** 455–461.

Green, B., Merchant, J., and Newgrain, K. (1987). Milk composition in the eastern quoll, *Dasyurus viverrinus* (Marsupialia: Dasyuridae). *Aust. J. Biol. Sci.* **40,** 379–387.

Green, S. W., and Renfree, M. B. (1982). Changes in the milk proteins during lactation in the tammar wallaby, *Macropus eugenii. Aust. J. Biol. Sci.* **35,** 145–152.

Gregory, M. E., Kon, S. K., Rowland, S. J., and Thompson, S. Y. (1955). The composition of the milk of the blue whale. *J. Dairy Res.* **22,** 108–112.

Gregory, M. E., Rowland, S. J., Thompson, S. Y., and Kon, V. M. (1965). Changes during lactation in the composition of the milk of the African black rhinoceros (*Diceros bicornis*). *Proc. Zool. Soc. London* **145,** 327–333.

Griffiths, M. (1978). "Biology of the Monotremes." Academic Press, New York.

Griffiths, M., McIntosh, D. L., and Coles, R. E. A. (1969). The mammary gland of the echidna, *Tachyglossus aculeatus,* with observations on the incubation of the egg and on the newly-hatched young. *J. Zool. (London)* **158,** 371–386.

Griffiths, M., Green, B., Leckie, R. M. C., Messer, M., and Newgrain, K. W. (1984). Constituents of platypus and echidna milk, with particular reference to the fatty acid complement of the triglycerides. *Aust. J. Biol. Sci.* **37,** 323–329.

Grigor, M. R., Allan, J. E., Carrington, J. M., Carne, A., Geursen, A., Young, D., Thompson, M. P., Haynes, E. B., and Coleman, R. A. (1987). Effect of dietary protein and food restriction on milk production and composition, maternal tissues and enzymes in lactating rats. *J. Nutr.* **117,** 1247–1258.

Gross, R., and Bollinger, A. (1959). Composition of milk of the marsupial *Trichosurus vulpecula*. *Am. J. Dis. Child.* **98,** 768–775.

Hall-Martin, A. J., Skinner, J. D., and Smith, A. (1977). Observations on lactation and milk composition of the giraffe *Giraffa camelopardalis. S. Afr. J. Wildlife Res.* **7,** 67–71.

Hanrahan, J. P., and Eisen, E. J. (1970). A lactation curve for mice. *Lab Anim. Care* **20,** 101–104.

Hurni, H., and Montalta, J. (1980). Katzenmilch und ihr Ersatz für die kunstliche Aufzucht. *Z. Versuchstierk.* **22,** 32–35.

Ivanova, G. M. (1965). Chemical composition and nutritive value of elks' milk. *Dairy Sci. Abstr.* **27,** 556. [Abstract]

Iverson, S. J., Kirk, C. L., Hamosh, M., and Newsome, J. (1991). Milk lipid digestion in the neonatal dog: The combined action of gastric and bile salt stimulated lipases. *Biochim. Biophys. Acta* **1083,** 109–119.

Iverson, S. J., Sampugna, J., and Oftedal, O. T. (1992). Positional specificity of gastric hydrolysis of long chain n-3 polyunsaturated fatty acids of seal milk triglycerides. *Lipids* **27,** 870–878.

Iverson, S. J., Bowen, W. D., Boness, D. J., and Oftedal, O. T. (1993). The effect of maternal size and milk energy output on pup growth in grey seals (*Halichoerus grypus*). *Physiol. Zool.* **66,** 61–88.

Iwata, H., and Ishii, T. (1946). Studies on fur animals. I. On the milk of the raccoon dog. *Jpn. J. Zootech. Sci.* **17,** 99–101. [In Japanese]

Jaoen, J. C., and Mens, P. (1981). Change in the composition of goat milk during milking. *Dairy Sci. Abstr.* **43,** 376. [Abstract]

Jenness, R. (1974). The composition of milk. *In* "Lactation: A Comprehensive Treatise" (B. L. Larson and V. R. Smith, eds.), pp. 3–107. Academic Press, New York.

Jenness, R., and Patton S. (1959). "Principles of Dairy Chemistry." Wiley, New York.

Jenness, R., and Sloan, R. E. (1970). The composition of milks of various species: A review. *Dairy Sci. Abstr.* **32,** 599–612.

Jenness, R., and Studier, E. H. (1976). Lactation and milk. *In* "Biology of Bats of the New World Family Phyllostomatidae, Part I" (R. J. Baker, J. K. Jones, and D. C. Carter, eds.), pp. 201–218. Museum of Texas Tech University, Special Publications, Lubbock, TX.

Jenness, R., and Odell, D. K. (1978). Composition of milk of the pygmy sperm whale (*Kogia breviceps*). *Comp. Biochem. Physiol. A* **61,** 383–386.

Jenness, R., Regehr, E. A., and Sloan, R. E. (1964). Comparative biochemical studies of milks—II. Dialyzable carbohydrates. *Comp. Biochem. Physiol.* **13,** 339–352.

Jenness, R., Erickson, A. W., and Craighead, J. J. (1972). Some comparative aspects of milk from four species of bears. *J. Mammal.* **53,** 34–47.

Jones, D. R. (1931). "Factors for Converting Percentages of Nitrogen in Foods and Feeds into Percentages of Proteins. Circular No. 183." U. S. Department of Agriculture, Washington, DC.

Jones, J. B. (1991). "Kjeldahl Method for Nitrogen Determination." Micro–Macro Publishing, Athens, GA.

Keen, C. L., Lönnerdal, B., Clegg, M. S., and Hurley, L. S. (1981). Developmental changes in composition of rat milk: Trace elements, minerals, protein, carbohydrate and fat. *J. Nutr.* **111,** 226–236.

Keen, C. L., Lönnerdal, B., Clegg, M. S., Hurley, L. S., Morris, J. G., Rogers, Q. R., and Rucker, R. B. (1982). Developmental changes in composition of cats' milk: Trace elements, minerals, protein, carbohydrate and fat. *J. Nutr.* **112,** 1763–1769.

Knight, C. H., Maltz, E., and Docherty, A. H. (1986). Milk yield and composition in mice: Effects of litter size and lactation number. *Comp. Biochem. Physiol. A* **84,** 127–133.

Koch, F. C., and McMeekin, T. L. (1924). A new direct Nesslerization micro-Kjeldahl method and a modification of the Nessler–Folin reagent for ammonia. *J. Am. Chem. Soc.* **46,** 2066–2069.

König, B., Riester, J., and Markl, H. (1988). Maternal care in house mice (*Mus musculus*): II. The energy cost of lactation as a function of litter size. *J. Zool. (London)* **216,** 195–210.

Kretzmann, M. B., Costa, D. P., Higgins, L. V., and Needham, D. J. (1991). Milk composition of Australian sea lions. *Neophoca cinerea:* Variability in lipid content. *Can. J. Zool.* **69,** 2556–2561.

Kretzmann, M. B., Costa, D. P., and LeBoeuf, B. J. (1993). Maternal energy investment in elephant seal pups: Evidence for sexual equality? *Am. Natl.* **141,** 466–480.

Kunz, T. H., Stack, M. H., and Jenness, R. (1983). A comparison of milk composition in *Myotis lucifugus* and *Eptesicus fuscus* (Chiroptera: Vespertilionidae). *Biol. Reprod.* **28,** 229–234.

Kunz, T. H., Oftedal, O. T., Robson, S. K., Kretzmann, M. B., and Kirk, C. (1995). Changes in milk composition during lactation in three species of insectivorous bats. *J. Comp. Physiol. (B)* **164;** 543–551.

Lascelles, A. K., and Lee, C. S. (1978). Involution of the mammary gland. *In* "Lactation: A Comprehensive Treatise" (B. L. Larson, ed.), pp. 115–177. Academic Press, New York.

Laxa, O. (1930). Le lait de renard-argente. *Annls Falsif. Expert. Chim.* **23,** 404.

Lemon, M., and Barker, S. (1967). Changes in milk composition of the red kangaroo, *Megaleia rufa* (Desmarest), during lactation. *Aust. J. Exp. Biol. Med. Sci.* **45,** 213–219.

Lhuillery, C., Martinet, L., Demarne, Y., and Lecourtier, M. J. (1984). Food intake of captive leverets before weaning and the composition of the milk of the brown doe-hare (*Lepus europaeus*). *Comp. Biochem. Physiol. A* **78,** 73–76.

Linzell, J. L. (1967). The effect of very frequent milking and oxytocin on the yield and composition of milk in fed and fasted goats. *J. Physiol.* **190,** 333–346.

Linzell, J. L., and Fleet, I. R. (1969). Accuracy of the micromethod of estimating milk fat concentration by high-speed centrifugation in capillary tubes. *J. Dairy Sci.* **52,** 1685–1687.

Linzell, J. L., Peaker, M., and Taylor, J. C. (1975). The effects of oxytocin on milk secretion and on the permeability of the mammary epithelium in the rabbit. *J. Physiol.* **253,** 547–563.

Lochmiller, R. L., Hellgren, E. C., Grant, W. E. Greene, L. W., and Dill, C. W. (1985). Description of collared peccary (*Tayassu tajacu*) milk composition. *Zoo Biol.* **4,** 375–379.

Lockyer, C. (1984). Review of baleen whale (Mysticeti) reproduction and implications for management, *Rep. Int. Whal. Commission (Special Issue 6),* 27–50.

Lönnerdal, B., Keen, C. L., Glazier, C. E., and Anderson, J. (1984). A longitudinal study of Rhesus monkey (*Macaca mulatta*) milk composition: Trace elements, minerals, protein, carbohydrate, and fat. *Pediatr. Res.* **18,** 911–914.

Lönnerdal, B., Keen, C. L. Hurley, L. S., and Fisher, G. L. (1981). Developmental changes in the composition of beagle dog milk. *Am. J. Vet. Res.* **42,** 662–665.

Lowry, O. H., Rosebrough, N. J., Farr, A. L., and Randall, R. J. (1951). Protein measurement with the Folin phenol reagent. *J. Biol. Chem.* **193,** 265–275.

Lucas, A., Gibbs, J. A. H., Lyster, R. L. J., and Baum, J. D. (1978). Creamatocrit: Simple clinical technique for estimating fat concentration and energy value of human milk. *Br. Med. J.* **1978,** 1018–1020.

Luckey, T. D., Mende, T. J., and Pleasants, J. (1955). The physical and chemical characterization of rat's milk. *J. Nutr.* **54,** 345–359.

Luhtala, A., Rautiainen, A., and Antila, M. (1968). Die Zusammensetzung der Finnischen Rentiermilch. *Acta Chem. Fenn.* **B41,** 6–9.

Luick, J. R., White, R. G., Gau, A. M., and Jenness, R. (1974). Compositional changes in the milk secreted by grazing reindeer. I. Gross composition and ash. *J. Dairy Sci.* **57,** 1325–1333.

Macart, M., and Gerbaut, L. (1982). An improvement of the Coomassie blue dye binding method allowing an equal sensitivity to various proteins: Application to cerebrospinal fluid. *Clin. Chim. Acta* **122,** 93–101.

Maltz, E. (1979). "Productivity in the Desert: Bedouin Goat, Ibex and Desert Gazelle." Ph.D. thesis, Tel Aviv University. [In Hebrew, English summary].

Maltz, E., and Shkolnik, A. (1984). Lactational strategies of desert ruminants: The Bedouin goat, ibex and desert gazelle. *Symp. Zool. Soc. London* **51,** 193–213.

Marier, J. R., and Boulet, M. (1959). Direct analysis of lactose in milk and serum. *J. Dairy Sci.* **42,** 1390–1391.

Martin, R. D. (1984). Scaling effects and adaptive strategies in mammalian lactation. *Symp. Zool. Soc. London* **51,** 87–117.

McCullagh, K. G., and Widdowson, E. M. (1970). The milk of the African elephant. *Br. J. Nutr.* **24,** 109–117.

Meier, H., Hoag, W. G., and McBurney, J. J. (1965). Chemical characterization of inbred-strain mouse milk—I. Gross composition and amino acid analysis. *J. Nutr.* **85,** 305–308.

Merchant, J., Green, B., Messer, M., and Newgrain, K. (1989). Milk composition in the red-necked wallaby, *Macropus rufogriseus banksianus* (Marsupialia). *Comp. Biochem. Physiol. A* **93,** 483–488.

Merchant, J. C., and Libke, J. A. (1988). Milk composition in the northern brown bandicoot, *Isoodon macrourus* (Paramelidae, Marsupialia). *Aust. J. Biol. Sci.* **41,** 495–505.

Messer, M., and Kerry, K. R. (1973). Milk carbohydrates of the echidna and platypus. *Science* **180,** 201–203.

Messer, M., and Mossop, G. S. (1977). Milk carbohydrates of marsupials. I. Partial separation and characterization of neutral milk oligosaccharides of the eastern grey kangaroo. *Aust. J. Biol. Sci.* **30,** 379–388.

Messer, M., and Green, B. (1979). Milk carbohydrates of marsupials. II. Quantitative and qualitative changes in milk carbohydrates during lactation in the Tammar wallaby (*Macropus eugenii*). *Aust. J. Biol. Sci.* **32,** 519–531.

Messer, M., Gadiel, P. A., Ralston, G. B., and Griffiths, M. (1983). Carbohydrates of the milk of the platypus. *Aust. J. Biol. Sci.* **36,** 129–137.

Messer, M., FitzGerald, P. A., Merchant, J. C., and Green, B. (1987). Changes in milk carbohydrates during lactation in the eastern quoll, *Dasyurus viverrinus* (Marsupialia). *Comp. Biochem. Physiol. B* **88,** 1083–1086.

Messer, M., Crisp, E. A., and Newgrain, K. (1988). Studies on the carbohydrate content of milk of the crabeater seal (*Lobodon carcinophagus*). *Comp. Biochem. Physiol. B* **90,** 367–370.

Meyerson-McCormick, R., Cranford, J. A., and Akers, R. M. (1990). Milk yield and composition in the punare *Trichomys apereoides*. *Comp. Biochem. Physiol. A* **96,** 211–214.

Montgomery, R. (1961). Further studies of the phenol–sulfuric acid reagent for carbohydrates. *Biochim. Biophys. Acta* **48,** 591–593.

Morris, D. L. (1948). Quantitative determination of carbohydrates with Dreywood's anthrone reagent. *Science* **107,** 254–255.

Mover, H., Ar, A., and Hellwing, S. (1989). Energetic costs of lactation with and without simultaneous pregnancy in the white-toothed shrew *Crocidura russula monacha*. *Physiol. Zool.* **62,** 919–936.

Mueller, C. C., and Sadleir, R. M. F. S. (1977). Changes in the nutrient composition of milk of black-tailed deer during lactation. *J. Mammal.* **58,** 421–423.

Munks, S. A., Green, B., Newgrain, K., and Messer, M. (1991). Milk composition in the common ringtail possum, *Pseudocheirus peregrinus* (Petauridae: Marsupialia). *Aust. J. Zool.* **39,** 403–416.

Munson, L. S., and Walker, P. H. (1906). The unification of reducing sugar methods. *J. Am. Chem. Soc.* **28,** 663–686.

Naismith, D. J., Mittwoch, A., and Platt, B. S. (1969). Changes in the composition of rat's milk in the stomach of the suckling. *Br. J. Nutr.* **23,** 683–693.

Nelson, W. L., Kaye, A., Moore, M., Williams, H. H., and Herrington, B. L. (1951). Milking techniques and the composition of guinea pig milk. *J. Nutr.* **44,** 585–594.

Nicholas, K. R. (1988a). Control of milk protein synthesis in the marsupial *Macropus eugenii:* A model system to study prolactin-dependent development. *In* "The Developing Marsupial: Models for Biomedical Research" (C. H. Tyndale-Biscoe and P. A. Janssens, eds.), pp. 68–85. Springer-Verlag, Berlin, Germany.

Nicholas, K. R. (1988b). Asynchronous dual lactation in a marsupial, the tammar wallaby (*Macropus eugenii*). *Biochem. Biophys. Res. Commun.* **154,** 529–536.

Nicholas, K. R., and Hartmann, P. E. (1991). Milk secretion in the rat: Progressive changes in milk composition during lactation and weaning and the effect of diet. *Comp. Biochem. Physiol. (A)* **98,** 535–542.

Nicholas, K. R., Loughnan, M., Messer, M., Munks, S., Griffiths, M., and Shaw, D. (1989). Isolation, partial sequence and asynchronous appearance during lactation of lysozyme and α-lactalbumin in the milk of a marsupial, the common ringtail possum (*Pseudocheirus peregrinus*). *Comp. Biochem. Physiol. B* **94,** 775–778.

Nicholas, K. R., Messer, M., Elliott, C., Maher, F., and Shaw, D. C. (1987). A novel whey protein synthesized only in late lactation by the mammary gland from the tammar (*Macropus eugenii*). *Biochem. J.* **241,** 899–904.

Nishikawa, I., Kawanishi, G., Cho, F., Honjo, S., Hatakeyama, T., and Wako, H. (1976). Chemical composition of Cynomolgus monkey milk. *Exp. Anim.* **25,** 253–264.

Oftedal, O. T. (1980). Milk and mammalian evolution. *In* "Comparative Physiology: Primitive Mammals" (K. Schmidt-Nielsen, L. Bolis, and C. R. Taylor, eds.). Cambridge University Press, Cambridge.

Oftedal, O. T. (1981). "Milk, Protein and Energy Intakes of Suckling Mammalian Young: A Comparative Study", Ph.D. Thesis, Cornell University, Ithaca, NY.

Oftedal, O. T. (1984a). Milk composition, milk yield and energy output at peak lactation: A comparative review. *Symp. Zool. Soc. London* **51,** 33–85.

Oftedal, O. T. (1984b). Lactation in the dog: Milk composition and intake by puppies. *J. Nutr.* **114,** 803–812.

Oftedal, O. T. (1985). Pregnancy and lactation. *In* "The Bioenergetics of Wild Herbivores" (R. J. Hudson, and R. G. White, eds.), pp. 215–238. CRC Press, Boca Raton, FL.

Oftedal, O. T. (1993). The adaptation of milk secretion to the constraints of fasting in bears, seals and baleen whales. *J. Dairy Sci.* **76,** 3234–3246.

Oftedal, O. T., and Jenness, R. (1988). Interspecies variation in milk composition among horses, zebras and asses (Perissodactyla:Equidae). *J. Dairy Res.* **55,** 57–66.

Oftedal, O. T., Hintz, H. F., and Schryver, H. F. (1983). Lactation in the horse: Milk composition and intake by foals. *J. Nutr.* **113,** 2096–2106.

Oftedal, O. T., Boness, D. J., and Tedman, R. A. (1987a). The behavior, physiology, and anatomy of lactation in the Pinnipedia. *Curr. Mammalogy.* No. 1, 175–245.

Oftedal, O. T., Iverson, S. J., and Boness, D. J. (1987b). Milk and energy intakes of suckling California sea lion *Zalophus californianus* pups in relation to sex, growth, and predicted maintenance requirements. *Physiol. Zool.* **60,** 560–575.

Oftedal, O. T., Boness, D. J., and Bowen, W. D. (1988). The composition of hooded seal (*Cystophora cristata*) milk: An adaptation for postnatal fattening. *Can. J. Zool.* **66,** 318–322.

Oftedal, O. T., Alt, G. L., Widdowson, E. M., and Jakubasz, M. R. (1993a). Nutrition and growth of suckling black bears (*Ursus americanus*) during their mothers' winter fast. *Br. J. Nutr.* **70,** 59–79.

Oftedal, O. T., Bowen, W. D., and Boness, D. J. (1993b). Energy transfer by lactating hooded seals, and nutrient deposition in their pups, during the four days from birth to weaning. *Physiol. Zool.* **66,** 412–435.

Oftedal, O. T., Bowen, W. D., and Boness, D. J. (1995). Lactation performance and nutrient deposition in pups of the harp seal, *Phoca groenlandica,* on the ice floes off southeast Labrador. *Physiol. Zool.* **68** (in press).

Ohta, K., Watarai, T., Oishi, T., Ueshiba, Y., Hirose, S., Yoshizawa, T., Akikusa, Y., Sato, M., and Okano, H. (1955). Composition of fin whale milk. *Sci. Rep. Whales Res. Inst.* **10,** 151–167.

Ono, K. A., Boness, D. J., and Oftedal, O. T. (1987). The effect of a natural environmental disturbance on maternal investment and pup behavior of the California sea lion. *Behav. Ecol. Sociobiol.* **21,** 109–118.

Ota, K., Makino, Y., Kimura, M., and Suzuki, J. (1991). Lactation in the Japanese monkey (*Macaca fuscata*): Yield and composition of milk and nipple preference of young. *Primates* **32,** 35–48.

Parkash, S., and Jenness, R. (1968). The composition and characteristics of goat's milk: A review. *Dairy Sci. Abstr.* **30,** 67–87.

Peaker, M. (1977). The aqueous phase of milk: ion and water transport. *Symp. Zool. Soc. London* **41,** 113–134.

Peaker, M., and Goode, J. A. (1978). The milk of the fur seal, *Arctocephalus tropicalis gazella;* In particular the composition of the aqueous phase. *J. Zool. (London)* **185,** 469–476. [Correct species: *Mirounga leonina,* M. Peaker, personal communication]

Peddemors, V. M., de Muelenaere, H. J. H., and Devchand, K. (1989). Comparative milk composition of the bottlenosed dolphin (*Tursiops truncatus*), humpback dolphin (*Sousa plumbea*) and common dolphin (*Delphinus delphis*) from southern African waters. *Comp. Biochem. Physiol. A* **94,** 639–641.

Perrin, W. F., Brownell, R. L., and DeMaster, D. P., eds. (1984). Reproduction in whales, dolphins and porpoises. *In* "Reports of the International Whaling Commission, Special Issue 6." International Whaling Commission, Cambridge, England.

Perry, N. A., and Doan, F. J. (1950). A picric acid method for the simultaneous determination of lactose and sucrose in dairy products. *J. Dairy Sci.* **33,** 176–185.

Pervaiz, S., and Brew, K. (1986). Composition of the milks of the bottlenose dolphin (*Tursiops truncatus*) and the Florida manatee (*Trichechus manatus latirostris*). *Comp. Biochem. Physiol. A* **84,** 357–360.

Peters, J. M., Maier, R., Hawthorne, E., and Storvick, C. L. (1972). Composition and nutrient content of elephant (*Elephas maximus*) milk. *J. Mammal.* **53,** 717–724.

Pilson, M. E. Q., and Kelly, A. L. (1962). Composition of the milk from *Zalophus californianus,* the California sea lion. *Science* **135,** 104–105.

Pilson, M. E. Q., and Waller, D. W. (1970). Composition of milk from spotted and spinner porpoises. *J. Mammal.* **51,** 74–79.

Ponce de Leon, A. (1984). Lactancia y composicion cuantitativa de la leche del lobo fino Sudamericano, *Arctocephalus australis* (Zimmerman, 1783). Industria Lobera y Pesquera del Estado, Montevideo, Uruguay, Anales 1984, pp. 43–58.

Powers, G. F. (1933). The alleged correlation between the rate of growth of the suckling and the composition of the milk of the species. *J. Pediatr.* **3,** 201–216.

Quicke, G. V., Sowler, S., Berry, R. K., and Geddes, A. M. (1984). Composition of mammary secretion from the epauletted fruit bat, *Epomorphorus wahlbergi. S. Afr. J. Sci.* **80,** 481–482.

Rammell, C. G., and Caughley, G. (1964). Composition of thar's milk. *N. Z. J. Sci.* **7,** 667–670.

Riedman, M., and Ortiz, C. L. (1979). Changes in milk composition during lactation in the northern elephant seal. *Physiol. Zool.* **52,** 240–249.

Robbins, C. T., Podbielancik, N., Roberta, S., Wilson, D. L., and Mould, E. D. (1981). Growth and nutrient consumption of elk calves compared to other ungulate species. *J. Wildlife Management* **45,** 172–186.

Roberts, S. B., and Coward, W. A. (1985). Dietary supplementation increases milk output in the rat. *Br. J. Nutr.* **53,** 1–9.

Roberts, S. B., Cole, T. J., and Coward, W. A. (1985). Lactational performance in relation to energy intake in the baboon. *Am. J. Clin. Nutr.* **41,** 1270–1276.

Rüsse, I. (1961). Die Laktation der Hündin. *Zentbl. Vet. Med.* **8,** 252–281.

Scheurmann, E., Senft, B., and Rietschel, W. (1977). Untersuchungen über die Zusammensetzung der Gayalmilch (*Bibos frontalis* Lambert 1837). *Zool. Gart. (Jena)* **47,** 24–32.

Schmidt-Nielsen, K., and Schmidt-Nielsen, B. (1952). Water metabolism of desert mammals. *Physiol. Rev.* **32,** 135–166.

Schoknecht, P. A., Cranford, J. A., and Akers, R. M. (1985). Variability in milk composition of the domestic ferret (*Mustela putorius*). *Comp. Biochem. Physiol. A* **81,** 589–591.

Sedmak, J. J., and Grossberg, S. E. (1977). A rapid, sensitive, and versatile assay for protein using Coomassie brilliant blue G250. *Anal. Biochem.* **79,** 544–552.

Shaffer, P. A., and Somogyi, M. (1933). Copper-iodometric reagents for sugar determination. *J. Biol. Chem.* **100,** 695–713.

Sherbon, J. W. (1967). Rapid determination of protein in milk by dye binding. *J. Assoc. Off. Anal. Chem.* **50,** 542–547.

Simon, K. J. (1959). Preliminary studies on composition of milk of Indian elephants. *Indian Vet. J.* **36,** 500–503.

Smith, A. (1970). Melk van wilde soogdiere. *Proc. S. Afr. Soc. Anim. Prod.* **9,** 63–72.

Smolenski, A. J., and Rose, R. W. (1988). Comparative lactation in two species of rat–kangaroo (Marsupialia). *Comp. Biochem. Physiol. A* **90,** 459–463.

Somogyi, M. (1952). Notes on sugar determination. *J. Biol. Chem.* **195,** 19–23.

Stern, I., and Shapiro, B. (1953). A rapid and simple method for the determination of esterified fatty acids and for total fatty acids in blood. *J. Clin. Pathol.* **6,** 158–160.

Stewart, R. E. A., Webb, B. E., and Lavigne, D. M. (1983). Determining lactose content of harp seal milk. *Can. J. Zool.* **61,** 1094–1100.

Taylor, B. A., Varga, G. A., Whitsel, T. J., and Hershberger, T. V. (1990). Composition of blue duiker (*Cephalophus monticola*) milk and milk intake by the calf. *Small Ruminant Res.* **3,** 551–560.

Teahan, C. G., McKenzie, H. A., and Griffiths, M. (1991). Some monotreme milk "whey" and blood proteins. *Comp. Biochem. Physiol. B* **99,** 99–118.

Tedman, R. A. (1980). "Lactation in the Weddell Seal, *Leptonychotes weddelli.*" Ph.D. thesis, University of Queensland, Australia.

Tilden, C. D., and Oftedal, O. T. (1985). Milk composition reflects pattern of maternal care in prosimiam primates. *Am. J. Primat.* (in press).

Treadway, J. L., and Lederman, S. A. (1986). The effects of exercise on milk yield, milk composition, and offspring growth in rats. *Am. J. Clin. Nutr.* **44,** 481–488.

Treus, V., and Kravchenko, D. (1968). Methods of rearing and economic utilization of eland in the Askaniyanova Zoological Park. *Symp. Zool. Soc. London* **21,** 395–411.

Trillmich, F., and Lechner, E. (1986). Milk of the Galapagos fur seal and sea lion, with a comparison of the milk of eared seals (Otariidae). *J. Zool. (London)* **209,** 271–277.

Turton, J. A., Ford, D. J., Bleby, J., Hall, B. M., and Whiting, R. (1978). Composition of the milk of the common marmoset (*Callithrix jacchus*) and milk substitutes used in hand-rearing programmes, with special reference to fatty acids. *Folia Primat.* **29,** 64–79.

Ullrey, D. E., Schwartz, C. C., Whetter, P. A., Rajeshwar Rao, T., Euber, J. R., Cheng, S. G., and Brunner, J. R. (1984). Blue–green color and composition of Stejneger's beaked whale (*Mesoplodon stejnegeri*) milk. *Comp. Biochem. Physiol. B* **79,** 349–352.

Venkatachalam, P. S., and Ramanathan, K. S. (1964). Effect of protein deficiency during gestation and lactation on body weight and composition of offspring. *J. Nutr.* **84,** 38–42.

Volcani, R., Zisling, R., Sklan, D., and Nitzan, Z. (1973). The composition of chinchilla milk. *Br. J. Nutr.* **29,** 121–125.

Weidner, K., and Jakobsen, P. E. (1966). Rapid method for determination of protein in milk and practical experience with it use. *In* "Proceedings of the 17th International Dairy Congress," pp. B161–B168.

White, J. C. D. (1953). Composition of whales' milk. *Nature* **171,** 612.

White, R. G., and Luick, J. R. (1984). Plasticity and constraints in the lactational strategy of reindeer and caribou. *Symp. Zool. Soc. London* **51,** 215–232.

Widdowson, E. M., and McCance, R. A. (1955). Physiological under-nutrition in the newborn guinea-pig. *Br. J. Nutr.* **91,** 316–321.

Wilson, D. E., and Hirst, S. M. (1977). Ecology and factors limiting roan and sable antelope populations in South Africa. *Wildlife Monogr.* **54,** 1–111.

Young, E. G., and Grant, G. A. (1931). The composition of vixen milk. *J. Biol. Chem.* **93,** 805–810.

Zarrow, M. X., Denenberg, V. H., and Anderson, C. O. (1965). Rabbit: Frequency of suckling in the pup. *Science* **150,** 1835–1836.

Zöllner, N., and Kirsch, K. (1962). Über die quantitative Bestimmung von Lipoiden (Mikromethode) mittels der vielen naturlichem Lipoiden (allen bekanntes Plasmolipoiden) gemeinsamen Sulfophosphovanillin-Reaktion. *Gesamte Exp. Med.* **135,** 545–561.
Zurowski, W., Kisza, J., Kruk, A., and Roskosz, A. (1974). Lactation and chemical composition of milk of the European beaver (*Castor fiber* L.). *J. Mammal.* **55,** 847–850.

B. Philogenetic and Ecological Variation in the Fatty Acid Composition of Milks

SARA J. IVERSON
OLAV T. OFTEDAL

I. Introduction

Lipid is the most variable constituent of milk among species, ranging from 0.2% by weight in milk of the black rhinoceros (Gregory *et al.*, 1965) to 60.0% in the milks of some phocid seals (Oftedal *et al.*, 1988; Iverson *et al.*, 1993). However, in all species studied milk lipid is composed primarily of fatty acids esterified as triglycerides (typically 97–99% by weight; Davies *et al.*, 1983; Iverson *et al.*, 1991, 1992).

Since the time of early studies on milk fatty acids, two things have been apparent. First, milk fatty acid composition differs markedly among species, both in the degree of unsaturation and in chain length (e.g., Hilditch, 1956; Hilditch and Williams, 1964). Second, milk fatty acids originate, to varying degrees, directly from the diet. As early as the 1920s, studies using stained fat showed that some fatty acids appeared to remain intact through the process of digestion and subsequent secretion in milk in various species, but not in cows (Folley and McNaught, 1961). Over the next decades, researchers demonstrated that the degree to which milk fat originated via *de novo* biosynthesis of fatty acids by the mammary gland versus direct uptake of circulating fatty acids from diet or endogenous sources varied among species.

The knowledge that the balance between *de novo* synthesis and dietary fat contribution to milk fatty acid composition may vary with proximate factors, such as diet and lactation stage, has led to numerous, systematic, and detailed studies in the human and some dairy animals [e.g., reviewed in Jensen (1989) and Jensen *et al.* (1990)]. However, this knowledge has largely been ignored with respect to the study of the fatty acid composition of other mammalian milks. Many researchers still report "the milk fatty acid composition of a species" as if it were a fixed product, relying on single

HANDBOOK OF MILK COMPOSITION

samples of unknown lactation stage and without considering effects of diet or environment (e.g., whether the animals studied were from captive or wild populations). Studies of this type continue to be cited repeatedly even when the less powerful analytical techniques of the 1960s and 1970s were employed. The misconception that the fatty acid composition of many species is "already known" appears to have inhibited the initiation of more reliable and systematic studies that are both possible and necessary if phylogenetic and ecological trends are to be understood.

In this review we present the most reliable data available. Throughout, the composition of fatty acids is expressed as weight percentage of total fatty acids, unless otherwise stated, and fatty acids are designated as carbon chain length number:number of double bonds where *n x* denotes the position (*x*) of the last double bond relative to the terminal methyl carbon (IUPAC nomenclature). In examining the variability of milk fatty acid composition among and within species, we stress that accurate and representative data are quite limited. The factors which contribute to differing fatty acid patterns among taxonomic groups will be briefly reviewed, as will the analytical techniques necessary to accurately identify and quantify milk fatty acids. In the end, we hope it will be apparent that much work remains to be done in order to understand the fatty acid patterns of mammalian milks.

II. The Sources of Milk Fatty Acids among Species

The milk fatty acid composition of a species or individual is the result of varying contributions of (1) direct uptake of circulating fatty acids, (2) *de novo* synthesis of fatty acids by the mammary gland from metabolites (e.g., acetate and NADPH) which provide sources of carbon and energy, and (3) further modification of fatty acids within the gland (i.e., desaturation or elongation; chain shortening is not believed to occur to any significant extent).

In monogastric mammals, ingested lipid (primarily triglyceride) is hydrolyzed in the stomach and small intestine, transported across the small intestine, and reesterified into the circulating chylomicrons for transport to tissues (Borgstrom, 1977; Patton, 1981). Thus, dietary fatty acids essentially remain intact through digestion and may be carried directly via chylomicrons to the mammary gland (Scow *et al.*, 1975; Nelson, 1992). The other circulating fatty acids which are taken up by the mammary gland are unesterified (e.g., bound to albumin) or carried in the very low-density lipoproteins (VLDL). These may originate directly from adipose tissue stores (and thus ultimately from the diet) or from endogenous synthesis within the liver or adipose tissue. Hence, in monogastric mammals dietary fatty acids may directly influence the composition of milk fatty acids to the degree that circulating fatty acids are taken up by the mammary gland (e.g., Iverson, 1993). In species with extensive foregut fermentation, such as ruminants, the circulating fatty acids in chylomicrons or VLDL will not reflect those from the diet due to the extensive microbial fermentation which takes place in the foregut. However, dietary changes, such as high concentrate versus high roughage (low concentrate) diet, appear to cause changes in milk fatty acid composition of ruminants (e.g., Calderon *et al.*, 1984; Jensen *et al.*, 1990).

De novo synthesis (and subsequent modification) is the other source of milk fatty acids. The most noteworthy difference between milk lipid and other tissue lipids is the presence, and in fact abundance, of short- and medium-chain fatty acids in the milk fat of many species. Whereas fatty acid synthesis in liver and adipose tissue produces primarily palmitate (16:0), mammary tissue in a number of species (e.g., species of primates, elephants, rodents, rabbits) synthesizes large amounts of medium-chain fatty acids (8:0–12:0) due to the presence of an enzyme specific to mammary tissue (Dils *et al.*, 1977; Dils, 1983). Additionally, ruminants esterify short-chain fatty acids (4:0 and 6:0) from butyryl-CoA and hexanoyl-CoA into milk triglycerides, whereas in nonruminants these cofactors are believed to be used preferentially as primers for the synthesis of longer-chain fatty acids (i.e., 16:0; Dils, 1983).

The relative importance of *de novo* synthesis of milk fatty acids by the mammary gland has generally been regarded as a species-specific characteristic (e.g., Morrison, 1970; Dils *et al.*, 1977). If so, this would largely determine the proportion of fatty acids which may be affected by circulating fatty acids and thus by dietary intake. For instance, in the single study of milk samples from wild African elephants, medium-chain fatty acids accounted for 92% of total fatty acids (McCullagh *et al.*, 1969). If typical of the species, this would indicate that no more than a small fraction (8%) of elephant milk fatty acids could be contributed from circulating fatty acids. *De novo* synthesis is easily detectable when short- or medium-chain fatty acids are present since they do not originate from circulating lipids. However, other fatty acids may also be produced by *de novo* synthesis. Early *in vitro* studies with guinea pig mammary tissue showed that while the mammary gland does not contain the medium-chain enzyme, it does synthesize the long-chain fatty acids (16:0, 16:1, 18:0, and 18:1) characteristic of guinea pig milk (Strong and Dils, 1972). While elongation of fatty acids in mammary glands is believed to be rare, desaturation of 16:0 to 16:1 and 18:0 to 18:1 has been demonstrated in cow, goat, pig, and mouse mammary tissue (Dils *et al.*, 1977; Bauman and Davis, 1974; Dils, 1983). To our knowledge, *de novo* synthesis has not been studied in other taxonomic groups.

The relative contribution from *de novo* synthesis of longer-chain fatty acids versus their uptake from the circulation is poorly understood and can only be elucidated through direct studies. In cows and goats, arteriovenous and radioactive tracer studies have shown that between 50 and 80% of milk lipid may be derived by direct uptake from serum triglycerides (Patton and Jensen, 1976; Annison, 1983). We know little about the extent of *de novo* synthesis among most nondairy species and especially among species, such as carnivores and marine mammals, which produce milks devoid of medium-chain fatty acids. We also do not know whether patterns of mammary *de novo* synthesis in such species are affected by the amounts or types of dietary fatty acids, by other dietary constituents, or by plane of nutrition; however, by analogy to other tissues that synthesize fatty acids (Nelson, 1992), we might expect this to be the case. In species in which mothers leave their young for prolonged periods to feed (e.g., some cave dwelling bats, nest building carnivores, winter-dormant bears, seals), do rates of *de novo* synthesis change during intermittent fasting and feeding? Or does *de novo* synthesis take place at all if diet or adipose tissue stores are sufficient to maintain circulating lipid levels?

III. Considerations in Sampling and Analysis of Milk Fatty Acids

Interpretation of milk fatty acid data is often difficult simply due to the lack of information given. Diet may be a key factor affecting fatty acid composition for most nonruminant species. Additionally, sampling and the methods of extraction, transesterification, and gas–liquid chromatography (GLC) analysis will also greatly influence milk fatty acid data. Reports of fatty acid composition should provide details on the samples collected and the methods of analysis used.

A. Sample Collection and Storage

As in evaluations of proximate milk composition, the stage of lactation (days postpartum) can also be a significant factor in milk fatty acid composition, even when a consistent diet is maintained throughout. A number of studies have documented an increase in the proportion of medium-chain fatty acids (by up to two- or threefold) over the lactation period in species which synthesize these components (e.g., human, horse, rat, rabbit; Jensen, 1989; Doreau *et al.,* 1992; Bitman *et al.,* 1985; Hall, 1971). Changes in patterns of longer-chain fatty acids over lactation have also been reported in marsupials such as the red kangaroo and tammar wallaby (Griffiths *et al.,* 1972; Green *et al.,* 1983). Of course, if changes in feeding behavior coincide with changes in lactation stage, milk fatty acid patterns will likely be affected. For instance, the black bear begins the first 2 or 3 months of lactation during the fast of winter dormancy. During this time levels of 18:2n 6 and 18:3n 3 average about 6.2 and 0.3% of milk fatty acids, respectively, whereas in the summer, bears foraging on fruits and leafy vegetation produce milk containing levels of these components as high as 20.0 and 15.0%, respectively (Iverson and Oftedal, 1992). Similarly, in otariid seals (i.e., fur seals and seal lions) large changes in milk fatty acids occur between the first week postpartum when females fast and subsequent lactation stages when they feed on various prey (Iverson, 1993). In phocid seals that fast throughout the entire lactation period, changes may occur in milk fatty acids as blubber stores become depleted (Iverson, 1988, Iverson *et al.,* 1995).

Although sampling methods (e.g., degree of mammary evacuation, evaporative water loss, adherence of fat to collection apparatus or containers) can cause substantial bias in the determination of milk fat concentration (Oftedal and Iverson, Chapter 10A), these do not appear to significantly affect the analysis of fatty acid composition (e.g., Jensen, 1989). Fatty acids are generally expressed as percentage of total lipid and any sample of milk lipid globules would be expected to contain the complete pattern. Of course, milk samples collected for fatty acid analysis should be free of contamination by blood or other components and if taken from dead animals should be collected shortly after death to avoid postmortem changes or destruction of fatty acids.

Fatty acids may be susceptible to autooxidation during storage and, thus, it is preferable that samples be extracted or placed in lipid solvents as soon as possible after collection. Oxidative damage can also be minimized by storage at low temperatures (preferably −70°C) and by flushing the sample with nitrogen prior to sealing the storage container. While these precautions are recommended, a less rigorous approach may not necessarily result in damage. For example, no differences were detected in the fatty acid compositions of sea lion milks stored frozen with solvents and antioxidants versus frozen without solvents for 6 years at −20°C (Iverson, 1988). However, if logistical constraints prevent frozen storage of samples (e.g., at research field sites), milks should be placed in solvent with antioxidant (e.g., butylated hydroxytoluene, BHT) at the time of collection.

In some cases, the fatty acid composition of milk has been inferred from gastric samples of suckling neonates. Although such gastric samples cannot be used to indicate total milk fat concentration (due to dilution and differential passage of gross constituents) or neutral lipid composition (due to hydrolysis of triglycerides), they may be used to assess maternal milk fatty acid composition in species which do not secrete medium- or shorter-chain fatty acids (Iverson, 1993). In pinnipeds, milks collected from mothers were identical in fatty acid composition to those obtained by gastric intubation of their pups, regardless of time since ingestion (Iverson, 1988, 1993). However, this observation needs to be verified in other species and is not likely to be true for species which secrete medium- and short-chain fatty acids (12:0 and shorter) since these can leave the milk fat globule and are absorbed directly across the stomach mucosa (Patton and Jensen, 1976; Patton *et al.,* 1982; Nelson, 1992).

B. Extraction of Lipids and Preparation of Fatty Acid Esters

In general, the most reliable and complete methods of extracting lipid from tissues for fatty acid analysis use a mixture of 2:1 (v/v) chloroform and methanol as developed by Folch *et al.* (1957) and Bligh and Dyer (1959) (reviewed in Christie, 1982). However, the lipid extract obtained by the Roese–Gottlieb procedure, which is an official method for the determination of total milk fat content (AOAC, 1975), has also frequently been used for milk fatty acid analysis. While the Roese–Gottlieb method is appropriate for lipid analysis of most milk samples (Jensen *et al.,* 1985), it may lead to erroneous results if samples have undergone extensive hydrolysis, for instance during storage or in gastric contents. The addition of ammonium hydroxide in the Roese–Gottlieb method (for disruption of the milk fat globule) forms soaps with most free fatty acids which then remain in the aqueous phase and are not extracted. This may not only significantly underestimate total fat content (Iverson, 1988), but if the free fatty acids released during hydrolysis are not proportional to the overall composition (e.g., if specific fatty acids or fatty acids at specific positions on the triglyceride backbone are hydrolyzed) then the reported fatty acid composition may be biased. Regardless of the extraction method used, evaporation of solvent to isolate the lipid should take place at moderate temperatures and under a stream of nitrogen to avoid oxidative damage and loss of fatty acids.

Volatile fatty acid derivatives are prepared by transesterification prior to analysis by GLC. Methyl esters are most commonly prepared and are suitable for most nonruminant milk samples. Acid-catalyzed procedures for preparation of methyl esters (such as methanolic hydrogen chloride, sulfuric acid in methanol, or boron trifluoride in methanol) are probably more reliable than those that are base catalyzed, as the latter do not esterify free fatty acids and their imprudent use can cause alterations to fatty acids (Christie, 1982). Medium- and long-chain fatty acids appear to be accurately quantitated using methyl ester preparations, but since evaporative loss of methyl esters of medium-chain fatty acids can occur, evaporation steps should be performed with caution. Substantial or complete loss of short-chain fatty acid methyl esters may occur due to their high volatility. Preparing butyl (e.g., using boron trifluoride in *n*-butanol) rather than methyl esters and avoiding evaporation steps wherever possible appears to solve this problem (Jensen *et al.,* 1990; Iverson, unpublished data in Jensen and Newburg, Chapter 6B). The method of transesterification should be included in fatty acid reports.

C. Analysis of Fatty Acid Composition Using GLC

The accurate analysis of milk fatty acid composition requires efficient separation of all components, correct identification of peaks, and accurate quantitation of peak areas. GLC analysis is a sophisticated technique and should be used with both theoretical understanding and direct verification. Excellent reviews of the theory and practice of GLC analysis are given in Christie (1982), Ackman (1991), and Poole and Poole (1991). However, a few points are pertinent to the interpretation of milk fatty acid data. Columns are the essence of GLC analysis. Packed columns do not give the kind of complete or reliable separations of fatty acids that are possible with capillary columns. Many of the reported analyses for mammalian milks have been obtained using packed columns; these analyses should be repeated using capillary columns. Superior resolution and separations of all components and their isomers have been obtained by one of the authors (S.J.I.) using a 30 m × 0.25 mm i.d. column coated with 50% cyanopropyl polysiloxane. Other stationary phases have been reviewed by Ackman (1991) and Poole and Poole (1991).

Temperature programming appropriate to the sample is also essential to the separation of components. Low starting temperatures are required for milks containing short- and also medium-chain fatty acids. Longer programs, suitable ramping, and higher temperatures are required for milks containing long-chain polyunsaturated fatty acids (LC-PUFA).

Accurate identification of components is critical to the fatty acid report. Identifications are generally made on the basis of retention times of the peak associated with each component on the chromatogram. Retention time of a given component will be dependent upon the sample injection technique (manual or automated), the column (coating and length), and the operating conditions of the instrument (e.g., carrier gas flow rate, temperature program). Identifications can be calculated by using equivalent and fractional chain length methods (e.g., Ackman, 1991), but these must be verified. Misidentifications are common in the literature and often obvious to the experienced lipid analyst. Fatty acid identities can be verified using a number of methods, including the use of complex standard mixtures, silver nitrate chromatography, or chemical degradative techniques (e.g., Christie, 1982). It is also essential to identify erroneous chromatogram peaks which may represent contaminants introduced during collection or solvent extraction, or artifacts produced during chemical reactions associated with transesterification. In many cases spurious peaks can resemble short-chain fatty acid methyl esters and sometimes appear inexplicably in only one of a series of similar samples. This is one important reason among others why more than a single sample should be analyzed. Spurious peaks can often be elucidated through modified extraction and transesterification procedures and then removed from the report. Unknown peaks should be discussed or listed in published reports.

Once identified, peak areas must also be accurately quantitated. Since the detector response varies among fatty acids with chain length and degree of unsaturation, individual response factors should be applied to the area count obtained for each component. Theoretical relative response factors are reviewed in Ackman (1991). Once theoretical response factors are set up, quantitative mixtures of fatty acid standards can be run to verify or calibrate these factors to the individual GC, column, and operating conditions.

IV. Selection Criteria for the Milk Fatty Acid Table

In selecting data for our table, all reports were critically evaluated with regard to quality and completeness of the report, sampling method, and analytical procedure. Although we would have preferred to restrict the included data to studies based on multiple samples and clearly defined sampling regimes, this would have excluded the majority of species that have been studied. We also recognize that obtaining milk samples from exotic and rare species can be extremely difficult and has often been done on an opportunistic basis. Thus, we have used as the most significant criteria the quality of the analysis itself, particularly with regard to separation and identification of fatty acid peaks. The likelihood that the data are in fact representative of the species studied will depend on sample size, stage of lactation, and dietary effects and, thus, this information is included in the tables to the extent possible.

Some of the most common problems with milk fatty acid data need to be emphasized, as these form the bases for exclusion of many reports from our table. A number of studies of fatty acid patterns in milks were conducted in the 1960s and 1970s and, thus, were unable to resolve the diversity of components, especially LC-PUFA, that are now routinely identified using sophisticated capillary GLC methodology. In early studies using packed columns, components were often misidentified, and listed fatty acids were often incomplete, with as few as 7–10 components adding up to 100% of total fatty acids. Although the "basic" fatty acid composition of a milk can often be summarized by about 10–14 components, 50–70 fatty acids and their isomers can be routinely identified and quantified in the milks of some species using temperature-programmed capillary GLC (Ackman *et al.*, 1988; Iverson, 1988). While it may not be necessary to identify all minor components, it is important to indicate in the report that the analysis was incomplete and that "others" or unidentified components were found and totaled. If in fact no other components were detected, despite sufficiently sensitive analytical methods, this should be noted.

One fairly common problem of early analyses using packed columns is that fatty acids longer than 18:3 are not reported. This appears to have been due to a failure to resolve other LC-PUFA from 18:3 and thus the aggregate was reported as 18:3. This error is especially evident in reports of marine mammal milks. Recent analyses of milks from over 12 species of pinnipeds and cetaceans using capillary columns have found less than 1% of 18:3*n* 3 (Iverson, 1988, unpublished data), which is consistent with the observation that this dietary component is generally less than 1% in marine fish and other marine mammal prey species (e.g., Ackman, 1980). Milks of most marine mammals also contain 20–30% of other LC-PUFA. Hence, the finding of Ashworth *et al.* (1966) that milks from wild northern fur seals contain 17.4% 18:3 and virtually none of the important LC-PUFA appears to be due to inadequate analytical resolution. Dietary 18:3*n* 3 originates in plants and is not readily incorporated or stored in the tissue lipids of animals unless it is fed to the relative exclusion of other dietary fats (Nelson, 1992). High levels of 18:3 are found in fruits (including some oilseeds) and leafy vegetation, as well as in some freshwater algaes (e.g., Ahlgren *et al.*, 1992), and, thus, are likely to occur in significant amounts only in milks of those species which consume these foods directly or indirectly (Iverson and Oftedal, 1992). Unfortunately, in the widely cited initial study of Glass *et al.* (1967), encompassing milks from more than 50 mammalian species, it also appears that 18:3 was not resolved adequately from other LC-PUFA. Although Glass *et al.* list significant amounts of unidentified long-chain components in some species, their values of 5–13% 18:3 in milks from wild-caught seals and the African lion, an obligate carnivore, are not reliable. In the absence of further information, it is not possible to evaluate whether the data for other species are similarly biased; thus, we have decided to exclude these reports from our tables. In temperature-programmed capillary GLC, the retention time of 18:3*n*-3 is distinct and usually about 6–12 min earlier than the most significant LC-PUFA, making it nearly impossible to misidentify. The later study by Glass and Jenness (1971) also used a packed column for analyses, but the updated method of transesterification may have been more reliable. The values for 18:3 and the coinciding list of other fatty acids appear to be consistent with more recent studies. Thus, we have included values from this report when they are the only data available for particular orders or families.

Artifacts or erroneous peaks in the chromatograms appear to be present in other published studies, such as in the report of large amounts of short-chain fatty acids in polar bear milk (e.g., 14% 4:0; Baker *et al.*, 1963) and Yeso brown bear milk (Ando *et al.*, 1979). Some transesterification procedures may produce erroneous peaks if not used with caution and in fact traces of basic transesterifying agent accidentally injected onto polyester columns can produce spurious peaks which can be mistaken for short-chain fatty acids (Christie, 1982). Short-chain fatty acids do not normally occur in nonruminant milk, as discussed previously (Section II), and have not been found in other studies of the milks of bears or other carnivores (Glass and Jenness, 1971; Jenness *et al.*, 1972; Iverson *et al.*, 1991; Iverson and Oftedal, 1992; Wamberg *et al.*, 1992), nor were they found in a later study of polar bear milk by Baker and colleagues (Cook *et al.*, 1970a). Apparent misidentification of components and/or reports of extremely large amounts of saturates of both odd-chain and very long-chain fatty acids, which are normally found in only trace amounts in most animal lipids (Christie, 1982), were present in many of the reports of Baker and colleagues using base-catalyzed transesterification methods [e.g., Arctic wolf, Lauer *et al.* (1969b); harp seal, Cook and Baker (1969); Van Horn and Baker (1971); beluga whale and fin whale, Lauer and Baker (1969); moose, Cook *et al.* (1970c); mountain goat, Lauer *et al.* (1969a); Dall sheep, Cook *et al.* (1970b)]; hence, these data were excluded from the table. Other milk fatty acid reports that were too incomplete to allow interpretation were also excluded. Some reports provide a column of "other" fatty acids which represents the sum of unidentified or unlisted fatty acids; although this is important information, if these sums equaled or exceeded 10% of total fatty acids, we did not include the data in our tables.

In Table I, data are listed by family within orders for a total of 82 species. Lactation stage, days postpartum, number of animals (and samples), diet, and type of column used in analysis are given wherever data were available (n/a indicates data were not available). Lactation stage is categorized as perinatal/early (first few days of lactation or early stage), mature (established lactation), and late (near weaning). We have included data on animals from wild populations wherever possible, but have also included data from captive animals when variations in diet are listed, when the species appear to synthesize most of their milk fatty acids *de novo,* or when no data are available from wild populations. The listed fat concentration of the milk is generally that measured in the same samples as were the fatty acids. However, when these data were not available, we were sometimes able to use fat values for a comparable stage of lactation as reported from other sources.

When more than one report was available for a species, the most complete and reliable data were included, even if these were in unpublished form; more than one report was included if it provided unique information with respect to diet or lactation stage. Most important fatty acids and their isomers are listed individually, but to economize on space, some minor components have been combined as designated. Isomers and antiisomers of saturates are included with "other." Fatty acid data are given as weight percentage of total milk fatty acids, except where (TG) indicates that only the triglyceride fraction of the milk lipid was analyzed. In most cases, the triglyceride fraction represents 97–99% of the total milk fatty acids. However, it should be noted that if the milk has undergone extensive hydrolysis such that triglyceride is only 70–80% of total lipid, a report based on the triglyceride fraction may not be representative of total milk fatty acids.

V. Patterns of Milk Fatty Acids among Taxonomic Groups

Based on the still relatively limited data which exist for many mammalian groups, some general patterns in milk fatty acid composition can be discerned along phylogenetic and ecological lines.

A. Monotremata and Marsupialia

The milks of monotremes and marsupials do not appear to contain short- or medium-chain fatty acids (Table I). These milks are composed predominantly of 14:0, 16:0, 16:1, 18:0, 18:1, and 18:2*n*6. The platypus, which consumes a diet mainly of aquatic freshwater invertebrates rich in 18:3*n*3 (10–13%) secretes a milk that is also quite high in 18:3*n*3 (8%). The highest levels of 18:3*n*3 (25–32%) are found in leaf-eating marsupials (e.g., brush-tail possum and koala). Although fatty acid data are listed for a large number of species, in most cases packed columns were used for analysis and only about 10 components were reported; LC-PUFA were analyzed in only three reports.

B. Insectivora and Chiroptera

No reliable data are available for insectivores and only a few reports exist for bats. The three bat species included are all insectivorous (Table I). Significant amounts of 12:0 (4%) occur in two species, suggesting some medium-chain *de novo* synthesis. High levels of 18:2*n*6 (12–22%) and 18:3*n*3 (2–11%) may reflect an insectivorous diet, similar to some of the insectivorous marsupials. Analyses were performed using packed columns and very few components, including LC-PUFA, were identified. An unpublished analysis of a single milk sample from *Myotis lucifigus* revealed that significant amounts (≥5%) of LC-PUFA (primarily 18:4*n*3, 20:4*n*6, 20:5*n*3, and 22:6*n*3) may be present in some species depending on diet (Iverson, personal communication).

C. Primates

Primates are somewhat better represented across families, but often sample size is low. All species appear to actively synthesize medium-chain fatty acids. Medium-chain synthesis may vary along phylogenetic lines, but conclusions must be considered tentative due to small sample sizes for most families. The milks of lemurs and lorises appear to contain virtually no 8:0, 1–6% 10:0, and relatively high amounts of both 12:0 (4–19%) and 14:0 (6–20%), whereas the milks of callitrichids may contain significant amounts of 8:0 (2%), high levels of 10:0 (8–22%), and generally higher amounts of 12:0 (8–17%) than do the lemurs (Table I). Species of Cebidae appear to secrete even higher proportions of both 8:0 (3 or 4%) and 10.0 (12–23%) than do the other families, with correspondingly less 12:0 and 14:0. Milks of species of Cercopithecidae may contain the highest proportions of 8:0 and the lowest proportions of 12:0 and 14:0. The other predominant fatty acids in primate milks were 16:0 to 18:2*n*6. Levels of 18:3*n*3 were less

TABLE I
Fatty Acid Composition of Mammalian Milks

Species	Lactation stage[a]	Days post-partum	No. of animals (samples)	Animal status[b] (diet)	Milk lipid (%)	GLC[c]	Weight % of total-fatty acids[d]											
							4:0	6:0	8:0	10:0	12:0	14:0	14:1	16:0	16:1	16:2–4 (all isomers)	15:0 and 17:0	15:1 and 17:1
Monotremata																		
Tachyglossidae																		
Common echidna (*Tachyglossus aculeatus*)	M	37–99	3 (4)	W^f	31.0	Pack					1.0	1.4	0.4	15.9	6.2		0.7	0.0
	M	37–99	2 (6)	C^g	31.0	Pack					tr	0.7	0.4	19.1	5.7		1.9	0.0
	M	37–99	1 (3)	C^h	31.0	Pack					0.5	6.1	0.5	28.4	3.6		0.5	0.0
Ornithorhynchidae																		
Platypus (*Ornithorhynchus anatinus*)	n/a	n/a	11 (11)	W^i	n/a	Pack						1.6	0.7	19.8	13.9		2.5	1.5
Marsupialia																		
Dasyuridae																		
Eastern quoll (*Dasyuridae viverrinus*)	M	98	3 (3)	C^j	15.0	Pack						1.7		24.6	4.0			
Tasmanian devil (*Sarcophilus harrisii*)	L	n/a	n/a	n/a	19.8	n/a						0.8	0.2	22.7	6.2		0.7	0.5
Myrmecobiidae																		
Numbat (*Myrmecobius fasciatus*)	L	na	5 (3)	W^f	12.0	Cap					0.1	0.9	0.2	14.1	3.4			
	L	n/a	2 (1)	C^k	n/a	Cap					1.2	5.2	0.4	24.9	2.2			
Peramelidae																		
Northern brown bandicoot (*Isoodon macrourus*)	M	39–56	7 (9)	C^l	18.0^m	Pack						2.9		27.3	5.0			
Thylacomyidae																		
Greater bilby (*Macrotis lagotis*)	L	n/a	n/a	n/a	n/a	n/a						1.9	0.3	28.1	3.7		0.9	0.4

TABLE I—*continued*

Species	18:0	18:1	18:2 n6	18:3 n3	18:4 n3	18:2–4 (other isomers)	20:1	20:2	20:3 n3,6	20:4 n3	20:4 n6	20:5 n3	22:1	22:5 n3	22:2–5 (other isomers)	22:6 n3	24:1	Others and/or unidentified	Total of fatty acids reported[e]	Reference
Monotremata																				
Tachyglossidae																				
Common echidna (*Tachyglossus aculeatus*)	3.9	61.2	5.1	0.8							0.5							3.0	100.0 (TG)	Griffiths *et al.* (1984)
	13.0	42.1	7.8	2.0							0.2							7.0	100.0 (TG)	Griffiths *et al.* (1984)
	7.3	42.9	7.7	0.5							0.3							1.7	100.0 (TG)	Griffiths *et al.* (1984)
Ornithorhynchidae																				
Platypus (*Ornithorhynchus anati*	3.9	22.7	5.4	7.6		0.6	2.2	0.5	0.2		2.4	4.5		4.2	0.4	tr	tr	5.3	100.0 (TG) (98% TG)	Gibson *et al.* (1988)
Marsupialia																				
Dasyuridae																				
Eastern quoll (*Dasyuridae viverrinus*)	8.7	36.1	15.4	2.2							1.6								94.2 (TG)	Green *et al.* (1987)
Tasmanian devil (*Sarcophilus harrisii*)	6.9	45.0	13.7	1.6															98.3	Green and Merchant (1988)
Myrmecobiidae																				
Numbat (*Myrmecobius fasciatus*)	7.0	59.7	7.9	0.1							0.2								93.6	Griffiths *et al.* (1988)
	9.5	40.0	8.8	0.6							1.1								93.9	Griffiths *et al.* (1988)
Peramelidae																				
Northern brown bandicoot (*Isoodon macrourus*)	10.1	36.3	12.4	1.2															95.2 (TG)	Merchant and Libke (1988)
Thylacomyidae																				
Greater bilby (*Macrotis lagotis*)	6.4	30.0	25.2	1.1															98.0	Green and Merchant (1988)

TABLE I—*continued*

Species	Lactation stage[a]	Days post-partum	No. of animals (samples)	Animal status[b] (diet)	Milk lipid (%)	GLC[c]	Weight % of total-fatty acids[d] 4:0	6:0	8:0	10:0	12:0	14:0	14:1	16:0	16:1	15:2–4 (all isomers)	15:0 and 17:0	15:1 and 17:1
Phalangeridae																		
Brushtail possum (*Trichosurus vulpecula*)	n/a	n/a	3 (3)	W[n]	n/a	n/a						1.0		18.9	1.3		0.8	
Potoroinae																		
Long-nosed potoroo (*Potorous tridactylus*)	M	84–126	3 (3)	C	9.0[m]	Pack						2.1	0.1	28.0	6.7		2.6	0.6
Macropodidae																		
Tammar wallaby (*Macropus eugenii*)	P/E	7	pooled	C[o]	2.0	Pack					tr	4.4	0.5	54.5	8.4		2.1	0.9
	M	126–182	? (12)	C[o]	4.0	Pack					tr	1.0	0.2	24.4	3.8		1.9	0.8
Wallaroo (*Macropus robustus*)	n/a	n/a	1 (1)	n/a	n/a	n/a						2.3		21.2	4.9		2.8	
Red kangaroo (*Macropus rufus*)	P/E	1–4	4 (4)	C[p]	0.9	Pack	tr	tr	tr	tr	0.6	3.3	1.2	51.4	7.6		1.5	0.0
	M	100–210	4 (5)	C[p]	5.4	Pack	tr	tr	tr	tr	tr	1.7	0.4	25.3	5.4		1.4	0.5
Ring-tailed rock wallaby (*Petrogale xanthopus*)	L	n/a	n/a	n/a	19.5[q]	n/a						1.8	1.1	17.4	4.4		4.5	1.8
Pademelon (*Thylogale billardierii*)	n/a	n/a	n/a	n/a	n/a	n/a						1.3		24.5	5.2		2.4	
Phascolarctidae																		
Koala (*Phascolarctos cinereus*)	n/a	n/a	n/a	n/a[r]	17.6[q]	n/a	0.0	0.0	0.0	0.0	0.1	3.3	0.2	24.4	4.3		1.8	0.8
Vombatidae																		
Naked-nosed wombat (*Vombatus ursinus*)	L	n/a	n/a	n/a	28.4[q]	n/a						1.2		22.7	3.8		1.1	0.7
Chiroptera																		
Vespertilionidae																		
Big brown bat (*Eptesicus fuscus*)	M	n/a	≥4 (4)	W[j]	16.4	Pack					3.9	1.5		21.5	8.3			

TABLE I—*continued*

Species	18:0	18:1	18:2 n6	18:3 n3	18:4 n3	18:2–4 (other isomers)	20:1	20:2	20:3 n3,6	20:4 n3	20:4 n6	20:5 n3	22:1	22:5 n3	22:2–5 (other isomers)	22:6 n3	24:1	Others and/or unidentified	Total of fatty acids reported[c]	Reference
Phalangeridae																				
Brushtail possum (*Trichosurus vulpecula*)	2.8	18.2	23.8	24 7														8.2	99.7 (TG)	Grigor (1980)
Potoroinae																				
Long-nosed potoroo (*Potorous tridactylus*)	5.6	36.5	13.9	1.4															97.4	Crowley *et al.* (1988)
Macropodidae																				
Tammar wallaby (*Macropus eugenii*)	2.8	16.6	6.0	0.3														3.0	100.0 (TG) (75% TG)	Green *et al.* (1983)
	16.3	40.3	5.6	2.2														3.1	99.7 (TG) (93% TG)	Green *et al.* (1983)
Wallaroo (*Macropus robustus*)	13.9	40.2	6.9															7.7	100.0 (TG)	Grigor (1980)
Red kangaroo (*Macropus rufus*)	2.2	15.6	11.1	0.7			0.2	0.9	0.8		1.0							1.9	100.0 (TG) (> 98% TG)	Griffiths *et al.* (1972)
	10.2	45.3	5.4	1.6			0.6	0.2	0.1		0.2							1.7	100.0 (TG) (> 98% TG)	Griffiths *et al.* (1972)
Ring-tailed rock wallaby (*Petrogale xanthopus*)	14.3	36.5	6.5	6.1															94.4	Green and Merchant (1988)
Pademelon (*Thylogale billardierii*)	15.0	36.4	7.2															7.8	99.8 (TG)	Grigor (1980)
Phascolarctidae																				
Koala (*Phascolarctos cinerus*)	5.2	16.8	10.7	32.5														0.0	100.0 (TG)	Parodi (1982)
Vombatidae																				
Naked-nosed wombat (*Vombatus ursinus*)	4.0	38.8	22.3	2.8															97.4	Green and Merchant (1988)
Chiroptera																				
Vespertilionidae																				
Big brown bat (*Eptesicus fuscus*)	7.4	37.9	15.7	2.1														1.9	100.0	Kunz *et al.* (1983)

TABLE I—*continued*

Species	Lactation stage[a]	Days post-partum	No. of animals (samples)	Animal status (diet)	Milk lipid (%)	GLC[c]	Weight % of total-fatty acids[d]											
							4:0	6:0	8:0	10:0	12:0	14:0	14:1	16:0	16:1	16:2–4 (all isomers)	15:0 and 17:0	15:1 and 17:1
Little brown bat (*Myotis lucifugus*)	M	n/a	≥13 (13)	W[j]	13.5	Pack					4.1	3.3		21.0	11.6			
Fringed myotis (*Myotis thysanodes*)	n/a	n/a	1 (1)	W[j]	n/a	Pack	0.0	0.0	0.0	0.0	0.3	0.9		18.7	7.5			
Primates																		
Lemuridae																		
Brown lemur (*Eulemur fulvus*)	M	40–75	1 (1)	C[s]	0.4	Cap			0.1	5.5	19.0	20.4	2.3	17.4	4.3		tr	0.5
Black lemur (*Eulemur macaco*)	M	40–75	1 (1)	C[s]	1.2	Cap			0.1	2.4	12.8	18.2	1.7	29.2	5.4		0.1	0.5
Mongoose lemur (*Eulemur mongoz*)	M	40–75	1 (1)	C[t]	0.3	Cap			0.0	0.7	2.2	5.3	0.3	26.3	6.5		0.3	0.4
Red-bellied lemur (*Eulemur rubriventer*)	M	40–75	1 (1)	C[s]	1.0	Cap			0.6	6.0	12.6	15.5	1.4	21.7	4.7		0.3	0.7
Gentle bamboo lemur (*Hapelemur griseus*)	M	40–75	1 (1)	C[u]	2.9	Cap			0.9	5.6	7.6	10.4	0.5	22.5	6.6		0.4	0.7
Ruffed lemur (*Varecia variegata*)	M	40–75	1 (1)	C[s]	0.9	Cap			0.0	1.1	4.3	7.8	0.5	24.3	6.2		0.3	0.5
Lorisidae																		
Slow loris (*Nycticebus coucang*)	M	32	1 (1)	C[v]	1.0	Cap			0.0	0.4	5.8	10.6	0.1	30.7	2.5		0.8	0.6
Greater galago (*Otolemur garnettii*)	P/E	4	1 (1)	C[w]	7.3	Cap			0.4	2.5	4.4	6.4	0.3	22.1	5.7		0.6	1.1
Callitrichidae																		
Common marmoset (*Callithrix jacchus*)	M	40–42	2 (1)	C[x]	5.4	Pack				8.0	8.5	7.7		18.1	5.5			
Golden lion tamarin (*Leontopithecus rosalia*)	P/E	3	1 (1)	C	5.8	Pack			2.2	22.2	17.5	9.9		14.7				
Cotton-top tamarin (*Saguinus oedipus*)	n/a	n/a	? (3)	C	n/a	Pack	0.0	0.0	2.4	14.7	15.7	12.1		21.5	2.2			

TABLE I—*continued*

Species	18:0	18:1	18:2 n6	18:3 n3	18:4 n3	18:2–4 (other isomers)	20:1	20:2	20:3 n3,6	20:4 n3	20:4 n6	20:5 n3	22:1	22:5 n3	22:2–5 (other isomers)	22:6 n3	24:1	Others and/or unidentified	Total of fatty acids reported[c]	Reference
Little brown bat (*lucifugus*)	5.4	24.7	12.0	11.0														7.0	100.0	Kunz *et al.* (1983)
Fringed myotis (*Myotis thysanodes*)	4.0	36.0	22.2	9.2														0.5	99.4	Glass and Jenness (1971)
Primates																				
Lemuridae																				
Brown lemur (*Eulemur fulvus*)	1.5	17.8	8.7	0.5			0.3	0.2	0.4		0.3	0.1	0.0	0.2		0.2	0.1	0.3	100.0 (TG) (70% TG)	Myher *et al.* (in press)
Black lemur (*macaco*)	1.7	16.9	8.3	0.5			0.4	0.4	0.3		0.6	0.1	0.0	0.3		0.3	tr	0.0	100.0 (TG) (75% TG)	Myher *et al.* (in press)
Mongoose lemur (*Eulemur mongoz*)	2.3	35.8	14.8	0.3			0.4	0.4	0.3		0.5	0.5	0.0	0.5		1.0	0.7	0.0	100.0 (TG) (82% TG)	Myher *et al.* (in press)
Red-bellied lemur (*Eulemur rubriventer*)	2.4	18.5	11.4	0.7			0.4	0.6	0.6		0.7	0.3	0.0	0.5		0.5	0.1	0.1	100.0 (TG) (82% TG)	Myher *et al.* (in press)
Gentle bamboo lemur (*Hapelemur griseus*)	1.5	25.9	9.8	1.1			0.3	0.3	0.3		0.4	0.1	0.0	3.1		1.3	0.4	0.2	100.0 (TG) (77% TG)	Myher *et al.* (in press)
Ruffed lemur (*Varecia variegata*)	2.4	44.8	5.8	0.3			0.4	0.2	0.1		0.1	0.2	tr	0.2		0.2	0.2	0.0	100.0 (TG) (83% TG)	Myher *et al.* (in press)
Lorisidae																				
Slow loris (*Nycticebus coucang*)	4.9	27.2	13.3	0.6			0.7	0.8	0.1		0.2	0.1	0.2	0.0		0.2	0.2	0.0	100.0 (TG) (92% TG)	Myher *et al.* (in press)
Greater galago (*Otolemur garnettii*)	2.8	38.0	9.9	0.8			1.0	0.4	0.6		1.0	0.2	0.2	0.5		0.6	0.6	0.0	100.0 (TG) (95% TG)	Myher *et al.* (in press)
Callitrichidae																				
Common marmoset (*Callithrix jacchus*)	3.4	29.6	10.9	0.9		0.4	1.2	0.1	0.4		0.1	1.3		0.8	0.1	2.5		0.7	100.0	Turton *et al.* (1978)
Golden lion tamarin (*Leontopithecus rosalia*)	2.0	15.5	15.9																100.0	Buss (1975)
Cotton-top tamarin (*Saguinus oedipus*)	3.4	19.6	8.0	6.0														0.0	99.6	Glass and Jenness (1971)

TABLE I—*continued*

Species	Lactation stage[a]	Days post-partum	No. of animals (samples)	Animal status[b] (diet)	Milk lipid (%)	GLC[c]	Weight % of total-fatty acids[d] 4:0	6:0	8:0	10:0	12:0	14:0	14:1	16:0	16:1	16:2–4 (all isomers)	15:0 and 17:0	15:1 and 17:1
Cebidae																		
Mantled howler (*Alouatta palliata*)	P/E	7	1 (1)	W[y]	3.4[z]	Cap	0.0	0.0	2.6	22.8	13.3	5.2	0.0	17.3	2.4	0.0	0.9	0.5
	M	30–150	4 (4)	W[y]	1.6[z]	Cap	0.0	0.0	2.2	11.9	5.7	2.9	0.0	23.4	2.2	0.0	1.2	0.6
Red howler (*Alouatta seniculus*)	M	30–150	5 (5)	W[y]	1.1[z]	Cap	0.0	0.0	4.0	20.8	9.3	3.8	0.0	20.3	3.5	0.0	0.3	0.3
Squirrel monkey (*Saimiri sciureus*)	M	95–153	13 (4)	C[aa]	5.1	Pack		0.4	4.3	7.9	5.7	4.6		20.0	2.4			
Cercopithecidae																		
Vervet (green monkey) (*Cercopithecus aethiops*)	n/a	n/a	1 (1)	C	n/a	Pack	0.0	0.8	6.4	9.6	2.7	1.6		25.1	4.4			
Crab-eating macaque (*Macaca fascicularis*)	M	44–119	8 (8)	C[ab]	n/a	Pack		1.1	8.1	7.9	1.8	2.1		22.1	5.4			
Talapoin monkey (*Mioopithecus talapoin*)	M	18–37	4 (5)	C[aa]	2.1	Pack		0.1	3.7	7.0	1.7	1.5		19.2	1.7		0.4	
Baboon (*Papio spp.*)	M	14–113	≤27 (48)	C[aa]	4.6	Pack		0.4	5.1	7.9	2.3	1.3		16.5	1.2		0.3	
Carnivora																		
Canidae																		
Dog (domestic, mixed) (*Canus familiaris*)	M	7–28	1 (8)	C[ac]	7.9	Cap	0.0	0.0	0.0	0.0	0.2	3.3	0.4	27.9	6.4		0.7	0.6
Ursidae																		
Black bear (*Ursus americanus*)	P/E	0–1	3 (3)	W[ad]	13.2	Cap	0.0	0.0	0.0	0.0	0.0	1.8	0.2	32.7	7.5	tr	0.2	0.3
	M	152–274	4 (4)	W[ae]	28.0	Cap	0.0	0.0	0.0	0.0	0.2	3.2	0.5	23.5	3.3	tr	0.8	0.5

TABLE I—*continued*

Species	18:0	18:1	18:2 n6	18:3 n3	18:4 n3	18:2–4 (other isomers)	20:1	20:2	20:3 n3,6	20:4 n3	20:4 n6	20:5 n3	22:1	22:5 n3	22:2–5 (other isomers)	22:6 n3	24:1	Others and/or unidentified	Total of fatty acids reported[e]	Reference
Cebidae																				
Mantled howler (*Alouatta palliata*)	3.0	15.6	5.8	9.0	0.0	0.0	0.1	0.0	0.0	0.0	0.5	0.0	0.0	0.0	0.0	0.0	0.0	1.3	100.0	Iverson and Oftedal (unpublished results)
	4.7	14.7	12.2	17.8	0.0	0.0	0.0	0.0	0.0	0.0	0.5	0.0	0.0	0.0	0.0	0.0	0.0	0.2	100.0	Iverson and Oftedal (unpublished results)
Red howler (*Alouatta seniculus*)	3.4	17.7	9.7	6.9	0.1	0.0	0.0	0.0	0.0	0.0	tr	0.0	0.0	0.0	0.0	0.0	0.0	tr	100.0	Iverson and Oftedal (unpublished results)
Squirrel monkey (*Saimiri sciureus*)	3.3	29.3	20.6	1.3															99.8	Buss and Cooper (1972)
Cercopithecidae																				
Vervet (green monkey) *Cercopithecus aethiops*)	7.2	34.6	7.6	0.0														0.0	100.0	Glass and Jenness (1971)
Crab-eating macaque (*Macaca fascicularis*)	4.9	28.9	15.8	0.8														1.1	100.0	Nishikawa *et al.* (1976)
Talapoin monkey (*Miopithecus talapoin*)	4.9	21.1	34.8	3.6															99.8	Buss and Cooper (1970)
Baboon (*Papio spp.*)	4.2	22.7	37.6	0.6															100.0	Buss (1969)
Carnivora																				
Canidae																				
Dog (domestic, mixed) (*Canus familiaris*)	4.3	40.7	11.9	0.5	0.3	0.0	0.6	0.5	0.2	0.0	1.2	0.0	0.0	0.0	0.0	0.0	0.0	0.5	100.0	Iverson *et al.* (1991)
Ursidae																				
Black bear (*Ursus americanus*)	3.0	41.2	6.2	0.3	0.0	0.0	0.4	0.5	0.5	0.0	2.8	0.3	0.0	0.2	0.0	0.2	0.0	1.5	100.0	Iverson and Oftedal (1992)
	5.8	39.9	15.7	5.1	0.2	0.0	0.3	0.1	0.1	0.0	0.3	tr	0.0	0.0	0.0	0.0	0.0	0.5	100.0	Iverson and Oftedal (1992)

TABLE I—*continued*

Species	Lactation stage[a]	Days post-partum	No. of animals (samples)	Animal status[b] (diet)	Milk lipid (%)	GLC[c]	Weight % of total-fatty acids[d] 4:0	6:0	8:0	10:0	12:0	14:0	14:1	16:0	16:1	16:2–4 (all isomers)	15:0 and 17:0	15:1 and 17:1
Procyonidae																		
Ringtailed coati (*Nasua nasua*)	n/a	n/a	1 (1)	C	n/a	Pack	0.0	0.0	0.0	0.0	0.0	2.4		24.0	6.3			
Mustelidae																		
American mink (*Mustela vison*)	M	22–25	6 (1)	C[af]	7.3[ag]	Cap				tr	0.2	2.1	0.1	27.0	3.4		0.1	0.4
	M	22–25	6 (1)	C[ah]	7.3[ag]	Cap				0.1	0.4	5.2	0.4	21.8	9.4		0.4	0.4
Felidae																		
Cat (domestic) (*Felis sylvestris catus*)	n/a	n/a	1 (1)	C	n/a	n/a	0.0	0.0	0.0	0.3	0.7	4.6	0.8	25.6	4.8		2.0	1.0
Jaguar (*Panthera onca*)	P/E	1	1 (1)	C	n/a	Pack	0.0	0.0	0.0	0.0	0.0	2.2		21.5	7.9			
Pinnipedia																		
Otariidae																		
Antarctic fur seal (*Arctocephalus gazella*)	P/E	0–9	3 (3)	W[ai]	n/a	Cap	0.0	0.0	0.0	0.0	0.2	4.7	0.2	17.9	8.7	1.0	0.6	0.3
	M	15–25	8 (8)	W[aj]	n/a	Cap	0.0	0.0	0.0	0.0	0.2	10.3	0.3	20.6	12.5	1.2	0.4	0.1
Northern fur seal (*Callorhinus ursimus*)	n/a	n/a	n/a	W	45.6	Pack					0.3	6.6	0.3	20.0	10.3		0.6	1.3
Australian sea lion (*Neophoca cinerea*)	M	90	1 (1)	W	25.4[ak]	Cap	0.0	0.0	0.0	0.0	0.2	7.0	0.2	18.2	6.9	1.0	1.8	0.7
California sea lion (*Zalophus californianus*)	P/E	0–7	3 (3)	W[ai]	32.4	Cap	0.0	0.0	0.0	0.0	0.2	3.9	0.2	19.0	8.4	1.1	0.9	0.3
	M	30–60	2 (2)	W[al]	31.0	Cap	0.0	0.0	0.0	0.0	0.1	4.5	0.1	18.1	6.5	0.8	1.0	0.2
	M	n/a	1 (1)	C[am]	n/a	Cap	0.0	0.0	0.0	0.0	0.2	6.8	0.2	14.9	5.8	0.3	0.6	0.2
Phocidae																		
Hooded seal (*Cystophora cristata*)	M	0–4	10 (10)	W[an]	61.0	Cap	0.0	0.0	0.0	0.0	0.1	4.4	0.2	11.7	13.4	1.0	0.3	0.2

TABLE I—*continued*

Species	18:0	18:1	18:2 n6	18:3 n3	18:4 n3	18:2–4 (other isomers)	20:1	20:2	20:3 n3,6	20:4 n3	20:4 n6	20:5 n3	22:1	22:5 n3	22:2–5 (other isomers)	22:6 n3	24:1	Others and/or unidentified	Total of fatty acids reported[e]	Reference
Procyonidae																				
Ringtailed coati (*Nasua nasua*)	5.2	43.2	18.4	0.0														0.5	100.0	Glass and Jenness (1971)
Mustelidae																				
American mink (*Mustela vison*)	10.2	39.3	9.2	0.9			1.2	0.4	0.1			0.9	0.3	0.3		3.4	0.1	0.5	100.0	Wamberg *et al.* (1992)
	3.5	18.2	4.8	1.3			7.0	0.4	0.0			6.4	3.8	0.8		13.8	0.7	1.1	100.0	Wamberg *et al.* (1992)
Felidae																				
Cat (domestic) (*Felis sylvestris catus*)	10.7	42.4	6.1	1.4														0.0	100.0	Parodi (1982)
Jaguar (*Panthera onca*)	6.0	44.4	14.4	0.0														4.2	100.6	Glass and Jenness (1971)
Pinnipedia																				
Otariidae																				
Antarctic fur seal (*Arctocephalus gazella*)	2.1	37.9	1.7	0.4	0.1	0.2	6.0	0.1	0.0	1.0	0.6	5.6	1.2	2.0	0.0	7.1	0.0	0.3	100.0	Iverson (1988, 1993)
	1.6	25.2	1.4	0.3	0.4	0.4	2.4	0.0	tr	0.6	0.5	11.6	0.5	2.2	0.4	6.3	0.0	0.7	100.0	Iverson (1988, 1993)
Northern fur seal (*Callorhinus ursimus*)	1.9	34.5	1.9	0.2			6.8	1.0			2.4	4.8	1.0	1.5		3.7		0.9	100.0	Dosako *et al.* (1982)
Australian sea lion (*Neophoca cinerea*)	2.5	19.0	1.7	1.0	1.1	0.2	6.1	0.3	0.0	2.3	1.2	5.6	2.3	2.5	0.2	16.2	0.2	1.3	99.4	Iverson (1988)
California sea lion (*Zalophus californianus*)	2.0	31.0	1.9	0.9	0.5	0.3	1.9	0.1	0.1	0.9	1.4	7.1	0.4	3.7	0.3	13.6	0.0	0.0	100.0	Iverson (1988, 1993)
	3.0	23.2	1.6	0.9	1.0	0.4	2.7	0.1	0.0	1.2	1.0	8.5	1.3	3.9	0.5	19.2	0.2	0.0	100.0	Iverson (1988, 1993)
	1.3	22.6	1.1	0.5	0.4	0.2	15.4	0.1	0.0	0.4	0.4	3.4	13.5	2.3	0.0	8.7	0.2	0.5	100.0	Iverson (1993)
Phocidae																				
Hooded seal (*Cystophora cristata*)	2.1	27.0	1.4	0.4	1.2	0.2	14.8	0.1	0.0	0.7	0.3	6.8	4.9	1.9	0.3	6.5	0.0	0.2	100.0	Iverson *et al.* (1995)

TABLE I—*continued*

Species	Lactation stage[a]	Days post-partum	No. of animals (samples)	Animal status[b] (diet)	Milk lipid (%)	GLC[c]	Weight % of total-fatty acids[d]											
							4:0	6:0	8:0	10:0	12:0	14:0	14:1	16:0	16:1	16:2–4 (all isomers)	15:0 and 17:0	15:1 and 17:1
Grey seal (*Halichoerus grypus*)	M	n/a	1 (1)	W[an]	≥32	Pack					tr	3.2	0.3	17.9	15.9	0.2	0.5	0.3
Weddell seal (*Leptonychotes weddelli*)	M	10–18	5 (5)	W[an]	39.7	Pack					0.2	8.7	0.7	13.3	11.8	0.7	0.3	0.3
Northern elephant seal *Mirounga angustirostris*)	L	25–26	3 (3)	W[an]	54.0[ao]	Cap	0.0	0.0	0.0	0.0	0.2	2.7	0.1	11.3	4.7	1.5	0.7	0.7
Harp seal (*Phoca groenlandica*)	M	4–9	5 (5)	W[an]	51.8	Cap	0.0	0.0	0.0	0.0	0.1	4.6	0.3	8.8	17.4	1.2	0.2	0.2
Cetacea																		
Delphinidae																		
Bottle-nosed dolphin (*Tursiops truncatus t.*)	L	n/a	1 (1)	W[ap]	19.0	Cap	0.0	0.0	0.0	0.0	0.3	3.2	0.2	21.1	13.3	1.1	0.8	0.7
Balaenopteridae																		
Fin whale (*Balaenoptera physalus*)	n/a	n/a	1 (1)	W[ap]	17.5	Cap	0.0	0.0	0.0	0.0	0.1	5.5	0.2	22.9	6.5	0.9	0.6	0.2
Balaenidae																		
Bowhead whale (*Balaena mysticetus*)	P/E	O[aq]	1 (1)	W	2.4	Cap	0.0	0.0	0.0	0.0	0.1	4.5	0.6	12.3	17.4	1 3	0.7	0.8
Sirenia																		
Trichechidae																		
West Indian manatee (*Trichechus manatus*)	L	≥365	1 (5)	C[ar]	12.7	Pack			0.6	3.5	4.0	6.3		20.2	11.6		1.5	
Proboscidea																		
Elephantidae																		
Indian elephant (*Elephas maximus*)	n/a	n/a	1 (1)	C	n/a	Pack	0.0	0.4	5.8	43.4	21.5	3.5		9.1	1.9			
African elephant (*Loxodonta africana*)	n/a	n/a	10 (10)	W	5.0[as]	Pack			9.7	64.5	17.4	1.2		2.6	0.5			

TABLE I—*continued*

Species	18:0	18:1	18:2 n6	18:3 n3	18:4 n3	18:2–4 (other isomers)	20:1	20:2	20:3 n3,6	20:4 n3	20:4 n6	20:5 n3	22:1	22:5 n3	22:2–5 (other isomers)	22:6 n3	24:1	Others and/or unidentified	Total of fatty acids reported[e]	Reference
Grey seal (*Halichoerus grypus*)	1.8	30.6	0.6	0.4	1.1	tr	2.7	tr	tr	0.4	1.4	8.8	0.5	3.5	0.8	8.1	0.0	0.8	99.8	Ackman and Burgher (1963)
Weddell seal (*Leptonychotes weddelli*)	2.0	38.8	1.7	0.4	1.0		7.5			0.4	0.2	4.7	1.6	0.9	0.3	4.1	0.2	0.2	100.2	Stull *et al.* (1967)
Northern elephant seal (*Mirounga angustirostris*)	3.1	37.9	1.5	0.5	tr	0.2	21.0	0.4	0.0	0.2	0.4	1.1	6.9	0.9	0.0	4.2	tr	0.2	100.0	Iverson (1988)
Harp seal (*Phoca groenlandica*)	1.6	23.0	1.1	0.5	0.8	0.2	17.2	0.1	0.0	0.3	0.4	6.1	5.9	3.2	0.2	6.5	tr	0.4	100.0	Iverson *et al.* (1992); Iverson (1988)
Cetacea																				
Delphinidae																				
Bottle-nosed dolphin (*Tursiops truncatus t.*)	3.3	23.1	1.2	0.2	0.2	0.2	9.0	0.2	0.4	0.4	1.4	6.0	2.8	2.0	0.3	6.4	0.1	2.1	100.0	Ackman *et al.* (1971)
Balaenopteridae																				
Fin whale (*Balaenoptera physalus*)	3.9	24.7	1.1	0.6	1.1	0.1	3.1	0.2	0.2	0.7	0.5	13.9	1.9	3.0	0.1	5.7	0.1	1.8	99.4	Ackman *et al.* (1968)
Balaenidae																				
Bowhead whale (*Balaena mysticetus*)	4.0	24.9	1.0	0.4	0.3	1.6	6.7	0.0	0.2	0.9	0.6	8.1	5.0	3.5	0.1	2.3	0.4	2.3	100.0	Iverson and Oftedal (unpublished results)
Sirenia																				
Trichechidae																				
West Indian manatee (*Trichechus manatus*)	0.5	47.0	1.8	2.2									0.4					0.4	100.0	Bachman and Irvine (1979)
Proboscidea																				
Elephantidae																				
Indian elephant (*Elephas maximus*)	0.5	9.8	2.3	0.5														1.4	100.0	Glass and Jenness (1971)
African elephant (*Loxodonta africana*)	tr	3.4	0.1	0.1				0.5											100.0	McCullagh *et al.* (1969)

TABLE I—*continued*

Species	Lactation stage[a]	Days post-partum	No. of animals (samples)	Animal status[b] (diet)	Milk lipid (%)	GLC[c]	Weight % of total-fatty acids[d]											
							4:0	6:0	8:0	10:0	12:0	14:0	14:1	16:0	16:1	16:2–4 (all isomers)	15:0 and 17:0	15:1 and 17:1
Perissodactyla																		
Equidae																		
Horse (*Equus caballus*)	M	28	5 (5)	C[at]	1.3	Cap			6.8	14.1	12.9	9.2		17.9	4.8			
	M	28	5 (5)	C[au]	1.1	Cap			8.0	17.1	14.3	8.7		15.4	4.1			
Tubulidentata																		
Orycteropodidae																		
Aardvark (*Orycteropus afer*)	n/a	n/a	1 (1)	C[j]	n/a	Pack	0.0	0.0	0.0	0.0	0.0	5.5		30.9	4.2			
Artiodactyla																		
Suidae																		
Pig (domestic) (*Sus scrofa*)	M	n/a	Pooled	C	5.5	Pack			0.0	0.1	0.4	3.1		27.6	9.2			
Tayassuidae																		
Collared peccary (*Pecari tajacu*)	M	n/a	3 (3)	C	6.7	Pack					tr	2.8	tr	29.4	6.0			
Camelidae																		
Dromedary camel (*Camelus dromedarius*)	M	n/a	> 55 (11)	C	3.6	Pack	tr	0.1	0.1	0.1	0.7	9.8	1.4	25.7	10.5		2.8	
Alpaca (*Lama pacos*)	n/a	n/a	1 (1)	C	n/a	Pack	0.0	0.0	0.2	0.3	0.3	5.8		24.0	15.6			
Cervidae																		
Mule deer (*Odocileus hemonius*)	n/a	n/a	1 (1)	W	n/a	Pack	7.8	2.1	0.3	0.8	0.7	12.7		35.6	1.1			
Giraffidae																		
Okapi (*Okapia johnstoni*)	n/a	n/a	1 (1)	C	n/a	Pack	0.6	1.4	1.9	7.0	1.1	14.0		30.5	4.4			

TABLE I—*continued*

Species	18:0	18:1	18:2 n6	18:3 n3	18:4 n3	18:2–4 (other isomers)	20:1	20:2	20:3 n3,6	20:4 n3	20:4 n6	20:5 n3	22:1	22:5 n3	22:2–5 (other isomers)	22:6 n3	24:1	Others and/or unidentified	Total of fatty acids reported[e]	Reference
Perissodactyla																				
Equidae																				
Horse (*Equus caballus*)	1.1	10.1	3.8	9.6															90.1	Doreau *et al.* (1992)
	1.2	8.3	6.1	4.4															87.5	
Tubulidentata																				
Orycteropodidae																				
Aardvark (*Orycteropus afer*)	7.5	35.8	10.1	1.3														4.8	100.0	Glass and Jenness (1971)
Artiodactyla																				
Suidae																				
Pig (domestic) (*Sus scrofa*)	5.5	32.0	13.0	0.6							1.5	0.2		0.4	0.8	0.1		3.2	97.7	Hrboticky *et al.* (1990)
Tayassuidae																				
Collared peccary (*Pecari tajacu*)	6.2	36.8	17.2	1.6															100.0	Brown *et al.* (1963)
Camelidae																				
Dromedary camel (*Camelus dromedarius*)	11.9	27.1	3.8	3.7				0.2			0.5	0.1				0.1	0.1	1.4	100.0	Sawaya *et al.* (1984)
Alpaca (*Lama pacos*)	7.7	35.8	6.2	0.0														4.0	100.0	Glass and Jenness (1971)
Cervidae																				
Mule deer (*Odocileus hemonius*)	15.0	14.7	1.4	2.1														5.5	99.8	Glass and Jenness (1971)
Giraffidae																				
Okapi (*Okapia johnstoni*)	11.8	20.9	3.5	0.0														2.8	100.0	Glass and Jenness (1971)

TABLE I—*continued*

Species	Lactation stage[a]	Days post-partum	No. of animals (samples)	Animal status[b] (diet)	Milk lipid (%)	GLC[c]	Weight % of total-fatty acids[d] 4:0	6:0	8:0	10:0	12:0	14:0	14:1	16:0	16:1	16:2–4 (all isomers)	15:0 and 17:0	15:1 and 17:1
Bovidae																		
Blackbuck antelope (*Antilope cervicapra*)	n/a	n/a	2 (2)	W	8.3	Pack	4.3	3.4	1.7	4.6	3.3	15.9		37.2	2.3			
Goat (domestic) (*Capra hircus*)	n/a	n/a	8 (≥8)	C[av]	3.4	n/a	2.6	2.6	3.1	9.8	5.2	9.9		27.3	2.2			
	n/a	n/a	8 (≥8)	C[aw]	3.0	n/a	2.6	2.9	3.5	11.5	6.5	10.5		28.1	1.9			
Blue duiker (*Cephalophus monticola*)	M	7–56	10 (21)	C[ax]	12.2	n/a	3.1	3.3	2.8	5.0	2.1	15.0		26.1	2.1			
Gazelle (*Gazella granti*)	n/a	n/a	1 (1)	C	n/a	Pack	3.4	3.1	2.7	5.0	3.1	15.5		33.2	2.7			
Sheep (domestic) (*Ovis aries*)	P/E	n/a	6 (6)	C[ay]	5.0	Pack		5.8	2.5	5.7	3.3	8.8		27.2	2.3			
Bighorn sheep (*Ovis canadensis*)	n/a	n/a	1 (1)	C	n/a	Pack	2.6	1.2	0.5	2.5	1.9	11.2		24.7	2.2			
Rodentia																		
Muridae																		
Gerbil (*Meriones unguiculatus*)	n/a	n/a	1 (1)	C	n/a	Pack	0.0	0.0	0.0	0.0	0.0	0.7		13.4	0.7			
Woodrat (pack rat) *Neotoma albuigula*)	n/a	n/a	1 (1)	C	n/a	Pack	0.0	tr	tr	6.7	9.4	9.4		10.6	0.7			
Deer mouse (*Peromyscus maniculatus bairdii*)	n/a	n/a	1 (1)	C	n/a	Pack	0.0	0.0	0.0	0.4	0.9	2.2		16.3	2.1			
Deer mouse (*Peromyscus melanophrys*)	n/a	n/a	1 (1)	C	n/a	Pack	0.0	0.0	0.0	0.5	5.1	13.7		22.8	2.6			
Slender-tailed cloud rat (*Phloeomys cumingi*)	n/a	n/a	1 (1)	C	n/a	Pack	0.0	0.0	0.0	0.0	0.0	4.2		29.4	4.4			
Norway rat (*Rattus norvegicus*)	M	13	9 (9)	C[az]	14.0	Cap			15.8	12.8	8.9	11.7		19.6	1.7			
	M	10	10 (10)[ba]	C[bb]	n/a	Pack				4.2	4.0	3.8	0.1	14.4	0.0			
	M	10	10 (10)[ba]	C[bc]	n/a	Pack				5.5	5.9	10.8	0.3	17.0	9.7	0 6		

TABLE I—*continued*

Species	18:0	18:1	18:2 n6	18:3 n3	18:4 n3	18:2–4 (other isomers)	20:1	20:2	20:3 n3,6	20:4 n3	20:4 n6	20:5 n3	22:1	22:5 n3	22:2–5 (other isomers)	22:6 n3	24:1	Others and/or unidentified	Total of fatty acids reported[c]	Reference
Bovidae																				
Blackbuck antelope (*Antilope cervicapra*)	6.8	19.4	1.6																100.3	Dill *et al.* (1972)
Goat (domestic) (*Capra hircus*)	8.0	22.2	3.3	0.9			0.1	tr	tr		0.2	0.1	tr			tr		2.0	99.6	Calderon *et al.* (1984)
	5.7	15.9	4.1	1.0			0.1	tr	tr		0.2	tr	tr			tr		5.1	99.8	Calderon *et al.* (1984)
Blue duiker (*Cephalophus monticola*)	16.6	20.3	3.7																100.0	Taylor *et al.* (1990)
Gazelle (*Gazella granti*)	5.8	15.8	6.4	0.0														3.3	100.0	Glass and Jenness (1971)
Sheep (domestic) (*Ovis aries*)	11.0	26.3	2.3	1.3															96.5	Leat and Harrison (1980)
Bighorn sheep (*Ovis canadensis*)	9.6	36.8	4.7	2.0														0.0	100.0	Glass and Jenness (1971)
Rodentia																				
Muridae																				
Gerbil (*Meriones unguiculatus*)	5.9	24.5	46.8	0.0														8.1	100.0	Glass and Jenness (1971)
Woodrat (pack rat) *Neotoma albuigula*)	2.6	14.6	44.0	0.7														1.0	99.7	Glass and Jenness (1971)
Deer mouse (*Peromyscus maniculatus bairdii*)	6.9	37.2	31.2	0.0														2.8	100.0	Glass and Jenness (1971)
Deer mouse (*Peromyscus melanophrys*)	6.7	31.4	15.9	0.0														1.2	100.0	Glass and Jenness (1971)
Slender-tailed cloud rat (*Phloeomys cumingi*)	3.8	46.6	7.9	0.0														3.7	100.0	Glass and Jenness (1971)
Norway rat (*Rattus norvegicus*)	3.2	12.3	9.4	0.8		0.2			0.5	0.4	0.6	tr		0.3	0.3	0.5		0.2	99.2	Mills *et al.* (1990)
	1.4	23.3	43.9	0.8		0.8	0.4	0.7	0.6		1.2	0.1		0.0	0.3	0.1			100.0	Yeh *et al.* (1990)
	2.3	12.6	5.6	1.8		0.1	1.2	0.1	0.2		1.0	11.0		2.8	0.6	7.4		3.2	99.7	Yeh *et al.* (1990)

TABLE I—*continued*

Species	Lactation stage[a]	Days post-partum	No. of animals (samples)	Animal status[b] (diet)	Milk lipid (%)	GLC[c]	Weight % of total-fatty acids[d]											
							4:0	6:0	8:0	10:0	12:0	14:0	14:1	16:0	16:1	16:2–4 (all iscmers)	15:0 and 17:0	15:1 and 17:1
Caviidae																		
Guinea pig (domestic) (*Cavia porcellus*)	M	5–10	3 (3)	C[bd]	5.7[ag]	Pack	0.0	0.0	0.0	tr	0.1	1.6		32.5	1.5			
Dasyproctidae																		
Acouchi (*Myoprocta pratti*)	n/a	n/a	1 (1)	C	n/a	Pack	0.0	0.0	0.0	0.0	0.0	4.9		54.9	8.6			
Lagomorpha																		
Leporidae																		
European hare (*Lepus europaeus*)	M	n/a	9 (9)	C	14.8[ag]	Cap		tr	6.8	13.1	4.8	5.2		27.4	5.5		0.8	0.4
Rabbit (*Oryctolagus cuniculus*)	M	n/a	7 (7)	C	15.2[ag]	Cap		tr	27.1	28.9	5.6	1.8		9.6	1.3		0.4	tr

TABLE I—continued

Species	18:0	18:1	18:2 n6	18 3 n3	18:4 n3	18:2–4 (other isomers)	20:1	20:2	20:3 n3,6	20:4 n3	20:4 n6	20:5 n3	22:1	22:5 n3	22:2–5 (other isomers)	22:6 n3	24:1	Others and/or unidentified	Total of fatty acids reported[c]	Reference
Caviidae																				
Guinea pig (domestic) (*Cavia porcellus*)	3.4	32.3	24.1	4.4															100.0	Smith *et al.* (1968)
Dasyproctidae																				
Acouchi (*Myoprocta pratti*)	2.4	21.9	6.8	0 6															100.0	Glass and Jenness (1971)
Lagomorpha																				
Leporidae																				
European hare (*Lepus europaeus*)	3.6	17.5	12.8	2.0							tr	tr		tr		tr		0.2	100.0	Demarne *et al.* (1978)
Rabbit (*Oryctolagus cuniculus*)	2.1	9.8	11.1	2.3							tr	tr		tr		tr			100.0	Demarne *et al.* (1978)

[a]Abbreviations used: P/E, perinatal period or early lactation; M, mature milk (mid- or peak lactation); L, late lactation; n/a, data not available.

[b]Abbreviations used: C, captive population; W, wild population.

[c]GLC, gas liquid chromatography: cap, using capillary column; pack, using packed column.

[d]Fatty acids are designated as carbon chain length; No. of double bonds and *n x* denotes the position (*x*) of the last double bond relative to the terminal methyl carbon. Where double bonds are present, if *n x* is not specified, this means the value is the sum of all isomers of that component. Blank spaces, values either not reported or not analyzed; 0.0, fatty acids not detected although capability existed to detect these components; tr, trace amounts (< 0.05%).

[e](TG) after value, only the triglyceride fraction of the milk was analyzed; (x% TG), the corresponding TG content of total lipid if reported.

[f]Diet: ants, termites.

[g]Diet: canned meat.

[h]Diet: powdered milk, eggs.

[i]Diet: freshwater insect larvae, shrimps, bivalve molluscs (diet estimated to contain 10–13% 18:3*n* 3).

[j]Diet: insectivorous.

[k]Diet: powdered milk, eggs, termites.

[l]Diet: puppy chow, bread, apples, access to plants and invertebrates.

[m]Fat content estimated from graphs.

[n]Diet: primarily leaves, also fruits and bark (Macdonald, 1984).

[o]Diet: grass, alfalfa, oats.
[p]Diet: alfalfa, oats, grasses, herbs.
[q]From Green (1984).
[r]Diet: leaves, primarily eucalypt leaves (Macdonald, 1984).
[s]Diet: dry bisquits, fruit.
[t]Diet: dry bisquits, vegetables, some fruit.
[u]Diet: dry bisquits, vegetables, some fruit, bamboo leaves.
[v]Diet: high-protein bisquits, fruit, mineral oil-soaked crickets regularly.
[w]Diet: high protein bisquits, fruit, mineral oil-soaked crickets 1×/week, occasional hard boiled egg.
[x]Diet: New world primate diet, tinned cat food, fruit.
[y]Diet: leaves (40–50% of diet), fruits.
[z]From Oftedal, Crissey, Glander and Rudran, unpublished results.
[aa]Diet: standard monkey bisquits (Wayne Monkey Diet or Purina Monkey Chow), largely corn-oil based.
[ab]Diet: pellet diet (vegetable based), apples, grapefruit.
[ac]Diet: Purina dog chow.
[ad]Fasting during winter dormancy.
[ae]Diet during postdormancy: omnivorous (fruits, nuts, acorns, succulent vegetation, scavenged meat, corn, human garbage, and handouts).
[af]Diet: 7.3% rendered lard + mixture of fish offal, fish, potato protein, wheat germ.
[ag]From Oftedal (1984).
[ah]Diet: 7.2% fish oil + mixture of fish offal, fish, potato protein, wheat germ.
[ai]Fasting during the initial perinatal period; diet prior to parturition unknown.
[aj]Diet: Antarctic krill (*Euphausia superba*).
[ak]From Kretzmann *et al.* (1991).
[al]Diet: Pacific mackerel, squid, anchovy, whiting.
[am]Diet: Atlantic herring.
[an]Fasting throughout all or most of lactation.
[ao]From Riedman and Ortiz (1979); fat content estimated from graphs.
[ap]Sample taken from North Atlantic waters.
[aq]Presuckling: fluid, white milk sample collected from pregnant female carrying 4-month near-term fetus, Alaska.
[ar]Captive for 1 week prior to sampling; diet: lettuce.
[as]From McCullagh and Widdowson (1970).
[at]High-roughage diet: 95% fescue hay, 5% concentrates (93% soybean meal).
[au]High-concentrate diet: 50% fescue hay, 50% concentrates (83% barley, 12% soybean meal).
[av]High-roughage diet: 60% alfalfa–oat hay mixture, 39% concentrates (barley, cottonseed meal).

[aw]High-concentrate diet: 20% ground alfalfa hay, 80% concentrates (70% barley, 8.5% cottonseed meal).
[ax]Diet: rabbit pellets, alfalfa hay.
[ay]Diet: chaffed hay, crushed oats, 25% concentrate nuts.
[az]Wistar Kyotc strain rats; diet: Ralston Purina Rat Chow Diet 5001.
[ba]Gastric samples from pups were analyzed and although data are informative, medium-chain components (8:0–12:0) are probably underestimated (see text, Sections III and V).
[bb]Sprague-Dawley strain rats; diet: Ralston Purina rat chow, 20% corn oil.
[bc]Sprague-Dawley strain rats; diet: Ralston Purina rat chow, 20% menhaden oil.
[bd]Diet: mixed crushed oats, rabbit pellets, cabbage.

than 1% in all milks except the leaf- and fruit-eating howler monkeys (at up to 18%). In the reports in which long-chain fatty acids were analyzed, these were present but in relatively minor amounts in most milks. However, in species which fed on biscuits or meat products, significant levels of LC-PUFA are reported.

D. Carnivora

Reliable data on fatty acid composition of carnivore milks are extremely limited. That which is available indicates no synthesis of short-chain fatty acids and only minor amounts of the medium-chain component 12:0 (Table I). The predominant fatty acids are 14:0–18:2. However, only three reports have analyzed LC-PUFA. Levels of 18:3 are low in most species (especially obligate carnivores) at 1% or less, except in the black bear which may have levels as high as 15% when feeding on fruits and leafy vegetation (Iverson and Oftedal, 1992). It appears that milk fatty acid patterns of carnivores are not only responsive to changes in diet but may also resemble those of the diet. For instance, species which include fruit and leafy vegetation in their diets are likely to have the highest proportions of 18:2n6 and 18:3n3, while species which include fish will have the highest proportions of LC-PUFA, at the expense of other fatty acids. In a controlled study of ranch mink, Wamberg *et al.* (1992) demonstrated that milk fatty acid composition was significantly altered when animals were changed from a lard-based diet to a fish oil-based diet, with the percentage of individual fatty acids being altered by up to sevenfold. In a natural habitat, fasting versus foraging and variable feeding habits during foraging appears to affect the milks of individual black bears (Iverson and Oftedal, 1992; see also pinnipeds). Hence, discussion of patterns of milk fatty acids among carnivores must be related to the dietary habits of individuals and species.

E. Pinnipedia, Cetacea, and Sirenia

More detailed work on milk fatty acids has been done with pinnipeds than with species of any other order (Table I). Pinnipeds and cetaceans, like the carnivores, do not secrete short- or medium-chain fatty acids and the composition of milk fatty acids is strongly directed by dietary fatty acids (Iverson, 1993). These milks are all rich in the marine LC-PUFA, containing up to 30% of these components, with relatively low levels of 18:2n6 ($\leq$2%) and 18:3n3 ($\leq$1%). Levels of 16:1 are relatively high and 18:0 relatively low compared to other orders. The direct influence of diet is perhaps best illustrated by comparing milk of a captive California sea lion (fed Atlantic herring) to that of animals from a wild population (Table I). The large amounts of 20:5n3 (8%) and 22:6n3 (19%) and low amounts of 20:1 and 22:1 (3 and 1%, respectively) in the wild animals correspond to those occurring in the natural diet, whereas the decrease in 20:5*n*3 and 22:6*n*3 coupled with the large proportions of 20:1 (15%) and 22:1 (13%) in the captive animal correspond with levels typically found in Atlantic herring (Iverson, 1993).

A sole report of milk from the manatee suggests that the mammary gland of this sirenean synthesizes medium-chain fatty acids (10:0 and 12:0), similar to the finding in other primarily herbivorous mammals. Other primary fatty acids are 14:0–18:1, but an analysis of LC-PUFA was not reported.

F. *Proboscidea and Perissodactyla*

Elephants appear to produce most of their milk fatty acids by *de novo* synthesis of medium-chain fatty acids. Proportions of 8:0–12:0 reportedly total 71 and 92% in Indian and African elephant milks, respectively. These proportions are greater than those that have been found in rabbit milk (70%; Smith *et al.*, 1968) or in the milks of some primates and rodents (40%; Table I). Levels of other fatty acids are comparatively minor.

Data from only a single species of the Perissodactyla (the horse) could be included in tables. The horse also secretes large proportions (34–39%) of medium-chain fatty acids (8:0–12:0) in milk. Levels of 18:2*n*6 and 18:3*n*3 are relatively high reflecting forage-based diets and appear to vary with low- versus high-concentrate diets (Doreau *et al.*, 1992). Even higher proportions of 18:3 ($>$30%) have been reported in another analysis of horse milk (Parodi, 1982), which may be related to dietary differences. Although reports from other species could not be included in Table I, these suggest that other members of the Perissodactyla, such as the tapir and the African black and Indian rhinos, also synthesize a large proportion of medium-chain fatty acids (Glass *et al.*, 1967; Klos *et al.*, 1974).

G. *Artiodactyla*

The artiodactyls are better represented than the perissodactyls and appear to be unique among mammals in the secretion of significant amounts of short-chain fatty acids (4:0 and 6:0). The proportions of short- and medium-chain fatty acids in milks are quite variable among species and it is possible that differences occur along family lines, although available data are not sufficiently robust to confirm this. The pig, peccary, and species of Camelidae appear to secrete very minor amounts of short- and medium-chain components (together totaling 1% or less), whereas the single cervid and most bovids analyzed secrete higher proportions of 4:0–12:0 (generally totaling 10% or more); goat milk may contain up to 27% 4:0–12:0. Few reports have included LC-PUFA analysis, but these are probably relatively minor components. The proportion of 16:1 is usually 2% or less in most artiodactyls, which is low compared to most other orders except rodents and rabbits, whereas levels of 18:0 (10–23%) are high compared to all other orders except some marsupials.

H. Rodentia and Lagomorpha

Unfortunately, data are available for very few families of rodents. Among the species for which there is information, rodents appear to be quite variable in their secretion of medium-chain fatty acids. These components are virtually absent in the milk of species, such as the gerbil, cloud rat, guinea pig, and acouchi, whereas they may reach 16–37% in the Norway rat and woodrat (Table I). Even among species of the same genus of deer mice (*Peromyscus*) medium-chain fatty acids may be absent or present in significant amounts. Short-chain fatty acids have not been found in rodent milks. In species studied, levels of 16:1 are generally low (about 2%) and 18:2n6 is quite high (up to 47%). LC-PUFA have been analyzed in one species, the Norway rat, and these appear to be present in only minor amounts in individuals fed on commercial pellets. However, in a controlled study in which rat dams were fed marine fish oil (20% of diet, Yeh *et al.*, 1990), the composition of LC-PUFA in milks (as analyzed by gastric contents of pups) increased dramatically to 11% 10:5n3 and 7% 22:6n3 compared to that in the pups' gastric samples of rat dams fed the same proportion of corn oil in their diet (Table I). The proportion of medium-chain fatty acids measured in gastric samples in this study (11% compared to 37% in mammary milk on a pelleted diet) was probably underestimated (see Section III,A); thus, it is not possible to evaluate whether a high-fat diet reduces *de novo* biosynthesis of medium-chain fatty acids by the mammary gland as it does in other tissues (e.g., Nelson, 1992).

The lagomorphs synthesize large proportions of medium-chain fatty acids. Demarne *et al.* (1978) found that the rabbit secreted higher levels of these components (62%) than did the hare (25%). However, these authors did not report whether both species were maintained on the same diet. Short-chain fatty acids have not been found in rabbit milk. Levels of 16:1 are low (1–5%) as in most rodents; 18:2n6 accounts for 11–13% of milk fatty acids which is lower than in most rodents.

VI. Conclusions

In conclusion, many questions remain about the relative contribution of mammary biosynthesis versus uptake of circulating fatty acids to the milk fatty acid composition in many species. This is an important area of research to the extent that patterns of fatty acid composition are to be understood from an ecological or phylogenetic perspective. Substantial gaps still exist in information for many mammalian groups, but even those that are represented in published reports usually have not been studied systematically with respect to diet, lactation stage, or other proximate factors which might affect the interpretation of milk fatty acid composition. It is clear that further interpretation of milk fatty acid data will require systematic study using statistically reliable sample sizes and sensitive analytical techniques.

Acknowledgments

The writing of this chapter was supported in part by an International Postdoctoral Fellowship to S. J. Iverson from the Natural Sciences and Engineering Research Council of Canada (NSERC). We thank W. D. Bowen for comments on an earlier draft of the manuscript. Valuable discussions with R. G. Ackman are also gratefully acknowledged.

References

Ackman, R. G. (1991). Application of gas–liquid chromatography to lipid separation and analysis: Qualitative and quantitative analysis. *In* "Analysis of Fats, Oils and Lipoproteins" (E. G. Perkins, ed.), pp. 270–300. American Oil Chemists' Society, Champaign, IL.

Ackman, R. G. (1980). Fish lipids, part 1. *In* "Advances in Fish Science and Technology" (J. J. Connell, ed.), pp. 86–103. Fishing News Books Ltd., Surrey, England.

Ackman, R. G., and Burgher, R. D. (1963). Component fatty acids of the milk of the grey (Atlantic) seal. *Can. J. Biochem. Physiol.* **41,** 2501–2505.

Ackman, R. G., Eaton, C. A., and Hooper, S. N. (1968). Lipids of the fin whale (*Balaenoptera physalus*) from North American waters. IV. Fin whale milk. *Can. J. Biochem.* **46,** 197–203.

Ackman, R. G., Eaton, C. A., and Mitchell, E. D. (1971). The bottle-nosed dolphin *Tursiops truncatus:* Fatty acid composition of milk triglycerides. *Can. J. Biochem.* **49,** 1172–1174.

Ackman, R. G., Ratnayake, W. M. N., and Olsson, B. (1988). The "basic" fatty acid composition of Atlantic fish oils: Potential similarities useful for enrichment of polyunsaturated fatty acids by urea complexation. *J. Am. Oil Chem. Soc.* **65,** 136–138.

Ahlgren, G., Gustafsson, I.-B., and Boberg, M. (1992). Fatty acid content and chemical composition of freshwater microalgae. *J. Phycol.* **28,** 37–50.

Ando, K., Mori, M., Kato, I., Yusa, K., and Goda, K. (1979). General composition and chemical properties of the main components of Yeso brown bear (*Ursus arctos yesoensis*) milks. *J. Coll. Dairying* **8,** 9–21. [In Japanese]

Annison, E. F. (1983). Metabolite utilization by the ruminant mammary gland. *In* "Biochemistry of Lactation" (T. B. Mepham, ed.), pp. 399–435. Elsevier, Amsterdam.

Ashworth, U. S., Ramaiah, G. D., and Keyes, M. C. (1966). Species differences in the composition of milk with special reference to the northern fur seal. *J. Dairy Sci.* **49,** 1206–1211.

Association of Official Analytical Chemists (AOAC) (1975). "Official Methods of Analysis" (W. Horwitz, A. Senzel, H. Reynolds, and D. L. Park, eds.). AOAC, Washington, DC.

Bachman, K. C., and Irvine, A. B. (1979). Composition of milk from the Florida manatee, *Trichechus manatus latirostris. Comp. Biochem. Physiol. A* **62,** 873–878.

Baker, B. E., Harington, C. R., and Symes, A. L. (1963). Polar bear milk. I. Gross composition and fat constitution. *Can. J. Zool.* **41,** 1035–1039.

Bauman, D. E., and Davis, C. L. (1974). Biosynthesis of milk fat. *In* "Lactation: A Comprehensive Treatise. Vol. II" (B. L. Larson and V. R. Smith, eds.), pp. 31–75. Academic Press, New York.

Bitman, J., Wood, D. L., Liao, T. H., Fink, C. S., Hamosh, P., and Hamosh, M. (1985). Gastric lipolysis of milk lipids in suckling rats. *Biochim. Biophys. Acta* **834,** 58–64.

Bligh, E. G., and Dyer, W. J. (1959). A rapid method of total lipid extraction and purification. *Can. J. Biochem. Physiol.* **37,** 911–917.

Borgstrom, B. (1977). Digestion and absorption of lipids. *In* "International Review of Physiology. Vol. 12. Gastrointestinal Physiology II" (R. K. Crane, ed.), pp. 305. Baltimore University Park Press, Baltimore, MD.

Brown, W. H., Stull, J. W., and Sowls, L. K. (1963). Chemical composition of the milk fat of the collared peccary. *J. Mammal.* **44,** 112–113.

Buss, D. H. (1975). Composition of milk from a golden lion marmoset. *Lab. Primate Newslett.* **14,** 17–18.

Buss, D. H. (1969). Fatty acid composition of baboon milk lipids. *Lipids* **4,** 152–154.

Buss, D. L., and Cooper, R. W. (1970). Composition of milk from talapoin monkeys. *Folia Primatol.* **13,** 196–206.

Buss, D. L., and Cooper, R. W. (1972). Composition of squirrel monkey milk. *Folia Primatol.* **17,** 285–291.

Calderon, I., De Peters, E. J., Smith, N. E., and Franke, A. A. (1984). Composition of goat's milk: Changes within milking and effects of a high concentrate diet. **67,** 1905–1911.

Christie, W. W. (1982). "Lipid Analysis." Pergamon Press, New York.

Cook, H. W., and Baker, B. E. (1969). Seal milk. I. Harp seal (*Pagophilus groenlandicus*) milk: Composition and pesticide residue content. *Can. J. Zool.* **47,** 1129–1132.

Cook, H. W., Lentfer, J. W., Pearson, A. M., and Baker, B. E. (1970a). Polar bear milk. IV. Gross composition, fatty acid, and mineral constitution. *Can. J. Zool.* **48,** 217–219.

Cook, H. W., Pearson, A. M., Simmons, N. M., and Baker, B.E. (1970b). Dall sheep (*Ovis dalli dalli*) milk. I. Effects of stage of lactation on the composition of the milk. *Can. J. Zool.* **48,** 629–633.

Cook, H. W., Rausch, R. A., and Baker, B. E. (1970c). Moose (*Alces alces*) milk. Gross composition, fatty acid, and mineral constitution. *Can. J. Zool.* **48,** 213–215.

Crowley, H. M., Woodward, D. R., and Rose, R. W. (1988). Changes in milk composition during lactation in the potoroo, *Potorous tridactylus* (Marsupialia: Potoroinae). *Aust. J. Biol. Sci.* **41,** 289–296.

Davies, D. T., Holt, C., and Christie, W. W. (1983). The composition of milk. *In* "Biochemistry of Lactation" (T. B. Mepham, ed.), pp. 71–117. Elsevier, Amsterdam.

Demarne, Y., Lhuillery, C., Pihet, J., Martinet, L., and Flanzy, J. (1978). Comparative study of triacylglycerol fatty acids in milk fat from two leporidae species: Rabbit (*Oryctolagus cuniculus*) and hare (*Lepus europaeus*). *Comp. Biochem. Physiol. B* **61,** 223–226.

Dill, C. W., Tybor, P. T., McGill, R., and Ramsey, C. W. (1972). Gross composition and fatty acid constitution of blackbuck antelope (*Antilope cervicapra*) milk. *Can. J. Zool.* **50,** 1127–1129.

Dils, R. R. (1983). Milk fat synthesis. *In* "Biochemistry of Lactation" (T. B. Mepham, ed.), pp. 141–157. Elsevier, Amsterdam.

Dils, R. R., Clark, S., and Knudsen, J. (1977). Comparative aspects of milk fat synthesis. *Symp. Zool. Soc. London* **41,** 43–55.

Doreau, M., Boulot, S., Bauchart, D., Barlet, J.-P. and Martin-Rosset, W. (1992). Voluntary intake, milk production and plasma metabolites in nursing mares fed two different diets. *J. Nutr.* **122,** 992–999.

Dosako, S., Taneya, S., Kimura, T., Ohmori, T., Daikoku, H., Suzuki, N., Sawa, J., Kano, K., and Katayama, S. (1982). Milk of the northern fur seal: Composition, especially carbohydrate and protein. *J. Dairy Sci.* **66,** 2076–2083.

Folch, J., Lees, M., and Sloane-Stanly, G. H. (1957). A simple method for the isolation and purification of total lipids from animal tissues. *J. Biol. Chem.* **226,** 497–509.

Folley, S. J., and McNaught, M. L. (1961). Biosynthesis of milk fat. *In* "Milk: The Mammary Gland and Its Secretion" (S. K. Kon and A. T. Cowie, eds.), pp. 441–482. Academic Press, New York.

Gibson, R. A., Neumann, M., Grant, T. R., and Griffiths, M. (1988). Fatty acids of the milk and food of the platypus (*Ornithorhynchus anatinus*). *Lipids* **23,** 377–379.

Glass, R. L, and Jenness, R. (1971). Comparative biochemical studies of milk—VI. Constituent fatty acids of milk fats of additional species. *Comp. Biochem. Physiol. B* **38,** 353–359.

Glass, R. L., Troolin, H. A., and Jenness, R. (1967). Comparative biochemical studies of milks—IV. Constituent fatty acids of milk fats. *Comp. Biochem. Physiol.* **22,** 415–425.

Green, B. (1984). Composition of milk and energetics of growth in marsupials. *Symp. Zool. Soc. London* **51,** 369–387.

Green, B., Griffiths, M., and Leckie, R. M. C. (1983). Qualitative and quantitative changes in milk fat during lactation in the tammar wallaby (*Macropus eugenii*). *Aust. J. Biol. Sci.* **36,** 455–461.

Green, B., and Merchant, J. C. (1988). The composition of marsupial milk. *In* "The Developing Marsupial. Models for Biomedical Research" (C. H. Tyndale-Biscoe and P. A. Janssens, eds.), pp. 41–54. Springer-Verlag, Berlin.

Green, B., Merchant, J., and Newgrain, K. (1987). Milk composition in the eastern quoll, *Dasyurus viverrinus* (Marsupialia: Dasyuridae). *Aust. J. Biol. Sci.* **40,** 379–387.

Gregory, M. E., Rowland, S. J., Thompson, S. Y., and Kon, V. M. (1965). Changes during lactation in the composition of the milk of the African black rhinoceros (*diceros bicornis*). *Proc. Zool. Soc. London* **145,** 327–333.

Griffiths, M., Friend, J. A., Whitford, D., and Fogerty, A. C. (1988). Composition of the milk of the numbat, *Myrmecobius fasciatus* (Marsupialia: Myrmecobiidae), with particular reference to the fatty acids of the lipids. *Aust. Mammalogy* **11,** 59–62.

Griffiths, M., Green, B., Leckie, R. M. C., Messer, M., and Newgrain, K. W. (1984). Constituents of platypus and echidna milk, with particular reference to the fatty acid complement of the triglycerides. *Aust. J. Biol. Sci.* **37,** 323–329.

Griffiths, M., McIntosh, D. L., and Leckie, R. M. C. (1972). The mammary glands of the red kangaroo with observations on the fatty acid composition of the milk triglycerides. *J. Zool. London* **166,** 265–275.

Grigor, M. R. (1980). Structure of milk triacylglycerols of five marsupials and one monotreme: Evidence for an unusual pattern common to marsupials and eutherians but not found in the echidna, a monotreme. *Comp. Biochem. Physiol. B* **65,** 427–430.

Hall, A. J. (1971). Fatty acid composition of rabbit (*Oryctolagus cuniculus*) milk fat throughout lactation. *Int. J. Biochem.* **2,** 414–418.

Hilditch, T. P. (1956). "The Chemical Constitution of Natural Fats," 3rd Ed. Wiley, New York.

Hilditch, T. P., and Williams, P. N. (1964). "The Chemical Constitution of Natural Fats," 4th Ed. Chapman and Hall, London.

Hrboticky, N., MacKinnon, M. J., and Innis, S. M. (1990). Effect of a vegetable oil formula rich in linoleic acid on tissue fatty acid accretion in the brain, liver, plasma, and erythrocytes of infant piglets. *Am. J. Clin. Nutr.* **51,** 173–182.

Iverson, S. J. (1988). "Composition, Intake and Gastric Digestion of Milk Lipids in Pinnipeds." Ph.D. thesis, University of Maryland, College Park, MD.

Iverson, S. J. (1993). Milk secretion in marine mammals in relation to foraging: Can milk fatty acids predict diet? *Symp. Zool. Soc. London* **66,** 263–291.

Iverson, S. J., Bowen, W. D., Boness, D. J., and Oftedal, O. T. (1993). The effect of maternal size and milk energy output on pup growth in grey seals (*Halichoerus grypus*). *Physiol. Zool.* **66,** 61–88.

Iverson, S. J., Kirk, C. L., Hamosh, M., and Newsome, J. (1991). Milk lipid digestion in the neonatal dog: The combined actions of gastric and bile salt stimulated lipases. *Biochim. Biophys. Acta* **1083,** 109–119.

Iverson, S. J., and Oftedal, O. T. (1992). Fatty acid composition of black bear (*Ursus americanus*) milk during and after the period of winter dormancy. *Lipids* **27,** 940–943.

Iverson, S. J., Oftedal, O. T., Bowen, W. D., Boness, D. J. and Sampugna, J. (1995). Prenatal and postnatal transfer of fatty acids from mother to pup in the hooded seal. *J. Comp Physiol. (B)* **931,** in press.

Iverson, S. J., Sampugna, J., and Oftedal, O. T. (1992). Positional specificity of gastric hydrolysis of long-chain n-3 polyunsaturated fatty acids of seal milk triglycerides. *Lipids* **27,** 870–878.

Jenness, R., Erickson, A. W., and Craighead, J. J. (1992). Some comparative aspects of milk from four species of bears. *J. Mammal.* **53,** 34–47.

Jensen, R. G. (1989). Lipids in human milk—Composition and fat-soluble vitamins. *In* "Textbook of Gastroenterology and Nutrition in Infancy" (E. Lebenthal, ed.), pp. 157–208. Raven Press, New York.

Jensen, R. G., Clark, R. M., Bitman, J., Wood, D. L., Hamosh, M., and Clandinin, M. T. (1985). Methods for the sampling and analysis of human milk lipids. *In* "Human Lactation: Milk Components and Methodologies" (R. G. Jensen and M. C. Neville, eds.), pp. 97–112. Plenum Press, New York.

Jensen, R. G., Ferris, A. M., Lammi-Keefe, C. J., and Henderson, R. A. (1990). Lipids of bovine and human milks: A comparison. *J. Dairy Sci.* **73,** 223–240.

Klos, Von H.-G., Jarofke, D., Langner, H. J., Siems, H., and Malek, E. (1974). Die chemische und mikrobiologische Zusammensetzung der Panzernashornmilch (Fortsetzung). *Zuchthygiene* **9,** 150–153.

Kretzmann, M. B., Costa, D. P., Higgins, L. V., and Needham, D. J. (1991). Milk composition of Australian sea lions, *Neophoca cinerea:* Variability in lipid content. *Can. J. Zool.* **69,** 2556–2561.

Kunz, T. H., Stack, M. H., and Jenness, R. (1983). A comparison of milk composition in Myotis lucifugus and Eptesicus fuscus (Chiroptera: Vespertilionidae). *Biol. Reprod.* **28,** 229–234.

Lauer, B. H., and Baker, B. E. (1969). Whale milk. I. Fin whale (*Balaenoptera physalus*) and beluga whale (*Delphinapterus leucas*) milk: Gross composition and fatty acid constitution. *Can. J. Zool.* **47,** 95–97.

Lauer, B. H., Blood, D. A., Pearson, A. M., and Baker, B. E. (1969a). Goat milk. I. Mountain goat (*Oreamnos americanus*) milk. Gross composition and fatty acid constitution. *Can. J. Zool.* **47,** 5–8.

Lauer, B. H., Kuyt, E., and Baker, B. E. (1969b). Wolf milk. I. Arctic wolf (*Canis lupus arctos*) and husky milk: Gross composition and fatty acid constitution. *Can. J. Zool.* **47,** 99–102.

Leat, W. M. F., and Harrison, F. A. (1980). Transfer of long-chain fatty acids to the fetal and neonatal lamb. *J. Dev. Physiol.* **2,** 257–274.

Macdonald, D. (1984). "The Encyclopedia of Mammals." Facts on File Publications, New York.

McCullagh, K. G., Lincoln, H. G., and Southgate, D. A. T. (1969). Fatty acid composition of milk fat of the African elephant. *Nature* **222,** 493–494.

McCullagh, K. G., and Widdowson, E. M. (1970). The milk of the African elephant. *Br. J. Nutr.* **24,** 109–117.

Merchant, J. C., and Libke, J. A. (1988). Milk composition in the northern brown bandicoot, *Isoodon macrourus* (Peramelidae, Marsupialia). *Aust. J. Biol. Sci.* **41,** 495–505.

Mills, D. E., Ward, R. P., and Huang, Y. S. (1990). Fatty acid composition of milk from genetically normotensive and hypertensive rats. *J. Nutr.* **120,** 431–435.

Morrison, W. R. (1970). Milk lipids. *Topics Lipid Chem.* **1,** 52–95.

Myher, J. J., Kuksis, A., Tilden, C., and Oftedal, O. T. (1995). A cross-species comparison of neutral lipid composition of milk fat of prosimian primates. *J. Lipid Res.* in press.

Nelson, G. J. (1992). Dietary fatty acids and lipid metabolism. *In* "Fatty Acids in Foods and their Health Implications" (C. K. Chow, ed.), pp. 437–471. Dekker, New York.

Nishikawa, I., Kawanishi, G., Cho, F., Honjo, S., Hatakeyama, T., and Wako, H. (1976). Chemical composition of Cynomolgus monkey milk. *Exp. Anim.* **25,** 253–264.

Oftedal, O. T. (1984). Milk composition, milk yield and energy output at peak lactation: A comparative review. *Symp. Zool. Soc. London* **51,** 33–85.

Oftedal, O. T., Boness, D. J., and Bowen, W. D. (1988). The composition of hooded seal milk: An adaptation to postnatal fattening. *Can. J. Zool.* **66,** 318–322.

Parodi, P. W. (1982). Positional distribution of fatty acids in triglycerides from milk of several species of mammals. *Lipids* **17,** 437–442.

Patton, J. S. (1981). Gastrointestinal lipid digestion. *In* "Physiology of the Gastrointestinal Tract" (L. R. Johnson, ed.), pp. 1123–1146. Raven Press, New York.

Patton, J. S., Rigler, M. W., Liao, T. H., Hamosh, P., and Hamosh, M. (1982). Hydrolysis of triacylglycerol emulsions by lingual lipase: A microscopic study. *Biochim. Biophys. Acta* **712,** 400–407.

Patton, S., and Jensen, R. G. (1976). "Biomedical Aspects of Lactation with Special Reference to Lipid Metabolism and Membrane Functions of the Mammary Gland." Pergammon Press, Oxford.

Poole, C. F., and Poole, S. K. (1991). "Chromatography Today." Elsevier, Amsterdam.

Riedman, M., and Ortiz, C. L. (1979). Changes in milk composition during lactation in the northern elephant seal. *Physiol. Zool.* **52,** 240–249.

Sawaya, W. N., Khalil, J. K., Al-Shalhat, A., and Al-Mohammad, H. (1984). Chemical composition and nutritional quality of camel milk. *J. Food Sci.* **49,** 744–747.

Scow, R. O., Blanchette-Mackie, E. J., Mendelson, C. R., Hamosh, M., and Zinder, O. (1975). Incorporation of dietary fatty acids into milk triglyceride: Mechanism and regulation. *Milk Lactation; Mod. Probl. Paediatr.* **15,** 31–45.

Smith, S., Watts, R., and Dils, R. (1968). Quantitative gas–liquid chromatographic analysis of rodent milk triglycerides. *J. Lipid Res.* **9,** 52–57.

Strong, C. R., and Dils, R. R. (1972). The fatty acid synthetase complex of lactating guinea-pig mammary gland. *Int. J. Biochem.* **3,** 369–377.

Stull, J. W., Brown, W. H., and Kooyman, G. L. (1967). Lipids of the Weddell seal, *Leptonychotes weddelli. J. Mammal.* **48,** 642–645.

Taylor, B. A., Varga, G. A., Whitsel, T. J., and Hershberger, T. V. (1990). Composition of blue duiker (*Cephalophus monticola*) milk and milk intake by the calf. *Small Ruminant Res.* **3,** 551–560.

Turton, J. A., Ford, D. J., Bleby, J., Hall, B. M., and Whiting, R. (1978). Composition of the milk of the common marmoset (*Callithrix jacchus*) and milk substitutes used in hand-rearing programmes, with special reference to fatty acids. *Folia Primatol.* **29,** 64–79.

Van Horn, D. R., and Baker, B. E. (1971). Seal milk. II. Harp seal (*Pagophilus groenlandicus*) milk: Effects of stage of lactation on the composition of milk. *Can. J. Zool.* **49,** 1085–1088.

Wamberg, S., Olesen, C. R., and Hansen, H. O. (1992). Influence of dietary sources of fat on lipid synthesis in mink (*Mustela vison*) mammary tissue. *Comp. Biochem. Physiol. A* **103,** 199–204.

Yeh, Y.-Y., Winters, B. L., and Yeh, S.-M. (1990). Enrichment of (n-3) fatty acids of suckling rats by maternal dietary menhaden oil. *J. Nutr.* **120,** 436–443.

C. Comparative Analysis of Milks Used for Human Consumption

BRENDA P. ALSTON-MILLS

I. Introduction

In this chapter, I describe the composition and some properties and uses for milks of domesticated mammals. Milk or milk products add to the quality of the human diet by providing highly digestible high-quality protein as well as other nutrients. In developing countries where rice or tubers are staples, milk protein is a supplement providing essential amino acids to the diet. Approximately 550 million metric tons of milk and milk products are consumed per year. Most of the milk is from cattle (88% *Bos taurus*). The second highest amount is from buffalo (*Bubalis bubalis.*). The species are listed in Table I.

For those countries where refrigeration is a problem, alternative methods of milk preservation must be employed. Often, vinegar is added to coagulate the milk. The curd is then shaped into flattened cakes and either dried in the sun or coated with salt. This forms a type of cheesecake that can be heated or toasted as desired. The cheesecake also provides a means of storage for later consumption or for barter or sale at the marketplace. Fermented milk is another common method for cheese manufacture. The curd is formed into balls and dried, while the residual whey milk, squeezed from the curd, can be consumed directly. The fat is churned and used as butter, flavoring, and can be sold as such. Because the fat is a source of essential fatty acids. it also provides needed calories in the human diet.

TABLE I
Species of Animals That Provide Milk for Human Consumption[a]

Alpaca	Camel	Goat	Musk ox
Ass	Caribou	Llama	Reindeer
Banteng (ox)	Cattle	Mare	Sheep
Buffalo	Elk	Moose	Yak

[a]Reproduced with permission from R. E. McDowell: "A Partnership for Humans and Animals" (1991). Kinnic Publishers, Raleigh, NC.

HANDBOOK OF MILK COMPOSITION

Because buffalo is a major source of milk in many countries, attention will be given to its composition compared to milk of the cow. Nine major breeds of buffalo are used for milk production. The average lactation period is approximately 285 days with a milk yield of about 1650 kg depending on feeding and management systems. There is a high yield variation among individuals as well as period of lactation. Buffalo milk is high in fat which provides it with rich flavor. In order to enhance the flavor of imported dry or skim milk powder from cattle, buffalo milk is often added. Much of the buffalo milk is used to prepare processed products such as cheese. Both fat and total solids are higher in buffalo milk than cow's milk as shown in a recent compilation of data comparing milks from cows and buffalo which is given in Table II. This information reflects analyses of milk specifically taken for human consumption. Total fat differs slightly from the overall proximate analyses listed in Table III.

It is important to emphasize again the influence of management practices, feeding regimen, and stage of lactation at which the sample was taken when comparing values. The chemical properties of buffalo milk and cow milk differ significantly. Cow milk contains β-carotene despite the fact that both types of milk possess similar vitamin A potency. Therefore, buffalo milk is much whiter in appearance than bovine cow's milk. Other chemical properties of buffalo milk are useful in the preparation of some cheeses and other processed foods. For example, the curd of buffalo milk withstands heating better than cow curd. Additionally, buffalo milk is easier to coagulate with rennet than milk from cows.

II. Chemical Properties of Milks

The composition of milks of different species of mammals was designed to provide the nutritional needs of their neonates. Specifically, only those mammals, e.g., cow, goat, sheep, horse, pig, which require passive

TABLE II
Percentage Composition of Milk between Two Breeds of Buffalo Compared to Two Breeds of Cow[a]

Species/breed	Fat	Total solids	Protein	Lactose
Carabao (water buffalo)	8.95	18.34	4.13	4.78
Buffalo	7.45	17.96	4.36	4.83
Cow				
Friesian	3.60	12.15	3.25	4.60
Native	4.97	13.45	3.18	4.59

[a]Personal communication from Dr. R. E. McDowell (1993).

TABLE III
Composition of Milks from Domesticated Mammals[a]

Order			g/dl					Reference
	Days of lactation	No. samples	Total solids	Protein	Lactose	Fat	Ash	
Perissodactyla								
Equus asinus (ass)	60–120	9	8.5	1.4	6.1	0.6	0.4	Anantakrishnan (1941)
Equus calabus (horse)	25–54	25	10.5	1.9	6.9	1.3	0.4	Oftedal *et al.* (1983)
	21	8	10.5	2.1	6.5	1.5	0.4	Alston-Mills and Larimore, manuscript in preparation
Artiodactyla								
Alces alces (moose)	> 2	15	21.5	8.5	3.0	10	1.5	Ivanova (1965)
Rangifer tarandus (reindeer)	21–30	6	26.3	9.5	3.4	10.9	1.3	Luhtale *et al.* (1968); Luick *et al.* (1974)
Rangifer arcticus (caribou)	?	3	23.6	7.6	3.7	11.0	1.3	Hatcher *et al* (1967)
Camelus bactrianus (bactrian camel)	23–91	30	15.2	4.3	—	4.3	0.9	Oftedal (1984)
Camelus dromedarius (dromedary)	?	15	13.6	3.6	5.0	4.5	0.7	Kheraskov (1961); Ohri and Joshi (1961); Khan and Appanna (1964); El-Bahay (1962)
	?	Pooled	11.9	2.5	4.7	3.9	0.8	Mehaia (1993)
Lama glama (llama)	—	1	16.2	7.3	6.0	2.4	—	Jenness (1974)
Bos taurus mature (cow)		> 2000	12.4	3.2	4.6	3.7	0.7	Macy *et al.* (1953)
Bos indicus (cow)	?	130	13.5	3.2	4.9	4.7		Basu *et al.* (1962)

Bos gruaniens (yak)		?	17.3	5.8	4.9	6.5	0.9	Vsakikh (1943); Markova (1956)
Ouibos moschatus (musk ox)	—	1	16.4	5.3	4.1	5.4	—	Tener (1956)
Bubalis bubalis (water buffalo)	30	42	16.8	4.3	4.9	6.5	0.8	Eltawil *et al.* (1976); Laxminaryan and Dastur (1968)
Cervus elaphus (elk)	14–77	28	19.0	5.7	4.2	6.7	1.3	Robbins *et al.* (1981)
Ovis aries (sheep)	13–35	20	18.2	4.1	5.0	7.3	0.8	Oftedal (1981)
Capra hircus (goat)	14–56	120?	12.0	2.9	4.7	3.8	0.8	Ronningen (1965)

[a]Compilation of tables from Jenness (1974) in "Lactation," Vol. III, Academic Press, New York; Oftedal (1984) in "Lactation Strategies," *Symp. Zool. Soc. London.*

immunity from colostrum contain β-lactoglobulin. All mammals producing lactose contain α-lactalbumin to modify galactosyl transferase in the lactose synthase system.

People have extended the use of milk beyond infancy as a nutrient source and, in doing so, consumed the milks of other mammals for socioeconomic reasons. Cow milk is the notable example. Consequently, the cow and her milk have entered into the human equation of life. Several aspects of human health are associated with the use of milks beyond the normal suckling period. As a high source of calcium, cow milk consumed in adolescence and early adulthood may help delay the onset of osteoporosis in women. Conversely, allergies to the foreign proteins, lactose intolerance, and technological problems associated with the production of by-products are encountered with the consumption of nonhuman milks and their products.

Cheese and fermented products made from milks of different domestic species vary greatly in body, texture, and flavor mainly because of the percentage composition of lipids and proteins, particularly the caseins. The body of fermented milk products is, of course, dependent on both lipid and casein composition and types but the α_{S1}-casein present in *Bos taurus* (western cattle) and *Bos indicus* (Zebu, Indian cattle) gives a strong curd in cheeses and yogurt. The remaining species of mammals mentioned in Table III do not contain α_{S1}-casein but are predominantly β-casein as in human milk. β-casein is responsible for the soft curd properties of goat's milk which can be made into low-moisture cheddar-type cheese. Although most cheeses produced lose whey protein during manufacture, yogurt and kefir do not. Therefore, they contain the nutritionally rich whey proteins high in cysteine and tryptophan.

One of the whey proteins, α-lactalbumin, is a strong calcium-binding protein. Electrophoresis of this protein in the absence of calcium shows that upon chelation of calcium, the protein migrates faster in an electric field because of the exposure of negative carboxylate groups when calcium is debound (Thompson *et al.* 1989). Typical R_m values of α-lactalbumin for several species are cow (1.14), horse (1.39), camel (dromedary, 1.21), goat (1.19), and sheep (1.17). The values indicate that although functionally alike, the protein is different in amino acid sequence and may differ in the amount of calcium bound to the molecule.

III. Uses for Milks of Domesticated Mammals

Although much of the milk produced by the domesticated mammals listed in Table I is consumed immediately in fluid form, numerous other uses have evolved over the centuries either as in home products or as a large-scale production commodity. The most notable example of the latter is Roquefort, a blue-veined cheese made from the milk of the ewe. In addition to the superb flavor, it provides a substantial income for the area of France. Goat's milk is extensively used worldwide in the manufacture of ice cream and yogurt as well as cheese. In California, it is spray dried to be reconstituted as a beverage or use in ice cream but at a premium price. The milk of water buffalo is desirable for the manufacture of mozzarella and provolone types of cheeses as well as yogurt and some ice creams. Mare's milk has been used in a variety of fermented milk products including kefir and yogurt-like products and is consumed primarily in the former Soviet Union and Eastern Bloc countries. Additionally, the use of mare's milk as a curative product is increasing in western countries (Solaroli *et al.*, 1993).

In recent years, milk from the camel has been used in the manufacture of soft cheeses and ice cream in locations such as Egypt and Saudi Arabia (Mehaia, 1993; Abu-Lehia *et al.*, 1989; El-Neshaway *et al.*, 1988). For centuries, the nomadic Bedouins used camel milk as a valuable source of nutrition.

IV. Summary

In countries where refrigeration and pasteurization are available, these methods to assure against microbial contamination are routinely used. Third world countries have had to devise methods to preserve milk for consumption. As indicated in this narrative, composition and chemical properties of milks vary according to species. Procedures used for milk preservation also vary depending on environmental constraints and processing methods. Thus, milk products to be used for immediate consumption or storage differ according to the geographic location. Milk and its products are also commercially important. Many of the natural cheeses made in the homes are sold in markets or offered in trade for other items. Moreover, we are fascinated by the fact that humankind has not only domesticated these animals but has creatively developed used for their milks as an important source of nutrition.

Acknowledgments

The author thanks Dr. Robert E. McDowell, Visiting Professor of Animal Science, North Carolina State University, Raleigh, North Carolina for his kind assistance in providing updated information for milk composition of water buffalo and cow as well as information found in the Introduction of this chapter. Appreciation is also given to Dr. Marvin P. Thompson, USDA (retired), who assisted in the organization and review of the manuscript.

References

Abu-Lehia, I. H., Al-Mohizea, I. S., and El-Behary, M. (1989). Studies on the production of ice cream from camel milk products. *Aust. J. Dairy Technol.* **44,** 31–36.

Anantakrishnan, C. P. (1941). Studies on ass's milk: Composition. *J. Dairy Res.* **12,** 119–130.

Basu, S. (1962). The influence of milk yield on the occurrence of postpartum oestrus in Murrah buffaloes. *Indian Vet. J.* **39,** 433–438.

Cook, H. W., Pearson, A. M., Simmons, N. M., and Baker, B. E. (1970). Effects of stage of lactation on the composition of the milk. *Can. J. Zool.* **48,** 629–633.

El-Bahay, G. M. (1962). Normal contents of Egyptian camel milk. *Vet. Med. J.* **8,** 7–18.

El-Neshaway, A. A., Farahat, S. M., Whabah, H. A. (1988). Production of salt cheese with low fat and salt contents. *Food Chem.* **28,** 219–224.

Eltawil, E. A., Moukhtar, S. A., Galat, E. S., and Khishin, E. S. (1976). Factors affecting the production and composition of Egyptian buffalo milk. *Trop. Anim. Health Prod.* **8**(2), 115–121.

Hatcher, V. B., McEwan, E. H., and Baker, B. E. (1967). Barren-ground caribou (*Rangifer tarandus gorenlandicus*): Gross composition, fat, and protein constitution. *Can. J. Zool.* **45,** 1101–1106.

Ivanova, G. M. (1964). Chemical composition and nutritive values of elks' milk. *Dairy Sci. Abstr.* **27**(4), 556.

Jenness, R. (1974). The composition of milk. *In* "Lactation" (B. L. Larson and V. R. Smith, eds.), Vol. III, pp. 3–107. Academic Press, New York.

Khan, Kh. U., and Appanna, T. C. (1964). Studies in camel milk: General composition. *Indian J. Physiol. Allied Sci.* **18,** 129–133.

Kheraskov, S. G. (1961). Composition, properties, and nutritive value of camel's milk. *Dairy Sci. Abstr.* **23,** 612.

Laxminaryana, H., and Dastur, N. N. (1968). Buffaloes' milk and milk products. *Dairy Sci. Abstr.* **30,** 177–186.

Luhtala, A., Rautiainen, A., and Antila, M. (1968). *Zussammensetzung Finnischen Rentiermilch. Suom. Kemistilihti B.* **41,** 6–9.

Luick, J. R., White, R. G., Gau, A. M., and Jenness, R. (1974). Compositional changes in the milk secreted by grazing reindeer. 1. Gross composition and ash. *J. Dairy Sci.* **57,** 1325–1333.

Macy, I. G., Kelly, H. J., and Sloan, R. E. (1953). "The Composition of Milks." National Academy of Science—National Research Council, Publication 254, pp. 70.

Markova, K. V. (1956). "Model Option Katal Russian Protsesson." Sb. Dokl. Vses. Soveshch. Moloch. Delu. p. 136.

Mehaia, M. A. (1993). Fresh soft white cheese (domiatai-type) from camel milk: Composition, yield, and sensory evaluation. *J. Dairy Sci.* **76,** 2845–2855.

Oftedal, O. T. (1984). Milk composition, milk yield, and energy output at peak lactation: A comparative review. *Symp. Zool. Soc. London* **51,** 33–85.

Oftedal, O. T. (1981). "Milk Protein and Energy Intakes of Suckling Mammalian Young: A Comparative Study." Ph.D. thesis, Cornell University, Ithaca, NY.

Oftedal, O. T., Hintz, H. F., and Schryver, H. F. (1983). Lactation in the horse: Milk composition and intake by foals. *J. Nutr.* **113,** 2096–2106.

Ohri, S. P., and Joshi, B. K. (1961). Composition of colostrum of camel. **13,** 604–607.

Robbins, C. T., Podbielancik, N., Robert, S., Wilson, D. L., and Mould, E. D. (1981). Growth and nutrient consumption of elk calves compared to other ungulate species. *J. Wildlife Management* **45,** 172–186.

Ronningen, K. (1965). Sommenhung mellon mengde, kjemisk innhold og smak, geitemjølk. *Rep. Inst. Animal Genet. and Breed Agric. Coll. Norway.* **213,** 118.

Solaroli, G., Pagliarini, E., and Peri, C. (1993). Composition and nutritional quality of mare's milk. *Ital. J. Food Sci.* **1,** 3–10.

Tener, J. S. (1956). Gross composition of musk-ox milk. *Can. J. Zool.* **34,** 569–571.

Thompson, M. P., Brower, D. P., Jenness, R., and Kotts, C. E. (1989). Phylogenetic variations in the calcium-dependent electrophoretic shift of α-lactalbumin. *J. Dairy Sci.* **72,** 3156–3165.

D. Infant Formulas

ROBERT G. JENSEN
SARAH C. COUCH
JAMES W. HANSEN
ERIC L. LIEN
KARIN M. OSTROM
UMBERTO BRACCO
ROGER A. CLEMENS

I. Introduction

Term infants at birth are usually able to suck, swallow, digest, and absorb nutrients from human milk and similar liquid foods. Human milk is believed to be the best source of nutrients and nonnutrient components for their optimal growth and development because it has been developed for this purpose through millennia of evolution (Jensen and Jensen, 1992). The synthesis of human milk is probably programmed by an interchange of information pre-and postnatally between mother and fetus or infant to produce the best milk for the infant. If, for whatever reason, breast-feeding was terminated or never started, the infant could not survive unless a suitable alternative food was fed. Since the shelf life of raw milk is very short, sweetened condensed and evaporated cow's milk were used. With the former, preservation is achieved by addition of sugar (high osmolality) and the latter by sterilization of the product by heat in a sealed can. Somewhat later, formulas for normal, preterm, and infants with special needs became available (Hansen *et al.*, 1988; Berry, 1988; Fomon, 1993; Tsang *et al.*, 1993).

II. Composition

A. Comparison between Human Milk and Formulas

The information presented earlier in this book shows the complex nature of human milk. The rationale for recommending human milk as the optimal food for infants is based on the nutrient balance which includes digestibility and absorbability, the growth-promoting substances therein,

HANDBOOK OF MILK COMPOSITION

and the host defense substances and mechanisms it contains (Garza *et al.*, 1993; Fomon, 1993). The physical organization of milk is noteworthy with the constituents distributed in solution, colloidal dispersions, emulsions, and cells. While formulas do not duplicate this complexity, they do fulfill classically recognized roles. The amounts of the major components are similar. Beyond this it would be very difficult or probably impossible to duplicate human milk.

B. Composition of Formulas

Data provided by the manufacturers of formulas are presented in Tables I–VII. It is not appropriate for us to comment on the possible merits of any individual formula. We make the following general statements (Hurrell *et al.*, 1988; Jensen and Jensen, 1992; Garza *et al.*, 1993; Fomon, 1993). The composition of human milk changes during a single feeding and as lactation progresses, while formulas remain uniform. Human milk contains oligosaccharides (see Chapter 8) and heat-sensitive proteins (see Chapters 5A and 5C) not found in formulas. Formulas usually contain more linoleic acid (18:2) and less 22:5n3, 22:6n3, and 20:4n6 than human milk (see Chapter 6A). The structure of triacylglycerols in formulas does not resemble that of human milk (see Chapter 6A). Unless an animal fat is added, formulas do not contain much cholesterol if made with milk proteins; none if they contain soy protein.

There are many other differences, but infants fed formula maintain growth and development within normal limits. There are some cases, e.g., preterm infants and those with allergies or metabolic problems, in which supplemented human milk or formulas are preferable. Also, the convenience of formula feeding is desirable. However, the long-term consequences of feeding formula compared to human milk are not known. A Framingham-type study of infant feeding practices would be desirable. In the Framingham study, the influence of life-style upon cardiovascular incidents has been studied for decades.

Readers seeking information on the relative merits of formulas should make their decisions with the advice of professionals. They should realize that the composition of formulas is changed as our knowledge of infant nutrition advances. The manufacturers of formulas are knowledgeable and will alter their products as needed. For those who want more information on infant nutrition, these books are recommended: Tsang and Nichols (1988); American Academy of Pediatrics (1993); Fomon (1993); and Tsang *et al.* (1993).

Acknowledgments

The generous provision of information by the formula companies and the assistance of their representatives is appreciated. Preparation of this chapter was supported in part by Federal funds made available through provision of the Hatch Act, Scientific Contribution No. 1543, Storrs Agricultural Experiment Station, Storrs, Connecticut.

References

Berry, H. K. (1988). Special and therapeutic formulas for inborn errors of metabolism. *In* "Nutrition During Infancy" (R. C. Tsang and B. L. Nichols, eds.), pp. 340–366. Hanley and Belfus, Philadelphia.

Committee on Nutrition (1993). "Pediatric Nutrition Handbook" (L. A. Barnes, ed.), 3rd Ed. American Academy of Pediatrics, Elk Grove Village, IL.

Fomon, S. J. (1993). Infant formulas. *In* "Nutrition of Normal Infants," pp. 424–442. Mosby, St. Louis, MO.

Garza, C., Butte, N. F., and Goldman, A. S. (1993). Human milk and infant formula. *In* "Textbook of Pediatric Nutrition" (R. M. Suskind and L. Lewinter-Suskind, eds.), pp. 33–49. Raven Press, New York.

Hansen, J. W., Cook, D. A., and Cordano, A. (1988). Human milk substitutes. *In* "Nutrition During Infancy" (R. C. Tsang and B. L. Nichols, eds.), pp. 378–398, Hanley and Belfus, Philadelphia.

Hurrell, R. F., Berrocal, R., Neeser, J.-R., Schweizer, T. F., Hilpert, H., Traitler, H., Colarow, L., and Lindstrand, K. (1989). Micronutrients in infant formulas. *In* "Micronutrients in Milk and Milk-Based Food Products" (E. Renner, ed.), pp. 239–303. Elsevier, New York.

Jensen, R. G., and Jensen, G. L. (1992). Specialty lipids for infant nutrition. I. Milks and formulas. *J. Pediatr. Nutr. Gastroenterol.* **15,** 232–245.

Tsang, R. C., and Nichols, B. L., eds. (1988). "Nutrition During Infancy." Hanley and Belfus, Philadelphia.

Tsang, R. C., Lucas, A., Uauy, R., and Zlotkin, S. (1993). "Nutritional Needs of the Preterm Infant." Williams & Wilkins, Baltimore, MD.

TABLE I
Cow's Milk-Based Infant Formulas: Casein-Predominant Formulas

	Lactogen (Nestle)[a]	Gerber (Gerber)[b]	Similac with iron (Ross)[c]	Lactofree (Mead Johnson)[d]
Energy (kcal)	670	680	676	676
Volume (ml)	1000	1000	1000	1000
Protein (g)	17	15	14.5	14.9
% kcal	10	8.8	9	8.8
Whey:casein	20:80	20:80	18:82	*
Source	Nonfat milk, demineralized whey	Nonfat milk	Nonfat milk	Milk protein isolate
Fat (g)	33.3	35.2	36.5	35.2
% kcal	46	48	48	49
Source	Butterfat, corn oil	Palm olein, soy, coconut, high-oleic sunflower oil	Soy, coconut oil	Palm olein, soy, coconut oil
Fatty acids (g) (%)[e]				
8:0	0.3 (0.8)	0.6 (1.6)	1.0 (2.8)	0.6 (1.6)
10:0	0.8 (2.4)	0.4 (1.2)	0.8 (2.4)	0.4 (1.2)
12:0	0.5 (1.6)	3.3 (9.3)	6.5 (18.8)	3.3 (9.3)
14:0	2.8 (8.6)	1.4 (4.1)	2.6 (7.5)	1.4 (4.1)
16:0	7.5 (23.3)	7.7 (22.0)	3.5 (10.0)	7.7 (22.0)
18:0	3.3 (10.2)	1.5 (4.3)	1.2 (3.5)	1.5 (4.3)
18:1	10.5 (32.6)	13.5 (3.8)	5.8 (16.6)	13.4 (38.0)
18:2	4.0 (12.4)	6.1 (17.2)	11.4 (33.0)	6.1 (17.2)
18:3	0.4 (1.1)	0.6 (1.8)	1.6 (4.6)	0.3 (1.8)
Saturated FA (g)[f]	17.4	15.4	16.0	15.6
Monounsaturated FA (g)	10.5	13.5	6.0	14.1
Polyunsaturated FA (g)	4.4	6.3	13.0	7.1
Carbohydrate (g)	74.0	73.0	72.3	70.0
% kcal	44	43	43	42
Source	Lactose	Lactose	Lactose	Corn syrup solids, citrate
Calcium (mg)	530	510	493	555
Phosphorus (mg)	440	390	380	370
Magnesium (mg)	53	41	41	54
Iron (mg)	8	12.2	12	12.2
Zinc (mg)	5	5.1	5.1	6.8
Manganese (μg)	47	34	34	101
Copper (μg)	400	610	610	510
Iodine (μg)	34	54	61	101
Selenium (μg)	*	*	15	*
Sodium (mg)	210	220	183	200

TABLE I—*continued*

	Lactogen (Nestle)[a]	Gerber (Gerber)[b]	Similac with iron (Ross)[c]	Lactofree (Mead Johnson)[d]
Potassium (mg)	720	730	710	740
Chloride (mg)	490	480	433	450
Vitamin A (IU)	2000	2000	2030	2000
Vitamin D (IU)	400	410	410	410
Vitamin E (IU)	8	13.6	20	13.5
Vitamin K (μg)	55	54	54	54
Vitamin C (mg)	54	61	60	81
Thiamine (μg)	400	680	680	540
Riboflavin (μg)	900	1020	1010	610
Vitamin B6 (μg)	500	410	410	410
Vitamin B12 (μg)	1.5	1.7	1.7	2
Niacin (μg)	5000	7100	7100	6800
Folic acid (μg)	60	102	100	108
Pantothenate (μg)	3000	3100	3040	3400
Biotin (μg)	15	30	30	20
Choline (mg)	50	*	108	81
Inositol (mg)	30	*	31.8	115
RSL (mosmol)	109	*	96.3	136
Osmolarity (mosm/liter)	287	*	270	*

[a]Corporate Communication, Nestle Research Centre, Lausanne, Switzerland (1993).
[b]Corporate Communication, Gerber Products Company, Fremont, Michigan (1993).
[c]Corporate Communication, Ross Products Division, Abbott Laboratories, Columbus, Ohio (1994).
[d]Corporate Communication, Bristol–Myers Squibb Company, Mead Johnson Research Center, Evansville, Indiana (1993).
[e]Grams fatty acid calculated by using a factor of 95% to allow for difference between weight of the triglyceride and the fatty acids released.
[f]Butyric acid; 4:0, 2.4% and 6:0, 0.8% omitted, but used for calculation of grams of saturated fatty acids.
*Data not available.

TABLE II
Cow's Milk-Based Infant Formulas: Whey-Predominant Formulas

	Enfamil-20 (Mead Johnson)[a]	SMA-20 (Wyeth)[b]	NAN (Nestle)[c]	Similac PM 60/40 (Ross)[d,e]	Good Start (Carnation)[f]
Energy (kcal)	680	676	670	676	100
Volume (ml)	1000	1000	1000	1000	148
Protein (g)	15	15	15	15.8	2.4
% kcal	9	9	9	9	9.6
Whey:casein	60:40	60:40	60:40	60:40	
Source	Nonfat milk, whey	Nonfat milk, demineralized whey	Whey, casein	Whey caseinate	Hydrolyzed whey protein
Fat (g)	36.1	36.0	32.3	37.6	4.9
% kcal	50	48	46	50	46.8
Source	Coconut, soy oils	Oleo, coconut, high-oleic safflower, soy oils	Butterfat, corn oil	Soy, coconut oils	Palmolein, soybean, coconut, high-oleic safflower oils
Fatty acids (g) (%)					
8:0	0.6 (1.6)	0.9 (2.6)	0.3 (0.8)	1.0 (2.8)	0.1 (1.6)
10:0	0.4 (1.2)	0.7 (2.0)	0.8 (2.4)	0.9 (2.4)	0.1 (1.6)
12:0	3.5 (9.3)	5.0 (13.8)	0.5 (1.6)	6.7 (18.8)	0.5 (9.5)
14:0	1.5 (4.1)	2.3 (6.5)	2.8 (8.6)	2.7 (7.5)	0.2 (4.1)
16:0	7.9 (22.0)	4.9 (13.5)	7.5 (23.3)	3.6 (10.0)	1.1 (23.3)
18:0	1.6 (4.3)	2.6 (7.1)	3.3 (10.2)	1.3 (3.5)	0.2 (4.2)
18:1	13.7 (38.0)	13.8 (38.3)	10.5 (32.6)	5.9 (16.6)	1.6 (33.0)
18:2	6.2 (17.2)	4.6 (12.8)	4.0 (12.5)	11.8 (33.0)	1.1 (20.1)
18:3	0.6 (1.8)	0.5 (1.3)	0.4 (1.1)	1.6 (4.6)	0.1 (2.1)
Saturated FA (g)	15.6	16.4	17.4	16	2.3
Monounsaturated FA (g)	13.7	14.6	10.5	6	1.6
Polyunsaturated FA (g)	6.8	5.1	4.4	14	1.0
Carbohydrate (g)	70.0	72.0	76.3	69	11.0
% kcal	41.0	43.0	45	41	43.6
Source	Lactose, citrate, nonfat milk, whey	Lactose	Lactose	Lactose	Lactose, maltodextrin
Calcium (mg)	530	420	420	380	64
Phosphorus (mg)	360	280	210	190	36
Magnesium (mg)	53	45	50	41	6.7

TABLE II—*continued*

	Enfamil-20 (Mead Johnson)[a]	SMA-20 (Wyeth)[b]	NAN (Nestle)[c]	Similac PM 60/40 (Ross)[d,e]	Good Start (Carnation)[f]
Iron (mg)	3.4[g]	1.5[h]	8.1	1.5	1.5
Zinc (mg)	5.3	5	5	5.1	0.8
Manganese (μg)	106	100	47	34	7
Copper (μg)	640	470	400	610	80
Iodine (μg)	41	60	34	41	8
Sodium (mg)	184	150	160	160	24
Potassium (mg)	730	560	660	580	98
Chloride (mg)	430	375	440	400	59
Vitamin A (IU)	2100	2000	2000	2030	300
Vitamin D (IU)	430	400	400	410	60
Vitamin E (IU)	13.6	9.5	8.1	20	2
Vitamin K (μg)	54	55	55	54	8.2
Vitamin C (mg)	55	55	54	60	8.0
Thiamine (μg)	530	670	400	680	60.0
Riboflavin (μg)	1020	1000	910	1010	135.0
Vitamin B6 (μg)	430	420	500	410	75.0
Vitamin B12 (μg)	1.6	1.3	1.5	1.7	2.2
Niacin (μg)	8500	5000	5000	7100	750.0
Folic acid (μg)	106	50	61	100	9.0
Pantothenate (μg)	3200	2100	3000	3040	450
Biotin (μg)	15.6	15	15	30	2.2
Choline (mg)	105	100	50	81	12
Inositol (mg)	31	27	30	160	18
RSL (mosmol)	98	91.4	90	96.3	99
Osmolarity (mosm/liter)	270	271	260	250	265

[a]Corporate Communication, Bristol–Myers Squibb Company, Mead Johnson Research Center, Evansville, Indiana (1993).
[b]Corporate Communication, Wyeth–Ayerst Research, Philadelphia, Pennsylvania (1993).
[c]Corporate Communication, Nestle Research Center, Lausanne, Switzerland (1993).
[d]Corporate Communication, Ross Products Division, Abbott Laboratories, Columbus, Ohio (1994).
[e]RTF formulation.
[f]Corporate Communication, Carnation Nutritional Products, Glendale, California (1993).
[g]12.8 with iron.
[h]12.0 with iron.

TABLE III
Soy Protein Formulations

	Alsoy (Nestle)[a]	Prosobee-20 (Mead Johnson)[b]	Isomil (Ross)[c,d]	Nursoy-20 (Wyeth)[e]	Gerber with soy (Gerber)[f]
Energy (kcal)	670	680	676	676	680
Volume (ml)	1000	1000	1000	1000	1000
Protein (g)	19	20	17	18	20
% kcal	11	12	10	11	12
Source	Soy protein isolate, met.	Soy protein isolate, met.	Soy protein isolate, L-met.	Soy protein isolate, L-met.	Soy protein isolate
Fat (g)	31.4	34.2	36.9	36.0	34.2
% kcal	45	48	49	48	48
Source	Palm, olein, soya oil, coconut oil	Coconut, soy oils	Soy, coconut oleo, oils	Coconut, high-oleic safflower, soy oils	Palm, olein, soy, coconut, high-oleic sunflower oils
Fatty acids (g) (%)					
8:0	0.5 (1.5)	0.5 (1.6)	1.0 (2.8)	0.9 (2.6)	0.5 (1.6)
10:0	0.5 (1.5)	0.4 (1.2)	0.8 (2.4)	0.7 (2.0)	0.4 (1.2)
12:0	2.8 (9.3)	3.2 (9.3)	6.5 (18.8)	5.0 (13.8)	3.2 (9.3)
14:0	1.1 (3.7)	1.4 (4.1)	2.6 (7.5)	2.3 (6.5)	1.4 (4.1)
16:0	6.5 (20.5)	7.5 (22.0)	3.5 (10.0)	4.9 (13.5)	7.5 (22.0)
18:0	1.2 (3.9)	1.4 (4.3)	1.2 (3.5)	2.6 (7.1)	1.5 (4.3)
18:1	10.7 (31.8)	13.4 (38.0)	5.8 (16.6)	13.8 (38.3)	13.0 (38.0)
18:2	7.8 (24.9)	6.8 (17.2)	11.4 (33.0)	4.6 (12.8)	5.9 (17.2)
18:3	0.8 (2.5)	0.6 (1.7)	1.6 (4.6)	0.5 (1.3)	0.6 (1.8)
Saturated FA (g)	12.8	14.4	21	16.4	14.5
Monounsaturated FA (g)	10.0	13.4	8	14.6	13.2
Polyunsaturated FA (g)	8.6	6.4	17	5.1	6.5
Carbohydrate (g)	74.0	68.0	69.6	69.0	68.0
% kcal	44	40	41	41	40
Source	Maltodextrin	Corn syrup solids, citrate	Corn syrup sucrose	Sucrose	Corn syrup sucrose
Calcium (mg)	600	640	710	600	640
Phosphorus (mg)	430	500	510	420	500
Magnesium (mg)	67	74	51	67	51
Iron (mg)	8.0	12.8	12	12.0	12.2
Zinc (mg)	6	5.3	5.1	5.0	5.1

TABLE III—*continued*

	Alsoy (Nestle)[a]	Prosobee-20 (Mead Johnson)[b]	Isomil (Ross)[c,d]	Nursoy-20 (Wyeth)[e]	Gerber with soy (Gerber)[f]
Manganese (μg)	270	170	200	200	200
Copper (μg)	1000	640	510	470	510
Iodine (μg)	50	69	100	60	102
Selenium (μg)	*	*	14	*	*
Sodium (mg)	230	240	300	200	320
Potassium (mg)	800	830	730	700	780
Chloride (mg)	490	560	420	375	600
Vitamin A (IU)	2000	2100	2030	2000	2000
Vitamin D (IU)	400	430	410	400	410
Vitamin E (IU)	8	13.6	20	9.5	20
Vitamin K (μg)	55	54	100	100	102
Vitamin C (μg)	108	55	60	55	61
Thiamine (μg)	400	530	410	670	410
Riboflavin (μg)	600	640	610	1000	610
Vitamin B6 (μg)	400	430	410	420	410
Vitamin B12 (μg)	1.5	2.1	3	2	3.1
Niacin (μg)	5000	8500	9130	5000	9200
Folic acid (μg)	60	106	100	50	102
Pantothenate (μg)	3000	3200	5070	3000	5100
Biotin (μg)	15	15.6	30	35	31
Choline (mg)	54	53	54	85	*
Inositol (mg)	16	32	34	27	*
ERSL (mosmol)	121	128	109.6	109	*
Osmolarity (mosmol/liter)	170	178	205	266	*

[a]Corporate Communication, Nestle Research Centre, Lausanne, Switzerland (1993).
[b]Corporate Communication, Bristol–Myers Squibb Company, Mead Johnson Research Center, Evansville, Indiana (1993).
[c]Corporate Communication, Ross Products Division, Abbott Laboratories, Columbus, Ohio (1994).
[d]RTF formulation.
[e]Corporate Communication, Wyeth–Ayerst Research, Philadelphia, Pennsylvania (1993).
[f]Corporate Communication, Gerber Products Company, Fremont, Michigan (1993).
*Data not available.

TABLE IVa
Formulations for Infants with Inborn Errors of Metabolism

	MSUD diet powder (Mead Johnson)[a]	Lofenalac (Mead Johnson)[a]	Phenylfree (Mead Johnson)[a]	Low Phe/Tyr diet powder (Mead Johnson)[a]	Low Met diet power (Mead Johnson)[a]
Energy (kcal)	470	460	400	460	520
Weight (g)	100	100	100	100	100
Protein (g)	9.9	15	20	15	15.5
% kcal	8	13	20	13	12
Source	Amino acids	Hydrolyzed casein, amino acids	Amino acids	Hydrolyzed casein, amino acids	Soy protein isolate
Fat (g)	19.0	17.1	6.5	17.1	26.6
% kcal	38.0	35.0	15.1	35.0	48.0
Source	Corn oil	Corn oil	Coconut, corn oil	Corn oil	Coconut, corn oil
Fatty acids (g) (%)					
8:0	—	—	0.2 (2.3)	—	0.4 (1.6)
10:0	—	—	0.1 (1.7)	—	0.3 (1.2)
12:0	—	—	0.9 (13.3)	—	2.5 (9.3)
14:0	—	—	0.4 (5.2)	—	1.1 (4.1)
16:0	2.2 (11.0)	2.0 (11.2)	0.7 (10.6)	2.0 (11.2)	6.0 (22.0)
18:0	0.4 (1.9)	0.4 (1.9)	0.2 (2.3)	0.4 (1.9)	1.1 (4.3)
18:1	5.3 (26.5)	4.7 (26.0)	1.4 (20.0)	4.8 (26.5)	10.1 (38.0)
18:2	11.9 (59.5)	10.8 (60.0)	2.9 (42.0)	10.7 (59.6)	4.6 (17.2)
18:3	0.1 (0.6)	0.1 (0.65)	—	0.1 (0.6)	0.5 (1.79)
Saturated FA (g)	2.6	2.4	2.2	2.4	11.4
Monounsaturated FA (g)	5.3	4.7	1.4	4.8	10.1
Polyunsaturated FA (g)	12.0	10.9	2.9	10.8	5.1
Carbohydrate (g)	63.3	60.0	66.0	60.0	51.0
% kcal	54.0	52.0	65.0	52.0	40.0
Source	Corn syrup solids, tapioca starch	Corn syrup solids, tapioca starch, citrate	Sucrose corn syrup solids, tapioca starch	Sucrose corn syrup solids, tapioca starch	Corn syrup solids, citrate
Calcium (mg)	490.0	430.0	510.0	430.0	480.0
Phosphorus (mg)	270.0	320.0	510.0	320.0	380.0
Magnesium (mg)	52.0	50.0	152.0	50.0	56.0
Iron (mg)	8.9	8.6	12.2	8.6	9.7
Zinc (mg)	3.7	3.6	7.1	3.6	4.0

TABLE IVa—*continued*

	MSUD diet powder (Mead Johnson)[a]	Lofenalac (Mead Johnson)[a]	Phenylfree (Mead Johnson)[a]	Low Phe/Tyr diet powder (Mead Johnson)[a]	Low Met diet power (Mead Johnson)[a]
Manganese (μg)	148.0	143.0	1020.0	144.0	129.0
Copper (μg)	440.0	430.0	610.0	430.0	480.0
Iodine (μg)	33.0	32.0	46.0	32.0	52.0
Sodium (mg)	185.0	220.0	410.0	220.0	185.0
Potassium (mg)	490.0	470.0	1370.0	470.0	630.0
Chloride (mg)	370.0	320.0	930.0	320.0	430.0
Vitamin A (IU)	1480.0	1430.0	1220.0	1440.0	1610.0
Vitamin D (IU)	300.0	290.0	152.0	290.0	320.0
Vitamin E (IU)	14.8	14.3	10.2	14.4	16.1
Vitamin K (μg)	74.0	72.0	102.0	72.0	80.0
Vitamin C (mg)	38.0	37.0	53.0	37.0	42.0
Thiamine (μg)	370.0	360.0	610.0	360.0	400.0
Riboflavin (μg)	440.0	430.0	1020.0	430.0	480.0
Vitamin B6 (μg)	300.0	290.0	910.0	290.0	320.0
Vitamin B12 (μg)	1.5	1.4	2.5	1.4	1.6
Niacin (μg)	5900.0	5800.0	8100.0	5800.0	6400.0
Folic acid (μg)	74.0	72.0	127.0	72.0	80.0
Pantothenic acid (μg)	2200.0	2200.0	3000.0	2200.0	2400.0
Biotin (μg)	37.0	36.0	30.0	36.0	40.0
Choline (mg)	63.0	61.0	86.0	61.0	40.0
Inositol (mg)	22.0	22.0	30.0	22.0	24.0
RSL (mosmol)	71.0	127.0	210.0	91.0	137.0

[a]Corporate Communication, Bristol–Myers Squibb Company, Mead Johnson Research Center, Evansville, Indiana (1993).

TABLE IVb
Formulations for Infants with Inborn Errors of Metabolism

	Calcilo XD (low Ca/vitamin D-free infant formula w/iron (Ross)[a]	Cyclinex -1 (Ross)[a]	Glutarex-1, I-Valex, Ketonex -1, Phenex -1 (Ross)[a]	Hominex -1, Propimex-1, Tyromex -1 (Ross)[a]
Energy (kcal)	513	515	480	480
Weight (g)	100	100	100	100
Protein equivalent (g)	11.4	7.5	15	15
% kcal	8.9	13.0	16.0	16.0
Source	Whey protein concentrate, sodium caseinate	L-Amino acids	L-Amino acids	L-Amino acids
Fat (g)	27.2	25.7	22.7	22.7
% kcal	50.4	47	45	45
Source	Corn, coconut oils	Palm, partially hydrogenated coconut, soy oils	Palm, partially hydrogenated coconut, soy oils	Palm, partially hydrogenated coconut, soy oils
Fatty acids (g) (%)				
8:0	0.5 (1.7)	0.9 (3.3)	0.7 (3.3)	0.7 (3.3)
10:0	0.4 (1.5)	0.6 (2.3)	0.5 (2.3)	0.5 (2.3)
12:0	3.3 (12.2)	4.5 (17.6)	4.2 (17.6)	4.2 (17.6)
14:0	1.4 (5.2)	1.9 (7.3)	1.7 (7.3)	1.7 (7.3)
16:0	3.1 (11.3)	6.3 (24.5)	5.5 (24.5)	5.5 (24.5)
18:0	1.1 (3.9)	2.2 (8.7)	2.0 (8.7)	2.0 (8.7)
18:1	5.7 (20.7)	5.9 (22.1)	5.0 (22.1)	5.0 (22.1)
18:2	11.6 (40.6)	3.2 (12.2)	2.8 (12.2)	2.8 (12.2)
18:3	0.7 (2.4)	0.3 (1.1)	0.3 (1.1)	0.3 (1.1)
Saturated FA (g)	9.8	16.3	14.6	14.6
Monounsaturated FA (g)	5.7	5.4	5	5
Polyunsaturated FA (g)	11.7	3.5	3.1	3.1
Carbohydrate (g)	52.3	52.0	46.3	46.3
% kcal	40.7	40	39	39
Source	Lactose	Hydrolyzed cornstarch	Hydrolyzed cornstarch	Hydrolyzed cornstarch
Calcium (mg)	< 50	650	575	575
Phosphorus (mg)	128	455	400	400
Magnesium (mg)	31	50	45	45
Iron (mg)	9.2	10	9	9
Zinc (mg)	3.8	9.5	8	8
Manganese (μg)	26	500	450	450
Copper (μg)	460	1,250	1,100	1,100
Iodine (μg)	31	100	95	95

TABLE IVb—*continued*

	Calcilo XD (low Ca/vitamin D-free infant formula w/iron (Ross)[a]	Cyclinex -1 (Ross)[a]	Glutarex-1, I-Valex, Ketonex -1, Phenex -1 (Ross)[a]	Hominex -1, Propimex-1, Tyromex -1 (Ross)[a]
Selenium (μg)	12	*	*	*
Sodium (mg)	108	215	190	190
Potassium (mg)	420	760	675	675
Chloride (mg)	292	400	325	410
Vitamin A (IU)	1,540	1,600	1,400	1,400
Vitamin D (IU)	0	400	350	350
Vitamin E (IU)	12.8	17	15	15
Vitamin K (μg)	41	60	50	50
Vitamin C (mg)	46	60	50	50
Thiamine (μg)	513	2,000	1,900	1,900
Riboflavin (μg)	770	1,000	900	900
Vitamin B6 (μg)	310	850	750	750
Vitamin B12 (μg)	1.28	5.6	4.9	4.9
Niacin (μg)	5,400	12,000	10,000	10,000
Folic acid (μg)	77	250	230	230
Pantothenate (μg)	2,300	7,800	6,900	6,900
Biotin (μg)	23	75	65	65
Choline (mg)	62	100	80	80
Inositol (mg)	123	50	40	40
RSL (mosmol)	69.3	70.5	94.7	97.1
Osmolarity (mosmol/liter)	193	220	273	273

[a]Corporate Communication, Ross Products Division, Abbott Laboratories, Columbus, Ohio (1993).

*Data not available.

TABLE V
Complete Special Formulations

	Pregestimil (Mead Johnson)[a]	Portagen (Mead Johnson)[a]	Nutramigen (Mead Johnson)[a]	Alimentum (Ross)[b]
Energy (kcal)	500	460	680	676
Amount	100 g	100 g	1000 ml	1000 ml
Protein (g)	14.1	16.5	19.0	18.6
% kcal	11.0	14.0	11.0	11
Source	Hydrolyzed casein, amino acids	Sodium caseinate	Hydrolyzed casein, amino acids	Casein hydrolysate cystine, tyrosine, tryptophan
Fat (g)	26.6	21.9	25.7	35.5
% kcal	48	40	35	48
Source	MCT oil, corn, high-oleic, safflower oils	MCT oil, corn, high-oleic, safflower oils	Corn oil	MCT oil, safflower, soy oils
Fatty acids (g) (%)				
8:0	10.7 (40.0)	13.2 (60.0)	—	9.7 (28.7)
10:0	5.9 (14.9)	5.3 (24.0)	—	6.5 (19.4)
12:0	0.1 (0.24)	0.1 (0.42)	0.1 (0.02)	0.1 (0.3)
14:0	0.01 (0.0)	0.04 (0.19)	0.1 (0.02)	0.03 (0.1)
16:0	1.1 (4.1)	0.4 (1.88)	2.8 (11.0)	1.3 (3.8)
18:0	0.3 (1.1)	0.1 (0.47)	0.6 (2.4)	0.5 (1.4)
18:1	4.8 (17.8)	0.9 (4.1)	6.7 (26.0)	2.6 (7.7)
18:2	5.4 (20.0)	1.8 (8.1)	14.9 (58.0)	12.4 (36.8)
18:3	0.3 (1.11)	0.01 (0.06)	0.5 (2.0)	0.4 (1.1)
Saturated FA (g)	16.1	19.1	3.6	18.2
Monounsaturated FA (g)	4.8	0.9	6.7	2.6
Polyunsaturated FA (g)	5.7	1.8	15.4	12.8
Carbohydrate (g)	51.0	54.0	91.0	68.9
% kcal	41.0	46.0	54.0	41
Source	Corn syrup solids, dextrose, citrate	Corn syrup solids, sucrose, citrate	Corn syrup solids, corn starch, citrate	Sucrose, modified tapioca starch
Calcium (mg)	470	440	640	710
Phosphorus (mg)	320	330	430	510
Magnesium (mg)	55	94	74	51
Iron (mg)	9.4	8.9	12.8	12
Zinc (mg)	4.7	4.4	5.3	5.1
manganese (μg)	155	590	210	200
Copper (μg)	470	740	640	510
Iodine (μg)	35	33	48	100

TABLE V—*continued*

	Pregestimil (Mead Johnson)[a]	Portagen (Mead Johnson)[a]	Nutramigen (Mead Johnson)[a]	Alimentum (Ross)[b]
Selenium (μg)	*	*	*	19
Sodium (mg)	195	260	320	300
Potassium (mg)	550	590	740	800
Chloride (mg)	430	410	580	540
Vitamin A (IU)	1900	3700	2100	2030
Vitamin D (IU)	380	370	430	305
Vitamin E (IU)	19	15	21	20
Vitamin K (μg)	94	74	106	100
Vitamin C (mg)	59	38	55	60
Thiamine (μg)	390	740	530	410
Riboflavin (μg)	470	890	640	610
Vitamin B6 (μg)	320	990	430	410
Vitamin B23 (μg)	1.5	3	2.1	3.0
Niacin (μg)	6300	9900	8500	9130
Folic acid (μg)	78	74	106	100
Pantothenate (μg)	2400	5000	3200	5070
Biotin (μg)	39	37	53	30
Choline (mg)	67	63	90	54
Inositol (mg)	24	22	32	34
RSL (mosmol)	122	*	124.4	123
Osmolarity (mosmol/liter)	141	126	290	330

[a]Corporate Communication, Bristol–Myers Squibb Company, Mead Johnson Research Center, Evansville, Indiana (1993).
[b]Corporate Communication, Ross Products Division, Abbott Laboratories, Columbus, Ohio (1993).
*Data not available.

TABLE VIa
Complex Modular Formulations

	RCF (Ross)[a,b]	AL110 (Nestle)[c]	Mono-/di Sacchr. Free Diet Powder (Mead Johnson)[d]	Protein Free Diet Powder (Mead Johnson)[d]	Similac Natural Care Human Milk Fortifier (Ross)[a]
Energy (kcal)	406	670	490	490	812
Amount	1000 ml	1000 ml	100 g	100 g	1000 ml
Protein (g)	20	19	22	0	22
% kcal	20	11	17.6	13	11
Source	Soy protein isolate, L-met.	Casein	Hydrolyzed casein, amino acids		Nonfat milk, whey
Fat (g)	34.2	31.4	31.4	21.9	41.9
% kcal	80	45	55.8	36	47
Source	Soy, coconut oils	Butter, corn oil	MCT oil, corn oil	Corn oil	MCT oil, soy, coconut oils
Fatty acids (g) (%)					
8:0	0.9 (2.5)	0.1 (0.8)	19.2 (60.0)	—	11.9 (28.5)
10:0	0.79 (2.2)	0.8 (2.3)	7.5 (24.0)	—	6.6 (15.8)
12:0	6.4 (17.7)	0.5 (1.6)	0.1 (0.42)	—	3.9 (9.5)
14:0	2.6 (7.1)	2.6 (8.2)	0.1 (0.19)	—	1.7 (4.0)
16:0	3.63 (10.1)	7.2 (22.8)	0.6 (1.88)	2.5 (11.2)	2.7 (6.5)
18:0	1.62 (4.5)	3.1 (9.9)	0.1 (0.47)	0.4 (1.9)	1.3 (3.0)
18:1	5.72 (15.9)	10.20 (32.5)	1.3 (4.10)	5.7 (26.0)	4.1 (9.9)
18.2	12.6 (34.9)	4.4 (14.1)	2.5 (8.10)	13.2 (60.0)	8.3 (19.9)
18:3	1.69 (4.7)	0.4 (1.2)		0.1 (0.6)	1.0 (2.4)
Saturated FA (g)[e]	16	15.5	27.6	2.9	25
Monounsaturated FA (g)	6	11.1	1.3	5.7	5
Polyunsaturated FA (g)	13	4.8	2.5	13.5	8
Carbohydrate (g)	0.04	74.0	33.0	72.0	86.1
% kcal	0	44.0	26.6	51.0	42.0
Source	Selected by physician	Maltodextrin	Tapioca starch, citrate	Corn syrup solids, tapioca starch, citrate	Lactose and polycose glucose polymers
Calcium (mg)	700	600	740	540.0	1,710
Phosphorus (mg)	500	400	500	300.0	850
Magnesium (mg)	50	67	86	63.0	100
Iron (mg)	1.5	8	15	10.8	3
Zinc (mg)	5	5	5	4.5	12.2

TABLE VIa—*continued*

	RCF (Ross)[a,b]	AL110 (Nestle)[c]	Mono-/di Sacchr. Free Diet Powder (Mead Johnson)[d]	Protein Free Diet Powder (Mead Johnson)[d]	Similac Natural Care Human Milk Fortifier (Ross)[a]
Manganese (μg)	200	47	240	180	100
Copper (μg)	500	400	740	540	2,030
Iodine (μg)	100	34	55	40	50
Selenium (μg)	14	*	*	*	15
Sodium (mg)	300	230	340	85	350
Potassium (mg)	730	800	860	340	1,050
Chloride (mg)	420	490	680	135	660
Vitamin A (IU)	2030	2000	3000	1800	5,520
Vitamin D (IU)	410	400	590	360	1,220
Vitamin E (IU)	20	8	30	18	32
Vitamin K (μg)	100	55	148	90	100
Vitamin C (μg)	55	54	92	47	300
Thiamine (μg)	410	400	610	450	2,030
Riboflavin (μg)	610	900	740	540	5,030
Vitamin B6 (μg)	410	500	500	360	2,030
Vitamin B12 (μg)	3.0	1.5	2.4	1.8	4.5
Niacin (μg)	9030	5000	9800	7200	40,600
Folic acid (μg)	100	60	123	90	300
Pantothenate acid (μg)	5020	3000	3700	2700	15,430
Biotin (μg)	50	15	61	45	300
Choline (mg)	52	50	105	77	81
Inositol (mg)	32	30	37	27	45
RSL (mosmol)	122	119	*	*	148.7
Osmolarity (mosmol/liter)	70	153	200	*	250

[a]Corporate Communication, Ross Products Division, Abbott Laboratories, Columbus, Ohio (1994).
[b]Values are for 1:1 dilution with water without added carbohydrate.
[c]Corporate Communication, Nestle Research Centre, Lausanne, Switzerland (1993).
[d]Corporate Commutation, Bristol–Myers Squibb Company, Mead Johnson Research Center, Evansville, Indiana (1993).
[e]Butyric; 4:0, 2.3% and 6:0, 0.8% omitted but used to calculate grams of saturated fatty acids. Palmitoleic acid; 16:1, 3.5% omitted but used to calculate grams of unsaturates.
*Data not available.

TABLE VIb
Complex Modular Formulations

	ProViMin (Ross)[a]	Pro-Phree (Ross)[a]	Enfamil Human Milk Fortifier (Mead Johnson)[b]
Energy (kcal)	313	520	370
Unit	100 g	100 g	100 g
Protein	73	Trace	18
% kcal	93	0	20
Source	Casein	None added	Whey
Fat (g)	1.3	29.5	1.0
% kcal	4.0	54	2.5
Source	Coconut oil	Palm, partially hydrogenated coconut, soy oils	Butterfat
Fatty acids (g) (%)			
8:0	*	1.02 (3.3)	0.01 (1.0)
10:0	*	0.71 (2.3)	0.03 (3.1)
12:0	*	5.46 (17.6)	0.04 (4.0)
14:0	*	2.26 (7.3)	0.12 (12.4)
16:0	*	7.56 (24.5)	0.28 (28.9)
18:0	*	2.76 (8.7)	0.13 (13.4)
18:1	*	6.82 (22.1)	0.01 (1.0)
18:2	*	3.75 (12.2)	0.01 (1.0)
18:3	*	0.34 (1.1)	0.03 (3.1)
Saturated FA (g)	*	20.00	0.64
Monounsaturated FA (g)	*	6.88	0.32
Polyunsaturated FA (g)	*	4.09	2.1
Carbohydrates (g)	2.0	60	70
% kcal	3.0	46	77.0
Source	Citrate	Hydrolyzed cornstarch	Corn syrup solids, lactose
Calcium (mg)	2,400	750	90.0
Phosphorus (mg)	1,700	525	45.0
Magnesium (mg)	200	63	1.0
Iron (mg)	40	11.9	—
Zinc (mg)	17	11	0.71
Manganese (μg)	200	620	4.7
Copper (μg)	2,100	1,450	62
Iodine (μg)	335	100	—
Selenium (μg)	70	*	*
Sodium (mg)	1,200	250	7
Potassium (mg)	3,300	875	15.6
Chloride (mg)	2,300	350	17.7

TABLE VIb—*continued*

	ProViMin (Ross)[a]	Pro-Phree (Ross)[a]	Enfamil Human Milk Fortifier (Mead Johnson)[b]
Vitamin A (IU)	6,740	1,800	950
Vitamin D (IU)	1,000	450	210
Vitamin E (IU)	67	19	4.6
Vitamin K (μg)	90	60	4.4
Vitamin C (mg)	200	70	11.6
Thiamine (μg)	2,240	2,100	151
Riboflavin (μg)	2,020	1,000	210
Vitamin B6 (μg)	1,350	970	114
Vitamin B12 (μg)	5.6	6.5	0.18
Niacin (μg)	24,000	14,000	3000
Folic acid (μg)	320	300	25
Pantothenate acid (μg)	10,100	7,000	730
Biotin (μg)	100	80	2.7
Choline (mg)	335	100	*
Inositol (mg)	105	50	*
RSL (mosmol)	493	43.2	*
Osmolarity (mosmol/liter)	182	150	*

[a]Corporate Communication, Ross Products Division, Abbott Laboratories, Columbus, Ohio (1994).
[b]Corporate Communication, Bristol–Myers Squibb Company, Mead Johnson Research Center, Evansville, Indiana (1993).
*Data not available.

TABLE VII
Premature/Low-Birth-Weight Infant Formulations

	Enfamil Premature Form-20 (Mead-Johnson)[a]	PreNAN (Nestle)[b]	Similac Special Care-20 (Ross)[c]	SMA Preemie-20 (Wyeth)[d]
Energy (kcal)	680	700	676	676
Volume (ml)	1000	1000	1000	1000
Protein (g)	20	20.4	18.3	20.0
% kcal	12	11.7	11	11.9
Source	Nonfat milk, whey	Nonfat milk, demineralized whey	Nonfat milk, whey	Nonfat milk, demineralized whey
Fat (g)	33.3	32.4	34.9	35.0
% kcal	44.0	42.9	47	46.7
Source	MCT oil, soy, coconut oils	MCT oil, butter-fat, corn, soy oils	MCT oil, soy, coconut oils	MCT oil, oleo, oleic, coconut, soy oils
Fatty acids (g) (%)				
8:0	10.1 (30.0)	7.5 (23.8)	9.9 (28.5)	*
10:0	4.0 (12.0)	5.1 (16.2)	5.5 (15.8)	*
12:0	3.2 (9,4)	0.5 (1.5)	3.3 (0.5)	*
14:0	1.2 (3.6)	1.3 (4.1)	1.4 (4.0)	*
16:0	2.0 (5.9)	4.0 (12.5)	2.3 (6.5)	*
18:0	0.8 (2.4)	1.7 (5.4)	1.1 (3.0)	*
18:1	3.7 (11.2)	6.2 (19.4)	3.4 (9.9	*
18:2	7.3 (22.0)	4.5 (14.1)	6.9 (19.9)	*
18:3	1.0 (3.1)	0.4 (1.3)	0.8 (2.4)	*
Saturated FA (g)	21.3	20.1	21	*
Monounsaturated FA (g)	3.7	6.2	4	*
Polyunsaturated FA (g)	8.3	4.9	7	*
Carbohydrate (g)	7.5	79.7	71.7	70.0
% kcal	4.4	45.4	42	41.5
Source	Corn syrup solids, lactose	Lactose, maltodextrin	Lactose, polycose, glucose polymers	Lactose, glucose polymers
Calcium (mg)	1,120	700	1,220	750
Phosphorus (mg)	560	450	610	380
Magnesium (mg)	46	80	81	70
Iron (mg)	1.7	10	2.5	3
Zinc (mg)	10.2	5.2	10.1	8
Manganese (μg)	43.0	49	80	134
Copper (μg)	850	630	1,690	700
Iodine (μg)	170	70	40	83
Selenium (μg)	*	*	13	*
Sodium (mg)	270	260	290	320

TABLE VII—*continued*

	Enfamil Premature Form-20 (Mead-Johnson)[a]	PreNAN (Nestle)[b]	Similac Special Care-20 (Ross)[c]	SMA Preemie-20 (Wyeth)[d]
Potassium (mg)	700	750	870	750
Chloride (mg)	580	400	550	530
Vitamin A (IU)	8,500	2,100	4,600	3,200
Vitamin D (IU)	1,840	700	1,010	510
Vitamin E (IU)	43	14	27	15
Vitamin K (μg)	54	84	81	70
Vitamin C (mg)	136	110	250	70
Thiamine (μg)	1,360	420	1,690	800
Riboflavin (μg)	2,000	940	4,190	1,300
Vitamin B6 (μg)	1,020	520	1,690	500
Vitamin B12 (μg)	1.7	1.5	3.7	2
Niacin (μg)	27,000	7,000	33,800	6,300
Folic acid (μg)	240	420	250	100
Pantothenate acid (μg)	8,200	3,100	12,840	3,600
Biotin (μg)	27	15	250	18
Choline (mg)	82	52	68	127
Inositol (mg)	116	31	37	30
RSL (mosmol)	*	123	124	128
Osmolarity (mosmol/liter)	176	238	210	242

[a]Corporate Communication, Bristol–Myers Squibb Company, Mead Johnson Research Center, Evansville, Indiana (1993).
[b]Corporate Communications, Nestle Research Centre, Lausanne, Switzerland (1993).
[c]Corporate Communications, Ross Products Division, Abbott Laboratories, Columbus, Ohio (1994).
[d]Corporate Communications, Wyeth–Ayerst Research, Philadelphia, Pennsylvania (1993).
*Data not available.

CHAPTER 11

Contaminants in Milk

A. Drugs and Contaminants in Human Milk

RUTH A. LAWRENCE
LINDA R. FRIEDMAN

I. Contaminants

The composition of human milk may include more than the nutrients for the infant that are well studied and well documented in other chapters in this text. Less well understood and less well studied is the composition of human milk as it reflects other ingestants of the mother that are absorbed into her bloodstream or stored in her bones or fat and reach the target organ, the breast, during active lactation. This pool of substances which could potentially reach the recipient nursling actually includes all substances that are ingested, inhaled, or absorbed through the skin or mucous membranes of the lactating woman. The materials and chemicals include herbs, spices, medications, recreational drugs, environmental contaminants including insecticides, heavy metals, and other toxins. Attention was first drawn to the environmental toxins when the food chain was contaminated through accidents and spills involving produce, livestock, and bovine milk for commercial use (Brillant *et al.*, 1978; Rogan *et al.*, 1980; Kimbrough, 1987; Ogaki *et al.*, 1987). The immediate clinical problem of an individual lactating woman who needs to take a medication while continuing to breast-feed has produced a need for study and understanding of how all compounds pass into the milk. Drugs can influence the composition of milk, stimulate or inhibit milk production or release, or pass into the milk and be absorbed by the nursing infant (Peterson and Bowes, 1983; Lawrence, 1989). This chapter reviews both the pharmacologic and the environmental substances that may reach the milk and discusses the mechanisms by which this occurs.

HANDBOOK OF MILK COMPOSITION

II. Chemical Constituents of Human Milk

The production of milk requires four major secretory processes: (1) exocytosis of major components, (2) secretion via the milk fat globule, (3) secretion in response to concentration gradients, and (4) pinocytosis and exocytosis of immunoglobulins. These events, which are synchronized in the alveolar cell of the lactating mammary gland, plus the paracellular pathway, are involved in the synthesis of this crucial secretion (Neville *et al.,* 1983).

Exocytosis accounts for the secretion into secretory vesicles of major milk components including proteins, lactose, calcium, phosphate, and citrate. The genetic information for milk proteins is specifically transcribed in the cell's nucleus and the proteins are synthesized by the translation of this mRNA. The sequestered proteins are transferred to the Golgi system and are ultimately moved to the apical portion of the cell and released into the lumen. Milk fat is secreted via the milk fat globule whose membrane is in constant flux. Proteins are partially excluded from this membrane suggesting that there is a selection and segregation of protein components at the site where the apical membrane engulfs the fat globule.

The third process is the secretion of ions and water across the apical membrane in response to gradients. There is an osmotic gradient set up by lactose and an electrochemical gradient. The pinocytosis and exocytosis of immunoglobulins represents the fourth pathway of milk manufacture. The paracellular pathway provides access for plasma components and leukocytes. It is more accessible early in lactation.

It is believed that the secretory activity of the alveolar cells is regulated such that the substances of nutritional significance replace in part the substances of protective or immunologic significance in the first week. The junctional complexes are more permeable to small ions and other plasma constituents at the onset of lactation than later.

The substances which are usually present in the composition of human milk have been investigated with reference to the mode in which they reach the milk and how their transport is regulated. The mammary gland synthesizes milk proteins, immunoglobulins, and milk sugars. The biochemical pathways involved in lipid synthesis are well described. The secretion of monovalent ions and water into milk, which is isoosmotic with plasma, is directly related to the transfer of solute. While lactose makes up most of the solute, monovalent ions and other osmotically active substances make up the rest.

The transfer of minor components and trace elements into milk needs considerable study. It is known, for example, that iron is tightly bound to lactoferrin and relatively little free iron is present in milk (Fransson and Lonnerdal, 1980). Zinc is abundant in human milk probably due to an active transport mechanism although the mechanism has not been studied in detail (Picciano and Guthrie, 1976). Iodine, on the other hand, is known to be actively transported resulting in high levels in the milk. In some cases there are maternal excesses which would suggest a pump that moves against a gradient (Gushurst *et al.*, 1984). Thus, the probable mechanism for transport of trace minerals and other elements, such as mercury and lead, except for zinc and iodine, is likely to be by binding to specific carrier proteins (Neville *et al.*, 1983).

It is obvious that much is known about the usual composition of milk. However, there is no such study of compounds not normally in human milk such as drugs and other chemicals. It is not possible to estimate the amount of a drug or nonnutrient that might be found in the milk purely by calculating from knowledge of milk production as it is now understood.

III. Pharmacokinetic Approach to Drug Transport Into Milk

When a woman is lactating, the breast may become an organ of excretion and possibly of receptor interaction. In order to study the transport of a drug or other nonnutrient substance into human milk, one has to study the amount of the drug that is free in the plasma. This is the net result of absorption of the substance into the maternal body, the state (bound to albumin or unbound) in the blood, the distribution within the tissues of the body, metabolic processes in which it could be involved, storage in bone, and pathways of excretion (see Figure 1) (Wilson *et al.*, 1980; Rivera-Calimlim, 1977, 1987). The concentration of a drug in the milk depends on the maternal plasma concentration and whether the drug is protein bound or free. The primary pharmacokinetics, applicable to any substance used therapeutically, are influenced by the size, route of absorption, and timing of the dose.

When a woman is exposed to a compound it may be through absorption from the gastrointestinal tract, the lungs (inhalation), the skin, or any portal of the body. The rate of entering the bloodstream differs if the compound is given intravenously, intramuscularly, or orally. After absorption, all substances are distributed through the bloodstream to various tissues or organs of the body, and in a lactating woman the breast becomes one of the compartments in which it is distributed (Figure 2). Lipid-soluble compounds deposit in fat and hydrophobic sites, highly protein-bound substances remain in the plasma, highly polar or charged substances are distributed in the "body water." The dilution of a substance in the body is dependent upon its chemical characteristics. The concentration of a substance is a function of the quantity of that substance and the volume of the space for distribution or the volume of distribution. It is complex, however, because the breast is one of many compartments of the body to which substances are distributed (see Figure 3) (Ellenhorn and Barceloux, 1988).

The volume of distribution (V_D) is not a physiologic state nor is it an actual space in the body. It is, however, a handy concept to utilize when

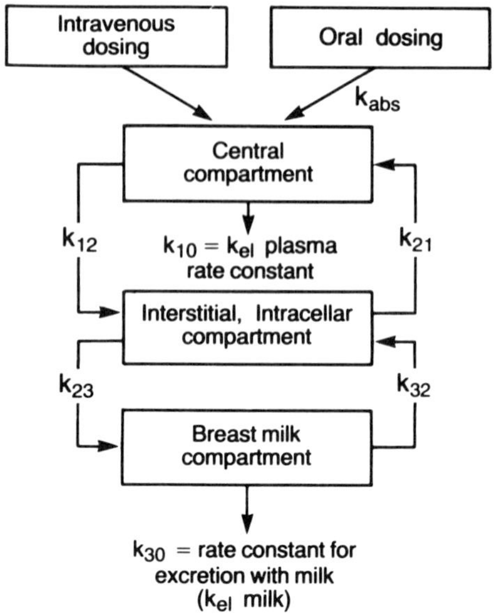

Figure 1 Three compartment model (Lawrence, 1994). (Modified from Wilson, 1983).

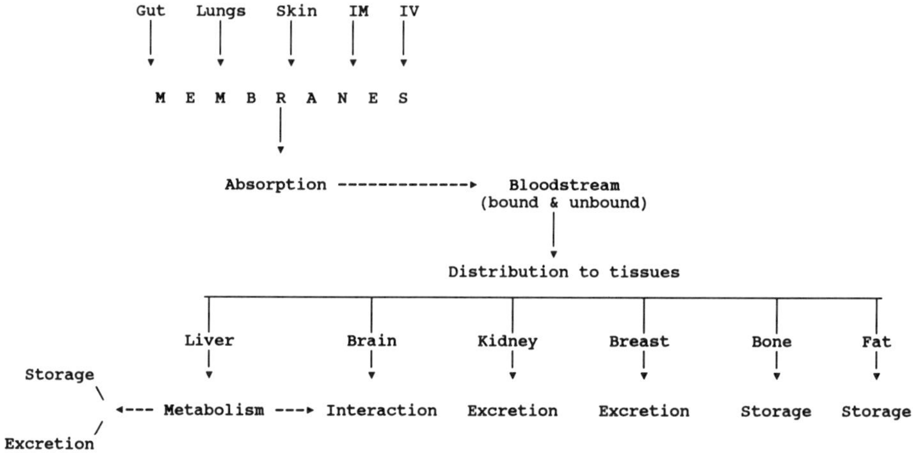

Figure 2 Diagram of drug distribution (modified from Rivera-Calimim, 1987).

estimating the amount of a drug in milk when there are no milk levels available with corresponding plasma levels. V_D is a theoretical term to describe "the space that would be occupied by the total body drug burden if it were distributed in the same concentration as present in plasma"

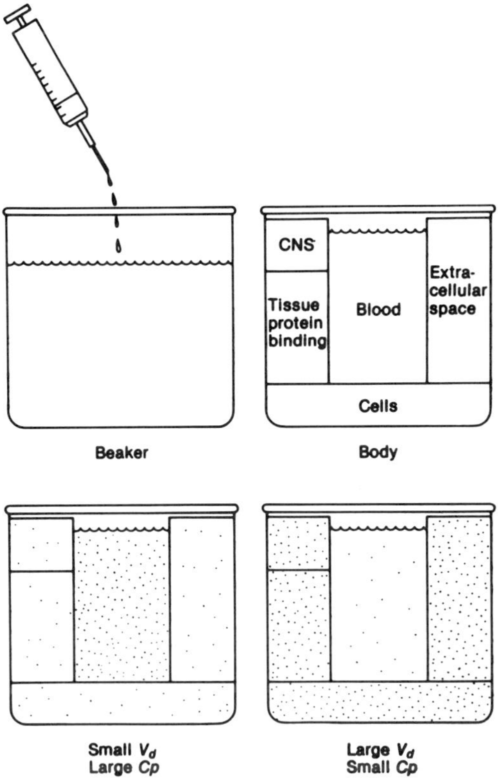

Figure 3 Beaker and body diagram (Ellenhorn and Barceloux, 1988).

(Ellenhorn and Barceloux, 1988). When a drug has a large volume of distribution (exceeds 1 liter per kilogram), the plasma levels are usually low and measured in nanograms per milliliter. Conversely, when drugs have a small volume of distribution, the concentration of the drug in the plasma (C_P) is large and measured in micrograms per milliliter. These drugs may be highly protein bound.

The distribution of the substance can be related to the concentration in the plasma. If the weight of the individual is known and the standard volume of distribution for this drug is known (see Table I), one can calculate the probable concentration of the drug of a given dose in the breast milk by substituting the formula (Peterson and Bowes, 1983):

$$C_B = \text{concentration in breast milk}$$

$$C_B = \frac{\text{Dose taken by mother}}{\text{Volume of distribution drug} \times \text{weight of mother}}$$

$$C_B = \frac{\text{mg}}{V_D \times \text{kg}}$$

TABLE I
Relevant Drug Information

Ratings AAAP[a], Scand[b]	M/P ratio	% Oral availability	Peak plasma time (hr)	% Protein bound	Volume of distribution
Alprazolam —	probably excreted into milk (Gilman *et al.*, 1990; PDR, 1992)				
		92 ± 17	1–2	71 ± 3	0.72 ± 0.12
Amiodarone —	plasma level in infant, 25% of mother's; concern that iodine may affect infant's thyroid (Briggs *et al.*, 1990; Gilman *et al.*, 1990; Somani *et al.*, 1989; Freedman and Somberg, 1991; Plomp *et al.*, 1992)				
	2.3–13	3–100	1.5–10	96–99	66 ± 44
Amitriptyline —	may cause galactorrhea (Dickey and Stone, 1975; Gilman *et al.*, 1990; Pittard and O'Neal, 1986; Reisner *et al.*, 1983; PDR, 1992)				
4, III	0.5–1.69	48 ± 11	4	94.8 ± 0.8	15 ± 3
Amoxapine —	less than 0.07% of maternal dose in milk (Gelenberg, 1979; PDR, 1992)				
4	0.21	99	1.5	90	
Amrinone	(Gilman *et al.*, 1990; Shammas and Dickstein, 1988)				
	?[c]	93 ± 12	0.5–2	35–49	1.3 ± 0.3
Atenolol —	one report of beta blocker effect in infant. Food decreases absorption (Gilman *et al.*, 1990; Mehvar *et al.*, 1990; AHFS, 1991; PDR, 1992)				
6, II	1.3–6.8	56 ± 30	1–4	5–16	0.95 ± 0.15
Atropine —	inhibits lactation (Gilman *et al.*, 1990; Kurz *et al.*, 1977; Rivera-Calimlim, 1987)				
6, III	*[d]	50		38.7 ± 1.1	3–3.3
				14–22	
Betaxolol	(Buckley *et al.*, 1990; Morselli *et al.*, 1989)				
	2.5–3.0	80–90	2–4	45–60	4.9–9.8
Bretylium	(Gilman *et al.*, 1990; Sietsema, 1989; Anderson, 1991)				
	*	12–37		0–8	5.9 ± 0.8
Captopril —	food in GI tract decreases absorption 30–40% (Gilman *et al.*, 1990; Reisner *et al.*, 1983; Roberts, 1984; Sietsema, 1989)				
6, II	0.006–0.6	63.5 ± 1.5	1 ± 0.5	30 ± 6	0.81 ± 0.18
Chlordiazepoxide —	in first weeks of life, may contribute to jaundice (Gilman *et al.*, 1990; Rivera-Calimlim, 1987)				
II	*	100		96 ± 5.8	0.3 ± 0.03
Chlorpromazine —	galactorrhea; drowsiness and lethargy in infant (Dahl and Strandjord, 1977; Gilman *et al.*, 1990; Wiles *et al.*, 1978)				
4, II	0.3–1.0	32 ± 19	2	95–98	9.8–35.3

TABLE I—*continued*

Ratings AAAP[a], Scand[b]	M/P ratio	% Oral availability	Peak plasma time (hr)	% Protein bound	Volume of distribution
Chlorthalidone —	may suppress lactation in first few months (Gilman *et al.*, 1990; White, 1984)				
6, III	0.03–0.05	64 ± 10		75 ± 1	3.9 ± 0.8
Clonidine	(Gilman *et al.*, 1990; Morselli *et al.*, 1989; PDR, 1992; Sietsema, 1989)				
	1.5–4.0	87 ± 12	3–5	20	2.1 ± 0.4
Desipramine —	metabolism varies widely; 5–10% of caucasians are poor metabolizers (Briggs *et al.*, 1990; Gilman *et al.*, 1990; Sietsema, 1989; Stancer and Reed, 1986; AHFS, 1991)				
4, II	0.4–1.2	33–68	4–6	82 ± 2	20 ± 3
Dexamethasone —	could suppress growth and interface with endogenous corticosteroid production (Brophy *et al.*, 1993; Gilman *et al.*, 1990)				
III	*	53 ± 40		68 ± 3	0.29–2.04
Diazepam —	lipid soluble; tends to be in higher concentrations in evening milk when the plasma levels are lower; neonatal lethargy, jaundice and weight loss possible (Briggs *et al.*, 1990; Gilman *et al.*, 1990; Speight, 1987; Syversen and Ratkje, 1985)				
4, III	0.1–2.7	100 ± 14		90–98	0.7–4.7
Diazoxide —	% protein bound decreases at higher concentrations (Gilman *et al.*, 1990)				
IV	?	86–96		94 ± 14	0.21 ± 0.02
Dicloxacillin	(Gilman *et al.*, 1990; Jusko *et al.*, 1975; Sietsema, 1989)				
IV	?	50–85 49 ± 11	1 ± 0.8	89–96	0.095 ± 0.026
Digitoxin -	In a child, there is an increase in the volume of distribution (Gilman *et al.*, 1990; Morselli, 1977; Sietsema, 1989)				
	?	84–93		90–97	0.54 ± 0.14
Digoxin —	0.06–0.3% of newborn dose in milk (Gilman *et al.*, 1990; Morselli, 1977; Morselli *et al.*, 1980; Rivera-Calimlim, 1987; Johnson *et al.*, 1991)				
6, II	0.45–1.0	40–100	1.5–3	20–40	5.17–7.35
Diltiazem —	2.4-fold decrease in clearance with multiple dosing (Hermann and Morselli, 1985; PDR, 1992; Sietsema, 1989)				
6	0.98–1.0	24–90	2–4	70–86	3–8
Disopyramide —	in some cases, drug is detected in infant's serum. % protein bound is dose dependent (Gilman *et al.*, 1990; MacKintosh and Buchanan, 1985; Mitani *et al.*, 1987; AHFS, 1991; Ribeiro *et al.*, 1987)				
6	0.4–1.07	83 ± 11 60–83	1–2	35–95	0.59 ± 0.15

TABLE I—*continued*

Ratings: AAAP[a], Scand[b]	M/P ratio	% Oral availability	Peak plasma time (hr)	% Protein bound	Volume of distribution
Doxepin —	may cause galactorrhea; report of a very sedated infant (Gilman *et al.*, 1990; Kemp *et al.*, 1985; Matheson *et al.*, 1985; Sietsema, 1989; Dickey and Stone, 1975)				
4	0.3–2.39	13–45			20 ± 8
Doxorubicin —	detectable in milk up to 72 hr (Egan *et al.*, 1985; Gilman *et al.*, 1990; Speth *et al.*, 1988)				
1	4.4	5	0.5	50–85	9–66
Ethambutol —	1.0–5.7% of therapeutic dose appears in milk (Gilman *et al.*, 1990; AHFS, 1991; Snider and Powell, 1984; Speight, 1987)				
6, II	1	77 ± 8	2–4	< 5–22	1.6 ± 0.2 2.5
Famotidine	(Campoli–Richards and Clissold, 1986; Gilman *et al.*, 1990; Echizen and Ishizaki, 1991)				
	0.41–1.78	45 ± 14	1–4	17 ± 7	0.94–1.42
Fenoprofen —	food delays and reduces absorption (Briggs *et al.*, 1990; Sietsema, 1989; Verbeeck *et al.*, 1983)				
	0.017	82.5 ± 2.5	1–2	99	0.08–0.1
Flecainide	(McQuinn *et al.*, 1990; PDR, 1992; Shammas and Dickstein, 1988; Tjandra-Maga *et al.*, 1989; AHFS, 1991; Speight, 1987)				
	0.8–4.6	60–95	0.5–6	40–60	5–13.4
Furosemide —	may decrease lactation; some reports say found in breast milk, some not (Gilman *et al.*, 1990; PDR, 1992; Roberts, 1984; Friedel and Buckley, 1991)				
I		11–90	1–2	91–99	0.07–0.35
Glutethimide	(Curry *et al.*, 1971; PDR, 1992; Atkinson *et al.*, 1988)				
	1	Erratic	1–6	47.3–59.3	
Haloperidol —	reported to cause galactorrhea (Gilman *et al.*, 1990; Whalley *et al.*, 1981)				
4, II	0.6–0.7	60 ± 18		92 ± 2	18 ± 7
Hydroxyzine	(Paton and Webster, 1985)				
IV	?		2 ± 0.9		19.5 ± 9.7
Imipramine —	may cause galactorrhea (Briggs *et al.*, 1990; Kirksey and Groziak, 1984; Morselli, 1976; Sietsema, 1989; AHFS, 1991)				
4, II	0.8–1.0	47 ± 21	1–2	85–95	20–40

TABLE I—*continued*

Ratings AAAP[a], Scand[b]	M/P ratio	% Oral availability	Peak plasma time (hr)	% Protein bound	Volume of distribution
Labetalol —	food increases bioavailability; extensive first pass hepatic metabolism; recovered in some infants, not all (Gilman *et al.*, 1990; Morselli *et al.*, 1989; Neuvonen and Kivisto, 1989; Donnelly and Macphee, 1991)				
6, II	0.4–2.6	10–80	0.33–2	45–55	2–16
Lidocaine	(Morselli, 1977; PDR; Sietsema, 1989; Zeisler *et al.*, 1986)				
6, II	0.4–1.1	34 ± 12	0.5	55–80	0.2–1.8
Lithium	(Gilman *et al.*, 1990; Schou and Amdisen, 1973; Sietsema, 1989)				
1, III	0.24–0.66	95 ± 5	2–4	0	0.79 ± 0.34
Lorazepam —	increased half-life in neonate (Gilman *et al.*, 1990; PDR, 1992; Sietsema, 1989; Summerfield, 1985)				
4, II	0.148–0.257	64–109	2	91 ± 2	1.3 ± 0.2
Lorcainide —	oral availability dose dependent; saturable first pass metabolism (Gilman *et al.*, 1990; Sietsema, 1989; Somani *et al.*, 1987)				
	?	35–65	1–2	85 ± 5	11.79 ± 7.15
Maprotiline	(Pinder *et al.*, 1977; Sietsema, 1989; Speight, 1987)				
III	1–1.5	36–67	9–16	88	22.6–52
Methicillin —	65% protein bound in neonate; half-life decreases with age (Gilman *et al.*, 1990; Roberts, 1984)				
	?	Negligible		39 ± 2	0.43 ± 0.1
Methyldopa —	galactorrhea; detected in plasma and urine of some infants (Gilman *et al.*, 1990; Myhre *et al.*, 1982; White *et al.*, 1985; Kwan *et al.*, 1976)				
6, III	0.19–0.34	8–62	1–4	1–16	0.28–1.40
Mexiletine —	lipid soluble (Gillis and Kates, 1984, Gilman *et al.*, 1990; Lewis *et al.*, 1981; Mitani *et al.*, 1987; PDR, 1992; Prescott *et al.*, 1977; Woosley, 1987)				
6, II	0.78–2	87 ± 13	2–4	50–75	4.9–9.5
Minoxidil —	rapidly excreted into milk (Gilman *et al.*, 1990; PDR, 1992; Valdivieso *et al.*, 1985)				
6	0.67–1.13	90	Within 1	0	2.7 ± 0.7
Moxalactam —	increased half-life in neonate (Gilman *et al.*, 1990; Miller *et al.*, 1984; AHFS, 1991)				
6, II	*	Negligible	0.5–2	45–67	0.25 ± 0.08
Nadolol	(Gilman *et al.*, 1990; Mitani *et al.*, 1987; AHFS, 1991)				
6, II	2–8	34 ± 5	2–4	16–30	1.9 ± 0.2

TABLE I—*continued*

Ratings AAAP[a], Scand[b]	M/P ratio	% Oral availability	Peak plasma time (hr)	% Protein bound	Volume of distribution
Oxazepam —	sparingly lipid soluble; less than 1/1000 of maternal dose in milk (Dusci *et al.*, 1990; Gilman *et al.*, 1990; Wretlind, 1987)				
II	0.1–0.33	97 ± 11		97.8 ± 2.3	0.6 ± 0.2
Prazepam —	drug is concentrated in milk (Committee on Drugs, 1989; Gilman *et al.*, 1990)				
4	*				14.4 ± 5.1
Prazosin —	small amounts found in milk; food does not affect bioavailability (Gilman *et al.*, 1990; AHFS, 1991; Marx, 1985; Vincent *et al.*, 1985)				
	*	43–82	1–3	95 ± 1	0.6 ± 0.13
Procainamide —	half-life three or four times longer in infant than adult. Mom may be a rapid or slow acetylator (Gilman *et al.*, 1990; Mitani *et al.*, 1987; Roberts, 1984; Pittard and Glazier, 1983)				
6	1.0–7.3	83 ± 16	0.66–2	16 ± 5	1.9 ± 0.3
Promethazine —	increase serum levels of prolactin (Paton and Webster, 1985; Sietsema, 1989)				
IV	?	25 ± 10	2.8 ± 1.4	76–80	13.4 ± 3.6
Propafenone —	oral bioavailability is dose dependent; good and poor metabolizers; bioavailability enhanced greatly by food (Gillis and Kates, 1984; Neuvonen and Kivisto, 1989; PDR, 1992; Libardoni *et al.*, 1991; Hii *et al.*, 1991)				
	*	5–50	2–3.5	96 ± 1	3 ± 1.4
Ranitidine —	decreases gastric acidity; large variation in amount in milk (Campoli-Richards and Clissold, 1986; Gilman *et al.*, 1990; AHFS, 1991)				
III	0.25–7	39–87	0.5–3	10–19	1.3 ± 0.4
Streptomycin —	very little absorbed through GI tract (Gilman *et al.*, 1990; Snider and Powell, 1984; Briggs *et al.*, 1990; Holdiness, 1984)				
6	0.12–1.0		1	48 ± 14	
Temazepam	(Gilman *et al.*, 1990; PDR, 1992; Motwani and Lipworth, 1991; Lebedevs *et al.*, 1992)				
	0.09–0.63	98.4 ± 15.6	1–3	96–97.6	1.06 ± 0.31
Tetracycline —	negligibly absorbed by infant; probably chelated by calcium in milk (Briggs *et al.*, 1990; Gilman *et al.*, 1990; Committee on Drugs, 1989; Sietsema, 1989; Beeley, 1986; Smilack *et al.*, 1991; Turner *et al.*, 1980)				
6, II	0.2–1.5	60–80	2	65–75	1.5 ± 0.08

TABLE I—*continued*

Ratings AAAP[a], Scand[b]	M/P ratio	% Oral availability	Peak plasma time (hr)	% Protein bound	Volume of distribution
Ticarcillin	(Briggs *et al.*, 1990; Gilman *et al.*, 1990; AHFS, 1991)				
6	Trace	0	0.5–1.25 after IM	45–65	0.21 ± 0.03
					0.34–0.42
Timolol —	when used for eye drops, 1.5 hr after dose, M/P is 6.02. Poor and extensive metabolizers exist (Fidler *et al.*, 1983; Gilman *et al.*, 1990; Johnsson and Regardh, 1976; Lustgarten and Podos, 1983; Sietsema, 1989)				
6, III	0.25–1.73	30–95	1–3	60 ± 3	2.4 ± 1.2
Tocainide —	food slows absorption but does not effect bioavailability (Gillis and Kates, 1984; Gilman *et al.*, 1990; Graffner *et al.*, 1980; Shammas and Dickstein, 1988; Sietsema, 1989; Wilson, 1988; AHFS, 1991)				
	2.14–3.04	89 ± 5	0.5–2	2–22	1.6–3.8
Trazodone —	food slows absorption (Gilman *et al.*, 1990; Verbeeck *et al.*, 1983; AHFS, 1991)				
4	0.09–0.21	81 ± 29	1–2	89–98	1.0 ± 0.3
Verapamil —	oral absorption > 90%; extensive first pass metabolism (AHFS, 1991; Briggs *et al.*, 1990; McAllister and Kirsten, 1982; Mitani *et al.*, 1987; Sietsema, 1989)				
6	0.23–0.94	10–27	0.5–2	90 ± 2	4.7 ± 2.5
Warfarin —	20–50% protein bound in milk (Gilman *et al.*, 1990; Orme *et al.*, 1977; PDR, 1992)				
6, II	< 0.2	93 ± 8	1–9	97–99	0.14 ± 0.06

[a]Rating of the American Academy of Pediatrics (Committee on Drugs 1989): 1, drugs that are contraindicated during breast-feeding; 2, drugs of abuse that are contraindicated during breast-feeding; 3, radiopharmaceuticals that require temporary cessation of breast-feeding; 4, drugs whose effect on nursing infant is unknown but may be of concern; 5, drugs that have caused significant effects on some nursing infants and should be given to nursing mothers with caution; 6, maternal medication usually compatible with breast-feeding; and 7, food and environmental agents and their effect on breast-feeding.
[b]Rating of the Swedish catalog of registered specialties (FASS) (Berglund *et al.*, 1984): I, does not enter breast milk; II, enters breast milk but is not likely to affect the infant when therapeutic doses are used. III, enters breast milk in such quantities that there is a risk of affecting the infant when therapeutic doses are used; and IV, not known whether it enters breast milk or not.
[c]Not known if it enters breast milk.
[d]Drug enters breast milk but M/P ratio is not known.
[e]Secreted into rat milk; not known if it gets into human milk.

If one has accessed the volume of distribution constant of a substance, an estimate can be made of the relative concentration in the milk even though no studies have been done to measure milk and plasma levels in a series of mothers who are taking the drug. For example:

Minoxidil: 0.5 mg to a 60-kg woman; V_D (minoxidil) = 12 liter/kg,

$$C_B = \frac{0.5 \text{ mg}}{(12 \text{ liter/kg})\,(60 \text{ kg})} = \frac{0.5 \text{ mg}}{720 \text{ liters}} = 0.0007 \text{ mg/liter} = 0.7 \text{ ng/ml}.$$

The infant would receive a negligible amount via the breast milk.

Phenytoin: 100 mg to a 60-kg woman; V_D (phenytoin) = 0.75 liter/kg,

$$C_B = \frac{100 \text{ mg}}{(0.75 \text{ liter/kg})\,(60 \text{ kg})} = \frac{100 \text{ mg}}{45 \text{ liters}} = 2.22 \text{ mg/liter} = 2.2\ \mu\text{g/ml}.$$

With feeding of 100 cc, the infant would receive 220 μg or 0.22 mg/feed.

IV. Properties of Substances That Influence Distribution in Milk

Materials which are highly protein bound usually do not pass into the milk or into other tissues or compartments. Free unbound drug has pharmacologic effects and the free drug concentration reflects most accurately the amount of active drug (Ellenhorn and Barceloux, 1988). Lipid solubility will influence distribution of a compound and its deposition into fat stores in the infant. The percentage of body fat among infants is significantly different (Lawrence, 1989). In term infants, it is 12%, premature infants, 3%, small-for-gestational-age infants, $< 12\%$, or large-for-gestational-age infants, 12–15%. Older children have 30% at 1 year which decreases to 18% by adulthood. As there are less body fat stores, more lipid-soluble substances will be available for distribution to the neonatal brain or other organs. Thus, the lipid solubility of a drug is important in the recipient infant. Extracellular fluid volume also varies with age. It is 50% at birth, 45% at 4 to 6 months, and 20 to 25% from 1 year to adult life. Water-soluble substances are distributed in the larger relative volume and result in a lower level in neonatal extracellular fluid than in plasma (Figure 4).

Other characteristics of the substance that impact the amount available in the breast milk are the compound's pH and pK_a which influence the affinity to cross the alveolar membrane into the slightly acidic milk. Except in the extremes, however, the pH has little practical applicability in determining the concentration in the milk according to Peterson and Bowes (1983).

Molecular size is of importance as items of large molecular size, such as insulin and heparin, do not cross the alveolar membrane. This barrier has more significance as the tight junctions close while lactation progresses over the first few days. Once a compound enters the milk, and the milk is

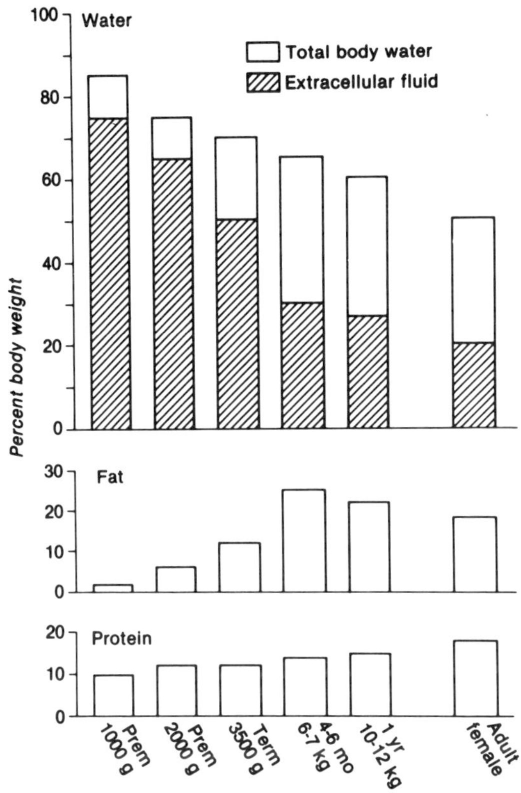

Figure 4 Comparative body composition of infants and adults (Lawrence, 1994).

ejected and received by the infant, it is out of the maternal system. When the peak level in the breast occurs without feeding and the compound is being rapidly excreted by the maternal kidneys, the compound may diffuse back equilibrating across the membrane. Thus, peak plasma times and peak milk times may differ.

The concentration of a drug reaches a maximum in the plasma depending upon the rate of administration which may be intravenous, intramuscular, oral, or cutaneous (Gilman *et al.*, 1990). Drugs given intravenously peak rapidly and higher than those given orally although the areas under the curve will be equal. Knowing the peak plasma time and the peak milk time can facilitate counseling a mother about her medications. Contrary to normal constituents of milk, such as proteins or minerals, the breast does not provide a standard level of a drug or a contaminant in every milliliter of milk.

The metabolism of the compound is an important consideration. Some substances are metabolized quickly by the liver into active metabolites or, in some cases, inactive metabolites. While the adult may handle a compound well, often the metabolism in a child, especially a newborn infant or premature, may be minimal. Materials that depend on the liver for metabolism may accumulate in the neonate because of immature liver function. Acetaminophen is rapidly metabolized in the adult and promptly excreted as an inactive metabolite (Ellenhorn and Barcelaux, 1988). Uniquely, acetaminophen is well handled by newborns who excrete it by the sulfhydryl pathway instead of a glucoronidyl pathway of adults. It appears to have no impact on the onset or severity of neonatal jaundice (Peterson and Bowes, 1983).

The oral bioavailability of a compound influences the amount absorbed into the maternal system, the rate of the absorption, and, thus, the peak plasma time. This may influence the level available to the breast at any given time especially if it is rapidly cleared by the maternal system. Of even greater significance for the infant is the oral bioavailability of the compound from the milk in the infant's stomach. Obviously, those drugs which must be given to the mother by injection because they are not orally bioavailable will not be active in the infant (see Table I). Heparin, insulin, and aminoglycocide antibiotics are well known for their lack of oral availability (Rivera-Calimlim, 1987). Furthermore, the absorption of many compounds is greatly diminished by the presence of food in the stomach and, therefore, less would be absorbed when presented to the infant in milk!

V. The Characteristics of the Infant

While the amount of some drugs in the milk may vary slightly in the early days of lactation as the tight junction intracellular spaces close, for practical purposes milk levels are considered to be the same at all stages of lactation. The recipient infant does change, however, and the impact of the same drug over time can change. The gestational age (GA) of the infant and its chronological age (CA) are important in estimating the infant's ability to absorb, detoxify, and excrete the compound. A 3-week-old CA 32-week premature (GA) has a conceptual age of only 35 weeks and would still not metabolize a drug as well as a full-term infant at birth (40 weeks). Maturing influences absorption, metabolism, and excretion as well as amount of body water and fat stores (Lawrence, 1989).

The amount of breast milk ingested, the possibility of other foods or fluids in the diet, and the frequency of feeds (intervals between feedings) all influence the impact of maternal drug on the absorption, metabolism, and excretion by the infant. The risk of a given drug to an infant will also depend on its ability to displace bilirubin from albumin (see Table II). When the drug binds to albumin, the bilirubin in the plasma is unbound and free to distribute in tissues including the brain. This is a consideration in jaundiced infants in the first month of life.

The drug that must be taken by the mother for a chronic disease, such as epilepsy, achieves a steady state and presents different risks than a drug taken for a week or 10 days, such as an antibiotic, or the drug taken episodically, such as one dose of acetaminophen for a headache. With a

TABLE II
Drugs That Displace Bilirubin from Serum Albumin[a]

Salicyclic acid
Sulfonamides
Fusosemide
Phenylbutazone

[a]Source: Ellenhorn and Barceloux (1988).

single dose, the risk of accumulation is absent and avoiding peak plasma time is a clinical remedial action to decrease exposure. Many active ingredients, often in combination of two or three, are available in over-the-counter medications to the public without a prescription. These medications are often considered by the public as not really "drugs." They are taken liberally. The opportunity for significant exposure has increased as more preparations are made available to relieve every symptom and discomfort. Some of these may alter milk production as well as get into the milk.

VI. Substances That Influence Milk Production

There are medications that may influence milk production or let down by enhancing or decreasing the action of prolactin or oxytocin. The most effective prolactin suppressant is bromocriptine and this is clinically utilized in the medical treatment of a pituitary prolactinoma. Bromocriptine is no longer utilized per FDA regulation in the management of postpartum lactation suppression. Across-the-counter decongestants, such as pseudoephedrine, pheniramine, and some antihistamines, dry secretions of the respiratory tree, may dry the eyes, and decrease milk production as well. Thus, it is recommended that a lactating woman avoid decongestants and seek relief of cold symptoms through local treatment such as nose drops. Diuretics may have a similar effect and were used at one time for suppressing lactation postpartum. A list of the drugs that have been associated with suppression of lactation include L-DOPA, clomiphene citrate (Clomid), monoamine oxidase inhibitors, and prostaglandin E and F_2. Ergot preparations also suppress lactation.

Some drugs are associated with breast enlargement and lactation is a side effect. Such a compound is chlorpromazine, which has been utilized to induce lactation but has other side effects that make this impractical. Metoclopramide (Reglan), a compound used to stimulate secretions, appears to act by sensitizing tissues to the action of acetylcholine (Souse *et al.*, 1975). It has been observed to enhance milk production in women who pump when the infant cannot go to the breast as in premature births (Ehrenkranz and Ackerman, 1986). The drug is excreted into milk. Milk production does increase but it diminishes if the drug is discontinued. Some find the side effects exceed the value of temporary increase in milk supply.

VII. Exposure to a "Recreational Drug"—Nicotine

Mothers who smoke cigarettes are at risk for a less successful lactation experience and for exposing their infant to carbon monoxide, nicotine, and cotinine in their milk. The effects of smoking on the duration of lactation have been shown in many studies in which smokers are noted to wean sooner. In a study by Lyon (1983), significantly more mothers who smoke choose to bottle feed their infants, but of those who did breast-feed, 70% had stopped by 6 weeks compared to 55% of nonsmoking mothers. The principal reason for stopping was inadequate milk and an "unsettled" baby. There was no difference in age, parity, or socioeconomic status of the mothers in either group. The association of smoking and early cessation of breast-feeding has been confirmed by others to be inversely related to the number of cigarettes smoked, but unrelated to social class, maternal age, or parity. The impact of smoking on the breast-feeding process was investigated to evaluate the consistent finding that smokers weaned early quite unrelated to socioeconomic and demographic factors (Whichelow, 1979; Woodward and Hand, 1988; Schwartz-Bickenbach *et al.*, 1987). Basal prolactin levels were reported to be significantly lower in smokers compared to nonsmokers, but increments in serum prolactin following suckling were not significantly different in smokers or nonsmokers (Anderson *et al.*, 1982). Likewise, oxytocin-linked neurophysin was measured and the response to suckling stimulation was not significantly different between smokers and nonsmokers. Serum nicotine and plasma adrenaline, but not dopamine or noradrenaline, increased significantly during smoking in these women. Heavy cigarette smoking was associated with the lowest baseline prolactin levels. It is not certain how this is related to early weaning and "insufficient milk syndrome."

The exposure of breast-feeding infants to nicotine and cotinine, the major metabolite, has been of considerable concern to many investigators (Andersen *et al.*, 1982; Schwartz-Bickenbach *et al.*, 1987; Luck and Nau, 1984, 1985, 1987; Labrecque *et al.*, 1989; Matheson *et al.*, 1985; Steldinger and Luck, 1988; Whichelow, 1979). It was first reported by Hatcher and Crosby in 1927. Nicotine reaches higher concentrations in milk than in serum because it is basic and the milk acidic. The milk plasma ratio is 2.9 ± 1.1. Not only did mother's milk contain nicotine (0.2 to 1.6 ng/l) and cotinine (3 to 30 ng/ml), but the urine of these infants also contained measurable amounts (Luck and Nau, 1985). The ratio of nicotine to creatinine ranged from 5.0 to 110 (median 14). The urine of breast-fed infants of nonsmokers did not contain these substances. The relative risks of nicotine via passive exposure (sidestream smoke) in all infants of smoking mothers and fathers has also been studied. Infants exposed to passive smoking had nicotine/creatinine ratios in the range of 4.76 to 218 (median 35). Other investigators made similar observations (Trundle and Skellern, 1983). The highest urinary excretion of cotinine, as expressed by ng cotinine/mg creatinine ratios, was observed in infants fully breast-fed by smoking mothers. Nursing, and to a lower degree, passive smoking, contribute to the exposure of infants to nicotine and its metabolite cotinine. There was a direct correlation between nicotine and cotinine levels in the mother's serum and in her milk. The levels of nicotine varied greatly over 24 hr but cotinine concentrations remained fairly constant and were a function of the number of cigarettes smoked (Labrecque *et al.*, 1989). The number of cigarettes smoked per day, and the individual smoking habits, like the time of smoking, the smoking frequency prior to nursing, and the time interval between nursing and the last cigarette, all influenced the nicotine level (Woodward *et al.*, 1986; Luck and Nau, 1987). As a result, early morning feeds had the lowest nicotine and levels rose throughout the day dropping again during the night. It is thought that most of the cotinine exposure comes from the mother's milk, whereas the highest exposure to nicotine comes from passive smoking. The half-life of nicotine in milk was reported to be 97 ± 20 min which was slightly greater than the half-life in serum, 81 ± 9 min (Luck and Nau, 1984). Cotinine remained consistent through the 4-hr interval. Thus, it has been recommended that mothers who must smoke extend the period following a cigarette, before nursing, to over an hour and a half. Decreasing the number of cigarettes or stopping smoking is an obvious solution.

The relationship of smoking, breast-feeding, and colic has been reported by Matheson and Rivrud (1989). Forty percent of infants breast-fed by smokers of five or more cigarettes per day had colic compared to 26% of infants of nonsmokers ($p < 0.005$). Bottle-fed infants were not influenced by maternal smoking but infants of smokers who breast-fed had a 50% chance of having colic.

VIII. Environmental Substances in Milk

In addition to the physiology of milk production, there are other factors which effect the body burden of contaminants and the amount excreted in the milk; the influence of maternal residence, urban or rural, industrial or nonindustrial, and proximity to unusual exposures, spills, or accidents. The relationship of increasing maternal age to increasing accumulation of contaminants in fatty tissues is generally accepted and levels correlate with studies on adipose tissue and blood fractions. Milk levels, however, have not provided clear correlations to maternal age (Dillon *et al.*, 1981). Discrepancies between studies may be explained by the narrow age range of mothers investigated and different exposure scenarios. Lower levels have been reported in multiparae than in primiparae which would be consistent with the fact that breast milk is one of the most important routes of excretions for persistent organohalogens like dichlorodiphenyl trichloroethane (DDT), polychlorinated biphenyls (PCBs), βHCH, HCB, PCDDs, PCDFs, and dieldrin (Jensen and Slorach, 1991). Thus, the multipara has lost organohalogens from her stores during multiple pregnancies and lactation. The body weight of the mother is usually related to higher levels in adipose tissue of thin individuals who have less fat in which to distribute the chemical. The observation as a corollary has been reported that thin mothers (weight less than 63 kg) have higher levels in their milk of DDT, dieldrin, heptachlor epoxide, and PCBs than heavier mothers (weight over 72 kg) (Polishuk *et al.,* 1977). Excessive weight loss can mobilize chemicals sequestered in the fat. Although black mothers have been reported to have higher DDT levels in both the United States and Brazil, it is thought to be a function of socioeconomic status, job, and residence, not of race (Davies *et al.,* 1972).

The diet may be the only source of contaminants for some individuals who live and work in a relatively clean environment. Those who eat more animal proteins and fats have higher residues of organohalogens. A greater intake of calories results in higher DDT concentrations (Kroger, 1972). When fish were contaminated, as in the Great Lakes in the 1970s, those who ate a lot of lake fish had higher levels of PCBs and polybrominated biphenyls (PBBs) (Drijver *et al.,* 1988; Schwarts *et al.,* 1983). Methylmercury levels are also noted to be higher in those who eat fish and/or live near the coast (Galster, 1976). In general, vegetarians had lower levels of these compounds in their systems (Noren, 1983).

Other variations are seen seasonally with higher levels in the summer than in winter. Smokers have higher levels of DDT and cadmium than nonsmokers.

Other characteristics of the milk itself influence the levels of contaminants in human milk. The concentration of fat in any given sample is most critical. The fat content of human milk is well known to vary from individual to individual, within the same individual during a single feeding, from feeding to feeding in a given day, and from day to day (Lawrence, 1989). Differences up to a factor of five times greater have been reported for PCBs by Mes and Davies (1978) during a single feeding and during the day. Colostrum fat is reported to have higher levels of residue than mature milk fat. There is a general decrease in levels between 6 to 12 months postpartum (Jensen and Slorach, 1991).

IX. Heavy Metal as Contaminants in Human Milk

Toxic industrial chemicals are a source of environmental contamination and also a source of serious exposure in certain workplaces. Breast milk may contain trace levels of most metals and other elements just as other body fluids do. Maternal exposure to heavy metals is another example of special risk groups whose body burden of a heavy metal may be a risk to the fetus or the breast-fed infant (Dabeka *et al.*, 1986; Jensen 1983; Perkins and Oski, 1976). While lead and mercury are most frequently studied because they have been the cause of infant poisonings through milk, other metals studied include arsenic, cadmium, chromium, and fluoride. They are found in higher concentrations in certain water supplies, cow's milk, and reconstituted formula than in human milk. In general, the amount of a heavy metal from maternal load that is passed via the placenta to the fetus far exceeds the amount passed via breast milk over a comparable period of time. Detailed tables of levels in human samples worldwide are available in the report of Jensen and Slorach (1991).

A. Lead

Lead is the single most common heavy metal pollutant of the environment. The relationship between heavy metals and nutrients has been recognized since the early part of this century, but analytical techniques have only recently made it possible to accurately track the problem (Dabeka and McKenzie, 1986). Nutrition is an important source of heavy metals in infants and children (Sternowsky and Wessolowski, 1985). About 50% of the body burden of lead has been accumulated by late childhood. Although many studies have looked at random samples of milk and serum of mother and infant (Dillon *et al.*, 1974; Ryu *et al.*, 1983; Walker, 1980; Wolff, 1983), few have looked at the impact of lead in breast milk over time (Sternowsky and Wessolowski, 1985). In general, urban dwellers are at greater risk than rural residents (Beattie *et al.*, 1975; Bryce-Smith and Stephens, 1982). The devastating effects of lead on brain development have been established (Smith *et al.*, 1963).

Higher values of lead in the milk of urban populations compared to rural ones, and higher values in colostrum compared to mature milk, have been reported. The calculated daily intake of a breast-fed infant weighing 5.5 kg and receiving 840 ml of milk per day in Hamburg, Germany would be 1.5–2.3 μg/kg/day of lead, while rural infants would receive only 0.9–1.3 μg/kg/day. Daily permissible intake according to the World Health Organization is 5 μg/kg/day. Concentrations of lead in samples collected randomly, by many investigators from 1933 to the present in many American cities and around the world, demonstrate no change in the range of 0 to 0.32 μg/ml with a mean of 0.01 to 0.27 μg/ml (Kovar *et al.*, 1984). Published maternal blood levels range from 0.21 to 0.87 μmol/liter. Children are reported to absorb 50% of the intake and retain 18% of dietary lead (adults, 10% absorbed and 5% retained) (Forbes and Reina, 1972). Higher lead levels were associated across Canada with age of house occupied, maternal exposure to heavy traffic for more than 5 years, and with high coffee consumption (Dabeka *et al.*, 1986). Except in situations of unusually high maternal lead levels, the breast-fed infant is at lower risk for increasing dietary lead than the bottle-fed infant whose formula is diluted with local water because the mother is an effective filter (Kroger, 1974).

B. Cadmium

Nutrition is the source of cadmium intake for infants and 50% of the body cadmium is accumulated by late childhood (Sternowsky and Wessolowski, 1985). Wide scatter in cadmium concentrations in milk have been reported and, as with lead, urban mothers have higher levels than rural mothers. Cadmium levels are higher in smokers and when the spouse smokes, but there are no correlations with any dietary factors (Dabeka *et al.*, 1986). Cadmium is also transferred via the placenta. The levels in breast milk and prepackaged formula are similar. The risk from breast milk is believed to be small (0.4 μg/liter) as reported by Kovar *et al.* (1984). Calculated daily intake of cadmium reported in 1985 in rural Germany, however, was 1.2–1.8 μg/kg/day, whereas in urban Hamburg, it was 1.6–2.2 μg/kg/day which is just above the adult values for daily permissible intake published by the World Health Organization of approximately 1 μg/kg/day. The authors did not obtain a history of cigarette smoking, although it was suggested that cadmium-containing fertilizers could be the link to the food chain. No known syndromes or problems have been associated with this level of exposure (Bhattacharyya, 1983) and the World Health Organization (1972) has not set levels for daily permissible intake for infants and children.

C. Organic and Inorganic Mercury

Inorganic and organic mercury exposures are usually tracked back to the food chain and isolated food accidents. Unexposed women reported by Pitken *et al.* (1976) in the United States had 0.9 ± 0.2 (SE) μg/liter total mercury in their milk compared to 3.6 μg/liter reported in Japanese women who consumed considerable fish which was probably contaminated (Fugita and Takabake, 1977). Inorganic mercury is approximately equally distributed in plasma and red blood cells (RBC) but organic mercury binds very efficiently to RBC and thus is less available for transfer into milk. The M/P of inorganic mercury is about 10 and for organic mercury 0.1 (Wolff, 1983).

Methylmercury is highly neurotoxic and has been responsible for many severe poisonings. Very high levels of methylmercury were found in mothers in Iraq during the catastrophic outbreak of methylmercury poisoning due to the ingestion of bread made with fungicide-treated grain (Bakir *et al.*, 1973). Their milk levels were 8.6% of the blood levels but the relationship was nonlinear at blood mercury levels below 50 ng/ml. Postnatal exposure from breast-feeding gave neonatal levels of 600 ng/ml. Intrauterine exposure had resulted in levels at birth higher in the infant than in the mother. Some infants were exposed only via the milk (mothers had not eaten contaminated bread during pregnancy) (Amin-Zaki *et al.*, 1974). Eight exclusively breast-fed infants were reported to have blood mercury levels in excess of 200 ppb, four were 1000 ppb, and one over 1500 ppb. At the time, they revealed no neurotoxic signs or symptoms (Amin-Zaki *et al.*, 1981). The safe intake level, utilizing a 10-fold safety factor, could be calculated from adult standards where 20 μg/liter blood mercury (30 μg/day intake or 0.4 μg/kg body wt) is considered toxic for an adult. Two micrograms/day might be safe for a 5-kg infant or a milk level of no more than 4 μg/liter (Fugita and Takabake, 1977; Woolf, 1983). Studies in mice have shown that methylmercury, given in an exclusively milk diet, was more efficiently absorbed than when given with other diets.

X. Insecticides

There are numerous chemicals used as insecticides worldwide but most belong to a few major families and a select number of these have been identified as a hazard because they have entered the food chain of which the human is the final link. The family of organohalide insecticides includes such well-known chemicals as endrin, aldrin, dieldrin, lindane, DDT, DDE (dichlorodiphenyl dichloroethene), chlordane, and heptachlor. It was discovered in 1948 that DDT was stored indefinitely in human tissues (Laug *et al.*, 1950, 1951) and, at that time, the general population in the United States had the six major organochlorine insecticides in their tissues (Jensen, 1983). These were DDT, BHC (benzene hexachloride), DDE, heptachlor, aldrin, and dieldrin. DDT was banned in 1972 and now all are banned except kethane, lindane, and chlordane, of which the latter is tightly controlled.

In mammals, absorption efficiency ranges from 75 to 96% with higher chlorinated congeners having the lowest absorption (Tanabe, 1981). The organohalogens bind to lipoproteins in the blood and are most abundant in the serum fraction (Birnbaum, 1985; Goo *et al.*, 1987; Skalsky and Guthrie, 1978). There is a shift in PCB distribution in rats as pregnancy advances. The higher-density lipoproteins give way to very low-density lipoproteins "which are not transported to the fetus but are a preferred source of lipid for mammary gland milk synthesis" (Jensen, 1991). The metabolism and elimination from the body of organohalogens is well known to be slow. Therefore, much is retained unchanged resulting in the gradual accumulation over time if exposure continues. Lactation is the most important route of elimination for females. In some species whose milk is high in fat, the body burden of chemicals in the fat stores can be transferred to the suckling offspring in a matter of weeks (Jensen, 1991).

This is true in the rat who has three times more milk fat than humans. The level of organohalogens in human milk is dependent upon the level in the fat, a reservoir of long standing, and not upon daily intakes during lactation. While blood levels are significantly increased by consuming a single meal of chemically laden fish, adipose tissue supplies most of the chemicals to the milk.

Human exposures occur through ingestion of contaminated foodstuffs or water, inhalation of vapors, and absorption through the skin (Morgan and Roan, 1974; Jensen and Slorach, 1991). Compounds were detected in the milk of women with excessive exposures, such as continual spraying of fields with aerial DDT vapor. Levels were reported at 120–770 parts per billion (Jensen, 1983). The average women in 1990 is not at risk and should not have measurable levels of organohalides in her body stores or her milk unless she has had an unusual personal exposure (Hofvander *et al.,* 1981; Rogan *et al.,* 1980; Rogan, 1983).

A. Organophosphate Insecticides

Organophosphates, parathion, malathion, and diazinon, have replaced the organohalogens as insecticides. While they are highly toxic with acute exposure, they have low chronic toxicity. The compounds reportedly reach the milk, but because of their rapid metabolism in humans, there is little potential for accumulation (Ellenhorn and Barceloux, 1988; Nolan *et al.,* 1984).

B. Herbicides

Paraquat, in acute exposures, is the most toxic herbicide known producing multisystem failure. Since its development in 1962, however, it has grown in popularity because of its rapid deactivation upon contact with the soil which is responsible for the compound's low chronic toxicity. Marijuana that has been sprayed with paraquat, however, presents a hazard as it remains toxic on the leaf. Marijuana itself is a hazard to the nursing infant. Paraquat has not been reported in human milk.

Dioxins are chlorophenoxy compounds used as herbicides. They have received considerable publicity because of the use of 2,3,7,8-tetrachlorodibenzo-*p*-dioxin (TCDD) and 2,4D as Agent Orange in Vietnam. They are known to be neurogenic, teratogenic, and possibly carcinogenic in animal models. TCDD is the most toxic of the 75 dioxin compounds known. Milk levels measured on pooled samples of women exposed to Agent Orange in Vietnam contained measurable dioxin. Individual exposures were not known nor were the levels measured in individual women (Schecter *et al.,* 1986). This same investigator, when testing pooled samples from high-risk women in this country, reported detectable amounts

(Schnecter and Gasiewicz, 1987). Unless a woman is known to have been exposed, however, there should be no risk of the chemical appearing in her milk. It is not practical to do mass milk screening on individual samples because of the complexity and expense of the test (Schnecter and Gasiewicz, 1987; Lindstrom *et al.*, 1988).

XI. Other Environmental Contaminants

A. Aflatoxins

Aflatoxins are fungal metabolites found as toxic contaminants of foods in the tropics and occasionally in the United States and other industrialized countries. They have been implicated in primary hepatocellular carcinoma (PHC).

To determine the risk to the fetus from transplacental exposure and to the breast-feeding infant from mother's milk, aflatoxins have been measured in breast milk, neonatal cord blood, and the serum of pregnant women (Lamplugh *et al.*, 1988). Infants in Africa are frequently exposed when the mother's level is elevated but there are significant seasonal variations with higher levels in the wet season. The development of an enzyme-linked immunosorbent assay for human milk has made possible the study of the relationship of aflatoxin and hepatitis B virus to PHC in early life (Wild *et al.*, 1987). Because of the close monitoring of foods in the United States, the risk of ingesting excessive amounts of aflatoxins is minimal except where homegrown crops, such as peanuts, are inadequately stored for home use.

B. PCB and PBB

PCB and PBB are the prime halogenated biphenyls that accumulate in human fat and are known to have caused symptoms due to serious chronic exposures that resulted from major accidental environmental exposures. In 1977, their manufacture was halted but removal of heat exchangers, transformers, and capacitors where they were used was not required. In the 1973 Michigan peninsula accident, PBBs got into cattle feed inadvertently, contaminated the cow milk and the beef entering the food chain, and caused human contamination (Wickizer *et al.*, 1981). Human milk was also found to contain PBB, the amount of which was relative to the mother's exposure. The incident was carefully studied (Wickizer and Brilliant, 1981). Symptoms were more closely related to mode of entry than to degree of exposure or serum PBB levels. Only chloracne and liver dysfunction have been reported in the adult victims. Not all women stopped breast-feeding when told their milk was contaminated (Hatcher, 1982). No infants who were breast-fed were reported to be symptomatic.

The greater the chlorine content of the compounds, the more resistant the compound is to metabolic degradation. They are magnified in the food chain reaching highest concentrations in fish, predators, and humans (Ellenhorn and Barceloux, 1988). While PCBs and PBBs clearly get into human milk, they are not a risk except in unusual exposures such as the Michigan peninsula accident or the Yusho (rice oil disease) incident in Japan (Rogan *et al.*, 1980; Wickizer *et al.*, 1981).

XII. Concluding Thoughts

The estimation of chemical hazards in breast milk from the workplace has been reviewed by Poitrast *et al.*, (1988). They concluded that, while the standard statement "to date, there is no evidence of harm to breastfeeding infants whose mothers are not exposed above a permissible limit (PEL)" is comforting, "no evidence of harm" is not the same as "evidence of no harm." The authors further suggest a systems approach to evaluating the hazard of transfer of occupationally encountered chemicals from mother to infants. The steps are: (1) evaluate the workplace to characterize the chemical exposure both qualitatively and quantitatively; (2) evaluate the probability of milk transfer from information on lipid solubility and elimination data; and (3) if a probability of significant hazard is determined to exist, and the mother still wishes to nurse, perform a quantitative analysis of her milk, determine an average daily intake, estimate an average daily chemical intake, and compare it to the recommended allowable daily intake.

References

American Hospital Formulary Service (1991). "AHFS Drug Information." American Society of Hospital Pharmacists, Bethesda, MD.

Amin-Zaki, L., Elhassani, S., Majeed, M. A., Clarkson, T. W., Doherty, R. A., and Greenwood, M. (1974). Intra-uterine methylmercury poisoning in Iraq. *Pediatrics* **54,** 587–595.

Amin-Zaki, L., Majeed, M. A., Greenwood, M. R., Elhassani, S. B., Clarkson, T. W., and Doherty, R. A. (1981). Methylmercury poisoning in the Iraqi suckling infant: A longitudinal study over five years. *J. Appl. Toxicol.* **1,** 210–214.

Andersen, A. N., Lund-Andersen, C., Larsen, J. F., Christensen, N. J., Legros, J. J., Louis, F., Angelo, H., and Molin, J. (1982). Suppressed prolactin but normal neurophysin levels in cigarette smoking breast-feeding women. *Clin. Endocrinol.* **17,** 363–368.

Anderson, P. O. (1991). Drug use during breast-feeding. *Clin. Pharmacokinet.* **10,** 594–624.

Atkinson, H. C., Begg, E. J., and Darlow, B. A. (1988). Drugs in human milk—Clinical pharmacokinetic considerations. *Clin. Pharmacokinet.* **14,** 217–240.

Bakir, F., Damluji, S. M., Amin-Zaki, L., Murtadha, M., Khalidi, A., Al-Rawi, N. Y., Tikriti, S., Dhahir H. I., Clarkson, T. W., Smith, J. C., and Doherty, R.A. (1973). Methylmercury poisoning in Iraq. *Science* **181,** 230–242.

Beattie, A. D., Moore, M. R., and Goldberg, A. (1975). Role of chronic low-level lead exposure in the aetiology of mental retardation. *Lancet* **1,** 589–592.

Beeley, L. (1986). Drugs and breastfeeding. *Clin. Obstet. Gynecol.* **13,** 247–251.

Berglund, F., Flodh, H., Lundborg, P., Prame, B., and Sannerstedt, R. (1984). Drug use in pregnancy and breastfeeding. *Acta Obstet. Gynecol. Scand.* **126,** 5–55.

Bhattacharyya, M. H. (1983). Bioavailability of orally administered cadmium and lead to the mother, fetus and neonate during pregnancy and lactation: An overview. *Sci. Total Environ.* **28,** 327–342.

Birnbaum, L. S. (1985). The role of structure in the disposition of halogenated aromatic xenobiotics. *Environ. Health Perspect.* **61,** 11–20.

Briggs, G. G., Freeman, R. K., and Yaffe, S. J. (1990). "Drugs in Pregnancy and Lactation: A Reference Guide to Fetal and Neonatal Risk," 3rd ed. Williams & Wilkins, Baltimore, MD.

Brilliant, L. B., Wilcox, K., Van Amburg, G., Eyster, J., Isbister, J., Bloomer, A. W., Humphrey, H., and Price, H. (1978). Breast-milk monitoring to measure Michigan's contamination with polybrominated biphenyls. *Lancet* **2,** 643–646.

Brophy, T. R., McCafferty, J., Tyrer, J. H., and Eadie, M. J. (1983). Bioavailability of oral dexamethasone during high dose steroid therapy in neurological patients. *Eur. J. Clin. Pharmacol.* **24,** 103–108.

Bryce-Smith, D., and Stephens, R. (1982). Lead and brain function. *Dev. Med. Child. Neurol.* **24,** 90–91.

Buckley, M. M. T., Goa, K. L., and Clissold, S. P. (1990). Ocular betaxolol: A review of its pharmacological properties, and therapeutic efficacy in glaucoma and ocular hypertension. *Drugs* **40,** 75–90.

Campoli-Richards, D. M., and Clissold, S. P. (1986). Famotidine: Pharmacodynamic and pharmacokinetic properties and a preliminary review of its therapeutic use in peptic ulcer disease and Zollinger–Ellison syndrome. *Drugs* **32,** 197–221.

Committee on Drugs (1989). Transfer of drugs and other chemicals into human milk. *Pediatrics* **84,** 924–936.

Curry, S. H., Riddall, D., Gordon, J. S., Simpson, P., Binns, T. B., Rondel, R. K., and McMartin, C. (1971). Disposition of glutethimide in man. *Clin. Pharmacol. Ther.* **12,** 849–857.

Dabeka, R. W., Karpinski, K. F., McKenzie, A. D., and Bajdik, C. D. (1986). Survey of lead, cadmium, and fluoride in human milk and correlation of levels with environmental and food factors. *Food Chem. Toxicol.* **24,** 913–921.

Dabeka, R. W., and McKenzie, A. D. (1986). Graphite-furnace atomic absorption spectrometric determination of lead and cadmium in food after nitric perchloric acid digestion and coprecipitation with ammonium pyrrolidine dithiocarbamate. *Can. J. Spectrom.* **31,** 44.

Dahl, S. G., and Strandjord, R. E. (1977). Pharmacokinetics of chlorpromazine after single and chronic dosage. *Clin. Pharmacol. Ther.* **21,** 437–448.

Davies, J. E., Edmundson, W. F., Raffonelli, A., Cassady, J. C., and Morgade, C. (1972). The role of social class in human pesticide pollution. *Am. J. Epidemiol.* **96,** 334–344.

Dickey, R. P., and Stone, S. C. (1975). Drugs that affect the breast and lactation. *Clin. Obstet. Gynecol.* **18,** 95–111.

Dillon, B. K., Wilson, D. J., and Schaffner, W. (1974). Lead concentrations in human milk. *Am. J. Dis. Child.* **128,** 491–492.

Dillon, J. C., Martin, G. B., and O'Brien, H. T. (1981). Pesticide residues in human milk. *Food Cosmet. Toxicol.* **19,** 437–442.

Donnelly, R., and Macphee, G. J. A. (1991). Clinical pharmacokinetics and kinetic-dynamic relationships of dilevalol and lavetalol. *Clin. Pharmacokinet.* **21,** 95–109.

Drijver, M., Duijkers, T. J., Kromhout, D., Visser, T. J., Mulder, P., and Louw, R. (1988). Determinants of polychlorinated biphenyls (PCBs) in human milk. *Acta Paediatr. Scand.* **77,** 30–36.

Dusci, L. J., Good, S. M., Hall, R. W., and Ilett, K. F. (1990). Excretion of diazepam and its metabolites in human milk during withdrawal from combination high dose diazepam and oxazepam. *Br. J. Clin. Pharmacol.* **29,** 123–126.

Echizen, H., and Ishizaki, T. (1991). Clinical pharmacokinetics of famotidine. *Clin. Pharmacokinet.* **21,** 178–194.

Egan, P. C., Costanza, M. E., Dodion, P., Egorin, M. J., and Bachur, N. R. (1985). Doxorubicin and cisplatin excretion into human milk. *Cancer Treatment Rep.* **69,** 1387–1389.

Ehrenkranz, R. A., and Ackerman, B. A. (1986). Metoclopramine effect on faltering milk production by mothers of premature infants. *Pediatrics* **78,** 614–620.

Ellenhorn, M. J., and Barceloux, D. G. (1988). "Medical Toxicology." Elsevier, New York.

Fidler, J., Smith, V., and DeSwiet, M. (1983). Excretion of oxprenolol and timolol in breast milk. *Br. J. Obstet. Gynaecol.* **90,** 961–965.

Forbes, G. B., and Reina, J. C. (1972). Effect of age on gastrointestinal absorption (Fe, Sr, Pb) in the rat. *J. Nutr.* **102,** 647–652.

Fransson, G. B., and Lonnerdal, B. (1980). Iron in human milk. *J. Pediatr.* **96,** 380–384.

Freedman, M. D., and Somberg, J. C. (1991). Pharmacology and pharmacokinetics of amiodarone. *J. Clin. Pharmacol.* **31,** 1061–1069.

Friedel, H. A., and Buckley, M. M. T. (1991). Torasemide: A review of its pharmacological properties and therapeutic potential. *Drugs* **41,** 81–103.

Fugita, M., and Takabake, E. (1977). Mercury levels in human maternal and neonatal blood, hair and milk. *Bull. Environ. Contam. Toxicol.* **18,** 205–209.

Galster, W. A. (1976). Mercury in Alaskan eskimo mothers and infants. *Environ. Health Perspect.* **15,** 135–140.

Gelenberg, A. J. (1979). Amoxapine, a new antidepressant, appears in human milk. *J. Nervous. Mental Dis.* **167,** 635–636.

Gillis, A. M., and Kates, R. E. (1984). Clinical pharmacokinetics of the newer antiarrhythmic agents. *Clin. Pharmacokinet.* **9,** 375–403.

Gilman, A. G., Rall, T. W., Nies, A. S., and Taylor, P., eds. (1990). "The Pharmacological Basis of Therapeutics." Peragon Press, New York.

Goo, Y. L., Emmett, E. A., Pellizzari, E. D., and Rohde, C. A. (1987). Influence of serum cholesterol and albumin on partitioning of PCB congers between human serum and adipose tissue. *Toxicol. Appl. Pharmacol.* **87,** 48–56.

Graffner, C., Conradson, T. B., Hofvendahl, S., and Ryden, L. (1980). Tocainide kinetics after intraveneous and oral administration in healthy subjects and in patients with acute myocardial infarction. *Clin. Pharmacol. Ther.* **27,** 64–71.

Gushurst, C. A., Mueller, J. A., Green, J. A., and Sedor, F. (1984). Breast milk iodide: reassessment in the 1980s. *Pediatrics* **73,** 354–357.

Hatcher, R. A., and Crosby, H. (1927). The elimination of nicotine in the milk. *J. Pharmacol. Exp. Ther.* **32,** 1–6.

Hatcher, S. L. (1982). The psychological experience of nursing mothers upon learning of a toxic substance in their breast milk. *Psychiatry* **45,** 172–181.

Hermann, P., and Morselli, P. L. (1985). Pharmacokinetics of diltiazem and other calcium entry blockers. *Acta Pharmacol. Toxicol.* **57**(Suppl. 2), 10–20.

Hii, J. T. Y., Duff, H. J., and Burgess, E. D. (1991). Clinical pharmacokinetics of propafenone. *Clin. Pharmacokinet.* **21,** 1–10.

Hofvander, Y., Hagman, U., Linder, C. E., Vaz, R., and Slorach, S. A. (1981). WHO collaborative breast feeding study: I. Organochlorine contaminants in individual samples of Swedish human milk, 1978–1979. *Acta Paediatr. Scand.* **70,** 3–8.

Holdiness, M. R. (1984). Clinical pharmacokinetics of the antituberculosis drugs. *Clin. Pharmacokinet.* **9,** 511–544.

Jensen, A. A. (1983). Chemical contaminants in human milk. *Res. Rev.* **89,** 1–128.

Jensen, A. A. (1991). Transfer of chemical contaminants into human milk. *In* "Chemical Contaminants in Human Milk" (A. A. Jensen and S. A. Slorach, eds.), pp. 10–19. CRC Press, Boca Raton, FL.

Jensen, A. A., and Slorach, S. A. (1991). Factors affecting the levels of residues in human milk. *In* "Chemical Contaminants in Human Milk" (A. A. Jensen and S. A. Slorach, eds.), pp. 199–208. CRC Press, Boca Raton, FL.

Johnson, B. F., Wilson, J., Johnson, J., and Flemming, J. (1991). Digoxin pharmacokinetics and spirapril, a new ace inhibitor. *J. Clin. Pharmacol.* **31,** 527–530.

Johnsson, G., and Regardh, C. G. (1976). Clinical pharmacokinetics of β-adrenoreceptor blocking drugs. *Clin. Pharmacokinet.* **1,** 233–263.

Jusko, W. J., Mosovich, L. L., Gerbracht, L. M., Mattar, M. E., and Yaffe, S. J. (1975). Enhanced renal excretion of dicloxacillin in patients with cystic fibrosis. *Pediatrics* **56,** 1038–1044.

Kemp, J., Ilett, K. F., Booth, J., and Hackett, L. P. (1985). Excretion of doxepin and N-desmethyldoxepin in human milk. *Br. J. Clin. Pharmacol.* **20,** 497–499.

Kimbrough, R. D. (1987). Human health effects of polychlorinated biphenyls (PCBs) and polybrominated biphenyls (PBBs). *Annu. Rev. Pharmacol. Toxicol.* **27,** 87–111.

Kirksey, A., and Groziak, S. M. (1984). Maternal drug use: Evaluation of risks to breast-fed infants. *World Rev. Nutr. Diet* **43,** 60–79.

Kovar, I. Z., Strehlow, C. D., Richmond, J., and Thompson, M. G. (1984). Perinatal lead and cadmium burden in a British urban population. *Arch. Dis. Child.* **59,** 36–39.

Kroger, M. (1972). Insecticide residues in human milk. *J. Pediatr.* **80,** 401–405.

Kroger, M. (1974). General environmental contaminants occurring in milk. *In* "Lactation" (B. L. Larson and V. R. Smith, eds.), Vol. III, pp. 135–157. Academic Press, New York.

Kurz, H., Mauser-Ganshorn, A., and Stickel, H. H. (1977). Differences in the binding of drugs to plasma proteins from newborn and adult man. I. *Eur. J. Clin. Pharmacol.* **11,** 463–467.

Kwan, K. C., Foltz, E. L., Breault, G. O., Baer, J. E., and Totaro, J. A. (1976). Pharmacokinetics of methyldopa in man. *J. Pharmacol. Exp. Ther.* **198,** 264–277.

Labrecque, M., Marcoux, S., Weber, J. P., Fabia, J., and Ferron, L. (1989). Feeding and urine cotinine values in babies whose mothers smoke. *Pediatrics* **83,** 93–97.

Lamplugh, S. M., Hendrickse, R. G., Apeagyei, F., and Mwanmut, D. D. (1988). Aflatoxins in breast milk, neonatal cord blood, and serum of pregnant women. *Br. Med. J.* **296,** 968.

Laug, E. P., Kunze, F. M., and Prickett, C. S. (1951). Occurrence of DDT in human fat and milk. *Arch. Ind. Hygiene Occup. Med.* **3,** 245–246.

Laug, E. P., Prickett, C. S., and Kunze, F. M. (1950). Survey analyses of human milk and fat for DDT content. *Fed. Proc.* **9,** 294.

Lawrence, R. A. (1994). "Breastfeeding: A Guide for the Medical Profession," 4th ed. C.V. Mosby, St. Louis, MO.

Lebedevs, T. H., Wojnar-Horton, R. E., Yapp, P., Roberts, M. J., Dusci, L. J., Hackett, L. P., and Ilett, K. F. (1992). Excretion of temazepam in breast milk. *Br. J. Clin. Pharmacol.* **33,** 204–206.

Lewis, A. M., Patel, L., Johnston, A., and Turner, P. (1981). Mexiletine in human blood and breast milk. *Postgrad. Med. J.* **57,** 546–547.

Libardoni, M., Piovan, D., Busato, E., and Padrini, R. (1991). Transfer of propafenone and 5-OH-propafenone to foetal plasma and maternal milk. *Br. J. Clin. Pharmacol.* **32,** 527–528.

Lindstrom, G. U. M., Sjostrom, M., Swanson, E., Furst, P., Kruger, C., Meemken, H. A., and Groebe, W. (1988). Multivariate statistical approach to a data set of dioxin and furan contaminations in human milk. *Bull. Environ. Contam. Toxicol.* **40,** 641–646.

Luck, W., and Nau, H. (1984). Nicotine and cotinine concentrations in serum and milk of nursing smokers. *Br. J. Clin. Pharmacol.* **18,** 9–15.

Luck, W., and Nau, H. (1985). Nicotine and cotinine concentrations in serum and urine of infants exposed via passive smoking or milk from smoking mothers. *J. Pediatr.* **107,** 816–820.

Luck, W., and Nau, H. (1987). Nicotine and cotinine concentrations in the milk of smoking mothers: Influence of cigarette consumption and diurnal variation. *Eur. J. Pediatr.* **146,** 21–26.

Lustgarten, J. S., and Podos, S. M. (1983). Topical timolol and the nursing mother. *Arch. Opthalmol.* **101,** 1381–1382.

Lyon, A. J. (1983). Effects of smoking on breast feeding. *Arch. Dis. Child.* **58,** 378–380.

MacKintosh, D., and Buchanan, N. (1985). Excretion of disopyramide in human breast milk. *Br. J. Clin. Pharmacol.* **19,** 856–857.

Marx, C. M. (1985). Drug use by nursing mothers. *In* "Manual of Neonatal Care" (Clohety and Stark, eds.), pp. 559–599. Little, Brown & Co., Boston/Toronto.

Matheson, I., Pande, H., and Alertsen, A. R. (1985). Respiratory depression caused by N-desemethyldoxepin in breast milk. *Lancet* **2,** 1124.

Matheson, I., and Rivrud, G. N. (1989). The effect of smoking on lactation and infantile colic. *J. Am. Med. Assoc.* **261,** 42–43.

McAllister, R. G., and Kirsten, E. B. (1982). The pharmacology of verapamil. IV. Kinetic and dynamic effects after single intravenous and oral doses. *Clin. Pharmacol. Ther.* **31,** 418–426.

McQuinn, R. L., Pisani, A., Wafa, S., Chang, S. F., Miller, A. M., Frapell, J. M., Chamberlain, G. V. P., and Camm, A. J. (1990). Flecainide excretion in human breast milk. *Clin. Pharmacol. Ther.* **48,** 262–267.

Mehvar, R., Gross, M. E., and Kreamer, R. N. (1990). Pharmacokinetics of atenolol enantiomers in humans and rats. *J. Pharm. Sci.* **79,** 881–885.

Mes, J., and Davies, D. J. (1978). Variation in the polychlorinated biphenyl and organochlorine pesticide residue during human breastfeeding and its diurnal pattern. *Chemosphere* **7,** 699–706.

Miller, R. D., Keegan, K. A., Thrupp, L. D., and Brann, J. (1984). Human breast milk concentration of moxalactam. *Am. J. Obstet. Gynecol.* **148,** 348–349.

Mitani, G. M., Steinberg, I., Lien, E. J., Harrison, E. C., and Elkayam, U. (1987). The pharmacokinetics of antiarrhythmic agents in pregnancy and lactation. *Clin. Pharmacokinet.* **12,** 253–291.

Morgan, D. P., and Roan, C. C. (1974). The metabolism of DDT in man. *In* "Essays in Toxicology" (W. J. Hayes, Jr., ed.), Vol. 5, pp. 39–97. Academic Press, New York.

Morselli, P. L. (1976). Clinical pharmacokinetics in neonates. *Clin. Pharmacokinet.* **1,** 81–98.

Morselli, P. L. (1977). Cardiovascular agents. *In* "Drug Disposition during Development" (P. Moselli, ed.), pp. 393–429. Spectrum Press, New York.

Morselli, P. L., Franco-Morselli, R., and Bossi, L. (1980). Clinical pharmacokinetics in newborns and infants. *Clin. Pharmacokinet.* **5,** 485–527.

Morselli, P. L., Boutroy, M. J., Bianchetti, G., and Thenot, J. P. (1989). Pharmacokinetics of antihypertensive drugs in the neonatal period. *Dev. Pharmacol. Ther.* **13,** 190–198.

Motwani, J. G., and Lipworth, B. J. (1991). Clinical pharmacokinetics of drugs administered buccally and sublingually. *Clin. Pharmacokinet.* **21,** 83–94.

Myhre, E., Rugstad, H. E., and Hansen, T. (1982). Clinical pharmacokinetics of methyldopa. *Clin. Pharmacokin.* **7,** 221–233.

Neuvonen, P. J., and Kivisto, K. T. (1989). The clinical significance of food–drug interactions: A review. *Med. J. Aust.* **150,** 36–40.

Neville, M. C., Allen, J. C., and Watters, C. (1983). The mechanism of milk secretion. *In* "Lactation: Physiology, Nutrition and Breastfeeding" (M. C. Neville and M. R. Neifert, eds.), pp. 49–102. Plenum Press, New York.

Nolan, R. J., Rick, D. L., Freshour, N. L., and Saunders, J. H. (1984). Chlorpyrifos. Pharmacokinetics in human volunteers. *Toxicol. Appl. Pharmacol.* **73,** 8–15.

Noren, K. (1983). Organochlorine contaminants in Swedish human milk from the Stockholm region. *Acta Paediatr. Scand.* **72,** 259–264.

Ogaki, J., Takayama, K., Miyata, H., and Kaskimoto, T. (1987). Levels of PCDDs and PCDFs in human tissues and various foodstuffs in Japan. *Chemosphere* **16,** 2047–2056.

Orme, M. L., Lewis, P. J., DeSwiet, M., Serlin, M. J., Sibeon, R., Baty, J. D., and Breckenridge, A. M. (1977). May mothers given warfarin breastfeed their infants? *Br. Med. J.* **1,** 1564–1565.

Paton, D. M., and Webster, D. R. (1985). Clinical pharmacokinetics of H1-receptor antagonists. *Clin. Pharmacokinet.* **10,** 477–497.

Perkins, K. C., and Oski, F. A. (1976). Elevated blood lead in a 6 month old breast-fed infant: The role of newsprint logs. *Pediatrics* **57,** 426–427.

Peterson, R. G., and Bowes, Jr., W. A. (1983). Drugs, toxins and environmental agents in breast milk. *In* "Lactation: Physiology, Nutrition and Breastfeeding" (M. C. Neville and M. R. Neifert, eds.), pp. 367–403. Plenum Press, New York.

Physicians Desk Reference (1992). Medical Economics Company, Oradell, New Jersey.

Picciano, M. F., and Guthrie, H. A. (1976). Copper, iron, and zinc contents of mature human milk. *Am. J. Clin. Nutr.* **29,** 242–254.

Pinder, R. M., Brogden, R. N., Speight, T. M., and Avery, G. S. (1977). Maprotiline: A review of its pharmacological properties and therapeutic efficacy in mental depressive states. *Drugs* **13,** 321–352.

Pitkin, R. M., Bahns, J. A., Filer, L. A., and Reynolds, W. A. (1976). Mercury in human maternal and cord blood, placenta and milk. *Proc. Soc. Exp. Biol. Med.* **151,** 565–567.

Pittard, W. B., and Glazier, H. (1983). Procainamide excretion in human milk. *J. Pediatr.* **102,** 631–633.

Pittard, W. B., and O'Neal, W. (1986). Amitriptyline excretion in human milk. *J. Clin. Psychopharmacol.* **6,** 383–384.

Plomp, T. A., Vulsma, T., and de Vijlder, J. J. M. (1992). Use of amiodarone during pregnancy. *Eur. J. Obstet. Gynecol. Reprod. Biol.* **43,** 201–207.

Poitrast, B. J., Keller, W. C., and Elves, R. G. (1988). Estimation of chemical hazards in breast milk. *Aviation Space Environ. Med.* **59,** A87–A92.

Polishuk, Z. W., Ron, M., Wasserman, M., Cucos, S., Wassermann, D., and Lemesch, C. (1977). Organocholine compounds in human blood plasma and milk. *Pestic. Monit. J.* **10,** 121–129.

Prescott, L. F., Pottage, A., and Clements, J. A. (1977). Absorption, distribution and elimination of mexiletine. *Postgrad. Med. J.* **43**(Suppl.), 50–55.

Reisner, S. H., Eisenberg, N. H., Stahl, B., and Hauser, G. J. (1983). Maternal medications and breast-feeding. *Dev. Pharmacol. Ther.* **6,** 285–304.

Ribeiro, C., and Longo, A. (1987). Procainamide and disopyramide. *Eur. Heart. J.* **8**(Suppl. A), 11–19.

Rivera-Calimlim, L. (1977). Drugs in breast milk. *Drug Ther.* **7,** 59–63.

Rivera-Calimlim, L. (1987). The significance of drugs in breast milk. *Clin. Perinatol.* **14,** 51–70.

Roberts, R. J. (1984). "Drug Therapy in Infants." Saunders, Philadelphia.

Rogan, W. J. (1983). Persistent pesticides and polychlorinated biphenyls. *Annu. Rev. Public Health* **4,** 381–384.

Rogan, W. J., Bagniewska, A., and Damstra, T. (1980). Pollutants in breast milk. *N. Engl. J. Med.* **302,** 1450–1453.

Ryu, J. E., Ziegler, E. E., Nelson, S. E., and Fomon, S. J. (1983). Dietary intake of lead and blood lead concentration in early infancy. *Am. J. Dis. Child.* **137,** 886–891.

Schecter, A., Ryan, J. J., and Constable, J. D. (1986). Chlorinated dibenzo-p-dioxin and dibenzofuran levels in human adipose tissue and milk samples from the north and south of Vietnam. *Chemosphere* **15,** 1613–1620.

Schecter, A., and Gasiewicz, T. A. (1987). Health hazard assessment of chlorinated dioxins and dibenzofurans contained in human milk. *Chemosphere* **16,** 2147–2154.

Schou, M., and Amdisen, A. (1973). Lithium and pregnancy—III, Lithium ingestion by children breast-fed by women on lithium treatment. *Br. Med. J.* **2,** 138.

Schwartz, P. M., Jacobson, S. W., Fein, G., Jacobson, J. L., and Price, H. A. (1983). Lake Michigan fish consumption as a source of polychlorinated biphenyls in human cord serum, maternal serum and milk. *Am. J. Public Health* **73,** 293–296.

Schwartz-Bickenbach, D., Schulte-Hobein, B, Abt, S., Plum, C., and Nau, H. (1987). Smoking and passive smoking during pregnancy and early infancy: Effects on birth weight, lactation period, and cotinine concentration in mother's milk and infant's urine. *Toxicol. Lett.* **35,** 73–81.

Shammas, F. V., and Dickstein, K. (1988). Clinical pharmacokinetics in heart failure—An updated review. *Clin. Pharmacokinet.* **15,** 94–113.

Sietsema, W. K. (1989). The absolute oral bioavailability of selected drugs. *Int. J. Pharmacol. Ther. Toxicol* **27,** 179–211.

Skalsky, H. L., and Guthrie, F. E. (1978). Binding of insecticides to human serum proteins. *Toxicol. Appl. Pharmacol.* **43,** 229–235.

Smilack, J. D., Wilson, W. R., and Cockerill, F. R. (1991). Tetracyclines, chloramphenicol, erythromycin, clindamycin, and metronicazole. *Mayo Clinic Proc.* **66,** 1270–1280.

Smith, H. D., Boehner, R. L., Charney, T., and Majors, W. J. (1963). The sequelae of pica with and without lead poisoning. *Am. J. Dis. Child.* **105,** 609–616.

Snider, D. E., and Powell, K. E. (1984). Should women taking antituberculosis drugs breastfeed? *Arch. Intern. Med.* **144,** 589–590.

Somani, P. (1989). Basic and clinical pharmacology of amiodarone: Relationship of antiarrhythmic effects, dose and drug concentrations to intracellular inclusion bodies. *J. Clin. Pharmacol.* **29,** 405–412.

Somani, P., Fraker, T. D., and Temesy-Armos, P. N. (1987). Pharmacokinetic implications of lorcainide therapy in patients with normal and depressed cardiac function. *J. Clin. Pharmacol.* **27,** 122–132.

Sousa, P. L. R., Barros, F. C., Pinheiro, G. N. M., and Gazalle, R. V. (1975). Re-establishment of lactation with metoclopramide. *J. Trop. Pediatr. Environ. Child Health* **21,** 214–215.

Speight, T. M., ed. (1987). "Avery's Drug Treatment." ADIS Press Limited, Auckland, Australia.

Speth, P. A. J., Van Hoesel, Q. G. C. M., and Haanen, C. (1988). Clinical pharmacokinetics of doxorubicin. *Clin. Parmacokinet.* **15,** 15–31.

Stancer, H. C., and Reed, K. L. (1986). Desipramine and 2-hydroxydespiramine in human breast milk and the nursing infant's serum. *Am. J. Psych.* **143,** 1597–1600.

Steldinger, R., and Luck, W. (1988). Half lives of nicotine in breast milk of smoking mothers: Implications for nursing. *J. Perinat. Med.* **16,** 261–262.

Sternowsky, J. H., and Wessolowski, R. (1985). Lead and cadmium in breast milk. *Arch. Toxicol.* **57,** 41–45.

Summerfield, R. J. (1985). Excretion of lorazepam into breast milk. *Br. J. Anesthesia* **57,** 1042–1043.

Syversen, G. B., and Ratkje, S. K. (1985). Drug distribution within human milk phases. *J. Pharm. Sci.* **74,** 1071–1074.

Tanabe, S., Nakagawa, Y., and Tatsukawa, R. (1981). Absorption efficiency and biological half-life of individual chlorobiphenyls in rats treated with Kanechlor products. *Agric. Biol. Chem.* **45,** 717–726.

Tjandra-Maga, T. B., Verbesselt, R., Van Hecken, A., Mullie, A., and De Schepper, P. J. (1986). Flecainde: Single and multiple oral dose kinetics, absolute bioavailability and effect of food and antacid in man. *Br. J. Clin. Pharmacol.* **22,** 309–316.

Trundle, J. I., and Skellern, G. G. (1983). Gas chromatographic determination of nicotine in human breast milk. *J. Clin. Hosp. Pharm.* **8,** 289–293.

Turner, E. S., Greenberger, P. A., and Patterson, R. (1980). Management of the pregnant asthmatic patient. *Ann. Intern. Med.* **6,** 905–918.

Valdivieso, A., Vlades, G., Spiro, T. E., and Westerman, R. L. (1985). Minoxidil in breast milk. *Ann. Intern. Med.* **102,** 135.

Verbeeck, R. K., Blackburn, J. L., and Loewen, G. R. (1983). Clinical pharmacokinetics of non-steroidal anti-inflammatory drugs. *Clin. Pharmacokinet.* **8,** 297–331.

Vincent, J., Meredith, P. A., Reid, J. L., Elliott, H. L., and Rubin, P. C. (1985). Clinical pharmacokinetics of prazosin—1985. *Clin. Pharmacokinet.* **10,** 144–154.

Walker, B. (1980). Lead content of milk and infant formula. *J. Food Protect.* **43,** 178–179.

Whalley, L. J., Blain, P. G., and Prime, J. K. (1981). Haloperidol secreted in breast milk. *Br. Med. J.* **282,** 1746–1747.

Whichelow, M. (1979). Breastfeeding and smoking. *Arch. Dis. Child.* **54,** 240–241.

White, W. B. (1984). Management of hypertension during lactation. *Hypertension* **6,** 297–300.

White, W. B., Andreoli, J. W., and Cohn, R. D. (1985). Alpha-methyldopa disposition in mothers with hypertension in their breastfed infants. *Clin. Pharmacol. Ther.* **37,** 387–390.
Wickizer, T. M., and Brilliant, L. B. (1981). Testing for polychlorinated biphenyls in human milk. *Pediatrics* **68,** 411–415.
Wickizer, T. M., Brilliant, L. B., Copeland, P., and Tilden, R. (1981). Polychlorinated biphenyl contamination of nursing mothers' milk in Michigan. *Am. J. Public Health* **71,** 132–137.
Wild, C. P., Pionneau, F. A., Montesano, R., Mutiro, C. I., and Chetsanga, C. J. (1987). Aflatoxin detected in human breast milk by immunoassay. *Int. J. Cancer* **40,** 328–333.
Wiles, D. H., Orr, M. W., and Kolakowska, T. (1978). Chlorpromazine levels in plasma and milk of nursing mothers. *Br. J. Clin. Pharmacol.* **5,** 272–273.
Wilson, J. H. (1988). Breast milk tocainide levels. *J. Cardiovasc. Pharmacol.* **12,** 497.
Wilson, J. T., Brown, R. D., Cherek, D. R., Dailey, J. W., Hilman, B., Jobe, P. C., Manno, B. R., Manno, J. E., Redetzki, H. M., and Stewart, J. J. (1980). Drug excretion in human breast milk: Principles, pharmacokinetics and projected consequences. *Clin. Pharmacokinet.* **5,** 1–66.
Wolff, M. S. (1983). Occupationally derived chemicals in breast milk. *Am. J. Ind. Med.* **4,** 259–281.
Woodward, A., and Hand, K. (1988). Smoking and reduced duration of breastfeeding. *Med. J. Aust.* **148,** 477–478.
Woodward, A., Grgurinovich, N., and Ryan, P. (1986). Breast feeding and smoking hygiene: Major influences on cotinine in urine of smokers' infants. *J. Epidemiol. Community Health* **40,** 309–315.
Woosley, R. L. (1987). Mexiletine and tocainide: A profile of two lidocaine analogs. *Rational Drug Ther.* **21,** 1–7.
World Health Organization (WHO) (April 1972). Evaluation of certain food additives and the contaminants mercury, lead, and cadmium. *In* "Technical Report Series 505." 16th report of the Joint FAO/WHO Expert Committee on Food Additives GENF.
Wretlind, M. (1987). Excretion of oxazepam in breast milk. *Eur. J. Clin. Pharmacol.* **33,** 209–210.
Zeisler, J. A., Gaarder, T. D., and DeMesquita, S. A. (1986). Lidocaine excretion in breast milk. *Drug Intell. Clin. Pharm.* **20,** 691–693.

B. Contaminants in Bovine Milk

ROBERT G. JENSEN

I. Introduction

A large number of environmental contaminants have been detected in bovine milk. These compounds, termed xenobiotics by Patton (1986), enter the cow as residues of pesticides, herbicides, etc., on feedstuffs or as drugs given to the cow orally, by injection, or as intramammary infusions for the treatment of mastitis. Contaminants also enter milk from equipment after milking. Many of these compounds are fat-soluble, log octanol/water partition coefficient > 5 (Patton, 1986) and will be stored in adipose tissue or secreted in milk fat. Less lipophilic compounds and their metabolites may be excreted in urine. The concentrations of these compounds have been reported as ppm (mg/liter), ppb (μg/liter), or ppt (ng/liter). The general types of compounds which have been or are found in milk are chlorinated pesticides, organophosphates, herbicides, fungicides, fasciolicides (compounds for the control of liver flukes), antibiotics and sulfonamides, detergents and disinfectants, polychlorinated and polybrominated biphenyls (PCBs and PBBs), dioxins, mycotoxins, heavy metals and other trace elements, nitrates, and somatropin (IDF, 1991). Some of the information I present was taken from the NDC Status Report (1991).

The presence of some of these compounds in bovine milk was reviewed by Kroger (1974), radionuclides by Lengemann *et al.* (1974), and contaminants in general by IDF (1991). The latter is a comprehensive review of milk contaminants in Europe. Unfortunately, a recent review of the status in the United States is not available.

A computer search showed that much of the recent research in the United States has been devoted to the development of new and improved methods for the analysis of environmental contaminants, e.g., a liquid chromatographic method for the analysis of sulfonamides in milk (Smedley and Weber, 1990).

I will not present a comprehensive review, but will give examples of several types of environmental contaminants.

II. Chlorinated Pesticides and Related Compounds: PCBs, PBBs, and Dioxins

A. DDT

DDT is the classic example of an environmental contaminant in milk. It came into general use immediately after World War II and was widely employed until its accumulation in fatty tissues and transfer into milk was observed. In 1972, DDT was banned (Lawrence and Friedman, Chapter 11A). There was concern about neurological disturbances and liver metabolism. The quantity of DDT and metabolites recently detected in milk is about 1.0 μg/kg (1.0 ppb), an amount well below the accepted daily intake of 20 μg/kg body wt (Roos and Tuinstra, 1991). The amount of DDT above is an average from milks from several unspecified European countries and dates were not given. Information on DDT and other pesticides is listed in Table I. Very small quantities will continue to be detected because DDT resists degradation, is stored in adipose tissue, and the methods used for detection are extremely sensitive, e.g., gas–liquid chromatography with an electron-capture detector.

B. Heptachlor

Milk contaminated with heptachlor was traced to cattle feed accidentally mixed with the pesticide (NDC, 1991). Based on the FDA action level of 0.1 ppm (fat basis) then in effect, the milk was removed from the market (FDA, 1990). Milk from the affected dairy cows could not be sold until shown to be safe. The pesticide is a potential carcinogen. Its use was banned in 1978. Data on heptachlor can be seen in Table I.

C. PBBs and PCBs

Livestock feed sold in Michigan in 1973 was accidentally mixed with PBB, a toxic fire retardant (NDC, 1991). Large quantities of milk and its products and meats were contaminated and destroyed. The FDA's action level of 0.3 ppm (fat basis) in milk was revoked. While there were apparently no long-term health effects, PPBs are very persistent. This was apparently the only incident involving PBBs in milk, but closely related compounds, PCBs, have been detected (Kadis, 1991). They are extremely stable and have been used in paints, resins, hydraulic fluids, dietectric fluids in transformers, copier paper, etc. They are no longer made in the United States and other countries as of 1972, but 1.5 billion pounds were manufactured in the United States alone. Their stability ensures that they will be found in the environment for years. About 0.8 μg/kg has been detected in milk (see Table I). The compounds may be deleterious, but the amounts are slowly decreasing. It is difficult to assess the affects of chronic toxicity so monitoring should continue (Kadis, 1991). However, there are many different

TABLE I
Amounts and Accepted Intakes of Some Halogenated Contaminants in Bovine Milk[a]

Compound	Median (μg/kg milk)	Accepted daily intake (μg/kg body wt)
DDT and metabolites[b]	1.0	20
Heptachlor, including epoxide[b]	< 0.3	0.5
Polychlorinated biphenyls[c]	0.8	1.0[d]
Dioxins (ng/kg on a fat basis)	1.5[e]	< 35[f]

[a]Based on milk containing 3.3% fat except for dioxins.
[b]Roos and Tuinstra (1991).
[c]Kadis (1991).
[d]Jensen and Slorach (1991).
[e]Expressed as TCDD equivalents. To obtain amount/kg milk, multiply by 3.3% = 50 pg/kg (Overstrom, 1991).
[f]Tolerable weekly intake of TCDD equivalents, < 35 pg/kg body wt (Overstrom, 1991).

PCBs in the mixtures that were used. Theoretically, up to 209 different PCBs could be formed during the manufacturing process, but in reality, fewer exist. Nevertheless, they are difficult to analyze because some of the compounds which contain less chlorine are potentially toxic because their structures are similar to 2,3,7,8-tetrachlorobenzo-*p*-dioxin (TCDD) which may cause health hazards (Bohm *et al.*, 1993). Unfortunately, these compounds are near the routine limit of detection of 0.1 to 0.5 μg/kg fat for the methods applied

D. Dioxins

The dioxins are some of the most toxic and carcinogenic compounds tested (Overstrom, 1991). They are found as impurities in pesticides and herbicides (Agent Orange, used in Vietnam). They are extremely persistent and their presence in the environment is ubiquitous. The most toxic and investigated compound in this group is TCDD. The amounts found are low, 1.5 ng/kg fat in Swedish milks. Although the amounts are small, the compounds accumulate in the body. Nevertheless, milk and dairy products do not appear to pose health problems. A tolerable weekly intake has been suggested as <35 pg TCDD equivalents /kg body wt/week. In another study, the estimated average daily intake for humans was 50 pg/day with 27% of this derived from milk (Travis and Hattemer-Frey, 1987).

E. Summary

When first used the pesticides decimated insect populations, reducing the spread of disease and the loss of food. Somewhat later, insects developed resistance to the pesticides and their accumulation in fatty tissues including milk fat was observed. The use of the common chlorinated pesticides has been banned in most countries. However, the compounds are degraded slowly and they will remain in fatty tissues for years. Their presence in milk is a result of the mobilization of fat in adipose tissue that occurs during lactation with feed as the ultimate source.

Constant monitoring of foods is needed. In the United States the FDA enforces tolerances and obtains data on the incidence and levels of pesticide residue and all foods other than meat, poultry, and some egg products, which are the responsibility of the USDA. Imported and domestic foods that are shipped interstate are tested. Animal feeds are also monitored. The FDA publishes an annual report describing their results. For example, in 1990, residues were not detected in 590 samples of milk and dairy products analyzed. No samples were found with levels over tolerance (FDA, 1991).

III. Veterinary and Other Drugs

A. Introduction

In order to obtain the greatest quantities of milk from their cattle, dairy producers use selective breeding, feed for optimum production, and maintain animal health. Mastitis—inflammation and infection of the mammary gland—reduces milk production, alters the composition of milk, and can irreversibly damage milk secretory tissue. Therefore, much of the veterinary treatment of dairy cattle involves intramammary infusion of antibiotics to control mastitis. The antibiotics can carry over into the milk unless stipulated procedures are followed. Other drugs are given to control endoparasites (trematodes, nematodes, etc.), ectoparasites (flies, lice, ticks, etc.), and several illnesses, and to increase milk production (somatotropin). Some of these carry over into the milk.

The concern about carryover of the veterinary drugs into milk, especially the antibiotics, is their potential for harmful effects on human health. Low-level doses of antibiotics for long periods could result in drug-resistant microorganisms. Consequently, in the United States, the FDA and state agencies routinely monitor milk and other foods for drug residues. The antibiotics can also destroy or hinder the growth of desirable microorganisms in cultures used to produce cheeses, yogurt, and other fermented dairy products with substantial economic losses.

B. Antibiotics

The antibiotics approved by the FDA for intramammary treatment of mastitis in lactating dairy cattle are listed in Table II. The information in this table was provided in a personal communication from Shotwell & Carr (1994). I have listed this source in the references so that readers can obtain the information as needed. All of the data except tolerance levels in milk are in the pamphlet by Boeckman and Carlson (1993). The pamphlet, sponsored by the American Veterinary Association and the National Milk Producers Association, is a producer manual containing a protocol for the prevention of dairy residues in milk and dairy beef. It is concise and full of useful information. Data on approved drugs for injection and oral use are provided in Tables III and IV. Drugs approved for topical application and as additives for feeds have not been listed because withholding times for milk were not listed (Boeckman and Carlson, 1993).

If the milk withholding times are applied, there should be no detectable residues in milk. The amounts decrease rapidly. For example, the amount of amoxicillin found in milk 36 hr after an intramammary application of 200 mg was 1.0 μg/dl (Heeschen and Blutgen, 1991). Data for excretion of other antibiotics are given in this reference. The Milk Safety

TABLE II
FDA-Approved Drugs for Intramammary Use in Lactating Cows[a]

Drug	Type[b]	Withdrawal (hr)	Tolerance (milk ppm)[c]
Amoxicillin trihydrate	Rx	60	0.01
Cephapirin (sodium)	OTC	96	0.02
Cloxacillin (sodium)	Rx	48	0.01
Erythromycin	OTC	36	0
Hetacillin (potassium)	Rx	72	0.01
Novobiocin	OTC	72	0.1
Penicillin G (procaine)	OTC	60–84	0
Pirlimycin	Rx	36	0.4
Salicylic acid[d]	OTC	48	0

[a]Adapted from a list of drugs approved for lactating cattle in a personal communication from Shotwell & Carr (1994). Most of the information is available from Boeckman and Carlson (1993). All except salicylic acid are antibiotics used for mastitis.
[b]Rx, prescription; OTC, over the counter.
[c]Level of drug tolerated in milk.
[d]Time milk must be discarded after treatment.

TABLE III
FDA-Approved Drugs for Administration to Lactating Dairy Cattle by Injection[a]

Drug	Type[b]	Withdrawal (hr)[c]	Tolerance (milk ppm)[d]
Amoxicillin trihydrate (antibiotic)	Rx	96	0.01
Ampicillin (antibiotic)	Rx	48	0.01
Chlorsulon, ivermectin (fasciolicide, nematocide)	OTC	—	1.0 0.025
Furosemide diethanolamine (diuretic)	Rx	48	None
Penicillin G (procaine) (antibiotic)	Rx	48	0
Sulfadimethozine (antimicrobial)	OTC	60	0.01
Tripelennamine HCl (antihistamine)	Rx	24	None

[a]From a list of drugs approved for lactating dairy cattle in a personal communication from Shotwell & Carr (1994). Some drugs with no withholding time have not been listed (Boeckman and Carlson, 1993).
[b]Rx, prescription; OTC, over the counter.
[c]Time milk must be discarded after treatment.
[d]Level of drug tolerated in milk.

TABLE IV
FDA-Approved Drugs for Oral Use in Lactating Dairy Cattle[a]

Drug	Type[b]	Withdrawal (hr)[c]	Tolerance (milk ppm)[d]
Chlorosulon (fasciolicide)	OTC	—	1.0
Chlorothiazide (diuretic)	Rx	72	None
Dexamethasone (anti-inflammatory)	Rx	72	None
Furosemide (diuretic)	Rx	48	None
Sulfabromomethazine (antimicrobial)	OTC	96	0.01
Sulfaethoxypyridazine (antimicrobial)	Rx	72	0
Thiabendazole (dewormer)	OTC	96	0.05
Trichloromethiazide, dexamethasone (diuretic, anti-inflammatory)	Rx	72	None

[a]Adapted from a list of drugs approved for lactating cattle in a personal communication from Shotwell & Carr (1994).
[b]OTC, over the counter; Rx, prescription.
[c]Time milk must be discarded after treatment.
[d]Level of drug tolerated in milk.

Branch of the FDA (1992) issued a memorandum in which they summarized a survey of state drug residue sampling done by the industry during January–June 1992. The data are summarized in Table V. The data suggest that the amount of milk containing detectable, antibiotic residues is small. However, some nonapproved drugs were used, i.e., gentamycin and ivermectin. Ivermectin is not an antibiotic. The temptation for noncompliant use is attractive because the financial losses caused by mastitis are substantial. An increase in monitoring appears to be indicated. According to Heeschen and Bluthgen (1991), in countries where regular testing for residues is not done, the percentages of residue-positive samples (herd milk) may reach 1 to 10%. In countries where regular testing is done the number is 1 to 5%. The latter is probably too high and is dependent on the tolerance level and the sensitivity of the methods used for detection.

The methods employed for detection of antibiotic residues in milk utilize the inhibition of growth of several microorganisms and other more specific tests (Bishop *et al.*, 1992). There are several cowside and tank screening procedures available (Boeckman and Carlson, 1993) which will detect antibiotics at the low ppb level. This increase in sensitivity requires regular reevaluations of the withholding times and milk tolerance. A "zero" tolerance established by a less-sensitive method may be invalidated by a more sensitive procedure which will detect much lower levels of residues.

TABLE V
Summary of State Survey for Antibiotic Residues in Milk[a]

	Total		Positive residues		
Milk product tested	*n*	Milk lbs	*n*	Milk lbs	%
Raw milk, farm[b] tank trucks January–June, 1992	1,828,020	59.5 billion	1505	45,610,408	0.08
Grade A[c] finished product 1991	107,381	—	24	—	0.02
January–June, 1992	52,618	—	4	—	0.008

[a]Milk Safety Branch, FDA (1992).
[b]Milk tested by industry for penicillin or β-lactam residues. Zero tolerance.
[c]Tested by state agencies. Residues not identified but were presumably penicillin. Sulfonamides (7/26818), tetracyclines (7/9196), gentamycin (1/1556), and ivermectin (1/23) detected in a small number of samples. (*n* positive/*n* tested).

According to Heeschen and Bluthgen (1991), regular monitoring resolves, on the international level, the problems of producing fermented milk products, yogurt, and cheeses, and a health hazard for humans can be virtually excluded. Determination of, adherence to, and monitoring of withholding times are paramount.

C. *Endoparasiticides*

Heeschen and Bluthgen (1991) listed 12 antihelmintics which have been used to control trematodes (liver flukes) and nematodes in dairy cattle. In the United States, a mixture of chlorosulon and Ivermectin is approved for injection (Table III) and chlorosulon and thiabendazole orally (Table IV). The precautions discussed for antibiotics are applicable here.

D. *Ectoparasiticides*

A large number of ticks, flies, mites, etc. torment dairy cattle (Heeschen and Bluthgen, 1991). Again, many compounds, essentially pesticides, have been employed. See Section II on pesticides for more information.

E. *Somatotropin*

Bovine somatotropin (BST) or bovine growth hormone is a protein produced by the pituitary gland in all cattle (Heeschen and Bluthgen, 1991).

Recently, a virtually identical recombinant BST has been produced by genetic engineering and is available at relatively low cost. When BST is injected into dairy cattle, a unit of milk is produced with less feed and protein supplement and with less animal excreta, i.e., manure, urine, and methane (Bauman, 1992). The FDA has recently approved the use of BST for this purpose, a move which has resulted in anti-BST furor by uninformed and partially informed groups.

The results from many studies on BST have been summarized by Bauman (1992) as follows: (1) BST is a protein which is digested when consumed orally; (2) BST is not human ST (HST) and does not elicit any of the biological responses in humans caused by HST; (3) trace amounts of BST occur naturally in milk (< 1 mg/ml) and are not increased above normal levels by dosage with BST; and (4) the overall nutrient composition of milk is not changed by BST treatment, nor is the response to processing treatments (see Chapter 5I).

Since they could not or would not use the information above, the anti-BST groups have focused on a purported increase in the incidence and severity of mastitis in cows that are "forced" to produce more milk. The rationale is that since more antibiotics will be used to treat the mastitis, there will be more drug residues in milk. The basic premise of more mastitis has been disproven (Bauman, 1992). Those proposing this sequence of events either did not know or have forgotten that a primary goal of dairy husbandry has been to increase production ever since the first cow was domesticated. Milk production has doubled during the past 40 years, while the incidence of mastitis has decreased. This was due in part to improved milking practices, i.e., cleansing and disinfection of the cow's teats and of the milking machines to prevent the entry and spread of microorganisms. It is very unlikely that the 15% or so increase in production that would be caused by use of BST would produce an epidemic of drug residues in milk.

IV. Detergents and Disinfectants

A. Introduction

Dairy equipment must be rinsed and cleaned after each use and disinfected prior to the introduction of milk. The ultimate purposes are to remove bacteria which may be adhering to surfaces or in milk residues and to destroy any which may be present. These processes are accomplished usually by pumping solutions of appropriate compounds through the equipment or cleaning in place (Palmer, 1991). Residues are removed by rinsing with clean water or by small quantities of milk which is then discarded. Detergent mixtures containing alkaline or acid compounds,

Ca-sequestering agents, and surface-active agents are employed with special formulations for specific equipment. Some of the residues contain denatured proteins and precipitated Ca complexes (milk stone) and are difficult to remove. Disinfection is done with compounds which usually contain chlorine or iodine or with quaternary ammonium compounds.

B. Significance

The amount of detergent/disinfectant residues found in milk will be minimal if proper procedures are used. Concentrations are usually less than 2 ppm. This is well below the lethal dosages of 0.5 to 3.0 g, but long-term effects of consuming milks with low levels are unknown. Sanitizers can be detected by taste. This occurs when the equipment has not been properly rinsed with water or when the length of the milk rinse is insufficient to remove all of the sanitizer. When the sanitizers are detectable by taste the milk will not be consumed. Remedies are informed employees, use of proper procedures, and regular monitoring, i.e, tasting the first several containers of milk emerging from packaging machines.

V. Mycotoxins

A. Introduction

Mycotoxins are potent hepatocarcinogens produced during the growth of molds on animal feedstuffs, and to a much lesser extent, cheeses (NDC, 1991; van Egmond, 1991). Aflatoxin B1, which is the direct precursor of the aflatoxin M-1 found in milk, can be produced by the molds *Aspergillus flavus* and *Aspergillus parasiticus* in feedstuffs. It can usually be detected in milk within 12 hr after contaminated feedstuffs are consumed by the cow.

B. Amounts and Significance

In the United States, the FDA has set action levels on aflatoxins of 20 ppb (20 μg/kg) in animal feed and 0.5 ppb (0.5 μg/kg) in milk (NDC, 1991). Materials containing amounts in excess of these are discarded. Levels in various studies ranged from 0.05 to 0.50 μg/kg milk (van Egmond, 1991). Some aflatoxins have been detected in mold-ripened cheeses and in cheese accidentally contaminated with molds, but the amounts were negligible and believed to be of little risk.

The state of Arizona established a monitoring program for aflatoxin in milk (Park, 1993). In 1978, almost 910,000 lbs of milk were dumped with levels of aflatoxin M1 as high as 10 ppb. All cottonseed (a potential growth medium for molds elaborating aflatoxins) produced in the state is tested. Those lots of cottonseed containing over 20 μg aflatoxin/kg were usually treated with ammonia to reduce aflatoxin levels. In 1989, 13% of 800 milk samples tested contained 0.2 to 0.5 μg/liter of aflatoxin M1. None contained more than 0.5 μg/liter, the action level.

There is a general correlation between the incidence of liver cancer in humans in certain areas of Asia and Africa and the amounts of aflatoxins in the diet (Park, 1993). However, a direct causal relationship has not been established. Nevertheless, the FDA acted upon the reasonable premise that an animal carcinogen should be considered to be a human carcinogen until proven otherwise. Levels of aflatoxins in foods should be controlled at the lowest practical level. These levels, 0.5 μg/kg milk and 20 μg/kg animal feed, are usually dictated by the sensitivity and ease of applicability of the analytical methods, in this case, immunoaffinity procedures (van Egmond, 1991). The aflatoxins are potential threats and monitoring should continue. However, in the United States and other similar countries where animal feedstuffs are used promptly before extensive mold growth occurs and mixed diets are consumed, the threat posed by aflatoxins is minimal.

VI. Metals

A. Introduction

The metals of concern are lead (Pb), cadmium (Cd), and mercury (Hg) which at sufficient levels in milk and dairy products could cause problems (Carl, 1991). Others of interest are arsenic (As), chromium (Cr), and Nickel (Ni). These metals, often designated as heavy because they were originated in the effluents from heavy industries, are found in fertilizer and sludges applied to fields, feedstuffs for cattle, and in postmanufacturing exposure such as the solder (Pb and Sn), used to seal cans of evaporated milk and the stainless steel in dairy equipment which contains chromium and nickel as well as steel.

B. Contents and Significance

The normal contents of the metals in milk are presented in Table VI along with acceptable limits in μg/kg of milk. Provisional tolerable weekly intakes for adults have been established by FAO/WHO as follows (μg/kg): lead, 50; cadmium, 7; and mercury, 3.3 (Carl, 1991). The much smaller quantities of Pb and Cd reported by Carl (1991) compared to the other data in Table II are probably the result of differences in analytical procedure. More recent methods were reported by Carl and by the IDF (1992).

TABLE VI
Amounts of Possible Toxic Metals in Bovine Milk

	Median (μg/kg)[a] (range)	Acceptable limits (μg/kg)[a]	Median (μg/liter)[b] (range)
Lead (Pb)	2.0–3.0 (1–5)	10–150	40 (30–60)
Cadmium (Cd)	0.5[c] (0.2–0.8)	5–50	— (1–30)
Mercury (Hg)	< 0.07 (0.07)	2–20	TR
Arsenic (As)			— (30–60)
Chromium (Cr)			15 (5–80)
Nickel (Ni)			— (0–30)

[a]Carl (1991). Acceptable limits from several European countries.
[b]Jenness (1989).
[c]IDF (1992).

Lead is a cumulative poison which behaves like calcium salts. It inhibits hemoglobin synthesis and has toxic effects on nervous tissue with possible permanent impairment of function. Most ingested leads goes into bony tissue. Cadmium binds with sulfhydryl groups of enzymes and interferes with oxidative phosphorylation. It can replace zinc in the metalloenzymes. Mercury has an affinity for the sulfhydryls in enzymes and has toxic neurological effects (Carl, 1991). Methylmercury is extremely toxic and has been a major problem in fish.

Chromium may be an essential trace mineral for humans (Anderson, 1988). It is involved in lipid and carbohydrate metabolism as an insulin potentiator and in nucleic acid metabolism. Deficiency signs and symptoms have been reported. A range of Cr intakes of 50 to 200 μg/day has been suggested, but the actual amounts consumed are probably less than 50 μg (NRC, 1989). The universal exposure of foods to stainless steel during processing and cooking ensures that some quantities will be present with larger amounts in acidic foods.

Stainless steel contains Cr and Ni. Trivalent chromium, the chemical form found in diets, is of low toxicity and does not present any problems. It is included here only because it is found in industrial effluents. Those planning to determine the amounts of Cr and Ni in foods must avoid contamination from processing and cooking equipment, surgical instruments, milking machines, laboratory mixers, etc.

Arsenic is a well-known classic poison but has also been a component of many medicines. It may be an essential nutrient, but in sufficient quantities, 0.76 to 1.95 mg As/kg body wt of a 70-kg human, is lethal (Nielsen, 1988). Since the amount in milk is small, it is unlikely to ever present a problem even with excess environmental contamination. Nickel may also be an essential nutrient (Nielsen, 1988). It has been estimated that a daily dose of 250 mg of soluble nickel would cause toxic symptoms in humans. Again, the amount in milk is far below this level.

Arsenic, Cr, and Ni are mentioned only because they accompany the more potentially toxic Pb, Cd, and Hg in industrial effluents. It is highly improbable that the amounts found in milk consumed in a mixed diet could cause problems, but again milk is at the end of a food chain. These compounds accumulate and milk should be monitored.

VII. Radionuclides

A. Introduction

Many of these elements are present naturally, but the types, some artificial, and amounts in the environment have increased following the testing of nuclear weapons, the operation of reactors, and the application of atomic energy (Lengemann, 1974). A more recent and not unexpected source is the reactor accident of the type that occurred in Chernobyl in 1986 (NDC, 1991). This resulted in excessive radioactive fallout in Sweden (Bruce and Slorach, 1987) and elsewhere.

B. Significance

Milk and dairy products are a major contributor to the accumulation of radionuclides believed to be hazardous to man; ^{90}Sr, ^{137}Cs, and ^{131}I (Lengemann *et al.*, 1974). Milk is thought to be the only major pathway for ^{131}I. This nuclide, although having a half-life of only 8 days, accumulates on forage and enters milk when the feeds are consumed by the cow. It then enters the thyroid gland where it is concentrated. The ^{90}Sr has a half-life of 28 years and behaves similarly to Ca in the body eventually locating in bone. ^{137}Cs has a half-life of about 30 days. It appears to be metabolized and distributed throughout the body in the same mode as K, becoming widely distributed in soft tissue.

Soon after the Chernobyl disaster, ^{131}I and ^{137}Cs appeared in components of the food chain. One week after the accident, and based on measurements of ^{131}I in Northern Italy, the consumption of leafy vegetables was halted in Italy. Consumption of milk was restricted for 3 weeks in pregnant women and in children less than 10 years of age (Rosen and Sinaiko, 1989). Both ^{137}Cs and ^{131}I were detected in breast milk in Italy and Austria. The radioactive fallout was also serious in some areas of Sweden (Bruce and Slorach, 1987). In these areas, ^{137}Cs activity concentrations were in some instances many-fold greater than the amounts found in low-fallout areas. Cows were not allowed to graze as usual to keep the levels of radionuclides in milk, especially ^{131}I, low.

The amounts of potentially dangerous radionuclides in milk and other foods are now very low. Hopefully, the use of nuclear weapons and Chernobyl-type accidents will never occur again. Prevention can be effective.

VIII. Summary

Contaminants in milk are caused by accidents, carelessness, and overzealous use of antibiotics. All could have and can be prevented by long-term planning, training of personnel, and monitoring with economic penalties. Nevertheless, since milk contains fat the persistent, lipophilic contaminants will be found for some time. Further, when the enormous quantity of milk produced is considered, the potential health hazard posed by the residues is almost nonexistent. Nevertheless, monitoring accompanied by regular reevaluation of the acceptable levels must continue, but with the realization that some residues will probably always be found in very low quantities.

Acknowledgments

Preparation of the manuscript was supported in part by an NIH contract and by federal funds made available through provision of the Hatch Act, Scientific Contribution, Storrs Agricultural Experiment Station, Storrs, Connecticut. The author appreciates the assistance given by Lynn S. Hinckley, Department of Pathobiology, University of Connecticut.

References

Anderson, R. A. (1988). Chromium. *In* "Modern Nutrition in Health and Disease, 7th Ed." (M. E. Shils and V. R. Young, eds.), pp. 268–273. Lee & Ferbiger, Philadelphia.

Bauman, D. E. (1992). Bovine somatotropin: Review of an emerging technology. *J. Dairy Sci.* **75,** 3432–3451.

Bishop, J. R., Senyk, G. F., and Duncan, S. E. (1992). Detection of antibioticl drug residues in milk and dairy products. *In* "Standard Methods for the Examination of Dairy Products, 16th Ed." (R. T. Marshall, ed.), pp. 347–395. American Public Health Association, Washington, DC.

Boeckman, S., and Carlson, K. R. (1993). "Milk and Dairy Beef Residue Prevention protocol, 1994 Producer Manual." Program of American Veterinary Medicine Association and National Milk Producers Federation. Available from Dairy Quality Assurance, 801 Shakespeare, Box 497, Stratford, IA, 50429.

Bohm, V., Schulte, E., and Thier, H-P. (1993). Polychlorinated biphenyl residues in food and human milk: Determination of coplanar and mono-*ortho* substituted congeners. *Z. Lebensm. Unters. Forsch.* **196,** 435–440.

Bruce, A., and Slorech, S. A. (1987). Dietary implications of radioactive fallout in Sweden following the accident at Chernobyl. *Am. J. Clin. Nutr.* **45,** 1089–1093.

Carl, M. (1991). Heavy metals and other trace elements. *In* "Residues and Contaminants in Milk and Milk Products, Special Issue 9101," pp. 112–119. International Dairy Federation, Brussels.

Food and Drug Administration (FDA) (1990). Food and drug administration monitoring program. *J. Assoc. Off. Anal. Chem.* **73,** 127A–146A.

Food and Drug Administration (FDA) (1991). Food and drug administration monitoring program. *J. Assoc. Off. Anal. Chem.* **74,** 121A–141A.

Heeschen, W. H., and Bluthgen, A. (1991). Veterinary drugs and pharmacologically active compounds. *In* "Residues and Contaminants in Milk and Milk Products, IDF Special Issue 9101," pp. 16–70. International Dairy Federation, Brussels.

International Dairy Federation (IDF) (1991). "Residues and Contaminants. Special Issue 9101." International Dairy Federation, Brussels. Available in the United States from USA–IDF, 464 Central Avenue, Northfield, IL 60093.

International Drug Federation (IDF) (1992). "Trace Elements in Milk and Milk Products." Bulletin 278, IDF, Brussels.

Jenness, R. (1988). Composition of milk. *In* "Fundamentals of Dairy Chemistry, 3rd Ed." (N. P. Wong, ed.), p. 11. Van Nostrand–Reinhold, New York.

Kadis, V. W. (1991). Polychlorinated biphenyls (PCBs). *In* "Residues and Contaminants in Milk and Milk Products," pp. 146–163. International Dairy Federation Special Issue 9101, Brussels. USA–IDF, 464 Central Avenue, Northfield, IL 60093.

Kroger, M. (1974). General environmental contaminants occurring in milk. *In* "Lactation: A Comprehensive Treatise, Vol. 3" (B. L. Larson and V. R. Smith, eds.), pp. 135. Academic Press, New York.

Lengemann, F. W., Wentworth, R. A., and Comar, C. L. (1974). Physiological and biochemical aspects of the accumulation of contaminant radionuclides in milk. *In* "Lactation: A Comprehensive Treatise, Vol. 3" (B. L. Larson and V. R. Smith, eds.), pp. 160–215. Academic Press, New York.

Milk Safety Branch, FDA (1992). National drug residue sampling survey, M-1-92-7, 200 C St., NW, Washington, DC 20204.

National Dairy Council (NDC) (1991). "Scientific Status Report 10: Food Safety/ Contaminants." NDC, Rosemont, IL.

National Research Council (NRC) (1989). Chromium. *In* "Recommended Dietary Allowances, 10th Ed." pp. 241–243. National Academic Press, Washington, DC.

Neilsen, F. H. (1988). Ultratrace minerals. *In* "Modern Nutrition in Health and Disease, 7th Ed." (M. E. Shils and V. R. Young, eds.), pp. 278–291. Lea & Ferbiger, Philadelphia.

Overstrom, H. (1991). Polychlorinated dibenzo-p-dioxins and dibenzofurans. *In* "Residues and Contaminants in Milk and Milk Products: Special Issue 9101," pp. 164–172. IDF, Brussels. USA–IDF, 464 Central Ave., Northfield, IL 60093.

Palmer, J. (1991). Detergents and disinfectants. *In* "Residues and Contaminants in Milk and Milk Products, Special Issue 9101," pp. 173–189, International Dairy Federation, Brussels. USA–IDF, 464 Central Avenue, Northfield, IL 60093.

Park, D. L. (1993). Controlling aflatoxin in food and feed. *Food Technol.* **47,** 92–96.

Patton, J. S. (1986). Cellular pathways in the movement of lipophilic xenobiotics from GI tract to breast milk. *In* "Human Lactation 2: Maternal and Environmental Factors" (M. Hamosh and A. S. Goldman, eds.), pp. 475–497. Plenum Press, New York.

Rosen, W., and Sinaiko, A. R. (1989). Environmental contaminants in breast milk. *In* "Textbook of Gastroenterology and Nutrition in Infancy, 2nd Ed." (E. Lebenthal, ed.), pp. 229–237. Raven Press, New York.

Ross, A. H., and Tuinstra, L. C. M. Th. (1991). Pesticides. *In* "Residues and Contaminants in Milk and Dairy Products, IDF Special Issue 9101," pp. 84–111. IDF, Brussels.

Shotwell & Carr (1994). "Research Drugs Approved for Lactating Cattle." Personal communication. Shotwell & Carr, Inc., 3003 LBJ Freeway, Suite 100 Dallas, TX, 75234-7755.

Smedley, M. D., and Weber, J. D. (1990). Liquid chromatographic determination of multiple sulfonermide residues in bovine milk. *J. Assoc. Off. Anal. Chem.* **73,** 875–879.

Travis, C. C., and Hattemer-Frey, H. A. (1987). Human exposure to 2,3,7,8TCCD. *Chemosphere* **16,** 2331–2342.

van Egmond, H. P. (1991). Mycotoxins. *In* "Residues and Contaminants in Milk and Milk Products, Special Issue 9101," pp. 131–145. IDF, Brussels. USA–IDF, 464 Central Avenue, Northfield, IL 60093.

CHAPTER 12

Summary

ROBERT G. JENSEN

It is not possible to summarize the enormous amount of data presented in this book except to state the obvious, milk is very complex. These fluids may contain about 100,000 compounds, contained and overlapping into the systems described in Chapter 2A.

The authors have attempted, and I believe succeeded, in providing the most reliable data available with appropriate commentary. Areas that should be investigated further are indicated or implicit throughout the book. The data bases are deficient in some areas. Most notable is the almost total lack of information obtained by modern methods on processed bovine milk to be consumed and on dairy products in general. There are many specific examples. There is almost no information on the contents of water-soluble vitamins in human and bovine milks that was determined by current methods.

Analyses of composition are a continuous process caused by development of newer and more sensitive analytical procedures. Only then can further information on the nutritive and nonnutritive roles of the constituents be determined. It is clear that these efforts must continue.

HANDBOOK OF MILK COMPOSITION

Index

Page number followed by *t* denotes table

FOOD SCIENCE AND TECHNOLOGY

International Series

Maynard A. Amerine, Rose Marie Pangborn, and Edward B. Roessler, *Principles of Sensory Evaluation of Food*. 1965.
Martin Glicksman, *Gum Technology in the Food Industry*. 1970.
Maynard A. Joslyn, *Methods in Food Analysis,* second edition. 1970.
C. R. Strumbo, *Thermobacteriology in Food Processing,* second edition. 1973.
Aaron M. Altschul (ed.), *New Protein Foods*. Volume 1, *Technology, Part A*—1974. Volume 2, *Technology, Part B*—1976. Volume 3, *Animal Protein Supplies, Part A*—1978. Volume 4, *animal Protein Supplies, Part B*—1981. Volume 5, *Seed Storage Proteins*—1985.
S. A. Goldblith, L. Rey, and W. W. Rothmayr, *Freeze Drying and Advanced Food Technology*. 1975.
R. B. Duckworth (ed.), *Water Relations of Food*. 1975.
John A. Troller and J. H. B. Christian, *Water Activity and Food*. 1978.
A. E. Bender, *Food Processing and Nutrition*. 1978.
D. R. Osborne and P. Voogt, *The Analysis of Nutrients in Foods*. 1978.
Marcel Loncin and R. L. Merson, *Food Engineering. Principles and Selected Applications*. 1979.
J. G. Vaughan (ed.), *Food Microscopy*. 1979.
J. R. A. Pollock (ed.), *Brewing Science,* Volume 1—1979. Volume 2—1980. Volume 3—1987.
J. Christopher Bauernfeind (ed.), *Carotenoids as Colorants and Vitamin A Precursors: Technological and Nutritional Applications*. 1981.
Pericles Markakis (ed.), *Anthocyanins as Food Colors*. 1982.
George F. Stewart and Maynard A. Amerine (eds.), Introduction to Food Science and Technology, second edition. 1982.
Malcolm C. Bourne, *Food Texture and Viscosity: Concept and Measurement*. 1982.
Hector A. Iglesias and Jorge Chirife, *Handbook of Food Isotherms: Water Sorption Parameters for Food and Food Components*. 1982.
Colin Dennis (ed.), *Post-Harvest Pathology of Fruits and Vegetables*. 1983.
P. J. Barnes (ed.), *Lipids in Cereal Technology*. 1983.
David Pimentel and Carl W. Hall (eds.), *Food and Energy Resources*. 1984.
Joe M. Regenstein and Carrie E. Regenstein, *Food Protein Chemistry: An Introduction for Food Scientists*. 1984.

Maximo C. Gacula, Jr. and Jagbir Singh, *Statistical Methods in Food and Consumer Research.* 1984.
Fergus M. Clydesdale and Kathryn L. Wiemer (eds.), *Iron Fortification of Foods.* 1985.
Robert V. Decareau, *Microwaves in the Food Processing Industry.* 1985.
S. M. Herschdoerfer (ed.), *Quality Control in the Food Industry,* second edition. Volume 1—1985. Volume 2—1985. Volume 3—1986. Volume 4—1987.
F. E. Cunningham and N. A. Cox (eds.), *Microbiology of Poultry Meat Products.* 1987.
Walter M. Urbain, *Food Irradiation.* 1986.
Peter J. Bechtel, *Muscle as Food.* 1986.
H. W.-S. Chan, *Autoxidation of Unsaturated Lipids.* 1986.
Chester O. McCorkle, Jr., *Economics of Food Processing in the United States.* 1987.
Jethro Jagtiani, Harvey T. Chan, Jr., and William S. Sakai, *Tropical Fruit Processing.* 1987.
J. Solms, D. A. Booth, R. M. Dangborn, and O. Raunhardt, *Food Acceptance and Nutrition.* 1987.
R. Macrae, *HPLC in Food Analysis,* second edition. 1988.
A. M. Pearson and R. B. Young, *Muscle and Meat Biochemistry.* 1989.
Dean O. Cliver (ed.), *Foodborne Diseases.* 1990.
Marjorie P. Penfield and Ada Marie Campbell, *Experimental Food Science,* third edition. 1990.
Leroy C. Blankenship, *Colonization Control of Human Bacterial Enteropathogens in Poultry.* 1991.
Yeshajahu Pomeranz, *Functional Properties of Food Components,* second edition. 1991.
Reginald H. Walter, *The Chemistry and Technology of Pectin.* 1991.
Herbert Stone and Joel L. Sidel, *Sensory Evaluation Practices,* second edition. 1993.
Robert L. Shewfelt and Stanley E. Prussia, *Postharvest Handling: A Systems Approach.* 1993.
R. Paul Singh and Dennis R. Heldman, *Introduction to Food Engineering,* second edition. 1993.
Tilak Nagodawithana and Gerald Reed, *Enzymes in Food Processing,* third edition. 1993.
Dallas G. Hoover and Larry R. Steenson, *Bacteriocins.* 1993.
Takayaki Shibamoto and Leonard Bjeldanes, *Introduction to Food Toxicology.* 1993.
John A. Troller, *Sanitation in Food Processing,* second edition. 1993.
Ronald S. Jackson, *Wine Science: Principles and Applications.* 1994.
Harold D. Hafs and Robert G. Zimbelman, *Low-fat Meats.* 1994.
Lance G. Phillips, Dana M. Whitehead, and John Kinsella, *Structure–Function Properties of Food Proteins.* 1994.
Yrjö H. Roos, *Phase Transitions in Foods.* 1995.
Robert G. Jensen, *Handbook of Milk Composition.* 1995.